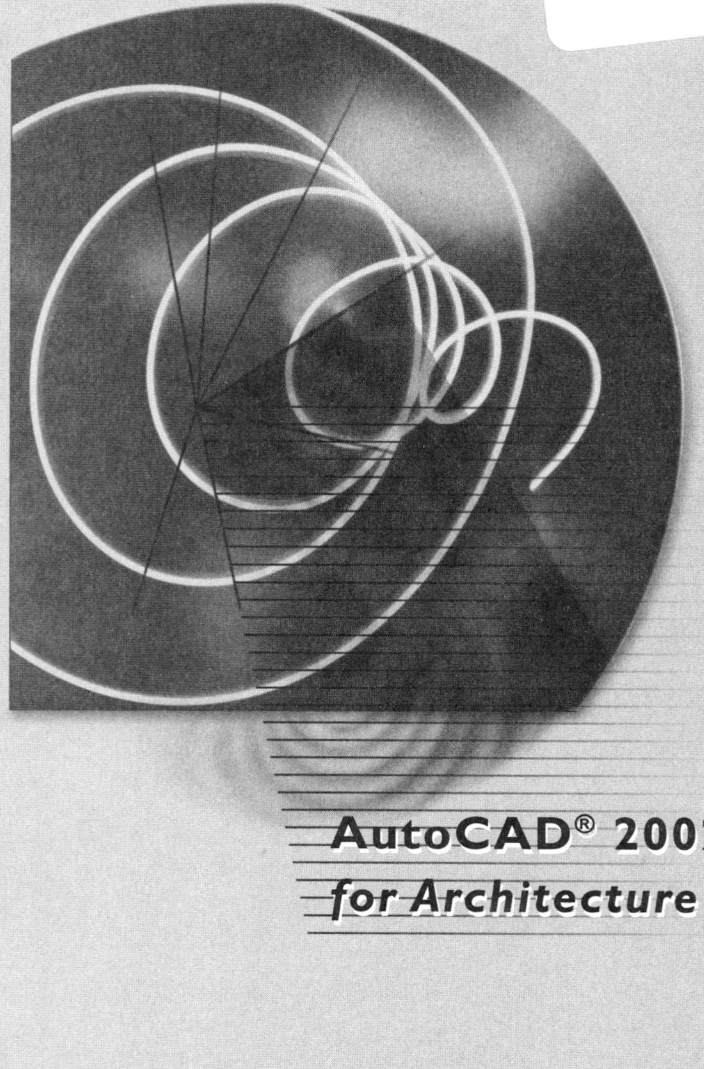

AutoCAD® 2002 *for Architecture*

AutoCAD® 2002 for Architecture

ALAN JEFFERIS
MICHAEL JONES
TEREASA JEFFERIS

autodesk® press

THOMSON LEARNING™

Australia • Canada • Mexico • Singapore • Spain • United Kingdom • United States

THOMSON LEARNING™

autodesk® press

AutoCAD 2002® for Architecture

Alan Jefferis Michael Jones Tereasa Jefferis

Business Unit Director:
Alar Elken

Executive Editor:
Sandy Clark

Acquisitions Editor:
James DeVoe

Development Editor:
John Fisher

Editorial Assistant:
Jasmine Hartman

Executive Marketing Manager:
Maura Theriault

Channel Manager:
Mary Johnson

Marketing Coordinator:
Karen Smith

Executive Production Manager:
Mary Ellen Black

Production Manager:
Larry Main

Production Editor:
Stacy Masucci

Art/Design Coordinator:
Mary Beth Vought

Printed in Canada
1 2 3 4 5 XXX 05 04 03 02 01

For more information contact Autodesk Press,
3 Columbia Circle, PO Box 15015,
Albany, NY 12212-5015.

Or find us on the World Wide Web at
www.autodeskpress.com

Library of Congress Cataloging-in-Publication Data
Jefferis, Alan.
AutoCAD 2002 for architecture / Alan Jefferis, Michael Jones, Tereasa Jefferis ; artwork by Tereasa Jefferis.
p. cm.
ISBN 0-7668-3847-1
1. Architectural drawing--Computer-aided design--Programmed instruction. 2. AutoCAD. I. Jones, Michael, 1964- ii. Jefferis, Tereasa. III. Title.

NA2728 .J4383 2001
720'.28'402855369--dc21 2001032502

CONTENTS

CHAPTER 7 CONTROLLING DRAWING ACCURACY

CHAPTER 8 DRAWING DISPLAY OPTIONS

SECTION 2

CHAPTER 9 BASIC METHODS OF SELECTING AND MODIFYING DRAWING OBJECTS

CHAPTER 11 POLYLINES

CHAPTER 12 SUPPLEMENTAL DRAWING COMMANDS

CHAPTER 13 CONTROLLING LINES AND LAYERS

CHAPTER 15 INQUIRY COMMANDS

SECTION 3

CHAPTER 16 PLACING TEXT ON A DRAWING

SECTION 4

CHAPTER 19 CREATING BLOCKS AND WBLOCKS

CHAPTER 21 OBLIQUE AND ISOMETRIC DRAWINGS

CHAPTER 22 WORKING WITH MULTIPLE DRAWINGS

CHAPTER 23 COMBINING DRAWINGS USING XREF

CHAPTER 26 AUTOCAD AND THE INTERNET

PREFACE

AutoCAD 2002 for Architecture is a practical comprehensive text and workbook that is easy to use and understand. The contents are designed to teach the skills needed to master the 2D drawing commands used on most construction-related drawings. The content can be used as presented to gain an understanding of the use of AutoCAD. This text can also be used with older releases of AutoCAD, but it is best suited for use with AutoCAD 2000, 2000i, and 2002. All concepts are presented using architectural examples so that each command can be correlated to a specific skill that will be required of architects, engineers, and drafters. Commands and drawing skills are presented in a sequence that is easy to understand and follow with each chapter building on the skills presented in the previous chapter. Other specific features of this text include the following:

- How a skill will relate to office practice is explained.
- Professional drawings by architects, engineers, and residential designers are used as illustrations.
- Chapter exercises review chapter contents in addition to reinforcing previous material.
- Each drawing chapter contains drawing problems that will reinforce the chapter concepts.

Individual chapters will provide knowledge to master the following drawing skills:

- Drawing lines and geometric shapes
- Using varied linetype and width
- Editing basic geometric shapes
- Combining skills to produce drawings such as floor and foundation plans, elevations, details, and section
- Editing an entire drawing to produce other similar drawings
- Understanding dimensioning methods and techniques
- Placing text within a drawing
- Creating varied text and dimensioning styles
- Creating symbol libraries to increase drawing speed and efficiency

This text provides a user-friendly environment to master skills that will lead to a productive career using tools that have long since replaced the T-square, parallel bar, or drafting machine. In addition, the authors have written an *Instructor's Guide* that contains supplemental drawing exercises, which are also available on disk. For more information, instructors can contact Delmar Publishers.

INTRODUCTION TO DRAWING COMMANDS

AutoCAD 2002 is a very powerful program that contains hundreds of commands to meet a variety of drawing needs. In this text, only the 2D command options that are typically used within the construction field will be explored. Several methods are available to execute each drawing command. Data entry can be made by using toolbars, menus, shortcut menus, and the keyboard. The selection method will depend on individual preference. Each command will be presented initially through the method that provides the fastest access. For many commands this will use of the keyboard, with available options used to supplement the basic procedure. With the keyboard presented as the primary option, the entire command sequence can be presented, displaying the many options that are available within a specific command.

Commands, options, or values that must be typed at the keyboard are listed in **BOLDFACE** capital letters. Once the command is typed, pressing the enter or return key (ENTER) is required and will be represented by the symbol (ENTER) throughout the text. Command sequences will be presented in the following manner:

Command: **L** (ENTER) *(Or click the Line icon on the Draw toolbar.)*

LINE Specify first point: **4,8** (ENTER)

Specify next point or [Undo]: **8,8** (ENTER)

Specify next point or [Undo]: (ENTER)

When input is required to complete a command, and the selected point can be any point, it will be presented using italics in the following manner:

Command: **L** (ENTER) *(Or click the Line icon on the Draw toolbar.)*

LINE Specify first point: *(Select a point.)*

Specify next point or [Undo]: *(Select a point.)*

Specify next point or [Undo]: (ENTER)

Command:

PREREQUISITES

Although this text can serve as an excellent reference for experienced professionals, no knowledge of architecture or construction is required to learn the command structure of AutoCAD. Some of the advanced projects do require a knowledge of basic residential architectural standards.

ONLINE COMPANION

The Online Companion™ is your link to AutoCAD on the Internet. We have compiled supporting resources with links to a variety of sites. Not only can you find out about training and education, industry, and the online community, but you can also find pointers to valuable archives compiled for AutoCAD users from various Web sites. In addition, there are pages specifically for users of *AutoCAD 2002 for Architecture* These include an owner's page with updates, a swap bank where you can share your drawings with other AutoCAD students, and a page where you can send us your comments. You can find the Online Companion at

http://www.autodeskpress.com/resources/olcs/index.html

When you reach the Online Companion page, click on the title **AutoCAD 2002 for Architecture**.

ABOUT THE AUTHORS

Mike Jones is the CAD Systems Manager and a drafting instructor at Clackamas Community College, which is an Autodesk Premier Training Center in Oregon City, Oregon. Mike has introduced new users and experienced professionals to the commands and options of AutoCAD for over 19 years. His training center experience is enhanced by 21 years of professional drafting and CAD system management experience in private industry.

Tereasa Jefferis is a graduate of Clackamas Community College and the Mechanical Systems Institute. She has been an architectural drafter for 5 years and has experience in architectural drafting, commercial plumbing drawings, and with HVAC systems. She designed several of the homes that are shown throughout this text. She currently works as an independent contractor with multiple mechanical engineering firms.

Alan Jefferis has been a drafting instructor at Clackamas Community College for 23 years. His professional experience includes 8 years of drafting experience with structural engineers, and 19 years as a professional building designer. He is currently the owner of Residential Designs, and is co-author of *Architectural Drafting and Design*, *Print Reading for Architectural and Construction Technology*, *Commercial Drafting and Detailing*, *AutoCAD for Architecture Release 12*, *AutoCAD for Architecture Release 13*, *AutoCAD for Architecture Release 14*, *AutoCAD for Architecture Release 2000*, and *Fundamentals of Mechanical Drafting and AutoCAD 2002*. All are available through Delmar Publishers.

ACKNOWLEDGMENTS

The quality of this text has been greatly enhanced by the donations of many professional offices and individuals within private and municipal agencies. My special thanks to the following:

Kenneth D. Smith
Kenneth D. Smith Architects, A.I.A.
El Cajon, California

Tom Sponsler, Director
Department of Environmental Services
City of Gresham, Oregon

F.D. Lee
Lee Engineering, Inc.
Oregon City, Oregon

G. Williamson Archer, A.I.A.
Archer & Archer P.A.
Meridian, Mississippi

Ron Lee, A.I.A.
Architects Barrentine, Bates & Lee, A.I.A.
Lake Oswego, Oregon

Thomas J. Kuhns, A.I.A.
Michael & Kuhns Architects, P.C.
Portland, Oregon

Ned Peck,
Peck, Smiley, Ettlin Architects

Havlin G. Kemp, P.E.
Van Domelen/Looijenga/McGarrigle/Knauf Consulting Engineers
Portland, Oregon

David Rogencamp
kpff Consulting Engineers
Portland, Oregon

Walt Cheever
South Central Technical College
N.Mankato, Minnesota

Shelly Fry
CPCC Corporate Center
Charlotte, North Carolina

Aaron Rumple
St. Louis Community College
St. Louis, Missouri

Dean Urevig
Dunwoody Instuite
Minneapolis, Minnesota

Sylvia Sullivan
College of the Canyons

Joseph Yamello
Greater Lowell Regional Vocational
Technical School

Robert Kamenski
Lenape A.V.T.S.

Walter Eric Lawrence
Gwinnett Technical Institute

All of these people have made substantial contributions to the success of this text. We're also especially thankful for the work and efforts of Gail Taylor of Taylor Wordsmiths, and John Fisher and Sandy Clark of Delmar Publishers. No one has been more helpful and supportive than Janice Ann. Thanks for the help and support. But most importantly, thanks to 'he who is able to keep you from falling, and to present you before his glorious presence without fault and with great joy.'

TO THE STUDENT

AutoCAD 2002 for Architecture is designed for students in architectural and engineering programs. The contents are presented in an order that has been tested within the classroom. Information is based on common drafting practices from the offices of architects, engineers, and designers. To make the best use of the information within the text, students should follow these few helpful hints.

Read the Text.

I know it sounds so simple, but if you'll read a chapter prior to sitting at a computer, you'll find your drawing time will be much more productive. Read through the chapter and make note of things that you're not quite sure of. With the reading complete, now try to mentally work through the commands that you've been reading about.

Study the Examples.

One of the many features of this text are the many illustrations and examples of how to complete each command sequence. Carefully study and compare the command sequence with the illustration of how to complete the command.

Practice at a Workstation.

Once you've read the chapter and studied the illustrations, complete the command sequences at a computer workstation. Practice each command several times and don't be afraid to explore all of the options within a command. Because each chapter is based on the previous chapters, don't hesitate to reread past material. Above all else, don't be afraid to experiment. It's very common for new computer users to be filled with a fear of breaking something. If you save your drawings often, there is very little that can go wrong by experimenting. Relax and have fun!

SECTION 1

DRAWING CONSTRUCTION

EXPLORING AUTOCAD TOOLS AND DISPLAYS

CREATING DRAWING AIDS

DRAWING AND CONTROLLING LINES

WORKING WITH DRAWING FILES

DRAWING ORGANIZATION

DRAWING GEOMETRIC SHAPES

CONTROLLING DRAWING ACCURACY

DRAWING DISPLAY OPTIONS

CHAPTER 1

Exploring AutoCAD Tools and Displays

This chapter will introduce methods for controlling the work environment, starting AutoCAD, and the tools and displays of AutoCAD. These include

- Drawing area
- Title bar
- Menu bar
- Command line
- Status bar
- Scroll bars
- User coordinate icons
- Cursors

Command selection methods are introduced:

- Toolbars
- Menus
- Keyboard entry
- Dialog boxes
- Shortcut menus
- Function Keys
- Control Keys

Commands to be introduced include

- **Line**
- **Arc**

- **Cursorsize**
- **Erase**
- **Filedia**
- **Assist**
- **Help**
- **Save**

You're about to start exploring the most popular computer-aided drafting program on the market. If you're like many new AutoCAD users, it will only take you a few minutes of exploring to realize that AutoCAD offers lots of options, and the task of learning the program might seem a bit overwhelming. Relax! It's really not that bad. CAD drafting, AutoCAD in particular, tends to become addictive. AutoCAD may seem frustrating for the first few weeks, but I think you'll love it.

An understanding of the tools you'll be using will help to eliminate the natural frustration that comes with learning. This chapter introduces key components of the hardware and the software that you'll be using to create and control your drawings.

CONTROLLING THE WORK ENVIRONMENT

Although becoming a proficient AutoCAD user will provide many exciting opportunities, spending long hours at a workstation also has its pitfalls. You might start to notice problems with your eyes, back, and wrist. Fortunately, most physical problems arising from computer use can be minimized or eliminated by proper workstation configuration, good work habits, and good posture.

EQUIPMENT

You might have little control over your workstation at school, but if you're setting up your own workstation, it should meet the following requirements:

- Your chair should be adjusted to a height that allows your feet to be supported on an inclined footrest or to rest on the floor. The chair should provide a tilt adjustment so that your back is firmly supported as you work. It should also have an armrest to provide support for your wrists. The height of the chair should allow your wrist to remain level as you type.
- The monitor should be placed so that the top of the monitor is level with your line of eyesight and between 18 and 30 inches from your eyes. Lighting should be placed so that it will not produce a glare on the display screen.
- Use a desk with a keyboard drawer that allows your wrist to remain level.
- Use an ergonomic keyboard, a foam wrist support, or forearm supports.

WORK HABITS

You can have a state-of-the-art workstation, but if you don't maintain good work habits, your body will wear out from fatigue. To keep your productivity high and maintain your sanity, you'll need to control your work habits and posture.

- Avoid looking at a monitor for long periods of time. To ease eyestrain, several times an hour look away from the monitor and force your eyes to focus on an object several feet away from your workstation.
- Avoid resting your hand on the mouse when you're not actually using it. When you are using the mouse, avoid resting your hand on the desk. Get in the habit of keeping your wrist level, or adjust the armrest of your chair to provide support.
- Take time to stretch your legs, neck, shoulders, back, and wrist several times each hour. In addition to reducing tension throughout your body, a stretch provides an excellent opportunity to stop focusing your eyes on the monitor.
- Most importantly, take time to walk away from your workstation at least once an hour. It's easy to get caught up with a deadline and convince yourself that you just don't have time for a break, but short breaks will help keep you productive and help to eliminate drawing errors.

AN INTRODUCTION TO CONTROLLING AUTOCAD

A workstation consists of neat stuff, but these gadgets can do nothing without software. Software consists of programs such as AutoCAD that are loaded into the computer to tell it what to do. The platform coordinates the computer and the software so that each can function. The operating systems that coordinate the software programs in your computer include Microsoft Windows 2000 Professional, Windows Millennium Edition, or Windows 98. If you've recently purchased a personal computer (PC), it will come loaded with Windows ME or 98 as the controlling platform. When a computer is booted using Windows ME or Windows 98, a screen similar to Figure 1.1 will be displayed.

MAKING YOUR SELECTIONS

No matter which Windows platform you use to run AutoCAD, they each have common features that must be understood to control AutoCAD. Before AutoCAD can be efficiently used, it's important that you understand the basics of Windows—including buttons, clicking, double-clicking, and dragging—before you tackle AutoCAD.

Icon Buttons

Pictures called icons are displayed to identify what program or command in a program is to be activated. Windows and AutoCAD use an abundance of icons as buttons that

Figure 1.1 *Although the display will vary greatly depending on what programs are loaded on the machine, the program icons are displayed as Windows is entered. Icon buttons are used throughout Windows and AutoCAD to represent specific programs or components within a program.*

cause actions when they are selected with the left mouse button. The buttons allow greater productivity and are used in place of written commands. Buttons are also used within AutoCAD to identify specific portions of the programs to be executed. Buttons that represent programs on your workstation desktop are shown in Figure 1.1. Figure 1.2 shows the buttons on the Standard toolbar of AutoCAD. Later portions of this chapter will discuss the use of each of the buttons.

Figure 1.2 *Buttons on the AutoCAD* **Standard** *toolbar can be used to perform specific functions.*

Clicking and Double-Clicking

A mouse is the pointing device used to enter commands with Windows applications. On a two-button mouse, the left button is primarily used for the select button. The right button is used to terminate a command or to select a shortcut menu to alter a command. Later chapters will introduce additional functions of each button. The mouse can be used to select commands from menus that are displayed beside, above, or inside the drawing area, or to select starting and ending points while in a command. Most often, you will be asked in this text to select a command from a menu. This is the process of going to a menu, placing the cursor on the desired option, and clicking the left button once as you would a telephone button. A single click of a mouse button is referred to as a *click* or a *pick*.

Menu items are typically selected with a single click. If a menu is displayed with six options, selecting one of the options with a single click will highlight that menu option. Figure 1.3 shows the **Draw** menu with the **Line** command highlighted. Clicking once again will start the command sequence.

Clicking a mouse button twice in rapid succession is referred to as double-clicking. If a double click is entered too slowly, it will be interpreted as two single clicks. A double-click is used to select a program, folder, or file for an immediate use.

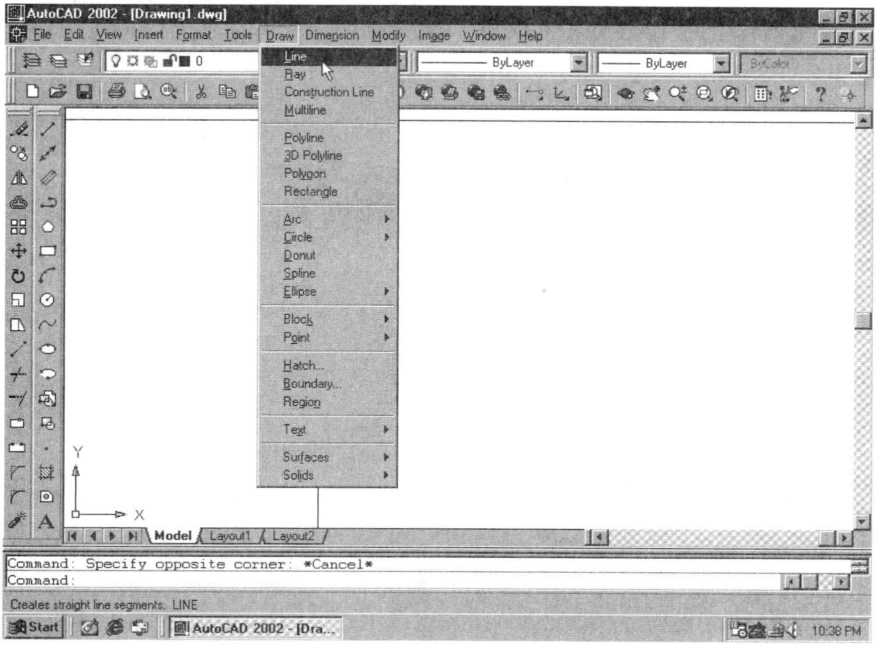

Figure 1.3 *Selecting* **Draw** *on the menu bar displays the* **Draw** *menu. As the cursor is moved to* **Line** *command, the command is highlighted for selection.*

Double-clicking the AutoCAD icon button of the original Windows display shown in Figure 1.1 will remove that display and start the program.

If you find yourself unsure about when to click or double-click, start with a single click. If the desired results are not achieved with a single click, use a double click.

Most commands and options are executed by clicking the left mouse button to make selections in Windows and AutoCAD. The right mouse button can produce several responses depending on what you're doing. Clicking a blank portion of the screen will typically reveal a hidden menu. The menu will be altered depending if you are in the middle of a command sequence, or if the right button is pressed before entering a command. Variations of the right button, or shortcut menus, will be discussed in later chapters as each command is discussed. The selections will be introduced with the command they affect. Clicking with the left button will remove the menu.

STARTING AUTOCAD

The easiest method of starting AutoCAD is to double-click the **AutoCAD** button from the desktop. (See Figure 1.1.) You can also start AutoCAD by double-clicking the program button in the **Programs** menu of Windows Explorer, or the **Run** menu will open AutoCAD. Each of these methods can be accessed by selecting the **Start** button. Double-clicking the **AutoCAD** button will produce the AutoCAD 2002 banner, followed by the AutoCAD drawing screen shown in Figure 1.4. Three areas are displayed on the screen as AutoCAD is opened: the Active Assistant, the TODAY window, and the drawing area.

The Active Assistant provides automatic or on-demand context-sensitive help. Depending on the settings of your computer, it might not be displayed as AutoCAD is entered. It will be further explored later in this chapter.

The TODAY window can be used to access drawings or the Autodesk web site. Key portions of this window include the following:

- Access to Autodesk Point A provides personalized content that is designed to help you accomplish your daily work with AutoCAD. Point A allows you to access the Internet while working in AutoCAD, giving you the ability to access industry news, links, and resources, utilize tips and techniques, new feeds, live updates, and content posted by your CAD manager.
- The Bulletin Board allows your CAD manager to post messages to the drawing team via the company's Intranet. Project information, new software patches, or drawings components to complete your project can be posted on the bulletin board.
- The tools for accessing drawings will be discussed in the next portion of this chapter. The benefits and the tools for connecting to Autodesk Point A will be discussed in Chapter 26.

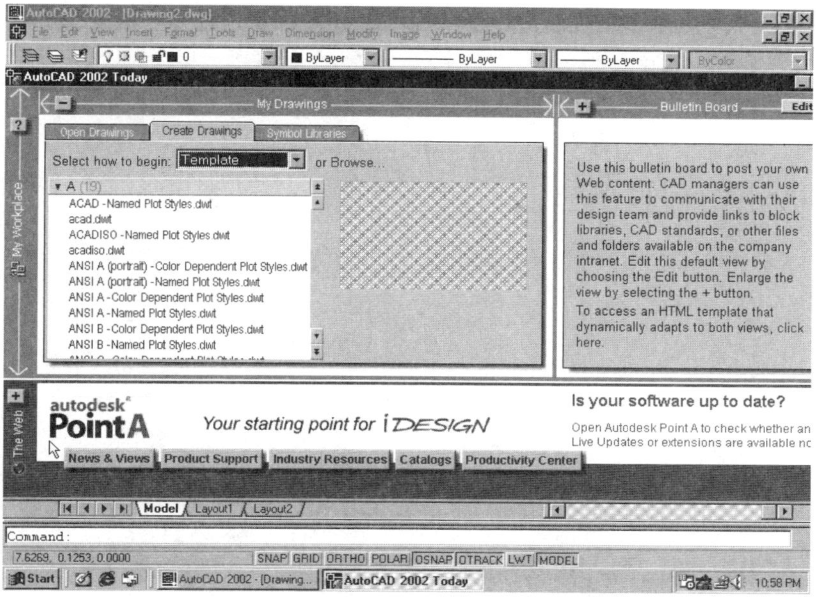

Figure 1.4 *Double-clicking the AutoCAD button will produce the default AutoCAD Startup window. In the default setting, the drawing window and the TODAY window are displayed.*

The drawing area is used to create and display drawings. Each portion of the display will be the subject of the balance of this chapter.

The exact settings displayed as AutoCAD is opened will vary depending on how the program in your workstation has been configured. Settings can be altered using the **General Options** portion of the **Systems** tab of the **Options** dialog box—this process will be explored later in this chapter. Settings for how AutoCAD will appear when opened include the following:

Show TODAY startup window—This option can be seen in Figure 1.4.

Show the traditional startup window—This option can be seen in Figure 1.5.

Do not show a startup dialog—Figure 1.6 shows this display option.

Most users will prefer one of the first two options. If you're using the TODAY startup window, click the **Create Drawings** tab. Notice that **Template** is now displayed in the **Select how to begin** list box. Selecting the arrow on the right side of the list box provides the options of **Template, Start from Scratch**, and **Wizards**. These same options, as well as **Open an Existing Drawing**, are available if the traditional **Startup** dialog box is used to access AutoCAD.

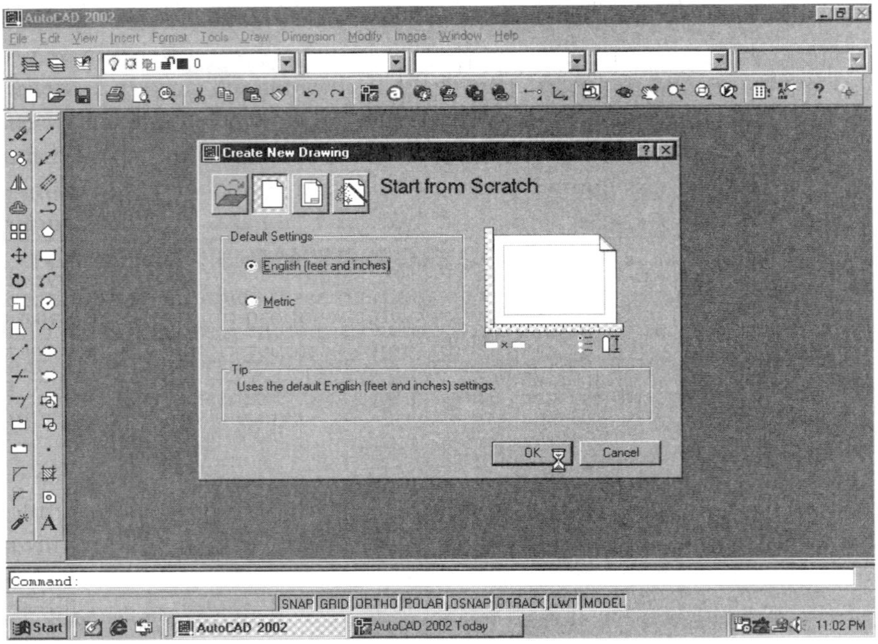

Figure 1.5 *The result of selecting the traditional startup window.*

EXPLORING THE STARTUP OPTIONS OF AUTOCAD

The drawing access options of AutoCAD include the following:

Open a Drawing—Opens an existing drawing. This option provides a list of the four most recently opened AutoCAD drawings.

Start from Scratch—Starts a new drawing with the expectation that you will provide all required measurement and drawing parameters.

Use a Template—Opens an existing drawing template for the creation of a new drawing. This option allows you to begin a drawing session using an existing template, which contains preset measurement and drawing parameters.

Use a Wizard—Creates a new drawing with drawing parameters set by the use of either the **Quick Setup** wizard or the **Advanced Setup** wizard.

Chapter 2 will introduce each of these tools for controlling the parameters of a drawing session. For now, click the **Close** button (the X in the upper right corner) to remove the TODAY window. The drawing area shown in Figure 1.6 is now displayed. If your machine displays the traditional startup display, press ESC, or click the **OK** button or the **Close** button to remove the **Startup** dialog box. This will allow access to

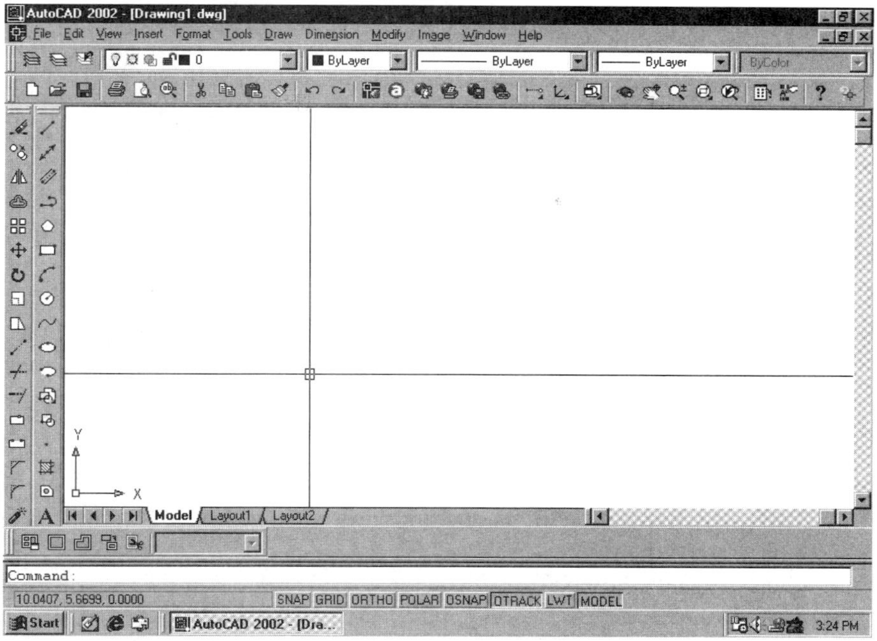

Figure 1.6 *The result of selecting* **Do not show a startup dialog**.

the default version of AutoCAD, displaying a drawing area of approximately 12" x 9" with measuring units set to four-place decimal inches. AutoCAD will provide either a light gray or black drawing screen, depending on how your machine has been configured. The exact size of the drawing area will also vary, depending on how the program is configured. Later chapters will explore how these settings can be changed to match your preferences.

The drawing area contains several important tools for controlling AutoCAD. To use AutoCAD effectively, you need to understand each of the methods of interacting with the software. This includes knowledge of the drawing area displays, toolbars, types of menus available, common components of a dialog box, function keys, and control keys. Don't let all of the different menus, boxes, and buttons confuse you. Each will become quite easy to use if you are willing to explore the program.

Tip: Push a few keys, explore some options. As long as you keep liquid away from the keyboard, there is very little you can do to hurt the hardware, software, or yourself.

EXPLORING THE DRAWING AREA TOOLS AND DISPLAYS

AutoCAD offers the user many tools and displays to enter information into the program, and to determine the status of the current drawing file. These tools include the drawing area, title bar, menu bar, command line, status bar, scroll bars, user coordinate icon, and cursors.

DRAWING AREA

Central to each of the AutoCAD tools is the drawing area, where the material that you enter is displayed. The size of the drawing area will vary depending on how many toolbars or dialog boxes are displayed. In the default setting of AutoCAD, it is the white area, surrounded by gray areas. Methods of altering the colors of each portion of the display will be introduced in later chapters.

Multiple Windows

One of the features that made Microsoft Windows so popular is that many programs can be opened at the same time. One of the features of AutoCAD is the ability to open several drawing projects in one AutoCAD session. Figure 1.7 shows an example of a drawing session with two drawings displayed. AutoCAD refers to the side-by-side arrangement as vertical tiles. As with other windows, the arrangement and size of the drawing window can be altered. The size of a window is changed by using the **Minimize/Maximize** buttons or by activating the window, and then dragging the edge to the desired size. Selecting **Window** from the menu bar will provide options for altering the arrangement of each window. Figure 1.8 shows the drawings arranged in a horizontal display. The arrangement of windows can be altered at any point during the drawing session. Later chapters explore the process for opening and working with multiple files.

Model and Paper Space Layouts

At the bottom of the drawing area are the Model and Layout tabs. Drawings are created in model space at full scale. If a wall is to be 6" wide, the lines that represent the walls are drawn 6" apart. If you're drawing a building that is 300' long, the drawing area can be adjusted so that the entire structure can be seen. Figure 1.9 shows a floor plan assembly displayed in the drawing area. Paper space is used to plot drawings that have been reduced to a scale such as 1/4"=1'–0". With 6" wide walls, the lines will be 1/8" apart. Paper space allows a 300' long building to fit on a 24" x 36" sheet of paper. Figure 1.10 shows the plan from Figure 1.10 in paper space. This is how the drawing will appear when plotted.

AutoCAD allows model and multiple paper space layouts to be displayed in a single drawing. When multiple sheets are combined in a single drawing, the number of individual drawings needed to produce and easily generate different views of a project can

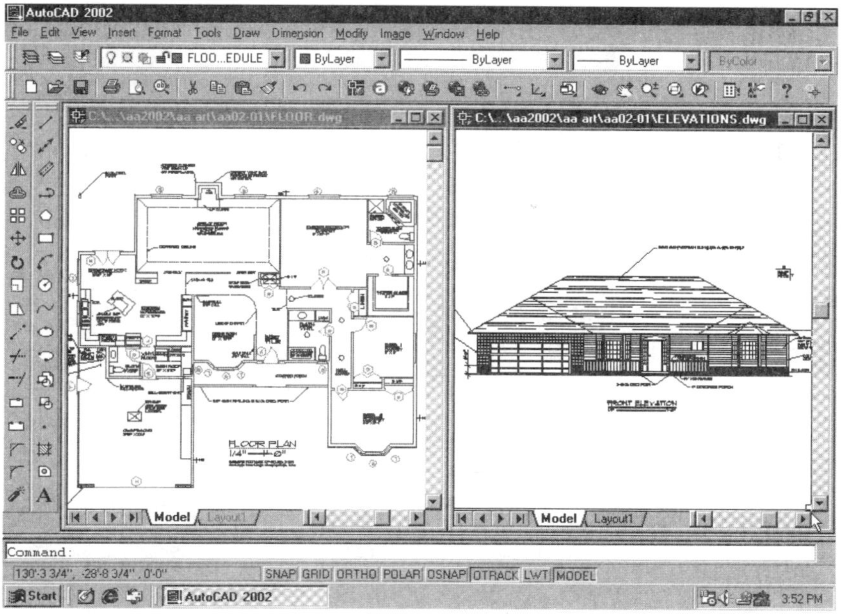

Figure 1.7 *AutoCAD allows multiple drawings to be opened. The name of each drawing is listed in the title bar of each window. The active drawing (Elevations) is displayed in a blue title bar, and the inactive drawing (Floor) has a gray title bar.*

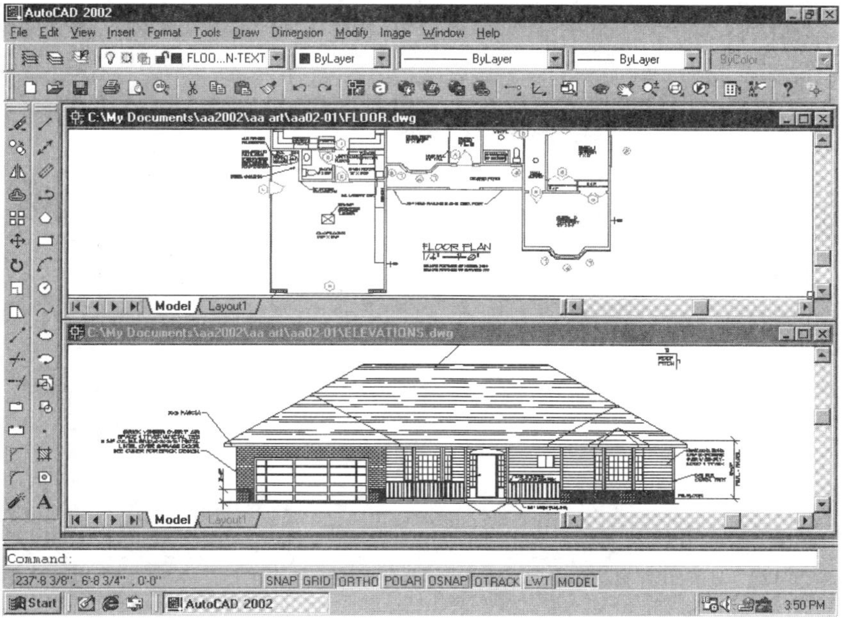

Figure 1.8 *AutoCAD drawings can be arranged in horizontal window arrangements.*

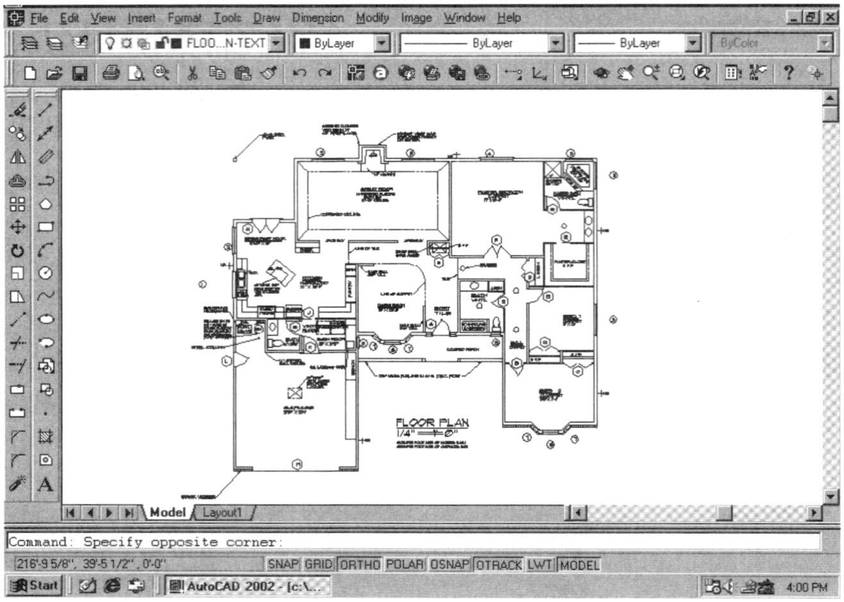

Figure 1.9 *Drawings are created using model space at full scale. Enlarging the drawing area will show the entire floor plan.*

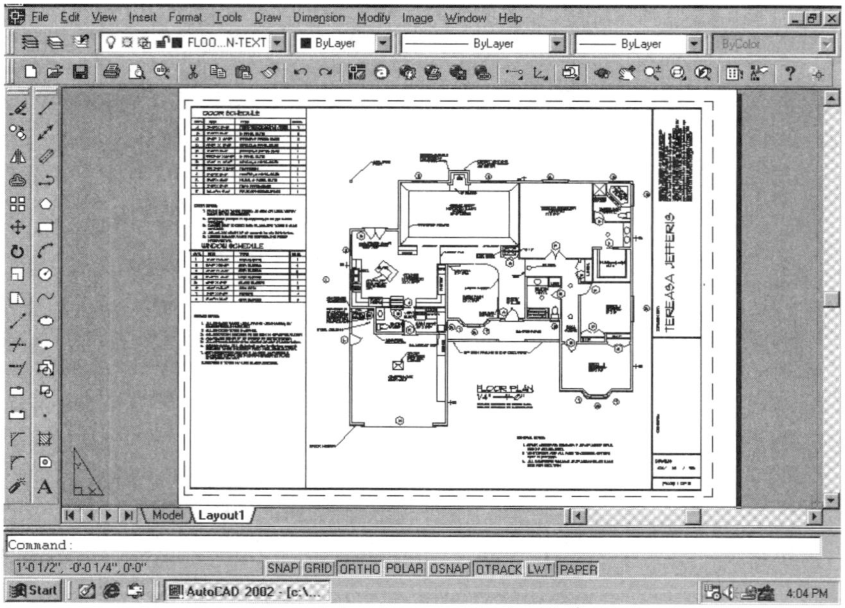

Figure 1.10 *The floor plan from Figure 1.9 is shown in paper space. Paper space scales a drawing to fit within the limits of a sheet of paper.*

be consolidated. Multiple layouts can be used to aid in the layout and the plotting of an object. Uses for model and paper space layouts will be further discussed through the balance of this text.

TITLE BAR

A title bar is located across the top of every window. It displays the icon and name of the current program as well as the name of the current file. The title bar serves as a handle to drag the window to a different location on the screen. Notice that in Figure 1.7, two drawings are opened and displayed in two windows within the AutoCAD program. The AutoCAD title bar is at the top of the drawing screen, and each drawing window has its own title bar. If multiple drawings are open, the title bar of the active window is highlighted with a blue bar, while inactive windows are displayed in another color. The name of each drawing session is listed in the title bar. Notice in Figure 1.7 that the active drawing is named Elevations and the inactive drawing is named Floor. At the far right end of each title bar are three buttons that can be used to control the size of the current window. These buttons are the **Minimize**, **Maximize**, **Restore**, and **Close** file buttons. The **Minimize** and **Close** buttons will always be displayed, but only the **Maximize** or the **Restore** button is displayed, depending on the current window setting. You'll notice that sometimes there are two rows of buttons. Although they function in a similar manner, the buttons at the end of each title bar control the display of the current program window.

The Minimize Button

Selecting the **Minimize** button (at A in Figure 1.11) with the select (left) button removes the current program or drawing from the display screen and places the program icon in the taskbar at the bottom of the drawing area. To restore the window, use a single click of the select button to select the program icon in the taskbar. (The task bar will be discussed later in this chapter.)

TIP: It's important to remember that if you select the **Minimize** button and remove the file from the screen, the file still exists—it's just been placed out of your view. It hasn't been saved or destroyed, just moved for convenience so you can do something else.

The Maximize Button

Selecting a window corner or border of a window with the cursor allows the size of the window display to be altered. When the size of a window has been reduced, selecting the **Maximize** button (at B in Figure 1.11) restores the current window to its largest size. Other windows that might be open are still open; they've just been covered by the current window. Notice that as soon as the window is maximized, the **Maximize** button is replaced by the **Restore** button.

The Restore Button

The **Restore** button, shown at C in Figure 1.11, is displayed only when a window is maximized. Selecting the **Restore** button allows the window to be reduced to its previous size.

The Close Button

Selecting the **Close** button (at D in Figure 1.11) closes the current program. When the **Close** button is selected while in a program such as AutoCAD, a box will be displayed with options to save or not to save the current file, or to cancel the **Close** command.

MENU BAR

The menu bar is located below the title bar and is used to display menus specific to the open program. The menu bar for AutoCAD can be seen in Figure 1.10. Each word of the menu bar represents a separate menu with options related to the key word. Moving the cursor to one of these words and clicking will display the corresponding pull-down menu. Figure 1.12 shows the **Draw** menu for AutoCAD. Some commands such as **Arc** have a solid black triangle on the right side. The triangle indicates that moving the mouse to this command will produce another menu of options related to that command. Moving the mouse to the **Arc** listing will produce the menu shown in Figure 1.13. Select the **Format** menu with a left click of the mouse. Notice that the **Plot Style** listing appears different from other listings. This option is inactive and is not currently available for selection. Inactive options become active only as other options are activated.

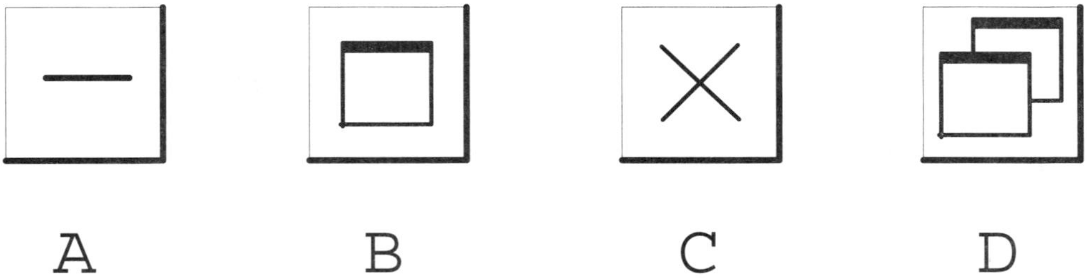

Figure 1.11 *Four buttons for controlling windows are located on the right end of the title bar. Selecting the* **Minimize** *button (A) will remove a window or drawing from the screen and place the title in the task bar. Selecting the* **Maximize** *button (B) will increase the window to its largest possible size. The* **Restore** *button (C) is displayed only when a window is maximized. Selecting the* **Restore** *button will return the window to its previous size. Selecting the* **Close** *button (D) will close a window and remove the program icon from the task bar.*

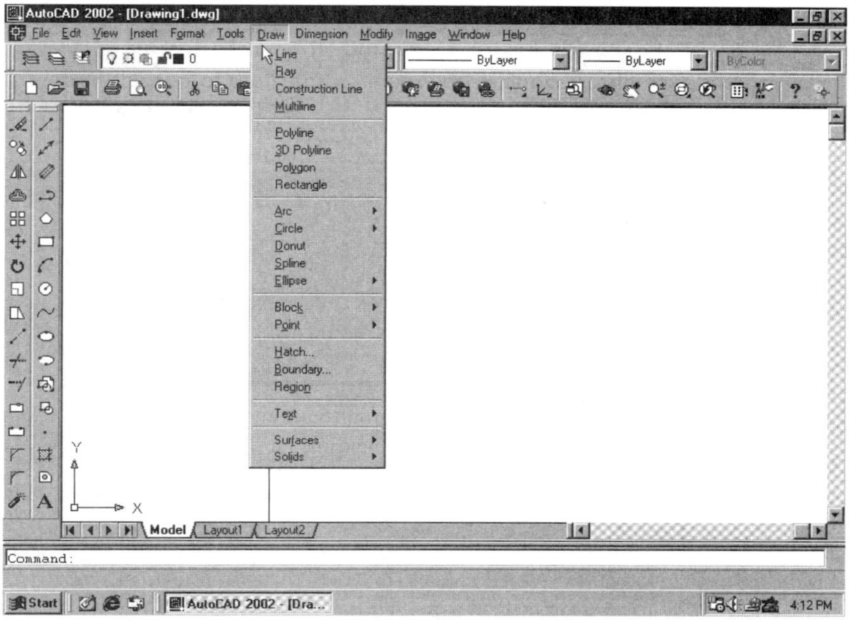

Figure 1.12 *The* **Draw** *menu of AutoCAD.*

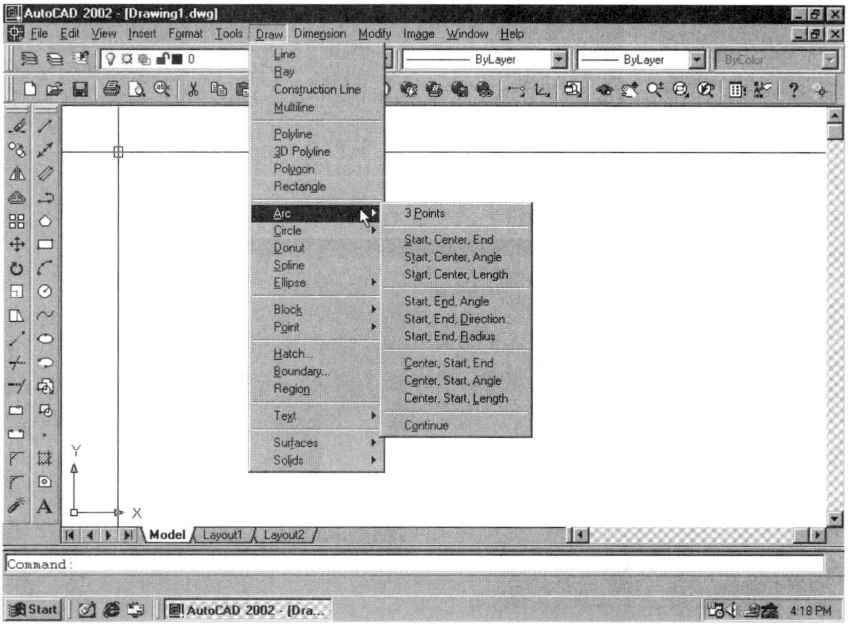

Figure 1.13 *Listings in a menu followed by a triangle will produce a cascading menu. The* **Arc** *cascading menu is displayed when Arc is highlighted.*

COMMAND LINE

The command line displays the prompt **Command:** and allows you to enter your request into AutoCAD using the keyboard. Command entry methods will be discussed later in this chapter. As you explore AutoCAD, it's important to remember that the command line displays prompts for completing commands. Even if you select commands by using a toolbar or menu, AutoCAD displays prompts and a command history in the command window. The command line also shows the result of selecting a command by clicking a button. If the **Line** command is started by clicking the **Line** button,

_line Specify first point

is displayed at the command line.

TIP: Users new to AutoCAD often fail to read the Command prompt and miss AutoCAD's attempt to communicate with them.

STATUS BAR

The status bar is located below the command line. Key elements of the status bar can be seen in Figure 1.14 and include the coordinate display and toggle buttons that can be used to control a drawing aid. The drawing aids in the status bar affect the placement or display of your drawings. Options are in the ON position when the button appears to be pressed in. Buttons include the following:

SNAP—Toggles an invisible grid ON/OFF for placing lines at precise intervals. Snap increments can be set using the **Snap and Grid** tab of the **Drafting Settings** dialog box. Display the dialog box by selecting **Drafting Settings** from the **Tools** menu.

GRID—Toggles a visible grid ON/OFF for placing modular objects. Grid increments can be set on the **Snap and Grid** tab of the **Drafting Settings** dialog box.

ORTHO—Toggles the ability to draw lines that are vertical and horizontal (ON) or at any angle (OFF).

Figure 1.14 *The status bar contains the coordinate display and eight buttons for controlling drawing aids.*

POLAR—Toggles the polar tracking feature that will describe the position of the cursor in distance and angle relative to another point. By default, the tracking angle is set to a 90° increment angle. The angle can be altered using the **Polar Tracking** tab of the **Drafting Settings** dialog box.

OSNAP—Toggles on the ability to select drawing objects such as endpoints, intersections, or eleven other options. AutoCAD automatically snaps to the endpoints of existing lines as the cursor is moved toward those lines. Other modes can be set by selecting the desired button on the **Object Snap** toolbar, or by selecting the desired mode from the **Drafting Settings** dialog box. Modes can also be selected by holding down SHIFT and the right mouse button at the same time.

OTRACK—Toggles the tracking feature that will allow specific angles to be tracked along a projected path based on an object endpoint to increase drafting precision.

LWT—Toggles the display of lineweight. When ON, assigned lineweights are displayed. When OFF, all lines are shown using the same lineweight.

MODEL/PAPER—Toggles between model and paper space. In model space, click MODEL to change to paper space.

Each button will be explained in detail in later chapters.

SCROLL BARS

The scroll bars consist of a vertical bar located on the right side of the drawing area and a horizontal bar located below the drawing area. The vertical bar consists of an up arrow at the top of the bar, a down arrow at the bottom of the bar, and a slide box that changes its position to reflect the current location's position. Scroll bars are used to alter the material viewed in the drawing area. They are also used in dialog boxes that are displayed throughout AutoCAD to alter program settings. For the drawing area, the scroll bars are one of many ways to alter the display in the drawing area. There are better, faster ways than scroll bars to alter the displayed view. In a dialog box, scroll bars offer a quick method of moving through a long list of options. Dialog boxes will be discussed later in this chapter.

USER COORDINATE SYSTEM ICON

The User Coordinate System icon describes the current coordinate system and orientation of the drawing area. AutoCAD drawings are imposed over an invisible coordinate system based on X, Y, and Z coordinates. AutoCAD uses both a fixed world coordinate system (WCS) and a movable user coordinate system (UCS). Each method will be explained in later chapters. In its standard setting, as you draw in AutoCAD, you're working in model space. When a drawing is to be plotted, it is plotted using paper space. Model and paper space will be discussed in detail in later chapters. Figure 1.15 shows the model and paper space UCS icons.

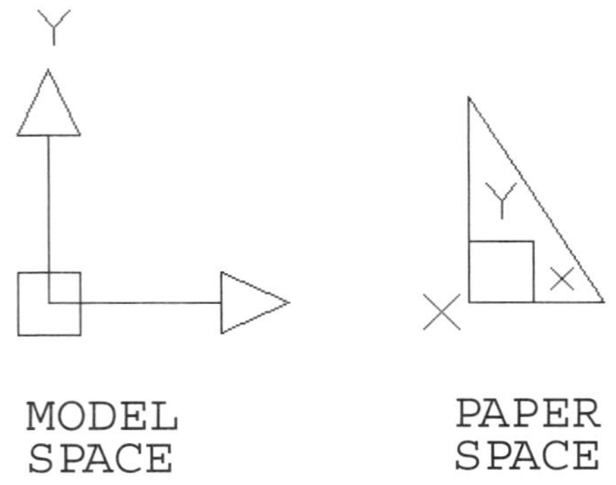

Figure 1.15 *The user coordinate system (UCS) icons for model and paper space.*

TIP: As you explore AutoCAD, the important thing to remember is that the icon is a reminder of what mode you are currently using to create your drawing. Unless your instructor tells you otherwise, work in model space as you create a drawing, and work in paper space to plot a drawing.

CURSORS

The term *cursor* refers to the pointing device that is placed on the screen to indicate the current mouse position. As seen in Figure 1.13, the cursor for the AutoCAD drawing area is crosshairs. As the mouse is moved, the location of the crosshairs is moved. The cursor provides a method of entering the location for drawing objects. When commands such as **Line** are started, the cursor is the primary method of selecting the beginning and ending points for a line.

Altering the Crosshairs Size

Depending on the complexity of the drawing being completed, some operators like to adjust the size of the cursor. Crosshairs that extend across the screen will aid in projecting objects from one view to another. Projecting heights from one elevation to another is much easier with enlarged crosshairs. The size of the cursor can be altered using either of the following methods:

Using a dialog box:

1. Select **Options** from the **Tools** menu. This will display the **Options** dialog box.
2. Select the **Display** tab. Use the slider bar in the **Crosshairs size** edit box to alter the cursor size.

3. When it is adjusted, click the **OK** button to return to the drawing area.

From the command line:

- Type **CURSORSIZE** (ENTER) at the Command prompt. This will produce the prompt:

  Command: **CURSORSIZE** (ENTER)

  Enter new value for CURSORSIZE <5>:

No matter how the command is entered, the default value is displayed as 5. The 5 represents the percentage of the screen the cursor will occupy. Select and hold the slider bar with the left mouse button and move it to the right to increase the extent of the cursor. If you're working at the command line, type the desired value and press (ENTER). Changing the value to 10 will double the size of the crosshairs. Entering a value of 2.5 will decrease the size of the cursor to half the original size. Entering a value of 100 will cause the lines forming the crosshairs to cross the entire screen.

Alternate Cursors

Just as with other Windows programs, the cursor changes shapes depending on where it is located in AutoCAD. When the cursor moves outside the drawing area, the crosshairs turn into an arrow. It can also appear as an hourglass, pointing finger, flashing vertical line, double arrow, or four-sided arrow. Each cursor can be seen in Figure 1.16.

- The cursor changes to an hourglass (at B in Figure 1.16) each time AutoCAD or Windows is processing information. While the hourglass is present, the current program is unable to process additional information. Once the command is executed, the hourglass reverts to the previous cursor, and new commands can be processed.

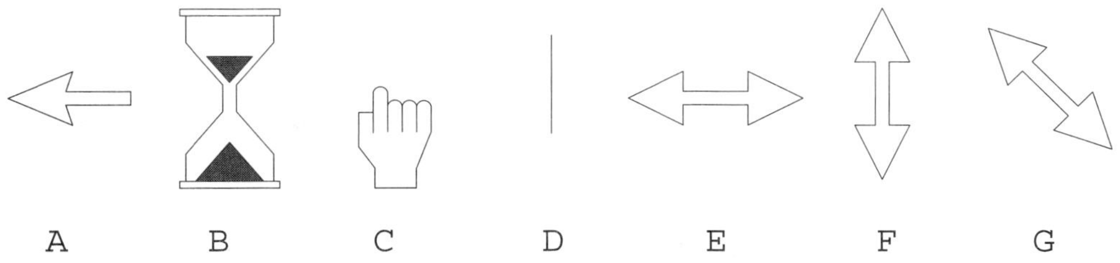

Figure 1.16 *The cursors of a Windows platform include (A) the arrow—used to make selections, (B) the hourglass—indicating a command is being processed, (C) the pointing finger—for making a selection, (D) the flashing bar—for entering text, (E) the horizontal double arrow—for altering a window size in the left/right direction, (F) the vertical double arrow—for altering a window in the top/ bottom direction, and (G) the diagonal double arrow—for altering a window in two directions in one movement.*

- The pointing finger (at C in Figure 1.16) is the politically correct symbol used to indicate that there is additional information available for a specified command or option. Clicking the select button will display additional information related to the designated topic.
- The arrow cursor changes to a flashing bar cursor (at D in Figure 1.16) when the cursor is moved into a portion of the screen where text input is required. Any text that is entered at the keyboard will be entered just to the left of the cursor bar.
- As the cursor is placed on the border of a window, it will change to a double arrow (at E, F, and G in Figure 1.16). The direction the arrow points will depend on the placement of the cursor. If the select button is pressed and held, the size of the window can be enlarged or decreased depending on the movement of the mouse. Using the arrow to move a side border will alter the horizontal size. Selecting the bottom or top border allows the vertical size to be altered. If the pointer is placed in the corner of a window and the button is held, the size of the window can be altered in both directions in one movement. The size of a window can be altered throughout a drawing session.

COMMAND SELECTION METHODS

AutoCAD contains more than 250 commands related to drawing or editing a drawing. Five options are available for entering commands, including selecting buttons from toolbars, menus, keyboard entry, dialog boxes, and shortcut menus. Consideration will also be given to using function and control keys to alter drawing settings or alter commands. Each method has its advantages. Throughout the balance of the text, commands will be presented in the order that tends to be the fastest method of entry. For most commands, the fastest entry method is by selecting a button from a toolbar or by keyboard entry. Remember that the key word is fastest, not best. As you're learning each command, take the time to explore each option. You might decide the keyboard is the fastest method of command entry.

As you begin to work with AutoCAD menus, you will see common terms used repeatedly. You will need to understand these so you can progress through AutoCAD. Common terms include the following:

Click—To use a pointing device to make a selection from a toolbar, menu, or in a dialog box.

Command—An instruction to be carried out by the computer, such as **Line** or **Arc**.

Default—A value that will remain constant until a new value is entered. Default values are shown on the command line in angle brackets: < 3 >.

Option—A portion of a command that requires a selection.

Pick—To use a pointing device to make a selection in the drawing.

Enter—ENTER is used to denote pressing the appropriate key (key names vary, depending on the keyboard manufacturer) after typing a request at the command line.

Escape—ESC is used to exit a command.

Select—To make a choice from a menu of commands or options.

Specify—To enter a value, such as a coordinate on the command line.

TOOLBARS

A toolbar contains buttons that represent commands. In its default display, the **Standard** and **Object Properties** toolbars are displayed above the drawing area just below the menu bar, and the **Modify** and **Draw** toolbars are displayed on the left side of the drawing screen. Key features of these four toolbars are as follows:

- The **Standard** toolbar contains the names of frequently used buttons. In addition to common methods for controlling AutoCAD, the toolbar contains standard Microsoft buttons such as **Open**, **Save**, **Print**, **Cut**, **Copy**, and **Paste**.
- The **Object Properties** toolbar can be used to set the properties of objects such as color, lineweight, linetype, and the layer an object is displayed on.
- The **Draw** toolbar contains buttons to represent each of the major drawing commands, including **Line**, **Polygon**, **Rectangle**, **Arc**, **Circle**, **Ellipse**, and **Point**.
- The **Modify** toolbar contains buttons to represent commands that can be used to edit drawing objects, such as **Erase**, **Copy**, **Mirror**, **Offset**, **Array**, **Move**, and **Rotate**.

Because of their ease of use and availability, for most AutoCAD users the toolbars will become the primary method of selecting commands. AutoCAD offers 24 different toolbars that can be displayed to meet various drawing needs. Toolbars can be displayed in two ways:

- Select **Toolbars** from the **View** menu.
- Type **TOOLBAR** (ENTER) at the Command prompt.

Each method will produce the toolbar list shown in Figure 1.17. A similar toolbar list can be displayed by moving the cursor to any open toolbar and right-clicking. Selecting **Dimension** in the menu will produce the **Dimension** toolbar shown in Figure 1.18. Each of the toolbars listed in Figure 1.17 will be discussed as specific commands are introduced throughout the text. As you get started using AutoCAD, it is more important that you feel comfortable using the toolbars that are currently displayed.

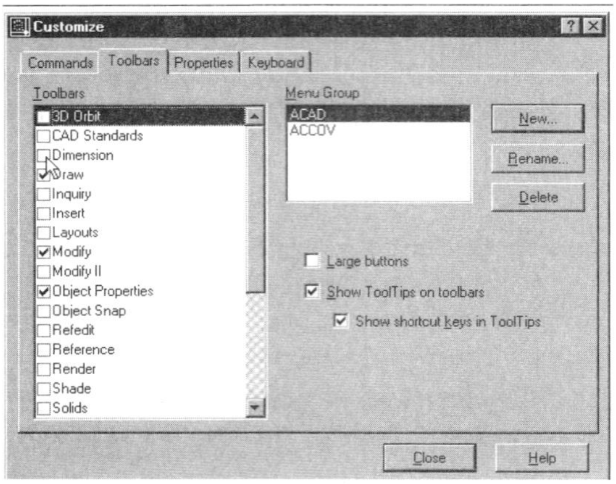

Figure 1.17 *Additional toolbars can be displayed by selecting* **Toolbars** *from the* **View** *menu.*

Figure 1.18 *The* **Dimension** *toolbar is displayed by making the selection started in Figure 1.17.*

To remove the **Customize** dialog box, click the **Close** button at the bottom of the dialog box or click the **Close** button in the upper right corner. The **Dimension** toolbar can also be removed from the screen by clicking the **Close** button in the toolbar title bar.

As the cursor is moved from the drawing screen to a toolbar, it changes from crosshairs to the arrow cursor. The cursor can now be used to select the desired button. You'll also notice as the cursor passes over, or rests on a button on the toolbar, that the button takes on a 3D appearance. Move the cursor to the button in the left corner beside the drawing window. As the cursor is placed above the button, the title of the button will be displayed below the cursor, similar to the display in Figure 1.19. The

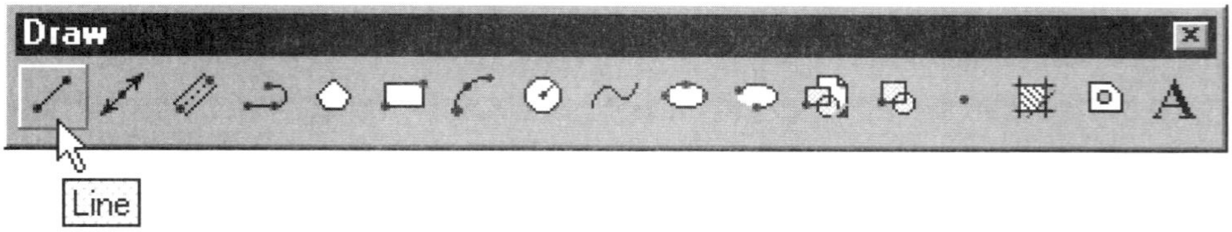

Figure 1.19 *Resting the cursor over a button provides a written description of the button's function called a tooltip. Notice that below the command line a brief description is also displayed.*

title is referred to as a tooltip. Tooltips provide a written display of the results that will occur if the button is selected. In this case, the cursor is over the **Line** button. Select the **Line** button with a single-click to start the **Line** command. This will display the following prompt at the command line:

Command:_line Specify first point:

The program is now waiting for you to select a starting point for a line. Move the cursor and select an endpoint for a line by clicking the select button of the mouse. As the endpoint is selected, the Command prompt is changed to request the next point of the line. The prompt will display the following:

Specify next point or [Undo]:

AutoCAD is waiting for you to select an endpoint for a line. Move the cursor to a new location and pick a point. The prompt will continue to request additional points after each new point is entered until you choose to stop drawing lines. As you draw one line, the command will continue to allow you to add to the existing line. To terminate the **Line** command, press SPACEBAR or ENTER, or click the right mouse button, which will produce the shortcut menu shown in Figure 1.20. Selecting the **Enter** or **Cancel** option will end the **Line** command. The other options of this menu will be explored in Chapter 3 as the **Line** command is explored further.

To start another line in a different location, press (ENTER), which automatically restarts the last command sequence. Repeating this process will allow an unlimited number of line segments to be drawn without your having to select the **Line** button for each line. Take time to enter AutoCAD and draw several different continuous line segments. Once you've drawn several lines, end the command, and then reenter the **Line** command. Now draw several individual line segments. This process can be used with each of the buttons on the **Draw** toolbar.

Although the commands will be discussed in detail in later chapters, take time to explore the **Arc**, **Circle** and **Rectangle** commands on the **Draw** toolbar. As you're exploring, you might also want to investigate the **Erase** command. Selecting the **Erase**

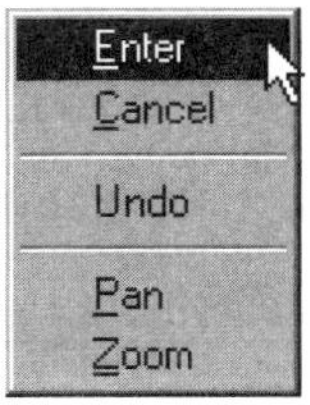

Figure 1.20 *Right-clicking while in the middle of the* **Line** *command will produce a command-mode menu. Selecting* **Enter** *or* **Cancel** *will terminate the* **Line** *command at the last point that was specified.*

button on the **Modify** toolbar will allow you to select lines and remove them from the drawing screen.

As you're exploring, notice that several of the buttons on the **Standard** toolbar have a small triangle in the corner of the button. Clicking and holding down the cursor on a button with a small black triangle in the lower right corner will produce a flyout menu, which is a submenu of the selected menu. Figure 1.21 shows the flyout menu for the **Zoom Window** button.

Altering Toolbars

Toolbars are used most often to enter commands because they are extremely flexible. In addition to being able to display multiple toolbars on the screen, you can move, dock, float, resize, and alter the contents of toolbars to allow for more convenient usage.

Moving Toolbars

A toolbar can be moved to any convenient area of the drawing screen. Placing the cursor in the title bar of the menu and then selecting and holding the title bar moves the toolbar. Pressing the select button and moving the mouse alters the location of the toolbar.

Floating Toolbars

A toolbar is referred to as a floating toolbar once it is moved from the border of the drawing display. When a toolbar is floating, it can be resized or docked, or its contents can be altered. When the **Dimension** toolbar is activated (see Figure 1.18), it is placed in the middle of the drawing screen, making it a floating toolbar. Selecting the title bar and holding the select button allows the toolbar to be moved to any desired location. Move the cursor to the top border of the **Draw** toolbar. Press and hold the select button and move the toolbar into the drawing area.

Resizing Toolbars

If the toolbar interferes with the drawing display, the size of a toolbar can be altered. Placing the cursor anywhere on the border of the toolbar and dragging it in the direction you want the bar to be resized alters the shape of a toolbar.

Figure 1.21 *Clicking and holding a button that contains a triangle will produce a flyout menu.*

Docking Toolbars
A toolbar is docked by selecting and holding down the select button on the toolbar title bar and then moving the toolbar to a position above, below, or on either side of the drawing screen. An alternative to docking a toolbar is to totally remove it. A toolbar can be removed from the screen while it is floating by clicking the **Close** button in the upper right corner. To restore the window, select the check box of the desired toolbar from **Toolbars** in the **View** menu.

Refloating Docked Toolbars
A docked toolbar can be refloated by placing the cursor anywhere on the border of the toolbar and then dragging the box to its desired position on the drawing screen.

Altering Toolbar Content
It might seem unimportant now, but one of the nice features of AutoCAD is that it allows you to modify existing toolbars or to create a new one with your favorite buttons on it. You can change the size or location of toolbar buttons. You can also add or delete buttons from a toolbar. Use the following method to modify a toolbar:

1. Select **Toolbars** from the **View** menu, and then activate the desired toolbar.
2. Change the size of the buttons by activating the **Large button** option. Keep in mind that this will reduce the size of your drawing area.
3. While the **Customize** dialog box is still displayed, select the **Commands** tab. This will display the different categories of commands featured on toolbars.
4. Select the desired category that you would like displayed in the image box.
5. Change the location of the buttons within the toolbar by selecting the desired button and dragging it to the new location in the toolbar.
6. Add a button to a toolbar by choosing the desired button from the image box and then dragging it to the desired location in the toolbar.
7. Remove a button from a toolbar by choosing the desired button and then dragging and it outside the toolbar.

   Create a new toolbar by choosing the desired button from the image box and then dragging it into the drawing area.

MENUS

As the cursor is moved into the menu bar, a key method of interacting with AutoCAD is presented. The menu bar displays the names of eleven menus used to control AutoCAD. As with other Windows menus, selecting a specific name will produce a display of that pull-down menu. Selecting a menu displays a complete listing of related commands. Move the cursor to the **Draw** menu and select **Draw** to display the menu shown in Figure 1.22. If you accidentally select a wrong menu, move to the

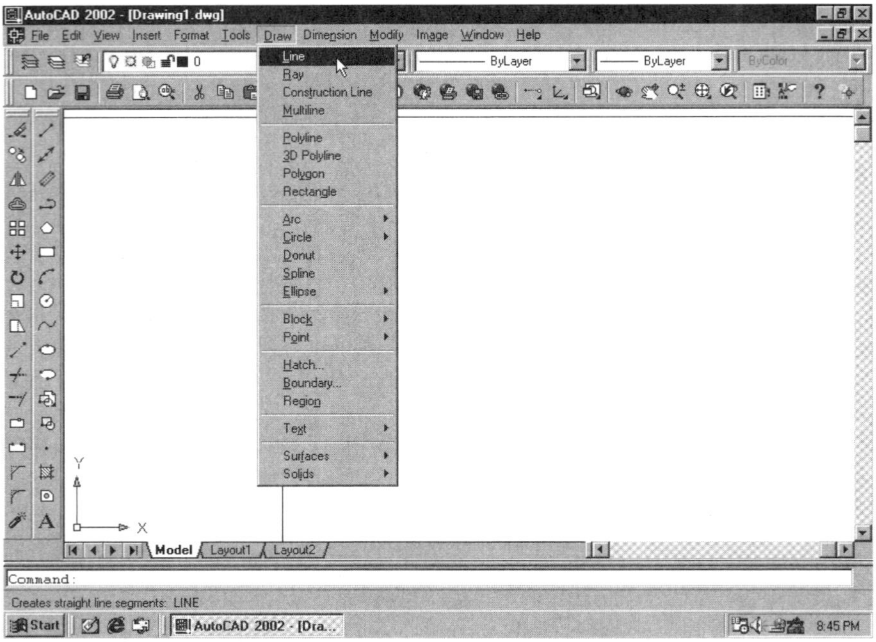

Figure 1.22 *Moving the cursor to the menu bar and selecting* **Draw** *will display the* **Draw** *menu.*

desired menu. Menus automatically open and close as the mouse is moved over the menu item.

Once in the **Draw** menu, you'll notice that three different types of commands are listed:

- Typical commands where the command is selected and executed
- Commands with a cascading menu of related commands or options
- Commands that produce a dialog box

Selecting one of the first eight options from the **Draw** menu removes the menu and returns the crosshairs to the screen so that the selected command can be executed. Select **Line** from the menu. This will remove the menu from the screen and start the **Line** command. Notice that the command line has now changed, to display

_line Specify first point:

AutoCAD is waiting for you to select a point. Draw a few lines. You'll notice that the command works the same as when you entered the command using the **Line** button on the **Draw** toolbar.

Exit the **Line** command by pressing (ENTER) and return to the **Draw** menu. Select Arc from the **Draw** menu to display the menu shown in Figure 1.23. Select the **3Points** option. This will remove the menu, return the crosshairs, and allow the selected command option to be executed.

Once the **Arc** option is selected, the command sequence to draw a three-point arc is as follows:

> Command: *(Select Arc from the Draw menu.)*
>
> _arc Specify start point of arc or [Center]: *(Select a start point for the arc with the mouse.)*
>
> Specify second point of arc or [Center/End]: *(Select a mid point for the arc.)*
>
> Specify end point of arc: *(Select the end point of the arc.)*

As the third point of the arc is selected, the arc will be displayed in the drawing area and the Command prompt awaits your next command.

Open the **Draw** menu. A third type of display is found in this menu. Notice that two of the options are followed by an ellipsis (…). Selecting an option such as **Hatch…** with a single click will produce the dialog box shown in Figure 1.24. Using dialog boxes

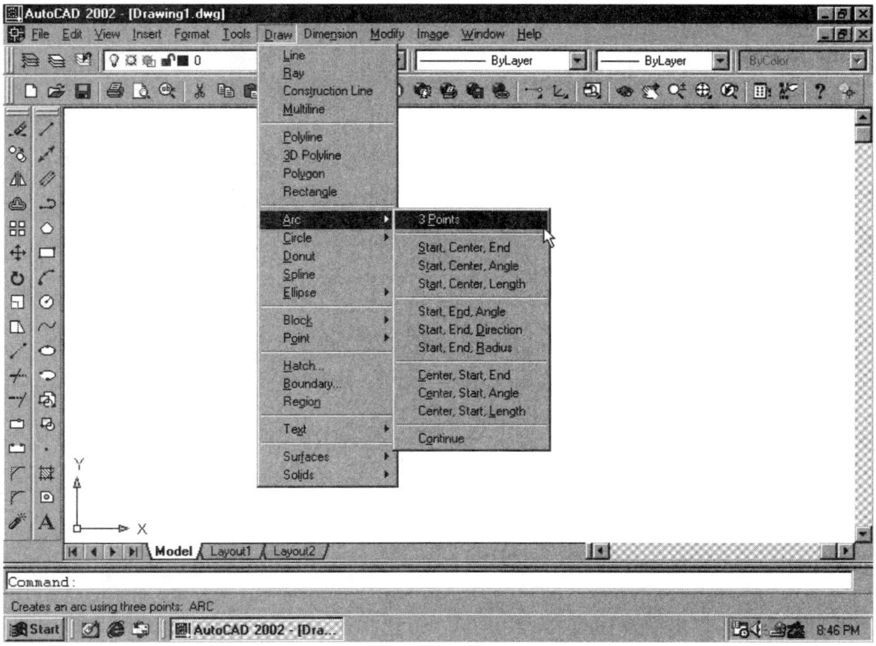

Figure 1.23 *The submenu produced when the* **Arc** *command is selected.*

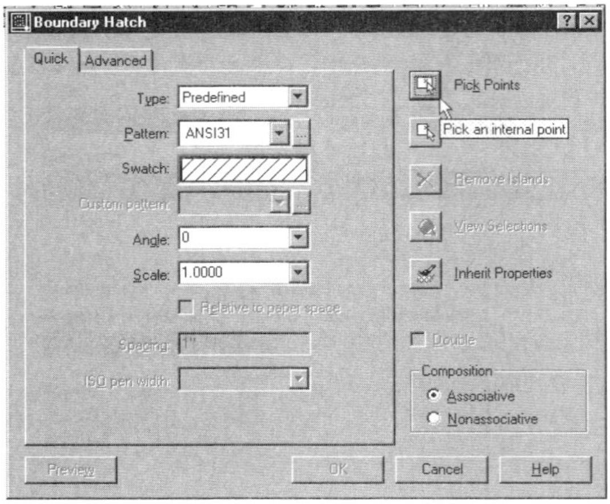

Figure 1.24 *Menu options followed by an ellipsis (…) will produce a dialog box. Dialog boxes allow the functions performed by certain commands to be adjusted.*

will be introduced later in this chapter and explained throughout the balance of the text as specific commands are introduced. To exit the **Boundary Hatch** dialog box, move the arrow to the **Cancel** button, and click the select button to return to the drawing screen.

KEYBOARD ENTRY

The command line allows you to enter your request to AutoCAD using the keyboard. AutoCAD will provide prompts on the command line to complete the command. Even with all of the icons and toolbars provided by Windows and AutoCAD, the keyboard is the fastest and easiest method of executing many commands. This is true even if you're a two-finger typist. Throughout this text, commands will be discussed by providing examples of toolbar button and keyboard entry. To enter a command by keyboard, type the name of the command and press (ENTER). To draw a line, type **Line** (ENTER) at the Command prompt. The text will automatically be entered at the Command prompt. For most commands, an alias can be used to reduce typing. For the **Line** command, an alias of **L** can be used to start the command. Once the command alias is typed, pressing (ENTER) will activate the specified command. Notice that once (ENTER) is pressed, the command line display will be altered. The command sequence is as follows:

Command: **L** (ENTER)

LINE Specify first point:

Use the cursor to select a starting point for the line just as you did when the command was started from the toolbar or menu.

Note: AutoCAD is not case sensitive. You can use all caps, all lowercase or a combination.

Expanding the Command Line

By default, the command window displays three lines of text. Many AutoCAD users prefer to have more lines of text displayed so that an entire command sequence can be seen at one time. Additional lines of command text can be viewed by selecting the top edge of the window and dragging it to a new location. Selecting **Options** from the **Tools** menu can also alter the size of the command window. This will produce the **Options** dialog box displayed in Figure 1.25. Select the **Display** tab—this will produce six groupings for controlling the screen display. (See Figure 1.26.) Notice that the box in the upper left corner is titled **Window Elements** and allows the size of the command line to be altered. (Other options for controlling the display will be discussed in later chapters.) To change the command line from three to four lines in this dialog box, use the following sequence:

1. Move the cursor to the text box for **Text lines in the command line window.**
2. Highlight the 3 by placing the cursor beside the 3, and then moving the cursor over the 3 while pressing the select button. This will place the 3 in a blue box, indicating that the 3 can be altered.

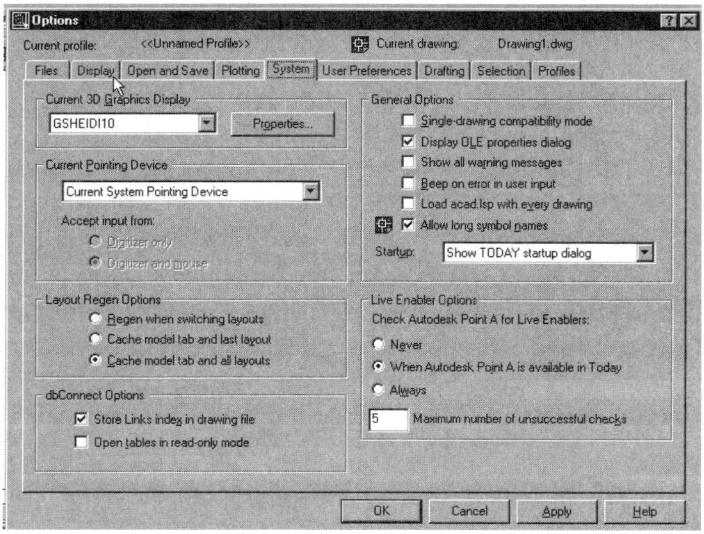

Figure 1.25 *Selecting* **Options** *from the* **Tools** *menu displays the* **Options** *dialog box.*

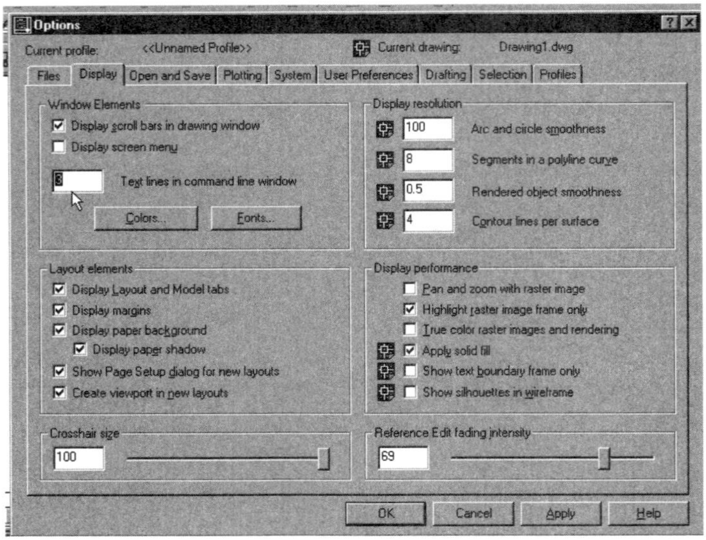

Figure 1.26 *Select the* **Display** *tab of the* **Options** *dialog box to adjust the size of the command window.*

3. Enter the desired number using the keyboard, in this case 4.
4. Select the **OK** button at the bottom of the dialog box. This will close the **Options** dialog box and return the drawing area. The command line will now display four lines of text.

As the size of the window is increased, a scroll bar is added to the window to allow viewing of previous commands. Up to 100 lines of text can be displayed, but a display of over 35 lines will totally fill the drawing area. The scroll bar on the right edge of the display can be used to show additional lines of text. In addition to displaying information about the current command sequence, AutoCAD can display a list of commands used throughout the drawing. This can be done with the F2 function key. (The function keys are located across the top of the keyboard; the purpose of each function key will be discussed later in this chapter.) Pressing F2 will produce a display of the AutoCAD Text window that contains the drawing history. The window can be used to display all the commands used to create the current drawing file. You can remove the display by pressing F2 again. Like other windows, the drawing history display can be moved to a new location or resized.

Note: You've just used **Options** from the **Tools** menu to change the size of the command window. One of the most useful tools for AutoCAD users who share a workstation is to be able to have the settings that control the work environment saved for future work sessions. The **Options** dialog box can be used to set and save user profiles for future drawing sessions. Although these options will not be introduced until Chapter 4, it is important to remember your favorite settings can be saved as you're exploring AutoCAD's options in this chapter. If you can't wait until Chapter 4, see the **Help** menu.

DIALOG BOXES

Although not a menu of commands, dialog boxes provide a convenient way to adjust certain command parameters. You encountered your first dialog box when AutoCAD opened and the TODAY or the **Startup** dialog box filled the screen. You've also opened the **Customize** and **Options** dialog boxes. A dialog box provides options for controlling commands, options, or portions of the drawing program. Figure 1.27 shows the dialog box for controlling layers. Access this dialog box by clicking the **Layers** button on the **Object Properties** toolbar, or by selecting **Layer** from the **Format** menu. Dialog boxes can also be opened from the command line by typing the name of the desired box to be displayed. Layers will be discussed in later chapters. For now, it is important to understand the features found in dialog boxes.

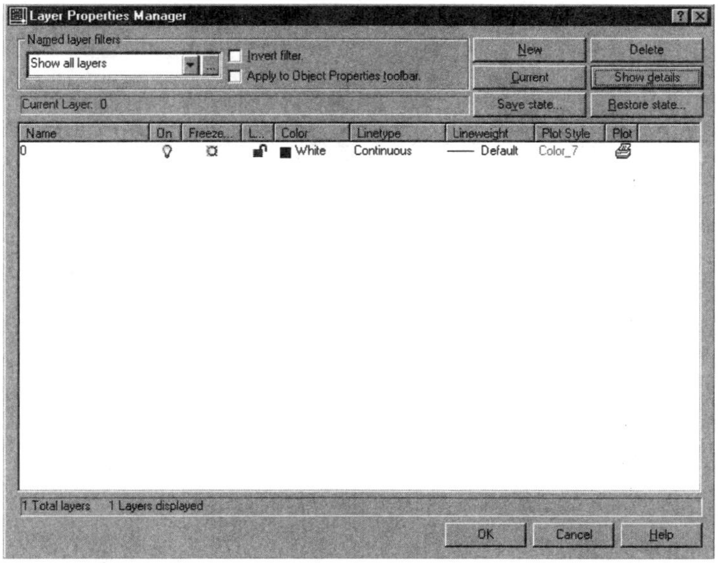

Figure 1.27 *A dialog box provides options for controlling commands, options, or portions of the drawing program. Display the* **Layer Properties Manager** *dialog box by selecting the* **Layers** *button on the* **Object Properties** *toolbar, or by selecting* **Layer** *from the* **Format** *menu.*

WARNING: AutoCAD provides system variables to control the display of dialog boxes. If dialog boxes are not displayed, type **FILEDIA** (ENTER) at the command line. The default value is 1. Options include

0 — does not display dialog boxes.

1— displays file dialog boxes.

Other system variables will be introduced as the commands they affect are introduced.

When a dialog box is selected, the crosshairs of the drawing area changes to an arrow. The arrow can be used to select an option by placing the arrow in the box or button by the desired option and pressing the select button. More specific information regarding the actual use of each dialog box will be discussed when each controlling command is introduced. Common components of a dialog box are seen in Figure 1.28; they often include **OK** and **Cancel** buttons, scroll bars, buttons, radio buttons, check boxes, edit boxes, image tiles, and alerts.

OK/Cancel

As you display a dialog box, you will notice that the cursor no longer responds to the drawing area. To return to the drawing and remove the dialog box, select either the

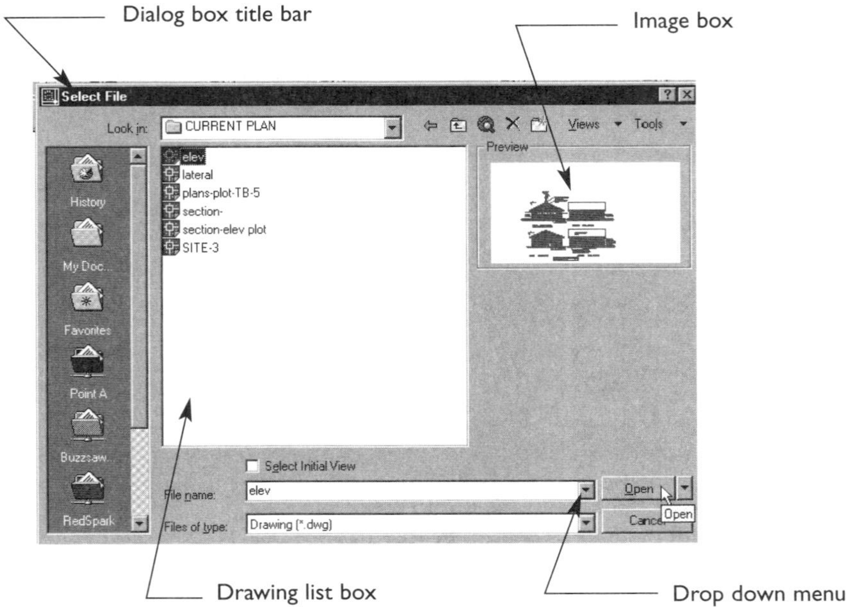

Figure 1.28 *Common components of a dialog box.*

OK or other similar action button, or the **Cancel** button. If you don't want to save the changes in a dialog box, close the box using the **Cancel** button.

- The **OK** option updates the dialog settings and removes the box from the drawing area.
- The **Cancel** option removes the box from the screen, but it will also delete any changes you might have made. Pressing esc has the same effect as clicking Cancel.

Buttons

Several buttons often are included in a dialog box in addition to **OK** and **Cancel**. Examples are seen in Figure 1.29. The **Plot** dialog box can be displayed by clicking the **Plot** button on the **Standard** toolbar or by typing **PLOT** (ENTER). Clicking one of these buttons will cause the selected option to be performed.

- Buttons with a wide border are default options.
- A button with a name followed by … will display another dialog box that must be addressed before proceeding.
- A button with a name followed by < indicates that an action needs to be performed in the graphics area before proceeding. This might include selecting an object or point in the drawing field. When an action button is selected, the dialog box is removed temporarily to allow for selecting an object from the drawing. Selecting the **Windows<** button provides prompts for selecting the area to be plotted.

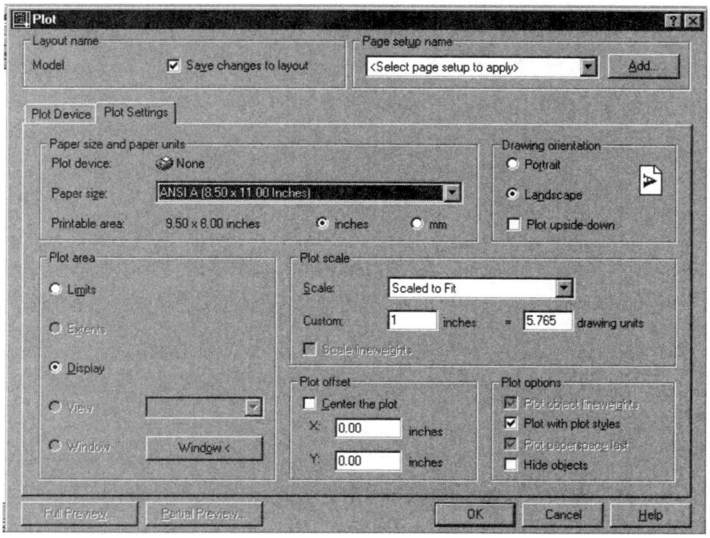

Figure 1.29 *Check boxes and radio buttons in a dialog box serve as toggle switches.*

Radio Buttons

Radio buttons are buttons that list mutually exclusive options. Examples can be seen in Figure 1.29. **Limits, Extents**, **Display**, **View**, and **Window** are each radio buttons found in the **Plot** dialog box. Each will present a different area to be plotted, but only one of the four options can be used per plot. Notice in Figure 1.29 that the **Extents, View**, and **Window** options are not displayed with black text. Gray options are inactive options and can't be selected until another option has been completed. Select one of the active or black options by moving the cursor to the desired button and clicking the select button.

Check Boxes

Check boxes serve as toggle switches. Display the **Drafting Settings** dialog box by selecting **Drafting Settings** from the **Tools** menu. Once it is open, select the **Object Snap** tab to produce the display shown in Figure 1.30. In the **Drafting Settings** dialog box, each of the modes is represented by a check box. Each mode is independent of the other listings. Selecting the **Endpoint** mode will allow the cursor to jump to the nearest endpoint of existing lines as the cursor is moved. Each of these options, as well as many others, will be presented throughout the text. For now, remember that if a box has an "x" or check in it, that option is enabled. If the box is empty, the option is disabled.

Edit Boxes

An edit box is a text entry box that allows information to be entered by keyboard. Figure 1.28 shows an example of an edit box. The **Select File** dialog box is displayed

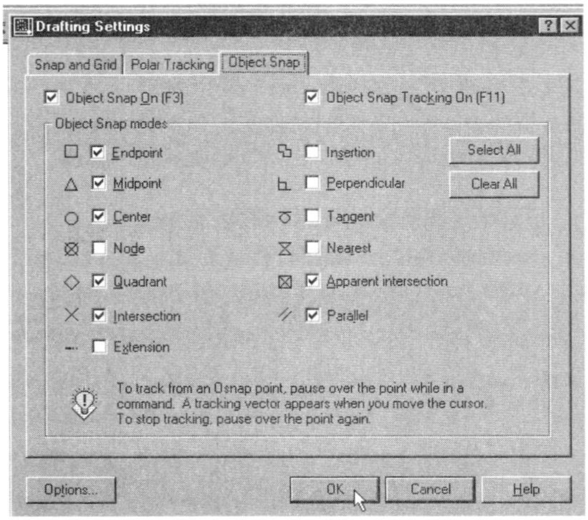

Figure 1.30 Object Snap *check boxes are displayed in the* **Object Snap** *tab of the* **Drafting Settings** *dialog box.*

by selecting **Open** from the **File** menu. When the edit box is selected, it contains a vertical line (a cursor) that flashes slowly. The flashing cursor means AutoCAD is waiting to be told the name of the file. Information can be typed into the box, in this case the **File name** edit box. If errors are made in typing, the cursor can be moved using the left and right arrow keys. Activate the text in the box by pressing (ENTER).

Edit Keys

Several keys can be used to aid input in a dialog box. These include text keys, arrow keys, and control keys. These functions include the following:

> TEXT—Symbols from the keyboard are inserted at the cursor. As a letter is inserted, the cursor and existing text move to the right. When the cursor reaches the right side limit, the cursor remains and text is moved to the left.
>
> LEFT ARROW—Moves the cursor to the left without changing the existing text.
>
> RIGHT ARROW—Moves the cursor to the right without changing the existing text.
>
> DEL—Deletes the character at the cursor and moves all remaining text to the left. The cursor will remain stationary.
>
> BACKSPACE—Deletes the character to the left of the cursor while the cursor and all remaining text is moved to the left.

Image Boxes

An image box is a portion of a dialog box that displays an image of the object or pattern that is available. Figure 1.31 shows the image box used to show the pattern that will be placed if this hatching option is selected. Display this box by selecting **Hatch** from the **Draw** menu.

Alerts

An alert is a message that is displayed in the lower left corner of a dialog box, or in a separate box that is superimposed over the dialog box. Alerts typically warn the user that they have made a mistake by displaying a message such as "Invalid filename." Another common type of alert is a warning that a file is about to be changed, for example:

The specified file already exists.
Do you want to replace it?
Yes   No   Cancel

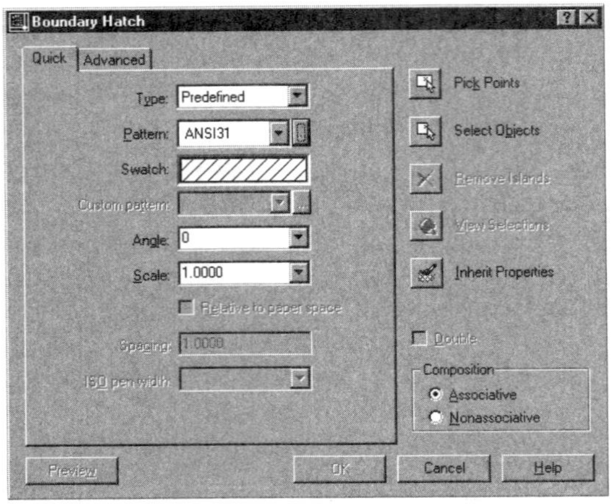

Figure 1.31 *An image box is a portion of a dialog box that displays an image of the selected item.*

SHORTCUT MENUS

Similar to most Microsoft Windows programs, AutoCAD contains menus that provide rapid access to options for tasks that are underway. Click the right mouse button to display these shortcut menus. As the name implies, the contents of the menu will change depending on when the menu is selected. If you're in the middle of a drawing command and right-click, the menu will be different than if you right-click while trying to edit a drawing. Five basic shortcut menus are available within AutoCAD. Each can be customized. The default settings of these menus are as follows:

- The default menu is displayed when a right-click is performed with the cursor in the drawing area prior to starting a command sequence.
- The command-mode menu appears when a right-click is performed in the middle of a command sequence. Options specific to the command will be displayed.
- The edit-mode menu is displayed with a right-click if objects have been selected, but no command is in progress.
- The dialog-mode menus appear if a right-click is performed while in a dialog box.
- Other menus will be displayed when right-clicking based on the cursor location. Right-clicking with the cursor over a toolbar produces a list of all toolbars. Right-clicking while over the command line, layout tabs, or status bar will each produce different context-specific menus.

FUNCTION KEYS

Twelve function keys are located across the top of the keyboard or to the left of the keyboard. The actual layout depends on the keyboard manufacturer. Ten function keys are used by AutoCAD to provide access to commands. Each is used as a toggle from ON to OFF and will be explained further in Chapter 5. Other keys can be programmed to perform specific commands. The following function keys are used by AutoCAD:

- **F1**—Help. Selecting this key clears half of the screen and displays the Help window. Using the Help options of AutoCAD will be discussed later in this chapter. Select the Cancel button or the Close button to remove the display from the screen.
- **F2**—Flip Screen. Pressing F2 in the middle of a drawing will remove the drawing from the screen and display a listing of the commands that were used to generate the drawing. Press F2 a second time and the text window will be removed.
- **F3**—Object Snap. Snap allows the cursor to be moved to specified points on drawing objects such as an endpoint or midpoint of a line. Later chapters will explore how to set the Object Snap settings. Pressing F3 will toggle Object Snap ON/OFF.
- **F4**—Tablet mode. This function key serves as a toggle if a digitizing tablet is used. Some professional offices use a digitizer tablet menu to control command options specific to their company. A digitizing tablet can be used as an alternative to a mouse for selecting objects from screen menus and controlling the crosshairs. If you're working at a station with a tablet, you'll see the following prompt:

  CALibrate the TABLET before turning it ON <off>

  If you're at a machine with no tablet, ignore the prompt and continue exploring.
- **F5**—Isoplane crosshairs mode. This is used when creating isometric drawings. This button toggles between the top, left, right, and bottom planes of an isometric square. These planes will be discussed as isometric drawings are explored.
- **F6**—Coordinate display. The coordinate display can be seen on the status line in Figure 1.8. This function key will toggle the coordinate display ON or OFF.
- **F7**—Grid display. In later chapters you will learn how to create a visible grid to aid in drawing layout. Once you have selected the size of the grid to be displayed, display or remove the grid by pressing the F7 key. You can also toggle grids ON/OFF by selecting the GRID button on the status bar.

- **F8**—ORTHO. ORTHO can be used to control the movement of the cursor. When it is ON, only vertical and horizontal lines can be drawn. When OFF, lines at any angle can be drawn. These options will be explored in Chapter 5. By switching ORTHO OFF, angular features can be drawn.
- **F9**—Snap mode. This key activates the ability to move the crosshairs within the drawing area at specific intervals. Snap will be explained in Chapter 5. Snap can also be toggled ON/OFF by selecting the SNAP button on the status bar.
- **F10**—Polar Tracking. This key toggles Polar Tracking ON/OFF.

Note: Remember that function keys can be used at any time throughout the life of a drawing including in the middle of a command. Although ORTHO in the ON setting will make rectangular features such as a floor plan much easier to draw, it can be quickly toggled OFF to allow for lines representing an angled wall to be drawn, and then reset to the ON position to finish other perpendicular features.

CONTROL KEYS

Like many other software programs, AutoCAD uses control keys to perform common tasks. These jobs are achieved when the control key is pressed in conjunction with a specified letter key. Many of these tasks duplicate the jobs performed by the function keys. The control keys are located in the lower corners of the keyboard and may be labeled CTRL. The control key functions areas follows:

CTRL + A—Selects all objects for editing.

CTRL + B—Snap mode toggle ON/OFF.

CTRL + C—**Copyclip** command copies selection to the clipboard.

CTRL + D—Coordinate display of status line toggle ON/OFF.

CTRL + E—Crosshairs in isoplane position toggle left/top/right.

CTRL + F—OSNAP mode toggle ON/OFF.

CTRL + G—Grid toggle ON/OFF.

CTRL + H—Toggles the **PICKSTYLE** value between 0 and 1.

CTRL + J—Same as using ENTER.

CTRL + K—**Hyperlink** command.

CTRL + L—ORTHO mode toggle ON/OFF.

CTRL + M—Same as using ENTE.

CTRL + N—Executes the **New** command.

CTRL + O—**Open** command, used to open a new file.

CTRL + P— Displays the **Plot** dialog box.

CTRL + R—Toggles between viewports.

CTRL + S— **Save** command.

CTRL + T—Tablet mode toggle ON/OFF.

CTRL + U—Polar mode toggle ON/OFF.

CTRL + V—**Pasteclip** command, pastes clipboard locations to a specified location.

CTRL + W—Toggles Object Snap Tracking ON/OFF.

CTRL + X—**Cutclip** cuts selected objects to the clipboard.

CTRL + Y—**Redo** command.

CTRL + Z—**Undo** command.

CTRL + 1—Toggles the display of the Properties window.

CTRL + 2—Toggles AutoCAD DesignCenter.

CTRL + 6—Toggles dbConnect Manager.

GETTING HELP

One of the best features of AutoCAD is that it offers free advice with no ridicule. You're going to be exposed to hundreds of commands, options, and menus, and sometimes it's easy to feel lost or overwhelmed. If you can avoid panic in the first 30 seconds you'll probably do fine. Your chances for success are even better if you remember that there has been an Active Assistance window as well as a **Help** button in every dialog box you've looked at.

USING ACTIVE ASSISTANCE

The Active Assistance window provides command related on-demand help. Depending on how your program is configured, the Active Assistance window will be displayed. Options for display include the following:

- **All commands**
- **New And enhanced commands**
- **Dialogs only**
- **On Demand**

The Active Assistance window can be relocated, enlarged, or reduced just as any other window.

If you've never used AutoCAD, the All Commands option will be helpful. In this setting, each time a command is started, the window displays information related to that command. Type L (ENTER) to start the **Line** command. As (ENTER) was pressed, the display changed to display the information shown in Figure 1.32. Typing **C** (ENTER) will change the window to show information about the Circle command.

As you gain experience with AutoCAD, you might find the window distracting. Display options for Active Assistance can be altered by moving the mouse anywhere in the assistance window and right-clicking. Selecting **Settings** from the menu will produce the display shown in Figure 1.33. Two important options alter the display:

- The **New and enhanced commands** radio button will display the Assistance box only as new or altered AutoCAD commands are started. This will allow you to draw lines without assistance but receive help when a command new to this release of AutoCAD is entered.
- The **Show on start** check box, if cleared, will make the default option inactive, and the window will not be displayed the next time an AutoCAD session is started.

If you decide you want to alter the Active Assistance display, type **ASSIST** (ENTER) at the Command prompt. This will display the Active Assistance window shown in Figure 1.34. Changes can be made by right-clicking in the window and selecting Settings.

Three other options that you should investigate are **Dialogs Only**, **Hover Help** and **On Demand**.

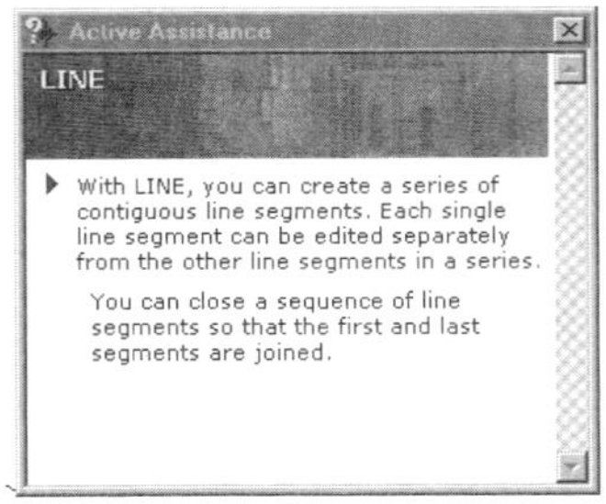

Figure 1.32 *The Active Assistance window provides command related on-demand help.*

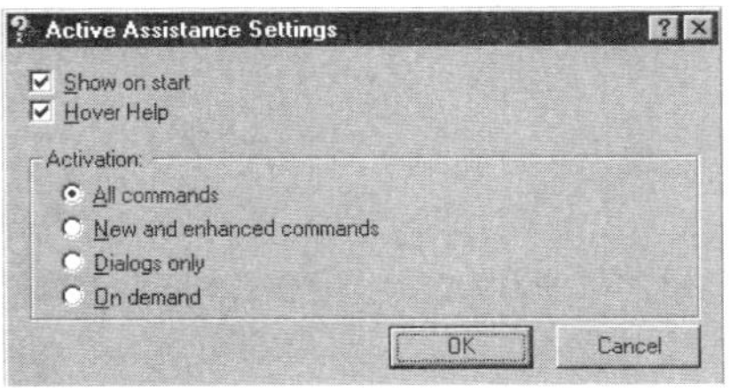

Figure 1.33 *The Active Assistance window automatically displays information about the active command.*

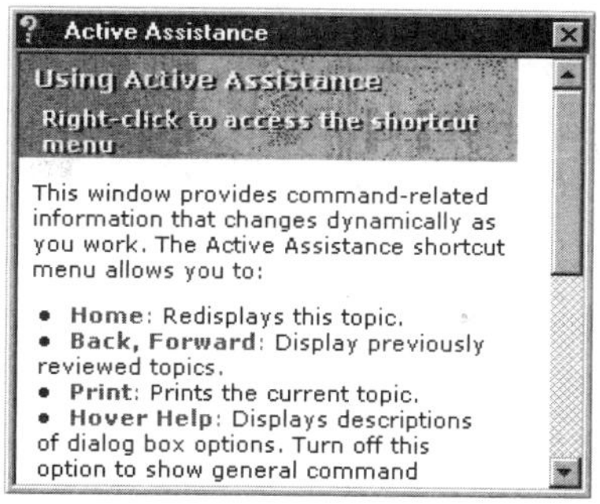

Figure 1.34 *The Active Assistance window can be used to determine when and how the Assistance box will be displayed.*

Dialogs only—When this radio button is selected, the assistance window is only displayed when a dialog box is displayed.

Hover Help—This check box controls how assistance will be displayed when a dialog box is open. With the box selected (the check mark), whenever the cursor is over an option in a dialog box, the Active Assistance window displays information relative to that option.

On Demand—With this radio button selected, the Active Assistance window is displayed only when the Active Assistance button is clicked. Clicking the Active Assistance button on the Standard toolbar or typing **ASSIST** (ENTER) at the Command prompt will produce the Active Assistance window.

USING THE HELP COMMAND

Additional help is available for all areas of AutoCAD by selecting the **Help** command. Access the **Help** command by pressing F1, by selecting **Help** from the menu bar, by clicking the **?** button on the **Standard** toolbar, or by typing **?** or **HELP** (ENTER) at the Command prompt. Each method will produce the AutoCAD Help: User Documentation dialog box shown in Figure 1.35. The **Help** command is a transparent command, one that can be completed in the middle of another command sequence. Start a transparent command by typing an apostrophe before entering the command. To get help in the middle of the **Line** command, make the following entry:

Command: **L** (ENTER) *(Or click the Line button on the Draw toolbar.)*

LINE Specify first point: *(Select desired starting point.)*

Specify next point or [undo]: **'HELP** (ENTER)

The F1 key can also be used in the middle of another command. When the **Help** command is accessed in the middle of a command, the **Help** command automatically provides command-related help. Asking for help in the middle of the **Line** command will produce the Help display shown in Figure 1.35, specific to the command.

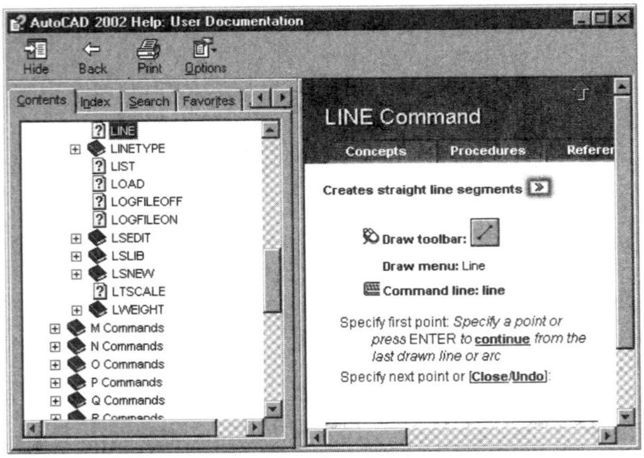

Figure 1.35 *Typing* **HELP** *(ENTER) at the command line, clicking the* **Help** *button on the* **Standard** *toolbar, or pressing the F1 key will produce the Help window.*

The left portion of the display is used for finding information in the Help window. The tabs at the top of the left screen provide the methods for using the Help window and the right side of the window is where the help text will be displayed. In the top left corner of the Help window is the **Hide** button. Selecting this button will hide the left half of the Help window. Once the desired material has been displayed, the window can be reduced. Selecting **Show** will restore the left portion of the Help window, allowing additional use of the Help window. Selecting the **Back** button will restore the previous help screen. The Help window offers help in the areas of **Contents**, **Index**, **Search**, **Favorites** and **Ask Me**.

CONTENTS

The **Contents** folder of the Help window can be used to obtain information about the Help program, or individual portions of the AutoCAD program. Notice that 10 listings are displayed in the **Contents** list, and each listing is preceded by a closed book. The **Contents** folder contains a How to Guide and a Users Guide to provide instruction on using each aspect of the program. The Command Reference portion of the folder provides information on each command used by AutoCAD. As you scroll through the listing of **Contents**, each listing will be highlighted. Clicking a highlighted menu item will provide a brief description of the selected item on the right side of the window. Double-clicking a menu item will provide a submenu of the contents. Use the following sequence to receive information about the **Line** command:

1. Display the Help window by selecting F1 or by any one of the other methods introduced in this chapter.

2. Double-click the Command Reference listing; it provides a list of commands presented in alphabetical order. To find information on the **Line** command, select the L Commands listing from the display.
3. Select LINE from the listing of L Commands. This will produce the display shown in Figure 1.35, and provide a brief description of the **Line** command.

To obtain a detailed description of the command, click the **>>** (**Click for more** button). Figure 1.36 shows the LINE command description. Clicking the **<<** (**Click to hide** button) will return to the original menu. Click the **Close** button in the Help title bar to close the Help window and return to the drawing screen.

INDEX

Selecting the **Index** tab will produce a menu that is like the index found at the end of a book. Look up a key term, and the book index will list several pages where information can be found. With the Help index, you don't even have to know the word you're looking for. If you enter the first few letters of a command or option, the Help window will display a list of commands and sources of information about that command. Entering **line** in the edit box will produce the display shown in Figure 1.37. This display box can be used to select information about the command. Double-clicking LINE command will produce a listing of related options. Double-clicking LINE [command Reference:ARC] will produce the help display for the **Line** command. You can also display the text by highlighting the desired option and clicking the **Display** button.

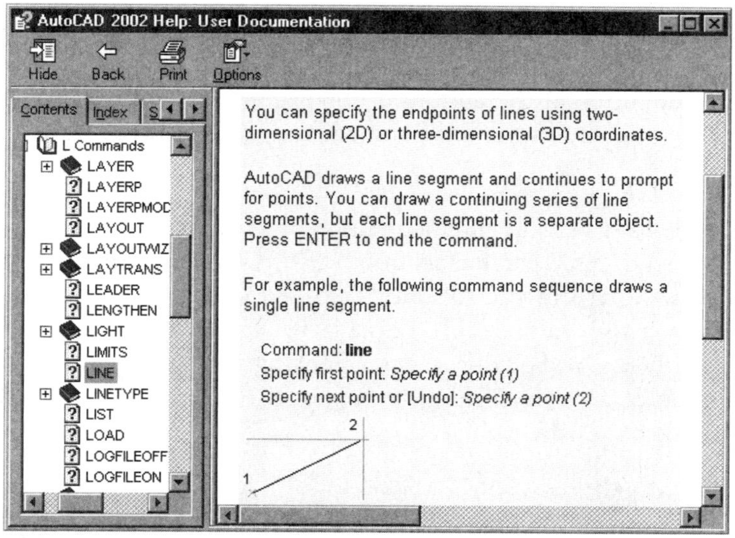

Figure 1.36 *Clicking the >> button in the LINE display produces a detailed description of the command.*

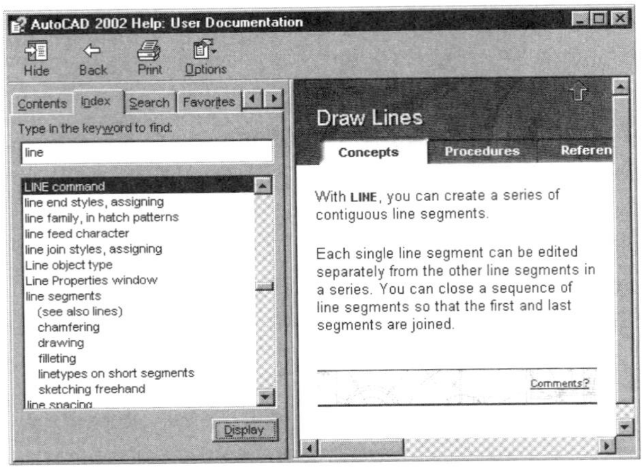

Figure 1.37 *Selecting the* **Index** *tab will produce a menu similar to an index at the end of a book. As commands are typed in the* **Keyword** *edit box, the display is altered to reflect the entry.*

Search

The **Search** option of Help produces a search of the Help based on words entered in the search box. Entering the word **LINE** in the search box restricts the search to the LINE command, as opposed to other line options. The search field can be restricted with the three options at the bottom of the display. These include **Search previous results, Match similar words**, and **Search titles only**.

Favorites

The **Favorites** tab allows you to assemble a list of favorite topics for quick access. Place a help topic selected with the **Content, Index,** or **Search** tab on the Favorites list by clicking the **Add** button at the bottom of the page. Remove topics from the list by highlighting the subject to be deleted and clicking the **Remove** button.

Ask Me

The **Ask Me** tab allows you to enter a question or a word in the **Topics** box. Once it is entered, select the search parameters from the **List Of Components to search** edit box. Enter **draw** a line in the top edit box, and select the Users Guide in the **Components to search** edit box. The display will be altered to list common solutions to the specified problem. For this search, Drawing Linear Objects is the first recommended resource. Double-clicking this option produces a display of related drawing options related to **Draw**.

AUTOCAD LEARNING ASSISTANCE

In addition to the Help window, help can be obtained through the AutoCAD Learning Assistance tutorial and the AutoCAD Support Assistant. AutoCAD provides additional help through a tutorial-based learning tool named the AutoCAD 2000

Learning Assistance (ALA). The ALA is provided on a CD separate from AutoCAD and must be installed before it can be accessed throught the Help window The ALA includes lessons for learning many of the new features of AutoCAD. The AutoCAD Support Assistant (ASA), available through Web access, contains technical documentation, solutions to common support issues and troubleshooting guidance for common AutoCAD error messages. AutoCAD users can periodically update the ASA through Web access.

EXPLORING THE SAVE COMMAND

Throughout this chapter you've explored several methods of controlling commands. The **Line** command was used as methods of using toolbars, menus and keyboard entry for command entry were explored. The lines you've created thus far will be destroyed if you end your drawing session without saving them. Chapter 4 will explore how and where files can be saved. This chapter will introduce the procedure for saving a drawing file.

If you have not already done so, enter the drawing area and draw several lines. Because this is a chapter for exploring, use the **Draw** toolbar to draw a circle and a rectangle. You have now created a work of art that must be saved for future generations to enjoy. Probably not, but it will create a file that can be used in Chapter 2 as you explore methods for opening an existing drawing. Use the following procedure to save a file to the hard drive:

1. Select **Save** from the **File** menu. This will display the **Save Drawing As** dialog box shown in Figure 1.38.

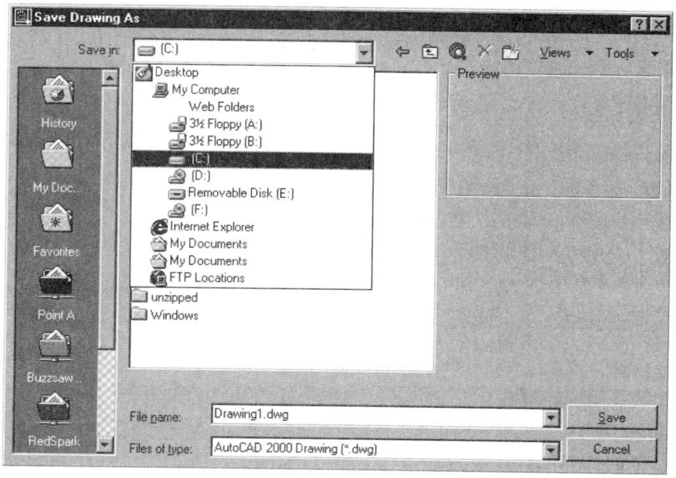

Figure 1.38 *Display the* **Save Drawing As** *dialog box by selecting* **Save** *from the* **File** *menu.*

2. Select the arrow in the **Save in** box. This display is similar to the display in Windows Explorer and allows the storage destination for the file to be selected. Select the C: drive for storage on a hard drive or A: for storage on a diskette. (Only save to a diskette at the end of a drawing session.)
3. The name of Drawing1 is currently displayed. To alter the name, click in the edit box, move the cursor so that Drawing 1 is highlighted (in a blue box.), and type the desired name, in this case, **EXE1-1**.
4. Select the type of file to be used if the drawing is to be saved to be compatible with older releases of AutoCAD
5. Click the **Save** button. The dialog box will be removed from the screen, the drawing will be saved, and the drawing area will be restored.

ENDING A DRAWING SESSION

You can continue to add objects to this drawing for as long as you want. When you want to end the drawing session, save the drawing again by selecting the Save button to update the file on your hard drive. To exit the drawing session, select **Exit** from the **File** menu or click the **Close** button on the title bar. For now, continue in AutoCAD and explore the Help window.

WHAT IT ALL MEANS

By now, you might be feeling overwhelmed by all the bells and whistles of AutoCAD. In this chapter you've been introduced to several methods of entering a command and explored a multitude of buttons, boxes, and controls for altering the commands and options. Each has been introduced with the **Line** command as an example. By the time you explore each of these controls in relation to the hundreds of commands, options, and buttons, your brain might be on overload. The good news is that you don't have to memorize every command, option, and control method. You only have to remember your favorites and the ones you use on a regular basis. Use the Help window (and this book) to help keep track of options that you don't use on a regular basis. The other key point of this chapter is that there are usually several ways to do one thing. Generally, one method is as good as another, but don't limit yourself to the first method you find. The key to becoming a successful AutoCAD operator is to explore. The following quiz will require you to do just that. Many of the questions will deal with information not introduced in this chapter. This is your first real opportunity to explore AutoCAD.

CHAPTER 1 QUIZ

1. List three sources for obtaining help with AutoCAD.
2. Select the **My Computer** button from the startup display. Select the **Printer** button and list which printers or plotters are configured to the current workstation.
3. Explain the difference between a click and a double click.
4. List the four screen areas displayed after you have entered AutoCAD.
5. Is a click or a double click required to start AutoCAD using the **Programs** menu?
6. Describe key functions of the left and right mouse buttons.
7. List three different methods that can be used for selecting commands and tell how they are accessed.
8. You've displayed a dialog box with 40 listings. Describe two methods that can be used to move through the list to the end.
9. List three sources for obtaining help with AutoCAD.
10. Explain the procedure to float the **Draw** toolbar.
11. Explain the following terms:

    Default

    Pick

    Command
12. List in order the first eight commands on the **Draw** menu.
13. You have just finished drawing a line and would like to draw an arc. Explain how to get the arc options using the menu.
14. What is the last **Arc** option listed in the menu?
15. List the options in the **File** menu.
16. You've displayed the **Insert** menu, but you wanted the **View** menu. How can you get **View** to be displayed?
17. List the tabs that are seen on the status line.
18. Is **Ellipse**, on the **Draw** menu, a command or an option? How can you be sure?
19. A button showing a paintbrush is on the **Standard** toolbar. What is the button, and what does it do?
20. List the four toolbars shown as AutoCAD is loaded.
21. What key will execute each of the following requests?

    Display grid

    ORTHO

    Snap mode

22. Someone has asked you what the last three commands were that you used on your drawing. How could you get a record if you don't remember?
23. List five items that can be found in a dialog box.
24. Search through each menu and list the location of these items:

    Erase

    Tablet

    Toolbars
25. What letter key would be needed with the control key (CTRL) to complete the following options?

    Grid toggle

    Save

    Backspace

    Coordinate display
26. List and describe five shortcut menus of AutoCAD.
27. How many sides will be drawn if the default value is selected for a polygon?
28. The **Draw** toolbar has been removed from the screen. Explain the procedure to display the **Draw** toolbar.
29. Describe the process to display a hidden toolbar.
30. You've discovered, after terminating the command, that you actually wanted to draw one more line. What is the quickest way to reenter the **Line** command?

CHAPTER 2

Creating Drawing Aids

This chapter will introduce methods for

- Entering the Drawing Environment, including the TODAY and traditional Startup window
- Starting a new drawing from within AutoCAD, including Wizards, Templates, and Start From Scratch
- Controlling drawing parameters from within a drawing

Volo View Express will also be briefly examined.

Commands to be introduced include

- **New**
- **Open**
- **Partialopen**
- **Partiaload**
- **Units**
- **Limits**
- **Zoom**
- **Grid**
- **Snap**
- **Pan**

In Chapter 1 you opened AutoCAD and were exposed to the TODAY window and the traditional **Startup** dialog box. To quickly get yourself drawing, you closed the TODAY and Active Assistance windows and went directly to the drawing screen. It is the equivalent to taking a piece of vellum with no other tools and starting to draw. This system worked well in kindergarten but is not suitable for the construction industry. This chapter will introduce methods of starting a drawing session as well as explore methods for controlling the drawing environment.

ENTERING THE DRAWING ENVIRONMENT

The TODAY window and the traditional **Startup** dialog box offer four alternatives to help set up a drawing file. Each can be seen in Figure 2.1. The decision as to which to use will depend on how you usually manage a drawing session and on personal preference. Both the TODAY window and the **Startup** dialog box will get you to the same point of a drawing session. AutoCAD offers two initial options for opening a drawing session using the TODAY window. These include the **Open Drawings** and **Create Drawings** tabs. If the **Create Drawings** tab is selected, options for starting a drawing session are given for **Template**, **Start from Scratch**, and **Wizards**. If the traditional **Startup** dialog box is used, options include **Open** Drawing, **Start from Scratch**, **Use a Template**, and **Use a Wizard** to organize drawings. Each startup method has advantages for drawing setup.

Open Drawings—Opens a drawing file created in a previous drawing session and allows editing or additions to be made to the existing drawing.

Wizards—Starts a drawing session when you know the scale or the paper size that will be needed to plot the project. This will allow you to use the setup assistance from AutoCAD. This is an excellent option to use for the design stage of a drawing. Drawing wizards offer helpful prompts to guide in drawing setup but do not provide title blocks. Later chapters will help you provide a title block to drawings created without a template. **Wizards** offers the **Quick Setup** and **Advanced Setup** options for determining how the new drawing environment will be established. Each option will be explored in the next section of this chapter.

Template—Allows predetermined drawing parameters to be used to control the drawing environment. This is an excellent design tool to use when you have planned the drawing requirements and can choose the space required to display the needed views. As you gain an understanding of the options that are available, you can create your own templates. These templates can include the company logo or similar items that can be created and saved for use on each drawing. Once you've created and stored your own template, the drawing parameters will be preset to meet your standards each time you start a new drawing. Chapter 4 will help you create your own template.

Start from Scratch—Starts a drawing with minimal drawing controls. This option is the equivalent of grabbing a blank sheet of paper with no title block. It is best suited for users who like to plan as they go. Use **Start from Scratch** for sketching, line drawings that will not be drawn to scale, or for drawings that do not require a high level of accuracy. This option also works

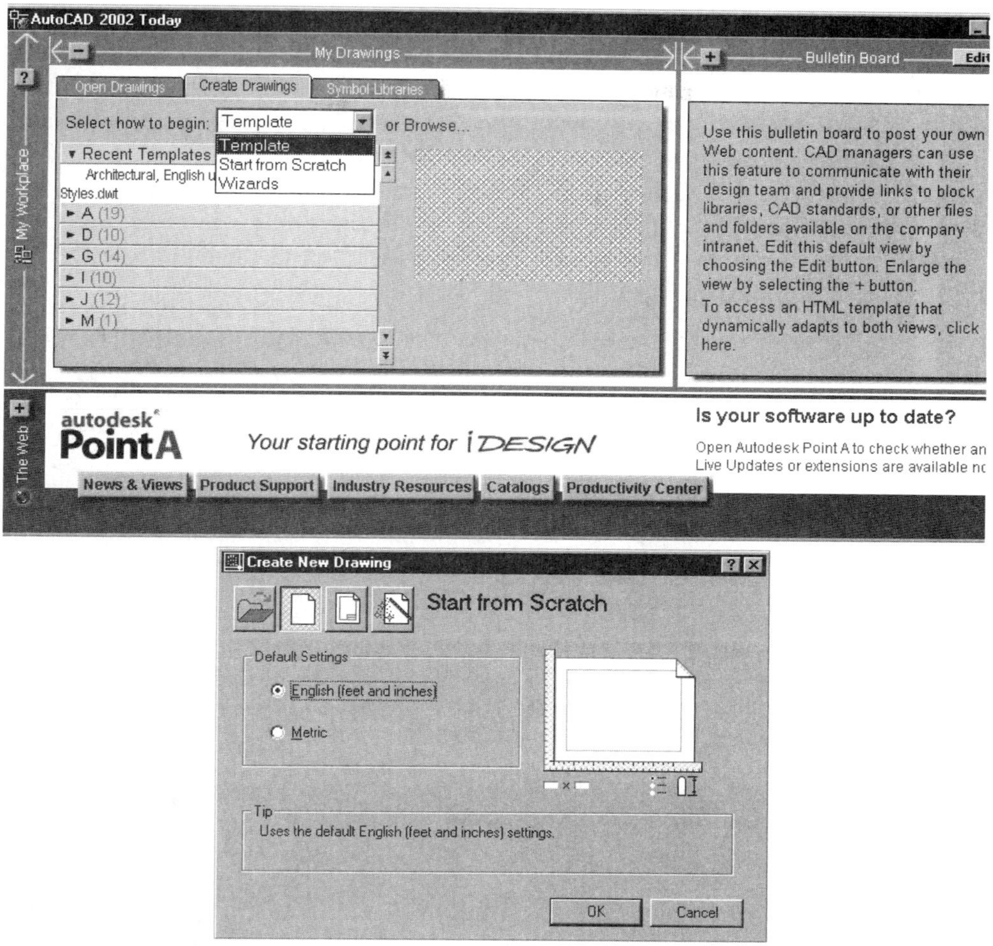

Figure 2.1 *The TODAY window or the traditional* **Startup** *dialog box will be displayed as AutoCAD is entered to provide help in setting a drawing parameter.*

well when you want to establish the drawing's parameters as they are needed. Although this option provides some basic drawing controls, you will need to add many other controls and variables to increase your drawing efficiency.

Note: If the **Startup** dialog box is not displayed as you enter AutoCAD, someone has hidden it from view. When it is hidden, you will automatically enter AutoCAD using the method that was last used to access the drawing area. You display the **Startup** dialog box or hide it from view by selecting **Options** from the **Tools** menu, and then selecting the **Systems** tab of the **Options** dialog box. Use the **Show Startup Dialog** button in the **General Options** box to control the display of the **Startup** menu. This can be a setting that you adjust and save as you create your own profile in Chapter 4.

OPEN AN EXISTING DRAWING

Much of your time in an office will be spent adding to or editing existing drawings. The **Open Drawing** option of the TODAY window is slightly different from the traditional method. In the default setting shown in Figure 2.2a, the names of drawings created in the last week are displayed. Moving the cursor over a file name will display the contents of the file in the image display box and list the full file name and the date and time of the last drawing session. The display of drawings to be opened can be altered using the **Select how to begin** list box. Display options include the following:

Most Recently Used—Provides a list of the drawings most recently used. Because you've only worked on drawing EXE 1-1, your list is restricted to this single listing. As you work on more drawings, this list will be expanded to include up to nine drawing names. More names can be displayed by using the slide bar on the right side of the listing.

History (by Date)—Displays a list of drawings for a specified date. The list is based on the most recent week. As the size of the list of drawing increases, the list is expanded to reflect drawings created in the last four weeks.

History (by Filename)—Displays a list of drawings based on the drawing file name and the number of drawings in that file. A listing of numbers and letters are displayed representing the starting number or letter of current file names. Clicking on a number or letter will display the names of files that have recently been used. Double-clicking a letter or number will close the listing.

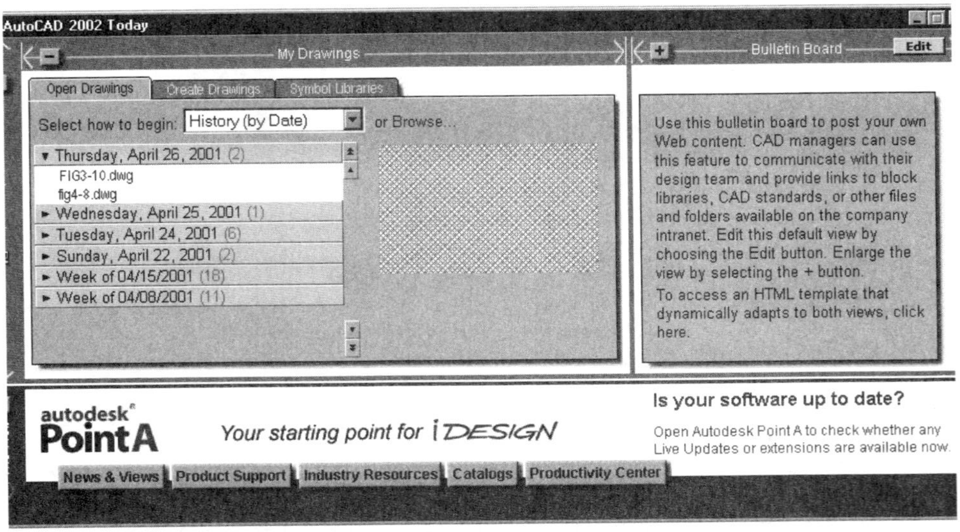

Figure 2.2a *Selecting* **Open Drawings** *in the TODAY window starts a search of the most recently used drawings. The search for existing drawings can also be defined by the file name or location.*

History (by Location)—Displays a list of drawings based on the drawing storage location. Letters are displayed to represent each of the drives and folders that contain drawings that have recently been used.

Browse—Displays a list of drawings similar to Windows Explorer using the location of the last drawing that was opened. Altering the location to be examined in the **Look in** list box will allow other folders and drives to be examined.

Selecting a file name with a single click will automatically open the file, placing the drawing in the drawing area.

If the traditional **Startup** dialog box is used, this option will display a startup menu similar to Figure 2.2b. The **Select a File** list box will display a list of the last four, and up to nine drawings that have been opened. In your dialog box, select the file created in Chapter 1, titled EXE1-1. Selecting the **OK** button will remove the dialog box and open the drawing that was created in Chapter 1. As you develop more than four drawings, the drawing file to be opened might not be displayed in the **Select a File** list box. Select the **Browse** button to find an existing drawing file that is not listed in the **Startup** dialog box. Selecting **Browse** will produce a **Select File** dialog box similar to Figure 2.3. (The display will vary depending on the number of drives and the file contents.)

The **Select File** dialog box can be used to list and select a drawing file from any of your storage locations. By default, files will be selected from the Acad2002 folder. Use the following steps to open an existing file that is not listed in the Select a File listing:

1. Select the arrow beside the **Look in** box to select the desired drive or folder that contains the drawing file to be opened. As the arrow is selected, a

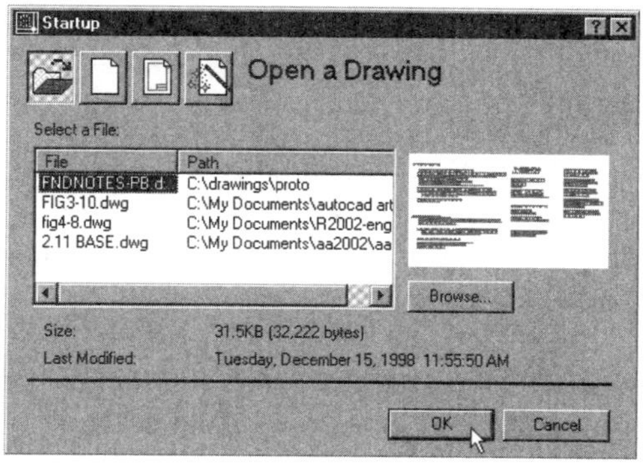

Figure 2.2b *Selecting the* **Open an Existing File** *button in the traditional* **Startup** *dialog box will display this dialog box. The last four drawings that you have worked on will be displayed in the* **Select a File** *edit box. Click the* **Browse** *button to search for files not listed.*

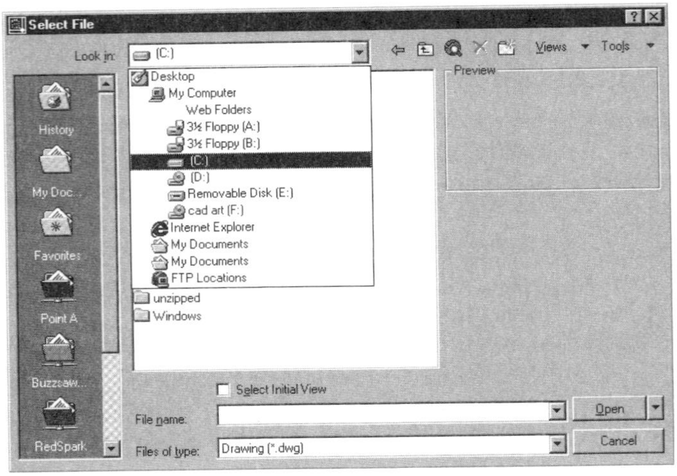

Figure 2.3 *The* **Look in** *list is used in a manner similar to using Windows Explorer.*

drop-down list is displayed, reflecting the contents of Windows Explorer. (See Figure 2.4.)

2. Move to the desired drive just as you would using Windows Explorer and select the file to be opened.
   - Click the desired drive to be searched to display a list of the drawing folders contained in the selected drive.
   - Double-click a folder to display a list of files contained in that folder. Highlighting a file with a single click will display an image of the file in the **Preview** box.
3. If this is the file you want to open, click the **Open** button to remove the dialog box and open the selected drawing on the drawing area, similar to Figure 2.5. (If this is not the file you expected, highlight another filename. Once you've found the desired file, select the **Open** button to bring the file into the drawing area.) You can also open a drawing by double-clicking the file name.

Note: Remember AutoCAD allows multiple drawings to be opened. An existing drawing can also be opened in the middle of another drawing session by selecting the **Open** button on the **Standard** toolbar, by selecting **Open** from the **File** menu, or by typing **OPEN** (ENTER) at the Command prompt. Each method will produce the dialog box shown in Figure 2.3.

CREATING NEW DRAWINGS

The **Create Drawings tab** of the TODAY window and the remaining three buttons of the traditional Startup dialog box offer three major options for creating a new drawing. These options include **Wizards, Start from Scratch**, and **Template**.

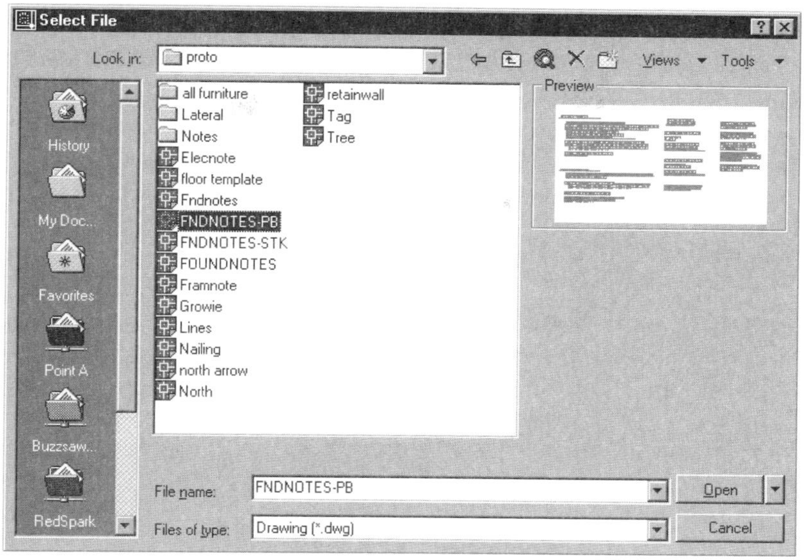

Figure 2.4 *Once in the desired drive, select the desired file to be opened. An image of the selected file will be displayed in the* **Preview** *box.*

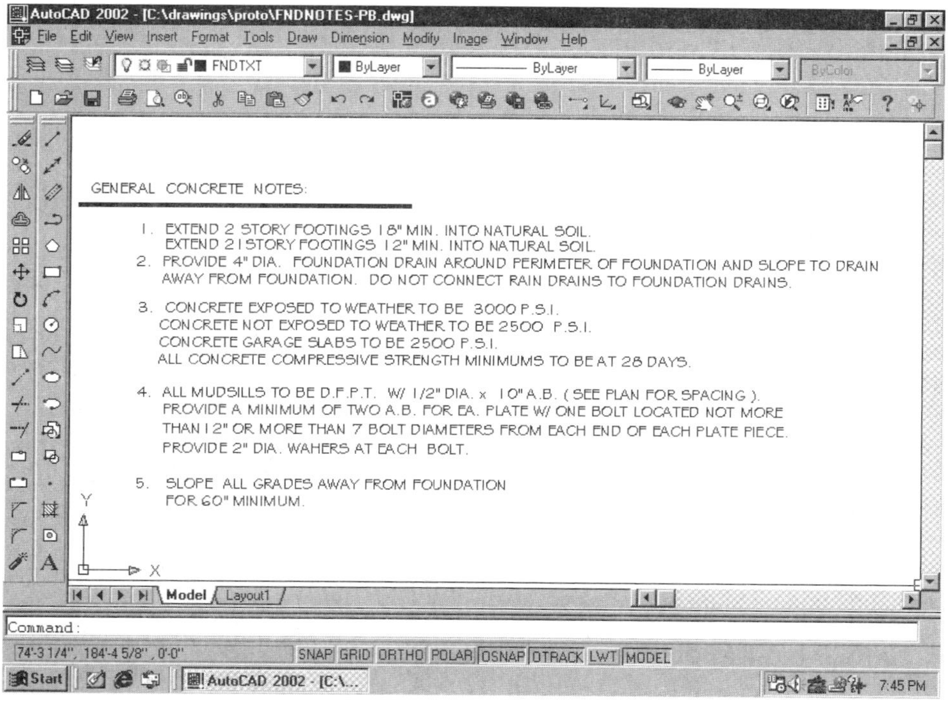

Figure 2.5 *The result of selecting the file in Figure 2.4.*

DRAWING SETUP USING A WIZARD

Using a wizard is an excellent method for starting a new drawing session if you know the size of paper the project will be plotted on. AutoCAD provides the **Quick Startup** wizard and the **Advanced Setup** wizard. The method chosen will depend on how much help you would like. Choosing **Quick Startup** will assist you in setting the drawing units and the drawing area. The **Advanced Setup** wizard will provide six start-up prompts.

Quick Setup

The **Quick Setup** option produces a drawing area in model space. This option works well for drawing, but it requires some adjustment for plotting. Selecting **Quick Setup** will produce the display shown in Figure 2.6 and allow the selection of the drawing units and the drawing area.

Setting the Drawing Units

One of the unique features of AutoCAD is that as you work in model space you will be drawing at full scale. The drawing will not be scaled until you change to paper space to plot your drawing. If you want to draw a structure that is 300' long, you will draw it full size, and then choose a scale as you plot, using AutoCAD's paper space option. (This will be covered in later chapters.) Units allow you to set the measuring method to be used to establish the drawing. In its default settings, your computer is set to measure lines in four-place decimal inch units. You can confirm this by looking at the coordinate display in the lower left corner of the status bar. To describe the

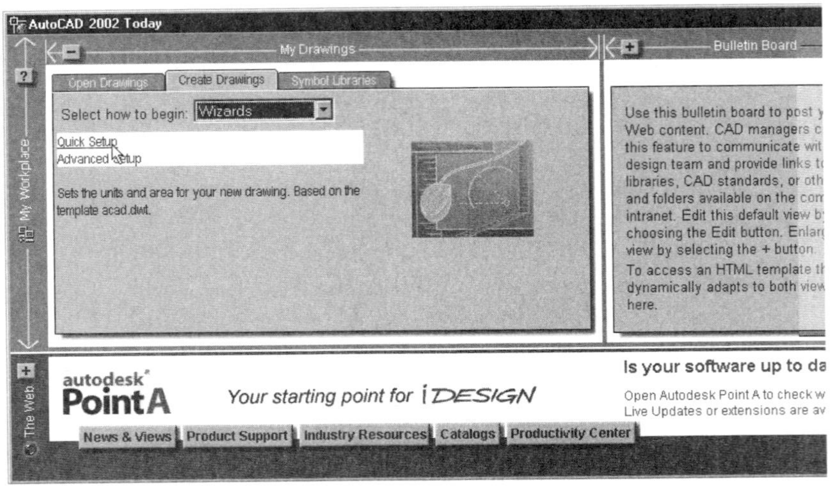

Figure 2.6 *Selecting* **Wizards** *from the* **Create Drawings** *tab of the TODAY window or selecting the* **Use a Wizard** *button in the traditional* **Startup** *dialog box provides two choices for drawing setup.*

current location of the crosshairs, two four-place decimal numbers are used. This would be fine for a drawing such as a beam connector shown in Figure 2.7. Most architectural projects are better suited to the Architectural option. This will measure objects in feet and inches. Options for unit settings and their display include the following:

> **Decimal – 15.0000**—This option can be used to create drawings in decimal or metric units. Metric units might be needed on government construction projects. If decimal dimensions are used, they should be expressed as two-place decimals unless the project requires a higher degree of accuracy.

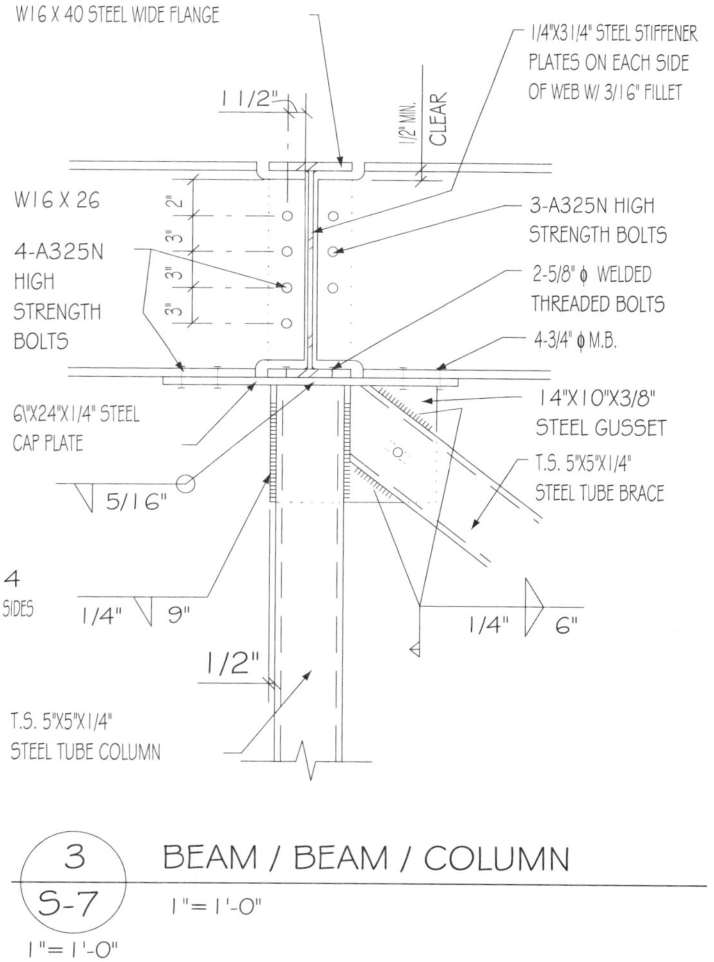

Figure 2.7 *Although some drawings can be drawn using Decimal or Engineering units, most drawings are created using Architectural units.*

Engineering – 1'–3.5000"—These units are best suited to civil drawings such as site plans, topography drawings, bridge and dam construction or other drawings dealing with large land areas.

Architectural - 1'–3 1/2"—Architectural units express distance in feet and inches and are used for drawing structures for architectural and structural clients. Architectural units are used for drawing most structures.

Fractional – 15 1/2—This option can be used on drawings that express measurements in fractions such as cabinet drawings.

Scientific – 1.5500E+01—This option is suitable for drawings used in chemical engineering. This unit of measurement is suitable when very large or very small measurements must be expressed on a drawing. Numbers are expressed using a base number and a multiplier. The E+01 represents a base number multiplied by 10 to the first power.

The units can be set by placing the cursor in the desired setting and pressing the select button. This will place a black dot in the button, indicating the selection is activated. Before proceeding, select the type of units that are appropriate for an architectural drawing such as a floor plan.

Setting the Drawing Area

Select the **Next>** button to continue the drawing setup. This will produce the display shown in Figure 2.8, showing a default value of 1' x 9". The drawing area should be based on what you expect to draw. For now, accept the default value and select the **Finish** button. This will display the AutoCAD drawing screen shown in Figure 2.9.

Advanced Setup

Selecting this option will display the dialog box shown in Figure 2.10. The **Advanced Setup** wizard will help you set six controls related to the units of measurement,

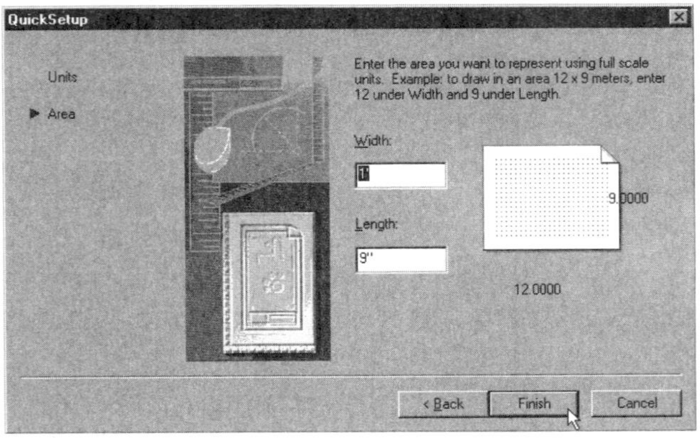

Figure 2.8 *Setting the area defines how much space is displayed on the screen. Set the area slightly larger than what you think you will need to complete the drawing.*

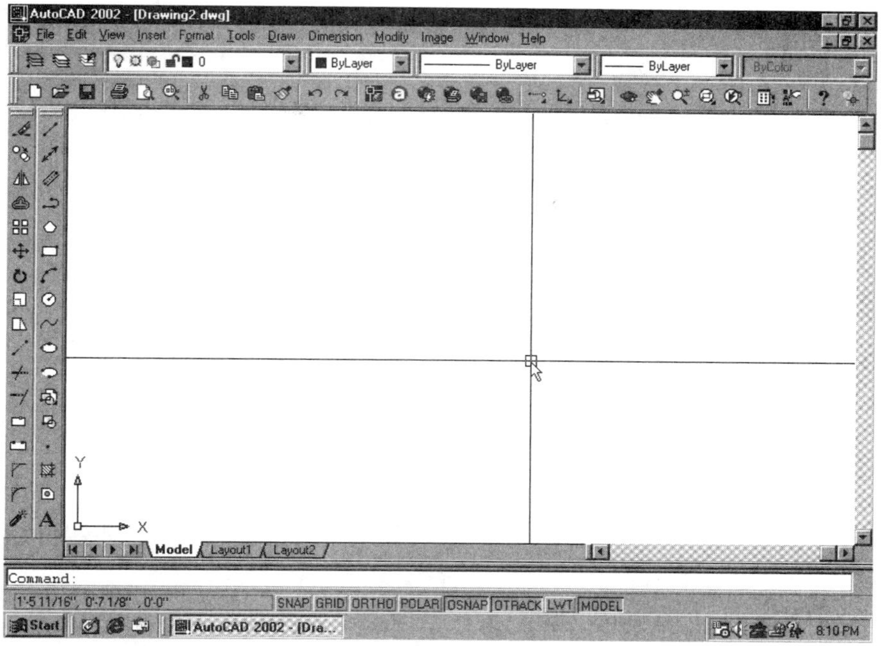

Figure 2.9 *Accepting the default value of 12" x 9" and selecting the* **Finish** *button will display the drawing screen. The grid is displayed in the drawing area by selecting the GRID button on the status bar.*

angle of measurement, direction of angle, and area. Select the **Next>** button to start the wizard.

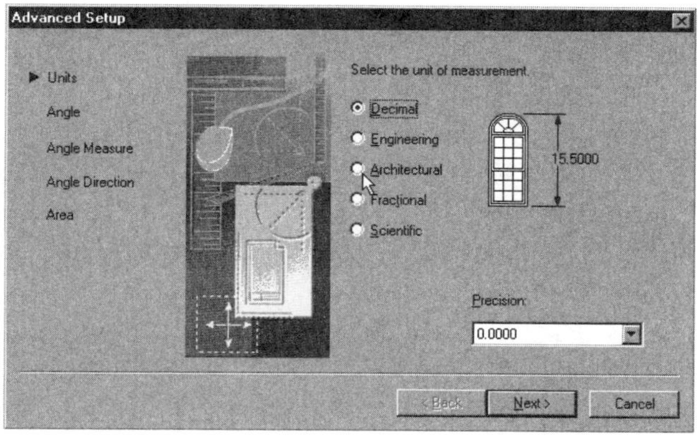

Figure 2.10 *Selecting the* **Advanced Setup** *wizard option will aid in setting six drawing aids. Notice that unlike the* **Quick Setup** *wizard, this method allows the precision of the units to be measured.*

Units

The **Units** control folder shown in Figure 2.11 is displayed as the **Advanced Setup** is opened. The default value is decimal. Choose the value that best meets the need of the drawing to be started using the methods described using the **Quick Setup** wizard.

Unit Precision

The **Precision** edit box is located at the bottom of the **Units** dialog box. For Architectural units, the default precision is set at 0'–0 1/16". This will work well for drawing a small detail, but will be of little help for drawing a 300' long structure. The drawing precision can be changed throughout the drawing, but should be kept as large as possible.

To change the setting, select the arrow on the right side of the box. This displays the **Precision** list shown in Figure 2.12. Selecting the down arrow will scroll through the list, allowing the degree of accuracy to be selected. Settings range from whole units of 0'–0" through 1/256". For now, select the 0'–0" setting by highlighting it with the mouse. This will return you to the **Units** box and allow you to continue the drawing setup. Notice, as the unit precision was changed from 1/16" to 0'–0", the display of units was also changed. The units will now display 1'–4". If you're satisfied with the units, select the **Next>** button to continue the drawing setup. The units can be changed at any point during the life of the drawing. Clicking the **Cancel** button will exit the wizard, discard any changes to the unit settings, and place the drawing area on the screen.

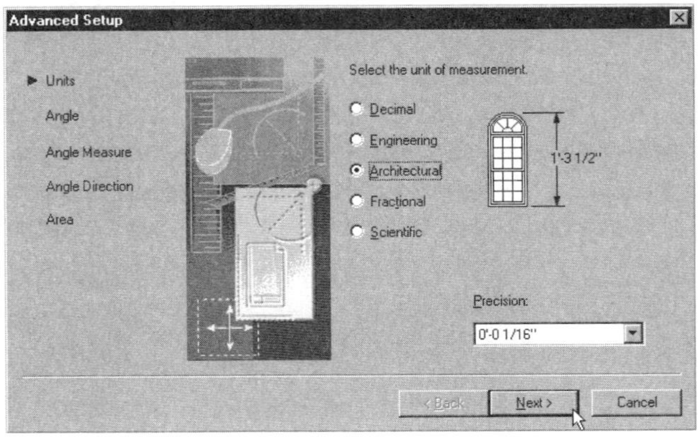

Figure 2.11 *The first step in using the* **Advanced Setup** *wizard is to set the drawing units. Notice this method allows the precision of the units to be set.*

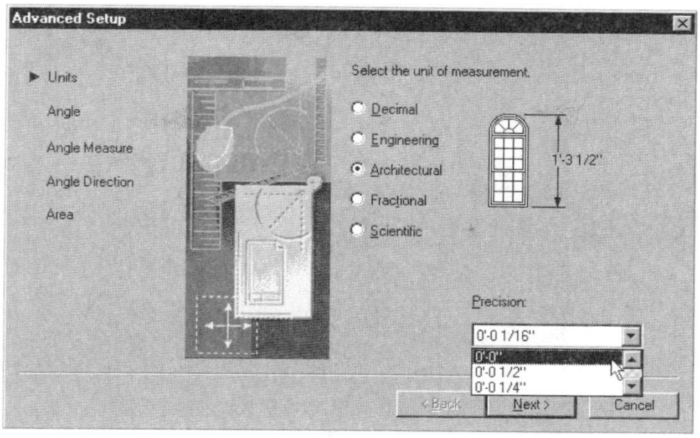

Figure 2.12 *Selecting the* **Precision** *arrow button displays the* **Precision** *list. Accepting the default value displays units with an accuracy of 1/16". Selecting 0'–0" will measure units in whole inches.*

Angles

Once the **Units** are selected, the **Angle** display for controlling angles shown in Figure 2.13 is displayed. Most architectural drawings can be completed easily using decimal angles. Options for measuring angles include the following:

> **The Decimal Degrees – 90.00**—This option is best suited for plan views such as the drawings shown in Figure 2.14. It will display angles in degrees and decimal parts of a degree.
>
> **Deg/Min/Sec – 90d00'00"**—This option is best suited for land measurements, such as site plans. This option will display angles in degrees, minutes, and seconds.
>
> **Grads – 100.00g**—Grads is the abbreviation for a gradient. One hundred grads equal one-quarter of a circle. This option of angle measurement is rarely used with architectural drawings.
>
> **Radians – 1.57r**—A radian is an angular unit measurement where $2\pi = 360°$ and a 90° is $\pi/2$ radians. This option of angle measurement is rarely used with architectural drawings.
>
> **Surveyor – N0d00'00"E**—This option presents surveyor's units, which are used to represent land, street, or sewer layouts, with angles displayed as bearings displayed as <N/S> <angle> <E/W>. The angle is based on north or south and will always be less than 90°. An example of common bearings is listed in Figure 2.15.

The method of angle measurement can be set throughout the life of the drawing with the methods that will presented at the end of this chapter. The method of angle measurement and precision is set using similar methods that were used to set the drawing units.

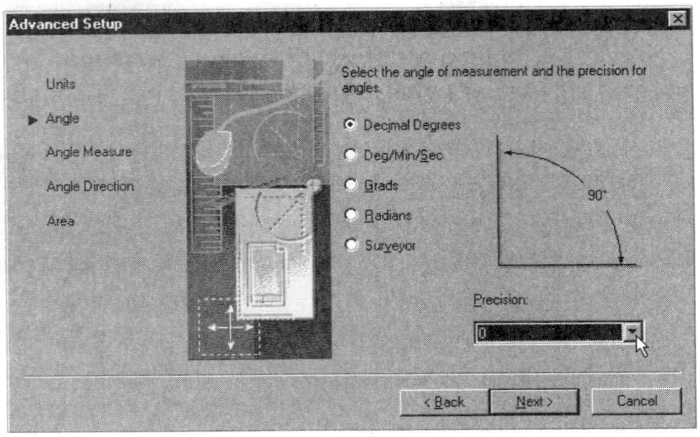

Figure 2.13 *The* **Advanced Setup** *menu allows the units for measuring angles and the precision to be selected.*

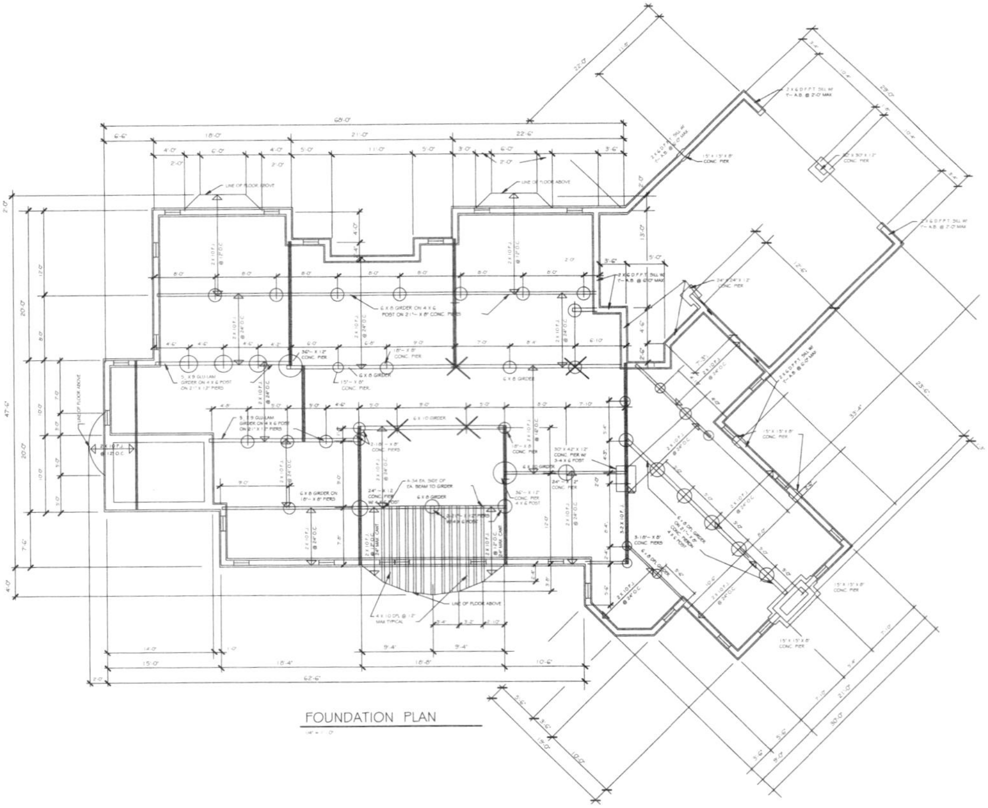

Figure 2.14 *Angles on a drawing such as this foundation plan can be measured conveniently using the Decimal Degrees angle option.*

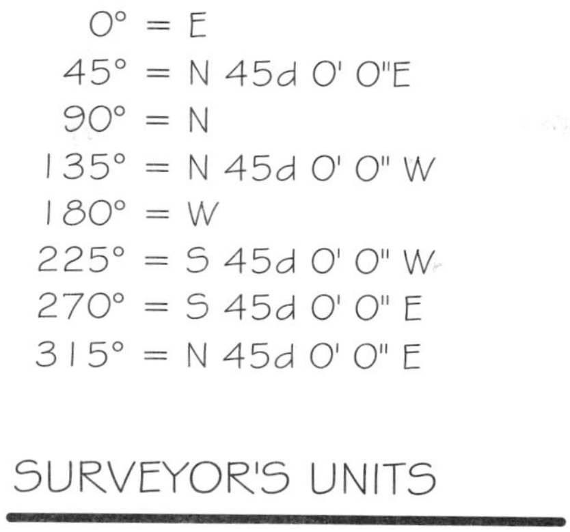

Figure 2.15. *Bearings are useful measuring units on site-related drawings. All angles are based on looking either north or south and turning either east or west.*

For now, accept the default option of Decimal Degrees. The precision for measuring angles can be set using the same method used to set the unit precision. Selecting the down arrow will provide options for zero to eight decimal places, with the default of 0. Normally, two-place accuracy is sufficient for architectural drawings while civil drawings might require a greater degree of accuracy. Accept the default by clicking the **Next>** button to proceed to the next page of the wizard.

Angle Measurement

The **Advanced Setup** wizard allows the starting point of angle measurement to be selected. The default option is east. Options are shown in Figure 2.16. Figure 2.17 shows the visual representation of the angles. Standard AutoCAD layout of angles places 0° at the three o'clock position and then moves counterclockwise.

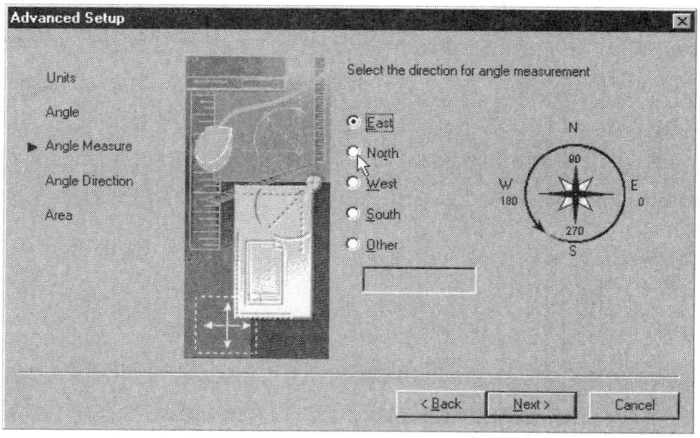

Figure 2.16 *The default starting point for measuring angles is east. You can select one of the other three quadrants, or any point in between, by choosing* **Other** *and providing an angle.*

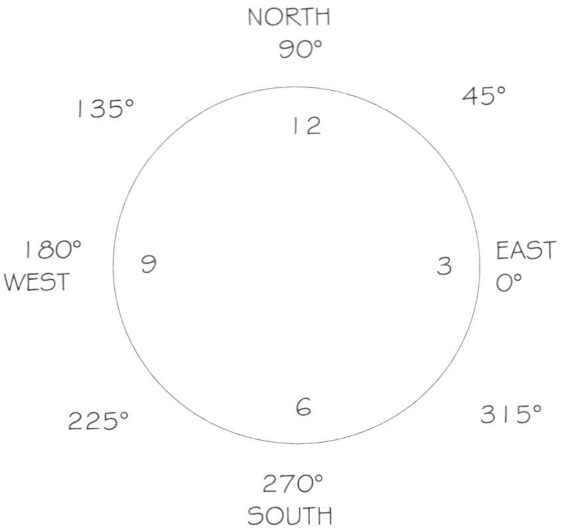

Figure 2.17 *A visual representation of the angle menu comparing compass and angular measurement.*

You can specify that angle measurement start in any location. To change the starting point of angles to north, select the **North** radio button. If you select north as the new starting direction, 0° will now be placed at north, 90° will be west, 180° will be south, and east will be at 270°. An angle other than one of the four compass points can be selected by choosing the **Other** button, and then entering the desired starting angle in the edit box. Selecting **Other** and entering a starting angle of 45° will place 0° halfway between north and east. When the desired starting point has been selected, clicking **Next>** will allow the wizard to continue.

Angle Direction

Figure 2.18 shows the **Advance Setup** wizard for selecting the directions that angles will be measured. The default setting for angle measurement is counterclockwise. Selecting the **Clockwise** button will change the direction of angle measurement. For now, accept the default. Remember that the defaults are there because they fit the needs of most users. The direction, like each of the other variables, can be altered at any time throughout the life of the drawing. Once the direction is selected, click the **Next>** button to proceed.

Area

The final setting of the **Advanced Setup** wizard allows the drawing area to be specified using the display shown in Figure 2.19. In the default setting, the drawing will be 1' x 9". Enter the values for the width and length of the area of the intended drawing to be completed in the appropriate edit box. Enter the desired value, and click the

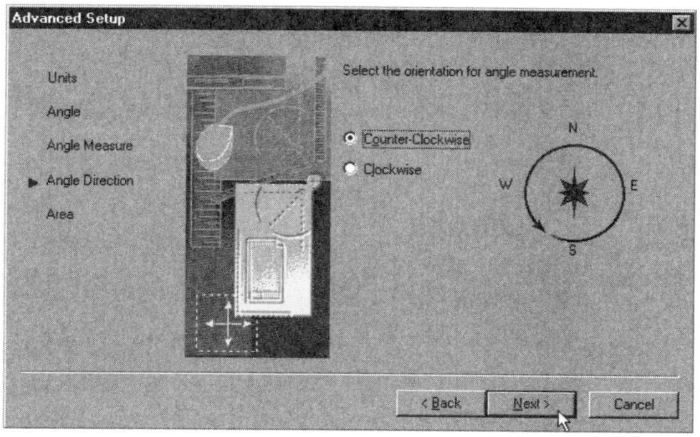

Figure 2.18 *The* **Advanced Setup** *wizard allows the direction for angle measurement to be entered. For most drawings, the default setting of counterclockwise works well.*

Finish button. Clicking the **Finish** button closes the wizard, establishes each of the parameters that were entered, and returns you to the drawing area.

Note: New CAD users are often fearful of making a wrong decision while using a wizard. With each setting it was stated that the setting could be adjusted throughout the life of the drawing. Don't make light of that fact. There are no wrong settings, only inconvenient settings. If the current setting is troublesome, change it, don't fight it.

DRAWING SETUP USING A TEMPLATE

Drawing templates with predetermined drawing values can be used to start a drawing. Each template contains values for drawing units, limits, and other values that can be altered throughout the life of a drawing. A template supplied by AutoCAD can be customized to meet specific requirements and saved for future use. Select **Template**

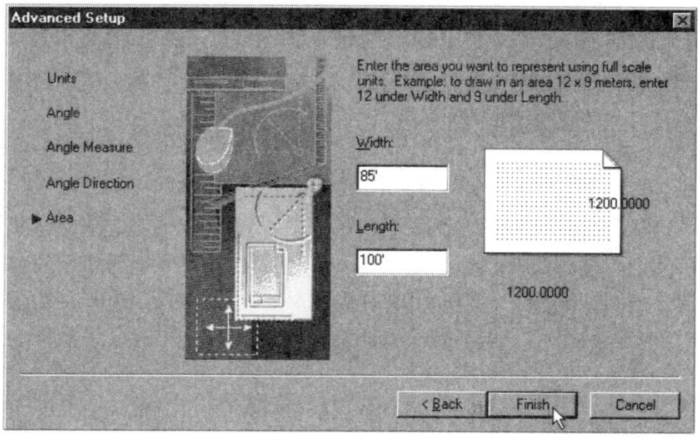

Figure 2.19 *Entering drawing units of 100',85' will allow a floor plan to be started.*

from the **Create Drawing** tab of TODAY to use a template to start a drawing session. Selecting **Template** will produce the display shown in Figure 2.20. The templates that start with A are listed. Scroll through the list using the arrows on the side. Double-clicking the A will display a listing of the letters representing the names of various templates contained by AutoCAD. Letters and their contents include the following:

A—ACAD templates

D—Din templates

G—Generic templates

I—ISO templates

J—JIS templates

M—Metric layout templates

Beside each category designation is the number of templates in parentheses for that category. Click on the category to display a listing of templates in that category. Moving the mouse over a listing displays an image of the selected template. Clicking the **Browse** button will display the **Select File** dialog box (see Figure 2.21). This box contains a listing of the templates contained in the default AutoCAD Drawing Template folder. This dialog box can be used to search for templates stored in other folders, other drives, other computers on the network, or on the Internet.

Scroll through the list of templates and highlight the Architectural, English units-Color Dependent Plot Styles.dwt. This will produce the display shown in Figure 2.22. Clicking the **OK** button will produce a drawing display as shown in Figure 2.23. The

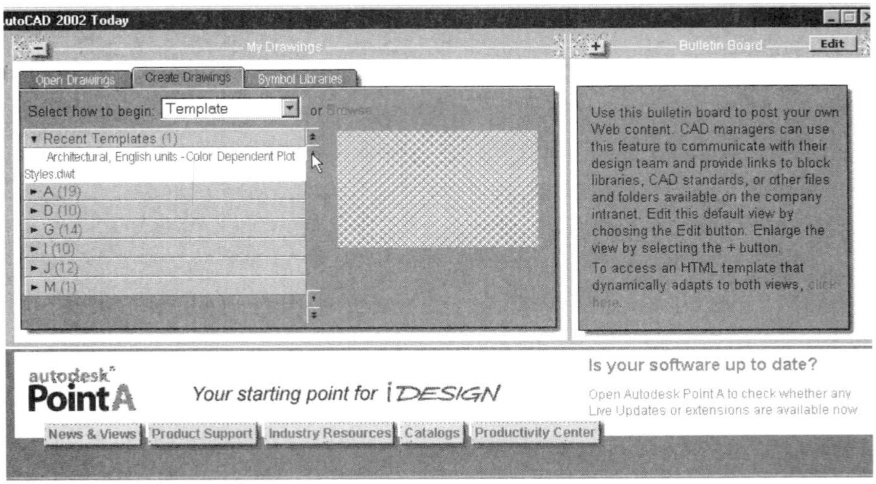

Figure 2.20 *Selecting the* **Template** *option will display a listing of the letters representing the names of various templates contained by AutoCAD.*

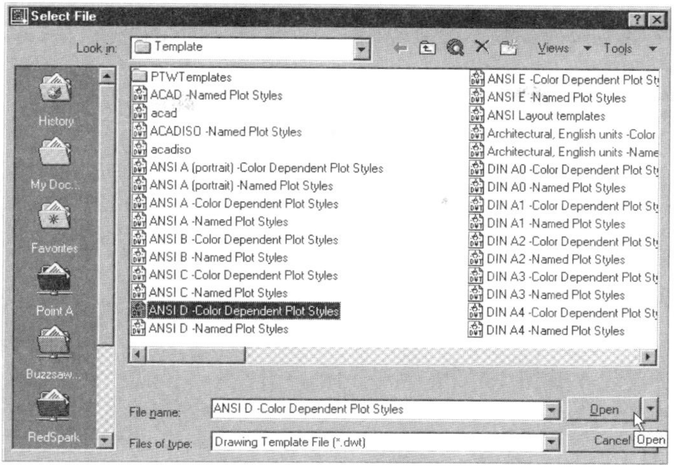

Figure 2.21 *Selecting* **Browse** *will display the* **Select File** *dialog box containing a listing of the templates contained in the default AutoCAD Drawing Template folder. This dialog box can be used to search for templates stored in other folders, other drives, other computers on the network, or on the Internet.*

drawing area will be approximately 36" x 24" with units measured in architectural units with 1/16" precision. The limits can be adjusted as needed.

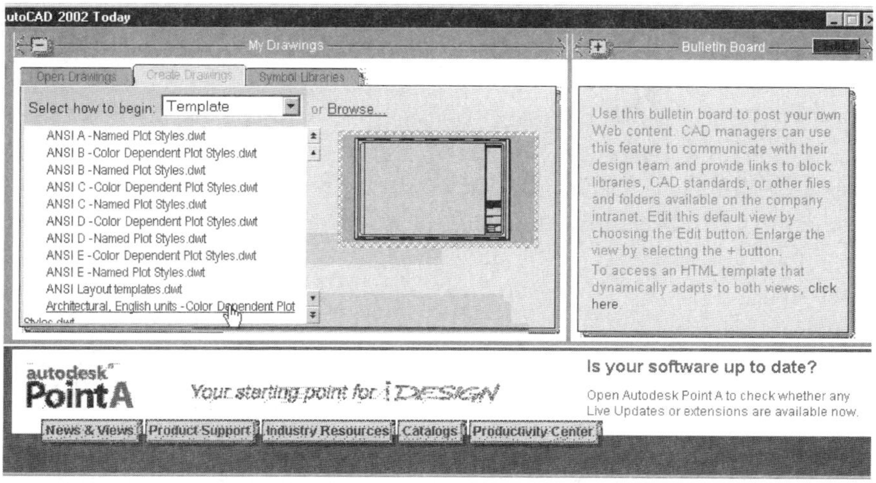

Figure 2.22 *Selecting the* **Use a Template** *button allows a drawing with preset drawing parameters to be used. Selecting* **Browse** *allows user-created templates to be accessed. Scroll through the list of templates and highlight the Architectural, English units-Color Dependent Plot Styles.dwt.*

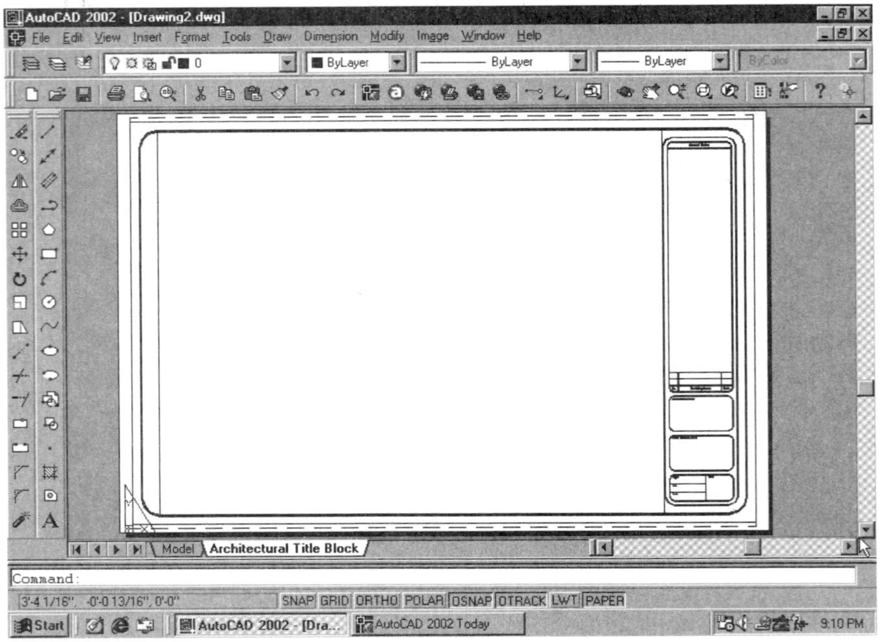

Figure 2.23 *The Architectural, English units-Color Dependent Plot Styles.dwt. template is suitable for basic architectural or engineering projects. Chapter 4 will introduce options for creating and saving your own title block.*

Selecting a Template

The template options are listed in groupings indicating the drafting standard used, the size of the title block in the preset layout, and the plot style. Listed drafting standards include ACAD, Architectural, ANSI, Din (based on the German Institute of Standardization), Generic, ISO (International Organization for Standardization), and JIS (Japanese Industry Standard). Architectural templates are based on decimal measurements. ACAD templates are also based on feet and inch measurements but do not contain a title block. ANSI and generic templates provided with AutoCAD are based on decimal units of measurement. Din, ISO, and JIS templates are based on metric measurements.

Rather than use the templates supplied with AutoCAD, you can develop your own template to meet specific needs of a project, your class or your employer. In addition to each of the drawing controls that were introduced in this chapter, your template could include common text fonts, linetypes, layers, and other related features. Each of these components will be introduced throughout the text. As each is introduced, features that you find useful should be added to your template. Methods for creating and saving templates will be covered in Chapter 4. For now, if you choose to use a template to enter the drawing area, accept the default and click **OK**.

Note: As the Architectural.dwt template is entered, you enter the drawing session in paper space. To create a drawing, select the Model tab at the bottom of the drawing area. This will remove the title block and produce a blank drawing screen. Now draw a few lines.

Select the Architectural Title Block tab and the lines that you drew will be displayed on the template. Click the PAPER button in the status bar (not the tab at the bottom of the drawing area) to display the drawing of the template in model space. If your drawing does not fit into the template, click the **Pan Realtime** button (the hand) on the **Standard** toolbar. Pan allows the drawing to be moved to the desired location. After you select **Pan**, a hand will be displayed in the drawing area. Press the select button, and move the cursor in the direction you would like to move the drawing.

When you're satisfied with the location, press ENTER to exit the **Pan** command. If the drawing is too big to be seen in the template, click the **Zoom** button (the magnifying lens) on the **Standard** toolbar. After you select **Zoom**, a magnifying lens will be displayed in the drawing area. Press the select button, and move the cursor down to reduce the size of the object. When you're satisfied with the location, press ENTER to exit the **Zoom** command. The full power of these commands will be introduced later.

START FROM SCRATCH

Selecting **Start from Scratch** from the TODAY window or the **Start from Scratch** button from the **Startup** dialog box allows a drawing to be started with a minimum of predetermined values. Setup is allowed in English or Metric units. English units are displayed in four-place decimal units, with drawing limits of 12" x 9". Rather than being prompted by a wizard or restricted by a template, **Start from Scratch** allows drawing controls to be set as needed using a dialog box or from the Command prompt. Selecting English will produce a display similar to Figure 2.9.

STARTING A NEW DRAWING FROM WITHIN AUTOCAD

In addition to the four methods you've just explored, AutoCAD allows a drawing to be started from inside an AutoCAD drawing. Use one of the methods introduced earlier in this chapter to start a new drawing. Once the drawing is started, draw a triangle so that the screen resembles Figure 2.24a. A new drawing can be started without harming the current drawing. To open another drawing window, select the **New** button at the far left end of the **Standard** toolbar, or select **New** from the **File** menu. You can also type **NEW** (ENTER) at the Command prompt. Each method will hide the existing drawing file and begin the process of starting a new file. Start the new drawing using any one of the four startup methods you prefer. Your original drawing will be removed from the drawing area, and a new drawing screen will be displayed. Use the **Line** command to draw a rectangle so that your drawing area resembles Figure 2.24b. Click the **Minimize** button and the original drawing will be displayed. Click the **Minimize** button for this drawing, and the display will resemble Figure 2.24c. The drawing tabs of both the open drawings are displayed. Several new drawings can be started, or you can open several existing files in one AutoCAD session. Although this might not seem important now, several of AutoCAD's features will make this very important: AutoCAD allows you to transfer objects, settings, and other types of information from one drawing to another. AutoCAD allows drawing attributes to be transferred between drawings using methods similar to those used with Windows Explorer. Drawing features can be moved between a drawing stored on your hard drive, a drawing on a network, or a drawing located across the Internet. These features will be addressed throughout the balance of the text. For now, remember that two or more drawings can be worked on in the same drawing session.

DISPLAYING MULTIPLE WINDOWS

As with other Windows programs, AutoCAD provides choices on how open windows and drawings will be displayed. Individual program or drawing windows within AutoCAD can be arranged in horizontal, vertical, or cascading tiles. You can display multiple layouts by selecting **Window** from the menu bar, as shown in Figure 2.25. The lower portion of the menu will display a listing of the current open drawing layouts. Selecting the name of the desired view will remove the current drawing and dis-

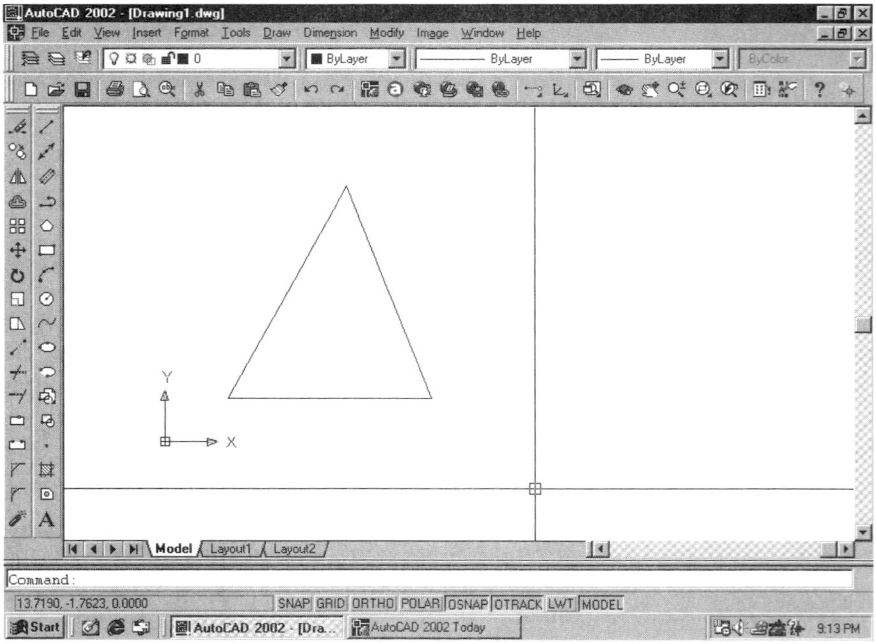

Figure 2.24a *A new feature of AutoCAD allows a drawing to be started from inside an active AutoCAD drawing. Use one of the methods introduced earlier in this chapter to start a new drawing. Once the drawing is started, draw a triangle.*

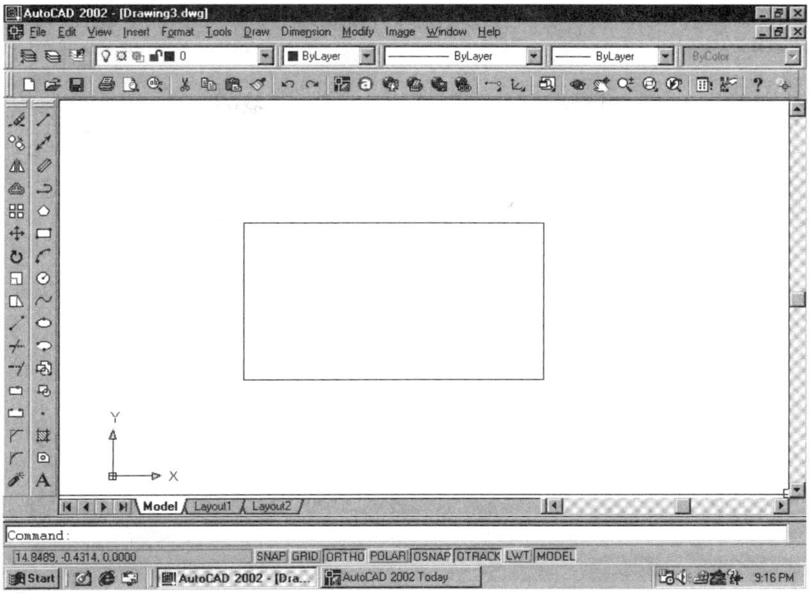

Figure 2.24b *A new drawing can be started without harming the current drawing. Another drawing window can be opened by selecting the* **New** *button on the* **Standard** *toolbar, by selecting* **New** *from the* **File** *menu, or by typing* **NEW** *(ENTER) at the Command prompt. Each method will remove the existing drawing from view and begin the process of starting a new file.*

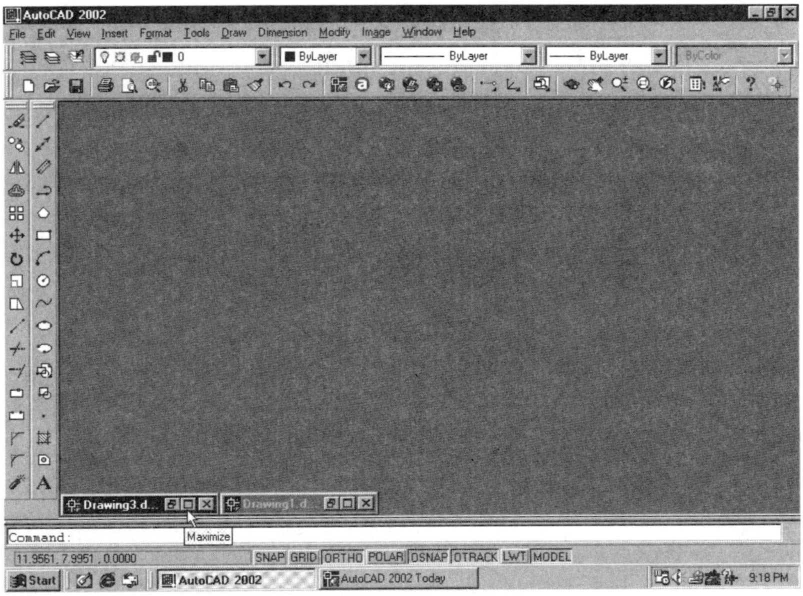

Figure 2.24c *Click the* **Minimize** *button, and the current display will be minimized to display the drawing tabs for all current drawings.*

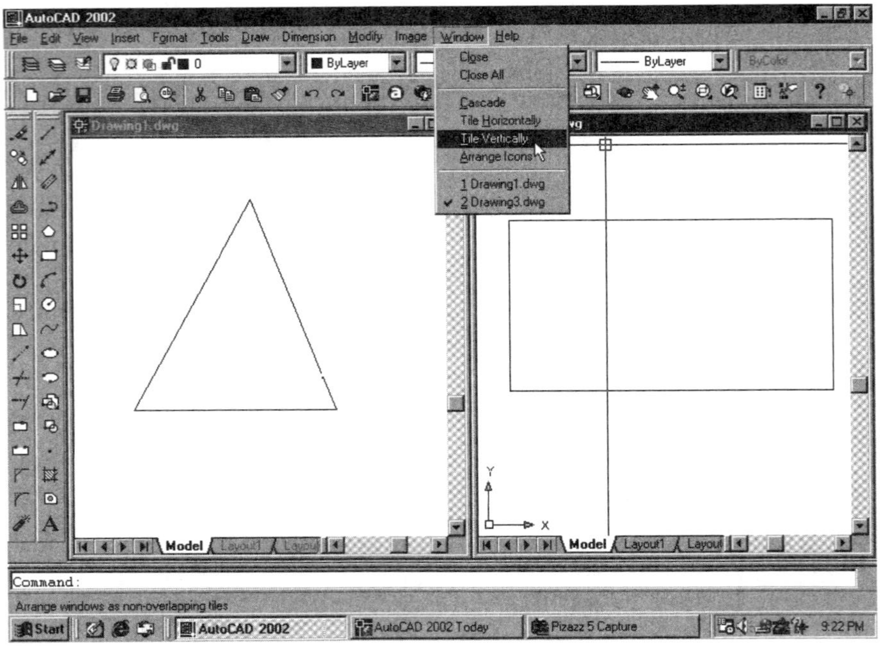

Figure 2.25 *Multiple layouts can be displayed from the* **Window** *menu.*

play the selected drawing. Choose a display for multiple windows by selecting one of the options from the top portion of the menu. A cascading or tiled display is often used when two or more windows are to be displayed.

Cascading Windows

Selecting **Cascade** from the **Window** menu displays the drawings similar to those shown in Figure 2.26a. Cascading drawings are stacked over each other in descending order. Drawings are displayed in the order that they are selected. To display one of the lower drawings, click the title bar of the desired drawing to move it to the top of the stack. Clicking the title bar and holding the select button down allows the window to be dragged anywhere on the screen.

Tile Displays

The **Tile** options allow drawings to be arranged in horizontal (Figure 2.26b) or vertical displays (Figure 2.26c). Tiled windows permit each window to be viewed at the same time. Just as with other programs, if you're unhappy with the size of a window, alter the size by selecting and dragging the edge to the desired size.

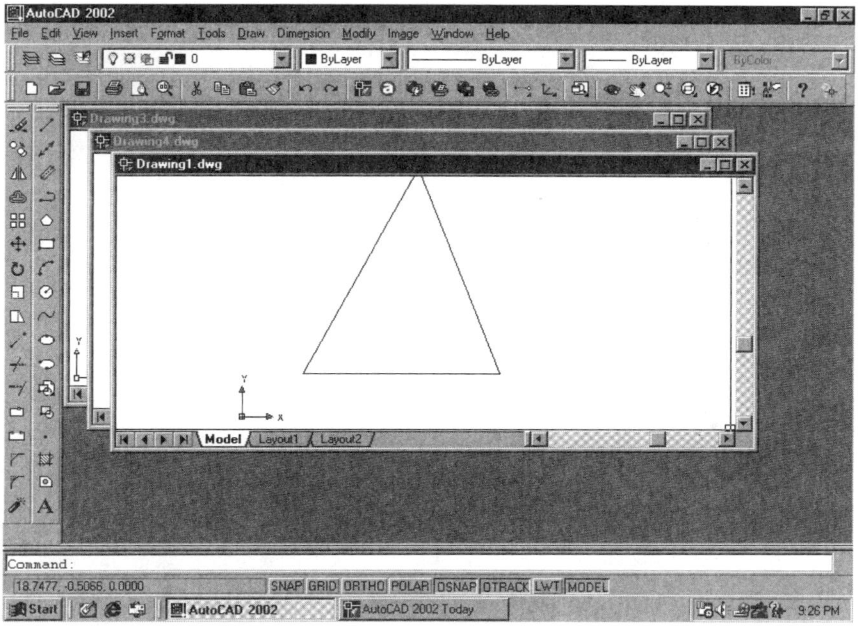

Figure 2.26a *Selecting* **Cascade** *from the* **Window** *menu displays drawing windows that are stacked over each other in descending order.*

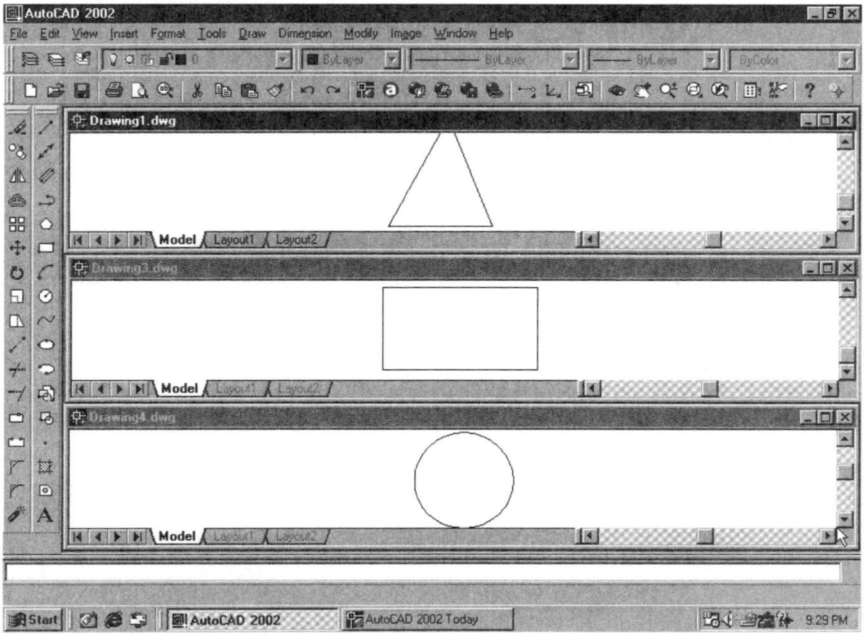

Figure 2.26b *The* **Tile Horizontal** *option displays drawing windows in horizontal rows.*

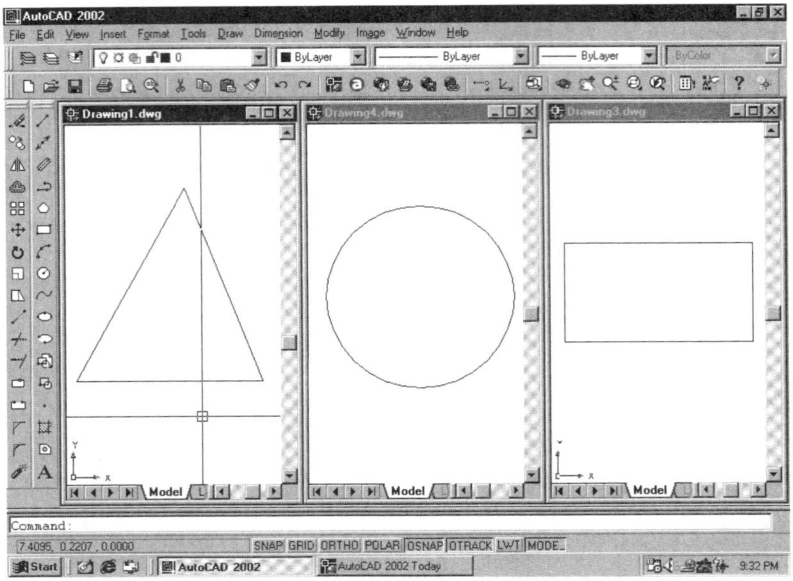

Figure 2.26c *The* **Tile Vertical** *option displays drawing windows in vertical columns*

WORKING WITH PARTIAL FILES

AutoCAD offers two additional methods of working with drawing files. Although each method is beyond your current needs as you explore AutoCAD, you should at least be aware of the **Partialopen** and **Partiaload** commands. Both commands can be used when you work with large files to allow only a specific portion of the drawing to be accessed. The projects that you work on throughout this text will typically be too small to benefit from **Partialopen**. On extremely detailed drawings, **Partialopen** will allow specific information to be viewed in the drawing area. If you're working on a complex floor plan, the plan can be divided into zones, and then each zone can be saved by what AutoCAD refers to as a view. **Partialopen** allows you to open a specific zone of the floor plan and display the framing information, while hiding the electrical or plumbing information. See Chapter 4 for an explanation of how each command can be used to start a drawing session.

Note: You now have several methods that can be used to start a drawing session. Just as important as how you start a drawing session is where the file that you want to open is located. When you're working with existing files, always use files stored on your hard drive to open the drawing session. AutoCAD needs work space as you work on a drawing session. If the drawing and the work space exceed the storage space on the disk, you might encounter unwanted results. If the drawing is stored on a network, you'll typically have no shortage of work space, but your hard drive is generally faster. If the most current file is saved on a diskette or in a network folder, copy the file to the hard drive prior to opening the file.

WORKING WITH FILES WITHOUT AUTOCAD

When Volo View Express is loaded, Autodesk now provides the option of being able to view AutoCAD drawings from the office network or from a Web site without having to use AutoCAD. Volo View Express is supplied with AutoCAD and can be used to view, review, and print AutoCAD drawings without loading AutoCAD. This feature will allow consultants to review AutoCAD drawings away from the work environment. The viewer will be especially useful in reviewing a project using the **Markup** without opening AutoCAD. To review a drawing, a client can be given a copy of the base drawing to be reviewed. Posting a drawing on a secure Web site (see chapter 26) will allow access by multiple consultants. With one of the options from the **Tools** or **Markup** tab of Volo View, text or a sketch can be added to the drawings. Once the review is complete, the suggested edits can be saved to a file written in redline markup language format (.rml) that is linked to the original file. The original drawing remains unaltered, giving you the final say over which suggested changes get added to the drawing base. Volo View Express can also be used to view drawings created using Architectural Desktop 2.0, Inventor, and Mechanical Desktop 4.0 if object enablers are downloaded from the Web site of the desired application.

CONTROLLING DRAWING PARAMETERS FROM WITHIN A DRAWING

No matter what method is used to begin a drawing, the drawing parameters can be altered at any time. Parameters can be altered through dialog boxes, menus, or the keyboard.

SETTING UNITS

The **Drawing Units** dialog box (see Figure 2.27) can be used to change the drawing units. The **Drawing Units** dialog box can be displayed at any point of a drawing session using one of the following methods:

- Select **Units** from the **Format** menu.
- Type **UN** (ENTER) at the Command prompt.

The drawing units are set by selecting the desired style from the **Type** list. Selecting the arrow on the right side of the **Precision** edit box will display the drop-down list. Select the desired precision by choosing the desired value from the **Precision** list. The dialog box can also be used to select the method for measuring angles and for setting the angle precision. Angle precision accuracy is set using the same methods as for changing the unit precision. Selecting the **Direction** button will display the **Direction Control** dialog box (see Figure 2.28). This dialog box allows the angle starting point and the angle direction to be set, by selecting the appropriate radio button or highlighting the desired accuracy. Once it is selected, the accuracy will be displayed in the edit box and the menu will be closed.

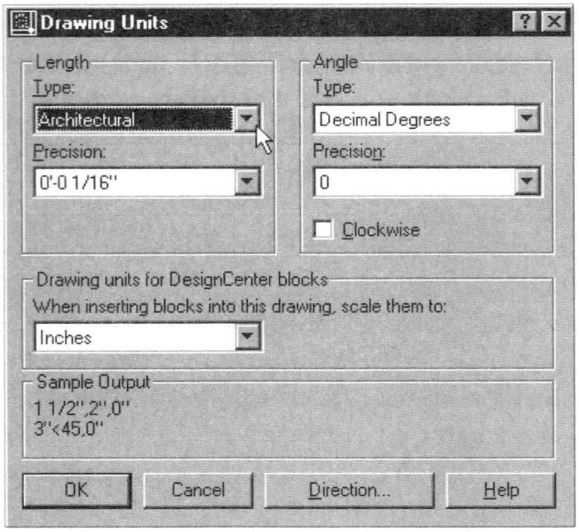

Figure 2.27 *Drawing units can be altered at any time using the* **Drawing Units** *dialog box. Display the dialog box by selecting Units from the Format menu, or by typing* **UN** *(ENTER) at the Command prompt.*

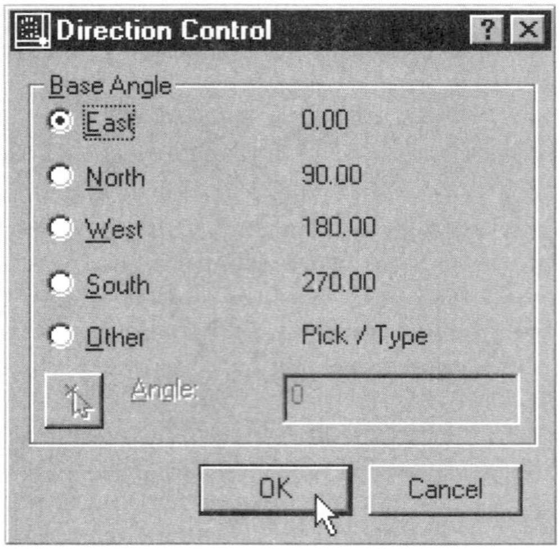

Figure 2.28 *Selecting the* **Direction** *button in the* **Drawing Units** *dialog box allows the orientation for angle direction to be altered.*

SETTING LIMITS

Changing the limits alters the size of the drawing displayed on the screen. Before you change the limits, draw a square inside the drawing area, similar to Figure 2.29a. This will help you to visualize the change that occurs when limits are set.

To change the size of the screen that can be viewed, select **Drawing Limits** from the **Format** menu or type **LIMITS** (ENTER) at the Command prompt. The drawing limits can also be altered in the middle of another command if you type **LIMITS** (ENTER) at the Command prompt. Each method of entering the command will produce the display:

Reset Model space limits:

Specify lower left corner or [ON/OFF] <0'-0",0'-0">:

Acceptable responses are ON, OFF, and (ENTER). ON will keep the current values (0'–0",0'–0") and activate limit checking. Limit checking will not allow objects to be drawn that are outside the limits. Think of limit checking as a border on your drawing paper to keep your project on the paper. OFF will disable limit checking but retain the limit values for future use of limit check.

Pressing (ENTER) retains the current location of the lower left drawing limit of (0'–0",0'–0"). This will be the most typical response. Pressing (ENTER) produces the prompt:

Specify upper right corner (1'-0", 0'-9"):

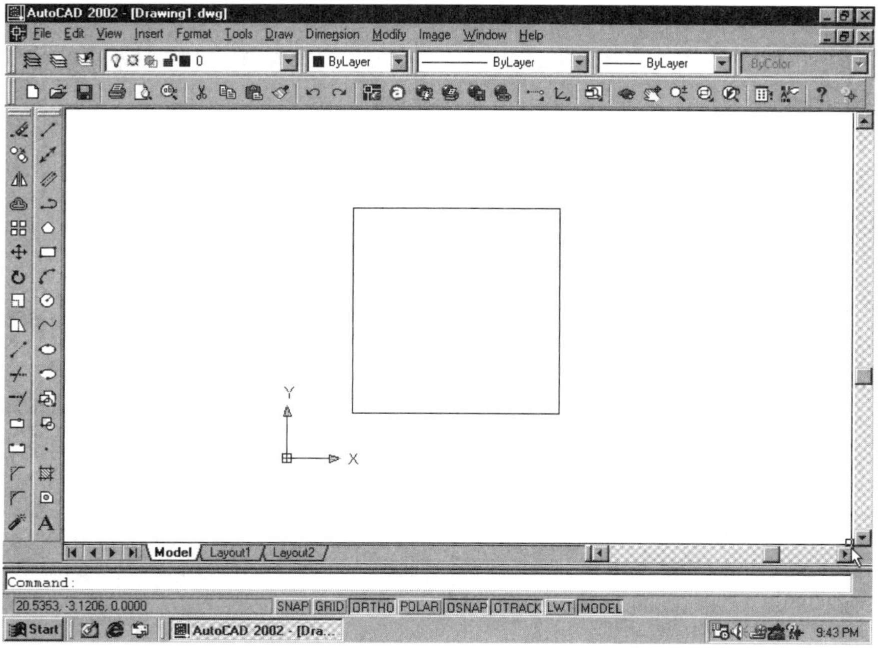

Figure 2.29a *With the drawing limits set at the default units of 1'–0",0'–9", an object might appear quite large.*

The default value is 1'–0",0'–9", with the first number representing the horizontal dimension and the second number representing the vertical dimension. This is your opportunity to change the size of the drawing area to fit your current project. Type **15',12'** (ENTER), and the Command prompt will return with no apparent change. Type **Z** (ENTER) at the keyboard, then **A** (ENTER) (the A represents all). Now the effect of your new limits can be seen in Figure 2.29b. Remember that the box is still the same size but it appears smaller because the area to be viewed has increased.

SETTING GRIDS

Another helpful tool for drawing layout is a grid. This will provide the equivalent of the nonrepro blue grid that is preprinted on drafting vellum. The **Grid** command will produce a visible grid of dots at any desired spacing. The grid size can be adjusted throughout the life of the drawing, depending on the size of the object being designed. The grid also can be displayed or removed to aid in visual clarity, but it will not be produced when the drawing is printed or plotted, even if it is displayed on the screen. Figure 2.30 shows an example of a drawing with the grid display.

The grid can be set by dialog box or from the command line. The current status is displayed on the status bar. Once the values have been set, the grid can be toggled

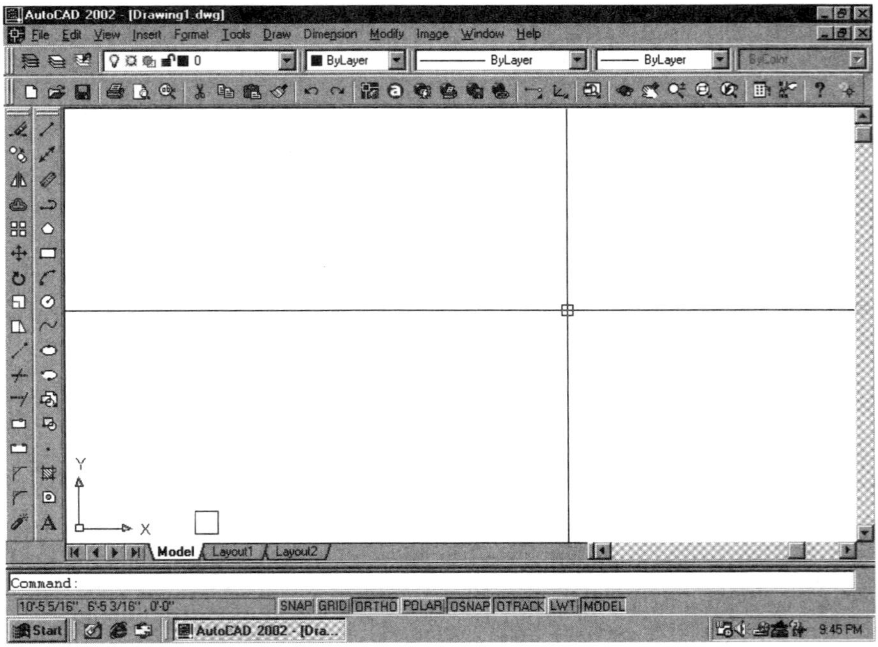

Figure 2.29b *By changing the location of the upper right corner and pressing* ENTER, *you can allow viewing of a larger area. In this example, the limits are set to 15',12'.*

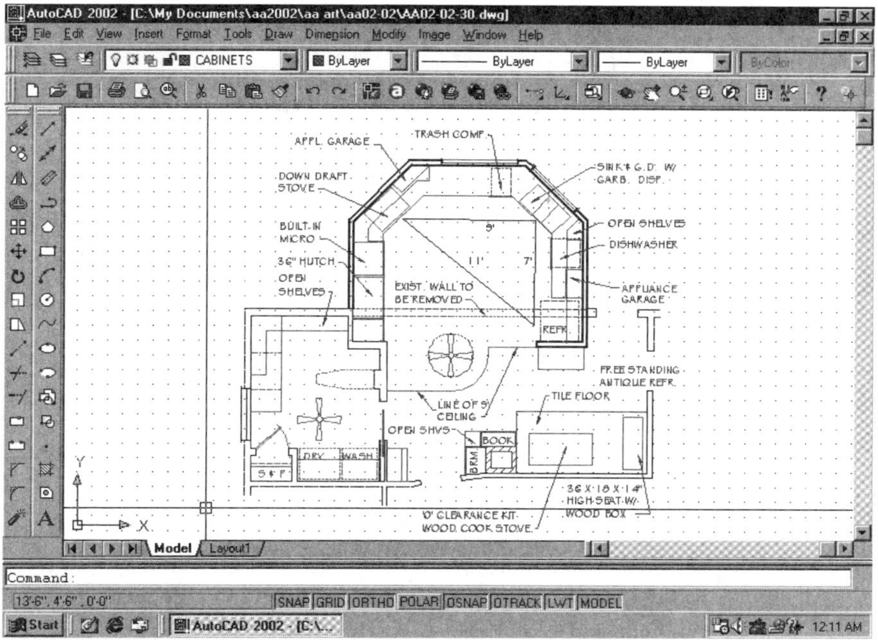

Figure 2.30 *Drawing limits of 15',12' with a grid value of 6".*

ON/OFF by selecting the GRID button on the status bar. The grid can also be toggled ON/OFF in the middle of another command by using the F7 key or by pressing Ctrl+G.

Assigning Grid Values Using a Dialog Box

Grid values can be set using the **Drafting Settings** dialog box, accessed by selecting **Drafting Settings** from the **Tools** menu. The dialog box, shown in Figure 2.31, can also be displayed by moving the cursor to the GRID or SNAP button on the status bar, right-clicking, and then choosing Settings from the shortcut menu, or by typing **RM** (ENTER) at the Command prompt. Once the **Drafting Settings** dialog box is displayed, grids are set using the following procedure:

1. Select the **Snap and Grid** tab.
2. Enter the desired value for the X (left-to-right) grid spacing.
3. Enter the desired value for the Y (top-to-bottom) grid spacing. If you want the Y value to match the X value, click the select button. Pressing (ENTER) will also match the X and Y values and close the dialog box. To enter a different Y value, enter the desired value.
4. Click the OK button.

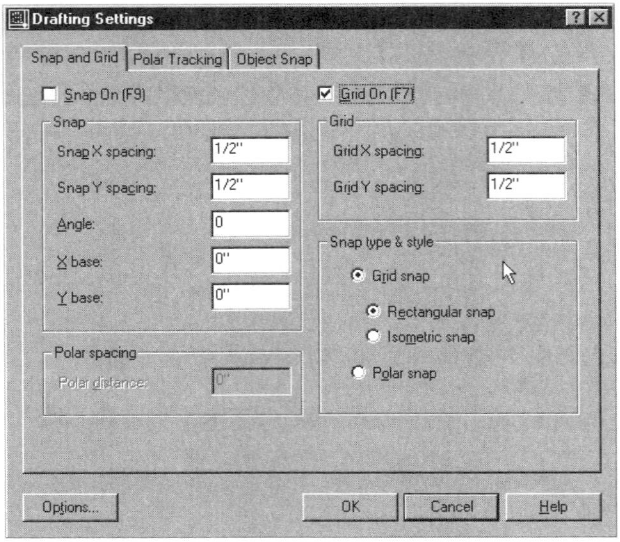

Figure 2.31 *Values for the grid can be altered on the* **Snap and Grid** *tab of the* **Drafting Settings** *dialog box by right-clicking the GRID button on the status bar and then selecting* **Settings**. *Grids can also be altered by selecting* **Drawing Aids** *from the* **Tools** *menu or by typing* **RM** *(ENTER) at the Command prompt and then selecting the* **Snap and Grid** *tab.*

The dialog box also offers methods of altering the angle used to display the grid. Type the desired angle for the grid in the **Angle** edit box. Figure 2.32 shows an example of a floor plan drawn with a grid rotated to 45°. Selecting the **Isometric Snap** setting will establish a grid suitable for isometric drawings. Isometric drawings will be discussed later in this text.

Setting Grids at the Command Line

The grid values can be established, altered, and displayed from the Command prompt. The controls for the grid are accessed at the Command prompt by typing **GRID** (ENTER). The command line will now read

Command: **GRID** (ENTER)

Specify grid spacing (X) or [ON/OFF/Snap/Aspect]
<0'-0">:

Type **12** (ENTER). This will return the Command prompt and display the grid field with dots at 1'–0" spacing on the drawing screen. Grid should be set in units common to the object to be drawn. For instance, when you draw a foundation plan, the grid might be set at 6".

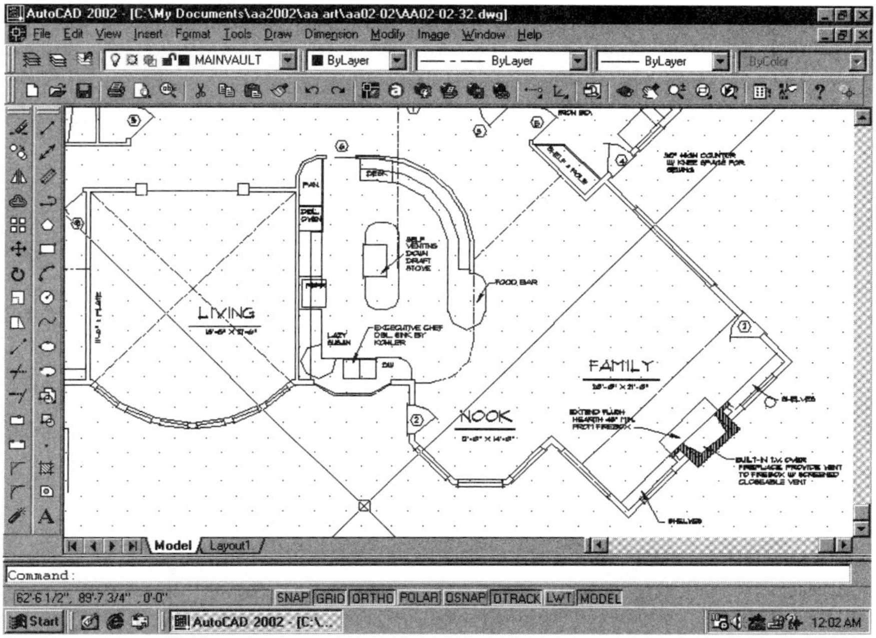

Figure 2.32 *The grid can be rotated to ease layout of an irregular shape.*

SETTING SNAP

When working on a modular component, it is helpful to be able to control the accuracy of lines entered with the mouse. The **Snap** command sets up an imaginary rectangular grid that controls placement of the drawing cursor in precise intervals. The distance between cursor movements is defined as snap resolution. Snap intervals can be set at any interval and can be changed throughout the life of the drawing. A change in the snap grid affects only the points entered after the change in resolution. Points entered prior to the change in resolution might no longer line up with the new snap location. The size of the grid setting and the size of the object to be drawn will affect the snap value. If the grid is set at 2', a snap value of 6" or 12" will be helpful in producing accuracy. If a detail of a beam to post is drawn, a snap of 1/2" might be used to accurately move through the drawing. If a 200' long structure is to be drawn, a 24" or 36" snap will be more helpful than the 1/2" accuracy used for a detail.

Snap can be set using the **Snap and Grid** tab of the **Drafting Settings** dialog box. It can also be set by right-clicking the SNAP button on the status bar and selecting **Settings** from the shortcut menu. Snap is adjusted using the same methods that were used to adjust the grid.

Values for the X and Y spacing can be entered in the appropriate edit box. Once the values have been set, snap can be toggled ON/OFF by double-clicking the SNAP button on the status bar. Snap can also be toggled ON/OFF in the middle of another command by using the F9 key or by pressing Ctrl+B. Entering an angle in the **Snap Angle** edit box controls the display angle for snap and grid. Entering a value in the X or Y **Base** edit boxes allows the base point for the snap pattern to be established.

Setting Snap at the Command Line

The snap values can be established, altered and displayed from the Command prompt. The controls for snap can be accessed at the Command prompt by typing **SN** (ENTER). This will produce the prompt:

> Specify snap spacing or [ON/OFF/Aspect/Rotate/Style/Type] <0'-0">:

Entering a numeric value will set the spacing of the grid for entry points. Type **6** (ENTER). The command line returns and allows for drawing. Type **L** (ENTER) and draw a line. You'll notice that the crosshairs jump across the screen in 6" intervals rather than moving smoothly as they had before. In addition to ON/OFF, settings include Aspect, Rotate, Style and Type.

Aspect

Some projects are set up where different values for the horizontal (X) and vertical (Y) directions are used. To do this, type **SN** (ENTER). Type **A** (ENTER) for Aspect and the prompt will display the additional listings:

> Specify horizontal spacing <0'-0">: **6** (ENTER)
>
> Specify vertical spacing <0'-0">: **12** (ENTER)
>
> Command:

Rotate

Because not all projects are rectangular, such as the floor plan in Figure 2.32, it might be helpful to rotate the snap grid to match the drawing. Type **SN** (ENTER) and then type **R** (ENTER) for Rotate at the prompt. This will produce the additional display:

> Specify snap spacing or [ON/OFF/Aspect/Rotate/Style/Type] <A>: **R** (ENTER)
>
> Specify base point <0'-0", 0'-0">:

AutoCAD is asking for the point of the drawing where you would like the snap grid to be rotated. To select a location, move the crosshairs to the desired location in the drawing and press the select button. This will now provide the prompt:

> Rotation angle <0>:

A number between −90 and 90 can be entered. Using a negative angle will rotate the snap grid counterclockwise. Using a positive angle will rotate the snap grid clockwise. Type a value and press (ENTER). The snap grid will be rotated around the base point just selected.

Style

The format of the snap grid is adjusted by using the Style option. Type **S** (ENTER) for Style at the snap prompt, and the prompt will display

> Specify snap spacing or [ON/OFF/Aspect/Rotate/Style/Type] <A>: **S** (ENTER)
>
> Enter Snap grid style [Standard/Isometric] <S>:

For orthographic drawings the style will not need to be set. For isometric drawings type **I** (ENTER).

Type

This setting toggles between polar and grid snap. Polar snaps align with polar angles relative to the Polar Tracking point. Grid snaps along the grid. Type **T** (ENTER) for Type at the snap prompt, and the prompt will display

> Specify snap type [Polar/Grid]<Grid>:

Note: Coordinating the grid and snap will provide great accuracy and speed. If the grid value is set at zero, the grid spacing automatically adjusts to the snap resolution. To specify the grid spacing as a multiple of the snap resolution, type **X** (ENTER) after the grid value. For instance, if the snap value is 2", typing **6X** (ENTER) at the Command prompt produces a grid of 12".

CHAPTER 2 QUIZ

1. What command will set an invisible grid to aid layout?
2. List the five options of the units.
3. What is the default unit for angle measurement?
4. What is the default for angle direction?
5. Your boss would like a drawing of a street and sewer layout based on the surveyor's notes. Describe the process for setting up the proper method of two-place angle measurement by keyboard.
6. List three methods to toggle between **Grid** ON/OFF.
7. List three methods to toggle between **Snap** ON/OFF.
8. What is the main difference between **Grid** and **Snap**?
9. You've just entered AutoCAD and don't want to establish any drawing parameters. How can you get to the drawing screen?
10. When would **Start from Scratch** be a useful method of entering a drawing?
11. What does **Area** control?
12. A site plan must be drawn with angles measured in a format such as N42° 30'-30" E. How can this be done if you forgot to adjust the angle measurement as you set up the drawing?
13. A friend wants to draw a house but has no idea how big it will be. What would you recommend setting for drawing limits as the drawing is started?
14. How can you use the **Advanced Setup** wizard to start a drawing?
15. What command is used to access the **Drafting Settings** dialog box from the Command prompt?

CHAPTER 3

Drawing and Controlling Lines

This chapter will introduce methods for

- Entering the **Line** command, including toolbars, keyboard, and the **Draw** menu
- Controlling the placement of lines, including absolute coordinates, relative coordinates, and polar coordinates
- Controlling lines with the tools of the status bar and for editing lines

Commands to be introduced include

- **Line**
- **Close**
- SNAP
- GRID
- ORTHO
- POLAR
- OSNAP
- **Erase**
- **Oops**
- **Undo**
- **Redo**

Now that you're able to set up drawing units and limits, basic drawing components can be explored. The main components of the drawings you'll be working with will consist of lines. Chapter 1 introduced the **Line** command using the **Draw** toolbar, the keyboard, and the **Draw** menu. This chapter introduces methods for making effective use of the **Line** command, as well as methods for locating and altering lines.

LINES

The basic components of every drawing are lines, which are used to represent topography, grades, pipelines, easements, walls, roads, property lines, and an endless list of other construction materials. Figure 3.1 shows the lines required to represent a portion of a framing plan for a multilevel steel structure. Although many of these lines are represented with different line patterns or thickness, each can be drawn using the **Line** command. Lines are used to represent different types of information within a drawing. Establishing and controlling varied linetypes will be introduced in later chapters.

ENTERING THE LINE COMMAND

The **Line** command is accessed by toolbar, keyboard, or menu. Each method was introduced in Chapter 1. To start the command, select the **Line** button on the **Draw** toolbar. This will produce the prompt:

> Command: *(Select the Line button.)*
>
> _line Specify first point:

Move the cursor to where you would like to start a line and click the select button. The "first point" will become the beginning point of a line segment. As the first point is selected, the prompt displays

> Specify next point or [Undo]:

Move the cursor to the desired location and click the select button to specify the endpoint. As the cursor is moved, a rubber-band line is displayed from the first point to the crosshairs location. This can be seen in Figure 3.2. The rubber-band line will help you to visualize the placement of the line to be drawn and to experiment with its placement. The rubber-band line will remain as the crosshairs are moved until the next point is selected or the **Line** command is terminated. There are five options for ending the **Line** command. Each method will draw the indicated line and terminate the command:

> Press (ENTER).
>
> Press (SPACEBAR).
>
> Press (ESCAPE).
>
> Right-click and select **Enter**.
>
> Right-click and select **Cancel**.

COMMAND LINE INPUT

A second method of entering commands is by keyboard. Enter the **Line** command by typing **L** (ENTER) at the Command prompt. The sequence for entering the **Line** command is as follows:

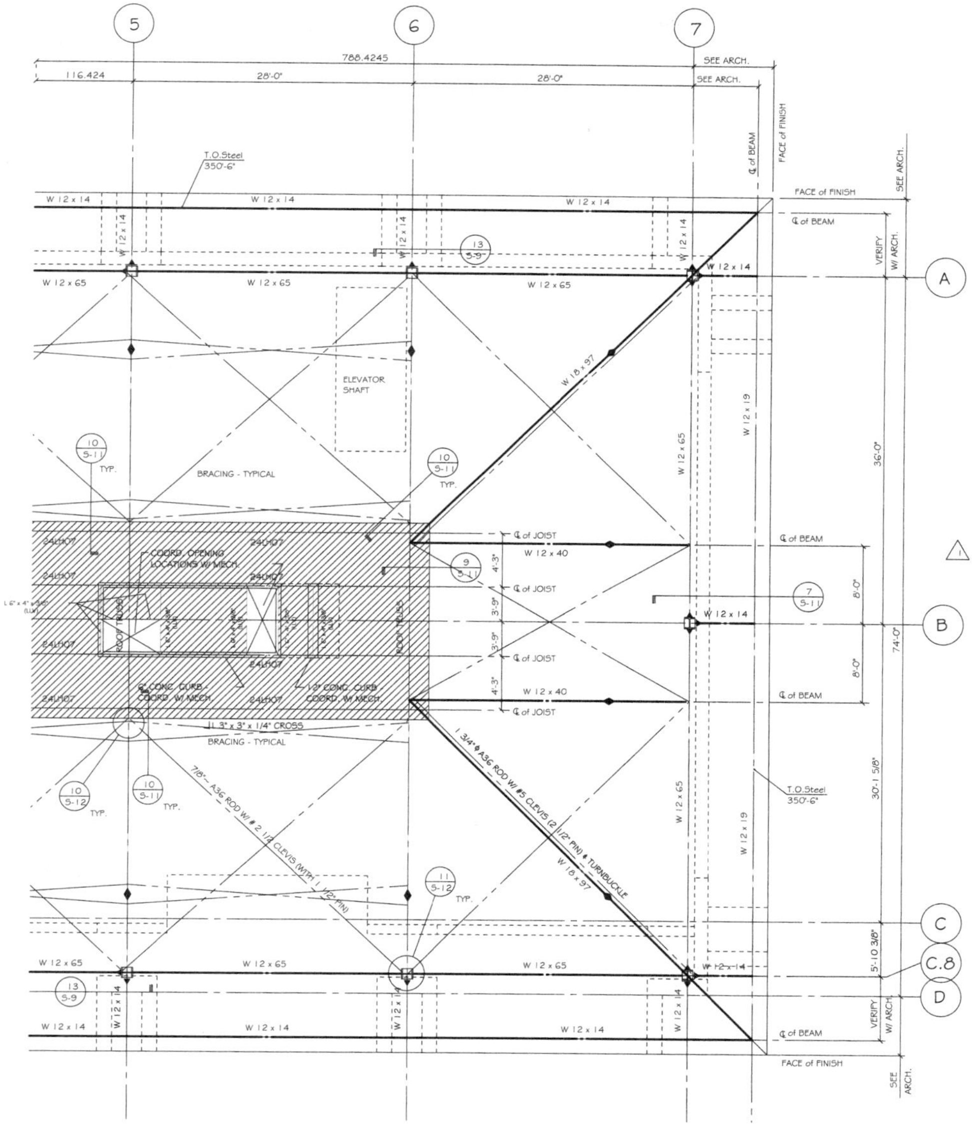

Figure 3.1 *Lines of varied widths and patterns are used on most construction drawings to represent varied materials. Methods of changing linetype and line width will be discussed in later chapters. (Courtesy Van Domelen/Looijenga/McGarrigle/Knauf Consulting Engineers.)*

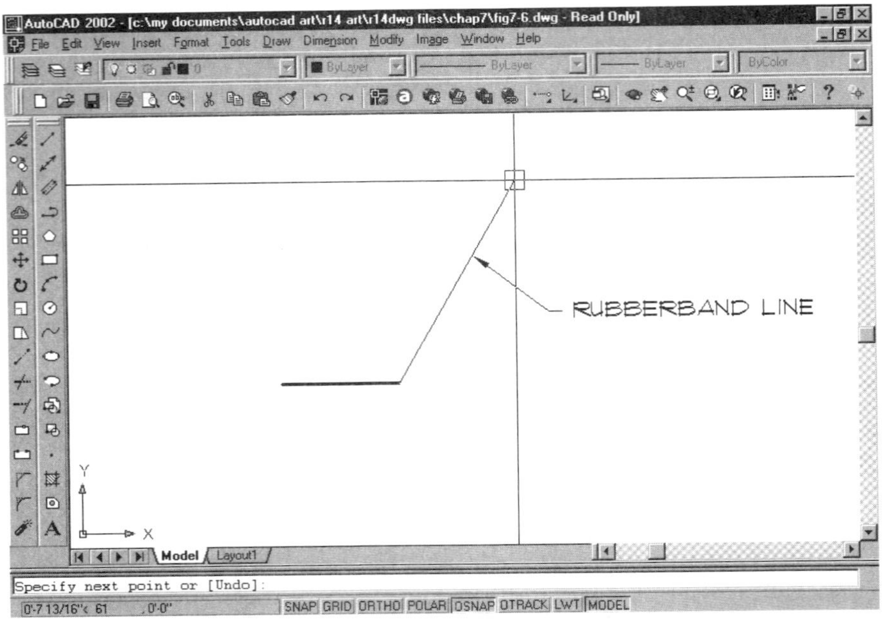

Figure 3.2 *When a "next point" is provided, a rubber-band line extends from the "first point" to the present crosshairs location. This line indicates the line that will be drawn if the current crosshairs location is used.*

Command: **L** (ENTER)

LINE specify first point: *(Specify an end point.)*

Specify next point or [Undo]: *(Specify an end point.)*

The **Line** command functions exactly the same no matter how it is entered.

MENU ENTRY

A third method for entering a command is to select **Line** from a menu. When you select **Line** from the **Draw** menu, the menu will disappear and the Command prompt will show

Command: *(Select Line from the Draw menu)*

_line Specify first point:

Notice that four other options for lines are presented in the **Draw** menu—**Ray**, **Construction Line**, **Multilines**, and **Polylines** commands. A fifth method of drawing lines is **Sketch**. The **Sketch** command allows freehand drawing methods to be reproduced. Each of these commands will be introduced in later chapters.

No matter which method is used to start the command, a request for the "first" and "next" points is displayed at the command line, and can be entered with the mouse, or by keyboard. Specific coordinates can be entered by keyboard rather than by selecting the points with the mouse. Methods of selecting line and point by keyboard will be discussed later in this chapter. As you start to experiment with drawing lines, you should draw objects combining all entry methods.

Note: You've entered the **Line** command using three different methods. As you start to experiment with drawing lines, you should draw objects combining all entry methods. Each method works well to initially start the command. The easiest way to repeat a command is to press (ENTER) or the SPACEBAR; both will repeat the previous command.

LOCATING LINES USING COORDINATE ENTRY METHODS

As you start to complete drawings in this chapter, it will be important to understand the user coordinate system (UCS) of AutoCAD. The UCS is based on a Cartesian coordinate system, dividing space into four quadrants. Figure 3.3 shows the division of space based on the Cartesian coordinates. Points in these quadrants are located in

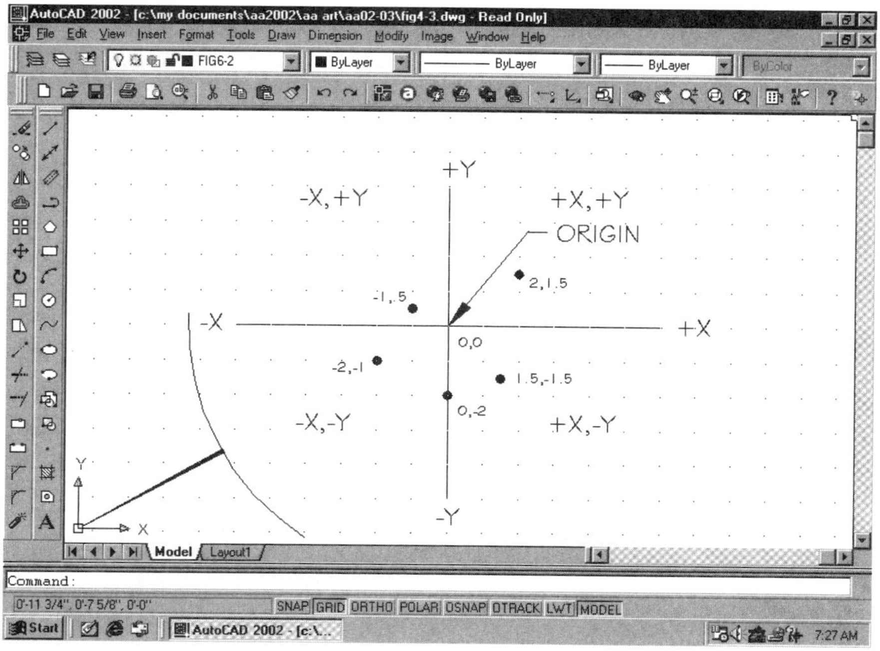

Figure 3.3 *The division of space using the Cartesian coordinate system. The intersection of the X and Y coordinates is assumed to be 0,0, which is the lower left corner of the screen. The grid has been moved to the center of the screen to represent negative X and Y locations.*

turn by using their points in a horizontal (X) and vertical (Y) direction along the plane. Points are always specified by the X coordinate followed by the Y coordinate.

Figure 3.4 shows the relationship of the drawing screen to the Cartesian coordinates. As you begin to draw, AutoCAD will ask you to specify a "first point." As you move to the "next point," the coordinate display at the left end of the status line will reflect the crosshairs movement. To activate coordinate display, press F6 or CTRL+D.

Because the fields of engineering and architecture deal with very specific locations, a means of controlling the point placement is desirable. This can be done by using absolute, relative, or polar coordinates for point entry methods.

ABSOLUTE COORDINATES

The absolute coordinate system locates all points from an origin assumed to be 0,0. The 0,0 origin point is assumed to be the same 0,0 point that the limits, snap, and grid are based on, although it can be changed. The axes for the X and Y coordinates intersect at 0,0. Each point on the screen is located by a numeric coordinate based on the original axis. Objects are drawn by entering the coordinates of X and Y separated by a comma. In the example in Figure 3.5, a line is drawn between 2,2 and 2,4. The next

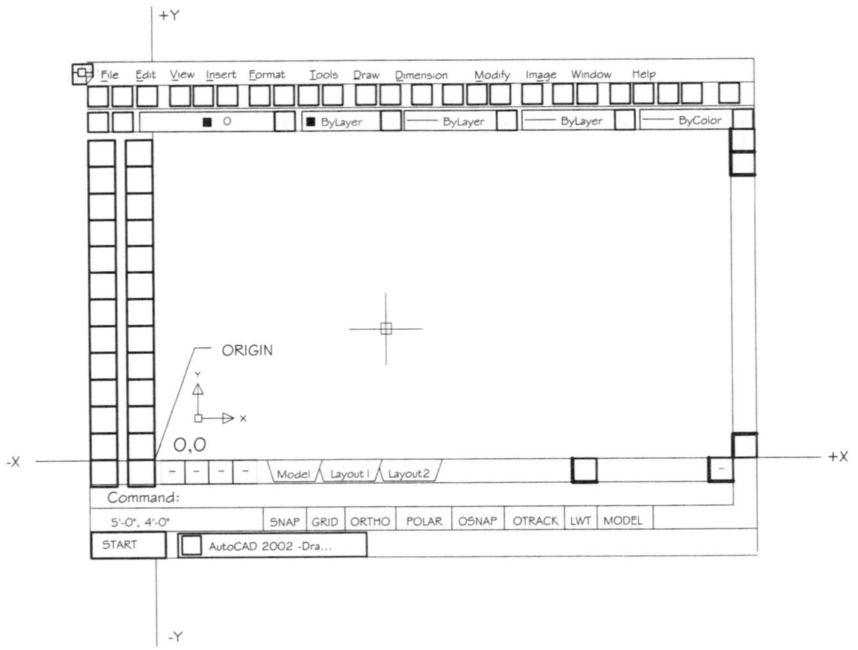

Figure 3.4 *The relationship of the drawing screen to Cartesian coordinates with the limits set at the lower left = 0,0, and the upper right values set as two positive values.*

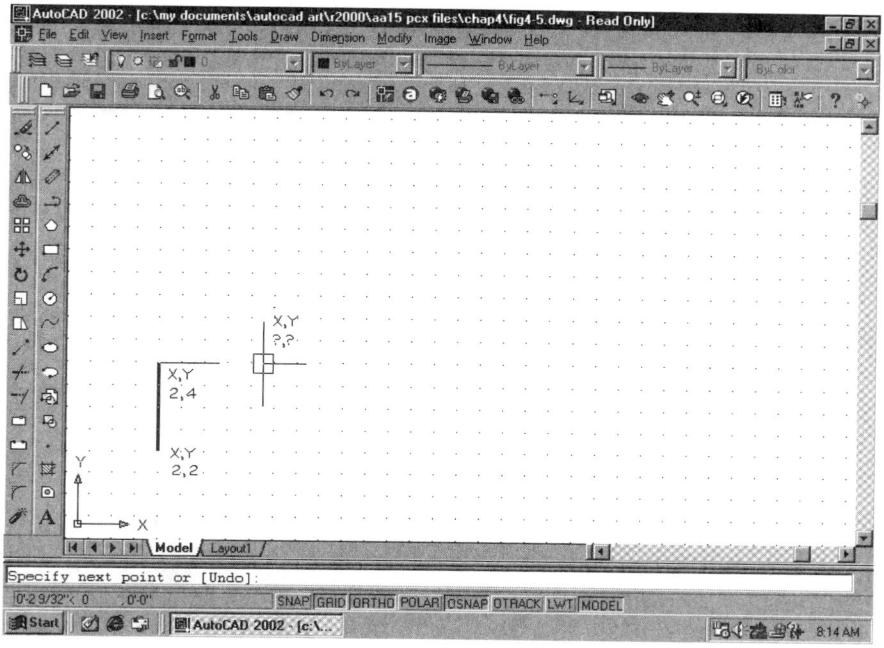

Figure 3.5 *With absolute coordinates, and the 0,0 location as the lower left corner of the drawing screen, the line drawn starts at 2" over (+X) and 2" up (+Y) and extends to a location 2" over (+X) and 4" up (+Y).*

"to point" is 4,4, based on the cursor location and the current coordinate display. To complete the drawing in Figure 3.6, use the following command sequence:

> Command: **L** (ENTER) *(Or click the Line button on the Draw toolbar.)*
>
> LINE Specify first point: **4,1.5** (ENTER)
>
> Specify next point or [Undo]: **6,1.5** (ENTER)
>
> Specify next point or [Undo]: **6, 3.5** (ENTER)
>
> Specify next point or [Close/Undo]: **5,5** (ENTER)
>
> Specify next point or [Close/Undo]: **4,3.5** (ENTER)
>
> Specify next point or [Close/Undo]: **C** (ENTER)
>
> Command:

Notice that **C** (ENTER) was typed as the last command, rather than numeric coordinates. C represents **Close. C** (ENTER) will end the **Line** command sequence at the point where the line sequence was begun. With Object Snap activated (click the OSNAP button on the status bar), moving the cursor to the desired endpoint will also allow easy closure.

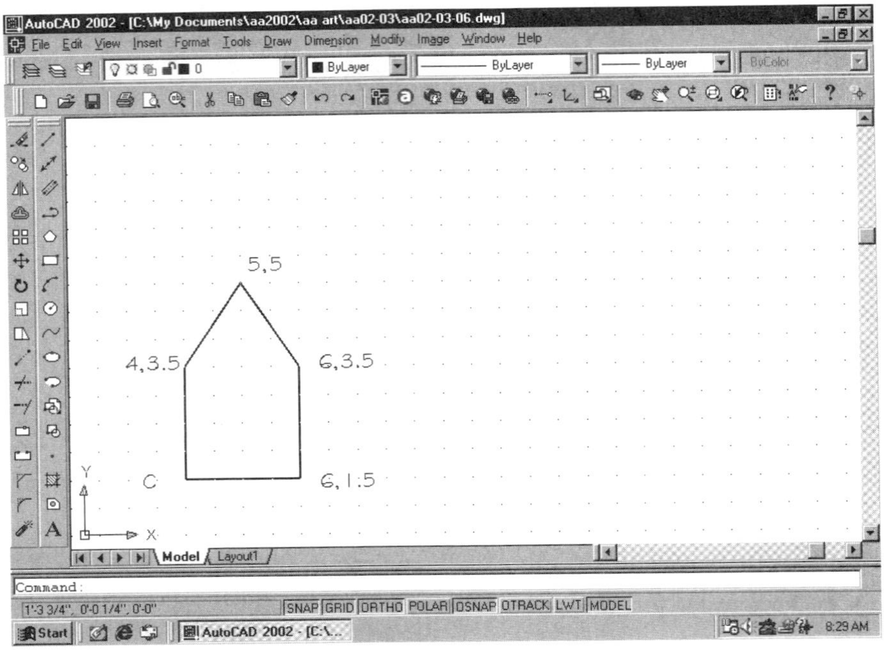

Figure 3.6 *A simple shape and the absolute coordinates required to describe it.*

Note: An alternative to using the **C** option to close the object is to use the Object Snap (OSNAP) option of AutoCAD. As the cursor is moved near the end of the original line a yellow box is placed around the end of the original line. (The color of the box can be altered in the **Options** dialog box by selecting the **AutoSnap Settings** area of the **Drafting** tab.) The box indicates that the nearest endpoint has been selected to become the next point. Object Snap will be discussed in detail in Chapter 7.

Because absolute coordinates are based on an origin of 0,0, they are often not convenient for most architectural and engineering projects. Many CAD users find relative coordinates more suited to a construction project.

RELATIVE COORDINATES

Relative coordinates locate "to points" based on the last point instead of the 0,0 origin. Relative coordinates are entered using a similar method as absolute coordinates. Absolute coordinates are entered as X,Y. Relative coordinates are entered as @ X,Y. In addition to preceding the coordinate value with the @ symbol, with relative coordinates, you need to be mindful of positive and negative values.

Figure 3.7 shows an object drawn using relative coordinates. Positive coordinates will move the crosshairs to the right or above the origin point. Negative coordinates will move the crosshairs down or to the left of the last entry point. The object in Figure 3.7 was drawn using the following command sequence:

Command: L (**ENTER**) *(Or click the Line button on the Draw toolbar.)*

LINE Specify first point: **4,1.5** (ENTER)

Specify next point or [Undo]: **@2,0** (ENTER)

Specify next point or [Undo]: **@0,2** (ENTER)

Specify next point or [Close/Undo]: **@–1,1.5** (ENTER)

Specify next point or [Close/Undo]: **@–1,–1.5** (ENTER)

Specify next point or [Close/Undo]: **@–1,–1.5** (ENTER)

Specify next point or [Close/Undo]: (ENTER)

Command:

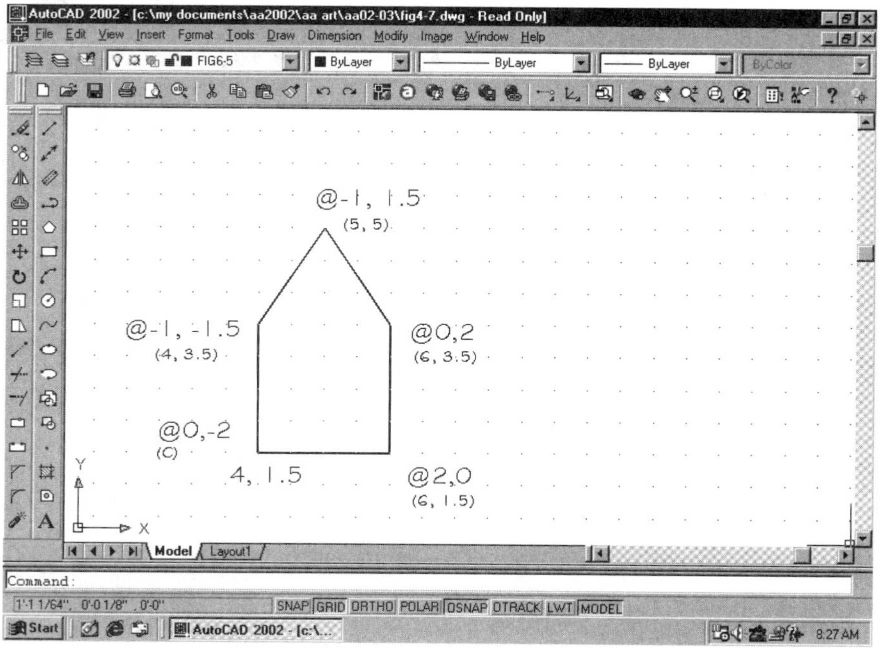

Figure 3.7 *Relative coordinates describe the location of the next point based on the last point. Relative coordinates must be preceded by the @ symbol. The value* **@2,0** *would extend a line 2" to the right with no vertical rise. Relative coordinates are a common entry choice for many projects.*

The first point was entered using absolute coordinates, with the following coordinate points entered by relative coordinates. The final coordinates could have been omitted and the object closed using Object Snap or by typing **C** (ENTER).

POLAR COORDINATES

Polar coordinates use the last entry point as the origin and combine a distance and an angle. This is useful on drawings such as sections for drawing a roof, or site plans with angled property lines. Polar coordinates are entered as

@3<27.5 (ENTER)

This would produce a line 3" long at a 27.5° angle, as shown in Figure 3.8. To draw the opposite side of this roof truss would require typing **@3<332.5** (ENTER). Remember, polar coordinates are based on the current origin of the crosshairs. In Figure 3.9 the location of line 2 at 332.5° is determined by subtracting 27.5° from 360°.

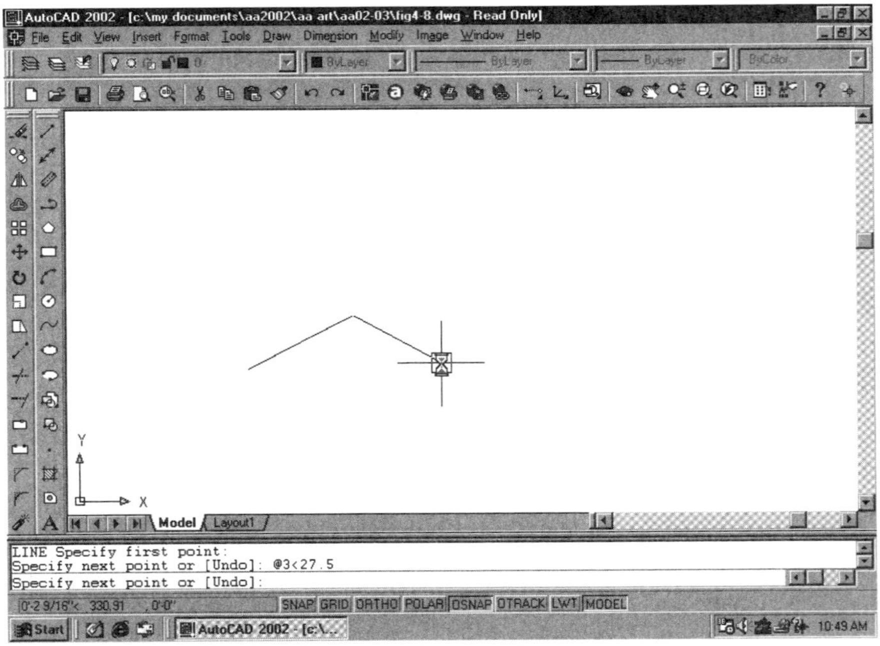

Figure 3.8 *Polar coordinates describe a point relative to the last position using a length and an angle. The @ symbol precedes the length, and the < symbol precedes the angle specification in degrees. Polar coordinates are useful in drawing angled sides of a parcel of land.*

To complete the triangle started in Figure 3.9, use the Object Snap feature of the cursor or type **C** (ENTER). Figure 3.10 shows an object drawn using polar coordinates. The required command sequence would be as follows:

Command: **L** (ENTER) *(Or click the Line button on the Draw toolbar.)*

LINE Specify first point: **4,1.5** (ENTER)

Specify next point or [Undo]: **@2<0** (ENTER)

Specify next point or [Undo]: **@2<90** (ENTER)

Specify next point or [Close/Undo]: **@1.8<124** (ENTER)

Specify next point or [Close/Undo]: **@1.8<236** (ENTER)

Specify next point or [Close/Undo]: *(With OSNAP ON, move the cursor to the origin point and press the select button or type* **C** *(ENTER).)*

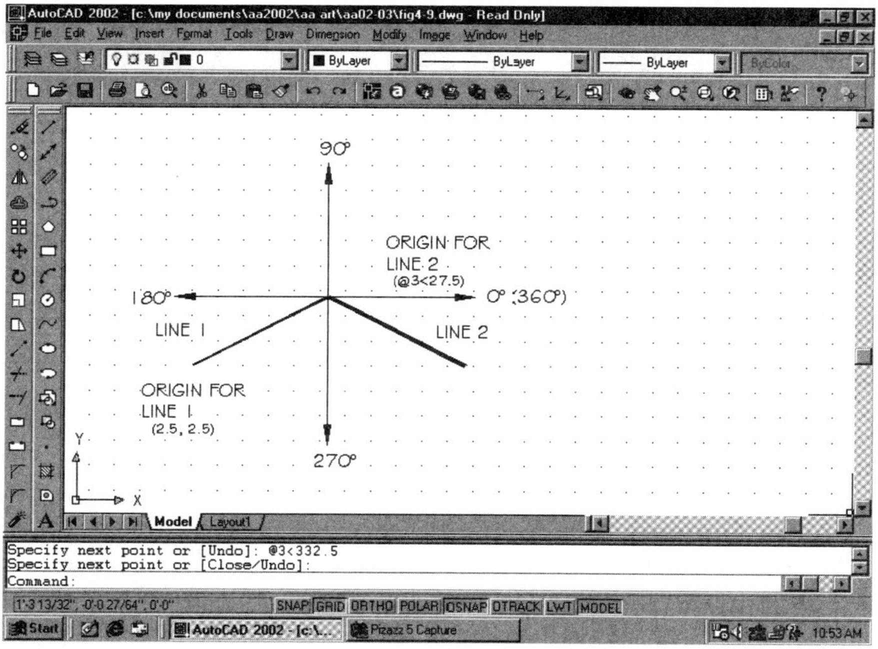

Figure 3.9 *Specifying angles using polar coordinates.*

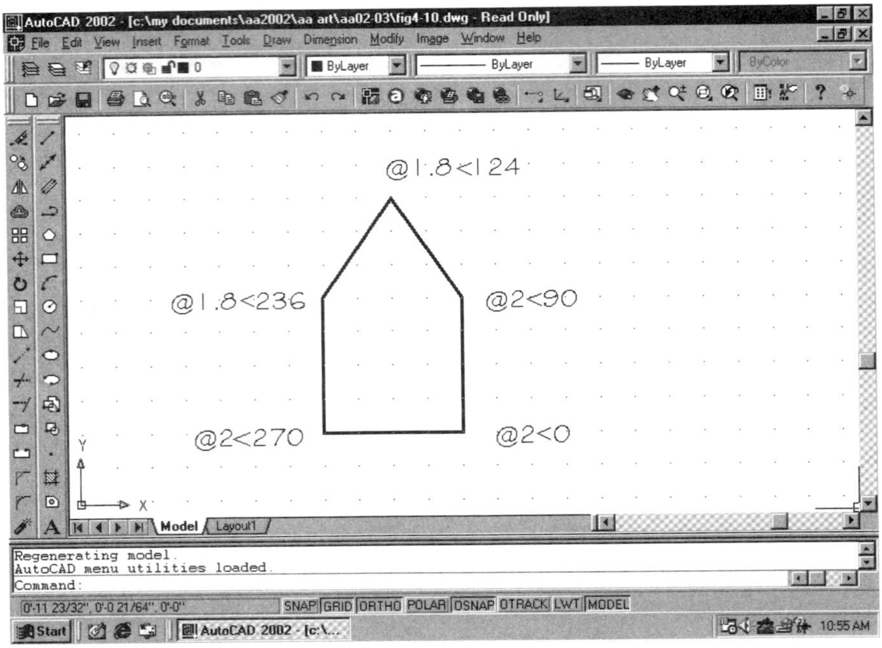

Figure 3.10 *The use of polar coordinates to establish a simple shape.*

CONTROLLING LINES

In addition to the aids for placing lines accurately, AutoCAD offers help for placing lines easily. Line continuation, closure, and the tools of the status bar can each be used to control the placement of lines.

LINE CONTINUATION

Once the **Line** command has been terminated, it can be resumed from the end of the last line segment that was drawn. This can be done even if other commands were used. The horizontal line segment drawn in Figure 3.11 was drawn prior to the circle. The circle was drawn by clicking the **Circle** button on the **Draw** toolbar. It could also be drawn by typing **C** (ENTER) at the Command prompt and selecting a center or radius point. This command will be further explored in the Chapter 6. If **L** (ENTER) is typed and then (ENTER) is pressed for the "first point," the new line will begin at the end of the last line segment.

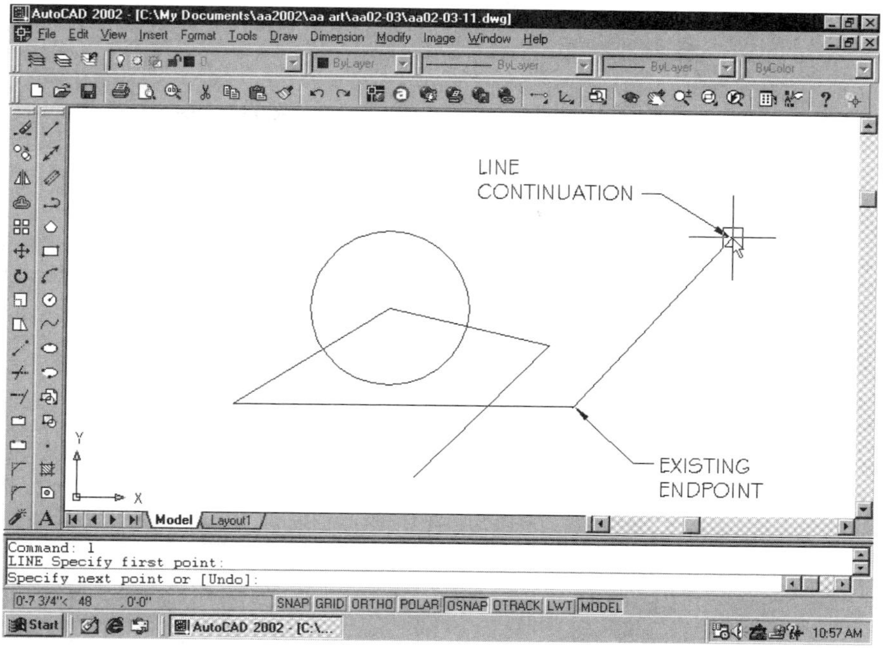

Figure 3.11 *A line can be continued from the end of the last line drawn by pressing* ENTER *for the "first point."*

LINE CLOSURE

If a sequence of lines will form a closed polygon, the last line can be drawn without providing endpoints or coordinates for the termination point. You've already used the Object Snap **Endpoint** mode to select an object. Rather than struggling to pick the exact endpoint of the first line, typing **C** (ENTER) at the "next point:" will close the polygon precisely. This can be seen in Figure 3.12 and in the following command sequence:

Command: **L** (ENTER) *(Or click the Line button on the Draw toolbar.)*

LINE Specify first point: **3,3** (ENTER)

Specify next point or [Undo]:**@6,0** (ENTER)

Specify next point or [Undo]: **@0,2** (ENTER)

Specify next point or [Close/Undo]: **@–6,0** (ENTER)

Specify next point or Close/[Undo]: **C** (ENTER)

Command:

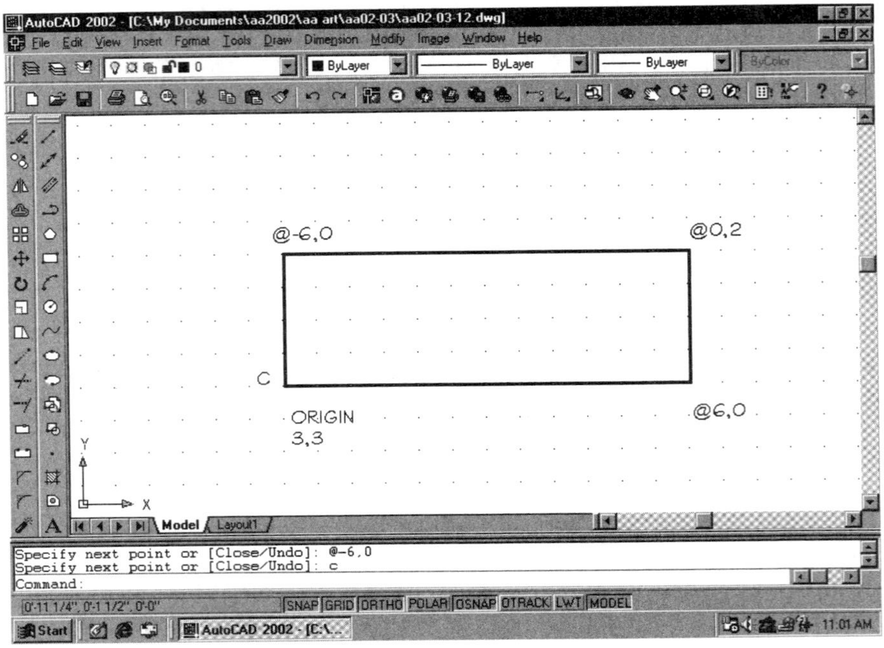

Figure 3.12 *Typing* **C** *(ENTER) for the last "to point" will close the polygon.*

Typically, the polygons you will draw for projects will be much more complex. It's not uncommon to have to interrupt the drawing of the shape to execute another command, such as **Erase**. Once the polygon has been interrupted, **C** (ENTER) cannot be used. This can be seen in Figure 3.13 and in the following command sequence:

Command: **L** (ENTER) *(Or click the Line button on the Draw toolbar.)*

LINE Specify first point: **4,2** (ENTER)

Specify next point or [Undo]: **@2,0** (ENTER)

Specify next point or [Undo]: **@0,2** (ENTER)

Specify next point or [Close/Undo]: **@2,2** (ENTER)

Specify next point or [Close/Undo]: **@–2,2** (ENTER)

Specify next point or [Close/Undo]: **@–2,–2** (ENTER) *(Wrong coordinate entered and a line drawn.)*

Specify next point or [Close/Undo]: (ENTER)

Command: E (ENTER) *(Or click the Erase button on the Modify toolbar.)*

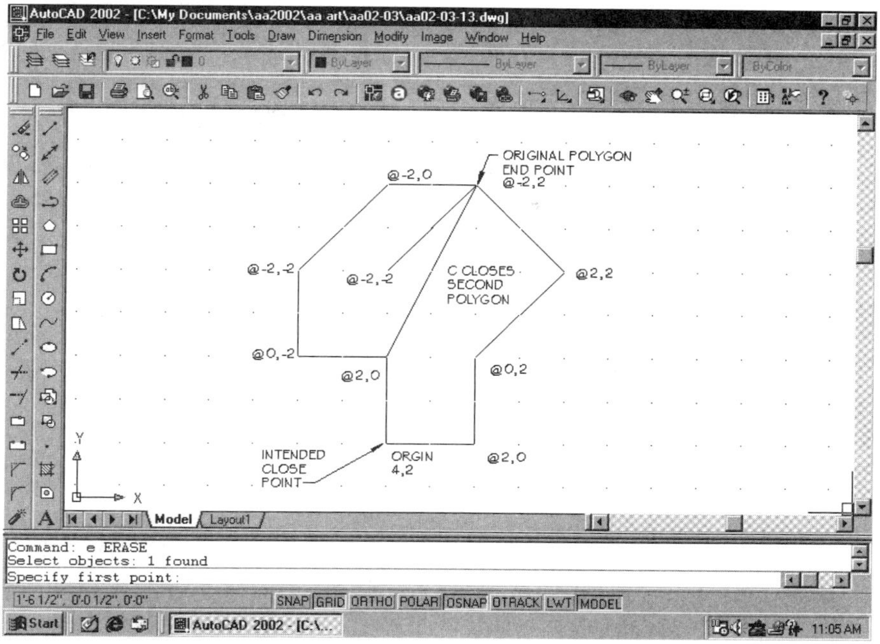

Figure 3.13 *If the drawing of the polygon has been interrupted,* **C** (ENTER) *will terminate the line at the original endpoint rather than at the close of the total polygon. The desired close point can be easily selected with OSNAP activated by moving the cursor near the desired point. As the cursor nears the endpoint, a box will be placed around the crosshairs, indicating an endpoint will be selected.*

ERASE

Select objects: (ENTER)

Command: **L** (ENTER) *(Or click the Line button on the Draw toolbar.)*

Specify next point or [Undo]: (ENTER) *(This uses the endpoint of the last line drawn as the new endpoint.)*

Specify next point or [Undo]: **@–2,0** (ENTER)

Specify next point or [Undo]: **@–2,–2** (ENTER)

Specify next point or [Close/Undo]:**@0,–2** (ENTER)

Specify next point or [Close/Undo]: **@2,0** (ENTER)

To point: **C** (ENTER)

Command:

Although the intended close is at the origin point of 4,2, the polygon will close at the start of the second sequence of lines that were drawn. Chapter 7 will offer other drawing tools to select exact points of a drawing entity as alternatives to C (ENTER)

In Figure 3.13 an error was made and the Line command was exited to correct the mistake. A better way to correct the mistake would have been to use the Undo command when the "next point: @–2,–2 (ENTER)" was entered. This would have returned the origin point to the end of the last segment drawn (–2,–2) and the polygon could have been continued, as shown in Figure 3.14.

CONTROLLING LINES WITH THE STATUS BAR TOOLS

The status bar was introduced in Chapter 2. This discussion will help you make practical application of these tools. Snap, grid, and Polar Tracking can each be set using the **Drafting Settings** dialog box. Access the dialog box by selecting **Drafting Settings** from the **Tools** menu, and then selecting the desired tab.

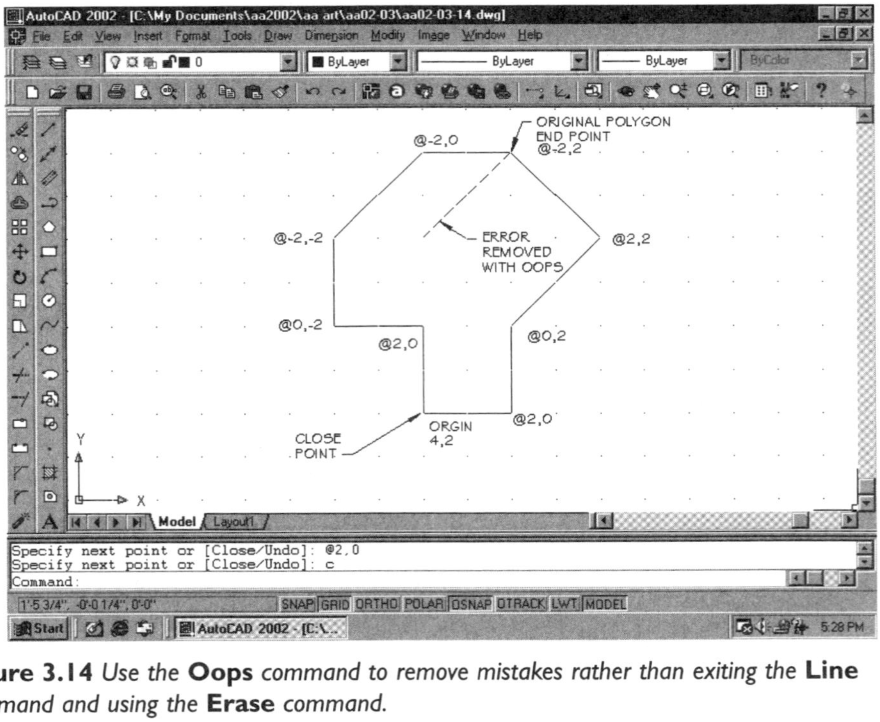

Figure 3.14 *Use the* **Oops** *command to remove mistakes rather than exiting the* **Line** *command and using the* **Erase** *command.*

SNAP

Snap was introduced in the last chapter. Although the snap grid is invisible, it can be a valuable tool for placing lines and other drawing elements. Snap should be adjusted to a modular size of the object being drawn. When you are working on a floor plan, a snap setting of 12" would be helpful for the initial layout of the walls. A setting of 4" allows lines to be easily placed as the plan is being fine-tuned. Snap will be most helpful if you adjust the value throughout the life of the drawing to meet specific needs. Keep in mind that when snap is ON, you might have a difficult time drawing angled lines. Snap can be toggled ON/OFF using the SNAP button on the status bar at any point of a command sequence.

GRID

The visible grid introduced in Chapter 2 should be used to plan and lay out a drawing. Setting the grid value as a multiple of the snap setting will greatly aid in placing drawing objects. The grid serves as a visual reference as well as an aid in placing lines. A grid of 24" and a snap of 6" will greatly aid in the layout of most plan views. The 24" spacing of the grid will provide a reference to keep track of object sizes, and the snap will provide a useful grid for placing lines. Grid can be toggled ON/OFF using the GRID button on the status bar at any point of a command sequence.

ORTHO

Most of the structures that you will draw will be composed of perpendicular lines. Type **L** (ENTER) at the Command prompt and draw a few lines. You'll notice that as a line is drawn from the "first point" to the "next point," a line is extended to the crosshairs and moves as the crosshairs move. The ORTHO mode will allow you to place the end of the rubber-band line in an exact horizontal or vertical position based on the current snap or grid pattern. If the snap and grid are rotated, ORTHO will respond by producing perpendicular lines based on the new grid. Toggle ORTHO ON/OFF by clicking the ORTHO button on the status bar. The residence shown in Figure 2.32 has a grid set to 45° for the right edge of the structure. As ORTHO was used on the right half of the structure, lines were automatically set to 45°, 135°, 225°, and 315°.

POLAR

Earlier in this chapter you were introduced to using polar coordinates to place a line. Activating Polar Tracking by clicking the POLAR button on the status bar will produce a display at specified intervals to aid in placing lines. The display will show the distance and the angle of the line in progress. Figure 3.15 shows an example of a polar display for placing a line. The display can be a helpful alternative to entering coordinates at the keyboard. Settings can be adjusted on the **Polar Tracking** tab of the

Drafting Settings dialog box (choose **Drafting Settings** from the **Tools** menu). The **Polar Tracking** tab can be seen in Figure 3.16. Key elements of the dialog box include the following:

Increment angle—This edit box allows the desired angle to be used with the tracker to be set. Default angles include 5, 10, 15, 18, 22.5, 30, 45, and 90 degrees. Selecting 10 will display a tracking indicator as the cursor is moved in 10° increments.

Additional angles—Click the **New** button to see this dialog box, which allows up to 10 additional angles to be added to the default list. If you're drawing an elevation with a roof at a 6/12 pitch, selecting an angle of 27.5° would aid in the layout of the roof. The setting could be entered as follows:

1. Click the **New** button.
2. Enter 27.5 in the edit box.
3. Click the **OK** button.

Settings can be removed from the default list by highlighting the setting to be removed and then clicking the **Delete** button.

Object Snap Tracking Settings—This option selects options for object track settings. Options include orthogonal and polar angle tracking settings.

Track orthogonally only—With this option active, the object tracker displays only horizontal or vertical object snap tracking paths for acquired object snap points when Object Snap Tracking is on.

Track using all polar angles—Allows the cursor to track along any polar angle tracking path when Object Snap Tracking is on.

Polar Angle Measurements—Sets the method to be used to measure polar angles. Options include **Absolute** and **Relative.**

Absolute—Displays polar tracking angles based on the current UCS.

Relative to last segment—Displays polar tracking angles on the most recent object.

Options—Clicking this button displays the **Options** dialog box. This dialog box can be used to adjust the color and size of the object snap marker. The dialog box also provides a convenient method of altering the size of the aperture box.

OSNAP

Activating Object Snap provides one of the easiest and fastest means of accurately drawing in AutoCAD. With this option ON (click OSNAP on the status bar), a marker will be placed on the crosshairs as it is moved near an endpoint, midpoint, or intersection of lines. With the marker displayed, the "next point" of a line will be placed

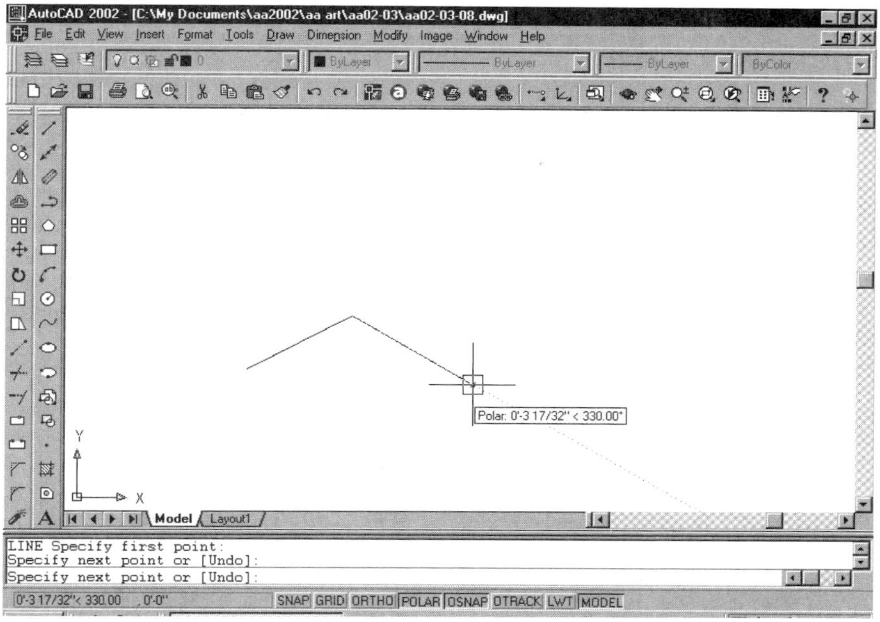

Figure 3.15 *Polar display for tracking a line. The display can be a helpful alternative to entering coordinates at the keyboard. Settings can be adjusted on the* **Polar Tracking** *tab of the* **Drafting Settings** *dialog box.*

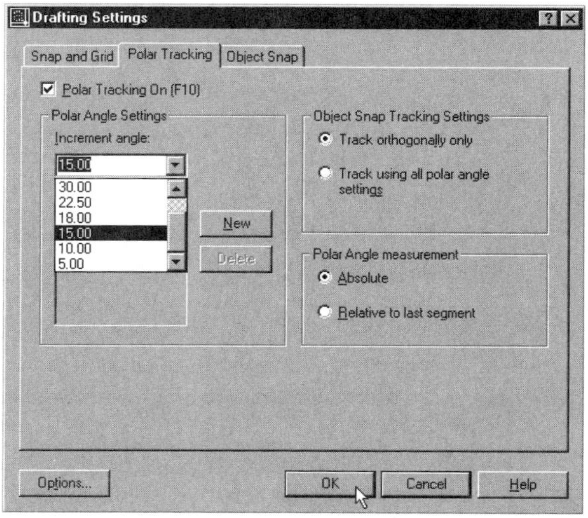

Figure 3.16 *The* **Polar Tracking** *tab of the* **Drafting Settings** *dialog box can be used to control the display used with Polar Tracking.*

exactly at the endpoint. Figure 3.17 shows an example of the display as the cursor is moved near the endpoint of an existing line. Rarely will you want to turn this feature off. Chapter 7 will explore methods of altering the default settings of Object Snap. If you want to explore on your own, select **Drafting Settings** from the **Tools** menu and select the **Object Snap** tab to view the options. Use the **Help** menu if you just can't wait until Chapter 7.

AN INTRODUCTION TO EDITING LINES

This section will introduce three methods of removing drawing errors from a drawing file. These methods include the **Erase**, **Oops**, and **Undo** commands. The **Redraw** command will also be introduced as a method of cleaning the drawing display. Chapter 8 will present methods for using these commands more effectively.

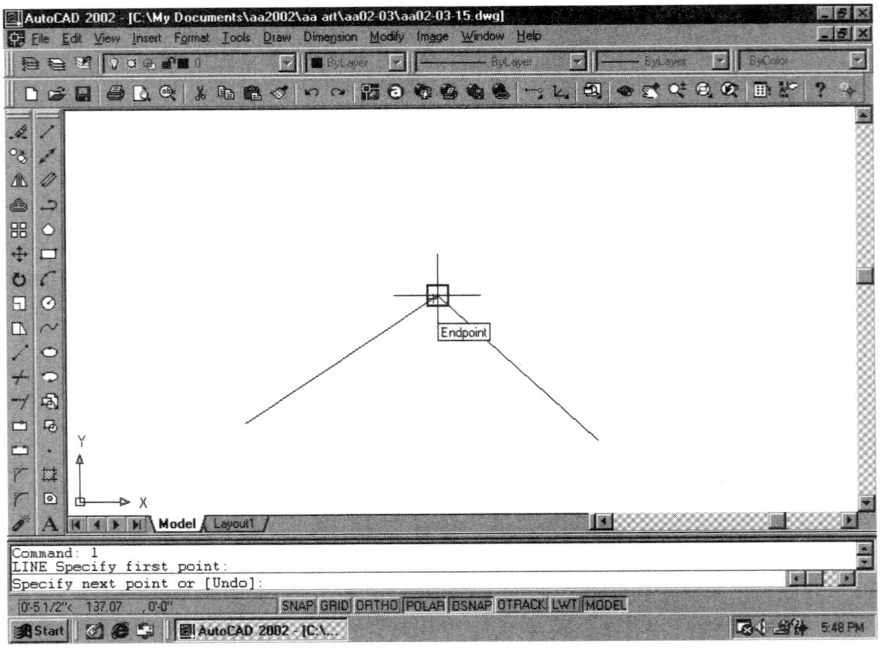

Figure 3.17 *Object Snap is an excellent method for increasing drawing accuracy. With the OSNAP button on the status bar active, a square will be added to the crosshairs as it moves near the endpoint of a line. The "next point" will automatically snap to the endpoint of the selected line.*

ERASE

The **Erase** command will allow you to remove unwanted drawing objects. To remove objects using the **Erase** command, select the **Erase** button on the **Modify** toolbar, type **E** (ENTER) or select **Erase** from the **Modify** menu. This will display the prompt:

> Command: **E** (ENTER) *(Or click the Erase button on the Modify toolbar.)*
>
> Select objects:

As the prompt is displayed, the crosshairs turn into a pick box. The pick box can be used to select objects to be removed. Place the box on the object to be removed and left-click. In Figure 3.18 the column centerline has been selected for removal. Once selected, the selected line will now appear highlighted. An unlimited number of objects to be erased can be selected. The command line will continue to show "Select objects" until (ENTER) is pressed. Pressing (ENTER) will remove the selected lines and restore the Command prompt, as seen in Figure 3.19.

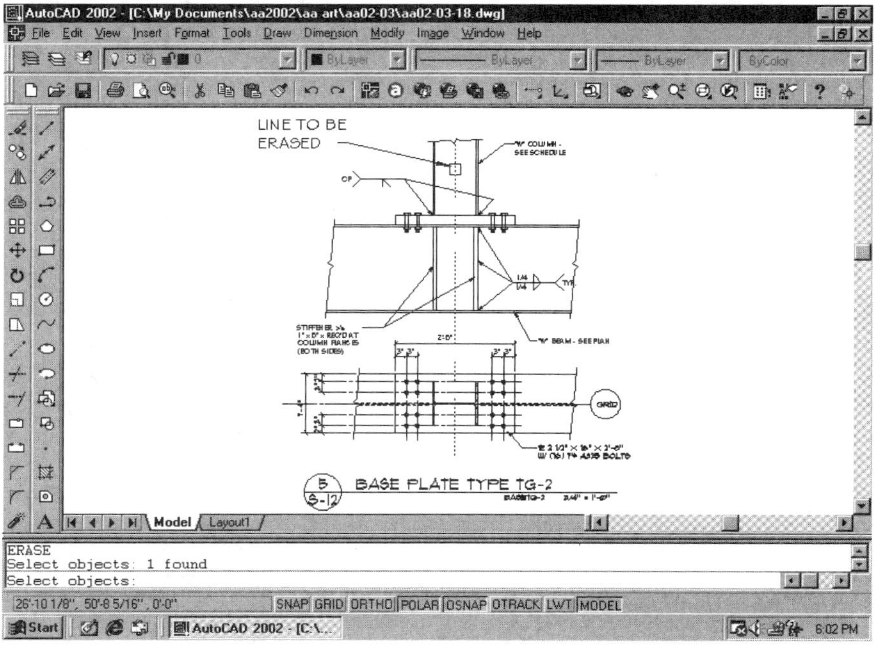

Figure 3.18 *As the* **Erase** *command is entered, the crosshairs will be replaced by a pick box. Placing the pick box on the desired item to be erased and selecting it will select that object for removal. (Courtesy Van Domelen/Looijenga/McGarrigle/Knauf Consulting Engineers.)*

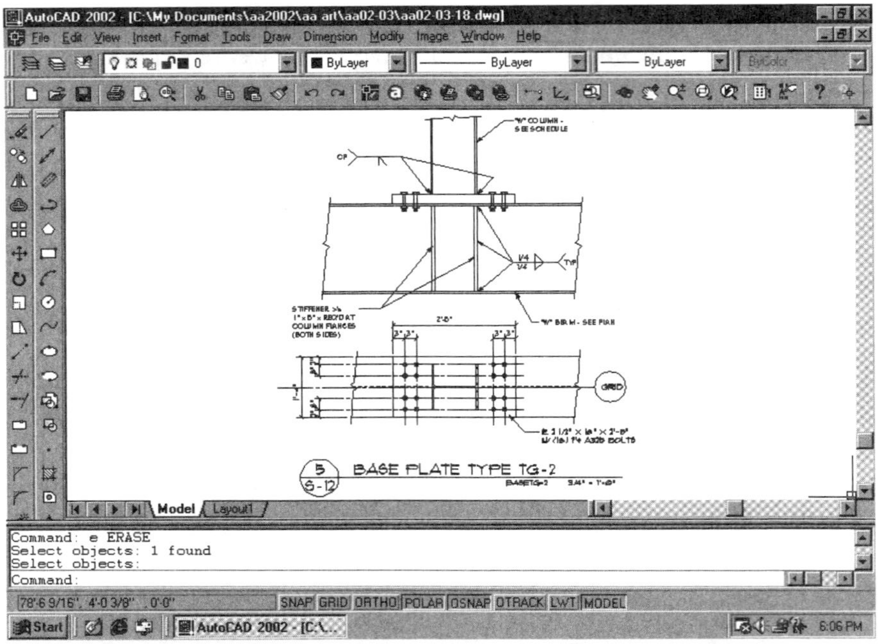

Figure 3.19 *Pressing* ENTER *or right-clicking will remove the selected object from the drawing screen. (Courtesy Van Domelen/Looijenga/McGarrigle/Knauf Consulting Engineers.)*

Erase allows for the last drawing object to be erased without your having to select the object. This can be done by selecting the **Erase** button or typing **E** (ENTER) and then **L** (ENTER). The Command prompt is as follows:

Command: **E** (ENTER) *(Or click the Erase button on the Modify toolbar.)*

Select object: **L** (ENTER)

1 found:

Select objects: (ENTER)

Command:

As the command line is restored, the last object drawn is eliminated. If you would like to remove the objects drawn with the last drawing sequence, type the following:

Command: **E** (ENTER) *(Or click the Erase button on the Modify toolbar.)*

ERASE

Select objects: **L** (ENTER)

Select objects: *(Select object.)*

1 found

Select objects *(Continue to select objects or press* (ENTER) *to end sequence.)*
(ENTER)

Command:

Selecting objects to be erased can be continued indefinitely until the entire drawing is removed, but this process is not the most efficient way to remove multiple objects. Later chapters will provide in-depth coverage of **Erase** and other editing options, as well as more efficient methods of selecting objects to be edited.

OOPS

Occasionally an object will be erased by mistake. The **Oops** command will restore the object removed by a previous **Erase** command. Chapter 10 will provide options for modifying drawing objects.

UNDO

An alternative to **Erase** or **Oops** is the **Undo** command. **Undo** will remove the command that was just executed. This command will remove the most recent line and then allow you to continue with the **Line** command from the previous line. This command can be used as many times as you wish, and it will continue to move you back through the drawing sequence one step at a time. Select **Undo** on the **Standard** toolbar, from the **Edit** menu, or by keyboard. When the toolbar is used to enter the command, simply select the button when you wish to remove the command that has just been executed. Selecting the button in the middle of a command sequence is an excellent method for removing an unwanted line, without ending the **Line** command. If the **Undo** button is selected after the command is ended, the entire line sequence will be removed. The **Undo** command can be entered by keyboard in the middle of the **Line** command using the following sequence:

Specify next point or [Close/Undo]: **@–2,–2** (ENTER)

Specify next point or [Close/Undo]: **U** (ENTER)

Specify next point or [Close/Undo]: **@–2,0** (ENTER)

If you decide that you really do want the line after you have used the **Undo** command, the **Redo** command can be used to correct your change of mind. **Redo** will reverse the effects of **Undo**. Perform a **Redo** by selecting the **Redo** button on the **Standard** toolbar or by typing **REDO** (ENTER) at the Command prompt.

CHAPTER 3 EXERCISES

Set limits, units, snap, and grid to best suit each project. Save each exercise to a diskette and to a folder on the hard drive or network, as per your instructor's directions.

1. Draw the following outline using the coordinates that are provided.

   Point of beginning: 1,1

point b	@4.2426<45
point c	@3.1623<342
point d	@2.5<0
point e	@4.0311<120
point f	@4.2720<159
point g	@3.9051<230

   C (ENTER)

   Save your drawing with a file name of E-3-1.

2. Draw the following outline using the coordinates that are provided.

   Point of beginning: 2,1

point b	@2.5<N80dE
point c	@3.5<N38d23'15"E
point d	@2.5<S40d20'40"E
point e	@1,1
point f	@1,–1
point g	@4<N
point h	@–6,1
point i	@3.5<S39d7'W

   back to the true point of beginning.

   Save your drawing with a file name of E-3-2.

3. Draw the following outline using the coordinates that are provided. Save the drawing as file name E-3-3.

   Point of beginning: 9,9

point b	@–5,–2.5
point c	@0,–1.5
point d	@4<S39d48'W
point e	@–1.5,–2

point f	@4.5,1.5
point g	@1.5<90
point h	@2<335
point i	@2.5<120
point j	@5.5<330
point k	@1,1
point l	@–2.5,1
point m	@–.5,–.5
point n	@3<N61d25'10"W
point o	@3.8125<E

@4.5<N to the true point of beginning

4. Use the following coordinates to lay out the required shape. Save as file name E-3-4.

   Starting point: 7.75, 7.95

   @1.91<127

   @–2,0

   @0,–2

   @–1.50, 1.75

   @–.96,0

   @1.62<137°

   @.46<227°

   @1.22<311°

   @–.92,0

   @1.62,–.5

   @-1.11,-4.13

   @3.5,0

   @1.62,1.62

   @1.75<308

   @3.32<80.73°

   back to the point of beginning

5. Use the attached drawing to show a 16" deep steel beam resting on a 3/4" x 12 steel top plate. Support the top plate on a 6" x 6" x 3/8" T.S. column. Provide a 3/4" x 16" steel end plate. Save the drawing with the file name E-3-5.

6 Use the attached drawing to draw the roofline of the structure. Assume all hips and valleys are drawn at 45° angles. Save the drawing as file name E-3-6ROOF.

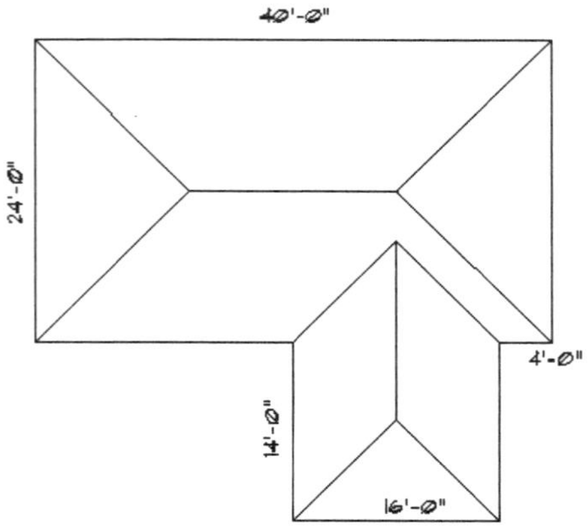

7. Use the attached drawing to draw a section of a one-story footing for a wood floor system. Assume the bottom of the footing extends 12" into the grade, and the top of the stem wall extends 8" above grade. Save the drawing as file name E-3-7-1STFTG.

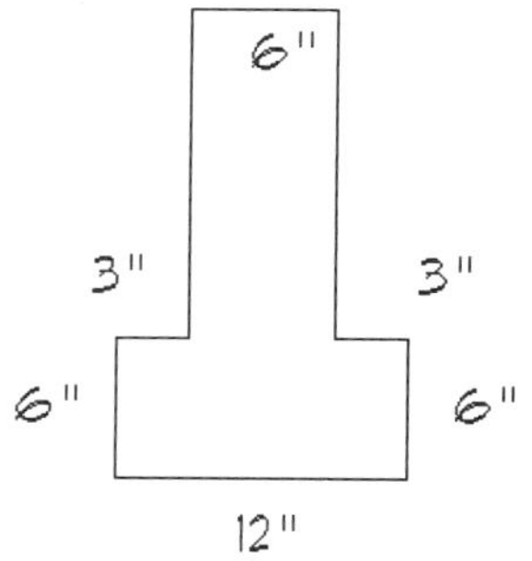

8. Use the attached drawing to draw a section of a one-story footing for an on-grade concrete floor system. Save the drawing as file name E-3-8-CONCFTG.

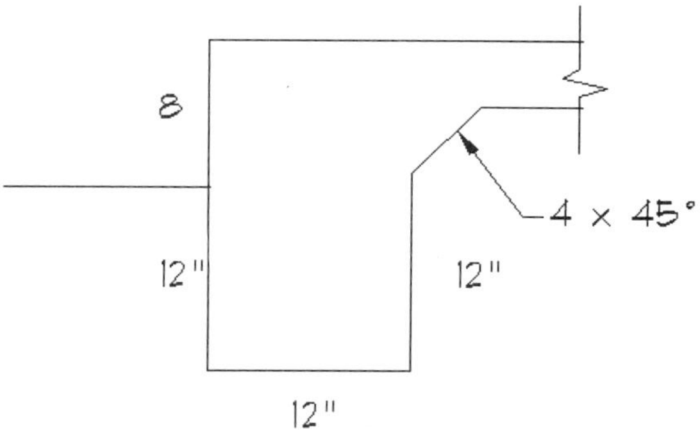

9. Use the attached drawing to complete a section through a 6" wide concrete panel, resting on a concrete footing 24" wide x 12" deep. Support the wall panel on 1" deep grout with beveled edges. Show a 3" x 6" pocket in the footing for reinforcing. Save the drawing as file name E-3-9CONCWALL.

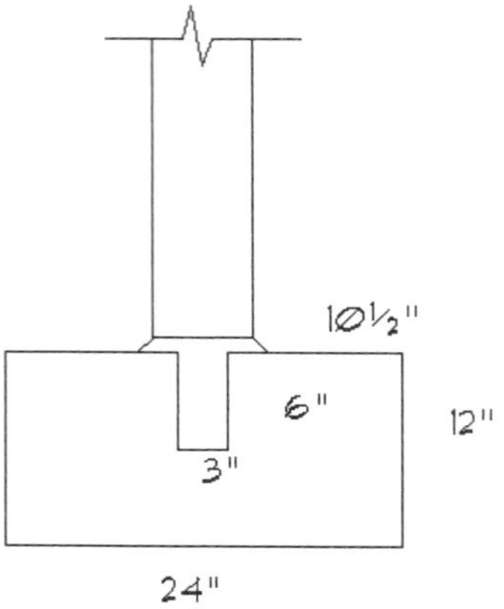

10. Use the attached drawing to complete a section view of a floor truss intersecting a 6" wide concrete tilt-up wall. Assume the truss to be 24" deep with 2" deep top and bottom. Save the drawing as file name E-3-10TRUSSWALL

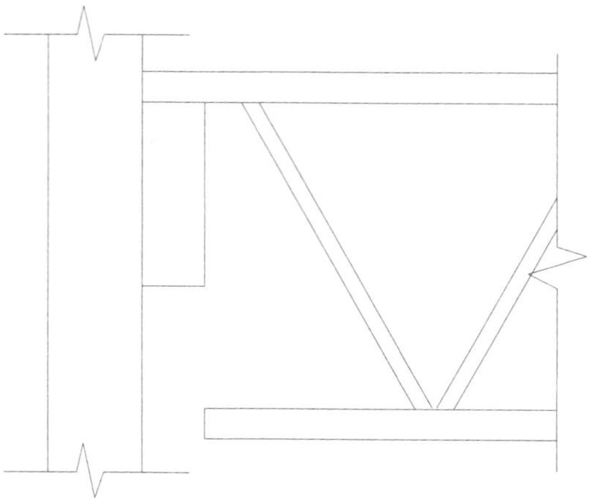

CHAPTER 3 QUIZ

1. Use the drawing below to provide absolute coordinates. Grid=.5", Snap=.25"

a. 2.5,2	e. ________	i. ________
b. ________	f. ________	j. ________
c. ________	g. ________	k. ________
d. ________	h. ________	l. ________
		a. ________

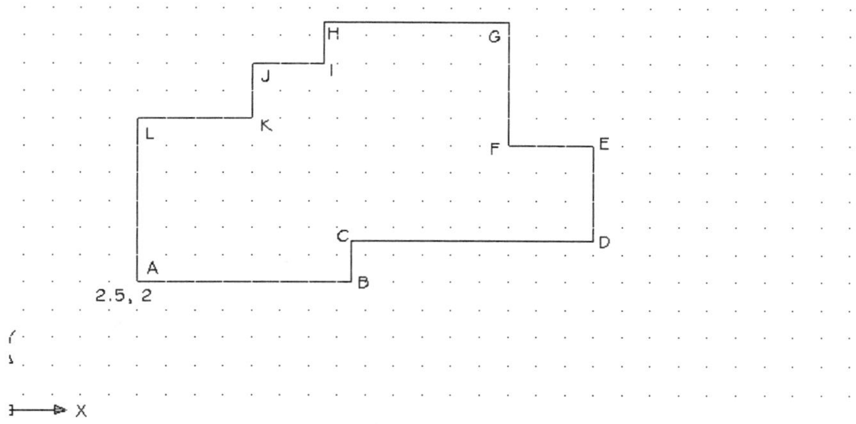

2. Use the drawing from Question 1 to provide relative coordinates.

a. 2.5,2	e. ________	i. ________
b. ________	f. ________	j. ________
c. ________	g. ________	k. ________
d. ________	h. ________	l. ________
		a. ________

3. Use the drawing from Question 1 to provide polar coordinates.

a. 2.5,2	e. ________	i. ________
b. ________	f. ________	j. ________
c. ________	g. ________	k. ________
d. ________	h. ________	l. ________
		a. ________

4. Use the drawing below to provide relative coordinates.

a. 1.5,1.5	e. ________	i. ________
b. ________	f. ________	j. ________
c. ________	g. ________	a. ________
d. ________	h. ________	

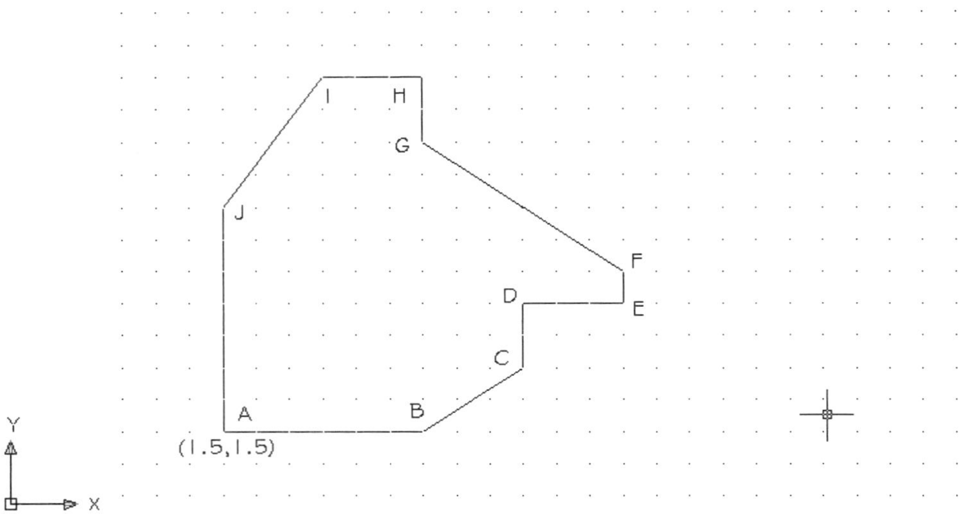

5. Use the drawing from Question 4 to provide absolute coordinates.

a. 1.5,1.5	e. ________	i. ________
b. ________	f. ________	j. ________
c. ________	g. ________	a. ________
d. ________	h. ________	

6. Use the drawing from Question 4 to provide polar coordinates. Use relative coordinates where insufficient information is available to determine polar entries.

a. 1.5,1.5	e. ________	i. ________
b. ________	f. ________	j. ________
c. ________	g. ________	a. ________
d. ________	h. ________	

7. What entry method uses the X,Y format to enter coordinates?
8. What entry method uses the @X<# format to enter coordinates?
9. What entry method uses the @X,Y format to enter coordinates?
10. Several entry coordinates are listed below. Circle the coordinates that will not work.

    a. 3,4
    b. –3,–4
    c. 3,4
    d. @3,<45
    e. @–3,–5
    f. –3,<27
    g. @1'–0",<27.5
    h. @ 27'6<n3615'e
    i. @ 36'<27.5
    j. N27d36'27"e,15'
    k. 27'6
    l. 25',13'6
    m. 14'<W30N
    n. 20'4,S27d36,15E
    o @3'6,7'4

11. Can an angled line be drawn with SNAP and GRID both ON?
12. Describe the line that is formed between 2,0 and 4,0.
13. Describe the line that is formed between 2,0 and 4,2.
14. List two methods for activating the coordinate display.
15. What settings must be changed to use absolute, relative, and polar coordinates on the same drawing?
16. How can the **Line** command be accessed from the menu?
17. List four methods for accessing the **Line** command.
18. What steps are required to erase a line?
19. If a line has been erased by mistake, how can it be restored?
20. What effect does **Erase** have on the cursor?
21. Describe the difference between **Oops** and **Undo**.
22. What effect will typing **C** (ENTER) have on a line segment?
23. How does the snap setting affect the use of the **Line** command?
24. What type of coordinate entry is best suited for drawing E-3-5PLAN?

    Use the **Help** command and list four methods of accessing the **Erase** command.

CHAPTER 4

Working with Drawing Files

This chapter will introduce methods for

- Creating and saving drawing templates
- Saving user preferences
- Saving drawing files
- Managing files and folders

Commands and major options to be explored in this chapter include

- Paper space
- Model space
- Viewports
- **Units**
- **Limits**
- **Pan**
- **Options**
- **Save**
- **Saveas**
- **Qsave**
- **Savetime**
- **Quit**
- **Exit**
- **Close**

- **Closeall**
- **Audit**
- **Recover**

In Chapter 1 you were introduced to the AutoCAD drawing environment and how to save a drawing. In Chapter 2 you explored methods of entering the AutoCAD drawing area when a new drawing file is created with the wizard or a template. In Chapter 3 you started to create great works of art. This chapter will introduce you to methods for creating a template drawing, explore methods to save drawing properties for future use, expand your methods for saving and managing drawing files, and explore methods to merge existing drawings.

CREATING AND SAVING DRAWING TEMPLATES

In Chapter 2 you were introduced to the templates that AutoCAD provides to help make each new drawing that you start. The template can be created and saved using the TODAY window or the Startup dialog box, and then used each time a drawing requiring those parameters is needed. Several useful drawing parameters were introduced in Chapter 2. Other useful values such as linetype, lettering size and styles, and dimensioning styles and format will be introduced in chapters that follow. Necessary values can be added to the drawing template for future use. Once created and saved, they can be opened to form the base for a new drawing file and save time.

USING AN EXISTING TEMPLATE

AutoCAD provides several templates to ease drawing setup. The default template is the ACAD-named plot styles. Scroll through the A listing and you'll find two architectural templates. Both templates will provide a drawing screen showing a 24" x 36" sheet of paper with an architectural style border that is suitable for most construction drawings. You'll also notice that most of the templates are either "named plot styles" or "color dependent." Plot styles alter the way a plotted drawing will appear. Each object in a drawing has a plot style property. These styles can be assigned by color or by named groups. With a change in the plot style, the color, linetype, and lineweight of an object can be altered on the print. This will allow objects within the drawing to be highlighted. The benefits of plotting using assigned styles will be addressed in detail in Chapter 25 as plotting is explored. For now, select the **Template** tab in the TODAY window or use the traditional **Startup** dialog box and select the **Use a Template** option. Select the Architectural, English units-Color Dependent Plot Style template. This will produce the template shown in Figure 4.1. Although the title implies that this template is for architectural drawing, it should not be used without some basic changes. The term *Architectural* is applied for the shape of the title block, rather than the current settings. Three areas of the template should be examined before you complete a drawing. These areas are paper space, model space, and viewport.

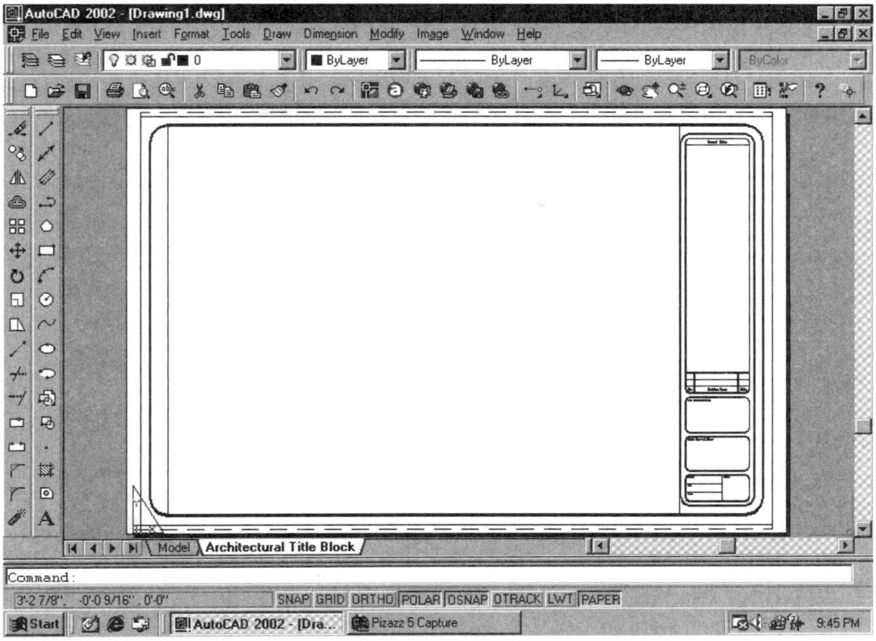

Figure 4.1 *The Architectural template can be used for a base layout that will meet most of the needs for construction related projects. Display the template by selecting the* **Use a Template** *option from the* **Startup** *dialog box and selecting the Architectural, English units option.*

Paper Space

As the template is displayed, you're working in paper space (notice the paper space icon in the lower left corner) with the Architectural Title Block tab active. The paper is 24" x 36" with a drawing area of approximately 2'–5 1/2" x 1'–10 1/2". In effect, you're now working on a blank sheet of paper. Move the cursor and you'll notice that it can be moved to any point in the display area. Paper space will allow you to start a line in the drawing area and finish the line at any point outside the drawing border.

Note: Paper space will be a great aid for plotting the drawing, but it should not be used to create a drawing.

Model Space

Model space is the area of the template where the drawing should be completed. Model space allows you to work in "real" space. Select the Model tab below the drawing area to switch from paper to model space. This will hide the template and display a blank screen with the model space icon in the lower left corner of the display. With model space active, you'll have a drawing area of approximately 20" x 11". The size will vary depending on the number of command lines displayed, and which toolbars you have displayed. The limits of this area can be adjusted to any size, at any time. Change the drawing limits to be more appropriate for an architectural drawing using values such as 150' x 80'. The size of the drawing space is altered using the **Limits** command that was introduced in Chapter 2. By changing the size of the space, you can display a large structure on the screen in its entirety. Before you explore new areas, enter the Architectural template, select the Model tab, and use the **Line** command to draw a box that is 90' x 50'. This box will be useful in visualizing how model space and the viewport work together. Notice that the drawing can be seen, but the template title block is not displayed. You're still working in the template, but in model space you're displaying only the space between the borders.

Viewports

The viewport is a 'hole' in paper space that allows the drawing in model space to be viewed. A viewport allows you to place a 100' long part on a 36" long sheet of paper. Figure 4.2 shows a representation of how the viewport and the template and model space work together. Switch back to paper space by selecting the Architectural Title Block tab. Select the PAPER button on the status bar. The PAPER button will now change to read MODEL, and the drawing will change to display the drawing shown in Figure 4.3. The template appears similar to Figure 4.1, but it has changed

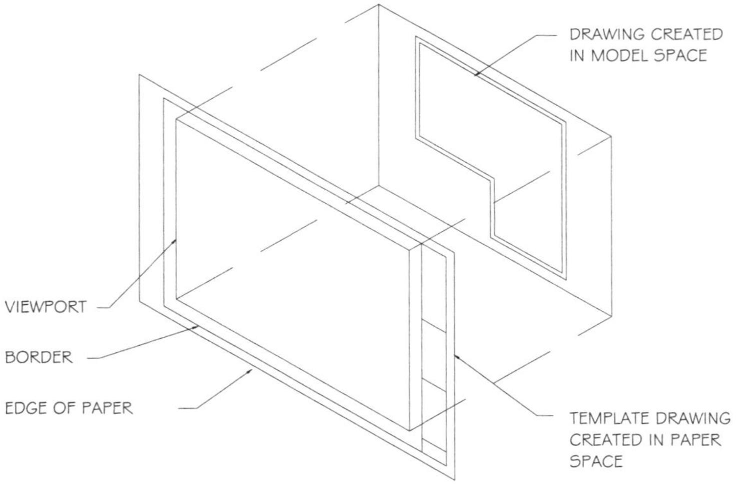

Figure 4.2 *A viewport is provided in each drawing template created in paper space. Additional layouts can be created to allow for different plotting displays.*

to reflect model space inside the template borders. The bold black line represents the edge of the viewport. The UCS icon has changed to reflect model space and has been placed inside the viewport. Move the crosshairs to the border of the template. As soon as the crosshairs touch the border, it changes to an arrow. You're not allowed to draw outside the viewport. You're now working with approximately a 12" x 9" drawing area or viewport. Nice looking border, but not many projects will fit into the drawing area.

Note: Even though the viewport display is in model space, your drawing should not be created or edited here. This area should only be used to adjust the relationship of the completed drawing (created in model space using the model tab) to the border.

Adjusting the Viewport Size

The size of the viewport must be adjusted based on the desired scale to be used when the drawing is plotted. This chapter will help you adjust the size of the viewport so that a scale of 1/4"=1'–0" can be used. Chapter 24 will explain how to adjust the viewport for varied scales.

If you haven't already done so, open the Architectural template and select the Architectural Title Block tab. Use the methods described in Chapter 2 to set the lim-

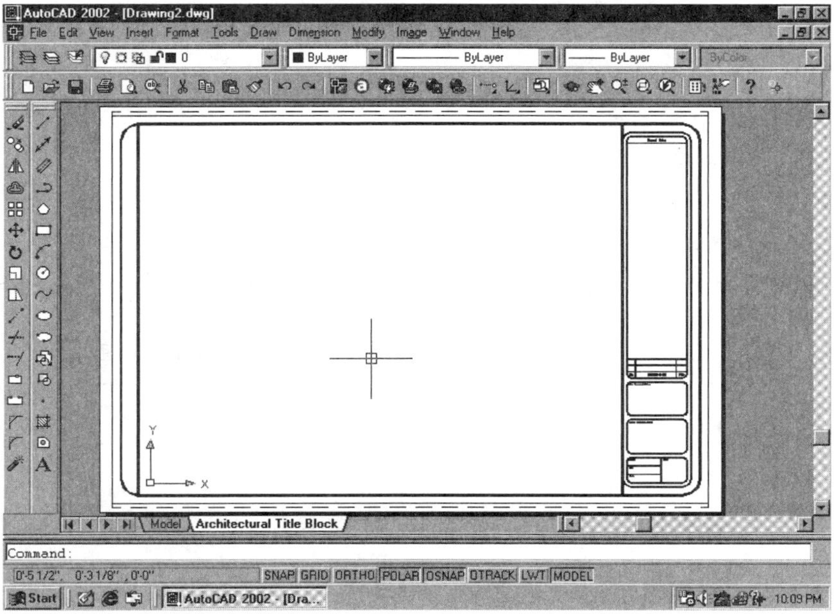

Figure 4.3 *Selecting the PAPER button on the status bar will toggle the display to a viewport with model space. Notice that the UCS icon is now inside the drawing area.*

its as 116',88'. Because many plan views in architecture are plotted at a scale of 1/4"=1'–0", this example will establish limits that allow a plan view to fit within the drawing area of the template. This size was established because at 1/4"=1'–0", each inch in paper space represents 4 feet in model space. Multiplying the drawing area size 29" (the horizontal distance between the template borders in paper space) x 4 (the number of feet per inch) and 22" (the vertical distance between the template borders in paper space) x 4 provides the basis for how the limits were determined. Methods for establishing drawing units for other drawing scales will be presented later in this chapter as you develop your own templates. Use the following procedures to update the template to architectural standards:

Type **UNITS** (ENTER) at the Command prompt. Use the Drawing Units dialog box to adjust the units to the desired precision.

Type **LIMITS** (ENTER) at the Command prompt. Use the following sequence to alter the drawing limits:

Command: **LIMITS** (ENTER)

Reset Model space limits:

Specify lower left corner [ON/OFF]: <0'-0",0'-0">: (ENTER)

Specify upper right corner <1'-0",0'-9">: **116',88'** (ENTER)

Command: **Z** (ENTER) *(Or select the Zoom button on the Object Properties toolbar.)*

Specify corner of window, enter a scale factor (nX or nXP), or

All/Center/Dynamic/Extents/Previous/Scale/Window/
<Realtime>: **A** (ENTER)

Command:

You should be familiar with the effects of using the **Limits** command. As you typed **Z** (ENTER), you started the **Zoom** command. Think of zoom as a telephoto lens on a camera. Zoom alters the area of a drawing to be displayed. (The **Zoom** command will be explored in detail in Chapter 8.) If you enter the **Zoom** command and type **A** (ENTER), you've selected the **All** option. You've asked AutoCAD to display all of the display area. This process will alter the size of the drawing area from 29" x 22" to 116' x 88'. Now you have a drawing area that is suitable for construction drawings. Later chapters will show you how to add other elements to the template to ease drawing setup.

Comparing Model Space, Paper Space, and the Template

Now that you've explored each aspect of a template, you can combine those features to create a useful drawing tool for future drawings. As you've explored each aspect of a template, a box has been drawn in model space, and the limits that have been set are for the viewport. Select the Model tab to restore the view of the 90' x 50' box. The box that was drawn earlier as you examined model space can be seen, but the template title block is not displayed. You're still working in the template, but in the model space, you're displaying only the space between the borders.

Selecting the Architectural Title Block tab displays the template with the title block and border, but all of the box can't be seen. Click the PAPER button on the status bar, and the viewport is displayed. See Figure 4.4a. The viewport can now be used to adjust the location of the drawing to the border using the **Pan** command. Select the **Pan** button (the hand) on the **Standard** toolbar. As the command is activated, a hand button will be placed in the drawing area. Pressing and holding the select button allows the drawing to be relocated in the viewport. When the drawing is in the desired location, press (ENTER) to end the **Pan** command. Chapter 8 will explore the **Pan** command in detail. Figure 4.4b shows the completed process.

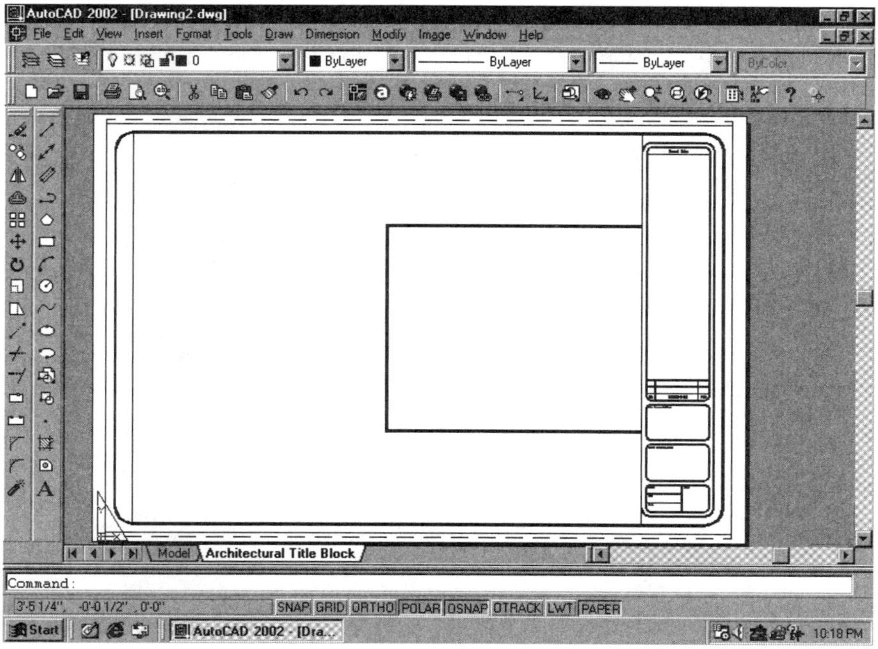

Figure 4.4a *Switching from model space to paper space displays the box (in model space) partially displayed in the template. Selecting the PAPER button in the status bar allows the display to be altered.*

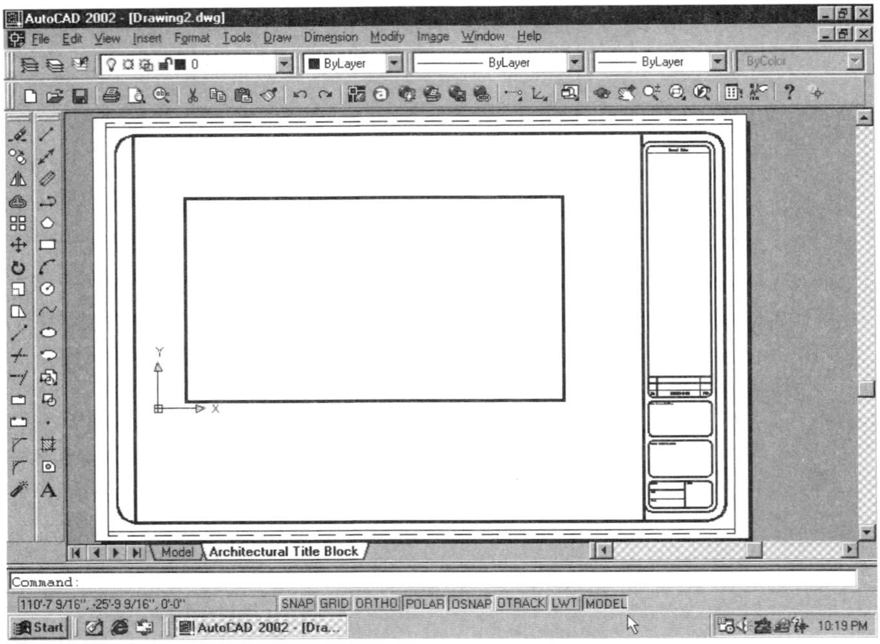

Figure 4.4b *The location of objects drawn in model space can be adjusted relative to the template using the* **Pan** *command. This drawing is now ready for plotting.*

Note: As a new AutoCAD operator, you might find this a bit overwhelming. If nothing else, keep in mind that a template is a great place to store drawing tools. It's like the junk drawer we all have in a desk. All the good stuff goes in the drawer for future use. A template is a drawing base where you will begin to store the settings that you use over and over. No one should be setting the precision of units or the drawing limits each time a drawing is started.

The second thing to remember is that you want to work in model space to create objects and paper space to plot them. Each of the preceding chapters has reminded you of this key point. It's important.

CREATING YOUR OWN TEMPLATE

Any drawing can be saved and used as a template. An existing AutoCAD template can be modified and saved with a new name or a template drawing can be created using the methods used to create a drawing file. Open a drawing and establish the drawing units, limits, snap, and grid values. Special consideration will need to be given to the drawing limits as you establish a drawing template. In Chapter 2, as the limits were established, common paper sizes were introduced. The size of paper that the completed project will be printed or plotted on must be considered as the drawing template is

DRAWING SCALE FACTORS								
ARCHITECTURAL SCALE FACTORS				ENGINEERING SCALE FACTORS				
SCALE	LIMITS		TEXT SCALE	SCALE	LIMITS		TEXT SCALE
	18 X 24	24 X 36			18 X 24	24 X 36	
1" = 1'-0"	18 X 24	24 X 36	12	1"= 1'-0"	18 X 24	24 X 36	12
3/4"= 1'-0"	24 X 32	32 X 48	16	1"= 10'-0"	180 X 240	240 X 360	120
1/2" = 1'-0"	36 X 48	48 X 72	24	1"= 100'-0"	1800 X 2400	2400 X 3600	1200
3/8"= 1'-0"	48 X 64	64 X 96	32	1"= 20'-0"	360 X 480	480 X 720	240
1/4" = 1'-0"	72 X 96	96 X 144	48	1"= 200'-0"	3600 X 4800	4800 X 7200	2400
3/16" = 1'-0"	96 X 128	128 X 192	64	1"= 30'-0"	540 X 864	864 X 1080	360
1/8" = 1'-0"	144 X 192	192 X 288	96	1"= 40'-0"	720 X 960	960 X 1440	480
3/32" = 1'-0"	192 X 256	256 X 384	128	1"= 50'-0"	900 X 1200	1200 X 1800	600
1/16" = 1'-0"	288 X 384	384 X 576	192	1"= 60'-0"	1080 X 1440	1440 X 2160	720

NOTE: ALL LIMITS ARE GIVEN IN FEET.

Figure 4.5 *The scale that a drawing will be plotted at must be considered as a template is constructed. The drawing limits will be affected by the scale to be used and the paper size.*

established. Although the drawing limits of model space can be altered at any time throughout the life of the drawing, establishing limits of the viewport that match standard paper sizes will aid in the use of the template.

To establish a template for use with drawings completed at 1/4"=1'–0", drawing values will need to be multiplied by a factor of 48 to determine the size to be used in the template. Determine the scale factor for 1/4"=1'–0" or 0.25"=12" by dividing 12 by 0.25. If the drawing is to be placed on D-size material, limits would need to be 1,152" x 1,728" or 96' x 144'. You can also determine the limits by multiplying the desired paper size by the number of feet in 1 inch of paper space. At a scale of 1/4"=1'–0", 1 inch of paper will contain 4 feet of drawing, resulting in the same 96' x 144' limits. Other common limits for D material are shown in Figure 4.5. Once the desired settings are established, the template can be saved and recalled for future use. Methods of saving a template will be introduced later in this chapter.

SAVING USER PREFERENCES

A drawing template works well for saving drawing settings that must remain constant for each member of the drawing team. There are many features of the drawing environment that do not affect the final project, but that affect the productivity of each team member.

Drawing profiles allow each user to adjust and save the drawing settings to meet personal preferences. Drawing profiles can be especially helpful if you're working at a computer that you share with several other people. If you like a black background and you share your computer with someone who likes a white background, AutoCAD allows these differences to be stored so that they can be changed quickly. Items that might be included in a profile include the **Startup** dialog box display, crosshairs size, display resolution of objects, intervals between automatic saves, text display format, screen colors, and plotter or printer information. Using a profile, you'll be able to adjust, save, and retrieve drawing settings the next time you use the workstation.

To set user profiles, select **Options** from the **Tools** menu. This will display the **Options** dialog box, with nine tabs to choose from. The tab that was used last is displayed by default. Each of these tabs will be explored in detail as commands are introduced throughout the text. The goal of this chapter is to introduce the **Options** dialog box and to help you save a few of your favorite settings. As new features are introduced, they can be added to your profile. Highlights of the **Options** tabs include the following:

Files—This tab is used to control the directories where AutoCAD will search for driver, menu, and other files. At this point of your AutoCAD career, only a few of the options need to be considered. Both points will be discussed later in this chapter.

Display—This tab allows the display of AutoCAD to be customized. Several of the options of this tab have been discussed in previous chapters. When the values are set and saved in a profile, they will automatically be reset when the profile is selected as the active profile. Take time to evaluate each portion of this box. If in doubt, click the **Help** button at the bottom of the tab for an explanation. Items you should adjust include the following:

Text Lines in the Command Line Window—Specifies the number of lines displayed in the docked window (two or three are usually adequate).

Colors—Allows the color of each aspect of the display screen to be altered. Selecting this option will display the **Color Options** dialog box and allow you to set the color of each element in the drawing area.

Crosshairs size—Allows the size of the crosshairs to be altered by entering a number that represents the percentage of the screen to be filled (100 allows easy projection from one object to another).

Display Resolution—The settings in this box affect the quality of the display of objects in the monitor. Do you want to display a circle as a circle, or as a series of straight line segments that appear to make a circle? The effects of this area will be discussed in Chapter 6 (100 is adequate for most drawings).

Open and Save—This tab will affect how and where drawing files are saved. This tab will be covered in detail later in this chapter.

> **SaveAs**—This portion of the dialog box can be used to determine the format used when saving your drawings. If you work at multiple workstations, this is your opportunity to save files to older drawing formats.
>
> **Automatic Save**—This option controls the time interval between automatic saves. This option is one of the greatest safety measures you can have to protect your drawing files (15 minutes is adequate for most drawings).
>
> **Create Backup Copy with Each Save**—This option toggles the creation of backup files. Typically, having a backup copy is a good safety feature. If your machine is short of storage space, deactivating this option will save your file over the existing copy of the file. If you decide to delete backup files, make sure you make multiple copies of your files on the hard drive, multiple diskettes, or a Zip drive.

Plotting—This tab controls all of the options that affect plotting. Components of this tab will be explored in the next chapter and discussed in detail in Chapter 25.

System—This tab controls how the system will operate. Take time to examine the **General Options** box. These settings control general options that relate to drawing operation. Key elements to consider:

> **Single Drawing Compatibility Mode**—One of the features of AutoCAD is that you can truly work on two drawings at once. This only holds true if you have this setting in the inactive mode.
>
> **Show Startup Dialog**—This setting will control the display of the **Startup** dialog box. Most new users prefer to have the dialog box displayed until they have developed their own work habits. Keep this setting activated in your profile.
>
> **Beep on Error in User Input**—In the active setting, AutoCAD will sound an alarm (a small beep) when an invalid response is given. If you're in the middle of the Line command, and AutoCAD is expecting a "to point" to be provided, the alarm will sound if you try and use the **Circle** command. As a new user, keep this activated. When it is deactivated, many new users get frustrated when the desired results are not achieved. The alarm often serves to bring you back to reality, reminding you to look at the Command prompt so that the desired information can be provided.

User Preferences—This tab affects options that optimize your performance in AutoCAD. One feature that experienced users of AutoCAD might want to explore is the **Right-click Customization** box. Some experienced users are so accustomed to using the right-click button in place of (ENTER) that they might be willing to forsake the new shortcut menus. If you're a new user, leave the menus active. Other options will be addressed in later chapters.

Drafting—Most of the options in this box have been introduced in previous chapters and will be covered in detail in later chapters. Options that you should address as you set your own profile include the following:

AutoSnap Marker Color—This control can be used to alter the color of the snap box.

AutoSnap Marker Size—This control can be used to alter the size of the snap box. As the size of the box is increased, it gets easier to see. As the drawing complexity increases, a large snap box can lead to inaccurate point selection.

Aperture Size—This slide bar can be used to control the size of the aperture box. The aperture box is the AutoCAD selection tool when a command is started.

Selection—These features control how AutoCAD will select drawing objects. This dialog box will be explored in detail in Chapter 9. A feature that you might want to set in your initial profile is the pick box size.

Profiles—This tab controls the creation and use of profiles. The **Profiles** tab can be seen in Figure 4.6. The next section of this chapter will help you explore this tab and create your first profile. As you gain experience and discover new settings that you would like to include in the profile, you can update the profile using this tab.

CREATING AND SAVING A PROFILE

To create a user profile, select **Options** from the **Tools** menu. Select the **Profiles** tab to display the dialog box shown in Figure 4.6. Go through each of the nine tabs and set the options to meet your preferences. Once you have selected the desired settings, follow this procedure:

1. Select the **Add to List** option. This will display the **Add Profile** dialog box shown in Figure 4.7.
2. Enter a profile name in the **Profile name** edit box. Use a name that will quickly describe the user or drawing type.
3. Provide a short description in the **Description** edit box that will quickly describe the contents.
4. Click the **Apply & Close** button to accept the settings. When the button is clicked, the dialog box will be closed and the profile name you provided will be added to the **Available profiles** list.

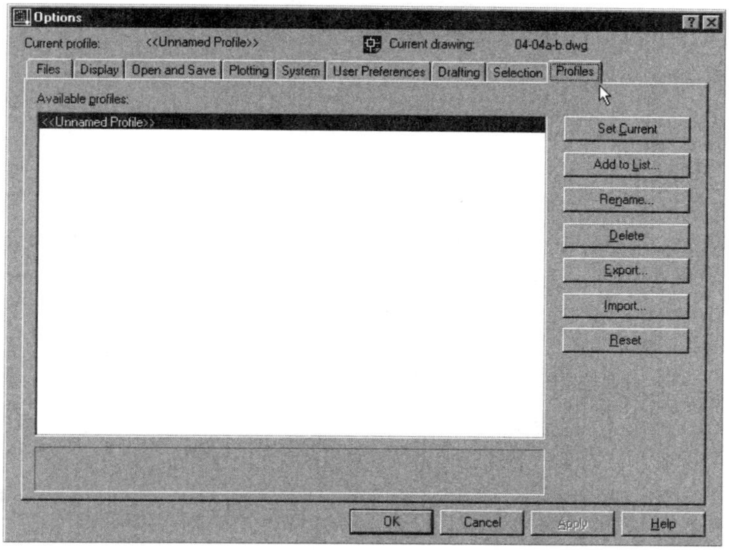

Figure 4.6 *The* **Profiles** *tab controls the creation and use of profiles. As you gain experience and discover new settings that you would like to include in the profile, the profile can be updated using this tab.*

Once more than one profile has been created, you can quickly select the desired profile from the **Profiles** tab of the **Options** dialog box by highlighting the listing of the desired profile and then clicking the **Set Current** button. Other buttons in this menu that should be considered include the following:

> **Rename**—This option allows an existing profile to be renamed. Clicking this button will produce the **Change Profile** dialog box shown in Figure 4.8. Alter the name as desired and then click the **Apply & Close** button to alter the profile name and close the dialog box.
>
> **Delete**—This option allows an existing profile to be removed from the **Available profiles** list. Highlight the profile to be deleted and click the **Delete** button. This will produce a warning to verify that you really want to delete the profile. Selecting the **No** button will terminate the delete process. Selecting the **Yes** button will delete the selected profile.
>
> **Export**—Selecting this option displays the **Export Profile** dialog box. (See Figure 4.9.) This dialog box can be used to export a profile as a file with an extension of .ARG. The profile can be saved and then shared with another computer. If you're working on a network, save your profile to your secure folder. Profiles should also be saved to a diskette.

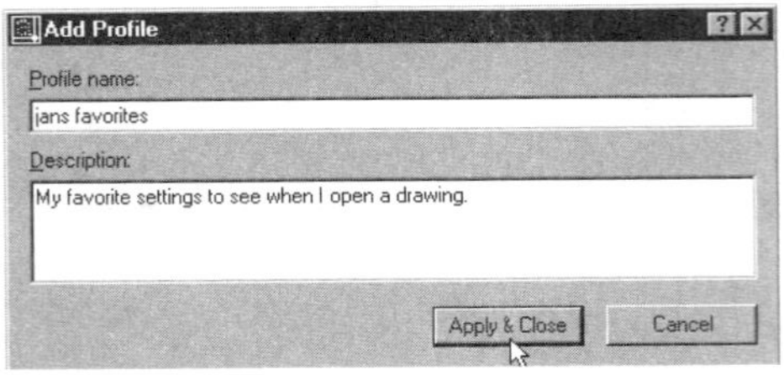

Figure 4.7 *Selecting the* **Add to List** *option will display the* **Add Profile** *dialog box. The box is used to name and describe the profile to be created.*

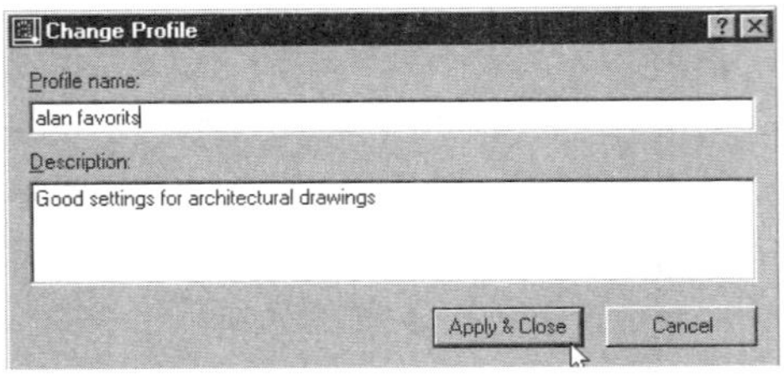

Figure 4.8 *Selecting the* **Rename** *button will produce the* **Change Profile** *dialog box. The box is used to alter the name of existing profiles.*

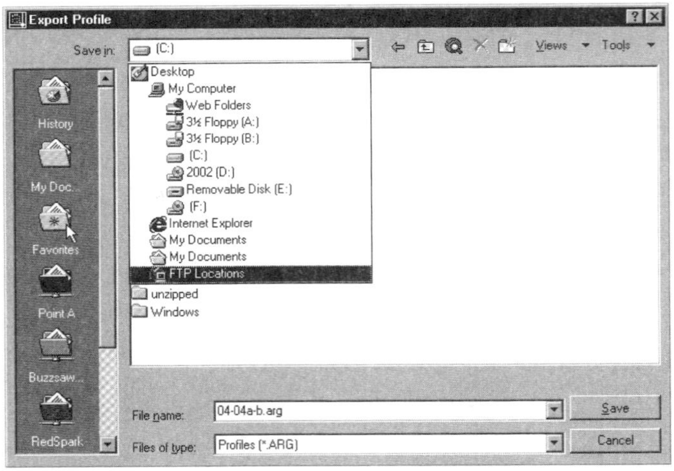

Figure 4.9 *The* **Export Profile** *dialog box can be used to export a profile as an .ARG file. The profile can be saved and then shared with another computer.*

Import—Selecting this option will display the **Import Profile** dialog box. Similar to the **Export Profile** dialog box, the display can be used to import a profile that was saved using Export.

Reset—Selecting this option will reset the computer to the AutoCAD default settings.

SAVING DRAWINGS

Chapter 1 introduced the **Save** command, and Chapter 3 gave you practice using it. This chapter will help you expand the use of the command. Once you have created your drawing, it will need to be saved. This means saving your drawing in a folder on the hard drive, on a diskette, or in your own folder on a network. Check with your instructor to see if saving on the hard disk is allowed. If so, set up a folder so that all of your drawings can easily be accessed. If not, select the letter of the appropriate drive so files will be saved to the diskette. Generally you should avoid saving a file to a diskette, except when the drawing session will be ended. Typically, a file should be saved to the hard drive throughout the drawing session to save time and avoid a possible problem caused by a full diskette.

You might be saving on the hard drive, tape drives, Zip drives, as well as on diskettes. Because working on the hard drive is so much faster, you will be saving to the hard drive during the drawing session, and saving on the diskette and the hard drive at the end of the drawing session. Many professionals save on the hard drive and make two diskette backups, which are each stored in different locations. The open drawing file should be saved to the hard drive approximately every 15 minutes. When the drawing session is done, the file should again be saved to update the existing file on the hard drive. The file should also be saved on another source, such as a diskette. This is especially true for students who must transport their drawing files between home and school.

No matter where you save your drawing, the most important thing to remember is to save often. Saving every 15 to 20 minutes will help you avoid losing your work if the power supply to your computer is interrupted. As you start saving your drawings, you will have three basic options: **Save**, **Saveas**, and **Qsave**.

THE SAVE COMMAND

Using the **Save** command will store your drawing in a selected location but still leave it displayed on the screen. This will allow you to continue working on the drawing. If your computer should crash, only work done after the last save will be lost. To save a drawing, select the **Save** button on the **Standard** toolbar, select **Save** from the **File** menu, or type **SAVE** (ENTER) at the Command prompt. Each will produce a similar result:

- The first time you save the drawing, the **Save Drawing As** dialog box shown in Figure 4.10a is displayed.
- The next time the drawing is saved, and all subsequent times, the save will occur automatically, with the name that was assigned during the initial save.

During the initial save, selecting the **Save in** edit box arrow will display the contents of Windows Explorer. When the destination drive has been selected, the contents of

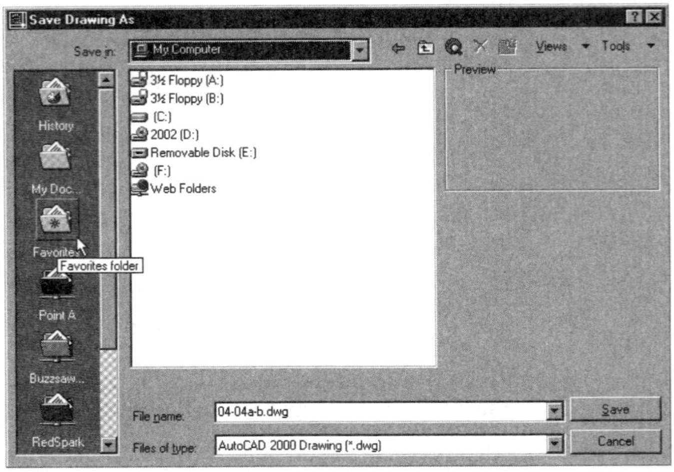

Figure 4.10a *Display the* **Save Drawing As** *dialog box by selecting the* **Save** *button on the* **Standard** *toolbar, selecting* **Save** *or* **Save As** *from the* **File** *menu, or by typing* **SAVE** *(ENTER) or* **SAVEAS** *(ENTER) at the Command prompt. The dialog box is displayed only the initial time the drawing is saved with the* **Save** *command. The second time the drawing is saved, the save will occur automatically using the name that was assigned during the initial save.*

the specified drive will be listed. Enter the desired file name in the **File name** edit box. The top five icons down the left side of the window can be used to provide quick access to common folders on your computer. Buzzsaw, RedSpark, and FTP icons farther down the left side, allow quick access to the Internet for saving files, posting information, and conducting meetings. Use the **Save as Type** edit box to control the drawing type to be used. Figure 4.10b shows a list of alternative drawing types that can be used. Once the type has been selected, click the **Save** button. (Guidelines for file names will be given later in the chapter.) The .DWG extension is added to the drawing file and it is saved to the designated drive. This will be the desired method of saving for most AutoCAD users.

File Names

As you save drawings, you might misname a drawing. If you haven't pressed (ENTER), just use BACKSPACE to remove the unwanted name. If the name is entered, the file can be renamed using Windows Explorer. Occasionally you will assign a drawing file a name that already has been used. In this case you will see a display similar to the one in Figure 4.11. If you proceed, you will destroy the existing file. The name will remain in the folder and your present drawing will be saved using the old file name, but the existing drawing will vanish. This is a fine way to update an old drawing, but be sure you can live without the old drawing.

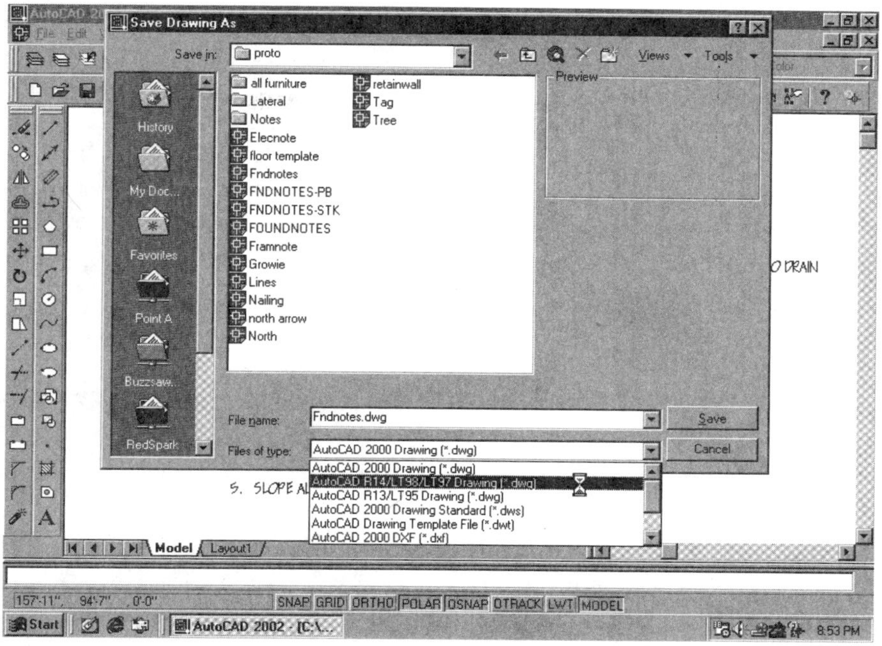

Figure 4.10b *The* **Save as type** *edit box allows drawings to be saved and be compatible with older versions of AutoCAD.*

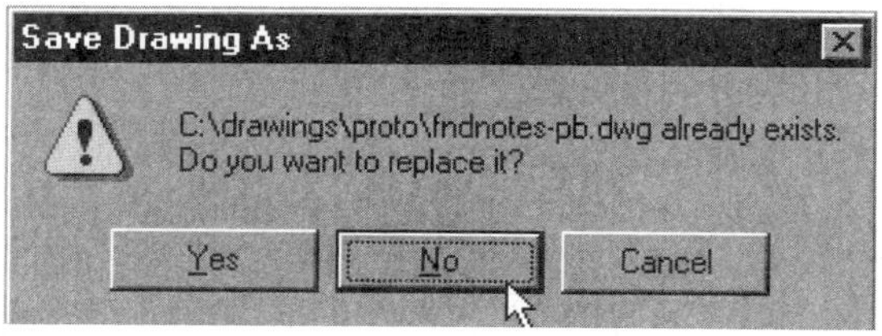

Figure 4.11 *Attempting to save a drawing using a name that has been assigned to another file will result in this alert message. If the* **Yes** *option is accepted, the drawing will be saved using the selected name, and the existing drawing will be overwritten.*

Backup Files

Earlier you discovered that when AutoCAD saves a file it adds an extension of .DWG or .BAK. The .DWG stands for a drawing file. The .BAK is the extension for a backup file. These are created automatically by AutoCAD each time you end a drawing. When a drawing is terminated using **Save**, the existing .DWG file becomes the .BAK file and a new .DWG file is created. The .BAK file might not contain all of your recent changes, but it should be only 15 minutes out of date if you've been saving at proper intervals. What a motivator to keep saving your drawings! Backup files are also great if your disk has been damaged. The .BAK file can be renamed to become a .DWG file using Windows Explorer and then reloaded. If the .BAK file cannot be converted to a .DWG file, or if it was scrambled, this will serve as one last reminder to save your files in more than one location. If you have a limited storage space, or if you save to multiple locations, you might not want to create backup files. If you decide you really don't want .BAK files created, you can deactivate the option by clearing the **Create backup copy with each save** check box on the **Open and Save** tab of the **Options** dialog box

THE SAVEAS COMMAND

The **Saveas** command is similar to the **Save** command.

- With **Save**, the **Save Drawing As** dialog box is displayed only the first time the drawing is saved.
- With **Saveas**, the dialog box is displayed each time the **Saveas** command is accessed.

Saveas allows the name of the file, the file destination, and the file type to be altered. **Saveas** can be accessed only from the **File** menu or from the Command prompt. Select **Save As** from the File menu, or type **SAVEAS** (ENTER) at the Command prompt. This will produce the **Save Drawing As** dialog box, seen in Figure 4.10a. The default name in the **File name** edit box will be the name you selected the last time the drawing was saved. If you like that name, click the **Save** button. If you want to change the name of the drawing file, enter the new name in the **File name** edit box.

Be sure that the correct drive destination is current. Note that the destination is listed in the **Save in** edit box. To change the current drive, use the arrow beside the **Save in** edit box to display the desired drive. Clicking the **Save** button will save the drawing to the specified drive. When a drawing is saved, it will be placed in the current drive. AutoCAD will assign the extension .DWG following the file name to help identify the contents of your directory.

Another benefit of **Saveas** is that it allows the file type to be altered. Examine Figure 4.10b again: it shows seven other alternatives for saving a file. These alternatives allow drawings created in AutoCAD 2002 to be saved so they will be compatible with different versions of AutoCAD. These options can be especially useful for students whose home machines are not current with those at school or for architectural and engineering firms with different versions of the software. To save a drawing created in AutoCAD 2002 to be used on a machine running R14, select the AutoCAD R14/LT98 Drawing (*.DWG) option. When a drawing is saved using a version other than the default, AutoCAD remembers the file format that was used, and uses the specified format in subsequent **Save**, **Qsave**, or **Autosave** operations that do not call for a dialog box.

Note: You can set the type of drawing format to be used for saving using the Open and Save tab of the Option dialog box and then saving the setting as part of your profile.

Open a drawing session, draw a few lines, and then save the file with a name of JUNK. As you assign a name, you do not have to type the .DWG at the end of the drawing file because it will be added automatically by AutoCAD. Once the drawing is saved, add a few more lines to the drawing and save the drawing again. If you save the drawing using the **Save** command, the **Save Drawing As** dialog box will not be displayed; the current file of JUNK.DWG is automatically updated. If you save the drawing using the **Saveas** command, the **Save Drawing As** dialog box will be displayed, allowing a new drawing name to be assigned to the drawing. Draw a few more lines and save the file using the **Saveas** command. When the **Save Drawing As** dialog box is displayed, change the drawing name from JUNK to TRASH, using the **File name** edit box. You will now have files named JUNK and TRASH. When you're done with the drawing session, be sure to delete these files using Windows Explorer.

QSAVE

Qsave, or quick save, will probably be your most frequently used method of saving a drawing. Selecting the **Save** button on the **Standard** toolbar or typing **QSAVE** (ENTER) at the Command prompt will save the current drawing using its given name without displaying the **Save Drawing As** dialog box. However, if **Qsave** is used on an unnamed drawing file, the **Save Drawing As** dialog box of Figure 4.10a will be displayed.

Note: Save, Saveas, and **Qsave** will each save a file for future use. Use **Qsave** for saving a drawing that you want to continue working on. It's the fastest method of saving because you only need to select the **Save** button on the **Standard** toolbar. **Qsave** saves named drawings without requesting a file name. Use **Saveas** when you want to save a drawing by giving it a new name or to save the drawing to a new destination. **Saveas** renames the current drawing and saves it to a file name. Save does the same thing as **Qsave** but is slightly slower to access because you need to select both the **File** menu and **Save**. **Save** saves the current drawing to a file name.

AUTOMATIC SAVING

By default, the existing drawing will be saved every 120 minutes. You can alter the time between automatic saves by using the **Open and Save** tab of the **Options** dialog box (select **Options** from the **Tools** menu). Use the following procedure to alter the interval between saves:

1. Display the **Options** dialog box.
2. Select the **Open and Save** tab.
3. Highlight the current value in the **Minutes between Saves** edit box.
4. Enter the desired value.
5. Click the **OK** button.

The save time can also be adjusted through the **Savetime** command. Type **SAVETIME** (ENTER) at the Command prompt. This will produce the prompt:

Command: **SAVETIME** (ENTER)

Enter new value for SAVETIME <120>: **15** (ENTER)

AutoCAD allows a value between 1 and 120 minutes to be entered. The value entered will start the timer when a change is made to the current drawing file. The commands **Save**, **Saveas**, and **Qsave** will reset the timer. With a value of 20, your drawing file will be automatically saved every 20 minutes with a name of AUTO.SV$. If this file needs to be accessed, Windows Explorer can be used to rename the file from a .SV$ file to a .DWG file. You can disable **Savetime** by entering a value of 0, but this could prove to be a huge mistake if your computer ever crashes.

SAVING A TEMPLATE

Earlier in this chapter you learned how to create a template. Once each of the values for the template is established, the drawing can be saved, to be reused whenever suitable values are required. Save the template using **Saveas** with the Drawing Template File (*.DWT) option selected. (See Figure 4.10b.) Enter the name of the drawing just as you would do with any other drawing. A name that describes the scale or paper size such as 14DSIZE or CSIZE will be helpful for sorting templates. Once the name has been entered, click the **Save** button. Provide a brief description of the template, such as "template for plan views." This description is displayed whenever you select this template from the **Create New Drawing** dialog box. The final step to save the template is to select the **OK** button. The template can now be reused when a new drawing is opened.

LEAVING AUTOCAD

Once a drawing is saved, you might want to leave the drawing and start another project. AutoCAD allows you to exit the program using the **Close** button or by using **Quit**, **Exit**, **Close**, and **Closeall** commands.

The Quit and Exit Commands

The **Quit** and **Exit** commands will exit the drawing and close AutoCAD. Execute the **Quit** command by selecting the **Close** button on the AutoCAD title bar or by typing **QUIT** (ENTER) at the Command prompt. If you attempt to leave a drawing with unsaved changes, you will see the alert box shown in Figure 4.12. If you click the **Yes** button, AutoCAD will save the changes to the default file. Clicking **No** will exit the drawing, close AutoCAD, and return you to other opened programs—any changes made since the last save will be destroyed. **Cancel** will terminate the **Quit** command and return you to the drawing. **Exit** functions in the same manner as **Quit**. It must be entered from the keyboard by typing **EXIT** (ENTER).

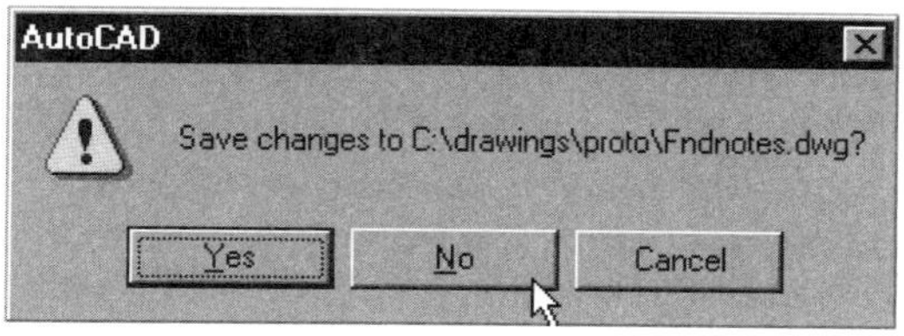

Figure 4.12 *An alert message will be displayed if you attempt to open a new drawing without saving the existing drawing.*

The Close Command

The **Close** command can be used to close the current drawings but remain in AutoCAD. The command must be entered from the Command prompt by typing **CLOSE** (ENTER) at the prompt. If there have been no changes, the drawing will be closed automatically but AutoCAD will remain open. If changes have been made since the drawing was last saved, you will be prompted to save or discard the changes.

The Closeall Command

If you are working in multiple drawings, the **Closeall** command can be used to close all open drawings. The command must be entered from the Command prompt by typing **CLOSEALL** (ENTER) at the prompt. If there have been no changes, the drawings will be closed automatically but AutoCAD will remain open. If changes have been made since each drawing was last saved, you will be prompted to save or discard the changes.

MANAGING FILES AND FOLDERS

No office could function if the drafters took their finished drawings and threw them into a room. If a drawing were needed, some poor drafter would have to wade in and hunt through thousands of sheets of paper to find the specific project. Architectural and engineering offices typically use well-organized filing systems to keep track of each project for quick, easy retrieval.

FILES

No matter what method is used to save a drawing, information that is to be kept on your disk must be stored in some orderly manner to allow quick, easy retrieval. Disks are sorted in files, folders, and paths. Your drawings will be stored as files and will need names to distinguish them from other files. File names for AutoCAD can contain up to 255 characters. Other guidelines for naming files include the following:

- Layer names can contain letters, digits, spaces, and special characters such as $ (dollars), - (hyphen), and _ (underscore).
- Special characters used by Microsoft, Windows, or AutoCAD cannot be used.

Do not use the following symbols:

* (asterisk)	< > (less-than and greater-than symbols)
‘ (back quote)	? (question mark)
: (colon)	“ ”(quotation marks)
, (comma)	; (semicolon)
= (equal sign)	| (vertical bar)
/ \ (forward and back slashes)	

FOLDERS

Files should be organized in folders to provide efficient storage. Folders are the equivalent of dividers in a file cabinet drawer. By default, your drawings will be stored in the Acad2002 folder. Information about files such as file size, time of creation, and the time of last file revision is automatically kept as files are stored. Folders can be created using Windows Explorer to organize information based on client names, class requirements, or any other similar criteria. Folder naming should follow the same guidelines used to name drawing files. Names should be used that will quickly define the contents such as DRAWINGS, MAPS, REPORTS, or CLIENTS. Folders can be placed inside other folders to better organize information. Subfolders can be created and named using the same guidelines used to create a folder. Figure 4.13 shows an example of how folders are set up within folders created by Windows Explorer in a hierarchical method. This use of paths in saving material will help in retrieval.

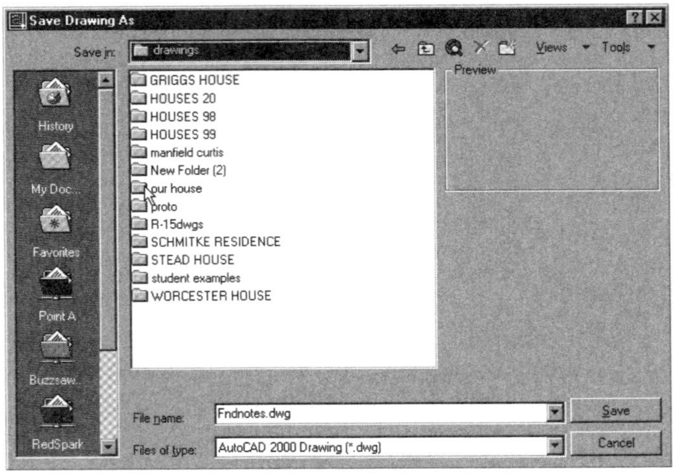

Figure 4.13 *Folders should be created to contain similar drawing files to ease retrieval. Notice, in the C drive, that all drawings are stored in the DRAWING folder. The DRAWING folder contains subfolders named for clients to ease retrieval.*

OFFICE PROCEDURE

Many small residential offices keep work for each client in separate folders or disks and label the disk by the client name. Drawings are then saved by contents such as floor, foundation, elevation, sections, specs, or site. Some offices assign a combination of numbers and letters to name each project. Numbers are usually assigned to represent the year the project is started, as well as a job number with letters representing the type of drawing. For example, 0153fl would represent the floor plan for the fifty-third project started in 2001. A few offices will save an entire house plan in one file. The drawings are separated by layers, which will be discussed in Chapters 5 and 13. The drawback to saving large amounts of information in one file is that more time is needed to load and process the information. Later chapters will introduce External Reference as a method of saving drawing files and methods of posting drawings on the Internet.

AIA Guidelines for Naming Files

Because it is rare that only one person in an office will use a file, most architectural and engineering offices use file names based on the AIA (American Institute of Architects) guidelines. The AIA guidelines provide a uniform file naming system that will be recognized by the various consultants that work on each project. Two methods of naming files are often used by architectural systems, one based on page numbering and the other based on the discipline.

One common practice in architectural firms is to name drawing files by the page number it will occupy in the drawing set based on AIA page numbering guidelines.

Common Page Numbers and File Names

A0.01—Index, symbols, abbreviations, notes, and location maps

A1.01—Demolition, site plans, and temporary work

A2.01—Plans, and schedules such as room material, door, or windows and keyed drawings

A3.01—Sections and exterior elevations

A4.01—Detailed, large-scale floor plans

A5.01—Interior elevations

A6.01—Reflected ceiling plans

A7.01—Vertical circulation drawings such as stair, elevator, and escalator drawings

A8.01—Exterior details

A9.01—Interior details

When this system is followed, a drawing file named A4.01 would represent a detailed floor plan. If plans for a five-level building were to be drawn, each floor could be saved as a separate drawing file using drawing names such as

A2.01—level 1

A2.02—level 2

A2.03—level 3

A2.04—level 4

A2.05—level 5

A second method of naming files is based on a letter that represents the discipline of the drawing originator, and a two-letter code that represents drawing types common to all disciplines. A number that represents the sheet sequence number is sometimes added to the file name. With this system, drawing files would have names such as A–FP01 or S–DT13. Some of the most common disciplines and drawing codes include the following:

Common Discipline Codes

A	Architectural	M	Mechanical
C	Civil	P	Plumbing
E	Electrical	Q	Equipment
F	Fire Protection	S	Structural
I	Interiors	T	Telecommunications
L	Landscape		

Common Drawing Codes

FP	Floor Plan	DT	Detail
SP	Site Plan	SH	Schedules
DP	Demolition Plan	3D	Isometric/3D drawings
QP	Equipment Plan	DG	Diagrams
XP	Existing Plan	EL	Elevation
SC	Section		

CSI Guidelines for Naming Files

Another popular method of naming construction drawing files is to use the guidelines provided by the Construction Specifications Institute (CSI). These guidelines recommend titles that start with a letter to represent a discipline code, a sheet designator, and a page number. Page numbers in file names are designated sequentially from 01 through 99. Discipline codes are the same as those used by the AIA. With the CSI format, in a file name such as A-501, A is a drawing by the architectural team, **5** represents details, and **01** represents sheet 1 of the detail section.

Sheet Type Designators

0—General symbols, abbreviations, notes, and location maps

1—Plans, and schedules such as room material, door, or windows and keyed drawings

2—Exterior elevations

3—Sections

4—Detailed, large-scale floor plans

5—Details

6—Schedules and diagrams

7—User defined

8—User defined

9—3D views such as isometrics, perspectives, and photographs

STORAGE PROBLEMS

One of the great thrills for a new CAD operator is seeing the message, "DISK FULL." It's not a serious problem, but it is discouraging. As a drawing is started, AutoCAD examines the disk to find a suitable storage space. Because AutoCAD creates space for several files, it will need to perform expected drawing commands and there might not be enough space to create the intended drawing. In the middle of a command you might find that a DISK FULL error is displayed. Your drawing will be terminated, but all work up to that point will be saved. You will typically encounter this problem only if you save to a diskette in the middle of a drawing session.

Occasionally a full disk will cause FATAL ERROR to flash across the screen just before your machine destroys two hours worth of drawing. Sure, you know that you should save every 15 minutes, but sometimes you've got to learn the hard way. The unfortunate aspect of a fatal error is that it might disrupt or destroy other files on your diskette. You'll find that you only have FATAL ERROR when you're 99 percent done with a project.

One Last Reminder: As you are working on a drawing file, save the drawing to the hard drive. Only save to a diskette when you are done with the drawing session and are closing the file. Working from a diskette will slow the drawing session and can lead to storage problems.

Diskette-Handling Procedures

Diskettes should serve only to back up projects that have been saved on the hard disk or network. Because diskettes can be damaged easily, extreme care must be taken with them. The biggest destroyers of information on a diskette are dust, smoke, magnets, and bending.

Dust and smoke interfere with the head in the disk drive as it attempts to interpret the diskette. Magnetic fields from such common items as a stereo speaker or a telephone can scramble the information on a disk into an unusable form. In addition to destroying information, a bend can hinder retrieval of a diskette from the drive unit. Disk labels that are not affixed well to the diskette also can cause the diskette to jam in the drive.

Guidelines for Safe Diskette Handling

- ALWAYS keep the disk labels in the outer corners of the disk.
- ALWAYS label diskettes clearly, indicating the contents.
- ALWAYS store diskettes in a protective carrying case and in a cool, dry location.
- ALWAYS store diskettes between 50° and 125° Fahrenheit (10° to 52° centigrade).
- NEVER place diskettes on top of a monitor.

Protecting Disks

Diskettes can be protected so that no one can accidentally enter information or change the original contents. There is a notch in the top right corner of the disk when you look at it from the front. Turn the diskette over and slide the black tab to its upper position. This will allow you to see through the notch and make the diskette write protected.

Correcting File Errors

No matter how carefully you follow the guidelines for handling a disk, they will occasionally become corrupted. AutoCAD provides the **Audit** and **Recovery** commands that attempt to make corrupted drawing files useable. Each are located in **Drawing Utilities**, on the **File** menu. The **Purge** command can be used to clean unneeded information from the drawing base and will be introduced in Chapter 15.

Examining a Drawing File with AUDIT

The **Audit** command can be used to examine the current drawing file for errors and to correct any errors that exist. These errors are not spelling mistakes or drawing errors, but errors that occasionally are made by the software. This command will detect errors, generate extensive descriptions of the problems, and recommend actions to correct the errors. Use **Audit** by selecting **Audit** from **Drawing Utilities** on the **File** menu or by typing **AUDIT** (ENTER) at the Command prompt. The command sequence is as follows:

Command: **AUDIT** (ENTER)

Fix any errors detected? [Yes/No] <N>

Responding with **N** (ENTER) will produce a report of the file and list the errors but will not repair the errors. The final listing of the report resembles

Total errors found 3 fixed 0

Responding (ENTER) at the prompt will produce a report of the file and list the errors. The listing resembles

19 Blocks audited		
Pass 1	1793	objects audited
Pass 2	1793	objects audited
Pass 3	3800	objects audited
Total errors found 2		fixed 2

With the **Auditctl** system variable set to 1, an Audit report file will be created using the file name with .ADT as the extension. This file will be placed in the directory containing the original drawing file and will list all of the functions required to fix file errors.

Recovering a Drawing

If errors are discovered that cannot be repaired by **Audit**, the **Recover** command can be used to try and salvage the file. Access **Recover** by selecting **Recover** from **Drawing Utilities** on the **File** menu or by typing **RECOVER** (ENTER) at the Command prompt. AutoCAD will automatically attempt to repair damaged drawings as they are opened. If the drawing can't be opened, a message will be displayed that the drawing must be recovered. Once the file to be recovered has been selected, AutoCAD will begin a series of diagnostic procedures. When the process is complete, a listing of problems and the corrective measures taken by the program will be displayed. Generally, the recovery process will open a damaged drawing. If the **Recover** command fails, it's time to get out the backup file. This is one last opportunity to remind you that no matter how fast and good computers have become, they occasionally fail, and generally only at times of great stress such as the end of a school term or the due date of a big project. Backup files can greatly reduce the stress when you have a file that can't be recovered.

CHAPTER 4 EXERCISES

1. Start a new drawing. You will be using this exercise as a base for a drawing of a small subdivision. The drawing will be drawn at a scale of 1"=30' and plotted on D-size paper. North will be at the top of the paper (twelve o'clock). Set all applicable defaults to two-place decimals.

   What unit setting should be used for this drawing? ______________

   Set the limits at the proper value. What option number will be required?

   ____

   Many of the property lines are described with angles measured in degrees, minutes, and seconds. Set the angle measurement system in the appropriate system. What default number was selected? ______________

2. Start a new drawing. Draw a square, a rectangle, and a triangle. Save the drawing using the name E-4-2.

3. Start a new drawing called EXER4-3 that has the same drawings contained on E-4-2. Add another circle and a rectangle. Save as E-4-3.

4. Edit drawing E-4-3 by adding any figure that you would like. Save file as E-4-4.

5. Get a directory listing of your diskette; list the following:

   The contents

   The time your first exercise was finished

   The date of your last exercise

   The number of bytes used

   The number of bytes free

6. Open the ARCHITECTURAL.DWT template.

   What is the current snap resolution? X _____ Y _____

   What is the current grid setting? X _____ Y _____

   What layer will new lines be added to? ______________

   You will be using this exercise to set up a drawing for a plan view of a room that will be drawn at 1/4"=1'–0" and plotted on D-size paper. Two-place decimal angles are desired.

   Set the units at the proper value. What option number will be required? ______________ What was the default? ______________

   The smallest fraction to be worked with will be 1/4". What denominator should be used? ______________

   Give the current system for measuring angles. ______________

   Will the current setting allow for two-place decimals? ______________

In which direction will angles be drawn? __________

What are the current drawing limits? __________

What limits will be needed to achieve the listed parameters of this exercise?

____

What is the current snap value? __________

Set the snap value to 6".

What is the current grid default value? __________

Set the grid to 12".

List two other methods of listing the 12" value.

CHAPTER 4 QUIZ

1. You're working on the fifth level of a multilevel project. Suggest three different names that can be used to save the file.
2. What command will store the drawing on the hard disk but leave the drawing on the screen?
3. How many characters can be used for a file using AutoCAD 2002? How many characters can be used if the file is to be compatible with an R14 file?
4. What command will store the drawing on a disk and terminate the drawing?
5. List three different methods for accessing the Save command.
6. You have a drawing file labeled SECTION. You draw a roof detail and start to save it as SECTION. What will AutoCAD respond and how will your drawings be affected?
7. Give six forms of information listed when the dialog box is used to save a drawing.
8. What two extensions are assigned to a drawing file?
9. What does **Quit** do differently from **Save**?
10. What is a fatal error?
11. Give two reasons for backing up a drawing in multiple locations.
12. List two reasons why you might receive an alert message while saving a drawing.
13. You've opened an existing drawing named FLOOR.DWG, made some changes, and now want to save the drawing while keeping FLOOR.DWG intact. How can this be done?
14. You've saved a file named LOST.DWG somewhere on your hard drive, but you're not sure where. How can you find the file?
15. Explain the difference between **Save** and **Saveas**.

    What is the default value for **Savetime**, and is this value adequate?

CHAPTER 5

Drawing Organization

This chapter will introduce

- Common linetypes used on construction drawings
- Methods of loading varied linetypes into a drawing
- Methods of controlling lineweight
- Methods of assigning color to drawing objects
- The use of layers for controlling lines, lineweight and color
- Making a check print

Commands to be introduced include

- **Linetype**
- **Properties**
- **Lineweight**
- **Color**
- **Layer**
- **Plot**

Now that you're able to set up drawing units and limits, basic drawing components can be explored. The main component of the drawings you will be working with will be lines. This chapter will introduce you to various linetypes used on construction drawings and to methods for assigning and controlling varied linetypes. As linetypes are added to a drawing, a method of managing these linetypes must also be considered. Color can be assigned to lines to help distinguish types of building components. Most professional AutoCAD users sort linetypes by the use of layers. The **Linetype**, **Color**, and **Layer** commands will be introduced in this chapter, with a detailed discussion presented in later chapters. In addition to exploring the methods of controlling lines, this chapter will explore the process of making a non-scaled check print. Plotting a drawing to scale will be explored in later chapters.

AN INTRODUCTION TO LINETYPES

Varied linetypes are used in every type of drafting and consulting firm involved in the construction process. This chapter will introduce you to common linetypes used to make construction drawings. It will explain how to load linetypes to be used in a drawing session and how to alter and control linetypes that are part of a drawing.

COMMON LINETYPES IN CONSTRUCTION DRAWINGS

Although each office may have its own standard, guidelines established by the American National Standards Institute (ANSI) are used throughout most of the construction industry to ensure drawing uniformity. Methods of altering line width will be introduced in this chapter and discussed later in Chapter 25 when the drawing is plotted. Common linetypes include object, hidden, center, cutting plane, section, break, phantom, extension, dimension, and leader. Examples of most of these types of lines can be seen in Figure 5.1. Specific uses for each linetype will be discussed throughout the text.

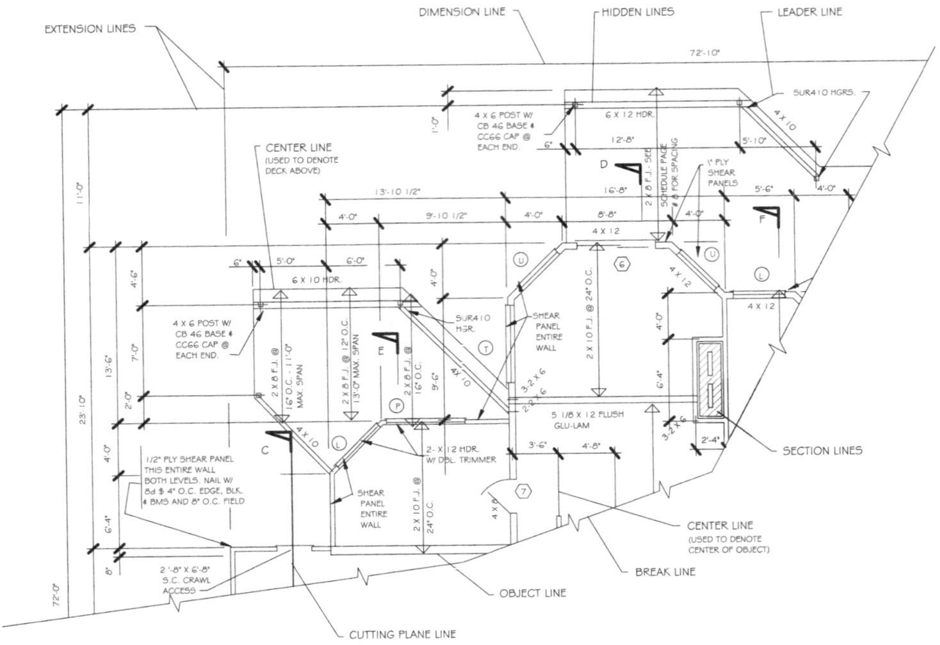

Figure 5.1 *Common linetypes used on construction drawings.*

Object Lines

Object lines are continuous lines used to describe the shape of an object or to show changes in the surface of an object. Object lines can be thick or thin depending on how they are being used. The common line width for thick lines is 0.6 mm; use a width of 0.3 mm for thin lines. On a drawing such as a floor plan, thick object lines are used to represent walls, and thin object lines are used to provide contrast when representing doors, windows, cabinets, or appliances. Figure 5.2 shows both uses of object lines. Uses of object lines include but are not limited to the following:

- Site plans

  Thick: building footprint

  Thin: flatwork, street and sidewalk edges, decks, balconies, or patios

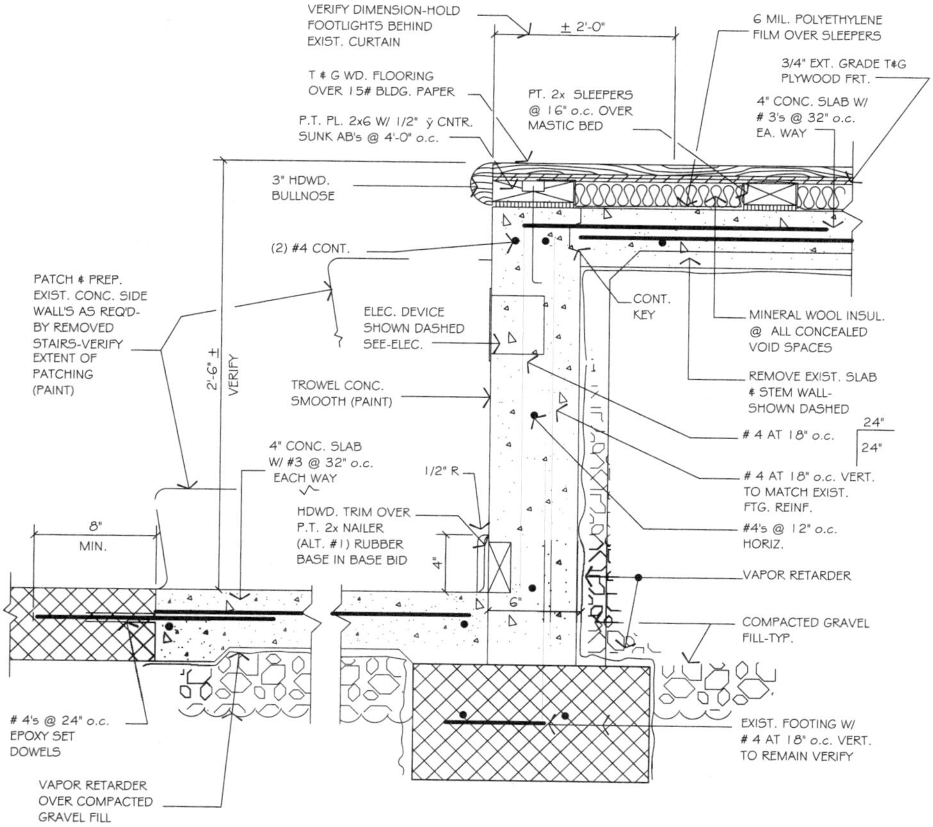

Figure 5.2 *Object lines are continuous lines used to describe the shape of an object or to show changes in the surface of an object.*

- Floor and framing plans

  Thick: walls and fireplace outlines

  Thin: doors, windows, cabinets, appliances, stairs, railings, plumbing fixtures, shelving, and framing members

- Elevations

  Thick: finish grades, outlines of objects, and shading

  Thin: windows, doors, trim, and wall and roof outlines

- Sections and details

  Thick: outline of structural material cut by the cutting plane such as beams, plates, sills, and girders; concrete shapes such as piers or footings; steel reinforcing; finish grades; and natural grades

  Thin: repetitive structural materials such as studs, posts, columns, joists, and rafters that lie beyond the cutting plane; structural materials such as sheathing and non-structural materials such as sheet rock, and materials beyond the cutting plane

Hidden Lines

Dashed or *hidden lines* are thin lines used to represent a surface or object that is hidden from view. ANSI specifies that the lines should be 1/8" long with 1/16" space between lines. The size of the line can be controlled using the **Ltscale** command. AutoCAD will automatically control the spacing, based on the line length. Figure 5.3 shows an example of hidden lines. In addition to hidden features, uses of hidden lines for drawings include but are not limited to the following:

Site plan—Easements, sewer or water lines, and structures to be removed

Floor plan—Appliances such as dishwashers that are under counters, refrigerators that are not in the contract, and knee spaces in cabinets

Framing plan—Beams and headers

Foundation plan—Outlines of footings, piers, and girders

Elevations—Outline of footings

Sections and details—Interior of tube steel columns, and steel reinforcing

Centerlines

A *centerline* is composed of thin lines that create a long-short-long pattern. The short line should be 1/8" long, the space should be 1/16", and the long segment must be between 3/4" and 1" long. Although the length of the long segment can vary from drawing to drawing, all long segments in a drawing should be consistent. AutoCAD will control the length and space of each portion of the line, or you can create your own pattern using the methods presented in Chapter 13. Centerlines are used to locate the

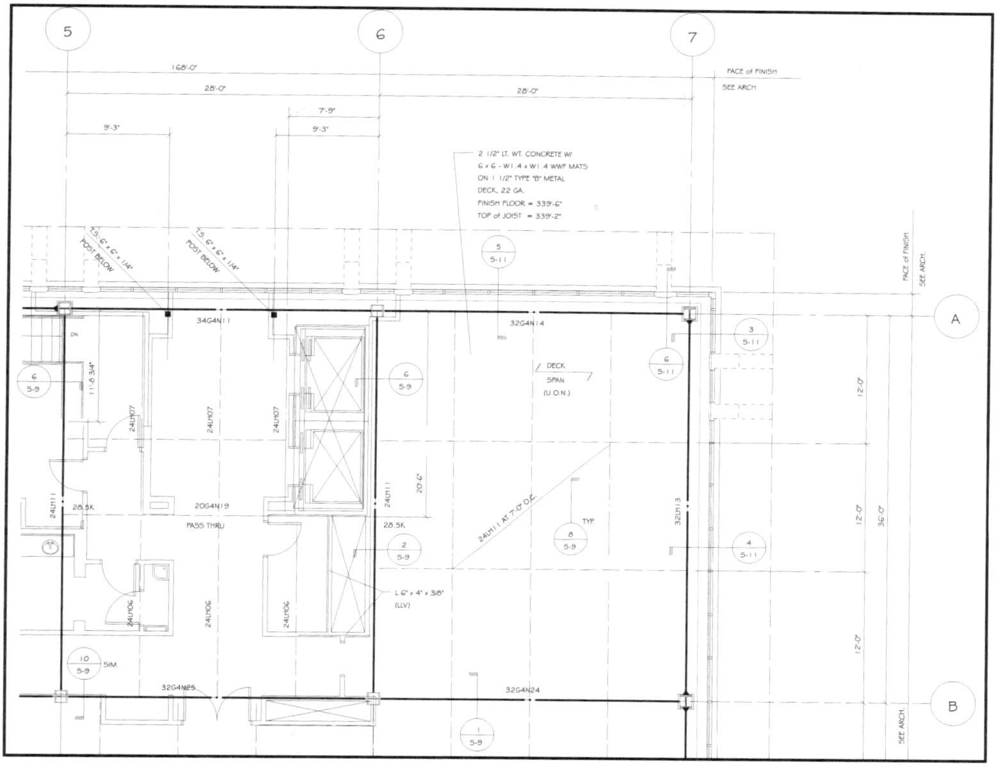

Figure 5.3 *Dashed or hidden lines are thin lines used to represent a surface or object that is hidden from view.*

center axis of circular features such as drilled holes, bolts, columns, and piers. Figure 5.4 shows the use of centerlines on a foundation plan. Uses for centerlines include but are not limited to the following:

Site plans—Center of easements, streets

Floor and framing plans—Upper levels or projections above the current floor, grid lines

Elevations—Finish floor or ceiling levels

Sections and details—Finish floor, ceiling of roofing levels, centers of steel columns and chevrons

Cutting Plane Lines

A *cutting plane line* is placed on the floor or framing plan to indicate the location of a section and the direction of sight when viewing the section. Cutting planes are a long-short-short-long line pattern. Cutting plane lines should be thicker than object

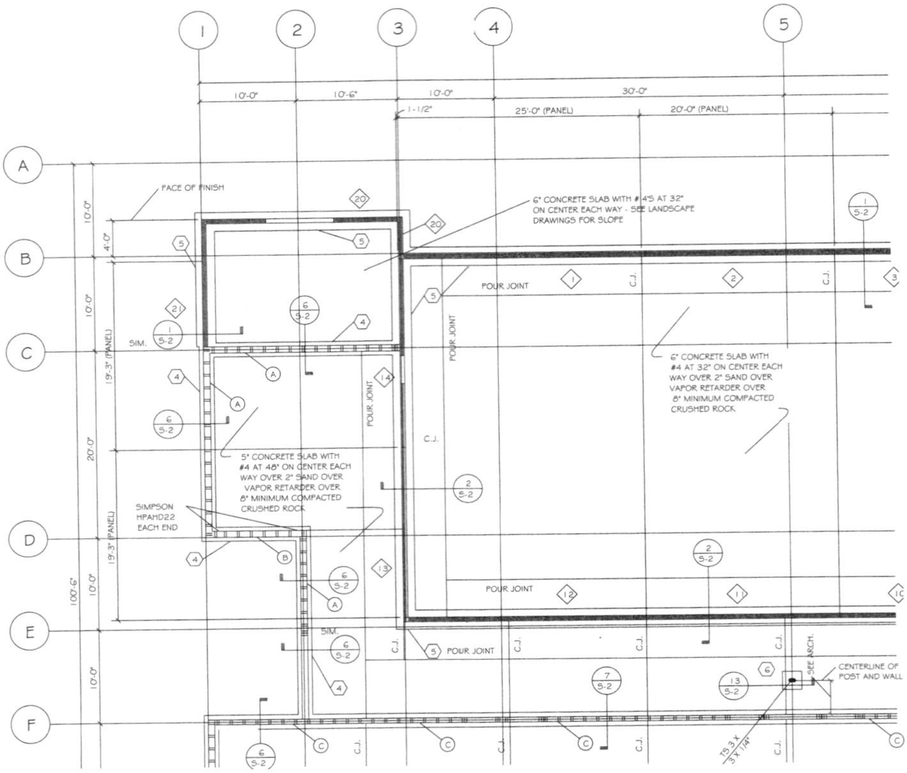

Figure 5.4 *A centerline is composed of thin lines that create a long-short-long pattern. Centerlines are used to locate the center axis of circular features such as drilled holes, bolts, columns, and piers. Centerlines are also used to show the girders on a foundation plan.*

lines. The long line segment should be between 3/4" and 1" long. The short lines should be 1/8" long, and the space between lines should be 1/16". Terminate the line with arrows to indicate the viewing direction. Place a letter at each end of the line using 1/4" high text to indicate the view or a detail reference bubble. Figure 5.5 shows examples of cutting planes and line terminators.

Section Lines

When an object has been sectioned to reveal its cross section, a pattern is placed to indicate the portion of the part that has been sectioned. In their simplest form, *section lines* are thin parallel lines used to indicate what portion of the object has been cut by the cutting plane. Section lines are usually drawn at a 45° angle but should not be parallel or perpendicular to any part of the object that has been sectioned. If an angle other than 45° is used, it should be between 15° and 75°. If two adjacent parts have

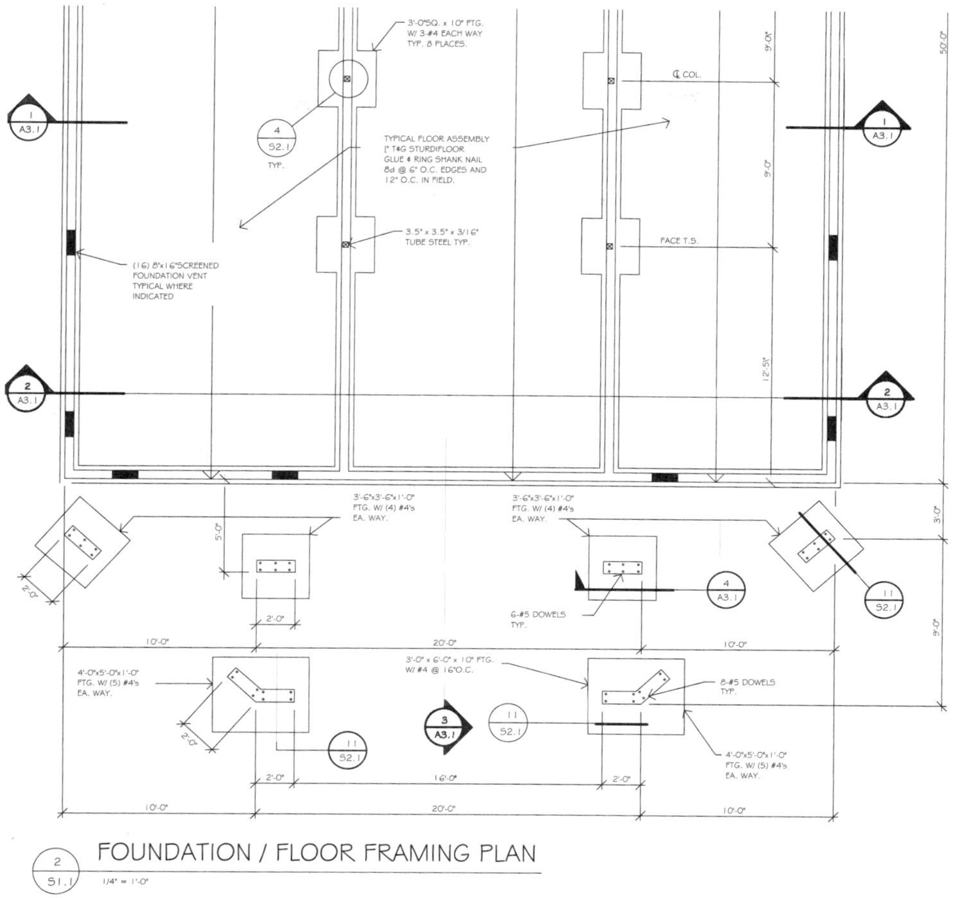

Figure 5.5 *A cutting plane line is used to indicate the location of a section and the direction of sight when viewing the section. The line is thicker than object lines and must terminate with arrows to indicate the viewing direction or a detail reference bubble.*

been sectioned, section lines should be placed at opposing angles (see Figure 5.6). Spacing between section lines can be altered, depending on the size of the part to be sectioned. Typically, a spacing of 1/8" should be provided between section lines. Section lines should be placed using the **Bhatch** command.

Section lines are generally used only to represent masonry in plan or sectioned views. Most materials have their own special pattern. Figure 5.7 shows common patterns used to indicate that an object has been sectioned. AutoCAD refers to these patterns as hatch patterns. Chapter 14 will discuss representing sectioned material, guidelines for placement of section lines and hatch patterns, and methods of using the **Bhatch** command.

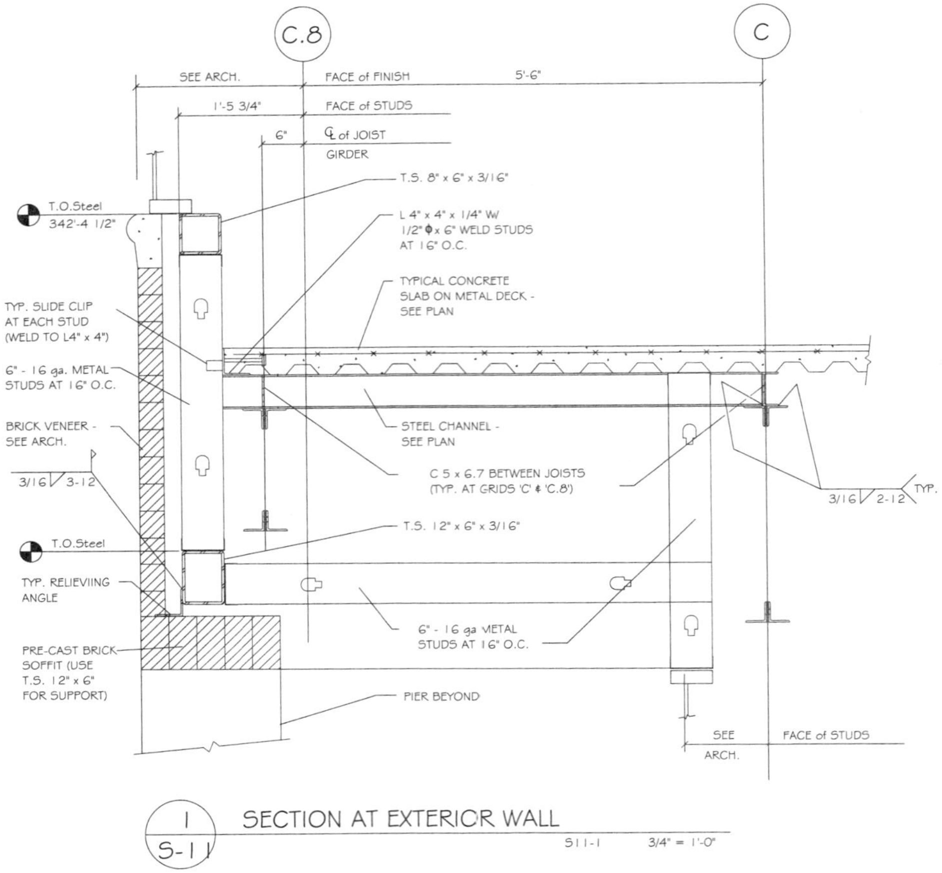

Figure 5.6 *Section lines are thin parallel lines drawn at a 45° angle to indicate what portion of the object has been cut by the cutting plane. Section lines are generally used only to represent masonry in plan or sectioned views.*

Break Lines

Break lines are used to remove unimportant portions of an object from a drawing so that it will fit into a specific space. If you consider a 10' long steel column that supports a beam and rests on a concrete footing, the column is the same between the beam and the bottom plate. If you need to draw a 10' steel column at full scale in a 5' space, break lines can be used to remove 5' of the column. The three common break lines used on construction drawings are the short, long, and cylindrical break lines. A short break line is a thick jagged line placed where material has been removed (see Figure 5.8 at A). A long break line is a thin line with a zigzag shape or inverted S shape inserted

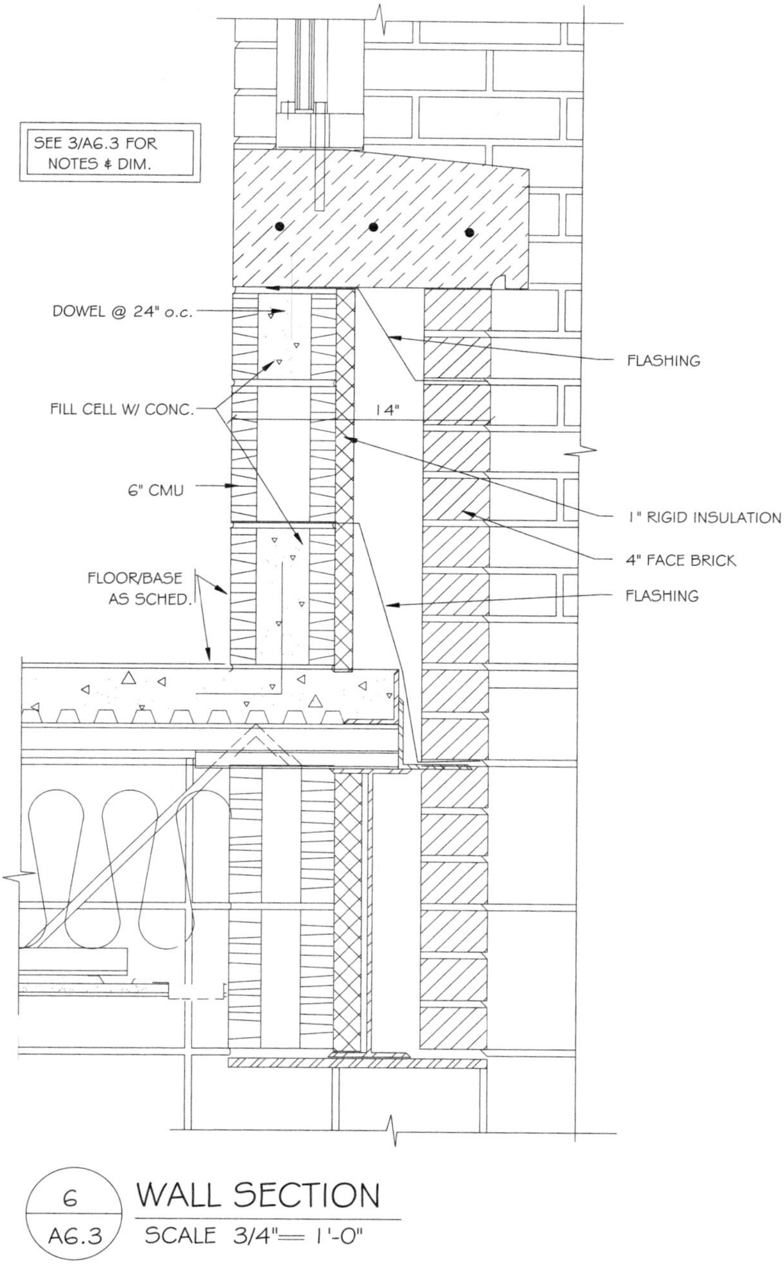

Figure 5.7 *Common patterns used to indicate that an object has been sectioned. Most construction materials have their own special pattern. AutoCAD creates these patterns through the* **Bhatch** *command.*

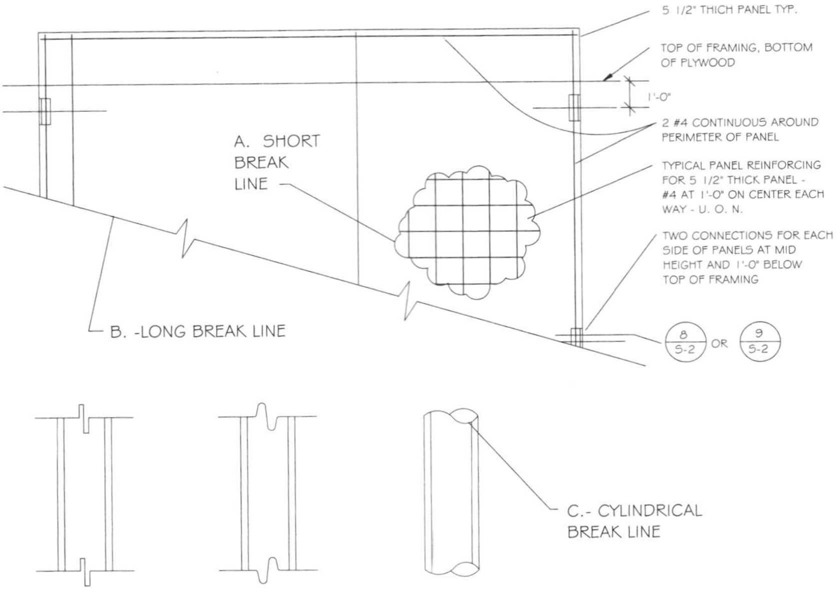

Figure 5.8 *Common break lines of construction drawings: a short break line (A) is a thick jagged line placed where material has been removed; a long break line (B) is a thin line with a zigzag shape or inverted S shape inserted into the line at uniform intervals; a cylindrical break line (C) is a thin line resembling a backward S.*

into the line at intervals (see Figure 5.8 at B). A cylindrical break line is a thin line resembling a backward S (see Figure 5.8 at C).

Phantom Lines

Phantom lines are thin lines in a long-short-short-long pattern that are used to show motion, or to show an alternative position of a moving part. They can also be used in place of centerlines to represent upper levels or projections of a structure, or on site plans to represent easements or utilities. The long line segment should be between 3/4" and 1" long but remain a constant length within each drawing. The short lines should be 1/8" long. The space between lines should be 1/16".

Extension Lines

Extension lines are thin lines used to relate dimension text to a specific surface of a part (see Figure 5.9). Provide a 1/16" to 1/8" space between the object being described and the start of the extension line. Extend the extension line 1/8" past the dimension line. An extension line can cross other extension lines, object lines, center, hidden, and any

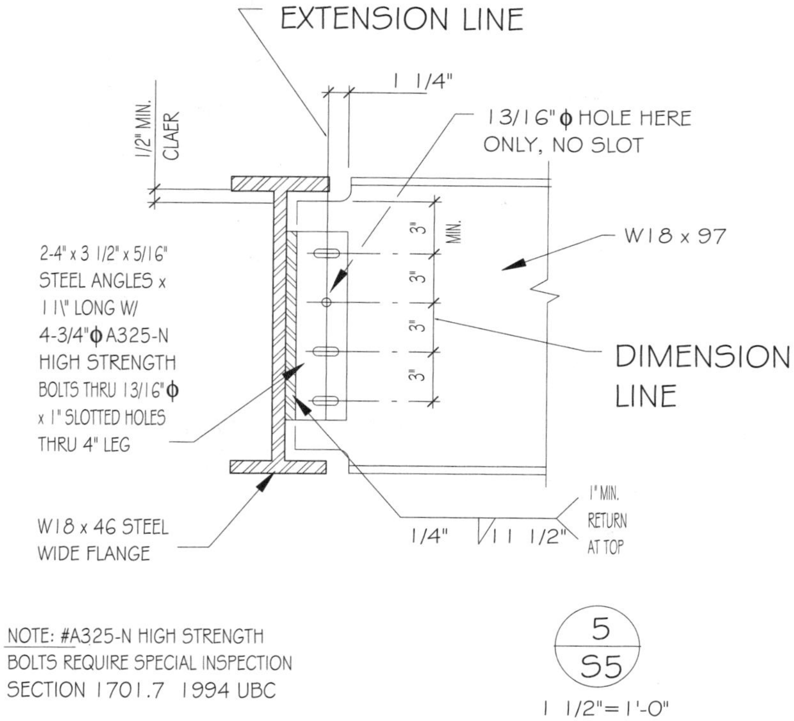

Figure 5.9 *Dimension lines are thin lines used to relate dimension text to a specific part of an object.*

other line, with one exception: An extension line can never cross a dimension line. Never! Chapters 17 and 18 will introduce other guidelines for placing extension lines using the various commands for placing dimensions.

Dimension Lines

Dimension lines are thin lines used to relate dimension text to a specific part of an object (see Figure 5.9). Dimension lines extend from one extension line to another. Dimension lines require a line terminator where the line intersects the extension line. The terminator is usually a tick mark or a solid arrowhead. Dimension text is typically 1/8" high text. Dimension text should be placed centered over the dimension line, although it might need to be placed off center to aid clarity when several lines of text are near each other. The space between the dimension text and the dimension line should be equal to half the height of the dimension text. Chapters 17 and 18 will provide specific uses of dimension lines and dimension text as the various commands for dimensioning are introduced.

Leader Lines

Leader lines are thin lines used to relate a dimension or note to a specific portion of a drawing. A solid arrowhead should be used at the object end of the leader line. A horizontal line approximately 1/8" long should be used at the note end of the leader line. Leader lines can be placed at any angle, but angles between 15° and 75° are most typical. Vertical and horizontal leader lines should never be used. The arrow end of the leader line should touch the edge of the part it describes. When used to describe a circle, the arrow should end at the edge of the circle, but if extended, the line would pass through the center of the circle. The leader line should extend far enough away from the object it is describing so that the attached note is a minimum of 3/4" from the outer surface of the object. See Figure 5.10 and Chapters 17 and 18.

AN INTRODUCTION TO CREATING LINETYPES

You've been introduced to the varied types of lines that are required to produce construction drawings. AutoCAD contains many of the linetypes found in the construction industry in the Standard linetype library. Access these linetypes by selecting **Linetype** from the Format menu or by typing **LINETYPE** (ENTER) or **LT** (ENTER) at the Command prompt. Each method will produce the **Linetype Manager** dialog box shown in Figure 5.11. To display the available linetypes in the **Load or Reload Linetypes** dialog box, click the **Load** button. Figure 5.12 shows the list of linetypes.

LINETYPE CHARACTERISTICS

The **Linetype** command will allow you to vary the linetype pattern for each object you draw or for the entire layer. Most offices assign a specific linetype to a specific layer. As you start to assign linetypes to drawing objects, you need to remember the following two key elements regarding linetypes:

Effects

Objects that will be affected by a change of linetype include line, circle, arcs, polygons, donuts, ellipse, hatch, and 2D polylines. Any other drawing object will be displayed as a continuous line. That will be good news as you start to add dimensions or text to a drawing and you're working on a layer drawn with centerlines or another line pattern as the current linetype.

Figure 5.10 *Leader lines are thin lines used to relate a dimension or note to a specific portion of a drawing. A terminator should be used at the object end of the leader line. Leader lines should extend a minimum of 3/4" from the object being described.*

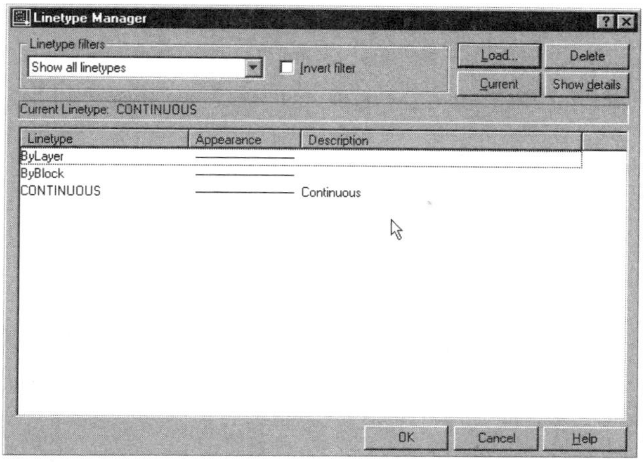

Figure 5.11 *The* **Linetype Manager** *dialog box can be used to change and load linetypes. Access it by selecting* **Other** *from the* **Linetype Control** *menu on the* **Object Properties** *toolbar, or by selecting* **Linetype** *from the* **Format** *menu.*

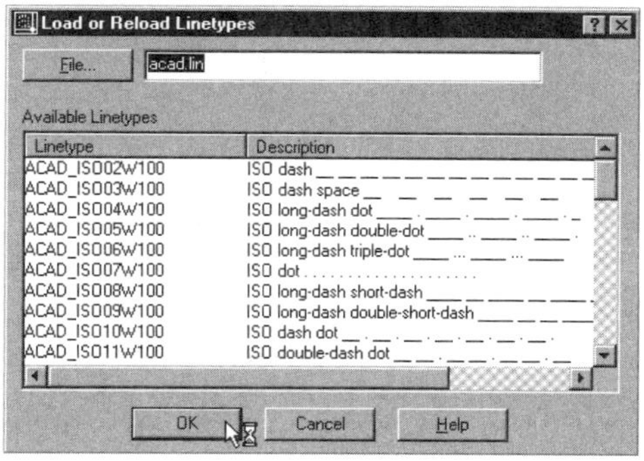

Figure 5.12a *Selecting the* **Load** *button will display the* **Load or Reload Linetypes** *dialog box with a list of AutoCAD linetypes. Use the scroll bar to see the balance of the list.*

Loading

A second concept to keep in mind as you work with linetypes is that you can't use a linetype unless it has been loaded into your drawing. The linetypes shown in Figure 5.12 exist in a library file stored in AutoCAD. They will be loaded automatically as a function of the **Linetype Manager** dialog box when the **Linetype** command is used.

LOADING LINETYPES INTO A DRAWING

As you display the **Linetype Manager** dialog box, you'll notice that three listings are displayed: ByLayer, ByBlock, and Continuous. Additional linetypes are loaded by using the **Load** button in the **Linetype Manager** dialog box. Use the following procedure to load the center, dashed, hidden, and phantom linetypes.

1. Display the **Linetype Manager** dialog box by selecting **Other** from the **Linetype Control** box on the **Object Properties** toolbar, by selecting **Linetype** from the **Format** menu, or by typing **LT** (ENTER) at the Command prompt.
2. Click the **Load** button. This will display the **Load or Reload Linetypes** dialog box shown in Figure 5.12.
3. Select the name, appearance, or description of the line to be loaded into the drawing to highlight the indicated linetype. To load more than one linetype, press and hold CTRL while selecting the names of the desired linetypes with the select button of the mouse.
4. When you're through selecting the linetypes to be loaded into a drawing, click the **OK** button.

The dialog box will be removed, and the selected linetypes will be displayed in the **Linetype** window of the **Linetype Manager** dialog box (see Figure 5.13). The lines are also shown in the **Linetype Control** box on the **Object Properties** toolbar when the display arrow is selected. Figure 5.14 shows the linetypes currently loaded into the drawing base.

MAKING A LINETYPE CURRENT

Remember that all you've done so far is to load linetypes into the drawing. To use the linetype, you must select one of the loaded linetypes as the current linetype. The following process will allow you to alter the lines that are used to create a drawing. Later in this chapter you'll be introduced to methods of assigning linetypes to a specific layer. Most professionals find linetypes easier to control when they are associated to a specific layer rather than scattered throughout one layer.

By default, you've been drawing with a continuous black line. Use the following procedure to alter the current linetype:

1. Select the linetype that will become the current linetype in the **Linetype Manager** dialog box. Select Center for now.
2. Click the **Current** button.
3. Click the **OK** button.

This process will make the selected linetype the current linetype to be used, close the dialog box, and return you to the drawing screen. Notice the name and pattern of the selected line will now be displayed in the **Linetype Control** box on the **Object Properties** toolbar. See Figure 5.15. The **Linetype Control** box can also be used to alter the linetype used to create objects. Click the arrow to display a list of current line-

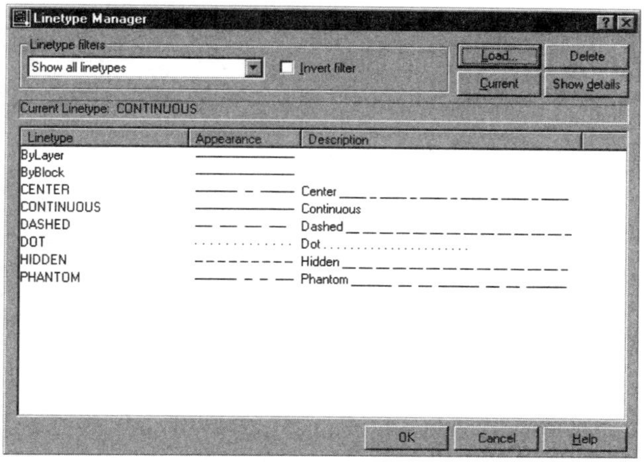

Figure 5.13 *Once a linetype has been selected to load, it will be displayed in the Linetype list.*

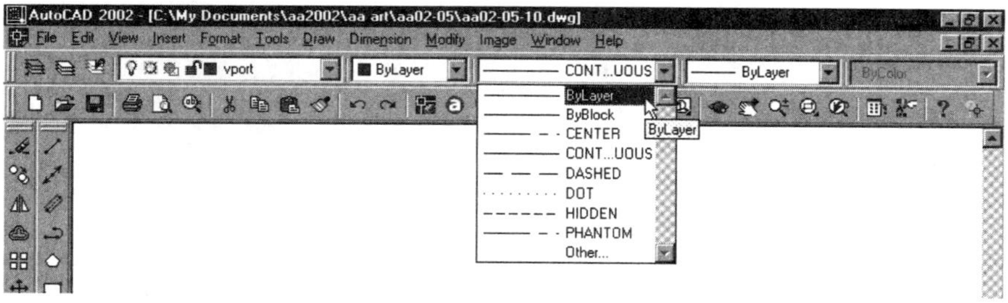

Figure 5.14 *The linetypes currently loaded into the drawing base can be found in the* **Linetype Control** *box on the* **Object Properties** *toolbar.*

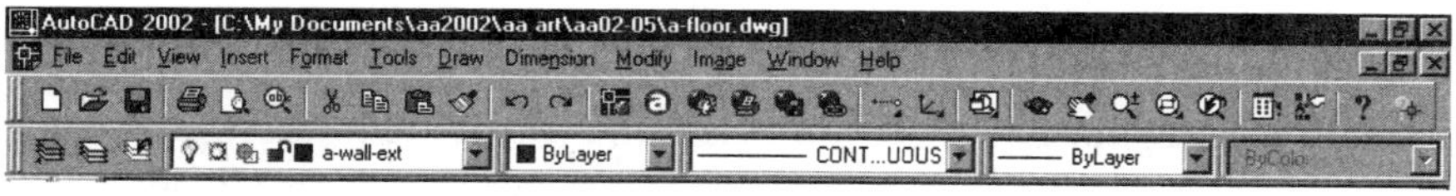

Figure 5.15 *The current linetype is displayed on the* **Object Properties** *toolbar. The* **Linetype Control** *box can be used to alter the linetypes that will be drawn.*

types. Select the name of the desired linetype to be used. This will close the list and make the selected linetype the current linetype.

The linetypes that are loaded into a drawing can also be displayed by selecting the **Properties** button on the **Standard** toolbar, by selecting **Properties** from the **Modify** menu or by typing **PROPERTIES** (ENTER) at the Command prompt. Each method

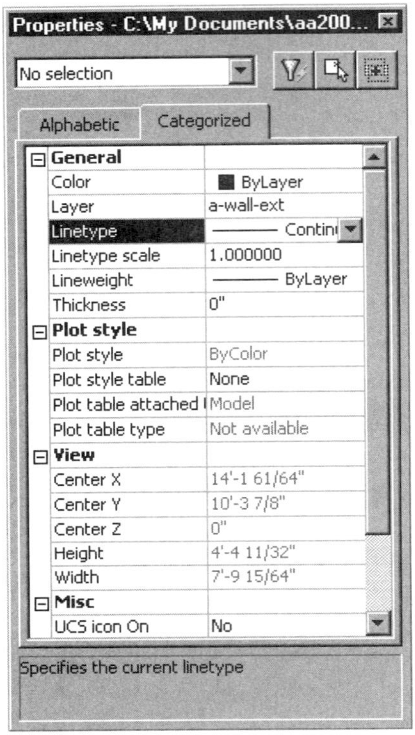

Figure 5.16a *The linetypes that are loaded into a drawing can be displayed by selecting the* **Properties** *button on the Standard toolbar or by selecting* **Properties** *from the* **Modify** *menu.*

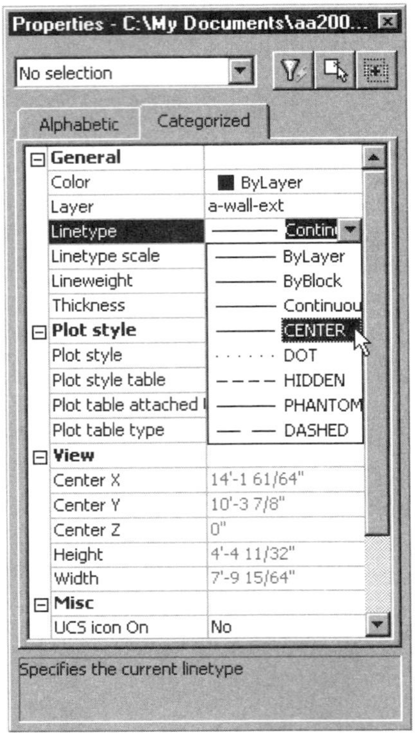

Figure 5.16b *Current linetypes can be displayed or altered by selecting the* **Linetype** *arrow.*

will produce the **Properties** dialog box. Selecting the current linetype will display a list arrow similar to Figure 5.16a. Selecting the arrow will display a listing of the current linetypes similar to Figure 5.16b.

Note: New users of AutoCAD often expect the linetype command to work miracles. Once a linetype has been made the current linetype, it will affect only drawing objects that are added after the linetype is changed. Objects drawn prior to the selection will not be affected. Objects drawn prior to the selection can be transformed to a different linetype using **Match Properties** or the **Change** command. Each will be discussed in Chapter 13.

ALTERING THE LINETYPE SCALE

Start the **Line** command and draw several lines. The new linetype should now be reflected on the drawing screen. If the line still appears as a continuous line, check the **Object Properties** toolbar to verify that you did select the current button. If the name and linetype show the desired linetype, most likely the linetype does not appear altered because you've altered the drawing limits from the original 12" x 9" screen. Use the following steps to alter the linetype scale.

1. Redisplay the **Linetype Manager** dialog box and click the **Show Details** button. This will produce a dialog box similar to Figure 5.17.
2. Alter the **Global scale factor**. If you're working on a template set for 1/4"=1'–0" plotting, change the **Global scale factor** to 48.
3. Click the **OK** button.

As you return to the drawing screen, lines should now reflect the selected pattern, as shown in Figure 5.18.

Note: Although it will be meaningless now, open your drawing template and type **PSLTSCALE** (ENTER) at the Command prompt. Change the default value from 1 to 0. This will ensure proper display of linetypes as drawings are plotted in the future. This process will need to be repeated for each layout that is created with a drawing.

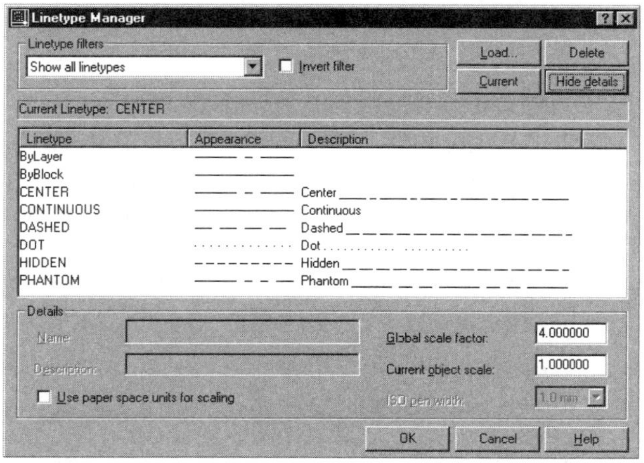

Figure 5.17 *Selecting the* **Show Details** *button in the* **Linetype Manager** *dialog box will display controls for linetype scaling. If you've altered the limits of the drawing screen from the 12" x 9", a line drawn with a centerline may appear to be drawn as continuous.*

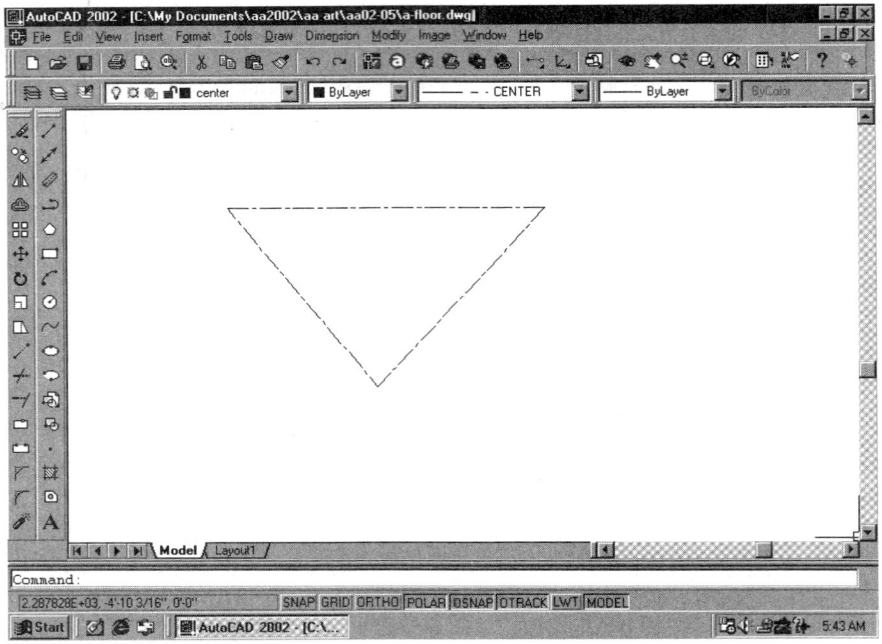

Figure 5.18 *The result of setting the* **Global scale factor**. *If you're working on a drawing that will be plotted at a scale of 1/4" = 1'–0", the scale factor is 48. Once the linetype scale is altered, the object will now be displayed using centerlines.*

AN INTRODUCTION TO LINEWEIGHT

You've been introduced to common linetypes and line widths earlier in this chapter. Varied line widths will help you and the print reader keep track of various components of a complex drawing. As you assign lineweight to drawings, remember that you're using thick and thin lines to add contrast, not represent thickness of an object. Chapter 11 will introduce an alternative method of using a thickness to accurately represent objects.

EXPLORING LINEWEIGHTS

Lineweights can be assigned to a layer by selecting **Lineweights** from the **Format** menu, or by typing **LW** (ENTER) or **LWEIGHT** (ENTER) at the Command prompt. Each method will produce the **Lineweight Settings** dialog box. Once you've changed some basic settings, lineweights can be altered using the **Lineweight Control** box on the **Object Properties** toolbar. For now, display the **Lineweight Settings** dialog box by selecting **Lineweights** from the **Format** menu. This will produce a display similar to Figure 5.19. Common lineweights and their width in millimeters and inches, point size, and pen size can be seen in Figure 5.20.

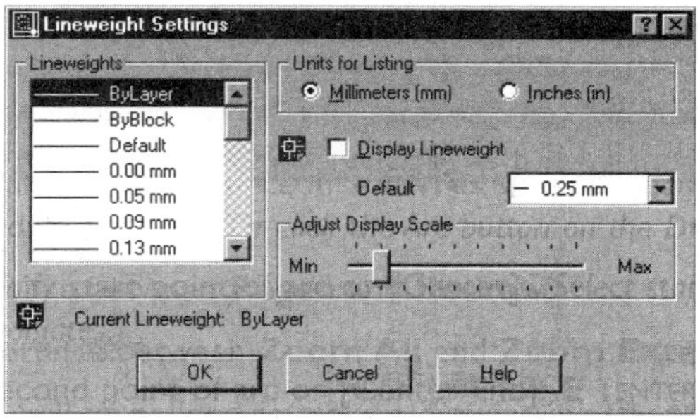

Figure 5.19 *The* **Lineweight Settings** *dialog box can be used to control drawing lineweights. Access the dialog box by selecting* **Lineweights** *from the* **Format** *menu or by typing* **LWEIGHT** *(ENTER) at the Command prompt.*

LINEWEIGHT	MILLIMETERS	INCHES	POINTS	PEN SIZE
	.25	.010	3/4 pt.	000
	.30	.012		00
	.35	.014	1 pt.	0
	.50	.020		1
	.53	.021	1-1/2pt.	
	.60	.024		2
	.80	.031		3
	1.00	.039		3-1/2
	1.20	.047		4

Figure 5.20 *Common lineweights and their width in metric and inches, point size and pen size.*

ASSIGNING LINEWEIGHTS

The **Lineweight Settings** dialog box can be used to set the units for listing line width, the default lineweight, the display scale, and as a display of the current settings. As you display the dialog box, notice that the **Display Lineweight** check box is not selected. In its current state, you can draw using varied lineweights, but the width will be reflected only on a print or plot and not in the drawing display. For most users, this should be the first adjustment you make. Select the **Display Lineweight** check box to make it active, and varied lineweights will be displayed on the screen. Most users are also more comfortable working in inches rather than metric units. Make this change by choosing the **Inches** radio button. With this change, you'll notice that the lineweights are now displayed in inches. To assign a lineweight, scroll down the list of lineweights, find the desired lineweight, and click the **OK** button. As you use the **Line** command, any line added to the drawing will now reflect the current linetype and lineweight.

Once the unit and display settings have been set, you can alter lineweights by selecting the **Lineweight Control** arrow on the **Object Properties** toolbar (see Figure 5.21). Lineweights can also be toggled ON/OFF by clicking LWT, the lineweight display control button on the status bar. A third method to control lineweight display, already discussed, is to activate the **Display Lineweight** check box in the **Lineweight Settings** dialog box. This will be helpful in the future, as you enlarge areas of the drawing using the **Zoom** command.

DISPLAYING LINEWEIGHTS

In addition to activating **Display Lineweight** in the **Lineweight Settings** dialog box, you also must consider the difference in lineweight display between model and paper space. In model space, lineweights are displayed in relation to pixels. Lineweights in paper space are displayed in the exact plotting width. For work in model space, lineweights should be selected that are proportional to their pixel value and not their real world width. Chapter 13 will explore the display scale for lineweights. For now, use the following guide for determining lineweight. Examples are all related to a floor plan.

Figure 5.21 *Once the unit and display settings have been set, lineweights can also be altered by selecting* **Lineweight Control** *on the* **Object Properties** *toolbar.*

mm	In.	Description	Use
0.25	0.010	very, very, thin	Very detailed areas
0.30	.0.012	very thin	Window symbols and door swings
0.35	.0.014	thin	Cabinets and plumbing fixtures
0.50	.0.020	moderate	Doors
0.60	.0.024	thick	Outline of major components, such as walls
0.80	.0.031	very thick	Underlining room titles
1.20	0.047	very, very thick	Borders

AN INTRODUCTION TO COLOR

Up to this point, all of your drawings have been done with black lines on a white background, or white lines on a black background. The choice of screen background can be altered by the following procedure:

1. Select Options from the **Tools** menu. This will display the **Options** dialog box.
2. Select the **Display** tab in the **Options** dialog box.
3. Click the **Colors** button from the **Window Elements** portion of the **Display** tab. This will produce the **Color Options** dialog box shown in Figure 5.22.
4. The box allows the color of each of the drawing window displays to be altered. To change the color of the drawing area, select the desired color tile.
5. Click the **Apply & Close** button to close the **Color Options** dialog box.
6. Click the **OK** button in the **Options** dialog box. The dialog box will close and the screen will display the selected background color.

Figure 5.22 *The* **Colors** *button on the* **Display** *tab will produce the* **Color Options** *dialog box.*

AutoCAD allows the color of each 2D drawing object to be altered using one of 255 colors. A name or an AutoCAD Color Index (ACI) number ranging from 1 through 255 can identify the color for objects. Rendered 3D drawings use the same 255 colors but can display millions of colors by altering the tint. The actual number of colors that are displayed will depend on third-party software and how the computer is configured.

COLOR NUMBERS AND NAMES

Each drawing object or layer can be assigned a color. The seven basic colors used by AutoCAD are listed in Figure 5.23. AutoCAD identifies colors using numbers from 1 through 255, with the selection based on the color displayed in the color palette. Multiple objects or layers can have the same color number. Objects of the same color can be assigned to a specific pen during the plotting process. For instance, all red objects can be assigned to a thick pen, and all blue objects can be assigned to a thin pen to produce good line contrast on the plotted drawing. Plot styles can be used to control plotting and colors, and they will be explored later.

SETTING OBJECT COLOR

The **Color** command can be used to assign colors to an entire layer or to individual objects on a layer. Setting colors for layers is the default method and will be explained as you proceed through layer options later in this chapter. The easiest method for assigning color is to select the **Color** button in the middle of the **Object Properties** toolbar. You can select the color tile, the color name, or the down arrow. Each method will display the list shown in Figure 5.24. Selecting one of the listed colors will close the list and return the drawing display. Now, if a line is drawn, it will be drawn using the selected color. Using the current settings, the line would be a red line drawn with a center linetype. Select the **Color** button again to redisplay the color list. Notice the last option is **Other.** Selecting this option will display a dialog box simi-

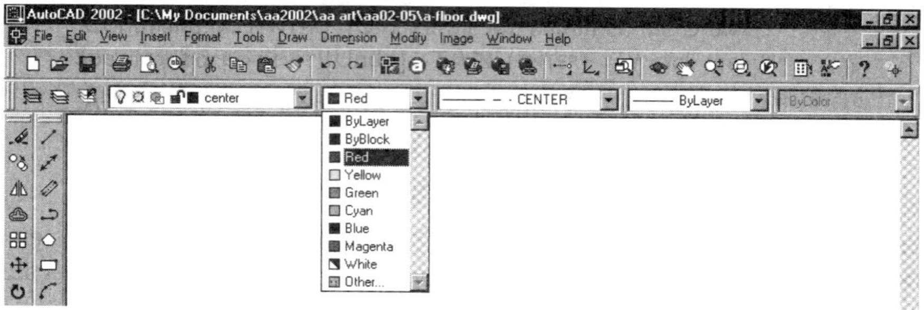

Figure 5.23 *The eight major colors of AutoCAD. The default color is 7. Depending on the background color, 7 will produce either a black or white line.*

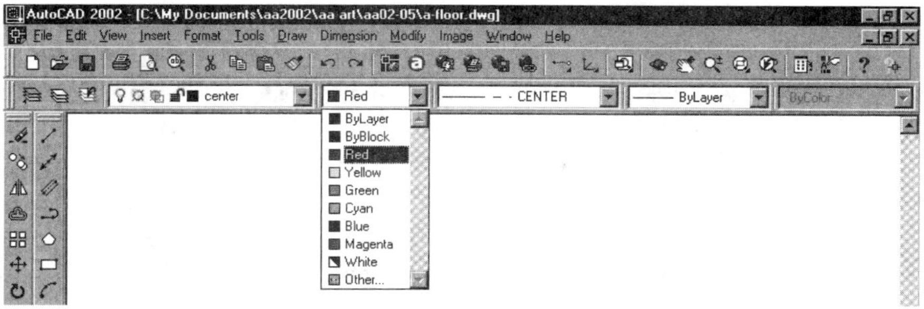

Figure 5.24 *Selecting the* **Color Control** *arrow on the* **Object Properties** *toolbar will display the color list and allow the color of the current layer to be changed.*

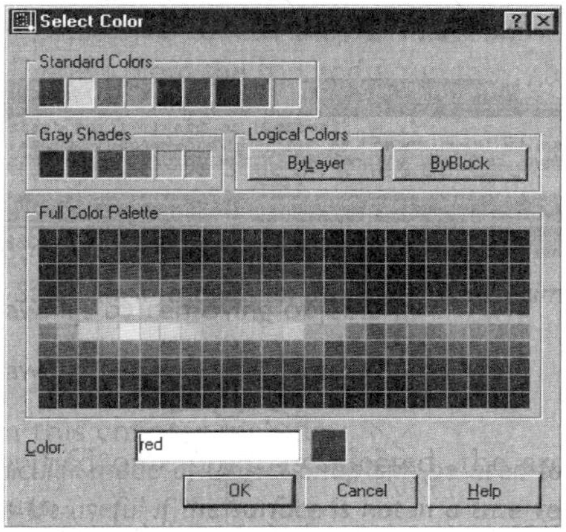

Figure 5.25 *Selecting* **Other** *from the* **Color** *list displays the color palette and allows a wide variety of color tints to be selected for drawing objects.*

lar to Figure 5.25. The **Select Color** dialog box can also be displayed by selecting **Color** from the **Format** menu or by typing **COLOR** (ENTER) at the Command prompt. The **Select Color** dialog box allows shades of the standard colors to be selected. Notice the current color is listed as red. Selecting any of the color tiles will alter the current color display tile and list the color by number. Clicking the **OK** button will close the display and make that the current color. Using the **Line** command now will create lines using the selected color.

Notice that the **Select Color** dialog box contains buttons for **ByLayer** and **ByBlock**. The **ByBlock** option will be discussed in Chapter 14. If **ByLayer** is accepted as the default, the color will be assigned once each layer has been defined. Most offices set colors using the **ByLayer** option rather than assigning colors to individual objects.

AN INTRODUCTION TO LAYERS

The use of layers in a CAD drawing is similar to overlay drafting procedures of manual design. Overlay drawing will have a base drawing, such as the valve assembly seen in Figure 5.26, on one sheet of Mylar®, and have overlay sheets of notes and dimensions on separate sheets of Mylar, as seen in Figure 5.27. Depending on who needed what, different sheets could be attached to the base and various prints made. Layers allow all the flexibility with none of the alignment problems of overlay drafting procedures. Until now you have been drawing on the 0 layer. AutoCAD will allow you to create different layers, assign each layer a different name, linetype, lineweight, and color, and allow some layers to be made invisible.

THE LAYER COMMAND

Start the **Layer** command by selecting the **Layers** button on the **Object Properties** toolbar, by selecting **Layer** from the **Format** menu, or by typing **LA** (ENTER) at the Command prompt. Each will display the **Layer Properties Manager** dialog box shown in Figure 5.28.

CREATING A NEW LAYER

Once the **Layer Properties** Manager dialog box is displayed, press (ENTER) or click the **New** button. Each method will add the highlighted name of Layer 1 in the **Name** edit box (see Figure 5.29a). With Layer 1 highlighted, a new layer name can be created by keyboard. Click the select button to accept the new layer name. If you would like to add multiple layers, press (ENTER), which adds the previously changed name to the drawings and adds a new layer titled Layer1. In Figure 5.29b, Layer1 is assigned the name of COOLSTUFF. (Follow the guidelines from Chapter 4 for naming files when assigning layer names.) Additional guidelines for naming layers will be

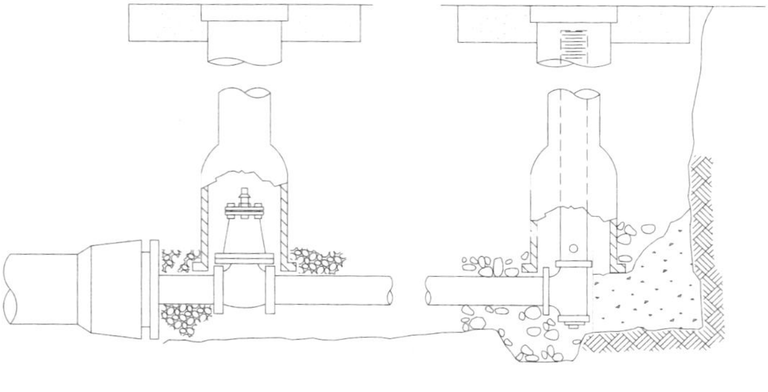

Figure 5.26 *The use of layers is similar to placing information on several sheets of paper in overlay drafting. This drawing shows the basic drawing with no supplemental information provided. (Courtesy Department of Environmental Services, City of Gresham, Oregon.)*

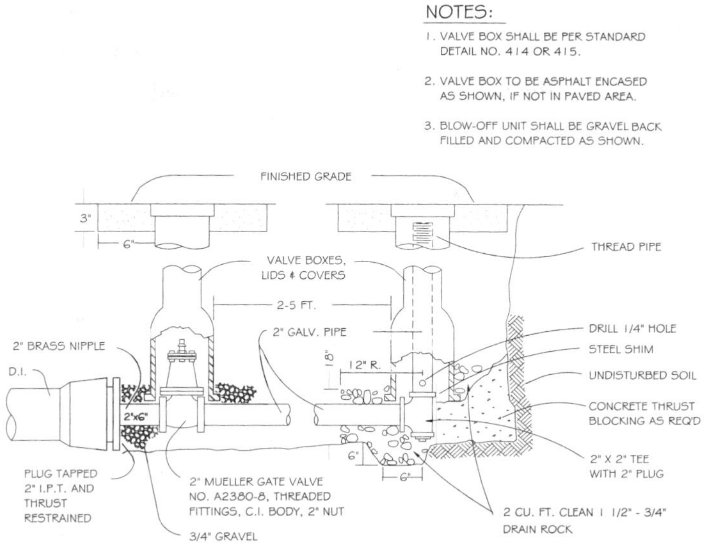

Figure 5.27 *Supplemental information added to the base drawing. (Courtesy Department of Environmental Services, City of Gresham, Oregon.)*

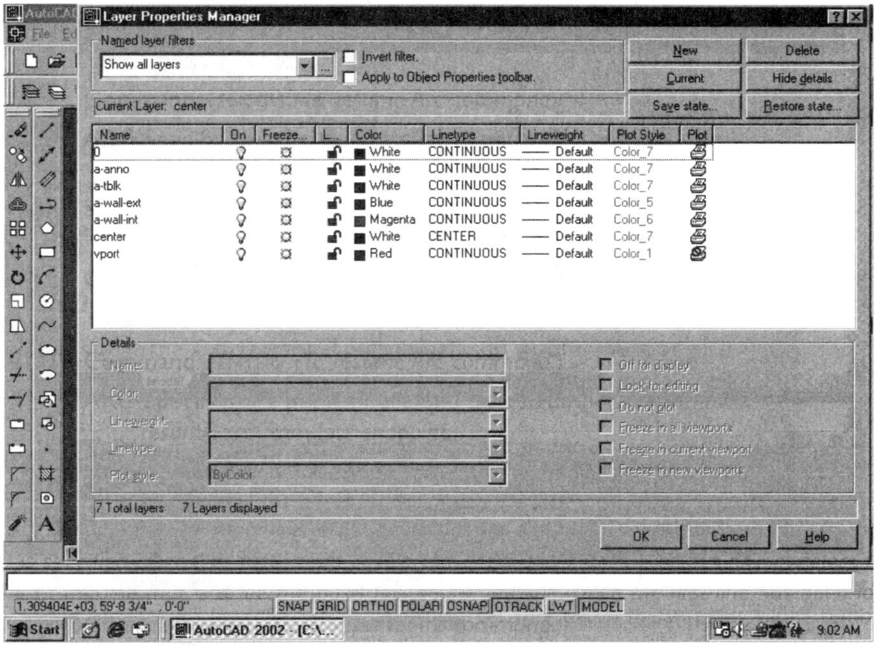

Figure 5.28 *Clicking the* **Layer** *button on the* **Object Properties** *toolbar or typing* **LA** *(ENTER) will display the* **Layer Properties Manager** *dialog box. The box can be used to create and control the properties of each drawing layer.*

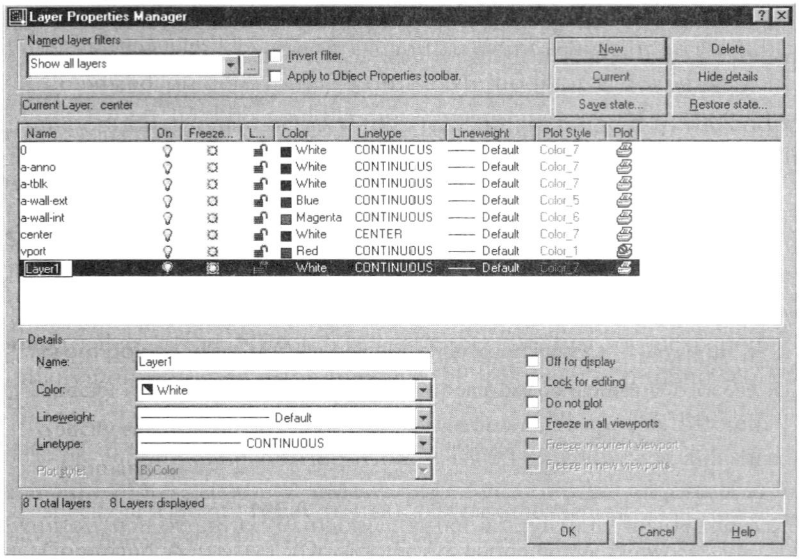

Figure 5.29a *Pressing* ENTER *or clicking the* **New** *button will display a new layer titled Layer1.*

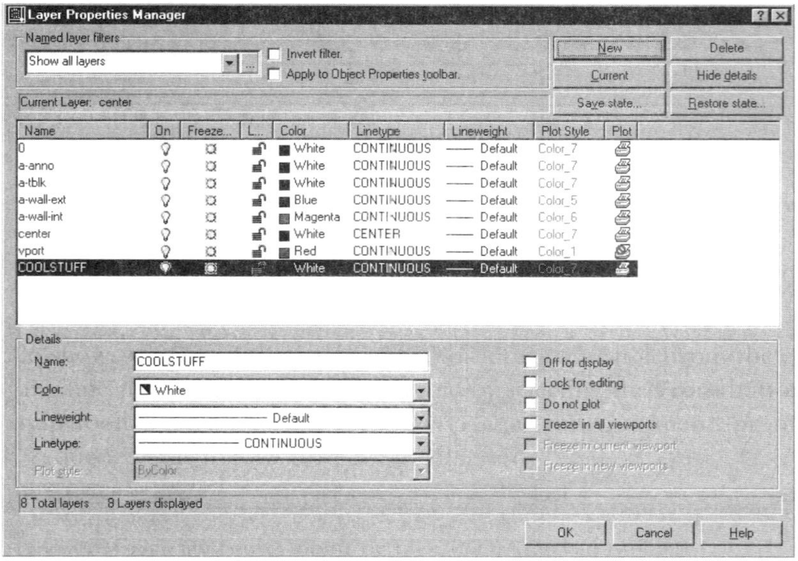

Figure 5.29b *The selected layer name in Figure 6.29a can be altered to reflect the desired layer name. Guidelines for naming layers are the same as those used to name a drawing file.*

discussed in Chapter 13. For now, keep the names short, simple, and descriptive, using names that will identify the contents. Once the name has been supplied, click the **OK** button to save the layer and return to the drawing area. Just as with **Linetype**, all you've done is to create a new layer. The current layer is still the default layer of 0.

MAKING A LAYER CURRENT

Close the **Layer Properties Manager** dialog box and reenter the drawing area and add a few lines. The lines will be the same linetype and color that was drawn before you created the COOLSTUFF layer. Redisplay the **Layer Properties Manager** dialog box by selecting the **Layers** button on the **Object Properties** toolbar. The dialog box displays 0 and COOLSTUFF as the only layers in the drawing. The box surrounds the 0 layer, indicating that it is the current layer. Use the following steps to make COOLSTUFF the current layer:

1. Select COOLSTUFF.
2. Click the **Current** button.
3. Click **OK**.

Clicking the **OK** button will make the selected layer the current layer, close the **Layer Properties Manager** dialog box, and return you to the drawing area.

Note: You can also make a layer current by displaying the **Layer Properties Manager** dialog box and double-clicking the name of the layer to be made current.

Notice that just to the right of the **Layers** button on the **Object Properties** toolbar is an edit bar that describes the current layer status. (See Figure 5.30a.) You'll see icons that describe the current status of the layers, with COOLSTUFF listed as the current layer. Clicking the down arrow will create a display of all the layers contained in the drawing, similar to Figure 5.30b. In this drawing, the only other option is the 0 layer, but this is a very convenient method for altering the current layer. To alter the current layer, select the down arrow, and then select the name of the layer you want as the current layer.

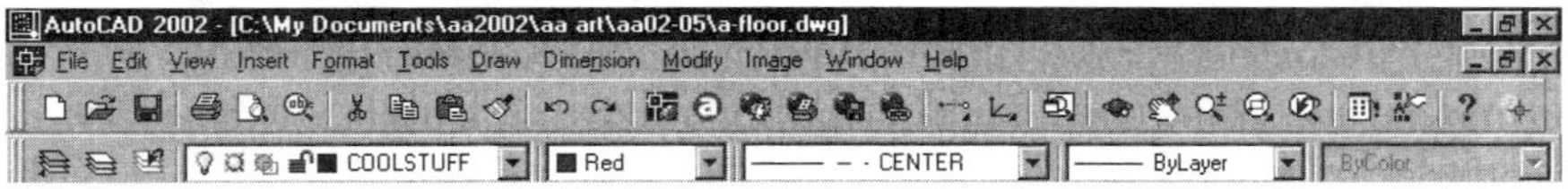

Figure 5.30a *The* **Layer** *box on the* **Object Properties** *toolbar describes the properties of the current layer.*

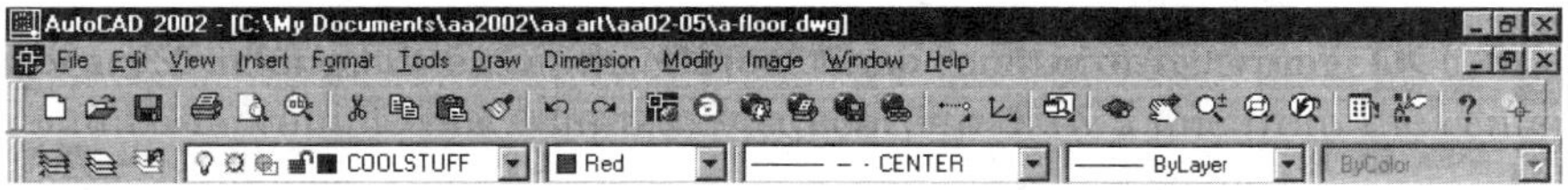

Figure 5.30b *Selecting the arrow displays the status of each drawing layer.*

ALTERING LAYER QUALITIES

Redisplay the **Layer Properties Manager** dialog box, select the COOLSTUFF layer and then click the **Current** button. COOLSTUFF has the same color, linetype, and lineweight qualities as the 0 layer. Any object created on the COOLSTUFF layer will appear the same as it would if it were drawn on the 0 layer. Layers become a powerful drawing tool when varied linetypes, lineweights, and colors are assigned to specific layers. Creating a layer with thin, blue, hidden lines for appliances can be a great way of easily separating them from walls that are drawn with thick, continuous red lines.

Linetype qualities can be assigned to all objects created, when this layer is the current layer, by selecting the current Linetype display. Use the following steps to alter the linetype of the COOLSTUFF layer:

1. Display the **Layer Properties Manager** dialog box.
2. Highlight the layer that you wish to alter.
3. With the layer highlighted, select the current linetype for that layer. This will produce the **Select Linetype** dialog box shown in Figure 5.31a. Only linetypes that have been loaded into the drawing will be displayed.
4. Select the desired linetype to be used.
5. Click the **OK** button to close the **Select Linetype** dialog box.

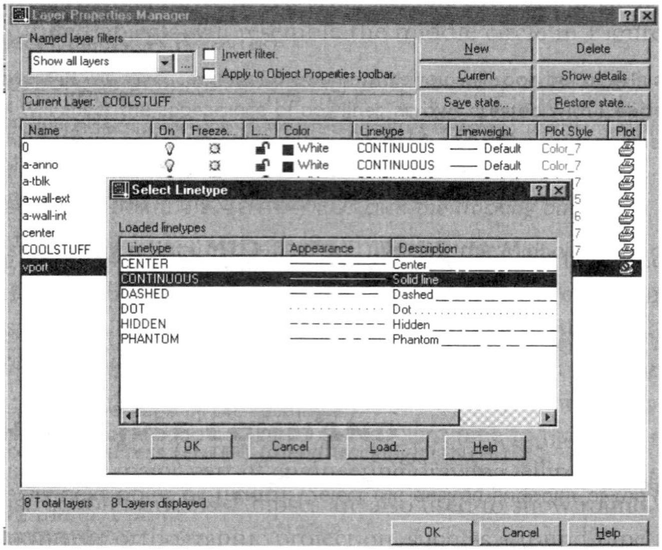

Figure 5.31a *Select the current linetype for the layer will produce the* **Select Linetype** *dialog box. Only linetypes that have been loaded into the drawing will be displayed. You can load additional linetypes into the drawing by selecting the* **Load** *button and following the guidelines presented early in this chapter.*

6. Click **OK** to close the **Layer Properties Manager** dialog box and return to the drawing area.

Selecting one of the linetypes will use that linetype for all drawing objects created using that layer. The lineweight and color can be altered using the same procedure.

Selecting the color display for COOLSTUFF will produce the **Select Color** dialog box shown in Figure 5.31b. Selecting the lineweight display for COOLSTUFF will display the **Lineweight** dialog box similar to Figure 5.31c. Once the desired qualities of a layer have been assigned, click the **Current** button if you want to use that layer as you return to the drawing editor. Clicking the **OK** button will close the **Layer Properties Manager** dialog box and return you to the drawing screen.

Note: Assigning colors, linetypes, and lineweights will affect all objects drawn on the layer that is altered, even objects drawn prior to making the changes.

CONTROLLING THE VISIBILITY OF LAYERS

Nine icons can be displayed by each layer name. These buttons represent various button controls for the use of each layer. The number of buttons displayed will vary depending on how much space is assigned for each button. The lightbulb button is a toggle for the ON/OFF setting of a layer. For the drawing shown in Figure 5.26, the Text layer was toggled OFF, and for Figure 5.27, the Text layer was toggled ON. Clicking the lightbulb will change the bulb from yellow to gray. The yellow bulb represents ON, the gray OFF.

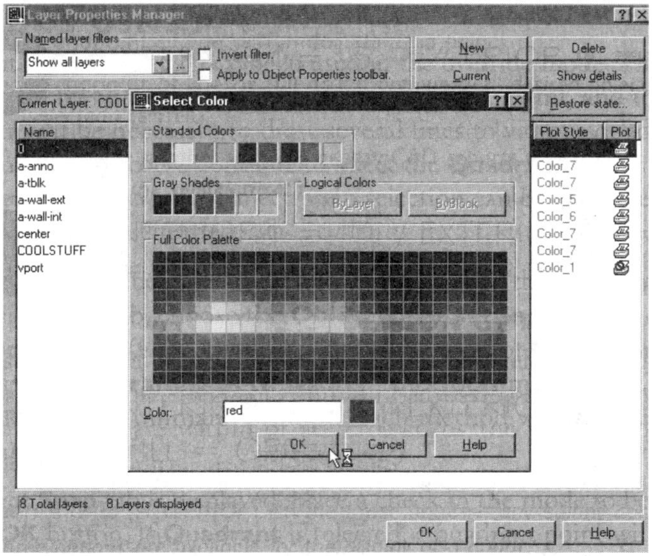

Figure 5.31b *The color of a layer can be adjusted by selecting the current color display for that line. The* **Select Color** *dialog box allows the color to be altered.*

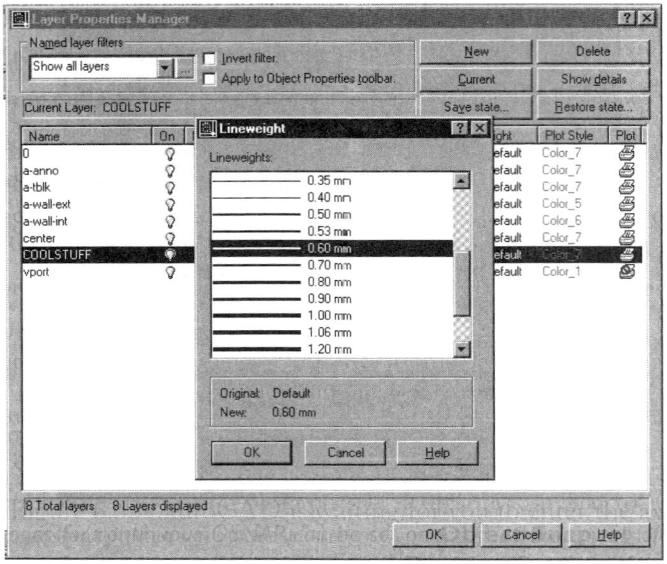

Figure 5.31c *The lineweight of a layer can be adjusted by selecting the current lineweight display for that line. The* **Lineweight** *dialog box allows the line width to be altered.*

The second button, a sun, can also be used to control the visibility of a layer—it is a toggle between Thaw/Freeze. When the sun is displayed, the contents of that layer will be displayed. Selecting the sun button will change it to a snowflake, which represents a frozen layer. The contents of a frozen layer will not be displayed. The benefits of using the ON/OFF and Thaw/Freeze options, as well as each of the other layer controls, will be discussed in Chapter 13. To remove the contents of a layer from the screen display, select the desired layer name and then select the lightbulb or the sun button and then select the **OK** button.

The visibility of layers can also be controlled from the **Layer Control** box on the **Object Properties** toolbar. Select the down arrow to display a list of current layers and select the lightbulb or sun button of the layer to be altered.

AN INTRODUCTION TO PRINTING

Once you've created a drawing, you'll need some method of passing the contents of the file on to those who will put it to use. This might include sending the drawing electronically over the Internet to another firm or sending the drawing by diskettes. For many, it also means making a hard copy such as a paper print or plot. No matter how technology advances, for most AutoCAD operators, a paper copy is still a common method for checking drawings. You've already been introduced to the layout templates as you explored drawing setup. Layouts provide a preview of how the drawing will appear prior to plotting. Each layout displays a border and the drawing as it will appear at the current lineweight and scale factor. Although it will be a bit overwhelming now, AutoCAD will allow you to establish multiple plotting layouts of the same drawing project. The same drawing can be plotted at different scales or configured for different plotting devices by using layouts. The settings to repeat the plotting process can be saved with the drawing, providing a valuable saving of time. The purpose of this chapter is not to explore all of the plotting functions, but to help you produce a quick check print of your drawing. This chapter will introduce the process of making a paper print using the system printer configured using your Windows platform. Later chapters will provide additional information on choosing additional printers, configuring a plotter, and controlling the many variables associated with plotting.

EXPLORING THE PLOTTING PROCESS

Start the print process by selecting the **Print** button on the **Object Properties** toolbar, by selecting **Print** from the **File** menu, or by typing **PLOT** (ENTER) at the Command prompt. Each selection method will produce the **Plot** dialog box shown in Figure 5.32. The dialog box contains the **Plot Device** and **Plot Settings** tabs. The **Plot Device** tab will not be considered in this chapter. See Chapter 25 for further discussion of altering the plotter.

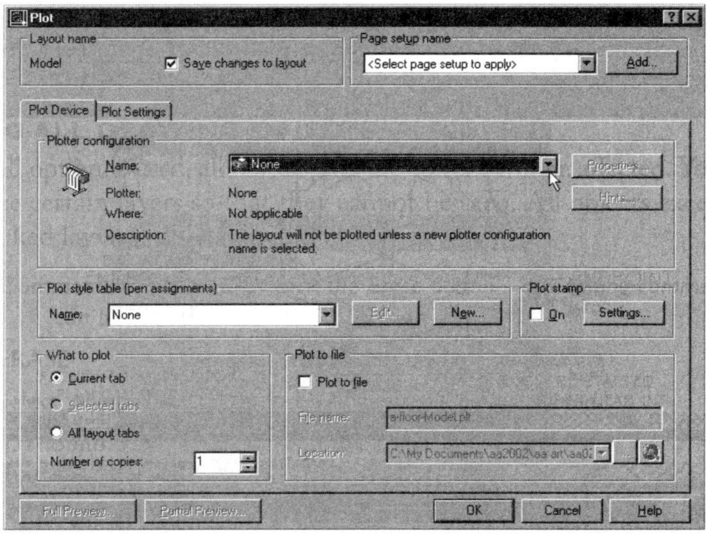

Figure 5.32 *Selecting the* **Plot** *button on the* **Standard** *toolbar, selecting* **Plot** *from the* **File** *menu, or typing* **PLOT** *(ENTER) at the Command prompt will display the* **Plot** *dialog box. The dialog box can be used to control each aspect of creating a plot.*

LAYOUT AND PAGE SETUP NAMES

The **Plot Settings** tab contains eight major areas that must be considered to produce a print. The **Layout name** and **Page setup name** portions of the **Plot** dialog box are used to specify layout and plotting device settings. The name of the current layout (Model, Layout1, or Layout2) is displayed on the left. Options for listing and choosing page setups for plotting are listed on the right. Each option will be explored as plotting is explored in later chapters. Chapter 24 will introduce methods of creating new page setups.

PAPER SIZE AND PAPER UNITS

The **Paper size and paper units** portion of the **Plot Settings** tab displays the current printing device, the paper size the plot will be made on, and the size of the printable work area. Selecting **Options** from the **Tools** menu and selecting the **Plotting** tab in the **Options** dialog box can also control printer and plotter selections. For the current discussion, the default printer will be used. The box also displays the plotting units in inches millimeters. Most construction projects will use the default setting of Inches for the plotting standard.

PLOT AREA

The **Plot** area provides options for describing what part of the current drawing will be plotted. Because often the entire drawing does not need to be plotted, five options are given to determine what portion of the drawing will be plotted.

Limits

Using this option will plot all of the drawing objects that lie within the current drawing limits, defined during the drawing setup.

Extents

Choosing this option will plot the current drawing with the extent of the drawing objects as the maximum limits that will be displayed in the plot.

Display

The default setting is to plot only the material that is currently displayed on the screen. If you have zoomed into a drawing, only that portion of the drawing will be plotted.

View

This option is currently not active. Methods of creating views will be introduced in Chapter 8 and the use of this option will be introduced in Chapter 24.

Window

This option will allow a specific area of a drawing to be defined for plotting. Initially the **Window** radio box is inactive. When the **Window<** button is clicked, the **Plot** dialog box will be removed from the screen, allowing you to select what is to be plotted. The command line will show the prompt:

> Command:_plot:
>
> *(Select the Window< button)*
>
> Specify window for printing
>
> Specify first corner:

AutoCAD is waiting for you to define a window or box that will surround what you would like to plot. To create the plotting window, move the cursor and select a point. Once the first corner of the window is selected, AutoCAD will provide a prompt to specify the second corner. Move the cursor to the opposite corner. As the cursor is moved, a box (the window) will be formed. Figure 5.33 shows the process for defining the window. Anything that lies within the window will be plotted. Once the window is defined, the **Plot** dialog box will be redisplayed, allowing you to continue defining the plot to be created. The **Window** radio button will be activated, allowing other options to be set. Each option will be discussed in later chapters.

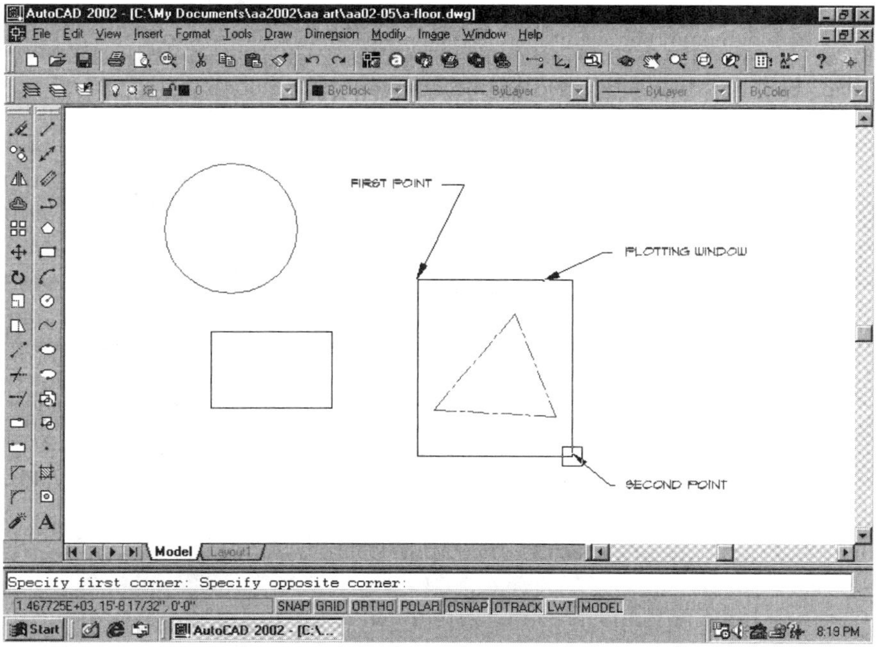

Figure 5.33 *When objects are selected to be plotted using the* **Window** *option, the size of the window can be set by selecting the window corners with the cursor. Selecting the Window< button will remove the dialog box and return the drawing to the screen. Once the corners of the select window are located, the* **Plot** *dialog box will be displayed again so that other print options can be set.*

DRAWING ORIENTATION

The **Drawing orientation** portion allows the drawing orientation to be specified. The current orientation of the box in Figure 5.34 shows that the drawing will be printed in the normal orientation for construction drawing, with the long paper edge horizontal. The drawing will be read looking from the top to bottom, and the title block will be along the right side of the print. This setting is referred to as Landscape. When the print is rotated so that the long edge of the paper is vertical, the print is referred to as a Portrait print. Adjust the orientation of the drawing to the paper by selecting the appropriate radio button. The **Plot upside-down** option plots the drawing in reverse of the selected position from the 0,0 origin. If **Landscape** is the selected position, and **Plot upside-down** is active, the top right corner of the drawing will be placed in the 0,0 position rather than the lower left corner of the drawing.

PLOT SCALE

The **Plot scale** portion controls settings for the scale of the plot. Its options include **Scale** and **Custom**.

Selecting a Scale

With the default setting, **Scaled to Fit**, in the **Scale** edit box, when a plot is made, it will be reduced in size to fit on the current printer or plotter paper. Unless you'll be plotting a small detail, most architectural drawings can't be printed at full size on a

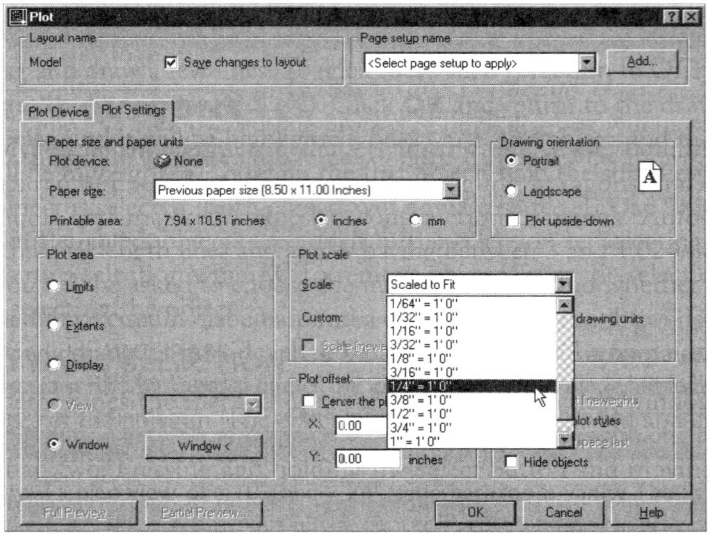

Figure 5.34 *To plot a drawing to a scale, select the arrow beside the edit box to reveal the scale list. Select the desired scale. Once the scale is selected, the list will close to allow you to continue setting the plot parameters.*

printer. Rarely can you get the floor plan for a structure to fit on A-size material. You can, however, print the entire floor plan on A-size material if the scale is reduced. The **Scaled to Fit** option reduces the selected material to be printed to an appropriate size to fill the current paper. You can print portions of a drawing to scale by selecting the desired scale and using the window selection method. To print a drawing to a scale, select the arrow beside the edit box to reveal the scale menu. (See Figure 5.34.) Select the desired scale, and the list will close to allow you to continue setting the print parameters.

Custom Scale

You've just altered the scale from **Scaled to Fit** to 1/4"=1'–0". The **Custom** window now has been altered to reveal 1 in the inches box, and 48 in the drawing units box. These units reflect the plot scale factor of 1/48 for a drawing to be scaled to 1/4"=1'–0" when printed. The **Custom** edit box allows you to print a drawing to a scale that you create. Figures 5.35a and 5.35b compare two different scaling options.

Note: Keep in mind that the goal of this chapter is to help you make a quick, non-scaled, check print so that you can evaluate your work. Stick with **Scaled to Fit** or one of the common architectural scales until plotting is discussed in detail.

Plot Offset

The **Plot offset** portion of the **Plot** dialog box can be used to specify an offset of the plotting area from the lower left corner of the paper. The default setting for plotting origin is X=0, Y=0. Printers typically specify the origin from the upper right corner. Specifying various locations for the origin can be a useful technique for altering the drawing location on the paper. With the **X** or **Y** origin box selected, new values can be entered. Entering an X value of 12 and a Y value as 10 would place the new origin 12 units to the right and 10 units above the original plotting origin.

Plot Options

The **Plot options** portion of the **Plot** dialog box can be used to specify options for lineweights, plot styles, and the current plot style table. By default, the **Plot with lineweights** and the **Plot with plot styles** options are active. Each plot option will be discussed in later chapters.

PREVIEWING

Selecting a preview allows the outcome of the current plot settings to be viewed prior to making the print. Two options are available for previewing the print: include **Partial Preview** and **Full Preview**.

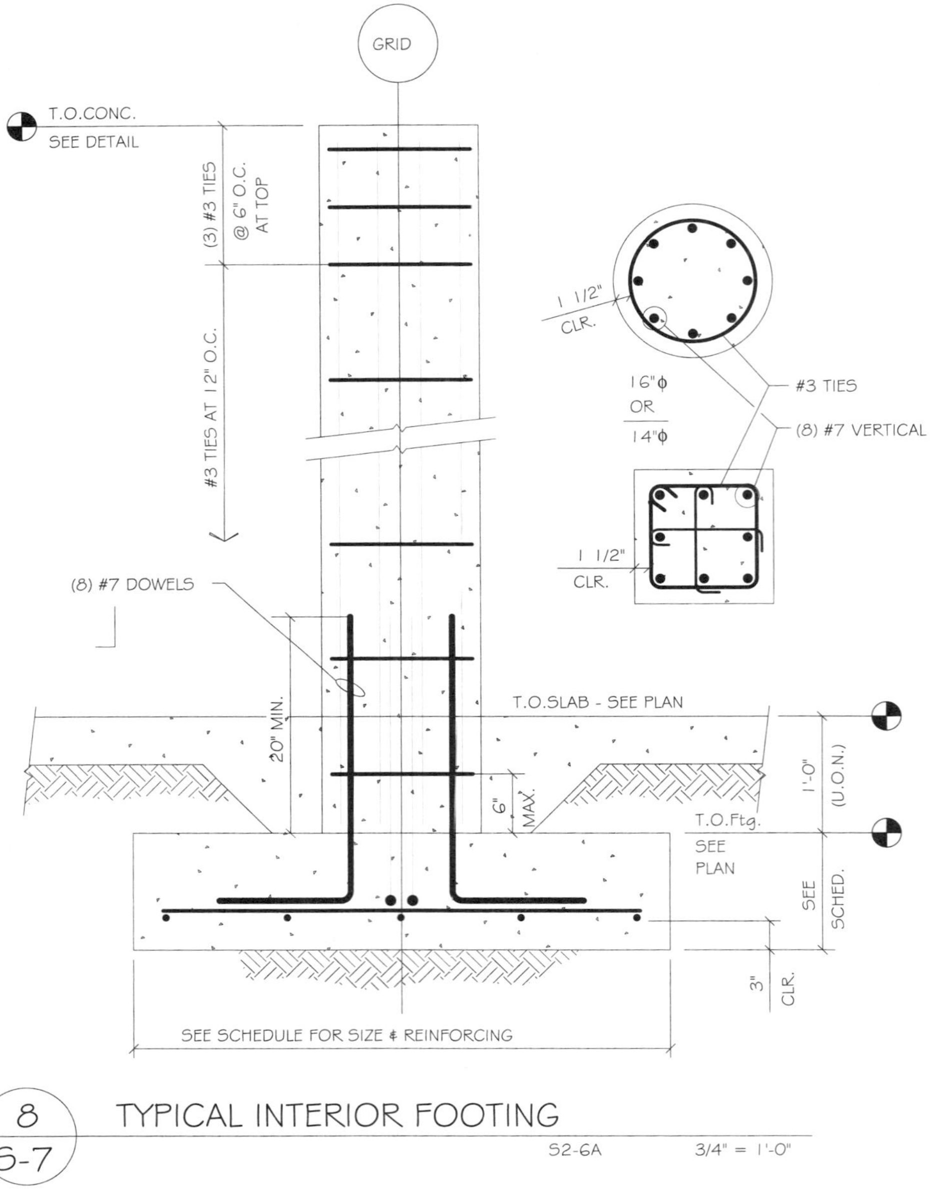

Figure 5.35a *A detail printed using the* **Scaled to Fit** *option. Most professionals use drawings printed with the* **Scaled to Fit** *option for check prints prior to plotting the drawing to scale.*

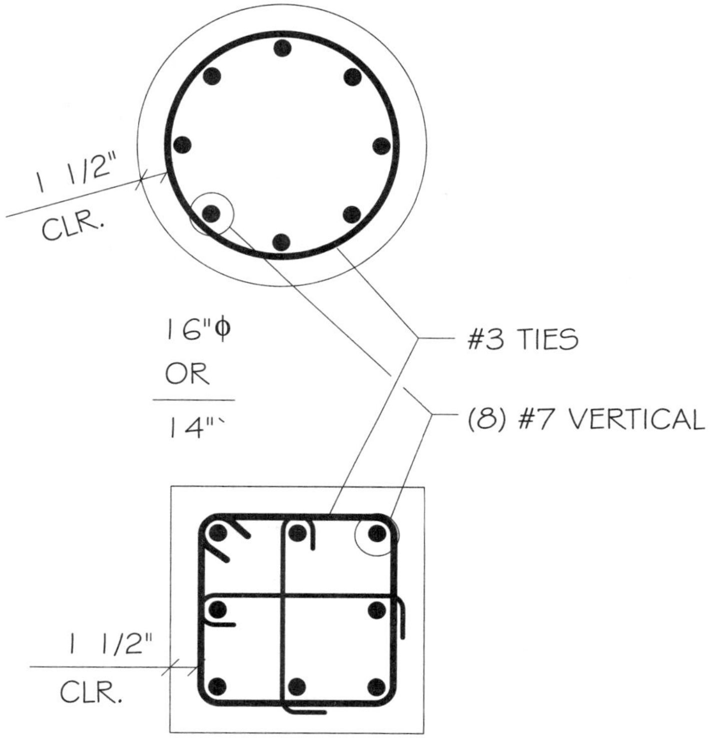

Figure 5.35b *A portion of the detail from Figure 6.35a printed using the scale of 3/4"=1'–0".*

Partial Preview

Clicking the **Partial Preview** button will produce a display similar to the one seen in Figure 5.36. A white box represents the paper size and the limits of the effective printable area are specified on the screen by a blue box. A warning also will be given of any problems that might be encountered during the plot. Common warnings displayed in the warning box include the following:

Effective area too small to display

Origin forced effective area off display

Plotting area exceeds plotter maximum

A small triangular Rotation Icon is displayed in the upper left corner of the paper image to represent a landscape rotation. As the rotation is set to **Portrait**, the icon is moved to the lower left corner of the paper image.

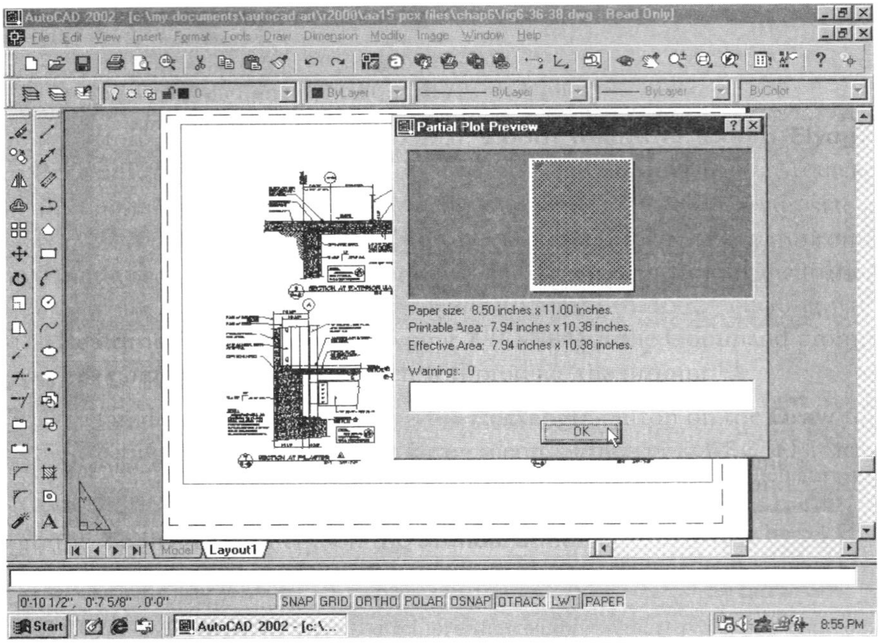

Figure 5.36 *A partial preview of a sheet of details to be plotted using the scale of 3/4"=1'–0"*

Full Preview

Clicking the **Full Preview** button will produce a display of the drawing as it will appear when printed or plotted. Figure 5.37 shows an example of a full preview. An outline of the paper size is drawn on the screen, which displays the portion of the drawing to be printed. The cursor is changed to show a magnifying glass with a + and – sign beside it. This represents the **Zoom** command and allows the image for viewing to be enlarged or reduced. It will not affect the plotted image. Press and hold the select button to alter the size of the image. Moving the mouse toward the top of the screen enlarges the image, and moving the mouse toward the bottom of the screen reduces the size of the image. Right-clicking will display the shortcut menu shown in Figure 5.38, allowing other options to be selected. Selecting the **Pan** option allows the image to be shifted, but the magnification of the image is not altered. Each of the options will be further discussed in Chapter 24. To continue with the plot process, select the **Exit** option to remove the menu.

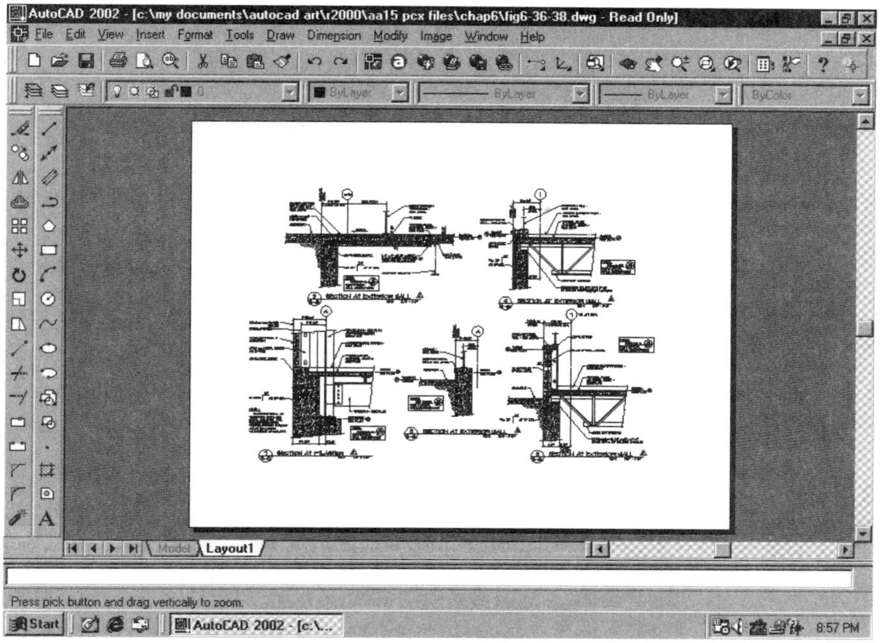

Figure 5.37 *A full preview will show the limits of the paper and the drawing to be plotted.*

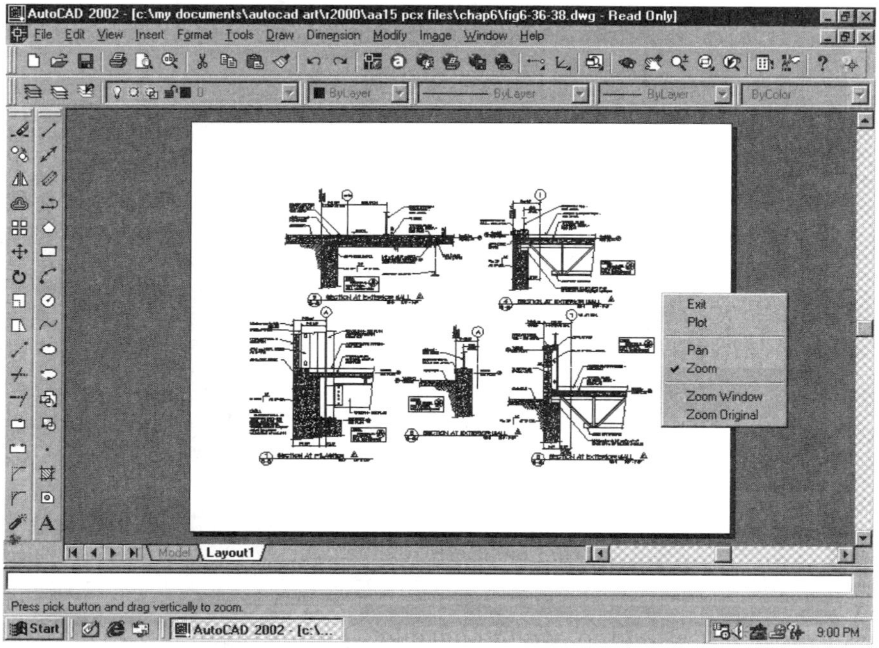

Figure 5.38 *Right-click while using a full preview to display the shortcut menu. The options can be used to control the display to be plotted.*

PRODUCING THE PRINT

Seven areas of the **Plot** dialog box have been examined. If you change your mind on the need for a print, click the **Cancel** button; the dialog box will be terminated, and the drawing will be redisplayed. If you are satisfied with the plotting parameters that have been established, click the **OK** button. The dialog box will be removed from the screen and the print will be produced. Figure 5.39 shows the results of the selected plot settings.

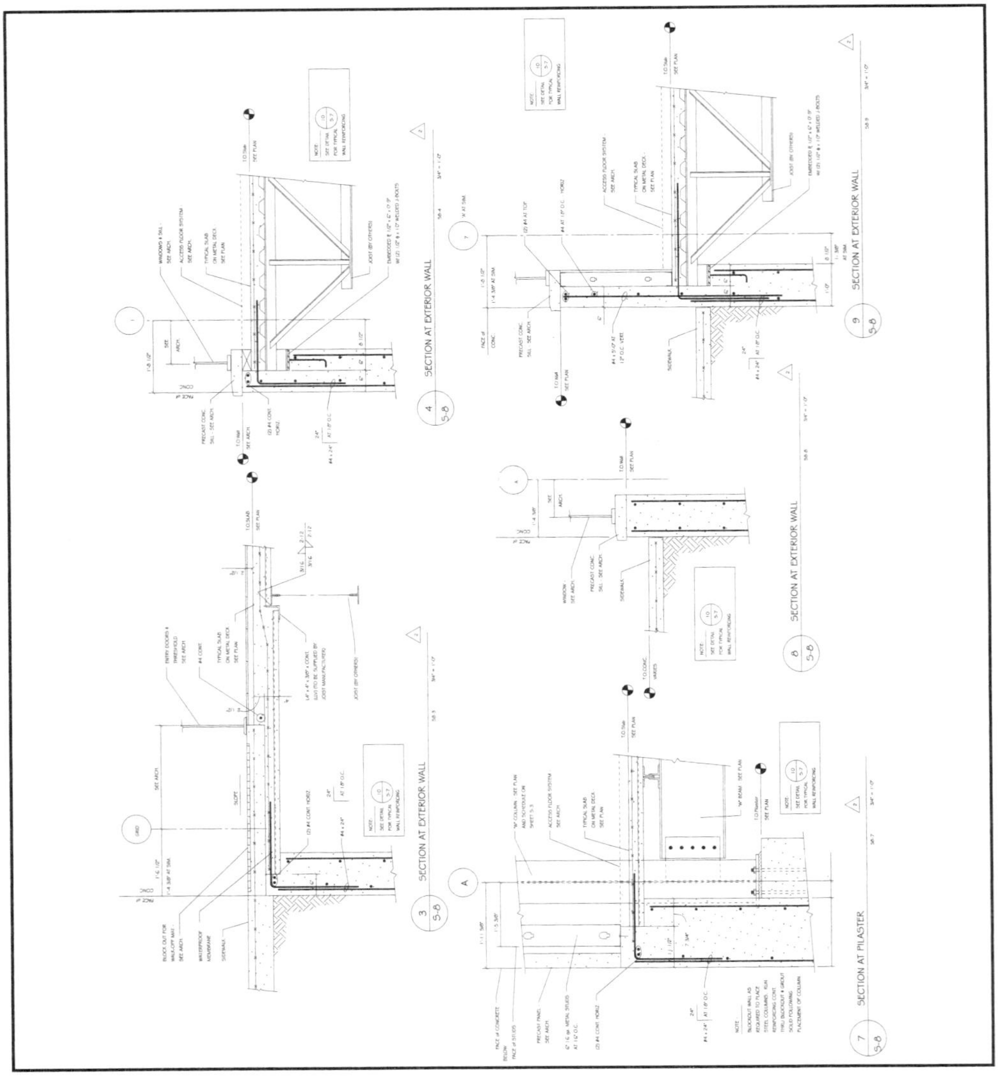

Figure 5.39 *The result of the previews shown in Figures 5.36 and 5.37. (Courtesy Van Domelen/Looijenga/McGarrigle/Knauf Consulting Engineers.)*

CHAPTER 5 EXERCISES

Set limits, units, snap, and grid to best suit each project. Use appropriate linetypes and lineweights that best suit each project. Dimensions, hatch patterns, and text do not need to be placed at this time. Save the following exercises in safe locations.

1. A parcel of land is to be developed. Use the following metes and bounds supplied by the surveyor and draw the site plan and the indicated utility easement. Set limits to 200',150'. Use a phantom line to represent the property line and use a centerline to represent the easement. Select different colors and layers for each linetype.

   SITE. Beginning at a point to be described as the northwesterly corner, 100.00' due south, thence 37.52' N89°37'E, thence 96.85' N58°40'52"E, thence 57.25' N3°50'40"W, and then back to the true point of beginning.

   EASEMENT. The easement is approximately an inverted L-shaped easement. The long leg of the "L" is 10' wide; the short leg is 15' wide. Lay out the easement beginning at a point that lies on the southerly property line, 20'–0" from the southwesterly corner of the property. Thence northerly 70'–0" along the center line of said easement to a point at N7°32'15"W, thence a 15' wide easement lying along a center line extending to the westerly property line at N50°55'04"W. Save the drawing with the file name E5-1 SITE.

2. Create a template drawing containing appropriate values for an architectural drawing. Include settings for each of the options discussed in previous chapters. Set limits and paper sizes assuming 24" x 36" plotting paper. Load linetypes and lineweights for lines that might be necessary on a plan view. Create a layer titled FLRWALLS. Save the file as a PLAN TEMPLATE drawing.

3. Use the drawing at right to complete a side view of a steel beam resting on a 6"x12"x3/4" steel plate supported on a steel column. Assume the beam to be 18" deep, and show the thickness of each plate as 7/8" thick. Provide a 10" long support plate, and an 18" high stiffener plate at the end of the beam. The support column is a TS 6"x6"x3/8". Provide a 4"x4"x3/8" T.S. brace @ 45° and a 4"x12"x12"x3/8" gusset plate. Provide appropriate break lines where needed. Save the drawing as E-5-3.

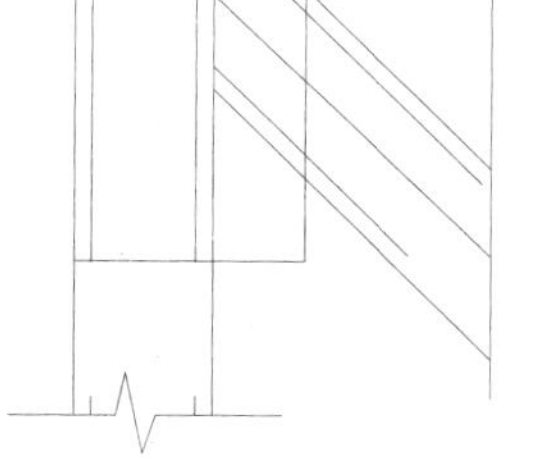

4. Use the drawing below to complete a side view of a steel beam resting on a steel plate supported on a steel column. Assume the beam to be 16" deep, and show the thickness of each plate as 5/8" thick. Provide a 14" long support plate. The support column is a TS 6"x6"x3/8" with a TS 6"x6"x3/8" chevron at a 45° angle. Provide centerlines for each TS. Provide appropriate break lines where needed. Save the drawing as E-5-4.

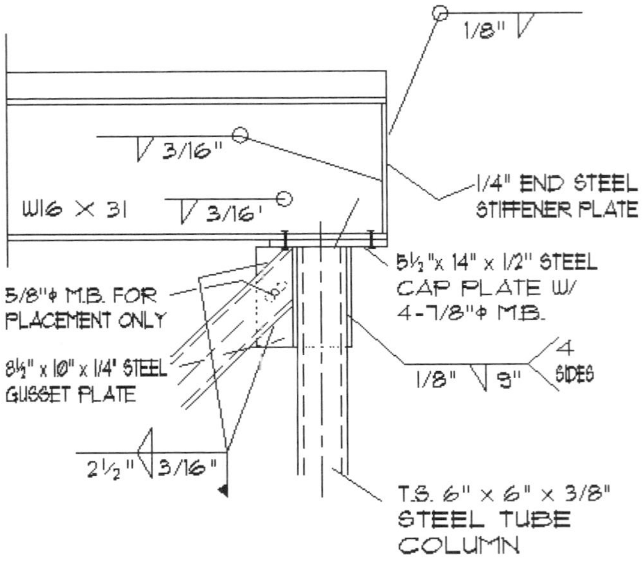

5. Use the drawing below to complete a side view of a steel column resting on a steel plate that is supported on 1" mortar over a 24" x 24" concrete pedestal. Assume the thickness of the plate as 5/8" and show 3/4" x 14" long anchor bolts 1 1/2" from each edge of the plate. Provide a 14"x10"x3/8" support plate. The support column is a TS 6"x6"x3/8" with a TS 6"x6"x3/8" chevron at a 45° angle. Provide centerlines for each TS. Provide appropriate break lines where needed. Save the drawing as E-5-5.

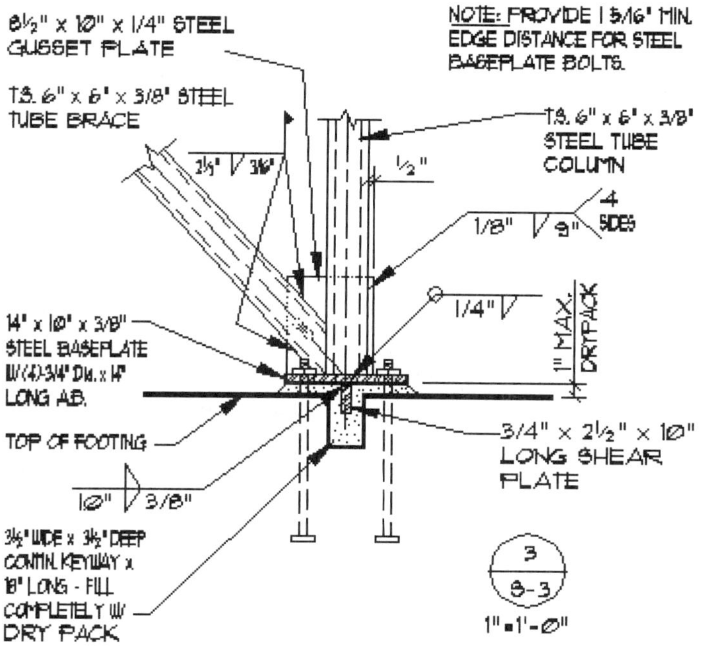

6. Use the drawing below to draw a partial foundation plan and place a detail marker. Show the outline of the footing with thin dashed lines and show the outline of the slab with a thick, continuous line. Draw a side view for a concrete slab foundation. Save the drawing as E-5-6.

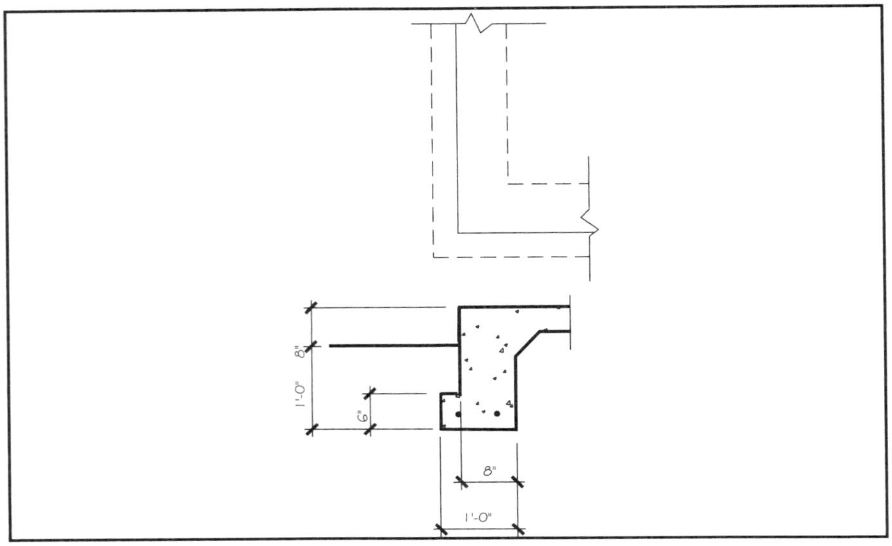

7. Use the drawing from Exercise 6 to obtain the needed sizes to draw a partial foundation plan and place a detail marker. Using the drawing below to show the outline of the footing with thin dashed lines and show the outline of the 8" wide stem wall with thick, continuous lines. Save the drawing as E-5-7.

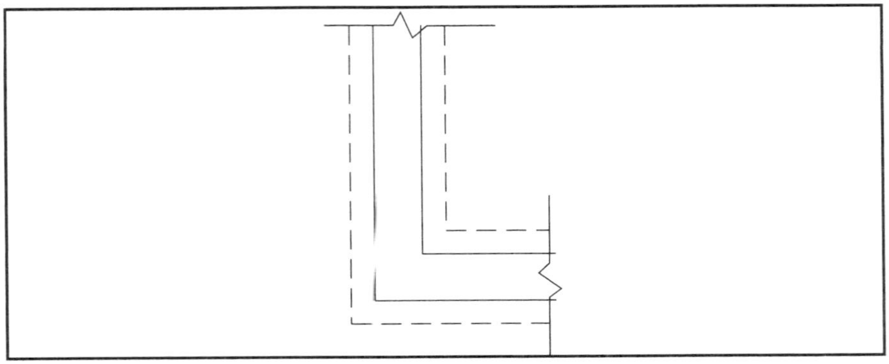

8. Use the drawing below to draw a site plan. Create your own north arrow. Save the drawing as E-5-8.

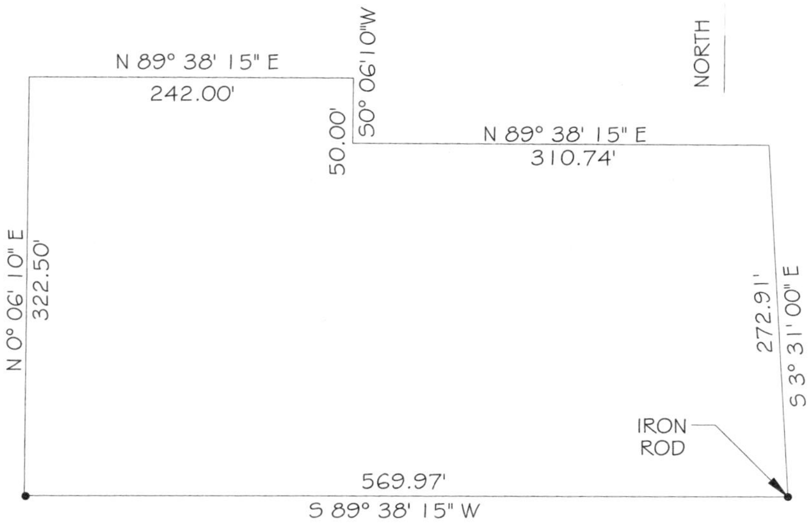

9. Use the drawing below to draw a section of a wall constructed of 8" concrete blocks. Use thin lines to represent the block. Show a #5 rebar centered in the wall using a thick dashed line. The individual blocks and hatch pattern do not need to be shown at this time. Save the drawing as E-5-9.

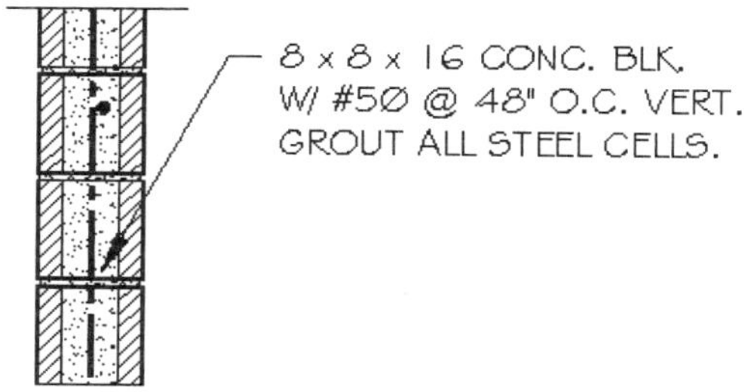

10. Use the drawing below to draw a 24' wide x 26' high concrete wall panel with a 10' x 12' door and a 3' x 7' door. Place each door 2' from the panel edge. Use thin lines to represent the wall outline and openings. Show the finish floor 12" above the bottom of the panel. Show a portion of the panel broken out to reveal #5 @ 12" o.c. horizontal and #5 @ 15" o.c. vertical. In addition to the standard reinforcing, draw lines to represent #5 rebar in the following locations:

    2" in from each edge of the panel

    2" clear from each edge of each opening

    Extend the vertical bars to within 2" of the top and bottom of the panel. Extend the horizontal door steel 12" past each opening. Save the drawing as E-5-10.

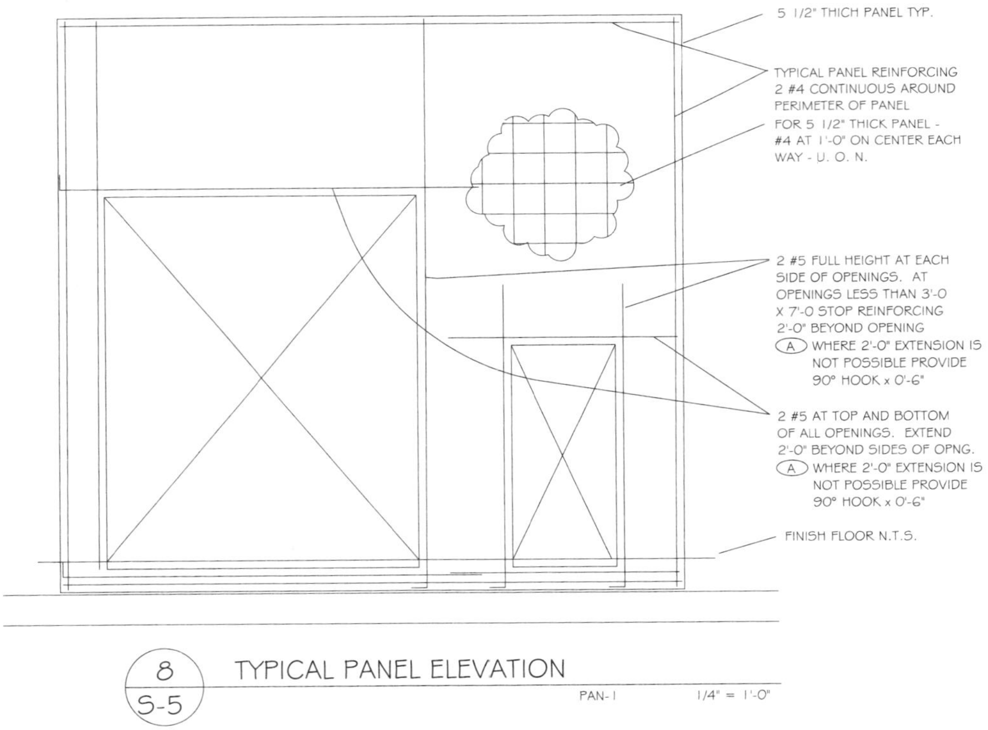

CHAPTER 5 QUIZ

1. List the methods for accessing linetype. Which is your favorite method? Which is your least favorite method? Explain each answer.
2. List two possible uses for an object line.
3. Describe the length and thickness of each line used in a centerline pattern.
4. List examples of lines that should be thick on a floor plan.
5. Describe the length and thickness of each line used in a hidden line pattern.
6. Describe the use of a section cutting plane line.
7. A note must be placed to describe a feature. Name and describe the type of line to be used.
8. A length of 24'–6" must be placed to describe the length of a room. What kind of line is used to reference the text to the drawing? In addition to this line, what else is needed to complete the correct placement?
9. Sketch how long and short break lines should be drawn.
10. List nine types of lines often found on construction drawings.
11. Describe the differences between a section line and a section cutting plane line.
12. A portion of a beam needs to be removed so that it will fit in a drawing. Name and describe the line used where the portion of the object has been removed.
13. Describe the options for a plot preview.
14. What command will allow the line pattern to be altered?
15. Explain the difference between plotting using the **Display** and **Window** options.
16. List four options for controlling the visibility of a layer.
17. You've just loaded five linetypes and have entered the **Line** command. Only continuous lines are being drawn. List two possible problems.
18. What is the difference between assigning a color to an object or a layer?
19. List four options for entering the **Layer** command.
20. What is the current layer name, and how can a new layer be named?
21. What are the limitations to assigning layer names?
22. A friend needs help to load five linetypes into a drawing. How can this be done quickly?
23. Centerline has just been made the current linetype. You return to the drawing and nothing is different. What's happened and why did it happen?
24. You've used the **Layer Properties Manager** dialog box to make centerline the current linetype. You return to the drawing. What's happened and why did it happen?

25. Use the **Help** menu and explore how the global scale factor will affect linetypes.
26. You've used **Lineweight Control** on the **Object Properties** toolbar to alter the lineweight. You return to the drawing area and draw several lines, but no change has been made to the lineweight. Why is there no change, and what must be done to see varied lineweights?
27. List the suggested lineweight for thick, moderate, and thin lines.
28. You want to draw one red line. How can it be done?
29. You want all of your lines to be red. What is the best way to do this?
30. You want to plot a drawing at 3/8"=1'–0". How can this be done?

CHAPTER 6

Drawing Geometric Shapes

Lines and points may be the base of all drawings, but it's difficult to imagine a large engineering or architectural project without curved features. This chapter will introduce drawing methods for creating common geometric shapes, including circles, arcs, ellipses, polygons, and rectangles.

Commands to be introduced include

- **Dragmode**
- **Circle**
- **Arc**
- **Donut**
- **Fill**
- **Regen**
- **Ellipse**
- **Polygon**
- **Rectangle**

DRAWING CIRCLES

Circles are found in a variety of projects throughout the construction world. Round windows, concrete piers and columns, steel tubing, and conduit lines—circular features will be prevalent throughout your drawings. Figure 6.1 uses circles to represent concrete piers for a structure. Start the **Circle** command by selecting the **Circle** button on the **Draw** toolbar, by selecting **Circle** from the **Draw** menu (as seen in Figure 6.2), or by keyboard. To draw a circle, select the **Circle** button on the **Draw** toolbar, or type **C** (ENTER) at the command prompt. Entering the **Circle** command on the toolbar or by keyboard will display the prompt:

Command: **C** (ENTER) *(Or click the Circle button on the Draw toolbar.)*

CIRCLE Specify center point for circle or [3P/2P/Ttr (tan tan radius)]:

AutoCAD is waiting for you to select a center point for the origin point of the circle. Once the center point is selected, you're prompted for the edge of a circle. The process can be seen in Figure 6.3.

Circles can be drawn by entering seven different combinations of information about the circle. Six of the circles are listed in the **Circle** submenu shown in Figure 6.2. The seventh, the **Donut** command on the **Draw** menu, produces a circle with varied line width and will be discussed later in this chapter.

Notice that once the center point is selected, a circle will appear on the screen, but the size will be altered as the mouse is moved. AutoCAD refers to this as drag or rubber band. Moving the mouse and changing the circle size is called dragging the circle into position. This drag image is typically helpful in determining the final locations of lines and other geometric shapes. If you find it distracting, type **DRAGMODE** (ENTER) at the command prompt and select OFF to disable it.

Note: As you draw circular features, you might want to review the **Display** tab of the **Options** dialog box. The **Arc and circle smoothness** setting can be used to control the smoothness of the display of arcs, circles, and ellipses. The default value is 100, with a range of 1 through 20,000. A higher number will produce a smoother display but will require longer regeneration times on larger drawings. A lower number will display circular features with a series of straight lines. The true shape of objects will be displayed on plots, regardless of the resolution.

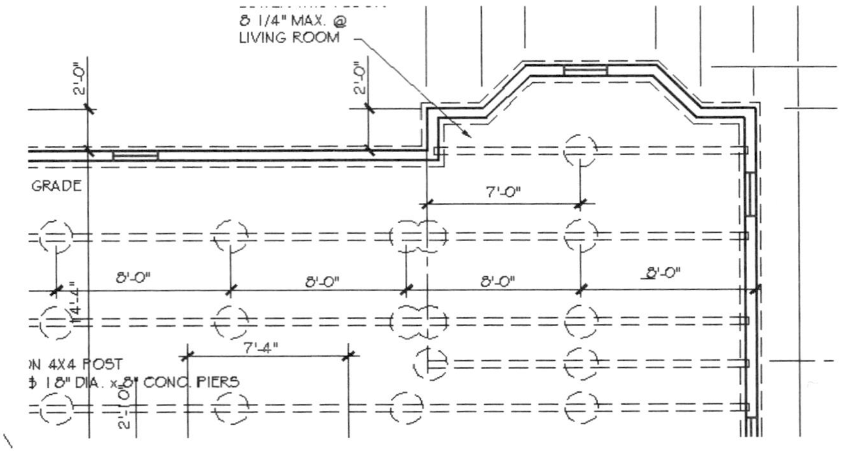

Figure 6.1 *Circles are used throughout architecture to represent many building components. (Courtesy Residential Designs.)*

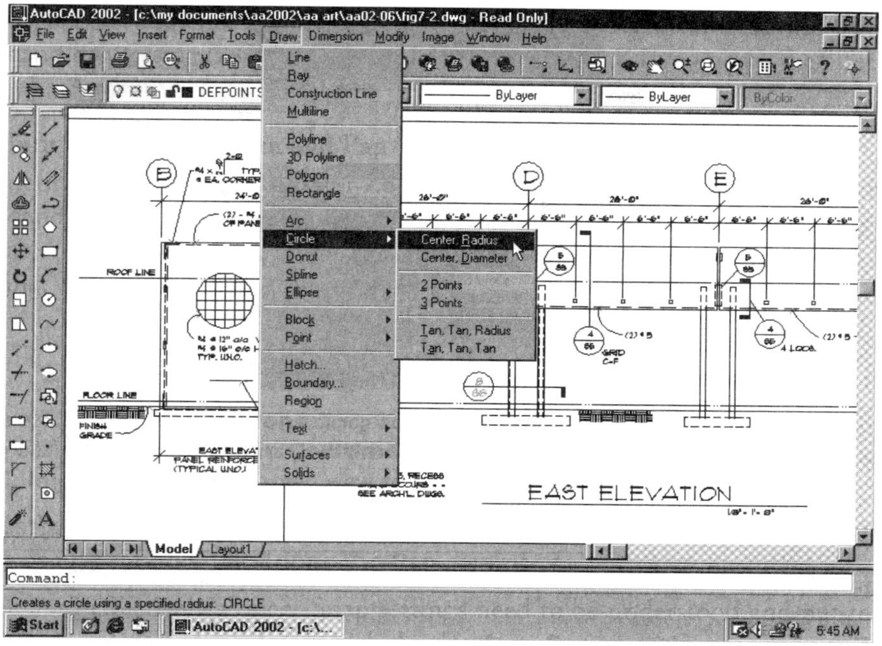

Figure 6.2 *To draw a circle, select the* **Circle** *button on the* **Draw** *toolbar, select* **Circle** *from the* **Draw** *menu, or type* **C** *(ENTER) at the command prompt.*

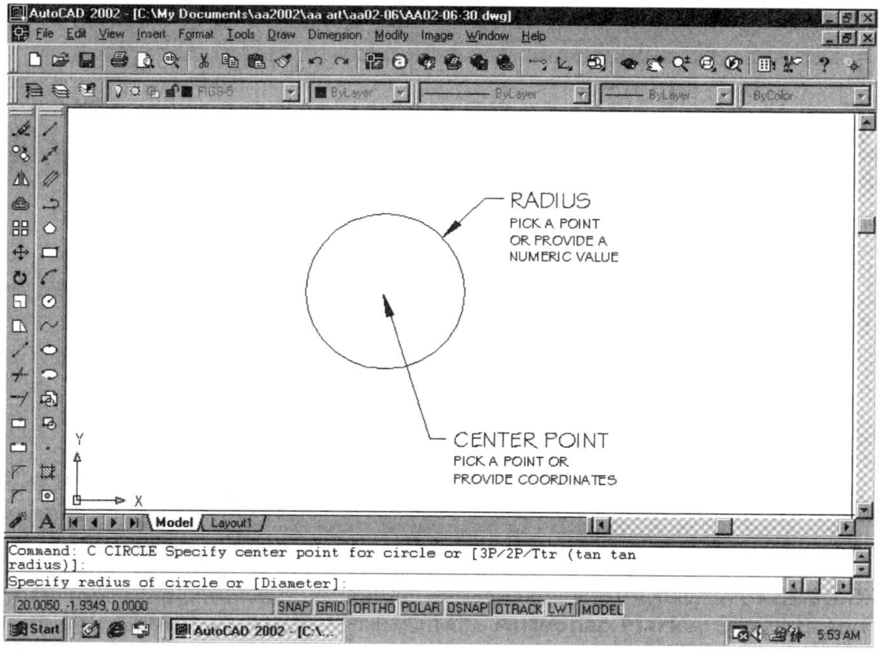

Figure 6.3 *Control the size of a circle by entering a value for the radius or the diameter.*

CENTER AND RADIUS

The default method of drawing a circle is to enter the center point and a radius. As the **Circle** command is started, enter the center point by typing coordinates or by selecting with the pointing device. In Figure 6.4 three circles have been drawn, and a fourth has been dragged into position. Notice that the rubber-band line extends from the chosen center point to the radius indicated by the crosshairs. Move the crosshairs to change the size of the circle. Type a numeric value and press (ENTER) to draw a circle with the desired radius. To draw a 24" diameter circle, the sequence is as follows:

Command: **C** **(**ENTER) *(Or click the Circle button on the Draw toolbar.)*

CIRCLE Specify center point for circle or [3P/2P/Ttr (tan tan radius)]
(Select center point.)

Specify radius of circle or [Diameter] <0'-0">: **12** **(**ENTER)

Command:

Note: To draw another circle, continue the command by pressing (ENTER) or SHIFT, or right-clicking and then selecting **Repeat CIRCLE**.

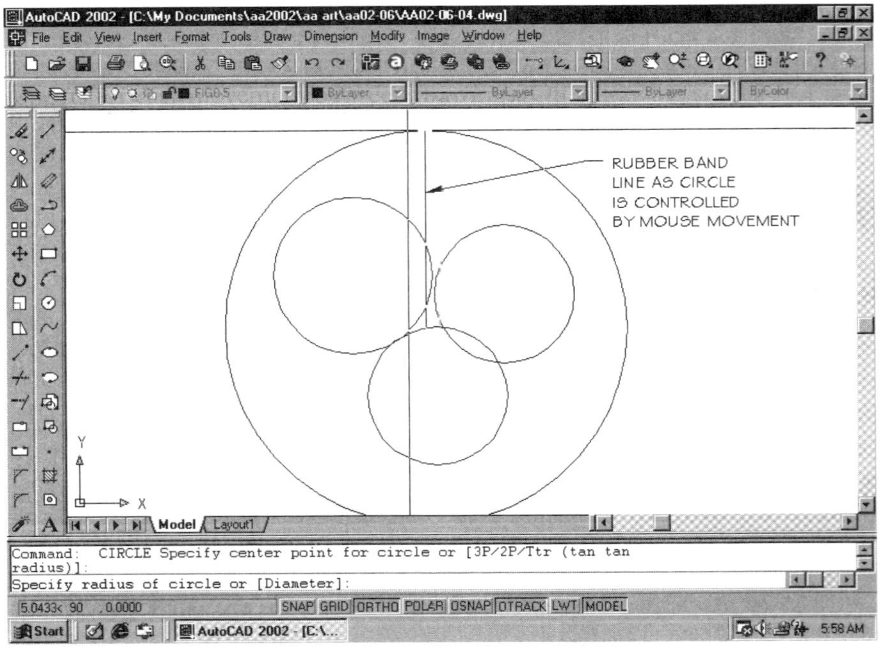

Figure 6.4 *Rather than entering a specific size, use the rubber-band line to see the effect on the circle to be drawn.*

CENTER AND DIAMETER

Sizes of circular features such as piers, pipes, and windows are typically given as a diameter. At the CIRCLE prompt, type **D** (ENTER) to switch the prompt from radius to diameter. When entered by toolbar or keyboard, the command sequence is as follows:

> Command: **C** (ENTER) *(Or click the Circle button on the Draw toolbar.)*
>
> CIRCLE Specify center point for circle or [3P/2P/Ttr (tan tan radius)]
> *(Select center point.)*
>
> Specify radius of circle or [Diameter] <0'-0">: **D** (ENTER)
>
> Specify diameter of circle:<2'-0"> *(Enter a diameter value or move the crosshairs to the desired location.)*
>
> Command:

The command sequence is similar when **Center, Diameter** is selected from the **Circle** submenu. AutoCAD simplifies the process with the menu so that the Diameter option is automatic.

Once the circle center is selected, the circle will be dragged across the screen but will not line up with the crosshairs. Figure 6.5 shows an example of a circle being created using the diameter. Notice that the length of the line extending from the center point is the diameter of the circle to be drawn. The line will disappear once the diameter is selected. The diameter entered will become the default setting for the next circle drawn.

TWO-POINT CIRCLES

Drawing a two-point circle can be useful if the diameter is known, but the center point is not. Draw the two-point circle by selecting a point and then selecting a point on the opposite side of the circle, as seen in Figure 6.6. The points can be selected with the crosshairs, or coordinates can be entered. The command sequence is as follows:

> Command: **C** (ENTER) *(Or click the Circle button on the Draw toolbar.)*
>
> CIRCLE Specify center point for circle or [3P/2P/Ttr (tan tan radius)]: **2P** (ENTER)
>
> Specify first end point of circle's diameter: *(Enter coordinates or select a point.)*
>
> Specify second end point of circle's diameter: *(Enter coordinates or select a point.)*
>
> Command:

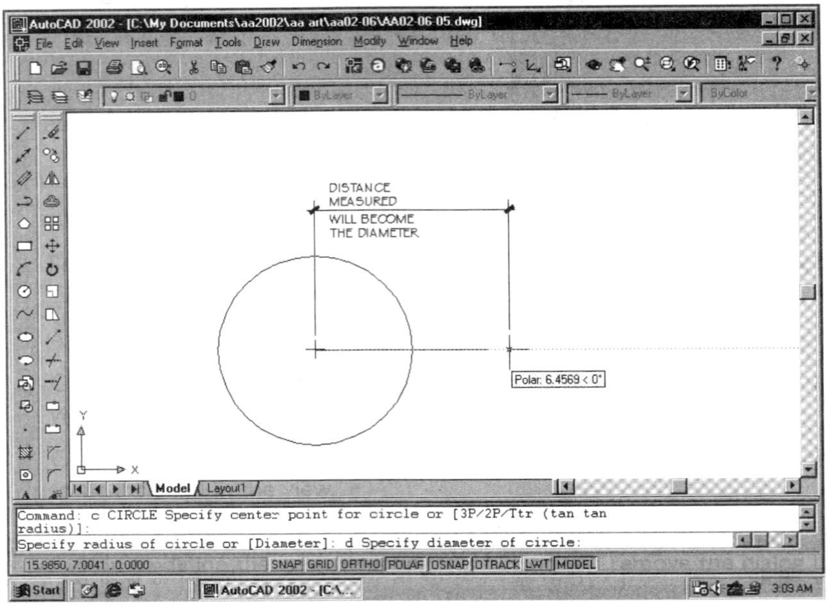

Figure 6.5 *With the Diameter option, the distance that the cursor is moved from the center point will become the diameter of the circle.*

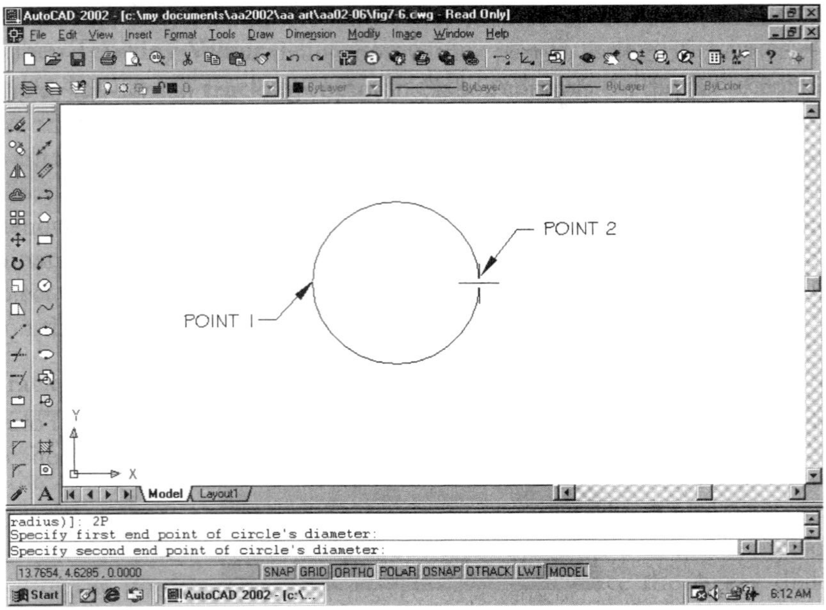

Figure 6.6 *A circle can be drawn by specifying the endpoints of a line that would be the diameter.*

THREE-POINT CIRCLES

If certain points of a circle are known, a circle can be drawn by entering any three points. Figure 6.7 shows an example of a three-point circle. The points can be selected with the crosshairs, or coordinates can be entered. The process for drawing a three-point circle is as follows:

> Command: **C** (ENTER) *(Or click the Circle button on the Draw toolbar.)*
>
> CIRCLE Specify center point for circle or [3P/2P/Ttr (tan tan radius)]: **3P** (ENTER)
>
> Specify first point on circle: *(Enter coordinates or select a point.)*
>
> Specify second point on circle: *(Enter coordinates or select a point.)*
>
> Specify third point on circle: *(Enter coordinates or select a point.)*
>
> Command:

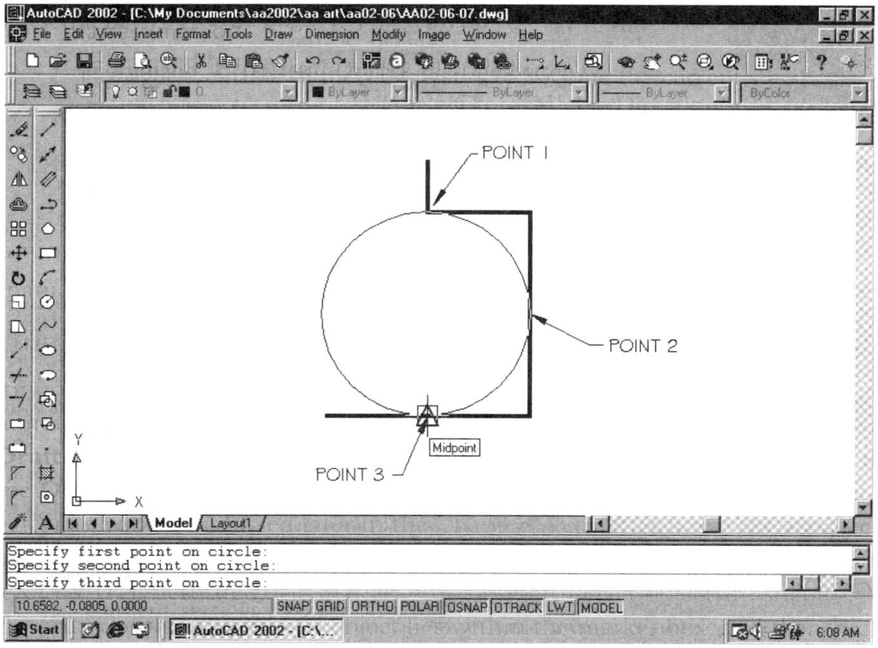

Figure 6.7 *The size of a circle can be determined by specifying three tangent points.*

TANGENT, TANGENT, RADIUS

An object is tangent if it intersects another object at one and only one point. A circle can be drawn tangent to existing features by selecting two features that the circle will be tangent to and then specifying a radius. This can be seen in Figure 6.8. The command sequence is as follows:

Command: **C** (ENTER) *(Or click the Circle button on the Draw toolbar.)*

CIRCLE Specify center point for circle or [3P/2P/Ttr (tan tan radius)]: **TTR** (ENTER)

Specify point on object for first tangent of circle: *(Select a line or circle.)*

Specify point on object for second tangent of circle: *(Select a line or circle.)*

Specify radius of circle: <2'-0"> *(Enter a numeric value or select a point.)* (ENTER)

Specify second point: *(Enter a numeric value or select a point.)* (ENTER)

Command:

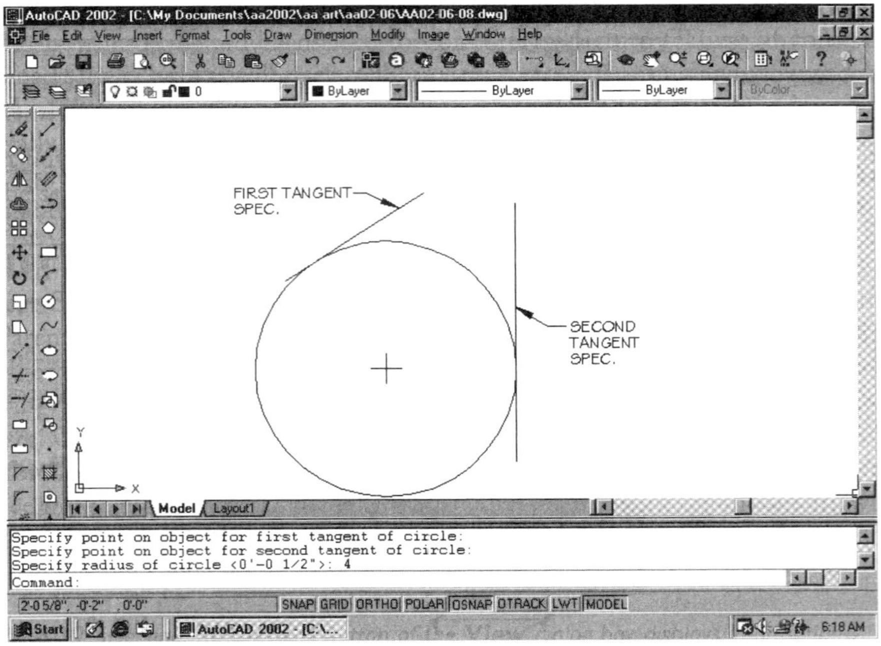

Figure 6.8 *A circle can be drawn by specifying two tangent surfaces and a radius.*

In Figure 6.9 each line was selected as a tangent point and a radius was entered. Because the radius is smaller than the distance between the lines, the first line does not touch the circle. If the line were extended, it would be tangent to circle.

TANGENT, TANGENT, TANGENT

This option works well for drawing a circle that is tangent to three objects. This option can be accessed only through **Tan, Tan, Tan** on the **Circle** submenu. The command sequence is as follows:

Command: *(Select Tan, Tan, Tan from Circle on the Draw menu.)*

_circle Specify center point for circle or [3P/2P/Ttr (tan tan radius)]: _3p

Specify first point on circle: _tan to *(Select the first object the circle is to be tangent to.)*

Specify second point on circle: _tan to *(Select the second object the circle is to be tangent to.)*

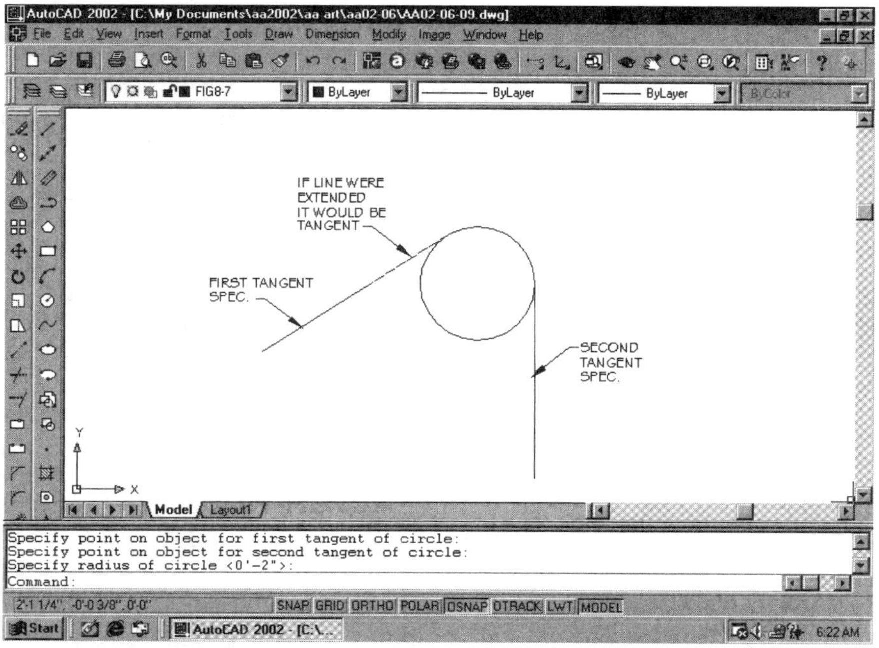

Figure 6.9 *If a radius is selected that is not large enough for the specified circle to touch each line, a circle is drawn tangent to one line and tangent to where the other line would be if it were extended.*

Specify third point on circle: _tan to *(Select the third object the circle is to be tangent to.)*

Command:

Figure 6.10 shows the development of a circle tangent to three other objects.

CIRCLE@

This option allows a circle to be drawn using the last endpoint of the last line drawn for the center point of a new circle. The option is best used for drawing concentric circles associated with pipes and circular columns. The command sequence is as follows:

Command: **C** (ENTER) *(Or click the Circle button on the Draw toolbar.)*

CIRCLE 3P/2P/Ttr/<Centerpoint>: @ (ENTER)

Diameter/<radius> <1.000> *(Specify desired diameter.)*

Command:

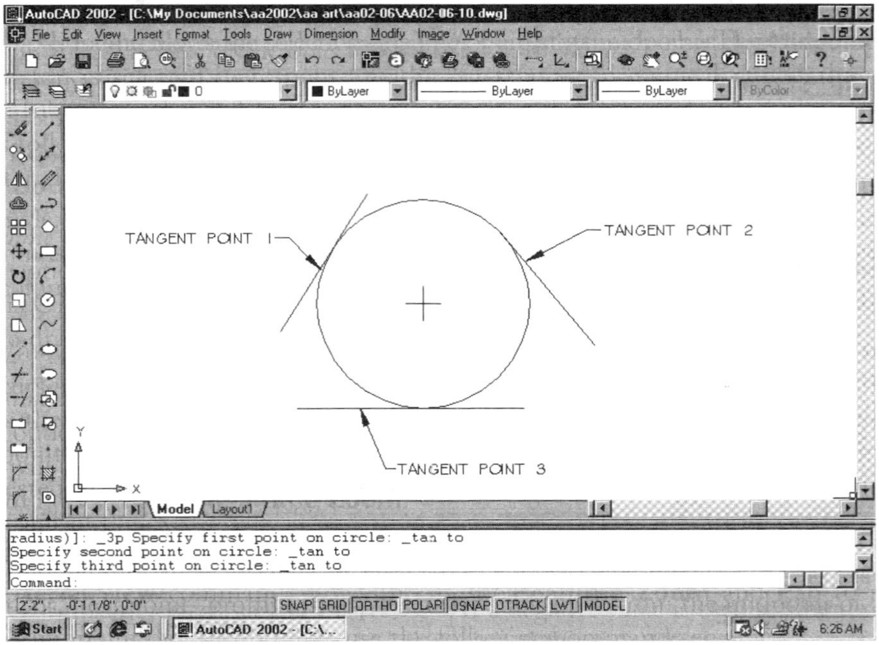

Figure 6.10 *Tan,Tan,Tan can be used to draw circles that are tangent to three lines.*

DRAWING ARCS

Arcs are used where a full circle is not required. Curved components are found throughout the structure seen in Figure 6.11. Start the command by selecting the **Arc** button on the **Draw** toolbar or by selecting **Arc** from the **Draw** menu. You can also start the command by typing **A** (ENTER) at the Command prompt. The menu can be seen in Figure 6.12. Eleven different options are available for drawing arcs, depending on what information is available or on personal preference. Selecting **Arc** on the toolbar will produce the prompt:

Command: _arc Specify start point of arc or [Center]:

The default setting of the toolbar and keyboard is the three-point arc. With the exception of the three-point arc, you can specify each arc drawing method by typing a letter and pressing (ENTER) or spacebar when prompted for a point of the arc. Letters and the function of the arc they represent are as follows:

A: included Angle

C: Center

D: starting Direction

E: Endpoint

L: Length of chord

R: Radius

Figure 6. 11 *Arcs are found throughout many structures (Courtesy Residential Designs.)*

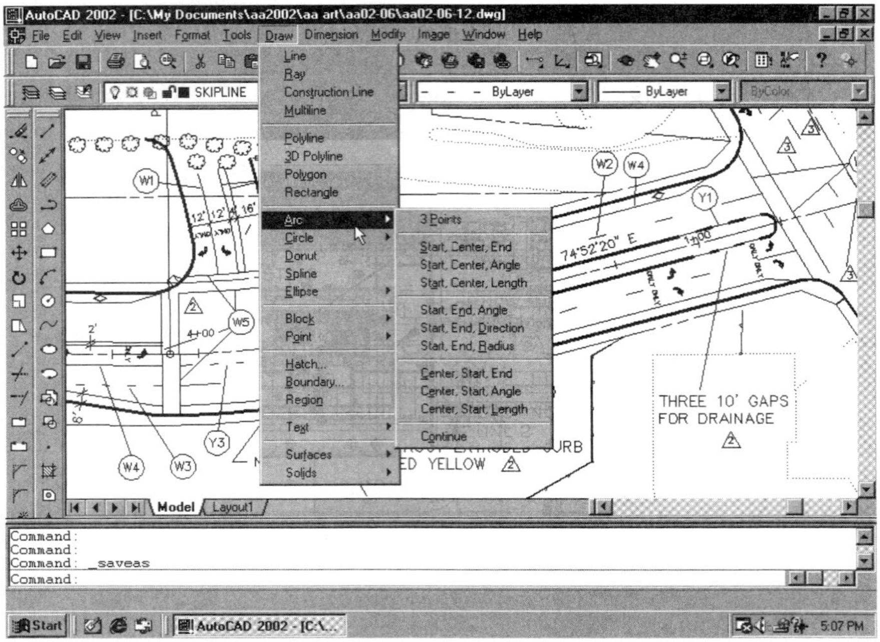

Figure 6.12 *The* **Arc** *menu. (Courtesy Lee Engineering, Inc.)*

Some of the terms used to describe an arc can be seen in Figure 6.13 Notice that "C" describes the center of the circle that the arc is part of, not the center of the arc. When the **Arc** command is started from the menu and the method is specified, prompts are automatically provided to match the option selected.

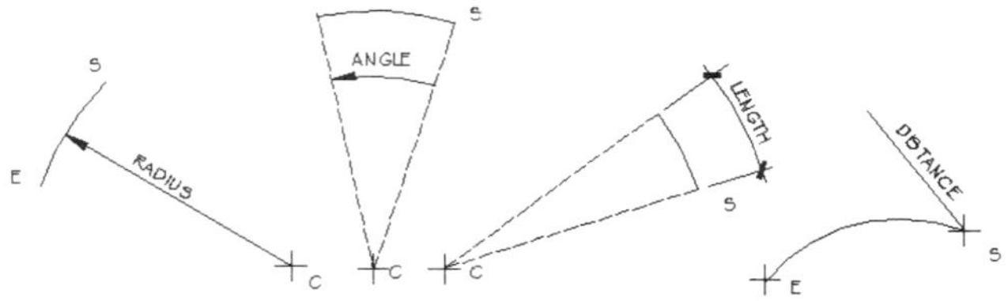

Figure 6.13 *Terms used to describe arcs are Start (S), Center (C), End (E), Angle, Length, Radius, and Distance.*

THREE-POINT ARCS

Drawing a three-point arc is similar to drawing a 3P circle and can be seen in Figure 6.14. The first and third points are the arc endpoints; the second point is any point on the arc. The command sequence for a three-point arc is as follows:

> Command: **A** (ENTER) *(Or click the Arc button on the Draw toolbar.)*
>
> ARC Specify start point of arc or [Center]: *(Select any point or enter coordinate.)*
>
> Specify second point of arc or [Center/ENd]: *(Select second point.)*
>
> Specify end point of arc: *(Select the arc's endpoint or enter coordinates.)*
>
> Command:

Notice that once the second point is selected, the arc will drag across the screen with the crosshairs. The arc will continue to change size until the endpoint is selected.

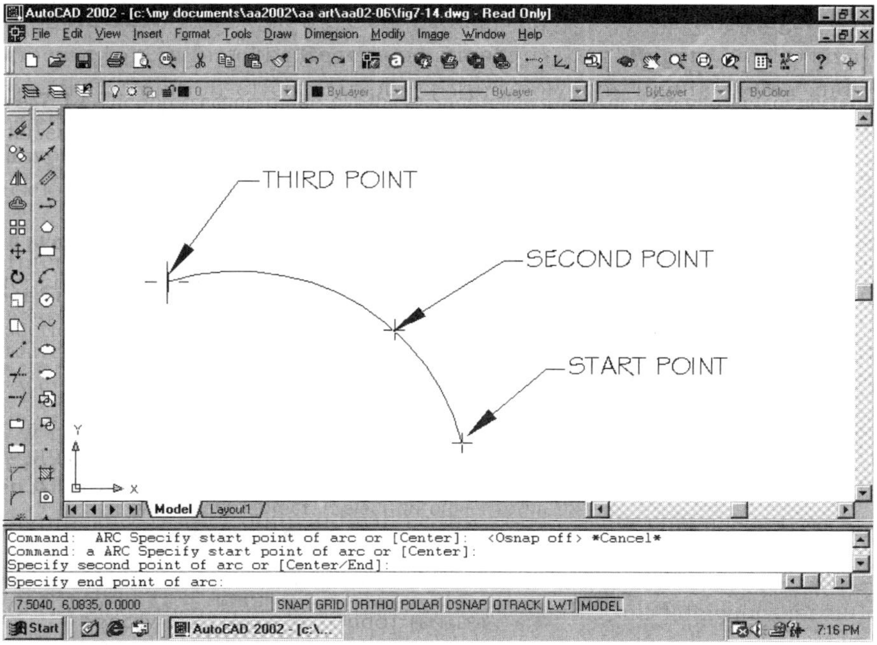

Figure 6.14 *Drawing a three-point arc is similar to drawing a three-point circle.*

START, CENTER, END ARCS

When the ends and center of an arch are known, the Start, Center, End arc option can be used. The stair railing in Figure 6.15 provides an example of where this arc option could be used. The command sequence is as follows:

> Command: **A** (ENTER) *(Or click the Arc button on the Draw toolbar.)*
>
> ARC Specify start point of arc or [Center]: *(Select any point or enter coordinate.)*
>
> Specify second point of arc or [Center/End]: **C** (ENTER)
>
> Specify center point of arc: *(Select second point.)*
>
> Specify end point of arc [Angle/chord Length]: *(Select the arc's endpoint or enter coordinates.)*
>
> Command:

As the last point is entered, the arc will be drawn and the command prompt will be returned. The process can be seen in Figure 6.16. Notice that as the last point is entered, the arc cannot end at the point entered. The endpoint that was specified is used only to determine the angle at which the arc will end. The actual radius was determined by the start and center points.

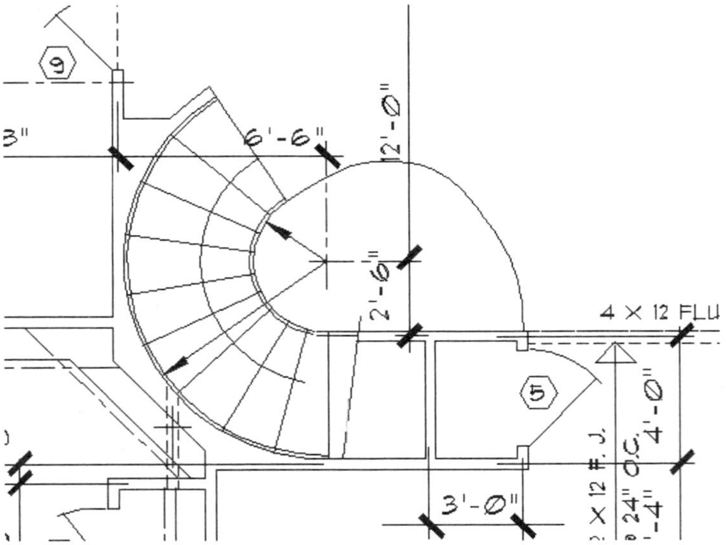

Figure 6.15 *A stair railing drawn using the Start, Center, End option. (Courtesy Residential Designs.)*

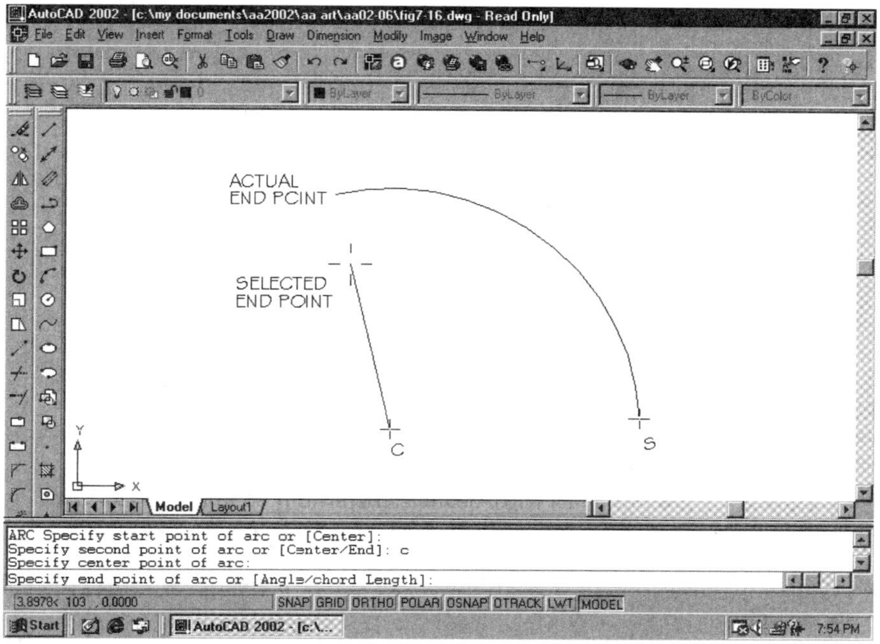

Figure 6.16 *The selected endpoint is used to determine the angle of the specified circle. The actual size of the circle is determined by the distance from the start to the center point.*

START, CENTER, ANGLE ARCS

In road and utility layouts for a site plan, it often is necessary to draw an arc using lengths or angles. Drawing an arc using the start, center point, and the included angle is ideal for this type of work. If a positive value is used for the angle, the angle will be measured in a counterclockwise direction. If a negative value is used, the angle will be measured in a clockwise direction. The effects of positive and negative values can be seen in Figure 6.17. The command sequence for a Start, Center, Angle arc is as follows:

Command: **A** (ENTER) *(Or click the Arc button on the Draw toolbar.)*

ARC Specify start point of arc or [Center]: *(Select any point or enter coordinate.)*

Specify second point of arc or [Center/End]: **C** (ENTER)

Specify center point of arc: *(Select center point of arc.)*

Specify end point of arc [Angle/chord Length]: **A** (ENTER)

Specify included angle: **30** (ENTER)

Command:

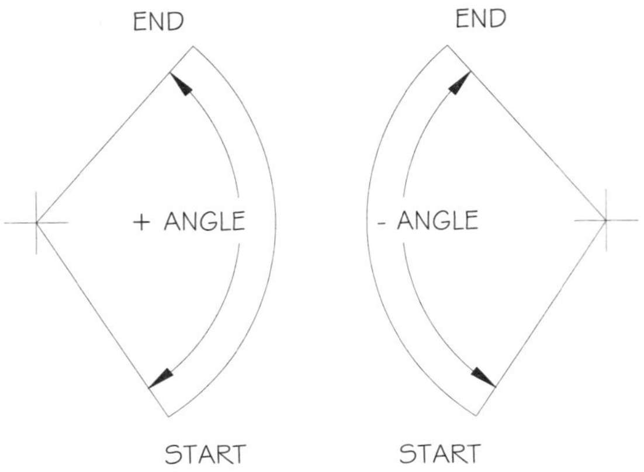

Figure 6.17 *The direction of layout is controlled by entering a positive (counterclockwise) or negative (clockwise) angle when drawing arcs with the Start, Center, Angle arc option.*

When the desired angle is entered, the arc will be drawn and the command line will return.

START, CENTER, LENGTH ARCS

A chord is the straight line connecting the endpoints of an arc. Subdivision maps and drawings relating to land often include arcs that require use of the chord length. Figure 6.18 shows an example of arcs that could be drawn using Start, Center, and Length of chord. The arc was drawn by specifying the start and center points and then supplying the chord length. A positive or negative value must be entered with the length. Figure 6.19 shows the difference between a chord of +3" or –3". The command sequence for a Start, Center, Length arc is as follows:

Command: **A** (ENTER) *(Or click the Arc button on the Draw toolbar.)*

ARC Specify start point of arc or [Center]: *(Select start point or enter coordinates.)*

Specify second point of arc or [Center/ENd]: **C** (ENTER)

Specify center point of arc: *(Select any point or enter coordinates.)*

Specify end point of arc [Angle/chord Length]: **L** (ENTER)

Specify length of chord: **3** (ENTER)

Command:

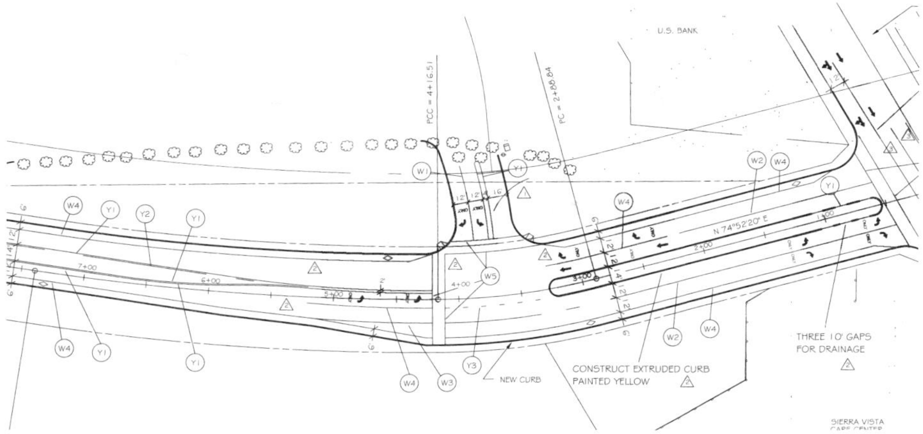

Figure 6.18 *The Start, Center, Length arc option is useful when parcels of land are drawn. (Courtesy Lee Engineering, Inc.)*

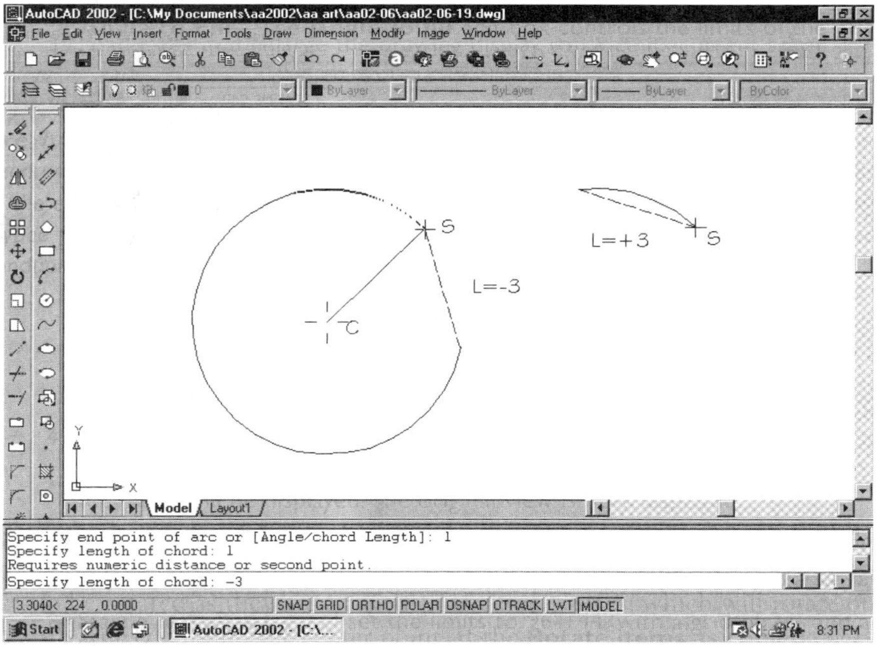

Figure 6.19 *Entering a positive or negative chord length determines which portion of the arc will be drawn using the Start, Center, Length arc option.*

When a length is specified or a point is selected, the arc will be drawn and the command line will return.

START, END, ANGLE ARCS

An arc drawn using Start, End, and included Angle is drawn counterclockwise from the start point to the endpoint if a positive angle value is used. If a negative angle is used, the angle will be drawn clockwise. The results of coordinate entry can be seen in Figure 6.20. The command sequence is as follows:

> Command: **A** (ENTER) *(Or click the Arc button on the Draw toolbar.)*
>
> ARC Specify start point of arc or [Center]: *(Select start point or enter coordinates.)*
>
> Specify second point of arc or [Center/End]: **E** (ENTER)
>
> Specify end point of arc: *(Select the desired endpoint or enter coordinates.)*
>
> Specify center point of arc [Angle/chord Length]: **A** (ENTER)
>
> Specify Included angle: *(Select desired angle.)*
>
> Command:

When a length is specified or a point is selected, the arc will be drawn and the command line will return.

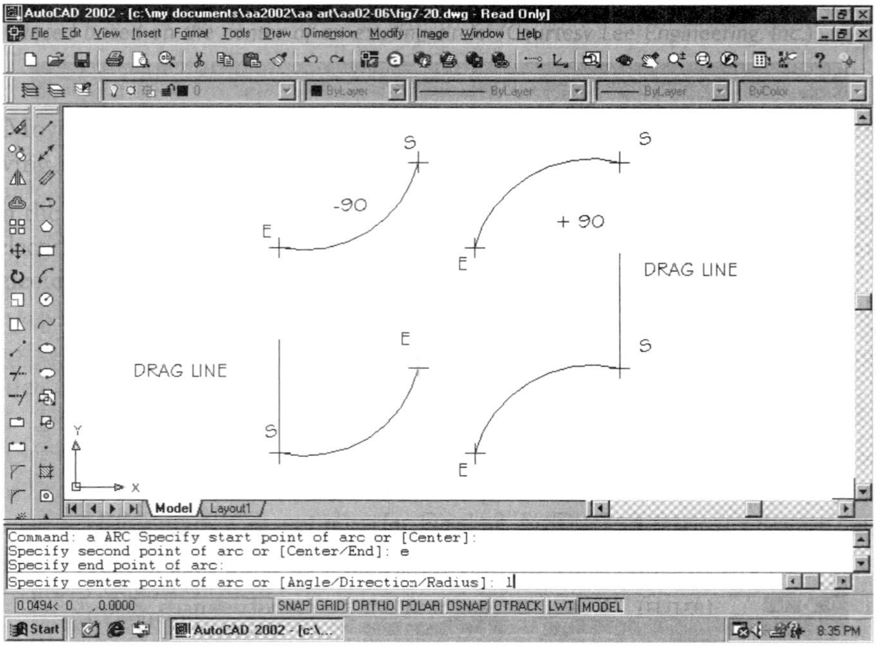

Figure 6.20 *Entering a positive or negative angle determines which direction an arc will be drawn using the Start, End, Angle arc option.*

START, END, DIRECTION ARCS

The Start, End, Direction option allows the direction of the arc to be selected. Major or minor arcs can be drawn in any orientation to the start point. Effects can be seen in Figure 6.21. The command sequence is as follows:

> Command: **A** (ENTER) *(Or click the Arc button on the Draw toolbar.)*
>
> ARC Specify start point of arc or [Center]: *(Select start point or enter coordinates.)*
>
> Specify second point of arc or [Center/End]: **E** (ENTER)
>
> Specify end point of arc: *(Select the desired endpoint or enter coordinates.)*
>
> Specify center point of arc or [Angle/Direction/Radius]: **D** (ENTER)
>
> Specify tangent direction for start of arc: *(Select desired direction.)*
>
> Command:

When a length is specified or a point is selected, the arc will be drawn and the command line will return.

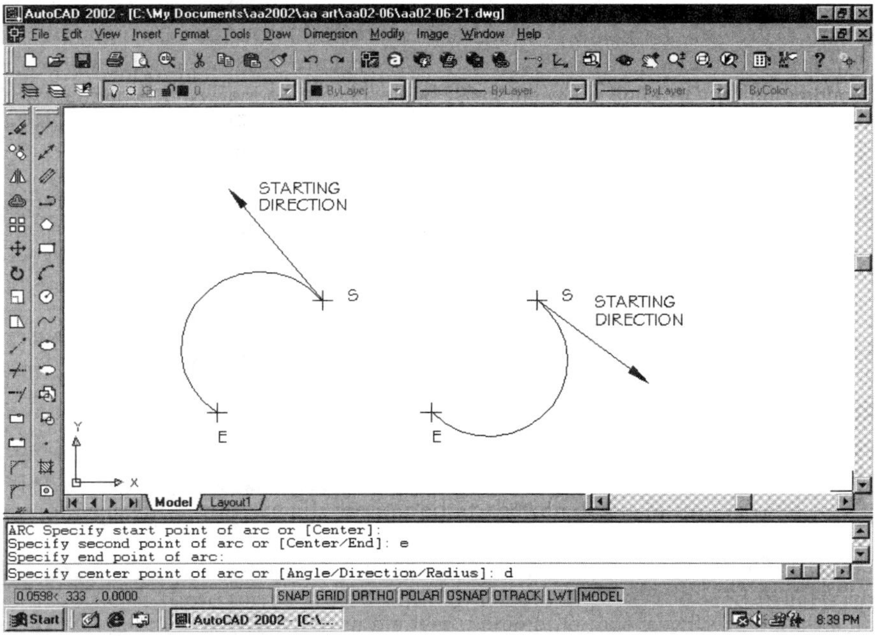

Figure 6.21 *The Start, End, Direction arc option allows the direction of the arc to be selected.*

START, END, RADIUS ARCS

Start, End, Radius arcs can only be drawn counterclockwise. Because the variables can produce different options, the arc is drawn counterclockwise from the Start point. Entering a positive or negative value for the radius will determine if the major or minor arc will be drawn. The effects of the value can be seen in Figure 6.22. The command sequence is as follows:

> Command: **A** (ENTER) *(Or click the Arc button on the Draw toolbar.)*
>
> ARC Specify start point of arc or [Center]: *(Select start point or enter coordinates.)*
>
> Specify second point of arc or [Center/End]: **E** (ENTER)
>
> Specify end point of arc: *(Select the desired endpoint or enter coordinates.)*
>
> Specify center point of arc or [Angle/Direction/Radius]: **R** (ENTER)
>
> Specify radius of arc: *(Select desired radius or enter a point.)*
>
> Command:

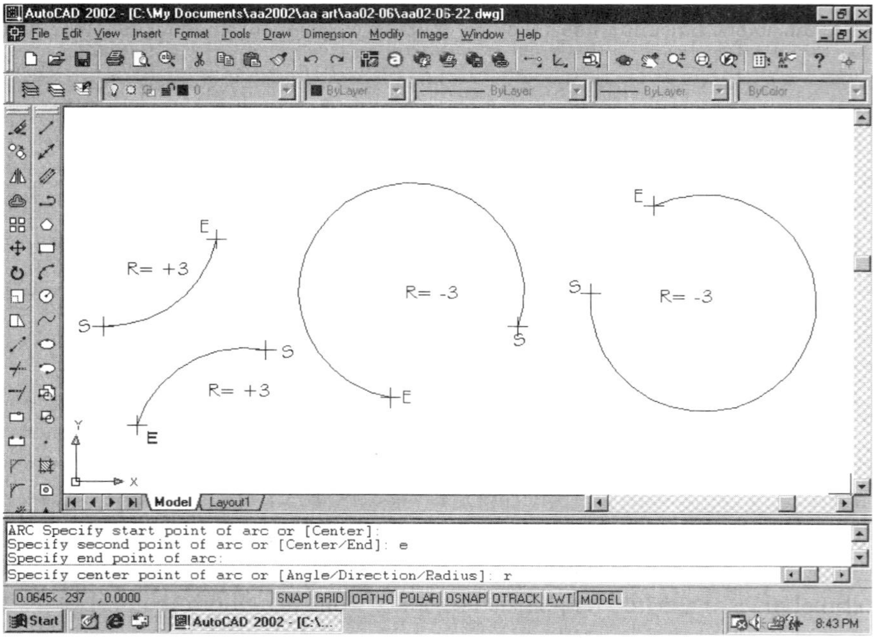

Figure 6.22 *Entering a positive radius determines which portion of the arc will be drawn using the Start, End, Radius arc option. The direction will always be counterclockwise.*

When a length is specified or a point is selected, the arc will be drawn and the command line will return (see Figure 6.23).

CENTER, START, END ARCS

The Center, Start, End option is similar to the Start, Center, End option. The effects of each command can be seen in Figure 6.24. The command sequence is as follows:

Command: **A** (ENTER) *(Or click the Arc button on the Draw toolbar.)*

ARC Specify start point of arc or [Center]: **C** (ENTER)

Specify center point of arc: *(Select center point or enter coordinates.)*

Specify start point of arc: *(Select the start point or enter coordinates.)*

Specify end point of arc or [Angle/chord Length]: *(Select the desired endpoint or enter coordinates.)*

Command:

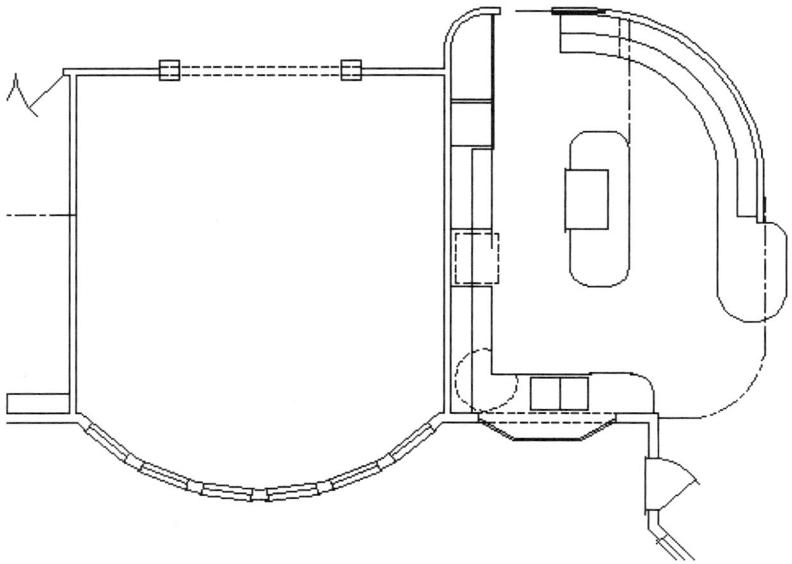

Figure 6.23 *Walls drawn using the Start, End, Radius arc option. (Courtesy Residential Designs.)*

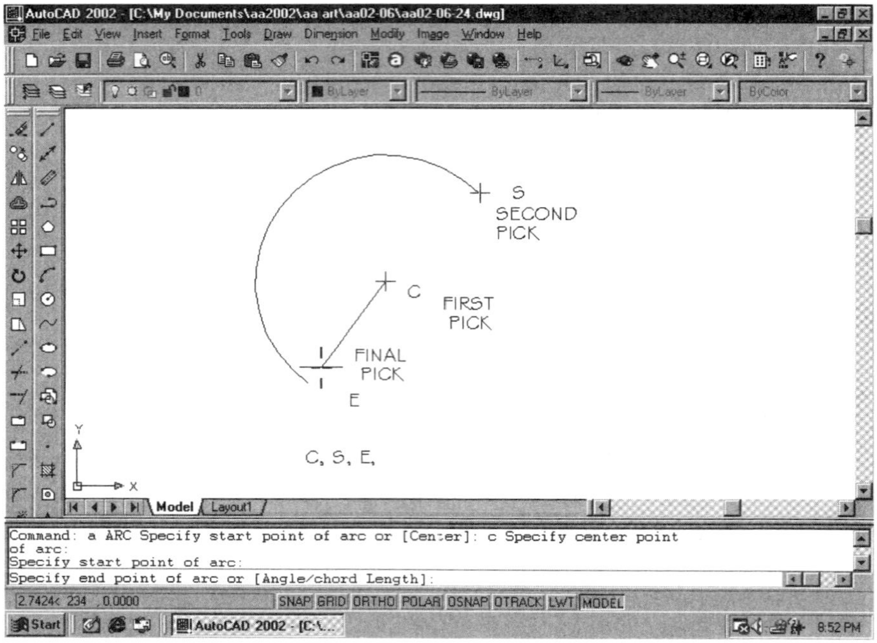

Figure 6.24 *The Center, Start, End arc option is similar to the Start, Center, End option.*

When a length is specified or a point is selected, the arc will be drawn and the command line will return.

CENTER, START, ANGLE ARCS

The Center, Start, Angle option is similar to the Start, Center, Angle option. The effects of each command can be seen in Figure 6.25. The command sequence for this arc is as follows:

Command: **A** (ENTER) *(Or click the Arc button on the Draw toolbar.)*

ARC Specify start point of arc or [Center]: **C** (ENTER)

Specify center point of arc: *(Select center point or enter coordinates.)*

Specify start point of arc: *(Select the start point or enter coordinates.)*

Specify end point of arc or [Angle/chord Length]: **A** (ENTER)

Specify included angle: *(Select the desired angle.)*

Command:

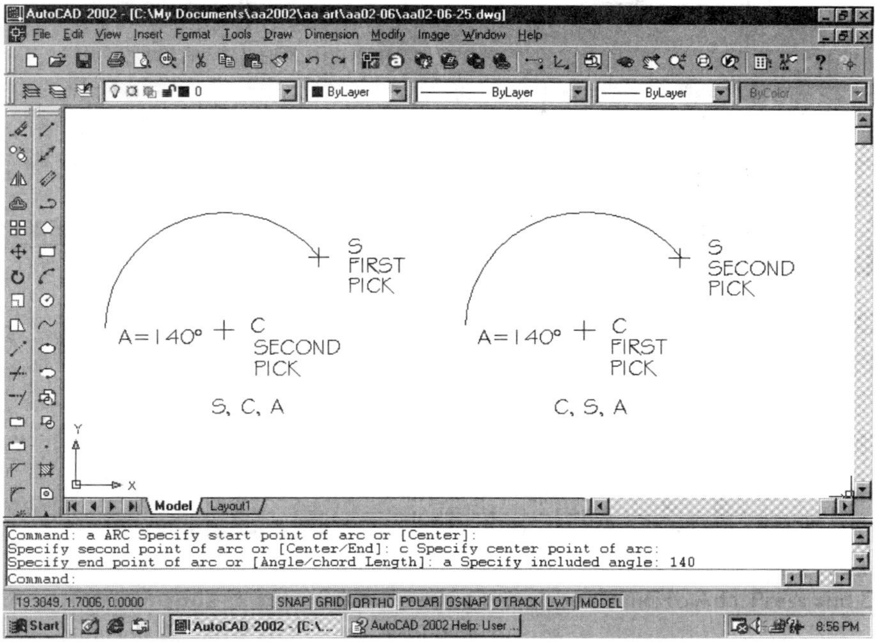

Figure 6.25 *The Center, Start, Angle arc option produces results similar to the Start, Center, Angle option.*

When a length is specified or a point is selected, the arc will be drawn and the command line will return.

CENTER, START, LENGTH OF CHORD ARCS

The Center, Start, Length of chord option is similar to the Start, Center, Length option. The effects of each command can be seen in Figure 6.26. The command sequence is as follows:

Command: **A** (ENTER) *(Or click the Arc button on the Draw toolbar.)*

ARC Specify start point of arc or [Center]: **C** (ENTER)

Specify center point of arc: *(Select center point or enter coordinates.)*

Specify start point of arc: *(Select the start point or enter coordinates.)*

Specify end point of arc or [Angle/chord Length]: **L** (ENTER)

Specify length of chord: *(Select desired length.)*

Command:

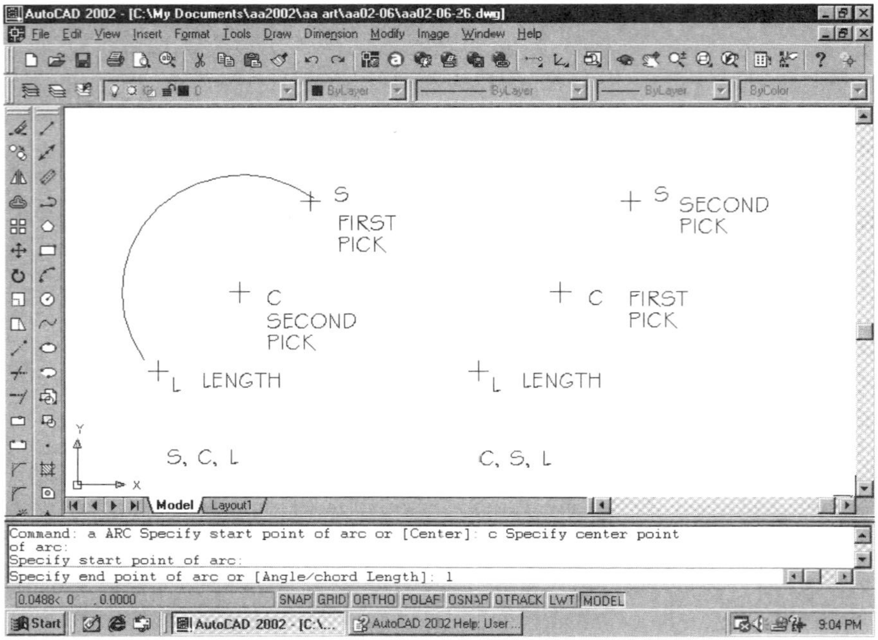

Figure 6.26 *The Center, Start, Length arc option produces results similar to the Start, Center, Length arc option.*

When a length is specified or a point is selected, the arc will be drawn and the command line will return.

CONTINUE

An arc can be continued from the previous arc. The **Arc** command can be continued by pressing (ENTER) or (SPACEBAR), or by selecting **Repeat ARC** from the shortcut menu in lieu of selecting the first arc entry point. When continued from a previous arc, the current arc's start point and direction are taken from the endpoint and ending direction of the last arc. Figure 6.27 shows an example of four continuous arcs. The command sequence is as follows:

ARC 1

Command: **A** (ENTER) *(Or click the Arc button on the Draw toolbar.)*

ARC Specify start point of arc or [Center/End]: *(Select any point or enter coordinate.)*

Specify second point of arc or [Center/End]: *(Select second point.)*

Specify end point of arc: *(Select the arc's endpoint or enter coordinates.)*

Command:

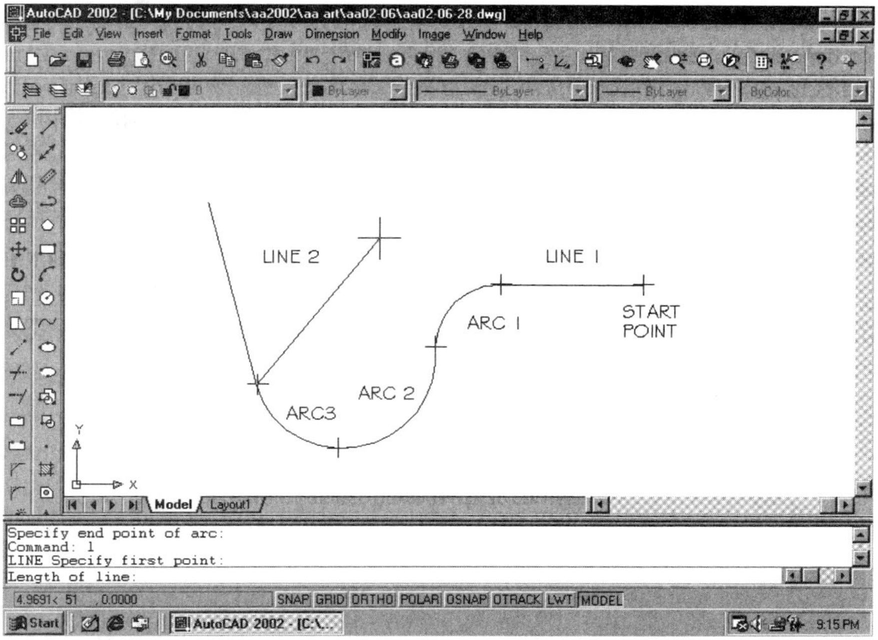

Figure 6.27 *The Continue option of the* **Arc** *command allows for multiple arcs to be drawn.*

ARC 2

Command: (ENTER) *(To resume Arc command.)*

ARC Specify start point of arc or [Center/End]: (ENTER) *(To resume arc from last endpoint.)*

Specify end point of arc: *(Select the arc's endpoint or enter coordinates.)*

Command:

ARC 3

Command: (ENTER) *(To resume Arc command.)*

ARC Specify start point of arc or [Center/End]: (ENTER) *(To resume arc from last endpoint.)*

Specify end point of arc: *(Select the arc's endpoint or enter coordinates.)*

Command:

ARC 4

Command: (ENTER) *(To resume Arc command.)*

ARC Specify start point of arc or [Center/End]: (ENTER) *(To resume arc from last endpoint.)*

Specify end point of arc: *(Select the arc's endpoint or enter coordinates.)*

Command:

The Continue option also can be used to extend an arc tangent to the endpoint of a line. Figure 6.28 shows an example of continuous arcs drawn from a straight line. The command sequence is as follows:

LINE 1

Command: **L** (ENTER) *(Or click the Line button on the Draw toolbar.)*

From point: *(Select starting point or enter coordinates.)*

Specify next point or [Undo]: *(Select point or coordinates.)*

Specify next point or [Undo]: enter

Command:

ARC 1

Command: **A** (ENTER) *(Or click the Arc button on the Draw toolbar.)*

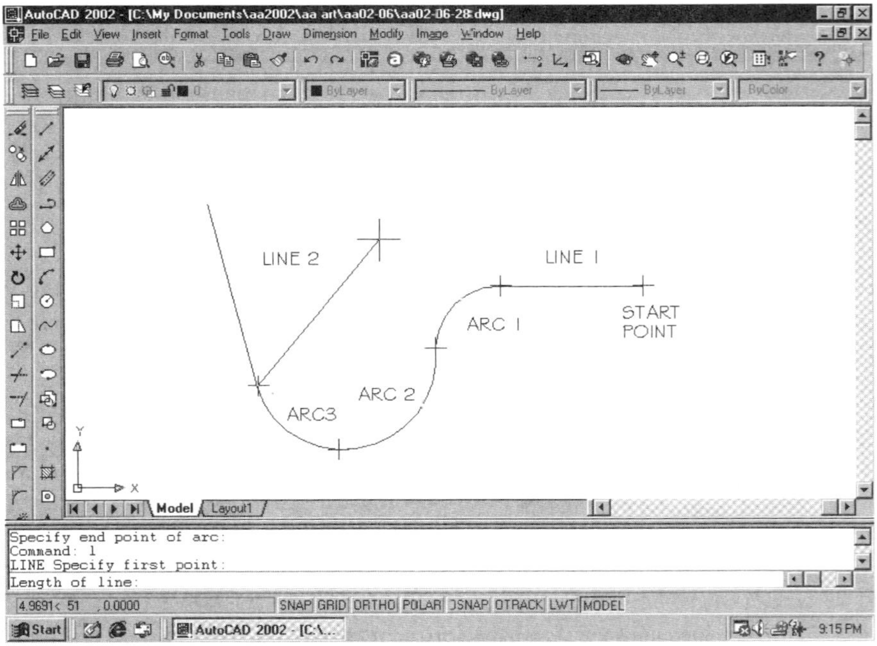

Figure 6.28 *Continue can be used to extend an arc tangent to the last line segment.*

ARC Specify start point of arc or [Center]: (ENTER) *(This will start the arc command at the line endpoint.)*

Specify endpoint of arc: *(Select the arc's endpoint or enter coordinates.)*

Command:

ARC 2

Command: (ENTER) *(This will resume the Arc command.)*

ARC Specify start point of arc or [Center]: (ENTER) *(This will start the arc at the arc endpoint.)*

(ENTER) *(This will start at the end of the last line drawn.)*

Specify endpoint of arc: *(Select the arc's endpoint or enter coordinates.)*

Command:

ARC 3

Command: (ENTER) *(This will resume the Arc command.)*

ARC Specify start point of arc or [Center]: (ENTER) *(This will continue the arc from the endpoint of the last arc.)*

Specify endpoint of arc: *(Select the arc's endpoint or enter coordinates.)*

Command:

LINE 2

Command: *(Select the Line button or type)* **L** (ENTER)

LINE Specify first point: (ENTER)

Specify next point or [Undo]: *(Select point or coordinates.)*

Specify next point or [Undo]: (ENTER) *(To end drawing sequence.)*

Command:

Notice in Figure 6.28 that line 2 was drawn tangent to arc 3 because the Continue option was used. To place a line that continues from an arc but is not tangent to the arc will require the arc endpoint to be selected as the line "next point," using the methods described in Chapter 8.

DRAWING DONUTS

A donut in AutoCAD is a filled circle or ring. Start the command by typing **DO** (ENTER) at the Command prompt, or by selecting **Donut** from the **Draw** menu. Each can be seen in Figure 6.29. Donuts are great when you are drawing the end view of reinforcing steel in detail. The command sequence to create a donut is as follows:

Command: **DO** (ENTER)

DONUT

Specify inside diameter of donut <0'-0?">**0** (ENTER)

Specify outside diameter of donut <0'-1">**.75** (ENTER)

Specify center of donut or <exit>: *(Select the desired location for the donut.)*

Specify center of donut or <exit>: *(The command will continue indefinitely; press* (ENTER) *to end the command.)*

Command:

Three different options are available for drawing donuts: a ring, a filled circle that is solid black, or a circle filled with lines. Each can be seen in Figure 6.30. The fill material is determined by the status of the **Fill** command. Typing **FILL** (ENTER) at the Command prompt will allow for a choice to be made for **Donut** and several other options. If OFF is entered, filled objects will resemble the bottom circle in Figure 6.30. With **Fill** ON, filled objects will resemble the middle circle in Figure 6.30.

Note: Fill will affect several of the commands you will soon be using. In addition to donuts, **Fill** will affect the appearance of polylines, arrows created with **Leader**, and the display of hatch patterns. In older, slower versions of AutoCAD, many users set **Fill** to OFF to a gain speed when working with large drawings. Unless you're trying to reproduce hollow objects, leave **Fill** in the ON setting.

Another thing to remember about **Fill** is that the effect of changing the setting is usually not be seen immediately. To see the effect of **Fill**, enter the command, alter the setting, and press (ENTER). No change shows until you type **REGEN** (ENTER) at the Command prompt. Now the effect of **Fill** can be seen.

USING THE ELLIPSE COMMAND

Although not a major part of engineering drawing, ellipses are sometimes used for construction symbols. Ellipses are also used to show round objects that are not being seen in true orthographic projection, such as a round pipe that is not perpendicular to the viewing plane of the drawing. To start the **Ellipse** command, select the **Ellipse** button on the **Draw** toolbar, select **Ellipse** from the **Draw** menu, or type **EL** (ENTER) at the command prompt. Menu options for **Ellipse** can be seen in Figure

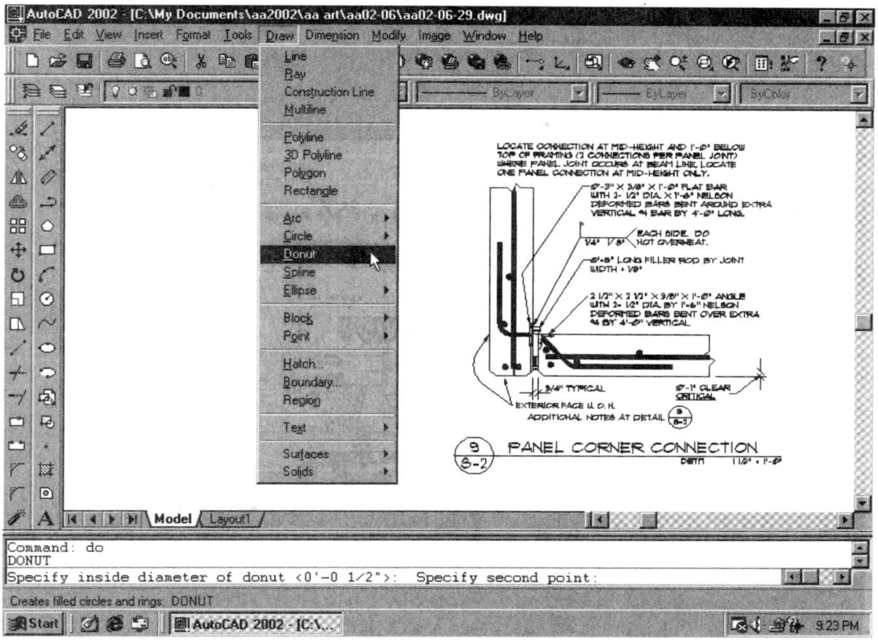

Figure 6.29 *Donuts can be drawn by selection from the* **Draw** *menu or by typing* **DO** *(*ENTER*) at the Command prompt.*

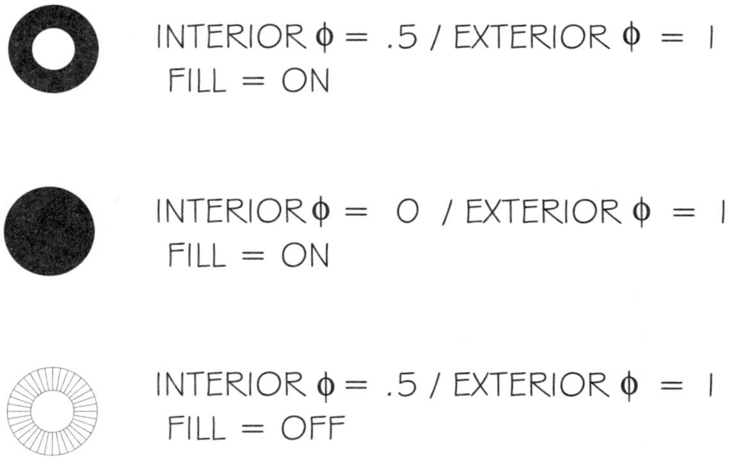

Figure 6.30 *Three options are available for the display of a donut.*

6.31. Figure 6.32 shows the major and minor axes that will be referred to in laying out an ellipse. The command sequence is as follows:

> Command: **EL** (ENTER) *(Or click the Ellipse button on the Draw toolbar.)*
>
> ELLIPSE
>
> Specify axis endpoint of ellipse or [Arc/Center]: *(Select a point or enter coordinates for the major axis endpoint.)*
>
> Specify other endpoint of axis: *(Select a point or enter coordinates for the opposite major axis endpoint.)*
>
> Specify distance to other axis or [Rotation]: *(Select a point or enter coordinates for the minor axis endpoint.)*
>
> Command:

An ellipse is now shown on the screen between the two selected points. The process can be seen in Figure 6.33. AutoCAD is waiting for you to enter the axis or rotation.

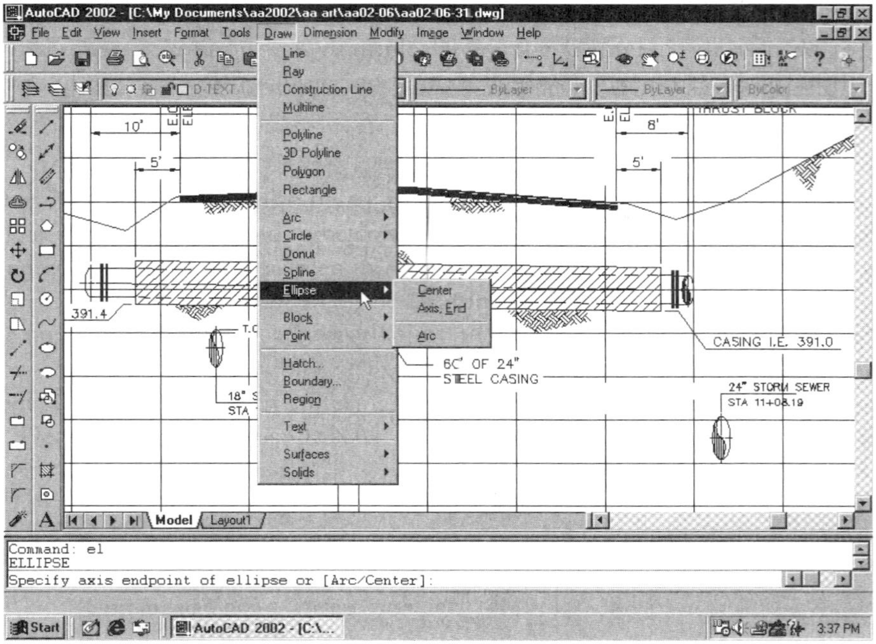

Figure 6.31 *The menu options for Ellipse. (Courtesy Lee Engineering, Inc.)*

If the angle is not known, the crosshairs can be moved until the ellipse represents the desired shape.

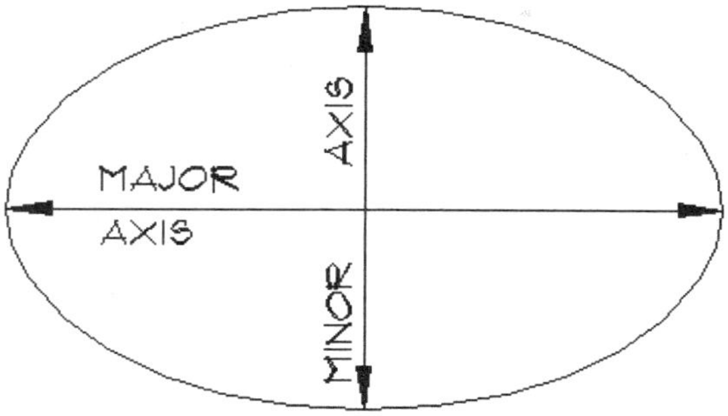

Figure 6.32 *An ellipse is defined by the major and minor axes.*

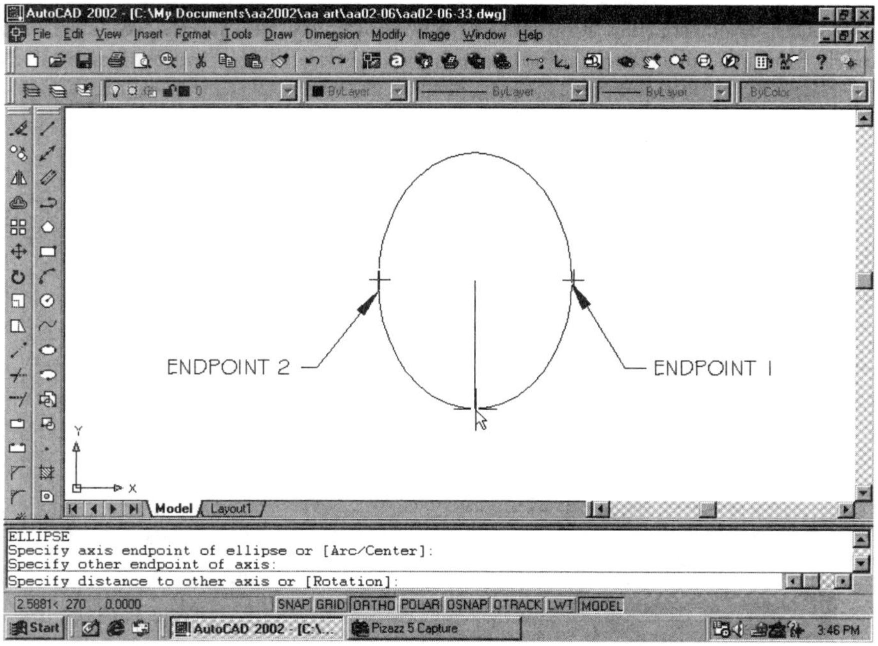

Figure 6.33 *Once the two endpoints are specified, the ellipse will change as the cursor is moved.*

If you know the ellipse represents a pipe that is at a 60° angle to the viewing plane, type **R** (ENTER). The command sequence is as follows:

> Command: **EL** (ENTER) *(Or click the Ellipse button on the Draw toolbar.)*
>
> ELLIPSE
>
> Specify axis endpoint of ellipse or [Arc/Center]: *(Select a point or enter coordinates for the major axis endpoint.)*
>
> Specify other endpoint of axis: *(Select a point or enter coordinates for the opposite major axis endpoint.)*
>
> Specify distance to other axis or [Rotation]: **R** (ENTER)
>
> Specify rotation around major axis: *(Enter desired angle of rotation.)* **60** (ENTER)
>
> Command:

The command sequence will produce the ellipse shown in Figure 6.34. Figure 6.35 shows the effect of rotating a circular pipe at 10° increments. An ellipse with a 90° rota

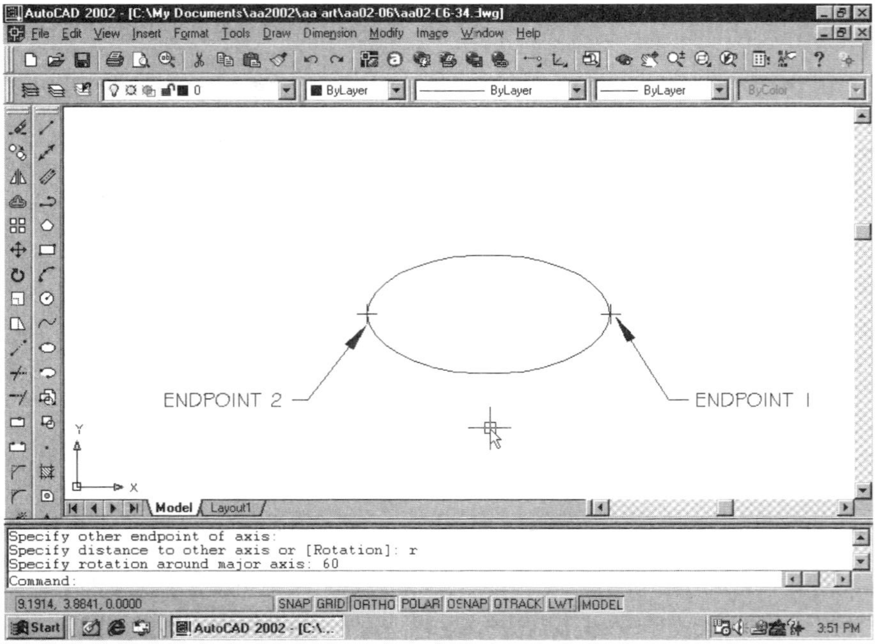

Figure 6.34 *Specifying the angle of rotation of the viewing plane will produce an ellipse. Using a 60° rotation produces the ellipse shown.*

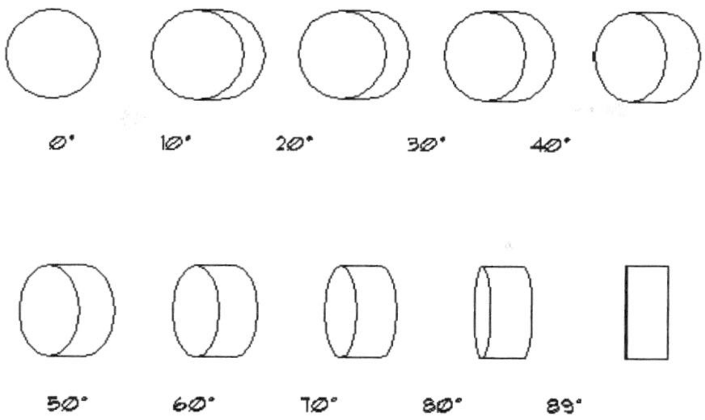

Figure 6.35 *The effects of rotating a circular object in 10° increments.*

tion cannot be drawn using the **Ellipse** command because it would produce a single straight line.

An ellipse also can be drawn by locating the center point and then specifying an endpoint for each axis. This process can be seen in Figure 6.36. The command sequence is as follows:

Command: **EL** (ENTER) *(Or click the Ellipse button on the Draw toolbar.)*

ELLIPSE

Specify axis endpoint of ellipse or [Arc/Center]: **C** (ENTER)

Specify center of ellipse: *(Select a point or enter coordinates for the major axis center point.)*

Specify endpoint of axis: *(Select a point or enter coordinates for the endpoint of major axis.)*

Specify distance to other axis or [Rotation]: *(Select a point or enter coordinates for the minor axis endpoint.)*

Command:

When the second axis endpoint is entered, the command prompt is returned and the ellipse is drawn on the screen.

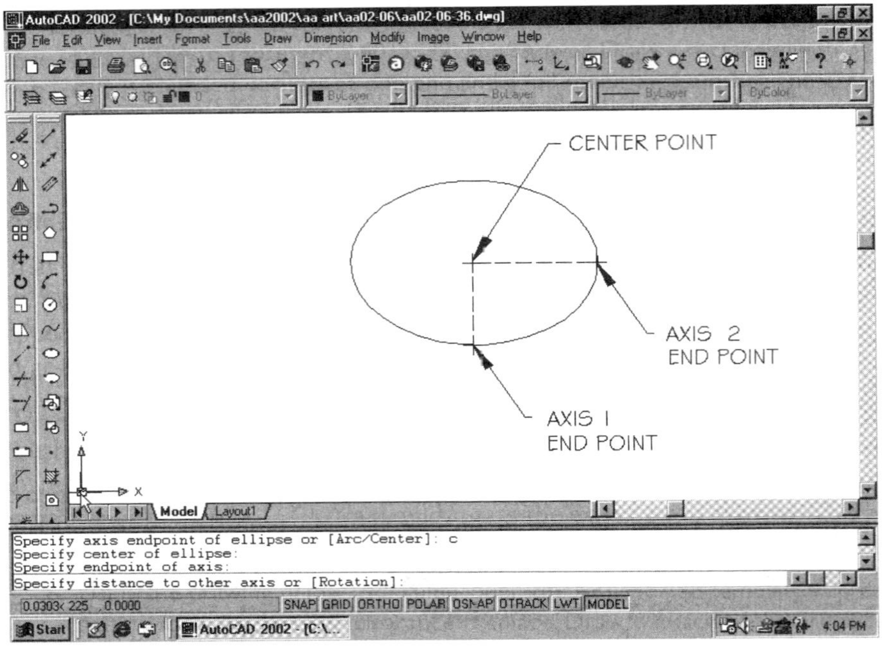

Figure 6.36 *An ellipse can be drawn by locating the center point followed by an endpoint for the major and minor axes.*

DRAWING POLYGONS

A polygon is any geometric shape lying on one plane, bound by three or more straight lines. Triangles, squares, pentagons, hexagons, and octagons are used as symbols and material shapes in construction drawings. AutoCAD draws polygons with equal length lines and angles having between 3 and 1,024 sides. To draw a polygon, you will be asked to choose the number of sides and whether the object is to be inscribed in or circumscribed around a circle. When the polygon is inscribed in a circle, the circle diameter provides the distance across the polygon points. When the polygon is circumscribed around a circle, the circle diameter provides the distance across the flat surfaces of the polygon. The effects of a hexagon inscribed in and circumscribed around a 2.5" diameter circle can be seen in Figure 6.37.

Start the **Polygon** command by selecting the **Polygon** button on the **Draw** toolbar, by selecting **Polygon** from the **Draw** menu, as shown in Figure 6.38, or by typing **POL** (ENTER) at the Command prompt. The command sequence is as follows:

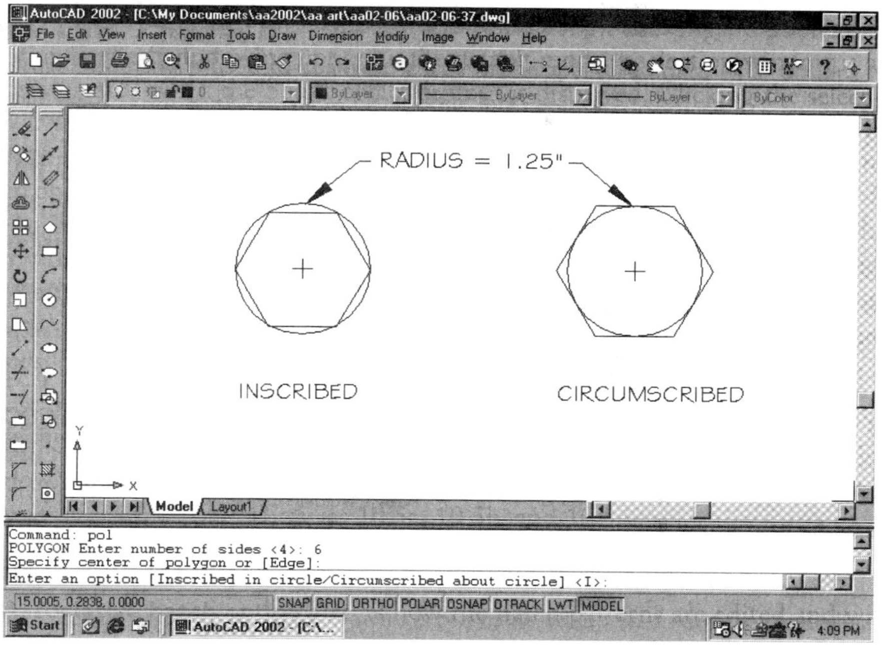

Figure 6.37 *Polygons are drawn based on a circle. Once the number of sides has been specified, you will be asked to specify whether the polygon will be placed inside or outside the circle.*

Command: **POL** (ENTER) *(Or click the Polygon button on the Draw toolbar.)*

POLYGON

Enter number of sides <4>: *(Select the desired number of sides.)* **8** (ENTER)

Specify center of polygon or [Edge]: *(Select a point or enter coordinates for the center point.)*

Enter an option [Inscribed in circle/Circumscribed about circle] <I>: *(Accept I or select C depending on your need and press* (ENTER)*)*

Specify radius of circle: *(Enter desired radius.)* **2.75** (ENTER)

Command:

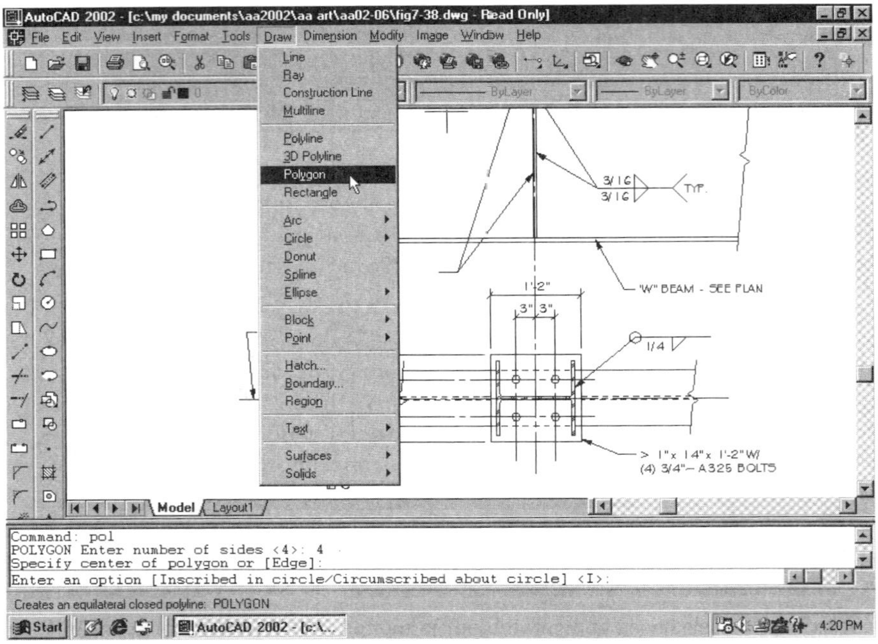

Figure 6.38 *The* **Polygon** *command on the* **Draw** *menu. The command is useful for drawing bolt heads. (Courtesy Lee Engineering, Inc.)*

The result of the sequence can be seen in Figure 6.39. The setting of ORTHO will affect how the polygon will be placed. When ORTHO is ON, the edges of the polygon will be parallel or perpendicular to the drawing screen. When ORTHO is OFF, the edges of the polygon will rotate based on the movement of the mouse. To activate ORTHO, click the ORTHO button on the status bar or press F8.

A polygon also can be drawn based on the location of one edge, rather than the location of the center point. Once the number of sides has been provided, you will be asked for an edge or center point. When **E** (ENTER) in typed at the prompt, a polygon similar to the one in Figure 6.40 can be drawn. The command sequence is as follows:

Command: **POL** (ENTER) *(Or click the Polygon button on the Draw toolbar.)*

POLYGON

Enter number of sides <8>: **6** (ENTER)

Specify center of polygon or [Edge]: **E** (ENTER)

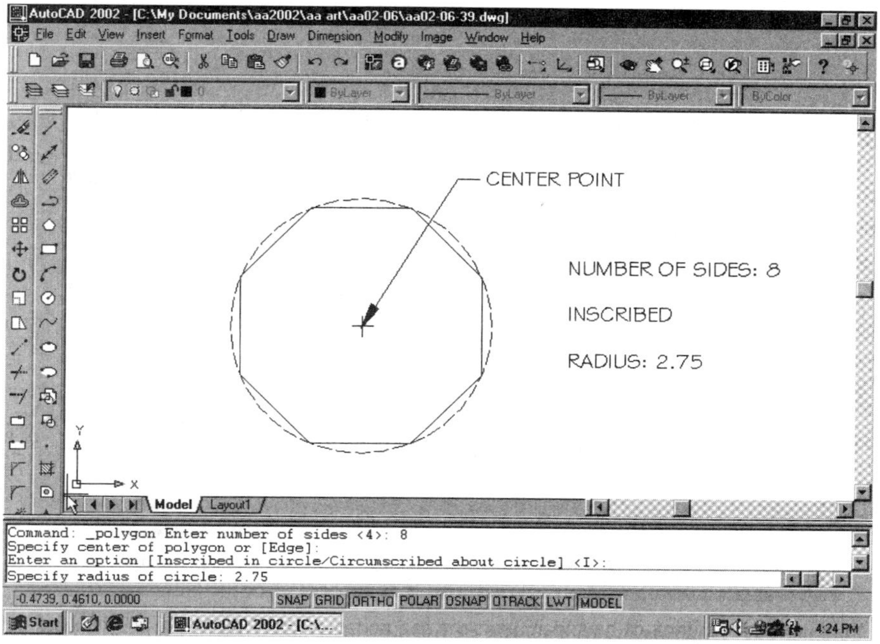

Figure 6.39 *The result of drawing an inscribed octagon with a 2.75" radius.*

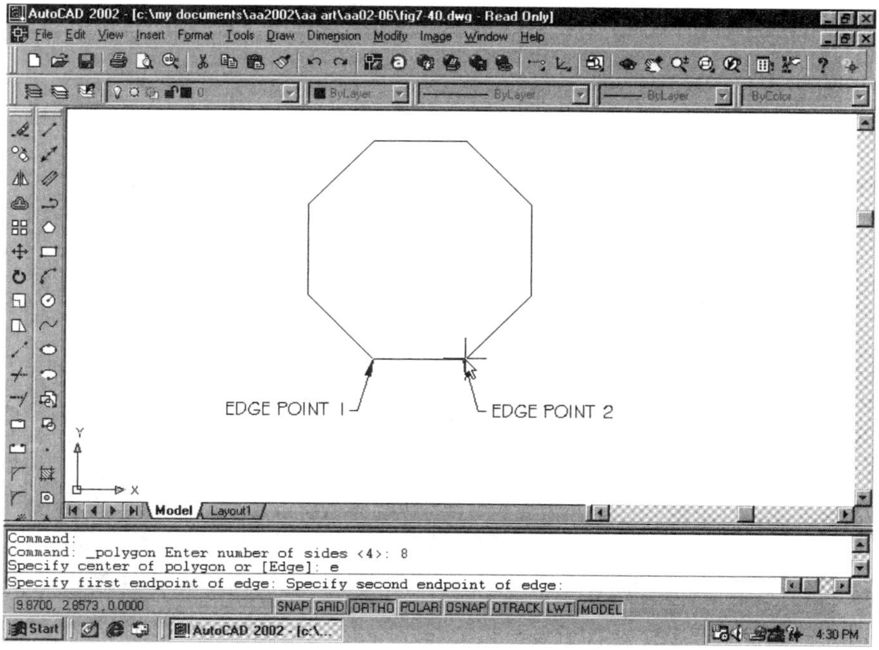

Figure 6.40 *A polygon also can be drawn by specifying the points that define one edge.*

Specify first endpoint of edge: *(Select a point or enter coordinates for the first endpoint.)*

Specify second endpoint of edge: *(Select a point or enter coordinates for the second endpoint.)*

Command:

As the first point is entered, a polygon will be displayed that will vary in size as the mouse is moved to the second point. As the second point is entered, the command prompt is returned and the polygon is drawn. The setting of ORTHO will affect the placement of a polygon using the Edge option.

DRAWING RECTANGLES

A rectangle is drawn by selecting the **Rectangle** button on the **Draw** toolbar, by selecting **Rectangle** from the **Draw** menu, or by typing **REC** (ENTER) at the command prompt. (A square can be drawn using the **Polygon** command.) The command sequence for drawing a rectangle is as follows:

Command: **REC** (ENTER) *(Or click the Rectangle button on the Draw toolbar.)*

RECTANGLE

Specify first corner point [Chamfer/Elevation/Fillet/Thickness/Width]: *(Select a corner of the rectangle with the mouse.)*

Specify other corner point or [Dimensions]: *(Select the opposite corner of the rectangle.)*

Command:

As the cursor is moved between the first and second corner, a rectangle will be displayed on the screen allowing the exact size to be determined. The sequence can be seen in Figure 6.41. Selecting either the Chamfer or Fillet option allows the corners of the rectangle to be altered. Selecting the Width option allows the width of the line to be altered. The other options affect only a rectangle drawn in 3D and will not be discussed in this chapter.

CHAMFER

A chamfer is an angled or mitered corner. Drawing a rectangle with the Chamfer option active will place a chamfer on each corner of the rectangle. Once the Chamfer option is selected, prompts will be given for the first and second chamfer distances. If the distances are equal, the result will be as shown in Figure 6.42. Providing different values will produce a rectangle similar to the example on the right in Figure 6.42. The command sequence to draw a rectangle with a chamfer is as follows:

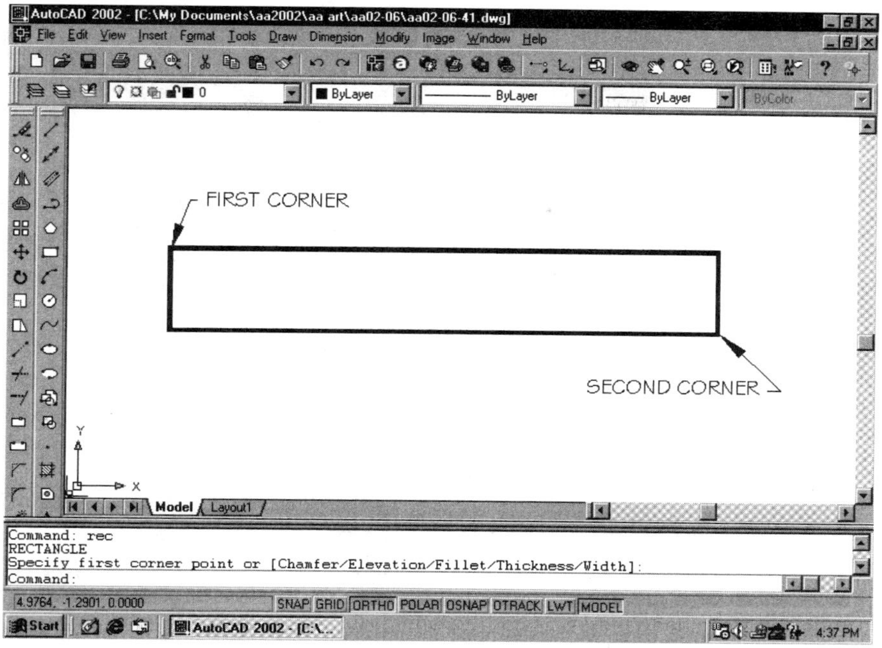

Figure 6.41 *Rectangular shapes can be drawn as one object using the* **Rectangle** *command on the* **Draw** *toolbar, the* **Draw** *menu, or by typing* **REC** *(ENTER) at the Command prompt.*

Command: **REC** (ENTER) *(Or click the Rectangle button on the Draw toolbar.)*

RECTANGLE

Specify first corner point or [Chamfer/Elevation/Fillet/Thickness/Width]: **C** (ENTER)

Select first chamfer distance for rectangles <0'-0">: *(Enter the desired size.)* **.5** (ENTER)

Second chamfer distance for rectangles <0'-01/2 ">: (ENTER)

Specify first corner point or [Chamfer/Elevation/Fillet/Thickness/Width]: *(Select a corner of the rectangle.)*

Specify other corner point or [Dimensions]: *(Select the opposite corner of the rectangle or enter a dimension to define the length.)*

Command:

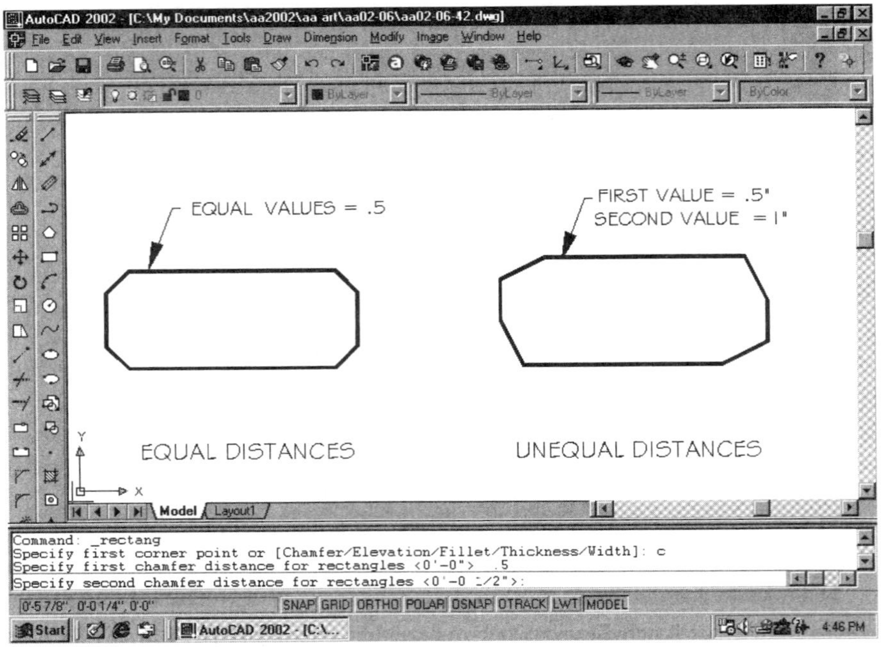

Figure 6.42 *The Chamfer option of the* **Rectangle** *command can be used to provide a chamfer at each corner of a rectangle. The rectangle on the left shows equal values for the chamfers while the rectangle on the right shows unequal values.*

FILLET

A fillet is a rounded corner. Drawing a rectangle with the Fillet option active will place a fillet on each corner of the rectangle. Once the Fillet option is selected, a prompt will be given to provide a radius for the fillet. Figure 6.43 shows an example of a rectangle with filleted corners. The command sequence to draw a rectangle with a fillet is as follows:

Command: **REC** (ENTER) *(Or click the Rectangle button on the Draw toolbar.)*

RECTANGLE

Specify first corner point or [Chamfer/Elevation/Fillet/Thickness/Width]: **F** (ENTER)

Select fillet radius for rectangles <0'-01/2">: *(Enter the desired size.)* (ENTER)

Specify first corner point or

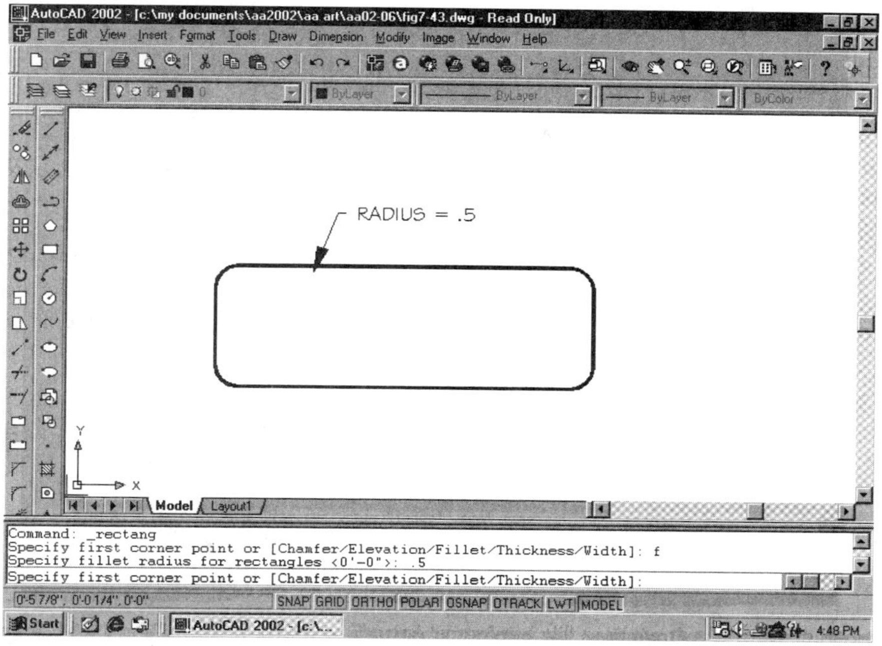

Figure 6.43 *The Fillet option of the* **Rectangle** *command can be used to provide a fillet at each corner of a rectangle. The value used for the radius will remain constant for future use until it is altered.*

[Chamfer/Elevation/Fillet/Thickness/Width]: *(Specify a corner of the rectangle.)*

Specify other corner point or [Dimensions]: *(Select the opposite corner of the rectangle.)*

Command:

Notice that the default radius value is 0'–0 1/2". The value that was provided for the chamfer will become the default value for all future chamfers and fillets until a new value is provided.

WIDTH

Selecting the Width option allows the line thickness to be altered. Because the fillet option was just used, setting the thickness for a rectangle will result in a rectangle with 1/2" radius corners to be drawn with a thick line. To draw a rectangle with square corners, enter the Rectangle command and alter the value for Fillet to 0, and then set the thickness, or vice versa. The command sequence is as follows:

Command: **REC** (ENTER) *(Or click the Rectangle button on the Draw toolbar.)*

RECTANGLE

Specify first corner point or
[Chamfer/Elevation/Fillet/Thickness/Width]: **W** (ENTER)

Specify line width for rectangles <0'-0">: **.25** (ENTER)

Specify first corner point or
[Chamfer/Elevation/Fillet/Thickness/Width]: **F** (ENTER)

Select fillet radius for rectangles <0'-01/2">: **0** (ENTER) *(The desired size entered.)*

Specify first corner point or
[Chamfer/Elevation/Fillet/Thickness/Width]: *(Specify a corner of the rectangle.)*

Specify other corner point or [Dimensions]: *(Select the opposite corner of the rectangle.)*

Command:

The results of this command sequence can be seen in Figure 6.44. The width will remain the current width for all future rectangles until the value is altered. Methods of providing thickness to other entities will be discussed as polylines are introduced.

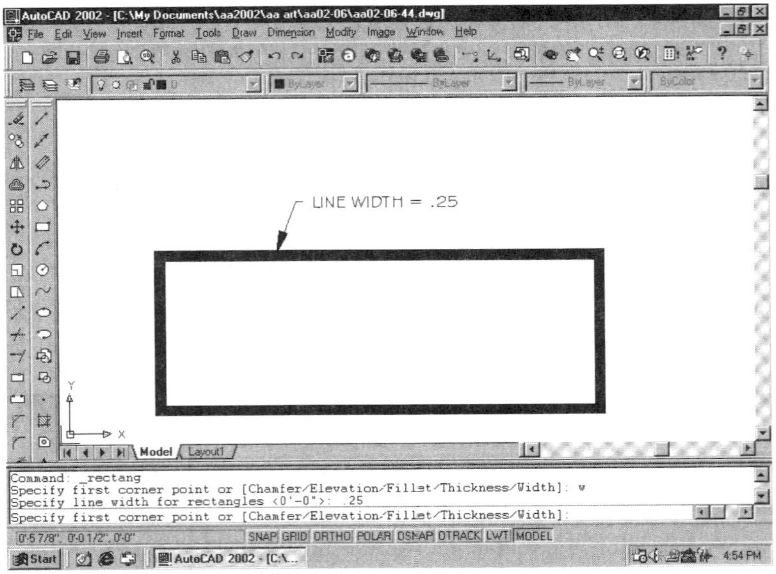

Figure 6.44 *The Width option can be used to assign a line width to the* **Rectangle** *command.*

CHAPTER 6 EXERCISES

Save the following exercises using the indicated file names.

1. Draw a 3" diameter circle, a circle with a 1" radius, and a 1" diameter circle that is tangent to two 1" long perpendicular lines. Make the smallest circle green with continuous lines. Make the largest circle red with hidden lines. Give one of the lines a width of 0.065". Save the drawing as E-6-1.
2. Draw two parallel lines that are 1" long and 1" apart, and provide an arc at each end of the lines to form a slotted hole. Draw these objects on a layer named SLOTS, using dashed blue lines, with a thin lineweight. Save the drawing as E-6-2.
3. Draw a 4" long line. Use the continuous option to place a 1" diameter circle at each end of the line. Save the drawing as E-6-3.
4. Draw a box formed by three 1" long lines. Draw an arc to close the polygon. Save as drawing E-6-4.
5. Draw an arc with a start point that is 2" to the right of the endpoint with an included angle of 232°. Save as drawing E-6-5.
6. Draw an arc with an endpoint that is @2,2 from the center point with an included angle of 45°. Save as drawing E-6-6.
7. Draw an ellipse with a 3" long axis at a 42° rotation. Save as drawing E-6-7.
8. Draw an ellipse with a center point that is 2.577 inches from the end of the major axis and 1" from the minor axis. Draw a four-sided polygon using the center of the ellipse as the center of the polygon. Make two edges of the square touch the ellipse. Save as drawing E-6-8.
9. Draw the following polygons:

3 sides	inscribed	2" diameter
3 sides	circumscribed	2" diameter
4 sides	inscribed	0.5" radius
4 sides	circumscribed	0.5" radius
6 sides	inscribed	0.75" radius
6 sides	circumscribed	0.75" radius
8 sides	inscribed	1.25" radius
8 sides	circumscribed	1.25" radius

Save the drawing as E-6-9.

10. Draw a donut with an inside diameter of 0.25", outside diameter of 0.75", and **Fill** set at ON. Draw another donut with an inside diameter of 0.0" and an outside diameter of 0.375" with **Fill** set at OFF. Draw a donut with an inside diameter of 0 and an outside diameter of 0.5" with **Fill** set to ON. Save the drawing as E-6-10.
11. Draw a rectangle with filleted corners using a 2"radius. Save the drawing as E-6-11.
12. Draw a rectangle with chamfered corners using a 1" distance for both settings. Save the drawing as E-6-12.
13. Draw a rectangle with chamfered corners using a 1.5" distance for the first distance and a 2.5 value for the second setting. Assign a line width of 0.065 and save the drawing as E-6-13.
14. Start a new drawing and draw at least three connected combinations of lines and arcs. Save the drawing as E-6-14.
15. Draw two perpendicular lines that are 2" long and form a corner. Draw a 3" diameter circle that is tangent to each of the lines. Draw three lines that are not connected and are neither parallel nor perpendicular to each other. Draw a circle that is tangent to all three line segments. Save this drawing as E-6-15.

CHAPTER 6 QUIZ

1. What is the maximum number of sides that can be drawn for a polygon?
2. When the **Arc** command is used, what do the following letters represent?

St	Ce	D
Len	E	R
Ang		

3. What is the default method for drawing arcs from the toolbar?
4. What is the command sequence to draw three continuous arcs?
5. What command determines whether the interior of a donut will be black or hatched with lines?
6. What is the difference between inscribed and circumscribed, and what are the effects of each on a polygon?
7. What four components can be used to construct an ellipse?
8. When would TTR be used for drawing a circle?
9. What are the options available for drawing circles?
10. What command controls the rubber-band effect of geometric shapes?
11. What options must be entered before a donut can be drawn?
12. What is the largest angle rotation that can be used with an ellipse?
13. What shape would a polygon with 500 sides appear as?
14. How does ORTHO affect the drawing of a polygon?
15. Use the **Help** menu and determine the use of Elevation as it relates to a rectangle.

CHAPTER 7

Controlling Drawing Accuracy

This chapter will introduce methods of controlling drawing accuracy by

- Controlling the aperture box
- Controlling drawing settings using the **Drafting Settings** dialog box
- Effectively using Object Snap options

Commands and options introduced in this chapter include

- **Aperture**
- **Dsettings**
- **OSNAP**

Up to this point, you've been creating drawing objects with little regard to drawing accuracy. In previous chapters, you were introduced to using the tools on the status bar to make the crosshairs move at specific intervals. This included the default settings of Object Snap. As you begin to combine geometric shapes into drawings, it is often helpful to join the shapes at exact locations. Joining objects by eye is difficult and should rarely be done. Objects that appear to be tangent might overlap when the **Zoom** command is used (see Figure 7.1). Perfect intersections can be achieved by using the Object Snap modes that were introduced in Chapter 3.

CONTROLLING THE APERTURE BOX

The normal cursor for the drawing area is the crosshairs surrounded by a box. If you enter a drawing command such as **Line**, the pick box is removed from the crosshairs as you select the first and next points. As OSNAP is activated, an aperture box is added to the crosshairs, as seen in Figure 7.2. The marker is displayed as the cursor is moved near one of the active Object Snap points such as an endpoint or midpoint. The shape of the marker will vary depending on the mode being used. Default settings were introduced in earlier chapters, and various Object Snap modes will be intro-

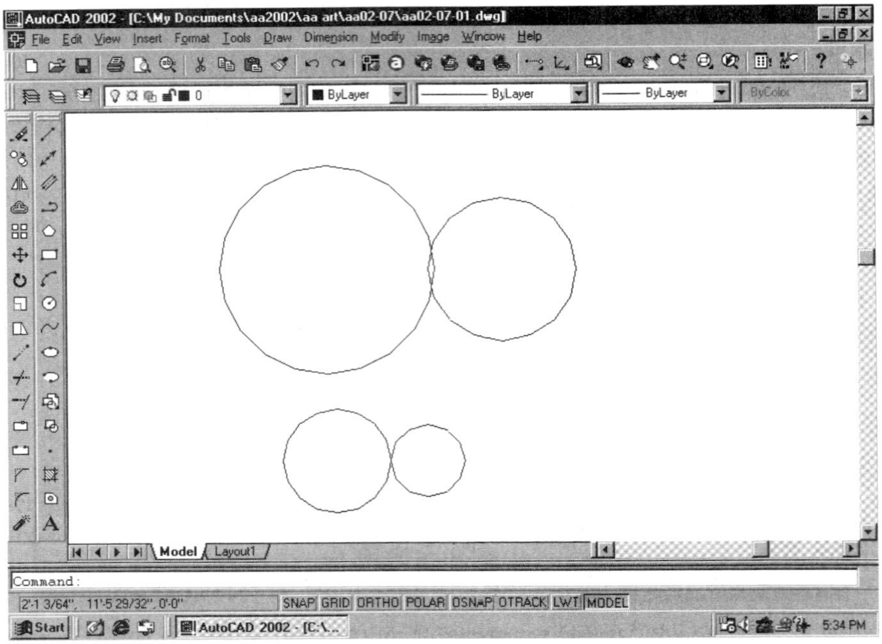

Figure 7.1 *Objects such as the lower circles that are placed by eye often appear to be perfectly placed. As the drawing is magnified using the Zoom command, circles that appear to be tangent are crossing.*

duced later in this chapter. Any object that lies within the target or aperture box is subject for the drawing or editing command being used. Although objects can be added or deleted from within the aperture box, that is time consuming. To obtain higher accuracy in object selection, you can adjust the size of the aperture box.

Note: The aperture box will not be displayed unless the **Display AutoSnap aperture box** check box is activated. Even if it is not displayed, an invisible aperture box will still function as you draw. The aperture box is displayed on the **Drafting** tab of the **Options** dialog box. To display the dialog box, select **Options** from the **Tools** menu.

ALTERING THE SIZE

The size of the aperture box can be adjusted using the **Drafting** tab of the **Options** dialog box or by keyboard. Access the **Options** dialog box by selecting **Options** from the **Tools** menu and select the **Drafting** tab. This will produce the dialog box shown in Figure 7.3. Select the slide bar in the **Aperture Size** area and hold the select button down. Dragging the bar to the left will decrease the size. Moving the bar to the right will increase the size. The size will be indicated in the Aperture Size window display.

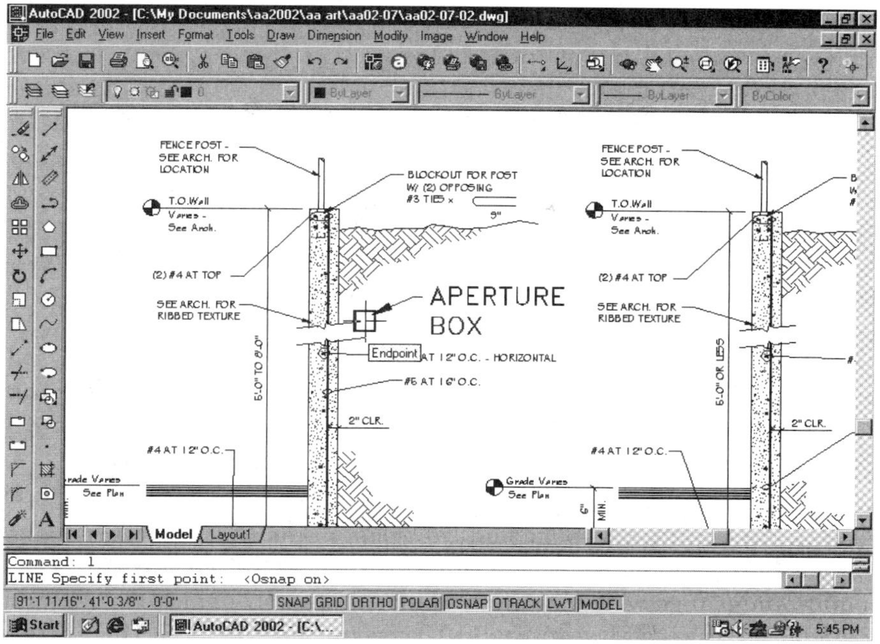

Figure 7.2 *As Object Snap is activated, an aperture box is added to the crosshairs. The box is used for selecting existing drawing objects to be used as a selection point for a new line, circle, arc, or ellipse. (Courtesy Van Domelen/Looijenga/McGarrigle/Knauf Consulting Engineers.)*

To adjust the size of the aperture box at the Command prompt, type

Command: **APERTURE** (ENTER)

Object snap target height (1-50 pixels) <10>:

A numeric value can now be entered. The larger the number of pixels selected, the larger the box. Figure 7.4 shows the effect of changing the value. The value entered will now be the default for all future Object Snap commands until a new value is entered. Changing the default for the Object Snap aperture box will not affect the size of the select box used for **Erase**.

ALTERING THE APERTURE COLOR

In addition to adjusting the size of the box, you can adjust the aperture color. Select **Options** from the **Tools** menu and click the **Display** tab of the **Options** dialog box. Select the **Color** button. This will display the **Color Options** dialog box similar to Figure 7.5. Select the list arrow beside the **Window Element** display. Selecting the **Model Tab Pointer** option will display a color menu. Once the desired color has been selected, clicking the **Apply & Close** button will change the crosshairs and box to reflect the selected color.

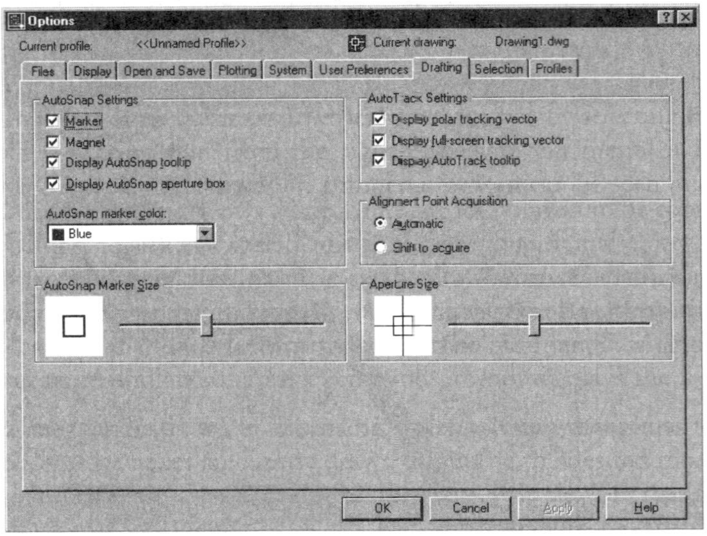

Figure 7.3 *The* **Options** *dialog box is accessed by selecting* **Options** *from the* **Tools** *menu. The dialog box can be used to adjust the aperture box, the marker size and color, AutoSnap settings, and AutoTrack settings.*

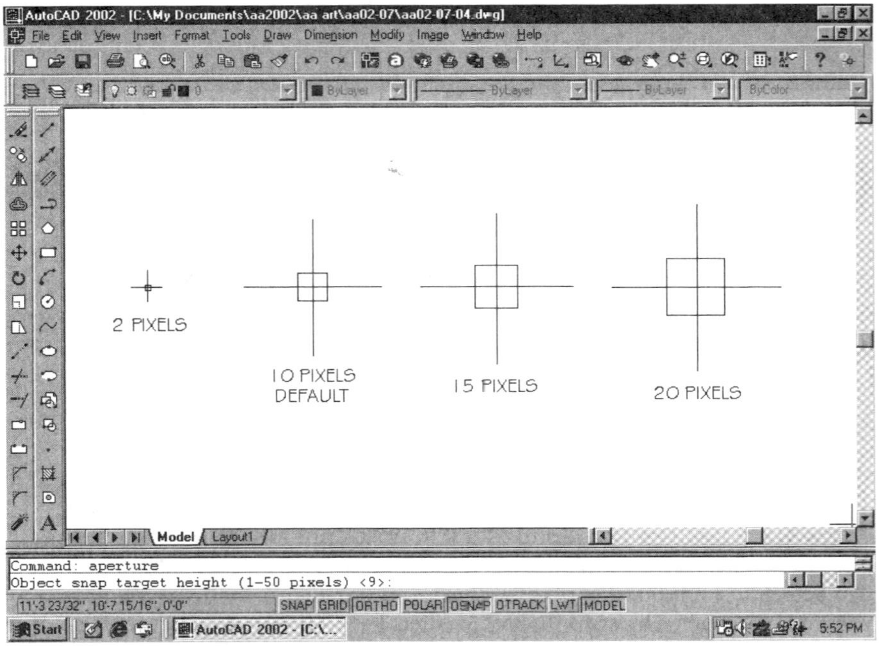

Figure 7.4 *As the value for the height of the aperture box is altered, the height is changed.*

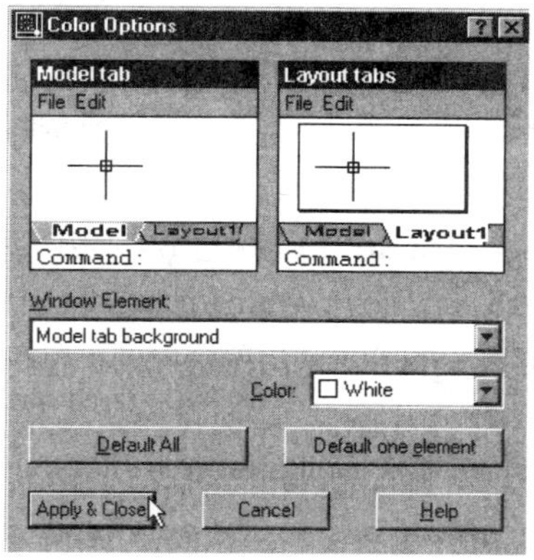

Figure 7.5 *The* **Color Options** *dialog box can be used to adjust the marker color. Access the dialog box by choosing* **Options** *from the* **Tools** *menu, selecting the* **Display** *tab, and clicking the* **Color** *button.*

CONTROLLING DRAWING SETTINGS

To be used effectively, Object Snap settings need to be understood and put to use. Major areas to be considered include the **AutoSnap settings** and **AutoTrack Settings**. Each can be set using the Options dialog box shown in Figure 7.3.

AUTOSNAP SETTINGS

The AutoSnap settings control the settings for each Object Snap mode. Settings include the marker, magnet, and tooltip.

Marker

As an Object Snap command is used, the marker for the current Object Snap mode will be displayed by the aperture box. AutoCAD uses a different geometric shape to represent each Object Snap mode. Specific markers will be introduced later in this chapter. With the **Marker** check box activated, each time the cursor is moved to a point controlled by Object Snap, the appropriate Object Snap marker will be displayed. For example, when you use the **Line** command, each time the cursor is moved toward the end of a line, the Endpoint marker is added to the cursor to indicate that the endpoint will be selected. If you deactivate this setting, the marker box will not be displayed, but the current Object Snaps will be unaffected. AutoCAD has it ON by default for a reason. It is usually very helpful for new users to be reminded of the current settings.

Altering the Marker Size

The size of the Object Snap marker can be adjusted using the **Drafting** tab of the **Options** dialog box. Select the slider bar in the **AutoSnap Marker Size** area and hold the select button down. Dragging the bar to the left will decrease the marker size. Moving the bar to the right will increase the marker size. The size will be indicated in the **AutoSnap Marker Size** window display.

Altering the Marker Color

Selecting the arrow in the **AutoSnap marker color** edit box will display a menu of the seven basic drawing colors. Selecting a color name will automatically change the marker to that color and close the color menu.

Magnet

The **Magnet** check box is a toggle for the ON/OFF setting. By default, AutoCAD magnet is ON. As you've used a drawing command, the cursor often moved toward the end of a line. That is the effect of having the magnet ON. The magnet automatically moves the cursor to lock it onto the nearest Object Snap point as the cursor is moved. Keeping the magnet on will increase your drawing speed.

Tooltip

The **Display AutoSnap tooltip** check box is a toggle for the ON/OFF setting. It is ON By default. Each time the cursor is moved near an active Object Snap mode, the

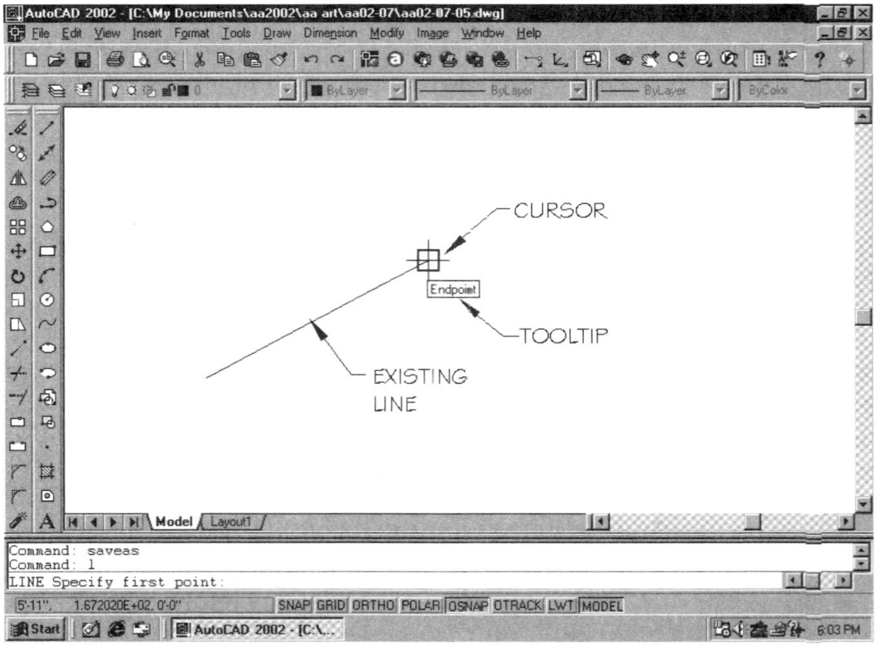

Figure 7.6 *Each time the cursor is moved near an active Object Snap mode, a tooltip below the cursor shows the Object Snap mode to be used.*

Object Snap mode to be used will be displayed in a tooltip below the cursor. Figure 7.6 shows the tooltip for the Endpoint Object Snap mode. On simple drawings, the tooltip might not seem important. On complex drawings with several Object Snap modes active, the tooltip can be helpful in determining what Object Snap mode will be used. Once you recognize each of the Object Snap markers, you might find the tooltips unnecessary.

AUTOTRACK SETTINGS

The AutoTrack settings control the settings that can enhance each Object Snap mode. Settings include the polar tracking vector, full-screen tracking vector, and AutoTrack tooltip displays. Each option is an ON/OFF toggle switch. By default, each is set to ON.

Polar Tracking Vector

If Polar Tracking is ON, the **Display polar tracking vector** check box, when activated, sets Polar Tracking so that lines can be drawn along angles relative to the drawing "first" or "next" points. Angle settings must be 90° divisors, such as 15°, 30°, and 45°. Tracking increments are set on the **Polar Tracking** tab of the **Drafting Settings** dialog box (see Figure 7.7a). You can access the dialog box by selecting **Drafting Settings** from the **Tools** menu or by typing **OS** (ENTER) or **DSETTINGS** (ENTER) at the Command prompt. With this option ON, tracking angles will be displayed similar to Figure 7.7b. The polar tracking display can be toggled ON/OFF using the POLAR button on the status bar.

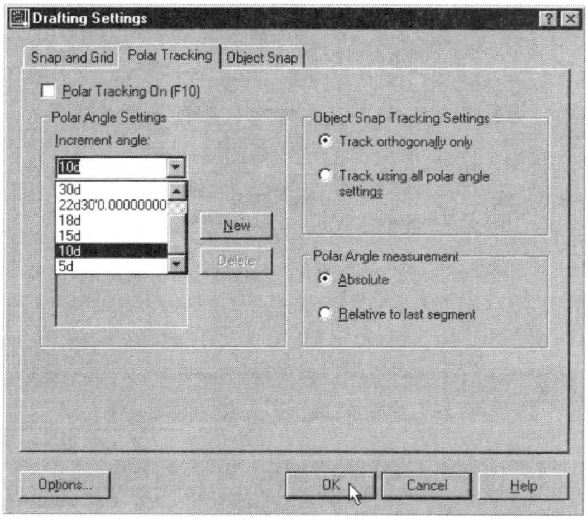

Figure 7.7a *Tracking increments can be set using the* **Increment angle** *edit box on the* **Polar Tracking** *tab of the* **Drafting Settings** *dialog box.*

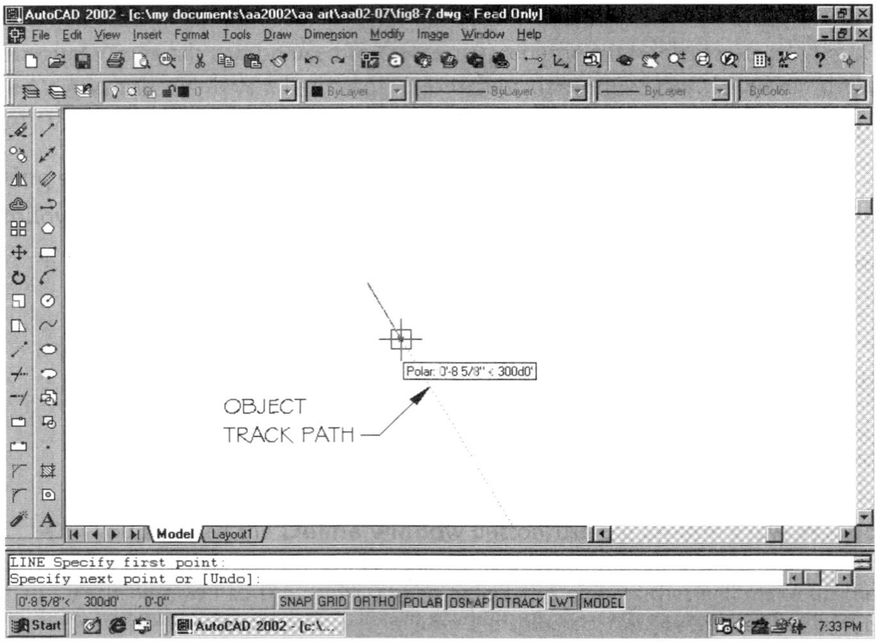

Figure 7.7b *If the Polar Tracking option is ON, tracking angles will displayed below the cursor.*

Full-Screen Tracking Vector

The **Display full-screen tracking vector**check box controls the display of tracking vectors. Tracking vectors are construction lines that can be used as a base, enabling objects to be drawn at a specific angle (see Figure 7.7b). They can also be used to draw objects at a specific relationship to other objects. In their default setting, tracking vectors extend to the edge of the drawing display area. Alignment paths can be shortened or eliminated. When this check box is not selected, the path is displayed only from the Object Snap point to the cursor. Controls for setting Object Snap tracking are found on the **Polar Tracking** tab of the **Drafting Settings** dialog box.

AutoTrack Tooltip

The AutoTrack tooltip is similar to the AutoSnap tooltip. The **Display AutoTrack tooltip** check box controls the display of the tracking coordinates.

OBJECT SNAP

Object Snap modes can be used to select an exact start or endpoint for connecting lines, rather than hoping you're close. Object snaps allow a line or other object to be snapped to a specific point. For instance, Object Snap will allow a line to be started exactly at the midpoint of an existing line, with absolutely no guessing. Object Snap has two modes of operation: single-point override and continuous. The continuous mode allows modes to be set and used throughout the drawing session. The override mode allows you to select an Object Snap mode to be used once in the middle of a command.

By default, Endpoint, Midpoint, Center, Intersection, and Extension are set on the ON position. The desired override Object Snap can be activated by selecting its button on the **Object Snap** toolbar, by selecting its button from the **Temporary Tracking** flyout menu of the **Object Properties** toolbar, by pressing SHIFT and right-clicking, or by keyboard. Display the toolbar by selecting **Object Snap** from **Toolbars** on the **View** menu. Figure 7.8 shows a listing of each Object Snap mode found on the **Object Snap** toolbar and the marker that will be displayed with each mode. An Object Snap mode can be used in the middle of a command by entering three letters to represent the mode for the "next" point of a command sequence. For instance, to draw a line that is perpendicular to an existing line would require the following sequence:

Command: **L** (ENTER) *(Or click the Line button.)*

LINE

Specify first point: *(Select any point for the starting point of the new line).*

Specify next point or [Undo]: **PERP** (ENTER) *(Or click the Perpendicular button.)*

to: *(Select an endpoint for the starting point of the new line).*

Specify next point or [Undo]:

Continuous Object Snap allows the Object Snap mode or modes to be selected for continuous use. Continuous Object Snap options are set on the **Object Snap** tab of the **Drafting Settings** dialog box. (Choose **Drafting Settings** from the **Tools** menu or type **OS** (ENTER) at the Command prompt to access the dialog box.) Continuous Object Snap will be discussed later in this chapter. No matter how they are accessed, using the Object Snap commands will save time and greatly improve your accuracy. Once OSNAP is activated, the marker box is used to select objects to snap to. Move the crosshairs so that the desired object lies within the marker box, and then click the

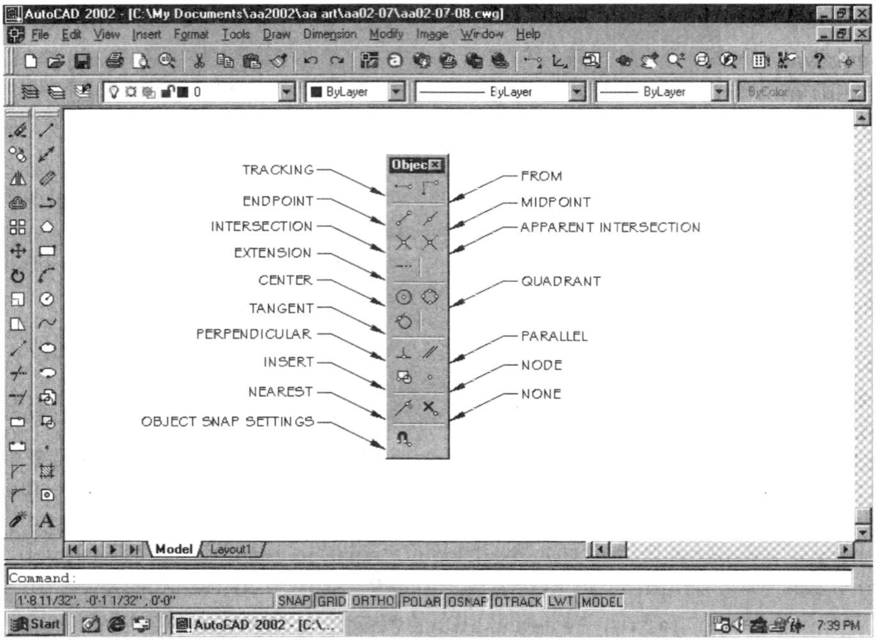

Figure 7.8 *The* **Object Snap** *toolbar and the marker box for each mode.*

select button. If more than one object lies within the marker box, the closest object will be selected.

OBJECT SNAP MODES

Object Snap tools include ENDpoint, MIDpoint, INTersection, EXTension, APParent Intersection, CENter, NODe, QUAdrant, INSertion, PERpendicular, TANgent, NEArest, PARallel, QUIck, and NONe. These modes can be used for drawing lines or arcs that connect at certain points to existing lines, circles, arcs, or ellipses. Object Snap modes also can be used any time a point needs to be specified, and for **Copy**, **Move**, and **Insert** commands, which will be discussed in later chapters. Notice in the modes list that the first three letters of some of the modes are written in capital letters. Typing these letters at the prompt for start point or endpoint will activate the desired Object Snap mode.

ENDPOINT

The Endpoint mode allows an arc, line, or circle to be drawn to or from the endpoint of a previous line or arc by snapping to the desired endpoint. Start the process by selecting the **Endpoint** button on the **Object Snap** toolbar when you are prompted for a "first point." To add a line that extends between the endpoints of two existing lines, the command sequence would be as follows:

> Command: **L** (ENTER) *(Or click the Line button.)*
>
> LINE Specify first point: *(Select the Endpoint button.)*

Now move the cursor near the end of the line you would like the new line to connect to. The cursor does not have to touch the end of the line, but it must be between the midpoint and endpoint so the correct end of the line is selected. As the cursor touches a line, the square marker box is displayed around the nearest endpoint. (See Figure 7.9.) Click the select button, and the new line segment will be connected to the selected endpoint. Select the **Endpoint** button again at the prompt for the "next point," to connect the new line to the end of the second line. The new line will automatically end at the selected end of the existing line segment.

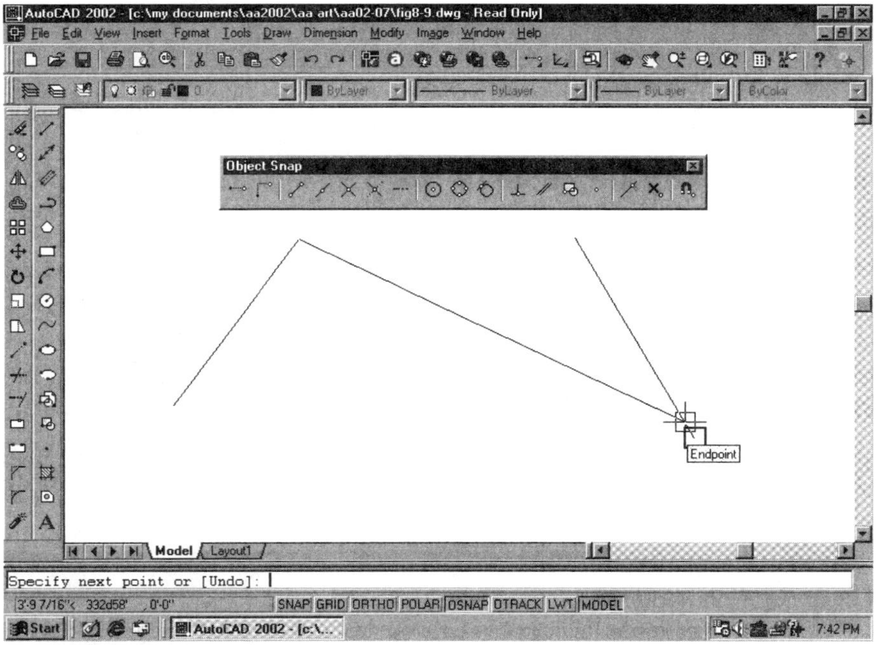

Figure 7.9 *When the Endpoint mode of Object Snap is selected, a marker is placed on the endpoint of the selected line to indicate the selected "first" or "next" point.*

The endpoint can also be selected at the keyboard, by typing **END** (ENTER) as the "first point" of the desired command. Move the marker box and select the desired line. To draw a line from the endpoint of an existing line to the endpoint of a second line, use the following command sequence:

Command: **L** (ENTER) *(Or click the Line button.)*

LINE Specify first point: **END** (ENTER) *(Or click the Endpoint button.)*

_end of *(Select desired line.)*

Specify next point or [undo]: **END** (ENTER) *(Or click the Endpoint button.)*

of *(Select desired line.)*

Specify next point or [undo]: (ENTER)

Command:

The command sequence can be seen in Figure 7.10. If you're thinking, "This is extra work," you're right. By default, AutoCAD automatically draws lines from endpoint to endpoint. However, the process will work wonders as different Object Snap modes are entered.

MIDPOINT

Midpoint allows the selection of a line or arc at its midpoint. The **Midpoint** button is selected in response to the "first point" prompt. To add a line that extends between the midpoints of two existing lines, the command sequence would be as follows:

Command: **L** (ENTER) *(Or click the Line button.)*

LINE

Specify first point: *(Select the Midpoint button on the Object Snap toolbar.)*

Now move the cursor near the line that you would like the new line to connect to. As the cursor touches a line, the marker box is displayed around the nearest midpoint. Click the select button, and the new line segment will be connected to the selected midpoint of the existing line. Select the **Midpoint** button as the "next point" is requested at the prompt and the new line will automatically end at the selected midpoint of the existing line segment.

The midpoint for an arc or line can also be selected by typing **MID** (ENTER) as the "first point" or "next point" prompt is displayed. To draw a line from the midpoint of a line to the midpoint of a second line, use the following command sequence:

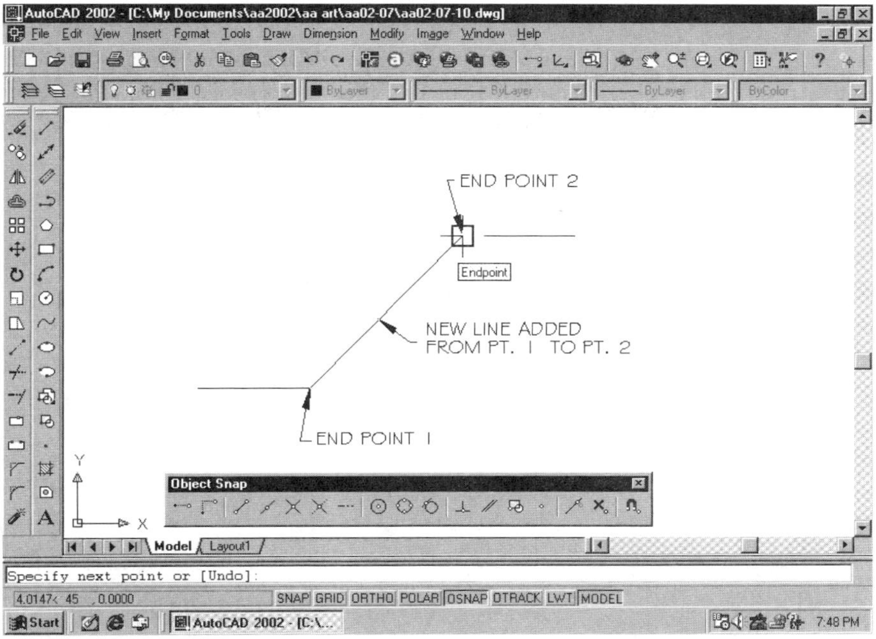

Figure 7.10 *To draw a line from the endpoint of one line to the endpoint of another, click the* **Endpoint** *button or type* **END** *(ENTER) when the "first point "and the "next point" prompts of the* **Line** *command are displayed.*

Command: **L** (ENTER) *(Or click the Line button.)*

LINE

Specify first point: **MID** (ENTER) *(Or click the Midpoint button.)*

_mid of *(Select desired line.)*

Specify next point or [Undo]: **MID** (ENTER) *(Or click the Midpoint button.)*

of *(Select desired line.)*

Specify next point or [Undo]: (ENTER)

Command:

The command sequence can be seen in Figure 7.11.

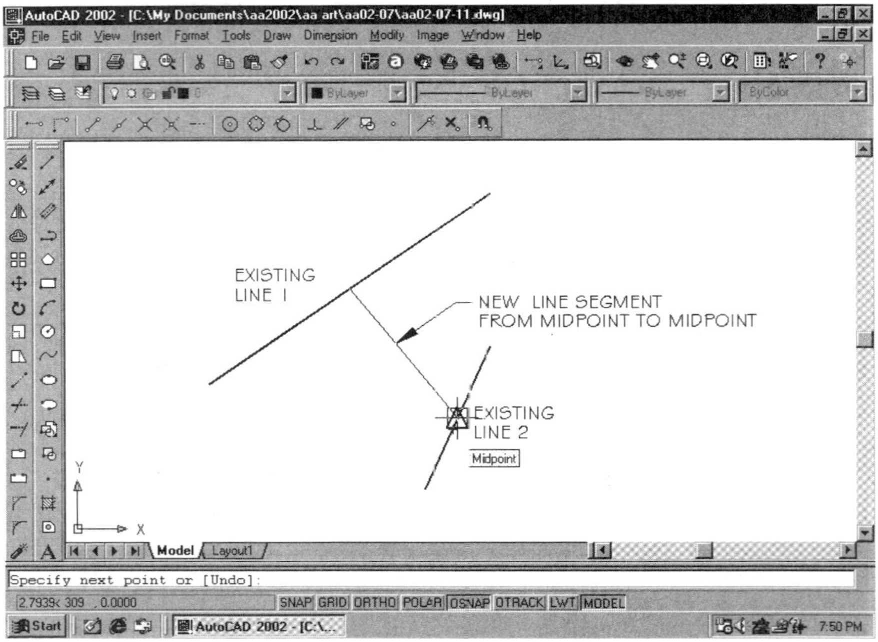

Figure 7.11 *The Midpoint mode can be used to draw a line from the midpoint of one line to the midpoint of another by selecting the* **Midpoint** *button or by typing* **MID** *(ENTER) for the "first" and "next" point prompts of the* **Line** *command.*

INTERSECTION

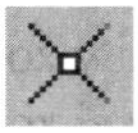

As its name implies, Intersection allows a line or arc to be extended to or from the intersection of any lines, arcs, or circles. An example of the use of the Intersection mode can be seen in Figure 7.12. The command sequence is similar to other Object Snap command entries and can be seen in Figure 7.13. By default, this option is ON. To draw a line from the intersection of two lines, move the cursor so the desired intersection lies within the target box. This will cause the cursor to snap to the intersection, allowing the next line to be drawn based on the selected intersection.

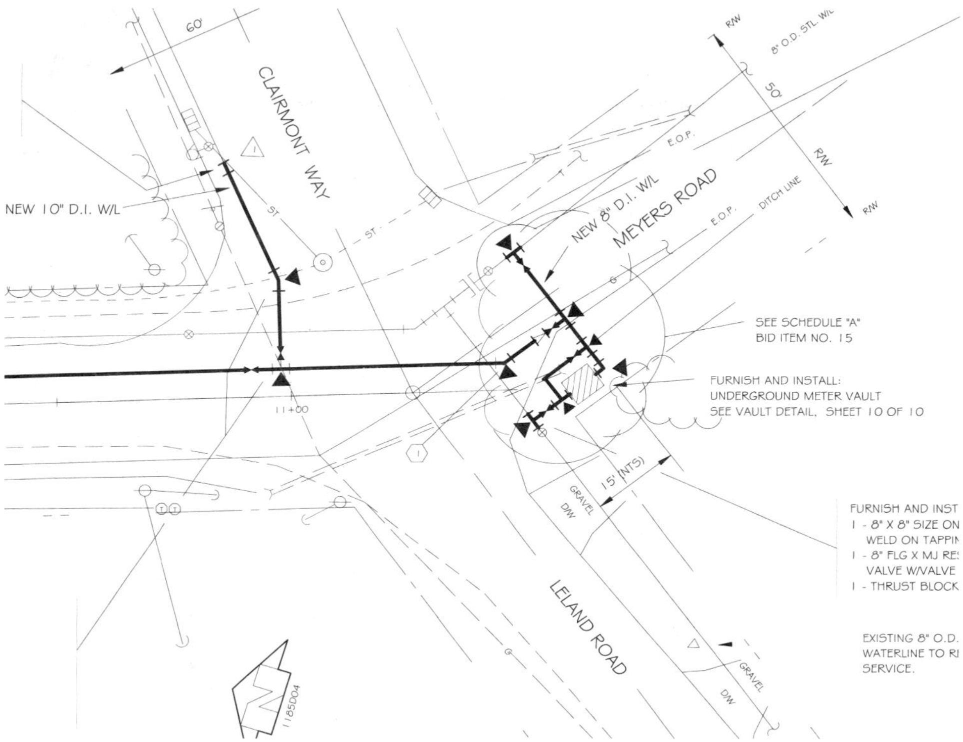

Figure 7.12 *The Intersection mode of Object Snap allows lines and arcs to be connected to the intersection of other geometric shapes. (Courtesy Lee Engineering, Inc.)*

The Intersection mode can also be used by selecting the **Intersection** button on the **Object Snap** toolbar or by entering INT (ENTER) as a "first point" or "next point," the intersection can be snapped to. The command sequence is as follows:

Command: **L** (ENTER) *(Or click the Line button.)*

LINE

Specify first point: **INT** (ENTER) *(Or click the Intersection button.)*

_int of *(Select desired to point.)*

Specify next point or [Undo]: *(Select desired to point.)*

Specify next point or [Undo]: (ENTER)

Command:

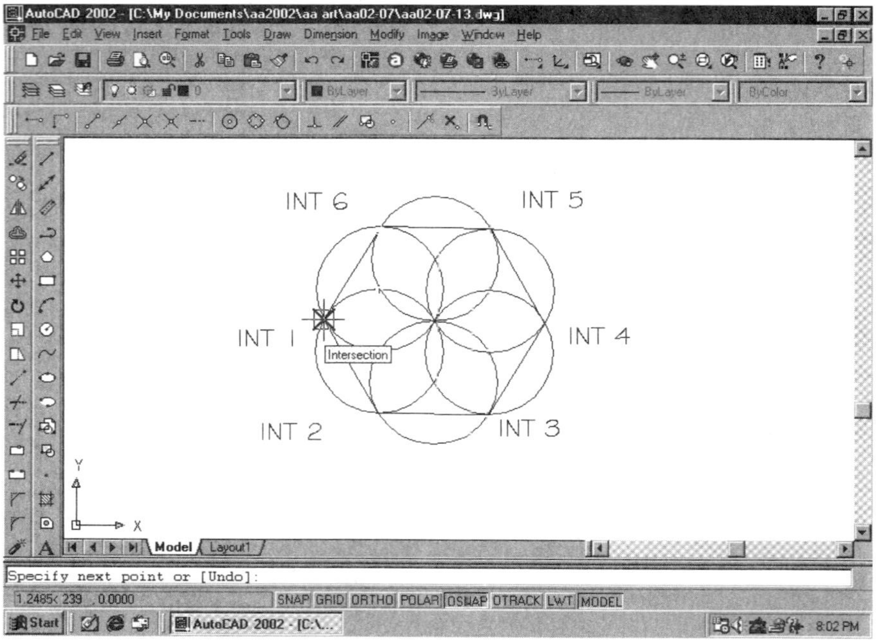

Figure 7.13 *The Intersection mode of Object Snap is used to draw a line from the intersection of each circle.*

EXTENSION

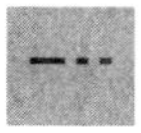

The Extension mode of Object Snap will use a point that lies along an extension path of a line as a first, next, or center point for an object. By default, this option is ON. This mode has two methods of operation. The first method will extend a line from an existing line. (See Figure 7.14.) This method can be used to lengthen a line. To extend a line from the end of an existing line, use the following procedure:

Command: **L** (ENTER) *(Or click the Line button.)*

LINE

Specify first point: **EXT** (ENTER) *(Or click the Extension button.)*

_ext of *(Select desired end point.)*

Specify next point or [Undo]: *(Select desired to point. Be sure OTRACK is ON.)*

Specify next point or [Undo]: (ENTER)

Command:

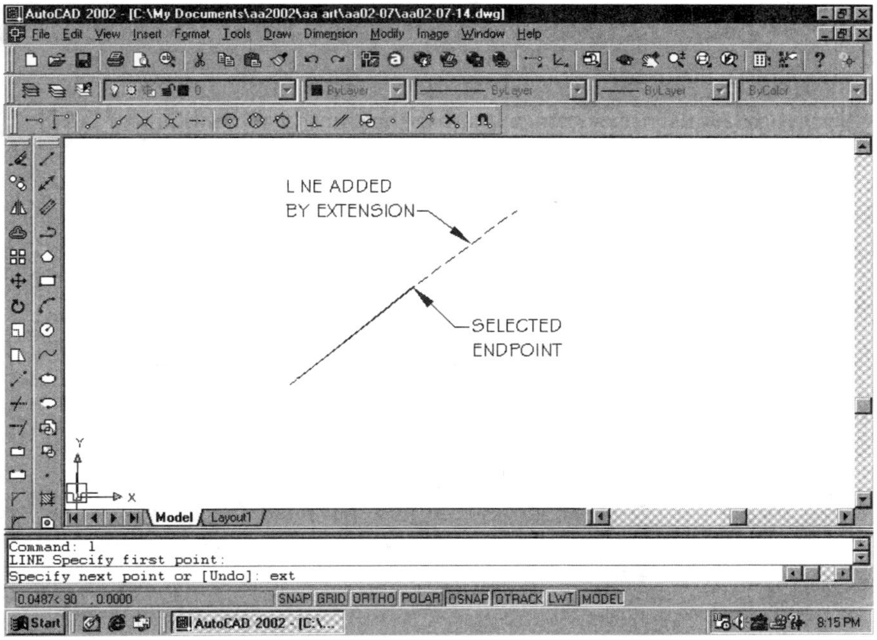

Figure 7.14 *The Extension mode of Object Snap will use a point that lies along an extension path of a line as a "next" point. When the endpoint of an existing line is selected for the extension point, the existing line will lengthened along the projected path to the "next" point.*

The second method will use a point that lies on the path as the line is extended from the existing line. This method will work well to place a point lying on a projected path. (See Figure 7.15.) Use the following procedure to select a point that lies on an extension of an existing line:

Command: **L** (ENTER) *(Or click the Line button.)*

LINE

Specify first point: **EXT** (ENTER) *(Or click the Extension button.)*

_ext of *(Move the cursor over the endpoint of the existing line but do not pick the endpoint. Extension will automatically select the indicated endpoint and project a path. Select a point that lies on the projected path.)*

Specify next point or [Undo]: *(Select desired to point.)*

Specify next point or [Undo]: (ENTER)

Command:

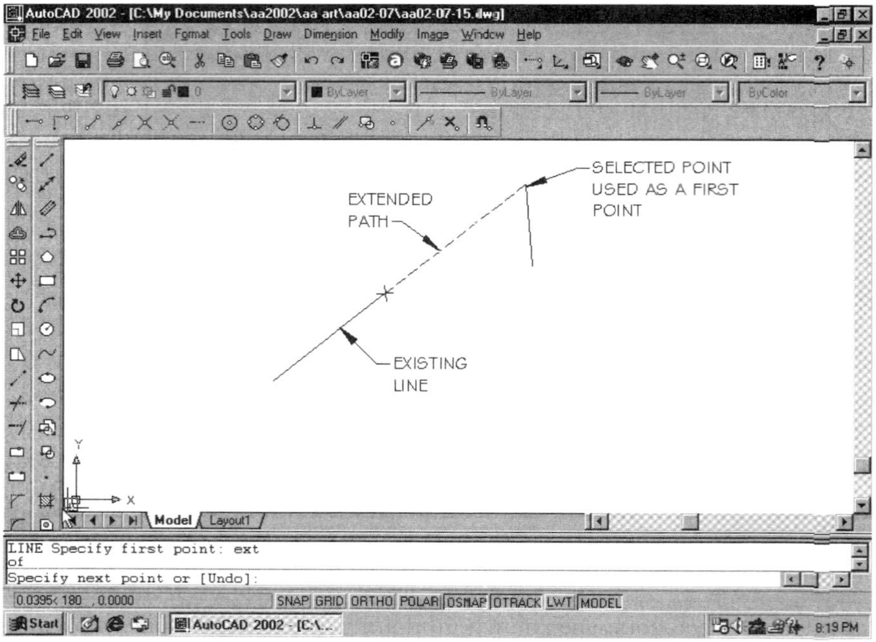

Figure 7.15 *Passing the EXT marker over the endpoint, but not selecting it as an endpoint, will use that point as a projection point for the extension path. Extension will use a point that lies on the path as the line is extended from the existing line.*

APPARENT INTERSECTION

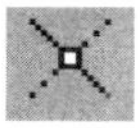

The Apparent Intersection mode finds the apparent intersection, or where two lines would intersect if one or both were extended. To find the apparent intersection, select the **Apparent Intersection** button or use the keyboard to type **APP** (ENTER) at the "first" or "next" prompt. Because the line will be generated from where two lines will intersect, two "first" points will need to be selected. The prompts will read " _appint of" and "and" to remind you to select two lines. The new line will be placed where the selected lines would meet. From the Command prompt, use the following command sequence:

Command: **L** (ENTER) *(Or click the Line button.)*

LINE

Specify first point: **APP** (ENTER) *(Or click the Apparent Intersect button.)*

_appint of *(Select first line to be projected.)*

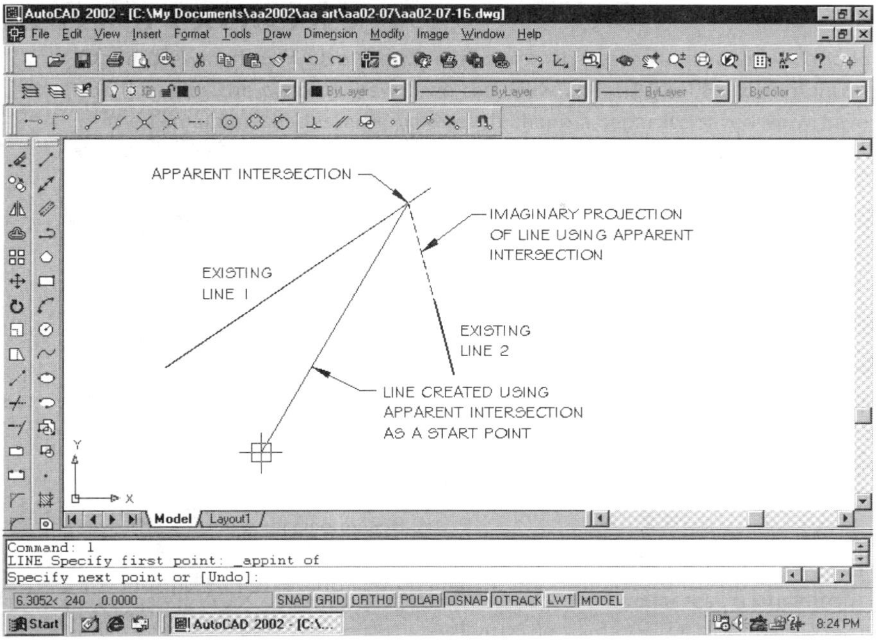

Figure 7.16 *The Apparent Intersection mode of Object Snap can be used to extend a line between two lines that would intersect. The line was started by using the Endpoint mode to select the end of line 2. Then the* **Apparent Intersect** *button was selected, or* **APP** *(ENTER) was typed for the "to" point. The marker box was used to select line 1 as the line to intersect. The endpoint of the original line must be selected to display the new line segment.*

and *(Select second line to be projected.)*

Specify next point or [Undo]: *(Select desired to point.)*

Specify next point or [Undo]: (ENTER)

Command:

Selecting (ENTER) will extend the line. The command sequence can be seen in Figure 7.16. Chapter 10 will present two alternatives to this option.

CENTER

By default, the Center mode is ON, allowing an object to be drawn based on the center of a circle. A line or arc often needs to be drawn from the center of an existing circle, or several circles might need to be drawn around one center point, as seen in Figure 7.17. This drawing could be easily drawn using the Center mode of Object Snap. To draw a line from the center of a circle, enter the **Line** command, and move the cursor over the edge of the circle. This will highlight the center of the cir-

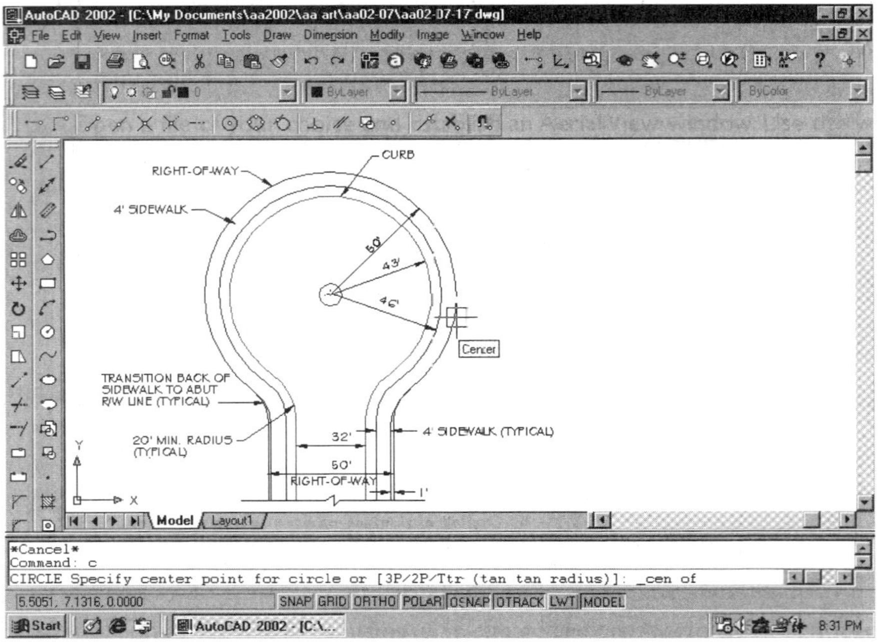

Figure 7.17 *The Center mode of Object Snap can be used whenever a drawing object needs to be placed based on the center of an existing circle. (Courtesy Lee Engineering, Inc.)*

cle. Move the cursor to the center point, and the cursor will snap to the center, allowing the center point to be used as the first point of the line to be drawn.

With the option OFF, a line or arc can be extended from the center point of a circle by selecting the **Center** button on the **Object Snap** toolbar. The Center mode can also be selected by keyboard using the following command sequence:

Command: **L** (ENTER) *(Or click the Line button.)*

LINE

Specify first point: **CEN** (ENTER) *(Or click the Center button.)*

_cen of: *(Select a circle.)*

Specify next point or [Undo]: *(Select the desired endpoint.)*

Specify next point or [Undo]: (ENTER)

Command:

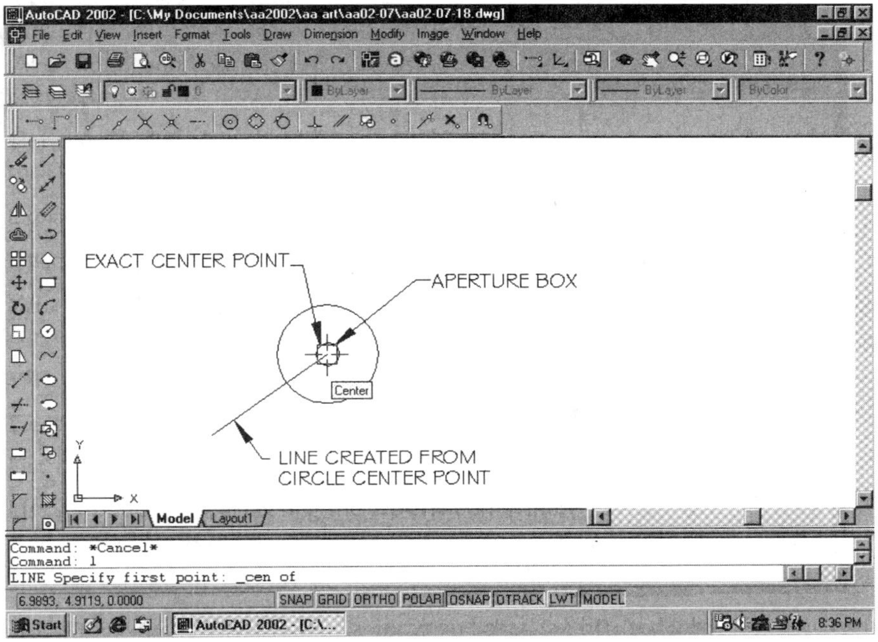

Figure 7.18 *To extend a line from the center of a circle, enter the* **Line** *command. When prompted for a "first" point, select the* **Center** *button or type* **CEN** *(ENTER) at the Command prompt. This will produce the center marker that can be used to select the desired circle. When you place the marker anywhere on the circle and select the circle, AutoCAD will automatically use the center point of that circle as the "first" point for a line.*

Touch any part of the circle, and a rubber-band line now extends from the exact center of the circle until a "next" point is selected. This process can be seen in Figure 7.18.

QUADRANT

The Quadrant mode will allow a line or arc to be snapped to the 0°, 90°, 180°, and 270° positions of a circle or arc. Start this mode by selecting the **Quadrant** button on the **Object Snap** toolbar or by typing **QUA** (ENTER) as the "first" or "next" point. Place the aperture box on the circle or arc near the quadrant point, and the new object will automatically be joined at the quadrant point, as seen in Figure 7.19. To draw a line from the quadrant point of an existing circle, use the following command sequence:

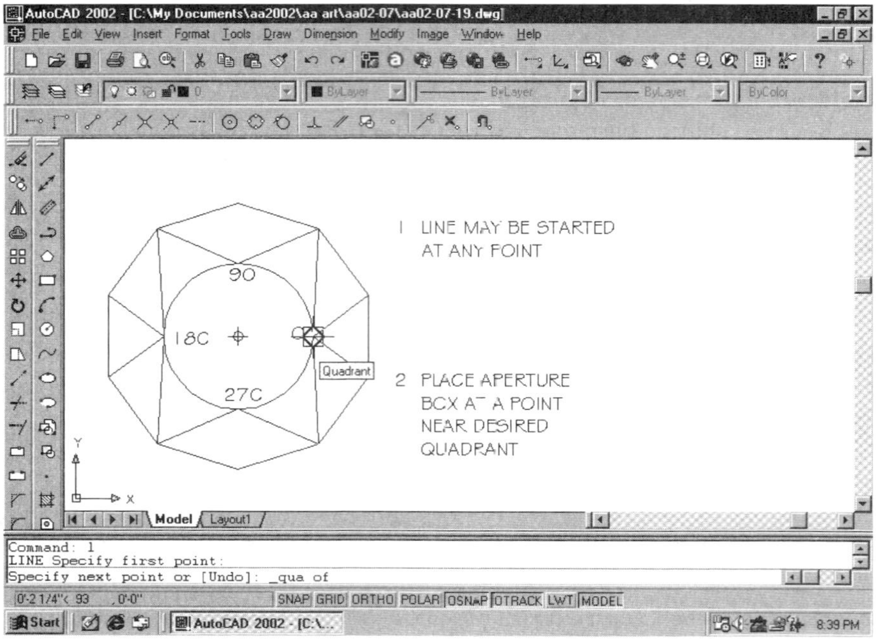

Figure 7.19 *The Quadrant mode of Object Snap can be used to attach a line or arc to the quadrant points of an existing circle.*

Command: **L** (ENTER) *(Or click the Line button.)*

LINE

Specify first point: **QUA** (ENTER) *(Or click the Quadrant button.)*

_qua of *(Select desired circle.)*

Specify next point or [Undo]: *(Select desired to point.)*

Specify next point or [Undo]: (ENTER)

Command:

TANGENT

The Tangent mode will allow a line to be snapped tangent to a circle or an arc. This very common combination of geometric shapes can be seen in Figure 7.20. Start this mode by selecting the **Tangent** button on the **Object Snap** toolbar or by typing **TAN** (ENTER) at the "first" or "next" point prompt. Figure 7.21 shows an exam-

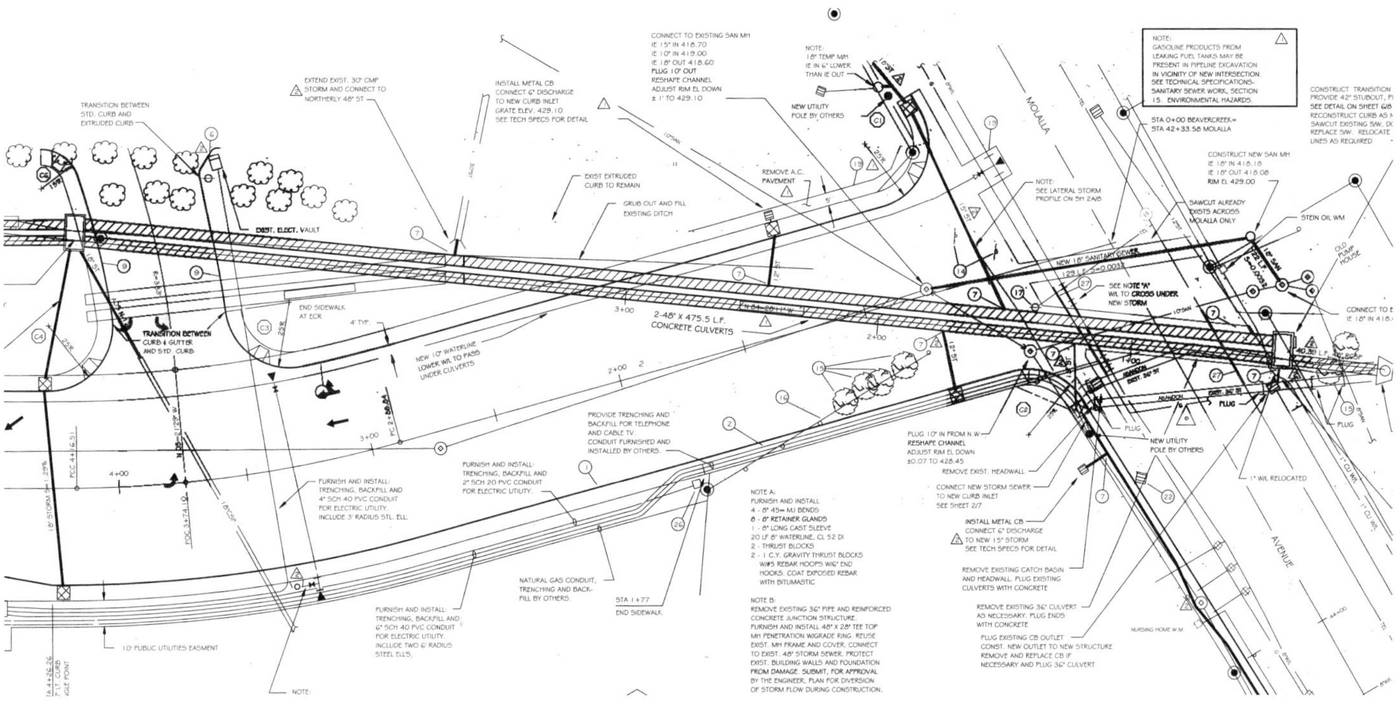

Figure 7.20 *A common drawing requirement is to draw tangent surfaces. Tangent Object Snap will allow a line or arc to be drawn tangent to an existing arc or circle. (Courtesy Lee Engineering, Inc.)*

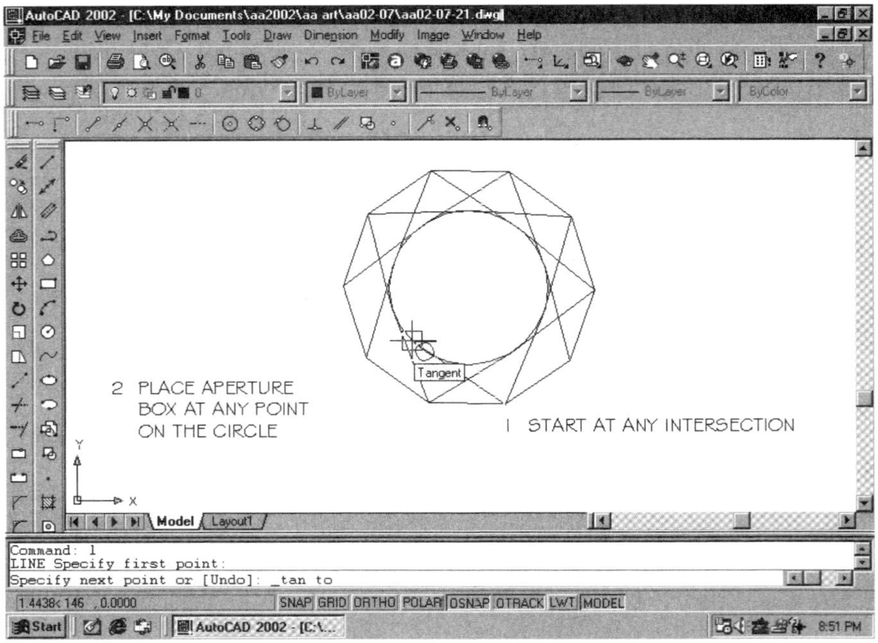

Figure 7.21 *Tangent is started by selecting the* **Tangent** *button or by typing* **TAN** (ENTER) *for the "first" or "next" point of the new line or arc.*

ple of how Tangent can be used. To draw a line tangent to an existing circle, use the following command sequence:

> Command: **L** (ENTER) *(Or click the Line button.)*
>
> LINE
>
> Specify first point: *(Select desired point.)*
>
> Specify next point or [Undo]: **TAN** (ENTER) *(Or click the Tangent button.)*
>
> _tan to *(Select desired arc or circle.)*
>
> Specify next point or [Close/Undo]: (ENTER)
>
> Command:

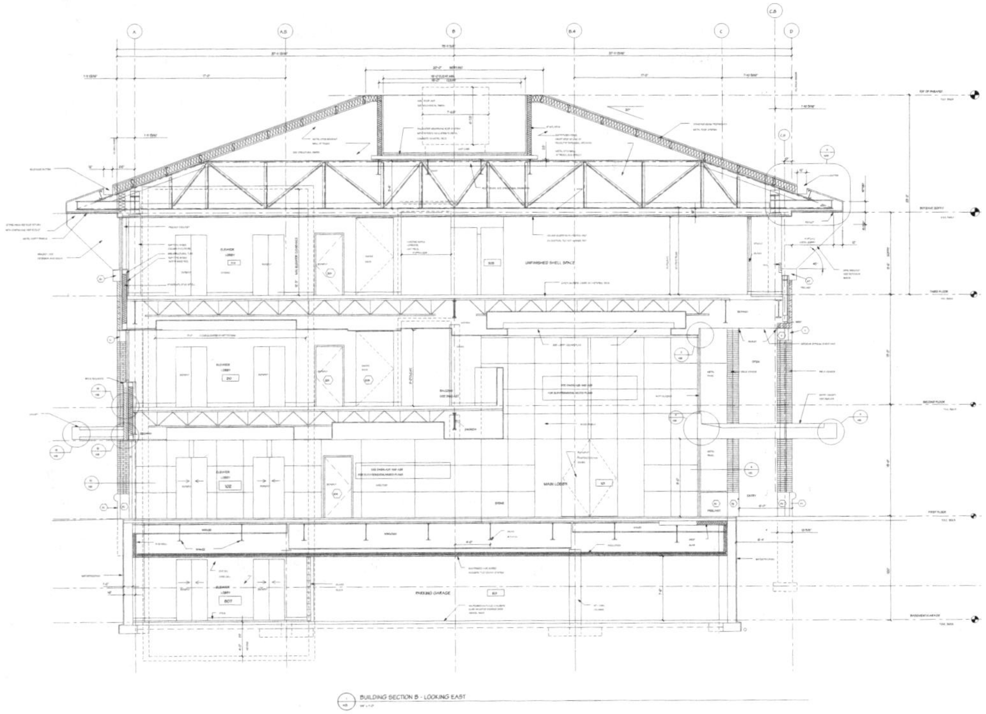

Figure 7.22 *The Perpendicular mode of Object Snap can be used to draw a line perpendicular to another surface. This can be useful if the surface is not in a true vertical or horizontal alignment. (Courtesy Peck, Smiley, Ettlin Architects.)*

PERPENDICULAR

As seen in Figure 7.22, perpendicular lines are a basis for most construction drawings. Previously introduced Object Snap modes could be used to bring a line or an arc toward or away from an object. The Perpendicular mode is used to draw a line that is perpendicular to an existing line. The process can be seen in Figure 7.23. To draw a line from any point perpendicular to an existing line, select the **Perpendicular** button on the **Object Snap** toolbar when prompted for a "first" or "next" point, or type **PER** (ENTER) at the Command prompt and use the following command sequence:

Command: **L** (ENTER) *(Or click the Line button.)*

LINE

Specify first point: *(Select desired point.)*

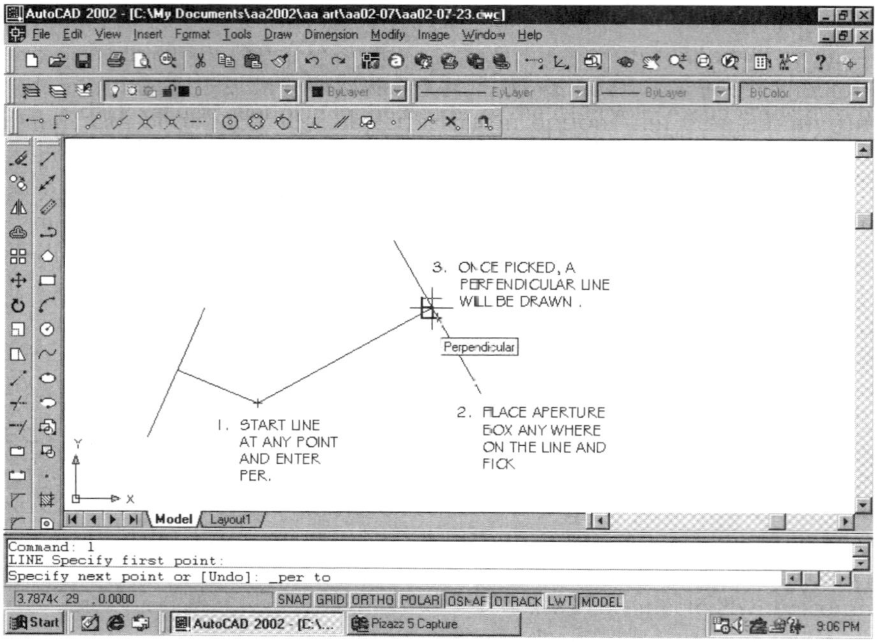

Figure 7.23 *The Perpendicular mode can be used to project a new line so that it will be at a 90° angle to an existing line.*

Specify next point or [Undo] **PER** (ENTER) *(Or click the Perpendicular button.)*

_per (Select desired line.)

Specify next point or [Undo]: (ENTER)

Command:

PARALLEL

The Parallel mode will draw a line parallel to an existing line. Unlike other modes, Parallel is specified after the "first" point is selected. Chapter 9 will explore an alternative method of placing parallel lines using the Offset command.

Command: **L** (ENTER) *(Or click the Line button.)*

LINE

Specify first point: *(Select desired to point.)*

Specify next point or [Undo]:**PAR** (ENTER) *(Or click the Parallel button.)*

_par to *(Move the cursor over, but do not select the endpoint of the desired line. As the aperture box moves over the endpoint, the point is marked and cursor snaps to a parallel path.)*

Specify next point or [Undo]: (ENTER)

Command:

The command sequence can be seen in Figure 7.24.

INSERTION

The Insertion mode will snap to the insertion point of a shape, text, attribute, or block. This might seem like meaningless stuff now, but combined with the information in later chapters, it will allow information to be inserted at an exact point.

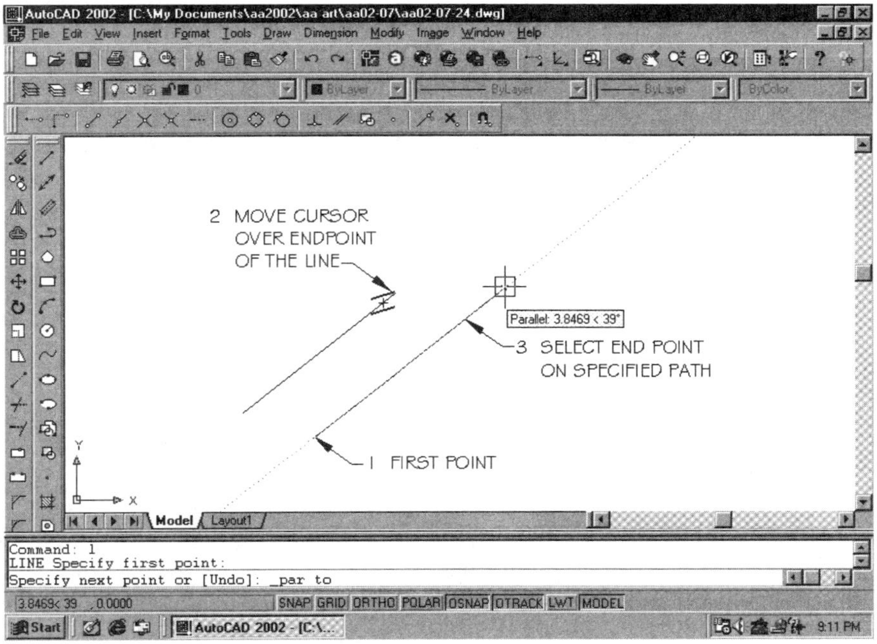

Figure 7.24 *The Parallel mode will draw a line parallel to an existing line. Unlike other modes, Parallel is specified after the "first" point is selected. Move the cursor over, but do not select the endpoint of the line that is to be the basis for the second line. Once the first line has been identified, specify the "next" point for the new parallel line.*

NODE

The Node mode can be used to snap to a point—lines or arcs can be extended from or to predetermined points. Methods of drawing a point will be introduced in Chapter 12. This could be used, for instance, to lay out points on a route survey. To activate this mode, select the **Node** button on the **Object Snap** toolbar or type **NOD** (ENTER) at either the "first " or "next" point.

NEAREST

The Nearest mode can be used to snap to the circle, line, or arc that is nearest to the crosshairs. To activate this mode, select the **Nearest** button on the **Object Snap** toolbar or type **NEA** (ENTER) at either the "first" or "next" point. To draw a line from any point of an existing line, use the following command sequence:

Command: **L** (ENTER) *(Or click the Line button.)*

LINE

Specify first point: **NEA** (ENTER) *(Or click the Nearest button.)*

_nea of *(Select desired line.)*

Specify next point or [Undo]: *(Select desired to point.)*

Specify next point or [Undo]: (ENTER)

Command:

The command sequence can be seen in Figure 7.25.

FROM

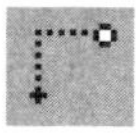

The From mode can be used to establish a point for a line or other drawing objects based on a known base point. The base point can be located by selecting a point with the cursor, or by providing polar or relative coordinates for the point. The From mode could be useful on site-related drawings where objects are to be located based on a specific point.

Figure 7.26 shows the use of the From mode for locating a circle from the endpoint of an existing line. The command can be entered by selecting the **From** button on the **Object Snap** toolbar or by typing **FRO** (ENTER) at the Command prompt. The command sequence is as follows:

Command: **C** (ENTER) *(Or click the Circle button.)*

CIRCLE Specify center point or [3P/2P/Ttr (tan tan radius)] **FRO** (ENTER) *(Or click the From button.)*

Base point <offset>: **END** (ENTER) *(Or click the Endpoint button.)*

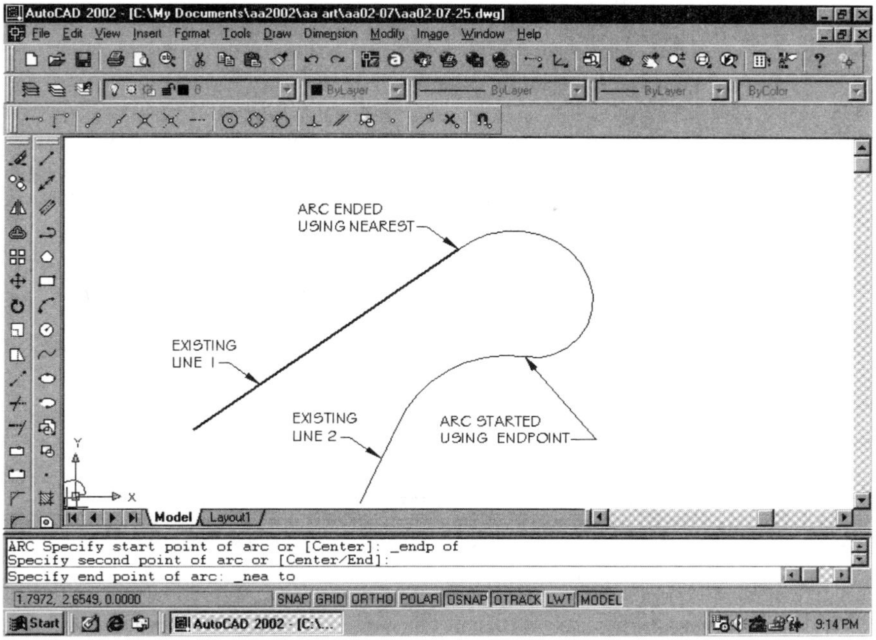

Figure 7.25 *The Nearest mode can be used to select the nearest drawing object for a "first" or "next" point when lines, arcs, circles, and other geometric shapes are placed.*

of: *(Select desired line endpoint.)*

of <offset>: **1.5** (ENTER) *(Select location or provide coordinates from base point.)*

Specify radius of circle or [Diameter]: *(Pick radius point or provide coordinates.)*

Command:

TEMPORARY TRACKING

The Temporary Tracking mode can be used to place a first or next point that lies along an implied projection of a line. The option is also useful for finding the center of a rectangle. The option functions similarly to the Apparent Intersection mode. By selecting the midpoint on the bottom side of the rectangle, and then selecting the midpoint of one of the sides, you can draw a line from the imaginary intersection of these two lines. Temporary Tracking can be started by selecting the **Temporary Tracking** button on the **Object Snap** toolbar or by typing **TK** (ENTER) at the Command prompt. The command sequence to draw a line from the center of a rectangle is as follows:

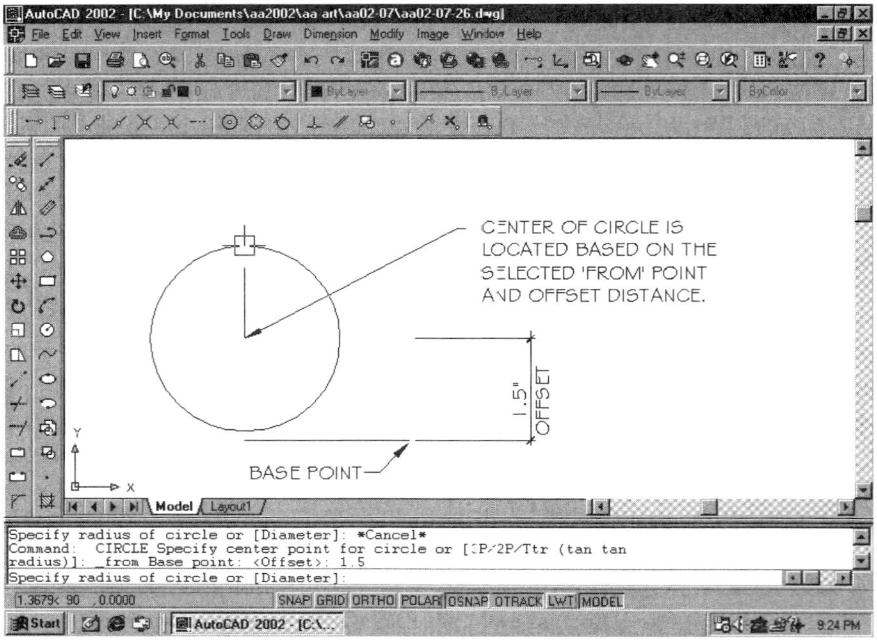

Figure 7.26 *The From mode of Object Snap allows a line, circle, or arc to be drawn from a base point by selecting the* **From** *button or by typing* **FRO** *(ENTER for one of the selection points for the drawing object.*

Command: **L** (ENTER) *(Or click the Line button.)*

LINE

Specify first point: **TK** (ENTER) *(Or click the Tracking button.)*

First tracking point: **MID** (ENTER) *(Or click the Midpoint button.)*

of *(Select the bottom edge of the rectangle.)*

Next point (Press (ENTER) to end tracking): **MID** (ENTER) *(Or click the Midpoint button.)*

of *(Select the edge of the rectangle.)*

Next point *(Press* (ENTER) *to end tracking)*: (ENTER)

Specify next point or [undo]: *(Select the desired endpoint of the line.)*

Command:

As the tracking sequence is ended, the rubber-band line will now be centered in the rectangle and the prompt for a "To" point is given. The command sequence can be seen in Figure 7.27.

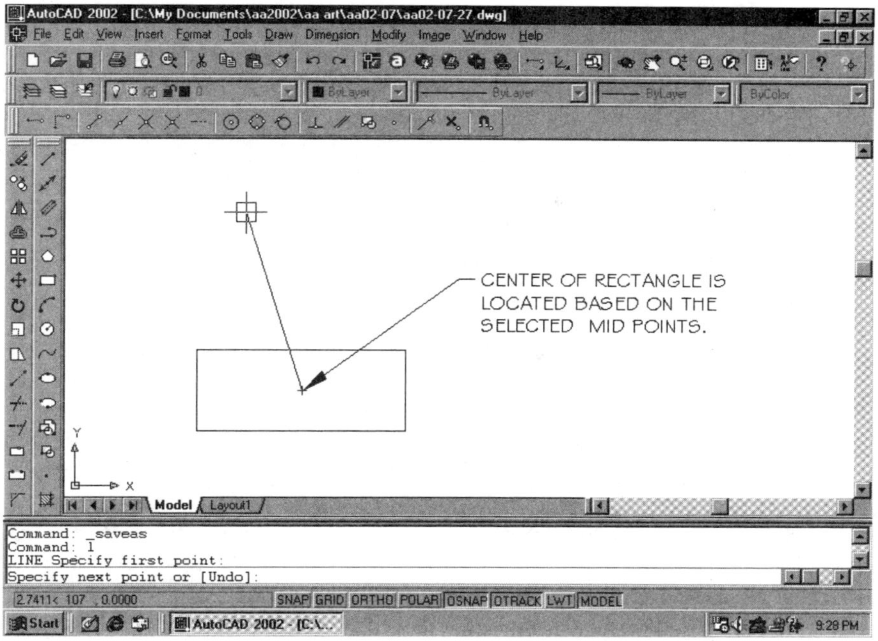

Figure 7.27 *Tracking can be used to determine the center point of a rectangle. After selecting the Tracking mode, select the midpoint of the bottom and then select the midpoint of one side of the rectangle. A line or arc can now be projected from the center of the rectangle.*

CONTINUOUS OSNAP

Up to this point, Object Snaps have been used in the override mode. The Object Snap mode was selected by button or keyboard for one specific point, and then the option was ended. It might be necessary to draw several lines to various quadrants of existing circles. Although you could continue to select the **Quadrant** button each time you draw a line, there is a faster way. Object Snap selection can be set to be in effect until the command is no longer needed.

Continuous OSNAP can be set on the **Object Snap** tab of the **Drafting Settings** dialog box. To display the tab, choose the **Object Snap Settings** button on the **Object Snap** toolbar or type **OS** (ENTER) at the Command prompt. You can also display the dialog box by selecting **Drafting Settings** from the **Tools** menu or by typing **DSETTINGS** (ENTER) at the Command prompt. Each method will produce a dialog box similar to Figure 7.28.

Once the **Object Snap** tab is displayed, place a check in the mode to be activated, and then click the **OK** button. If **Quadrant** is selected, a quadrant point will automatically be selected if a circle is selected for a "first" or "next" point each time a line or arc is

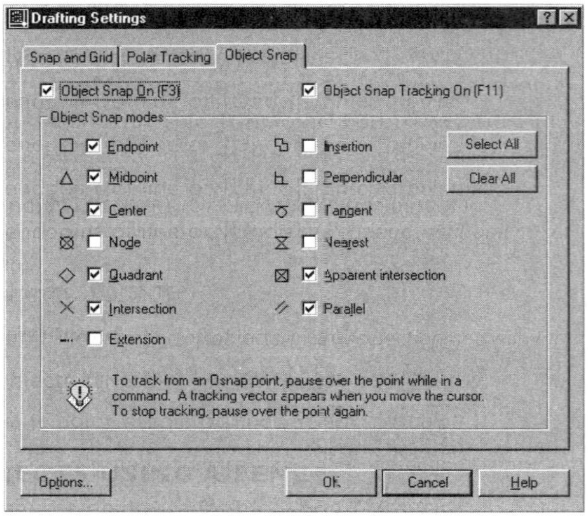

Figure 7.28 *Modes for continuous OSNAP can be set on* **Object Snap** *tab of the* **Drafting Settings** *dialog box. Access the dialog box by selecting* **Drafting Settings** *from the* **Tools** *menu or by typing* **OS** *(ENTER) at the Command prompt.*

drawn. Figure 7.29 shows an example where the Center mode is used to place columns in the center of each pier, and lines to represent beams that start at the center of each column. The selection can be ended or altered by returning to the dialog box and altering the active modes.

Continuous OSNAP can also be set to combine multiple modes. This is useful, for instance, if several lines need to be constructed from the intersection of one object to the center of other objects, as seen in Figure 7.30. Activate multiple modes by selecting them on the **Object Snap** tab of the **Drafting Settings** dialog box. There is no limit to the number of mode check boxes that can be selected. A temporary override of continuous OSNAP can be achieved by selecting the **Snap to None** button on the **Object Snap** toolbar, or by entering **NON** (ENTER) at the prompt for an object snap. Deactivate continuous OSNAP by clicking the OSNAP button on the Status bar or by pressing F3.

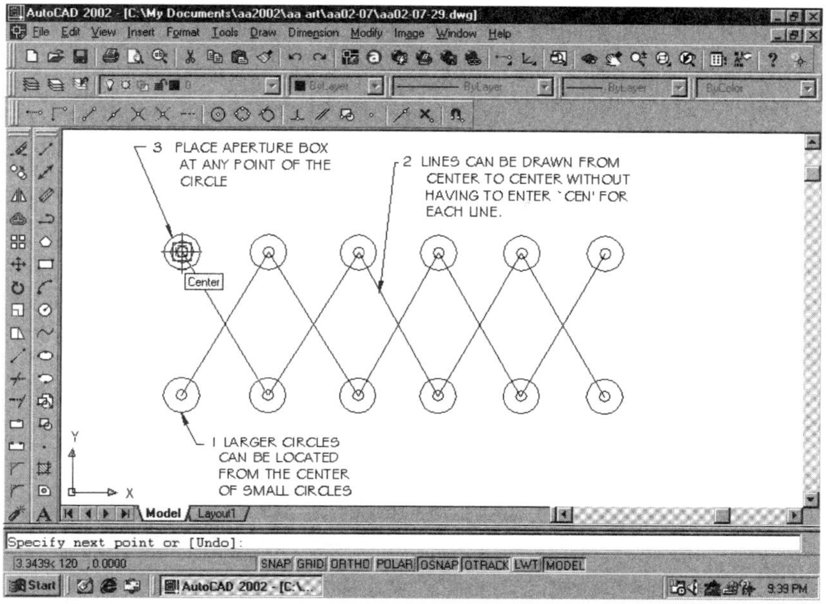

Figure 7.29 *An Object Snap mode can be used repeatedly if it is selected on the* **Object Snap** *tab of the dialog box. Active modes can be deactivated by clicking the OSNAP button on the status bar or by pressing F3.*

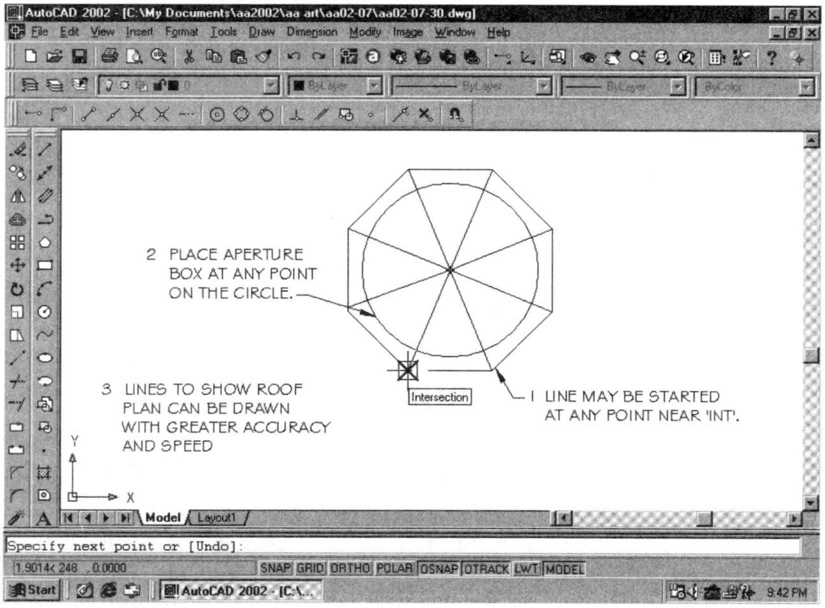

Figure 7.30 *Continuous OSNAP can be set to combine multiple options for "first" and "next" points.*

CHAPTER 7 EXERCISES

1. Draw an equilateral triangle inscribed in a 4" diameter circle. Use Object Snap Midpoint to form a second triangle inside the first. Draw a third triangle inside the second. Save the drawing as E-7-1.
2. Draw the base to a chimney in plan view that is 60" x 32" with 8" wide walls. Draw an 8" wide support wall in the center of the chimney. Save the drawing as E-7-2.
3. Use the drawing below to draw a 1" diameter circle, a 2.25" diameter circle, a four-sided polygon, and a triangle. Complete the drawing using the CEN, MID, INT, and TAN modes. Save the drawing as E-7-3.

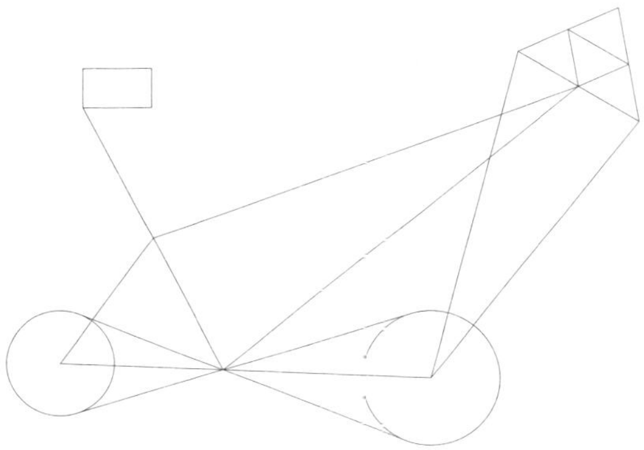

4. Set the limits and units to draw a 100' x 75' lot. Assume north to be at the top of the page. The north and south property lines will be 100' long. Five trees are located on the property. Each is located from the southwest corner with coordinates and diameters listed as follows:

	Coordinates	Trunk diameter	Branch diameter
a.	15',12'-9"	9" diameter	6' diameter
b.	30',17'	13" diameter	7.5' diameter
c.	42'-6", 19'-3"	19" diameter	9.5' diameter
d.	50', 28'-2"	15" diameter	8.5' diameter
e.	62'-0", 5'-3"	26" diameter	14' diameter

Locate each tree with a point. Use an "X" to locate each tree. Use the appropriate Object Snap modes to draw the diameter of the trunk and the branch structure around each tree center. Draw a line from the southwest property corner through the center of the trees to the southeast corner. Draw another line from the southeast property corner to the northern limits and tangent to the branch diameter of each tree and ending at the southwest corner. Save the drawing as E-7-4.

5. Use the drawing below as a guide to draw a 4' x 4' window with a half-round window above. Draw all window dividers as 1/2" wide. Save the drawing as E-7-5.

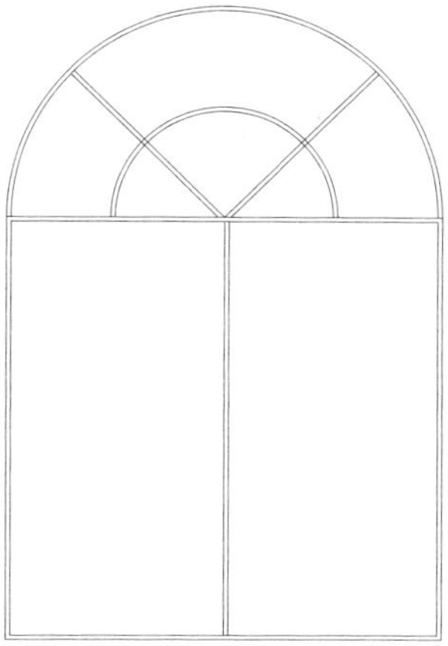

6. Using the drawing below as a guide, open the ARCHBASE template and draw a 32" x 21" double sink. Show the outline of the sink with rounded corners using a 2" radius. Divide the sink into two equal portions. Draw a 2" diameter circle in the center of the right portion of the sink. Save the drawing as E-7-6SNK.

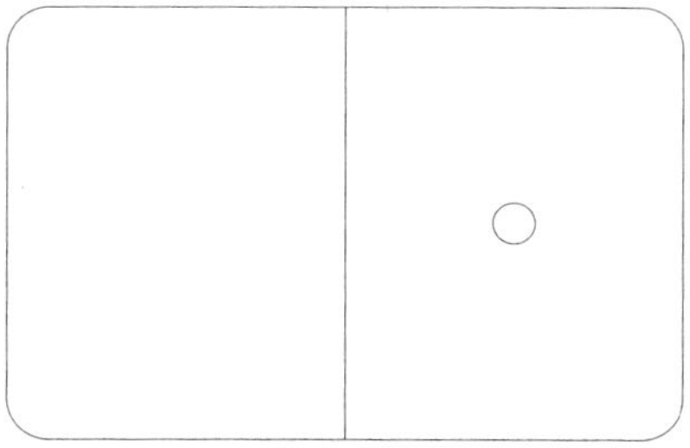

7. Using the drawing below as a guide, open the ARCHBASE template and draw a rectangle to represent a 60" x 32" tub/shower. Draw a rectangle with 3" radius rounded corners at one end and a full radius at the other end. Draw a triangle at the squared end to represent a shower head. Save the drawing as E-7-7TUB.

8. Using the drawing below as a guide, open the ARCHBASE template and draw a 48" wide x 24" deep, 0' clearance fireplace. Draw the face of the fireplace as 8" wide on each side, with the interior edges 20" long, set at a 15° angle. Save the drawing as E-7-8FIRE.

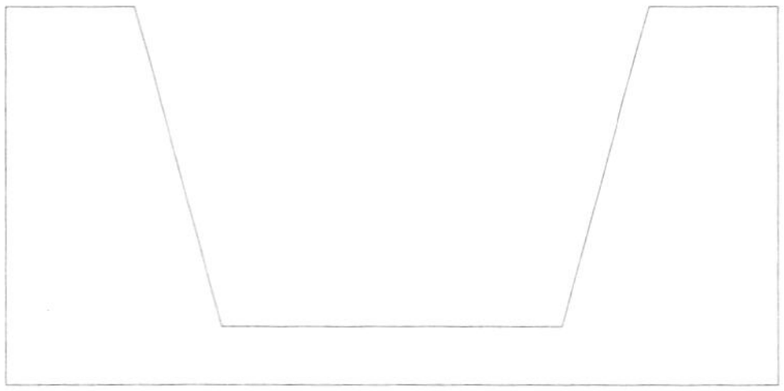

9. Open the ARCHBASE template and draw a 42" x 36" rectangle using continuous lines. Draw diagonal lines from opposite corners using hidden lines and show a 3" diameter circle at the intersection of the interior crossing lines. Save the drawing as E-7-9SHOW.

10. Draw a plan view of a 10' long x 24" wide bathroom vanity. The vanity is to have a 30" wide knee space in the center that should be represented by hidden lines. Place a 18" x 16" oval sink centered on each side of the knee space. Save the drawing as E-7-10VAN.

CHAPTER 7 QUIZ

1. What command is used to change the size of the aperture box?
2. What is the size range for the aperture box?
3. Type the command sequence to provide running snap to an endpoint.
4. List three Object Snap modes for connecting a line to a circle.
5. What can be done to save time when using Object Snap on a huge drawing?
6. List three methods to stop continuous OSNAP.
7. What steps would be needed to draw a perpendicular line to the midpoint of an existing line?
8. List the letters that are used to activate the Object Snap mode used to connect geometric shapes.
9. How can you get the dialog box for Object Snap to be displayed?
10. With the Quadrant mode of Object Snap, at what degree will lines be attached to a circle?

CHAPTER 8

Drawing Display Options

In the last chapter you were introduced to drawing accurately by snapping to specific points of an object. You can improve your drawing accuracy by changing the size of the view of the drawing and changing the drawing resolution. You were introduced to the **Zoom** and **Pan** commands in Chapter 2. This chapter will explore

- Options of the **Zoom** command
- Other commands that can be used to alter or move through the display

Commands to be explored in this chapter include

- **Zoom**
- **Pan**
- **View**
- **Aerial View**
- **Redraw**
- **Regen**
- **Regenall**
- **Viewres**

ZOOM

Similar to the zoom lens of a camera, AutoCAD's **Zoom** command has a potential zoom ratio of 10 trillion to 1. "Zooming in" on a drawing can magnify a portion of a drawing to allow for better visual quality. Zoom can be used in the middle of another command sequence, which can greatly improve accuracy on detailed drawings. As the apparent size of the object is increased, the area of the drawing that can be seen is reduced. This can be seen in Figures 8.1 and 8.2. The opposite holds true: when you reduce the apparent size of an object, more of the drawing can be seen. It is important to keep in mind that the actual size of the object is not changing, only the magnification.

Zoom options can be accessed from the **Zoom** toolbar and by selecting one of three different buttons on the **Standard** toolbar, **Zoom Realtime**, **Zoom Flyout**, and **Zoom Previous**; these buttons will be discussed throughout the chapter. You can also access the **Zoom** command from the **View** menu, by the right-click shortcut menu, or by keyboard. Figure 8.3a shows the **Zoom** toolbar and the **Zoom** flyout menu found on the **Standard** toolbar. Zoom is also activated by selecting **Zoom** from the **View** menu. This will produce the selections seen in Figure 8.3b. Right-clicking will produce the **Realtime** option of **Zoom**. Typing **Z** (ENTER) at the Command prompt will also start the command sequence. Each will produce the prompt:

Command: **Z** (ENTER) *(Or click the Zoom button.)*

[All/Center/Dynamic/Extents/Previous/Scale/Window]<Real time>:

Each of the options listed at the Command prompt and in the menu can be selected from the **Zoom** flyout menu on the **Standard** toolbar.

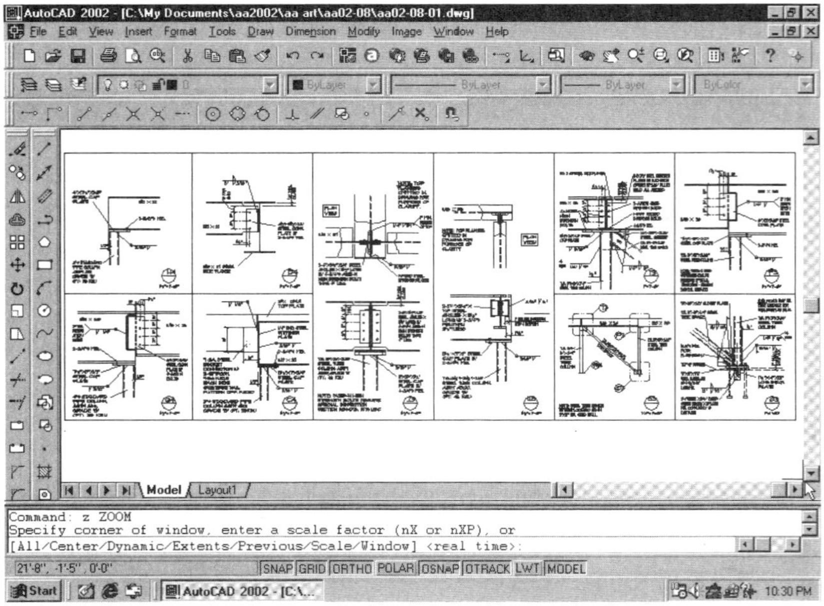

Figure 8.1 *As the drawing limits are enlarged, many features of the drawing might not be legible. (Courtesy Kenneth D. Smith Architects & Associates, Inc.)*

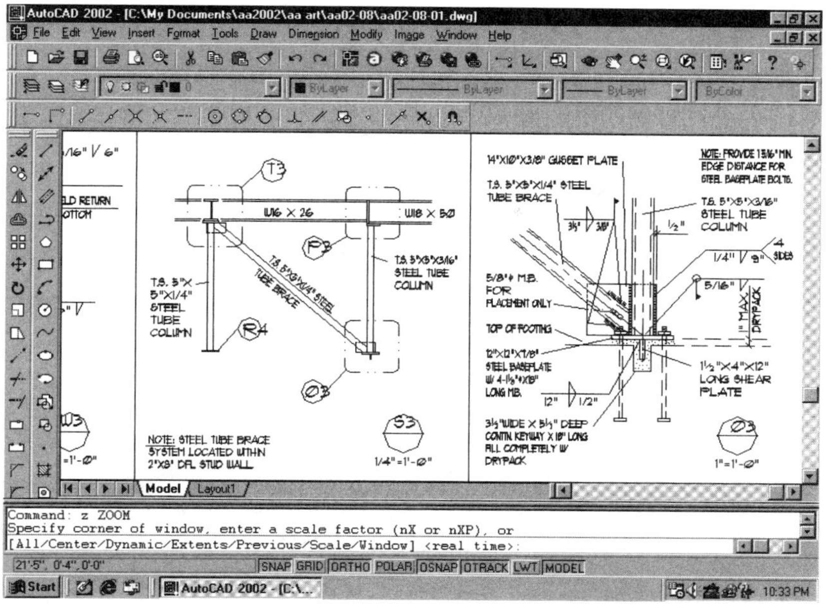

Figure 8.2 *The* **Zoom** *command alters the position of a drawing to be viewed, giving the sense that the portion of the drawing being viewed has been enlarged. (Courtesy Kenneth D. Smith Architects & Associates, Inc.)*

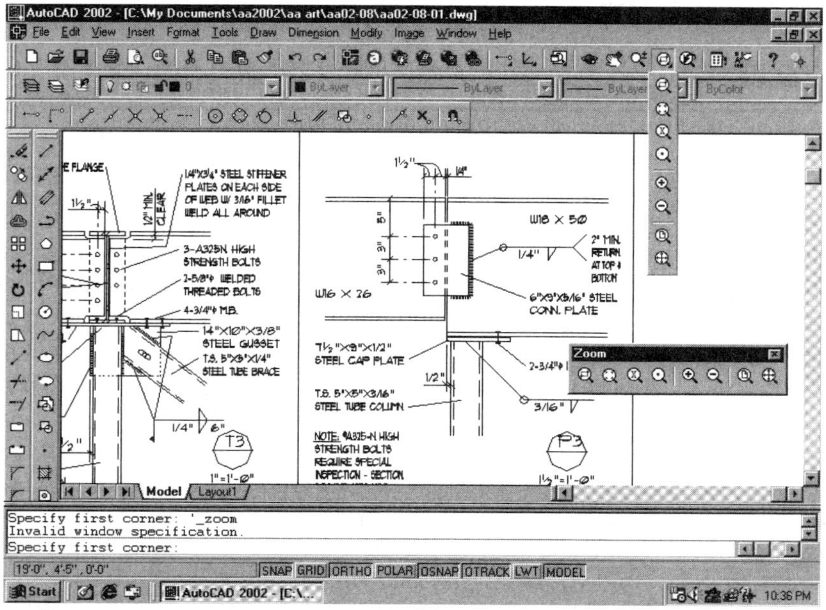

Figure 8.3a *The* **Zoom** *toolbar and the* **Zoom** *flyout menu on the* **Standard** *toolbar can be used to select* **Zoom** *options.*

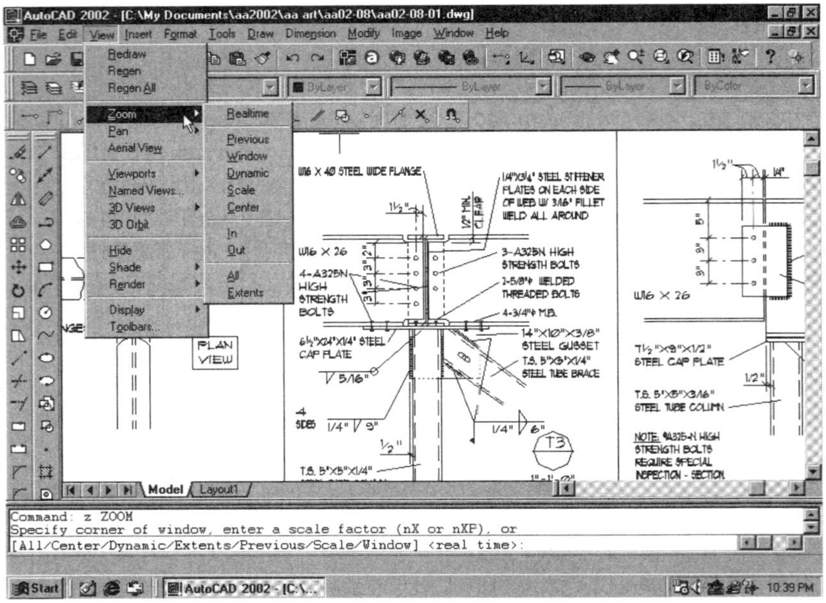

Figure 8.3b *The* **Zoom** *command can be selected by typing* **Z** *(ENTER) at the Command prompt or by selecting the command from the* **View** *menu.*

REALTIME COMMAND

Realtime zooming means that as you move the cursor, you see the results. Start the **Zoom Realtime** command by the right-click shortcut menu, by selecting the **Zoom Realtime** button on the **Standard** toolbar, from the **Zoom** submenu of the **View** menu, or by typing **RTZOOM** (ENTER) at the Command prompt. As the command is activated, the cursor will change to a magnifying glass with a + and – symbol.

Move the cursor around on the screen and nothing happens. As you press and hold the select button, moving the cursor toward the top of the screen, the magnification of the drawings is enlarged. Placing the magnifying glass in the middle of the screen and moving to the top will enlarge the display 100%. Pressing and holding the select button and moving from the middle to the bottom of the screen will decrease the magnification 100%. Placing the magnifying glass at the bottom of the screen and moving to the top will enlarge the display 200%. Moving from the top to the bottom of the screen will decrease the magnification 200%. Each command can be repeated indefinitely, until the desired magnification is achieved. If you have trouble making **Realtime Zoom** alter the drawing size, remember that you must hold down the select button as the cursor is moved.

The command is ended by pressing ESC, SPACEBAR, ENTER, or right-clicking to display the shortcut menu and selecting **Exit** (see Figure 8.3c). Selecting **Exit** will end

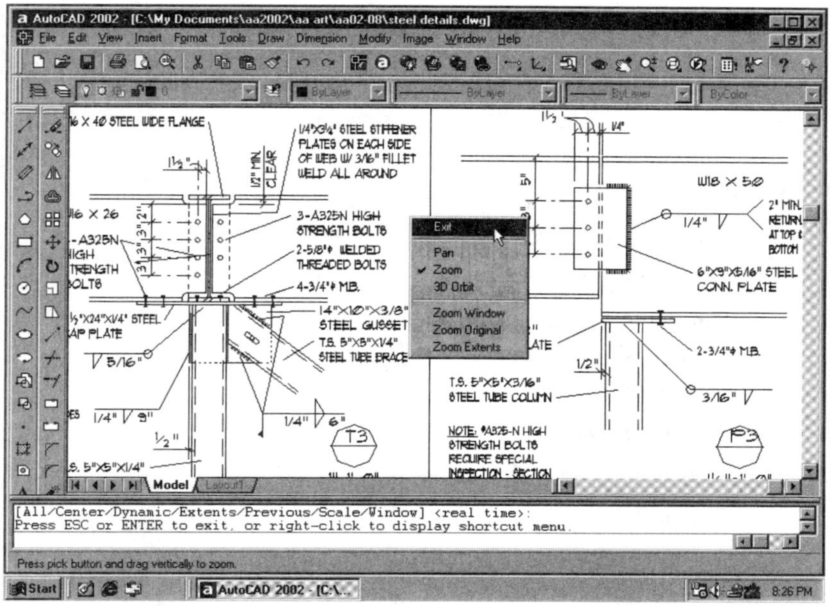

Figure 8.3c *Right-clicking while in the* **Zoom** *command will produce the* **Zoom** *shortcut menu.*

Zoom and return the Command prompt. The menu also allows other viewing options to be selected. By alternating between the **Pan** and the **Zoom** commands, you can enlarge a drawing while keeping the desired portion on the screen. Each of these options will be discussed throughout the balance of this chapter.

ZOOM WINDOW

The Window option of **Zoom** is a common option for zooming into a drawing. This option lets you select the area you wish to view by providing two opposite corner locations that will form the viewing window. Select the size of the view box by entering coordinates or by using the select button to select locations on the screen. Two methods are available for using this option of the **Zoom** command. By default, the command is waiting for you to enter two points that will form the corners of a rectangle. The command sequence is as follows:

Command: **Z** (ENTER) *(Or click the Zoom button.)*

[All/Center/Dynamic/Extents/Previous/Scale/Window]<Real time>:

Specify first corner: *(Pick a corner.)*

Specify other corner: *(Pick a corner.)*

Command:

This process can be seen in Figure 8.4a. Objects within the view box will be redisplayed on the screen as the new display. You select the first corner of the zoom display by moving the cursor to the desired location. Once a corner is selected, the crosshairs will switch to a box that is enlarged or reduced as the mouse is moved across the drawing area. Figure 8.4b shows the results of the selection made in Figure 8.4a. The process is similar when entering coordinates with a mouse.

A zoom window can also be created using the Window option from the **Zoom** command. The command process is as follows:

> Command: **Z** (ENTER) *(Or click the Zoom Window button.)*
>
> [All/Center/Dynamic/Extents/Previous/Scale/Window]<Real time>:
> **W** (ENTER)
>
> Specify first corner: *(Pick a corner.)*
>
> Specify other corner: *(Pick a corner.)*
>
> Command:

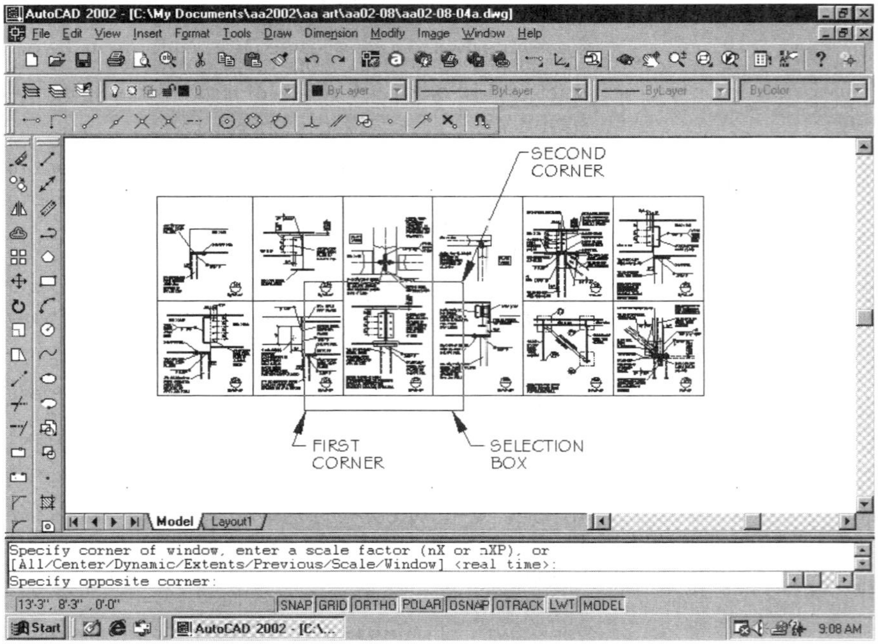

Figure 8.4a *The* **Zoom Window** *option allows the display area to be selected by specifying opposite corners of the "window." (Courtesy Kenneth D. Smith Architects & Associates, Inc.)*

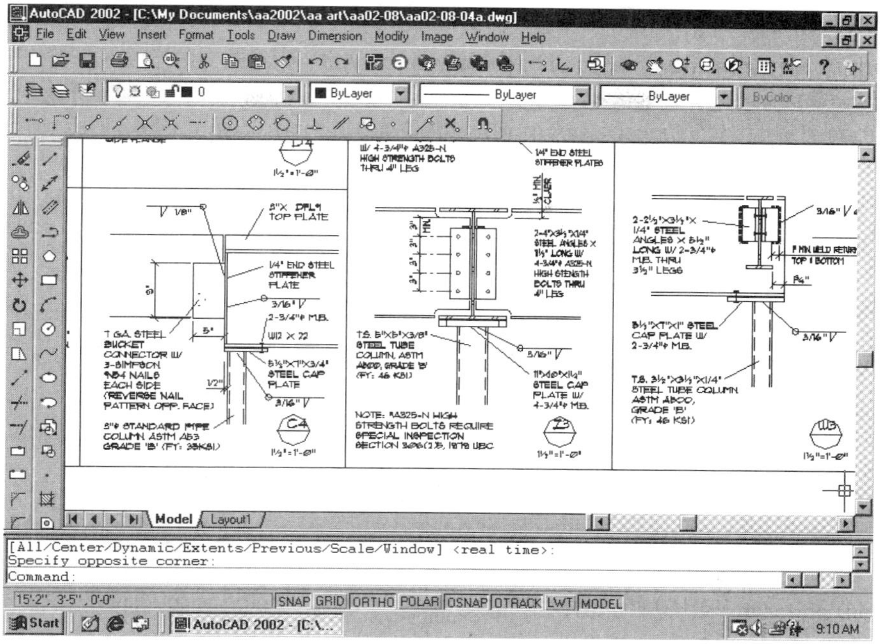

Figure 8.4b *The* **Zoom** *window created in Figure 8.4a now displays the selected area. (Courtesy Kenneth D. Smith Architects & Associates, Inc.)*

Typing **Z** (ENTER), **W** (ENTER), or selecting the **Zoom Window** button are the most efficient methods of starting the command.

TRANSPARENT ZOOM

One of the really great features of **Zoom** is that it is transparent, meaning it can be used in the middle of another command. If you start to draw a line and realize that you're straining to see the line, the screen image can be magnified using **Zoom** in the middle of a **Line** command sequence. Simply select the button of the desired Zoom option or enter the command by keyboard. To use a transparent command from the keyboard, type an **'** (apostrophe) prior to the command. The command sequence for using the window option of **Zoom** in the middle of the Line command would be as follows:

> Command **L** (ENTER) *(Or click the Line button on the Draw toolbar.)*
>
> LINE
>
> Specify first point: *(Select first line endpoint.)*
>
> Specify next point or [Undo]: **'Z** (ENTER)

_zoom

>>Specify corner of window, enter a scale factor (nX or nXP), or

[All/Center/Dynamic/Extents/Previous/Scale/Window] <Real time>: **W** (ENTER)

>>Specify first corner: *(Pick desired start point.)*

>>Specify opposite corner: *(Select opposite corner to be enlarged.)*

Resuming LINE command.

Specify next point or [Undo]: *(Select desired next point.)*

Specify next point or [Undo]: (ENTER)

Command:

Zoom can be used at any point during the **Line** command or any other command when a prompt is given. In the **Line** command, the **'Zoom** command could have been started at the prompt for "first point" or at the "next point" prompt. Start the **Line**, **Circle**, or **Arc** command and then use **'Zoom** to become familiar with this option of **Zoom**. Your eyes and back will love you as you zoom more and squint less.

ZOOM ALL

The **Zoom All** option will allow for an entire drawing to be viewed, based on the drawing limits. If the drawing extends beyond the drawing limits, the screen display will show the entire drawing. **Zoom All** should be used if you are zoomed in to a small area of a drawing, such as Figure 8.2. Using the All option returns all of the drawing to the screen so that the display resembles Figure 8.1. **Zoom All** can be selected from the **Zoom** flyout menu on the **Standard** toolbar, from the right-click menu, from the **Zoom** submenu on the **View** menu, or by typing **Z** (ENTER), **A** (ENTER) at the Command prompt. The command sequence is as follows:

Command: **Z** (ENTER) *(Or click the Zoom All button.)*

ZOOM

Specify corner of window, enter a scale factor (nX or nXP), or

[All/Center/Dynamic/Extents/Previous/Scale/Window] <Realtime>: **A** (ENTER)

Command:

ZOOM CENTER

The **Zoom Center** option allows a zoom to be specified by its desired center point. Once a center point is selected, a prompt will be given for the height of the display. If the current default is maintained, the drawing is redisplayed based on the new center point, but the magnification is not changed. If a smaller value for the height is selected, the display magnification will be increased. If a greater value for the height is selected, the display magnification will be reduced. The command sequence is as follows:

Command: **Z** (ENTER) *(Or click the Zoom Center button.)*

ZOOM

Specify corner of window, enter a scale factor (nX or nXP), or

[All/Center/Dynamic/Extents/Previous/Scale/Window] <Realtime>: **C** (ENTER)

Specify center point: *(Enter coordinates or select a point.)*

Enter magnification or Height < 17'-4">: **10** (ENTER)

Command:

ZOOM EXTENTS

The **Zoom Extents** option will display the entire drawing based on the size of the drawing—not the size of the limits that were used for **Zoom All**. If a small object is being examined with large drawing limits, Extents will enlarge the object to fill the drawing screen. This can be quite helpful during the initial stages of a drawing. **Zoom Extents** is also helpful if the extent of the drawing exceeds the drawing limits, because the option allows the entire drawing to be examined. The command sequence is as follows:

Command: **Z** (ENTER) *(Or click the Zoom Extents button.)*

ZOOM

Specify corner of window, enter a scale factor (nX or nXP), or

[All/Center/Dynamic/Extents/Previous/Scale/Window] <Realtime>: **E** (ENTER)

Command:

ZOOM PREVIOUS

Each of the other **Zoom** options is used to enlarge an area of a drawing for easier viewing. The Previous option will allow you to go backward, so that the previous drawing area can be seen. When you use this option, the drawing displayed prior to the zoom will be returned to the screen. A maximum of 10 previous views can be restored using the Previous option. The command sequence for **Zoom Previous** is as follows:

Command: **Z** (ENTER) *(Or click the Zoom Previous button.)*

ZOOM

Specify corner of window, enter a scale factor (nX or nXP), or

[All/Center/Dynamic/Extents/Previous/Scale/Window] <Realtime>: **P** (ENTER)

Command:

ZOOM SCALE

The default selection for the **Zoom** command is Scale or magnification factor (X). The scale can be set to be relative to the full view or to a current view.

Scale Relative to Full View

When you enter a number for the scale factor, the entire drawing will be affected. A scale factor of 1 will display the current size. A scale factor of 3 will make objects appear three times as large. By entering a number smaller than 1, you will decrease the size of the object. If you enter **.5**, the object will appear half as big as the full display. The command sequence is as follows:

Command: **Z** (ENTER) *(Or click the Zoom Scale button.)*

ZOOM

Specify corner of window, enter a scale factor (nX or nXP), or

[All/Center/Dynamic/Extents/Previous/Scale/Window] <Realtime>: **S** (ENTER)

Enter scale factor (nX or nXP): **.5** (ENTER)

Command:

Scale Relative to Current View

If you enter a numeric value followed by an "X", the scale of the zoom will be relative to the current view, rather than the entire drawing. The command sequence is as follows:

Command: **Z** (ENTER) *(Or click the Zoom Scale button.)*

ZOOM

Specify corner of window, enter a scale factor (nX or nXP), or

[All/Center/Dynamic/Extents/Previous/Scale/Window] <Realtime>: **S** (ENTER)

Enter scale factor (nX or nXP): **1.5X** (ENTER)

Command:

Scale Relative to Paper Space Units

AutoCAD thinks of space as model space and paper space. 2D drawings should be drawn using model space. Paper space is used to plot views of the model that have been created. **Zoom XP** can be used to scale each space relative to paper space units. This will be helpful, for instance, in printing drawings on the same sheet that are drawn at different scales. This topic will be covered in more detail in later chapters.

ZOOM IN

Zoom In automatically scales the current drawing screen display by a scale factor of 2. Selecting this Zoom option produces the same result as typing a scale factor of 2X (ENTER) from the command line. The **Zoom In** option is available on the **Zoom** toolbar, the **Zoom** flyout menu on the **Standard** toolbar, and from the **Zoom** submenu of the **View** menu.

ZOOM OUT

Zoom Out automatically enlarges the drawing area that is displayed on the screen by a factor of 2. The drawing is, in effect, reduced by a scale factor of .5X. The **Zoom Out** option is available on the **Zoom** toolbar, the **Zoom** flyout menu on the **Standard** toolbar, and from the **Zoom** submenu of the **View** menu.

ZOOM DYNAMIC

The **Zoom Dynamic** option will allow specific areas of the drawing to be selected for Zoom viewing. The command will seem complicated at first. Once you master it, you'll be able to select any portion of the drawing for the next view without requiring a **Regen**. The importance of this might seem lost on you as you compare a dynamic to a window zoom while working on the exercises. As you start to work on larger projects using hundreds of thousands of bytes of information, a **Zoom Dynamic** will be well worth the effort because of the time you save in not having to regenerate a drawing. The original view box can be enlarged or reduced and moved about the drawing screen to select the exact portion of the drawing to be zoomed. The command sequence is as follows:

> Command: **Z** (ENTER) *(Or click the Zoom Dynamic button.)*
>
> ZOOM
>
> Specify corner of window, enter a scale factor (nX or nXP), or
>
> [All/Center/Dynamic/Extents/Previous/Scale/Window] <Realtime>: **D** (ENTER)

The screen will be cleared, and the original view will be returned, surrounded by three color-coded boxes. Each box can be seen in Figure 8.5 and represents a different viewing area.

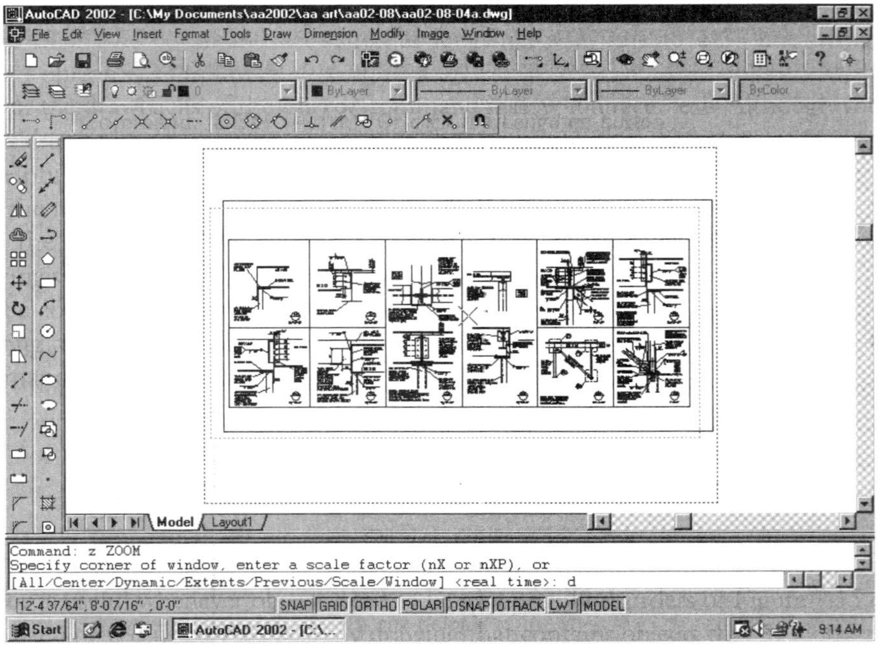

Figure 8.5 *Selecting the* **Zoom Dynamic** *option will display the original drawing area, as well as three view boxes.*

The first box that will be displayed is the Pan box. The Pan box allows you to drag the drawing across the screen to determine which portion of the drawing you would like to view. (See Figure 8.5.) Once the desired portion of the drawing is shown on the screen display, click the select button, and the zooming window, the view box, is displayed. The view box represents the current viewport (viewports will be discussed later). For now, you only have one viewport. The view box allows the size of the current viewport to be adjusted, either by shrinking or enlarging it, or by moving it at its current size around the screen. Adjust the size of the window and press (ENTER). Click the select button if you need to readjust the portion of the drawing to be displayed. It sounds complicated, but it's not. However, it might not be worth the effort to learn it either. **Zoom Dynamic** was a great method of moving through a drawing, until **Realtime Zoom** and **Pan** were added to AutoCAD. With these two options added to the **Standard** toolbar, **Zoom Dynamic** has been rendered almost obsolete.

PAN

You can best visualize the **Pan** command by thinking of your drawing as a large sheet of paper that is rolled up. As you pan across your drawing, it's just as if you were unrolling the drawing to view a different portion. The drawing magnification is not changed, just the portion of the drawing that is displayed. This can be helpful when

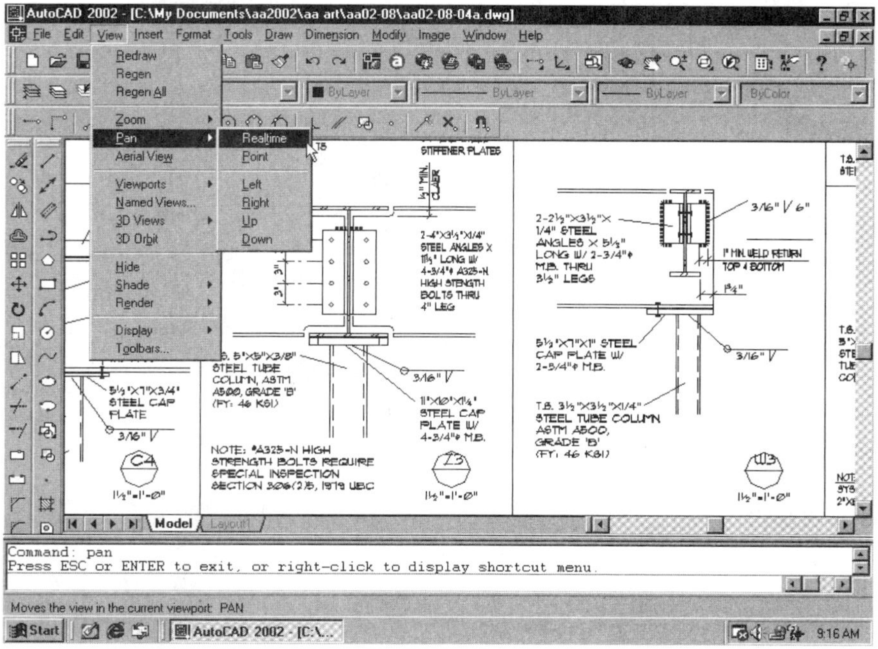

Figure 8.6 *Start the* **Pan** *command by selecting the* **Pan** *button on the* **Standard** *toolbar, by selecting* **Pan** *from the View menu, or by typing* **P** *(*ENTER*) at the Command prompt.*

you've zoomed in on a drawing, similar to the one in Figure 8.2, and you are looking at a specific detail. Rather than having to **Zoom All**, and then **Zoom Window** to a new detail, you can use **Pan** to bring in the desired detail. Start the **Pan** command by selecting the **Pan Realtime** button on the **Standard** toolbar, by selecting **Pan** from the **View** menu, by selecting **Pan** from the right-click shortcut menu or by typing **P** (ENTER) at the Command prompt. The submenu for **Pan** can be seen in Figure 8.6.

PAN REALTIME

The **Pan Realtime** command is similar to **Zoom Realtime**. The easiest method of starting the command is to select the **Pan Realtime** button on the Standard toolbar. No matter the access method, as the command is activated, the cursor will change to a hand cursor. The Command prompt displays the following:

> Press (ESC) or (ENTER) to exit, or right-click to display shortcut menu.

Move the cursor (the hand) to the desired pick point and then press and hold the select button. This point now becomes the displacement point of the **Pan** Command. As the cursor is moved while the select button is being held, the screen display is altered.

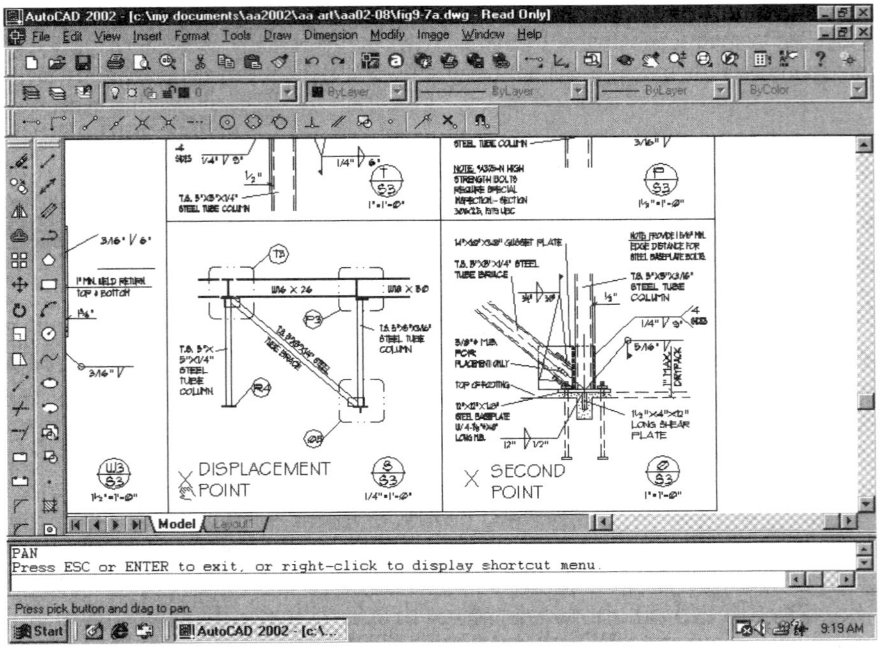

Figure 8.7a Pan *allows a drawing to be "scrolled" across the screen without the drawing magnification being changed. (Courtesy Kenneth D. Smith Architects & Associates, Inc.)*

Panning can be repeated indefinitely, until the desired display is achieved. End the command by pressing (ESC) or (ENTER). Figures 8.7a and 8.7b compare the effects of using the **Pan** Command.

Combining Realtime Zoom and Pan

By entering either **Pan Realtime** or **Zoom Realtime** using the appropriate button on the **Standard** toolbar or the shortcut menu, you can easily toggle between the two options and have a convenient method of moving between different portions of a large drawing. To experiment, try this:

1. Open a drawing and draw several lines and circles.
2. Click the **Zoom Realtime** button and use the select button to enlarge the drawing.
3. Now click the **Pan Realtime** button, and the cursor will change to the hand and allow the enlarged view to be panned.

Remember that when you right-click, a shortcut menu will be displayed that allows you to toggle between the **Zoom** and **Pan** commands.

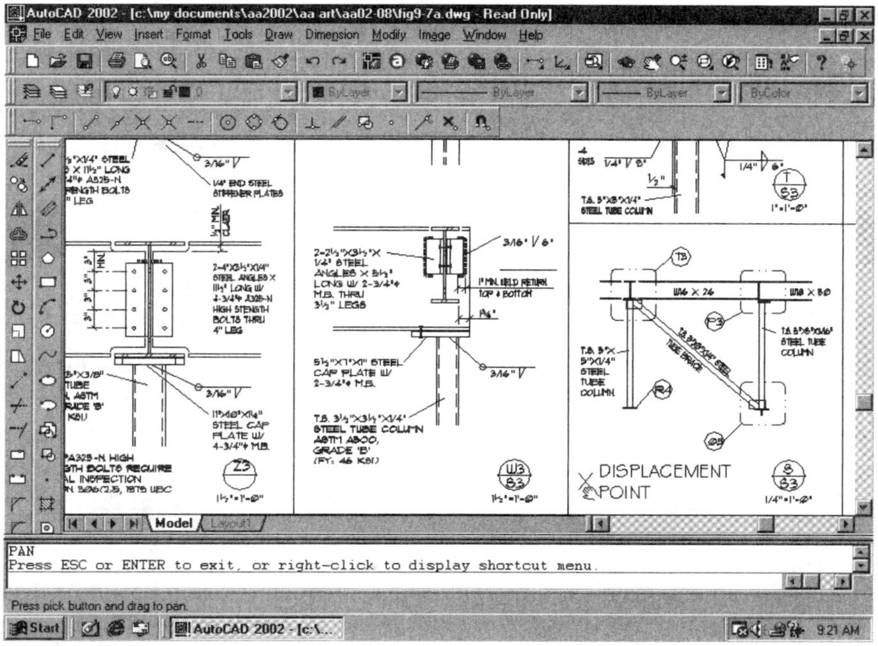

Figure 8.7b *The effects of the* **Pan** *that was started in Figure 8.7a. (Courtesy Kenneth D. Smith Architects & Associates, Inc.)*

Entering Realtime Pan from the Keyboard

You can begin **Realtime Pan** by typing **-P** (ENTER) at the Command prompt. Once started, the command will prompt you to select a point on the drawing screen and then ask, in effect, where would you like to move that point. You can enter a point by providing coordinates from the keyboard or by selecting a point on the screen. The drawing can be panned in greater increments by entering coordinates. Figure 8.7a shows the **Pan** process. The results can be seen in Figure 8.7b. The command sequence is as follows:

> Command: **-P** (ENTER) *(Or click the Pan Realtime button.)*
>
> Specify base point or displacement: *(Enter point or coordinates of point to be dragged.)*
>
> Specify second point: *(Enter point or coordinates to bring first point to.)*

In most cases, the **Pan Realtime** button is much quicker to enter and easier to use. The **–P** (ENTER) option is there to make users of older AutoCAD versions feel at home.

VIEW

An alternative to **Zoom** and **Pan** is the use of predefined drawing views. The **View** command allows views to be named and saved so that they can be restored later. By default, AutoCAD begins each drawing as one view. As the limits are set, you are deciding how big the view is. The screen can be divided into smaller viewports for easy movement between drawings. On a drawing similar to Figure 8.1, each detail box could be named as a view so that each could be revised easily. By naming each detail as a view, you have, in effect, predefined the areas to be viewed. Instead of using **Zoom** or **Pan** to select a view, you could select the desired predefined view. The View command might seem like a waste of time on small exercises. However, the **View** command offers great flexibility for quickly viewing areas of a large, detailed drawing if you need to constantly move from one place to another.

Begin defining views by selecting the **Named Views** button on the **Standard** toolbar, or by selecting the **Named View** button on the **View** toolbar. Selecting **Named Views** from the **View** menu or typing **V** (ENTER) at the Command prompt can also be used to work with named views. Each method will produce the **View** dialog box shown in Figure 8.8. Figure 8.9 shows the details that have been used throughout this chapter and the view names that will be used. The **View** dialog box contains the **Named Views** and the **Orthographic & Isometric Views** tabs. This chapter will explore only the **Named Views** tab.

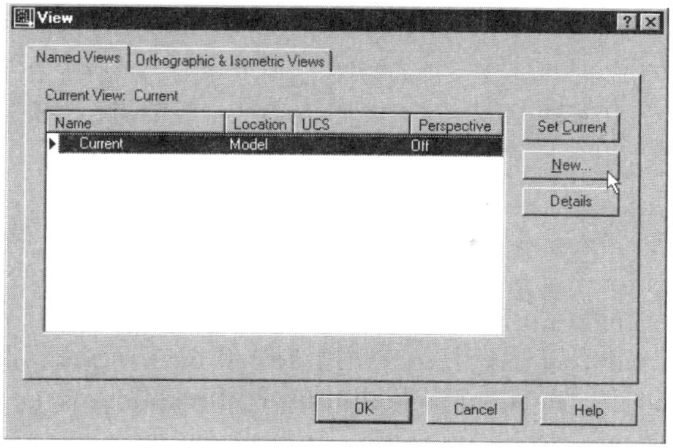

Figure 8.8 *The View command offers great flexibility for quickly viewing areas of a large, detailed drawing. Access the* **View** *dialog box by selecting the* **Named Views** *button on the* **Standard** *toolbar, by selecting the* **Named View** *button from the* **View** *toolbar, by selecting* **Named Views** *from the* **View** *menu or typing* **V** *(*ENTER*) at the Command prompt.*

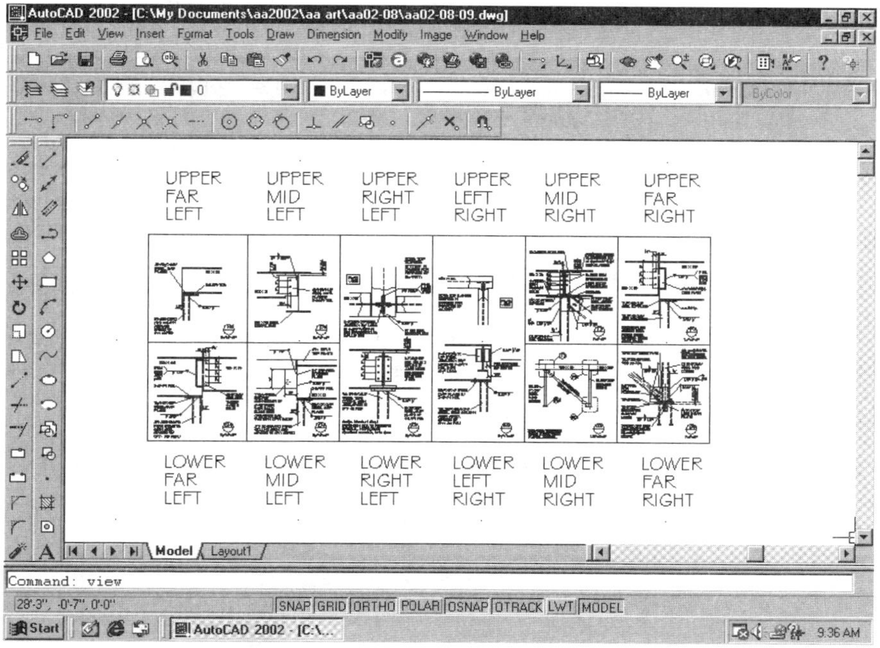

Figure 8.9 *The sheet of the details that have been used throughout this chapter and the view names that will be used.*

EXPLORING THE NAMED VIEWS TAB

The **Named Views** tab can be used to create, set, rename, and delete named views. The **Named Views** tab displays a listing of the current drawing views. In a new drawing, the only view listed is the Current view.

Creating and Saving a Named View

The following steps can be used to create a named view.

1. Select the **New** button on the **Named Views** tab of the **View** dialog box. This will display the **New View** dialog box shown in Figure 8.10.
2. Assign a view name in the **View name** edit box.
3. Choose the **Current display** radio button if the view on the screen is the view that you would like to have saved.

   Choose the **Define window** radio button if you need to select the objects to be included in the view.
4. If you selected the **Define window** button, use the **Define View Window** button to define the view. Selecting this button will remove the dialog box, returning you to the drawing area. Select the objects that will make up the view. Once the objects for the view are selected, the **New View** dialog box will be restored.
5. Click **OK**. The **View** dialog box will be restored, and the name of the selected view will be displayed.

Figure 8.11 shows the **View** dialog box with a partial listing of the twelve views described in Figure 8.9. The **New** option selects and saves the current screen display using a name that you supply. View titles can be up to 255 characters long, following the same guidelines used to name drawing files. Views are often named by their location in the job or by their specific function. If four rows of details will be drawn, each row could be given a letter and each column of details could be given a number to produce an easy grid to work with. Each room of a floor plan can also be turned into a view for easy movement through a project.

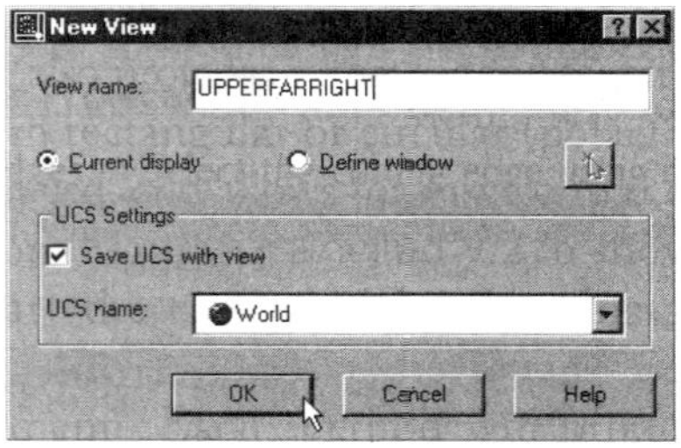

Figure 8.10 *The* **New View** *dialog box is used to assign drawing view names. Selecting the* **New** *button in the* **View** *dialog box displays this dialog box.*

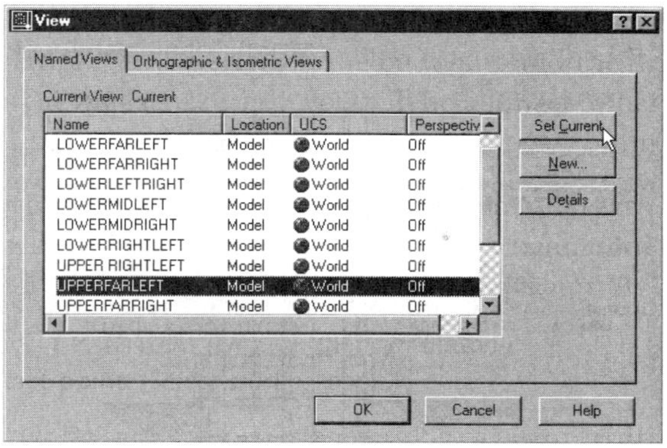

Figure 8.11 *The named views reflected in Figure 8.9 can be seen in the* **View** *dialog box. (Courtesy Kenneth D. Smith Architects & Associates, Inc.)*

Altering the Current View

The screen display can easily be altered between named views. To make the LOWER FAR LEFT view of Figure 8.9 the current view, use the following steps:

1. Display the **View** dialog box. Typing **V** (ENTER) at the Command prompt is the fastest method.
2. Highlight the name of the view that is to become the current view.
3. Click the **Set Current** button.
4. Click the **OK** button. The dialog box will be closed and the LOWER FAR LEFT view will now occupy the screen display area.

Renaming a Named View

The name of a view can be altered in the View dialog box. Use the following steps to change the existing name of a view:

1. Display the **View** dialog box.
2. Highlight the name of the view to be renamed.
3. Right-click to produce the shortcut menu and choose **Rename**.
4. Enter the desired name of the view.
5. Click the **OK** button to accept the new name, close the dialog box, and return to the drawing area.

An alternative to using the shortcut menu to rename a view is to use F2 once the name is highlighted. This will place the selected view name in an edit box and allow the new name to be entered.

Delete

This option will allow one or more views to be removed from the list of saved views. You're not removing the drawing, you're just removing the saved view from the list of named views. Delete views using a procedure similar to renaming them:

1. Display the **View** dialog box.
2. Highlight the name of the view to be deleted.
3. Right-click to produce the shortcut menu and choose **Delete**.
4. Click the **OK** button to close the dialog box and return to the drawing area.

View Details

Selecting the **Details** button in the **View** dialog box will display the **View Details** dialog box, similar to the display in Figure 8.12. This box provides information on the width and height of the selected view. Information provided includes the view name, the area of the view relative to the overall viewport, the coordinates of the view target location, and other information that is beyond the scope of entry-level discussion.

AERIAL VIEW

Aerial View is a useful method of viewing different portions of a large drawing file. An overall view of the entire drawing is displayed in a small window that is overlaid on the drawing screen. This is like the picture-in-picture feature of some TVs. Portions of the drawing can be selected for viewing from this small window and redisplayed while you can still see the entire drawing in the Aerial View window. This is especially helpful when you are working on a very large file. Many AutoCAD users often lose a sense of where they are in the drawing file when they have used the **Zoom** command several times in the drawing. The use of Aerial View provides that per-

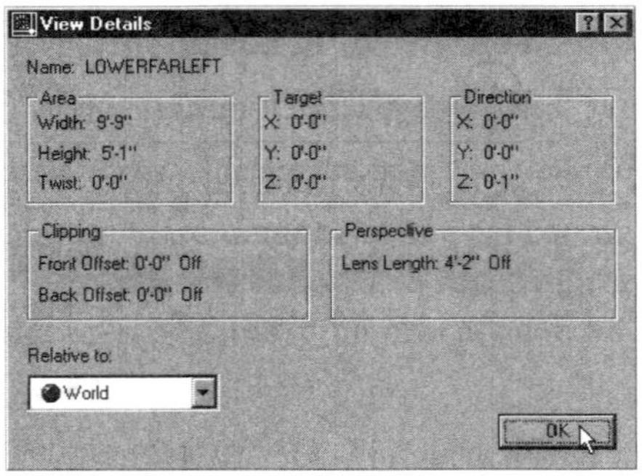

Figure 8.12 *Selecting the* **Details** *button of the* **View** *dialog box displays information including the view name, the area of the view relative to the overall viewport, and the coordinates of the view target location.*

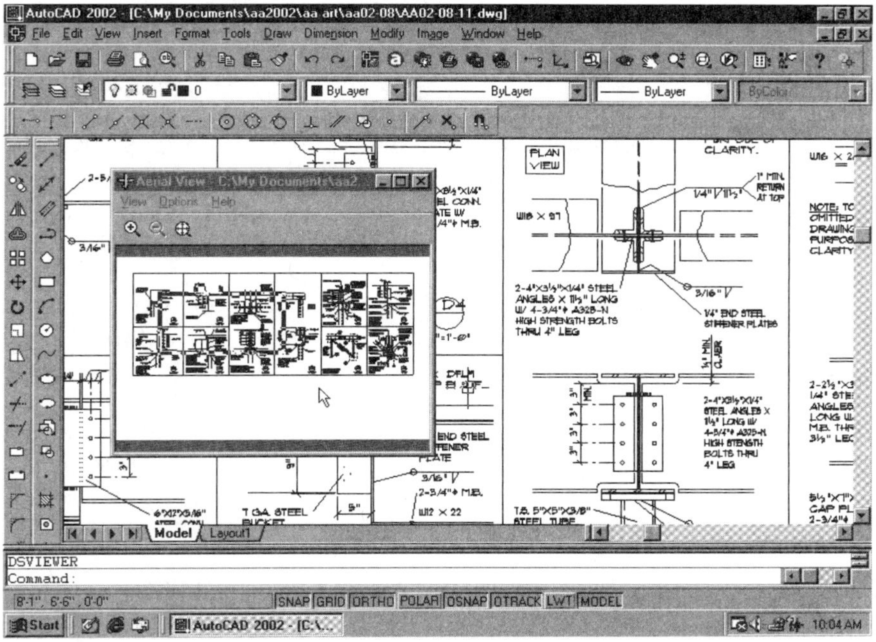

Figure 8.13 Aerial View *provides an option for moving between views of a drawing. The drawing extents are shown in the Aerial View window.*

spective by displaying the overall drawing in the smaller window. It also allows quick movement from one portion of the drawing to another, using **Realtime Zoom** and **Pan** options.

ACCESSING THE AERIAL VIEW WINDOW

Display the Aerial View window by typing **AV** (ENTER) at the Command prompt or by selecting **Aerial View** from the **View** menu. Each method will display the window seen in Figure 8.13. As the aerial view is displayed, it will consist of the current view surrounded by a thick black box. The size of the view can be altered using either **Zoom** or **Pan**.

AERIAL VIEW WINDOW MENUS

In addition to the standard **Help** menu, the Aerial View menus include **View** and **Options**.

View Menu

The options on the **View** menu include **Zoom In**, **Zoom Out**, and **Global**. If the entire drawing is displayed as **Aerial View** is started, the **Zoom In/Zoom Out** options will not be active. These options are used to change the magnification within the Aerial View window.

Zoom In

The **Zoom In** option of **Aerial View** is the same as the option found in the **Zoom** option of the **View** menu. The option increases the magnification of the image displayed in the Aerial View window.

Zoom Out

The **Zoom Out** option of **Aerial View** is the same as the option found in the **Zoom** option of the **View** menu. The option decreases the magnification of the image in the Aerial View window.

Global

Selecting this option will display the entire drawing and the current view in the Aerial View window.

Options Menu

The **Options** menu includes the options of **Auto Viewport**, **Dynamic Update**, and **Realtime Zoom**.

Auto Viewport

This option is a toggle switch between ON/OFF. When a check is displayed, the setting is ON. In the ON mode, switching to a different tiled viewport will automatically cause the new viewport to be displayed in the Aerial View window. In the OFF mode, a new viewport will not be displayed. Tiled viewports will be discussed in later chapters.

Dynamic Update

This option is a toggle switch between ON/OFF. When a check is displayed, the setting is ON. In the ON mode, the window will update the display after each change in the drawing.

Realtime Zoom

This option controls how the window will be updated. In the default setting, the window will be zoomed in real time. If you click this option, the view will be enlarged in **Zoom** rather than in **Realtime Zoom**.

USING THE AERIAL VIEW WINDOW

The **Aerial View** command allows the use of **Realtime Zoom** and **Pan** to move between areas of the drawings.

Zoom

The **Zoom** option of **Aerial View** changes the viewport center and view size by allowing points to be selected that will become the new corners of the current view. Use the following steps to alter the view with the **Zoom** option of the Aerial View window:

1. Display the Aerial View window.
2. Right-click and select **Realtime Zoom**. **Realtime Zoom** is active when the listing is preceded by a check. **Realtime Zoom** can also be activated from the **Options** menu of the Aerial View window.
3. In the Aerial View window, left-click to display the pan and zoom box. You can toggle between pan/zoom by clicking the select button. The Pan window will have an X in the middle of the box. The Zoom window will have an arrow on the right edge of the window.
4. With the **Zoom** option active, move the cursor to control the size of the window (see Figure 8.14).
5. Press ESC to end **Realtime Zoom**.
6. Left-click to toggle to **Realtime Pan**.

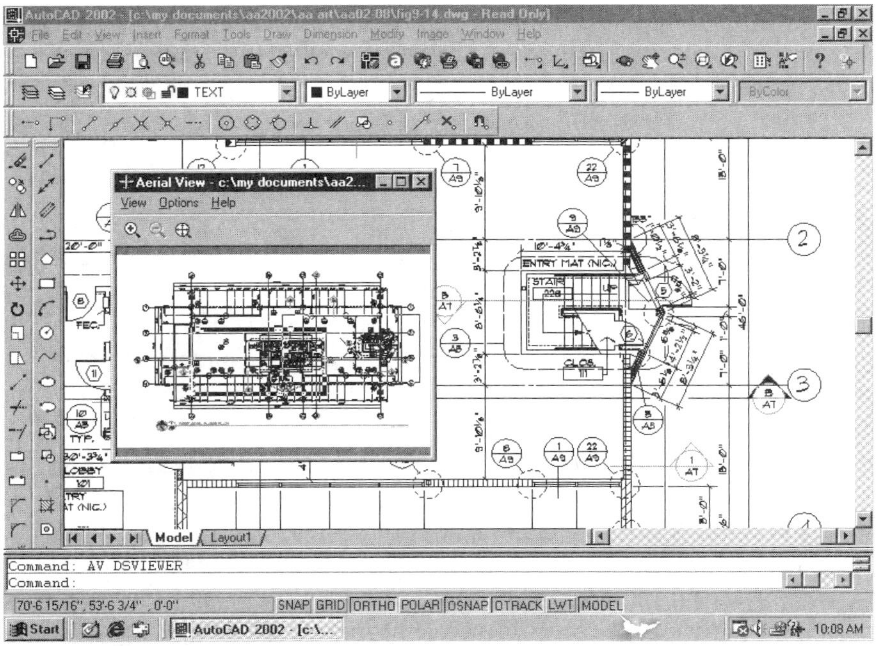

Figure 8.14 *With the* **Zoom** *option active, an Aerial View window can be created. As the cursor is moved, the window size is altered and then the area and size of the drawing are altered.*

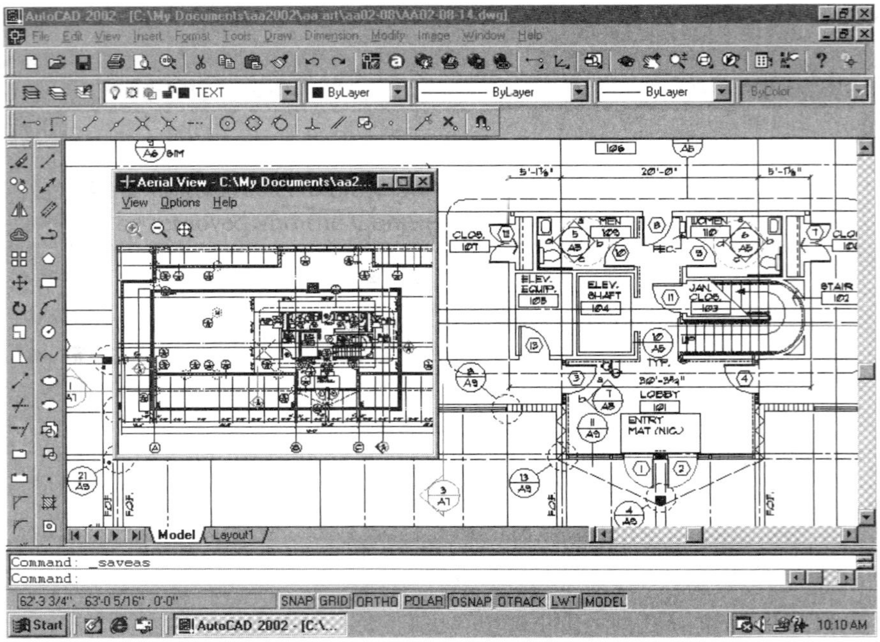

Figure 8.15 *Using the* **Pan** *option in the Aerial View window to pan the drawing. As the reference box is moved in the aerial view, the portion of the drawing displayed in the drawing area of the screen is altered.*

Pan

This option allows the display to be altered from side to side, or top to bottom, without changing the drawing magnification. The Aerial View window provides a point of reference by showing the entire structure. If the **Pan** option is selected, a reference box will be displayed in the Aerial View window to pan the drawing. As the reference box is moved in the Aerial View window, the portion of the drawing displayed in the drawing area of the screen is altered. Move the reference box, and the drawing will snap to show the area contained in the reference box. Without mouse buttons being held down, the drawing will drag in real time to reflect the movement of the reference box. Figure 8.15 shows an example of the effect of the **Pan** option.

REFRESHING THE SCREEN

This chapter has introduced methods of controlling the drawing display using the **Zoom**, **Pan**, **View**, and **Aerial View** commands. Each command alters the view that is placed in the drawing area. The final aspect of controlling the drawing view is adjusting the quality of the view that is displayed. The view quality can be controlled using redraw, regeneration, and view resolution commands.

REDRAW

The **Redraw** command removes the current screen display momentarily and then replaces the drawing. As the drawing is replaced, blips are removed from the drawing editor. Lines that might have been partially erased from view while other objects were edited will also be restored. Start the command by selecting the **Redraw All** button on the **Standard** toolbar or by typing **R** (ENTER) at the Command prompt. The same results can be achieved by selecting **Redraw** from the **View** menu. **Redraw** can be used while another command is in progress if it is entered at any non-text prompt. (Typing **R** (ENTER) in the middle of a command sequence where text is expected will confuse AutoCAD and frustrate you with a misspelled word.) On drawings similar in size to the exercises that you've been doing, **Redraw** will seem almost instantaneous. On larger drawings, such as a subdivision map or floor plan, **Redraw** can take slightly longer.

REGEN

Similar to **Redraw** is the slightly slower **Regen** command. AutoCAD clears the drawing screen and then regenerates the entire drawing. AutoCAD recalculates the current screen display with 14-place accuracy, keeping track of each pixel in the X and Y directions. If you have zoomed in on a specific part of a drawing, **Regen** is regenerating even the part of the drawing that cannot be seen. Start a screen regeneration by typing **REGEN** (ENTER) at the Command prompt or by selecting **Regen** from the **View** menu. To terminate the **Regen** operation, press ESC. **Regen** can also be used to improve the resolution of circular objects and polygons Figures 8.16a and 8.16b compare the effects of **Redraw** and **Regen** on drawing objects. You'll learn how to create viewports in later chapters. Using viewports will speed up each **Regen** operation because only the current viewport is regenerated.

REGENALL

In later chapters you'll learn how to divide the drawing screen into different viewports so that specific areas of the drawing can be viewed quickly. When multiple viewports are created, the **Regen** command will affect only the current drawing viewport. All other viewports will remain unaffected. The **Regenall** command will regenerate all of the views, so that blips in each view will be removed, and reindex the drawing database to increase the accuracy of objection selection. Start the command by selecting **Regen All** from the **View** menu or by typing **REA** (ENTER) at the Command prompt. The significance of the command will be discussed in later chapters.

REGENAUTO

AutoCAD allows the regeneration mode to be set. With a small drawing, an automatic regeneration might not even be noticed. As the drawing size increases, using **Regen**

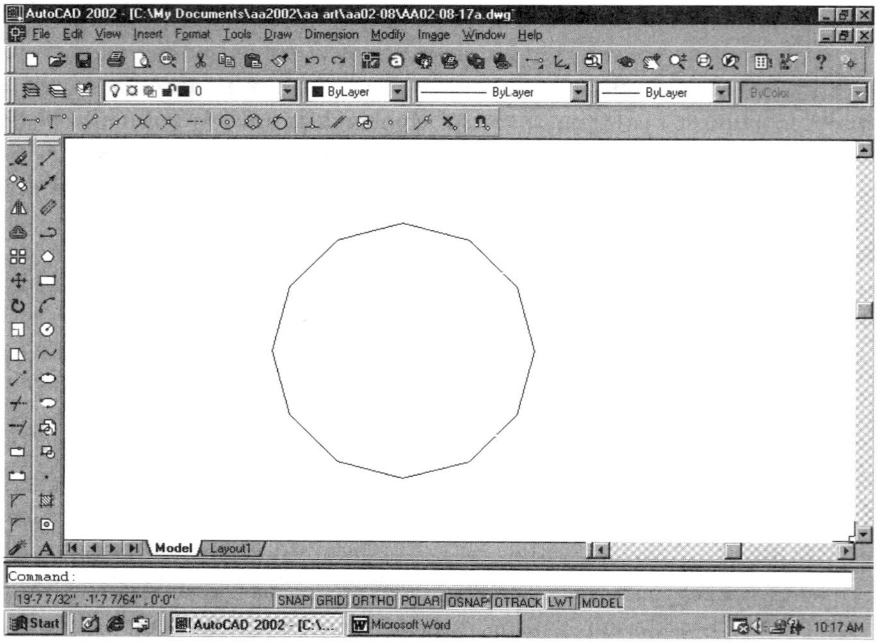

Figure 8.16a *Circular objects often appear as straight line segments when magnified or when a low view resolution is used.* **Redraw** *will not affect the smoothness of such features.*

might prove to be a frustration. AutoCAD allows a prompt to be given to remind you that the drawing is about to be regenerated. This aspect of **Regen** can be entered only by keyboard. The command sequence is as follows:

> Command: **REGENAUTO** (ENTER)
>
> Enter mode [ON/OFF] <ON>: *(Enter On or Off, or press* (ENTER)*)*

If the ON option is retained, AutoCAD will perform an automatic regeneration when needed while altering the screen display in commands such as **Zoom**. When OFF is the current mode and a regeneration is needed as the screen is altered, AutoCAD will display the prompt:

> About to Regen—proceed <Y> Enter Y or N, or press (ENTER)

If you accept the default, the drawing regeneration will proceed, allowing the **Zoom** to take place. If the NO option is selected, the **Zoom** will not be completed. Because of AutoCAD's increased ability to handle huge amounts of information quickly and easily, you really can leave this option in the "that's nice" file of your mind and explore other viewing options.

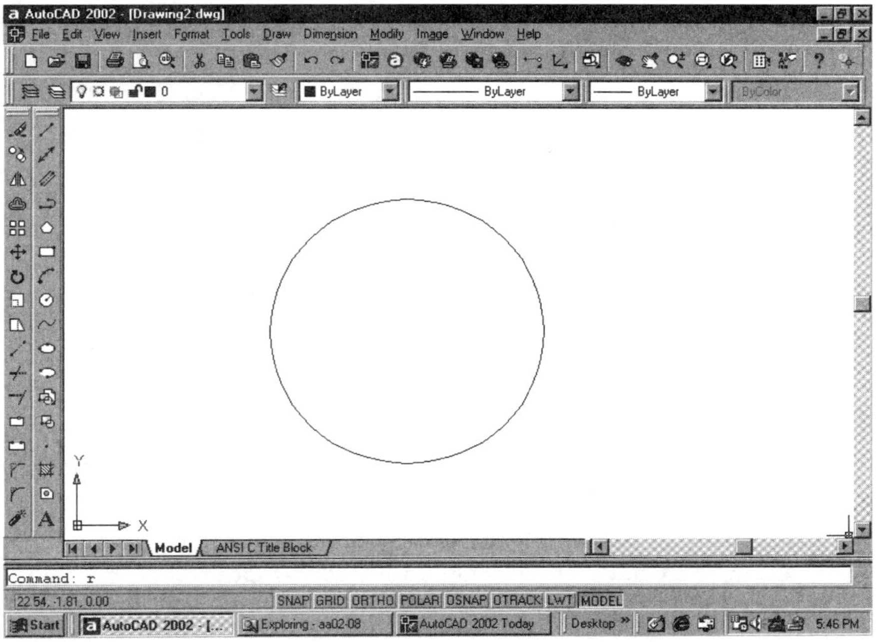

Figure 8.16b *The smoothness of circular objects will be improved with the* **Regen** *command.*

VIEW RESOLUTION

The speed of each **Zoom** or **Pan** can be enhanced if a **Redraw** is used, rather than a **Regen**. The speed of all view changes also can be influenced by the view resolution. Resolution refers to the amount of detail that is represented when arcs and circles are drawn. The higher the resolution, the more lines that are used to represent an object and the smoother the arc or circle will appear. Higher resolution also means that more time is needed for zooms and pans.

The resolution of arcs and circles can be controlled from the **Options** dialog box or by keyboard. To control the resolution from the dialog box, select **Options** from the **Tools** menu and then select the **Display** tab. The **Arc and circle smoothness** option in the **Display Resolution** portion can be used to alter the smoothness of an arc. The **Segments** in a polyline curve option will be considered in later chapters. The other two options in this box affect 3D drawings and will not be considered in this text.

Arc and Circle Smoothness

The **Arc and circle smoothness** option controls the smoothness of circles. A value from 1 through 20,000 can be provided in the edit box, in place of the default value of 100. With a low number such as 25 or 50, circles will appear as a series of flat lines.

As the value increases, the smoothness of the circle or arc is increased. Regeneration time on large drawings will slow slightly as the value is increased.

The resolution of a circular feature can also be adjusted using the **Viewres** command. The command is entered from the keyboard using the following sequence:

> Command: **VIEWRES** (ENTER)
>
> Do you want fast zooms? [yes/no] <Y> *(a Y or N is required.)* (ENTER)
>
> Enter circle zoom percent (1–20000) <100>: **150** (ENTER)
>
> Command:

In setting **Viewres**, the first decision to be made is for fast zooms.

- Responding **N** (ENTER) will cause all **Zoom**, **Pan**, and **View** restorations to be done using regeneration.
- Accepting the default of Yes will keep a large virtual screen and execute most **Zoom**, **Pan**, and **View** restorations at **Redraw** speed.

Zooming to the limits of a drawing might require regeneration, even though you have selected fast zooms. When a small circle is drawn with a low value, it will appear as a polygon. Using the **Regen** command will restore its circular shape.

CHAPTER 8 EXERCISES

1. Open drawing E-6-9. Use the **Zoom** command to make the view containing the triangles the current view. Save this view as TRIANGLES. Zoom into each of the remaining shapes, and make each shape a separate saved view. Save the entire drawing as E-8-1. List each view name in the space provided.

2. Using drawing E-8-1 as a base, make the view containing the squares the current view. Draw a square in the center of one of the existing squares using Midpoint Object Snap to form a diagonal square. Repeat this process four additional times, using **Zoom** as required. Inside the final square, draw a circle that is tangent to the smallest square. Save this view with the name TINY and save the entire exercise as E-8-2.

3. Load E-8-1 and enlarge the drawing display to twice its current size. Save this view as VIEW2X. Zoom out so that the edges of the largest triangle touch the edge of the screen. Name this view with the name of the command required to perform the zoom. Set the drawing so that screen markers will be produced and save the drawing as E-8-3. Write the name of the views created and the commands used in the space below.

   VIEWS COMMANDS

4. Load drawing E-8-3 and use it as a base for this exercise. Change the limits of this drawing to 36,18 and set the screen display so that the total content of the limits will be displayed. The original view is now displayed:

   Use **Zoom Dynamic** to make the object's original view fill the screen. Write the steps that were required to adjust the views. Save the drawing as E-8-4.

5. Start a new drawing and set the limits to 96',144' with a grid of 24" and a snap of 6". Set units to measure in architectural, with fractions set at 16. Set angles to be measured in degrees/minutes/seconds with two-place accuracy. Set the view resolution to 150. Divide the drawing into the following seven views: All, Upper Left, Upper Cen, Upper Right, Low Right, Low Cen, and Low Left. Set All as the current view. Save the drawing as a drawing template. This drawing can be used as a future base for drawings done at 1/4"=1'–0" scale. Save the drawing as E-8-5.

6. Open a file of your choice and establish a minimum of three named views. Provide names that quickly describe the area captured in the viewport. Save the file as E-8-6.
7. Open a file of your choice and establish an Aerial View window. Use the window to zoom a portion of the drawing and then save the file as E-8-7.

CHAPTER 8 QUIZ

1. What command controls the use of screen markers, and how should it be set to provide markers?
2. Compare and explain the difference in **Viewres** settings between 50 and 500 and the default value.
3. Explain the difference between **Regen** and **Redraw**.
4. Describe the difference between **Zoom All** and **Zoom Extents**.
5. What commands will be needed to make a drawing exactly twice as large as the current screen display?
6. What option is used to scale space relative to paper space units?
7. Describe the meaning of "displacement point" and "second point."
8. List the steps to enlarge a portion of a drawing using the **Zoom** command.
9. List six options for accessing the **Zoom** command.
10. Describe the process that allows the **Zoom** command to be used in the middle of another command.
11. What term is used to describe a command that can be used while another command is being executed?
12. What command would you use if you wanted to create a picture-in-picture of an AutoCAD drawing?
13. What must be typed at the Command prompt to start **Aerial View**?
14. Describe how to reduce the size of the Aerial View window.
15. Other than putting you to sleep, what's the point of all of the options presented in this chapter?

SECTION 2

ENHANCING DRAWINGS

CHAPTER 9

Basic Methods of Selecting and Modifying Drawing Objects

By now you've explored the basic components of a drawing. This chapter will explore methods for

- Selecting objects for editing, including the pick box, Window, Crossing window, automatic selection mode, Wpolygon, Cpolygon, Fence, All, and Last
- Modifying the selection process using Noun/Verb, Shift to Add, Press and Drag, Implied Window, and Object grouping
- Modifying drawings by removing objects
- Modifying drawings by adding objects

Commands explored in this chapter include

- **Pickbox**
- **Erase**
- **Copy**
- **Mirror**
- **Mirrtext**
- **Offset**
- **Array**

Each command will reproduce all or parts of a drawing. A second group of editing commands for refining drawings and two new methods of sorting objects will be explored in the next chapter.

SELECTING OBJECTS FOR EDITING

AutoCAD provides many options for selecting objects. Individual objects can be selected using a pick box. Groups of objects can be selected for editing using one of the following options: Window, Crossing, Auto, WPolygon, CPolygon, Fence, All, Last, Add, Remove, or Multiple. As different methods of selecting objects for editing are discussed, each selection method will be introduced using the **Erase** command, although other commands can be used. These object selection options can be used any time an object selection needs to be made within an editing command sequence. (Selecting a button on a toolbar or entering the command at the keyboard are generally the fastest methods.) No matter which method is used to access the command, a prompt will be displayed asking you to

Select objects:

Each of the following selection methods can be entered at the Select objects prompt for any of the editing commands.

SELECTING OBJECTS USING A PICK BOX

As you've edited your drawings to this point, you selected objects to be edited one at a time. One or more objects have been selected using the pick box and clicking the select button. As the prompt for the edit command is entered, the crosshairs will change to an object selection target or pick box and the object to be edited can be selected. Any object that is included in the pick box will be affected by the command. If the size of the pick box is altered, it will affect the accuracy of your selections. As the size of the box is enlarged, information that is not desired in the selection set might be added. With the box too small, you might have trouble selecting information.

The size of the box can be altered to enlarge or reduce the size of the selection area. Adjust the size of the pick box on the **Selection** tab of the **Options** dialog box (select **Options** from the **Tools** menu). As the slide bar is moved to the right, the size of the pick box is enlarged. Move the slide bar to the left to decrease the size of the pick box. The size can also be adjusted from the keyboard. Typing **PICKBOX** (ENTER) at the Command prompt will produce a prompt of

Command: **PICKBOX** (ENTER)

Enter new value for PICKBOX <3>: *(Enter desired value.)* (ENTER)

AutoCAD is now waiting for you to enter a numeric value from 0 to 50, to represent the size of the pickbox.

Selecting the Object

You can select objects to be edited by choosing any point on the object. Place the pick box on top of the object and click the select button. You can also select objects by entering coordinates for the location of the target. When you select objects with a thickness, it is important to place the pick box so that it touches an edge and not the center of the object. You also need to be careful not to select objects at their intersections with other objects.

SELECTING WITH A WINDOW

Rather than selecting objects to edit one at a time, you can select a group of objects by surrounding them completely in a window by responding **W** (ENTER) at the Select objects prompt. This procedure was used in Chapter 8 with the **Zoom** command and is shown in Figure 9.1a. Any object that lies entirely inside the selection window will be edited. Any object that is partially inside the window will not be edited. If the selection set in Figure 9.1a is accepted, the text that is within the selection box will be removed, and the leader lines that pass through the right side of the box will be not be affected by the edit. The result of the selection set can be seen in Figure 9.1b. The command sequence to use the **Erase** command with a window is as follows:

> Command: **E** (ENTER) *(Or click the Erase button.)*
>
> ERASE
>
> Select objects: **W** (ENTER)
>
> Specify first corner: *(Select point for first corner.)*
>
> Specify opposite corner: *(As the first point is selected, a window will*

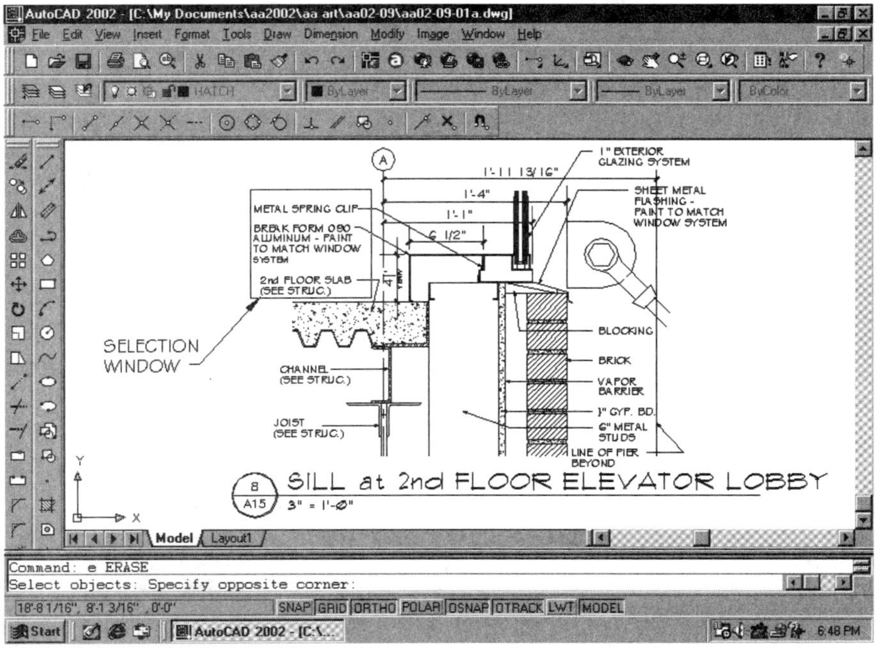

Figure 9.1a *If the Window option is used for selecting objects to be edited, all objects that lie entirely in the window will be edited. In this selection set, the text in the window will be removed, but the leader lines that cross the right side of the selection window will not be included in the selection set. (Courtesy Peck, Smiley, Ettlin Architects.)*

drag across the screen until a second point is selected.)

Select objects: *(Selected objects will be highlighted. Press* (ENTER) *when selection set complete.)*

When prompted to select objects, you now have an opportunity to alter the selection set by adding additional objects using another window, or by using the Add or Remove selection methods. If no changes need to be made, press (ENTER). Making changes to the selection set will be introduced later in this chapter.

SELECTING WITH A CROSSING WINDOW

Choosing an object to edit with the Crossing option is similar to the Window option. You can select groups of objects by surrounding them partially in a window by typing **C** (ENTER) at the Select objects prompt. The command process will be the same as the Window option. The crossing window will appear different from the Window option window. The crossing window is constructed using a dashed line, and a continuous line represents the Window option window. Any objects that lie entirely in the window will be edited. Any objects that are partially inside the window will also be edited. If the selection set in Figure 9.1a was selected using the Crossing option and accepted for editing, both the text and the leader lines would be changed by the specified editing command (see Figure 9.2).

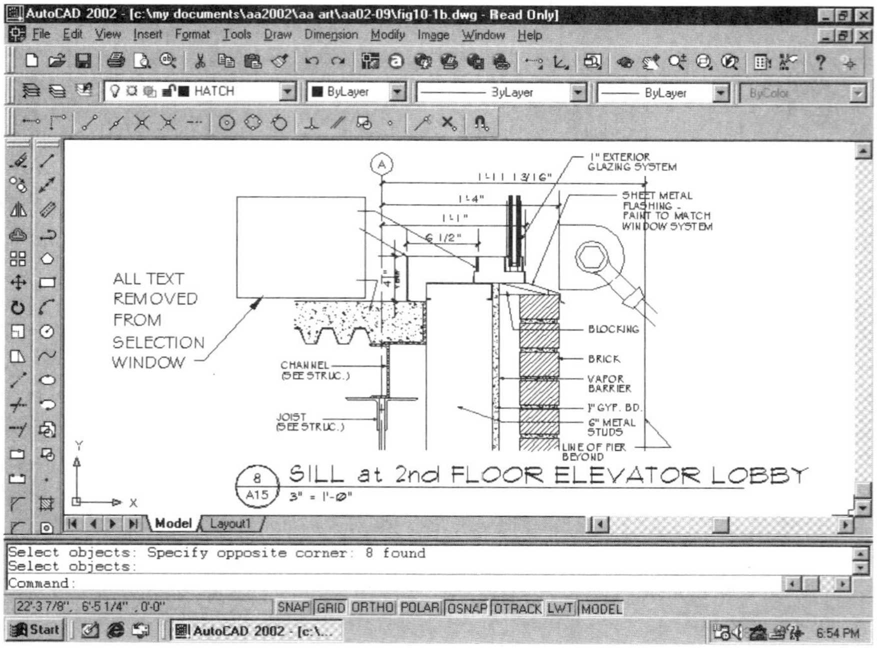

Figure 9.1b *The result of the* **Erase Window** *option started in Figure 9.1a.*

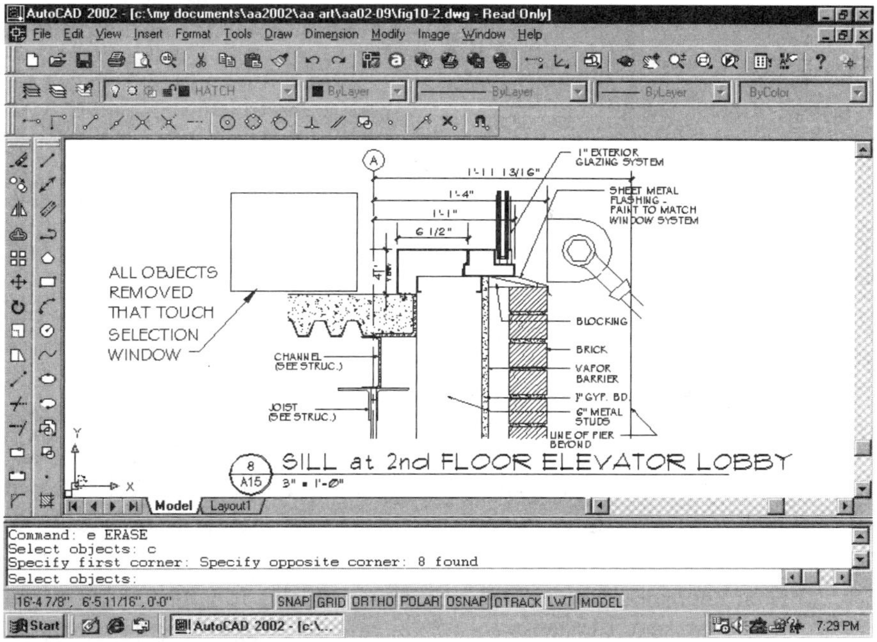

Figure 9.2 *The result of the* **Erase Crossing** *option started in Figure 9.1a.*

AUTO SELECTION MODE

The auto selection mode is similar to editing using a window. The default mode for selecting objects is with the pick box. If the select button is clicked after the editing command is selected, the window method of selecting objects will be used. When prompted for the object to be erased, click the select button. This will turn the pick box into the corner of the selection box. As the cursor is moved, the window will be formed. Any items that the window encloses will be selected. When you're satisfied with the size of the selection box, click the select button to accept the objects. To remove the selected objects from the drawing, press (ENTER). The command process is as follows:

Command: **E** (ENTER) *(Or click the Erase button.)*

ERASE

Select objects: *(Click the select button to select the first corner of the selection window.)*

Specify opposite corner: *(The window will drag across the screen until a second point is selected.)*

Select objects: (ENTER)

Command:

Implied Window Selection

The direction that is used to move the cursor as the selection window is defined will affect the selection process. If the cursor is moved from left to right when the selection set is picked, only objects completely enclosed in the window will be erased. When the selection window is created by moving from right to left, any object that the window crosses will be affected by the edit command. The results of implied window selection can be seen in Figure 9.3.

SELECTING OBJECTS WITH THE WPOLYGON OPTION

The WPolygon option is especially useful for selecting an irregularly shaped group of objects to edit. The option is similar to the Window option, but you are allowed to select points to form a polygon around objects to be edited. The polygon can be any shape except a shape that crosses itself, such as a figure eight. The command sequence is started once the editing option has been entered and you are prompted to select points to form the polygon. You can select points to define the polygon by using a pointing device or you can enter them as coordinates, as seen in Figure 9.4. Close the polygon by pressing ENTER. All objects lying totally inside the polygon will be edited. The process is completed using the following steps:

Command: **E** (ENTER) *(Or click the Erase button—any editing command can be used.)*

ERASE

Select objects: **WP** (ENTER)

First polygon point: *(Enter a point.)*

Specify endpoint of line or [Undo]: *(Enter a point.)*

Specify endpoint of line or [Undo]: *(Enter a point.)*

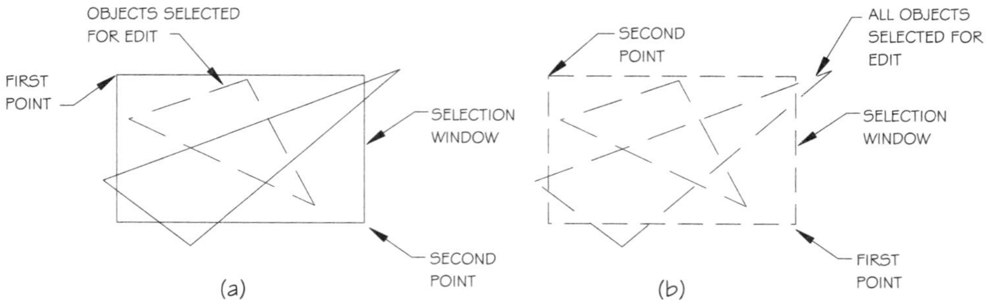

Figure 9.3 *If the cursor is moved from left to right, only objects completely enclosed will be in the selection set. When the selection window is created by moving from right to left, any object that the window crosses will be affected by the edit command.*

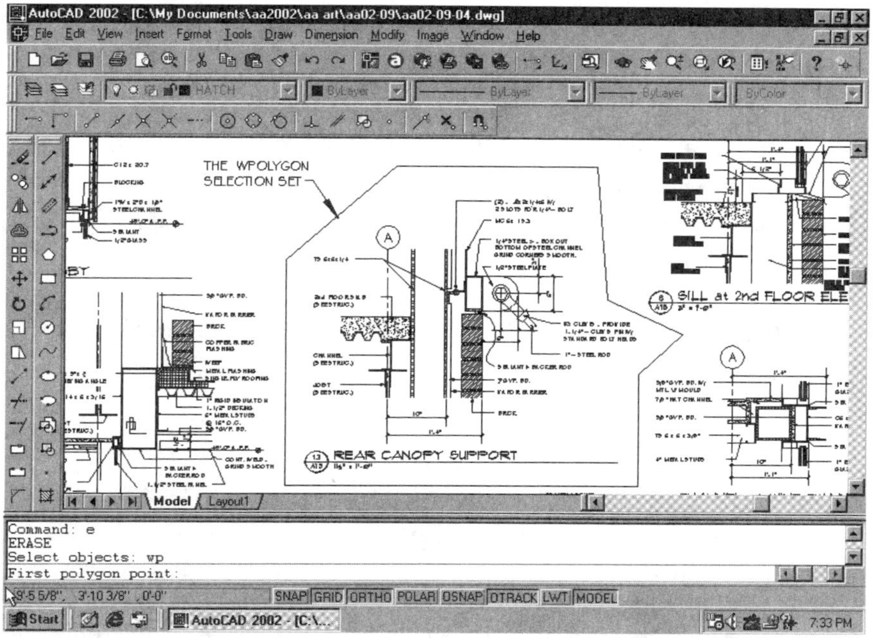

Figure 9.4 *The WPolygon option can be used as an alternative to the Closing option to select objects in an irregularly shaped group. Points for the polygon can be entered by selecting points or entering coordinates. (Courtesy Peck, Smiley, Ettlin Architects.)*

Specify endpoint of line or [Undo]: *(Enter a point.)*

Specify endpoint of line or [Undo]: *(Continue until desired objects are selected.)* (ENTER)

100 found *(The number will vary as AutoCAD reports the number of objects in the selection set.)*

Select objects: (ENTER)

Command:

SELECTION OBJECTS WITH THE CPOLYGON OPTION

Objects can also be edited by typing **CP** (ENTER) at the Select objects prompt. This option will be similar to the effects of combining Crossing and WPolygon, in that it will edit all objects that are within or are crossed by the polygon. The command sequence is as follows:

Command: **E** (ENTER) *(Or click the Erase button—any editing command can be used.)*

ERASE

Select objects: **CP** (ENTER)

First polygon point: *(Enter a point.)*

Specify endpoint of line or [Undo]: *(Enter a point.)*

Specify endpoint of line or [Undo]: *(Enter a point.)*

Specify endpoint of line or [Undo]: *(Enter a point.)*

Specify endpoint of line or [Undo]: *(Continue until object set is complete.)*
(ENTER)

100 found *(The number of objects in the selection set will vary.)*

Select objects: (ENTER)

Command:

SELECTING OBJECTS USING A FENCE

Objects can be selected for editing by Fence. This option is similar to WPolygon and CPolygon, except that it does not close the polygon. The fence might be a series of line segments or it could form a polygon that can cross itself. This allows long rows or columns to be edited in one command sequence. As seen in Figure 9.5, the selection fence edits only items that intersect or cross the fence. Fence points can be selected with a pointing device or by entering coordinates. The command sequence is as follows:

Command: **E** (ENTER) *(Or click the Erase button—any editing command can be used.)*

ERASE

Select options: **F** (ENTER)

First fence point: *(Enter a point.)*

Specify endpoint of line or [Undo]: *(Enter a point.)*

Specify endpoint of line or [Undo]: *(Enter a point.)*

Specify endpoint of line or [Undo]: *(Enter a point.)*

Specify endpoint of line or [Undo]: *(Continue until object set is complete.)* (ENTER)

5 found

Select objects: (ENTER) *(Selecting* (ENTER) *will execute the editing command.)*

Command:

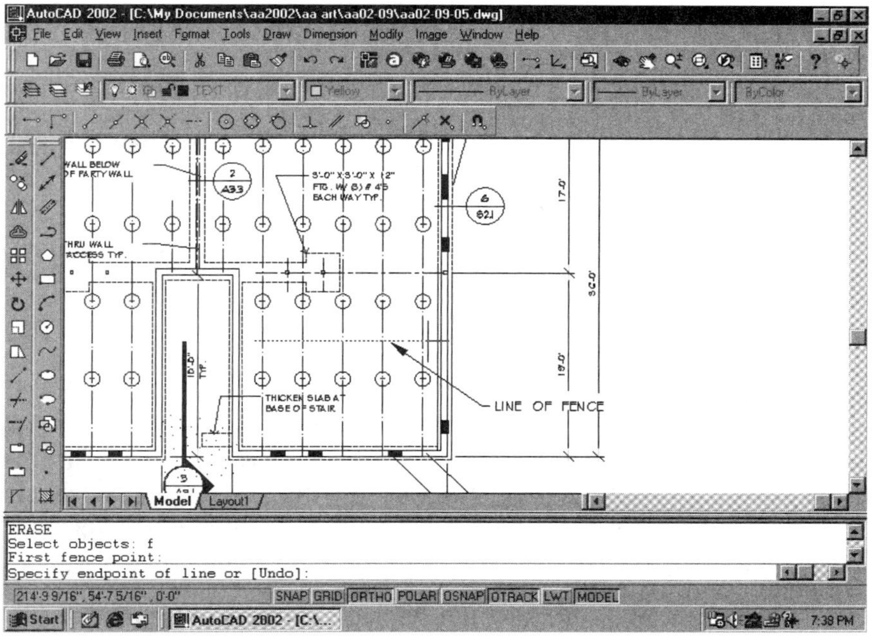

Figure 9.5 *The Fence option allows the use of straight lines to include objects in the selection set. Any objects that are touched by the fence will be included in the selection set. (Courtesy Scott R. Beck, Architect.)*

SELECTING ALL DRAWING OBJECTS

When the All option is used, all objects that are current will be affected. You've learned how to freeze certain layers so that they cannot be seen. All objects except those on frozen or locked layers will be edited.

Command: **E** (ENTER) *(Or click the Erase button—any editing command can be used.)*

ERASE

Select objects: **ALL** (ENTER) *(All items will be highlighted.)*

137 found

Select objects: (ENTER) *(All items will be removed.)*

Command:

SELECTING THE LAST OBJECT

The Last option allows the most recently created object to be selected for editing.

Command: **E** (ENTER) *(Or click the Erase button—any editing command can be used.)*

ERASE

Select options: **L** (ENTER)

1 found

Select options: (ENTER)

Command:

ADDING OBJECTS TO THE SELECTION SET

When selection sets are defined, you might not have selected all objects intended for editing. Typing **A** (ENTER), for Add, at the Select objects prompt will allow for additional objects to be added to the selection set. Once Add is entered, objects can be added to the selection set individually or by any other selection method. When you are satisfied with the selection set, press (ENTER) to edit the drawing. Objects are added to a selection set using the following steps:

Command: **E** (ENTER) *(Or click the Erase button—any editing command can be used.)*

ERASE

Select objects: **W** (ENTER) *(Any selection method can be used.)*

Specify first corner: *(Select first corner.)*

Specify opposite corner: *(Specify second point.)*

5 found

Select objects: **A** (ENTER)

Select objects: **C** (ENTER) *(Any selection method can be used.)*

Specify first corner: *(Select first corner.)*

Specify opposite corner: *(Specify second point.)*

20 found, 25 total

Select objects: (ENTER)

Command:

REMOVING OBJECTS FROM THE SELECTION SET

This option will allow for objects to be removed from the selection set. Once objects are selected for editing, typing **R** (ENTER) will allow objects to be taken away from the selected objects to edit. Objects can now be removed from the selection set individually or with any other selection method. When you are satisfied with the selection set, click the select button to edit the drawing. The command sequence is as follows:

Command: **E** (ENTER) *(Or click the Erase button—any editing command can be used.)*

ERASE

Select objects: **C** (ENTER) *(Any selection method can be used.)*

Specify first corner: *(Select first corner.)*

Specify opposite corner: *(Specify second point.)*

5 found

Select objects: **R** (ENTER)

Select objects: *(Select objects to be removed using any selection method.)*

1 found, 1 removed, 4 total

Remove objects: (ENTER)

Command:

MULTIPLE SELECTION SET

The Multiple selection option works well when drawing objects are stacked over each other. In Chapter 5 you learned to place parts of your drawing on different layers. The Multiple option will allow you to edit material on different levels at one time.

DEFINING THE SELECTION MODE

The **Selection** tab of the **Options** dialog box offers six modes that will affect the formation of the selection set. Choosing **Options** from the **Tools** menu opens the dialog box. An examination of this tab will reveal that three of the settings are active. Any combination of selection methods can be used.

NOUN/VERB SELECTION

Up to this point, as you have edited an object, you've used a process AutoCAD referred to as Verb/Noun. You've selected the verb (**Erase**) and then selected the object to be edited (the noun). With the **Noun/Verb selection** check box active, AutoCAD allows you to reverse the process. The object can be selected first, and then the editing command.

With **Noun/Verb selection** active, any object can be selected merely by choosing it with the pick box. Open a drawing and select a line. Notice the line is highlighted and a box appears in the middle and at each end of the line, but the Command prompt has not changed. Typing **E** (ENTER) will erase the selected line. The command sequence is as follows:

Command: *(Select object to edit using any method.)*

Command: **E** (ENTER) *(Or click the Erase button—any listed editing command can be used.)*

Noun/Verb selection can be used with the following commands to edit objects:

Array	**Ddchprop**	**Mirror**	**Wblock**
Block	**Dview**	**Move**	**Explode**
Change	**Erase**	**Rotate**	
Chprop	**Hatch**	**Scale**	
Copy	**List**	**Stretch**	

Some commands require that the edit command be executed before the objects can be selected. These commands include the following:

Break	**Divide**	**Fillet**	**Offset**
Chamfer	**Extend**	**Measure**	**Trim**

If you use these commands, they will ignore any selection set created prior to starting the command.

USING SHIFT TO ADD SELECTION

This selection method controls how objects are added to an existing selection set. Start a new drawing and draw a few lines. Toggle **Noun/Verb selection** to OFF (no X) and **Use Shift to add to selection** ON (X). Click **OK** and return to the drawing. Select an object to erase, and it will be highlighted. Select a second object, and the first object no longer will be highlighted. To retain the first object, hold the shift key down while the second object is selected. Both objects will now be edited. If **Use Shift to add to selection** is not enabled, objects are added to the selection by selecting them individually or using a selection window. To remove objects from the selection set, press SHIFT while selecting the object.

PRESS AND DRAG SELECTION

This mode determines the method used for drawing a selection window. With the **Press and drag** option active, the selection window can be made with one button selection rather than two. Pressing and holding the select button and moving the cursor diagonally creates the selection window. Releasing the select button completes the window when it is the size you want. When this option is not selected, objects can be selected for editing using the normal Window option.

IMPLIED WINDOWING

With the **Implied windowing** option in the default mode, a selection window is automatically created when the Select object prompt is displayed. Clicking the select button will automatically select the first corner point. A prompt will then be given requesting the other corner. If you draw the window from left to right, the window selects objects that lie entirely within the window. If the window is drawn from right to left, all objects in and touching the window will be selected. With the option off, the window must be set for each edit. In the OFF position, clicking the select button does nothing. A mode such as Window or Crossing needs to be entered to start the creation of a window.

OBJECT GROUPING

The **Object grouping** option determines if grouped objects will be recognized as individual objects or as a group. In the default setting of ON, grouped objects function as a group. When OFF, grouped objects can be edited individually. Grouping of objects will be discussed in later chapters.

ASSOCIATIVE HATCH

The **Associative hatch** option affects how the hatch patterns are edited and will be discussed in Chapter 14.

MODIFYING A DRAWING USING ERASE

The **Erase** command allows you to remove unwanted objects (mistakes) from a drawing. Because you're familiar with **Erase** by now, the command serves as a means of using some of the selection methods just covered. The selection prompts Point, Multiple, Window, Crossing, WPolygon, CPolygon, Fence, All, Last, Add, and Remove should start to seem more meaningful. Figure 9.6 shows a detail marker that has been selected for editing by automatic window. The selected object is highlighted, while AutoCAD waits for you to select other objects. When the selection process is complete, press (ENTER) to remove the object.

The Window option can be seen in Figure 9.7. Notice that at the Select object prompt, **W** was entered rather than **WINDOW**. Remember, an object must be entirely inside the window to be edited. Figure 9.8 shows the objects to be erased as highlighted. If you would like to Add or Remove objects from the selection set, enter the corresponding letter at the Select objects prompt. Figure 9.9 shows the effects of the Window **Erase** with no objects added or removed.

If you type **C** (ENTER) at the Select object prompt, objects that are within or are crossed by the window are selected to be erased, as seen in Figure 9.10. The result of a Crossing window selection can be seen in Figure 9.11. Because the engineer would like to keep the beams and remove the column, the Remove option was used to restore

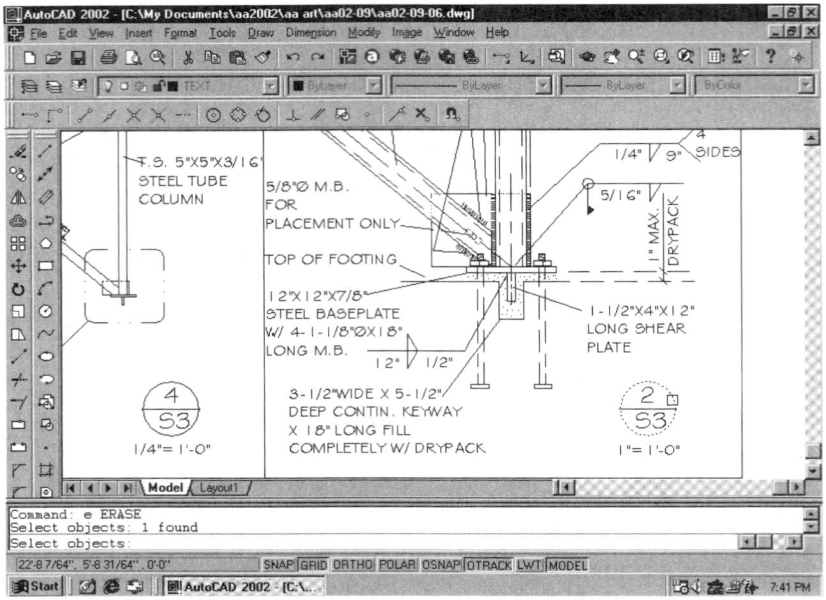

Figure 9.6 *The detail marker at the bottom of the drawing has been selected for* **Erase**. *(Courtesy Kenneth D. Smith Architect & Associates, Inc.)*

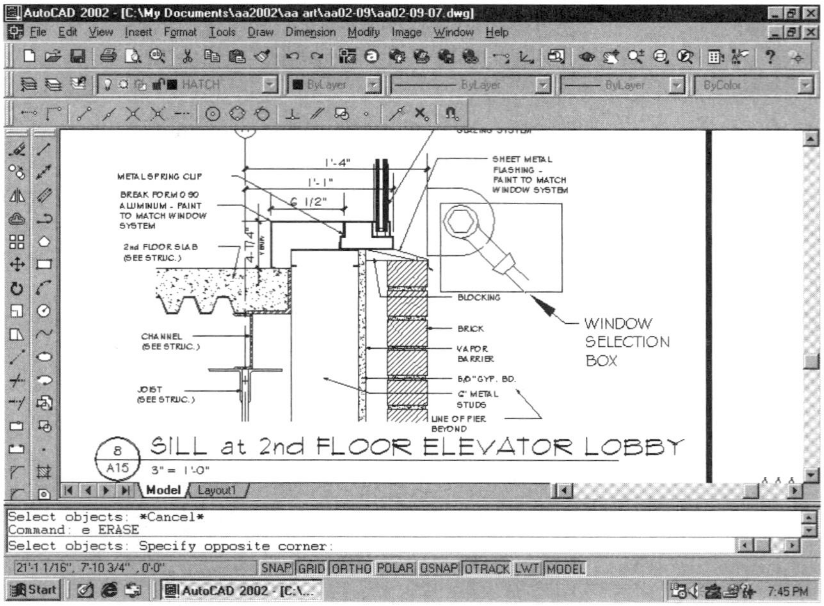

Figure 9.7 *Using automatic selection or the Window option to form a selection set for the* **Erase** *command. (Courtesy Peck, Smiley, Ettlin Architects.)*

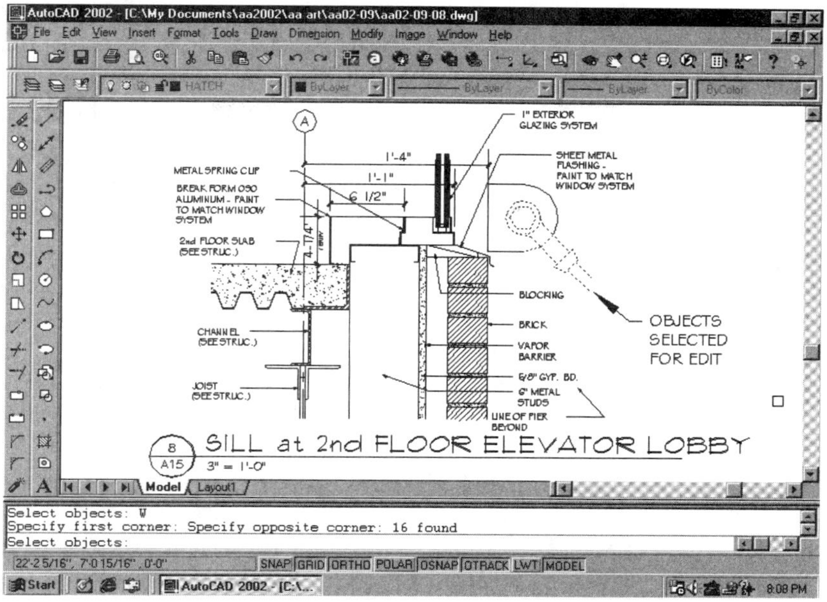

Figure 9.8 *Results of the Window selection set are now displayed to confirm the choice of objects to be included in the selection set. Notice that the number of objects included in the set is displayed above the Command prompt. (Courtesy Peck, Smiley, Ettlin Architects.)*

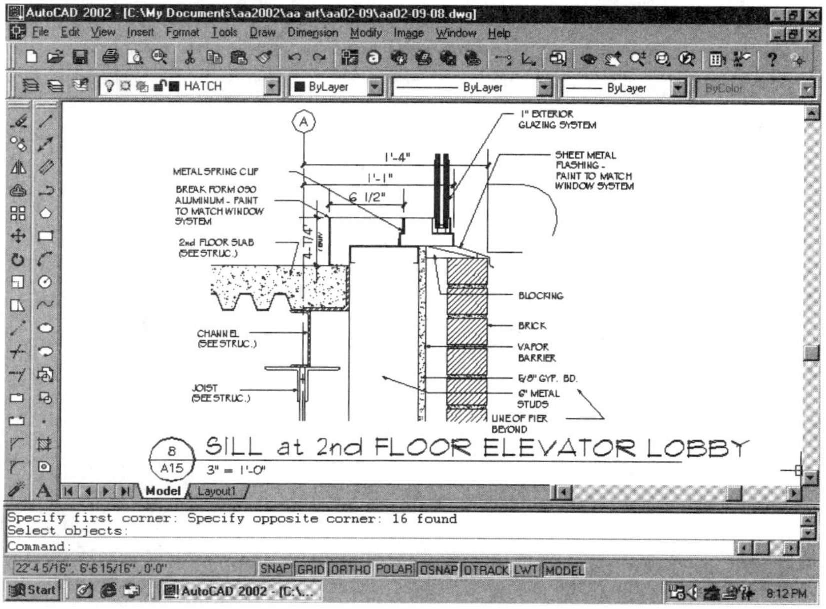

Figure 9.9 *The effect of the Window option used with the* **Erase** *command. (Courtesy Peck, Smiley, Ettlin Architects.)*

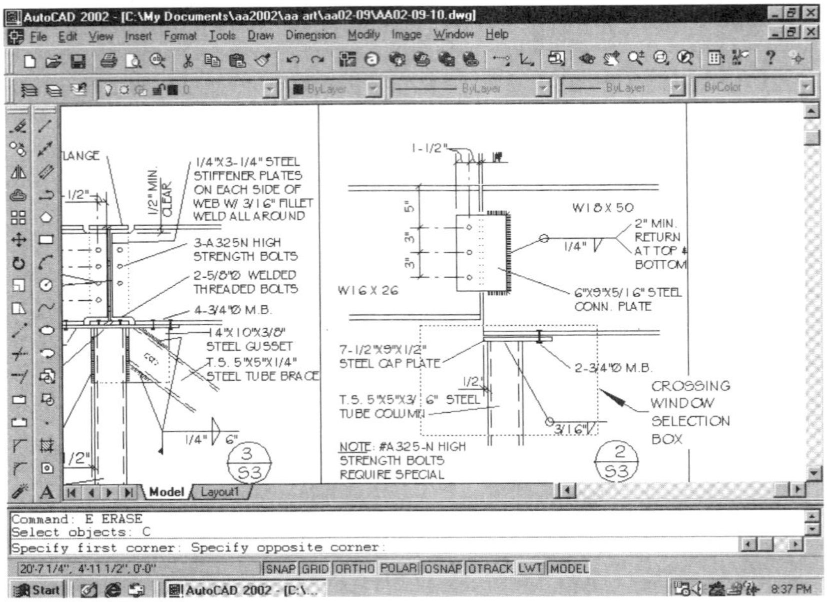

Figure 9.10 *Using the Crossing option to remove the column, support plate, and related information. Any object touched by the window will be edited. (Courtesy Kenneth D. Smith Architect & Associates, Inc.)*

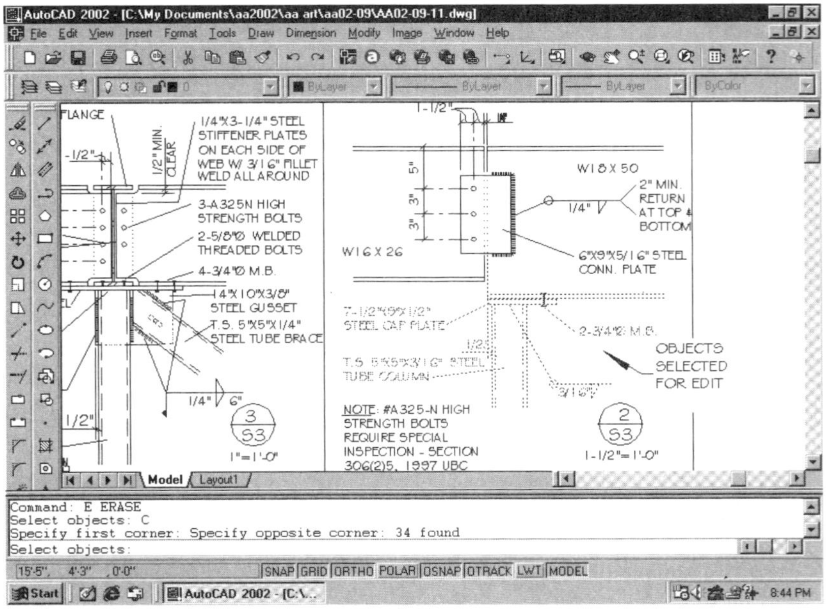

Figure 9.11 *Results of the Crossing window selection set are now displayed to confirm the choice of objects to be included in the selection set. (Courtesy Kenneth D. Smith Architect & Associates, Inc.)*

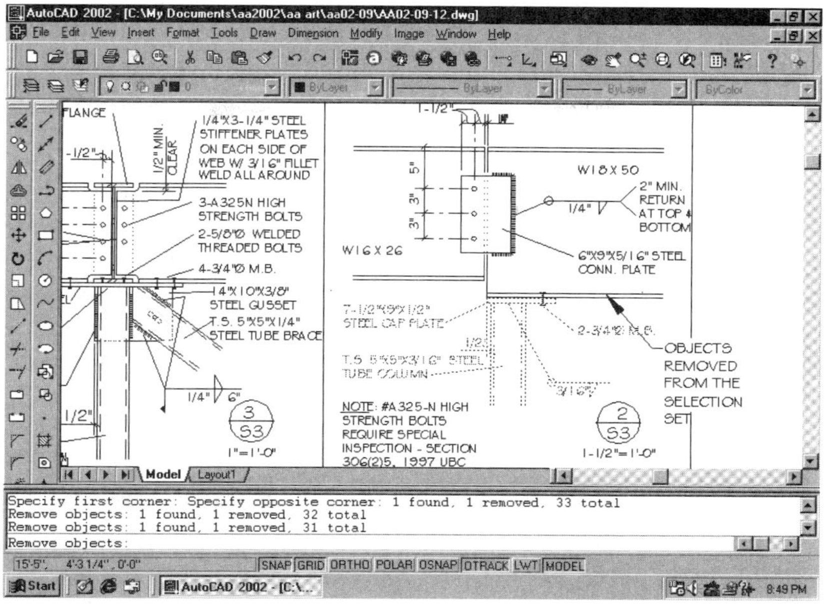

Figure 9.12 *If you would like to include more objects in the set, type* **A** *(***ENTER***) at the Select objects prompt. Typing* **R** *(***ENTER***) at the prompt will remove objects from the set. To accept the set, press* **ENTER***. (Courtesy Kenneth D. Smith Architect & Associates, Inc.)*

the selected lines of the beams. Figure 9.12 shows the final selection set. This drawing could now be used as the basis to form a new detail of two intersecting beams of different sizes with no support column.

MODIFYING A DRAWING BY ADDING OBJECTS

The balance of this chapter will introduce the skills needed to alter a drawing by adding objects to the drawing. Later chapters will help you edit a drawing by altering existing drawing components. The **Copy**, **Mirror**, **Offset**, and **Array** commands can each be used to modify a drawing by adding objects to the drawing base. Objects added using one of these commands can be selected with any of the selection methods introduced earlier in this chapter.

COPY

Whenever multiple objects need to be drawn, the **Copy** command is a convenient method of reproducing objects. **Array**, **Block**, and **Wblock** are also efficient means of reproducing objects (**Block** and **Wblock** will be covered in later chapters). Start the **Copy** command by selecting the **Copy** button on the **Modify** toolbar, by typing **CP** (ENTER) at the Command prompt, or by selecting **Copy** from the **Modify** menu. Figure 9.13 shows an example of a drawing created using the **Copy** com-

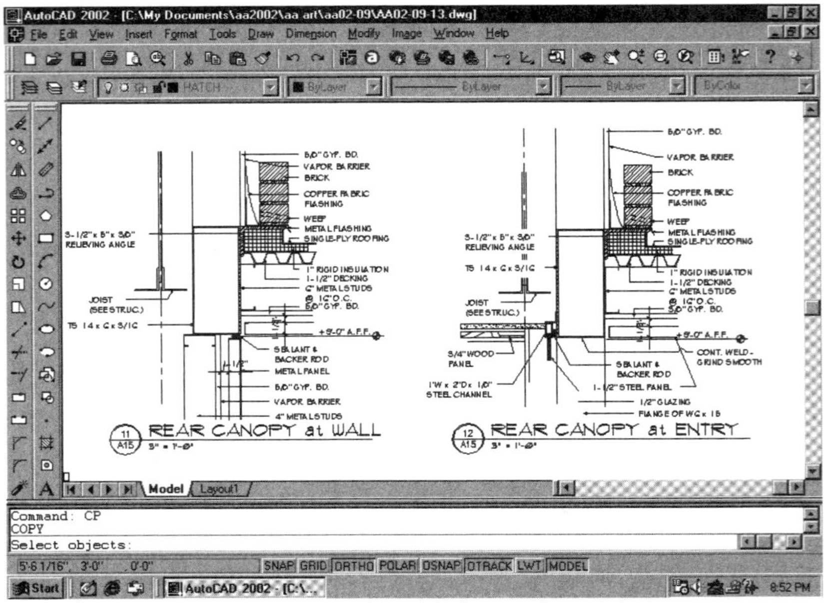

Figure 9.13 *The* **Copy** *command can be used to duplicate common features. The command can be typed, selected on the* **Modify** *toolbar, or selected from the Modify menu. (Courtesy Peck, Smiley, Ettlin Architects.)*

mand. The drawing on the left was drawn first and copied to produce the drawing on the right. Once copied, the text was edited slightly to produce the detail. The command sequence is as follows:

Command: **CP** (ENTER) *(Or click the Copy button.)*

Select objects: **W** (ENTER) *(Select objects by desired selection method.)*

Specify first corner: *(Select window corner.)*

Specify opposite corner: *(Select opposite window corner.)*

54 found

Specify base point or displacement or [Multiple]: *(Select base point or enter coordinates of object to be moved.)*

Specify second point of displacement or < use first point as displacement>: *(Select a point to relocate the base point to.)*

Command:

The Copy process can be seen in Figures 9.14a through 9.14d.

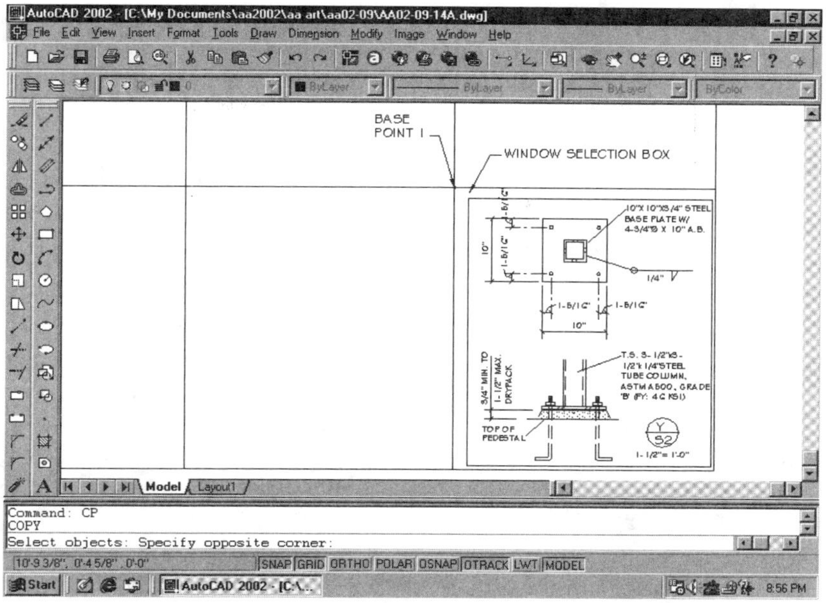

Figure 9.14a *The* **Copy** *command is started by defining the selection set. (Courtesy Kenneth D. Smith Architect & Associates, Inc.)*

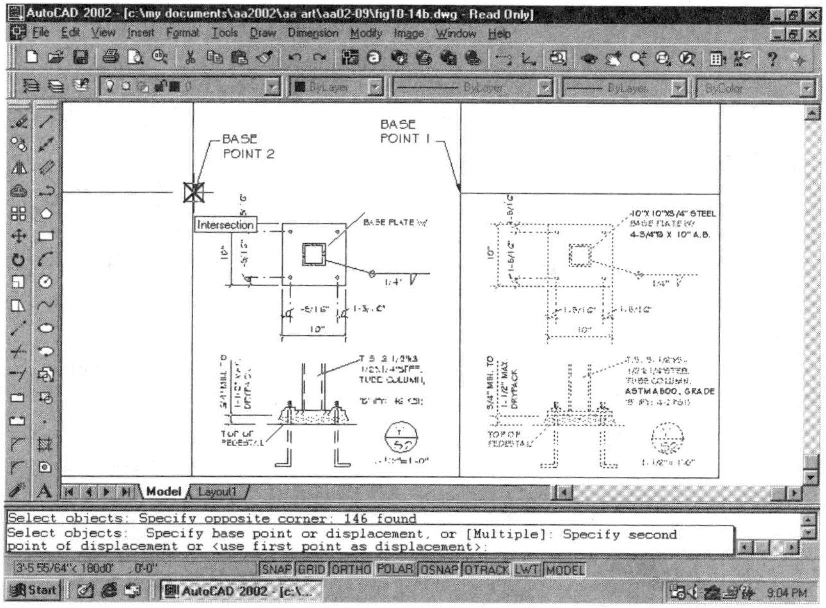

Figure 9.14b *The second step of the* **Copy** *command is to specify a base point and then specify a new location for the base point. (Courtesy Kenneth D. Smith Architect & Associates, Inc.)*

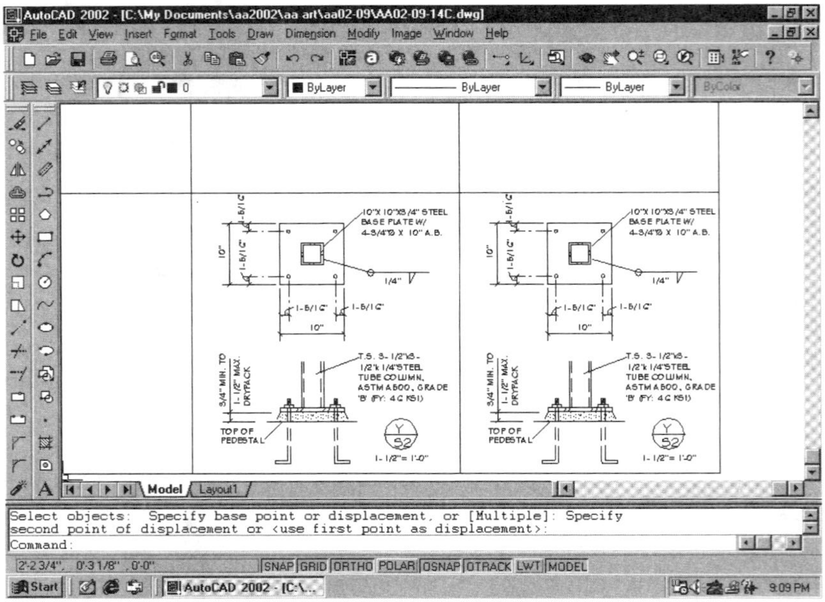

Figure 9.14c *Once* **ENTER** *is pressed for the new location, the objects in the selection set will be duplicated. (Courtesy Kenneth D. Smith Architect & Associates, Inc.)*

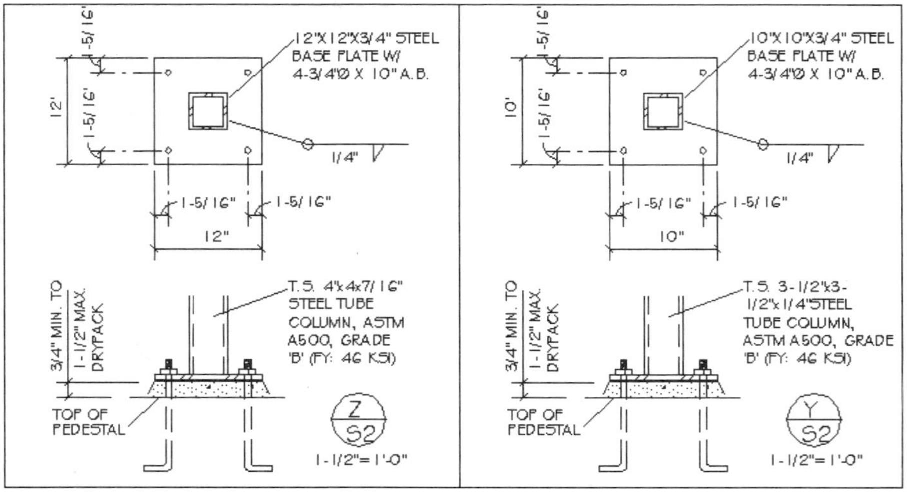

Figure 9.14d *Once copied, the detail can be edited to provide information about a similar column-plate intersection. Notice that the size of the plate, column, and the bolt locations have been edited. (Courtesy Kenneth D. Smith Architect & Associates, Inc.)*

Multiple Copies

It's not uncommon to need to reproduce multiple copies of an object throughout a drawing. Multiple copies of an object or group of objects can be made once the selection set is selected. Notice in the **Copy** command sequence that the alternative to the first point of displacement is Multiple. When you type **M** (ENTER) at the prompt, the selection group will be placed at an unlimited number of locations. The command sequence to make four copies of a group of objects is as follows:

Command: **CP** (ENTER) *(Or click the Copy button.)*

COPY

Select objects: **C** (ENTER) *(Select objects by desired selection method.)*

Specify first corner: *(Select window corner.)*

Specify opposite corner: *(Select opposite window corner.)*

25 found

Select objects: (ENTER)

Specify base point or displacement or [Multiple]: **M** (ENTER)

Specify base point: *(Select base point or enter coordinates of object to be moved.)*

Specify second point of displacement or < use first point as displacement>: *(Select a point to relocate the base point to for the first copy.)*

Specify second point of displacement or < use first point as displacement>: *(Select a point to relocate the base point to for the second copy.)*

Specify second point of displacement or < use first point as displacement>: *(Select a point to relocate the base point to for the third copy.)*

Specify second point of displacement or < use first point as displacement>: *(Select a point to relocate the base point to for the final copy.)*

Specify second point of displacement or < use first point as displacement>: (ENTER)

Command:

The process can be continued until all of the desired copies are drawn. To end the sequence, press (ENTER) at the displacement prompt.

Practical Use for Multiple Copy

In a practical application, Figure 9.15 shows the effect of the Multiple **Copy** command. One girder and one pier were drawn for the foundation plan. Multiple **Copy** was used to copy the three additional piers along the girder. The Object Snap Tracking mode was used for exact placement. Once the first girder was completed, Multiple **Copy** was used again to place the girders in the foundation plan.

MIRROR

Objects in a drawing often need to be reversed. Something as simple as reversing a door swing, or as complex as flipping a floor plan for an apartment building plan, (Figure 9.16) can be done using the **Mirror** command. The command also can be used any time objects are symmetrical, such as a fireplace or double doors. When you draw half the object, you can create the other half by using **Mirror.** Start the **Mirror** command by selecting the **Mirror** button on the **Modify** toolbar, by typing **MI** (ENTER) at the Command prompt, or by selecting **Mirror** from the **Modify** menu.

Once objects are selected to be mirrored (flipped), you will be asked to describe a "mirror line" by selecting each end point. The process can be seen in Figure 9.17. Think of the mirror line as a fold line between the old and the new objects to be mirrored. As you pick the first point of the mirror, drag the selection set into position and alter

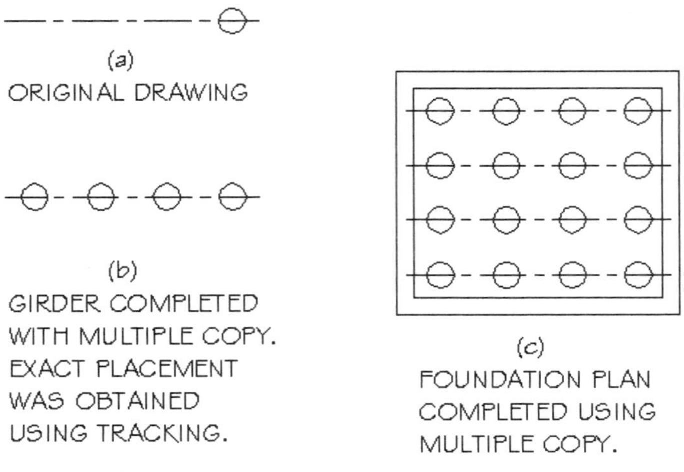

Figure 9.15 *The Multiple mode of* **Copy** *allows objects to be copied several times without the command having to be repeated. The drawing was started by drawing the girder and one pier. The required number of piers was copied, and finally the girder and its piers were copied to finish the drawing.*

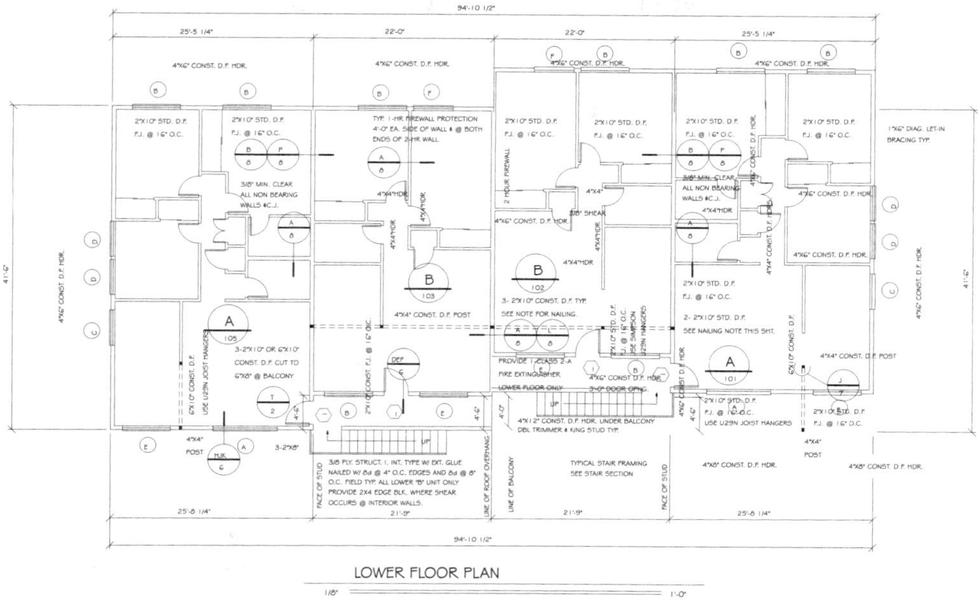

Figure 9.16 *The* **Mirror** *command can be used to flip existing objects. The command can be selected on the* **Modify** *toolbar, from the* **Modify** *menu, or entered by keyboard. (Courtesy Kenneth D. Smith Architect & Associates, Inc.)*

it as the cursor is moved. The use of ORTHO and Object Snaps can greatly aid in the placement of the mirrored image. The new placement can be seen in Figure 9.18.

You'll be given the option of keeping or discarding the original object prior to making the **Mirror** operation permanent. Your choice will depend on the use of the drawing. If the object is being completed by **Mirror**, don't delete the old object. The final result will resemble Figure 9.19. The command sequence is as follows:

Command: **MI** (ENTER) *(Or click the Mirror button.)*

Select objects: **C** (ENTER) *(Choose objects using the desired selection method or combination of methods.)*

Specify first corner: *(Select window corner.)*

Specify opposite corner: *(Select opposite window corner.)*

255 found

Select objects: (ENTER)

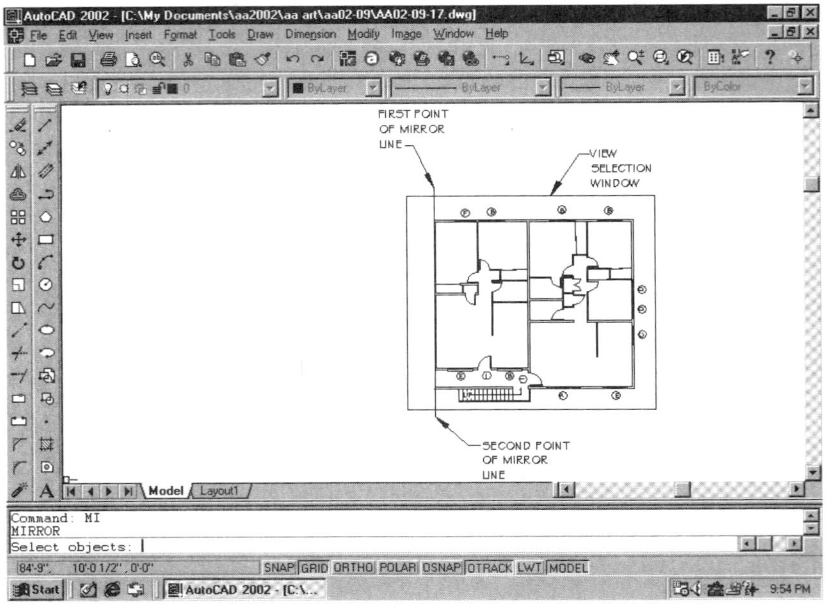

Figure 9.17 *Once the selection set has been defined, two points will be required to define the mirror line. (Courtesy Kenneth D. Smith Architect & Associates, Inc.)*

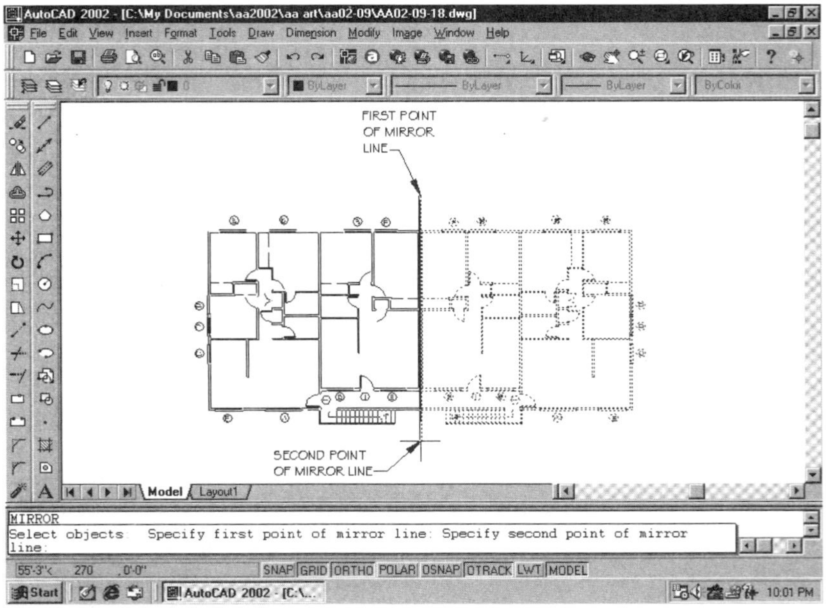

Figure 9.18 *As the second point is entered, the selection set will be mirrored into position. (Courtesy Kenneth D. Smith Architect & Associates, Inc.)*

Specify first point of mirror line: *(Select a point or enter coordinates.)*

Specify second point of mirror line: *(Select a point or enter coordinates.)*

Delete source objects? [Yes/NO] <N>: (ENTER)

Command:

If the object is being flipped, you will want to remove the old object, as in Figure 9.20. The command prompts would be similar to those just noted, until the last line is reached.

Delete source objects? [Yes/NO] <N>: Y (ENTER)

Typing **Y** (ENTER) will now delete the original object and display the mirrored copy of the original object.

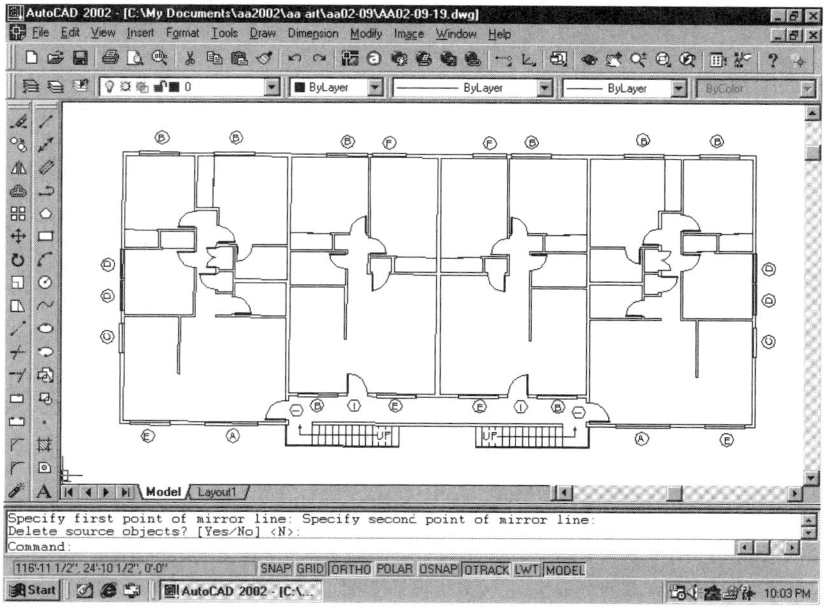

Figure 9.19 *Before the* **Mirror** *command is complete, an option will be given to "Delete source objects?" If you respond* **N** (**ENTER**)*, the selection set will be mirrored and the original objects will be retained. (Courtesy Kenneth D. Smith Architect & Associates, Inc.)*

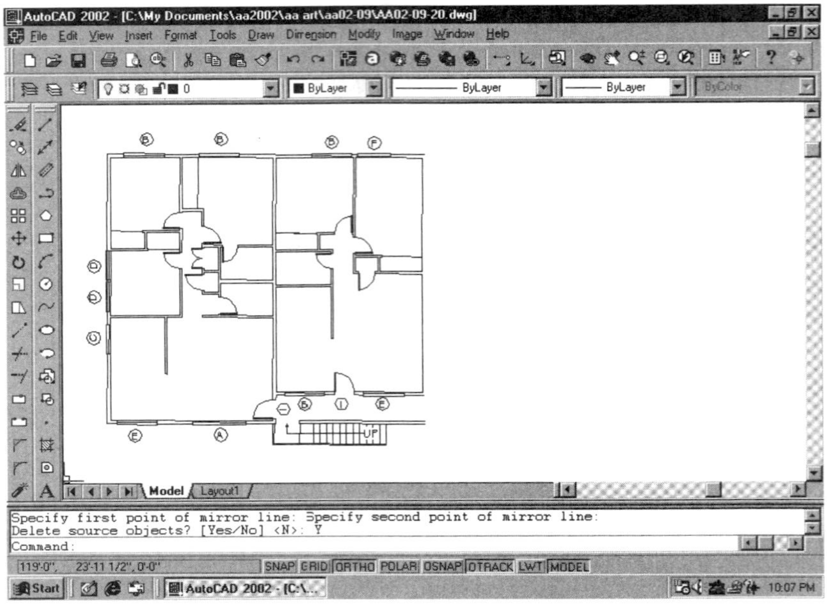

Figure 9.20 *If you respond* **Y** *(***ENTER***) to "Delete source objects?" the selection set will be mirrored and the original objects will be removed. (Courtesy Kenneth D. Smith Architect & Associates, Inc.)*

One of the drawbacks of the **Mirror** command is how it affects text. Although you will not add text to a drawing for a few more chapters, an additional feature of **Mirror** will be helpful soon. As AutoCAD is currently set, if you were to mirror a drawing with text, the text would be reversed (see Figure 9.21). To mirror the drawing without flipping the text, adjust the **Mirrtext** value to zero. With **Mirrtext** set to zero, the drawing will be flipped, but the text will remain readable, as seen in Figure 9.22. This will be covered again as text is explored, but you might want to set the value in your drawing template now so that when you do add text, you won't need the **Undo** or **Oops** command. The command sequence is as follows:

> Command: **MIRRTEXT** (ENTER)
>
> New value for MIRRTEXT <1>: **0** (ENTER)
>
> Command:

Practical Uses for Mirror

Mirror is useful whenever objects need to be grouped. The girders of Figure 9.15 could have been placed using **Mirror**. Any drawing that contains objects that lie along a centerline can be created using the **Mirror** command. Common symbols that could be created with **Mirror** for a floor plan include fireplaces, stairs, pairs of swinging doors, bay

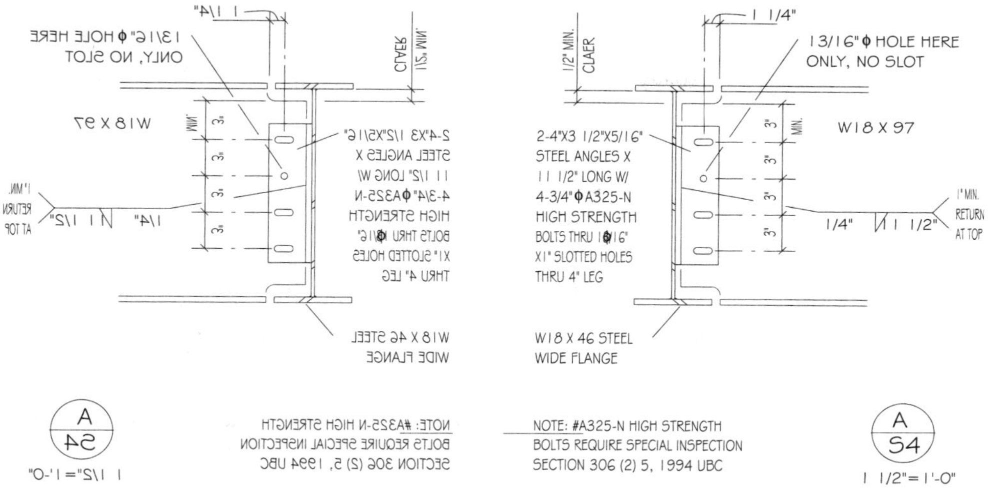

Figure 9.21 *The default setting of the* **Mirror** *command also will reverse text. (Courtesy Kenneth D. Smith Architect & Associates, Inc.)*

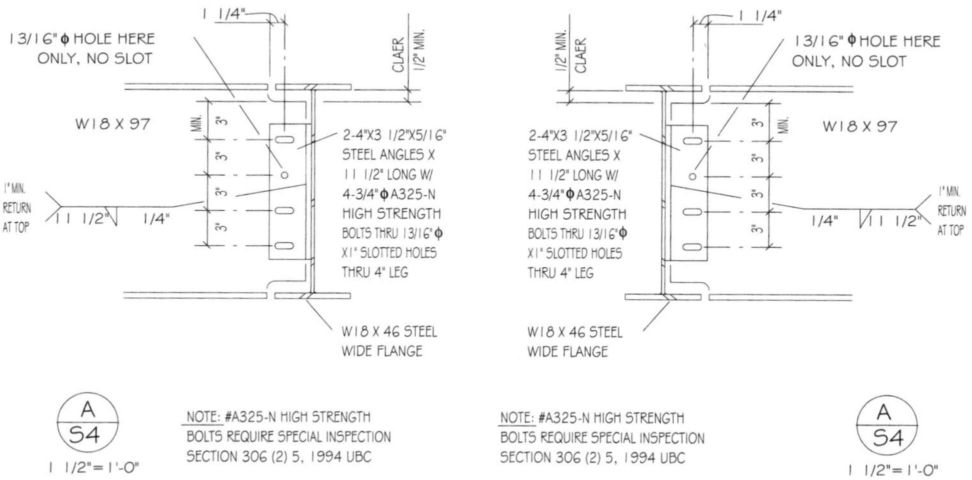

Figure 9.22 *If the* **Mirrtext** *variable is set to 0, text will not be mirrored. (Courtesy Kenneth D. Smith Architect & Associates, Inc.)*

windows, sinks, convenience outlets, toilets, and tubs. Drawing half of the symbol and then using **Mirror** to complete it will create each of these symbols. Later chapters will explain how to make and save each symbol so that it only needs to be created once. If you're really eager, explore Blocks in the **Help** menu.

OFFSET

The **Offset** command allows a line, circle, arc, or polygon to be copied and relocated parallel to and at a specific distance from the original object. The uses of this command are almost endless. A circle can be drawn and offset to represent a conduit or sewer pipe. Starting with a horizontal and a vertical line, an entire rectangular structure can be drawn by offsetting lines and then cleaned up using other editing commands, presented in later chapters.

Using **Offset**, lines can be drawn in the exact location without thought as to their exact length. Once all of the boxes are lightly drawn to outline the desired rooms, the lines to be retained are darkened. Use **Offset** with the same attitude. Get the location right without worrying that the line might be too long, and then edit the length with **Fillet**, **Stretch**, or **Trim**. (Each of these commands will be introduced in Chapter 10.) Figures 9.23a and 9.23b show the use of **Offset** to lay out a simple floor plan.

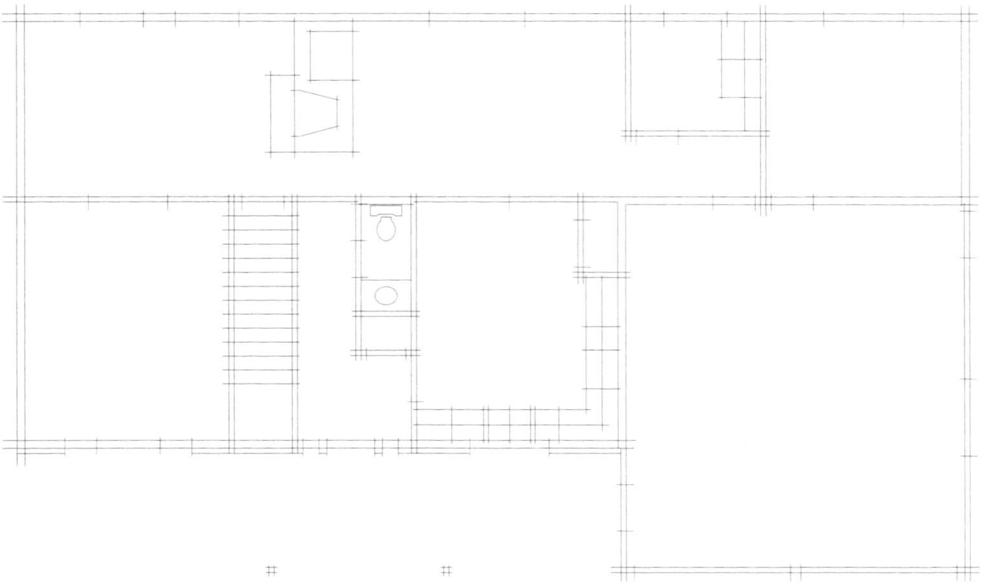

Figure 9.23a *With one horizontal and one vertical line, the* **Offset** *command can be used for drawing parallel lines at a specified distance apart to create a floor plan.*

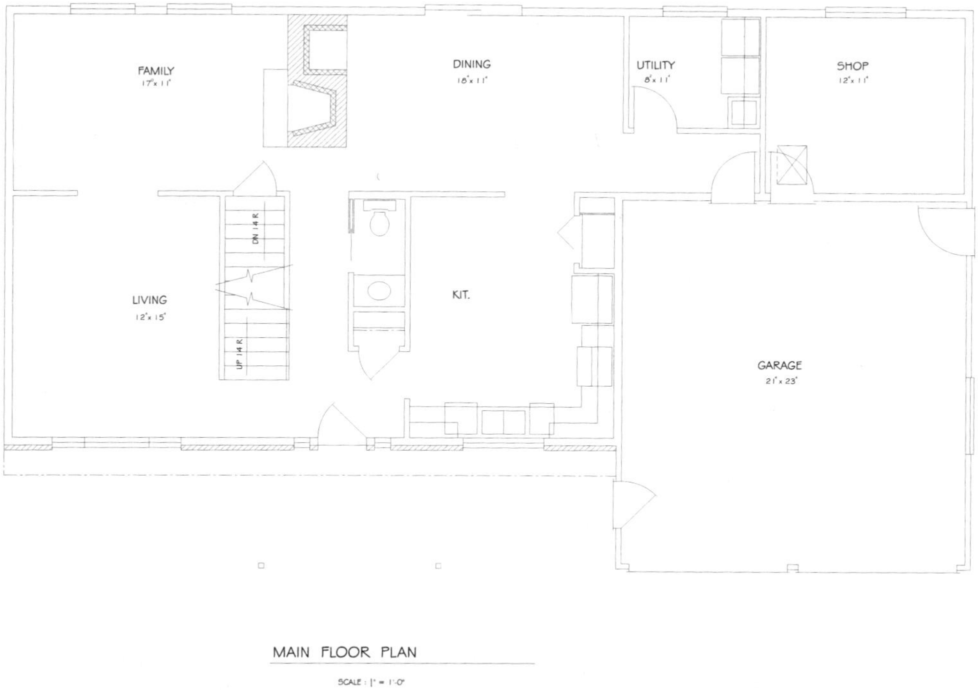

Figure 9.23b *Once the basic lines have been put in the correct position using* **Offset**, *other editing commands can be used to finish the drawing.*

Start the command process by selecting the **Offset** button on the **Modify** toolbar, by typing **O** (ENTER) at the Command prompt, or by selecting **Offset** from the **Modify** menu. The command sequence to offset a line 18" from the original line is as follows:

> Command: **O** (ENTER) *(Or click the Offset button.)*
>
> OFFSET
>
> Specify offset distance or [Through] <Through>: **1'6"** (ENTER)

A distance can be entered by providing a numeric value or by typing **T** (ENTER). When the distance is entered, a new prompt will be displayed that prompts

> Select the object to offset or <exit>:

Lines, arcs, circles, ellipses, xlines, rays, and polylines can be offset (the last three objects will be introduced in later chapters). Once selected, the object will be highlighted and the prompt will read

> Specify point on side to offset:

Move the cursor slightly to the desired side and click the select button to offset the object and redisplay the prompt:

Select the object to offset or <exit>:

The process can be continued indefinitely, with the offset distance that was entered originally remaining as the default. Terminate the **Offset** process by pressing (ENTER). Once the process is terminated, a new default distance can be entered or a new command sequence can be started. The command sequence can be seen in Figure 9.24.

Typing **T** (ENTER) at the original prompt allows a point to be entered with a pointing device, and the object to be offset will pass through the indicated point. The command sequence is as follows:

Command: **O** (ENTER) *(Or click the Offset button.)*

OFFSET

Specify offset distance or [Through] <1'-6">: **T** (ENTER)

Select the object to offset or <exit>: *(Select an object.)*

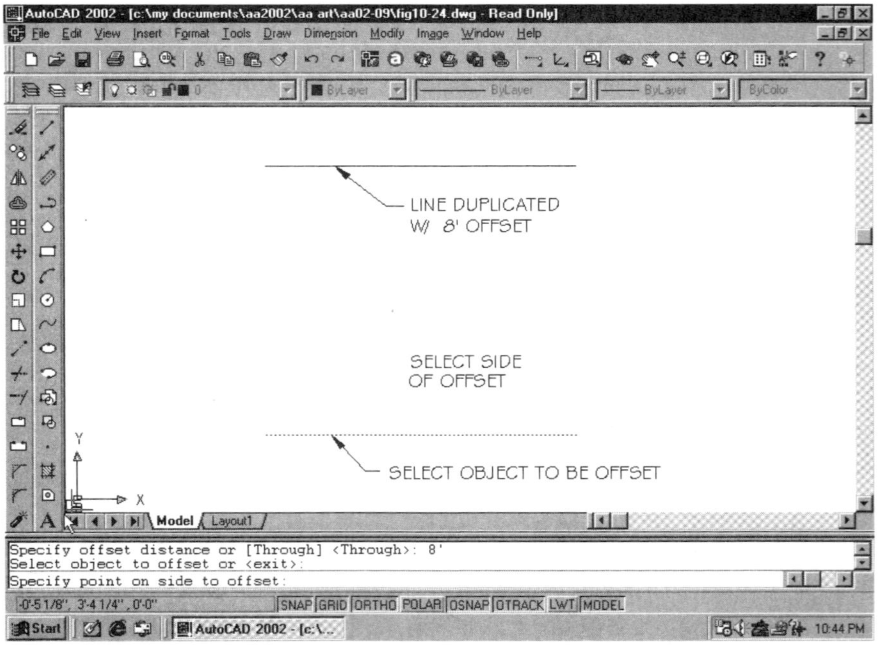

Figure 9.24 *Complete the* **Offset** *command by selecting the distance to be offset, the object to be offset, and the side of the original on which the offset will occur.*

Specify through point: *(Move cursor to the desired point and select.)*

Select the object to offset or <exit>: (ENTER)

Command:

The **Offset** process with the Through option can be seen in Figure 9.25.

Practical Uses for Offset

Open the ARCHBASE template. Now draw intersecting horizontal and vertical lines. Refer to the floor plan in Figure 9.23b. Many clients come to a designer with a floor plan that they like, along with a list of a thousand changes that they would like to make. The **Offset** command can be useful in the layout of a floor plan. Assume the two lines that have been drawn represent the outer edges of the walls in the upper right corner of the plan shown in Figure 9.23b. Offset each line 6" toward the inside of the house. Because the room in Figure 9.23b is 12'–4" wide, offset the vertical line you just created a distance of 12'–4" to the left. Offset this same line 4" to represent the wall thickness. Offset this line 8'–2", and then again 4". This will provide the wall on the left side of the Utility room.

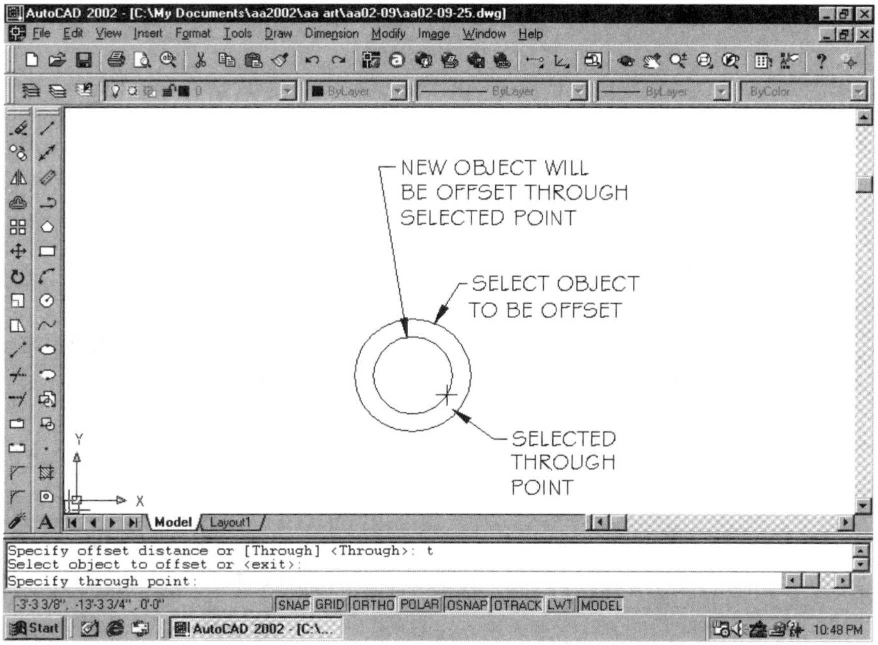

Figure 9.25 *The Through option of* **Offset** *allows the offset distance to be specified using the pointing device.*

Figure 9.26 *By drawing two perpendicular lines and using the* **Offset** *command, you can easily draw a floor plan.*

Now go back to the inner horizontal line that you drew. Offset this line 11'–4" toward the bottom of the drawing. Offset this line 6" toward the bottom of the drawing. These two lines will represent the wall between the shop and the garage. To place the wall between the utility room and the hall, offset the inner garage wall 42". Offset this new line 4". The entire process is shown in Figure 9.26. Save this crude drawing as FLOOR10.

ARRAY

Until now, to reproduce a drawing element, you have used the **Copy** command. The **Array** command will also allow for multiple copies of an object or group of objects to be reproduced—such as the beam, post, and columns of a post-and-beam foundation. The command also provides several added features not available with **Copy**. **Array** reproduces objects in rectangular or circular (polar) patterns, allows the object to be rotated as it is reproduced, and allows for easy control of the spacing of the object during the **Array** operation. Figures 9.27a and 9.27b show examples of a rectangular and a polar array. Start the command by selecting the **Array** button on the **Modify** toolbar, by typing **AR** (ENTER) at the Command prompt, or by selecting **Array** from the **Modify** menu. Each method will produce the **Array** dialog box shown in Figure 9.28. Once the dialog box has been accessed, the type of array pattern should be selected.

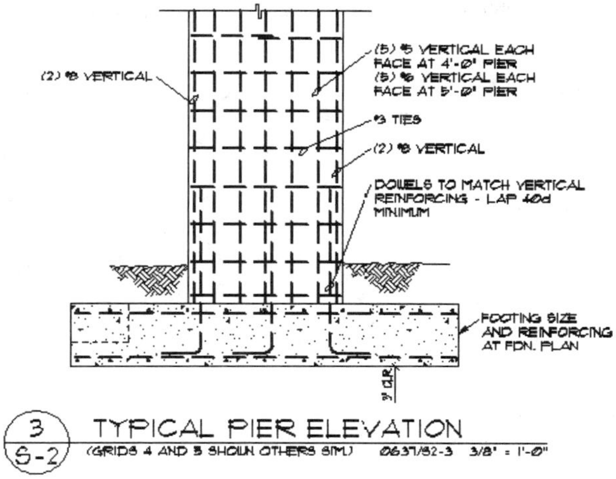

Figure 9.27a *The steel reinforcement was placed using the* **Array** *command. (Courtesy Van Domelen/Looijenga/McGarrigle/Knauf Consulting Engineers.)*

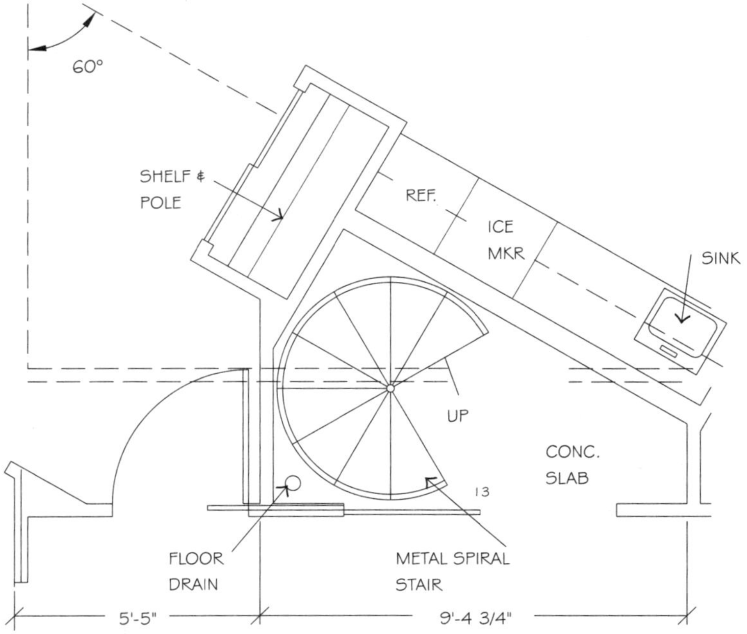

Figure 9.27b *The treads of this stair were placed using the Polar option of the* **Array** *command. (Courtesy Piercy & Barclay Designers, Inc.)*

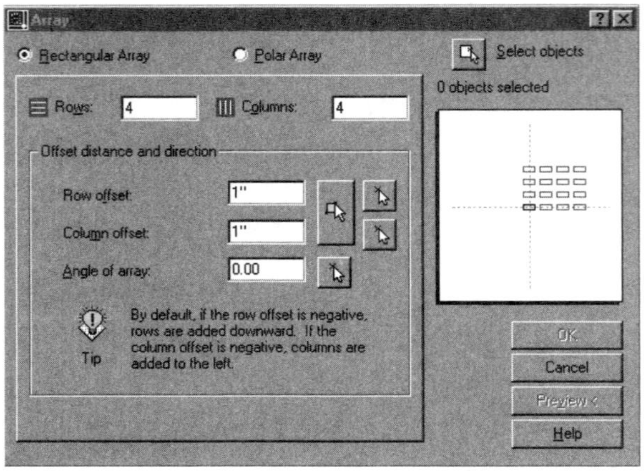

Figure 9.28 *The* **Array** *dialog box is used to recreate an object in rectangular and circular pattern. Access the* **Array** *dialog box by selecting the* **Array** *button or by typing* **AR** *(***ENTER***) at the Command prompt.*

Rectangular Array

A rectangular array will prove suitable for projects that are arranged in rows or columns. Selecting the **Rectangular Array** button starts the process. Six additional specifications must be provided to perform a rectangular array. These include selecting the object to be arrayed, the number of rows and columns, the row and column offsets, and the array angle. The elements can be seen in Figure 9.29.

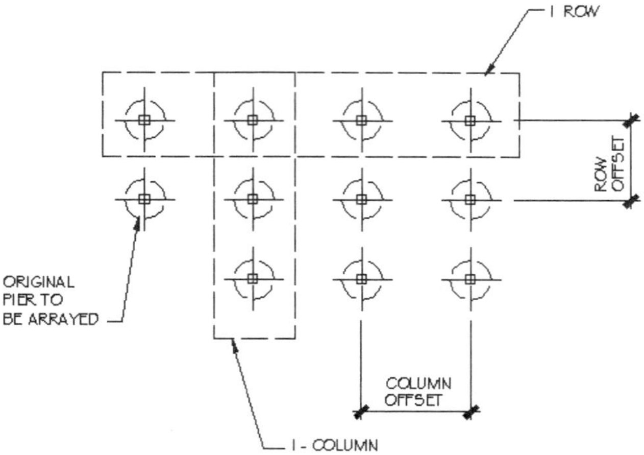

Figure 9-29 *The number of rows and columns and the spacing of each must be specified to use a rectangular array.*

Selecting the Object to Array
Selecting the **Select objects** button will remove the **Array** dialog box and allow objects to be selected for the array pattern. As the dialog box is removed, drawing objects can be selected using any of the drawing selection methods introduced earlier. When the selection set is complete, press (ENTER) or (SPACEBAR).

Specifying the Number of Rows
In the **Rows** edit box, specify the quantity of horizontal rows to be created. Any whole number can be entered for the value. Figure 9.30 shows an example of an array with two, three, and four rows.

Specifying the Number of Columns
In the **Columns** edit box, specify the number of vertical columns to be used. Any whole number can be entered for the value. If the value of 1 was used for the row value, a value other than 1 must be used for the column value. Otherwise, AutoCAD will respond with the prompt:

```
One-element array, nothing to do.
```

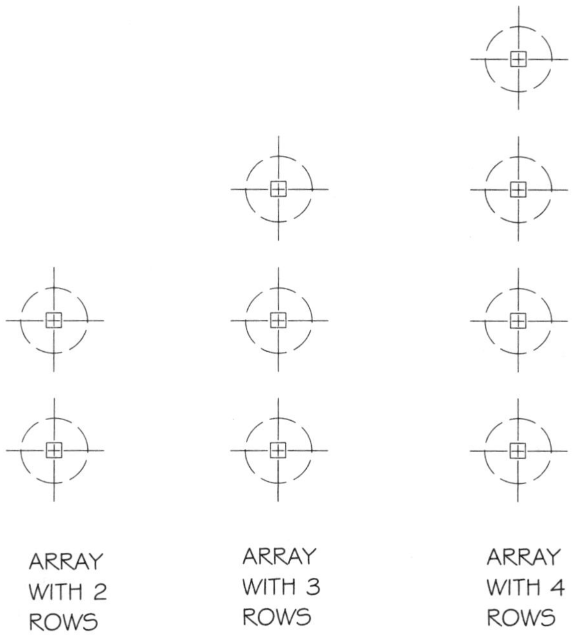

Figure 9.30 *Altering the number of rows of an array.*

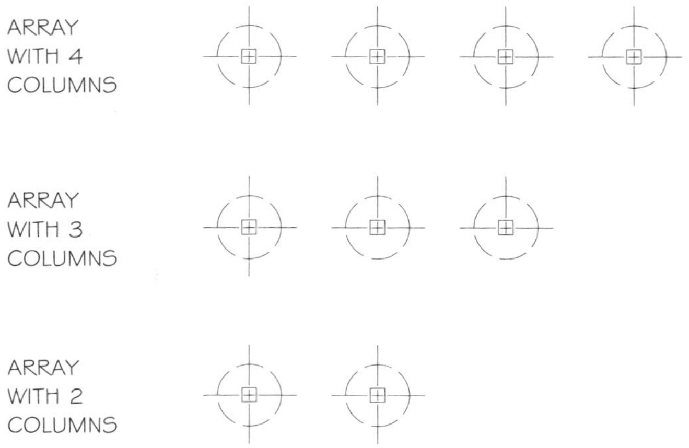

Figure 9.31 *Altering the number of columns in an array.*

Figure 9.31 shows the effect of an array using two, three, and four columns.

Specifying the Row Offset

The **Row offset** edit box allows the offset distance to be set for rows. The offset distance includes the size of the object to be arrayed. If you are arranging 16" wide chairs in rows, and would like 16" between rows, the spacing would require a 32" unit cell. This is shown in Figure 9.32a. You can set the distance between rows by selecting the **Pick Row Offset** button and then selecting a point with the mouse, or by selecting the **Pick Both Offset** button and then selecting the unit cell as two opposite points of a rectangle. The rectangle will provide information for the row spacing. If more than one column is to be arrayed, the rectangle will provide information for both the row and the column distances. The unit cell spacing can be seen in Figure 9.32b.

Specifying the Column Offset

The **Column offset** edit box for a rectangular array provides the distance between columns. The same guidelines for spacing rows apply to spacing column offsets. Figure 9.33 shows the results of an array pattern with three rows and eight columns.

Positive and Negative Offset Values

As values are provided for rows and columns, they can be positive, negative, or a combination of both. Using positive values for both will array the objects up and to the right of the original object, as seen in Figure 9.34. If a negative value is used for the row distance, new objects will be added below the original. If negative values are used for the column distance, new objects will be added to the left of the original object, as seen in Figure 9.35. Figures 9.36 and 9.37 show the effects of combining values.

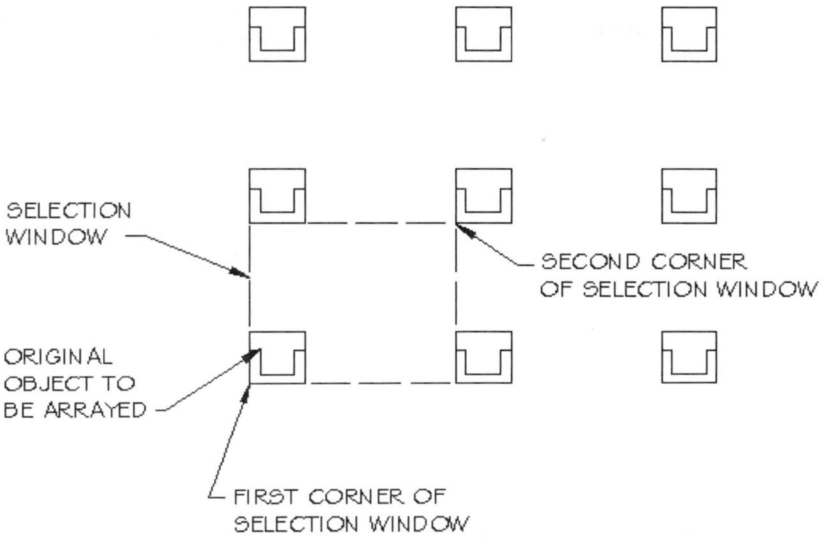

Figure 9.32a *The offsets for rows and columns can be specified by selecting the two opposite corners of a window.*

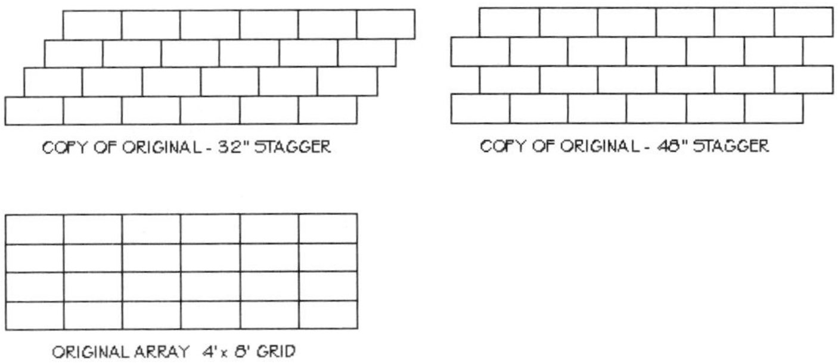

Figure 9.32b *The original array pattern at the bottom was created using four rows and six columns. The patterns at the top were altered by moving rows.*

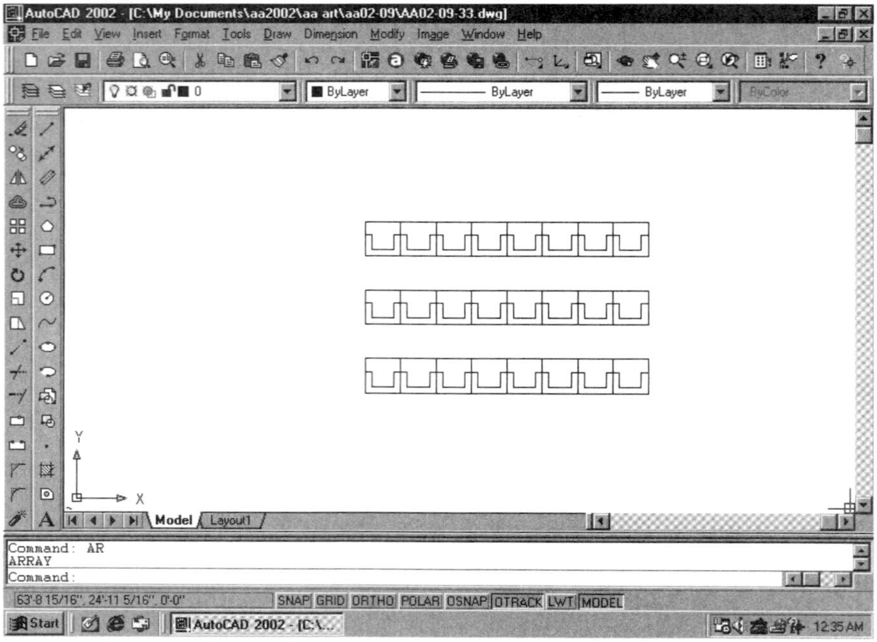

Figure 9.33 *The effect of row and column spacing in an array of three rows and eight columns.*

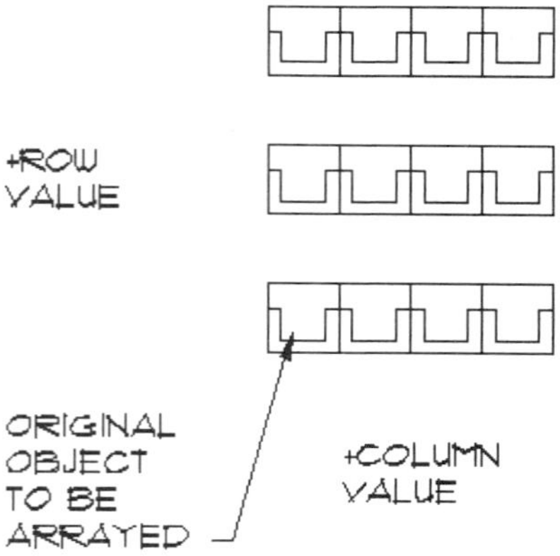

Figure 9.34 *An array based on entering two positive distances.*

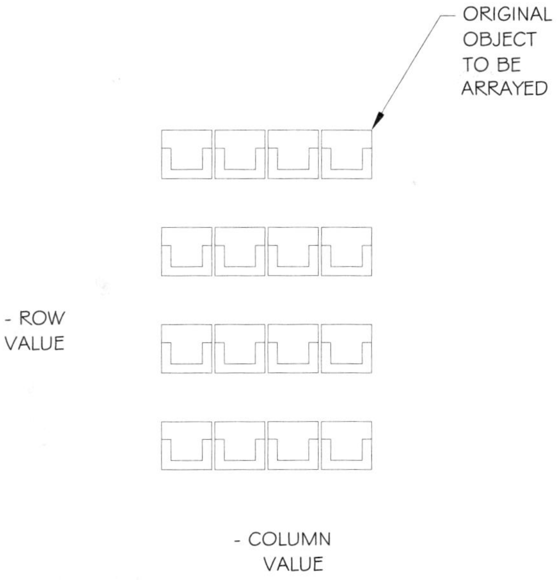

Figure 9.35 *An array based on entering two negative values.*

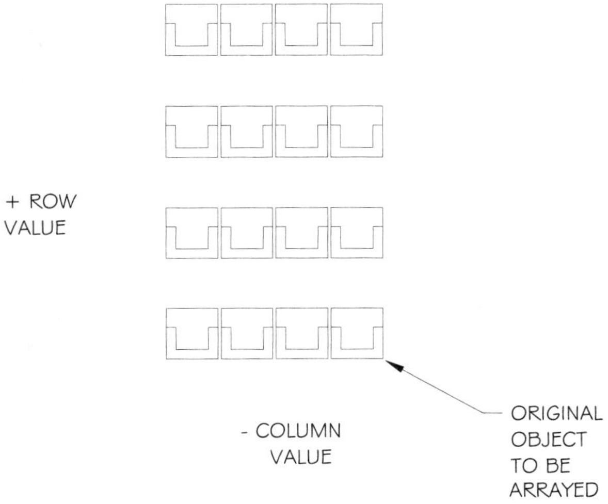

Figure 9.36 *An array based on a positive row value with a negative column value.*

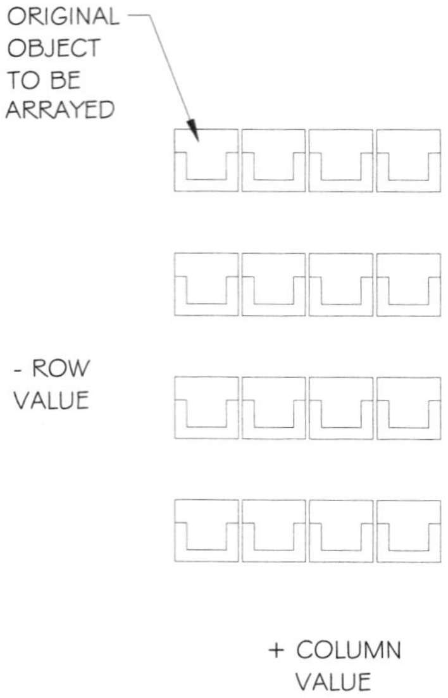

Figure 9.37 *An array based on a negative row value with a positive column value.*

Specifying an Angle for Rectangular Arrays

The array pattern is normally based on a horizontal or vertical baseline. Entering a numeric value in the **Angle of array** edit box will rotate the baseline angle. Selecting the **Pick Angle of Array** button will also allow the angle array to be altered. Selecting this button will prompt for two lines to be specified. The angle of the line drawn will become the angle for the array pattern. Figure 9.38 shows the effect of using an angled array.

Summary of Steps in the Rectangular Array Process

1. Display the **Array** dialog box by selecting the **Array** button on the **Modify** toolbar or by typing **AR** (ENTER) at the Command prompt.
2. Select the method of array to be used.
3. Select the objects to be arrayed using the **Select objects** button.
4. Specify the number of rows.
5. Specify the number of columns.
6. Specify the row offset.
7. Specify the column offset.
8. Specify the array angle if required.

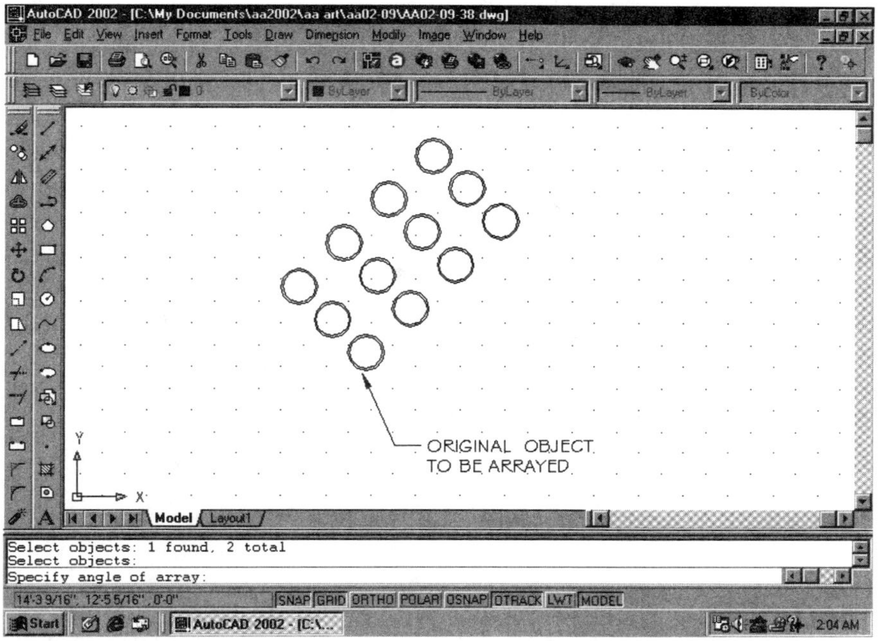

Figure 9.38 *A rotated array can be drawn by selecting the* **Pick Angle of Array** *button.*

9. Preview the array by selecting the **Preview** button. If the pattern is okay, click the **Accept** button. If the pattern is not what you expected, select the **Modify** button, and alter the pattern as needed.

10. Click the **Accept** button to accept the array as specified. If you decide not to complete the pattern, select the **Cancel** button.

Polar Array

A polar array is a circular layout of objects based on a center point. Selecting the **Polar Array** button starts the process. Once **Polar Array** is selected, the dialog box will be altered as shown in Figure 9.39. Six additional specifications must be provided to perform a polar array. These include selecting the center point of the array pattern, the method of array, the rotation of objects, and the object base point.

Selecting the Center Point

The center point is the point that the selected objects will be arrayed around. The center point can be selected by entering coordinates in the **Center point X** and **Y** edit boxes or by selecting the **Pick Center Point** button. Selecting the **Pick Center**

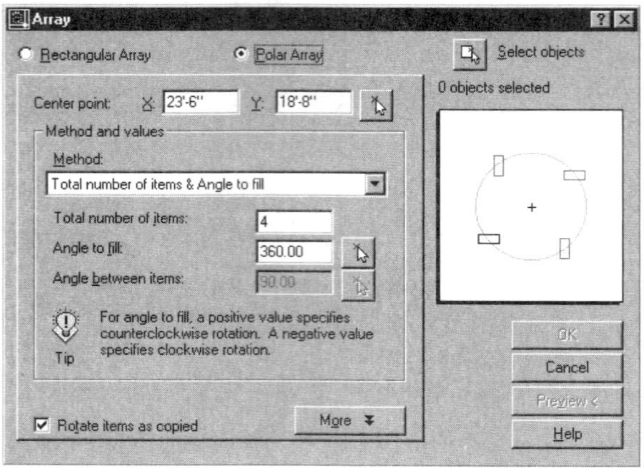

Figure 9-39 *Selecting the* **Polar Array** *button in the* **Array** *dialog box will alter the display of the dialog box, allowing objects to be created in a circular pattern.*

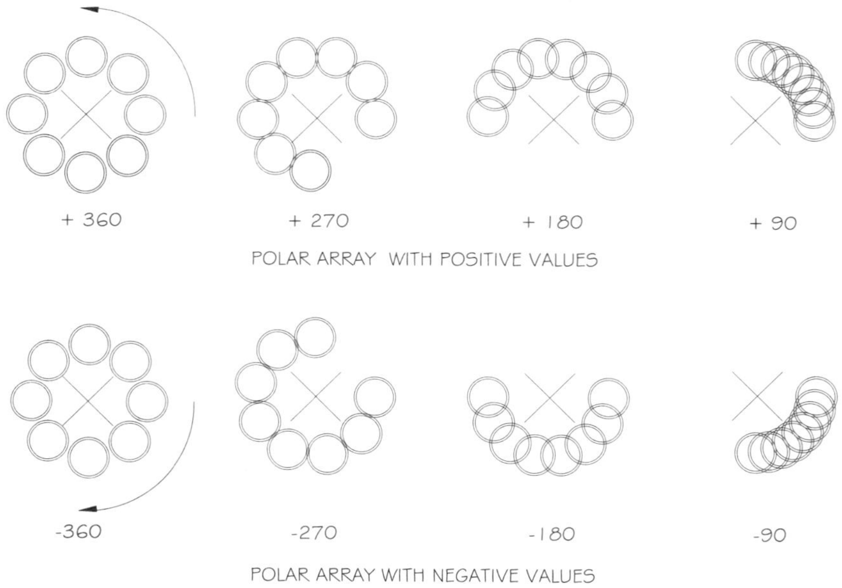

Figure 9.40 *The effect of adjusting the* **Angle to fill** *edit box for a polar array.*

Point button will temporarily remove the dialog box to allow a center point to be specified with the mouse.

Selecting the Method of Array

Three methods of completing a polar array pattern are available. Each method can be selected from the **Method** edit box. Options include the following:

- Total number of items & Angle to fill
- Total number of items & Angle between items
- Angle to fill & Angle between items

The option selected will determine which of the three prompts in the **Method and values** area will be active.

> **Number of Items**—The **Total number of items** edit box allows you to specify how many items, including the original, will be displayed around the center point.
>
> **Angle to Fill**—The **Angle to fill** edit box controls the limits of the display around the center point. The default of 360° will locate the specified number of items around a full circle. A number less than 360° will array the indicated number of objects around a selected center point in a specified degree range. The pattern will start with the original object and spread the balance of the pattern from its location. A positive value will produce a counterclockwise rotation. A negative value will cause a clockwise rotation. Examples of each pattern can be seen in Figure 9.40.
>
> **Angle Between Items**—The **Angle between items** edit box can be used to specify the angle between items. The number of items to be rotated, combined with the angle distance between each item, is often used to locate bolts or other similar features for drawing connectors (see Figure 9.41). Materials also can be arrayed over a specified degree range and intervals, such as showing pipes that intersect a cooling unit over a 75° area at each 15°, as seen in Figure 9.42.

Rotating Arrayed Objects

Selecting the **More** button will display this option. When active, this option allows objects to be rotated as they are arrayed. The default is yes, which will rotate objects as they are arrayed around the center point. If the **Rotate items as copied** check box is inactive, the objects will not be rotated and will appear similar to the right side of Figure 9.43.

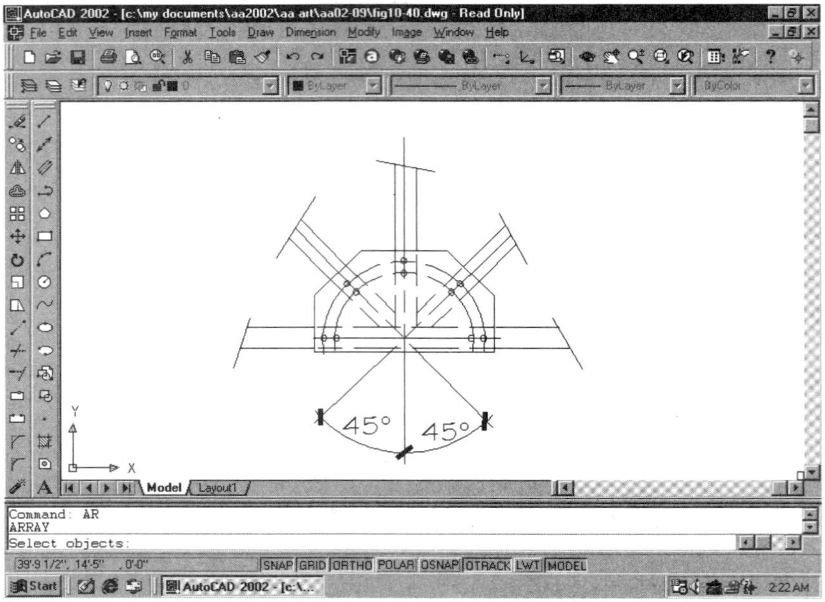

Figure 9.41 *Specifying the number of items to be rotated and the* **Angle between items** *is a common method for defining an array pattern.*

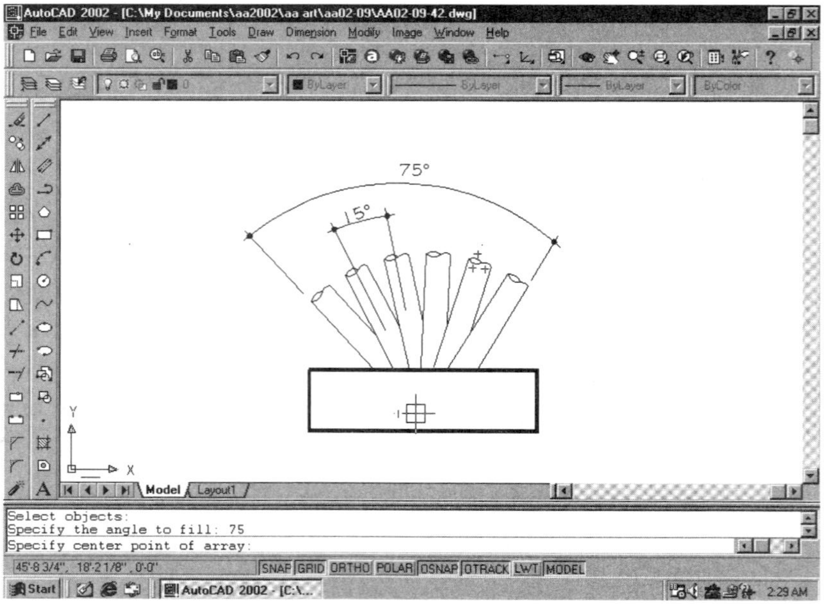

Figure 9.42 *Specifying the* **Angle to fill** *and the* **Angle between items** *is also used to define the array pattern.*

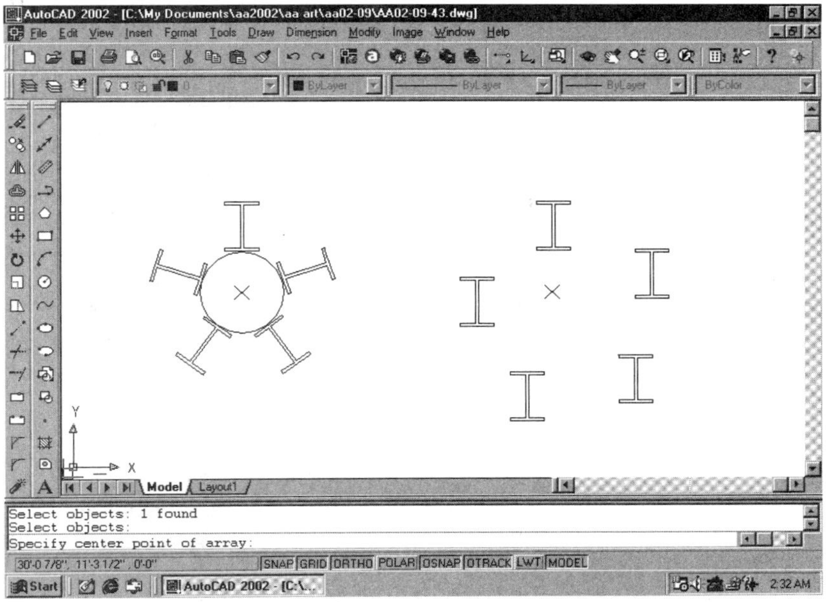

Figure 9.43 *Objects can be rotated as they are arrayed if the* **Rotate objects as copied** *check box is activated.*

Selecting an Object Base Point

The **Object base point** option specifies a new base point relative to the selection set. The base point will remain a constant distance from the center point of the array as the objects to be arrayed. The point used depends on the type of the object to be arrayed. Default base points are listed as follows:

Object type	**Default base point**
Arc, circle, ellipse	Center point
Polygon, rectangle	First corner
Donut, line, polyline, ray	Start point
Block, paragraph text, single-line text	Insertion point
Construction lines	Midpoint
Region	Grip point

Although some of these terms are unfamiliar, they will be introduced in later chapters.

Summary of Steps in the Polar Array Process

The entire process for a polar array of 50 segments around a 250° arc is as follows:

1. Display the **Array** dialog box by selecting the **Array** button on the **Modify** toolbar or by typing **AR** (ENTER) at the Command prompt.
2. Select the method of array to be used, **Polar Array**.
3. Select the objects to be arrayed using the **Select objects** button.
4. Select the center point of the array pattern.
5. Select the method. For this example, the Total number of items & Angle to fill option was used. Notice that the **Angle between items** edit box is inactive.
6. Enter 270 in the **Total number of items** edit box.
7. Enter 250 in the **Angle to fill** edit box.
8. Keep the **Rotate items as copied** check box active.
9. Preview the array by selecting the **Preview** button. If the pattern is okay, click the **Accept** button. If the pattern is not what you expected, select the **Modify** button, and alter the pattern as needed.
10. Click the **Accept** button to accept the array as specified. If you decide not to complete the pattern, select the **Cancel** button.

Practical Uses for Array

Array should be used whenever an object needs to be added to a drawing in a repetitive pattern. The girder and piers that were drawn in Figure 9.10 could easily have been placed using the **Array** pattern. Lines to represent parking spaces on a site plan, lines to represent siding on an elevation, rectangles to represent joist or rafters in longitudinal sections, and circles to represent piers on a foundation are just a few examples of how the **Array** command can be used to increase drawing efficiency.

CHAPTER 9 EXERCISES

1. Start a new drawing and draw a horizontal line. Draw polygons with three, four, five, six, and eight sides with the centerline of each polygon on the horizontal line. Draw all polygons inscribed in a circle with a 0.5" radius. Make copies of the polygons with three, five, and eight sides and place them directly below their counterparts. Make copies of the four-sided and six-sided polygons and place them in a third row directly below their counterparts. Mirror the entire drawing with a horizontal mirror line, and delete the old objects. Save the drawing as E-9-1.
2. Open the existing file E-8-5. Copy the window so that a total of three pairs of windows are drawn exactly 4" apart. Align the bottoms of the windows. Save the drawing as E-9-2.
3. Draw half of a W 8" x 28" (I-shaped) beam. The beam has a total height of 8.28". The top and bottom flanges are 0.25" thick, with a total width of 5.25". The web is a total of 0.28" thick. Use the **Mirror** command to complete the drawing. Save the drawing as E-9-3BM.
4. Using the following figure as a guide, draw an exterior swinging door that is 36" wide with a 12" wide sidelight. Use the **Mirror** command to draw a pair of doors with double sidelights. Save the drawing as E-9-4-DOOR.

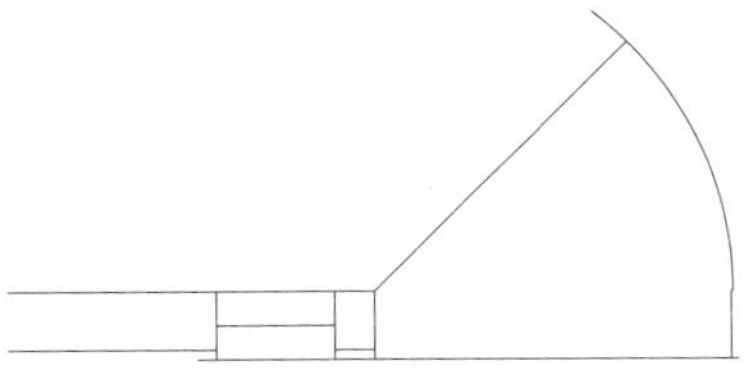

5. Using the following figure as a guide, draw a steel connector plate. The plate is 15 1/2" x 10" h. Draw the bolts as 3/4" diameter hexagons, 1 1/2" down and 1 1/2" in from the top and ends. Bolts are 3" O.C. Each end of the plate is symmetrical. Draw a 2 1/2" x 8" strap, centered in the bottom of the plate. Provide two bolts with similar spacing as the plate. Save the drawing as E-9-5.

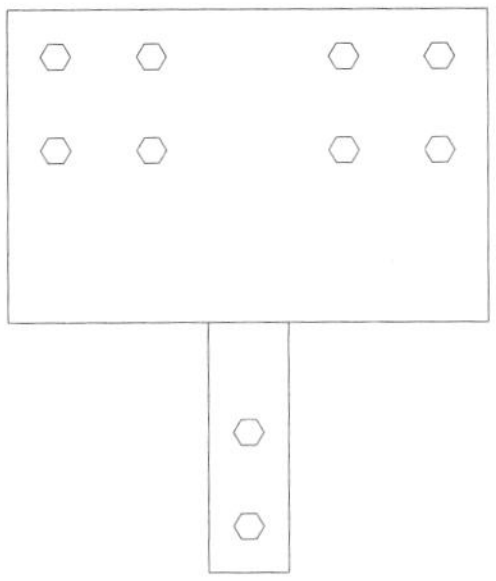

6. Use the sketch below as a guide and lay out the outline of the apartment floor plan. Draw the exterior walls as 6" wide. Use this unit plan and lay out four different building layouts with a minimum of four units per building. Save the drawing as APARTFLR.

7. Draw a post-and-beam foundation plan for a 24' x 16' structure. Exterior walls will be 8" thick. Piers are 18" diameter at 8' O.C. along a 4" wide beam. Repeat the beam and piers at 4' O.C. Save this exercise as PB-FND.
8. Start a new drawing that can be used as a base for shingle siding in elevation. Draw several lines approximately 36" long, at 10" O.C. Draw shingles similar to the sketch below with widths that vary from 4" through 12". Copy this row of shingles to form a minimum of five rows of shingles. Stagger each row with the rows above and below it, so that the shingle seams do not line up. Save the drawings as SHINGLE.

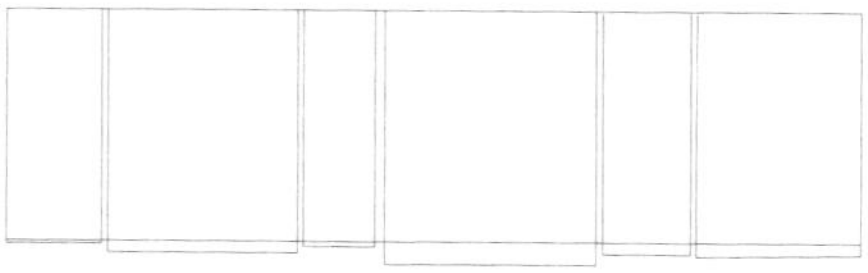

9. Start a new drawing and set the limits as needed to draw a 4" square centered in a 15" diameter circle. Draw an 8" and a 12" diameter circle with the same center point as the 15" circle. Make three additional copies of this object and move each so that the four 15" circles are tangent. Save the drawing as E-9-9.

10. Open drawing E-8-5. Your client thinks he would like a grid in each half but isn't really sure. Copy the drawing and draw a grid in one of the windows. Provide one vertical and three horizontal dividers. Mirror the grid into the other. Save the drawing as E-9-10.
11. Start a new drawing and draw a 60' wide x 44' deep structure. Use 8" wide walls. No doors or windows need to be shown. Create a layer to represent the suspended ceiling and draw the plan for a suspended ceiling using 24" x 48" panels. Make a second copy of the plan and show the panels laid out in the opposite direction. Save the drawing as E-9-11.
12. Use the base drawing started in Figure 9.11 and draw a roof framing plan. Freeze the suspended ceiling information and create a new layer to represent roof trusses at 32" O.C. Show a portion of the roof with 4' x 8' plywood over the trusses. Save the drawing as E-9-12.
13. Draw a column that is 10'–0" tall and 12" wide. Show a 20" x 4" and a 16" x 4" corbeling 4" down from the top and 4" up from the bottom of the column. Save the drawing as E-9-13.
14. A circular structure is to be built on a hillside. Use the drawing below as a guide and draw the pier plan. Save the drawing as E-9-14.

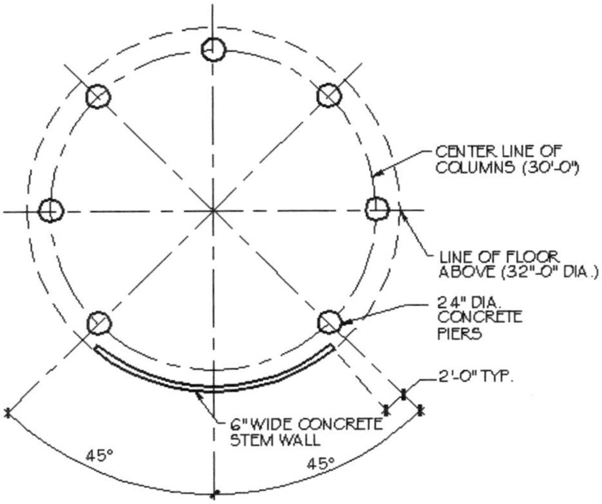

15. Draw a plan view of a 40' diameter water reservoir. The reservoir will be supported on (12) 6" diameter x 3/8" steel columns. The columns will be equally spaced, on a 36' diameter circle centered on an 18" diameter x 3/4" thick vertical inlet pipe in the center of the reservoir. Assume north to be at the top of the drawing, and locate a 12" diameter x 3/8" thick vertical outlet pipe at the outer edge of the tank, at 37° west of true north. Save the drawing as E-9-15.

CHAPTER 9 QUIZ

1. Describe the process used to draw a wall that is 6" thick.
2. Several copies of an object need to be made. What option is available so that the **Copy** command does not have to be repeated?
3. You have selected several objects for editing when you realize you have accidentally selected an object that you do not want changed. What can you do so that you do not have to start the edit sequence over again?
4. How will **Undo** affect objects that have just been moved?
5. List three methods suitable for selecting irregularly shaped objects for editing.
6. When you use the Fence option, what objects will it select for editing?
7. What editing mode will allow you to select objects and then decide how to edit them?
8. What process will display the dialog box that controls selection methods?
9. List the commands in sequence required to complete Exercise 9.
10. Explain when the Through option of **Offset** could be used.
11. Start the **Erase** command using each method and describe the advantages or disadvantages.
12. Experiment with the **Mirror** command. Describe the effects of ORTHO on the mirror line.
13. **Mirror** is being used to complete a symmetrical object. What options are available and how do they affect the **Mirror** sequence?
14. Where is **Erase** listed in a menu?
15. Describe how to change the size of the pick box.
16. What is the effect of base point on the **Copy** command, and how should it be selected?
17. List ways that the **Copy** command can be used to draw a floor plan.
18. What is the benefit of having so many different methods of selecting objects for editing?
19. What are the drawbacks to enlarging the pick box to twice its normal size, or decreasing it to half its normal size?
20. Use the **Help** menu to explain the difference between the normal setting for an editing command and the Single option.
21. Explain the difference between Noun/Verb and Verb/Noun selection.
22. Explain the meaning and the effect of the following letters for selection groups: W, C, F, WP, CP, R, A, and P.

23. What effect does Press and drag have on selection?
24. Use the **Help** menu to list the effects of the **Pickadd**, **Pickauto**, and **Pickdrag** variables.
25. How can Object Snap be used with editing commands?
26. Explain the difference between the two types of array patterns.
27. Twelve steel columns need to be drawn 22' on center in one horizontal line. What will the array sequence be?
28. Three horizontal lines of 10 steel columns need to be drawn 18' on center. The horizontal lines of columns will be 32' apart. What will the array sequence be?
29. Use the drawing for Exercise 9.14 and give the command sequence to array the supports.
30. Using an illustration as an example, define a unit cell.
31. How could an array pattern for objects aligned at 45° from horizontal be drawn?
32. Ten rows and three columns are to be arrayed below and to the left of the original object at 60" O.C. How can this be done?
33. Describe how to produce a clockwise rotation.
34. Give the combination of alternatives for describing a polar array.
35. If a positive column distance and a negative row distance are entered, where will the objects be arrayed in relation to the original object?

CHAPTER 10

Modifying the Position and Size of Drawing Objects

In Chapter 9 you were introduced to editing commands that can affect an entire drawing. This chapter will explore

- Commands that can be used to modify part of an object by altering the position or size of the object
- Commands that can be used to edit objects by altering their length
- Methods to select objects for editing using grips
- Methods for selecting objects for multiple edits
- Methods for naming object groups

Commands explored in this chapter include

- **Move**
- **Align**
- **Rotate**
- **Scale**
- **Stretch**
- **Lengthen**
- **Trim**
- **Extend**
- **Break**
- **Chamfer**
- **Fillet**
- **Select**
- **Group**

Mastering these commands will greatly increase your drawing ability. All of the commands can be selected on the **Modify** toolbar, entered by keyboard, or selected from the **Modify** menu. Each command can also be entered from the shortcut menu by selecting an object to edit, and then right-clicking to select the editing command.

ALTERING THE POSITION OR SIZE OF OBJECTS

The **Modify** menus offer six options for modifying a drawing by changing the position of an object or by changing the size of the object. These commands include **Move**, **Align**, **Rotate**, **Scale**, **Stretch**, and **Lengthen**.

MOVE

The **Move** command allows one or more objects to be selected and moved from their current location to a new one. Access the command by selecting the Move button on the **Modify** toolbar, by typing **M** (ENTER) at the Command prompt, or by selecting **Move** from the **Modify** menu. Figure 10.1a and 10.1b compare the effects of the **Move** command.

The command sequence for **Move** is as follows:

Command: **M** (ENTER) *(Or click the Move button.)*

MOVE

Select objects: *(Use any selection method.)*

Select objects: (ENTER)

Specify base point or displacement: *(Enter point or coordinates.)*

Specify second point of displacement or <use first point as displacement>: *(Enter point or coordinates.)*

Command:

The process can be seen in Figures 10.2a through 10.2c.

- Use any selection method to select the objects to be moved.
- The "base point of displacement" prompt is asking for a reference point for moving the selection set. Any point can be selected, but it might help you to visualize the results of moving by selecting a point in the drawing, such as an intersection or endpoint. This is an excellent time to make use of OSNAP.
- The "second point of displacement" prompt is asking where you would like to move the first point.

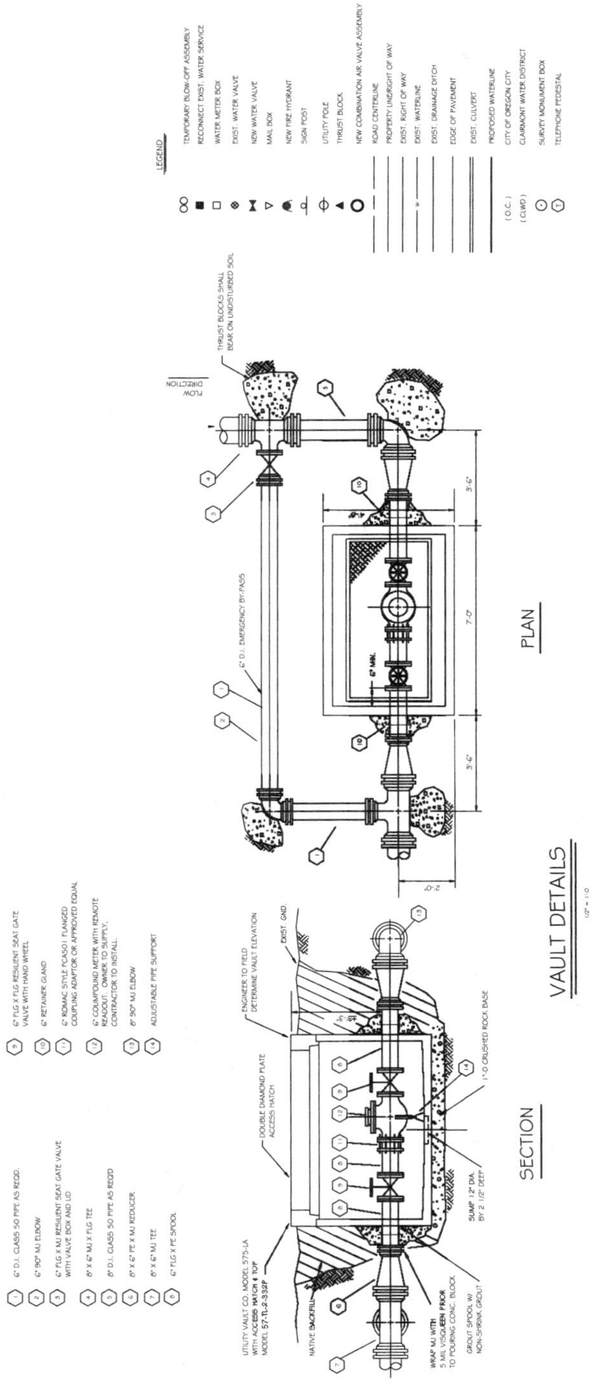

Figure 10.1a *The* **Move** *command can be used to alter the layout of a drawing. (Courtesy Lee Engineering, Inc.)*

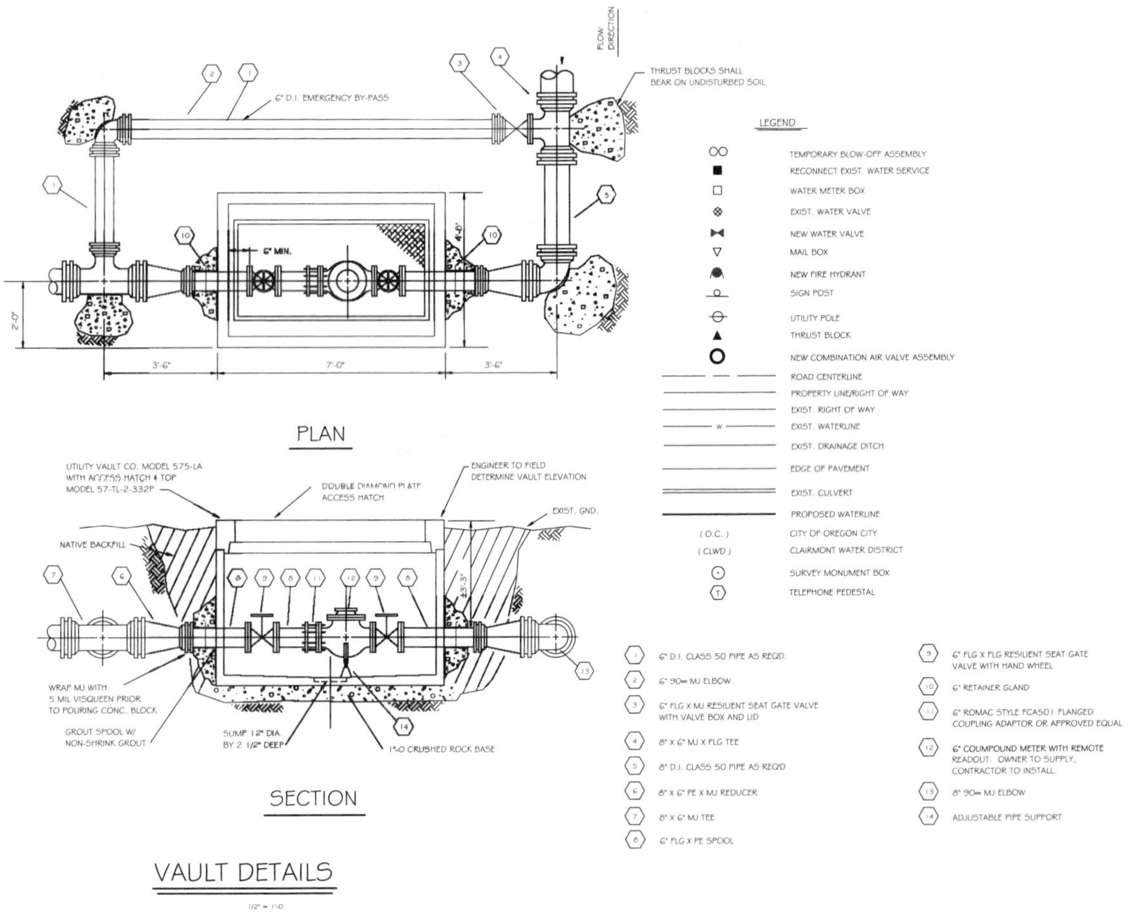

Figure 10.1b *The* **Move** *command was used to place the plan view of Figure 10.1a above the section drawing. (Courtesy Lee Engineering, Inc.)*

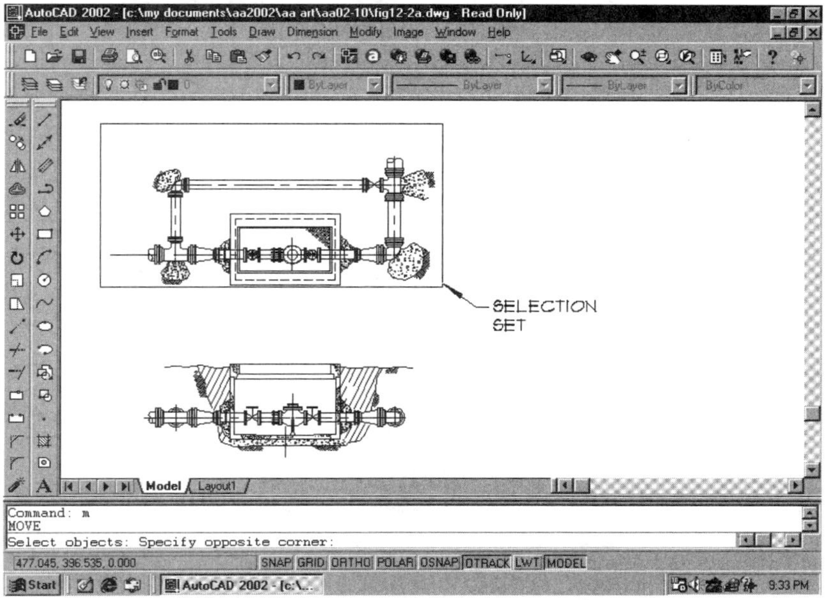

Figure 10.2a *The first step in the process for moving a drawing is to select the objects to be moved. (Courtesy Lee Engineering, Inc.)*

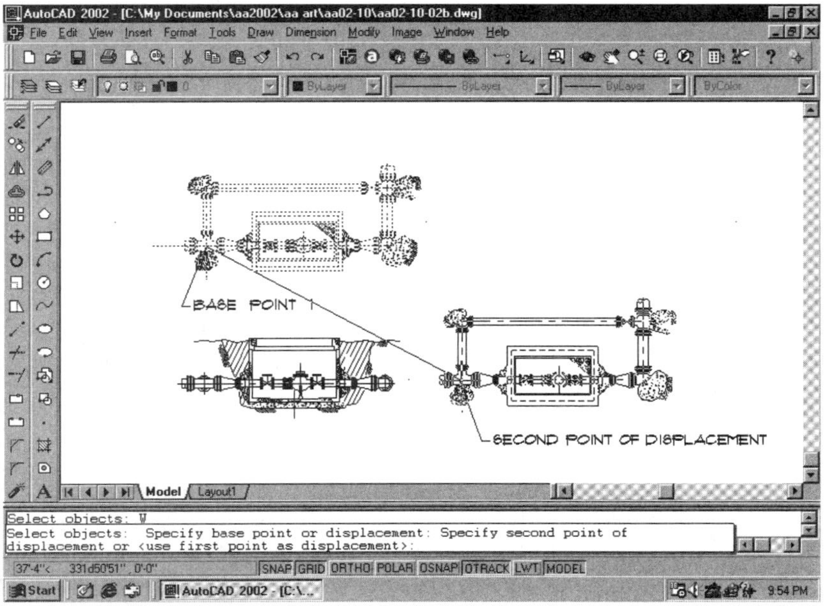

Figure 10.2b *The second step is to select the base point of displacement and the new location for this point. (Courtesy Lee Engineering, Inc.)*

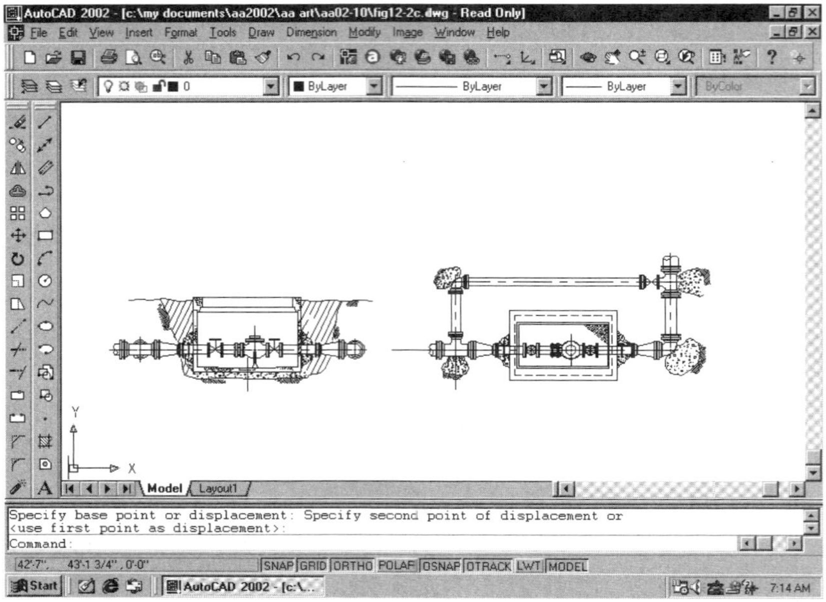

Figure 10.2c *Finally, the second point of displacement is selected, and the Move process is complete. (Courtesy Lee Engineering, Inc.)*

Each point can be entered by selecting a point with a pointing device or by entering coordinates. SNAP, ORTHO, OSNAP, and **Zoom** can be very helpful for aligning related objects as they are moved.

Practical Use: Move should give you an immense sense of freedom, but not a sense of reckless abandonment. You still need to plan your drawings, but the **Move** command offers you great flexibility in drawing arrangement. Space between views can be altered as the needs of the drawing change. On the other hand, you have to keep in mind that each of your drawings will need to fit on a sheet of paper of a set size. For most architectural projects, this means your drawing will need to fit on a sheet of paper that is 36" x 24". If you're going to try and place ten details on this sheet, there is only so much space where the details can be moved. AutoCAD still has not incorporated the CRAM command. Planning and wise use of the editing commands are the only hope for a good solution.

ROTATE

In Chapter 9 you learned how to array portions of a drawing. In a similar fashion, you can draw an object or group of objects and then rotate them around a base point to change their orientation. Start the command by selecting the **Rotate** button on the **Modify** toolbar, by typing **RO** (ENTER) at the Command prompt, or by

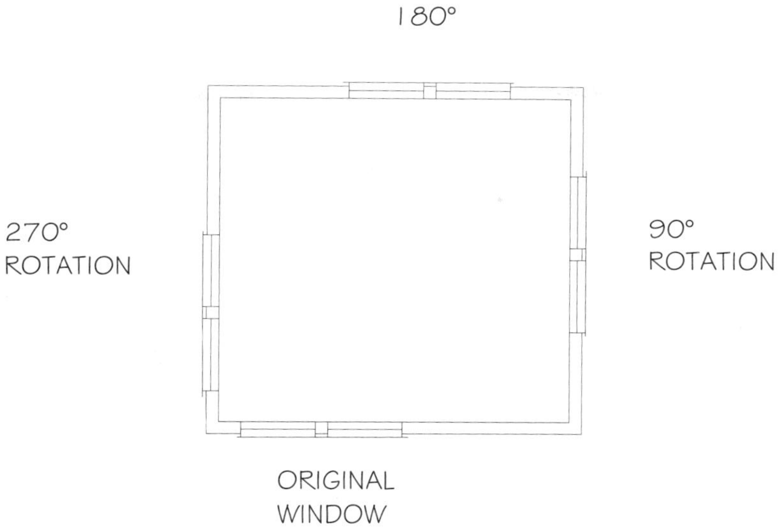

Figure 10.3 *The* **Rotate** *command allows objects to be drawn and rotated into a new position.*

selecting **Rotate** from the **Modify** menu. You can also start **Rotate** from the shortcut menu by picking the object, right-clicking, and then selecting **Rotate**. Figure 10.3 shows an example of a window that was drawn for a horizontal wall on a floor plan. The window was then copied and rotated 90° to be used in a vertical wall.

To complete the command, you will need to select objects to be rotated, a base point to rotate the object around, and the desired degree of rotation.

- Entering a positive rotation angle rotates the selected object in a counter-clockwise direction.
- Entering a negative rotation angle produces a clockwise rotation.

The base point can be anywhere on the drawing screen, but selecting one corner of the object as the base point can ease visualization of the rotation. The command sequence is as follows:

Command: **RO** (ENTER) *(Or click the Rotate button.)*

ROTATE

Current positive angle in UCS: ANGDIR=counterclockwise
ANGBASE=0

Select objects: *(Use any selection method.)*

8 found

Select objects: (ENTER)

Specify base point: *(Select any point such as an endpoint or midpoint.)*

Specify rotation angle or [Reference]: *(Enter angle or drag into position.)*

Command:

Note that instead of entering a precise angle, you can drag the selected objects into the new position as the cursor is moved. If ORTHO is toggled ON, the objects will be snapped at 90° intervals. With ORTHO off, the objects can be dragged into any desired angle. When you enter an angle, you can rotate objects to exact locations based on the current location. The process can be seen in Figure 10.4.

Objects also can be rotated based on absolute rotation angles. Suppose you've drawn the top chord of a truss at 22 1/2° for a 5/12 pitch. The owners change their minds and desire a 6/12 pitch (27 1/2°). You could rotate the existing top chord 5° and hope you get it close enough, or use the Reference option and place the roof exactly where it should be. The correct option uses the following command sequence:

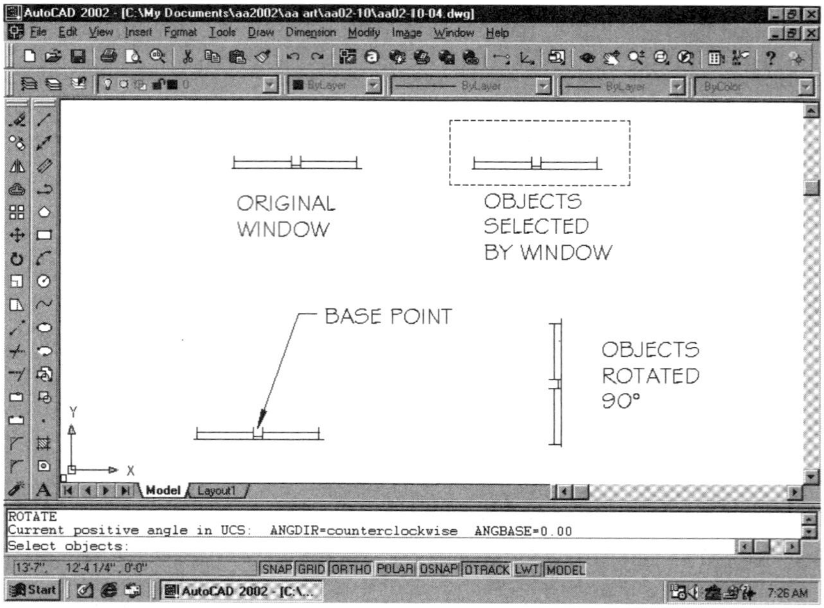

Figure 10.4 *Objects are rotated by defining the selection set, specifying a base point for the rotation, and providing the rotation angle.*

Command: **RO** (ENTER) *(Or click the Rotate button.)*

ROTATE

Current positive angle in UCS: ANGDIR=counterclockwise ANGBASE=0

Select objects: *(Use any selection method.)*

5 found

Select objects: (ENTER)

Specify base point: *(Select any point.)*

Specify rotation angle or [Reference]: R (ENTER)

Specify the reference angle: **22.5** (ENTER)

Specify the new angle: **27.5** (ENTER)

Command:

The process can be seen in Figures 10.5a, 10.5b, and 10.5c.

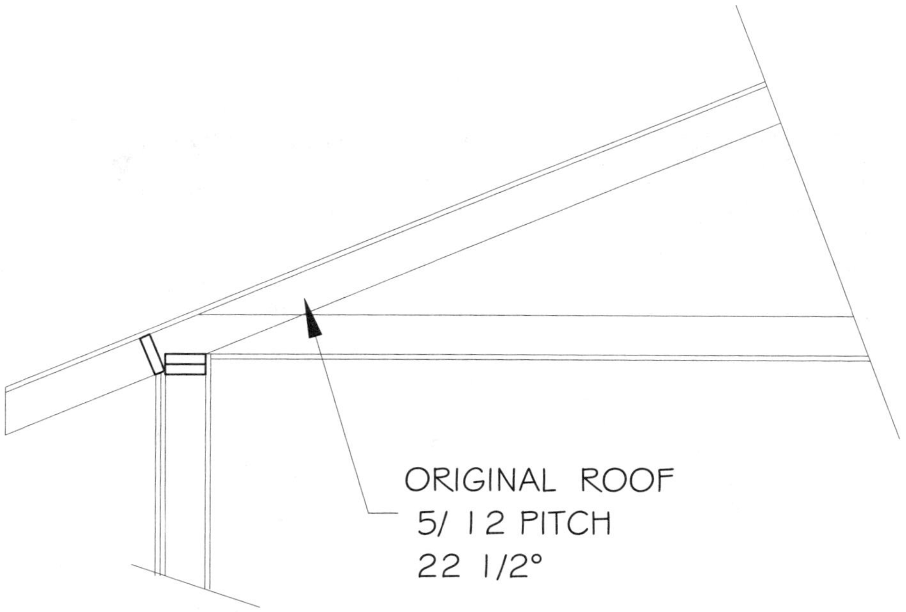

Figure 10.5a *The pitch of a roof can be altered by use of the* **Rotate** *command.*

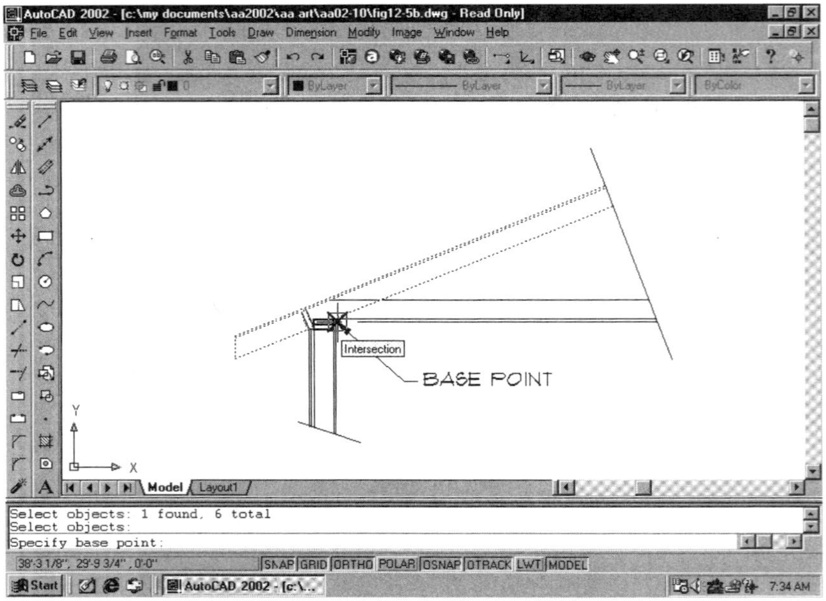

Figure 10.5b *The selection set is defined and the base point is specified.*

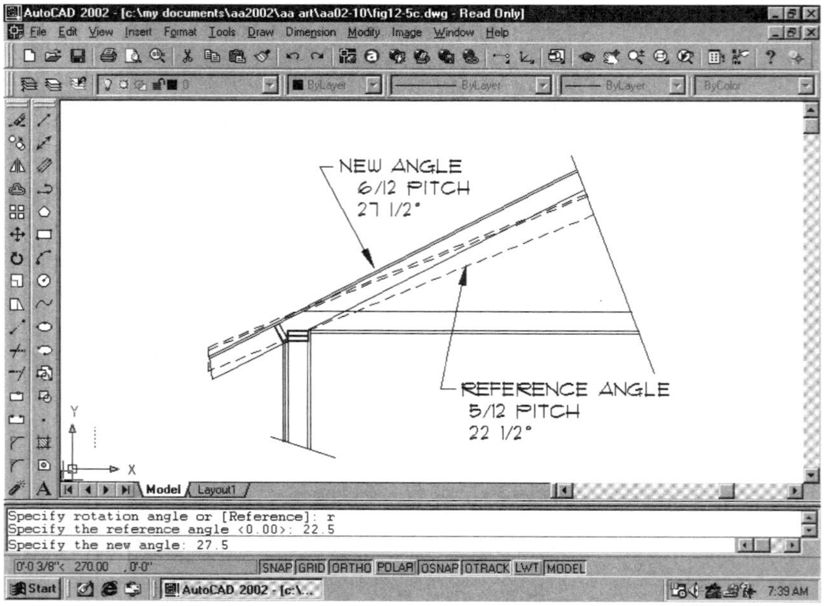

Figure 10.5c *With the reference angle specified (existing angle of 22.5°), the new angle of 27.5° is specified and the roof is rotated.*

ALIGN

Think of the **Align** command as a combination of the **Move** and **Rotate** commands. Although primarily a command for 3D drawing, it works well when an object or group of objects need to be moved and rotated. Start the command by typing **AL** (ENTER) at the Command prompt, or by selecting it from **3D Operation** on the **Modify** menu. The command will prompt you for two source points and two destination points. The source points will be used to define points on the original object in its original position. The destination points represent the locations where the object will be placed. The command can be used in two ways. If one pair of source and destination points is provided, the selected object will be moved in a fashion similar to when the **Move** command is used. The benefit of the **Align** command is in providing two pairs of points. If two pairs of source and destination points are provided, the selected object can be moved, rotated, and scaled to match other objects. This could be useful in moving, rotating, and aligning a room with another when redesigning a floor plan. Use the following command sequence to move and rotate a kitchen from one area of a floor plan to another:

Command: **AL** (ENTER)

ALIGN

Select objects: *(Select object using any selection method.)*

Specify first source point: *(Select point with mouse or enter coordinates.)*

Specify first destination point: *(Select point with mouse or enter coordinates.)*

Specify second source point: *(Select point with mouse or enter coordinates.)*

Specify second destination point: *(Select point with mouse or enter coordinates.)*

Specify third source point or <continue>: (ENTER)

Scale objects based on alignment points? [Yes/No] <No>: (ENTER)

When (ENTER) is pressed, the object will be moved and rotated. The selection process can be seen in Figure 10.6. The result of the command can be seen in Figure 10.7. If Y (ENTER) is typed at the last prompt, the objects being rotated will be scaled to match the distance indicated by the first and second destination points. Figure 10.8 shows the process of scaling the objects to meet the alignment points.

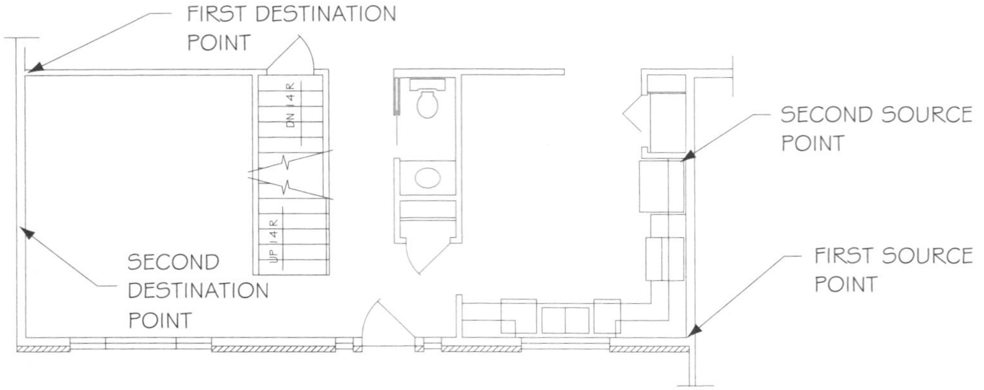

Figure 10.6 *The* **Align** *command allows portions of a drawing to be moved and rotated. The command will prompt you to provide source points, and then ask where you would like those points moved to.*

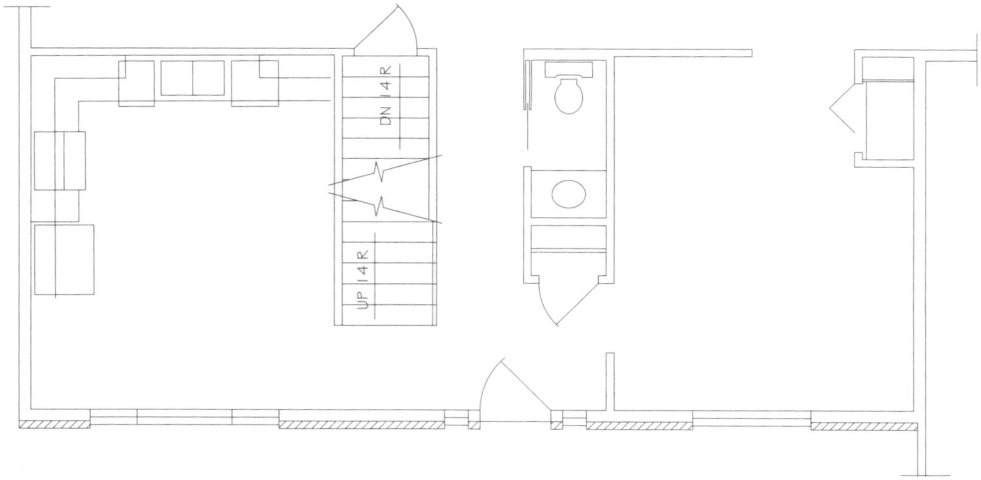

Figure 10.7 *By using the destination points shown in Figure 10.6, the kitchen was moved and rotated to a new position with the* **Align** *command.*

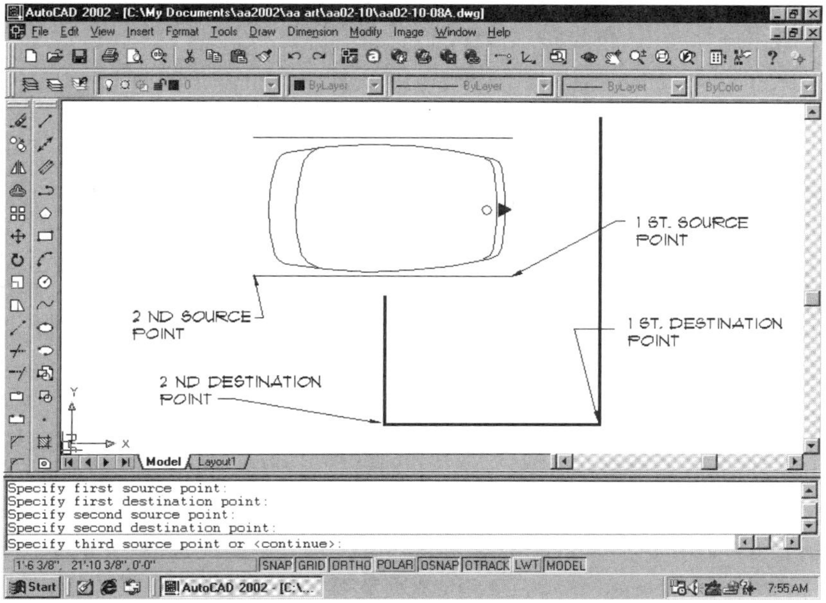

Figure 10.8a *If you choose to scale objects based on alignment points, an object or group of objects can be moved and scaled to match the selected destination points.*

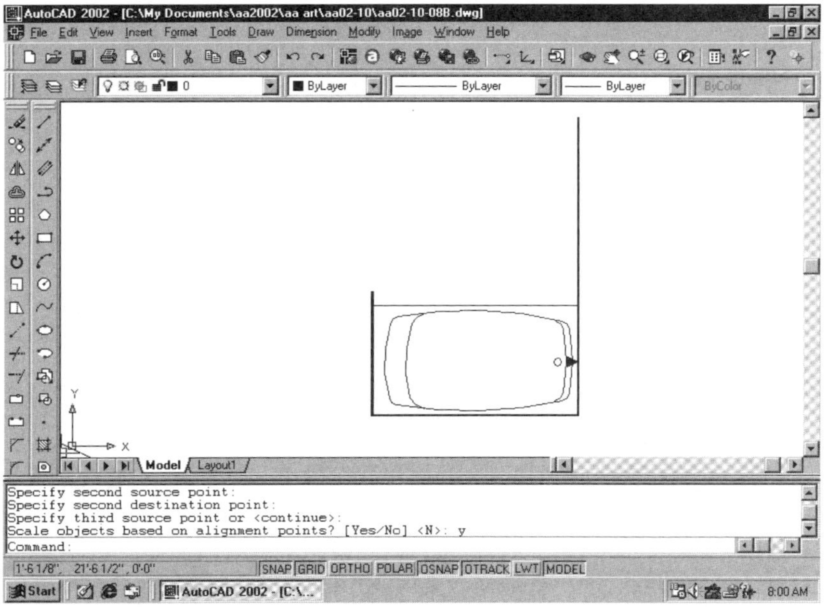

Figure11.8b *The 6' long tub is scaled to fit in a 5' space.*

SCALE

Up to this point, your drawings have been drawn at full scale. If you needed to draw a plot of land that is 100' square, you set the limits and units accordingly, zoomed out to the limits, and drew the plot of land. The whole process was done at real scale with the size of the zoom used to control the viewing screen. The **Scale** command will allow you to change the size of an existing object or an entire drawing, as seen in Figure 10.9. Start the command by selecting the **Scale** button on the **Modify** toolbar, by typing **SC** (ENTER) at the Command prompt, or by selecting **Scale** from the **Modify** menu. The command can be completed using the following process:

Command: **SC** (ENTER) *(Or click the Scale button.)*

SCALE

Select objects: *(Select objects using any selection method.)*

1 found

Select objects: (ENTER)

Specify base point: *(Select base point.)*

Specify scale factor or [Reference]:

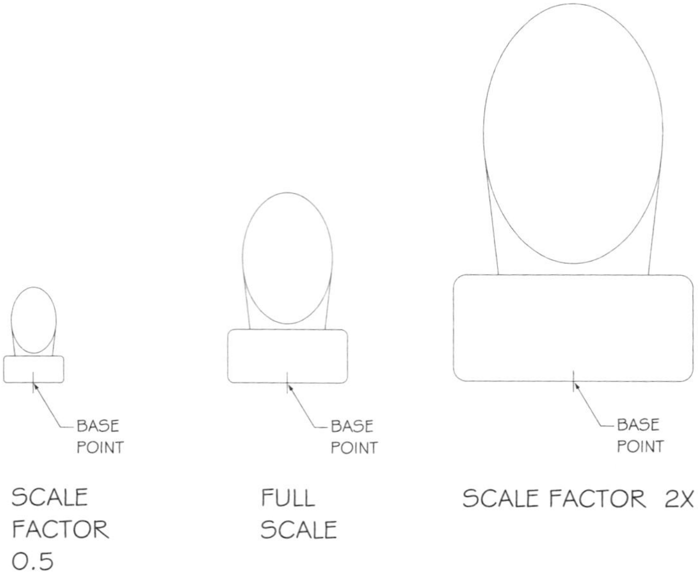

Figure 10.9 *The* **Scale** *command allows the size of objects to be altered. In this case, the toilet on the left is half the size of the one in the middle.*

Scale Factor

The default for entering the effect of scaling is to enter a numeric value. To enlarge an object to twice its existing size, type **2** (ENTER) at the prompt. Using whole numbers will enlarge the selected objects by that factor. Type **3** (ENTER) at the prompt to enlarge the objects to three times their existing size. Typing a fraction, such as **.5** (ENTER), will reduce the selected object to half its original size.

Reference Factor

This option will allow an object to be enlarged to an absolute length rather than a relative scale. For instance, a tub that is 5' long can be scaled to 6' long. Rather than figuring the proportion of enlargement, type **R** (ENTER) at the Command prompt. The process can be seen in Figure 10.10. The command can be completed using the following process:

> Command: **SC** (ENTER) *(Or click the Scale button.)*
>
> SCALE
>
> Select objects: *(Select objects using any selection method.)*
>
> 1 found

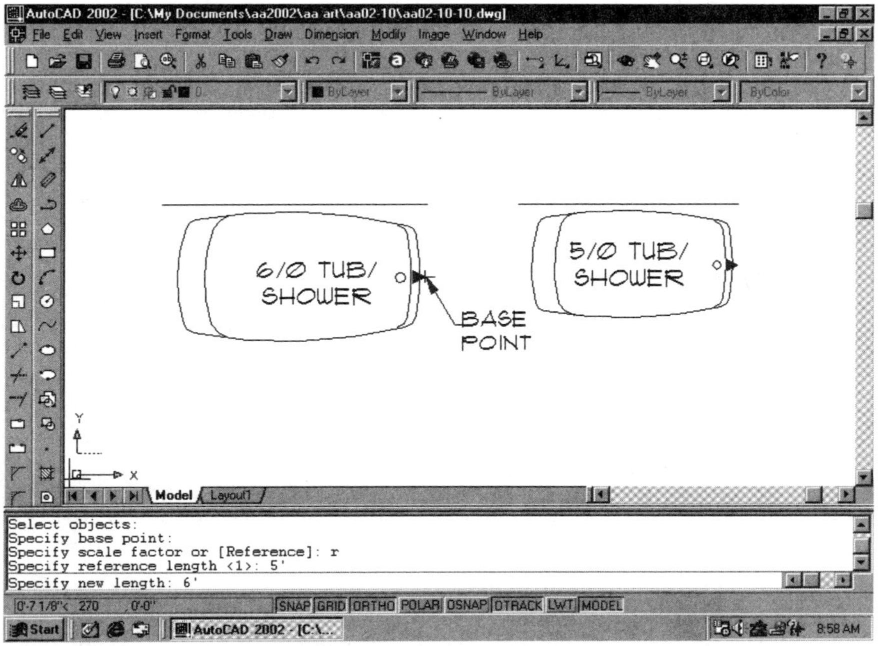

Figure 10.10 *Entering a reference factor allows an object to be scaled by an absolute length.*

Select object: (ENTER)

Specify base point: *(Select base point.)*

Specify scale factor or [Reference]: **R** (ENTER)

Specify reference length <1>: **5'** (ENTER)

Specify new length: **6'** (ENTER)

Command:

STRETCH

The **Stretch** command allows you to enlarge or shrink drawing objects. This can be especially helpful when editing the shape and size of a floor plan to meet the changing needs of a client, as seen in Figure 10.11. Drawings made with lines, arcs, solids (Chapter 14), and polylines (Chapter 11) can all be stretched. Start the command by selecting the **Stretch** button on the **Modify** toolbar, by typing **S** (ENTER) at the Command prompt, or by selecting **Stretch** from the **Modify** menu. The command sequence is as follows:

Command: **S** (ENTER) *(Or click the Stretch button.)*

STRETCH

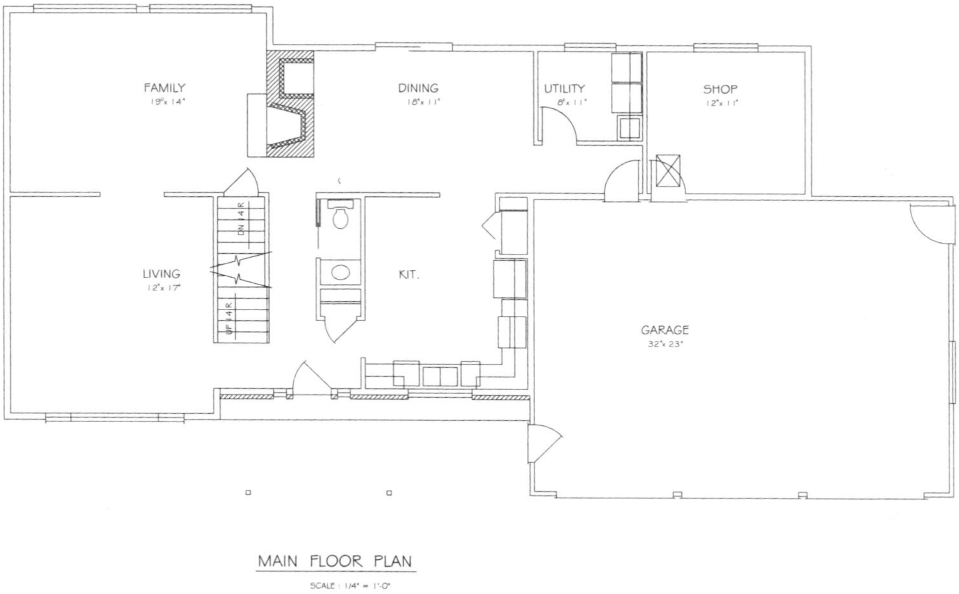

Figure 10.11 *The* **Stretch** *command can be used to elongate an object or group of objects. Compare the garage size with the garage in Figure 9.23b.*

Select objects to stretch by crossing-window or crossing-polygon...

Select objects: *(Select object to be stretched using automatic select window, window, crossing window, or a polygon options.)*

Specify first corner: *(Select a point for the corner of the window.)*

Specify opposite corner: *(Select the second point for the corner so that the object to be stretched is included in the window.)*

23 found

Select objects: (ENTER)

The automatic selection window only needs to cross a portion of the desired object because the default selection mode is Crossing. Objects can be added to the selection set by any of the methods described in Chapter 9. When the Window option is used to select the selection set for objects to be stretched, the desired effects are not always achieved. If the entire object is selected, **Stretch** will function like the **Move** command and move the entire object. Once the selection set is complete, the prompt will display the following:

Specify base point or displacement: *(Select a point to serve as a reference. Be sure OSNAP is toggled ON.)*

Specify second point of displacement: *(Select the location that you would like to stretch the base point to.)*

Command:

As the cursor is moved, the effect is displayed as the object is stretched between the base and new points. The process can be seen in Figures 10.12a and 10.12b.

Not only can **Stretch** be used to lengthen objects, but it also can be used to stretch some objects and move others, as seen in Figures 10.13a, 10.13b, and 10.13c.

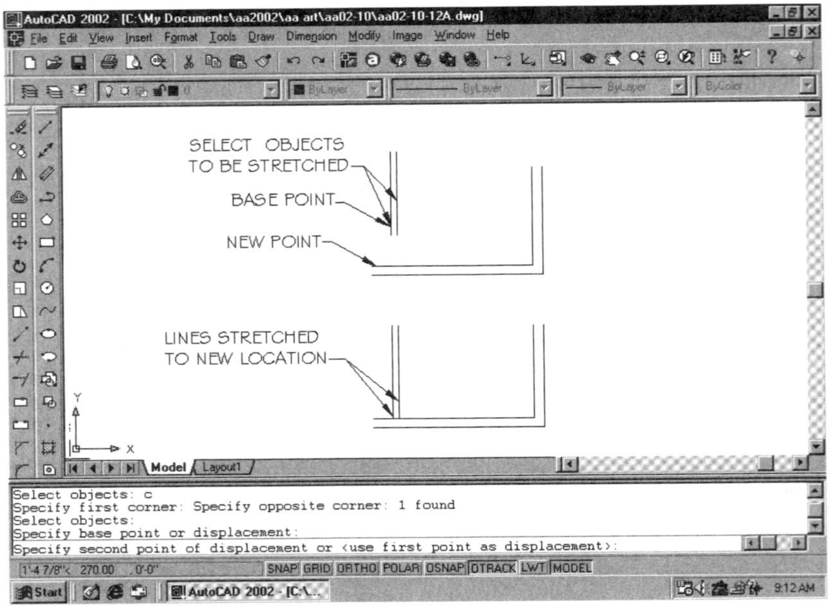

Figure 10.12a *The Window option is used to form the selection set of objects that are to be stretched. Once objects are selected, prompts will be displayed for the base point of the original objects and the new position they are to be stretched to.*

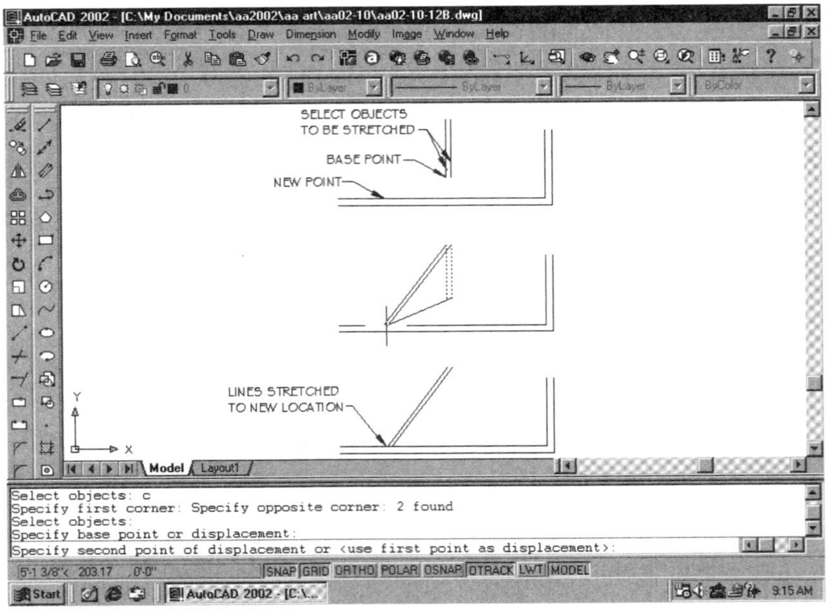

Figure 10.12b *In addition to adding length to line segments, the* **Stretch** *command can be used to alter the angle of the selected lines.*

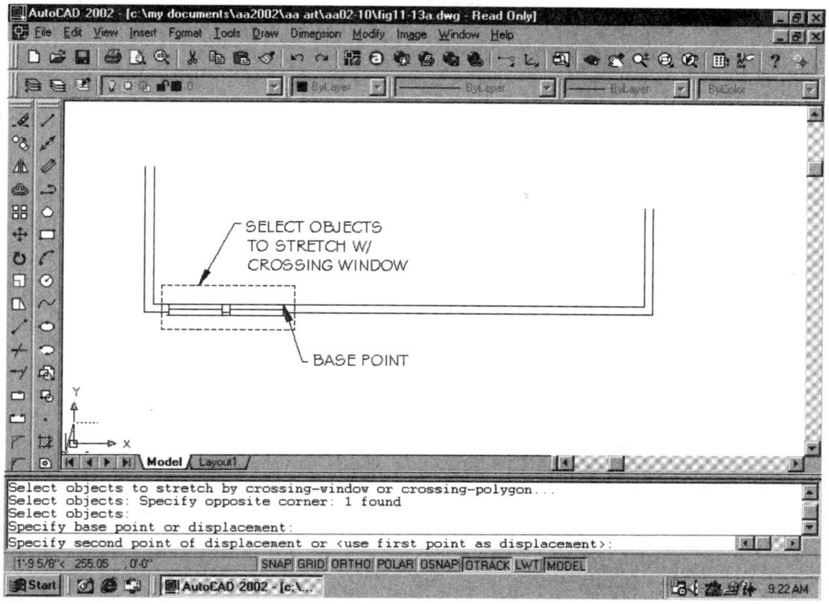

Figure 10.13a *The* **Stretch** *command can also be used to alter the location of objects.*

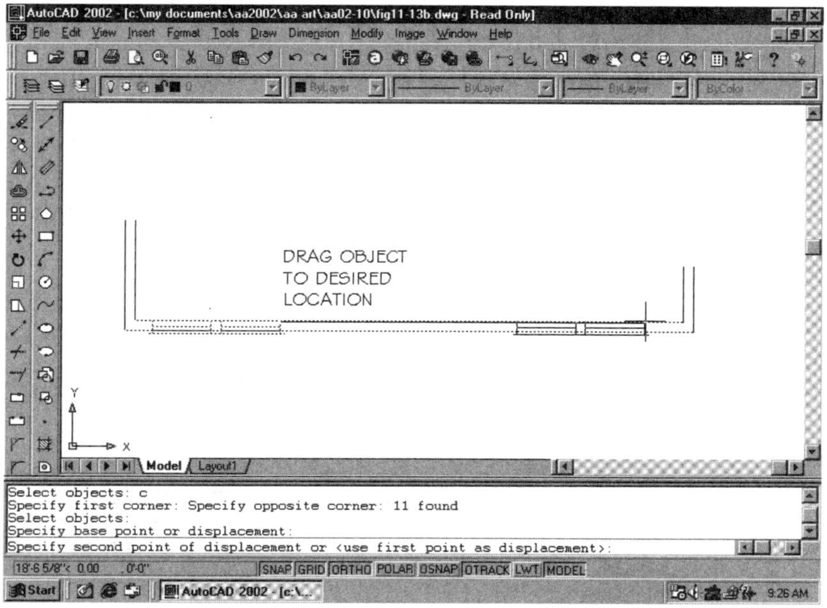

Figure 10.13b *The window and walls are stretched to alter the location of the window. Lines on the left side of the window are stretched, while the lines on the right side are shortened.*

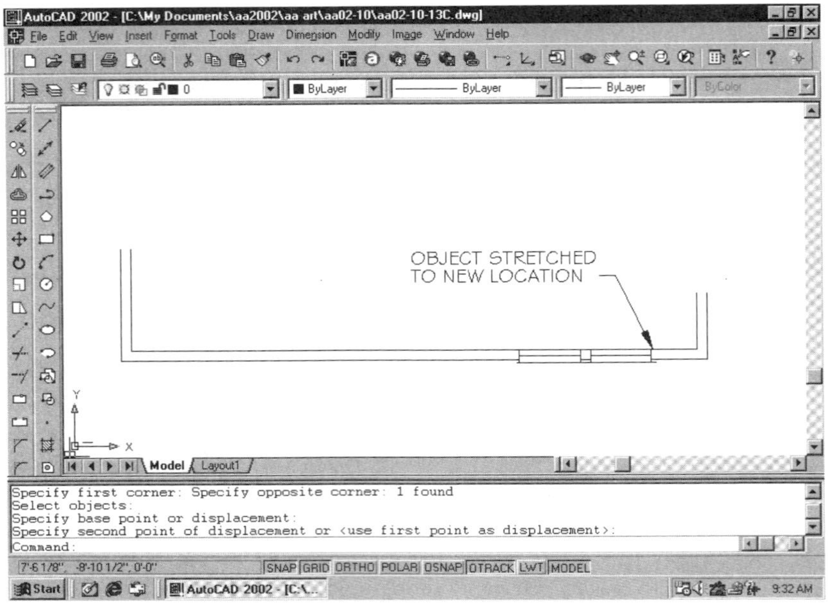

Figure 10.13c *The completed* **Stretch** *command. This is much more efficient than moving the window, stretching the left lines, and trimming the right lines.*

LENGTHEN

The **Lengthen** command is similar to the **Stretch** command. **Stretch** is used to edit the length of groups of objects and closed polygons. However, **Lengthen** is used to edit the length of objects, but not closed polygons. Start the **Lengthen** command by selecting the **Lengthen** button on the **Modify** toolbar, by typing **LEN** (ENTER) at the Command prompt, or by selecting **Lengthen** from the **Modify** menu. The command sequence is as follows:

Command: **LEN** (ENTER) *(Or click the Lengthen button.)*

LENGTHEN

Select an object or [DElta/Percent/Total/DYnamic]: *(Select an object to edit.)*

Current Length: 10'-0"*(Length will vary.)*

Select an object or DElta/Percent/Total/DYnamic <Select object>: **DE** (ENTER)

Using the Delta Option of Lengthen

Typing **DE** (ENTER) at the Select object prompt will produce the prompt:

Enter delta length or [Angle] <0'-0">: **5'** (ENTER) *(Enter the length to add.)*

Select an object to change or [Undo]: *(Select the object again.)*

Select an object to change or [Undo]: (ENTER)

Command:

The command sequence can be seen in Figure 10.14. Notice that when DElta was entered at the Select prompt, a choice could be made between length and angle. The default is to provide a length. Entering a positive number lengthens the line. Providing a negative response shortens the line. With each option of **Lengthen**, the end of the object that is selected will be the end that is edited.

Note: The length entered is the length that will be added to the selected line. The value will not become the length for the total line length.

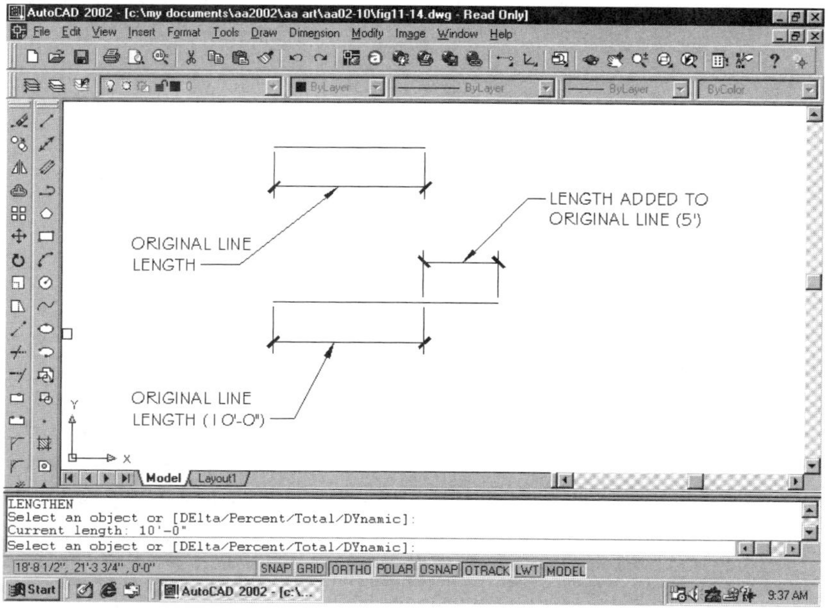

Figure 10.14 *The DElta option of the* **Lengthen** *command can be used to lengthen or shorten a line.*

The Angle option can be used to change the included angle of an arc. The effects of this command sequence can be seen in Figure 10.15. The command sequence is as follows:

Command: **LEN** (ENTER) *(Or click the Lengthen button.)*

LENGTHEN

Select an object or [DElta/Percent/Total/DYnamic]: *(Select an arc to edit.)*

Current Length: 10'-0", included angle: 90 *(Length and angle will vary.)*

DElta/Percent/Total/DYnamic <Select object>: **DE** (ENTER)

Enter delta length or [Angle] <0'-0">: **A** (ENTER)

Enter delta angle <0>: **60** (ENTER)

Select an object to change or [Undo]: *(Select arc to be edited.)*

Select an object to change or [Undo]: (ENTER)

Command:

Using the Percent Option of Lengthen

The Percent option allows the length of a line or arc to be altered by a percentage of the existing length. The existing line length represents 100%. To lengthen the line, provide a value greater than 100%. Providing a percentage of less than 100% will shorten the line. The command sequence to lengthen a 2" line to a 3" line using the Percent option is as follows:

Command: **LEN** (ENTER) *(Or click the Lengthen button.)*

LENGTHEN

Select an object or [DElta/Percent/Total/DYnamic]: *(Select an object to edit.)*

Current Length: 10'-0" *(Length will vary.)*

Select an object or [DElta/Percent/Total/DYnamic]: **P** (ENTER)

Enter percentage length <100.0000>: **150** (ENTER)

Select an object to change or [Undo]: *(Select line to be edited.)*

Select an object to change or [Undo]: (ENTER)

Command:

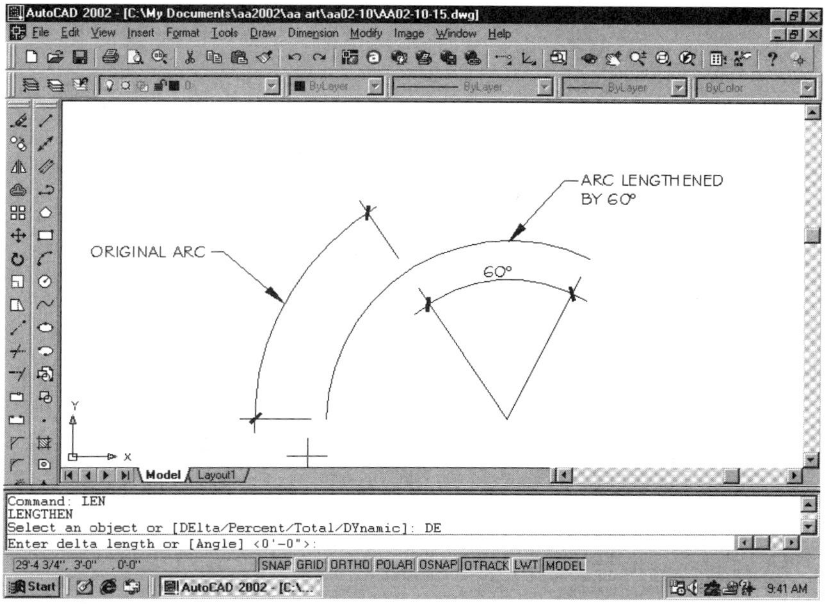

Figure 10.15 *The DElta and Angle options of the* **Lengthen** *command can be used to lengthen or shorten an angle.*

The effects of the Percent option can be seen in Figure 10.16. When used with an arc, the angle of the arc will be edited based on a percentage of the total angle of the specified arc.

Using the Total Option of Lengthen

The Total option allows the length of a line or angle of an arc to be edited by providing the desired ending result. The command sequence to alter an existing line to a 35' line would be as follows:

Command: **LEN** (ENTER) *(Or click the Lengthen button.)*

LENGTHEN

Select an object or [DElta/Percent/Total/DYnamic]: *(Select an object to edit.)*

Current Length: 10'-0" *(Length will vary.)*

Select an object or [DElta/Percent/Total/DYnamic]: **T** (ENTER)

Specify total length or [Angle]: **35'** (ENTER)

Select an object to change or [Undo]: (ENTER)

Command:

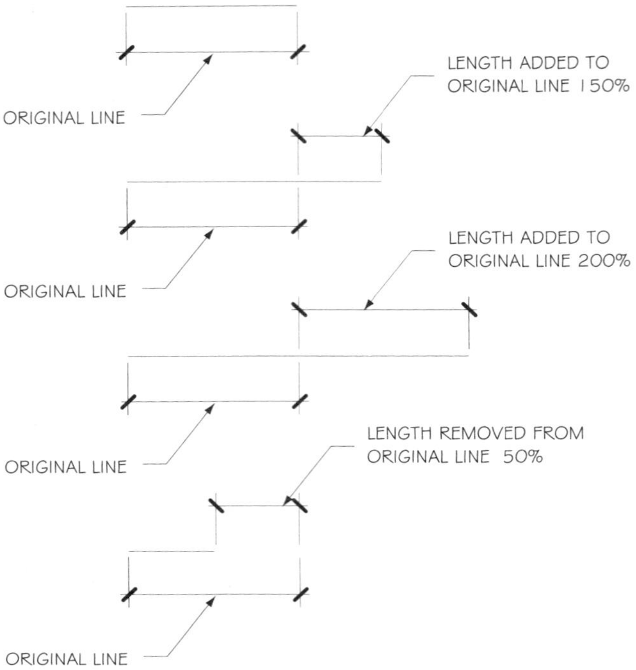

Figure 10.16 *The lengths of lines and arcs can be altered by entering a percentage of the original object. Percentages less than 100% will shorten the line or arc, and those greater than 100% will lengthen the line or arc.*

The result of this sequence can be seen in Figure 10.17. When first prompted to select an object, you might prefer to select an object before choosing the Total option. When a line is selected first, the length of the existing line will be displayed.

Using the DYnamic Option of Lengthen

The DYnamic option allows the selected line or angle to be dragged to the selected point to achieve the required length. The results of the DYnamic option can be seen in Figure 10.18.

Command: **LEN** (ENTER) *(Or click the Lengthen button.)*

LENGTHEN

Select an object or [DElta/Percent/Total/DYnamic]: *(Select an object to edit.)*

Current Length: 10'-0"*(Length will vary.)*

Select an object or [DElta/Percent/Total/DYnamic]: **DY** (ENTER)

Select an object to change or [Undo]: *(Select line to be edited.)*

Specify new endpoint: *(Select new endpoint for line to be edited.)*

Select an object to change or [Undo]: (ENTER)

Command:

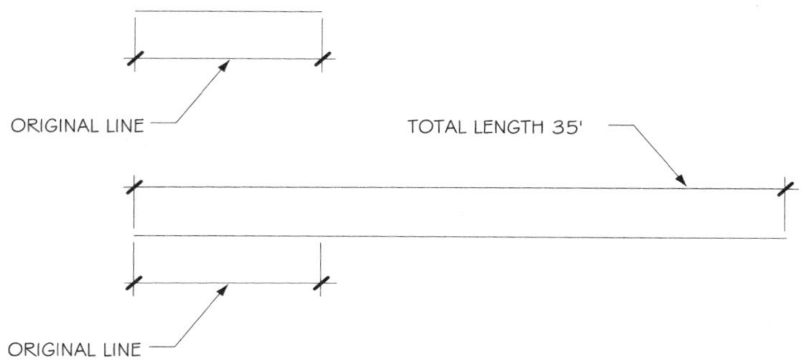

Figure 10.17 *The Total option of* **Lengthen** *allows a line or angle of an arc to be lengthened based on the desired total length or angle.*

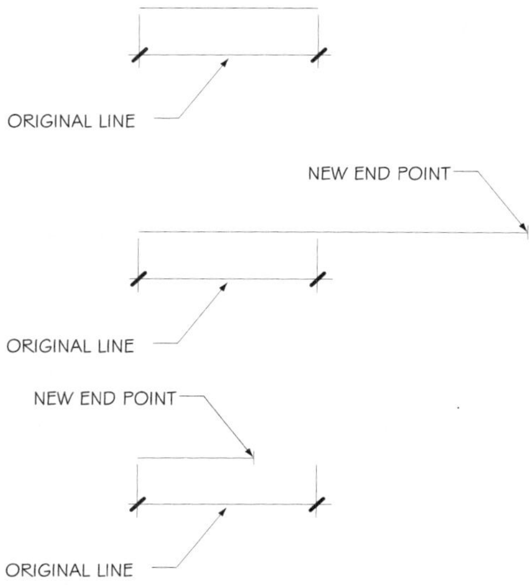

Figure 10.18 *The DYnamic option of* **Lengthen** *allows a line or angle of an arc to be dragged to the desired length by selecting a new ending point.*

MODIFYING OBJECTS BY ALTERING LINES

Up to this point, you've learned to modify objects by making additional copies and by altering the position or size of an object. The commands contained in this portion of the **Modify** menu will allow you to modify a drawing by altering intersections of lines, circles or arcs, or by removing a portion of an object. The commands include **Trim**, **Extend**, **Break**, **Chamfer**, and **Fillet**.

TRIM

The **Trim** command was introduced in previous chapters as aids to drawing layout were explored. **Trim** was used for its ability to clean the corners of intersecting lines, as seen in Figure 10.19, or to modify the size of objects. This chapter will expand on the command, allowing you to make fuller use of the command's power. Start the **Trim** command by selecting **Trim** on the **Modify** toolbar, by typing **TR** (ENTER), or by selecting **Trim** from the **Modify** menu. The command allows you to edit lines, arcs, circles, or polylines using a cutting edge. (Polylines will be introduced in the next chapter.) Any portion of the selected object extending past the designated cutting edge will be removed. Lines, arcs, circles, and polylines can be used as cutting edges. Once the cutting edges are defined, the command will continue until (ENTER) is pressed. Any object that extends past the cutting edge can be trimmed. Notice that

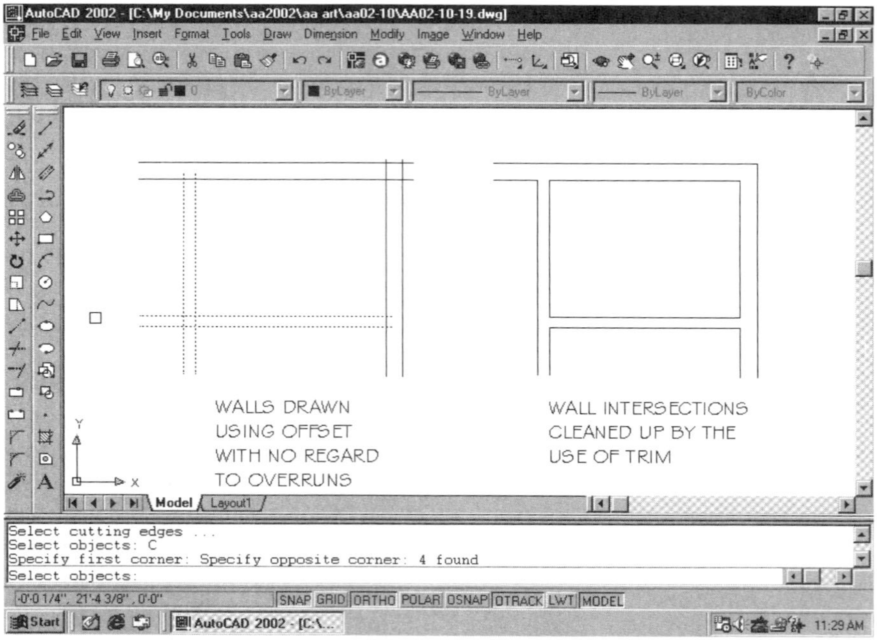

Figure 10.19 *The* **Trim** *command can be used to eliminate line overruns that were created using* **Offset**.

an object can be selected to be a cutting edge and it can also be an object to be trimmed.

Multiple Trim Selections

To select more than one cutting edge, hold SHIFT while selecting the cutting planes. The Window and Crossing window options work well for selecting multiple objects for cutting planes. Selecting the All option will use all drawing objects as trim edges. To use the All option, you must type **ALL** (ENTER) (not **A**).

Using the Trim Edge Option

The Edge option of **Trim** can be used to trim two objects that do not actually intersect, but would intersect if one of the objects were extended. AutoCAD refers to this as an implied intersection. For the Edge option, two objects must be selected. Once these edges have been selected, type **E** (ENTER) when prompted to select another object. This will now display the option

Enter an implied edge extension mode [Extend/No extend]
 <Extend>:

When Extend is selected, AutoCAD checks the cutting edge object to verify if it will intersect when extended. If it would touch, the implied intersection is used to trim the second object. This option is shown in Figure 10.20. The command sequence is as follows:

Command: **TR** (ENTER) *(Or click the Trim button.)*

TRIM

Current settings: Projection=UCS Edge=None

Select cutting edges... *(Select the cutting edge using any object selection method.)*

1 found

Select objects: (ENTER)

Select object to trim or shift-select to extend or
 [Project/Edge/Undo]: **E** (ENTER)

Enter an implied edge extension mode [Extend/No Extend]
 <Extend>: (ENTER)

Select object to trim or shift-select to extend or
 [Project/Edge/Undo]: *(Select object to trim.)*

Select object to trim or shift-select to extend or
 [Project/Edge/Undo]: (ENTER)

Command:

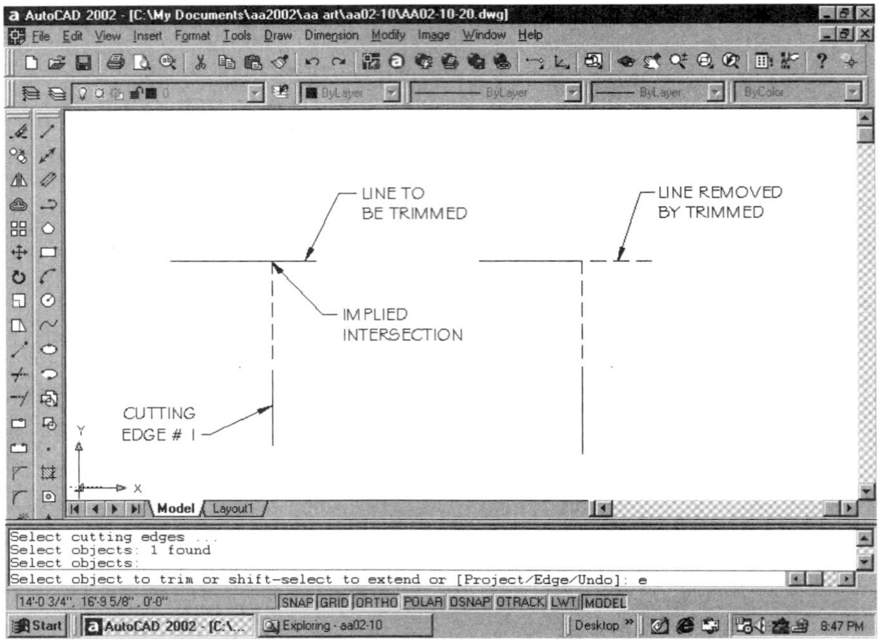

Figure 10.20 *Lines that would touch, if one were extended, can be trimmed by selecting the Extend mode of the* **Trim Edge** *option.*

Using the Trim Extend option

The Extend option of **Trim** can be used to extend objects that do not actually intersect, but would intersect if one of the objects were extended. AutoCAD refers to this as an implied intersection. To use the Extend option, select the line to be used as the destination. When prompted to select another object, press and hold SHIFT and select the line to be extended with the select button. The command sequence is as follows:

Command: **TR** (ENTER) *(Or click the Trim button.)*

TRIM

Current settings: Projection=UCS Edge=None

Select cutting edges... *(Select the cutting edge using any object selection method.)*

1 found

Select objects: (ENTER)

Select object to trim or shift-select to extend or [Project/Edge/Undo]: *(SHIFT~ + select object to extend.)*

Select object to trim or shift-select to extend or
[Project/Edge/Undo]: (ENTER)

Command:

Using the Trim Undo option

Occasionally you might trim an object and not be pleased with the results. The **Trim Undo** option will allow an unsatisfactory **Trim** to be undone while still in the command sequence. The command sequence is as follows:

Command: **TR** (ENTER) *(Or click the Trim button.)*

TRIM

Current settings: Projection=UCS Edge=Extend

Select cutting edges... *(Select the cutting edge using any object selection method.)*

1 found

Select objects: (ENTER)

Select object to trim or shift-select to extend or
[Project/Edge/Undo]: *(Select object to be trimmed.)*

Select object to trim or shift-select to extend or
[Project/Edge/Undo]: **U** (ENTER)

Command has been completely undone.

Select object to trim or shift-select to extend or
[Project/Edge/Undo]: *(Select object to be trimmed.)*

Select object to trim or shift-select to extend or
[Project/Edge/Undo]: (ENTER)

Command:

Using the Trim Project Option

The **Trim Project** option is primarily a 3D option and that aspect of the command will be discussed only briefly in this text. Refer to the **Help** menu for further information. The option can be used just as the Edge option as a means of trimming objects that extend past a projection of the selected cutting edge. The command sequence is as follows:

Command: **TR** (ENTER) *(Or click the Trim button.)*

TRIM

Current settings: Projection=UCS Edge=None

Select cutting edges... *(Select the cutting edge using any object selection method.)*

I found

Select objects: (ENTER)

Select object to trim or shift-select to extend or
[Project/Edge/Undo]: **P** (ENTER)

Enter a projection option [None/Ucs/View] <Ucs>: **N** (ENTER)

Select object to trim or shift-select to extend or
[Project/Edge/Undo]: *(Select object to trim)*

Select object to trim or shift-select to extend or
[Project/Edge/Undo]: (ENTER)

Command:

Practical Use of Trim

As you discovered earlier, combining the **Line**, **Offset**, and **Trim** commands will allow almost any object to be drawn easily. Draw two perpendicular lines to represent the corner of an object without regard to the exact length. The exact size of the object can be determined with **Offset** or **Trim**. Open the FLOOR10 file that was started earlier. With the use of **Trim**, this partial floor plan can be completed so that it more closely resembles the drawing in Figure 9.23b.

EXTEND

Start the **Extend** command by selecting the **Extend** button on the **Modify** toolbar, by typing **EX** (ENTER), or by selecting **Extend** on the **Modify** menu. The command allows an object to be lengthened to an exact boundary point. The command will ask you to designate boundary edges, and then select objects to be extended to that edge. The boundary edge can be a line, arc, circle, or polyline. The process can be seen in Figure 10.21. The command sequence is as follows:

Command: **EX** (ENTER) *(Or click the Extend button.)*

EXTEND

Current settings: Projection=UCS Edge=Extend

Select boundary edges... *(Select the edge using any object selection method.)*

I found

Select objects: (ENTER)

Select object to extend or shift-select to trim or
[Project/Edge/Undo]: *(Select object to be extended using any selection method.)*

Select object to extend or shift-select to trim or
[Project/Edge/Undo]: (ENTER)

Command:

As the object to be extended is selected, the line is extended to the other object that was selected as the boundary edge.

Using the Extend Edge Option

The **Extend Edge** option is similar to the Edge option presented with the **Trim** command. Edge will allow lines that do not presently touch but have an implied intersection to be extended. In the default setting, the command will automatically extend using an implied edge. If the Edge setting has been altered, the sequence to extend to an implied edge is as follows:

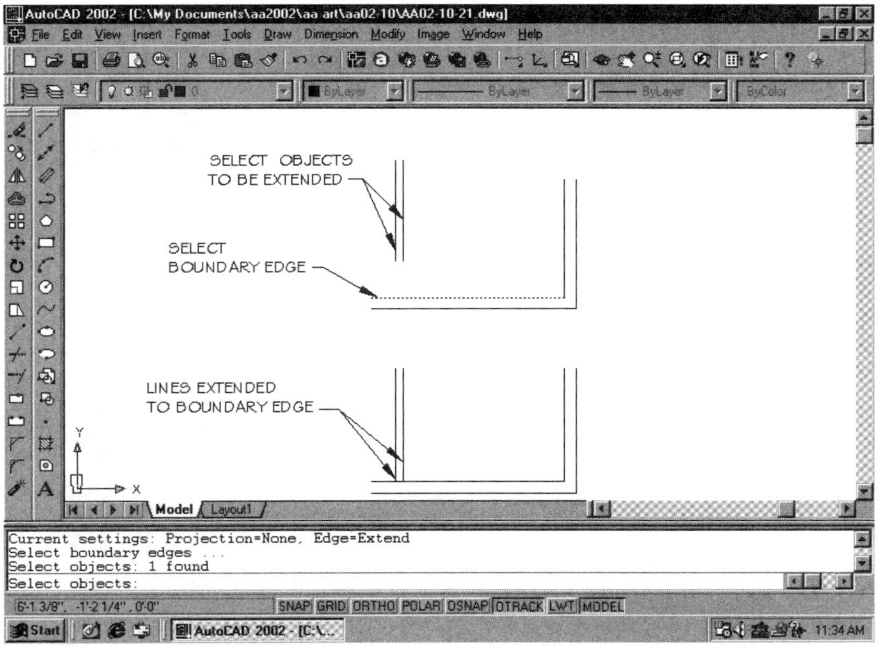

Figure 10.21 *The* **Extend** *command can be used to lengthen lines to a selected boundary edge.*

Command: **EX** (ENTER) *(Or click the Extend button.)*

EXTEND

Current settings: Projection=UCS Edge=Extend

Select boundary edges... *(Select the edge using any object selection method.)*

1 found

Select objects: (ENTER)

Select object to extend or shift-select to trim or [Project/Edge/Undo]: **E** (ENTER)

Enter an implied edge extension mode [Extend/No Extend] <Extend>: *(Select object to be extended using any selection method.)*

Select object to extend or shift-select to trim or [Project/Edge/Undo]: (ENTER)

Command:

The result of the command can be seen in Figure 10.22.

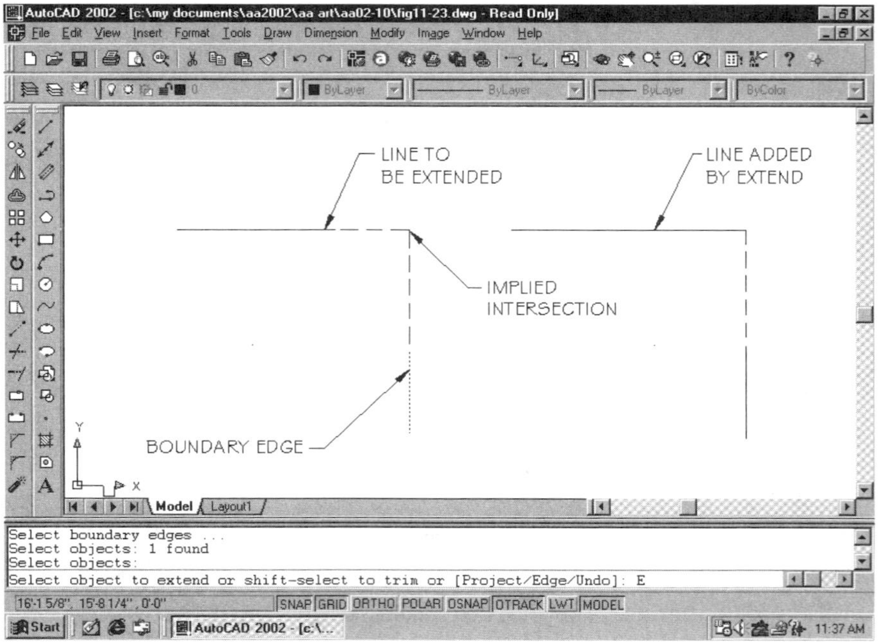

Figure 10.22 *Lines that would touch, if both were extended, can be extended by selecting the Edge option of* **Extend**.

Using the Extend Trim Option

The Trim option of **Extend** allows an object to be trimmed while within the **Extend** command. The command sequence to use the Trim option is as follows:

Command: **EX** (ENTER) *(Or click the Extend button.)*

EXTEND

Current settings: Projection=UCS, Edge=Extend

Select boundary edges... *(Select the edge using any object selection method.)*

1 found

Select objects: (ENTER)

Select object to extend or shift-select to trim or [Project/Edge/Undo]: *(SHIFT + Select object to be trimmed using any selection method.)*

Select object to extend or shift-select to trim or [Project/Edge/Undo]: (ENTER)

Command:

Using the Extend Undo Option

The **Extend Undo** option allows a sequence that has not produced satisfactory results to be undone while still in the command sequence.

Using the Extend Project Option.

This mode is used for 3D drawings and will not be discussed in this text. Use the **Help** menu for further information on this subject.

BREAK

The **Break** command can be used to remove a portion of a line, circle, arc, or polyline, (polylines will be introduced in later chapters). Start the **Break** command by selecting the **Break** button on the **Modify** toolbar, by typing **BR** (ENTER) at the Command prompt, or by selecting **Break** from the **Modify** menu. **Break** can be completed using two different methods, but their effects are similar.

Select Object, Second Break Point

The default option of **Break** allows you to remove a segment of an object by using two selection points. The first selection point selects the segment to be broken and specifies where the break will begin. The second point specifies where the break will end. The command is as follows:

Command: **BR** (ENTER) *(Or click the Break button.)*

BREAK Select object: *(Select an object to break.)*

Specify second break point or [First point]: *(Select a second point.)*

Command.

Once the second point is selected, the portion of the object from the first to the second point will be removed and the Command prompt will be returned. The sequence can be seen in Figure 10.23. If a circle is being edited, the break occurs in a counterclockwise direction from the first point to the second point, as seen in Figure 10.24. Type **F** (ENTER) when the second point is requested to allow the procedure to follow the Select Object, Two Break Points method.

Select Object, Two Break Points

This option allows you to remove a segment of an object by using three selection points. The first request selects the object to be broken. Once the segment has been defined for editing, the next selection point specifies where the break will begin and the final selection point specifies where the break will end. The command sequence is as follows:

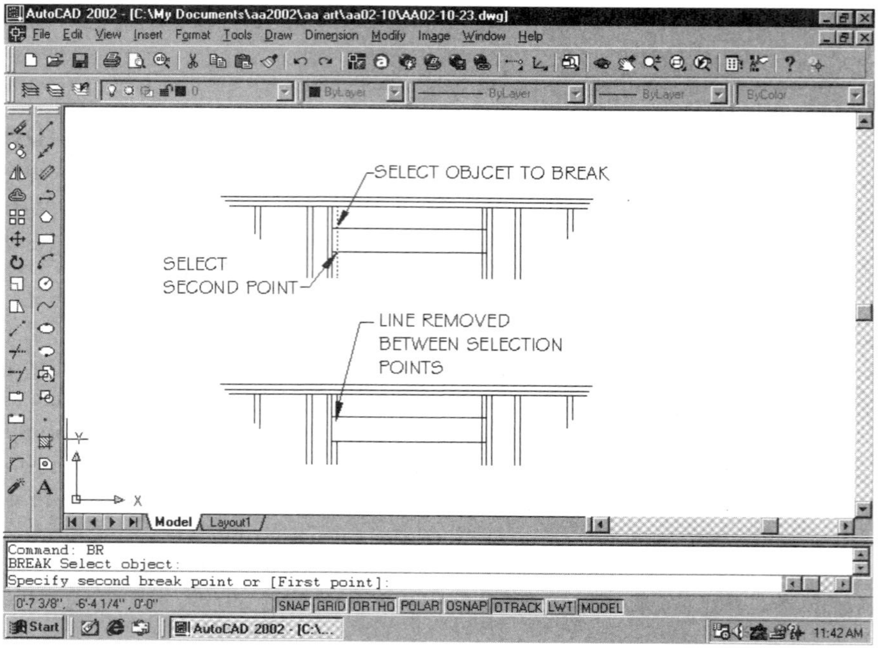

Figure 10.23 *The default option of* **Break** *allows for selecting the line to be broken and then specifying the segment to be removed.*

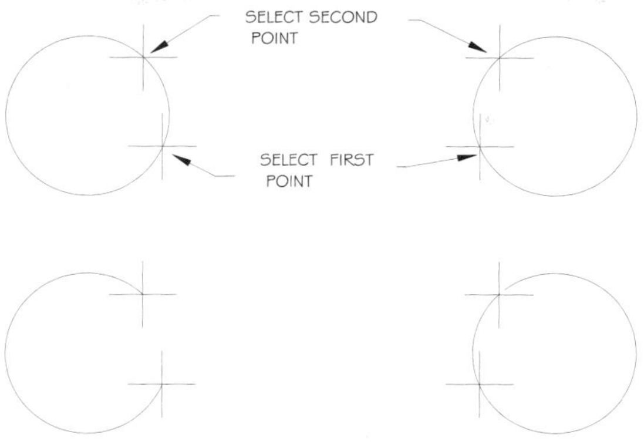

Figure 10.24 *When a break is specified for a circle, it will occur in a counterclockwise direction.*

Command: **BR** (ENTER) *(Or click the Break button.)*

BREAK Select object: *(Select an object to break.)*

Specify second break point or [First point]: **F** (ENTER)

Specify first break point: *(Select a point to start the break.)*

Specify second break point: *(Select a point to end the break.)*

Command:

The command sequence can be seen in Figure 10.25.

Break @

This option will break a segment, but no apparent change will take place. Break @ will divide an object into two portions, but no space will be left between the portions. The @ option is useful for erasing only a portion of an object by using **Break @** and then using the **Erase** command. The command sequence is as follows:

Command: **BR** (ENTER) *(Or click the Break at Point button on the Modify toolbar.)*

BREAK Select object: *(Select an object to break.)*

Specify second break point or [First point]: @ (ENTER)

Command.

The sequence is shown in Figure 10.26.

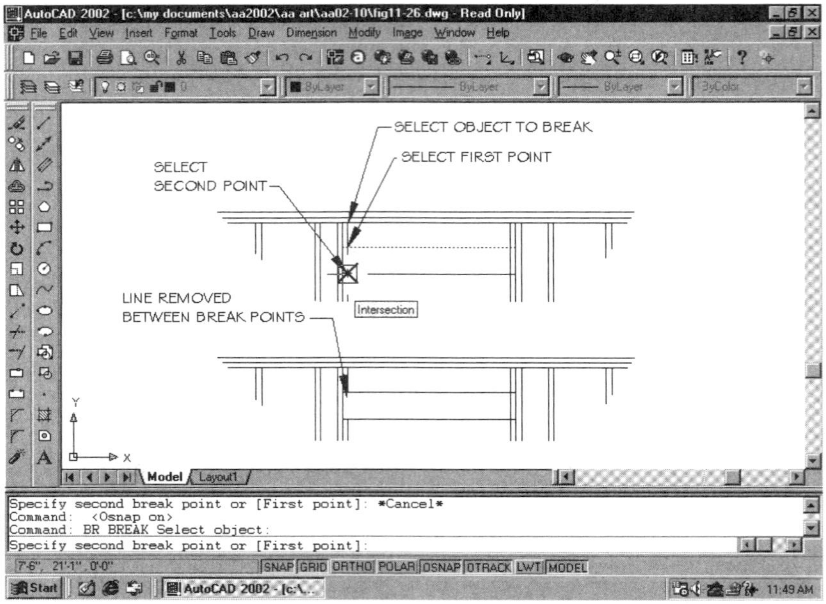

Figure 10.25 *The* **Break** *command's First option can be used to remove a segment from a line. Specified points are used to define which part of the line will be removed.*

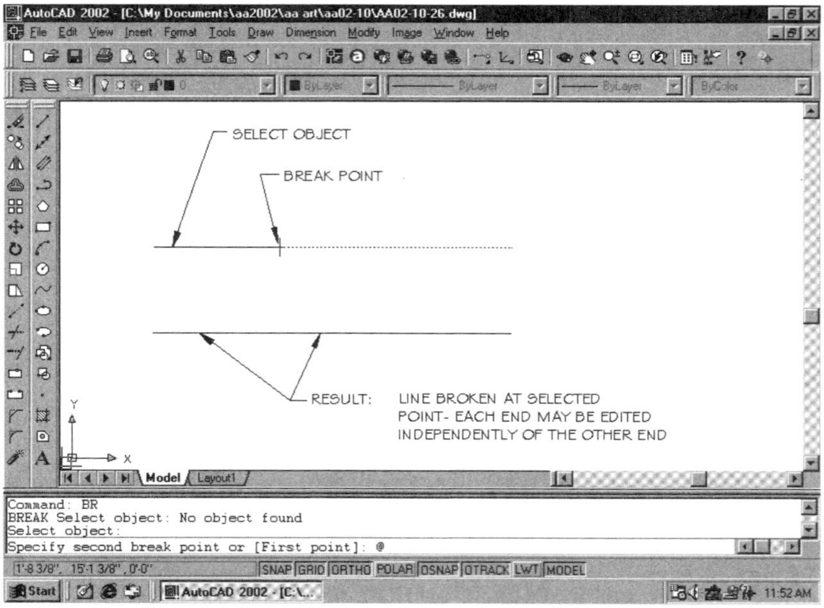

Figure 10.26 *The* **Break @** *command will break a line into two segments, but no visible change will occur.*

EDITING WITH THE CHAMFER COMMAND

A chamfer is an angled edge formed between two intersecting surfaces, typical of what might be found on the edge of the cabinet counter shown in Figure 10.27. The **Chamfer** command trims two intersecting lines and provides a mitered corner. You'll be asked to select two lines and two distances. This will form a new line segment between the two existing line segments. Start the **Chamfer** command by selecting the **Chamfer** button on the **Modify** toolbar, by typing **CHA** (ENTER) at the Command prompt, or by selecting **Chamfer** from the **Modify** menu. The command sequence to set the default values to 0", which can be used to trim corners, is as follows:

Command: **CHA** (ENTER) (Or click the Chamfer button.)

CHAMFER

(TRIM mode) Current chamfer Dist1 = 0'-0 1/2", Dist2 = 0'-0 1/2"

Select first line or [Polyline/Distance/Angle/Trim/Method]: **D** (ENTER)

Specify first chamfer distance <0'-0 1/2">: **0** (ENTER)

Specify second chamfer distance <0'-0">: (ENTER)

Select first line or [Polyline/Distance/Angle/Trim/Method]:

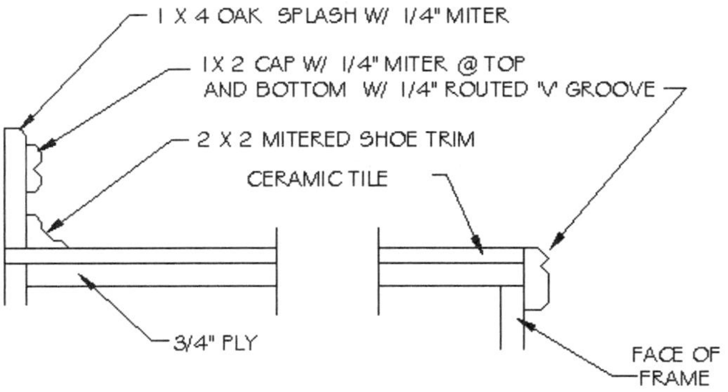

Figure 10.27 *The* **Chamfer** *command can be used to miter a corner.*

The first value will become the default for the second value. When the distances are equal, the results will produce a 45° miter. When the second value or (ENTER) is entered, the command sequence will continue. If the command is continued, objects will be chamfered to form square corners. Because no chamfer distance has been entered, the process and the results would resemble Figure 10.28.

To provide a mitered corner, the sequence would be similar. For the following example, a distance of 36" will be used—although any numeric value can be used, as long as the value is no longer than that of the line being chamfered. If the value exceeds the length, the following message will be displayed and the command terminated:

Distance is too large

Invalid

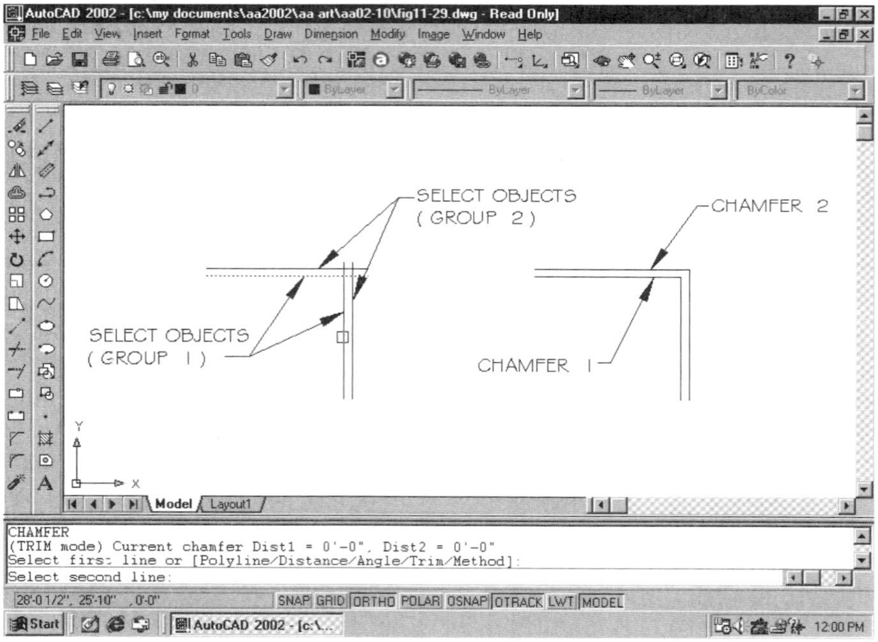

Figure 10.28 *With the value set at 0,* **Chamfer** *can be used to trim corners.*

The command sequence is as follows:

> Command: **CHA** (ENTER) *(Or click the Chamfer button.)*
>
> CHAMFER
>
> (TRIM mode) Current chamfer Dist1=0'-0", Dist2=0'-0"
>
> Select first line or [Polyline/Distance/Angle/Trim/Method]: **D** (ENTER)
>
> Specify first chamfer distance <0'-0 1/2">: **36"** (ENTER)
>
> Specify second chamfer distance <3'-0">: (ENTER)
>
> Select first line or [Polyline/Distance/Angle/Trim/Method]: *(Select first line to be chamfered.)*
>
> Select second line: *(Select second line to be chamfered.)*

As the second line is selected, the chamfer will be performed and the Command prompt returned. The effects of the command sequence can be seen in Figure 10.29.

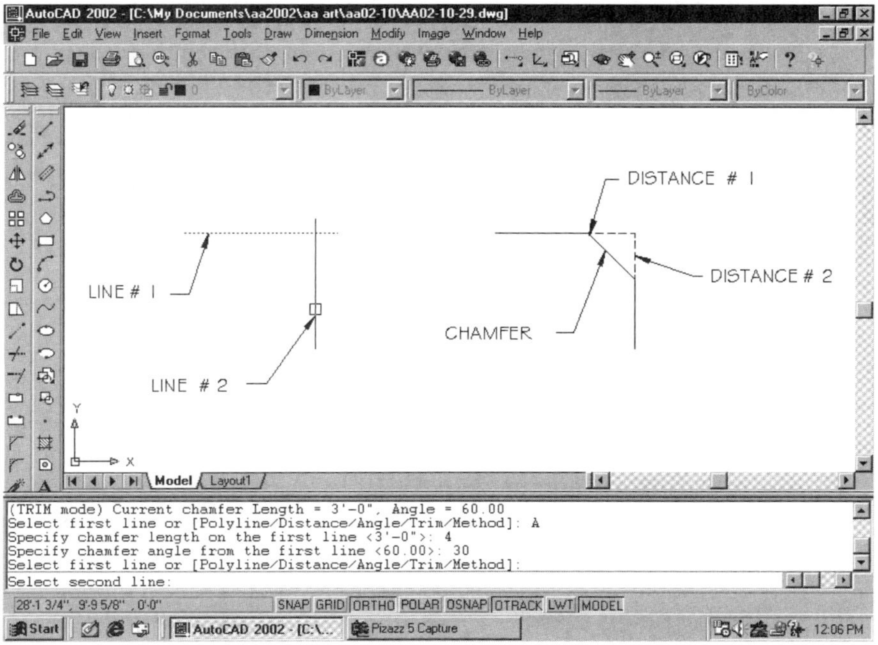

Figure 10.29 *In mitering a corner, prompts will be displayed for the "first distance," and the "second distance." Once the distances are provided, a new segment will be created between the two points.*

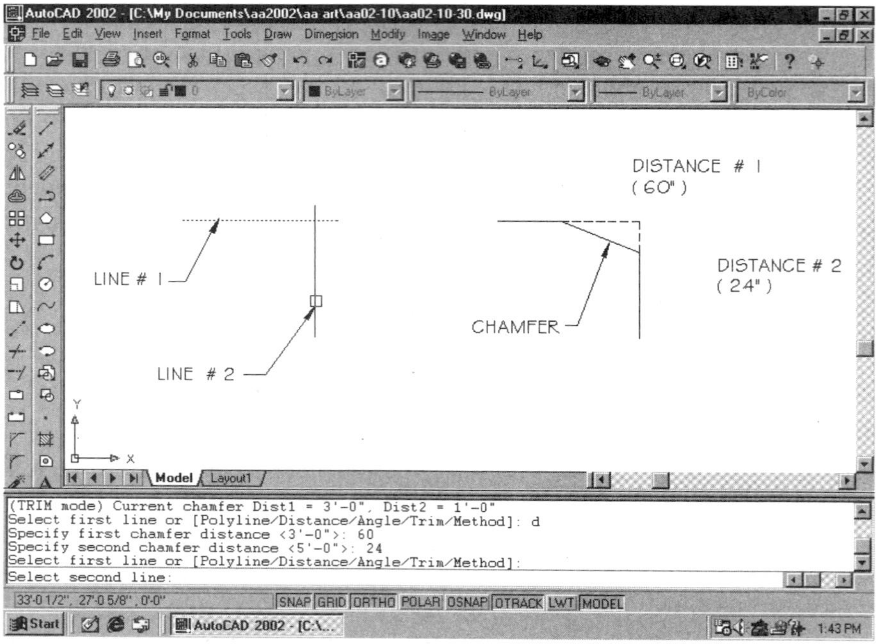

Figure 10.30 *Providing different values for the chamfer distances will affect the angle of the chamfer.*

Figure 10.30 shows an example of two different distance values. **Chamfer** can be used to square or chamfer two intersecting lines, as well as to square or chamfer two lines that do not touch.

Selecting the Chamfer Angle

Chamfer also allows the chamfer to be created using one distance and an angle. Start this option by typing **A** (ENTER) when prompted for the first line. The command sequence to place a 30° angle 4" from a corner is as follows:

Command: **CHA** (ENTER) *(Or click the Chamfer button.)*

CHAMFER

(TRIM mode) Current chamfer Dist1=3'-0", Dist2=3'-0"

Select first line or [Polyline/Distance/Angle/Trim/Method]: **A** (ENTER)

Specify chamfer length on the first line <0'-1">: **4"** (ENTER)

Enter chamfer angle from the first line <0>: **30** (ENTER)

Select first line or [Polyline/Distance/Angle/Trim/Method]: *(Select first line.)*

Select second line: *(Select second line.)*

Command:

The process can be seen in Figure 10.31.

Selecting the Chamfer Trim Mode

The common method of using the **Chamfer** command is to create the chamfer between two lines and remove a portion of the lines where the chamfer is created. The No Trim option allows for the choice of not removing the lines by the chamfer (by default, the lines are trimmed). The command sequence is as follows:

Command: **CHA** (ENTER) (Or click the Chamfer button.)

CHAMFER

(TRIM mode) Current chamfer Dist1=3'-0", Dist2=3'-0"

Select first line or [Polyline/Distance/Angle/Trim/Method]: **T** (ENTER)

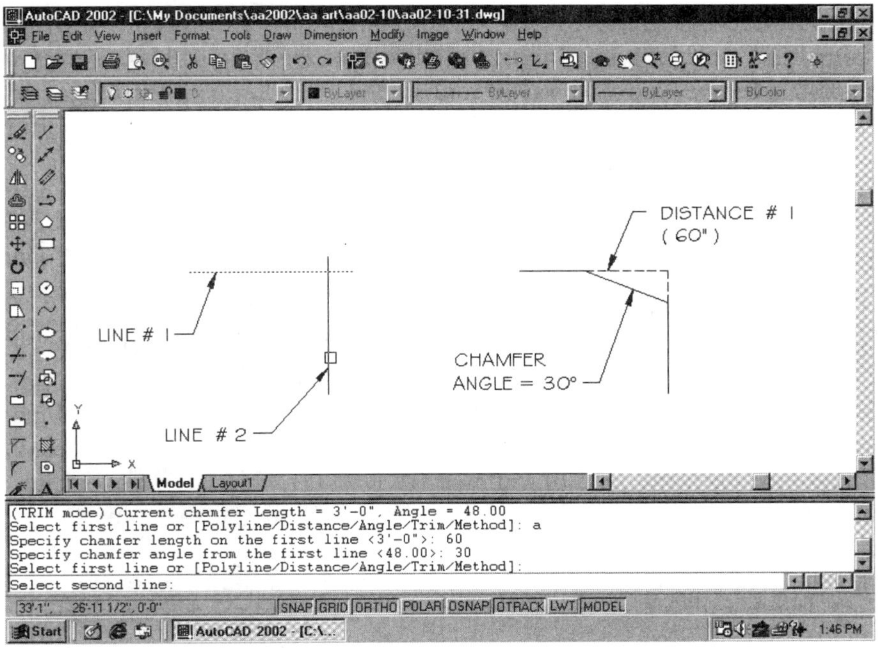

Figure 10.31 *Selecting the Angle option provides a prompt for one length and an angle. AutoCAD converts the angle to a length.*

Enter Trim mode option [Trim/No trim] <Trim>: **N** (ENTER)

Select first line or [Polyline/Distance/Angle/Trim/Method]: *(Select first line.)*

Specify second line: *(Select second line.)*

Command

Reminder: If you alter the trim mode of **Chamfer**, the chamfer will be performed, but no lines will be removed. The **Chamfer Trim** mode will also affect the **Fillet** command. (**Fillet** will be discussed in the next section of this chapter.) Most users find the Trim option (the default) more useful than the No Trim option.

Selecting the Chamfer Method

When a chamfer is placed using the Distance or Angle option, the values are kept as the defaults until you change them. Using the Method option, the values selected for one option will not affect the other option, and you can switch between using the Distance or Angle method. The command sequence is as follows:

Command: **CHA** (ENTER) *(Or click the Chamfer button.)*

CHAMFER

(TRIM mode) Current chamfer Dist1=3'-0", Dist2=3'-0"

Select first line or [Polyline/Distance/Angle/Trim/Method]: **M** (ENTER)

Enter Trim method [Distance/Angle] <Trim>: **D** (ENTER)

Select first line or [Polyline/Distance/Angle/Trim/Method]: *(Select first line.)*

Specify second line: *(Select second line.)*

Command: (ENTER)

EDITING WITH THE FILLET COMMAND

You were introduced to the **Fillet** command as a tool for basic drawing construction. The command is used connect two lines to form a corner, or allow two lines, arcs, or circles to be fitted together smoothly in rounded corners. Start the command by selecting the **Fillet** button on the **Modify** toolbar, by typing **F** (ENTER) at the Command prompt, or by selecting **Fillet** from the **Modify** menu. Once the command is started, set the desired radius using a process similar to setting the chamfer distance. Once the two objects are selected, the radius will be applied and the Command prompt will be returned. The **Fillet** process can be seen in Figure 10.32. If the Trim option is set to No Trim, the results of the **Fillet** will not be evident. Be sure the Trim option is set to Trim. The command can be used to form a radius between intersecting lines or two lines that do not touch.

For the command to function with crossing lines as you intend, you need to be careful which portion of the lines you select. Draw two more intersecting lines similar to those in Figure 10.32. Start the **Fillet** command and select the short ends of the lines. Once the **Fillet** command is completed, only the small portion of the lines will remain, as shown in Figure 10.33.

Altering the Fillet Radius

The **Fillet** radius can be altered using the Radius option. Once the first object is selected, use the Radius option to alter the radius of the fillet. Typing **R** (ENTER) at the first prompt allows a radius to be entered for rounded corners. To transform two existing perpendicular lines into a 60" radius corner, the command sequence is as follows:

Command: **F** (ENTER) *(Or click the Fillet button.)*

FILLET

Current settings: Mode=TRIM, Radius=0'-01/2"

Select first object or [Polyline/Radius/Trim/Method]: **R** (ENTER)

Specify fillet radius <0'-01/2">: **60"** (ENTER) *(or 5' (ENTER))*

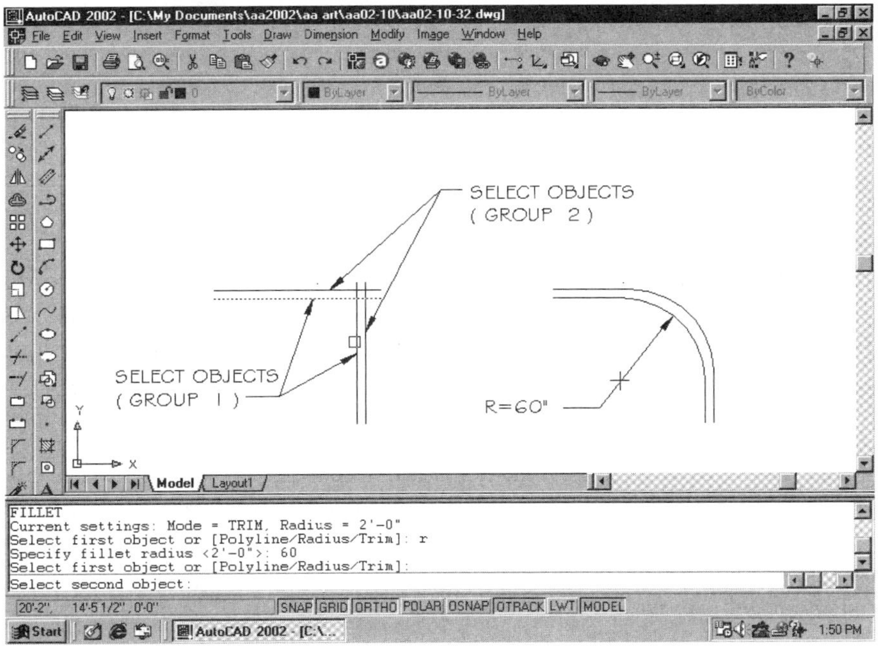

Figure 10.32 *By accepting the default or assigning a new value for the Radius, you can use the* **Fillet** *command to round square corners.*

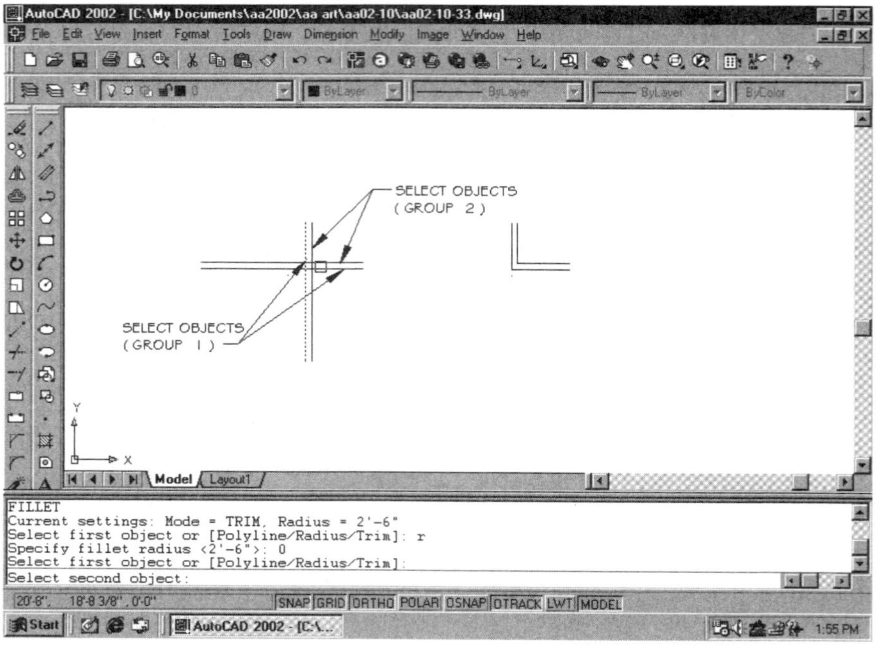

Figure 10.33 *Care must be taken when selecting objects to be edited by the* **Fillet** *command. Selecting the short ends of the lines removes the other portions of the lines.*

Select first object or [Polyline/Radius/Trim]: *(Select first line.)*

Select second object: *(Select second line.)*

Command:

As the second object is selected, the radius will be formed between the two selected lines, similar to Figure 10.32. The portion of the object that is selected will affect how the **Fillet** will be drawn. Examples of selections and their results can be seen in Figure 10.34. Don't speed past the Command prompt. If you've entered a radius value that exceeds the length of the lines, no change will be made to the drawing, and the following prompt will be displayed:

Radius is too large

Select second object:

Command:

A smaller radius value will need to be selected to make the command work.

One of the most common uses of **Fillet** is to form a corner with no radius. Setting the radius to 0 and then proceeding with the command will allow the command to be used to form corners. (See Figure 10.35.)

Practical Uses for Combining Offset, Fillet and Trim

Open the FLOOR10 drawing that you started in Chapter 9. Because you've seen the finished floor plan in Figure 9.23b, it's not too hard to visualize the lines in Figure 9.26 representing the rooms of the finished plan. Using a combination of **Offset** and **Fillet** will make it much easier to visualize. Figure 10.36 shows two examples of how **Fillet** can be used to clean up the floor plan. Notice that if **Fillet** is used at Fillet 3, the lower portion of the line will be removed and needs to be redrawn. This would be a good opportunity to use **Trim** instead of **Fillet**.

SELECTING OBJECTS WITH GRIPS

In Chapter 9 you were introduced to several methods of selecting objects for editing. Grips provide an additional method of selecting objects for editing, as well as a method of combining several of the editing procedures introduced in this chapter.

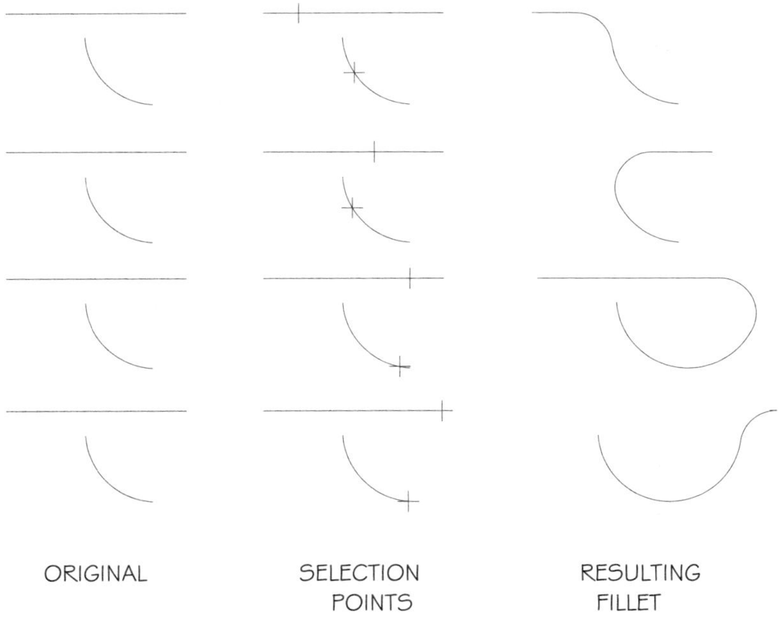

Figure 10.34 *Selecting different portions of the line segment will affect the results of the* **Fillet**.

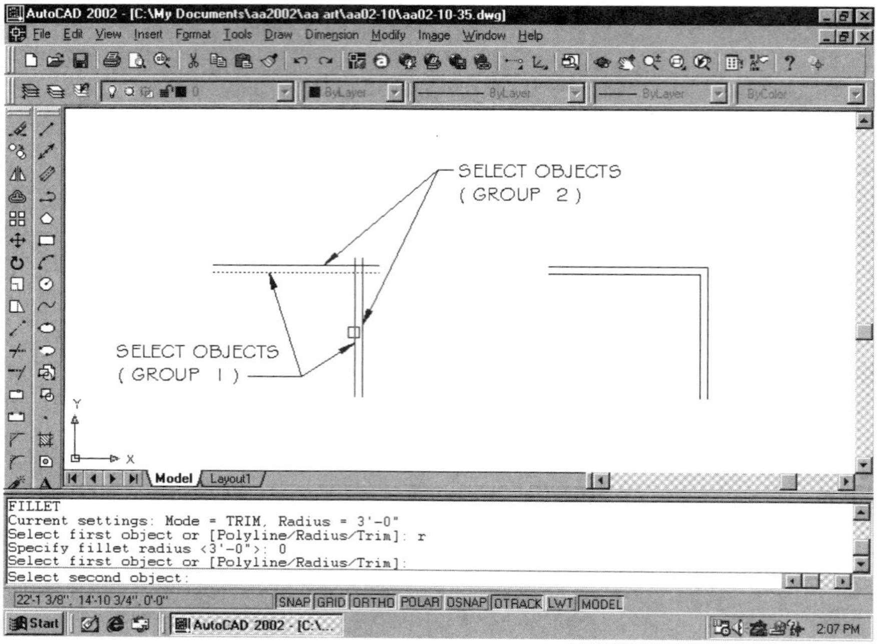

Figure 10.35 *With the value of the Radius set to 0, the* **Fillet** *command can be used to square corners of crossing lines.*

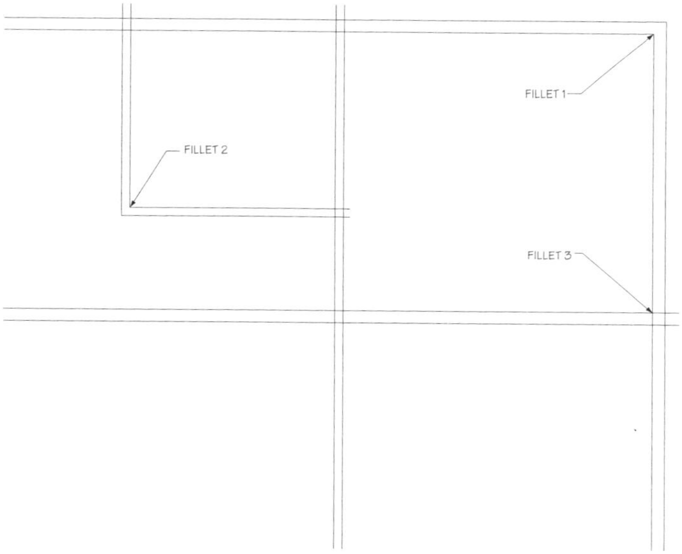

Figure 10.36 *The* **Fillet** *command was used to clean up the corners at Fillets 1 and 2. If* **Fillet** *had been used at Fillet 3, the lines that will be used to form the garage would have been removed.*

Grips are similar to moving a heavy object by its handles. You can move the object without using the handles, but they make life so much easier. Grips are handles placed on objects for easy manipulation, similar to the OSNAP points, with **Stretch**, **Move**, **Rotate**, **Scale**, and **Mirror**. The default setting for grips is ON. Grips can be toggled ON/OFF on the **Selection** tab of the **Options** dialog box. Select **Options** from the **Tools** menu to display the dialog box (see Figure 10.37). With the **Enable grips** check box selected, grips will be displayed when an edit or inquiry command is given; grip boxes will be displayed at various object-specific positions, as seen in Figure 10.38. When the cursor is moved over a grip, the cursor will snap to the grip for editing. This is sometimes faster that using an OSNAP location.

USING GRIPS

Enter AutoCAD and draw several lines, arcs, and circles. Move the cursor so that it is over one of the lines and click the select button. Three boxes or grips are now displayed. Move the cursor to the center grip, and click the select button. The grip has changed from a hollow box to a solid square. This grip is now said to be hot. Notice that the Command prompt has also changed to display the following prompt:

```
** STRETCH **
Specify stretch point or [Base point/Copy/Undo/eXit]:
```

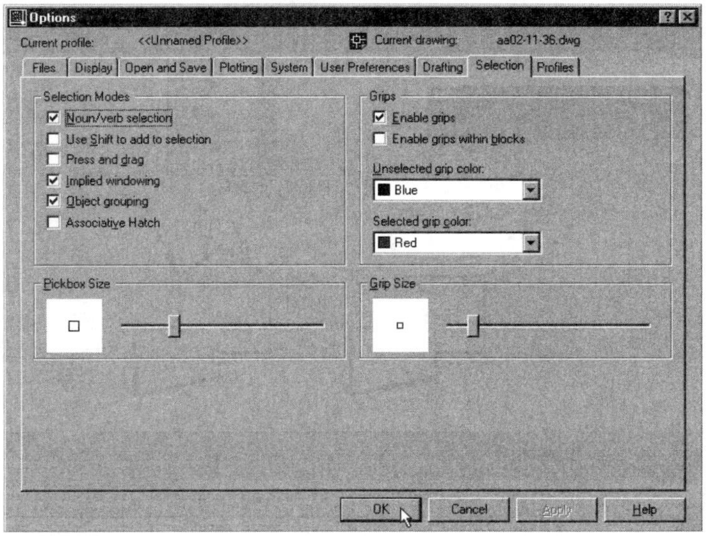

Figure 10.37 *Grips can be toggled ON/OFF on the* **Selection** *tab of the Options dialog box. Access the dialog box by selecting* **Options** *from the* **Tools** *menu or by typing* **DDGRIPS** *(***ENTER***) at the Command prompt.*

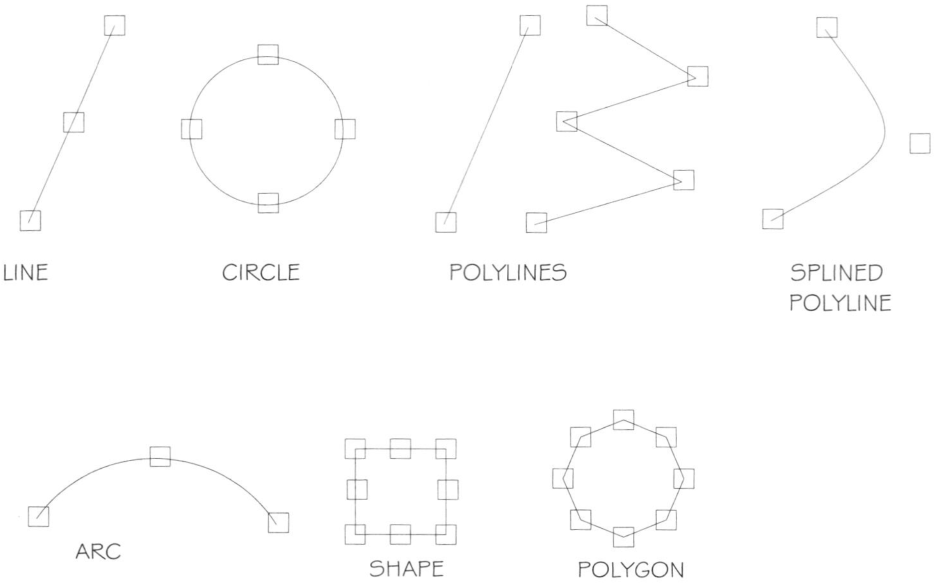

Figure 10.38 *With the default setting* **Enable grips** *check box toggled to ON, grips will be displayed for each geometric shape*

We'll come back to the prompt later. With the center grip hot, and the prompt reflecting **Stretch**, move the cursor slightly. The hot grip in the center of the line functions like the Move command. As the cursor moves, so does the line. Clicking the select button will move the line to this new location. Press ESC to cancel the command. Select the same line with the cursor and make one of the end grips hot. Now the grip at the other end remains a stationary point, and as the hot grip is moved, the selected line is stretched. The effect of **Stretch** on a line can be seen in Figure 10.39. Grips function in a similar manner with arc and circles. Figures 10.40a and 10.40b show the effect of stretching each. If you decide you don't want to edit a selected line, press ESC to deactivate the grips.

GRIP OPTIONS

Earlier you saw that the command displayed in grips has several options. These grip options can be used once the grips for an object have been activated. These options include Base point, Copy, Undo, and eXit.

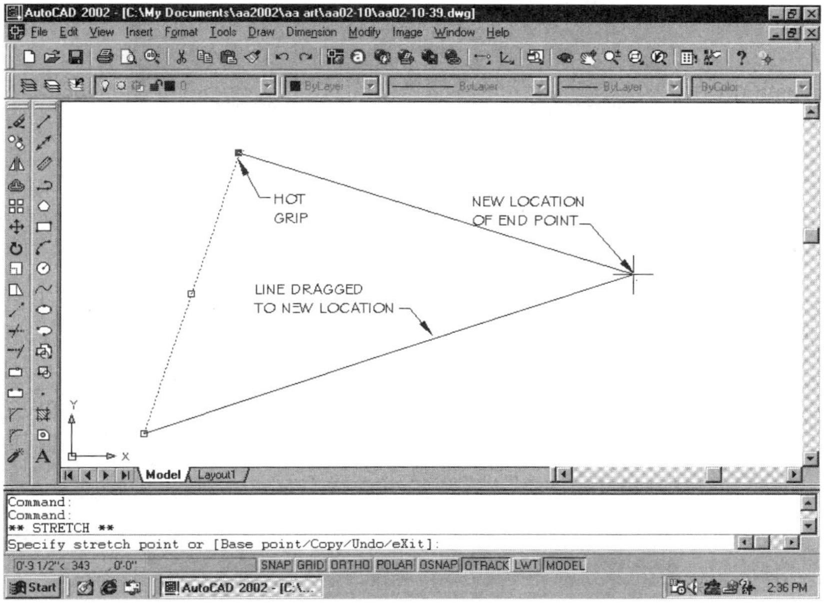

Figure 10.39 *Select a drawing object to display its grips. Moving to one of the grips and clicking the select button will make that grip hot. The object can now be edited by* **Stretch**, **Move**, **Rotate**, **Scale**, *and* **Mirror**.

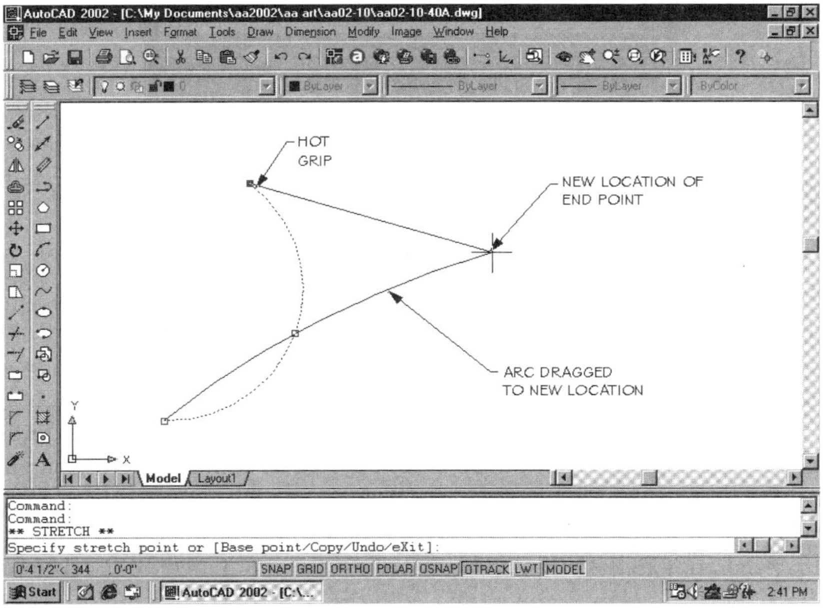

Figure 10.40a *The* **Stretch** *command applied to an arc using the hot grip.*

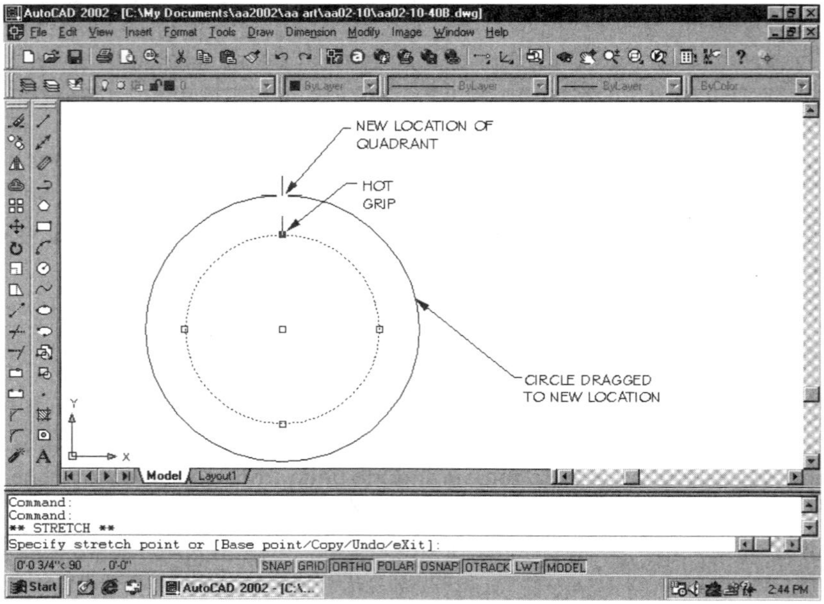

Figure 10.40b *The size of a circle can be altered using* **Stretch** *by selecting a hot grip and a new location.*

Base Point

Typing **B** (ENTER) at the prompt allows the hot box to be disregarded and a new base point to be selected.

Copy

Typing **C** (ENTER) at the prompt allows one or more copies of the selected object to be made.

Undo

Typing **U** (ENTER) at the prompt will undo the previous selection.

eXit

Typing **X** (ENTER) at the prompt terminates the grip edit and returns you to the Command prompt.

CHANGING GRIP OPTIONS

Once a grip box has been selected and made hot, right-clicking will produce the menu shown in Figure 10.41. Any one of the options can be selected to alter the function to be preformed by the grip.

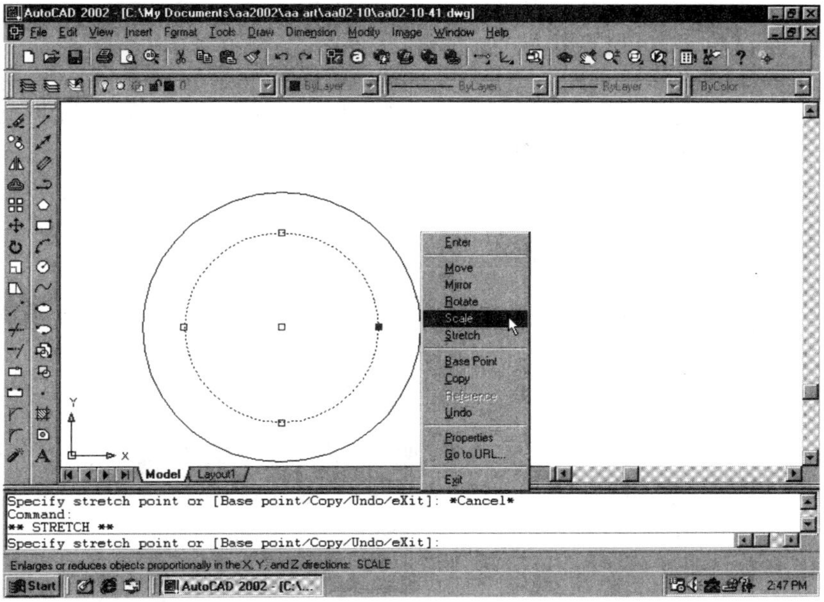

Figure 10.41 *Right-clicking produces this menu that allows the grip edit option to be altered.*

SELECTING OBJECTS FOR MULTIPLE EDITS

Often an object or group of objects might need to be edited with several different editing commands. If you were arranging units in a multi-building apartment complex, a unit may need to be mirrored, moved, and rotated to form a new unit. Up to this point, you would have to select the objects to be mirrored, use the **Mirror** command, select the objects to be moved, use the **Move** command, select the objects to be rotated, and use the **Rotate** command. It works, but you'll never impress the boss as worthy of the big bucks. The S**elect** command will allow an object or group of objects to be pre-selected and used for one or more editing functions. The command sequence to select is as follows:

Command: **SELECT** (ENTER)

Select objects: **W** (ENTER) *(Select objects using any of the selection methods.)*

First corner: Other corner: 100 found

Select objects: (ENTER)

Command:

The Command prompt is returned but no apparent changes have occurred. The objects that were selected have been grouped as a set and will be affected by any editing command. Start the **Mirror** command and when prompted to select objects, type **P** (ENTER) to automatically select the objects that were just placed in the set. The sequence to mirror, move, and rotate the objects is as follows:

Command: **MI** (ENTER) *(Or click the Mirror button.)*

MIRROR

Select objects: **P** (ENTER)

100 objects found

Select objects: (ENTER)

Specify first point of mirror line: *(Select first point.)*

Specify second point of mirror line: *(Select second point.)*

Delete source objects? [Yes/No]<N>: **Y** (ENTER)

Command: **M** (ENTER) *(Or click the Move button.)*

MOVE

Select objects: **P** (ENTER)

100 objects found

Select objects: (ENTER)

Specify base point or displacement or <use first point as displacement>: *(Select base point.)*

Specify second point of displacement or <use first point as displacement>: *(Select new location for base point.)*

Command: **RO** (ENTER) *(Or click the Rotate button.)*

ROTATE

Current positive angle in UCS: ANGDIR=counterclockwise ANGBASE=0

Select objects: **P** (ENTER)

100 objects found

Select objects: (ENTER)

Specify base point: *(Select base point.)*

Specify rotation angle or [Reference]: **45** (ENTER)

Command:

The **Group** command allows similar functions, but with more power.

Note: If you forget to use **Select** before the editing command, don't worry. The same results can be obtained by entering the commands in the following sequence: type **MI** (ENTER), select your objects, type **RO** (ENTER), and type **P** (for previous).

NAMING OBJECT GROUPS

While **Select** allows a group to be created, the **Group** command allows multiple groups to be created, named, and saved. For instance, the apartment unit that was just edited could be saved as a group called Base. It could be mirrored, and both units saved as a new group called Units (a group that is part of another group is referred to as nested). The Units group could also be mirrored and saved as a third group called BLDG1. Start the **Group** command by typing **G** (ENTER) at the Command prompt. This will produce the **Object** Grouping dialog box shown in Figure 10.42.

CREATING AND NAMING A GROUP

In Figure 10.42 you can see that this drawing has no named groups. To create a group, enter the desired name at the keyboard. As you type, the name will be entered in the **Group Name** box in the **Group Identification** area. Group names can be up to 225 characters long and include numbers, letters, or characters not used by Microsoft Windows or AutoCAD. Two separate words can't be used, but compound words can be joined by a dash or underline to create a name such as base_unit. As with the naming of drawing files, be sure to use names that will be easily recognized and that will clearly describe the contents of the group.

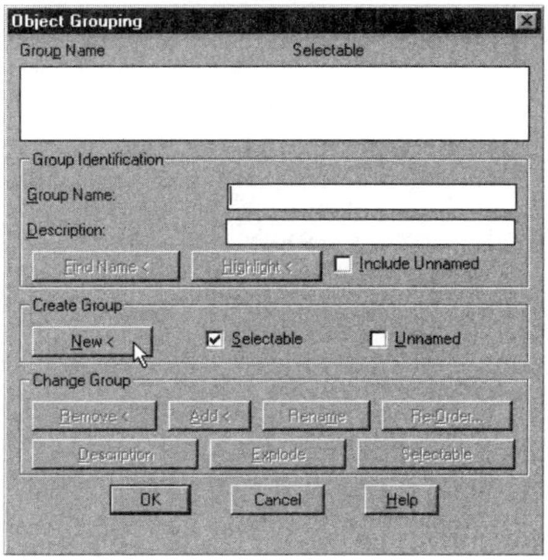

Figure 10.42 *Typing* **G** *(*ENTER*) at the command line will display the* **Object Grouping** *dialog box.* **Group** *can be used to select and save drawing objects in groups for editing.*

Description

A description can be created to help you remember the contents of a specific group. To create a description, click in the **Description** box and type a description.

New

Once the desired name and description have been entered, select the **New** button. This will remove the dialog box from the screen and prompt you to select objects. Select objects to be included in the group using any of the selection methods. When the selection process is complete, press (ENTER) to redisplay the dialog box. The group name will be listed in the **Group Name** box. As the group is created, the word *yes* now displayed beside the group name indicates that it is selectable. If a group is selectable, selecting one object in the group selects the entire group. Once the group is created, the other areas of the dialog box can be used to alter or use the group.

IDENTIFYING A GROUP

Key elements of the **Group Identification** area can be seen in Figure 10.43. It is used to identify named groups.

Find Name

Selecting this button removes the dialog box from the screen and provides the prompt:

Pick a member of a group

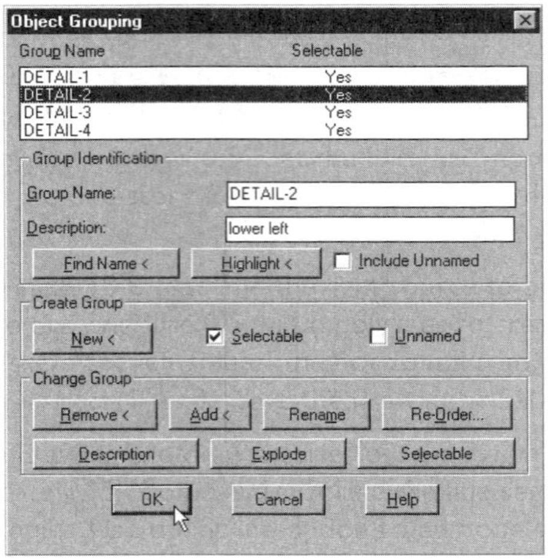

Figure 10.43 *Once groups are identified, selecting one of the groups will activate each portion of the dialog box.*

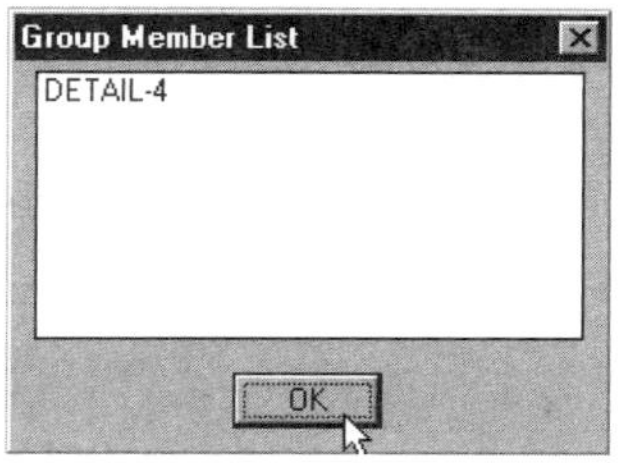

Figure 10.44 *Selecting* **Find Name** *will display the* **Group Member List** *box. Select an object on the drawing, and the group that it belongs to will be displayed.*

When an object is selected, a **Group Member List** dialog box similar to the one shown in Figure 10.44 is displayed. This box shows all groups where the selected object is a member. Clicking the **OK** button returns the **Object Grouping** dialog box. (If you select an unnamed group, the **Add<** button in the **Change Group** area allows a name to be provided for an unnamed group.)

Highlight

The **Highlight** button is not active as the dialog box is displayed. Clicking one of the named groups activates the button. Clicking the **Highlight** button removes the dialog box from the screen and shows the named group in dashed lines as well as the **Continue** button. This option is especially helpful on complicated drawings with multiple groups. Clicking the **Continue** button restores the **Object Grouping** dialog box.

Include Unnamed

Activating the **Include Unnamed** check box will display unnamed groups in the **Group Name** box. Unnamed groups can be renamed using the **Rename** button in the **Change Group** area.

CREATING A NEW GROUP

The key element of the **Create Group** area can be seen in Figure 10.43. The **New** box was covered previously in the section, "Creating and Naming a Group."

Selectable

Activating the **Selectable** check box determines the status of named groups as selectable. Remember that selecting one object in a named group selects the entire group.

Unnamed

If the **Unnamed** check box is active, AutoCAD will provide names for unnamed groups.

CHANGING A GROUP

Key elements of the **Change Group** area can be seen in Figure 10.43. All elements except **Re-Order** are inactive as the dialog box is displayed. Highlighting one of the named groups will activate all of the boxes in the **Change Group** area.

Add

The **Add** button allows drawing objects to be added to an existing group. Once this button is clicked, you will be prompted to select objects. If the selected object is part of an existing group, the group name and description boxes will display the current status of the object. Clicking the **OK** button completes the selection process.

Description

This **Description** button allows the description to be updated and saved. To alter a description, click in the **Description** box and type the desired material to be added. When you have finished, click the **Description** button to update the description and save it for future reference.

Explode

The **Explode** button is used to remove the selected group from the effects of Group select action. The objects that were in the group will not be altered, only the ability to select those objects as a group. Highlight the group to be removed in the **Group Name** list and then click the **Explode** button to explode the group.

Remove

Selection of the **Remove** button allows individual objects to be removed from the grouping. Highlight the group to be edited in the **Group Name** list and click **Remove** to display the Remove objects prompt. The dialog box will be removed and the objects in the group will be highlighted. Objects can be selected for removal using any of the selection methods. When finished removing objects from the set, press (ENTER) to return to the **Object Grouping** dialog box.

Rename

The **Rename** button can be used to rename an existing group. Highlight the name of an existing group so that the name and description are displayed in the **Group Identification** area. Edit the name as desired and click **Rename** to make the new name the current name.

Re-Order

Click the **Re-Order** button to produce the **Order Group** dialog box shown in Figure 10.45. Notice that the lower portion of the box includes three listings with numbered entries. Objects in a group are numbered in the same order that they were selected as the group was created. Re-ordering is typically used in CAD/CAM applications of manufacturing and is not often a part of the construction drawings. Options include the following:

Group Name
Displays the name of the group to be re-ordered.

Description
This selection displays the description of the selected group.

Remove from position (0–x):
For each listing, x will vary depending on the size of the objects in the group. This selection specifies the current numbered position of the object to be reordered.

Replace at position (0–x):
This selection specifies the new position the object will be moved to.

Number of objects (1–x):
This selection lists the range of numbers to be re-ordered.

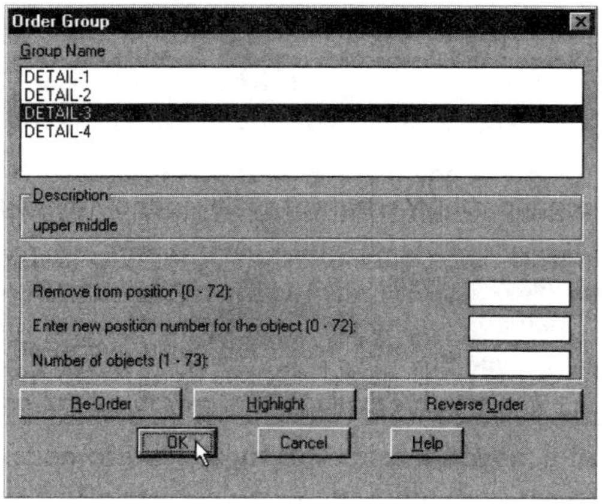

Figure 10.45 *The* **Order Group** *dialog box is displayed when the* **Re-Order** *button is selected. Re-ordering allows the order in which objects are selected for certain functions to be altered. It is not generally used with construction drawings.*

CHAPTER 10 EXERCISES

1. Open a template and draw a 38' long x 18 1/2" deep floor truss. Top and bottom chords will be made out of two 3x4s with 1" diameter aluminum webs at approximately 45°, equally spaced throughout the truss. Save the drawing as E-10-1.

2. Draw four sets of perpendicular crossing lines. On one set of lines, use the **Fillet** command to square the corners. On the second set of lines, provide a 12" fillet. On the third set of lines use a 12" chamfer. On the fourth set of lines use a 12" and 24" chamfer. Save the drawing as E-10-2.

3. Draw an 8" wide concrete block retaining wall extending 8' above a 4" thick concrete floor slab. Show the wall extending 8" above the finish grade. Thicken the slab to 8" thick at the edge resting on the footing. Use a 16" wide x 12" deep concrete footing, and show a 4" diameter drain on the soil side of the wall. Show a 2x6 top plate supporting 2x8 floor joists flush with the soil side of the wall. Reinforce the wall with #5 diameter @ 12" o.c. each way, 2" clear of the interior face. Use varied line weights for concrete, grade, sectioned wood, and steel. Adjust the widths to achieve good contrast. Save the drawing as E-10-3.

4. Open drawing E-9-4. Copy and rotate the pair of doors to provide doors for walls at 45°, 90°, 135°, 180°, and 270° rotations. Save the drawing as E-10-4.

5. Open drawing E-9-5 and make a copy of the column cap. Adjust the base strap so that it is 3" longer and 1" wider. Add one additional bolt. Each end of the side plate must be 2" longer (4" total) and 1 1/2" higher. Save the drawing as E-10-5.

6. Open drawing E-9-7. Enlarge the foundation plan 8' in each direction. Add required beams and piers to meet the original criteria. Place 4' x 4' diagonal corners in the exterior walls. Save the drawing as E-10-6.

7. Open drawing E-7-1 and make three additional copies. In the original drawing, provide a 6" line from each intersection and perpendicular to the opposite side of the triangle. Rotate the new triangles and place a point of each new triangle tangent to the existing triangle and perpendicular to the 6" lines. Draw a 10" diameter circle with a center point in the center of the original triangle. Trim all elements that extend beyond the circle. When complete, your drawing should resemble the attached drawing. Save the drawing as E-10-7.

8. Start a new drawing and draw a metal chimney cap that extends a total of 12" above a 4' wide wood chimney chase. Trim the chase with 3" wide trim and show 6" wide horizontal siding. The top of the chimney cap is 14" wide x 1". The main portion of the cap is 12" x 7". The chimney is 10" wide x 4" high. Use **Mirror**, **Offset**, **Trim**, and **Fillet** to complete the drawing. When complete, your drawing should resemble the drawing below. Save as E-10-8.

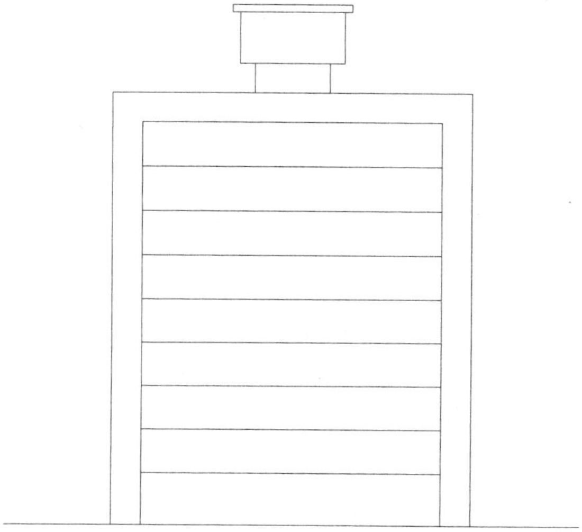

9. Start a new drawing and draw a section of a standard one-story foundation similar to the sketch below. Copy and edit the drawing to the proper size for a two-story footing. Save the drawing as Figure E-10-9.

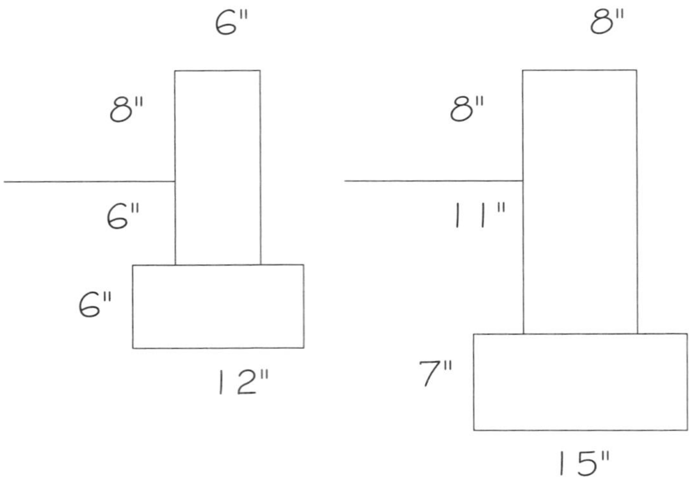

10. Start a new drawing and draw one half of a roof truss with a 5/12 pitch with a 26' span. All chords will be made of 6" wide material; all webs will be made from 4" material. Use a 24" overhang. The webs should be spaced equally at the center of the bottom chord. As a minimum, use **Offset**, OSNAP, and **Trim**. Use the **Mirror** command to complete the drawing. Save the drawing as E-10-10.

11. Use the floor plan shown in Figure 10.23b and draw the walls and cabinets for this residence. No doors, windows, plumbing fixtures, or appliances need to be drawn at this time. Assume the following:

   - All exterior heated walls and toilet walls are 6" thick.
   - All exterior unheated walls, interior walls are 4" thick.
   - The fireplace is 3'–4" wide. Determine the length by allowing for a 36" wide hallway.
   - The stairway, the hall between the stair and bathroom and the space between the stair and the front wall to be 42" wide.
   - The bathroom is 36" wide.

   Although door and windows symbols should not be drawn at this time, provide the opening in the wall for each door. Provide the outline of all cabinets. Assume 24" wide base and 12" wide upper cabinets. Do not draw appliances. Save this drawing as EFLOOR.

12. Open drawing EFLOOR. Assume north to be at the top of the drawing so that the front door is presently facing south. Make all changes so that the current wall between the family and living rooms remains straight for the entire length of the house. Alter the width of the wall as needed as it changes from interior/exterior/toilet wall.
    - Add 2'–0" to the family room in the north/south (n/s) direction.
    - Add 3'–6" to the living room in the east/west (e/w) direction. Add 18" in the n/s direction.
    - Enlarge the bathroom to 42" wide (+6" e/w).
    - Remove 24" e/w from the shop.
    - Remove the garage doors on the south face of the garage and place them on the east wall.
    - Remove the window from the east wall and place (2) 6'–0" wide windows on the south wall.
    - In addition to the 2'–8" personnel door on the east wall, enlarge the garage to accommodate (3) 8'–0" wide doors with a 12" minimum space.
    - Provide 24" space on each end of the garage between the door and the end of the wall.

    Save the drawing as E12FLOOR.
13. Draw five concentric circles with the smallest having a diameter of 6" and the other sizes increasing in 3" increments. Remove a 1" wide strip going across each of the circles in both the vertical and horizontal directions. The center of the strip to be removed should be at the center of each quadrant. Rotate the remaining portions of the second and forth circles (starting at the inside) so that the remaining portions of these circles have been revolved 45° while remaining concentric. Save the drawing as E-10-13.
14. Use EFLOOR as a base. Draw the plan view of a simple tree. Draw the trees in groups of two and three with each tree in the group to have a different size. Place at least four groups of trees around the house. Design walkways, patios, and a driveway.
15. Draw concentric shapes of three, four, five, six, seven, and eight sides that are tangent to one circle. Make copies of these objects that are 1/4 size, 1/2 size, full size, 2 x normal size, and 90% of the normal size. Save the drawing as E-10-15.
16. Open drawing E-10-4 and make two copies of the door. Scale one door to be 2'–6" wide and the other door to be 2'–8" wide. Leave the window beside the door as is on each option. Once the sizes are altered, make copies so that double doors are available for each option. A total of four door arrangements should be available when complete. Save the doors as E-10-16.

17. Open Drawing E-9-10 and make two additional copies of the window. Enlarge one window to 5'–0" wide. Reduce the other window to 75% of the original size. Save the drawing as E-10-17.

18. Draw a U-shaped kitchen with a 30" x 48" island. Provide a 32" x 21" sink, a 24" x 24" dishwasher, a 34" x 24" refrigerator, and 27" x 27" oven in the kitchen, with a 30" x 22" cook top in the island. Create a second copy of the kitchen that will be flopped and set at a 45° angle to the first kitchen. Save the drawing as E-10-18.

19. Using the drawing below as a guide, lay out the outline of the office floor plan. Use 8" wide exterior walls. Copy the plan. Save the drawing as E-10-19.

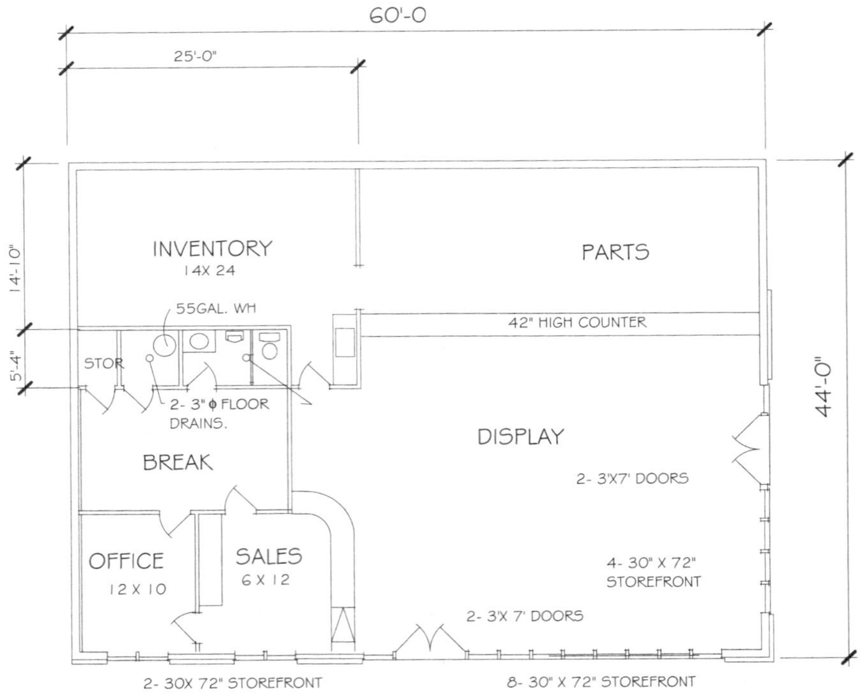

20. Draw a 3" diameter circle, a 6" line, and a polygon with eight sides that will fit in a 5 3/4" diameter circle. Save this portion of the drawing as E-10-20a. Make enough copies of each object and then use each option of each grips editing function to edit each object. Save the drawing as E-10-20b.

21. Use E-10-20a as a template drawing to complete this assignment. Make a second copy of these objects. Use **Select** to create a group consisting of the two circles and two polygons. Use **Mirror, Rotate**, and **Scale** to edit the group. Save the drawing as E-10-21.
22. Use Drawing E-10-20a as a template drawing to complete this assignment. Make a second copy of these objects and then mirror both groups so that you have a total of four circles, lines and polygons. Use **Group** to name and save at least six different groups. Save the file as E-10-22.
23. Using the drawing below as a guide, draw and array a pattern to represent Spanish clay tile roofing. Save the drawing as E-10-23.

24. Use a reference guide to determine the size of a W8x35 steel piling. Draw a plan for a 38' x 76' structure showing three rows of piling. Each of the two exterior rows will be inset 8" from the exterior face of the structure. The interior row of pilings will be 18' in from the front (the 76' long surface) of the structure. Pilings will be placed at 19' o.c., equally spaced from each exterior wall. Save the drawing as E-10-24.

CHAPTER 10 QUIZ

1. Explain the uses of the **Break** command.
2. Describe two different uses of the **Fillet** command.
3. List the command sequence for **Rotate**.
4. Explain the difference between **Move** and **Copy**.
5. List the command sequence to **Extend** a line to meet another line and sketch and label the process.
6. Two perpendicular lines cross at a corner. List four methods to edit the lines to form a non-crossing corner.
7. A line has been drawn short of its required destination. How can the error be corrected?
8. What two options are available for offsetting an object?
9. In the **Chamfer** command, what is the relationship between the distances and the selection of lines?
10. You are preparing to **Move** some objects, so you type **P** at the Select objects prompt. What will be moved?
11. Use the **Help** menu to determine what **Setvar** controls a preset **Fillet** value?
12. How can an object be rotated in a clockwise direction?
13. A door is drawn at a 45° angle. It should be 49°. How can this error be corrected?
14. Six objects need to be moved. As you select the objects, an extra object is included in the selection set, and an intended object is not included. Describe the command process and prompts to move these objects.
15. List and describe the three **Break** options.
16. Describe how each of the editing commands presented in this chapter could be used in drawing a floor plan.
17. What will happen if you try to place a 60" radius fillet in two intersecting lines that are 48" long?
18. Explain how **Chamfer** will affect two lines that do not touch if they are selected as the objects of this command.
19. What will be the effect of the Trim option of **Chamfer** on two intersecting lines?
20. What will be the effect of the **Trim Edge** option on two lines that do not intersect?
21. How will the Edge option of **Extend** affect two lines?
22. How will Window and Crossing window affect objects selected for **Stretch**?
23. What are the options that are given if the DElta option of **Lengthen** is selected?

24. A 6" long line is selected to be edited using **Lengthen.** What will be the result of entering 50% x 150% x 200%?
25. How many selection points will be required when you break an arc using **Break @**?
26. Provide an example of how the **Align** command could save time.
27. List the commands used to create E-10-9.
28. What must be done to display grips on a line?
29. What five editing functions can be executed after a grip is hot?
30. What is a hot grip and how do you create one?
31. You're exploring options and select **Gripblock**. What will this option do and what can you do if you don't want this option?
32. What advantage does **Select** have over grips?
33. Describe the process of selecting and naming a group of objects using **Group**.
34. What is a nested group?
35. What is a selectable group?
36. You've named a group, but you come to realize the name does not describe the drawing well. How do you fix the problem?
37. Save a group of objects without providing a name. How does AutoCAD keep track of this group?
38. You're exploring again and you use the Explode option of **Group**. What effect will this have on your computer? What effect will it have on the current drawing? What effect will it have on the selected group?
39. How do you make the options in the **Change Group** area of the **Object Groupin** dialog box active?
40. Use the **Help** menu and determine what variables affect grips. What are the options and their effects?

CHAPTER 11

Polylines

This Chapter will introduce drawing individual line segments that function as one line by the use of polylines. The chapter will explore the **Pline** command and examine methods of editing polylines.

Commands to be examined in this chapter include

- **Pline**
- **Pedit**
- **Explode**

One of the best ways to get objects to stand out in a complicated drawing is to vary the line thickness. In previous chapters you've learned how to add thickness to lines. Thickness can also be assigned easily with the **Pline** command, which creates polylines. In plan views, walls typically are drawn thicker than all other types of lines. On site drawings, polylines are often used to represent contour elevations at specific intervals, such as every five or ten feet. Polylines are used throughout structural drawings to represent various materials. Sections and details often show several line widths to help distinguish concrete, steel, and wood that has been cut by the cutting plane from materials that lie beyond the cutting plane. Figure 11.1 shows examples of polylines used to represent steel reinforcing in concrete details.

In addition to being able to display varied widths, a polyline has several other unique features. When the line is drawn using the Line command, each segment is a separate object. An irregularly shaped object can be drawn with polylines, yet each individual line is part of the whole. Notice that in Figure 11.2 the horizontal line, the arc, and the vertical line forming the L-shaped rebar have been selected to be erased. Because these features were drawn using the **Pline** command, they function as one object. This feature becomes important when the **Bhatch** command is used.

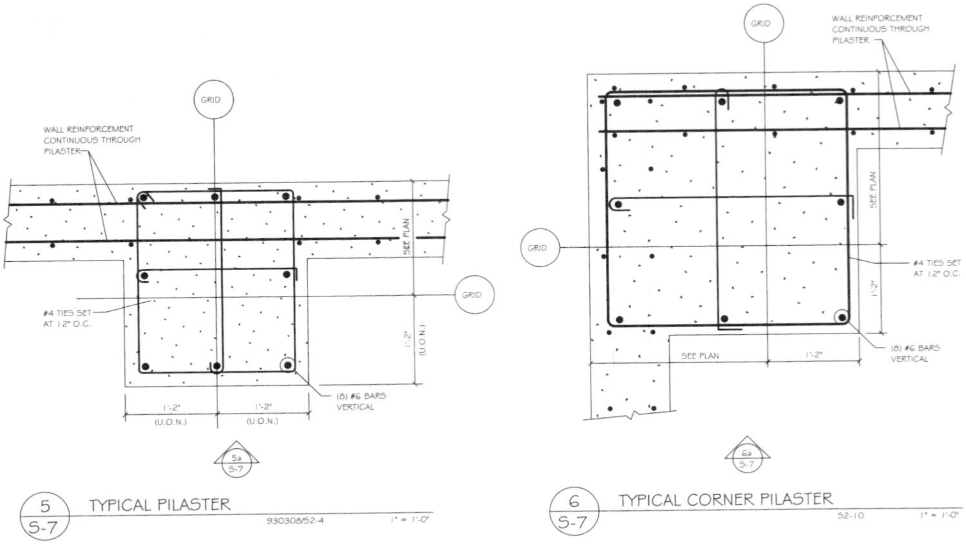

Figure 11.1 *The Pline command can be used to distinguish between various materials. Notice that the steel is drawn with a polyline and can be easily separated from the edge of the concrete. Not only can polylines be thicker, but they move as one object rather than as individual lines. (Courtesy Van Domelen/Looijenga/McGarrigle/Knauf Consulting Engineers.)*

EXPLORING THE PLINE COMMAND

Start the **Pline** command by selecting the **Polyline** button on the **Draw** toolbar, by typing **PL** (ENTER) at the Command prompt, or by selecting **Polyline** from the **Draw** menu. The command sequence is as follows:

> Command: **PL** (ENTER) *(Or click the Polyline button.)*
>
> PLINE
>
> Specify start point: *(Select any point.)*

Once a beginning point has been selected, a new prompt will be displayed:

> Current line–width is 0'-0"
>
> Specify next point or [Arc/Halfwidth/Length/Undo/Width]: *(Specify endpoint of line.)*

This prompt will continue to be displayed each time a new endpoint is specified. Each new line segment will continue from the end of the preceding line. The command will continue until (ENTER) is pressed.

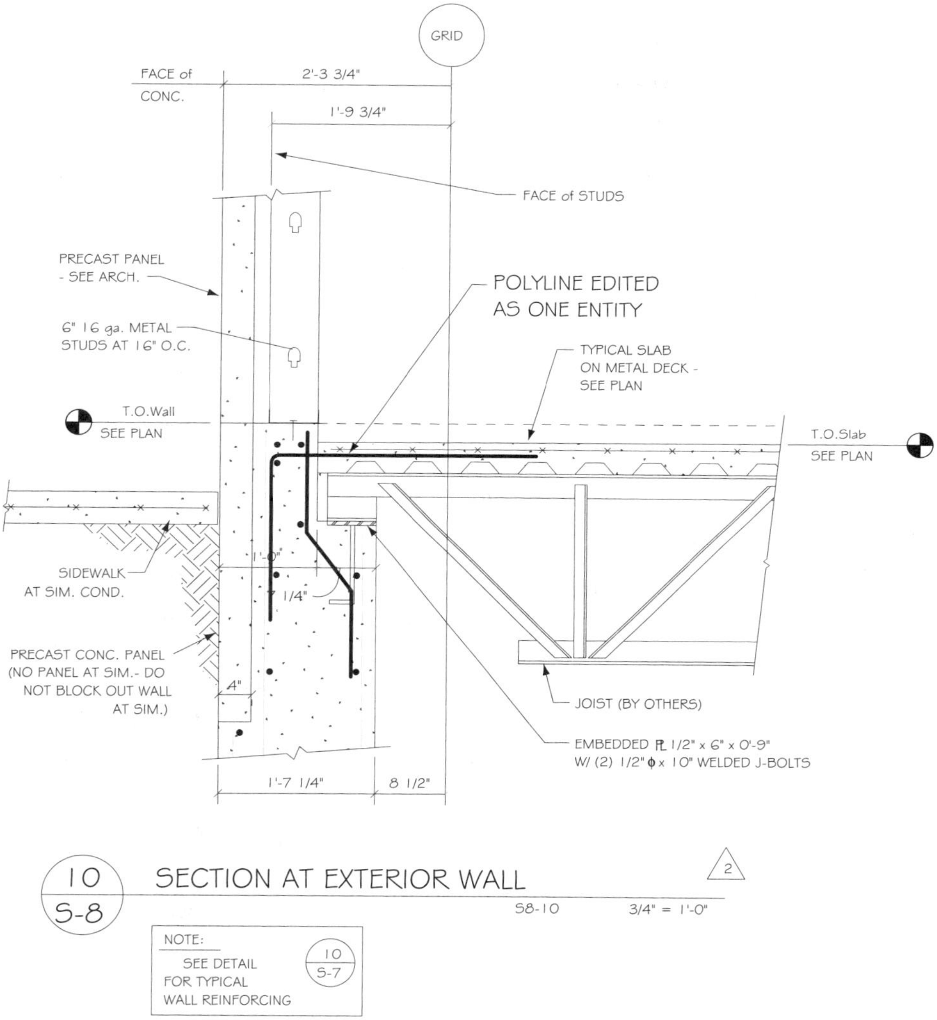

Figure 11.2 *Selecting the L-shaped rebar that was drawn using* **Pline** *at any point will select the horizontal line, the arc, and the vertical line as one object. (Courtesy Van Domelen/Looijenga/McGarrigle/Knauf Consulting Engineers.)*

CHANGING THE WIDTH

Enter the drawing area and draw a few polylines. Once you've drawn a polyline with several different drawing segments, select one of the segments to be erased. During editing is when you can really appreciate a polyline. Selecting one segment of a line selects the entire polyline for editing. A line can be made up of hundreds of segments, but if one is selected, they are all selected. With the line width set to "0", polylines look just like any other line. The 0 width works great if you want to outline an irregularly shaped area and find the area or the perimeter, where line width is not important. To change the width of a line, start the **Pline** sequence and select a start point. As the option prompt is displayed, type **W** (ENTER). This will allow information for a starting and ending width to be entered. The command sequence to draw a 1/8" wide line is as follows:

Command: **PL** (ENTER) *(Or click the Polyline button.)*

PLINE

Specify start point: *(Select any point.)*

Current line–width is 0'-0"

Specify next point or [Arc/Halfwidth/Length/Undo/Width]: **W** (ENTER)

Specify starting width <0'-0">: **.125"** (ENTER)

Specify ending width <0'-01/8">: (ENTER)

AutoCAD rounds the option to the nearest fraction, depending on how the units have been set. If you enter a fraction and the default returns as 0, exit **Pline** and enter **Units**, and set the fractional values as desired. For the examples used in this chapter, the units are set to architectural and the denominator for fractions is set to 1/8" (.125"). Notice that the prompt for the ending width shows 1/8" as a default value. By responding ENTER at the prompt, the option prompt will return allowing for selection of the endpoint. With the width selected, the command sequence is as follows:

Specify next point [Arc/Halfwidth/Length/Undo/Width]: *(Specify starting point.)*

With the starting and ending width the same, a line with uniform thickness will be drawn, as shown in Figure 11.3.

Drawing Tapered Lines

The **Pline Width** option allows the width of a polyline to be changed for each segment or from one end of a segment to the other. A tapered polyline with a starting width of 1/8" and an ending width of 1/4" can be drawn by using the following command sequence:

Command: **PL** (ENTER) *(Or click the Polyline button.)*

PLINE

Specify start point: *(Select any point.)*

Current line–width is 0'-0"

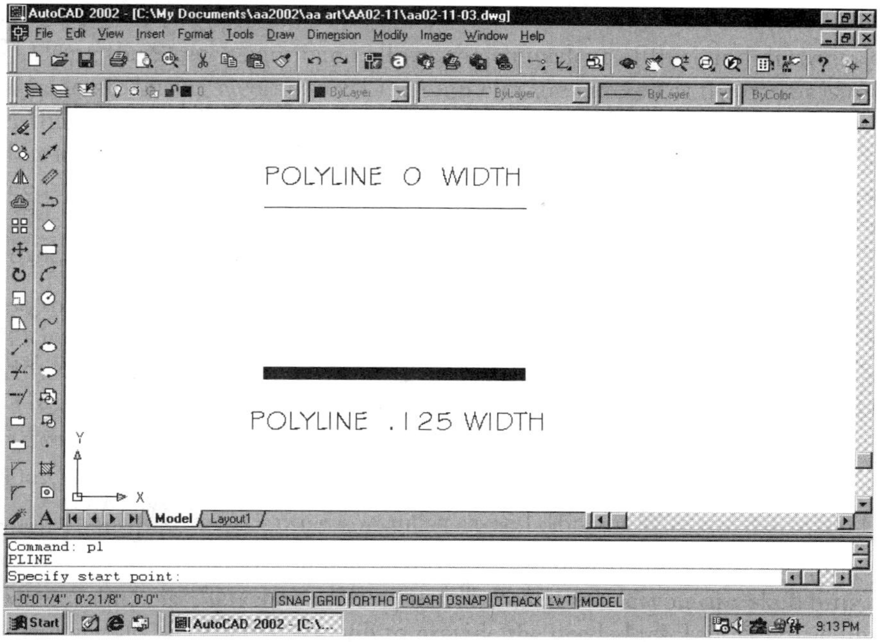

Figure 11.3 *If drawn with a width value of 0, the polyline will appear as any other line segment.*

Specify next point or [Arc/Halfwidth/Length/Undo/Width]: **W** (ENTER)

Specify starting width <0'-0">: .125" (ENTER)

Specify ending width <0'-01/8">: .25" (ENTER)

Specify next point [Arc/Halfwidth/Length/Undo/Width]: *(Select endpoint.)*

Specify next point [Arc/Close/Halfwidth/Length/Undo/Width]: (ENTER)

Command:

As the endpoint is entered, the line will be drawn and the option prompt will be returned to allow continuing a polyline. Notice that if the command is continued, the default for the next line segment is the ending width of the one just completed. Figure 11.4 shows an example of varied line width.

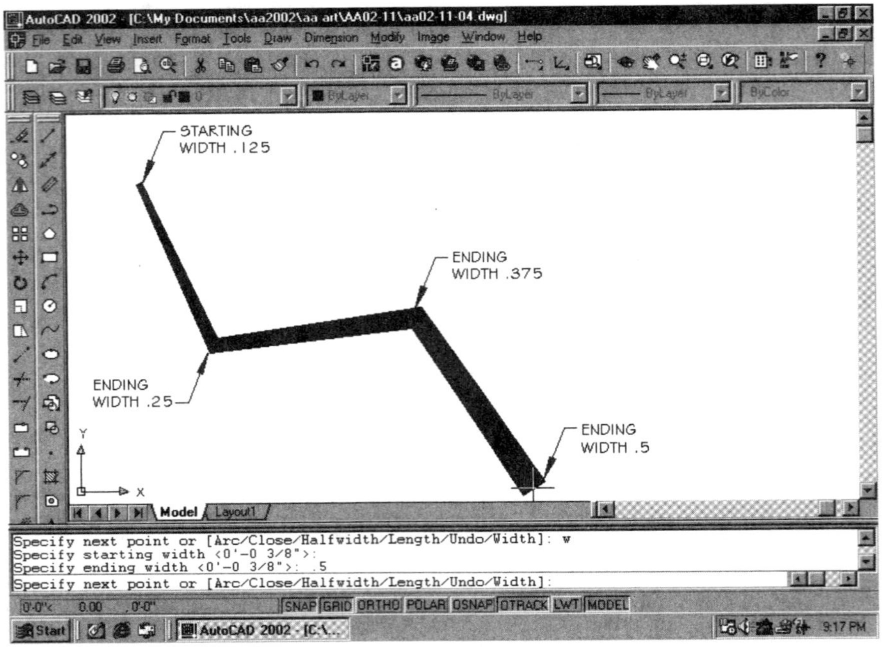

Figure 11.4 *Polylines can be used to draw a line with varied widths.*

Choosing a Line Width

When you experiment with **Pline**, the width selection is not terribly important. As you begin to work on more involved projects that will be printed or plotted, the width selection is critical. One of the considerations in selecting a line width is the scale at which the drawing eventually will be printed. A 1/4" polyline will look huge on the screen when you've zoomed in on a feature, but if printed or plotted at 3/8"=1'–0", it will appear as a 0 width line on a drawing plotted at 1"=30'–0".

It's also important to realize that line width can be controlled in the plotting process. This will be covered in a later chapter as layers and colors are discussed, and as you explore plotting options. For now, the key is to realize that polyline width can easily be altered.

Halfwidth Option

The Halfwidth option allows the width of a polyline to be set from the center of the polyline to an edge and is very similar to setting the width. It's like describing the size of a circle by referring to the radius instead of the diameter. The command sequence is as follows:

Command: **PL** (ENTER) *(Or click the Polyline button.)*

PLINE

Specify start point: *(Select any point.)*

Current line–width is 0'-01/4"

Specify next point or [Arc/Halfwidth/Length/Undo/Width]: **H** (ENTER)

Specify starting half–width <0'-01/8">: (ENTER)

Specify ending width <0'-01/8">: (ENTER)

Specify next point [Arc/Halfwidth/Length/Undo/Width]: *(Select end-point.)*

Specify next point [Arc/Close/Halfwidth/Length/Undo/Width]: (ENTER)

Command:

Length Option

This option will allow a polyline to be drawn to a specified length and parallel to the last polyline drawn. The new line will extend in the same direction as previous lines. To change the direction in which the line is drawn, enter a negative length. This process can be seen in Figure 11.5. If the last segment was an arc, a line drawn with **L** will be tangent to the arc.

Command: **PL** (ENTER) *(Or click the Polyline button.)*

PLINE

Specify start point: *(Select any point.)*

Current line–width is 0'-01/4"

Specify next point or [Arc/Close/Halfwidth/Length/Undo/Width]: **L** (ENTER)

Specify length of line: *(Enter desired length.)* 2 (ENTER)

Specify next point or [Arc/Close/Halfwidth/Length/Undo/Width]: (ENTER)

Command:

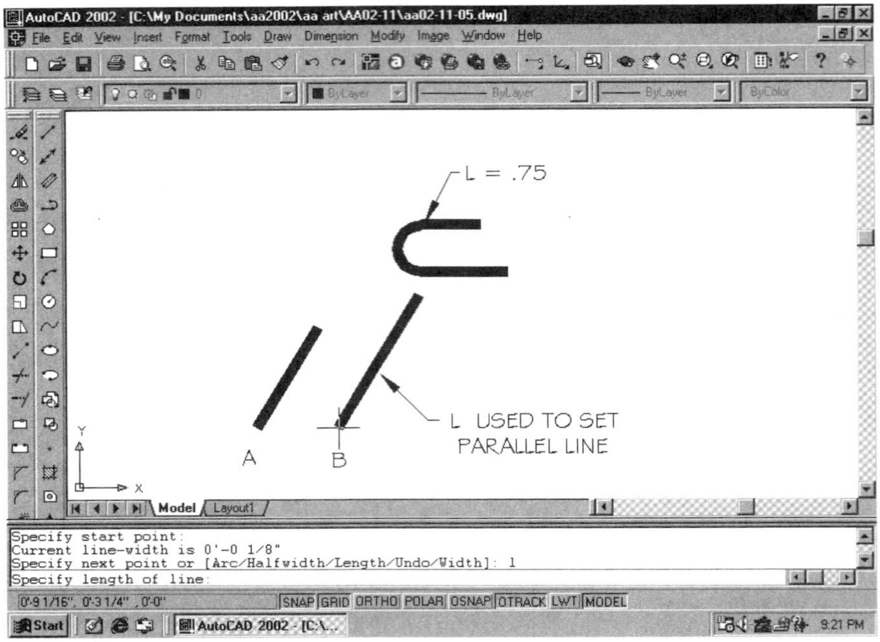

Figure 11.5 *The Length (L) option of* **Pline** *will draw a polyline parallel to another polyline at a specified length.*

POLYLINE ARCS

Arc is the first option listed in the **Pline** menu. When you select Arc by typing **A** (ENTER) at the prompt, AutoCAD starts the Arc submenu and displays the prompt:

Command: **PL** (ENTER) *(Or click the Polyline button.)*

PLINE

Specify start point: *(Select starting point.)*

Current width is 0'-01/4"

Specify next point or [Arc/Halfwidth/Length/Undo/Width]: **A** (ENTER)

Specify endpoint of arc or [Angle/CEnter/Direction/Halfwidth/Line/Radius/Second pt/Undo/ Width]: *(Select any point.)*

Specify endpoint of arc or [Angle/CEnter/CLose/Direction/Halfwidth/Line/Radius/Second pt/Undo/ Width]: ENTER

Command:

The Arc option will continue to prompt for arc endpoints until (ENTER) is selected.

As the Arc submenu is displayed, an arc is started at the first point that was selected and is dragged to the current location of the cursor. Several of the options for Arc are the same as the options for straight line polylines. The Width option is similar, with the width for an Arc ranging from 0 up to the diameter of the arc.

Continuous Polyline Arcs

If an arc is started from a previous polyline segment, the new arc will be tangent to the existing line by default. The arc will continue in the same direction as the most recent polyline segment. An example of how the arc will be constructed can be seen in Figure 11.6. An arc can also be constructed using the Angle, CEnter, CLose, Direction, Halfwidth, Line, Radius, and Second pt methods.

Polyline Arc Angles

This option will allow the start point, center, and angle that the polyline arc is to span (included angle). By default, the angle will be drawn counterclockwise. By entering a negative value, the angle will be drawn clockwise. The command process to draw an arc with a 45° included angle is as follows:

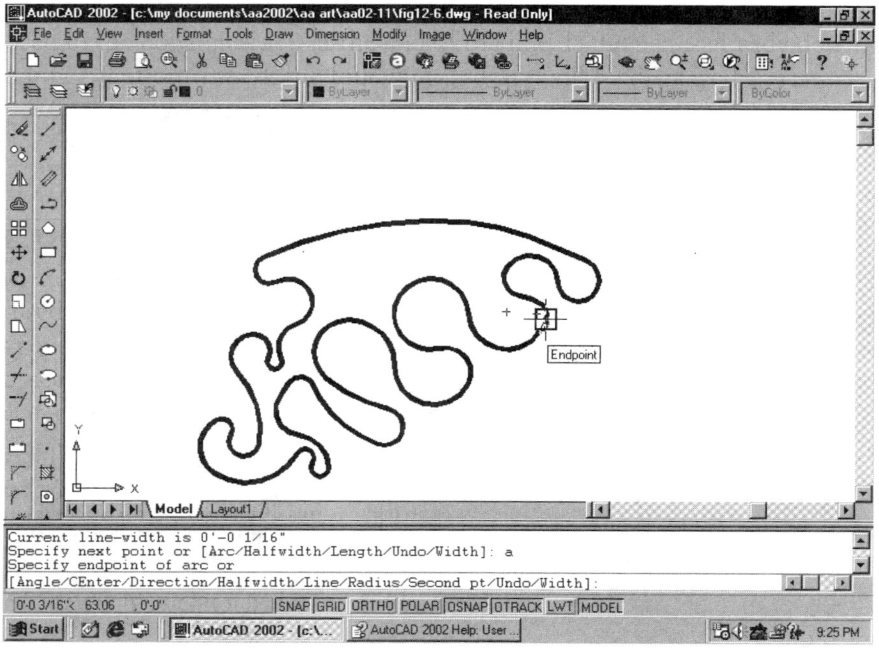

Figure 11.6 *Arcs can be constructed of polylines by using the Arc option of Pline.*

Command: **PL** (ENTER) *(Or click the Polyline button.)*

PLINE .

Specify start point: *(Select starting point.)*

Current width is 0'-01/4"

Specify next point or [Arc/Halfwidth/Length/Undo/Width]: **A** (ENTER)

Specify endpoint of arc or [Angle/CEnter/Direction/Halfwidth/ Line/Radius/Second pt/Undo/ Width]: **A** (ENTER)

Specify included angle: **45** (ENTER)

Specify endpoint of arc or [CEnter/Radius]: *(Select desired endpoint.)*

Specify endpoint of arc or [Angle/CEnter/CLose/Direction/ Halfwidth/Line/Radius/Second pt/Undo/ Width]: (ENTER)

Command:

As the endpoint is selected, the center is rotated as needed, based on the two endpoints. This process can be seen in Figure 11.7.

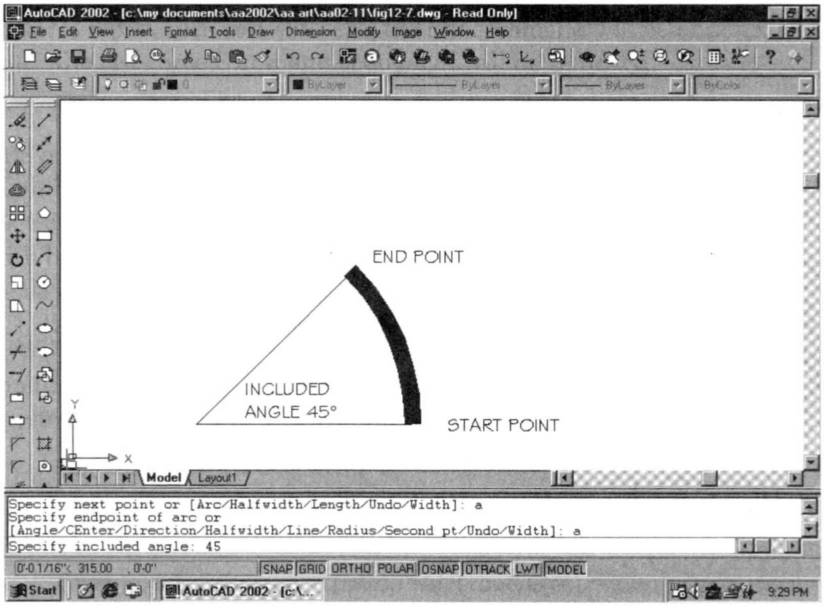

Figure 11.7 *Selecting the Angle (A) option of the Arc submenu will allow an arc to be drawn based on the angle provided.*

Arc Center

A polyline arc can be drawn by specifying the endpoint and the centerpoint. When you type **CE** (ENTER) at the last prompt, you will be allowed to define the first endpoint and the center point of the arc. The first steps of the command sequence are identical to other Arc commands. The command sequence for drawing a polyline arc by defining the center point is as follows:

Command: **PL** (ENTER) *(Or click the Polyline button.)*

PLINE

Specify start point: *(Select starting point.)*

Current width is 0'-01/4"

Specify next point or [Arc/Halfwidth/Length/Undo/Width]: **A** (ENTER)

Specify endpoint of arc or [Angle/CEnter/Direction/Halfwidth/Line/Radius/Second pt/Undo/ Width]: **CE** (ENTER)

Specify center point of arc: *(Select desired center point.)*

Specify endpoint of arc or [Angle/Length]: **A** (ENTER)

Specify included angle: **45** (ENTER)

Specify endpoint of arc or [Angle/CEnter/CLose/Direction/Halfwidth/Line/Radius/Second pt/Undo/ Width]: (ENTER)

Specify endpoint of arc or [Angle/CEnter/CLose/Direction/Halfwidth/Line/Radius/Second pt/Undo/ Width]: *(Select desired endpoint)*

Specify endpoint of arc or [Angle/CEnter/CLose/Direction/Halfwidth/Line/Radius/Second pt/Undo/ Width]: (ENTER)

Command:

Arc Radius

Typing **R** (ENTER) will allow for specifying the first endpoint and the radius to be selected. The command sequence is as follows:

Command: **PL** (ENTER) *(Or click the Polyline button.)*

PLINE

Specify start point: *(Select starting point.)*

Current width is 0'-01/4"

Specify next point or [Arc/Halfwidth/Length/Undo/Width]: **A** (ENTER)

Specify endpoint of arc or [Angle/CEnter/Direction/ Halfwidth/Line/Radius/Second pt/Undo/ Width]: **R** (ENTER)

Specify radius of arc: *(Enter desired radius.)* **3.25** (ENTER)

Specify endpoint of arc or [Angle]: **A** (ENTER)

Specify included angle: *(Specify desired angle)* **45** (ENTER)

Specify direction of chord for arc <30>: *(Select direction)*

Specify endpoint of arc or [Angle/CEnter/CLose/ Direction/ Halfwidth/Line/Radius/Second pt/Undo/ Width]: (ENTER)

Command:

If an angle for the chord (the line from endpoints of the arc) is entered, the arc will be drawn and the prompt will be redisplayed for continuing the arc command. Then, a new arc will rubber-band on the screen. Press (ENTER) to stop the command or enter the letter representing the desired option to continue.

Arc Direction

The Direction option allows for control of the bearing of the polyline arc chord. The default for **Pline** is to draw the arc tangent to the preceding segment. The Direction option will allow you to override this action and specify a starting direction for the arc that is not tangent to the previous line. The process can be seen in Figure 11.8. The sequence is as follows:

Command: **PL** (ENTER) *(Or click the Polyline button.)*

PLINE

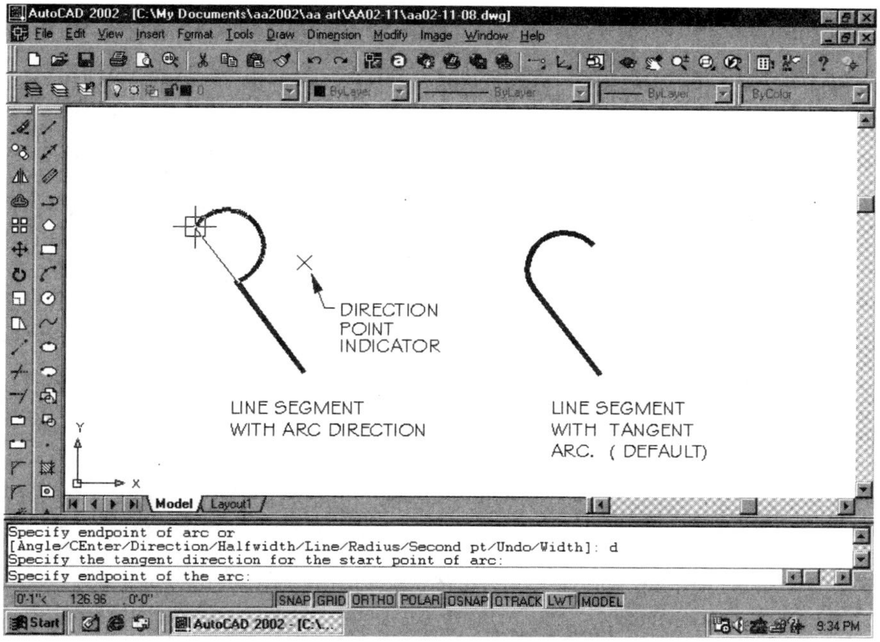

Figure 11.8 *The Direction (D) option of the Arc submenu allows an arc to be drawn that is not tangent to an existing polyline.*

Specify start point: *(Select starting point.)*

Current width is 0'-01/16"

Specify next point or [Arc/Halfwidth/Length/Undo/Width]: **A** (ENTER)

Specify endpoint of arc or [Angle/CEnter/Direction/Halfwidth/Line/Radius/Second pt/Undo/ Width]: **D** (ENTER)

Specify the tangent direction for the start point of arc: *(Select a point to indicate the desired arc direction.)*

Specify endpoint of arc: *(Specify desired endpoint)*

[Angle/CEnter/CLose/Direction/Halfwidth/Line/Radius/Second pt/Undo/ Width]: (ENTER)

Specify endpoint of arc or [Angle/CEnter/CLose/Direction/
Halfwidth/Line/Radius/Second pt/Undo/ Width]: (ENTER)

Command:

Three-Point Arc

A polyline arc can be drawn by entering a second point between each endpoint. This allows for better placement of the arc. The sequence can be seen in Figure 11.9. The command sequence is as follows:

Command: **PL** (ENTER) *(Or click the Polyline button.)*

PLINE

Specify start point: *(Select starting point.)*

Current width is 0'-01/16"

Specify next point or [Arc/Halfwidth/Length/Undo/Width]: **A** (ENTER)

Specify endpoint of arc or [Angle/CEnter/Direction/
Halfwidth/Line/Radius/Second pt/Undo/ Width]: **S** (ENTER)

Specify second point on arc: *(Specify desired second point for arc to pass through.)*

Specify endpoint of arc: *(Specify desired endpoint of arc.)*

Specify endpoint of arc or [Angle/CEnter/CLose/Direction/
Halfwidth/Line/Radius/Second pt/Undo/ Width]: (ENTER)

Command:

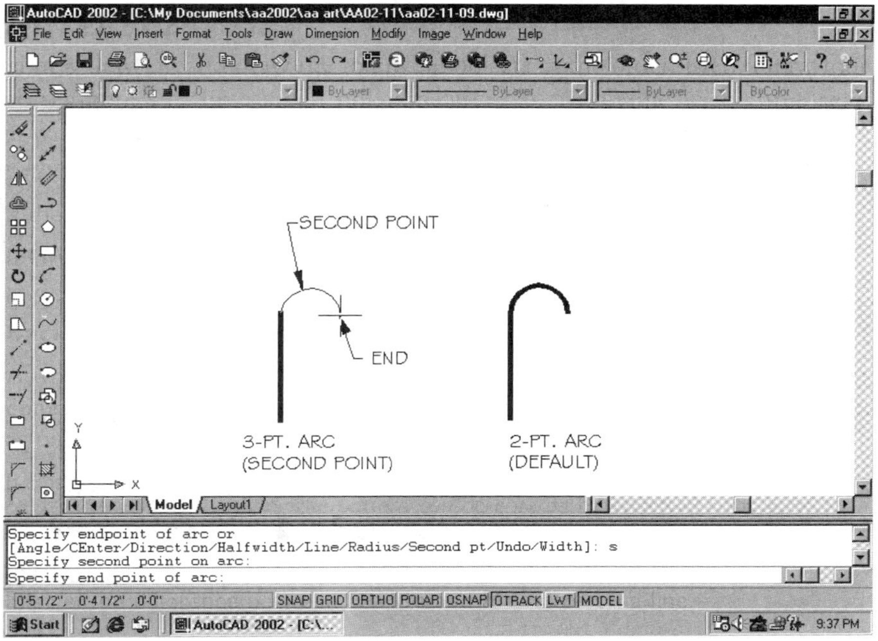

Figure 11.9 *Using the Second point (S) option of the Arc submenu draws an arc based on three points.*

CLOSING A POLYGON

Polygons are ended by using automatic OSNAP or by typing **C** (ENTER) the way you draw with the **Line** command. Each option provides a line segment from the present location, back to the original starting point. These two options are also available with the **Pline Arc** option. Typing **CL** (ENTER) at the prompt will also allow a polygon constructed of polylines to be closed. Remember that **CL** must be entered when in the arc submenu rather than **C** to distinguish between CLose and CEnter. The results can be seen in Figure 11.10. The final sequence to close a polyline is as follows:

Specify end point of arc or [Angle/CEnter/CLose/Direction/
Halfwidth/Line/Radius/Second pt/Undo/ Width]: **CL** (ENTER)

Command:

NOTE: Just because AutoCAD gives you an option doesn't mean you need to use it. Automatically snapping to an endpoint is a much faster method for closing a polygon.

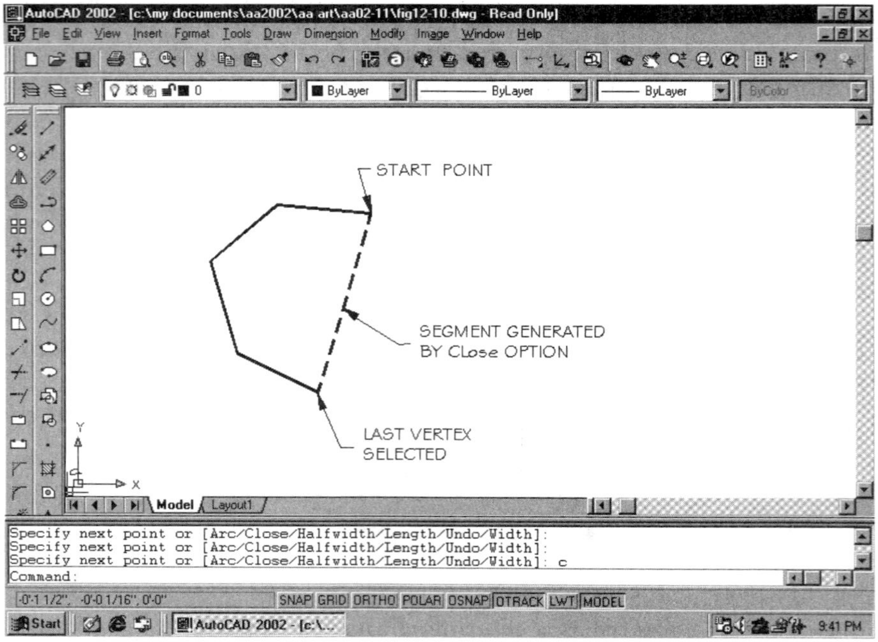

Figure 11.10 *Polygons drawn using polylines can be completed using the Close (CL) option. Automatic OSNAP and* **C** *(***ENTER***) can also be used to close a polygon constructed of polylines.*

UNDOING A POLYLINE

As with other drawing options, **Pline** allows for removing the previous segment or segments in the reverse order from the way they were drawn. This can be done with the Undo option by typing **U** (ENTER) at the prompt.

> Specify end point of arc or [Angle/CEnter/CLose/Direction/
> Halfwidth/Line/Radius/Second pt/Undo/ Width]: **U** (ENTER)

This will remove the last polyline segment that was drawn, as seen in Figure 11.11. Undo can be used repeatedly until just one point is left in the polyline. Separate from the Undo option, the **Undo** command can be used once the drawing command has been terminated. The **Undo** command removes an entire command sequence and will be covered in later chapters.

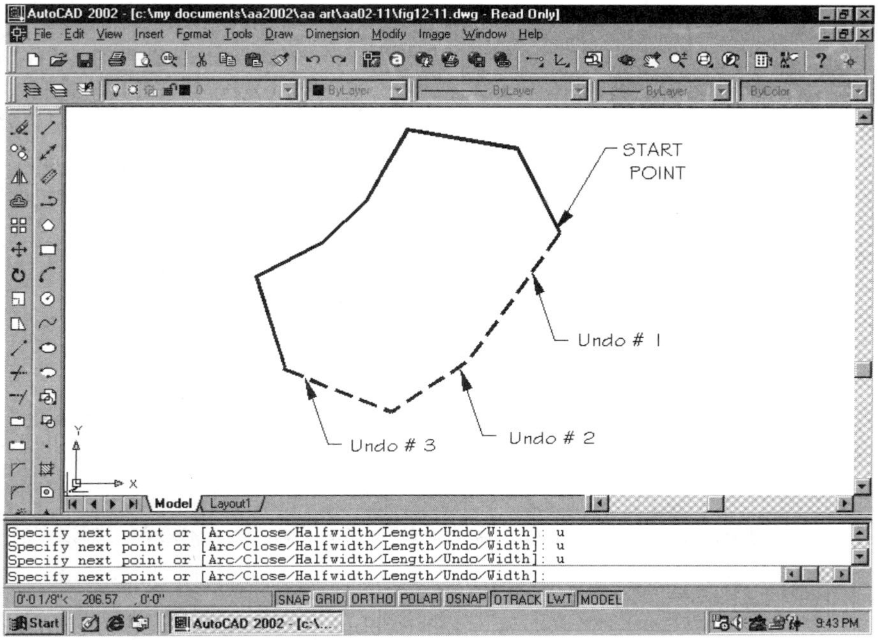

Figure 11.11 *The Undo (U) option will remove the last polyline segment drawn.*

EDITING A POLYLINE

Polylines can be edited with the commands introduced in previous chapters. Because polylines are segments drawn as a single object, they are edited as a single object. This can facilitate the selection process, because only one segment needs to be selected, rather than each of the segments that make up the entire object. This can be seen in Figure 11.12.

Most of the commands explained in Chapter 10 will modify a polyline in exactly the same way as another element. **Fillet** and **Chamfer** each have an added feature for a polyline.

ALTERING POLYLINES WITH FILLET

The **Fillet** command can't be used to fillet two different polylines in the same way you use it to fillet two lines. However, the command can be used on closed polygons constructed of a polyline, by the use of two different methods. The command sequence to **Fillet** one corner of a closed polygon with a radius of 3/8" is as follows:

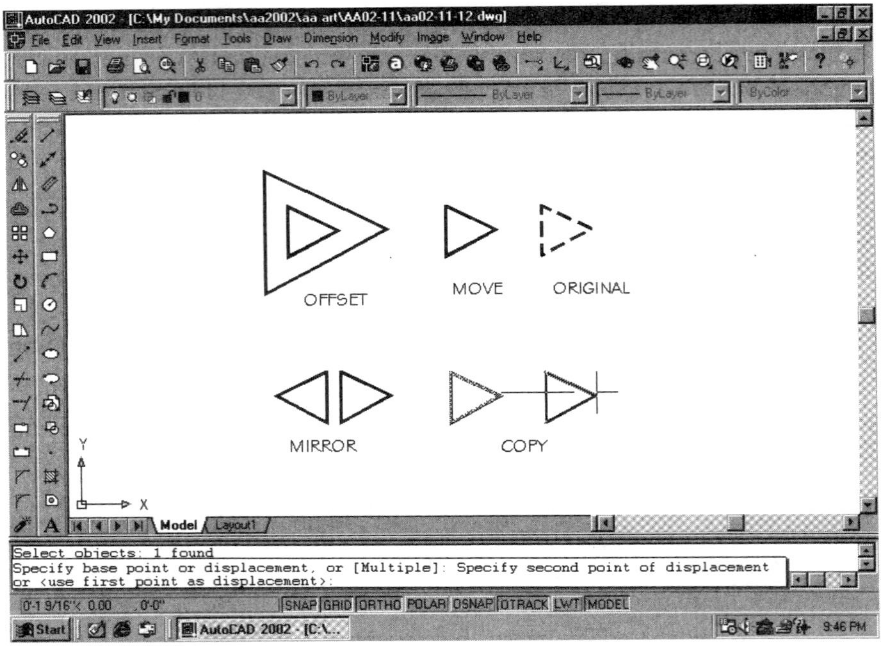

Figure 11.12 *When editing a polyline, remember that the entire polyline acts as one object.*

Command: **F** (ENTER) *(Or click the Fillet button.)*

FILLET

Current settings: Mode = TRIM, Radius = 0'-0"

Select first object or [Polyline/Radius/Trim]: **R** (ENTER)

Specify fillet radius <0'-0">: **.375** (ENTER)

Select first object or [Polyline/Radius/Trim]: *(Select edge of object to receive fillet.)*

Select second object: *(Select second edge to receive fillet.)*

Command:

The fillet can also be applied to all edges of the polygon in one command sequence. Once the radius is set, the command sequence is as follows:

Command: **F** (ENTER) *(Or click the Fillet button.)*

FILLET

Current settings: Mode=TRIM, Radius=0'-0 3/8"

Select first object or [Polyline/Radius/Trim]: **P** (ENTER)

Select 2D polyline: *(Select object to receive fillets.)*

4 lines filleted *(Quantity will vary with each object.)*

Command:

The effects of the **Fillet** command on polylines can be seen in Figure 11.13. The same effects can be achieved by selecting the first object instead of entering **P**.

ALTERING POLYLINES WITH CHAMFER

Editing polylines with **Chamfer** is similar to using the **Fillet** command. The command sequence for Chamfer is as follows:

Command: **CHA** (ENTER) *(Or click the Chamfer button.)*

CHAMFER

(TRIM mode) Current chamfer Dist1=0'-0 3/8", Dist2=0'-0 3/8"

Select first line or [Polyline/Distance/Angle/Trim/Method]: *(Select object to receive chamfer.)*

Select second line: *(Select second edge to be chamfered.)*

Command:

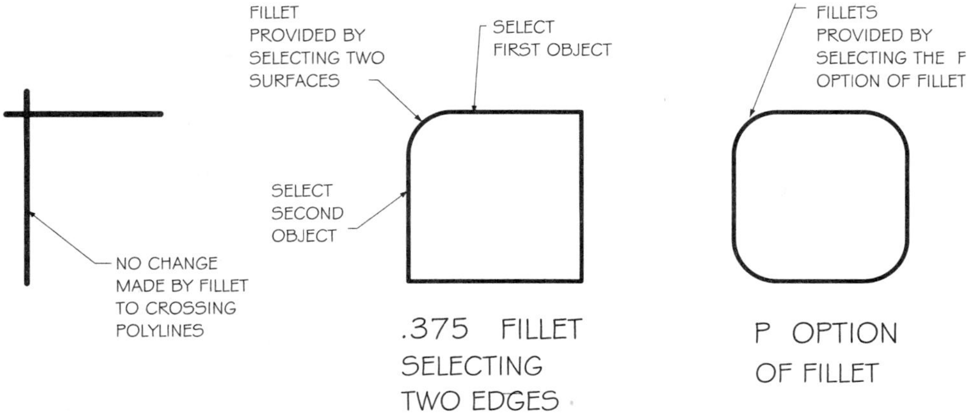

Figure 11.13 *Two separate polylines are not affected by the* **Fillet** *command. The* **Fillet** *command can be used to modify polygons constructed of polylines, or the Polyline option of* **Fillet** *can be used to fillet all edges of a polygon.*

Once the distances are set, the command will edit a polyline as two line segments, if the default option is selected. If the Polyline option is selected, all corners of the selected polyline will receive chamfers. The command sequence is as follows:

> Command: **CHA** (ENTER) *(Or click the Chamfer button.)*
>
> CHAMFER
>
> (TRIM mode) Current chamfer Dist1=0'-03/8", Dist2=0'-03/8"
>
> Select first line or [Polyline/Distance/Angle/Trim/Method]: **P** (ENTER)
>
> Select 2D polyline: *(Select object to receive chamfer.)*
>
> 4 lines were chamfered
>
> Command:

The effect of chamfering a polygon can be seen in Figure 11.14.

EXPLODING POLYLINES

One of the benefits of a polyline is also one of its drawbacks. Whether you've drawn 2 or 2,000 connected polyline segments, they all function as one object. That's great if you want to move 2,000 lines at once by selecting only one option, but a hindrance if you want to change just one of the segments. To overcome this obstacle, use the **Explode** command. This is not the long-awaited command you hope for when you're really frustrated with a computer. It explodes only groups of drawing objects, and it doesn't even make a loud noise. Access the command by selecting the **Explode** button (the stick of dynamite) on the **Modify** toolbar, by typing **X** (ENTER) at the Command prompt, or by selecting **Explode** from the **Modify** menu. The command sequence is as follows:

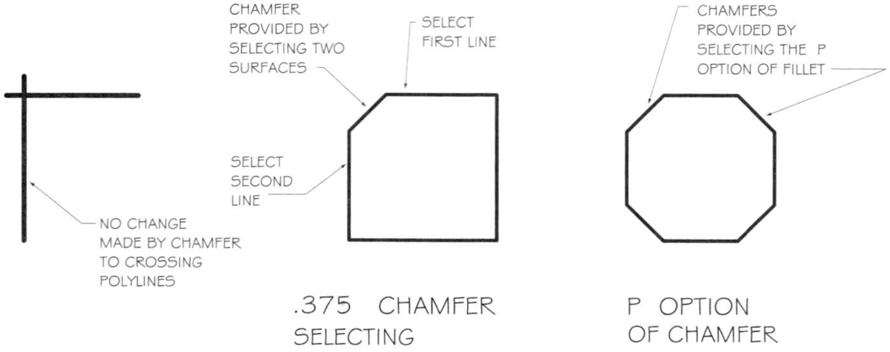

Figure 11.14 *Two separate polylines are not affected by the* **Chamfer** *command. The* **Chamfer** *command can be used to modify polygons constructed of polylines, or the Polyline option of* **Chamfer** *can be used to chamfer all edges of a polygon.*

Command: **X** (ENTER) *(Or click the Explode button.)*

EXPLODE

Select objects: *(Select objects to be edited.)*

1 found

Select object: (ENTER)

Once (ENTER) has been pressed, the polyline will be transformed into individual line or arc segments. If the polyline had a specified width, the following message will be displayed at the command line:

Exploding this polyline has lost width information.

The UNDO command will restore it.

Command:

The effects of **Explode** on a polyline can be seen in Figure 11.15. If you can tolerate the loss of line width, proceed. If the line width is important to the drawing, typing **U** (ENTER) at the Command prompt will restore the polyline to its unedited form. If you want the polyline to function as separate lines but retain their width, **Explode** the polyline and assign width using the **Lineweight** and the **Change** commands. See the **Help** menu or later chapters on editing.

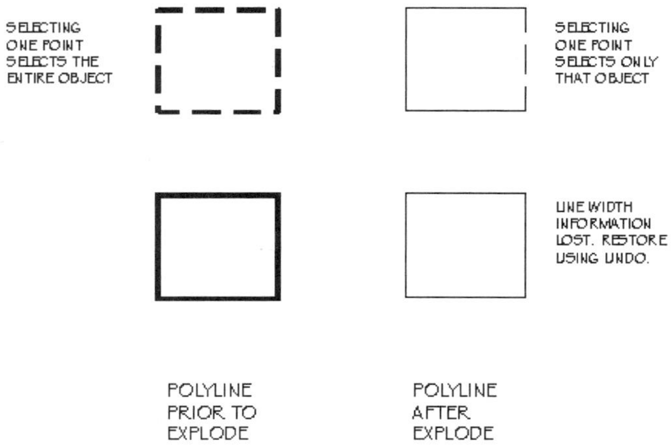

Figure 11.15 *The* **Explode** *command allows segments of a polyline to be treated as individual objects.*

ALTERING POLYLINES WITH PEDIT

In addition to using the standard editing options, you can edit polylines using the **Pedit** command. Access the command by selecting the **Edit Polyline** button on the **Modify II** toolbar, by typing **PE** (ENTER) at the Command prompt, or by selecting **Polyline** from the **Modify** menu. Options for the command will vary slightly depending on whether the polyline forms a closed shape or a line. Options for the command when a closed polyline is selected are as follows:

> Command: **PE** (ENTER) *(Or click the Edit Polyline button.)*
>
> PEDIT
>
> Select polyline or [Multiple]: *(Select polyline to be edited.)*
>
> Enter an option [Open/Join/Width/Edit
> vertex/Fit/Spline/Decurve/Ltype gen/Undo]:

If an open polyline is selected, the Open option is replaced with the Close option. Pressing (ENTER) will exit back to the command line. You can select objects by any of the selection methods discussed in previous chapters. Notice, when you are prompted to select objects, the cursor changes to a selection box to allow you to select polylines individually. The Window and Crossing methods are also excellent for selecting polylines.

Occasionally, segments might be selected for **Pedit** that are not polylines. When this happens, you will be given the following prompt:

> Object selected is not a polyline.
>
> Do you want to turn it into one? <Y>:

If the default is accepted, the segment will become a polyline and the **Pedit** prompt will be displayed, allowing options of the new polyline to be altered. A circle cannot be changed to a polyline, but it can be drawn using the 360° arc of **Pline** Arc option. A donut also could be used to draw a circle that would have qualities similar to a polyline. The editing options for a polyline include Open, Join, Width, Edit vertex, Fit, Spline, Decurve, Ltype gen, Undo, and eXit.

Open

The Open (O) option will remove the closing segment of a polyline. If a line was drawn back to the starting point without using the Close option, opening the polyline has no visible effect. The results can be seen in Figure 11.16. If **O** is entered for an open polyline, the reverse happens—it will be closed.

Close

The Close (CL) option creates a closing segment of the polyline while in the **Pline** command routine. Once you've moved on to another command sequence, the poly-

Figure 11.16 *A polyline segment removed with the Open option.*

line must be edited with **Pedit C** (Close) option. If you select **C** on a closed polyline, it will be opened. Figure 11.16 shows the effects of the close command.

Join

The Join (J) option adds lines, arcs, and other polylines that meet a selected polyline by typing **J** (ENTER) at the Command prompt. This option can be useful for joining two or more individual polylines so that they function as one polyline. This option can only be used with open polylines and cannot be used to join segments that do not touch the selected polyline. Objects that cross the polyline will not be joined.

Width

The Width (W) option will allow the current width of a polyline to be changed. Polylines that consist of segments with varying widths or tapers will be changed so that all segments have the new width. When (ENTER) is pressed at the option prompt, a new prompt will be given requesting the following:

Specify new width for all segments:

The width can be entered by keyboard or by selecting two points with the select button. Once the width is altered, the **Pedit** prompt will be displayed again, allowing for other editing to take place. Press (ENTER) to return to the Command prompt. Figure 11.17 shows the effect of the Width option.

Edit Vertex

Once a polyline has been formed, the shape might need to be altered with the Edit vertex (E) option. Altering the shape can be done by typing **E** (ENTER) at the option prompt. An X will be placed in the first vertex and a new prompt will be displayed with the following options:

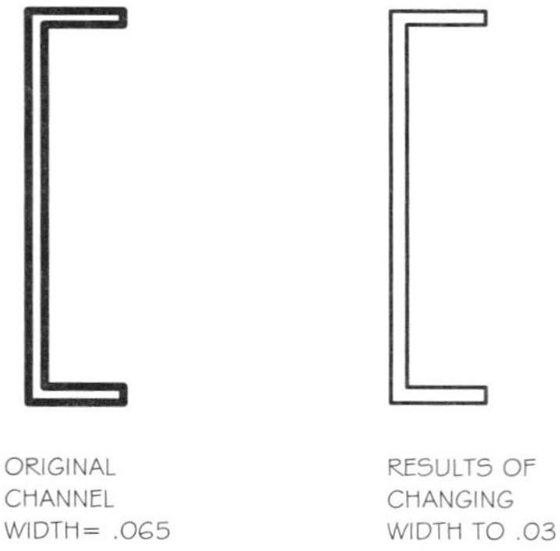

Figure 11.17 *The effect of changing the width of a polyline.*

Enter an option [Open/Join/Width/Edit
vertex/Fit/Spline/Decurve/Ltype gen/Undo]: **E** (ENTER)

Enter a vertex editing option [Next/Previous/Break/Insert/
Move/Regen/Straighten/Tangent Width/eXit] <N>:

Unless the X is placed in the location you would like to edit, use the Next or Previous options to move the X around the polyline to select the desired editing location.

Next
The Next (N) mode moves the X to the next vertex. By pressing (ENTER), you can move the X around the object, as seen in Figure 11.18.

Previous
The Previous (P) mode moves the X to the previous vertex. By pressing (ENTER), you can move the X around the object in the direction opposite from the one used by Next. The option can be seen in Figure 11.18.

Break
The Break (B) mode will remove a portion of a polyline. The portion to be broken is selected by the N or P option. Once the X is in the desired position, the prompt reads

Enter a vertex editing option [Next/Previous/Go/eXit] <N>:

As the X is now moved with N or **P**, any point between the original starting location and the present location will be removed. When the X is moved to the end of the segment to be used, type **G** (ENTER) for Go. This will remove the desired segment. If Break is used on a closed polygon, the closing section of the polygon will be removed

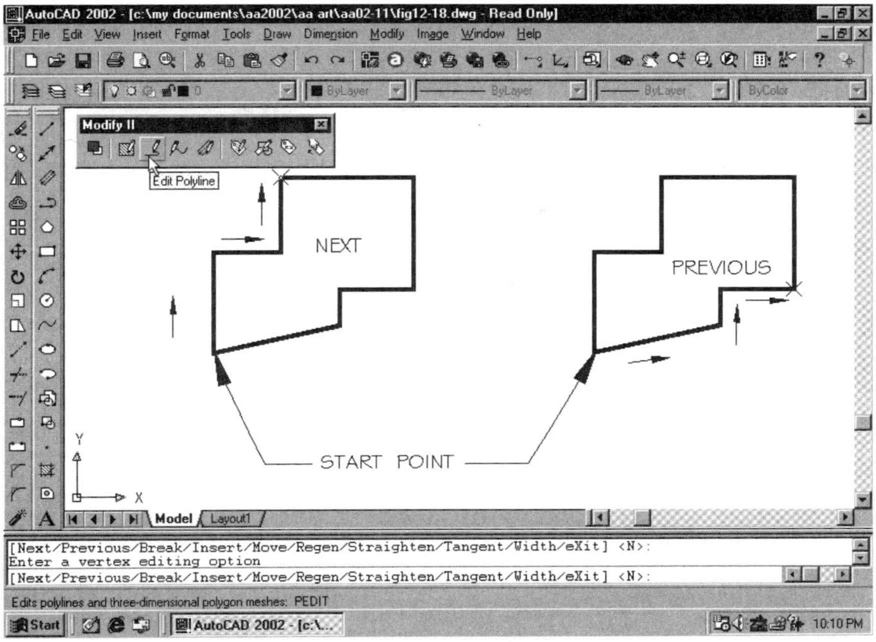

Figure 11.18 *The vertex of a polyline can be selected by moving the X using the Next or Previous option.*

along with the selected portion. The process can be seen in Figure 11.19. The command sequence is as follows:

Enter an option [Open/Join/Width/Edit vertex/Fit/Spline/Decurve/Ltype gen/Undo]: **E** (ENTER)

Enter a vertex editing option [Next/Previous/Break/Insert/Move/Regen/Straighten/Tangent Width/eXit] <N>: **B** (ENTER)

Enter an option [Next/Previous/Go/eXit] <N>: (ENTER)

Enter an option [Next/Previous/Go/eXit] <N>: **G** (ENTER)

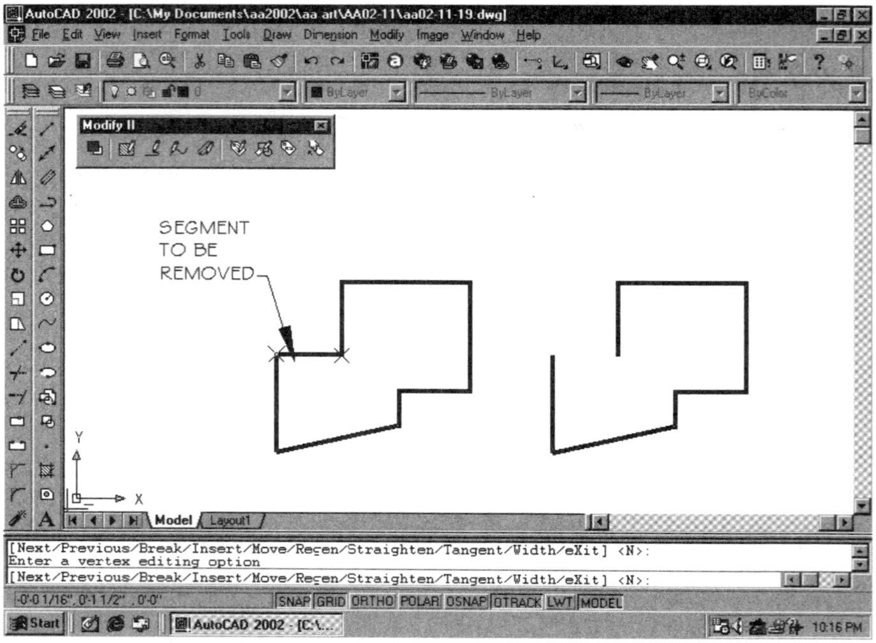

Figure 11.19 *Removing a portion of a polyline using the Break option.*

Insert

The Insert (I) mode will edit an existing polyline by adding a new vertex. The new vertex is added after the vertex marked by the X. The process can be seen in Figure 11.20. The command sequence is as follows:

> Enter an option [Open/Join/Width/Edit vertex/Fit/Spline/Decurve/Ltype gen/ Undo]: **E** (ENTER)
>
> Enter a vertex editing option [Next/Previous/Break/Insert/Move/Regen/Straighten/Tangent/Width/eXit] <N>: (ENTER) *(Move X to the desired location and press (ENTER).)*
>
> [Next/Previous/Break/Insert/Move/Regen/Straighten/Tangent/Width/eXit] <N>: **I** (ENTER)
>
> Specify location of new vertex: *(Select new location.)*

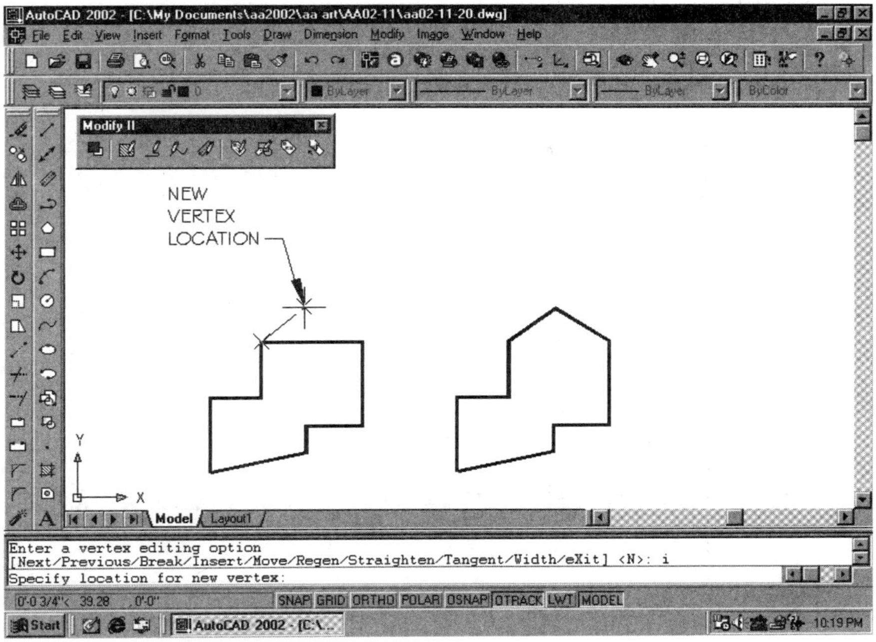

Figure 11.20 *Altering the shape of a polyline by inserting a new vertex.*

Move

The Move mode (M) allows for moving an existing polyline vertex to a new location. The process can be seen in Figure 11.21. The vertex to be moved must be the one currently marked by the X. The command sequence is as follows:

Enter an option [Open/Join/Width/Edit vertex/Fit/Spline/Decurve/Ltype/Undo/eXit]: **E** (ENTER)

Enter a vertex editing option [Next/Previous/Break/Insert/Move/Regen/Straighten/Tangent/Width/ eXit] <N>: (ENTER) *(Move X to the desired location and press (ENTER).)*

Enter a vertex editing option [Next/Previous/Break/Insert/Move/Regen/Straighten/Tangent/Width/eXit] <N>: **M** (ENTER)

Specify new location for marked vertex: *(Select new location.)*

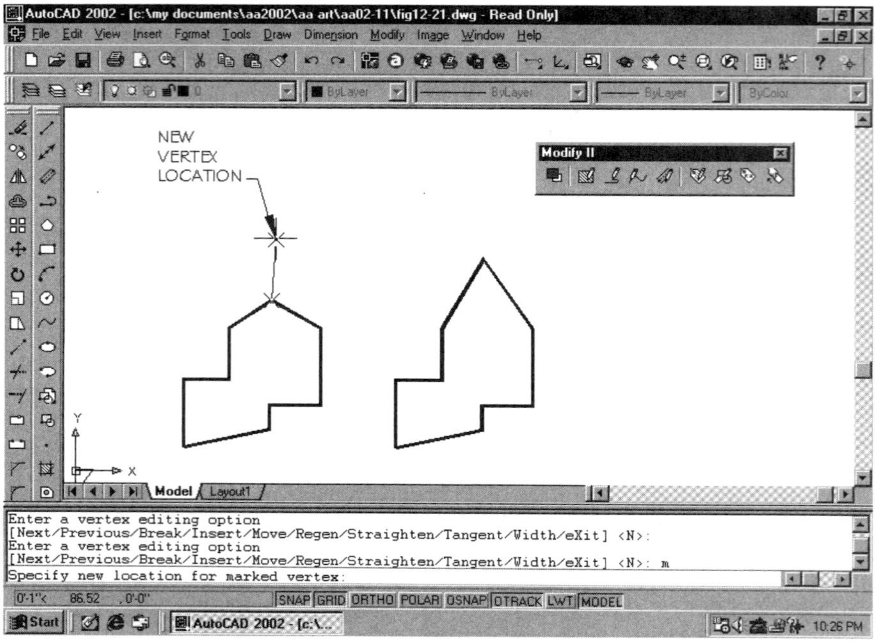

Figure 11.21 *Altering the shape of a polyline by moving an existing vertex to a new location.*

Regen

The Regen (R) mode will regenerate the edited polyline.

Straighten

The Straighten (S) mode will straighten existing polyline segments that lie between two selected points. Move the X to the desired location to mark the start of the edit. The X can then be moved to mark the end of the edit. Any arcs or segments between the two marks will be deleted and replaced by a straight segment. The process can be seen in Figure 11.22. The command sequence is as follows:

Enter an option [Open/Join/Width/Edit vertex/Fit/Spline/Decurve/Ltype/Undo/eXit]: **E** (ENTER)

Enter a vertex editing option [Next/Previous/Break/Insert/Move/Regen/Straighten/Tangent/Width/ eXit] <N>: (ENTER) *(Move X to the desired location and press* (ENTER)*.)*

Enter a vertex editing option

[Next/Previous/Break/Insert/Move/Regen/Straighten/Tangent/Width/
eXit] <N>: **S** (ENTER)

Enter an option [Next/Previous/Go/eXit] <N>: (ENTER) *(Continue pressing (ENTER) until the X is moved to the desired location)*

Enter an option [Next/Previous/Go/eXit] <N>: (ENTER)

Enter an option [Next/Previous/Go/eXit] <N>: (ENTER)

Tangent

The Tangent (T) mode will allow you to attach a tangent direction to the current vertex for later use in curve fitting. The command sequence is as follows:

Enter an option [Open/Join/Width/Edit
vertex/Fit/Spline/Decurve/Ltype/Undo/eXit]: **E** (ENTER)

Enter a vertex editing option [Next/Previous/Break/Insert/Move/
Regen/Straighten/Tangent/Width/ eXit]<N>: (ENTER) *(Move X to the desired location.)*

Enter a vertex editing option

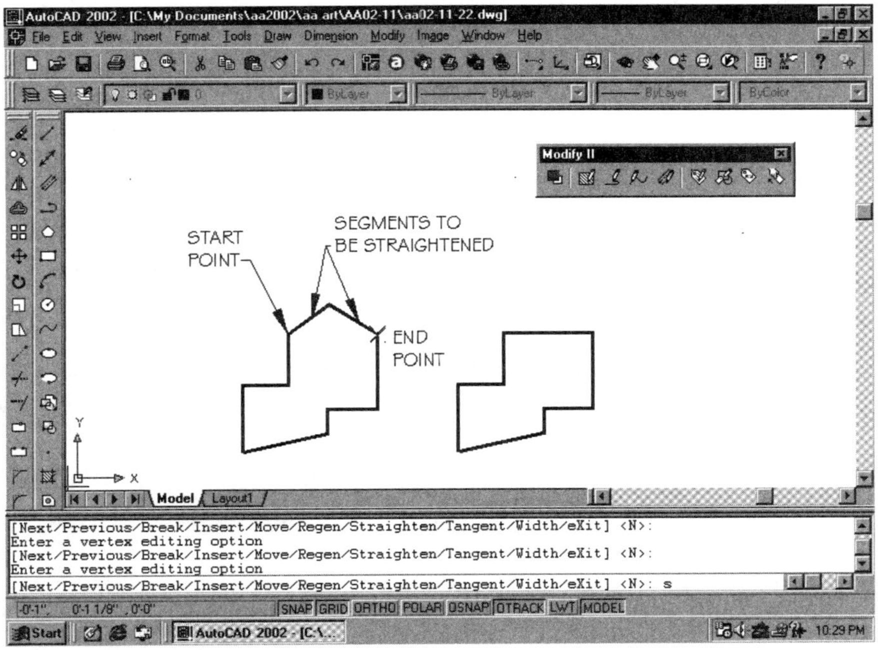

Figure 11.22 *Removing an existing polyline vertex using the Straighten mode of the Edit vertex option.*

[Next/Previous/Break/Insert/Move/Regen/Straighten/Tangent/Width/
eXit] <N>: **T** (ENTER)

Specify direction of vertex tangent: *(Specify a point or enter an angle.)*

Enter a vertex editing option

Either a specific tangent angle can be specified from the keyboard or a point can be selected to mark the direction from the currently marked (X) vertex.

Width

The Width (W) mode changes the starting and ending width of the segment following the marked vertex. You've explored the Width option of **Pedit**, which changes the entire polyline. The Width mode of the Edit Vertex option allows one segment only to be edited. The process can be seen in Figure 11.23. The command sequence is as follows:

Enter an option [Open/Join/Width/Edit
vertex/Fit/Spline/Decurve/Ltype/Undo/eXit]: **E** (ENTER)

Enter a vertex editing option [Next/Previous/Break/Insert/Move/
Regen/Straighten/Tangent/Width/ eXit]<N>: (ENTER) *(Move X to the desired location.)*

Enter a vertex editing option

[Next/Previous/Break/Insert/Move/Regen/Straighten/Tangent/Width/
eXit] <N>: **W** (ENTER)

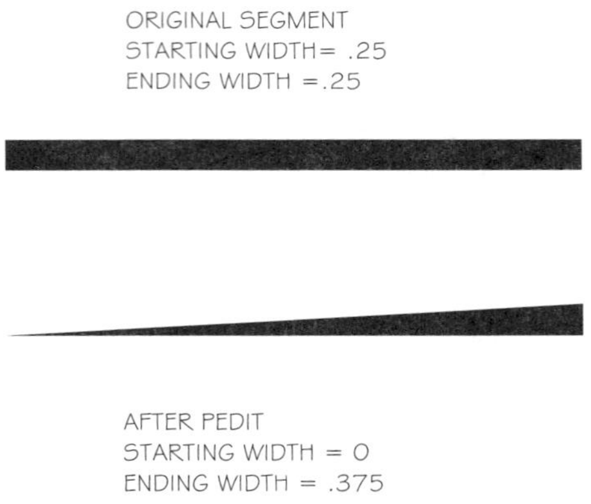

Figure 11.23 *Altering the width of an existing pline using the Width mode of the Pedit Edit vertex option.*

Specify starting width for next segment <0'–01/8">: **0** (ENTER) *(Enter desired width.)*

Specify ending for next segment width <0'-0">: **.375** (ENTER) *(Enter desired width.)*

Exit

The Exit (X) mode exits from Vertex editing and returns you to the **Pedit** prompt.

Fit

The Fit (F) option of **Pedit** allows straight line segments to be converted to curved lines. This can be especially helpful on a topography where lines are typically drawn from elevation to elevation. Fit can also be used to convert zigzag lines to the smooth curves often used to represent insulation in section view, as seen in Figure 11.24. The command process is as follows:

Command: **PE** (ENTER) *(Or click the Edit Polyline button.)*

PEDIT

Select polyline:

Enter an option [Close/Join/Width/Edit vertex/Fit/Spline/Decurve/Ltype gen/Undo]: **F** (ENTER)

As (ENTER) is pressed, Fit transforms the points of the original polyline into smoothed curves. The process can be seen in Figure 11.25.

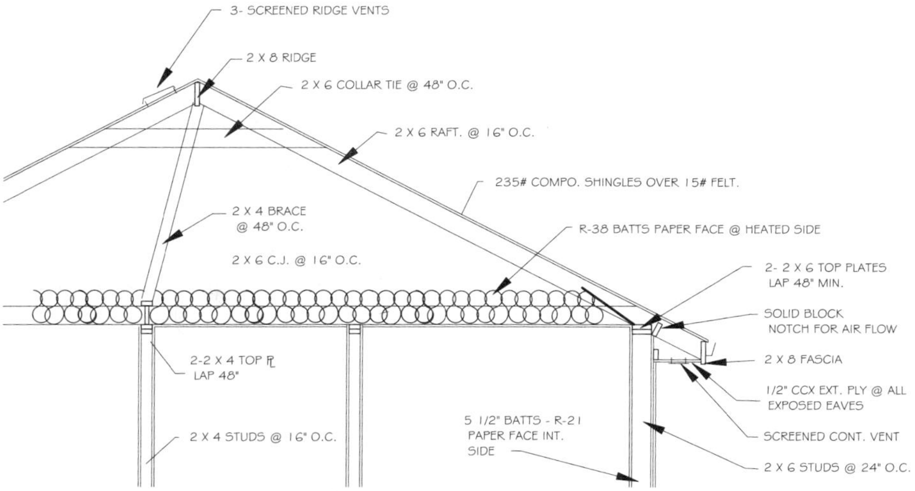

Figure 11.24 *The ceiling insulation is drawn using the Fit option of* **Pedit**.

Spline

Unlike the Fit curve that passes through all the vertices, a Spline (S) curve only passes through the first and last points of the polyline. In between, the curve will be close to each vertex, but it will not pass through them. The more control points (vertices) that you specify, the smoother the Spline curve will be. Figure 11.26 shows a Spline curve. The command process is as follows:

> Command: **PE** (ENTER) *(Or click the Edit Polyline button.)*
>
> PEDIT
>
> Select polyline:
>
> Enter an option [Close/Join/Width/Edit
>   vertex/Fit/Spline/Decurve/Ltype gen/Undo]: **S** (ENTER)

As (ENTER) is pressed, the Spline option transforms the points of the polyline into smoothed curves. The Fit and Spline options are not often used on construction drawings. For further information, remember to use the **Help** menu in AutoCAD. You may also want to research the commands **Splines**, **Splinetype**, **Splinesegs**, **Splinedit**, and **Nurbs** for related types of drawing objects.

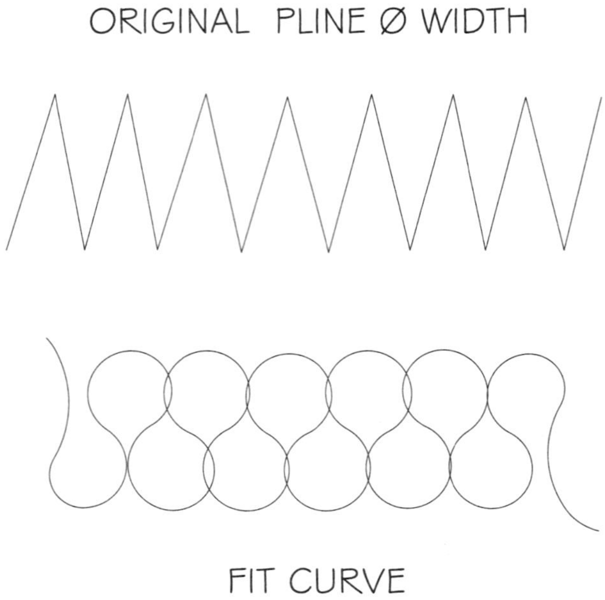

Figure 11.25 *Changing a polyline using the Fit option.*

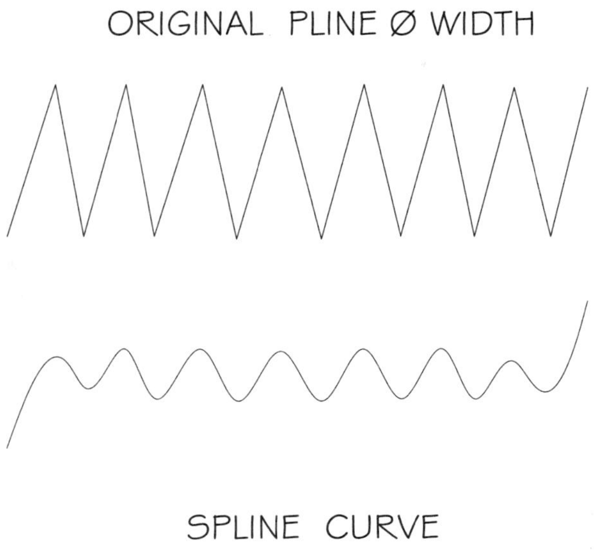

Figure 11.26 *Altering a polyline using the Spline option.*

Decurve

The Decurve (D) option can be used to remove any vertices that have been inserted by the Fit or Spline option and to straighten segments of a polyline. The effects can be seen in Figure 11.27. The command sequence is as follows:

Command: **PE** (ENTER) *(Or click the Edit Polyline button.)*

PEDIT

Select polyline:

Enter an option [Close/Join/Width/Edit vertex/Fit/Spline/Decurve/Ltype gen/Undo]: **D** (ENTER)

Ltype Gen

Figure 11.28 shows a polyline drawn with a centerline linetype. Notice that the pattern extends from vertex to vertex. Each segment is, in effect, a separate centerline. The polyline also can be drawn so that the pattern extends throughout the entire polyline, as seen in Figure 11.29. The Ltype gen (L) option of the Pedit command will adjust how the pattern is displayed. Toggled to the ON position, Ltype gen will draw line patterns in a continuous pattern from beginning to end with no consideration for vertices. In the OFF position, line patterns will be based on each vertex. Ltype gen does not affect polylines with tapered segments.

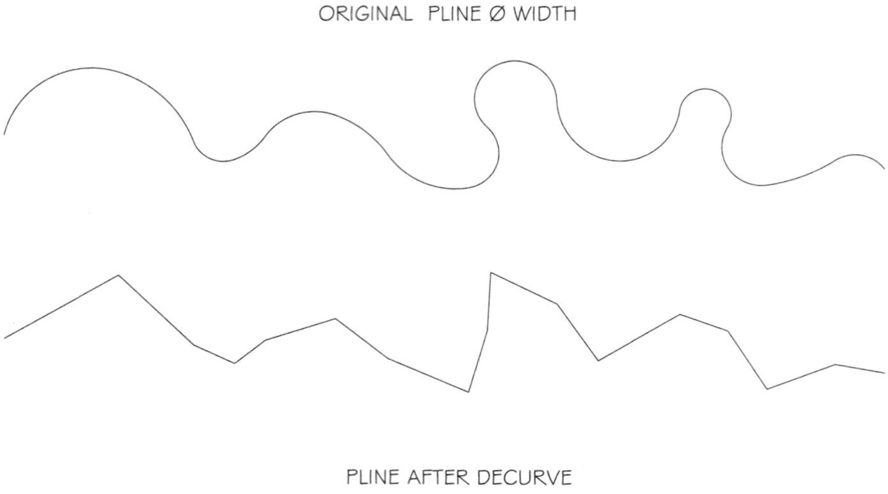

Figure 11.27 *Decurve will remove any vertices that were inserted with the Fit or Spline option.*

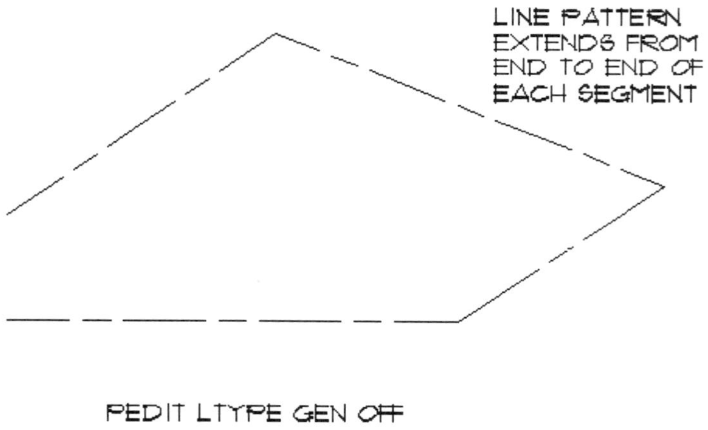

Figure 11.28 *Using the OFF setting of the Ltype gen option of* **Pedit** *will draw a linetype pattern from vertex to vertex.*

Undo

The Undo (U) option of Pedit will undo the most recent Pedit operation. Multiple entries of U will step back through the drawing, removing previous Pedit entries.

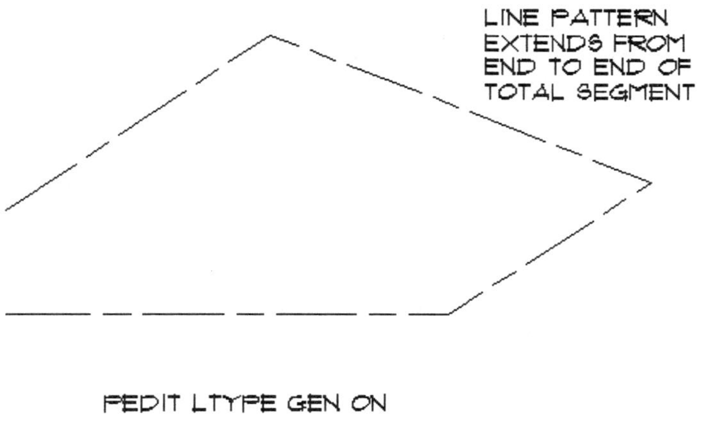

Figure 11.29 *Using the ON setting of the Ltype gen option of* **Pedit** *will draw a linetype pattern from end to end of the entire line, disregarding the vertex.*

Exit

The eXit (X) option is the default for Pedit. Pressing (ENTER) will also end the Pedit command and return to the Command prompt.

CHAPTER 11 EXERCISES

1. Start a new drawing and draw a polyline polygon consisting of five sides with a 0 width. Set the width of the polyline as 1/16" and draw a polygon with a minimum of four sides. Use the Close option to close the polygon. Change the width to .125" and draw three open polylines. Save the drawing as E-11-1.
2. Start a new drawing and draw a polyline with a beginning width of 0.125", a length of 1.5", and an ending width of 0.25". Continue from this segment with a line that is 1.5" long and 0.125" wide. Draw a third segment that will close a polygon with a starting width of 0.25" and ending with a 0 width. Save the drawing as E-11-2.
3. Start a new drawing and draw a circle with a 3" diameter and a thickness of 1/16". Use the three-point Arc option to draw a polyline circle with a width of 1/8" and a radius of 1.5". Save the drawing as E-11-3.
4. Start a new drawing and use a width of 0 to draw a 3" long line. Draw a 0.125" wide arc with a radius of 1.75" and an angle of 60° on the right end of the original line segment. Use the right end of the straight line segment as the center of a 45° arc that ends at the left end of the arc segment. Save the drawing as E-11-4.
5. Start a new drawing and draw a four-sided polygon using **Pline** with a width of 0.125". Copy the polygon so that there are a total of five polygons. Edit one of the polygons so that the line width is 0.25". Edit another polygon so that the width is 0.065". Fillet two corners with a 0.5" radius. Fillet the remaining two corners with a 0.25" radius. Edit a third polygon so that the width of all lines is 0.0 wide. Provide a 0.4" x 0.25" chamfer at each corner. On the fourth polygon, add a vertex at some point so that the polygon has five segments. Save the drawing as E-11-5.
6. Open Drawing E-10-9 and convert the lines that represent the footing to polylines. Set the width as 0.5". Set the line that represents the finish grade as 0.75". Copy the one-story footing and explode the drawing. Save the drawing as E-11-6.
7. Open Drawing E-10-8. Assume the light source to be in the upper left corner to shade the drawing. Use 0.5 width polylines to create shade on all overhanging materials. On the left side of the horizontal trim, draw a shadow that tapers from 0 up to 1" wide. Save the drawing as E-11-7.
8. Start a new drawing and draw a series of seven zigzag polyline 0.0 width lines at approximately 15° from vertical. Make the line segments 10" long. Make two additional copies. Use the Spline option to edit one set of lines and the Fit option to edit the other set. Save the drawing as E-11-8.
9. Open Drawing E-10-6 and adjust the limits as required to draw a typical wall section that is similar to the drawing on page 465. Show:

- 2x6 sill with a 1/2" x 10" anchor bolt.
- 2x8 floor joist w/ 2x8 rim joist.
- 3/4" plywood sub-floor
- 8' high studs with a double top plate and a single base plate
- 2x6 ceiling joist
- 2x6 rafters @ a 27 1/2° (5/12) pitch
- 2x8 fascia and solid eave blocking

Use 0.25" wide polylines to represent any materials that would be cut by the cutting plane, such as the plate, sill, fascia, and blocking. Save the drawing as E-11-9.

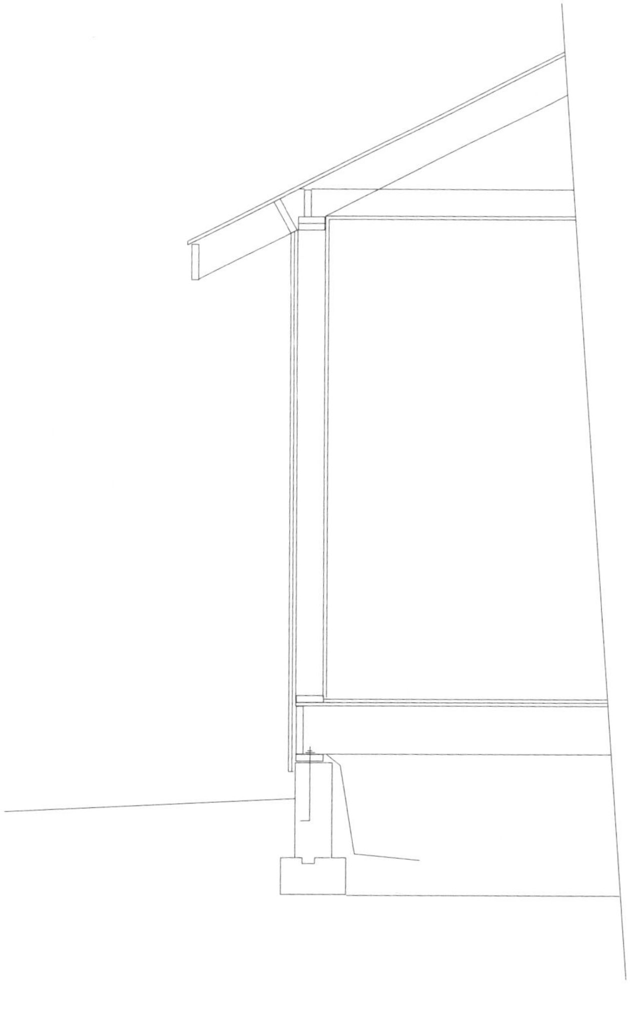

10. Start a new drawing and set the units and limits to draw the grading plan shown below. Lay out the grids and approximate the contour lines. Use a 6" wide polyline to represent 5' intervals and a 2" wide polyline to represent 1' intervals. Once all straight line segments have been drawn, edit the plines to provide the most accurate, smooth transitions possible. Save the drawing as E-11-10.

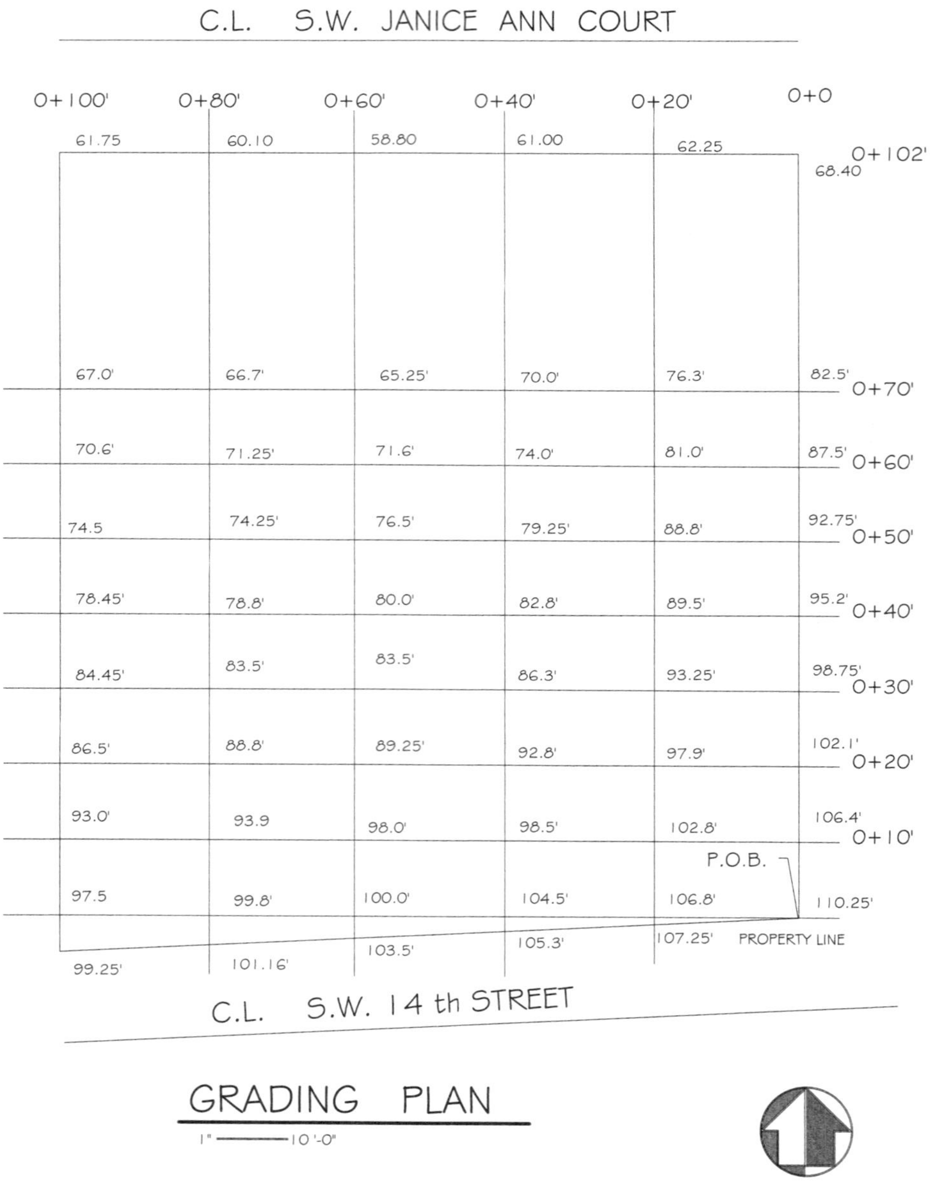

CHAPTER 11 QUIZ

1. Use the **Help** menu and determine what **Plinegen** controls, and what options are included.
2. Give the command sequence and show all options to draw a polyline with a width of 0.125".
3. List two options to control line width other than **Pline**.
4. Major drawings for a residence include the site, floor, and foundation plans, elevations, and sections. Find a set of professional drawings and list common ways polylines can be used to enhance these drawings.
5. What option would allow a 0.25" wide line to be drawn by entering **.125**?
6. Explain the difference between the Spline and Fit options.
7. Explain the difference between Open and Close **Pedit** options.
8. How does **Explode** affect a polyline?
9. To edit a vertex that is formed between the third and fourth lines drawn, what two options will be needed?
10. What option will allow an additional vertex to be added to a completed polyline?
11. A polyline has been drawn, but one vertex is 1/2" to the left of where it belongs. What option will fix the problem?
12. What **Pedit** option will remove any vertices that have been inserted by the Fit or Spline option?
13. What **Pedit** option will remove and convert arcs to straight line segments?
14. What option will display an X at the start of a polyline segment, and what three options affect the X?
15. What is the default setting for **Pedit**?
16. How does the Halfwidth option differ from the Width options?
17. A friend needs to draw a 2" long polyline parallel to an existing pline. How can this be done?
18. How can a circle be drawn using the **Pline** command?
19. How can a polyline arc be drawn in a clockwise direction?
20. How can the direction of an arc chord be controlled?

CHAPTER 12

Supplemental Drawing Commands

This chapter will introduce methods for marking and dividing drawing space and explore five new methods for drawing lines.

Commands to be introduced in this chapter include

- **Point**
- **Divide**
- **Measure**
- **Ray**
- **Xline**
- **Sketch**
- **U**
- **Undo**
- **Redo**

MARKING AND DIVIDING SPACE

Three commands are available in AutoCAD for marking and dividing space. Although these commands can't be used to draw lines, each can be used to mark space in a drawing and provide a location for placing lines.

PLACING POINTS

Often, a point needs to be marked on a drawing. For example, the loads on a column on the upper level of a structure will need to be carried down through several floors into the foundation. AutoCAD will allow you to mark these load locations on a drawing by drawing a point. To place a point in a specific location, select the **Point** button on the **Draw** toolbar, type **PO** (ENTER) at the Command prompt, or select **Point** from the **Draw** menu. If the command is selected from the **Draw** menu, the command will allow for multiple points to be located. Each method will produce a prompt:

> Command: **PO** (ENTER) *(Or click the Point button.)*
>
> POINT
>
> Current point modes PDMODE=0 PDSIZE=0'–0"
>
> Specify a point: *(Select a point location or enter coordinates.)*

The prompt is now waiting for coordinates to be specified. Points are usually specified with the select button.

Controlling the Point Display

The shape that is used to identify a point can be altered through the **Point Style** dialog box, shown in Figure 12.1. Display the dialog box by selecting **Point Style** from the **Format** menu or by typing **DDPTYPE** (ENTER) at the Command prompt. The dialog box allows the style and size of the point to be controlled. The active setting is shown in a black box. In this case, a point is used for the marker. The marker will be displayed on the screen but will not be plotted. Notice the second box from the left in the top row contains no marker. If this option is selected, the point will be marked with an invisible marker. Although the point can't be seen, it will be selected when the Node mode of Object Snap is active.

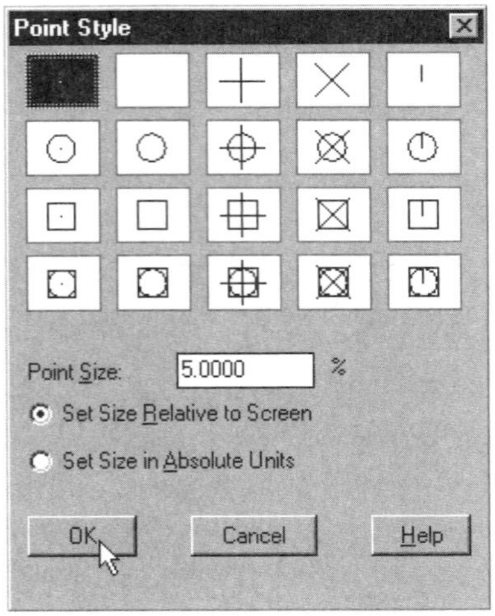

Figure 12.1 *Selecting* **Point Style** *from the* **Format** *menu or typing* **DDPTYPE** *(ENTER) at the Command prompt will produce the* **Point Style** *dialog box. The box can be used to alter the size and shape of the point marker.*

The current point size is listed as 5% of the display screen size. Point size can be altered to meet the needs of the drawing. Just as important as the size is the display method. In the default setting, the size of the point will be displayed relative to the size of the display area. As you zoom in and out of a drawing, points will remain a set percentage of the drawing display size. If you select the display to be in absolute units, points will disappear as you zoom out and become huge as you zoom in. Entering a positive value for the point size will display the point relative to the drawing area. Use the following procedure to display a point that is marked by crosshairs surrounded by a circle and a square, and is displayed at 10% of the screen size:

1. Select **Point Style** from the **Format** menu.
2. Select a marker from the **Point Style** dialog box.
3. Change the display in the **Point Size** edit box to read 10%.
4. Click the OK button.

Note: You'll notice as you work through this chapter that several of the buttons noted by commands are not found on the standard toolbars. You can find these command buttons by selecting **Customize** from the **Toolbars** menu. You can also right-click when the cursor is in any open toolbar, and then select **Customize**. Select the **Categories** edit box in the **Customize Toolbars** dialog box to display a listing of command categories. To display the **Point** button, select Draw from the **Categories** edit box. Review Chapter 1 and the **Help** menu to place buttons on a toolbar.

EXPLORING THE DIVIDE COMMAND

Although this command does not edit an object, **Divide** will allow an object to be divided into any number of segments of equal length for easy editing. This is done by placing markers along a line. This can be especially helpful when you are dividing space between two floors to locate where each stair tread and riser will be placed, or when locating stress points along a beam. **Divide** will enable you to select points on a line, arc, circle, or polyline.

Before using the **Divide** command to mark an object, you must decide how each mark will be indicated. The default marker for a point is a single point. The point style must be altered so that it can be seen on the line. The crosshairs, the X, and the crosshairs and circle, selected in the **Point Style** dialog box, make excellent markers (see Figure 12.1).

Once the marker is set, start the command sequence for **Divide** by typing **DIV** (ENTER) at the Command prompt, or by selecting **Divide** from the **Point** cascading menu on the **Draw** menu. The command sequence is as follows:

Command: **DIV** *(Or click the Divide button.)*

DIVIDE

Select object to divide: *(Select the objects.)*

This is your opportunity to select a single object to divide. If you select an object that cannot be divided, an error message will be displayed and the command will be exited. Once an object is selected, another prompt will be displayed.

> Enter the number of segments or [Block]: *(Enter number.)* **14** (ENTER)

Typing a number between 2 and 32,767 and pressing (ENTER) will cause the object to be divided into the number of segments specified. The object is not physically divided into separate segments—the markers are just placed so that exact points can be selected. The process can be seen in Figure 12.2.

Another useful feature of the **Divide** command is its ability to insert a group of objects called a block at repeated distances. Blocks will be covered in a later chapter, but the idea of inserting them is not difficult to remember. You can insert blocks by dividing a line into the desired number of segments and then inserting the block at each marker. This can be done if the Block option is selected. Figure 12.3 shows the effect of aligning blocks with the object. The command sequence to divide a circle and place a block named SLOT is as follows:

> Enter the number of segments or [Block]: **B** (ENTER)
>
> Enter name of block insert: *(Type name of desired block.)* **SLOT** (ENTER)
>
> Align block with object? [Yes/No]<Y>: (ENTER)
>
> Number of segments: *(Enter number.)* **14** (ENTER)
>
> Command:

EXPLORING THE MEASURE COMMAND

The **Measure** command is similar to the **Divide** command. **Divide** places a specified number of markers at equal distances along a segment. **Measure** places markers at specific distances along a line, arc, circle, or polyline. This command can be used whenever objects occur at a repetitive spacing, such as when you are placing studs at 16" o.c. along a plate or placing a manhole at every 527.25 feet along a sewer line. Figure 12.4 shows an example of how **Measure** could be used as a drawing aid. Start the command by typing **ME** (ENTER) at the Command prompt, or by selecting **Measure** from the **Point** cascading menu of the **Draw** menu. Be sure to change the point marker to something that will be visible when placed on a line. The command sequence to measure an object is as follows:

> Command: **ME** (ENTER) *(Or click the Measure button.)*
>
> MEASURE
>
> Select object to measure: *(Select object.)*

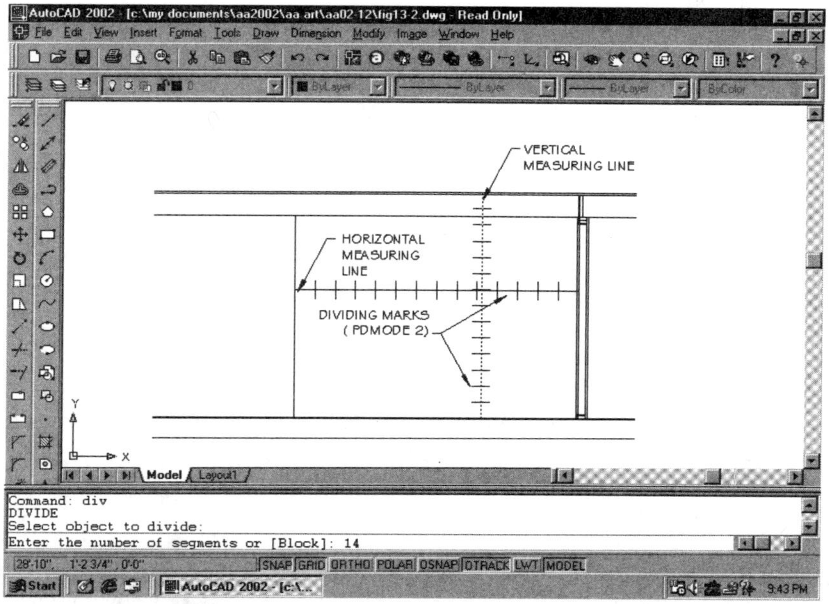

Figure 12.2 *Dividing a horizontal line into fourteen segments.*

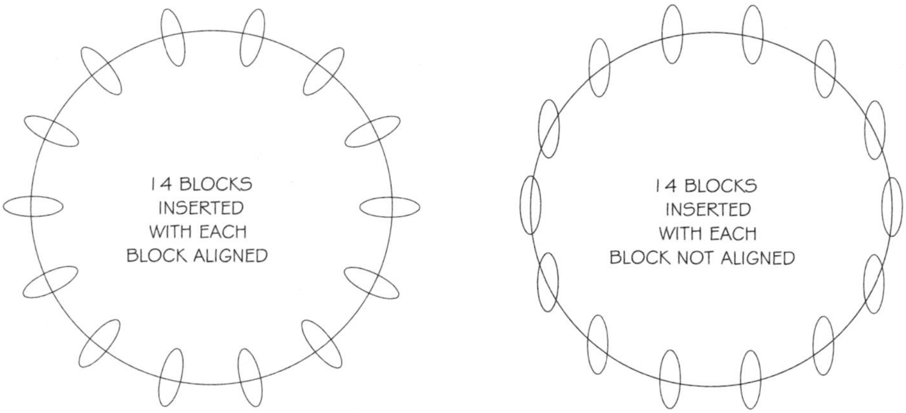

Figure 12.3 *Predetermined drawing components can be inserted in a drawing at each specified* **Divide** *marker using the Block option of* **Divide***.*

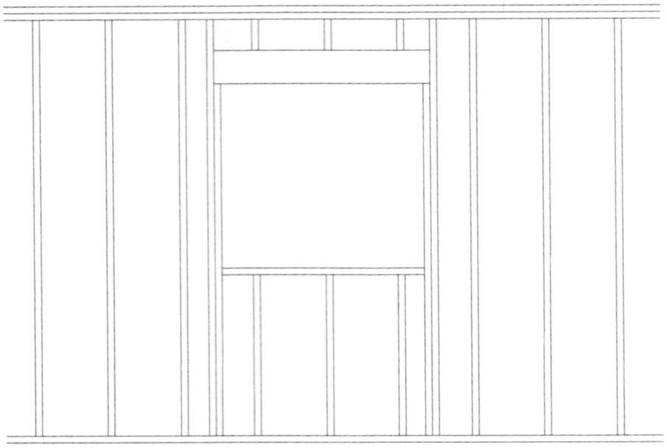

Figure 12.4 *The* **Measure** *command can be used to place markers at a specific distance on a line, arc, circle, or polyline, so that objects can be drawn.*

This is your opportunity to select a single object to measure. If you select an object that cannot be measured, an error message will be displayed and the command will be exited. Once an object is selected, another prompt will be displayed:

Specify length of segment or [Block]: *(Enter distance.)* **16** (ENTER)

Command:

You can enter any specific distance by keyboard or specify two points with the select button. The markers will be placed, starting at the selected end of the line of the object to be measured. This process can be seen in Figure 12.5.

To use the select button to enter the distances, move the cursor to the desired location and click the select button. This will produce the following command sequence:

Command: **ME** (ENTER) *(Or click the Measure button.)*

MEASURE

Select object to measure: *(Select object.)*

Specify length of segment or [Block]: *(Select a location.)*

Second point: *(Enter a second point or coordinates.)* (ENTER)

Command:

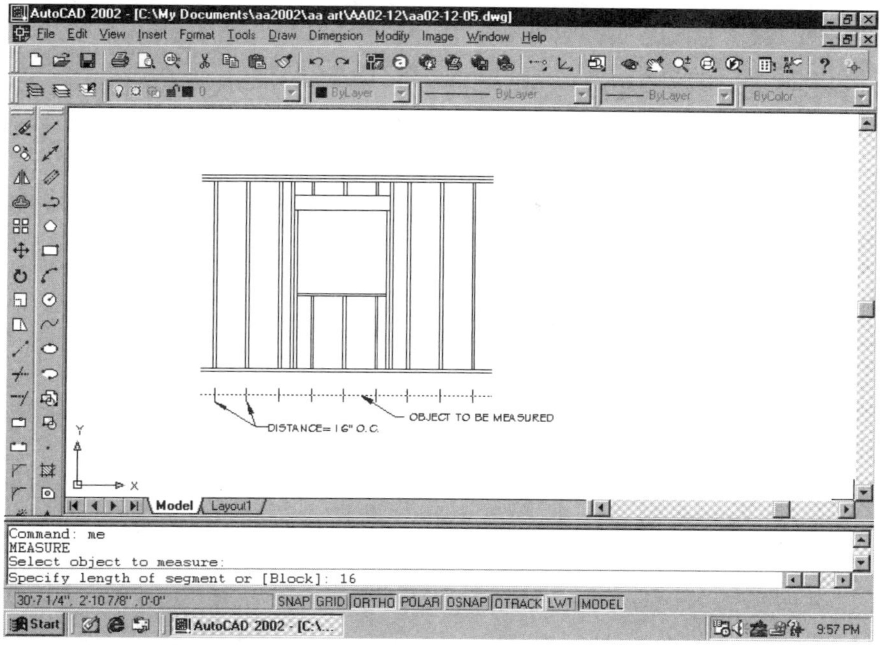

Figure 12.5 *Once a segment has been selected for measurement, the desired distance to be measured can be entered by keyboard or by selecting the distance with the select button.*

DRAWING LINES OF VARIED WIDTH AND LENGTH

Each command introduced in this section can be used as an alternative to the **Line** command. The **Ray** and **Xline** commands allow semi-infinite and infinite lines to be added to a drawing. The **Sketch** command allows freehand sketching to be inserted in a drawing.

CREATING LINES USING THE RAY COMMAND

If you've done manual drafting, you've used light lines to plan your drawing prior to placing dark lines in graphite or ink. Those light lines are called projection lines and are meant to be seen only by the drafter. AutoCAD allows a similar line to be drawn using the **Ray** command. A ray is a semi-infinite line with a defined starting point that extends into infinity without changing the drawing area. Sounds technical, but they are really very helpful.

Rays can be especially helpful in projecting features from one drawing to another. This often occurs when lines are projected from a floor, roof, or grading plan to develop exterior elevations. If you were to use a line to project from a floor plan to develop an elevation, it must have two endpoints. A zoom might be required to go from the "first" point to the "next" point. A transparent zoom (**Z**) could be used, but this is time consuming when compared to using a ray. A ray will provide a guideline that extends from the selected point indefinitely, no matter how many **Zoom** previous operations are required.

Start the **Ray** command by choosing **Ray** from the **Draw** menu, or by typing **RAY** (ENTER) at the Command prompt. The command sequence to draw a ray is as follows:

Command: **RAY** (ENTER) *(Or click the Ray button.)*

RAY

Specify start point: *(Select desired starting point of ray.)*

Specify through point: *(Select desired direction of ray.)*

Specify through point: (ENTER)

Command:

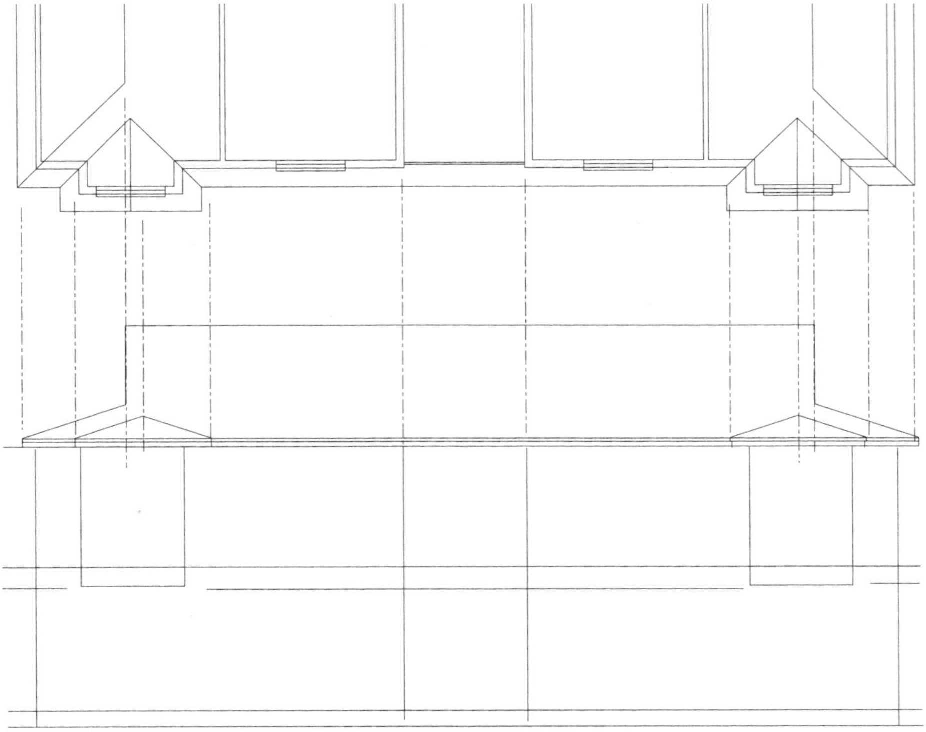

Figure 12.6 *Rays and xlines make excellent projection lines. Because of their length,* **Zoom** *and* **Stretch** *operations can be minimized.*

Figure 12.6 shows the layout of an elevation using the **Ray** command. A ray will show when the drawing is plotted. Rays should be kept on a layer separate from the drawing being created, so they can be easily removed from the drawing base by freezing the layer. They can also be erased easily using the Crossing window or Fence option of **Erase**.

CREATING LINES USING THE XLINE COMMAND

A line created using the **Xline** command is similar to a ray, except that it is infinite in length. With **Ray**, you selected a starting point, and the line is extended from that point. With **Xline**, the line extends through the point an infinite distance in both directions and offers more options than with a ray. Access the command by selecting the **Construction Line** button on the **Draw** toolbar, by typing **XL** (ENTER) at the Command prompt, or by selecting **Construction Line** from the **Draw** menu. Unlike the **Ray** command, **Xline** offers six options to place the line. The default setting is to select a point for the line to pass through.

Through Point

By selecting a through point, you can display the line through the point. Once the point is selected, the line will be rotated to the desired angle through the selected point. Once a line location is selected, the command sequence will allow for additional lines to be drawn through the point. The command sequence is as follows:

Command: **XL** (ENTER) *(Or click the Construction Line button.)*

XLINE

Specify a point or [Hor/Ver/Ang/Bisect/Offset]: *(Select desired option.)*

Specify through point: *(Select a direction for the line to extend.)*

Specify through point: (ENTER)

Command:

Hor Option

Choosing the Hor option will allow a horizontal line to be drawn through a specified point. The line will remain horizontal, regardless of the ORTHO setting. The command sequence is as follows:

Command: **XL** (ENTER) *(Or click the Construction Line button.)*

XLINE

Specify a point or [Hor/Ver/Ang/Bisect/Offset]: **H** (ENTER)

Specify through point: *(Select the point for the line to pass through.)*

Specify through point: (ENTER)

Command:

Ver Option

The Ver option will pass a vertical line through the selected point. The command functions in the same way as the horizontal option, once the Ver option is selected. The command sequence is as follows:

Command: **XL** (ENTER) *(Or click the Construction Line button.)*

XLINE

Specify a point or [Hor/Ver/Ang/Bisect/Offset]: **V** (ENTER)

Specify through point: *(Select the point for the line to pass through.)*

Specify through point: (ENTER)

Command:

Ang Option

The Ang option allows a specific angle for the line to be specified. The selected angle is relative to horizontal. The angle option can be useful in laying out roofs or other inclined surfaces. The command sequence is as follows:

Command: **XL** (ENTER) *(Or click the Construction Line button.)*

XLINE

Specify a point or [Hor/Ver/Ang/Bisect/Offset]: **A** (ENTER)

Enter angle of xline (0) or [Reference]: **30** (ENTER)

Specify through point: *(Select the point for the line to pass through.)*

Specify through point: (ENTER)

Command:

If the Reference option of the command is selected, a line can be drawn at an angle to any line. This option could be used to draw a line 30° from a roof drawn at a 6/12 pitch—even if you can't remember what angle 6/12 represents and you don't feel like using the **List** command. Once the reference angle is selected, a location for the line to pass through needs to be indicated. The command sequence is as follows:

Command: **XL** (ENTER) *(Or click the Construction Line button.)*

XLINE

Specify a point or [Hor/Ver/Ang/Bisect/Offset]: **A** (ENTER)

Specify angle of xline (0) or [Reference]: **R** (ENTER)

Select a line object: *(Select the line to be used as the reference.)*

Enter angle of xline <0>: **30** (ENTER)

Specify through point: *(Select the point for the line to pass through.)*

Specify through point: (ENTER)

Command:

Bisect Option

The Bisect option can be used to draw a construction line that bisects an existing angle. The line is created by selecting three points. The first point to be selected is the Angle vertex point. The second and third points are located on the lines forming the angle to be bisected. Figure 12.7 shows a construction line used to bisect an angle. The command sequence to bisect an angle is as follows:

Command: **XL** (ENTER) *(Or click the Construction Line button.)*

XLINE

Specify a point or [Hor/Ver/Ang/Bisect/Offset]: **B** (ENTER)

Specify angle vertex point: (ENTER) *(Select the point and press (ENTER).)*

Specify angle start point: *(Select the end of one of the lines to be bisected.)*

Specify angle end point: *(Select the end of the other line to be bisected.)*

Specify angle end point: (ENTER)

Command:

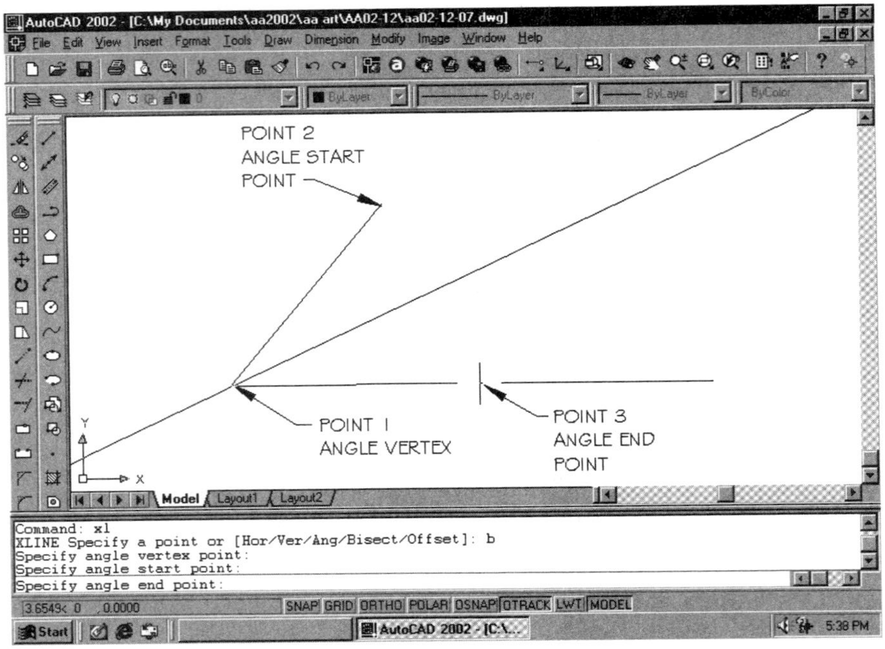

Figure 12.7 *The Bisect option of* **Xline** *can be used to divide an angle in half, by selecting three points.*

Offset Option

The Offset option draws construction lines that are parallel to the selected line, based on a specified offset distance or point. This option functions like the **Offset** command. You'll be prompted for a line, a distance, and a side to offset. Select an offset distance, and then you will be prompted for a line to be offset and for a through point. The command sequence to offset a line 12" is as follows:

Command: **XL** (ENTER) *(Or click the Construction Line button.)*

XLINE

Specify a point or [Hor/Ver/Ang/Bisect/Offset]: **O** (ENTER)

Specify offset distance or [Through] <Through>: **12** (ENTER)

Select a line object: *(Select the line to be offset.)*

Specify side to offset: *(Select the side of the offset.)*

Select a line object: (ENTER)

Command:

CREATING LINES WITH THE SKETCH COMMAND

Using the **Sketch** command allows the equivalent of freehand drawings to be generated using a computer. Line segments are entered as the cursor is moved, rather than providing a "next" point. The command is useful for entering signatures or other irregular material on a drawing. Because each tiny segment of the sketch is recorded as a line segment, drawing file size can become extremely large in a very short time. The use of **Sketch** should be kept to a minimum to avoid filling up drawing space on a disk. The size of the drawing is limited only by your storage media.

Access the **Sketch** command by selecting the **Sketch** button or typing **SKETCH** (ENTER) at the Command prompt. Best results usually occur when ORTHO and SNAP are toggled to OFF, allowing greatest flexibility in line placement. With ORTHO mode ON, the **Sketch** command will draw only horizontal and vertical lines. With the SNAP mode on, the record increment (segment length) will equal the snap value.

Segment Length

As the command is entered, you will be prompted to provide information on the *record increment*, which is the segment length to be used while you are sketching. The increment should be set small enough to provide the needed resolution. Exceeding the resolution will only result in wasted disk space by storing lines that aren't necessary. The default line length is <0'–0 1/8">. A new value can be entered by keyboard.

The increment that is chosen should be determined by the size of the reproduced drawing. If the drawing is to be reproduced at full scale on an 8 1/2" x 11" sheet of paper, 1/8" will not produce high-quality resolution. As the mouse is moved, a line segment will be recorded for each movement greater than 1/10 of an inch. In a 10" wide space, 100 segments could be created. By changing the length to .5", a 10" long line could be generated with 20 segments in the same space, which would greatly decrease the resolution. However, if the drawing is going to be reproduced at a scale of 1/4"=1'–0", a line segment of .5" would give extremely high resolution. A 1" long line segment might be more appropriate. Figure 12.8 compares four record increments. You can specify the width by providing a value by keyboard or by moving the cursor. To enter a value using the keyboard, type the value followed by (ENTER). The command sequence is as follows:

> Command: **SKETCH** (ENTER) *(Or click the Sketch button.)*
>
> Record increment <0'-01/8">: .065 (ENTER)
>
> Sketch. Pen eXit Quit Record Erase Connect.

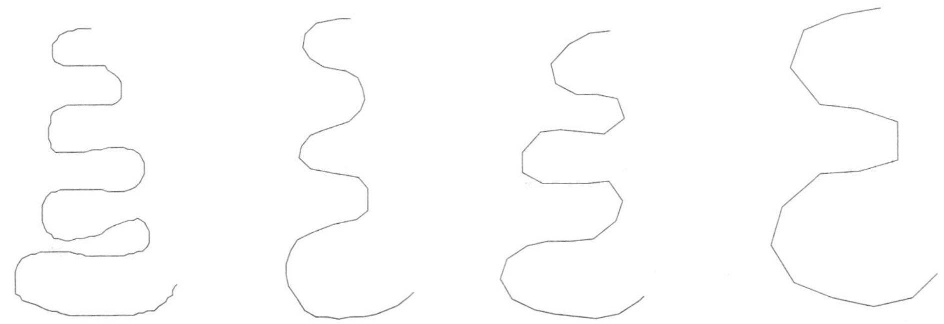

Figure 12.8 *The smaller the value for the segment length, the smoother the resulting curves will be. A scale should be selected that provides the desired accuracy without requiring unneeded lines.*

AutoCAD is now ready to sketch. To set the line increment using the cursor, click the select button when the prompt for the increment is provided. The sequence is as follows:

> Command: *(Select the Sketch button or type)* **SKETCH** (ENTER)
>
> Record increment <0'-01/8">: *(Click the select button.)*
>
> Second point: *(Move the cursor the desired increment distance and click the select button.)*
>
> Sketch. Pen eXit Quit Record Erase Connect.

For a drawing requiring great accuracy, enter the line increment value using the keyboard. Once the width is entered, several options will be listed in the following prompt:

> Sketch. Pen eXit Quit Record Erase Connect.

Once the segment length is set, move the cursor into the drawing area. You'll notice that nothing happens as the cursor is moved throughout the drawing area.

Pen (P)

Think of the **Sketch** command as sketching on paper with a pencil. The **Sketch** command treats the cursor as a "pen" and allows for lifting the pen and moving it from one location to another without drawing a line. As **Sketch** is started, the drawing pen is

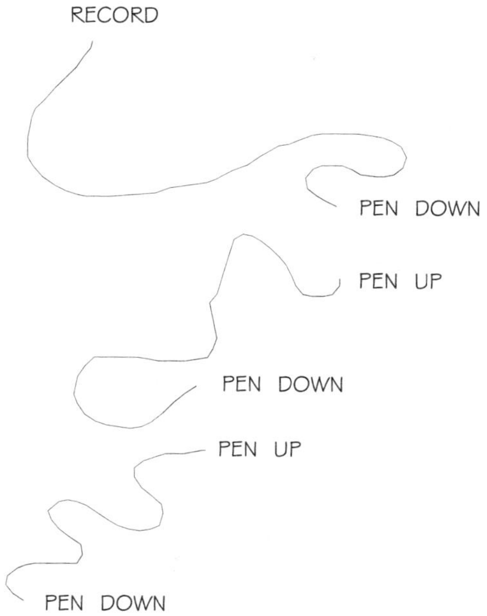

Figure 12.9 *Controlling the pen position of UP/DOWN allows breaks to be created between segments.*

in the "up" position, allowing for selection of a starting location. Once the start location is selected, the pen can be put down on the paper to allow for sketching. Click the select button or press P on the keyboard to raise and lower the pen. If you use the keyboard to control the pen, don't press (ENTER) after pressing P. Most users skip the keyboard once they've entered the command, and use the select button to control the pen.

Each time the select button is clicked, the pen is raised or lowered to allow for movement and drawing. The process can be seen in Figure 12.9.

Period (.) Line to Point

The period (.) option can be used to draw a straight line from the endpoint of the last sketched line to the current location of the cursor. Sketch a few segments and raise the pen. Move the cursor to a new location and press the period (.) key. This will produce a straight line from the end of the sketch segment to the new cursor location. The effect can be seen in Figure 12.10.

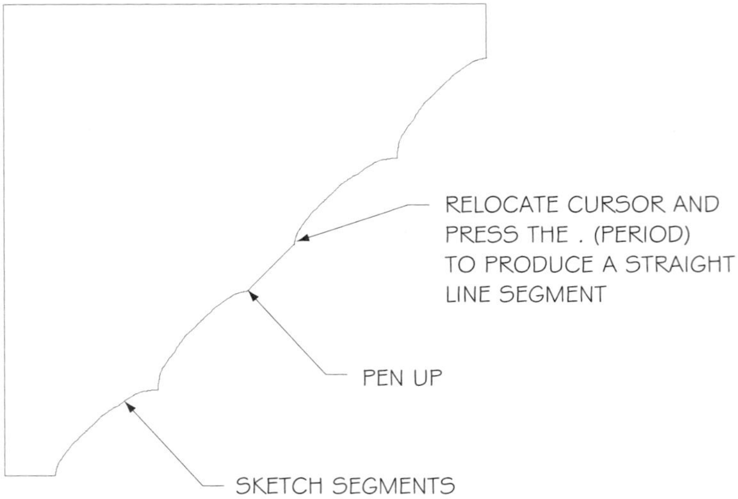

Figure 12.10 *The period (.) option of the* **Sketch** *command allows a straight line to be drawn between the end of the last sketch segment and the present cursor location.*

Record (R)

When sketches are created, they are not actually part of the drawing until they are recorded. Enter the drawing area and sketch a few lines. You'll notice that the lines are green (color will vary depending on configuration of AutoCAD), even though you've done nothing to change the drawing color. The Record (R) option records all of the temporary sketch lines as a part of the drawing base and converts them to the present drawing layer and color. Once recorded, sketch lines cannot be edited using the sketch options, although they can be edited using any of the edit commands. Pressing R does not affect the position of the pen, and you can continue sketching after you've used it. Pressing (ENTER) will record the sketch segments and exit to the Command prompt.

Record and Exit (X)

Using the X will record the sketch segments to the drawing base, return the Command prompt to the screen, and re-establish the Command prompt. Using this option will display the number of lines recorded just above the command line.

Quit (Q)

Pressing Q on the keyboard will function similarly to the **Erase** command. Q will remove all temporary sketch segments entered since the last Record and return you to the Command prompt. This option can be used if you sketch an object and then decide that you don't need it.

Erase (E)

This option allows for part of the sketch segments to be removed prior to recording. Start this option by pressing E; it will remove a portion of the sketch from the selected point to the end of the sketch segments. Once the option is started, a message is displayed:

Erase: Select end of delete. < Pen up >

The pen will be placed in the up position to allow you to select the end of a segment. Move the crosshairs to the portion of the line to be removed. The line will be displayed with the indicated portion of the line removed. If the removed section is acceptable, press P to erase the segment. If the selection is unacceptable, press E. The message "Erase aborted" will be displayed and the **Sketch** prompt will be returned.

Connect (C)

Typing **C** (ENTER) is similar to the Endpoint option of Object Snap when drawing lines, and will allow sketching to continue from the end of the last drawn sketch segment. When the option is selected, the following prompt will be displayed and the crosshairs will be displayed:

Connect: Move to endpoint of line

When the crosshairs are within one record increment of the end point, the pen will be placed in the down position at the end of the line. If the pen is already down when the Connect option is used, the following prompt will be displayed:

Connect Aborted

To terminate the Connect command, type **C** (ENTER) a second time.

THE U COMMAND

You've already used the Undo option to remove a drawing object. The U option has been used in the middle of a command routine to remove an object that might not be what you intended. The **U** command also can be used to backtrack through a command sequence once the sequence has been ended. Use the command carefully. Once it is used, objects are gone and will have to be redrawn if you want them back. Start a new drawing, draw a series of several lines, and press (ENTER). Draw a circle, a polygon, and an ellipse in the order listed. Now if you type **U** (ENTER) at the Command prompt, you will see

Command: **U** (ENTER)

ELLIPSE GROUP

The last item drawn, the ellipse, will be removed from the screen. Notice that the most recent command (ellipse) will be listed and the Command prompt will reappear. If the process is repeated, the polygon will be removed.

Command: **U** (ENTER)

POLYGON GROUP

By continuing to enter **U**, you can remove the entire drawing.

Command: **U** (ENTER)

CIRCLE

Command: **U** (ENTER)

LINE

UNDO

The **Undo** command allows several command sequences to be undone at one time and permits several operations to be carried out as the objects are undone. The **Undo** command should give you the freedom to know you can try anything, and if the outcome is undesirable, enter **Undo** at the Command prompt to restore the drawing to its original state. The command sequence is as follows:

Command: **UNDO** (ENTER)

Enter the number of operations to undo or
[Auto/Control/BEgin/Mark/Back] <1>:

If a number is typed, such as **1** (ENTER), the last command will be undone. If you type **3** (ENTER), the last three command sequences will be removed.

AUTO

The Auto option issues a prompt to toggle between ON and OFF. When Auto is ON, the default value, any group of commands that is used to insert an item, or group of items, is treated as one item and removed by **U**. This will be helpful later when you work with blocks, wblocks, and macros. If Auto is OFF, each command in a group of commands is treated as an individual one.

CONTROL

The Control option is another method to limit the **Undo** command; it also can be used to completely disable **Undo**. The command sequence is as follows:

Command: **UNDO** (ENTER)

Enter the number of operations to undo or
[Auto/Control/BEgin/Mark/Back] <1>: **C** (ENTER)

Enter an UNDO control option [All/None/One] <All>:

If All is selected, **Undo** will not be limited, and it will erase back to the last Mark. The None option will disable **U** and **Undo** commands so that your machine will function as if these commands have never been placed in AutoCAD. You can reactivate **Undo** and **U** by entering **Undo** and typing **A** (ENTER). Selecting the One option limits **U** and **Undo** to a single operation before returning to the command line.

BEGIN AND END

The Begin option of **Undo** groups a sequence of drawing operations. The group of drawing commands is defined by the use of the Begin option of **Undo**. The end of the group is marked by the End option. Within a specified group, AutoCAD will treat grouped commands as a single operation.

BACK

The Back option of **Undo** will remove everything in the entire drawing. Not some, not most, we're talking *everything*. Do you get the feeling you need to use this option carefully? AutoCAD will even help you ponder the use of this option by displaying the prompt:

This will undo everything. OK? <Y>

If (ENTER) is pressed, this will accept the default Yes value, and really clean up your drawing. Typing **N** (ENTER) will ignore the Back option. If you do use the Back option and then change your mind, use the **Redo** command to restore your drawing.

The Back option also can be used with the Mark option so that the entire drawing is not erased.

MARK

This option will limit the distance that Back will search through a drawing as it erases work. A mark can be placed in a drawing prior to executing several commands. If the outcome of those commands is not what you expected, **Undo Back** will remove all of the commands until the mark is reached. The command sequence is as follows:

Command: **UNDO** (ENTER)

Enter the number of operations to undo or
[Auto/Control/BEgin/Mark/Back] <1>: **M** (ENTER)

Command: **L** (ENTER) *(Now draw several features in an existing drawing.)*

Command: **UNDO** (ENTER)

Enter the number of operations to undo or
[Auto/Control/BEgin/Mark/Back] <1>: **B** (ENTER)

LINE

Mark encountered

Command:

NUMBER

The Number option of **Undo** will undo the specified number of drawing operations. The effect is the same as if the **U** command had been used the same number of times, but only a single regeneration is required.

REDO

Occasionally you will remove an object with **U** or **Undo** by mistake. The **Redo** command can be used to restore items that have been deleted. **Redo** will only restore objects, however, when used immediately after an **Undo** or **U**. If **U** or **Undo** is followed by a command such as **Line**, and then **Redo**, **Redo** will have no effect on the material that was deleted prior to the other command sequence. Start the command by selecting the **Redo** button on the **Standard** toolbar or by typing **REDO** (ENTER) at the Command prompt.

CHAPTER 12 EXERCISES

1. Start a new drawing and set the units and limits to be appropriate for this problem. Draw the outline of a 15' x 28' apartment unit. Make the 15' walls 6" thick, with the long walls 8" wide. Use 4" wide lines for interior walls and lay out a 12' x 15' bedroom, and a 5'–6" x 9' bathroom. No windows or doors are required at this time. Save the drawing as E-12-1.
2. Start a new drawing and draw a line 27'–6 1/2" long. Divide the line into five equal segments. Copy the original line and place markers at 16" o.c. Save the drawing as E-12-2.
3. A stairway is to be drawn between two floors of a townhouse with a distance of 9'–1 1/2" from finish floor to finish floor. Plan a set of stairs with 7 1/2" maximum rise and 10 1/2" minimum run per step. Determine the required number of stairs and the total run. Draw horizontal lines to represent each floor. The upper floor level will be 13 1/2" thick. Divide the space between floors into the required number of equal risers. Divide the total required run into the required individual steps. Using any editing process presented in the last two chapters, edit the grid that has been laid out to show the shape of the stairs. Draw a diagonal line that passes through the front edge of each step. Offset this line down 12" to determine the required depth of the stairs. Save the drawing as E-12-3.

   Required risers. Total run.
4. Use Figure 12.4 as a guide and draw an elevation showing an 8' high, 2x6 stud wall. Use (2) 2x6 top plates and a single base plate. Show an opening for a 6' wide x 4' high window. Use a 4x8 header over the window, and set the bottom of the header at 6'–8" above the floor level. Place the studs at 16" o.c. Save the drawing as E-12-4A. Make a copy of the elevation and show the wall at 9' high with the studs located at 24" o.c. Set the bottom of the header at 7'–10" above the floor. Save the elevation as E-12-4B.
5. Use the **Sketch** command and create a logo that can be used in the title block of a template drawing. Save the drawing as E-12-5.
6. Start a new drawing and draw three lines 10'–0" long. Set the point marker to show a marker with a circle with an X in the center. Measure the line into seven equal segments. Divide the second line into 16" long segments using a square around the crosshairs marker. Show the third line with perpendicular lines at 19.2" o.c. Save the drawing as E-12-6.
7. Use the **Sketch** command and draw four curved lines using 0.065", 0.125", 0.25", and 0.5" long line segments. In at least two of the lines, include a straight line segment between the sketched segments. Save the drawing as E-12-7.

8. Start a new drawing and draw a rectangle. Place a point using the crosshairs surrounded by a circle and a square in the center of the rectangle. Draw lines of infinite length that pass through the point in a horizontal, vertical and at 30° to the horizontal and vertical planes. Trim the lines that extend beyond the rectangle. Save the drawing as E-12-8.

9. Start a new drawing and draw a circle with a semi-infinite line extending outward from each quadrant. Draw lines that extend from the center point to the edge of the circle and are placed at 15° intervals for a 360° pattern. Save the drawing as E-12-9.

10. Draw two lines that intersect. One of the lines is to be horizontal and the second line is to be 67° above the first line. Draw a line that bisects the intersection of the two lines. Save the drawing as E-12-10.

CHAPTER 12 QUIZ

Start a new drawing and draw the following objects in the order listed. Draw four pairs of line segments, a three-sided polygon, a circle, a polyline with three segments with a width of 0.125, and an ellipse. Copy the circle, move the ellipse, and enlarge the polygon to twice its size.

1. If the **U** command is used after the polygon is enlarged, what will be affected?
2. If **Undo 2** is used after the polygon was drawn, what will be affected?
3. If a mark is placed after the circle is drawn, what will the effect be if the **Undo Back** command is used after the circle is copied?
4. A whole drawing sequence is an experiment that you might want to delete. List the command and subcommands that would allow the entire sequence to be removed as one object. Explain when these subcommands are used.
5. If 5 is used with the **Undo Number** option, what will the effect be?
6. In using the **Undo Number 5** option, one too many objects were removed. How can this be corrected?
7. Explain the difference between **Divide** and **Measure** on a line.
8. Explain the difference between a Trace and a polyline line segment.
9. What is the default setting for a sketch line if the units are set to architectural units with fractions of 1/16"?
10. What is the danger of having the sketch resolution too fine?
11. Explain how the correct option of **Sketch** will affect a drawing.
12. What is a ray and how could it be used?
13. What is the difference between a ray and an xline?
14. What is the effect of Hor on a ray if ORTHO is OFF?
15. Describe the difference between the two angle options of **Xline**.

CHAPTER 13

Controlling Lines and Layers

This chapter will examine methods for

- Controlling lines in AutoCAD
- Organizing drawing information by the use of layers
- Altering linetypes, colors, and layers

Commands to be introduced include

- **Ltscale**
- **Celtscale**
- **Layer**
- **Layer Previous**
- **Match Properties**
- **Properties**
- **Qselect**

CONTROLLING LINETYPES

A linetype is a specific pattern of lines of varied length and spacing presented in a uniform pattern. Varied linetypes traditionally have been used throughout the architectural and engineering communities. Chapter 5 introduced methods for loading and drawing with varied linetypes. This chapter will introduce methods for altering factors that control linetypes.

CHANGING THE LINE SEGMENT SIZE

The dashes that comprise the line pattern are currently defined as drawing units that might be unsuitable for your current drawing needs. A centerline drawn in a 12" x 9" drawing area will appear as a centerline until you zoom in on the line. If the same centerline is used in a drawing area that is 50' wide, the centerline will be displayed as a continuous line. To have the centerline appear as a centerline requires that the scale of the line be altered. The **Global scale factor** setting of the **Linetype Manager** dialog box can be used to control the length of line segments in a drawing. Access the dialog box by typing **LT** (ENTER) or by choosing **Linetype** from the **Format** menu. Select the **Show Details** button to display the current line scale values.

You can also alter the segment length of each pattern by typing **LTS** (ENTER) at the Command prompt. (LTS represents linetype scale.) No matter how the value is entered, by default AutoCAD thinks of the selected length as one unit. This unit could be in inches, feet, millimeters, kilometers, or any other unit of measurement. Altering the value will allow you to alter the selected unit length in a way similar to the way the **Scale** command alters the size of objects. The distances specified in the linetype definition will be multiplied by the **Ltscale** value to produce a new length for all patterns. The command sequence is as follows:

Command: **LTS** (ENTER)

LTSCALE Enter new linetype scale factor <1.00>: **2**
(ENTER)

If a value of 2 is entered, all line segments and spaces will be doubled in size. The effect of **Ltscale** on line segments can be seen in Figure 13.1.

The linetype scale is determined by the size the drawing will be plotted at. Using a linetype scale factor of 2 will make the line pattern twice as big as the line pattern created by a scale of 1. For a drawing to be plotted at 1/4"=1'–0", a dashed line that should be 1/8" when plotted, would need to be 3" long when drawn at full scale. This can be determined by multiplying .125 x 24 (the scale factor). You can also determine the linetype scale factor by using half of the scale factor. The drawing scale factor is always

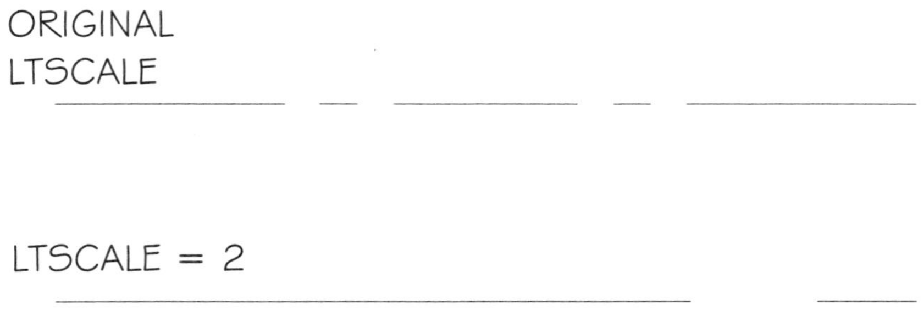

Figure 13.1 *What appears as a centerline on a 12" x 9" screen will appear as a continuous line when the limits are changed to 100' x 75'. The* ***Ltscale*** *command allows the length of line segments to be altered so that the pattern will still be seen.*

the reciprocal of the drawing scale. Using a drawing scale of 1/4"=1'–0" would equal .25"=12". Dividing 12 by .25 produces a scale factor of 48 with a linetype scale value of 24. Keep in mind that when the linetype scale is adjusted, it will affect ALL linetypes within the drawing. Usually this is not a problem, unless you are trying to create one unique line. Typically, the linetype scale factor can be preset as template drawings are set up. In most offices, template drawings are established based on the size of the paper or the scale the finished drawing will be plotted at. Common plotting scales and their respective scale factors include the following:

ARCHITECTURAL VALUES		ENGINEERING VALUES	
Drawing scale	**LTSCALE**	**Drawing scale**	**LTSCALE**
1"=1'–0"	6	1"=1'–0"	6
3/4"=1'–0"	8	1"=10'	60
1/2"=1'–0"	12	1"=100'	600
3/8"=1'–0"	16	1"=20'	120
1/4"=1'–0"	24	1"=200'	1200
3/16"=1'–0"	32	1"=30'	180
1/8"=1'–0"	48	1"=40'	240
3/32"=1'–0"	64	1"=50'	300
1/16"=1'–0"	96	1"=60'	360

CHANGING INDIVIDUAL LINE SEGMENT SIZES

Using **Ltscale** changes the values of all of the lines in a drawing. Using the **Celtscale** command allows the scale of existing lines to remain unaltered, but changes the scale of future lines. Typing **CELTSCALE** (ENTER) at the Command prompt will produce the prompt:

Enter new value for CELTSCALE <1.0000>:

The value that is assigned to **celtscale** is used in combination with the value assigned to the linetype scale. The **celtscale** acts as a multiplier of the linetype scale, producing a net scale effect on lines being drawn. With a linetype scale value of 2, and a celtscale of .25, the net scale factor would be 2 x .25=.50. Figure 13.2 shows the effects of using **Celtscale** to control the current line width.

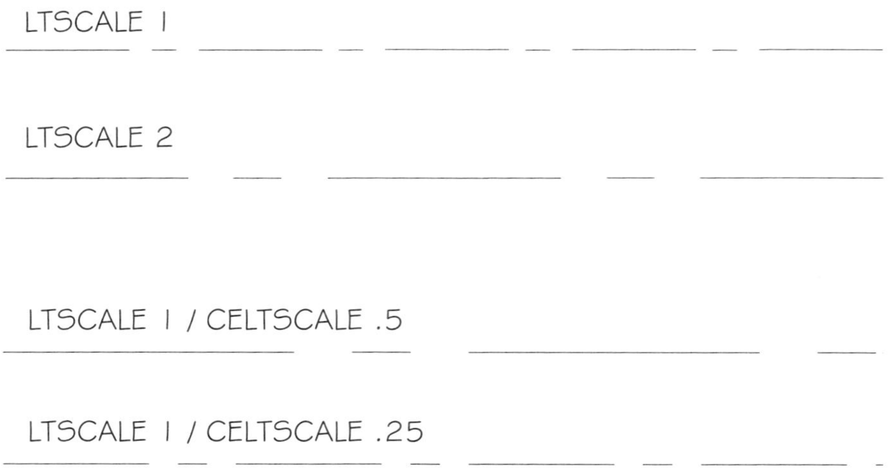

Figure 13.2 *The effects of* **Ltscale** *and* **Celtscale** *on line length.*

NAMING LAYERS

Layers were introduced in Chapter 5 along with methods to create, alter, and control a layer. As your drawing ability increases, it is crucial to fully understand the importance of using layers. The use of layers will be increasingly important as the techniques in the following chapters are introduced. Layers are used like separate sheets of paper to separate information within a drawing file. The base components of a floor plan can be placed on one layer, and the text, dimensions, framing information, and plumbing can be placed on separate layers to provide the information needed by various subcontractors. To help you control information, consideration must be given to layer names, controlling linetypes and colors by layer, as well as methods of displaying and hiding layer information.

LAYER NAMES

Names for layers have not become standardized throughout the architectural or engineering communities. Most offices have developed a standard method of assigning names, so that all members of the staff have easy access to information on a drawing. Even within a firm, names will vary depending on the type of project to be drawn.

As you set up names for layers, use the same guidelines for naming layers that were used to name linetypes. Use a name that will still be meaningful in six months. This might include layer names such as UWALLS, MWALLS, LWALLS, UBEAMS, MBEAMS, and LBEAMS. Up to 255 characters (including spaces) can be used in naming a layer but it helps to keep layer names short and descriptive.

AIA STANDARDIZED LAYER NAMES

As with other areas of architecture and engineering, there is no one standard method for naming and using layers on construction drawings. The second edition of the *AIA CAD Layer Guidelines* was released in September of 1997. Copies of the standard can be obtained from the AIA Web site at *www.aia.org*. The guidelines are intended to coordinate with ISO (International Organization for Standardization) and CSI (Construction Specifications Institute) standards. The AIA recommends a name comprised of a discipline code and a major group name. A minor group name and a status name can also be added to the layer name, depending on the complexity of the project. The guidelines can be used to label both layers and drawing files with components of file names also used in the layer names. Because name components are common to both file and layer names, care must be taken to avoid creating project specific references within a layer name. Project designations should not be included in folder names. Use the guidelines presented in Chapter 4 for naming project folders and the following guidelines for naming files and layers. Figure 13.3 shows examples of the new guidelines and how they can be applied to layers and file names.

Naming Model/Sheet Files

The guidelines that were introduced in Chapter 4 work well for projects that do not require input from several different subcontractors. When several firms are responsible for the project, one system that meets the needs of each discipline is required to name layers. The AIA CAD standard can be used to name layers and drawing files. Drawing files can be classified as either a model or sheet file.

- Model files are the components of a set of working drawings that are drawn full scale. These are the elements—such as walls, doors, and windows—that you've been drawing in model space. Chapter 4 introduced guidelines for naming model files.
- A sheet file is a completed drawing, such as a floor plan, that is prepared for plotting. A sheet file can contain one or more drawings that will be plotted on one sheet of paper with a title block.

Later chapters will introduce merging a model file into a template drawing using the **Insert** and **Xref** commands. Sheet files allow for plotting in paper space at a scale of 1"=1". Regardless of the type of file, the AIA CAD layer standards will provide a better means of tracking the contents of files and layers. The initial component of file and layer names is a discipline code.

Discipline Code

The discipline code is a character code that is used to identify the originator of the drawing. The discipline code will be the same for a model file, sheet file, or layer name. The code allows instant recognition of the originator of the drawing or layer. The code can be either one or two characters. If only one character is used, it is followed by a hyphen. The first character represents one of the following 16 disciplines:

Discipline Codes for the Drawing Originator

A	Architectural	M	Mechanical
C	Civil	P	Plumbing
E	Electrical	Q	Equipment
F	Fire Protection	R	Resource
G	General	S	Structural
H	Hazardous Materials	T	Telecommunications
I	Interiors	X	Other Disciplines
L	Landscape	Z	Contractor/Shop Drawings

The second character to be used should either be a hyphen or a user-defined modifier. Suggested modifiers might include G for graphics, K for kitchen, or V for audiovisual equipment. A layer or file with a discipline code of AG would represent architectural graphics, such as a rendering developed by the architectural firm.

Guidelines for Naming Model Files

Figure 13.3 shows a sample of a model file name. The discipline code is followed by the drawing-type code. The AIA guideline lists eleven codes that are common to all disciplines. The codes are listed below and preceded by an asterisk that represents the space for the discipline code.

Model File Drawing Type Codes

*–FP	Floor Plan	*–SC	Section
*–SP	Site Plan	*–DT	Detail
*–DP	Demolition Plan	*–SH	Schedules
*–QP	Equipment Plan	*–3D	Isometric/3D
*–XP	Existing Plan	*–DG	Diagrams
*–EL	Elevation		

In addition to drawing type codes that apply to all disciplines, specific codes are listed for architectural, civil, electrical, fire protection, interiors, mechanical, plumbing, structural, and telecommunications. A–CP for ceiling plans and A–EP for enlarged plans are examples of drawing types for the architectural field. S–FP for framing plans and S–NP for foundation plans are examples of structural drawing types. See the *AIA CAD Layer Guidelines* for a complete listing of discipline specific codes.

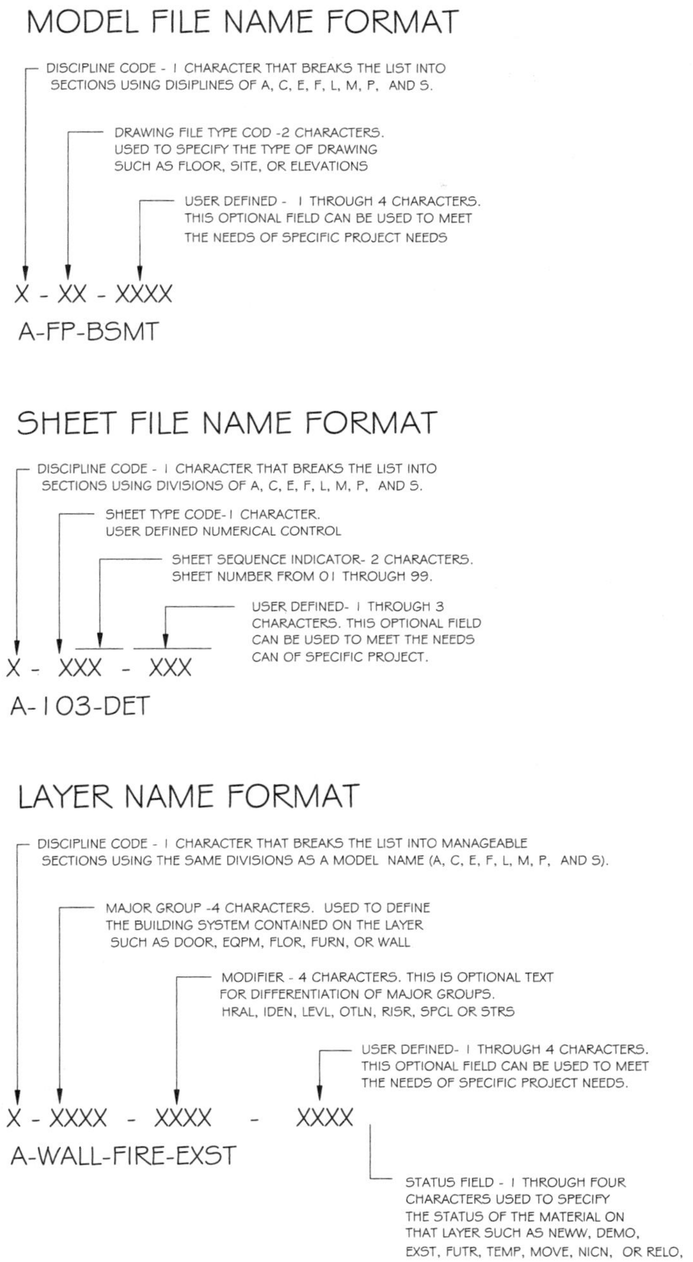

Figure 13.3 *The AIA CAD Layer Guidelines use abbreviations to name layers and drawing files. By using a uniform system to name layers and files, architects, engineers, and consultants working for several different firms can easily produce a coordinated set of drawings.*

User-Definable Codes

The user defines the final four digits of the model file name. Guidelines for naming files and folders that were presented in Chapter 4 can be used for the final four digits. The fifth-level floor plan for a structure could be saved with a name of A–FP–2.05.

Guidelines for Naming Sheet Files

Figure 13.3 shows an example of a sheet file name. The discipline code is followed by a one-digit sheet-type code. The AIA guideline lists nine common codes that apply to all disciplines.

Sheet Type Designators

0 General (symbols, legends, notes, etc.)
1 Plans
2 Elevations
3 Sections
4 Large scale drawings that are not details
5 Details
6 Schedules/Diagrams
7 User-defined
8 User-defined
9 3D views

A two-digit sheet sequence number ranging from 01 to 99 follows the sheet type designator. The last component of a sheet title is a three-digit or letter, user-defined code. A sheet file name of S–405CON could be used to represent a sheet of concrete details that have been assembled for plotting by the structural engineer. The sheet would be page 5 in the drawing set.

Guidelines for Naming Layers

The method for naming layers is similar to naming files. A layer name can be composed of the discipline and major group name. A minor name and status code can be added to the sequence. Figure 13.4a shows some alternatives for using components with layer names. The discipline code is the same as the one- or two-character code that is used for model and sheet file names.

Major Group Name

The major group code is a four-character code that identifies a building component specific to the defined layer. Major group layer codes are divided into the major groups of architectural, civil, electrical, fire protection, general, hazardous, interior, landscape, mechanical, plumbing, equipment, resource, structural, and telecommunication. Codes such as ANNO (annotation), EQIP (equipment), FLOR (floor), GLAZ (glazing), and WALL (walls) are examples of major group codes that are associated with the architectural layers. A complete listing of major codes for each group can be found in the *AIA CAD Layer Guidelines*.

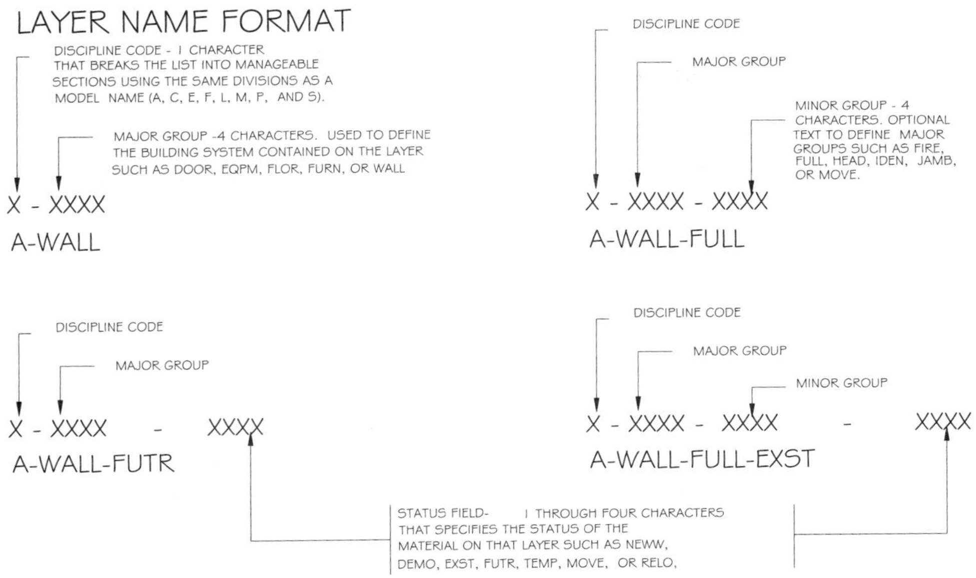

Figure 13.4a *Layer names fall into one of four categories when based on the AIA CAD Layer Guidelines.*

Minor Group Name

The minor group code is an optional four-letter code that can be used to define subgroups to the major group. The code A–FLOR (architectural–floor) might include minor group codes for OTLN (outline), LEVL (level changes), STRS (stair treads or escalators), EVTR (elevators), or PFIX (plumbing fixtures). A layer name of A–FLOR–IDEN would contain room names, numbers, and other related titles or tags. A complete listing of minor group codes specific to each discipline is listed in the AIA guideline.

Status Code

The status code is an optional four-letter code that can be used to define a subgroup of either a major or minor group. See Figure 13.4a. The code is used to specify the phase of construction. The layer names for the walls of a floor plan (A–WALL–FULL) could be further described using the status code of NEWW (new work), EXST (existing walls), or DEMO (demolition). A wall shown on the floor plan that is to be moved would be displayed on the A–WALL–FULL–MOVE layer. Figure 13.4b shows a listing of common layer names and their contents.

ARCHITECTURAL LAYER NAMES

ANNOTATION

A-ANNO-DIMS	DIMENSIONS
A-ANNO-KEYN	KEY NOTES
A-ANNO-LEGN	LEGENDS & SCHEDULES
A-ANNO-NOTE	NOTES
A-ANNO-NPLT	NON-PLOTTING INFORMATION
A-ANNO-NRTH	NORTH ARROW
A-ANNO-REVS	REVISIONS
A-ANNO-REDL	REDLINE
A-ANNO-SYMB	SYMBOLS
A-ANNO-TEXT	TEXT
A-ANNO-TTLB	BORDER & TITLE BLOCK

FLOOR PLAN

A-FLOR-EVTR	ELEVATOR CAR & EQUIPMENT
A-FLOOR-FIXT	FLOOR INFORMATION
A-FLOR-FIXD	FIXED EQUIPMENT
A-FLOR-HRAL	HANDRAILS
A-FLOR-NICN	EQUIPMENT NOT IN CONTRACT
A-FLOR-OVHD	OVERHEAD ITEMS (SKYLIGHTS)
A-FLOR-PFIX	PLUMBING FIXTURES
A-FLOR-STRS	STAIR TREADS
A-FLOR-TPIN	TOILET PARTITIONS

ROOF PLAN

A-ROOF	ROOF
A-ROOF-OTLN	ROOF OUTLINE
A-ROOF-LEVL	ROOF LEVEL CHANGES
A-ROOF-PATT	ROOF PATTERNS

ELEVATIONS

A-ELEV	ELEVATIONS
A-ELEV-FNSH	FINISHES & TRIM
A-ELEV-IDEN	COMPONENT IDENT. NUMBERS
A-ELEV-OTLN	BUILDING OUTLINES
A-ELEV-PATT	TEXTURES & HATCH PATTERNS

SECTIONS

A-SECT	SECTIONS
A-SECT-IDEN	COMPONENT IDENT. NUMBERS
A-SECT-MBND	MATERIAL BEYOND SECTION CUT
A-SECT-MCUT	MATERIAL CUT BY SECTION PLANE
A-SECT-PATT	TEXTURES AND HATCH PATTERNS

CIVIL LAYER NAMES

ELEVATIONS

C-BLDG	PROPOSED BUILDING FOOTPRINT
C-PKNG	PARKING LOTS
C-PKNG-CARS	AUTOMOBILES
C-PKNG-DRAN	DRAINAGE INDICATORS
C-PROP	PROP. LINES & BENCHMARKS
C-PROP-BRNG	BEARINGS & DISTANCE LABELS
C-PROP CONS	CONSTRUCTION CONTROLS
C-PROP-ESMT	EASEMENTS, RIGHTS- OF -WAY
C-ROAD	ROADWAYS
C-ROAD-CNTR	CENTER LINES
C-ROAD-CURB	CURBS
C-SSWR	SANITARY SEWERS
C-STRM	STORM DRAIN / CATCH BASINS
C-STRM-UNDR	UNDERGROUND DRAIN. PIPES
C-TOPO	PROPOSED CONTOURS & ELEV..
C-TOPO-BORE	TEST BORINGS
C-TOPO-RTWL	RETAINING WALL

STRUCTURAL LAYER NAMES

GENERAL LAYERS

S-ABLT	ANCHOR BOLTS
S-BEAM	BEAMS
S-COLS	COLUMNS
S-DECK	STRUCTURAL FLOOR DECKING
S-GRID	COLUMN GRIDS
S-GRID-DIMS	COLUMN GRID DIMENSIONS
S-GRID-EXTR	COLUMN GRIDS OUTSIDE OF BUILD.
S-GRID-IDEN	COLUMN GRID TAGS
S-JOIS	JOIST
S-METL	MISCELLANEOUS METAL
S-WALL	STRUCT. BEARING / SHEAR WALLS

FOUNDATION /SLAB PLAN LAYERS

S-FNDN	FOUNDATION
S-FNDN-PILE	PILES / DRILLED PIERS
S-FNDN-RBAR	FOUNDATION REINFORCING
S-SLAB	SLAB
S-SLAB-EDGE	EDGE OF SLAB
S-SLAB-REBAR	REINFORCING
S-SLAB-JOIN	SLAB CONTROL JOINTS

Figure 13.4b *Common layer names and their contents based on the AIA CAD Layer Guidelines.*

CONTROLLING LAYERS

Layers can be controlled by using the **Layer Properties Manager** dialog box or by using the **Layer Control** menu on the **Object Properties** toolbar. Access the **Layer Properties Manager** dialog box shown in Figure 13.5 by selecting the **Layer** button on the **Object Properties** toolbar, by typing **LA** (ENTER) at the Command prompt, or by selecting **Layer** from the **Format** menu. Previous chapters introduced each element of the layer and linetype properties. This chapter will introduce methods of setting and controlling each of the layer elements.

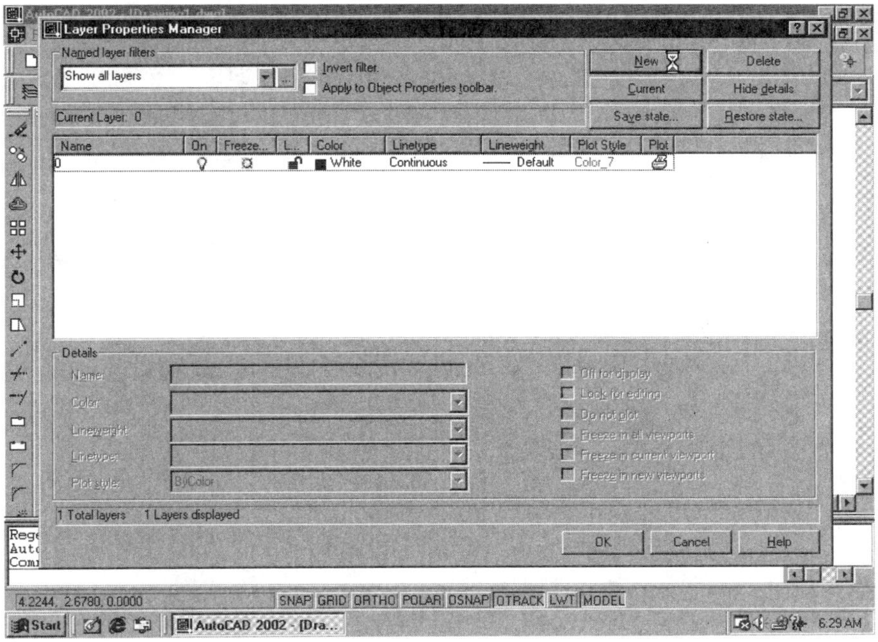

Figure 13.5 *The* **Layer Properties Manager** *dialog box can be used to control the characteristics of linetype and layers. It can be accessed by selecting the* **Layer** *button on the* **Object Properties** *toolbar or by typing* **LA** *(ENTER) at the Command prompt.*

ASSIGNING COLOR BY LAYER

Colors and linetypes can be assigned by layers after a layer has been created. See Chapter 5 for a review of how to create a layer.

Figure 13.6 shows an example of the dialog box with the addition of the MWALLS, UPWALLS, and BWALLS layers. Notice that the current layer is still 0, but new layers have been added. As new layers are created, they are initially placed at the bottom of the layer list. As the screen is regenerated, the names will be displayed in alphabetical order. Once a layer has been created, the qualities for that layer color and linetype can be assigned.

The MWALLS layer is currently listed as black. Any lines that are created on this layer will be drawn as black lines. To alter the color of a layer, select the layer so that the layer name is highlighted. You can alter the color of each layer by selecting the color button from the **Color list** or by choosing a color in the **Color** box in the **Details** area. (The **Details** area is not active until a layer is highlighted and the **Show Details** button is selected.) Selecting the color button for the selected layer will display the **Select Color** dialog box shown in Figure 13.7. As the desired color is selected from

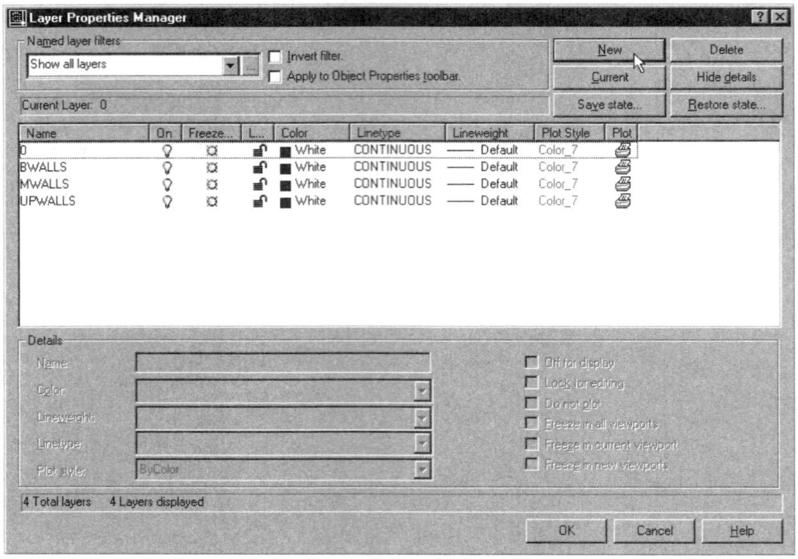

Figure 13.6 *To create new layers, click the* **New** *button. A layer with a named Layer 1 will be added. Highlight the name of Layer 1 and type a more descriptive name.*

the palette, the name of the color will be displayed in the **Details** area, and the color will be displayed. Clicking **OK** will remove the color palette. You can set the color for a layer by selecting the **Color** button in the **Color** box on the **Object Properties** toolbar or by selecting the arrow beside the **Color** box. Either method will produce the **Color** menu shown in Figure 13.8. You can specify the color by selecting a color button or by entering the name or number of the color. Selecting the Other option of the menu will display the **Select Color** dialog box shown in Figure 13.7. If you want to select another color, simply select another color chip, and the display will be updated. When you're satisfied with the color, click the **OK** button, and the dialog box will be removed and the **Layer Control box** redisplayed. You can alter the color of individual objects on a layer by selecting the **Color** button on the **Object Properties** toolbar. Selecting this button will allow the color of the layer to remain unchanged but allow all items drawn after you select the button to have a color different from the specified layer color.

Objects and layers can be assigned any color you desire. It may be helpful to assign like objects of different layers a different color. For instance, if all plumbing fixtures are blue, it might not be readily apparent if the UPLUMB fixtures are ON when only the lower-layer components are desired. Assigning a different color to the LPLUMB, MPLUMB, and UPLUMB layers might help you keep track visually of a drawing as different layers are set to ON/OFF for viewing or plotting. Color as it relates to plotting will be introduced in later chapters.

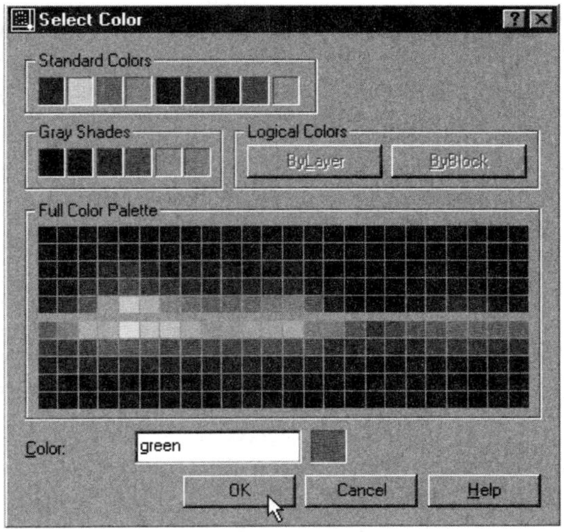

Figure 13.7 *Selecting the* **Color** *button in the* **Layer Properties Manager** *dialog box will display the* **Select Color** *dialog box, allowing the current layer color to be changed.*

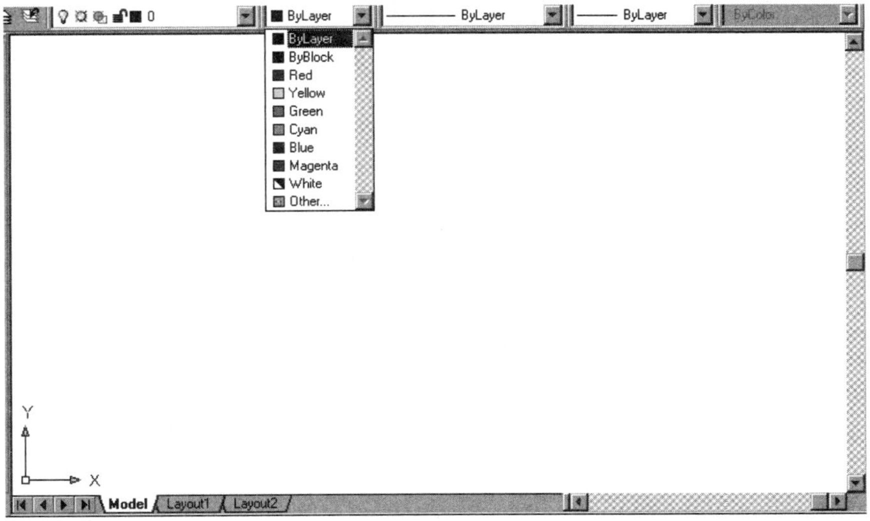

Figure 13.8 *You can alter the color for a layer by selecting the* **Color** *button on the* **Object Properties** *toolbar or the down arrow beside the* **Color** *box.*

ASSIGNING LINETYPES TO LAYERS

The current linetype default for each layer is a continuous line. In previous chapters, methods of drawing different linetypes were introduced. The Linetype command allows for setting and drawing different linetypes without regard to the layer. The Linetype option of **Layer** will allow all lines on a specific layer to be drawn as one type of line. Linetypes can now be assigned to a specific layer and all lines on that layer will be drawn with the selected linetype.

The linetype can be altered using methods similar to altering layer color. Highlighting the name of the layer to be altered and then selecting the linetype name will display the **Select Linetype** dialog box shown in Figure 13.9. The dialog box can be used to load and select linetypes. Selecting the **Load** button will display the **Load or Reload Linetypes** dialog box shown in Figure 13.10 and display a list of linetypes contained in AutoCAD. Highlight the names of the linetypes to load and click the **OK** button to remove the dialog box and restore the **Select Linetype** dialog box. As the box is restored, it will be updated to reflect the linetypes that were just loaded. One of these linetypes can now be selected from the **Select Linetype** dialog box and assigned to the layer that is highlighted in the **Layer Control** box. Assign a linetype by selecting the linetype and clicking the **OK** button. As the button is clicked, the **Layer Properties Manager** dialog box will be redisplayed, reflecting the selected linetype.

Note: To select multiple linetypes, hold down CTRL while you select the linetypes. To load a sequence of linetypes, hold down SHIFT and select the first and last listing. All linetypes that are listed between the first and last listings will be selected. These two selection methods can be used any time multiple objects are to be selected from a list.

SETTING THE CURRENT LAYER

Up to this point, you've created a layer with the name of MFLOOR and assigned the color red and a linetype of CENTER. If you exit the dialog box and start to draw, nothing will appear to have changed. To draw red centerlines, you need to set the current layer to MFLOOR. To make the MFLOOR layer current, highlight the name of the layer in the **Layer Properties Manager** dialog box and click the **Current** button. Clicking the **OK** button will keep this change and restore the drawing screen. As you start the **Line** command, segments will be drawn using the selected color and linetype that were just added. A new layer can also be made current by selecting the down arrow beside the **Layer** box on the **Object Properties** toolbar and selecting the desired layer.

SETTING LAYERS ON/OFF

All of the new layers you have created up to this point are set to ON. If you were to switch to a layer such as UPWALLS, and then draw, you would be able to see what is drawn. In a multilevel structure it is sometimes helpful to view the walls of several different levels to check alignment, but typically, you will want to have only one level

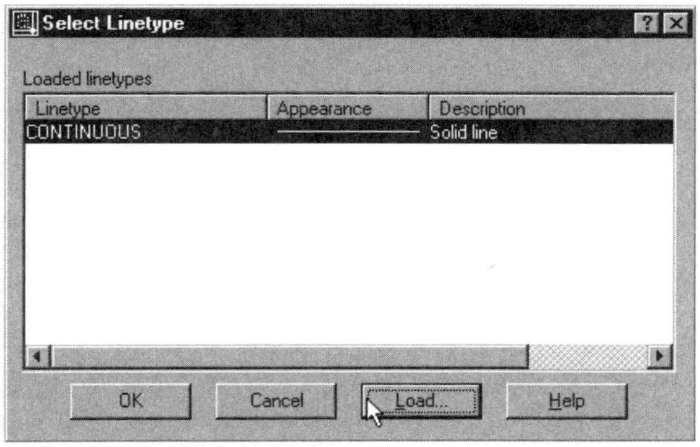

Figure 13.9 *Selecting the name of the current linetype in the* **Layer Properties Manager** *dialog box will display the* **Select Linetype** *dialog box, allowing for selection of one of the current linetypes.*

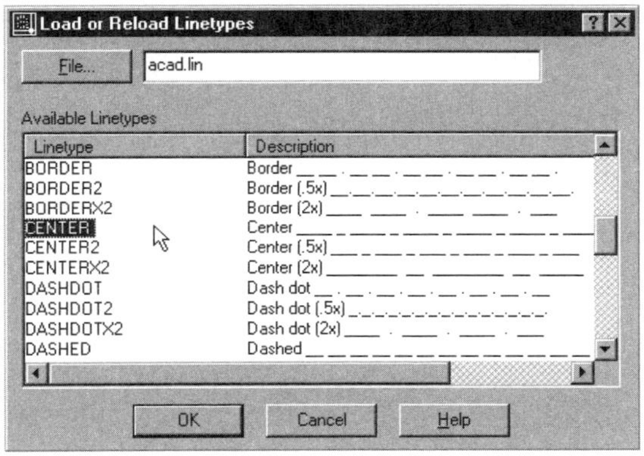

Figure 13.10 *You can add linetypes to the drawing base by highlighting the desired layer name and clicking the* **OK** *button.*

of a structure current. If a layer is set to OFF, objects drawn on that layer will be invisible. The information is retained, but not displayed or plotted.

Shown below are several drawings showing different layers in a drawing file for a simple residence and the listing for the layers contained in the drawing file. The drawing file contains the floor, framing, and electrical plans for each level as well as the foundation, roof, and site plans. Figure 13.11 shows the floor plan of a residence with all of the nonstructural material ON, and all of the structural material OFF. Figure 13.12 shows the same floor plan with the structural material layers toggled to ON, and the architectural information set to OFF. This particular drawing file contains information for the floor plan, framing plan, electrical plan, roof plan, foundation plan, and site plan. Through the ON/OFF option, each drawing can be displayed and edited. Each plan can be seen in Figure 13.11 through 13.16.

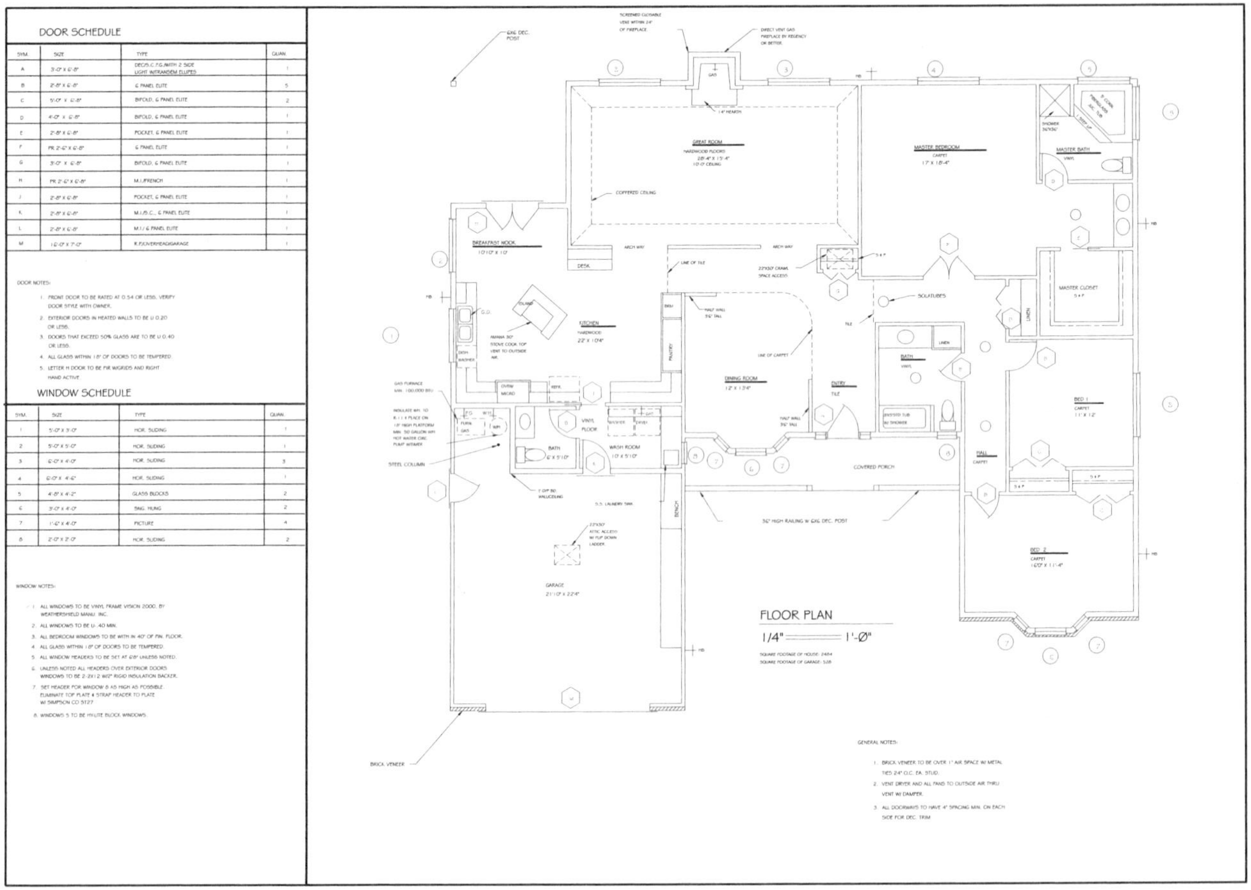

Figure 13.11 *All of the drawings seen in Figures 13.11 through 13.16 are contained in the same drawing file. Each drawing is stacked over others using the layers. By use of a layer matrix, layers can be easily selected to display the desired drawing. (Courtesy Tereasa Jefferis.)*

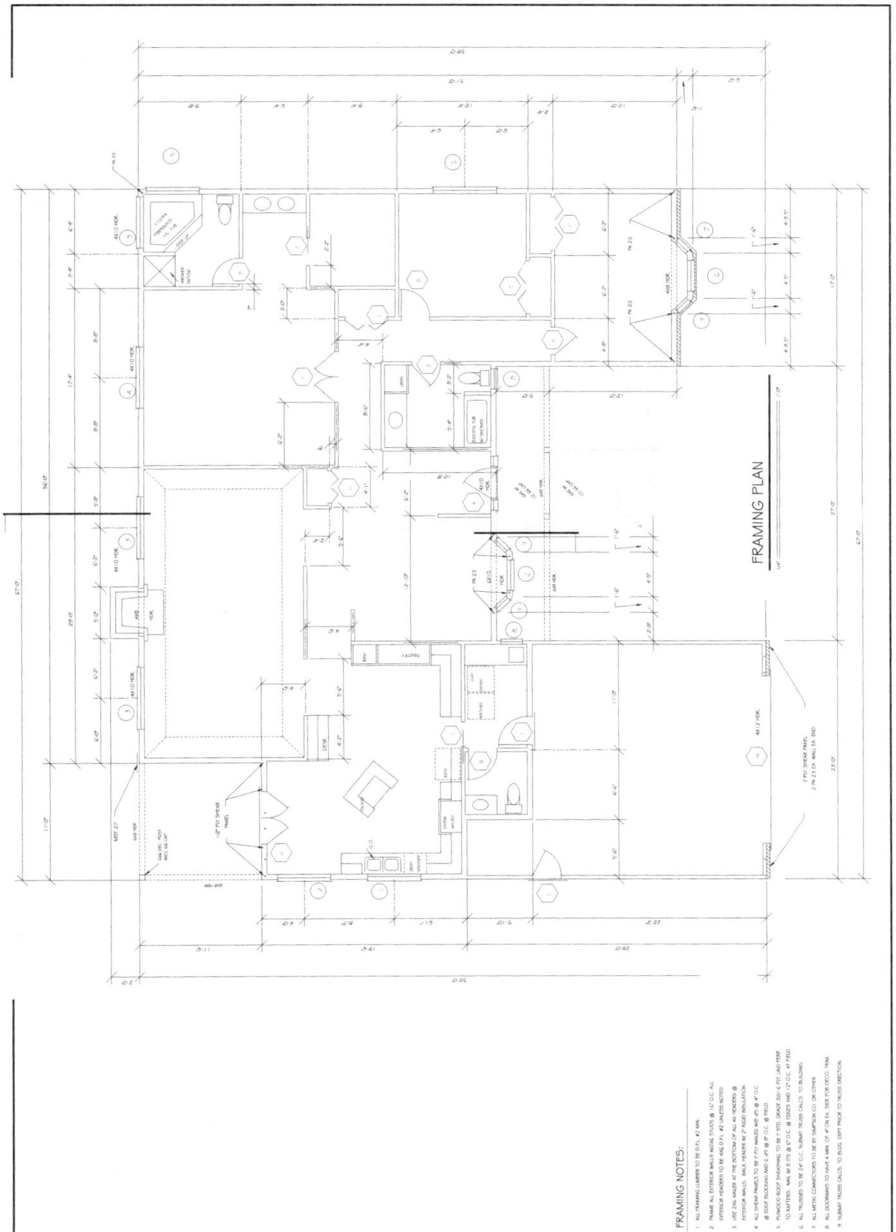

Figure 13.12 *The framing plan for the residence seen in Figure 13.11. (Courtesy Tereasa Jefferis.)*

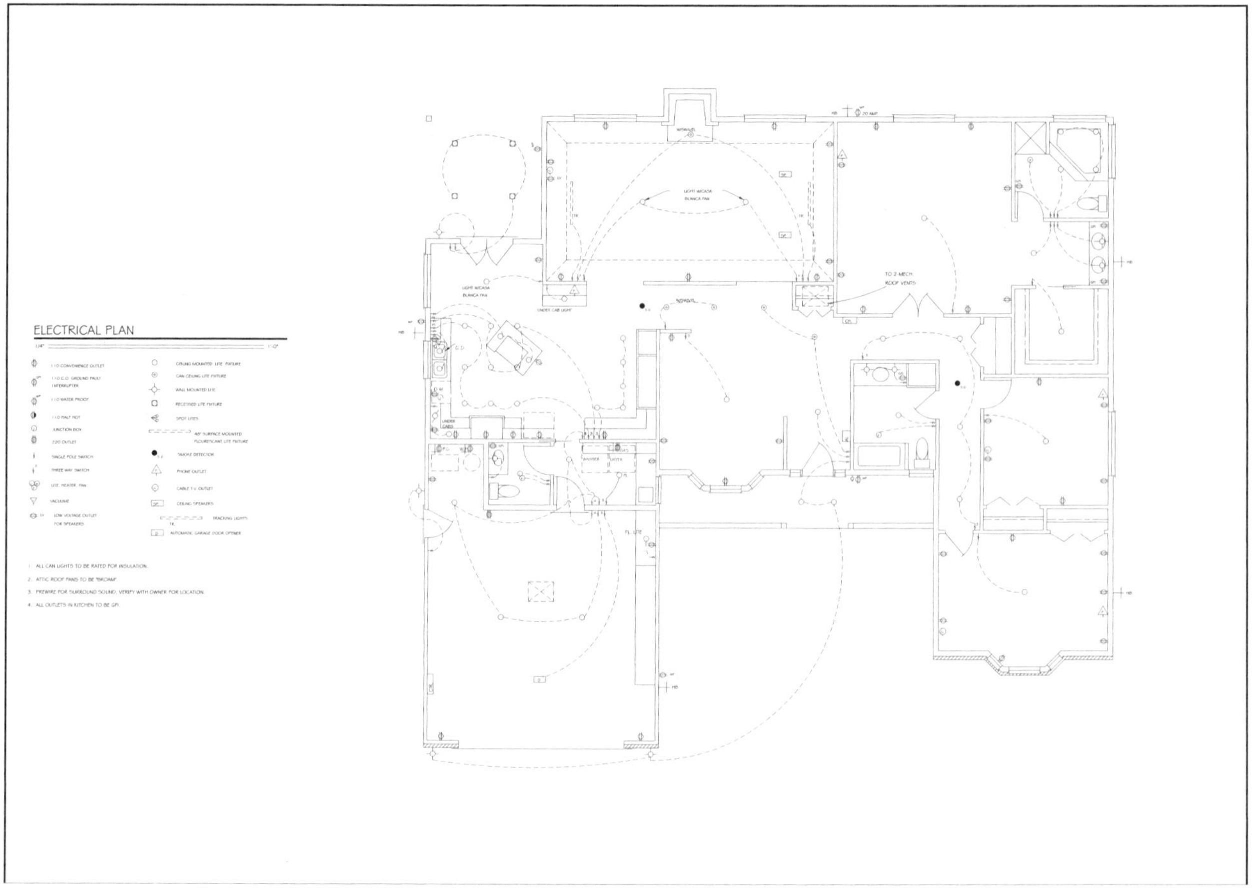

Figure 13.13 *An electrical plan uses the walls, windows, doors, and cabinets of the floor plan for a drawing base. (Courtesy Tereasa Jefferis.)*

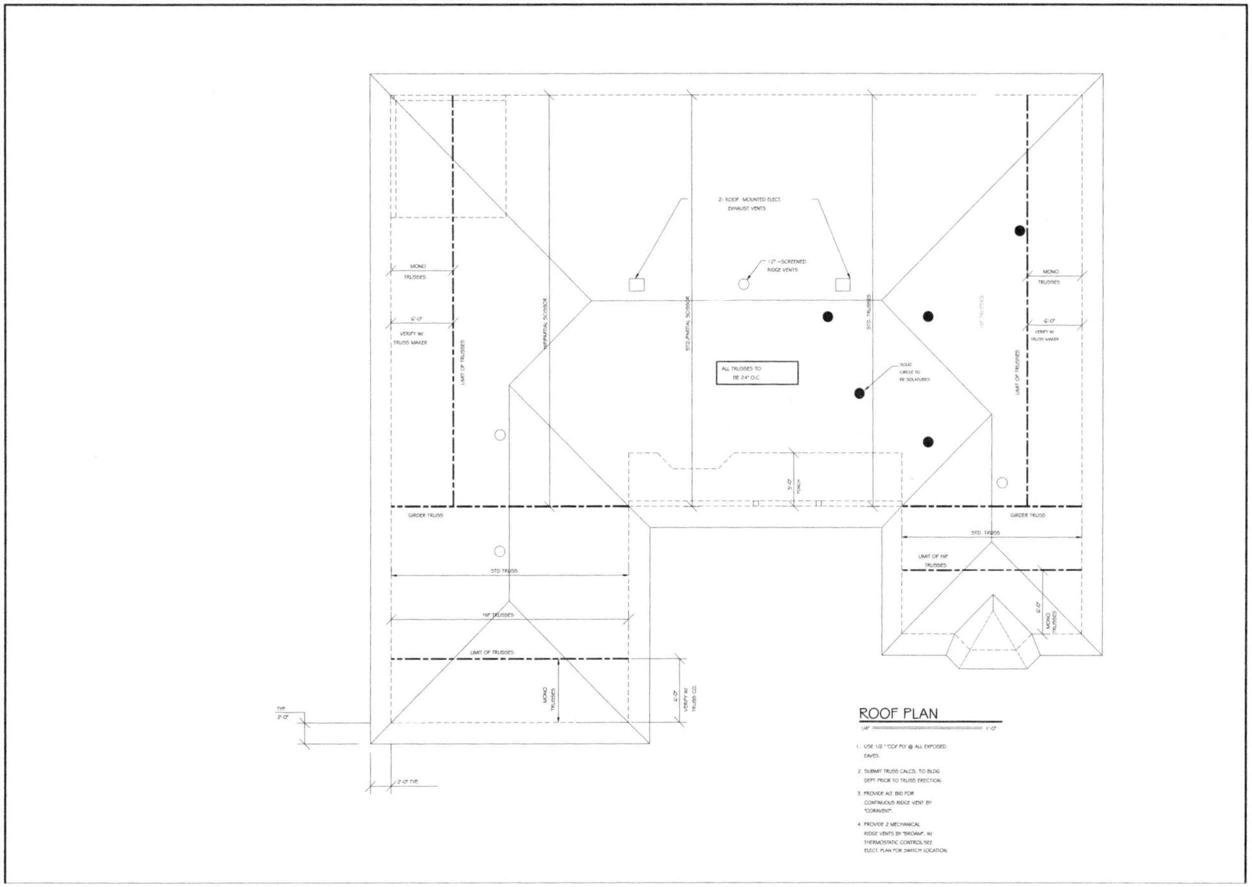

Figure 13.14 *None of the layers on the floor, framing, or electrical plans is used on the roof plan, but the walls serve as an outline for the roof. Using OSNAP and the* **Line** *command, with the walls thawed, the outline can be quickly drawn before all other layers are frozen . (Courtesy Tereasa Jefferis.)*

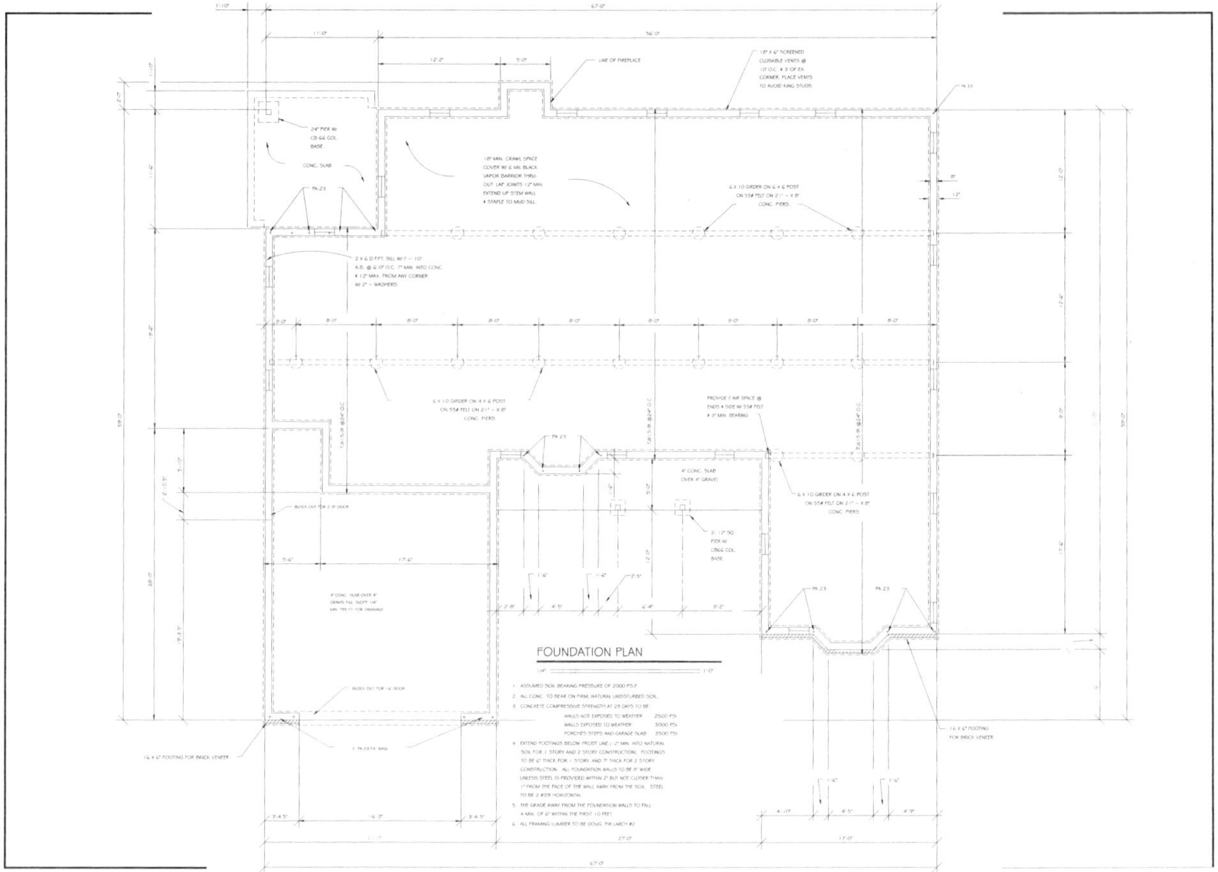

Figure 13.15 *The foundation can be drawn using the walls of the floor plan as a guide. With careful planning and good layer control, many of the dimensions on the floor plan can be used on the foundation. (Courtesy Tereasa Jefferis.)*

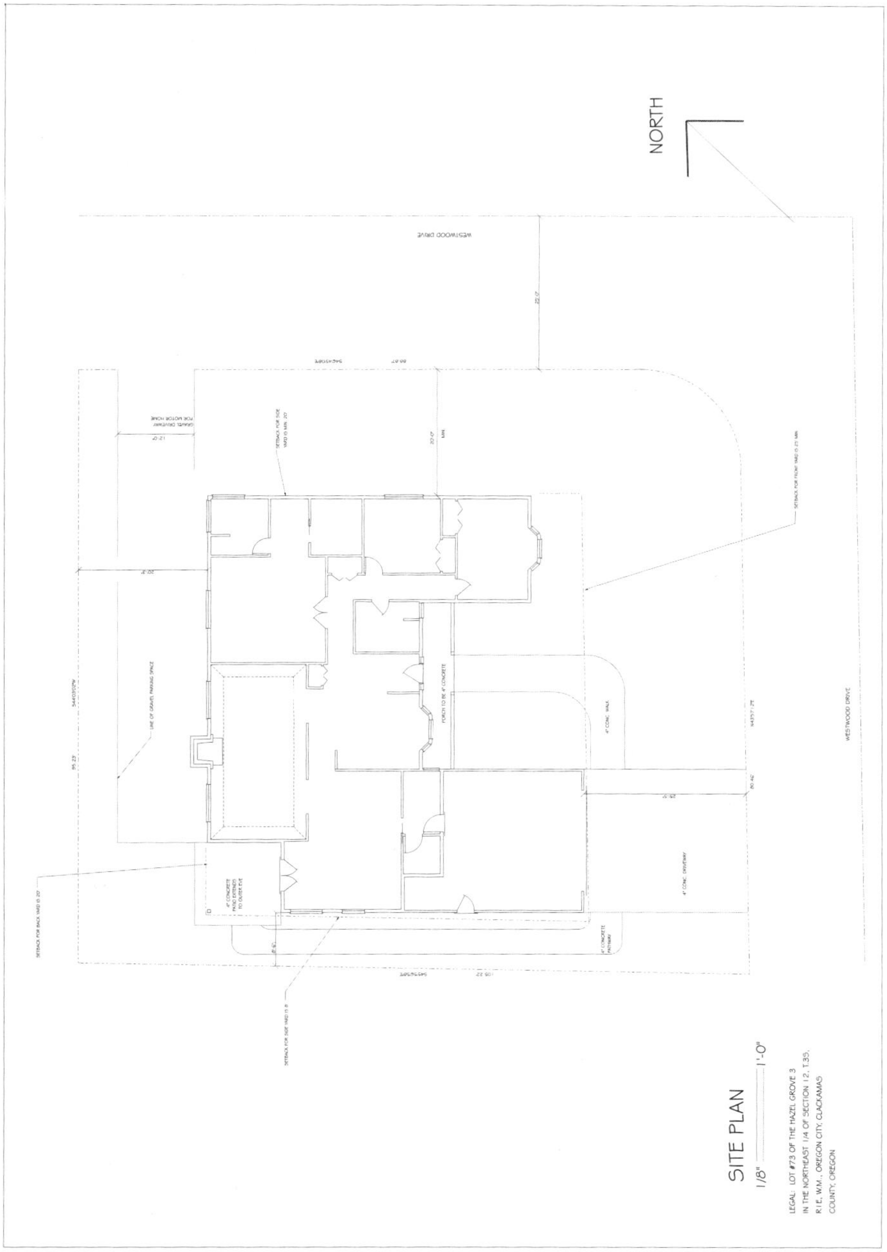

Figure 13.16 *A site plan often requires careful use of varied linetypes and Celtscale. (Courtesy Tereasa Jefferis.)*

Drawing Layers for Electrical, Floor, Foundation, Framing and Site Plans

A-Flor-anno-note	Continuous	E-Lite-flor	Continuous
A-Flor-cabs	Continuous	E-Lite-swch	Continuous
A-Flor-dims	Continuous	E-Lite-wall	Continuous
A-Flor-door-schd	Continuous	E-Soun	Continuous
A-Flor-door-symb	Continuous	E-symb	Continuous
A-Flor-fixt	Continuous	E-wire	Dashed
A-Flor-iden	Continuous	S-Fndn-anno-genr	Continuous
A-Flor-plmb	Continuous	S-Fndn-beam	Center
A-Flor-strs	Continuous	S-Fndn-dimn-extr	Continuous
A-Flor-ttlb	Continuous	S-Fndn-dimn-intr	Continuous
A-Flor-wall	Continuous	S-Fndn-foot	Hidden
A-Flor-wind-schd	Continuous	S-Fndn-jois	Continuous
A-Flor-wind-symb	Continuous	S-Fndn-iden	Continuous
A-Roof-beam	Center	S-Fndn-latl	Continuous
A-Roof-fram	Continuous	S-Fndn-rbar	Continuous
A-Roof symb	Continuous	S-Fndn-pier	Hidden2
A-Roof text	Continuous	S-Fndn-slab	Continuous
A-Roof-ttlb	Continuous	S-Fndn-symb	Continuous
A-Roof wall	Dashed	S-Fndn-ttlb	Continuous
C-anno	Continuous	S-Fndn-wall	Continuous
C-Dimn	Continuous	S-Fram-anno-genl	Continuous
C-dran	Continuous	S-Fram-anno-symb	Continuous
C-esmt	Phantom2	S-Fram-base-dimn	Continuous
C-prop	Phantom	S-Fram-beam	Center
C-roads-	Center	S-Fram-dimn-extr	Continuous
C-symb	Continuous	S-Fram-dimn-intr	Continuous
C-topo-exst	Dashed	S-Fram-jois	Continuous
C-topo-fin	Continuous	S-Fram-latl	Continuous
C-topo-wall	Continuous	S-Fram-latl-schd	Continuous

E-Anno-legn	Continuous	S-Fram-symb	Continuous
E-Anno-schd	Continuous	S-Fram-ttlb	Continuous
E-Lite-clng	Continuous		

Layer Off

To turn layers off, display the **Layer Properties Manager** dialog box. The ON/OFF button is the lightbulb—with yellow representing ON and black representing OFF. Select the ON/OFF button for the desired layers to be turned off, and click the **OK** button. The dialog box will be removed and any information that is stored on a layer that is OFF will be removed from the screen. If you attempt to set the current layer to be turned OFF you will be given the following prompt:

The current layer is turned off.

If you click the **OK** button, the dialog box is removed, and the items on the current layer are removed from the drawing screen. To restore the current layer, redisplay the **Layer Properties Manager** dialog box and restore the current layer. If you accept the current layer OFF, the current layer will be made invisible. Now if you draw objects, they will be drawn but not displayed. This situation occurs more by accident than by design. If you find yourself drawing a line but nothing is displayed, check the listing to see if the current layer has been accidentally switched OFF.

Layer On

Layers that have been made invisible by the OFF option can be restored by the ON option. Display the **Layer Properties Manager** dialog box and highlight the desired layers to be set to ON. Select the black lightbulb to turn it ON and click the **OK** button to restore the selected layers, remove the dialog box, and display the information on the selected layers.

THAWING AND FREEZING LAYERS

If you have set layers to OFF, you removed information from the display screen but not from the drawing base. The information merely became invisible. If you request a **Regen**, it would be performed with all of the information on the ON layers redisplayed. Information on layers that are set to OFF would also be processed, which will cause a slight slowdown on large drawings. Using the Freeze/Thaw options, you will be able to control the extent of information to be scanned during the **Regen** process. Layers that are in the Thaw mode are processed for a **Regen**; information that is frozen is not processed during a **Regen**, reducing the time required to **Regen**. This will be a factor only on huge drawing files.

Freeze

Display the **Layer Properties Manager** dialog box, select the layer that you would like to freeze, and select the **Thaw** (sun) button, showing the Thaw mode. It will be changed to a snowflake, showing the Freeze mode. Clicking the **OK** button will activate the selection, restore the drawing screen, and remove the objects on the selected layer from the display. Unlike the OFF option, the current layer cannot be frozen.

Thaw

The Thaw option of the **Layer** command will allow for specified frozen layers to be thawed. To thaw a layer that is frozen, display the **Layer Properties Manager** dialog box and select the **Freeze** (snowflake) button of the layer to be thawed. As the snowflake is selected, it will change to the sun button, indicating that the layer has been thawed. Clicking the **OK** button will activate the selection, restore the drawing screen, and display the material on the layer that is now thawed.

Safeguarding Layers

The final pair of options of the **Layer** command is Lock and Unlock. As the name implies, a lock is used to protect something. AutoCAD allows you to lock layers so that they cannot be edited accidentally. Information on a locked layer is still visible; it's just protected. To lock a layer, highlight the desired layer and select the **Lock** button followed by the **OK** button.

To unlock a layer, highlight the desired layer and select the **Unlock** button, and click the **OK** button. This process will reverse the consequences of the Lock. Remember that locking a layer is not a security device, but a method of protecting information on a layer from careless editing. Entities on a locked layer will not be included in a selection set of an edit command. Locking a layer will allow you to view the information on the layer, but it will not be affected it as you edit a drawing. You can continue to draw on a locked layer and objects can still be selected for OSNAP locations, thawed or frozen, or altered using the **Linetype** or **Color** button. Objects on locked layers can't be edited.

Filters

Filters provide the ability to display certain layer names in the listing box, based on a specific attribute. Layers can be filtered based on layer name, color, lineweight, contents, locked or unlocked status, freeze or thaw status, and plot styles. There are several other filter options that will not be considered until viewports are explored in later chapters. Filters are controlled in the **Layer Properties Manager** dialog box using the **Named layer filters** edit box arrow. Choosing the **Filter** edit arrow will produce the list shown in Figure 13.17. Selecting the [...] button will produce the dialog box shown in Figure 13.18 and allow the parameters of the filter to be set. In the example, the

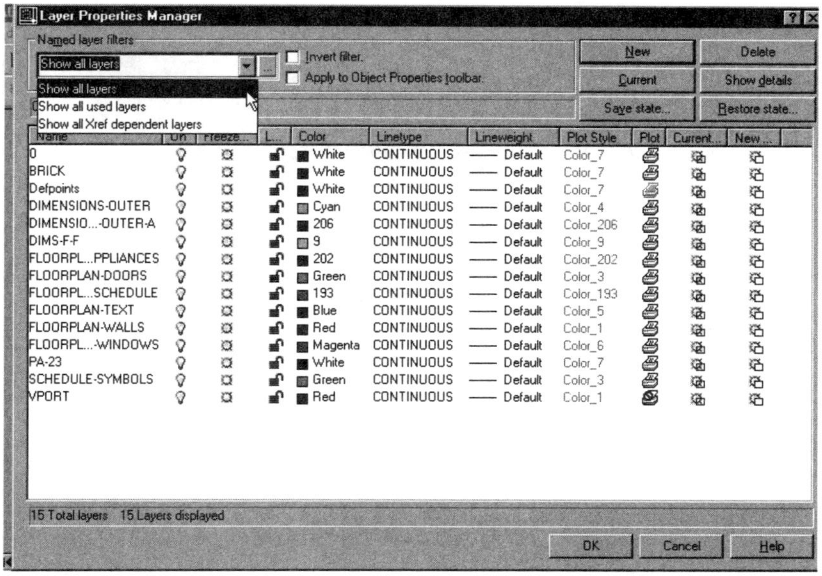

Figure 13.17 *Selecting the arrow beside the* **Named layer filters** *edit box will display this list, which can be used to select filters to display.*

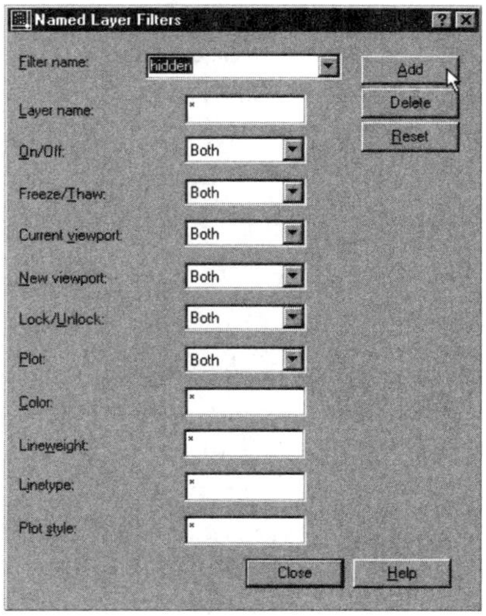

Figure 13.18 *Selecting the [...] button will produce the* **Named Layer Filters** *dialog box, allowing the scope of the filter to be defined.*

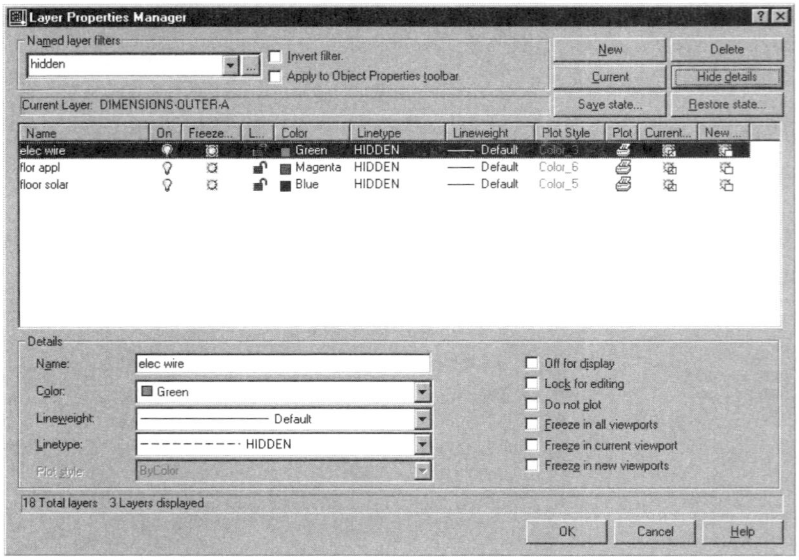

Figure 13.19 *Layers displayed based on the Hidden linetype.*

filter is based on the Hidden linetype. Clicking the **Add** button and then **Close** will shut the **Named Layer Filters** dialog box. Selecting the **Filter** edit arrow will display the list shown in Figure 13.17, with the Hidden filter now shown in the list. Selecting Hidden will produce a layer listing of layers created using the specified criteria. (See Figure 13.19.) All of the existing features are still displayed on the screen—they just won't be listed in the **Layer** display until the filter is altered. Opening the **Named layer filters** list and selecting the **Show All Layers** option will restore the total layer listing.

RENAMING LAYERS

The current name of a layer can be altered to better define the use of the layer at any time throughout the life of a drawing. Only the 0 layer and XREF layers can't be renamed. (XREF layers are layers that are created on a drawing that is attached to the current drawing, and they will be explained in later chapters.) Rename layers inside the **Layer Properties Manager** dialog box by selecting the name of the layer to be renamed, and then selecting it again. As it is selected the second time, the name will be highlighted and a box will be placed around the original name with a blinking cursor. Pressing BACKSPACE while the name is highlighted will remove the existing name and allow a new name to be entered. Pressing (ENTER) once will accept the name. Clicking **OK** removes the dialog box and restores the drawing area.

DELETING LAYERS

An unused layer can be deleted from the drawing base at any time throughout the life of the drawing—with the exception of the 0 layer or an XREF layer. Layers that are referenced to a BLOCK and a layer named DEFPOINTS cannot be deleted, even if the layer is empty. XREF, DEFPOINTS, and BLOCK will be explored in later chapters. Delete an unused layer from inside the **Layer Properties Manager** dialog box by selecting the name of the layer to be removed and then clicking the **Delete** button—the layer will be removed from the list.

ALTERING LINETYPE, COLOR, AND LAYERS

AutoCAD allows individual properties within a layer to be altered, as well as the properties of the entire layer to be changed. The **Layer Control** button on the **Object Properties** toolbar can be used to alter the qualities of an entire layer. Individual objects on a layer can be changed using the **Make Object's Layer Current** button, or by using the **Match Properties**, **Object Properties**, **Properties**, and **Qselect** commands.

LAYER CONTROL MENU

The **Layer Control** menu shown in Figure 13.20 can be used to alter the visibility of an existing layer. Display the menu by selecting the **Layer Control** arrow to the right of the box on the **Object Properties** toolbar. This menu can be used to toggle ON/OFF, Thaw/Freeze, Lock/Unlock, Plottable/Non-Plottable, and to set the current layer.

- Adjust the visibility of layers by selecting the ON/OFF button or the Thaw/Freeze button. As the button for each layer is selected, the layer will be displayed or removed, depending on the setting of the button.
- Lock or unlock layers by selecting the button of the desired layer.
- The Plottable/Non–Plottable button allows a layer to remain visible on the screen, but to be removed from the plotting set. In the default setting, all layers will be plotted. Selecting a layer and choosing this button will set the layer to non-plottable.
- Alter the current layer by selecting the name of the layer that you want to make current. As soon as the layer name is selected, the menu will be closed, the layer will be changed, and the drawing area will be restored.

Note: The Plottable/Non-plottable control can be a great asset for displaying information that you need to see, but others might not. A non-plottable layer could be created for placing questions to ask the engineer or instructor. The questions can be seen on the display, but they will not be plotted.

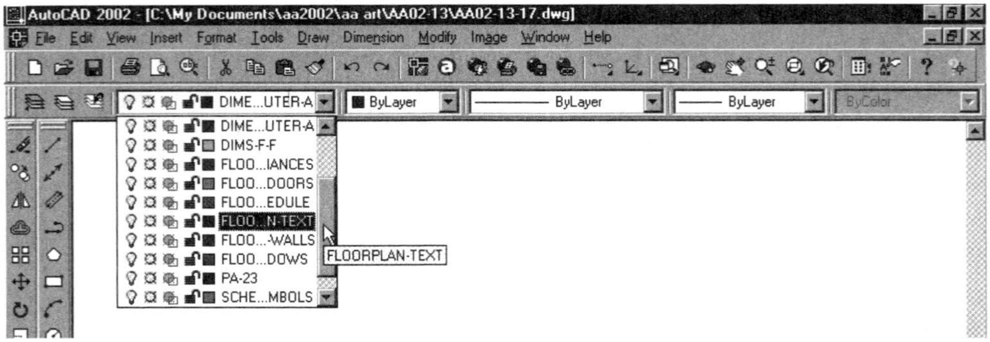

Figure 13.20 *Access the* **Layer Control** *menu by selecting the* **Layer Control** *arrow on the* **Object Properties** *toolbar. The menu can be used to toggle ON/OFF, Thaw/Freeze, Lock/Unlock, and Plottable/Non-plottable.*

MAKING AN OBJECT'S LAYER CURRENT

The **Make Object's Layer Current** button is located at the far left end of the **Object Properties** toolbar. The button can be used to alter the current layer. Selecting the button displays the following Command prompt:

Command: _ai_molc

Select object whose layer will become current:

The cursor will be changed to a pick box as the program waits for you to select a drawing object. Selecting a drawing object will make the layer that contains the selected object the current layer. The Command prompt will then display the name of the current layer, and drawing can continue.

ALTERING LAYERS USING LAYER PREVIOUS

The **Layer Previous** command can be used to undo changes made to layer settings. The command will affect settings such as On/OFF Freeze, Thaw, Lock, Color, Linetype, and Lineweight. **Layer Previous** will not undo the renamed layers, deleted layers, or added layers. Start the command by selecting the **Layer Previous** button on the **Object Properties** toolbar or by typing **LAYERP** (ENTER) at the Command prompt.

ALTERING THE DISPLAY WITH MATCH PROPERTIES

The properties of one layer object can be transferred to another object by the use of the **Match Properties** command. The **Match Properties** button is found on the **Standard** toolbar. Two other methods of executing the command include typing MA (ENTER) at the Command prompt or selecting **Match Properties**

from the **Modify** menu. The command will copy the properties of the first object to the second object, including the layer, the linetype, and the color. Each method will produce the following prompt:

Command: **MA** (ENTER) *(Or click the Match Properties button.)*

MATCHPROP

Select Source Object: *(Select the object with the properties that you would like to copy.)*

Current active settings: Color Layer Ltype Ltscale Lineweight Thickness

PlotStyle Text Dim Hatch

Select destination object(s) or [Settings]: *(Select the object that you would like to change.)*

Select destination object(s) or [Settings]: (ENTER)

Command:

Instead of selecting the object that you would like to change, typing **S** (ENTER) for Settings at the second prompt will produce the **Property Settings** dialog box shown in Figure 13.21. The box can be used to specify the qualities that will be copied to selected objects. One or more of the properties can be assigned including the **Color**, **Layer**, **Linetype**, **Linetype Scale**, **Lineweight**, and **Thickness**—as well as several other qualities that will be discussed in later chapters. Once the qualities have been selected, click the **OK** button and resume the command sequence by selecting the object to receive the new properties.

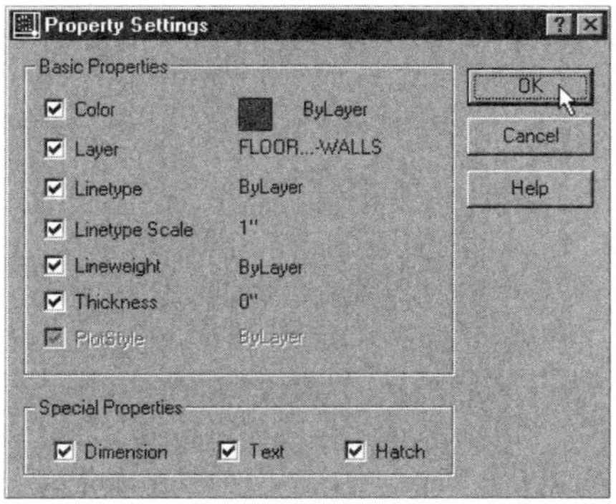

Figure 13.21 *Once a source object has been selected, typing* **S** *(*ENTER*) for Settings at the Command prompt will display the* **Property Settings** *dialog box. The dialog box allows specific properties of a layer to be altered without affecting other layer qualities.*

CHANGING OBJECT PROPERTIES

The **Properties** command can be used to control the properties of existing objects. Start the command by selecting the **Properties** button on the **Standard** toolbar, by typing **MO** (ENTER) or **CH** (ENTER) at the Command prompt or by selecting **Properties** from the **Modify** menu. Each method will display the **Properties** dialog box similar to Figure 13.22, showing the qualities of the current layer. Object properties can be listed in alphabetical order or categorized by groups depending on which tab is selected. The dialog box allows drawing properties including the Color, Layer, Linetype, Thickness, and Linetype scale to be altered. To alter a property for the selected object, choose either the name or value of the desired property. Each property will display a specific listing. Figure 13.23 shows the list for altering lineweights. Once the desired properties are altered, select the **Close** button to remove the dialog box and return to the drawing.

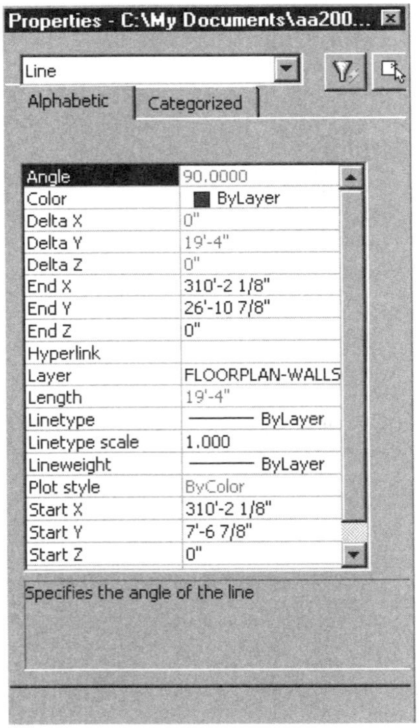

Figure 13.22 *The Properties dialog box can be used to control the properties of existing objects.*

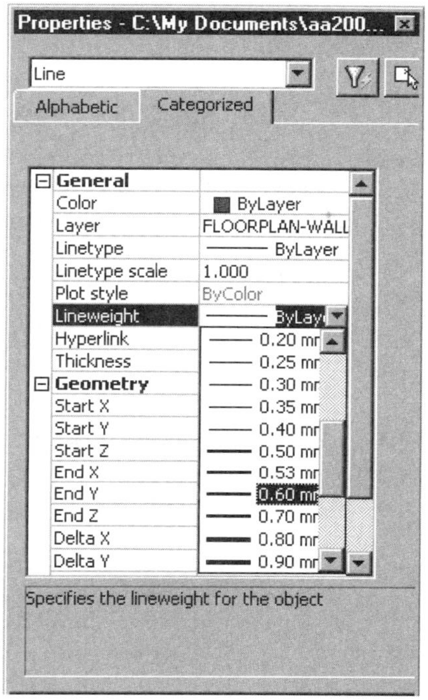

Figure 13.23 *To alter a property for the current object, choose either the name or value of the desired property. Each property will display a specific listing.*

CHANGING OBJECTS WITH QSELECT

The **Qselect** command can be used to create selection sets for altering the drawing objects. You can create selection sets to either include or exclude objects matching the specified object type and criteria. For instance, **Qselect** allows you to make a selection set made up of all objects that are green, or a selection set could be made to include everything except green objects. Start the command by right-clicking and choosing **Quick Select** from the shortcut menu, by selecting the **Quick Select** button in the **Properties** dialog box, or by selecting **Quick Select** from the Tools menu. You can also start the command by typing **QSELECT** (ENTER) at the Command prompt. Each method will produce the **Quick Select** dialog box shown in Figure 13.24. The dialog box contains seven key areas for adjusting the selection set:

> **Apply to**—If a selection set has been defined and is current, it will be selected automatically in the **Apply to** box. If no selection set has been defined, by default the only option is to apply changes to the entire drawing. Choosing the **Select Objects** button creates selection options.

Select Objects—The **Select Objects** button is located beside the **Apply to** edit box. Selecting the button will remove the **Quick Select** dialog box temporarily so that items can be selected to apply the selection filters to.

Object type—This box specifies the object type for the filter. Until a selection set is created, the **Object type** box displays a list of each shape included in the drawing, similar to Figure 13.25. Once a selection set is created, the list will include only the object types contained in the selection set.

Properties—The **Properties** list shows all searchable properties for the selected object to be used with the filter. The **Properties** display will vary depending on the object selected in the **Object type** edit box. If a circle is selected in the **Object type** edit box, a listing similar to Figure 13.26 will be displayed.

Operator—The **Operator** box controls the filter range and is dependent on the selected property. Operator options include: =Equals, < >Not Equal, >Greater than, and <Less than (see Figure 13.26).

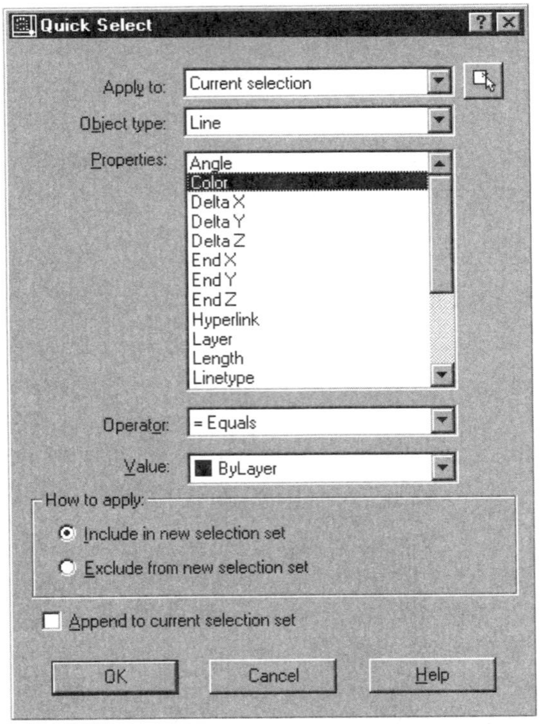

Figure 13.24 *Display the* **Quick Select** *dialog box by right-clicking and choosing* **Quick Select** *from the shortcut menu, by selecting the* **Quick Select** *button from the* **Properties** *dialog box, by selecting* **Quick Select** *from the* **Tools** *menu, or by typing* **QSELECT** *(ENTER) at the Command prompt.*

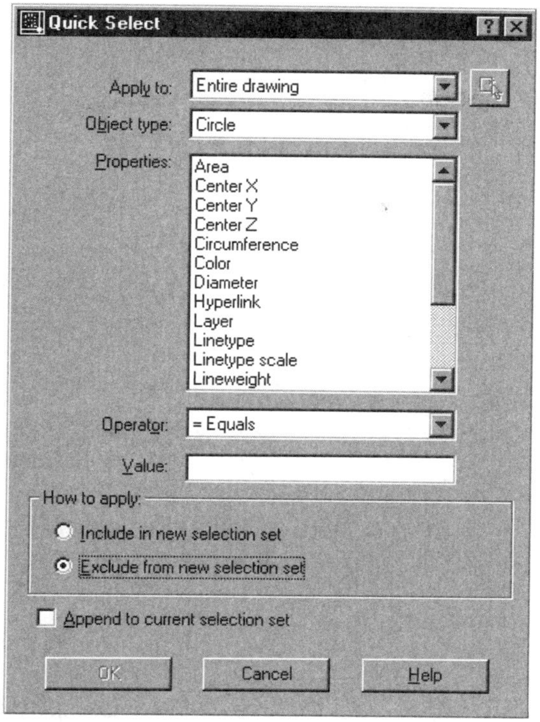

Figure 13.25 *The* **Object type** *box will only display a list of each shape included in the drawing. Once a selection set is created, the list will include only the object types contained in the selection set.*

Value—The **Value** edit box allows a value for the filter to be entered. For instance, if the property is a circle, a value of 24" could be used to include all circles equal to, not equal to, greater than, or less than 24" to define the selection set.

How to Apply—The **How to Apply** area contains radio buttons for including or excluding objects from the selection set.

- When **Include in new selection set** is active, a new selection set will be created composed of objects that match the filtering criteria.
- When **Exclude from new selection set** is active, a selection set will be created composed of objects that do not match the filtering criteria.

Append to current selection set—This check box determines if the selection set replaces or appends the current selection set.

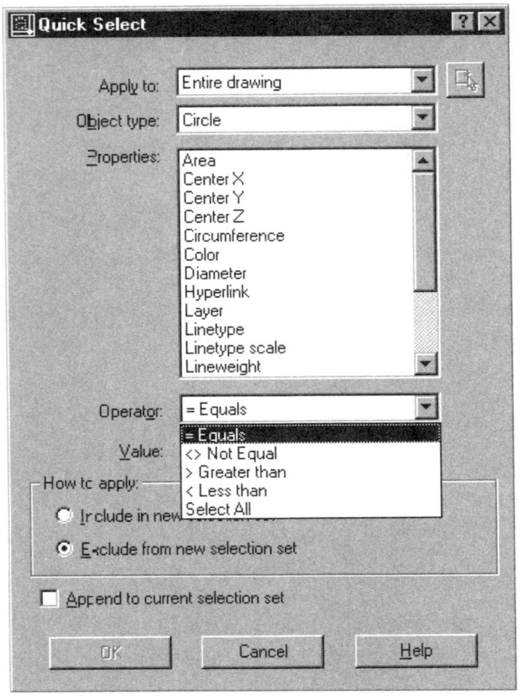

Figure 13.26 *The* **Properties** *display provides a list of all searchable properties for the selected object to be used with the filter. The display will vary depending on the object selected in the* **Object type** *edit box and the operator and value selected.*

Creating a Selection Set Using Qselect

A selection set for editing all green can be created using the following steps:

1. Display the **Quick Select** dialog box using one of the methods described earlier.
2. Using the **Apply to** edit box, select Entire drawing.
3. Select Multiple for the **Object type**.
4. Select Color in the **Properties** edit box.
5. Select=Equals in the **Operator** edit box.
6. Use the down arrow beside the **Value** edit box, and select Green.
7. Select **Include in new selection set** in the **How to apply** area.
8. Click the **OK** button to close the box.

AutoCAD will close the **Quick Select** dialog box and provide grips on each green object in the entire drawing. The Command prompt will show the number of items in the selection set, similar to Figure 13.27.

Excluding Objects from the Selection Set

A selection set excluding objects of a specific property can be created. Using the objects in Figure 13.27, a selection set excluding all circles larger than 24" could be created using the following sequence:

1. Display the **Quick Select** dialog box using one of the methods described earlier.
2. Using the **Apply to** edit box, choose Current Selection.
3. Select Circle for the **Object type**.
4. Select Diameter in the **Properties** edit box.
5. Select >Greater Than in the **Operator** edit box.
6. Enter a value of 24.
7. Select **Exclude from new selection set** in the **How to apply** area.
8. Click the **OK** button to close the box.

AutoCAD will close the **Quick Select** dialog box and remove from the selection set all circles with a diameter greater than 24".

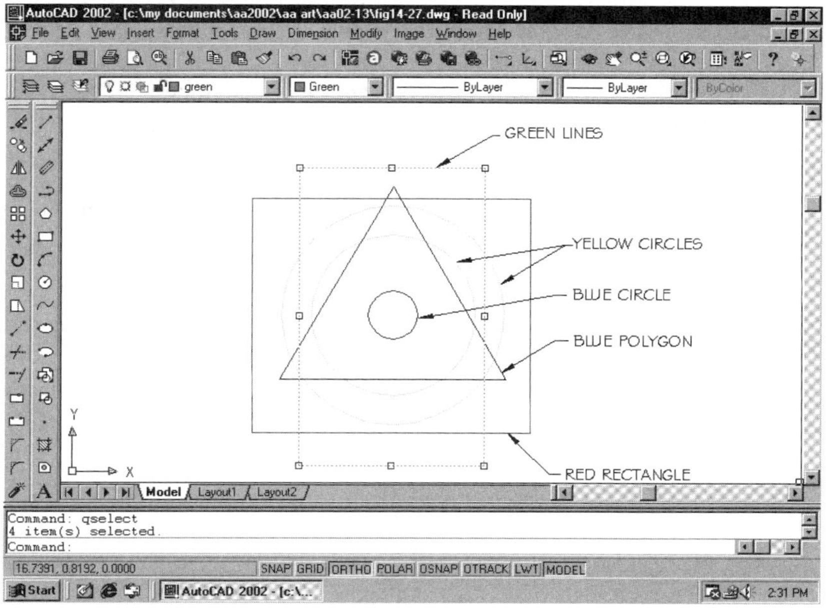

Figure 13.27 Qselect *can be used to create a selection set made up of all objects of a specific property description. In this example, all objects that are green have been selected for editing.*

CHAPTER 13 EXERCISES

1. Start a new drawing and load CENTER, CENTER2, CENTERX2, DASHED, DASHED2, and HIDDEN. Draw a line with each linetype, long enough so that the line pattern can be seen. Save the drawing as E-13-1.

2. Open E-11-4. Create layers for WALL with red lines and DOORWIN with green lines. Move the existing entities to the appropriate layer and save the drawing as E-13-2.

3. Open drawing E-7-4. Create layers for the specified layers and options:

PROPLINE	red	Centerx2	property line
TREES	Green	Dashed	tree shape
T-LAYOUT	7	Dot2	point/baseline

   Set the **Ltscale** appropriate for future plotting at a scale of 1"=10'–0". Save the drawing as E-13-3.

4. Open drawing E-7-1. Assign six different layers, each with a different linetype and color. Change each of the existing shapes to a different layer. Draw a different geometric shape on the remaining layers. Use **Pline** and OSNAP. Save the drawing as E-13-4.

5. Open the Architectural drawing template, set the **Ltscale**, and set the scale values appropriate to 1/4"=1'–0". Set the units to Architectural, with 1/16" fractional value. Establish layers that would be suitable for a two-level residential floor plan. Include the following: walls, dimensions, architectural notes, structural text, joist, beams, furniture, appliances, planting, plumbing, cabinets, section tags, doors and windows, decks at each level, electrical fixtures, and electrical wiring. No provisions need to be made for other drawings that typically are grouped with a floor plan. Assign colors to each layer. Assign linetypes appropriate to the layer contents. Save the drawing as QUARTER. This will become the basis for future architectural plan drawings that are to be plotted at 1/4"=1'–0".

6. Open drawing E-10-11 and create separate layers using an AIA long format so that walls, cabinets, fireplace, doors, windows, and plumbing fixtures can be separated from each other. Assign each layer a different color. Add center and hidden lines to the drawing base. Add a dishwasher to the kitchen, using hidden lines. Save the drawing as FLRPLN.

CHAPTER 13 QUIZ

1. How many linetypes are in the ACAD.LIN library?
2. List five commands that can be used to alter the appearance of a line.
3. List four objects that are affected by a change in linetype.
4. List the command sequence to draw with a centerline and a hidden line.
5. What would be the line scale value for a line that is to be plotted at a scale of 1"=20'?
6. When you describe the state of a layer, what qualities are considered?
7. List the four groups of information to be used in layer names recommended by the AIA.
8. What aspects of Change Properties apply to 2D drawing?
9. Describe guidelines for naming a model file using the AIA guidelines.
10. You are working on a drawing containing several different linetypes and would like to find out their names. What command and option should be used?
11. List the preferred linetype scale for the following scales.

    Full size

3"=1'–0"	1"=1'–0"
3/4"=1'–0"	3/8"=1'–0"
1/4"=1'–0"	1/8"=1'–0"
1/16"=1'–0"	1"=10'–0"
1"=20'–0"	1"=60'–0"

12. What will typing **DDLTYPE** at the Command prompt allow you to do?
13. Describe the difference between the effects of **Ltscale** and **Celtscale**.
14. A layer has a title of S-ANNO-SYMB. Who created the drawing and what do you think it would contain?
15. What information is in the minor group portion of a layer name?
16. You need to erase some information that is on a layer that is currently frozen. What is your plan?
17. You're about to start the plans for a one-level residence that will require a site plan, floor plan, framing plan, electrical plan, foundation plan, exterior elevations, and sections. How many drawing files do you expect to use, and how will you create layers for these files? (It's not in this chapter, you've got to search through your cranium.)

18. What is the advantage of using **Qselect**?
19. How can the **Properties** dialog box be accessed and what can be done with it?
20. What are the uses for the include/exclude buttons of **Qselect**?

CHAPTER 14

Placing Patterns in Drawing Objects

This chapter will introduce methods of representing various materials using different line or shape patterns referred to as hatch patterns.

Commands to be introduced in this chapter include

- **Bhatch**
- **Hatch**
- **Region**
- **Hatchedit**

HATCHING METHODS

Hatch patterns are repetitive patterns of lines, dots, and other symbols used to represent a surface or specific material. Concrete material is typically filled with a pattern of dots and triangles, soil as a series of perpendicular lines, and masonry as diagonal or crossing diagonal lines. In addition to the many patterns used in the construction industry, a solid pattern can be placed in an object. Figure 14.1 shows an example of a drawing with hatch and solid patterns added to define each material. The majority of hatch patterns are found in details and sections similar to Figure 14.2, where they are used to distinguish between materials. Hatch patterns can be used on elevations to represent siding and roofing materials as well as shades and shadows, as seen in Figure 14.3. Hatch patterns are also used on plan views to represent various materials and changes in floor or ceiling heights and to distinguish between existing and new construction. No matter the drawing, using the AIA guidelines, the pattern should be placed on a layer with a minor code of PATT, with a name such as A-FLOR-PATT or A-ELEV-PATT

Hatch patterns are placed using either the **Bhatch** (**Boundary Hatch**) or the **Hatch** command. Although the patterns will look exactly alike there is a major difference

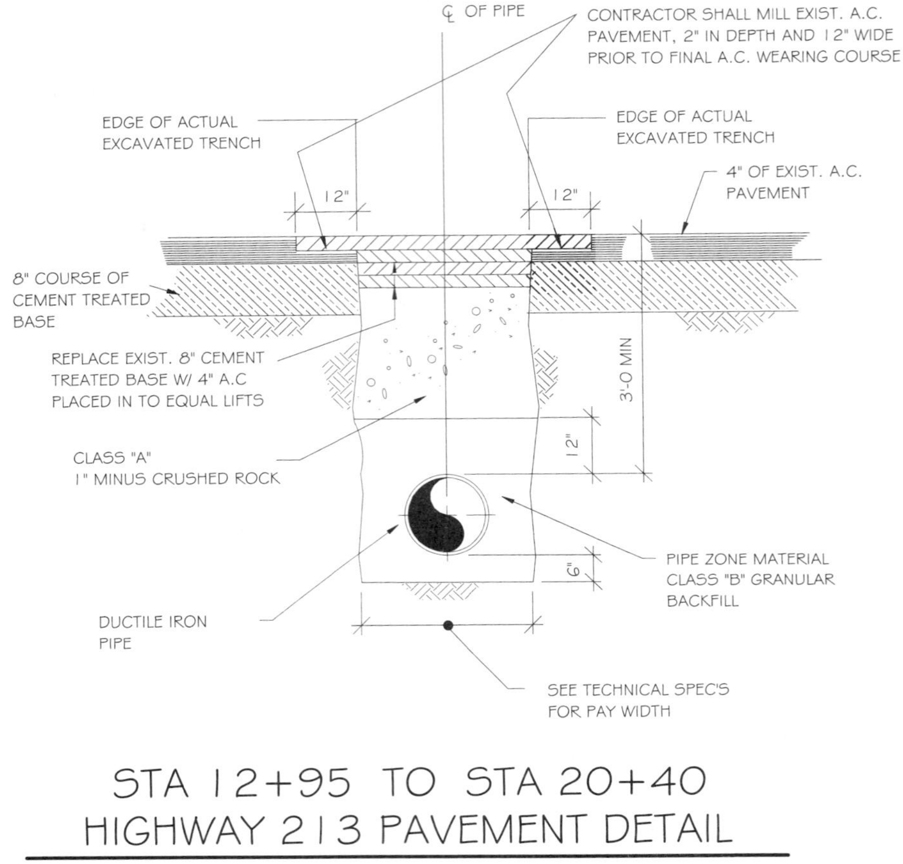

Figure 14.1 *Shading and drawing patterns to represent materials can be produced by using the* **Bhatch** *command. (Courtesy Department of Environmental Services, City of Gresham, Oregon.)*

between the two commands. **Bhatch** creates patterns that can be associative or non-associative, while **Hatch** creates only non-associative patterns.

- Associative patterns are linked to their boundaries. If the boundary is altered, the pattern is altered to match the boundary.
- Non-associative patterns are not affected by alterations to their boundaries.

The other major difference between the **Bhatch** and **Hatch** commands is in how they are added to the drawing. Boundary hatch patterns are applied using the **Boundary Hatch** dialog box. Display the **Boundary Hatch** dialog box by selecting **Hatch** on the

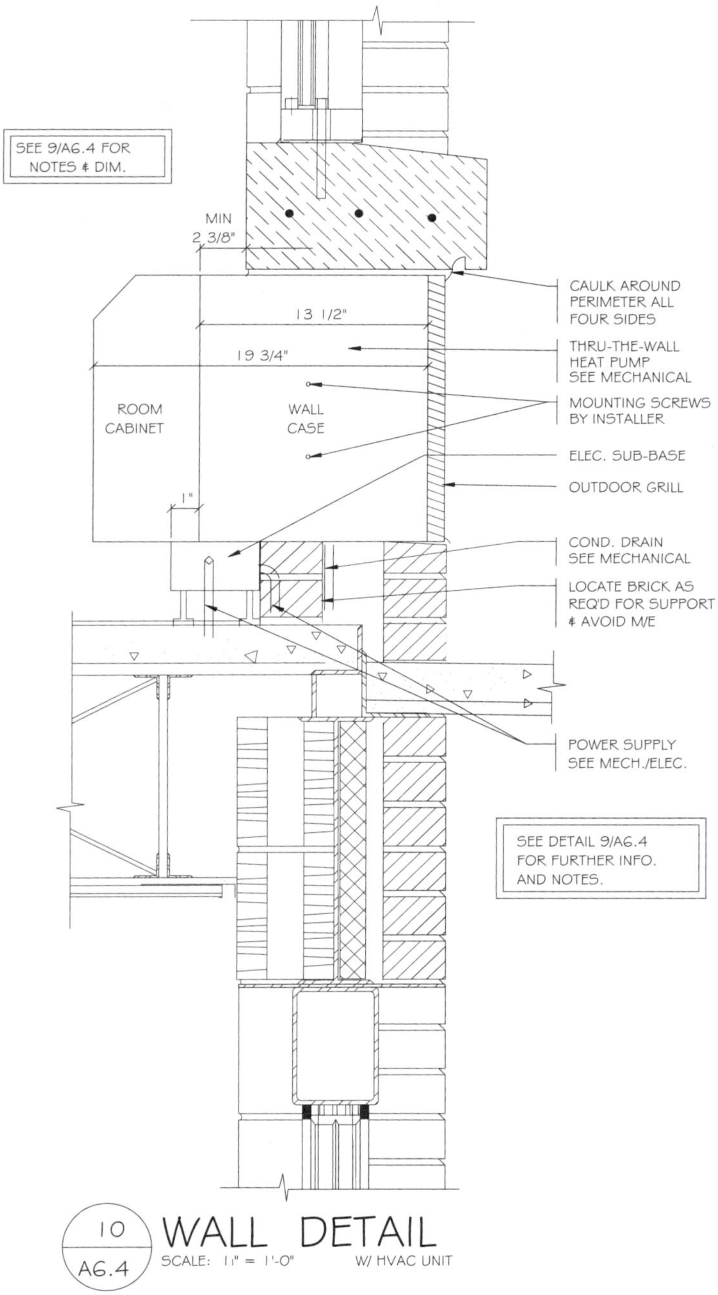

Figure 14.2 *Hatch patterns are used to distinguish various materials from one other. (Courtesy G. Williamson Archer, AIA of Archer & Archer, P.A.)*

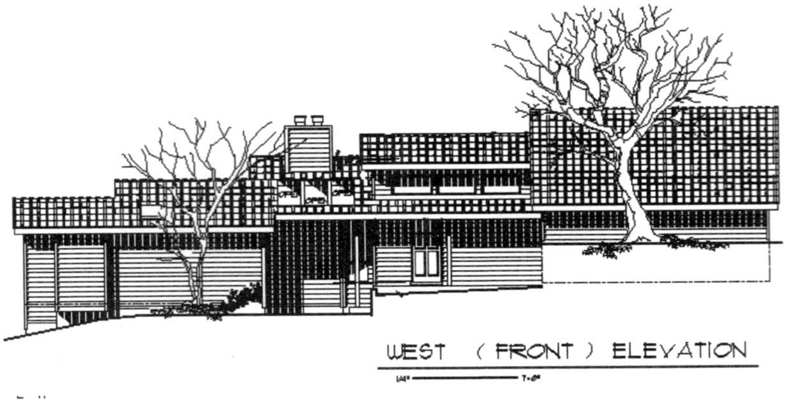

Figure 14.3 *Hatch patterns have been used to represent the tile roof, wood siding, and the shadows. (Courtesy Residential Designs.)*

Draw toolbar, by typing **BH** (ENTER) at the Command prompt, or by selecting **Hatch** from the **Draw** menu.

Hatch patterns are added to a drawing from the command line. **Hatch** will not be discussed in this text. Consult the **Help** menu if you're interested in using the slower, harder method.

USING THE QUICK TAB

The **Boundary Hatch** dialog box consists of the **Quick** and **Advanced** tabs. The **Quick** tab of the **Boundary Hatch** dialog box is shown in Figure 14.4. The **Quick** tab can be used for placing patterns for many common shapes encountered on construction drawings. As you prepare to hatch an area using **Bhatch**, you can take several steps to speed up the process. Zooming into an area rather than keeping the entire drawing in view will eliminate the number of lines that must be examined to determine the boundary set that will define the hatch area. Speed can also be gained if extra layers are turned OFF or Frozen so that objects on these layers are not examined as part of the selection set.

PATTERN TYPE

The **Type edit** box is used to set the pattern type. Options include Predefined, User defined, and Custom. The default value is to use a Predefined pattern. Selecting the down arrow displays the listings shown in Figure 14.5. Predefined patterns can be viewed in the **Hatch Pattern Palette** dialog box (shown later in Figure 14.7a). For most uses, the predefined patterns will be all that are ever needed. Selecting the User defined option creates a pattern based on the current linetype in the drawing. This

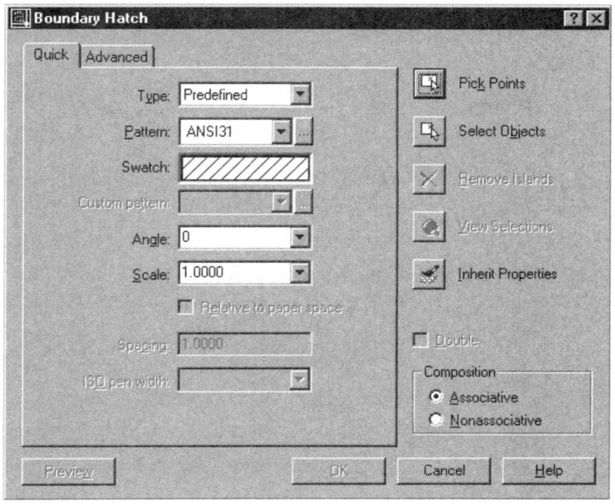

Figure 14.4 *Start The* **Bhatch** *command by selecting* **Hatch** *on the* **Draw** *toolbar, by typing* **BH** *(***ENTER***) at the Command prompt, or by selecting* **Hatch** *from the* **Draw** *menu. Each method will produce the* **Boundary Hatch** *dialog box.*

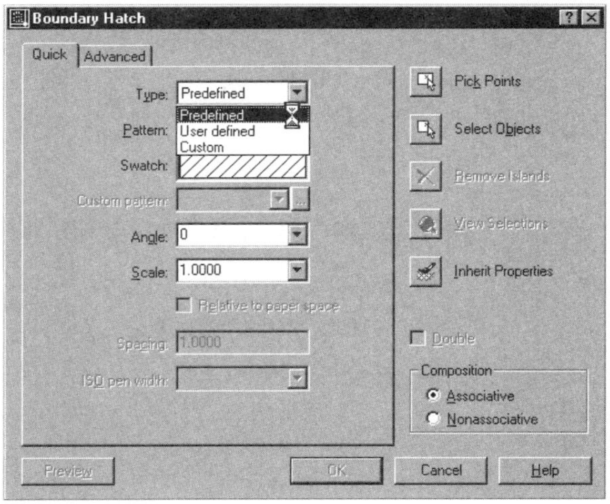

Figure 14.5 *Selecting the arrow beside the* **Pattern** *edit box will display options for selecting the pattern type.*

option allows the angle and spacing of the pattern to be altered. The Custom option allows a user created pattern to be to be added to the drawing base. Custom patterns allow a drawing pattern to be saved in the acad.pat file or in its own PAT file. Methods of creating patterns will be discussed later.

SELECTING A PATTERN

Notice that the ANSI31 hatch pattern is the default setting shown in the Pattern edit box. A different pattern can be selected for a **Bhatch** and it will remain the default until a new pattern is selected. You can select a new pattern using the **Pattern** edit box using three different methods. Selecting the pattern name in the edit box or the edit box arrow will display a listing of the hatch pattern names similar to the list shown in Figure 14.6. The slide bar can be used to scroll through the list of predefined patterns. Clicking the name in the pattern box a second time will close the list without making any changes. Selecting a name from the list will make that pattern the current pattern and close the list. This method of pattern selection works well if you are familiar with the names of patterns.

If you're a new user, names like ANSI31 or ANSI32 will be meaningless. Select the **Pattern** [...] button beside the edit box to see a display of each image in the **Hatch Pattern Palette** dialog box. See Figure 14.7a. The **ANSI** and **Other Predefined** (see Figure 14.7b) tabs will provide the patterns typically associated with construction drawings. The slide bar can be used to move through the listing to select the desired hatch pattern. Once a pattern is selected for use in a drawing, click the **OK** button to return to the **Quick** tab.

A third method of selecting a hatch pattern is to use the **Swatch** display. Clicking the **Swatch** pattern will display the most recently used tab of the **Hatch Pattern Palette** dialog box.

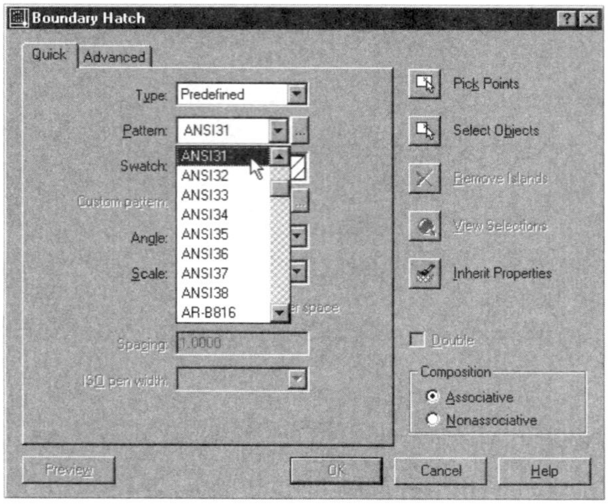

Figure 14.6 *The* **Pattern** *list can be used to select the pattern to be used for the Bhatch command.*

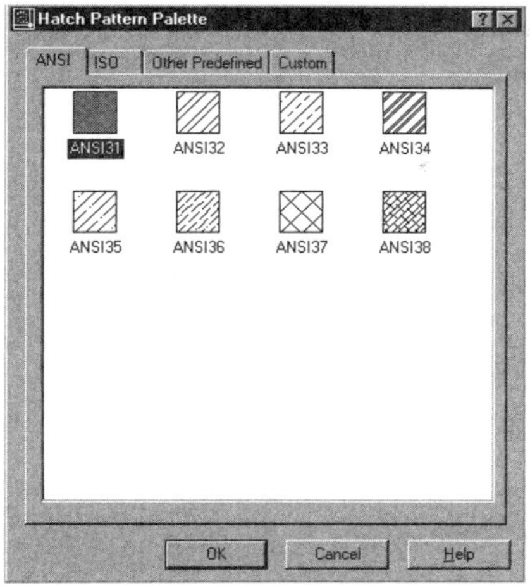

Figure 14.7a *Selecting the* **Pattern** *[...] button beside the* **Pattern** *edit box will display the* **Hatch Pattern Palette** *dialog box.*

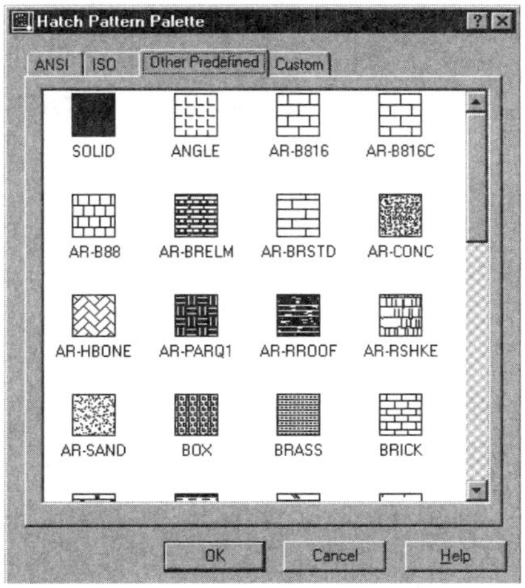

Figure 14.7b *The* **Other Predefined** *tab provides the patterns typically associated with construction drawings. The slide bar can be used to move through the palette to select the desired hatch pattern.*

Of the many interesting patterns found in the **Hatch Pattern Palette** dialog box, several hatch patterns are specific to architectural drawings.

ANSI Tab

Pattern	Common Use
ANSI31	Masonry in plan and section view, thin veneers in section views, shading for elevations
ANSI32	Steel plates in section view
ANSI37	Firebrick in section view

Other Predefined Tab

Pattern	Common Use
SOLID	General usage when a surface needs to be filled solid
AR–B816 &16C	Concrete masonry units in elevation
AR–CONC	Poured concrete in section view
AR–HBONE	Flooring or brick patterns in plan and elevations
AR–PARQ1	Flooring or brick patterns in plan and elevations
AR–RROOF	General roofing in elevation (non specific to material)
AR–RSHKE	Cedar shake in elevation (typically not used for roofing)
AR–SAND	EIFS, stucco or plaster in elevations and section
BRICK	Brick in elevation
DOTS	Shading in elevation
EARTH	Soil in section

PATTERN ANGLE

Once the pattern has been selected, the **Angle** edit box should be considered. By default, the pattern will be placed at an angle of 0. The pattern will be reproduced as shown in the **Swatch** display and in the **Hatch Pattern Palette** dialog box. You can enter a value by keyboard or select a value from the **Angle** list shown in Figure 14.8. Entering a value other than zero allows the hatch pattern to be rotated. By entering an angle of 45°, you can use pattern ANSI31 to represent a shadow, as seen in Figure 14.9.

PATTERN SCALE

The **Scale** option sets the size of the pattern elements with the default of one drawing unit. The relationship of the pattern can be controlled by the drawing scale factor. Adjust the scale value using the same method as used to adjust the angle value,

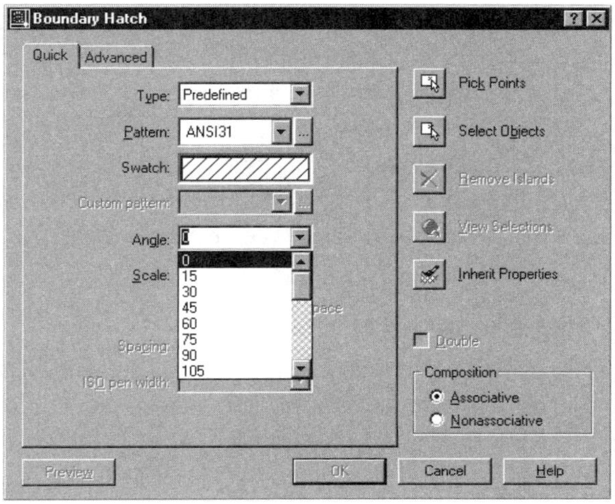

Figure 14.8 *A value for rotating the pattern can be entered by keyboard or by selecting a value from the* **Angle** *list.*

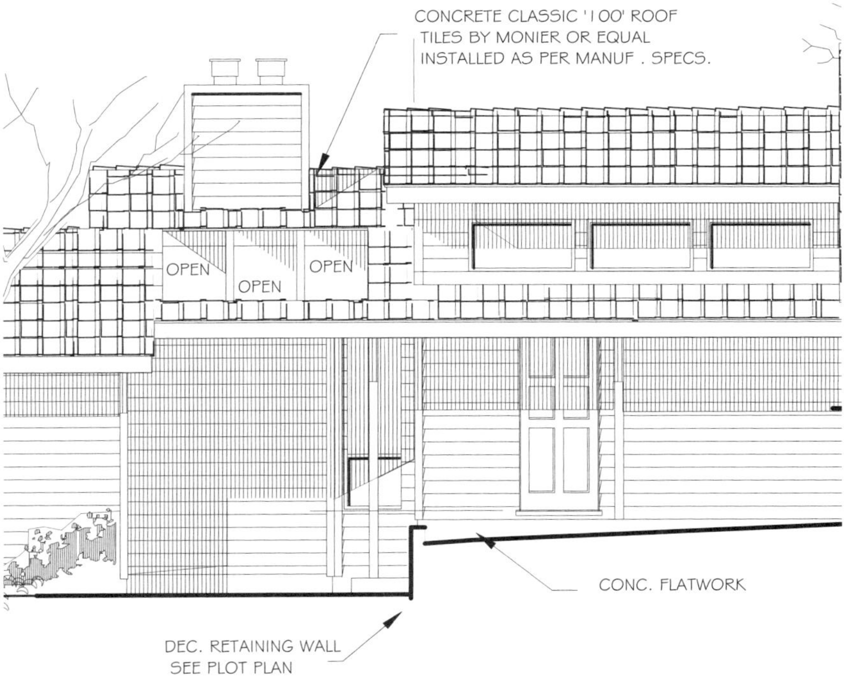

Figure 14.9 *The hatch pattern ANSI31 is drawn at a 45° angle by default. Providing a rotation angle of 45° will reproduce the pattern in a vertical position. (Courtesy Residential Designs.)*

either by keyboard entry or by choosing from the **Scale** list. Figure 14.10 shows an example of three differently scaled patterns. Selecting the largest scale pattern that is still visually appealing will save **Regen** time. Drawing a line the length of the desired pattern to be inserted can help to ease the scale selection. If you need a hatch pattern to represent a concrete block, draw a 16" long line below the area to receive the hatch. Later in the command process, you'll have a chance to preview how the pattern will appear. The 16" line will serve as a reference to verify that the scale selection is accurate.

SELECTING OBJECTS FOR HATCHING

Once the pattern, angle, and scale have been selected, use one of the methods from the right side of the dialog box to choose the area to be hatched. For now, the options are limited to **Pick Points** and **Select Objects**.

If the **Pick Points** button is selected, only one point will need to be selected to place the pattern.

If the **Select Objects** button is selected, any of the selection methods introduced in previous chapters can be used to define the area.

Other methods of selecting the boundary can be selected using the options on the **Advanced** tab, which will be discussed later in this chapter.

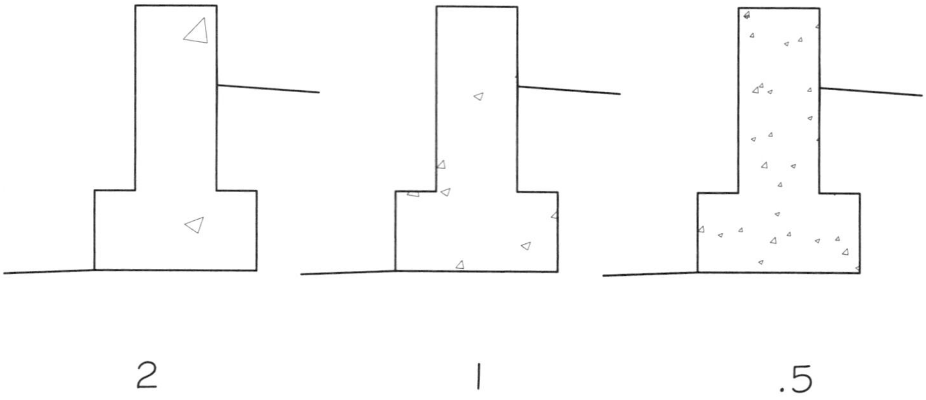

Figure 14.10 *The effect of altering the scale factor. Always use the largest size possible to save* **Regen** *time.*

Using the Pick Points Option

The **Pick Points** option can be used to hatch drawing objects that form an enclosed boundary. Once you've selected the **Pick Points** button, the dialog box will be removed and the drawing will be restored. A prompt is now displayed requesting "Select internal point." Select any point that lies within the area to be hatched. Once it is selected, the command line will display the prompt:

Select internal point: Selecting everything ...

Selecting everything visible…

Analyzing the selected data...

Analyzing internal islands…

Select internal point: (ENTER)

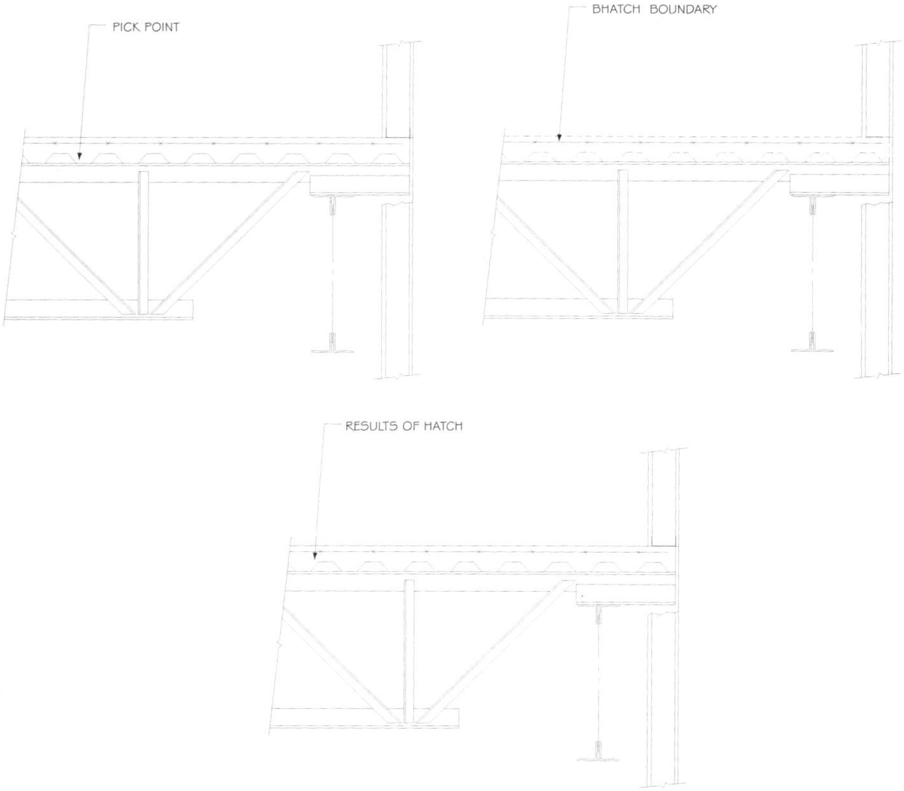

Figure 14.11 *Using* **Bhatch** *to add a pattern to a concrete floor slab. Once the area to receive the pattern is selected, the boundary will be highlighted. The AR-SAND pattern at a scale of .5 and an angle of 0 was selected. With this pattern, the angle is unimportant.*

Because the **Pick Points** button was selected, you can select the object to be hatched by selecting one point that lies within the object. As the point is selected, the boundary of the area to be hatched will be displayed as dashed lines. AutoCAD is waiting for you to select additional objects to be hatched. Press (ENTER) or right-click to redisplay the **Boundary Hatch** dialog box. Press (ENTER) a second time to apply the hatch pattern to the selected object. Figure 14.11 shows the resulting hatch pattern.

Using the Select Objects Option

Selecting the **Select Objects** button will remove the **Boundary Hatch** dialog box and prompt you to select objects. Figures 14.12a and 14.12b show the process for selecting an object to be hatched using the **Select Objects** option. In Figure 14.12a, the fireplace was selected using the Window selection method. Figure 14.12b shows the hatch pattern that would result. The object to be hatched does not need to form a closed boundary, or it can contain an internal area referred to as an island that will not be hatched. Figure 14.13 shows the results of selecting an object with an island using the Window selection method.

If the object to be hatched is surrounded by a polyline, the object can be selected by picking the object at any point, as seen in Figure 14.14. The pattern will be the same

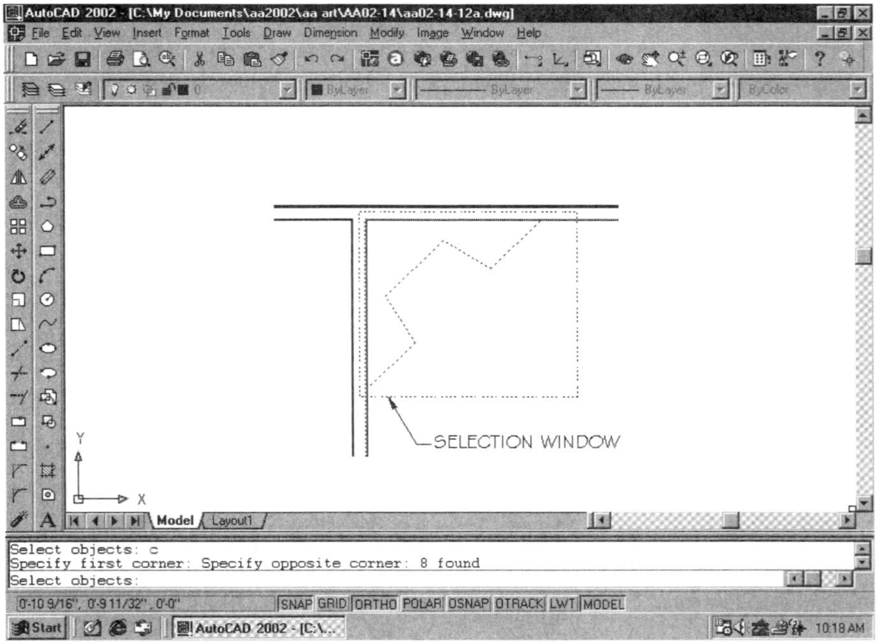

Figure 14.12a *With the* **Select Objects** *option, an object can be selected using any selection method, such as Window or Crossing.*

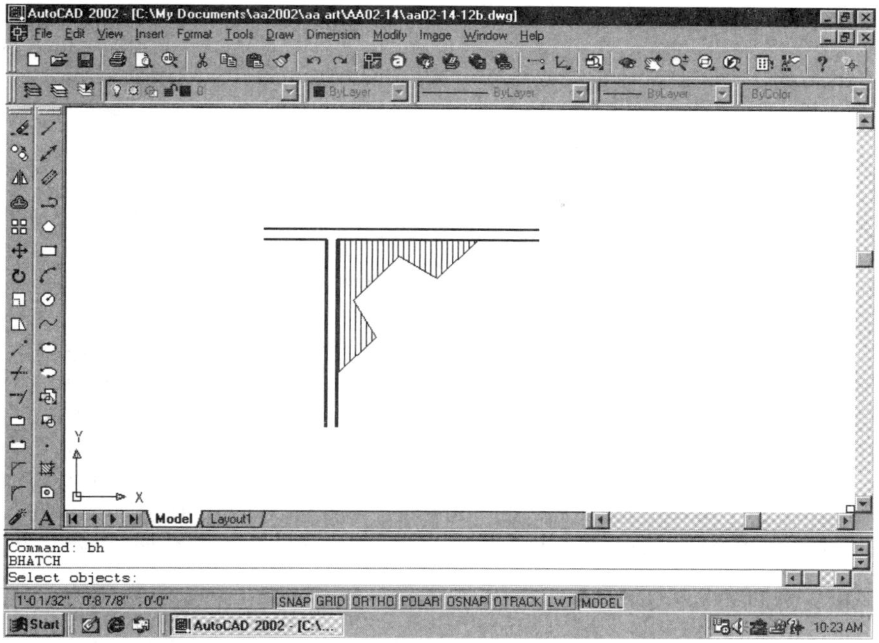

Figure 14.12b *The resulting hatch pattern for the selection set in Figure 14.12a.*

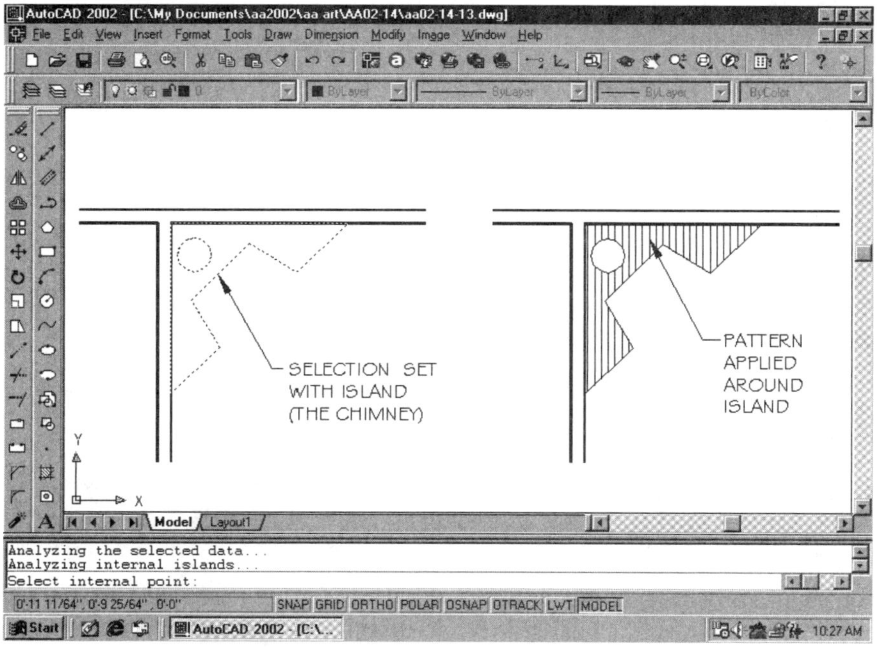

Figure 14.13 *The result of using the* **Bhatch Select Objects** *option on an object with an island. The placement of the pattern can be altered with the Style option.*

as the pattern in Figure 14.12b. Once the object to be hatched is selected, press (ENTER) to restore the **Boundary Hatch** dialog box. Pressing (ENTER) a second time will apply the pattern. Figure 14.15 shows the results of selecting an object with islands to be hatched by selecting an edge.

USING THE PREVIEW OPTION

If you're unsure of any of the pattern parameters, select the **Preview** button prior to applying the hatch pattern. The **Preview** option provides a display of how the pattern will appear based on the current settings. **Preview** is especially helpful when a large, irregularly shaped boundary is to be hatched. When the **Preview** option is selected, the dialog box is removed and the object with the hatch pattern is displayed as seen in Figure 14.16. Notice that the boundary is still dotted, which is a reminder that the hatch pattern has not yet been added to the drawing file. Press (ENTER) or right-click to restore the **Boundary Hatch** dialog box. If the pattern is acceptable, click the **OK** button or press (ENTER). If the preview did not display the intended results, any of the pattern options can be altered. In Figure 14.17, a new hatch pattern was selected by returning to the **Boundary Hatch** dialog box. The **Preview** option was selected.

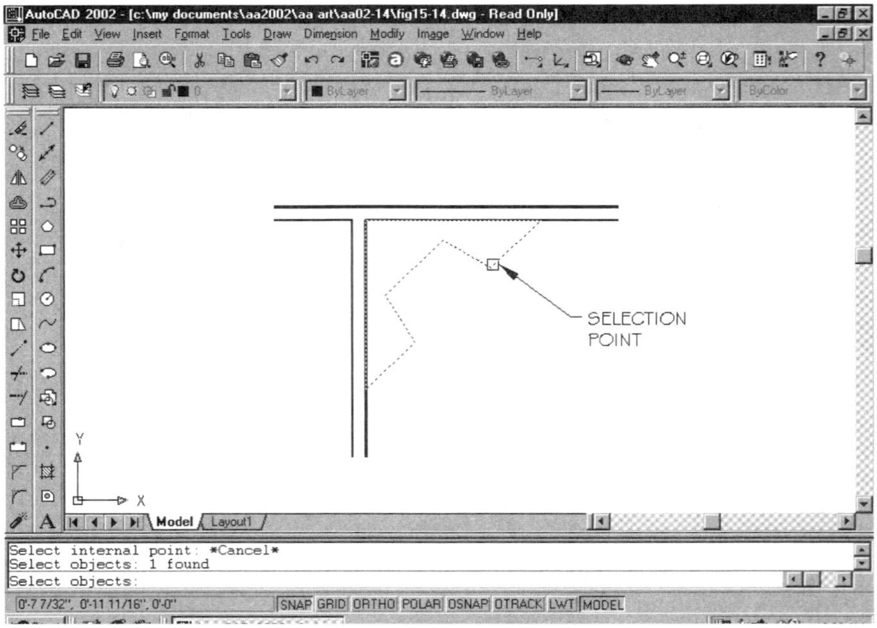

Figure 14.14 *When the boundary of an object is formed by a polyline, it can be selected by single point. The resulting pattern will be the same as the pattern in Figure 14.12b.*

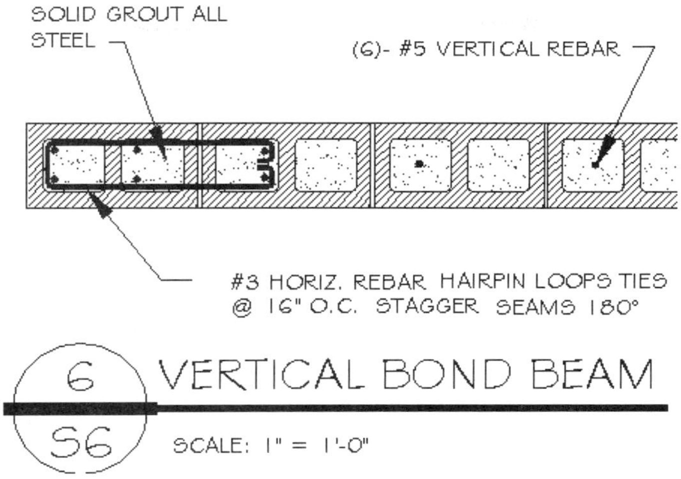

Figure 14.15 *The resulting hatch pattern for a selection set with islands.*

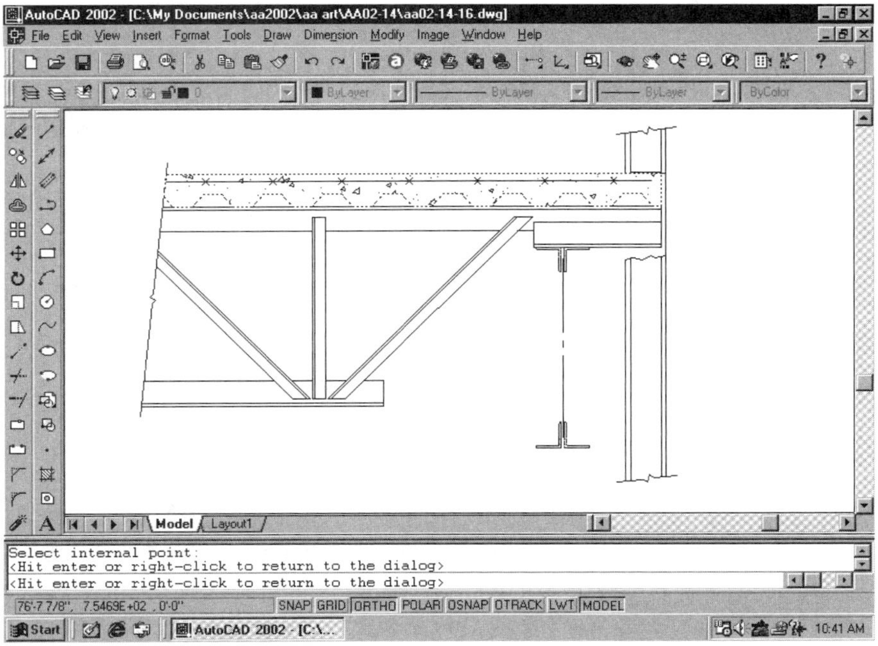

Figure 14.16 *You can display the proposed pattern by selecting the* **Preview** *button. If the pattern is not what you expected, the properties can be altered before you apply the pattern.*

Using **Preview** to test alternatives can go on indefinitely. The final change is to adjust the **Angle**. The results can be seen in Figure 14.18. Because this view was acceptable, (ENTER) was used to return to the **Boundary Hatch** dialog box, and the **OK** button was clicked.

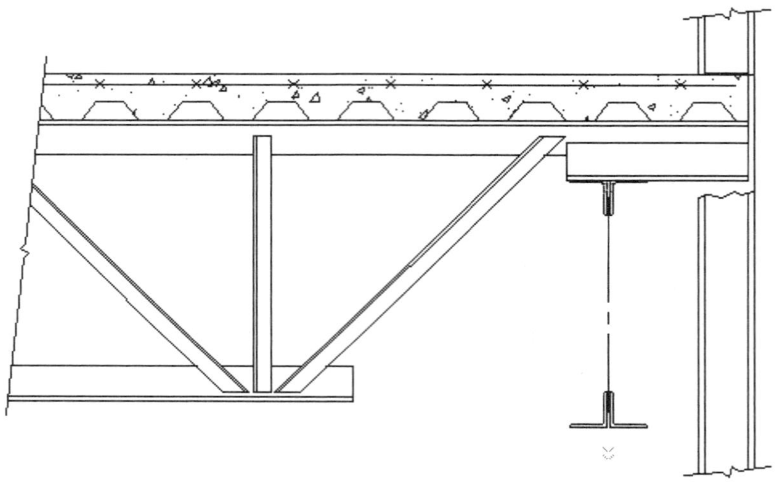

Figure 14.17 *If the* **Preview** *is acceptable, you can add it to the drawing base by clicking the* **OK** *button. You can select a new pattern by returning to the* **Boundary Hatch** *dialog box.*

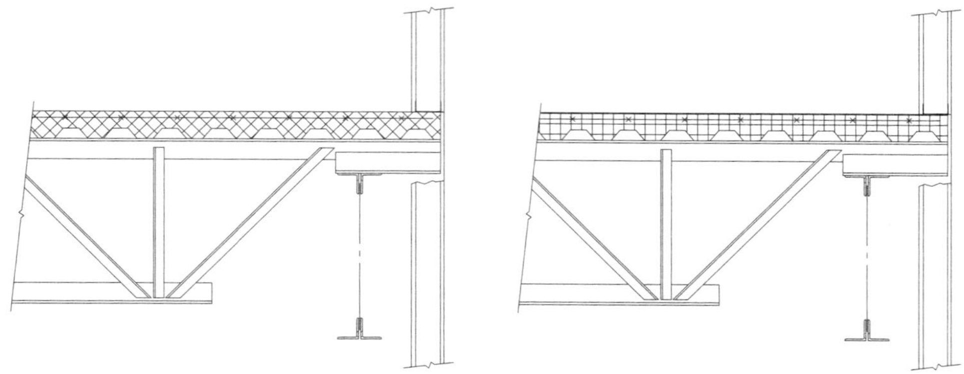

Figure 14.18 *The* **Preview** *option can be used to view the hatch pattern prior to adding it to the drawing base. If you are unhappy with the pattern, change the pattern, the scale, and the angle options and then reuse the* **Preview** *option.*

Summary of Steps in Placing a Hatch Pattern

To place the hatch pattern in Figure 14.18 would require the following steps.

1. Display the **Boundary Hatch** dialog box by selecting **Hatch** on the **Draw** toolbar, by typing **BH** (ENTER) at the Command prompt, or by selecting **Hatch** from the **Draw** menu.
2. Choose the **Quick** tab.
3. Select the [...] button to display the **Hatch Pattern Palette** dialog box.
4. Choose the **ANSI** tab.
5. Select the ANSI37 display and click **OK**.
6. Select the **Pick Points** button in the **Boundary Hatch** dialog box.
7. Specify an internal point in the object to be hatched.
8. Select the **Preview** button to display the proposed hatch pattern.

If the pattern is not acceptable, alter the hatch properties using the **Quick** tab, and then use **Preview** option again.

9. If the pattern is acceptable, click the **OK** button to apply the hatch pattern to the drawing.

REMOVING ISLANDS

Figure 14.19 shows a concrete floor slab with four islands. When the boundary is selected, each of the islands will automatically be excluded from the hatch. The hatch pattern was omitted from each of the cavities in the floor on the left side of Figure 14.19. Once the object to be hatched has been defined, selecting the

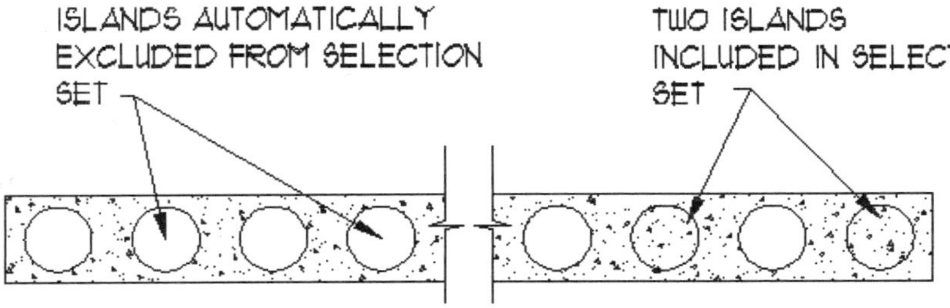

Figure 14.19 *The* **Remove Islands** *option can be used to control the placement of the pattern in islands within the boundary.*

Remove Islands button in the **Boundary Hatch** dialog box allows islands to be included and receive the hatch pattern. The example on the right side of Figure 14.19 shows the result of removing two of the islands from the boundary set.

VIEW SELECTION

With all of the options available for altering the boundary or patterns used in a boundary hatch, you might lose track of what options you've selected. The **View Selections** button will remove the dialog box and display the currently defined boundaries.

INHERIT PROPERTIES

This option is very useful when you are hatching an object using an existing pattern and you can't remember the qualities of the existing hatch pattern. Selecting the **Inherit Properties** button will close the **Boundary Hatch** dialog box and return the drawing area with the prompt:

Select associative hatch object: *(Select the hatched pattern that you would like to duplicate.)*

Inherited Properties: Name <ANSI31>, Scale <2.000>, Angle <0>

Select internal points: (ENTER)

Pressing (ENTER) will restore the **Boundary Hatch** dialog box.

COMPOSITION

The final portion of the **Boundary Hatch** dialog box to be examined is the **Composition** area, which contains toggles for **Associative** and **Nonassociative**. With the **Associative** button active, as an object containing a hatch pattern is edited, so is the hatch. With **Nonassociative** activated, if an object with a hatch pattern is altered, the object is altered, but the hatch pattern remains unaffected. Figure 14.20 shows the effect of stretching two steel beams and the effect of an associative pattern.

CONTROLLING PATTERN PLACEMENT

Although some hatch patterns such as soil, sand, and concrete can be placed without regard to the starting point of the boundary, other patterns such as brick or block should be placed relative to the hatch boundary. The **Snap** command can be used to control the relationship of the hatch pattern to the boundary. Use the following command sequence to make one of the hatch boundaries the starting point for the pattern.

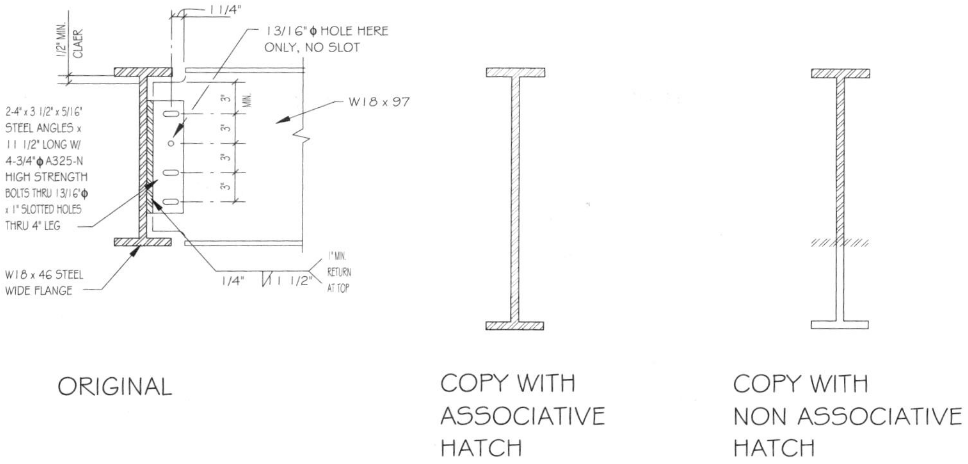

Figure 14.20 *With* **Associative** *active, stretching a steel beam will alter the hatch pattern. With the* **Associative** *box inactive, the beam is stretched, but the hatch pattern is unaltered.*

Command: **SNAP** (ENTER)

Specify snap spacing or [ON/OFF/Aspect/Rotate/Style/Type] <0'-0 1/2">: **R** (ENTER)

Specify base point <0'-0",0'-0">: *(Select a corner of the pattern boundary)*

Specify rotation angle <0>: (ENTER)

Command: **BH** (ENTER)

The pattern can now be placed using the **Bhatch** command.

DEFINING BOUNDARY HATCH USING THE ADVANCED TAB

Although the **Quick** tab can be used to hatch simple objects, the **Advanced** tab provides options to help increase the efficiency of selecting boundary hatch boundaries. Unless you're working on a very large drawing, you will probably not need this option. The **Bhatch** command evaluates all objects displayed on the screen when you select objects to be hatched. If you are working on a very large structure, the options on the **Advanced** tab can speed the boundary selection process. Key elements of the **Advanced** tab include controls for island detection, object type, boundary set, and island detection method. The **Advanced** tab can be seen in Figure 14.21.

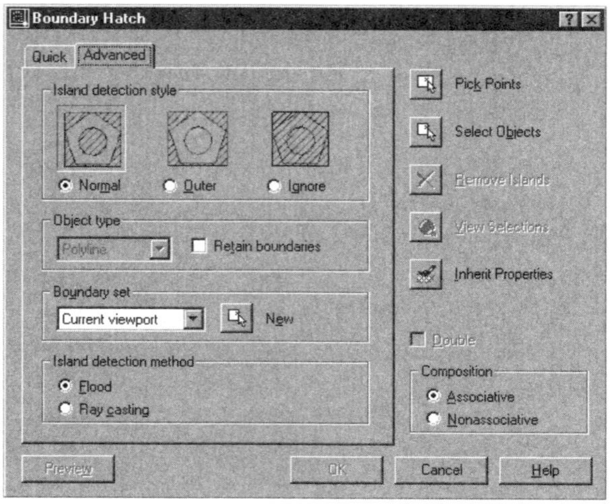

Figure 14.21 *The* **Advanced** *tab provides options to help place hatch patterns in complex shapes. Key elements of the tab include controls for island detection, object type, boundary set, and island detection method.*

ISLAND DETECTION STYLE

The **Island Detection Style** area of the **Advanced** tab allows three styles of island detection for displaying the hatch pattern. The **Normal** setting is the default and the image tile reflects how the pattern will be displayed. You can change the setting by selecting one of the other radio buttons.

Normal

Figure 14.22 at A shows how hatching sets made up of multiple objects will be hatched when the default of **Normal** style is selected and each of the objects is selected as a boundary. **Normal** hatching style works inward, starting at the area boundary, and proceeds until another boundary is found. The pattern is turned off until another boundary is discovered.

Outer

Figure 14.22 at B shows how hatching sets made up of multiple objects will be hatched when the **Outer** option is selected and each of the objects is selected as a boundary. The **Outer** hatching style works inward, starting at the area boundary and proceeds until another boundary is found. The pattern is turned off at the second boundary.

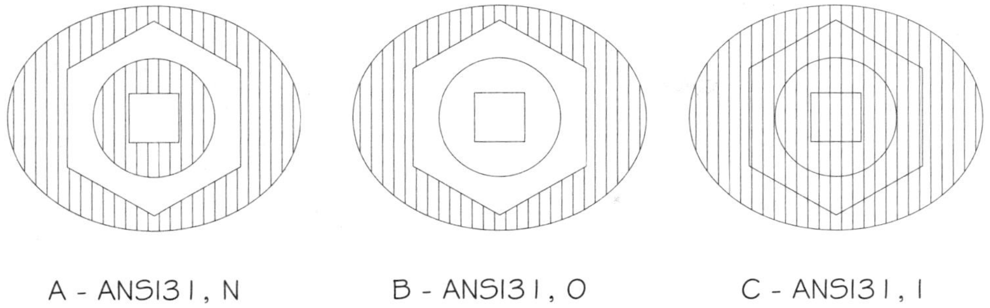

Figure 14.22 *The* **Island Detection Style** *area controls how the pattern will be displayed around multiple object sets. The* **Normal** *option works inward until another boundary is found. The pattern is not displayed until another boundary is found. The* **Outer** *option will display only the pattern in the outer boundary area. The* **Ignore** *option ignores all boundaries and hatches all areas within the outer boundary.*

Ignore

Figure 14.22 at C shows how multiple objects will be hatched when the **Ignore** option is selected. The **Ignore** hatching style works inward, starting at the area boundary, and passes through all objects within that boundary.

Note: With the **Normal** setting, if a hatch operation encounters another shape or text, the pattern will be placed around the text. Although text has not yet been added to the drawing, it will be important to remember that text inside a hatch boundary will still be legible (see Figure 14.23).

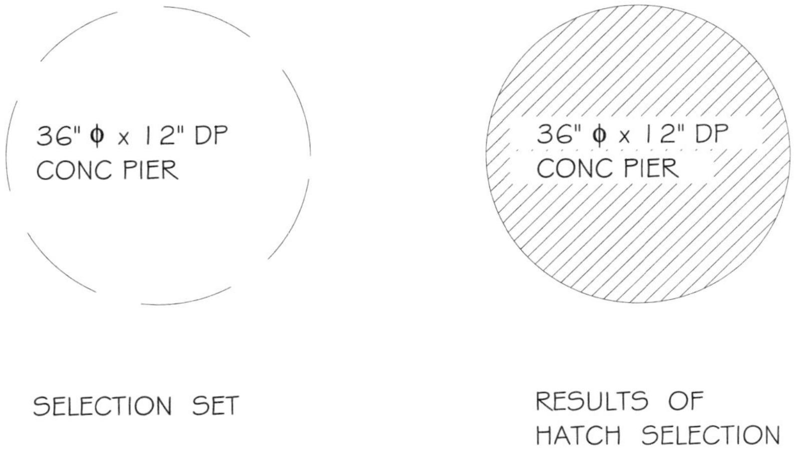

Figure 14.23 *Text will be treated like an island and automatically excluded from the area to be hatched.*

OBJECT TYPE

The **Object Type** area controls boundary options for applying the hatch pattern. The options for the boundary include the **Retain Boundaries** check box and the **Object Type** box. The **Retain Boundaries** option specifies whether or not a temporary boundary will be added to the drawing. Boundaries for hatch patterns can be specified by **Object Type**. These options are not active unless the **Retain Boundaries** option is active. **Object Type** controls the type of new boundary to be created. Options are Region and Polyline. You've explored polylines in a previous chapter. A closed shape comprised of a polyline can be used to form a boundary for hatch patterns.

Region

A region is an enclosed 2D area created from shapes called loops. Loops are a sequence of connected curves and lines that do not intersect and define an area on a plane. Sounds technical until you realize that loops are nothing more that the basic shapes that make up a drawing. Loops can be a combination of lines, polylines, circles, arcs, elliptical arcs, splines, and solids. Loops or regions form the area where the hatch will be applied. Create a region by selecting the **Region** button on the **Draw** toolbar, by selecting **Region** from the **Draw** menu, or by typing **REG** (ENTER) at the Command prompt. Regions are created using the following command sequence:

Command: **REG** (ENTER)

Select objects: **W** (ENTER) *(Select objects to form the region using any selection method and press (ENTER).)*

4 found

Select objects: (ENTER)

1 loop extracted.

1 Region created

Command:

The selected lines will now function as a polyline, forming a boundary for future hatch patterns.

BOUNDARY SET

This portion of the dialog box defines how the selection set for the hatch pattern will be selected when the **Pick Points** option is selected. By fine-tuning the selection set, you can have the **Bhatch** command function more quickly because AutoCAD has to examine fewer objects to define the selection set.

The default selection to define boundaries is Current viewport.

Selecting **New** will remove the dialog box from the screen and allow a new boundary to be selected. The existing boundary set will be discarded and you will be allowed to construct a new boundary set to define the area to be hatched.

Once the new boundary set is selected, the dialog box will be returned and the default setting is now Existing set. An unlimited number of boundary sets can be selected. The last set selected will remain the default until a new boundary set is selected. When Current viewport is selected, the selection set will be created from everything that is visible in the current viewport. Selecting this option when there is a current boundary set will discard the current selection and place the pattern in everything in the current viewport. The normal setting is to place the pattern based on the Existing boundary set.

ISLAND DETECTION METHOD

The final option of the **Advanced** tab is the **Island Detection Method** area. In its current state of **Flood**, internal features (islands) are selected to be part of the boundary set. When the **Ray casting** option is current, AutoCAD sends an invisible object called a ray to the nearest object it can find and then traces the boundary in a counterclockwise direction. You might be tempted to store this with other boring facts about AutoCAD that you wish you never knew, but it really is important when you use **Bhatch**. If the ray hits text or an internal boundary of an object, the boundary definition error is displayed.

EDITING HATCH PATTERNS

Features of boundary hatch patterns can be edited using the **Hatchedit** command. Start the command by selecting a hatch pattern to edit and the right–clicking in the drawing area, by selecting **Edit Hatch** on the **MODIFY II** toolbar, by typing **HE** (ENTER) at the Command prompt, or by selecting **Hatch** from the **Modify** menu. Each method will produce the prompt:

Command: **HE** (ENTER) *(Or click the Edit Hatch button.)*

Select hatch object: *(Select hatch pattern to be edited.)*

As the pattern is selected, a **Hatch Edit** dialog box similar to the one seen in Figure 14.24 will be displayed. (The exact display will vary depending on which pattern is selected.) The box is very similar to the **Boundary Hatch** dialog box. Each of the features of the pattern can be altered, and the options function as they do in the **Boundary Hatch** box.

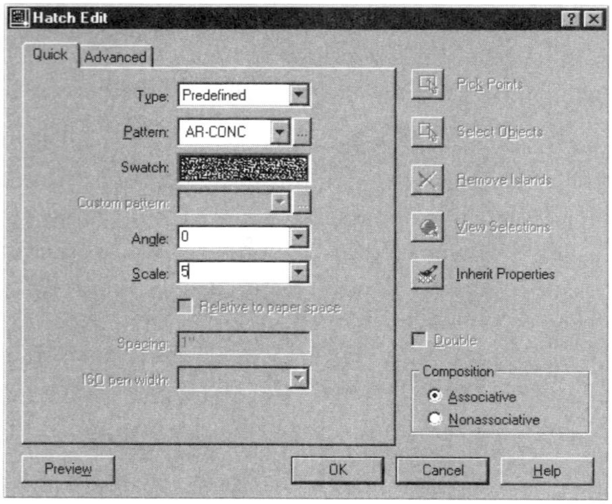

Figure 14.24 *The* **Hatch Edit** *dialog box can be used to edit the pattern once it has been added to the drawing base. Display the dialog box by selecting a hatch pattern to edit and then right-clicking.*

CHAPTER 14 EXERCISES

1. Start a new drawing and use the following coordinates to draw the required object.

   Start: 1,1

B. = @ 3,0	C. = @ –5,.25	D. = @ –.75,1	E.= @ –1,1
F. = @ .75,1	G. = @ .5,1	H. = @ 0,1	J. = @ .25,.25
K. = –2.25,0	L. = @ .25,–.25	M. = @ 0,–1	N. = @ .75,–1
P. = @ –1,–1	Q.= @ 0,–1	R. = @ 1,–.75	S. = C

   Make a copy of the completed shape. Use the SOLID pattern to fill one of the objects. Turn Fill to OFF. Use the ANSI31 hatch pattern and adjust the scale factor to a suitable scale to hatch the object so that it appears solid. Save the drawing as E-14-1.

2. Start a new drawing and use the drawing below as a guide to draw a 15" diameter circle. Draw a 12" diameter circle with a centerpoint 10" above the first centerpoint. Array the 12" circle so that there are a total of six circles. Copy the pattern so that you have a total of three circular patterns. Create layers to separate the object from the hatching. Use the **Bhatch** command to form three completely different hatching styles. Use different hatch patterns for each area to be hatched. Save the drawing as E-14-2.

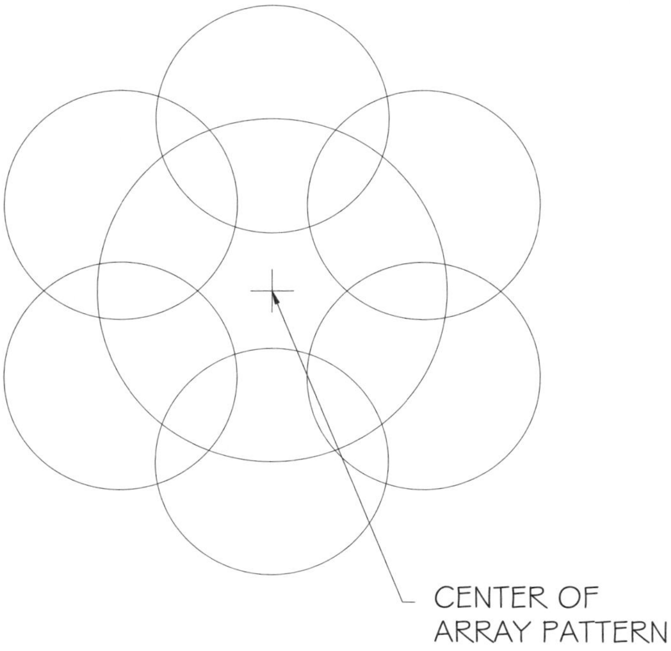

3. Start a new drawing and draw a red square, surrounded by a blue circle, surrounded by a green octagon, surrounded by a cyan triangle. Copy this pattern so that there are a total of three copies. Create a separate layer for hatching. Use ANSI34 and create a double hatching pattern. Provide examples of **Normal**, **Outer**, and **Ignore** styles of hatching. Save the drawing as E-14-3.
4. Start a new drawing and draw a one-story footing. Use a 6" wide stem wall extending from the natural grade 8". Support the wall on a 12" x 6" foundation. Show a 2x4 key, a 2x6 plate with a 1/2" diameter x 10" anchor bolt. Place wood, steel, and concrete on separate layers with different colors. Use varying widths of polylines to distinguish each material. Use EARTH to represent the soil and AR-CONC to represent the footing and stem wall. Copy the drawing and show a 2x10 floor joist with 3/4" plywood floor sheathing. Show the bottom of a 2x6 stud wall with single-wall construction. On the second copy, show a post and beam floor with 2" decking and a 6" deep beam. Save the drawing as E-14-4.
5. Draw the plan view of a 5' wide masonry fireplace. Use hatch patterns to distinguish between the firebrick of the firebox and normal brick of the chimney. Save the drawing as E-14-5.
6. Open drawing 13-3 and create a layer for oak (green), maple (red), and Dutch elm (yellow). Change the trees from their existing layer to these new layers. Tree E is an oak, A is a maple, and the other three are Dutch elm. Hatch each type of tree with a different pattern. Save the drawing as E-14-6.
7. Open drawing E-7-5 and create a white SHADE layer. Hatch the glass area with DOTS, rotated at 45°. Adjust the scale so that the glass appears gray rather than black or white. Save the drawing as E-14-7.
8. Open drawing PB-FND (E-9-7). Create layers for FNDWALLS, FNDFOOT (dashedx2), FNDPIER (hidden), FNDCONC, FNDTXT, and FNDBEAM (center). Assign each layer a different color. Assume the wall to be made from concrete masonry units and hatch the wall with an appropriate pattern. Offset the wall lines to form a 16" wide footing centered on the wall. Trim the footings as required and change the footing lines to the appropriate layer. Save the drawing as E-14-8.
9. Open drawing E-11-9 and complete the drawing by providing the section. Use approximate sizes if dimensions are missing. Provide suitable hatch material and save as E-14-9.

10. A residence is 24' x 36' wide. The ridge is 36' long and the roof is a 6/12 pitch with 24" overhangs, with a 2x6 fascia. The roof has a 12" overhang at the gable end walls. The finish floor is 18" above the finish grade and the plate height for the walls is 8'–0". You are to design an elevation of the 36' front side, which must include the following:

    - 3' x 6'–8" front door (show some type of appropriate decorative pattern)
    - 8' x 5' sliding/picture window with grids
    - 18" x 60" shutters each side of window
    - 12" wide x 6" deep stucco columns each side of window shutters with 6" horizontal siding above and below the window and shutters
    - A 24" wide x full-height area of brick on each side and above the front door (remember, brick should be approximately 8" long)

    Hatch the roof with AR-ROOF.

    The area (36") between the doorway brick and the stucco columns is to be AR-RSHKE (rows should be approximately 10" wide).

    All other areas are to have vertical siding @ 8" o.c.

    Show a shadow that extends 6" below the top of the door and window and show shade that would result from the sun being above and to the left of this gorgeous structure. (Hey! Be glad it's not in your neighborhood!)

    Create a separate layer for each material and save this work of art as E-14-10.

CHAPTER 14 QUIZ

1. A circle surrounds a square, which encloses a hexagon. What style is used to hatch the circle and hexagon?
2. In the object described in question 1, what effect will **Ignore** have on each shape?
3. What option should be used to fill an object with a hatch pattern?
4. List the dialog boxes that are associated with **Bhatch**.
5. How can the **Hatch Pattern Palette** dialog box be accessed?
6. How can the scale factor of the **Bhatch** pattern be controlled by the plotted scale factor?
7. What process is used to keep the hatch pattern from crossing text?
8. Explain the differences in using the **Pick Points** option and the **Select Objects** option.
9. What is direct hatching?
10. Describe the process to create a hatch pattern of vertical lines at 1/16" spacing.
11. What is the effect of using the ANSI32 pattern?
12. Use the **Help** menu to determine what effect the **Explode** command will have on a hatch pattern.
13. What does associative have to do with **Bhatch**?
14. You've defined a hatch pattern, but you're not sure how it will appear. List two methods to view the pattern.
15. What is an island?

CHAPTER 15

Inquiry Commands

As the drawings you're working with get more complex, you might find it hard to keep track of some of the drawing parameters you have established. This chapter will introduce methods for

- Determining information about drawing objects
- Determining information about drawing files
- Using the AutoCAD calculator

The **Inquiry** toolbar and the **Inquiry** menu on the **Tools** menu contain commands used to obtain information about drawing objects or an entire drawing. Commands to be introduced in this chapter include

- **Distance**
- **Area**
- **Massprop**
- **Region**
- **List**
- **Dblist (Data Base List)**
- **Locate Point (ID Point)**
- **Time**
- **Status**
- **Purge**
- **Cal**

DETERMINING DISTANCE

The **Distance** command measures the distance and angle between two selected points. This command can be especially helpful when used with SNAP or OSNAP locations to measure exact locations. **Distance** can be useful in determining the size of a room during the design stage, or checking an angle of an object from a baseline. Access the command by selecting the **Distance** button on the **Inquiry** toolbar, by selecting **Distance** from **Inquiry** on the **Tools** menu, or by typing **DI** (ENTER) at the Command prompt. The command sequence is as follows:

> Command: **DI** (ENTER) *(Or click the Distance button.)*
>
> DIST: Specify first point: *(Select a point.)*
>
> Specify second point: *(Select second point.)*

The resulting display for the points in Figure 15.1 would be

> Distance=14'-0" Angle in XY plane=225.0 Angle from XY plane=0
>
> Delta X=–9'-101/2"Delta Y=–9'-101/2"Delta Z=0'-0"
>
> Command:

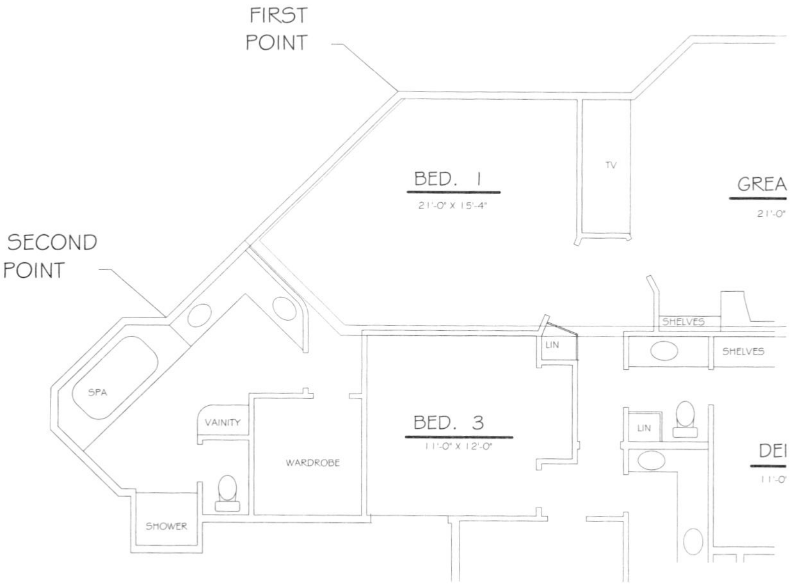

Figure 15.1 *The* **Distance** *command measures the distance and angle between two selected points. (Courtesy Residential Designs.)*

DETERMINING AREA

The **Area** command can be used to determine the area and perimeter of an object. If OSNAP is used, the **Area** command can be used to determine exact square footage sizes of structures and, at the same time, determine wall lengths. A running total can be kept and areas can be added to or deleted from the total. Access the command by selecting the **Area** button on the **Inquiry** toolbar, by selecting **Area** from **Inquiry** on the **Tools** menu, or by typing **AA** (ENTER) at the Command prompt. As the command is entered, four options are given:

Command: **AA** (ENTER) *(Or click the Area button.)*

AREA

Specify first corner point or [Object/Add/Subtract]: *(Specify a point or enter an option.)*

FIRST CORNER POINT

When you accept the **Area** "first corner point" option, you will be asked to supply a series of next corner points to specify the area to be defined. Points can be selected with the pointing device or by any of the other selection method. Using OSNAP will greatly increase the accuracy in computing the area. Figure 15.2 shows an example of computing the area of a retaining wall by using the point option. The command sequence is as follows:

Command: **AA** (ENTER) *(Or click the Area button.)*

AREA

Specify first corner point or [Object/Add/Subtract]: *(Specify first point.)*

Specify next corner point or press (ENTER) for total:

Continue to select points as required. In the example in Figure 15.2, press (ENTER) at Point 12. This will produce a display listing the area (in square inches and square feet when units are set for architectural units) and the perimeter. The same results will occur if Point 13 (Point 1) is entered as the ending point.

OBJECT

The Object option will compute the area of a circle, ellipse, spline, or an object composed of closed polylines. Selecting a circle will cause the area and the circumference to be displayed. When a closed polyline is selected, the area and perimeter will be displayed. If an open polyline is selected, the area is computed based on an imaginary line extending from each end of the polyline. The command sequence is as follows:

Command: **AA** (ENTER) *(Or click the Area button.)*

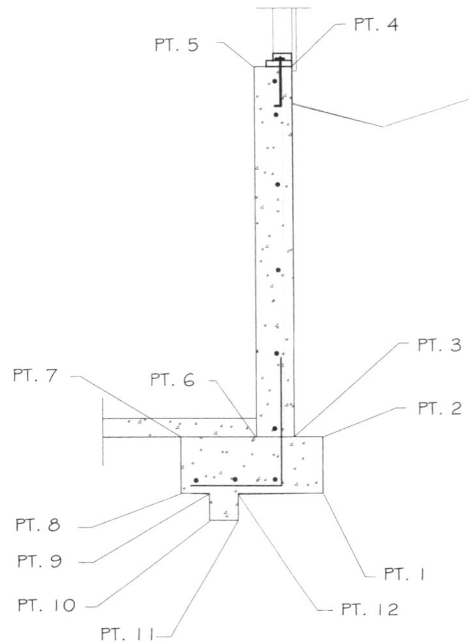

Figure 15.2 *The Area command can be used to compute the area of a specified shape. The area and the perimeter are given at the Command prompt.*

AREA

Specify first corner point or [Object/Add/Subtract]: **O** (ENTER)

Select object: *(Select object.)*

Area=27.63 square in. (0.1918 square ft.) Length 1'–15/8"

Command:

Once the object is selected, a display of the area and circumference will be displayed. If an open polyline is selected, AutoCAD will calculate the area as if a line were drawn between the endpoints. (See Figure 15.3.) The line that is projected by AutoCAD is not included in the calculation of the area or perimeter. The command will list the area confined by the polyline and the length of the line.

ADD AND SUBTRACT

Using the Add option will allow each area measured in this mode to be displayed and added to a running total. The running total will be displayed after each addition. To begin a running total, select the Add option before selecting the objects to be added. Using the Subtract option provides the opposite effect to that of Add. Each area measured in the Subtract mode will be displayed and then subtracted from the running total. With both the Add and Subtract options, the prompt will be altered to reflect

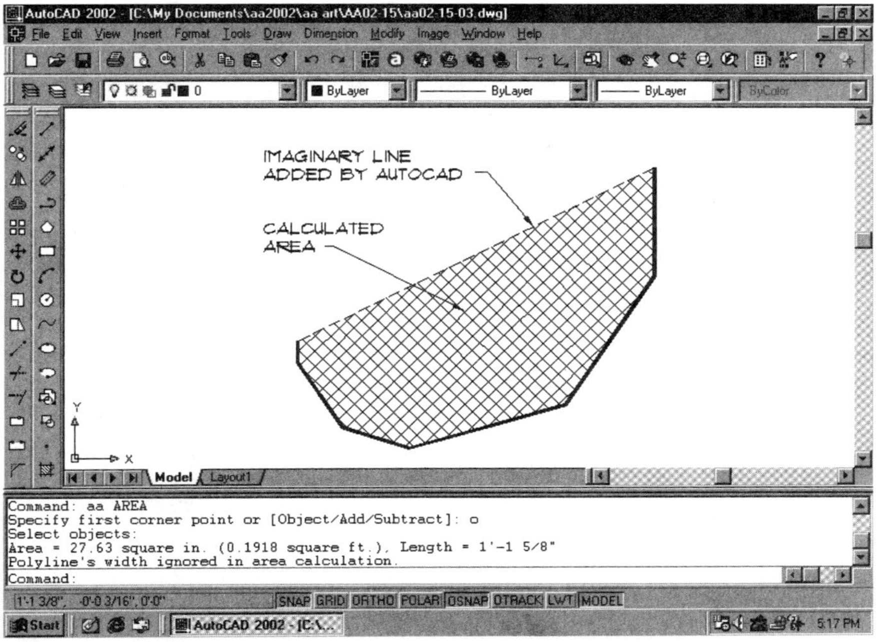

Figure 15.3 *The Object option of* **Area** *can be used to obtain information about an open or closed polyline. AutoCAD will provide the line required to close the polygon and calculate the area without displaying the line.*

the active mode. The selected option will stay in effect until the command is canceled. Figure 15.4 is an example of computing the usable area of a parcel of land with the Add and Subtract options. The area of the lot was determined using continuous OSNAP to select each corner of the lot, and then subtracting the area of the residence and a garage, which are both outlined by a polyline. The command sequence is as follows:

Command: **AA** (ENTER) *(Or click the Area button.)*

AREA

Specify first corner point or [Object/Add/Subtract]: **A** (ENTER)

Specify first corner point or [Object/Subtract]: *(Select point 1.)*

Specify first corner point or press (ENTER) for total (ADD mode):
(Select point 2.)

Specify first corner point or press (ENTER) for total (ADD mode):
(Select point 3.)

Specify first corner point or press (ENTER) for total (ADD mode):
(Select point 4.) (ENTER)

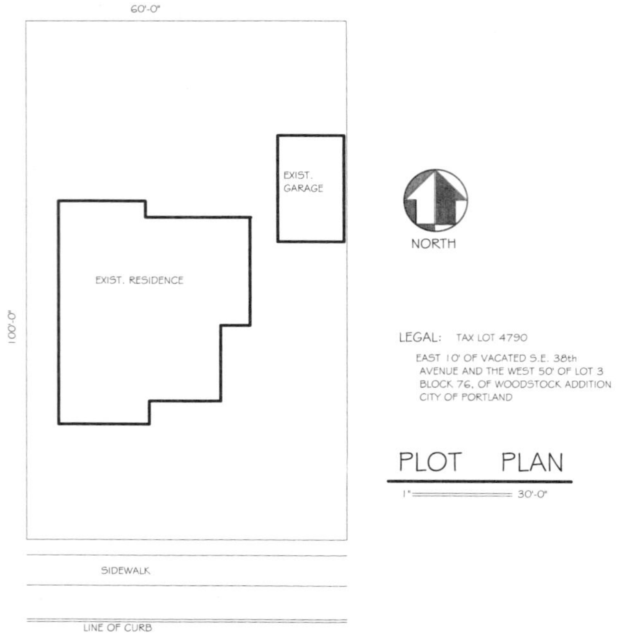

Figure 15.4 *The Add and Subtract options can be used with the* **Area** *command to add areas to or subtract areas from a running total. (Courtesy Residential Designs.)*

Area=864000.0 square in. (6000.000 square ft.), Perimeter=320'-0"

Total area=864000.0 square in. (6000.000 square ft.)

Specify first corner point or [Object/Subtract]: **S** (ENTER)

Specify first corner point or [Object/Add]: *(Select point 1.)*

Specify first corner point or press (ENTER) for total (SUBTRACT mode): *(Select point 2.)*

Specify first corner point or press (ENTER) for total (SUBTRACT mode): *(Select point 3.)*

Specify first corner point or press (ENTER) for total (SUBTRACT mode):

(Select point 4.) (ENTER)

Area=189154.2 square in. (1313.571 square ft.), Perimeter= 118'-31/2"

Total area=674845.8 square in. (4686.429 square ft.)

Specify first corner point or [Object/add]: *(Select point 1.)*

Specify first corner point or press (ENTER) for total (SUBTRACT mode): *(Select point 2.)*

Specify first corner point or press (ENTER) for total (SUBTRACT mode): *(Select point 3.)*

Specify first corner point or press (ENTER) for total (SUBTRACT mode):

(Select point 4.) (ENTER)

Area=36365.9 square in. (252.541 square ft.), Perimeter=65'-7"

Total area=638479.9 square in. (4433.888 square ft.)

Specify first corner point or [Object/Add]: (ENTER)

Command:

MASS PROPERTIES

The **Mass Properties** (**Massprop**) command can be used to calculate the properties of 2D and 3D objects. Although primarily a 3D command, **Mass Properties** can be used to determine the area, perimeter and several other features of 2D shapes. Start the command by selecting the **Mass Properties** button on the **Inquiry** toolbar, by selecting **Mass Properties** from **Inquiry** on the **Tools** menu, or by typing **MASSPROP** (ENTER) at the Command prompt.

To use the **Mass Properties** command with a 2D object, first you need to convert the 2D object to a region. If an object that is not a region is selected, the following prompt is displayed:

Command: **MASSPROP** (ENTER) *(Or click the Mass Properties button.)*

Select Objects: *(Select desired objects using any selection method.)*

1 found

No solids or regions selected.

Command:

To change an object to a region, select the **Region** button on the **Draw** toolbar or type **REG** (ENTER) at the Command line and select the objects to be converted. **Mass Properties** can now be used to display the properties of the selected region. The command sequence is as follows:

Command: **REG** (ENTER) *(Or click the Region button.)*

Select Objects: *(Select desired objects using any selection method.)*

104 found

Select objects: (ENTER)

1 loop extracted

1 Region created

Command:

Once objects are selected, AutoCAD will display the mass properties in the text window. The mass properties for the shape in Figure 15.5 are seen below.

Regions	
Area:	277.1929 sq in
Perimeter:	91.8271 in
Bounding box:	X: 23.5672 — 46.5362 in
	Y: -3.1626 — 18.0234 in
Centroid:	X: 39.2238 in
	Y: 9.3193 in
Moment of inertia:	X: 33838.1482 sq in sq in
	Y: 433392.8323 sq in sq in
Product of inertia:	XY: 97969.2639 sq in sq in
Radii of gyration:	X: 11.0487 in
	Y: 39.5412 in

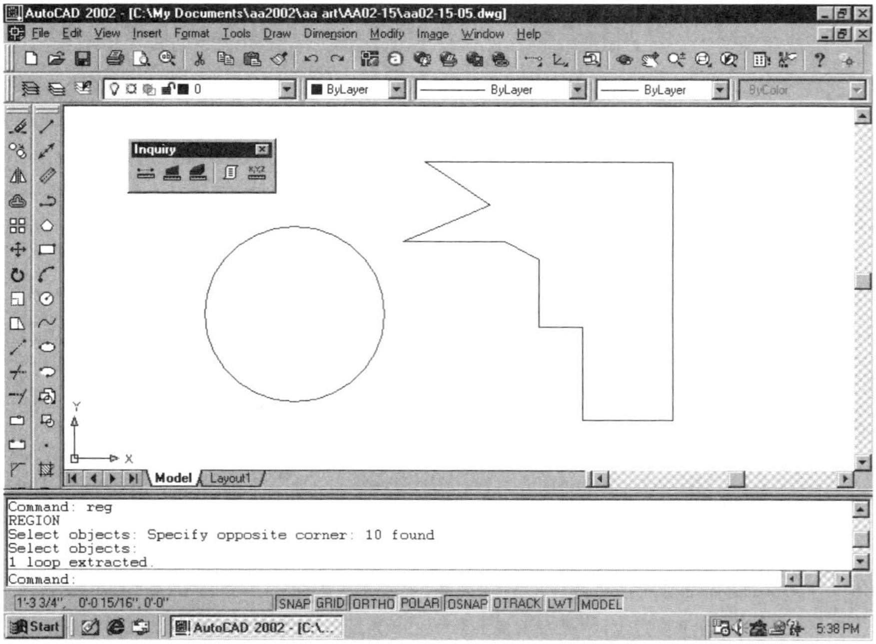

Figure 15.5 *The* **Mass Properties** *command can be used to calculate the properties of 2D objects, once they have been converted to regions. Once a region is selected, AutoCAD will display the mass properties in the text window.*

Principal moments (sq in sq in) and X-Y directions about centroid:

I: 4704.5411 along [0.5527 -0.8334]

J: 11989.3890 along [0.8334 0.5527]

Write analysis to a file? [Yes/No] <N>: (Press (ENTER) to continue)

Although **Mass Properties** gives more information than is typically needed for most 2D situations, the command can be used to get the area and perimeter of an object. Although much of the information provided by this command is beyond the scope of this text, other information includes the following:

Bounding box—The two coordinates that define the bounding box are the diagonally opposite corners that enclose the rectangle.

Centroid—The center of the region.

Moment of inertia—A value used in engineering formulas to compute the distributed loads.

Product of inertia—A property used to determine forces causing motion in an object.

Radii of gyration—A value used to determine the moments of inertia of a solid.

Principal moments—The properties derived from the products of inertia.

LISTING INFORMATION ABOUT AN OBJECT

The **List** command displays a listing of information about an object in the drawing. Access the command by selecting the **List** button on the **Inquiry** toolbar, by selecting **List** from **Inquiry** on the **Tools** menu, or by typing **LI** (ENTER) at the Command prompt. The command sequence is as follows:

Command: **LI** (ENTER) *(Or click the List button.)*

Select objects: *(Select an object.)*

One or several objects can be selected for listing, although the information might not fit on the screen all at once. To view additional pages, press (ENTER) to resume the output display. Pressing F2 will terminate the text box and return the Command prompt. Selecting the **Close** button in the title bar will also close the window.

Figure 15.6 shows a partial floor plan with a line selected for List and the resulting list for this line. The information that is displayed depends on the type of object that is selected—but the object's type, position relative to the current UCS, the layer location, and either model or paper space are always listed. The color and linetype are listed if either is not set by layer. Four values are always given to describe a line. The X and Y values specify the horizontal and vertical distances of the "first point" and the "next point," from the 0,0 point when the limits were established. The line in Figure

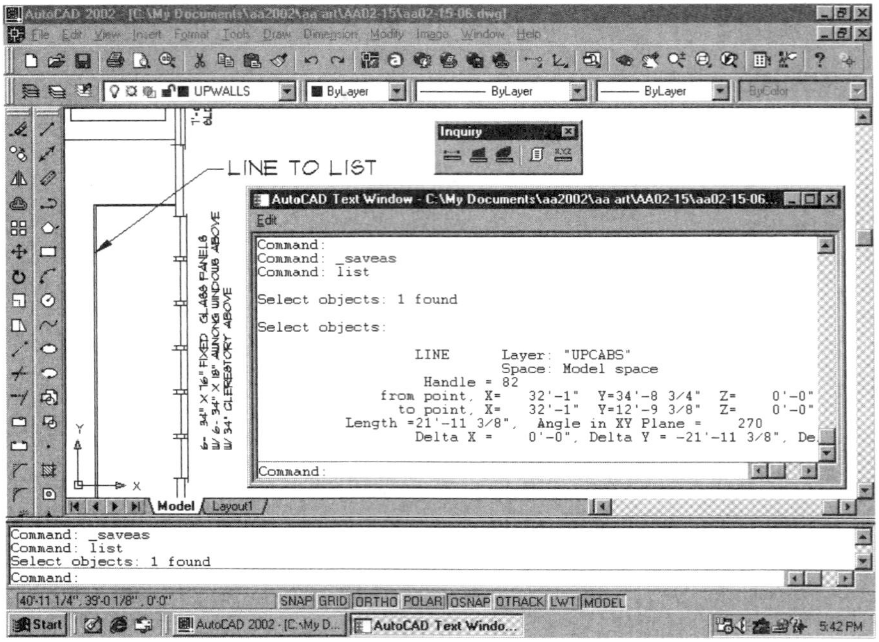

Figure 15.6 *The* **List** *command can be used to supply information about a drawing object.*

15.6 is 32'–1" to the right of the X axis. Notice that distances of the from point and the to point of the Y axis are 34'–8 3/4" and 12'–9 3/8" respectively. The difference between these two values is 21'–11 3/8", which is the specified length. Figure 15.7 shows an example of each of the values on a vertical and an inclined line.

Listing circles will provide information based on the center point and radius, the area, and the circumference. Listing a circle is an easy way to determine the area of a pier for doing concrete estimates. An example of a listing for the circle in Figure 15.8 is as follows:

CIRCLE	Layer: FNDFOOT
	Space: Model space
Handle	2D
Center point	X=119'-8" Y=103'-6" Z=0'-0"
radius	0'-9"
circumference	4'-89/16"
area	254.47 sq in (1.7671 sq ft)

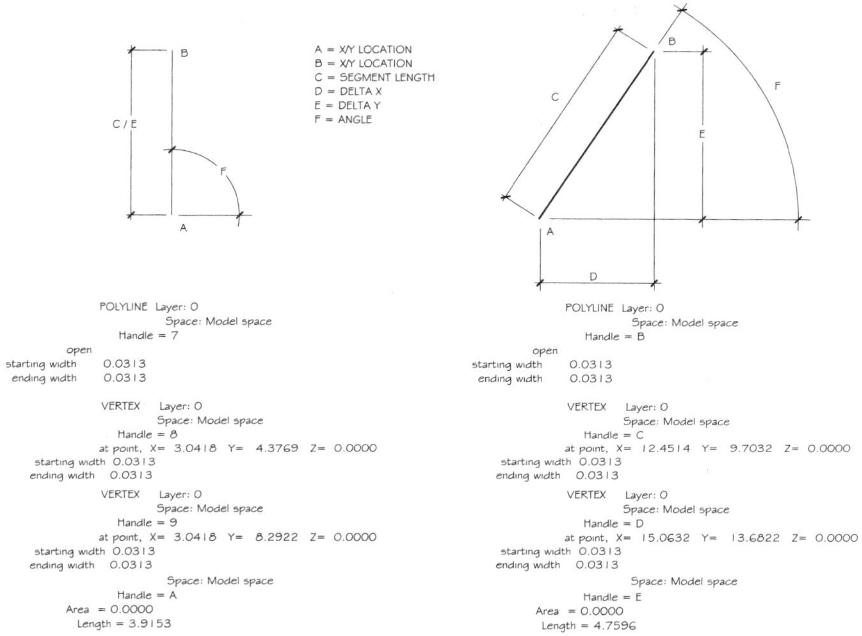

Figure 15.7 *List values for two lines.*

The **List** command can also be used to provide information about text in the drawing base. (Text will be introduced in the next chapter.) The listing to describe the text for the footing in Figure 15.8 is as follows:

TEXT	Layer: FNDTEXT
Space:	Model space
Handle	10
Style	ARCHSTYL Font file=archstyl
start point	X=27'-1 7/8" Y=17'-3 3/4" Z=0'-0"
height	0'-6"
text	THICKEN SLAB AT BASE POINT OF STAIR
rotation angle	0
width scale factor	1.000
obliquing angle	0
generation	normal

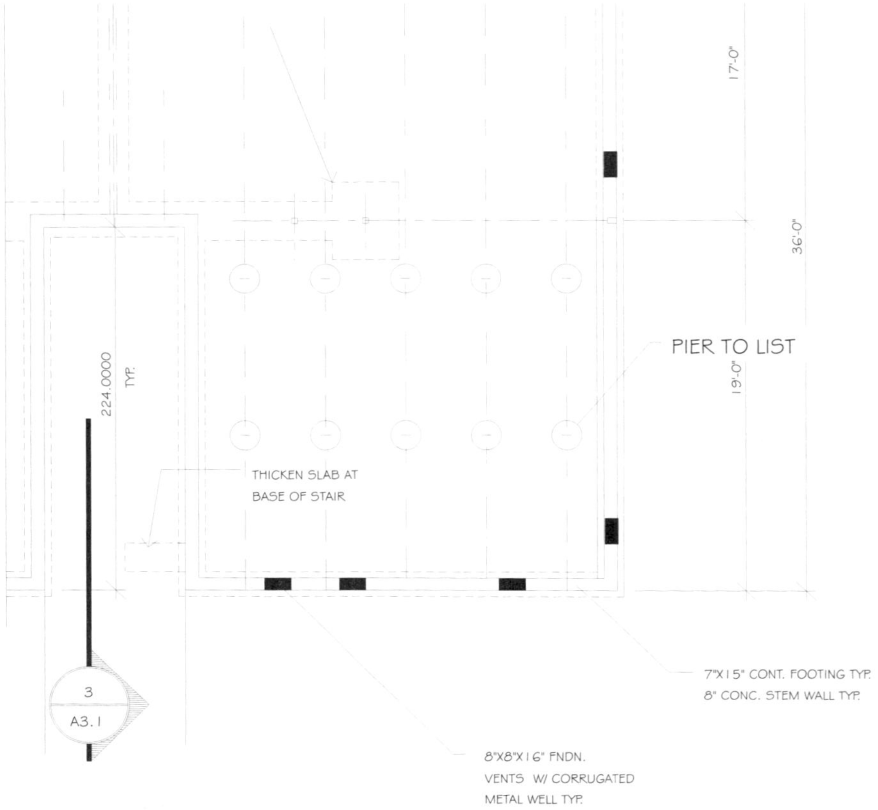

Figure 15.8 *The* **List** *command provides information based on the center point, the radius, and the diameter. (Courtesy Scott R. Beck, Architect.)*

OBTAINING INFORMATION ABOUT A DRAWING

The **List** command is used to obtain information about selected objects. The **Dblist** command will produce information about the entire drawing. This command can be useful in comparing drawing sequences or information contained in two similar drawings. Information is presented in the same order as used for **List**, one page at a time. The command can be accessed only from the command line. The command sequence is as follows:

Command: **DBLIST** (ENTER)

Pressing (ENTER) will advance through a listing of information for the entire drawing. Pressing ESC will terminate the scrolling of the listing, and pressing F2 will remove the text window and restore the drawing screen. You can print listings from both the **Dblist** and **List** commands if your terminal is connected to a printer, by pressing PRINTSCREEN.

IDENTIFYING A POINT

Information about a specific point in a drawing can be obtained by using the **ID Point** command. This command will allow you to determine the exact location of a specific point. Access the command by selecting **Locate Point** on the **Inquiry** toolbar, by typing **ID** (ENTER) at the Command prompt, or by selecting **ID Point** from **Inquiry** on the **Tools** menu. The sequence is as follows:

Command: **ID** (ENTER) *(Or click the Locate Point button.)*

Point: *(Select a point.)*

Entering a point will display the corresponding X, Y, and Z coordinates. This can be helpful in determining the dimension of an object.

The **ID Point** command also can be used to identify a specific location in the drawing. If **ID** is entered at the Command prompt, coordinates can be supplied and that point will be identified in the drawing. This can be useful with relative coordinates entry methods to start an object at a specific location.

The **ID Point** listing for the location of intersection of the line in Figure 15.6 is as follows:

Command: **ID** (ENTER) *(Or click the Locate Point button.)*

Specify point: *(Select desired point. For this example the handrail in Figure 15.6 was used.)*

X=31'-11"xY=34'-10"xZ=0'-0"

TRACKING DRAWING TIME

The current time is displayed at the bottom of the drawing screen on the status bar. The **Time** command uses the date and time maintained by the computer to keep track of the time related to the drawing session. Access the command by selecting **Time** from **Inquiry** on the **Tools** menu, or by typing **TIME** (ENTER) at the Command prompt. Before the command can function properly, the date and time must be set. Using **Date/Time** on the **Control Panel**, accessed from **My Computer**, can set an accurate date or time. If you reenter a drawing, AutoCAD can now provide the current time, drawing creation time, last update time, and time elapsed while editing the drawing. Each might seem unimportant in an educational setting, but the information is highly useful in keeping track of hours spent on a project for billing purposes and minor bookkeeping functions, such as determining profit margins.

TIME DISPLAY VALUES

Access the time display by typing **TIME** (ENTER) at the Command prompt. The display will resemble

Current time: 16 Aug, 2001 @ 11:15:25.510 AM

Times for this display:

Created: Tuesday, Jul 21, 2001 at 3:3106.970 AM

Last updated: Friday, Aug 10, 2001 at 2:2506.970 AM

Total editing time: 0 days 3:27:38.175

Elapsed timer (on): 0 days 1:59:51.510

Next automatic save in: 0 days 00:12:20.70

Enter option [Display/On/Off/Reset]:

Command:

Times are listed as day, month, year, hour, minute, seconds, and milliseconds.

- The Current time is the current date and time.
- Created time is marked when the drawing was initially started.
- The Last updated time lists the last editing session when either the **End** or **Save** command was used with the default file name.
- The Total editing time is the amount of time used editing the current drawing, since it was created until Save is used.

If **Quit** is used to end a drawing session, the time spent in that session is not added to the accumulated time.

Time spent printing and plotting is excluded from the Editing time

This time is updated continuously and cannot be reset or stopped. This timer is most helpful to supervisors in tracking efficiency and logging total hours for billing.

The Elapsed time can be reset and turned On or Off. This timer is best suited for individual use each day to keep track of your time if you will be working on several different jobs. The Next automatic save time indicates when the next save is scheduled to occur (see Chapter 3).

PROMPT OPTIONS

After the time display, four options are listed on the prompt line.

- The Display option will redisplay the Time display and update the current values.
- The On option will activate the User Elapsed timer. The default value is set to On when a drawing is started.
- The Off option will stop the User Elapsed timer.
- The Reset option clears the User Elapsed timer to zero.

The command line can be returned by (ENTER) or F2.

STATUS

The **Status** command will list the current value settings of a drawing. Access the command by selecting **Status** from **Inquiry** on the **Tools** menu or by typing **STATUS** (ENTER) at the Command prompt. The listings of the drawing shown in Figure 15.8 are as follows:

3931 objects in FOUND			
Model Space Limits are:	X: 0'-0"	Y: 0'-0" (off)	
	X: 275'-0"	Y: 250'-0"	
Model space uses	X: 45'-31/8"	Y: 48'-71/2"	
	X:194'-113/8"	Y:152'-73/8"	
Display shows	X: 89'-5"	Y: 83'-63/4"	
	X: 145'-05/8"	Y:123'-93/8"	
Insertion base is	X: 0'-0"	Y: 0'-0"	Z: 0'-0"
Snap resolution is	X: 0'-1" Y: 0'-1"		
Grid spacing is	X: 2'-0" Y: 2'-0"		
Current space:	Model space		
Current layer:	FNDFOOT		
Current color:	BYLAYER—1 (red)		
Current linetype:	BYLAYER—DASHED		
Current lineweight:	BYLAYER		
Current plot style:	ByLayer		
Current elevation:	0'-0" thickness: 0'-0"		

Fill on   Grid off   Ortho on   Qtext off   Snap off   Tablet off

Object snap modes: Center, Endpoint, Intersection, Node, Extension

Free dwg disk (C:) space: 2047.7 Mbytes

Free temp disk (C:) space: 2047.7 Mbytes

Free physical memory: 2.6 Mbytes (out of 31.5M)

Press (ENTER) to continue: (ENTER)

Free swap file space: 1996.6 Mbytes (out of 2016.5M).

Command:

Several elements in the list, such as color, linetype, Snap, Grid, and Ortho, should be very familiar by now. Other terms might seem a little foreign. An important value to keep track of is Free dwg disk. The value refers to the space that is available on the drive that contains the drawing. When this space is used up, AutoCAD will terminate the drawing session and save what has been done up to the point of termination. If you're working on the hard drive, this will rarely be a problem. Remember that much of the information provided by the **Status** command can be obtained using the **Properties** command.

PURGE

This command does not provide information about a drawing, but it can be a useful tool within a drawing file to eliminate unused items and minimize storage space used. Objects that have been named in a drawing session, such as Blocks, Dimstyles, Layer, Linetypes, Shapes, and Text styles, are examples of items that can be purged from a file. **Purge** can be used, for instance, if a template drawing is used as a base for a new drawing. The base drawing might contain the LAYER listings for a three-story structure with prefixes such as UPPER, MAIN, and LOWER. If only a two-level structure is to be drawn, all layers with the MAIN prefix could be purged.

Items to be purged can be eliminated at any time after entering the drawing file is opened. Start the **Purge** command by typing **PU** (ENTER) at the Command prompt or by selecting **Purge** from **Drawing Utilities** on the **File** menu. Each method will produce a **Purge** dialog box similar to Figure 15.9.

Four choices can be made regarding information to be deleted from the drawing base including:

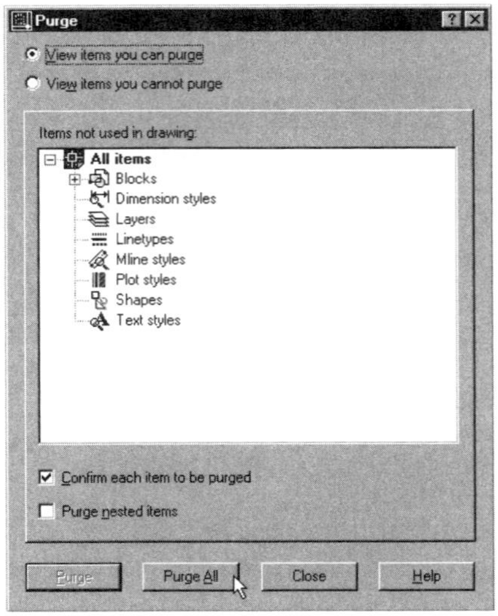

Figure 15.9 *The* **Purge** *dialog box can be used to delete unused objects and features from the drawings. Display the dialog box by typing* **PU** *(ENTER) at the Command prompt or by selecting* **Purge** *from* **Drawing Utilities** *on the File menu.*

View items you can purge—In the default setting, this radio button, and the **Items not used in drawing** display window are displayed. The window lists categories of named items that have been defined in the drawing but are not presently being used. Double-clicking any of the listed features will display a listing of individual named items for that category that have not been used. For instance, if you created a layer but placed nothing on that layer, the name of the layer would be listed as an item that is not in use. Notice in Figure 15.10 that layers Red and Hidden are listed. The **Purge** command will remove these and other unused items from the drawing. Execute the command by selecting the **Purge** button or the **Purge All** button.

View items you cannot purge—This radio button is OFF in the default setting. Notice in Figure 15.11 that the display window has been altered to list features that have been used in the drawing. The 0 and Wall layer are listed as layers that cannot be removed from the drawing. Notice that with the Wall layer highlighted, a warning is displayed in the dialog box: "This Layer cannot be purged if it contains objects."

Confirm each item to be purged—Once an option has been entered, you can decide if you want verification of each object that is to be purged. With this check box active, a **Confirm Purge** warning box will be displayed, allowing a YES or NO answer to be given for each item to be purged.

Purge nested items—A nested object is an object that is part of another object. In later chapters you'll learn how to save components in groups called blocks. A sink, stove and dishwasher are examples of items that could be saved as a block. These blocks could then be saved in bigger blocks that represent typical kitchen arrangements. The kitchen sink is said to be nested in the larger kitchen block. This option, if active, will purge drawing items even if they are contained within or referenced by other unused objects.

Once the desired settings have been entered, you can remove objects by selecting the **Purge** button or the **Purge All** button.

- Selecting the **Purge** button will remove multiple objects, one at a time. A warning will be displayed before each item is removed from the drawing base.
- Selecting the **Purge All** button still allows you to rethink the process. Selecting the **Yes** button will purge the item and continue prompting for items to purge. Selecting the **Yes to All** button will remove all items not used in the drawing.

If you decide that you've made a mistake, exit the command and select the **Undo** button to restore the purge.

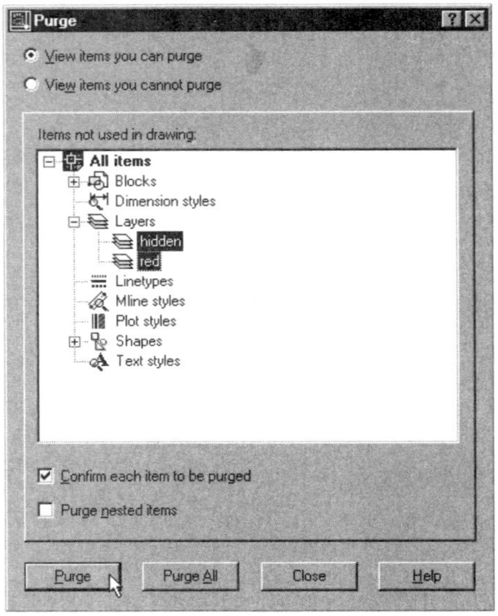

Figure 15.10 *With* **View items you can purge** *as the default setting, the* **Items not used in drawing** *list is displayed. The window lists categories of named items that have been defined in the drawing but are not presently being used. The layers Red and Hidden have been selected to be purged.*

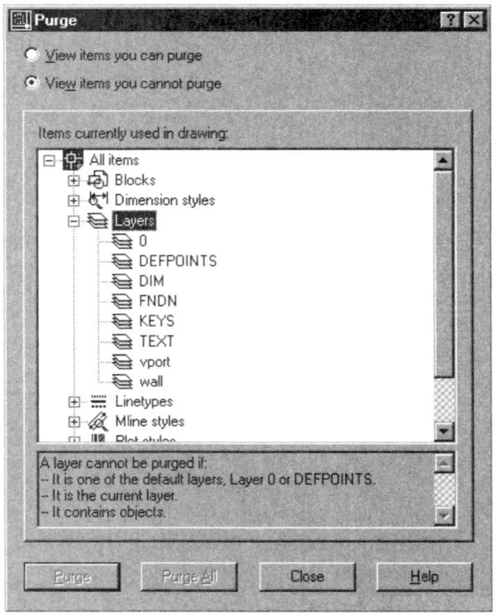

Figure 15.11 *If the* **View items you cannot purge** *radio button is ON, the window lists features that have been used in the drawing. The 0 and Wall layers are listed as layers that cannot be removed from the drawing.*

USING THE AUTOCAD CALCULATOR

The calculator of AutoCAD can be used just as a hand-held calculator and also as a means of accurately locating points in a drawing. Typing **CAL** (ENTER) at the Command prompt starts the geometry calculator that will compute standard mathematical formulas. The prompt at the command line is as follows:

Command: **CAL** (ENTER)

>> Expression:

The double angle brackets (>>) preceding the Command prompt indicate that AutoCAD us waiting for input.

PERFORMING MATHEMATICAL CALCULATIONS

Mathematical calculations can be completed by AutoCAD, but specific symbols must be included as the values are entered. Symbols recognized by AutoCAD include the following:

+Add	–Subtract
*Multiply	/Divide
^Exponents	[ ]Grouped expressions

- To add 4 and 4, type **4+4** (ENTER) at the prompt.
- To subtract 50 from 100, type **100–50** (ENTER) at the prompt.
- To multiply 50 by 50 would require **50*50** (ENTER). Dividing 100 by 50 would require **100/50** (ENTER).

Of course, none of these problems requires a calculator, but the command is very useful for performing advanced calculations without having to clean up your work area to find a pocket calculator. By typing **CAL** (ENTER) at the Command prompt, you can use the calculator transparently. If the command line is currently set to display only one line of text, the answer will not be displayed. Adjust the display of text as needed to display the problem and the solution.

In addition to performing simple calculations, you can use the calculator to execute advanced formulas. Parentheses are used to group symbols and values of complex mathematical equations into sets. **Cal** evaluates expressions based on standard mathematical procedures using

- Left to right for operations of equal precedence.
- Expressions in parentheses first, starting with the innermost set.
- Operations are performed by exponents first; followed by multiplication, division, addition, and subtraction.

Vector operators that can be used include the following:

+	Add vectors
–	Subtract vectors
*	Multiply a vector by a real number or scalar product of vector
/	Divide a vector by a real number
&	Vector product of vectors
[ ]	Signifies expression

Point/vector operators that can be used include the following:

[ ]	Encloses point or vector
,	Separates coordinate values
<	Precedes angle
@	Precedes distance
*	Denotes WCS (overrides UCS)

Entering Feet and Inches in the Calculator

Numbers are entered into the calculator just as they are at the Command prompt to enter coordinates. Ten feet three inches would be entered as **10'-3"** or **10'3"**. If you are curious about how many inches that would equal, press (ENTER). Feet/inch values will be converted to inch values. The process is as follows:

Command: **CAL** (ENTER)

>> Expression: **10'3"** (ENTER)

123.0

Command:

Entering Angles

The default units of angle values are decimal degrees. Angles are entered using a format of 45d30'30". The minute (') and second (") symbols can be omitted if the values are zero. If the value is less than one degree, enter the value as **0d40'10"**. Angles can be entered in radians if the value is followed by the letter r (5.5r), or in grads when followed by the letter g (15g). No matter how the values are entered, they will be converted to decimal degrees.

Converting Units of Measurement

As metric dimensioning becomes more important in the construction industry, the need to convert values will increase. The calculator can be used to convert units of measurement from one system to another. Enter the value in the following format:

CVUNIT (value, from unit, to unit). The process to convert 120 inches to millimeters is as follows:

Command: **CAL** (ENTER)

>> Expression: **CVUNIT**(120,inch,mm) (ENTER)

3048.0

Command:

You can find additional values accepted by the calculator in the **Help** menu by researching Standard Numeric Functions.

USING THE CALCULATOR TO CREATE OR EDIT POINTS

In addition to the normal functions of a calculator, the **Cal** command can be used to create or edit a drawing. Listed below are common functions.

Function	Calculation to be performed
CUR	Allows a point to be selected using a cursor.
DIST	(P1,P2) Distance between point P1 and P2.
DPL	(P,P1,P2) Distance between point P and line P1-P2.
ILL	(P1,P2,P3,P4) Gives the intersection between lines P1-P2 and P3-P4.
PLD	(P1,P2,Dist) Point on a line P1-P2, that is the specified distance from P1.
PLT	(P1,P2,T) Point on line P1-P2 defined by the parameter T.
RAD	Gives the radius of a selected circle.
ROT	(P,Origin,Angle) Rotates point P through an angle about the origin.
ROT	(P, P1ax, P2ax) Rotates point P through angle using line P1-P2 as the rotation axis.

In addition to the functions listed above, some commonly used functions have shortcuts that combine some of the previous functions with the Endpoint Object Snap mode. The shortcuts include the following:

Function	Shortcut for	Calculation to be performed
DEE	Dist(end,end)	Distance between two end points
ILLE	ill(end,end,end,end)	The intersection of two lines defined by four endpoints
MEE	(End+end)/2	Midpoint between two endpoints

NEE	Nor(end,end)	Unit vector in the XY plane and normal to two endpoints
VEE	Vec(end,end)	Vector from two endpoints
VEE1	Vec1(end,end)	Unit vector from two endpoints

To determine the length of a line, use the following sequence:

Command: **CAL** (ENTER)

>> Expression: **DEE** (ENTER)

>> Select one endpoint for DEE: *(Select desired endpoint.)*

>> Select another endpoint for DEE: *(Select the other endpoint.)*

10.2500 (Length given for specified line.)

Command:

To determine the intersection of two lines, use the following sequence:

Command: **CAL** (ENTER)

>> Expression: **ILLE** (ENTER)

>> Select one endpoint for ILLE: First line: *(Select first endpoint.)*

>> Select another endpoint for ILLE: First line: *(Select the other endpoint.)*

>> Select one endpoint for ILLE: Second line: *(Select first endpoint.)*

>>Select another endpoint for ILLE: Second line: *(Select the other endpoint.)*

(12.25 5.25 0.0) (Intersection given for specified lines in XYZ coordinates.)

Command:

To determine the radius of a circle, use the following sequence:

Command: **CAL** (ENTER)

>> Expression: **RAD** (ENTER)

Select circle, arc, or polyline segment for RAD function: (*Select circle.)*

1.13702 (Radius for specified circle.)

Command:

CHAPTER 15 EXERCISES

1. Open drawing E-14-9 and determine the area of one side of the cross-sectional view of the manhole cover. Assume all unmarked radii to be 0.25". Maintain a thickness of 7/8" at the cover support. Compute the following:

   Surface area of one side the cover ______________

   Sectional area ______________

   Perimeter of section ______________

   Manhole surface area ______________

   Circumference ______________

2. A steel fabricating company needs to order enough primer to cover (5,000) 21 1/2" x 8 1/2" steel plates. Each plate has (4) 3/4" diameter holes with the center points located 2" down, 2" from plate edge and 3" o.c. The plates will be made of 1/2" steel. Draw the top, front, and side views of the plate, and save the drawing as E-15-2. Use the drawing to determine the following information:

   Surface area 1 side ______________

   Surface area 2 of sides ______________

   Perimeter ______________

   Area of edges ______________

   Area of each hole ______________

   Total area of holes ______________

   Total area of 1 plate ______________

   Total area of 5,000 plates ______________

3. Open drawing E-15-2. Copy the listed times required to complete the project.

   Created ______________

   Last updated ______________

   Total editing time ______________

   Terminate the drawing session using **End** or **Save**.

4. Open drawing E-13-3 and provide the following information:

   Drawing name ______________

   Total objects ______________

   Limits ______________ ______________

   Limits ______________ ______________

Current layer ______________

Current linetype ______________

Free disk space ______________

Status of: Grid ______________ Snap ______________ Ortho ______________

Qtext ____________ Tablet ______________

Terminate the drawing with **Save**.

5. Open drawing E-7-5 and determine the surface area of the window.

   AREA ______________ PERIMETER ______________

   What is the distance between the endpoints of the inner side of the interior arch? ______________

   List the values that describe the outer arch at the top of the window.

   List the total editing time ______________

   Current drawing session time ______________

   Terminate the drawing using **Save** or **End**.

CHAPTER 15 QUIZ

1. Give four types of information that will be given if a circle is selected for a list.
2. Explain the difference between **List** and **Dblist**.
3. Write the name of the command and describe two different uses for identifying a point.
4. What are the two methods of selecting objects to determine their area?
5. What command is best suited to determine the length of a line between two points?
6. What is the process to set the internal clock of your computer?
7. What value of the status listing keeps track of available drawing space?
8. What is the meaning of DELTA X and DELTA Y?
9. You've opened a drawing, made a few changes, and then discarded the drawing. Will the time spent in this drawing be added to the accumulated time?
10. Explain each of the categories used to measure time.
11. Describe how the **Area** command can be used on an open polyline segment.
12. You need to determine the area of a rectangle. Explain the difference in selecting three and four points.
13. List the symbols used by calculator for add, subtract, multiply, divide, and to indicate a set.
14. Determine the following values for your workstation:

    Free drawing space

    Free temp disk space

    Free physical memory
15. Determine how many millimeters are in 500 inches, and how many feet are in 3000 mm. Explain the entries needed to determine the answer.
16. What is the circumference of a 4" diameter circle to the nearest 1/6"?
17. What function of the calculator should be used to determine the radius of an existing circle?
18. Research the operations required for the calculator to determine the following functions:

    The cosine of an angle

    The square root of a positive number
19. What will the ILLE calculator function do?

    Research and describe the calculator function to rotate a point about an axis.

SECTION 3

PLACING TEXT AND DIMENSIONING

PLACING TEXT ON A DRAWING

INTRODUCTION TO DIMENSIONS

PLACING DIMENSIONS ON DRAWINGS

CHAPTER 16

Placing Text on a Drawing

If you have previous drafting or design experience, you will still remember the long hours of practice required to develop your lettering skills. Throw away all the templates, lettering guides, and rub-on letters of manual drafting, because AutoCAD has a better way. In addition to general considerations for placing text on construction drawings, this chapter will introduce methods for

- Placing small amounts of text using the **Text** command
- Placing large amounts of text using the **Mtext** command
- Creating varied styles and fonts of text
- Editing text on a drawing
- Finding and editing text
- Placing text in template drawings

New commands to be examined in this chapter include

- **Text**
- **Style**
- **Ddedit**
- **Mirrtext**
- **Qtext**
- **Mtext**
- **Spell**

GENERAL TEXT CONSIDERATIONS

Before text is placed on a drawing, several important factors should be considered. Each time a drawing is started, consider who will use the drawing, the information the text is to define, and the scale factor of the text. Once these factors have been considered, the text values can often be placed in a template drawing and saved for future use, similar to the stock notes seen in Figure 16.1. Architectural and engineering offices often use two different styles of lettering. Many engineering offices that do civil, municipal or government projects tend to use a simple block lettering shape. Architectural and engineering offices dealing with construction projects typically use a style of lettering that features thin vertical strokes, with thicker horizontal strokes to give a more artistic flair to the drawing. Other common variations are compressed and elongated letter shapes. In addition to the basic shape of the letters, some offices use a forward or backward slant to make their office lettering more distinctive. You are unlikely to find all of these variations on one professional drawing. Figure 16.2 shows some of the variations that can be found on drawings.

WOOD:

1. MANUFACTURE OF THE STRUCTURAL GLU-LAMINATED TIMBER SHALL BE BY AN AITC APPROVED FACILITY AND SHALL BE IN CONFORMANCE WITH AMERICAN NATIONAL STANDARDS ANSI.AITC A190.1-1983. APPEARANCE GRADE SHALL BE INDUSTRIAL UNLESS NOTED OTHERWISE ON FRAMING PLANS.

2. ALL GLU-LAMINATED ADHESIVES SHALL BE FOR WET-USE CONFORMING TO ANSI. AITC A190.1 1983 SEC. 4.5.1.2 AND ASTM D2559-82..

3. ALL GLU-LAM BEAMS, JOIST, POST AND BLOCKING AT EDGES OF STRUCTURE SHALL BE PRESSURE TREATED W/ 0.05 PCF K-520 COPPER NAPHTHENATE IN LIGHT SOLVENT, IN ACCORDANCE WITH AWPA STANDARDS C-14, UNLESS NOTED. ALL CUTS AND BOLT HOLES SHALL BE MADE PRIOR TO TREATMENT, UNLESS NOTED. INADVERTENT FIELD DRILL HOLES AND CUTS SHALL BE REPEATEDLY FLOODED WITH 2% COPPER NAPHTHENATE, PRIOR TO FASTENER INSTALLATION IN CONFORMANCE WITH AWPA STANDARDS.

4. MINIMUM LUMBER GRADES SHALL BE AS FOLLOWS UNLESS NOTED ON PLANS:
GLU-LAM BEAMS-SIMPLE SPANS. . . . . COASTAL D-FIR 24F-V4
GLU-LAM CANTILEVERS . . . COASTAL D-FIR 24F-V8
STUDS, JOIST, AND BLOCKING . . . COASTAL D-FIR # 2 AND BETTER
4 x 6 x POST AND BEAMS . . . COASTAL D-FIR #1 AND BETTER
DECKING . . . CEDAR #1 OR BETTER

5. PRE-MANUFACTURED FRAMING CONNECTORS SHALL BE BY SIMPSON CO. OR APPROVED EQUAL. JOIST HANGER NAILS SHALL BE COMMON. CONNECTORS SHALL BE ATTACHED TO FRAMING MEMBERS WITH THE MAXIMUM SIZE AND NUMBER OF NAILS OR BOLTS AS SPECIFIED BY SIMPSON.

Figure 16.1 *A partial listing of standard construction notes. (Courtesy Residential Designs.)*

ARCHITECTURAL OFFICES USE MANY
STYLES OF LETTERING.
SOME TEXT IS ELONGATED.
SOME OFFICES USE A COMPRESSED STYLE.
NO MATTER THE STYLE THAT IS USED,
IT MUST BE EASY TO READ.

Figure 16.2 *Some common types of lettering found on construction drawings.*

TEXT SIMILARITIES

With all of the variation in styling, two major areas are typically uniform throughout architectural and engineering offices. Text height and capital letters are as close to an industry standard as architects and engineers get. Text placed in the drawing area is approximately 1/8" high. Titles are usually between 1/4" and 1" in height, depending on office practice. The other common feature is that text almost always comprises capital letters, although there are exceptions based on office practice. The letter "d" is always printed in lowercase when representing the pennyweight of a nail. A typical use would resemble

USE (3) -16d @ EA. RAFT. / PL CONNECTION

Occasionally manufacturers use lowercase letters to represent the size of special nails or fasteners specific to their products. With CAPSLOCK activated, all lettering will be capitalized unless SHIFT is pressed. Pressing SHIFT will make any letters typed while the key is depressed lowercase letters. If you have typing experience, it will take some time getting used to not having to press SHIFT at the start of each sentence.

Numbers are generally the same size as the text they are placed with. When a number is used to represent a quantity, it is generally separated from the balance of the note by a dash. This can be seen in the nailing specification above. Another common method of representing the quantity is to place the number in parentheses () to clarify the quantity and the size. This would resemble

USE (3) -3/4" DIA. M.B. @ 3" O.C.

Fractions are generally placed side by side (3/4") rather than one above the other so that larger text size can be used. When a distance such as one foot two and one half inches is to be specified, it can be done as 1'–2 1/2". The ' (foot) and " (inch) symbols are always used, with the numbers separated with a dash. If metric sizes are to be represented on drawings, the use of a comma as a number separator is somewhat different than with English units. Metric sizes of four digits are often written with no comma. Three thousand one hundred and fifty millimeters is written as 3150 mm. When a number five digits or larger must be specified, a space is generally placed where the comma would normally be placed. Fifty-six thousand, three hundred and forty-five millimeters is written as 56 345 mm. Other common methods of placing numbers will be presented as dimension methods are introduced in later chapters.

TEXT PLACEMENT

Text placement refers to the location of text relative to the drawing and within the drawing file. Care should be taken with the text orientation so that it is generally placed parallel to the bottom or parallel to the right edge of the page. When structural material shown in plan view is described, text is ideally placed parallel to the member being described. Figure 16.3 shows methods of placing text in plan views. On plan views such as Figure 16.4, text is typically placed within the drawing, but arranged so that it does not interfere with any part of the drawing. Text is generally placed within 2" of the object being described. A leader line is used to connect the text to the drawing. The leader line can be either a straight line or an arc depending on office practice.

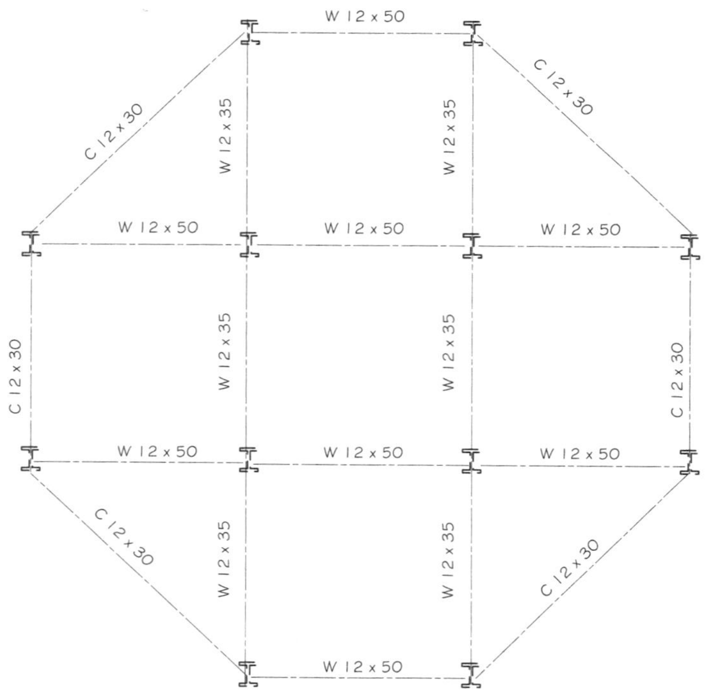

Figure 16.3 *Text should be placed so that it can be read from the bottom or right side of the drawing.*

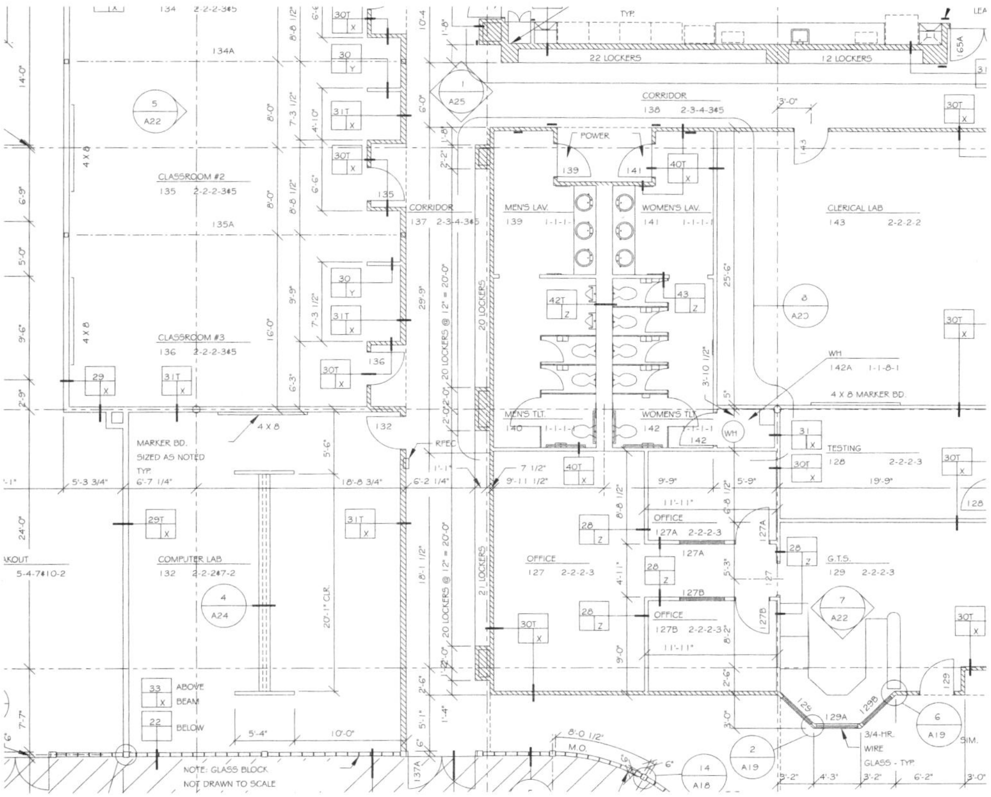

Figure 16.4 *Text can be placed within the drawing, but it should not interfere with drawing clarity. (Courtesy Tom Kuhns, Michael & Kuhns, Architects, P.C.)*

On details, text can be placed within the detail if large open spaces are part of the drawing. It is preferable to keep text out of the drawing. Text should be aligned to enhance clarity and can be either aligned left or right. Figure 16.5 shows an example of a detail using good text placement. The final consideration for text placement is to decide on layer placement. Text should be placed on a layer with a major group code of ANNO. Minor group codes include KEYN (key notes), LEGN (legends and schedules), NOTE, NPLT (non-plotted text), REDL (redline), REVS (revisions), SYMB (symbols) TEXT, and TTLB (border and title block).

TYPES OF TEXT

Text on a drawing is considered to be either a general or local note. General notes can refer to an entire project or to a specific drawing within a project. Local notes refer to specific areas of a project such as the notes in Figure 16.4. Small amounts of local

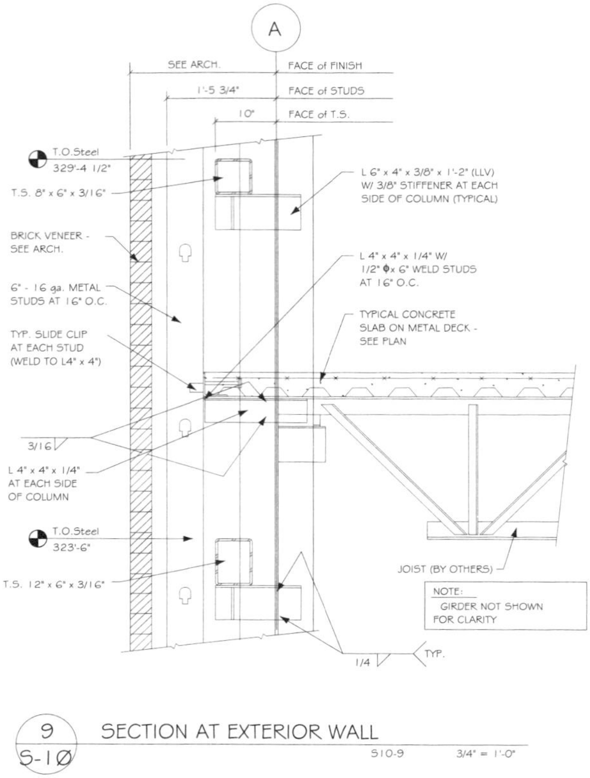

Figure 16.5 *Text should be aligned so that it can be quickly read. (Courtesy Van Domelen/Looijenga/McGarrigle/Knauf, Consulting Engineers.)*

notes should be placed using the **Text** command. The notes shown in Figure 16.1 are general notes that specify the materials of a framing plan. The **Mtext** command should be used to place large amounts of text. Practical uses for including each in template drawings will be discussed later in this chapter.

LETTERING HEIGHT AND SCALE FACTOR

One of the first prompts you'll be given as you place text will be for the height. Manually, the height might have been 1/8" high text, but in AutoCAD you're drawing in real (model) space. In previous chapters, the scale factor required to make line segments the desired size when plotted was considered. Similar factors must be applied to text height to produce the desired 1/8" high text when plotting is finished. Although text height can be adjusted before plotting, it is best to multiply the desired height (1/8") by the scale factor. The text scale factor is the reciprocal of the drawing scale. For drawings at 1/4"=1'–0", this would be 48. By multiplying the desired height of 1/8" (.125) x 48 (the scale factor), you see that the text should be 6" tall.

Quarter-inch-high lettering should be 12" tall. You'll notice that the text scale factor is always twice the line scale factor. Other common text scale heights include the following:

ARCHITECTURAL VALUES		**ENGINEERING VALUES**	
Drawing scale	**Text Scale**	**Drawing scale**	**Text Scale**
1"=1'–0"	12	1"=1'–0"	12
3/4"=1'–0"	16	1"=10'	120
1/2"=1'–0"	24	1"=100'	1200
3/8"=1'–0"	32	1"=20'	240
1/4"=1'–0"	48	1"=200'	2400
1/16"=1'–0"	64	1"=30'	360
1/8"=1'–0"	96	1"=40'	480
3/32"=1'–0"	128	1"=50'	600
1/16"=1'–0"	192	1"=60'	720

Working at a scale of 1"=20', 1/8" high lettering would be required to be drawn as 30" high text (.125" x 240).

USING THE TEXT COMMAND

The **Text** command creates dynamic text that can be seen on the screen as it is entered at the keyboard. Multiple lines of text can be placed even though the command name on the **Draw** menu implies otherwise. Text allows some basic editing to be performed and multiple lines of text to be entered while in the same command sequence. Start the **Text** command by selecting **Single-Line Text** from the **Draw** menu, by selecting the **Text** button on the **Text** toolbar, or by typing **TEXT** (ENTER) at the Command prompt. The command sequence is as follows:

Command: **TEXT** (ENTER) *(Or click the Text button.)*

Current text style: "Standard" Text height: 0'-0 3/16"

Specify start point of text or [Justify/Style]: *(Select text starting point.)*

Specify height: <0'-0 3/16">: **6"** (ENTER)

Specify rotation angle of text <0>: (ENTER)

Enter text: *(Enter desired text)*

Enter text: (ENTER)

Command:

Once you know the other Justify options, the **Text** command can be continued with any one of the options. Notice that the first display is the text style. The current style is referred to as Standard. Methods for creating different styles will be explored after you've explored the **Text** command. Because the "start point" is the first setting requiring input, it is a logical place to start.

START POINT

The name says it all. This is your opportunity to dictate where the start of the text will be placed. Indicating a point with the cursor will make that point the lower left corner of the text location. This type of text placement is referred to as *left justified.* As the prompt for text is displayed, the crosshairs will display the starting point at the lower left corner of the text. The relation of the text to the starting point can be seen in Figure 16.6. If you don't like the location, move the cursor to a new location and select that location with the select button. Two other options are available instead of placing a start point. Each will be introduced after the basic sequence has been explored. Once the start point has been selected, a prompt for the height will be displayed.

HEIGHT

The second option specifies the text height, with the default of 0'–0 3/16". If you'll be plotting this drawing at 1/4"=1'–0", type **6** (ENTER) for the height. You can also adjust text height by dragging the pointing device until the distance between the cursor and the insertion point indicates the desired text height. If you have not adjusted the initial drawing limits, any text you create is going to appear to be of gigantic proportions. If the height is set to zero, the text height will be controlled by the **Style** height value. (The **Style** command will be introduced later in this chapter.)

ROTATION ANGLE

The third option to be adjusted prior to entering text is the rotation angle of the text. The rotation angle can be visualized by thinking of a line extending from the start point at the specified angle, with the text above this line. The rotation angle can be entered at the Command prompt by entering X and Y coordinates, or by using the cursor. To use the cursor, drag the pointing device until the angle between the cursor and the

Figure 16.6 *The default placement with the* **Text** *command is to place text on the right side of the start point. This is called left justified.*

insertion point represents the desired text rotation angle. Figure 16.7 shows an example of text rotated at various angles. Although text is read looking from the bottom or right side of the drawing, this option will allow text to be aligned with specific objects of a drawing, as seen in Figure 16.8. You can accept the default of 0 rotation by pressing (ENTER), or you can enter a numeric value.

PLACING TEXT

The desired text can now be entered by keyboard. Be sure to have CAPSLOCK activated. Notice that as you type, the text is displayed on the screen. Depending on the zoom setting, you might be unable to read it. Either cancel the command and **Zoom** into the area where you will be adding the text or read the text at the **Text** prompt line that replaces the command line. If any special characters are used, they will show up on the screen, but not at the text prompt. When the text entry is completed, press (ENTER) to add the text to the drawing. To add additional lines of text, press (ENTER), and the new start position will be adjusted to be directly below the first line of text. The prompts for start point, height, and rotation angle will be skipped and the values for the first line of text will be reused. The spacing between lines of text will be half the height of the text. The command sequence is as follows:

> Command: **TEXT** (ENTER) *(Or click the Text button.)*
>
> Current text style: "Standard" Text height: 0'-0 3/16"
>
> Specify start point of text or [Justify/Style]: *(Select text starting point.)*
>
> Specify height: <0'-0 3/16">: **6"** (ENTER)

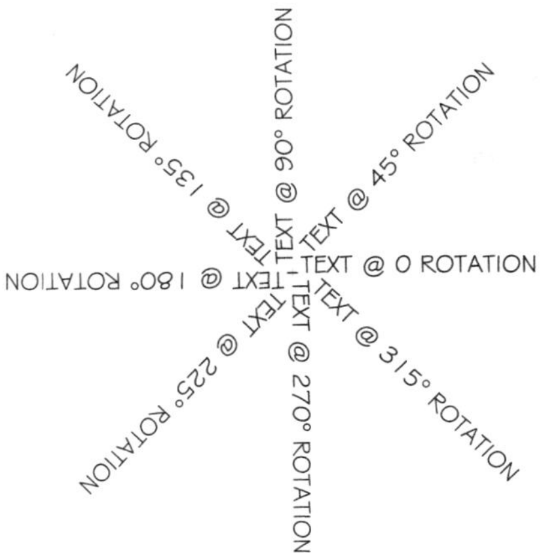

Figure 16.7 *Lines of text can be rotated so that they can be read from any angle.*

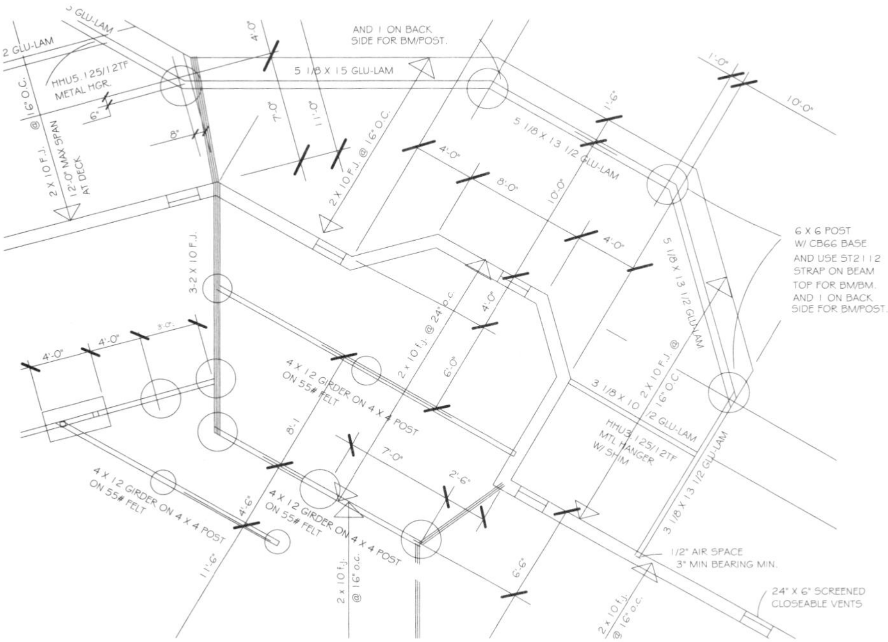

Figure 16.8 *Text should be rotated to be parallel to the structural member when a structure is an irregular shape. (Courtesy Residential Designs.)*

Specify rotation angle of text <0>: (ENTER)

Enter text: **FINISH FLOOR** (ENTER)

Enter text: **ELEV. 100.00'** (ENTER)

Enter text: (ENTER)

Command:

Figure 16.9 shows the result of this command sequence. As you work with multiple lines of text, each succeeding line will have all of the attributes of the original text. Once the desired text has been entered, press (ENTER) twice to exit **Text**. Pressing ESC will destroy any text entered during the active command sequence. The **Mtext** command, which will be discussed later in this chapter, should be used to create large bodies of text.

JUSTIFYING TEXT

As the prompt for the start point was displayed, the options of Justify and Style were given. To justify text is to determine where and how the text will be located within a defined space. The default for placing text is for left justified text, which places the left

Figure 16.9 *The results of the* **Text** *command using a specified start point, a height of 6", and a rotation angle of 0.*

edge of each new line of text directly in line with the line of text above. The lower left starting point is defined, and text will be placed to the right of the starting point. This method is used on construction drawings such as sections, and details where large amounts of written information can be neatly aligned. Figure 16.10 shows an example of left justified text.

Fourteen other options are available for aligning text. Each option can be accessed by the following sequence:

Command: **TEXT** (ENTER) *(Or click the Text button.)*

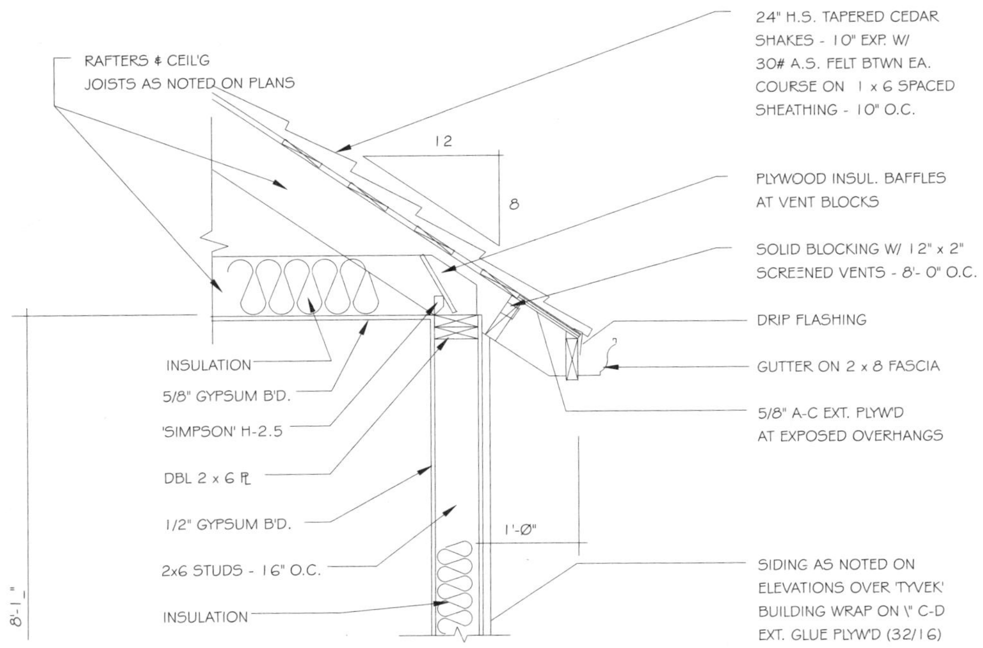

Figure 16.10 *Left justified text helps add to the clarity of a drawing. (Courtesy Piercy & Barclay Designers, Inc., A.I.B.D.)*

Current text style: "Standard" Text height: 0'-0 3/16"

Specify start point of text or [Justify/Style]: **J** (ENTER)

Align/Fit/Center/Middle/Right/TL/TC/TR/ML/MC/MR/BL/BC/BR: **R** (ENTER)

Once you feel comfortable with each option, the option can be accessed without displaying the Justify menu by typing the letter of the option at the Command prompt. The process to align text from the right can be started by the following sequence:

Command: **TEXT** (ENTER) *(Or click the Text button.)*

Current text style: "Standard" Text height: 0'-0 3/16"

Specify start point of text or [Justify/Style]: **R** (ENTER)

Specify right endpoint of text baseline:

Even though Right is not listed as one of the current options, **R** will be accepted and will change the prompt to the right endpoint prompt for the text baseline. This prompt will then be followed by the height and rotation prompts. As text is entered using the Right option, it will be placed just as it was when Left justified was used. When (ENTER) is pressed to place the text, it will be added to the drawing as right justified text.

Align

The Align option will allow two useful options to be accessed. The starting and ending point can be selected so that text can be aligned with an inclined object, similar to the way the rotation angle option is used. This method of selection is superior to using Rotation when the angle is not known. The second benefit of Align is the ability to select the start and endpoints of the line of text. The drawback to this option is that text height is assigned based on the amount of text to be placed between the start and end points. Figure 16.11 shows an example of aligned text. The command sequence is as follows:

Command: **TEXT** (ENTER) *(Or click the Text button.)*

Current text style: "Standard" Text height: 0'-0 3/16"

Specify start point of text or [Justify/Style]: **A** (ENTER)

Specify first endpoint of text baseline: *(Select starting point.)*

Specify second endpoint of text baseline: *(Select starting point.)*

Specify height <0'-0 3/16"> *(Enter desired text height.)*

Enter text: *(Enter desired text.)*

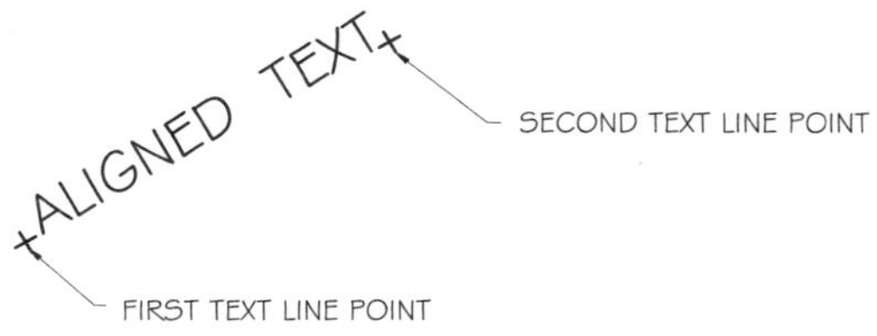

Figure 16.11 *Text can be aligned by specifying two points. The desired text will now be placed between these points. If more than one line of text is required, the height of each line may vary.*

Fit

The Fit option is very similar to the Align option. Using Fit, the text height will remain constant throughout several different lines of text, but the width will be varied. This option uses a command sequence similar to Align, but it allows for a height to be specified prior to placing the text. Figure 16.12 shows an example of text placed using the Fit option. The command sequence for Fit is as follows:

Command: **TEXT** (ENTER) *(Or click the Text button.)*

Current text style: "Standard" Text height: 0'-0 3/16"

Specify start point of text or [Justify/Style]: **F** (ENTER)

Specify first endpoint of text baseline: *(Select starting point.)*

Specify second endpoint of text baseline: *(Select starting point.)*

Specify height <0'-0 3/16"> *(Enter desired text height.)*

Enter text: *(Enter desired text.)*

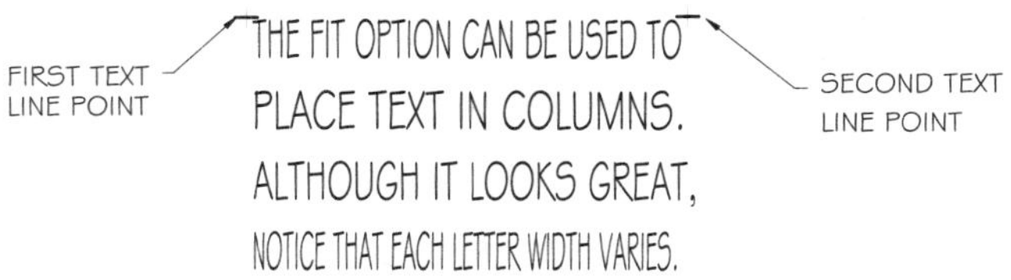

Figure 16.12 *The Fit option will place text between two specified points, but the height will remain constant if more than two lines of text are required.*

Center

The Center text option will place a line of text centered on a specified point, as seen in Figure 16.13. The command sequence is as follows:

Command: **TEXT** (ENTER) *(Or click the Text button.)*

Current text style: "Standard" Text height: 0'-0 3/16"

Specify start point of text or [Justify/Style]: **C** (ENTER)

Specify center point of text: *(Select desired center point.)*

Specify height <0'-0 3/16">: *(Enter desired text height.)*

Specify rotation angle of text <0>: *(Enter desired angle.)*

Enter text: *(Enter desired text.)*

To place several lines of text using the Center option, press (ENTER) to move the cursor to start the line below the first line. Type the second line of type and press (ENTER) when the line is complete. This process can be continued indefinitely. Remember that **Mtext** is better suited for entering large amounts of text.

Middle

This option is similar to the effects of Center, but text is centered vertically and horizontally on the start point, as seen in Figure 16.14. The command sequence is similar to the sequence used for the Center option, once the Middle point is selected.

Right

This option is the reverse of the default left justified option. Figure 16.15 shows an example of right justified text.

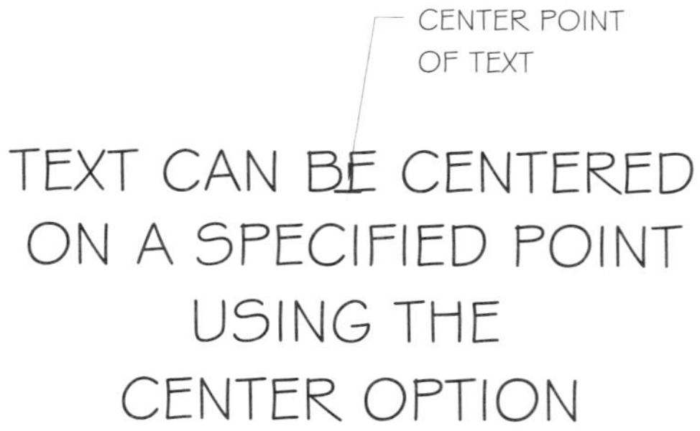

Figure 16.13. *The Center option allows text to be spaced equally around a specified center point.*

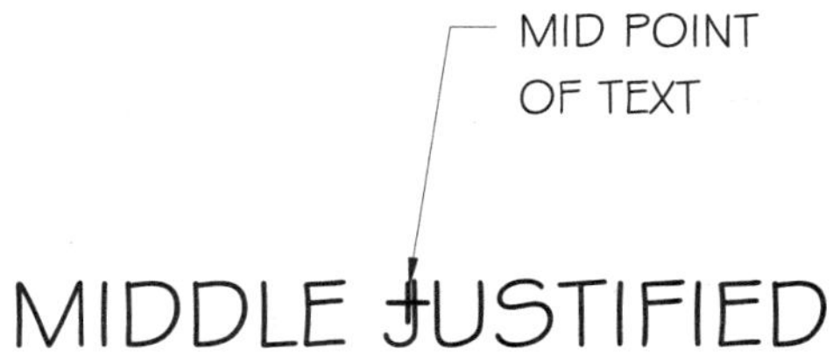

Figure 16.14 *The Middle justified option will center text along an imaginary line.*

Figure 16.15 *The Right justified option will align the right ends of lines of text.*

Letter Options

The remaining nine options can be entered by typing the designated letters at the Justify prompt:

TL-top left	TC-top center	TR-top right
ML-middle left	MC-middle center	MR-middle right
BL-bottom left	BC-bottom center	BR-bottom right

The effects of each can be seen in Figure 16.16. The command sequence is similar to the sequence used for left justified text, although the start point is referred to by the new location. For instance, the command sequence for BR is as follows:

Command: **TEXT** (ENTER) *(Or click the Text button.)*

Current text style: "Standard" Text height: 0'-0 3/16"

Specify start point of text or [Justify/Style]: **BR** (ENTER)

Specify bottom right point of text: *(Select desired start point.)*

Specify height <0'-0 3/16">: *(Enter desired text height.)*

Specify rotation angle of text <0>: *(Enter desired angle.)*

Enter text: *(Enter desired text.)*

The result of this sequence can be seen in Figure 16.17.

STYLE OPTION

The last option of **Text** to be considered is the Style option that will determine the appearance of the text characters. The option is selected by typing **S** (ENTER) at the first text prompt. The sequence and options include the following:

Command: **TEXT** (ENTER) *(Or click the Text button.)*

Current text style: "Standard" Text height: 0'-0 3/16"

Specify start point of text or [Justify/Style]: **S** (ENTER)

Enter style name or (?) <STANDARD>:

Standard will remain the default style until the **Style** command is used to create other styles. The **Style** command will be explored next in this chapter. If the question (?) option is selected, the following prompt will be displayed:

Enter text style(s) to list <*>: (ENTER)

Current text style: "STANDARD"

Specify start point of text or [Justify/Style]:

Figure 16.16 *The start points for each of the nine letter options for placing text.*

Figure 16.17 *BR (bottom right) justification is often used when notes must be placed at the left side of an object.*

WORKING WITH STYLE

The term *style* in CAD drafting describes the characteristics of a group of text such as the font, height, width, and the angle used to display the text. One style can be used for all drawing text, a second style used for the drawing titles, and a third style used for all information in the title block. The **Style** command can be used to add different fonts to your drawing as well as change the properties of the letter shapes. As styles are created, they should be stored in drawing templates for future use with all similar projects. Assigning a style name, selecting a font, and defining style properties are the three areas to be considered when setting a new style of text.

NAMING A STYLE

Start the command by selecting the **Text Style** button on the **Text** toolbar, by typing **ST** (ENTER) at the Command prompt, or by selecting **Text Style** from the **Format** menu. Each option will produce the dialog box shown in Figure 16.18, which can be used to name a style, assign a font, control the style options, and preview the results.

New styles can be created in the **Text Style** dialog box by selecting the **New** button. This will produce the **New Text Style** dialog box shown in Figure 16.19. The default new style is named Style1. Accept the name by clicking the **OK** button. The **New Text Style** dialog box will be closed and Style1 is now displayed as the current name in the **Style Name** edit box. Style1 can also be altered before it is accepted as a style name by using the same procedure as in renaming a drawing or layer name. While the name is highlighted, type the desired name. Type the name of **Titles** in the edit box. Clicking the **OK** button will remove the dialog box and change the name Style1 to the name provided. The name Style1 can also be renamed to something more meaningful by selecting the **Rename** button and using the **Rename Text Style** dialog box

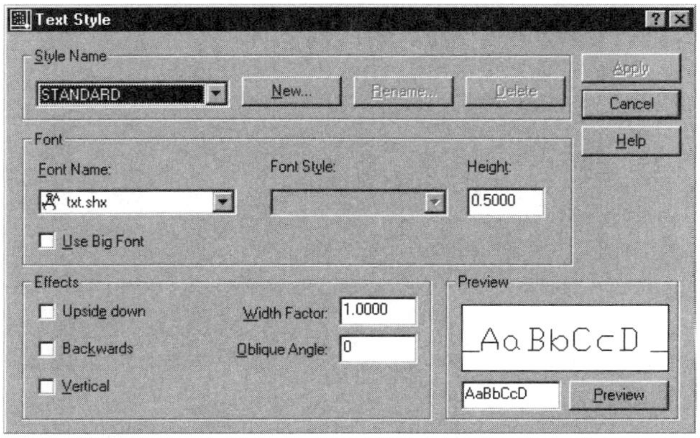

Figure 16.18 *Selecting the* **Text Style** *button on the* **Text** *toolbar, typing* **ST** *(ENTER) at the Command prompt, or selecting* **Text Style** *from the* **Format** *menu produces the* **Text Style** *dialog box that can be used to name and control text styles.*

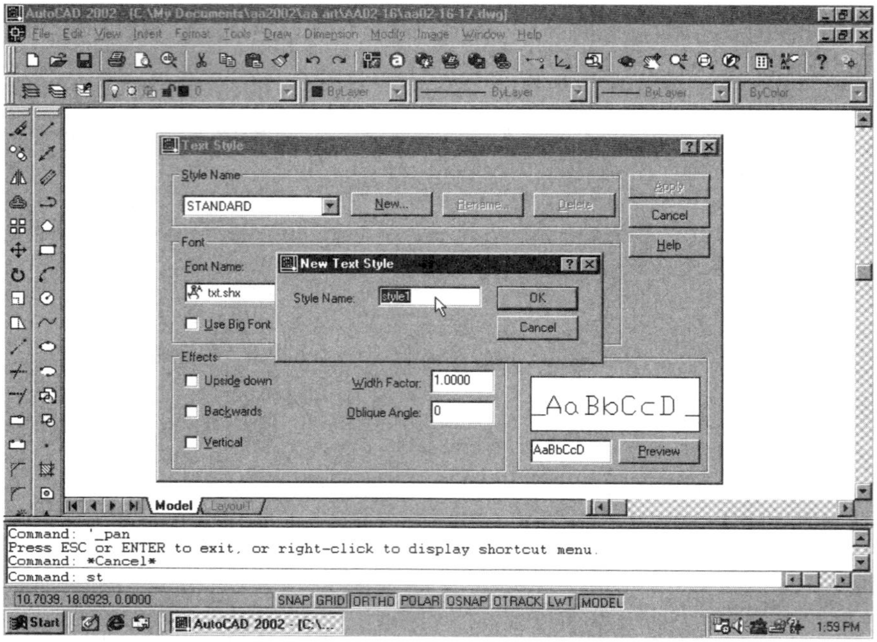

Figure 16.19 *Selecting the* **New** *button produces the* **New Text Style** *dialog box, allowing a new text style to be named.*

that can be seen in Figure 16.20. Selecting the Cancel button will remove the **Rename Text Style** dialog box and allow a different aspect of the **Style** command to be addressed, or allow the command to be terminated. The third option in the **Style Name** area is the **Delete** button. Selecting this button allows a style to be deleted from the current list of text styles.

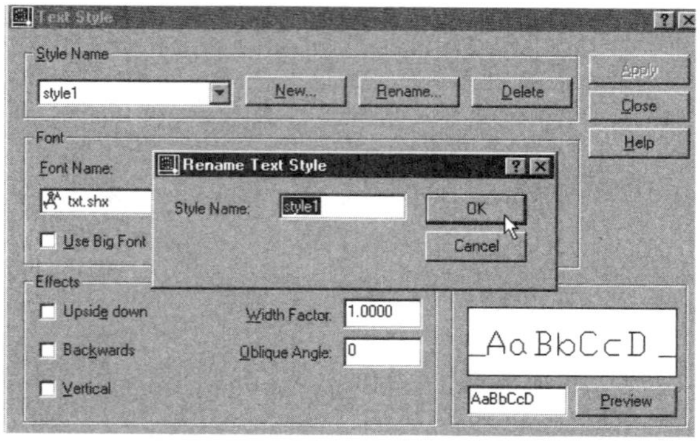

Figure 16.20 *Style1 can be renamed in the* **New Text Style** *dialog box before the name is assigned, or it can be altered by selecting the* **Rename** *button.*

Choosing the Style Name

The name of the text style to be created should reflect the use of the text. The name can contain up to 255 letters, numbers, spaces, and special characters. (Names longer than 31 characters will spill out of the edit box.) The same guidelines that were used to name linetypes and layers should be used to name text styles. Some offices use the name of the text font for the style name. Names such as titleblock, gennotes (general notes), fndnotes (foundation notes), or sitenotes are examples of names that could be used to describe different styles of text. These names could be abbreviated further as TB, GEN, FND, and ST to speed typing text selections. Names that include the text font and usage are helpful. Much of the lettering in this text was created using a font called ARCHSTYLE, which is created by a third-party vendor for use with AutoCAD. You can create styles in the template drawing such as, archtitl, archtxt, and italitxt. These would be used for creating titles and text using the font archstyle, and a text style using the font italic for drawing text.

LETTERING FONTS

The word *font* is written in an italic font. The term font is used to describe a particular shape of text that includes all lower and upper case text and numerals. The shape of the letters is one of the defining differences between architectural and mechanical lettering. Construction drawings usually contain text fonts that are more artistic than the gothic block lettering used on mechanical drawings. Most offices use different fonts throughout a drawing to distinguish between titles and text, or to highlight specific information.

AutoCAD allows all of the fonts loaded into your computer to be accessed by the **Style** command. Many of the fonts contained in AutoCAD can be seen in Figure 16.21. The illustrations in this chapter have been created using the default font of TXT, a simple font design, with letters consisting only of straight-line segments. Because of its simplicity, it is quick to regenerate and plot. Many engineering offices use fonts such as ROMANS, ROMANC, ROMANT, or ITALICC to letter drawings. In addition to these fonts, some architectural offices use fonts that more closely resemble freehand lettering, such as Stylus BT, City Blueprint and Country Blueprint for notes on a drawing and ROM, ROMB, or SASBO for titles. Examples of these fonts can be seen in Figure 16.22. It is important to select fonts that are easy to read and quickly reproduced by a plotter. The complexity of a font also affects the amount of storage space required, the speed of regens, and the plotter speed.

Lettering Programs

One of the benefits of AutoCAD is that many third-party suppliers have developed material to supplement the standard program. Many offices add text libraries to the basic program. You'll notice that much of the text used throughout the illustrations in this text is not found in the AutoCAD font files. This lettering is produced using

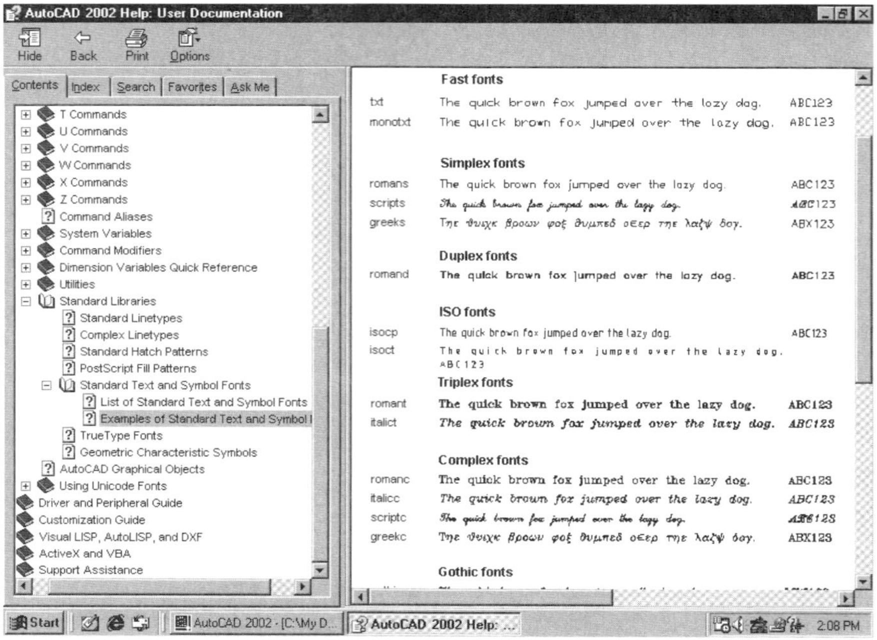

Figure 16.21 *Standard AutoCAD fonts.*

material created by third-party vendors. Consult advertisements in local trade magazines for other available text libraries. The **Text Style** dialog box can be used to select the font to be used with each lettering style. The box will be discussed in detail shortly, but it's important to remember that in addition to the fonts contained in AutoCAD, TrueType fonts contained in additional programs in your computer can be used in AutoCAD drawings. Methods of accessing these files will be discussed as ways of creating text styles are introduced.

Choosing a Font

The second aspect of defining a text style is choosing a font. Once the style name has been entered, selecting the down arrow beside the **Font Name** edit box will produce a list of all of the text fonts available to you. A sample of the list can be seen in Figure 16.23. The display will depend on the status of the **Use Big Font** check box. With the box inactive, all fonts in all programs on your computer will be displayed. With the box active, only the fonts contained in AutoCAD will be displayed in the list. The AutoCAD fonts are preceded by the Divider icon and followed by .SHX. The TrueType icon precedes other fonts available to AutoCAD. Scroll through the list and select STYLUS BT font or an architectural font of your choice. As the name of the font is selected in the list, the name will be displayed in the edit box, the list will be closed, and a sample of the font will be displayed in the **Preview** window.

DEFINING STYLE PROPERTIES

Once the style has been named and the font has been assigned, seven options are available to further define the style being created. These options included **Font Style**, **Height**, **Upside-down**, **Backwards**, **Width Factor**, **Oblique Angle**, and **Vertical**.

Font Style

So far you've named a style (TITLES) and assigned the STYLUS BT font to the text style. The **Font Style** edit box is activated once a font is named. This edit box will allow

THE STYLE 'STANDARD' IS COMPOSED OF THE TXT FONT.

THE STYLE 'NOTES 'IS COMPOSED OF THE ROMANS FONT.

THE 'TITLERESPONSE' STYLE IS COMPOSED OF THE ROMANC FONT.

THE TBLOCK STYLE IS COMPOSED OF THE FONT ROMANT.

THE STYLE 'GENERALNOTES' IS COMPOSED OF THE FONT ITALICC.

THE 'TITLEBLOCK' STYLE IS COMPOSED OF THE FONT ROMB.

THE STYLE 'ENGINEERSTAMP' IS COMPOSED OF THE SASBO FONT.

THIS TEXT IS A SAMPLE OF THE 'STYLUS BT' FONT

THE STYLE 'FNDNOTES' IS COMPOSED OF THE FONT COUNTRYBLUEPRINT

THE 'TEXT' STYLE IS COMPOSED OF THE CITYBLUEPRINT FONT.

THIS TEXT IS A SAMPLE OF TEXT BY A THIRD PARTY DEVELOPER . THE FONT IS 'ARCHSTYL'.

Figure 16-22 *Common fonts used in architectural and engineering offices.*

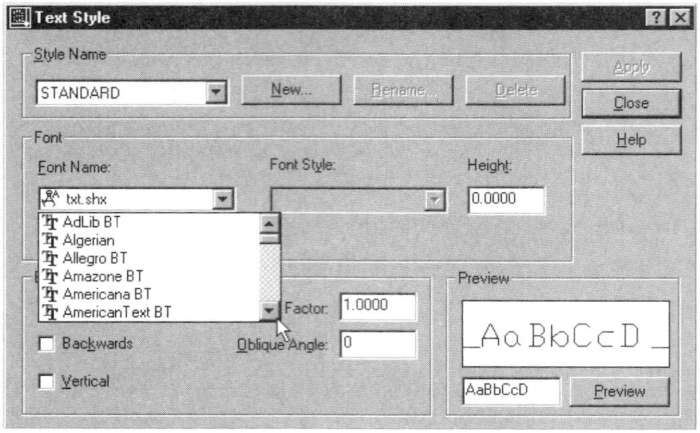

Figure 16.23 *Selecting the down arrow beside the* **Font Name** *edit box will produce a listing of the AutoCAD and the True Type font files. Scroll through the desired list and select the name of the font to be used with the style being created.*

characters to be formatted with bold, italic, oblique or regular text. Options will vary depending on the selected font.

Height

If a height is entered at this prompt, all text created using this style of text will be the same height. If all TITLES throughout the drawing are to be the same size, the height variable of the **Style** command can be set. When height is set with the **Style** command, the variable will not have to be set each time the font is used. The height of text is best controlled with the Height option of the command used to create the actual text. Entering a height now will delete the Height prompt in the **Text** and **Mtext** command sequences when this style is used. Responding with a 0 will allow the height to be determined each time this style is used.

Upside-Down

The default for this option will place text in the normal position. Activating this box will place text upside-down. Although this option is not typically used in construction drawings, an example of upside-down text can be seen in Figure 16.24.

Backwards

The default value for this option, inactive, will produce text that is read from left to right. Activating this box will produce text that is read from right to left or backwards. Although this option is not used often, it could be useful if a drawing is to be plotted on one side of a sheet of vellum or Mylar, and text on the other side. Figure 16.25 shows an example of text printed backwards.

Figure 16.24 *Text can be printed in both normal and upside-down orientation.*

Width Factor

Font characters are displayed and plotted based on a ratio of their height and width. No matter what the actual height is, the width is assigned a value of 1. The default value for the width factor is also 1, making the width equal to the height of the letter. If a value of .5 is entered, text will be half as wide as it is tall. A value of 2 will produce an elongated style of lettering. Figure 16.26 shows a comparison of various width factors.

Oblique Angle

This option will allow text to be tilted from true vertical. The default angle of 0 will display text in the normal position, with the vertical leg of each letter in a vertical position. Entering a positive value will slant text to the right. Entering a negative value will produce a backslant. Figure 16.27 shows examples of altering the obliquing angle.

Figure 16.25 *Text can be printed in both the normal and backward orientation.*

Figure 16.26 *Adjusting the* **Width Factor** *will affect the appearance of text.*

Figure 16.27 *The* **Oblique Angle** *box of the* **Text Style** *dialog box affects the angle of the vertical stroke of each letter.*

Vertical

This option is not active if a TrueType font has been selected. The option will display .SHX fonts in the vertical position, as shown in Figure 16.28.

Applying the Style

Once all of the parameters have been selected for the style, the options of **Apply** and **Close** remain to set the style. Choosing the **Apply** button will apply any style changes made in the dialog box to the text in that current style within this current drawing. Figure 16.29 shows an example of text created using the TXT.SHX font and the same

Figure 16.28 *Placing text in a vertical position might be required on some construction drawings.*

TEXT STYLES CAN BE CHANGED ONCE THEY HAVE
BEEN CREATED. THIS IS TXT.SHX FONT.

*TEXT STYLES CAN BE CHANGED ONCE THEY
HAVE BEEN CREATED. THIS IS ITALIC WITH
A WIDTH FACTOR OF 1.5.*

Figure 16.29 *Text created using the THX.SHX font was changed to ITALIC.SHX with a width factor of 1.5 by changing the appropriate values and then selecting the* **Apply** *button.*

text after the font was changed to ITALIC.SHX with a width factor of 1.5. The text was altered by changing the appropriate values, selecting the **Apply** button, and then selecting the **Close** button. If you choose the **Close** option without selecting **Apply** first, it will accept the current style values and remove the dialog box. Existing styles will remain unchanged, and new text will be added using the current settings.

SPECIAL CHARACTERS

In addition to the symbols on a keyboard, it is often necessary to print special symbols, such as a degree or diameter symbol. This can be done by typing a code letter preceded by two % (percent) symbols. Common symbols of AutoCAD include the following:

%%C=diameter symbol

%%D=degree symbol

%%O=overscore text

%%P=plus/minus symbol

%%U=underscore text

The symbols for overscoring and underscoring text are toggles. Type %%U preceding the text to be underscored to activate underscoring. The symbol must also be typed at the end of the phrase to turn off underscoring. An example of the sequence would be

%%UPROVIDE AN ALTERNATE BID FOR ALL ROOFING
MATERIALS%%U

As the text line is displayed at the command line, the %%U will be displayed in addition to the desired text. When (ENTER) is pressed, the %%U will be removed from the screen and the text will be underscored. Although a single percent symbol can be created by using the percent key (SHIFT + 5), when the percent symbol must proceed another special character, %%% must be used. To write 50°+/–5% you must enter the following sequence:

50%%D%%P5%%%

Remember that these symbols apply to the fonts in AutoCAD and do not apply to text created by third-party vendors. It's also good to remember there is an easier method of applying special text: Use the **Mtext** command.

EDITING EXISTING TEXT

Text can be edited by using the backspace key and entering corrections before making the text part of the drawing base. Text can be edited once it is part of the drawing base using the **Ddedit** command. Start the command by selecting the **Edit Text** button on the **Text** toolbar, by typing **ED** (ENTER) at the Command prompt, or by selecting **Text** from the **Modify** menu. Typing **ED** (ENTER) will produce the following prompt:

Command: **ED** (ENTER) *(Or click the Edit Text button.)*

DDEDIT

Select an annotation object or [Undo]:

Select the text to edit by moving the cursor to the line of text and select it with the select button. The error does not have to be selected, only the line of text containing the error. As the text is selected, the **Edit Text** dialog box is presented, containing the line of text. Figure 16.30a shows a line of text in a drawing and the line contained in the **Edit Text** dialog box. Because the line of text is too long to fit in the edit display, the left and right arrow keys can be used to alter the line of text that is displayed.

Once the proper portion of the text line is displayed, editing can be performed by moving the arrow to the desired location and using the select button. In the example, the following note is used:

aLL FRAMINGG LUMBER IS TOBE D.F.L. # 2 OR BETTER UNLESS-
NOTED.

The "a" of "aLL" should be in capital letters, "FRAMINGG" is misspelled, and a space should be placed between "TOBE" and "UNLESSNOTED." To fix the "a" move the arrow to the right side of the "a" and click the mouse select button. Use BACKSPACE to eliminate the "a". Without moving the cursor, the error can be fixed by typing an **A**.

Use the arrow key to move the cursor to the right side of "FRAMINGG" and click the select button. Press BACKSPACE to eliminate the last "G". Now if the cursor is placed between the "O" and "B" of "TOBE", you can insert a space by pressing SPACEBAR. The text should now match Figure 16.30b. "UNLESSNOTED" can be corrected in

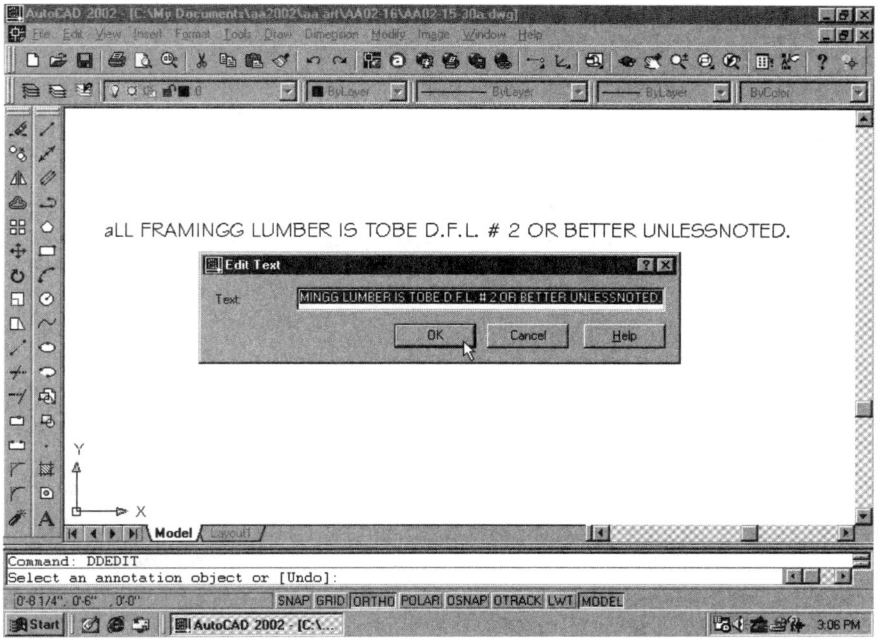

Figure 16.30a *The* **Ddedit** *command allows a string of text to be edited. Once selected, the text string will be placed in the* **Edit Text** *box.*

the same method by pressing END to scroll that area of text into the edit box. The corrected text will now read as it was intended (see Figure 16.30c.) If the edit is not what you desired, the **Undo** option can be used to restore the line of text to its previous status. Once you are satisfied with the text, click the **OK** button to return to the drawing. The **Ddedit** command is still active. If no other editing needs to be done, press the * key to return the command line.

TEXT AND THE MIRROR COMMAND

The **Mirror** command was introduced in Chapter 9. If you mirror an object that includes text, the text will also be reversed. To avoid the problems this will create, use the **Mirrtext** command. Start the command by typing **MIRRTEXT** (ENTER) at the Command prompt. This will produce the prompt:

> Command: **MIRRTEXT**
>
> Enter new value for MIRRTEXT <1>:

Pressing (ENTER) at the prompt will accept the default value and the text will be reversed as the object is mirrored. Typing **0** (ENTER) at the prompt will mirror the object and the text, but keep the text readable.

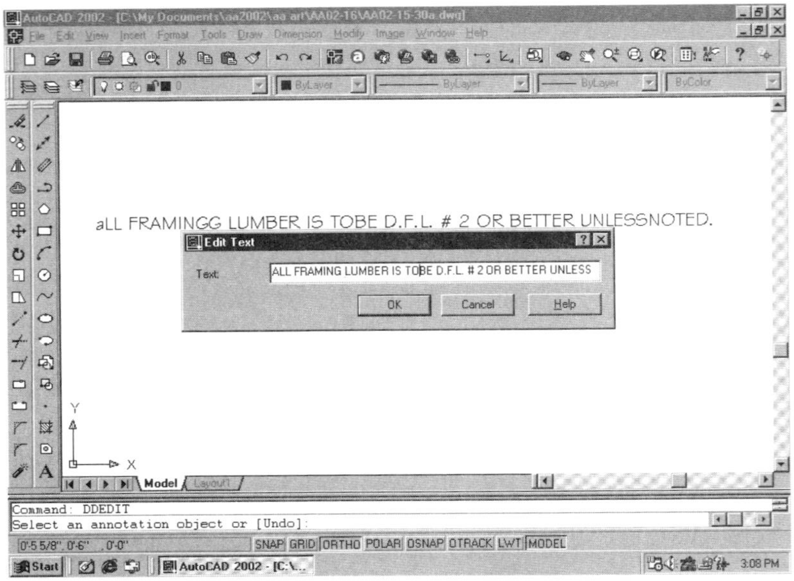

Figure 16.30b *The "a" was edited by moving the cursor to the right of the letter and clicking the select button. Once selected, the error can be removed with the backspace arrow, and the desired correction made at the keyboard. Even though the "A" and the word "FRAMING" have been corrected, these changes are not reflected in the drawing until* **OK** *is clicked.*

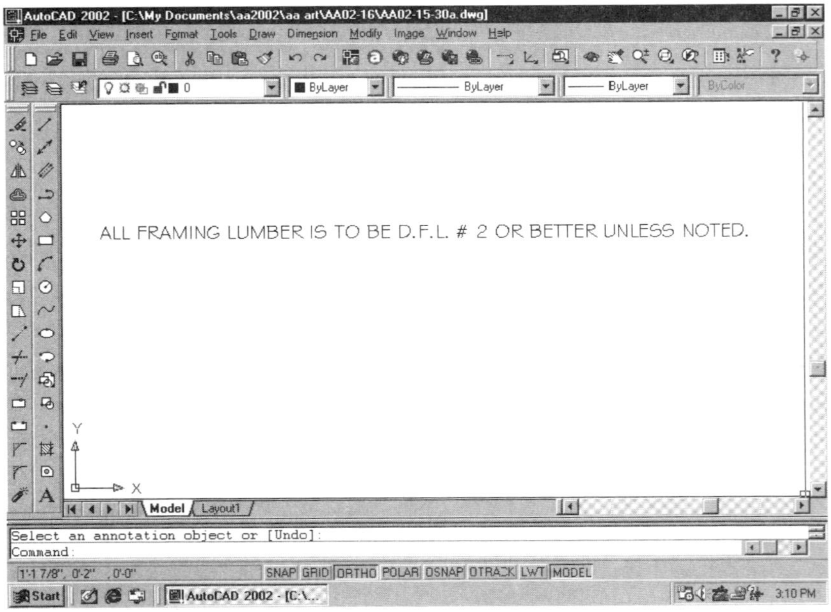

Figure 16.30c *The note shown in Figure 16.30a is now as it was intended.*

QUICK TEXT

Quick text is not an additional way to create text, but a very useful way to display it. As you're editing drawings containing large amounts of text, you might find that **Regen** times start to slow. The **Qtext** command will display lines of text as rectangles, which are much quicker to regenerate. The height of the rectangle is the same as the text height, but the length of the rectangle is typically longer than the line of text. The command sequence is as follows:

> Command: **QTEXT** (ENTER)
>
> Enter mode [ON/OFF] <OFF>:

With **Qtext** set to ON, text will retain its current setting until a **Regen** is performed. Figures 16.31a and 16.31b show an example of **Qtext** in both the ON and OFF settings.

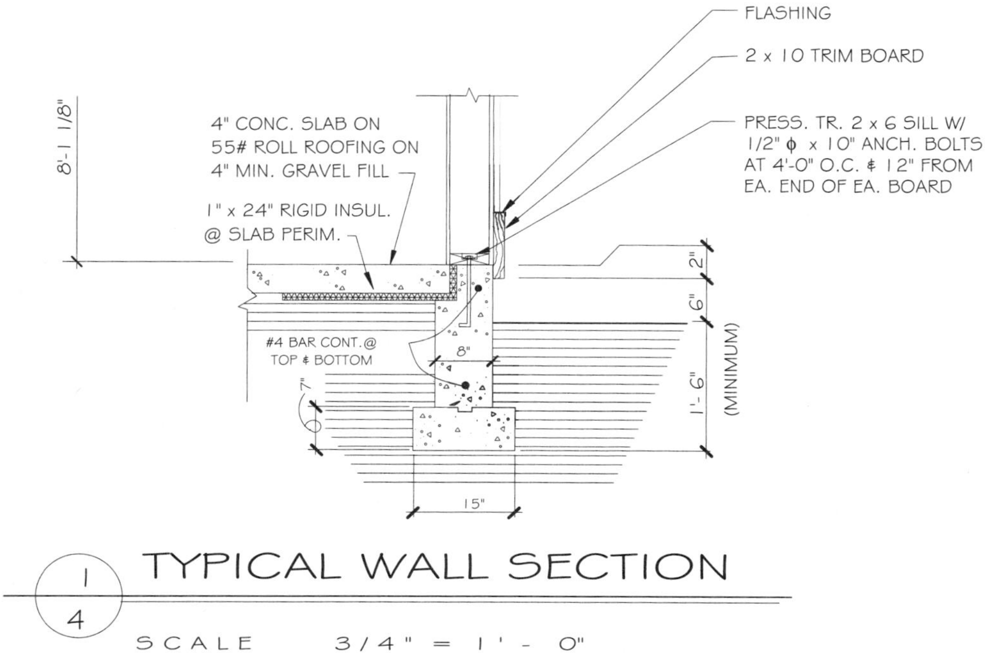

Figure 16.31a *An example of normal text. Figure 16.31b shows the same drawing with quick text. (Courtesy Piercy & Barclay Designers, Inc., A.I.B.D.)*

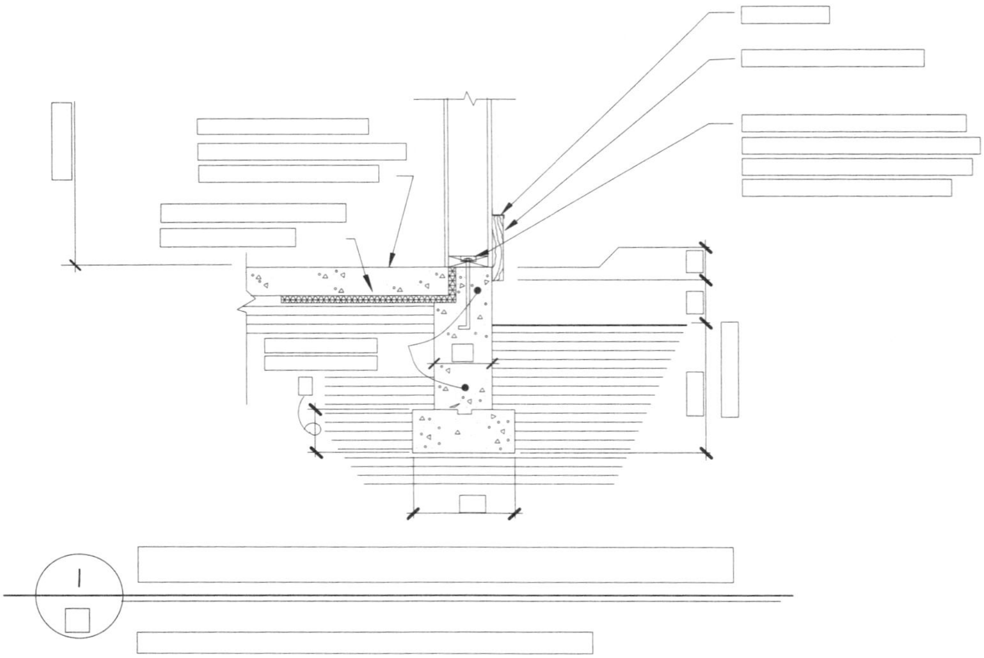

Figure 16.31b Qtext *can be used to speed the* **Regen** *of very large drawings. The effect of* **Qtext** *on the drawing in Figure 16.31a. (Courtesy Piercy & Barclay Designers, Inc., A.I.B.D.)*

CREATING TEXT WITH MTEXT

You might be feeling overwhelmed with text by now. You've worked with **Text**, **Style**, and **Qtext**, and you might be wondering why you need another way to enter text into a drawing. The answer is because each method offers more options than the previous method. Text is ideal for placing local notes in and around the drawing. However, paragraphs created in **Text** can be very frustrating to edit if they need to be moved and resized. **Qtext** is a tool for displaying text and offers no benefits for creating or editing text. **Mtext** (**Multiline Text**) is ideal for placing large areas of text on a drawing. Multiline text can be created in the **Multiline Text Editor** dialog box, which offers greater flexibility in creation and editing of the text. Using **Mtext** is similar to using a word processing program within AutoCAD. **Mtext** also offers advantages when editing the drawing. Lines of text placed with **Text** act as individual lines when they are moved. Text placed with **Mtext** functions as one object. Multiline text also allows text to be easily underlined and varied fonts, colors, and heights to be assigned throughout the text. **Mtext** is ideal for entering and editing large bodies of text such as those seen in Figure 16.32. A paragraph can be moved by selecting one character rather than the entire paragraph. Access the command by selecting the **Multiline Text** button on the **Draw** or **Text** toolbar, by typing **T** (ENTER) at the Command prompt or by selecting **Multiline Text** from **Text** on the **Draw** menu. Each will produce the following prompt:

Command: **T** (ENTER)

_mtext Current text style: "titles" Text height: 12"

Specify first corner:

By default, AutoCAD is waiting for you to select where you would like to place the text. Once you select the first corner, you will be prompted:

Specify opposite corner or [Height/Justify/Line spacing/
Rotation/Style/Width]:

Although you can respond to the second prompt by providing a value for one or more of the options, selecting the second corner to define the text window width will produce the **Multiline Text Editor** dialog box and allow each of these values and many more to be easily set. Notice in Figure 16.33, as the box is being dragged to indicate the second corner, an arrow is shown indicating the direction that excess text will flow out of the box. Methods of altering the direction will be introduced shortly. By

STEEL JOIST
1. STEEL JOISTS SHALL BE DESIGNED, FABRICATED AND ERECTED PER SJI STANDARD SPECIFICATION.
2. CAMBER ALL JOISTS WITH STANDARD CAMBERS PER ABOVE STANDARD SPECIFICATION UNLESS NOTED ON THE DRAWINGS.
3. PROVIDE BRIDGING AS CALLED FOR BY ABOVE SPECIFICATIONS. DO NOT USE SLOTTED HOLES IN BOLTED CONNECTIONS UNLESS APPROVED BY ENGINEER.
4. PROVIDE SHOP DRAWINGS AND CALCULATIONS THAT INDICATE JOIST SIZES, LOADING, MEMBER SIZES, PANEL POINT LOCATIONS, CAMBERS, BRIDGING AND ANY OTHER INFORMATION THAT MAY BE PERTINENT TO THE JOB. SHOP DRAWINGS SHALL BE STAMPED BY AN ENGINEER (THE PRODUCT ENGINEER) REGISTERED IN THE STATE OF OREGON.
5. PROVIDE INSPECTION REPORT PREPARED BY THE PRODUCT ENGINEER CERTIFYING THAT THE JOISTS HAVE BEEN FABRICATED ACCORDING TO THE DESIGN ASSUMPTIONS AND REQUIREMENTS. REPORT IS TO ACCOMPANY THE JOISTS ON DELIVERY TO THE PROJECT.
6. JOIST ERECTOR SHALL EXERCISE EXTREME CARE DURING ERECTION OF JOISTS TO PREVENT THE JOISTS FROM BUCKLING LATERALLY. USE SPREADER BARS FOR LIFTING JOISTS AND PROVIDE LATERAL BRACING AS NECESSARY. REMOVE ANY DAMAGED JOISTS FROM THE JOB SITE. DO NOT ATTEMPT TO REINFORCE DAMAGED JOISTS.

STEEL FLOOR DECK
1. STEEL FLOOR DECK TO BE VERCO 22 GAGE 'TYPE B' FORMLOK OR EQUIVALENT, FOR UNSHORED CONSTRUCTION.
2. WELD DECK TO SUPPORTS AT ALL VALLEYS AND TO MARGINAL MEMBERS AT 24" O.C. WITH 1/2" DIA. FUSION WELDS UNLESS OTHERWISE NOTED ON DRAWINGS.
3. BUTTON PUNCH SIDE SEAMS AT 24" ON CENTER MAXIMUM.

STEEL ROOF DECK
1. STEEL ROOF DECK TO BE VERCO 20 GAGE 'TYPE HSB-36' OR EQUIVALENT.
2. WELD DECK TO SUPPORTS AT ALL VALLEYS AND TO MARGINAL MEMBERS AT 24" O.C. WITH 1/2" DIA. FUSION WELDS UNLESS OTHERWISE NOTED ON DRAWINGS.
3. BUTTON-PUNCH SIDE LAPS AT 24" ON CENTER MAXIMUM.

Figure 16.32 Mtext *is ideal for creating large quantities of text.*

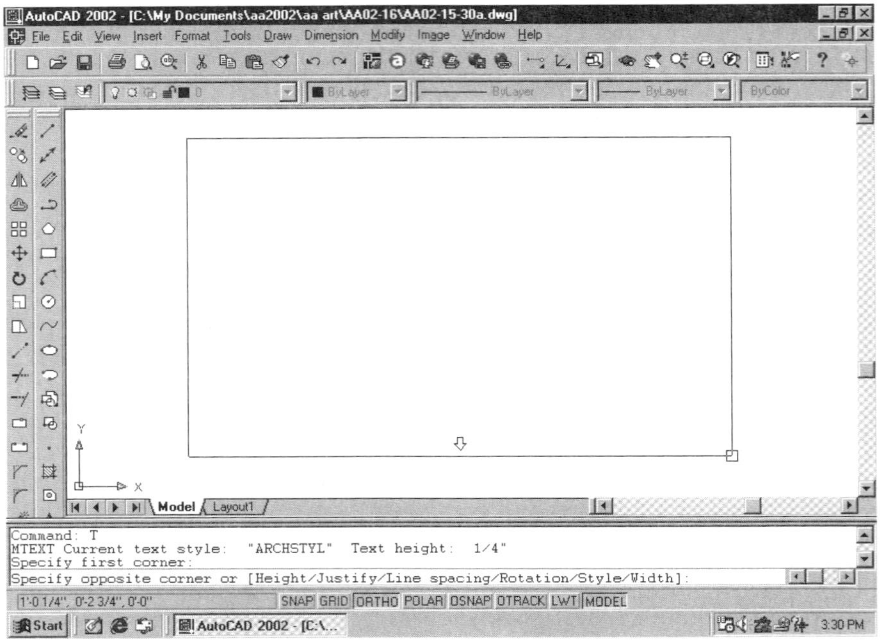

Figure 16.33 *Multiline text is started by selecting a box to define the width that the text will occupy. If the text does not fit into the box, it will overflow the bottom of the box in the current setting (the down arrow). The direction of overflow can be adjusted.*

prompting for a second point, AutoCAD is asking you to provide a window to place your text in. The width of the window you specify is very important. The width of the window will control the width of your paragraph. Extra text will spill out of the top or bottom of the window depending on the justification, so that the depth of the window is not critical. Once the other corner is selected, the **Multiline Text Editor** dialog box shown in Figure 16.34 will be displayed. Text can now be entered in the edit window using procedures similar to those used with other text editors. AutoCAD allows a third-party text editor to be used it you select the editor from the **File** tab in the **Options** dialog box (select **Options** on the **Tools** menu). See the AutoCAD User's Guide in the **Help** menu to select a third-party editor. The balance of this chapter will introduce the many capabilities of the AutoCAD program editor.

As text is entered at the keyboard, it will be displayed in the edit box of the text editor. Text will wrap to the next line as the desired width is reached, just as with any word processing program. Notice in Figure 18-34 that the editor contains tabs for controlling text features. The default tab controls characters and there are also tabs for **Properties**, **Line Spacing**, and **Find/Replace**.

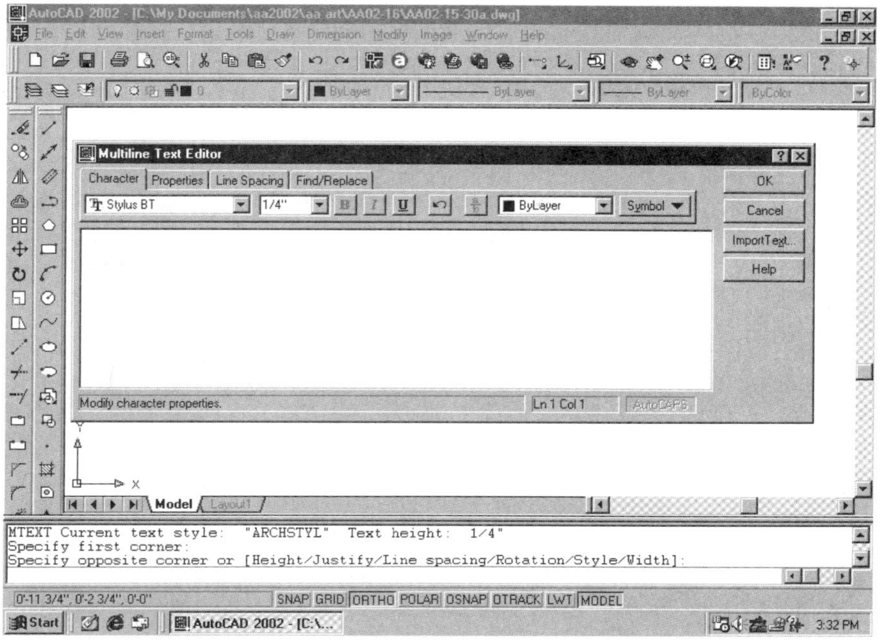

Figure 16.34 *The* **Multiline Text Editor** *is used to add large amounts of text to a drawing. By using the* **Character, Properties, Line Spacing,** *and the* **Find/Replace** *tabs, you can assign text various colors, and heights, underlining, and other alterations. In addition to using the AutoCAD text editor, you can use a third-part editor by selecting the desired editor using the* **File** *tab of the* **Options** *dialog box (select* **Options** *from the* **Tools** *menu).*

CONTROLLING TEXT CHARACTERS

As with **Text**, the qualities of **Mtext** can be easily altered. Qualities controlled on the **Character** tab include the font, height, bold, italic, underline, undo, stack, color, and special symbols. This dialog box also includes the **?** (question mark) button in the title bar. Selecting the **?** button will produce a question mark button beside the cursor. Now if any of the buttons are selected, an on-screen explanation will be displayed. Figure 16.35 shows an example of selecting the **Stack** button.

Font

Clicking the down arrow beside the **Font** edit box displays a list similar to the font list displayed with the **Style** command. The list allows a font for new text to be selected or the font of selected text to be changed. The current font for multiline text is the same as the current font for the **Style** command. With **Mtext**, the style can be altered at any time throughout the use of the command, including mid-sentence. Figure 16.36 shows an example of text that was altered to provide different fonts within the same command sequence.

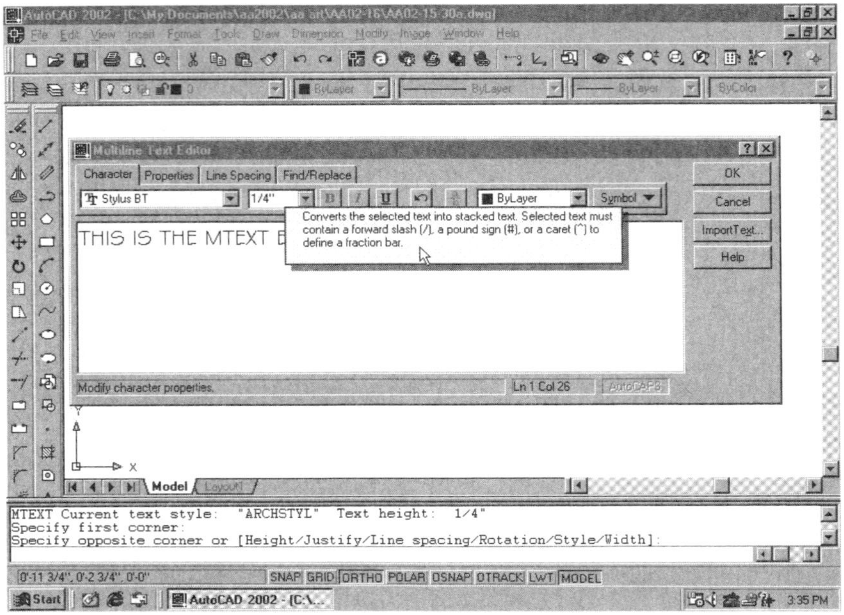

Figure 16.35 *Selecting the* **?** *button in the title bar and then selecting a box or button in the* **Multiline Text Editor** *will provide a brief explanation of the function.*

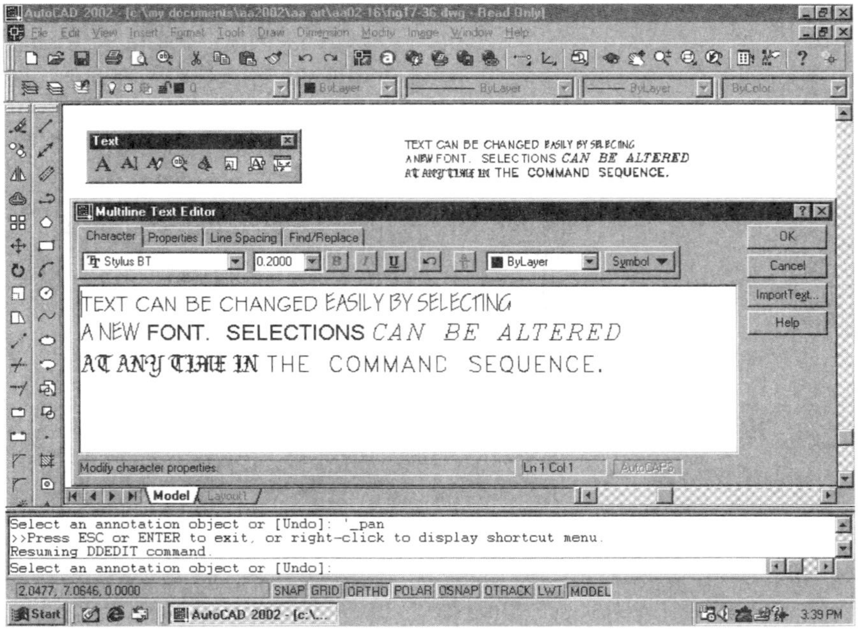

Figure 16.36 *The text font of one word or a string of words can be altered at any point of the drawing's life.*

Height

Just as with the **Text** command, this option controls the height of the text. The default value is the same as the current style. The value for text height can be changed at any point of the command, allowing words to be printed at varied heights. Selecting the down arrow will display the heights of all current styles. To provide a new height, select the **Height** edit box and click the cursor to highlight the current height. Remove the current height with the backspace arrow and then enter the desired height value. Text height can be altered before the text is added to the drawing by highlighting the text to be altered, and then providing the required height in the edit box. Figure 16.37 shows an example of how text height can be altered.

Bold

Selecting the **Bold** button toggles between bold and normal text. The options are active only for TrueType fonts. Bold text can be created by selecting the **Bold** button before text is entered. Existing normal text can be changed to bold by highlighting the desired text to be changed, and then clicking the **Bold** button. Figure 16.38 shows an example of text that was altered to provide bold text.

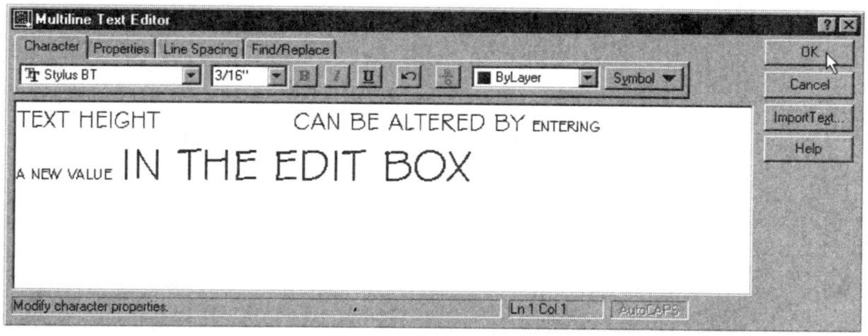

Figure 16.37 *Multiline text height can be altered at any time throughout the life of the drawing. Selecting the down arrow will display a choice of heights. You can also enter a new height by selecting the* **Height** *box and replacing the current value with the desired value.*

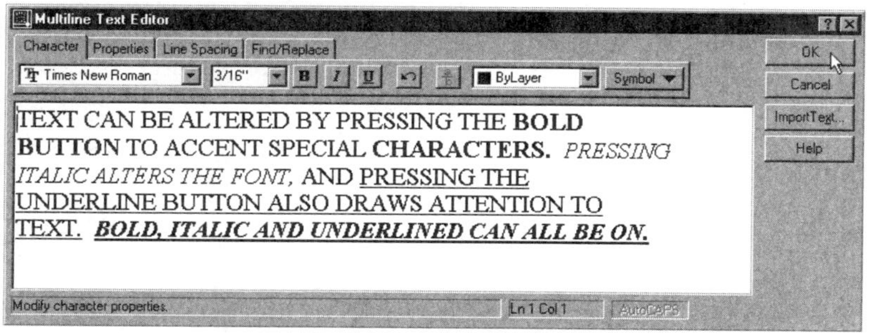

Figure 16.38 *Text can be highlighted by selecting the* **Bold**, **Italic**, *or* **Underline** *button. One or more of the features can be used in combination to affect the font.*

Italic

Selecting the **Italic** key toggles italic formatting for text. As with other options, **Italic** can be applied to text being created as well as used to modify existing text. The option is available only for TrueType fonts.

Underline

Selecting this button allows the underline function to be toggled ON and OFF. This option of **Mtext** is so much easier than %%something that is required with the **Text** command.

Undo

This option can be used to undo the last sequence of the text editor. Changes can be made to the text or to one of the formatting options.

Stack

The **Stack** option can be used to place a portion of text over other text. This is useful in writing stacked numbers such as $^{1}/_{4}$ rather than 1/4 or in writing equations. With the text written as 1/4, highlight the fraction by pressing and holding the select button and moving the cursor over the text. Once the desired text is selected, selecting the **Stack** button will stack the indicated text. See Figure 16.39.

Color

Selecting the **Color** button allows the color of the text to be altered. By default, multilinetext will be assigned the color of the current layer. To select a new text color, select the down arrow beside the **Color** edit box and select the desired color. Selecting the Other color option will produce the **Select Color** dialog box and allow for a full range of colors to be selected.

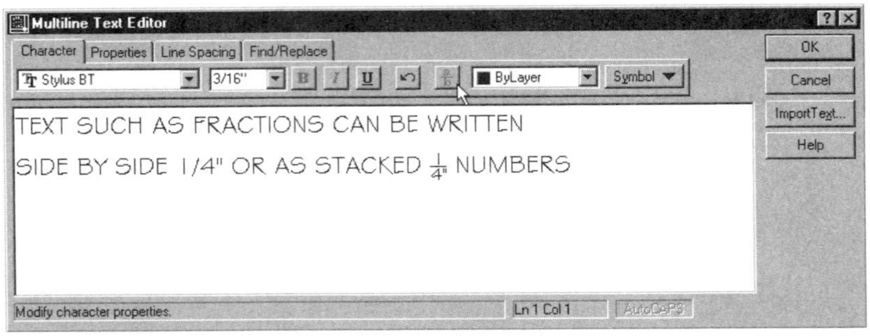

Figure 16.39 *Text such as fractions can be written side by side or stacked depending on office preference.*

Symbol

Selecting the down arrow beside the **Symbol** button displays the list shown in Figure 16.40. The degree, plus/minus, and diameter are often used on construction documents. Each can be inserted in multiline text by selecting the down arrow to display the list, and then selecting the desired symbol. If this seems too easy, the symbols can also be inserted using the special characters that were discussed as **Text** was introduced. Selecting the Other option will display the **Character Map** dialog box shown in Figure 16.41 and allow one of the symbols from the **Character Map** to be inserted. Symbols will vary depending on the current font.

Control Buttons

As with other dialog boxes, control buttons are provided to implement the desired settings. This box features the **OK**, **Cancel**, **Import Text**, and **Help** buttons. The results of the **OK**, **Cancel**, and **Help** buttons should be familiar by now. The **Import Text** button allows ASCII (American Standard Code for Information Interchange) and RTF (Rich Text Format) text files to be inserted in a drawing file. This might seem unimportant as you're learning AutoCAD now, but it means you can create schedules or large bodies of text up to 16 K in size using your second favorite software, and then

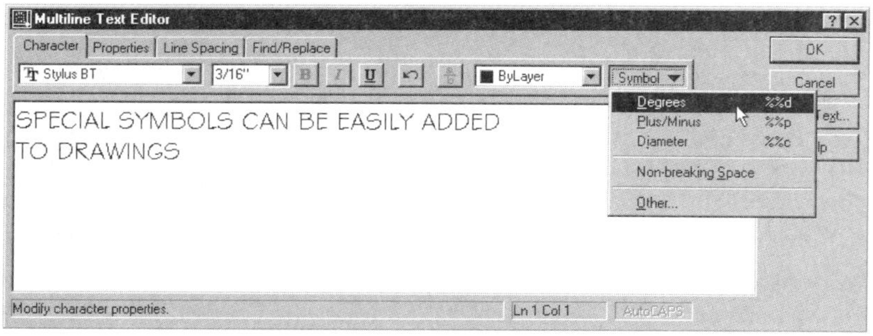

Figure 16.40 *Using* **Mtext***, special characters such as the degree, plus/minus, and diameter symbols can be added without using any %% special character.*

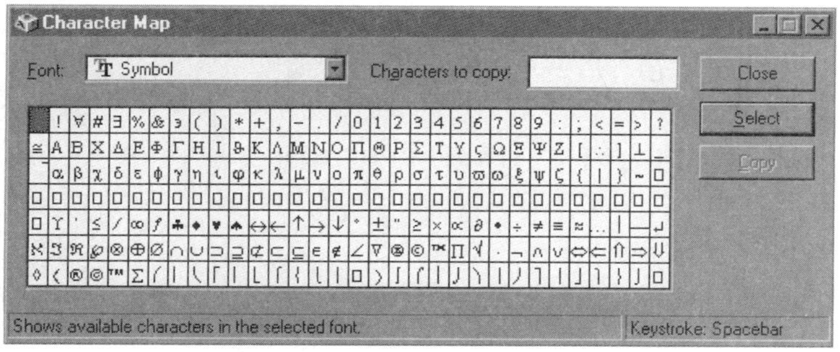

Figure 16.41 *Selecting the Other option from the* **Symbols** *list displays the character map.*

import the file into AutoCAD once you feel more comfortable with multiline text. Imported text retains it original character formatting and style properties, but it can be edited as if it had been created within **Mtext**.

Note: Files larger than 16K can be inserted into AutoCAD if they are not imported directly into AutoCAD. To insert a file larger than 16K, minimize AutoCAD, and then open the file to be inserted using the word processor that was used to create the file. Now save the text file to the Clipboard. Maximize AutoCAD and paste the file into the drawing.

ADJUSTING TEXT PROPERTIES

Selecting the **Properties** tab of the **Multiline Text Editor** will provide options for controlling text properties by changing the style, justification, width, and rotation. Figure 16.42 shows the **Properties** tab. The tab can be selected at any point of the **Mtext** operation.

Style

The **Style** option allows an existing text style to be applied to a new or selected text body. Character formats for font, height, bold, and italic can be overridden by applying a new style to existing multiline objects. Information can be typed in one style, and altered before the text is placed in the drawing base. To alter the current text style, select the **Properties** tab and then select the **Style** down arrow to display a listing of current text styles. Figure 16.43 shows an example of the **Style** list.

Justification

Justification with **Mtext** is similar to that of the **Text** command. Selecting the **Justification** option allows the alignment of text to be altered. The default setting places the cursor in the top left corner of the text box. Selecting the **Justification** down arrow will produce the list shown in Figure 16.44. Figure 16.45 shows examples of how text will be placed based on each setting.

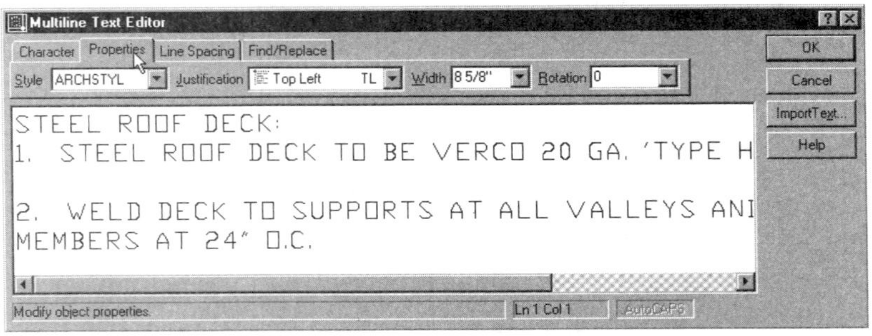

Figure 16.42 *Selecting the* **Properties** *tab displays the options for altering the style, justification, width, and rotation of multiline text.*

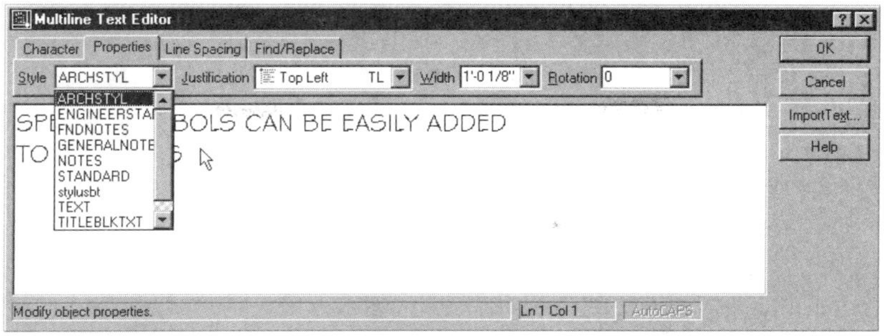

Figure 16.43 *Selecting the down arrow for* **Style** *displays a list of current text styles. Styles for multiline text are created using the same procedures as used to create a style for text.*

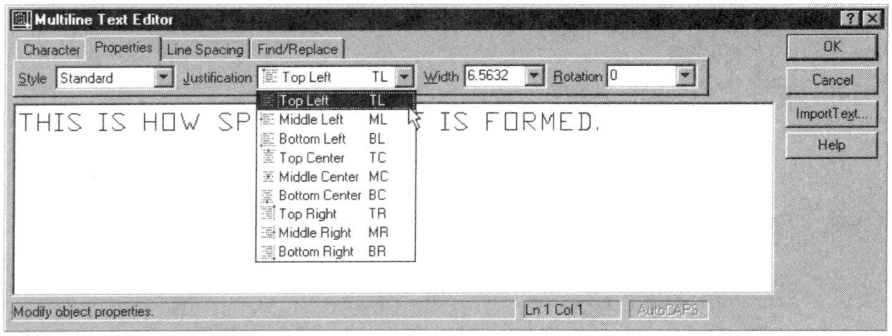

Figure 16.44 *Selecting the* **Justification** *down arrow will produce a list of options for arranging multiline text that are similar to the* **Text** *justification options.*

Width

This option allows the width of the text to be redefined. Selecting the down arrow beside the **Width** edit box will display a list of current multiline text widths. One of the current widths can be selected or a new value can be added to the edit box.

Rotation

This option functions just as the option did with the **Text** command, to set the text rotation angle. Selecting the **Rotation** down arrow displays a listing of angles in 15° increments. One of these options can be selected from the list, or any numerical value between 0 and 360 can be assigned. The option can be used with the current text before it is added to the drawing, or existing text can be selected and altered.

LINE SPACING

The **Line Spacing** tab includes two settings for controlling line spacing. These settings include the **At Least/Exactly** and the **Single (1.0x)** options.

Figure 16.45 *The options for justification of multiline text.*

At Least/Exactly

This setting specifies how the spacing between lines will be placed. The default setting is to allow AutoCAD to automatically add space between lines of text based on the height of the largest character in the line of text. The **Exact** option sets the spacing so that it will always be the same between lines of text. The **Exact** option maintains the exact spacing no matter the variation of individual letters in each line. The **Exact** option should be used when constructing tables like a window or door schedule. Because the spacing between lines of text will always be the same, text placed with this option will have a very orderly appearance.

Single (1.0x)

This menu can be used to control the spacing increment between lines of text multiline text. This setting is the equivalent to choosing between single, double or triple spaced text with a word processor. The spacing is the vertical distance measured from bottom to bottom of lines of text. The default distance is 1.66 times the height of the text. A setting of 1 represents single spacing, and 2x represents double-spacing.

USING FIND/REPLACE

Selecting the **Find/Replace** tab of the **Multiline Text Editor** displays the options shown in Figure 16.46. The options help create search patterns for words or specified strings of text. Options include the Find box and button, the **Replace with** box and button, **Match Case**, and **Match Word**.

Find

The **Find** box can be used to enter a word or string of text that you wish to find. This can be especially helpful in finding a key word in a long list of specifications. To find a word, select the **Find** box by placing the cursor in the box and then type the desired word. Figure 16.47 shows an example of using **Find** to locate the word JOISTS. Clicking the **Find** button to the right of the **Find** edit box starts the search for the text in the **Find** box. Press (ENTER) to find the next occurrence of the selected word. Figure 16.48 shows the result of selecting joists and replacing it with the word trusses.

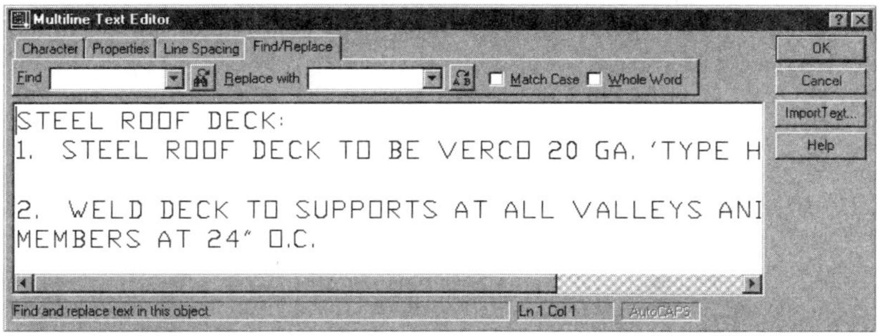

Figure 16.46 *Selecting the* **Find/Replace** *tab provides options for editing multiline text.*

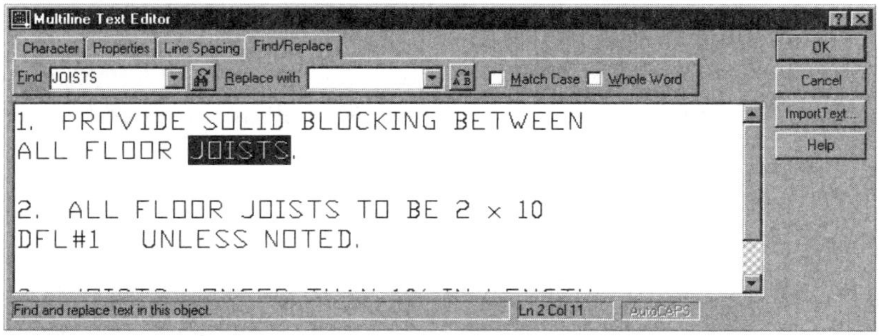

Figure 16.47 *Selecting the* **Find** *box allows a word or word string to be specified for a word search. The word JOISTS has been selected for the search. Clicking the* **Find** *button will highlight the first use of the specified word. Pressing* ENTER *will highlight the next occurence of the selected word or string.*

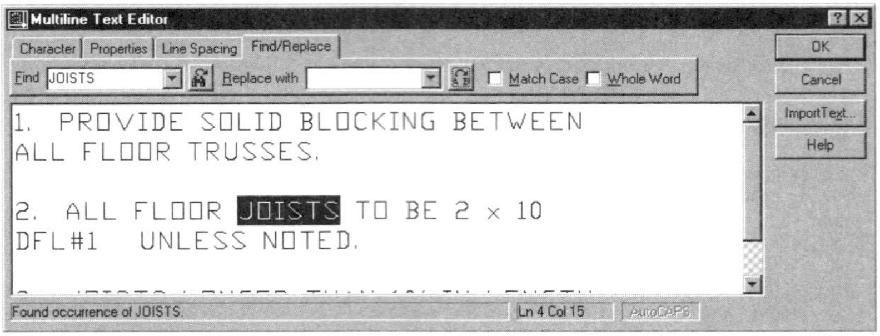

Figure 16.48 *With the word JOISTS as the active word for the search, the first ocurence of the word was altered to read TRUSSES.*

Replace With

The **Replace with** option can be used in conjunction with **Find** to alter a word or word string. Once a word has been found, enter a word in the **Replace with** box and click the Replace button.

Match Case

With this box checked, only the exact word with the same case will be identified. With the box inactive, placing Truss in the **Find** box would identify truss in the body of text. With the **Match Case** box active, truss would not be identified. Only uses of the word or string that are identical will be located.

Whole Word

With this box inactive, any use of the selected word will be identified with the **Find** option. With truss indicated in the **Find** box, Truss–joist would be identified in the document. Selecting this option will identify the word only if it is a single word. When the word is part of another text string, it will be ignored.

EDITING MULTILINE TEXT

A multiline text object can be edited using the **Copy**, **Erase**, **Grips**, **Mirror**, **Move**, and **Rotate** commands. The text in multiline text can be edited using either the **Ddedit** or **Properties** command or by using the **Text Style**, **Scale Text**, or **Justify Text** buttons on the **Text** toolbar.

USING DDEDIT

To use the **Ddedit** command with multiline text, you start the same way as when editing text created with the **Text** command. The quickest method of accessing the command is from the Command prompt, but you can also access the command by selecting **Edit Text** on the **Text** toolbar or by selecting **Text** from the **Modify** menu. The command sequence is as follows:

Command: **ED** (ENTER) *(Or click the Edit Text button.)*

Select an annotation object or [Undo]: *(Select the desired text to be edited.)*

When the paragraph to be edited is selected, the text will be displayed in the **Multiline Text Editor** dialog box. Each of the four tabs that were used to create the text can be used to alter the selected text. The mouse can also be used to move the cursor to the desired location. As the cursor is moved into the edit box, it changes to a blinking cursor. To edit text, move the cursor to highlight the text to be changed. With the text highlighted, any one of the features set using the text editor can be altered. When the changes are made, click the **OK** button to return to the drawing screen. As the screen is restored, the prompt for selecting additional annotation is still displayed. The program is waiting for you to select additional text to be edited. Press (ENTER) a second time to exit **Ddedit** and restore the Command prompt.

In addition to using the four tabs of the **Multiline Text Editor** dialog box to edit text, standard Windows control keys can be used to alter the text in the editor. The control keys include the following:

CTRL+A	Selects all text in the **Multiline Text Editor**.
CTRL+B	Applies or removes bold format for selected text.
CTRL+C	Copies selected text to the Clipboard.
CTRL+I	Applies or removes italic format for selected text.
CTRL+SHIFT+L	Converts selected text to lowercase.
CTRL+U	Applies or removes underline format for selected text.
CTRL+SHIFT+U	Converts selected text to uppercase.
CTRL+V	Pastes Clipboard contents to cursor location.
CTRL+X	Cuts selected text to the Clipboard.
CTRL+SPACEBAR	Removes character formatting in selected text.
(ENTER)	Ends the current paragraph and starts a new line.

EDITING WITH PROPERTIES

The **Properties** command can be used to alter the properties of multiline text. The command has been introduced earlier, but its power to alter text should not be forgotten. The properties of multiline text can be displayed using the following steps:

1. Select the body of text to be edited.
2. Right-click in the drawing area to display the shortcut menu.
3. Select Properties from the shortcut menu.

You can also start the command by selecting the **Properties** button on the **Standard** toolbar or by typing **MO** (ENTER) at the Command prompt. Each method will display the **Properties** dialog box similar to Figure 16.49. When the properties are listed by category, seven general properties affecting the text and nine specific properties are listed for multiline text.

General Qualities

Each of the seven general properties are used to display or set the specified property. For instance, if you select a block of multiline text, and then select **Color**, the color property will be highlighted, and the edit arrow will be displayed. Selecting the edit arrow will display the color palette, allowing the color of the text to be altered. Each of the other general properties of the text can be altered using the same procedure.

Specific Text Properties

Each of the multiline text features that are controlled in the **Multiline Text Editor** dialog box can be controlled using the **Properties** dialog box. To alter the current **Style**, select the text to be altered. With the text highlighted, selecting **Style** will display a listing of the current styles. Selecting the name of the new style will automat-

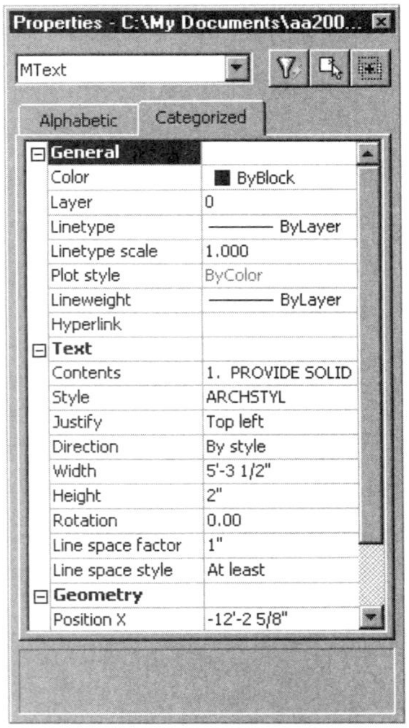

Figure 16.49 *The* **Properties** *dialog box can be used to edit multiline text. Select the* **Properties** *button on the* **Standard** *toolbar or type* **MO** *(ENTER) at the Command prompt to produce the* **Properties** *dialog box.*

ically update the selected text to the indicated style. Each of the other properties can be altered using the same procedure. Other text specific properties include the following:

Contents—The edit box specifies the selected text string. Only a portion of the text will be displayed for long text strings.

Style—Specifies the style name of the selected text body and allows a new style to be specified. Selecting the **Text Style** button on the **Text** toolbar can also alter the style of existing text.

Justify—Specifies the attachment point of the text body using the nine common justification options for placement.

Direction—Allows the horizontal or vertical direction of the selected multiline text to be specified.

Width—Allows the width of multiline text to be edited. With the text highlighted, a grip will be displayed at each corner of the text box. To alter the width of the text, select a grip to make it hot. Move the hot grip to the desired location and the text will be redisplayed to the specified width.

Height—Specifies and alters the height of the selected text. Text can be enlarged or reduced by altering the current value. Selecting the **Scale Text** button on the **Text** toolbar can also alter the scale of existing text.

Rotation—This option specifies the rotation angle of the text body. Enter the desired rotation angle by keyboard and press (ENTER). The text will be rotated around the current justification point.

Line space factor—Controls the line spacing factor of the text body. Enter the desired value by keyboard and press (ENTER). The text will be altered to display the selected spacing value.

Line space style—Specifies the spacing based on text height using the Exact and At least options.

Note: Keep in mind that the Properties dialog box can be displayed beside the drawing area and used to alter more than just multiline text. With the box open, the Properties listing will be altered as each new object is selected. The name of the listed object will be displayed in the object edit box located at the top of the dialog box.

USING THE SPELL CHECKER

In addition to using **Ddedit** and **Properties** to edit text, you can use the **Spell** command to correct spelling errors in text and multiline text. Access **Spell** by selecting **Spelling** from the **Tools** menu or by typing **SP** (ENTER) at the Command prompt. Each option will prompt you to select objects to be checked for spelling. Once a word or body of text is selected, the **Check Spelling** dialog box similar to Figure 16.50 will be displayed. If no spelling errors are contained in the selected text, you will see the display shown in Figure 16.51. Clicking **OK** or pressing (ENTER) will return you to the drawing. Pressing (ENTER) (ENTER) will remove the AutoCAD alert and reenter the **Spell** command so that additional text can be checked.

If errors are found in a selection, the line of text will be displayed in the **Context** box of the **Check Spelling** dialog box. The misspelled word will be displayed in the **Current word** box. A possible listing of alternatives can be found in the **Suggestions** box. If the current word is as you would like it, select the **Ignore** button, and the spell checker will move to the next word identified by the checker. If you would like to use one of the suggested words, move the cursor to highlight the word and then select **Change**. The editor will proceed through your text until the alert seen in Figure 16.51 is displayed.

CHANGING THE DICTIONARY

Selecting the **Change Dictionaries** button will produce the dialog box shown in Figure 16.52. The **Main dictionary** box will allow you to change the language that is stored in the dictionary. The **Custom dictionary** and **Custom dictionary words** boxes

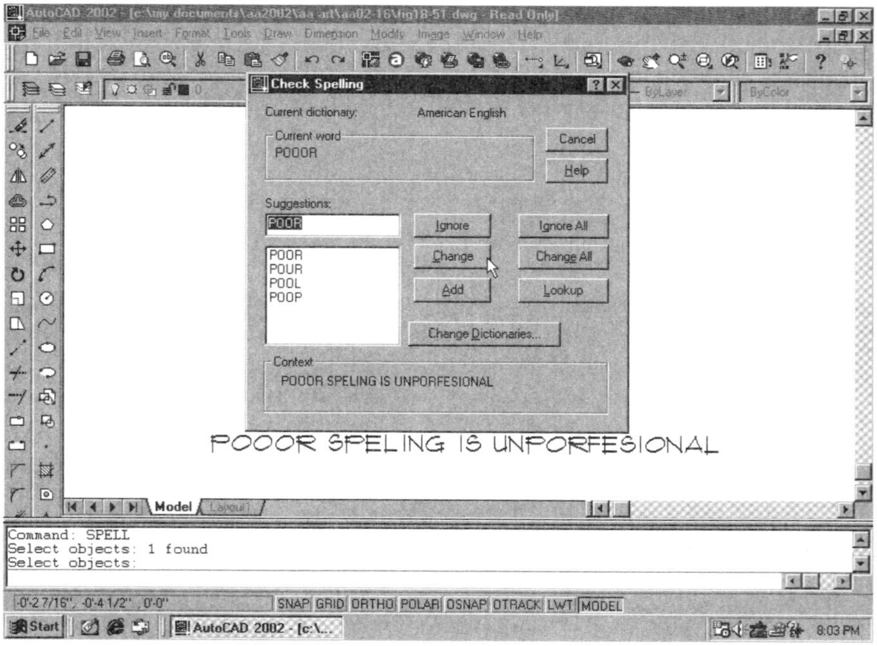

Figure 16.50 *The* **Spell** *command can be used to check text and multiline text. The current word "pooor," four suggestions, and the current context of the word are all displayed.*

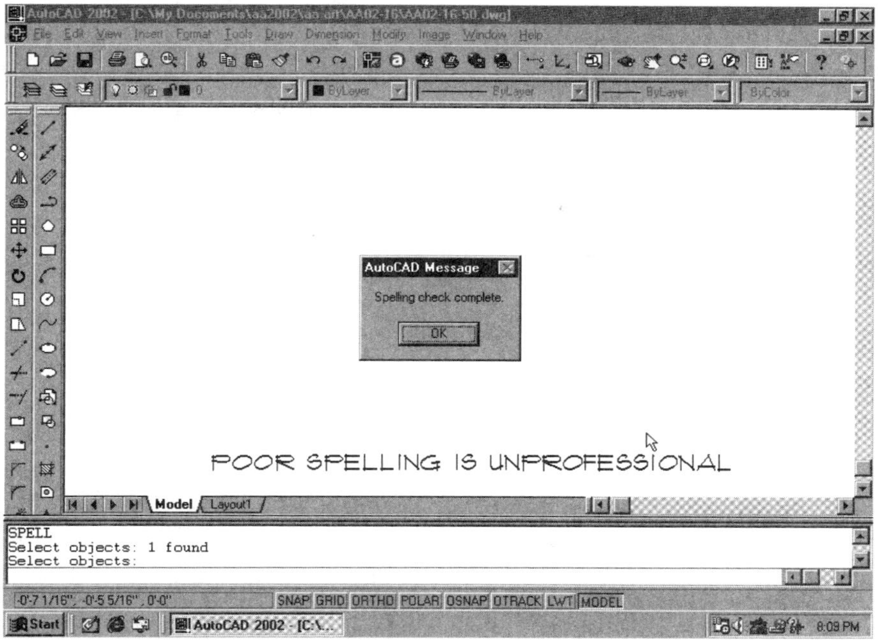

Figure 16.51 *If no errors are found, you will see this display.*

can be very useful for adding words, acronyms, client names or vendor names specific to your area of expertise to the dictionary. When creating a custom dictionary, use the same guidelines that were presented in Chapter 2 for naming drawing files. An extension of .CUS must be added after the name. To enter a custom name for an architectural dictionary, type ARCHWORD.CUS in the **Custom dictionary** box. If several custom dictionaries have been created, the **Browse** button can be used to select a dictionary.

To add words to a specific dictionary, select the **Custom dictionary words** box. Names of up to 63 characters can be entered in the box. Multiple names can be entered at one time when separated by a comma. Selecting the **Add** button will add the names to the existing list of names contained in the dictionary. Figure 16.53 shows a custom dialog box for the ARCHWORD.CUS dictionary with three words to be added to the existing list.

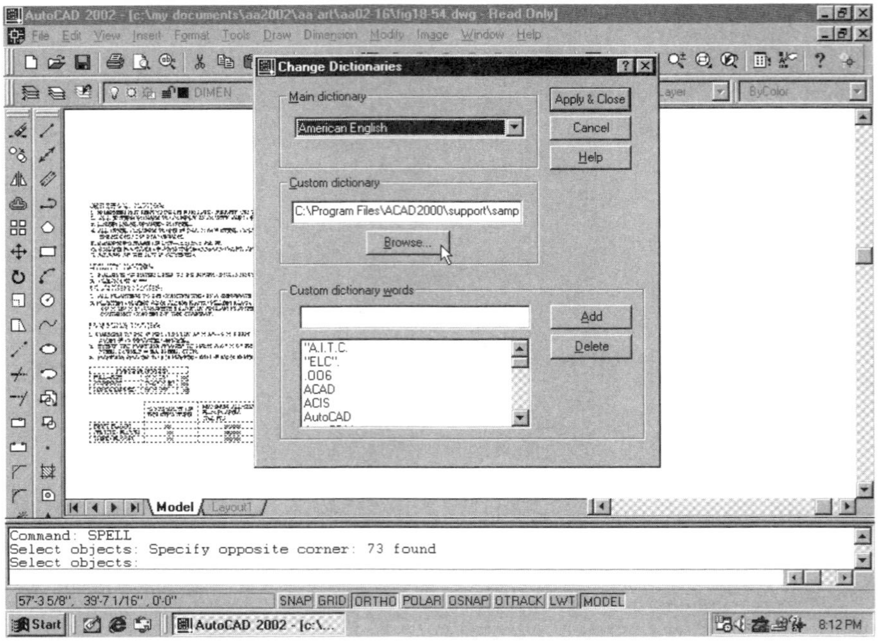

Figure 16.52 *A separate dictionary can be created to contain words specific to your field.*

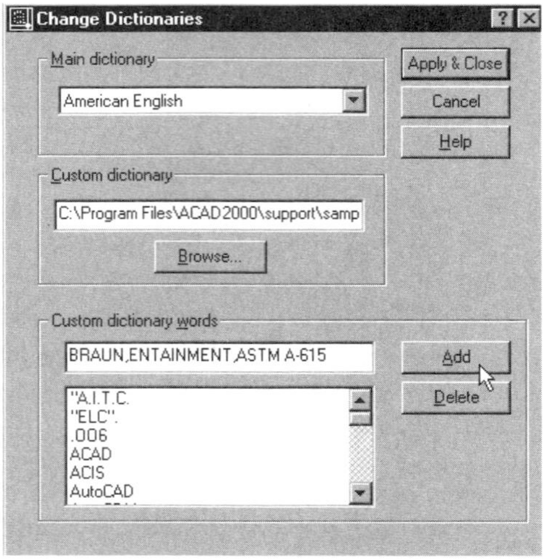

Figure 16.53 *Three words are to be added to the dictonary. A partial listing of the ARCH-WORDS dictonary is shown in the lower left corner of the dialog box.*

FINDING AND REPLACING TEXT

The **Find** command is another tool that can be used to edit text. Start the command by right-clicking and choosing **Find** from the shortcut menu, by selecting the **Find and Replace** button on the **Standard** toolbar, by selecting the **Find and Replace** button on the **Text** toolbar, or by selecting **Find** from the **Edit** menu. Each will produce the **Find and Replace** dialog box similar to Figure 16.54. The **Find and Replace** dialog box can be used to select, find, replace and zoom to text created by the **Text** or **Mtext** commands. The dialog box can also be used to edit dimension text, block attribute text, hyperlink descriptions, and hyperlinks. Each of these four types of text will be introduced in later chapters. The command can be used in a method similar to the **Find/Replace** tab of the **Multiline Text Editor** dialog box. To find text using the dialog box, enter the text or text string that you would like to find in the **Find text string** edit box. The search field can be the entire drawing or limited to the current text selection. Once the text string and the search field are defined, select the **Find** button. If the text is found, it will be displayed in the **Search results** window. Choosing the **Zoom to** button will display the selected text on the screen.

In addition to locating specific text in a drawing, the **Replace with** edit box can be used to replace a word or phrase. After entering the text to be found, enter a phrase to be used as the replacement text in the **Replace with** edit box. Choosing the **Replace** button will replace the specified text with the replacement text. Choosing the **Replace All** button will replace each usage of the selected text or phrase.

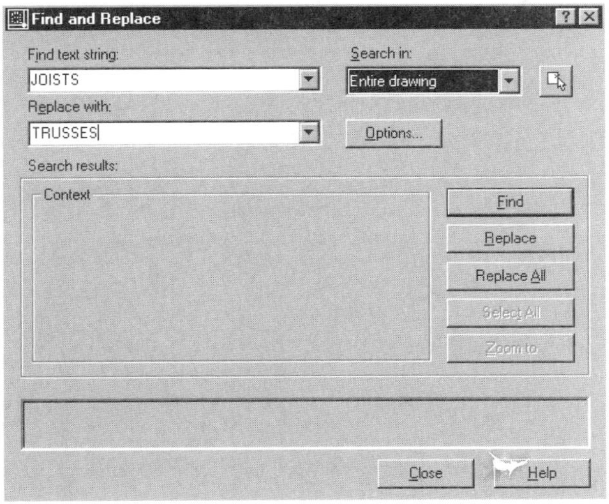

Figure 16.54 *The* **Find and Replace** *dialog box can be used to select, find, replace and zoom to text created by the* **Text** *and* **Mtext** *commands.*

INCORPORATING TEXT INTO TEMPLATE DRAWINGS

Much of the text used to describe a drawing can be standardized and placed in a template drawing. Text can also be saved as a wblock and inserted in drawings as needed. (Wblocks will be discussed in later chapters.) Figure 16.55 shows an example of notes that an office includes as standard notes for all site plans. These notes can be typed once and saved as a separate drawing so that they do not have to be retyped with each use. The wblock can then be either inserted in every site plan or nested in the

GENERAL NOTES:

1. ENGINEER NOT RESPONSIBLE FOR LAND SURVEY OR TOPOGRAPHY.
2. ALL EXTERIOR SIGNS TO COMPLY W/ COUNTY SIGN ORDINANCE.
3. LOWER LEVEL FRAMED IN STEEL.
4. ALL STEEL COLUMNS TO BE 3" DIA. X 60" STEEL COLUMN FILLED W/ CONC. EMBEDDED 24" INTO GRADE.
5. SQUARE FOOTAGE OF LOT--00,000 SQ. FT.
6. SQUARE FOOTAGE OF STRUCTURE--00,000 SQ. FT. APPROXIMATELY.
7. 00.00% OF THE LOT IS COVERED.

UTILITY NOTES:

1. BALANCE OF WATER LINES TO BE SHOWN ON FLOOR PLANS.
2. CLEAN OUT @ ****

PLANTER NOTES:

1. ALL PLANTERS TO BE CONSTRUCTED W/ A SEPARATE PERMIT.
2. PLANTER ON WEST SIDE ALONG FLOYD MILLER BLVD. TO BE 44' X 7.5' X 3' CONCRETE BLOCK W/ SIMILAR PLANTER IN THE. SOUTHWEST CORNER BY THE STAIRWAY.

PARKING NOTES:

1. PARKING TO BE 4" MIN. CONC. W/ 6" X 6"--4 X 4 WWM OVER 4" COMPACTED GRAVEL.
2. EVERY TWO PARKING SPACES TO HAVE A 6" X 6" WHEEL STOP W/ (2) 1/2" x 12" STEEL DOWELS @ EA. WHEEL STOP.
3. PARKING SPACES TO BE MARKED WITH 4" WIDE WHITE STRIPES.

PARKING SPACES		
FULL-SIZE	9' X 20'	xx
COMPACT	7'-6" X 15'	xx
HANDICAPPED	12' X 20'	xx

	OCCUPANCY OF THE STRUCTURE	MAXIMUM ALLOWABLE FLOOR AREA (SQ. FT.)	TOTAL OCCUPANTS (FOR BLDG.)
FIRST FLOOR	xx	xx,xxx	xx
SECOND FLOOR	xx	xx,xxx	xx
THIRD FLOOR	xx	xx,xxx	xx

Figure 16.55 *Many offices have standard notes that are used for all site plans. These notes can be typed once and saved as a separate drawing or stored as a template so that they do not have to be retyped with each use.*

template drawing on the SITE-ANNO layer. The layer can be frozen, thawed when needed, and then either the **Ddedit** or the **Properties** command can be used to make minor changes based on specific requirements of the job. Notice that several notes and each of the schedules include the text 'xxx'. These are notes that have been set up as blocks and stored with attributes that can be altered for each usage. Attributes will be discussed later.

Many drawings, such as sections, contain basically the same notes that are displayed as local notes. Although sections for a tilt-up structure will be radically different from a section for a steel rigid frame, a template sheet can be developed for each type of construction. The actual drawing might look different, but the notes to specify standard materials are similar for most buildings using the same type of construction. Local notes can also be placed in the template drawing as shown in Figure 16.56. These notes can also be stored in a template or as a wblock, moved to the desired drawing, thawed, moved into the needed position, and edited as seen in Figure 16.57. This can greatly increase drafting efficiency.

1/2" CD APA 32/16 PLY. ROOF SHTG. INT. GR. W/ EXT. GLUE. LAY FACE GRAIN PERP. TO JOISTS & STAGGER JOINTS. USE: 8d COM. NAILS @ 6" O.C. BOUNDARY & EDGES, 8d COM. NAILS @ 12" O.C. FIELD. SEE ROOF-FRAMING PLANS FOR SPECIAL NAILING.

'TRUS-JOIST' 16" TJI 35" @ 24" O.C.

MINERAL CAP SHEET OVER 2 LAYERS OF ASB. FELT AS PER. JOHN MANSVILLE SPEC. #406

PROVIDE 1/2" GYP. BD. DRAFTSTOPS FROM CEILING TO PLY ROOF SHTG. FOR EVERY 3,000 SQ. FT. MAX.

1 1/2" 100#/FT LT. WT. CONC. OVER 15# ASPH. SATURATED FELT W/ 6" x 6" W-4" x 4" WWM IN SLAB OVER PLY. SHTG.

3/4" CD APA 42/20 T&G PLY FLR. SHTG., INT. GRD. W/ EXT. GLUE. LAY FACE GRAIN PERP. TO JOISTS & STAGGER JOINTS. USE: 10d COM. NAILS @ 6" O.C. BOUNDARY & EDGES, 10d COM. NAILS O.C. FIELD. SEE FLOOR FRAMING PLANS FOR SPECIAL NAILING.

LINE OF SUSP. CEILING (CLASS 'A')

1 1/2" 100#/FT LT. WT. CONC.

3/4" CD APA 42/20 T&G PLY. FLOOR SHEATHING.

R-25 RATED INSULATION

PROVIDE 1/2" GYP. BD. DRAFT STOPS FROM CEILING TO PLY. FLOOR SHTG. FOR EVERY 1,000 SQ. FT. MAX.

2-2 x 4 DFL TOP PLATES LAP 48" MIN. W/ 10-16d COM.

5/8" TYPE "X" GYP. BD.

2 x 4 STUDS @ 16" O.C. UNLESS NOTED

2-2 x 6 DFL TOP PLATES LAP 48" MIN. W/ 12-16d COM.

R-21 RATED INSULATION

2 x 6 SOLE PLATES.

R-38 RATED INSULATION

Figure 16.56 *Standard notes to be used with sections showing light frame construction.*

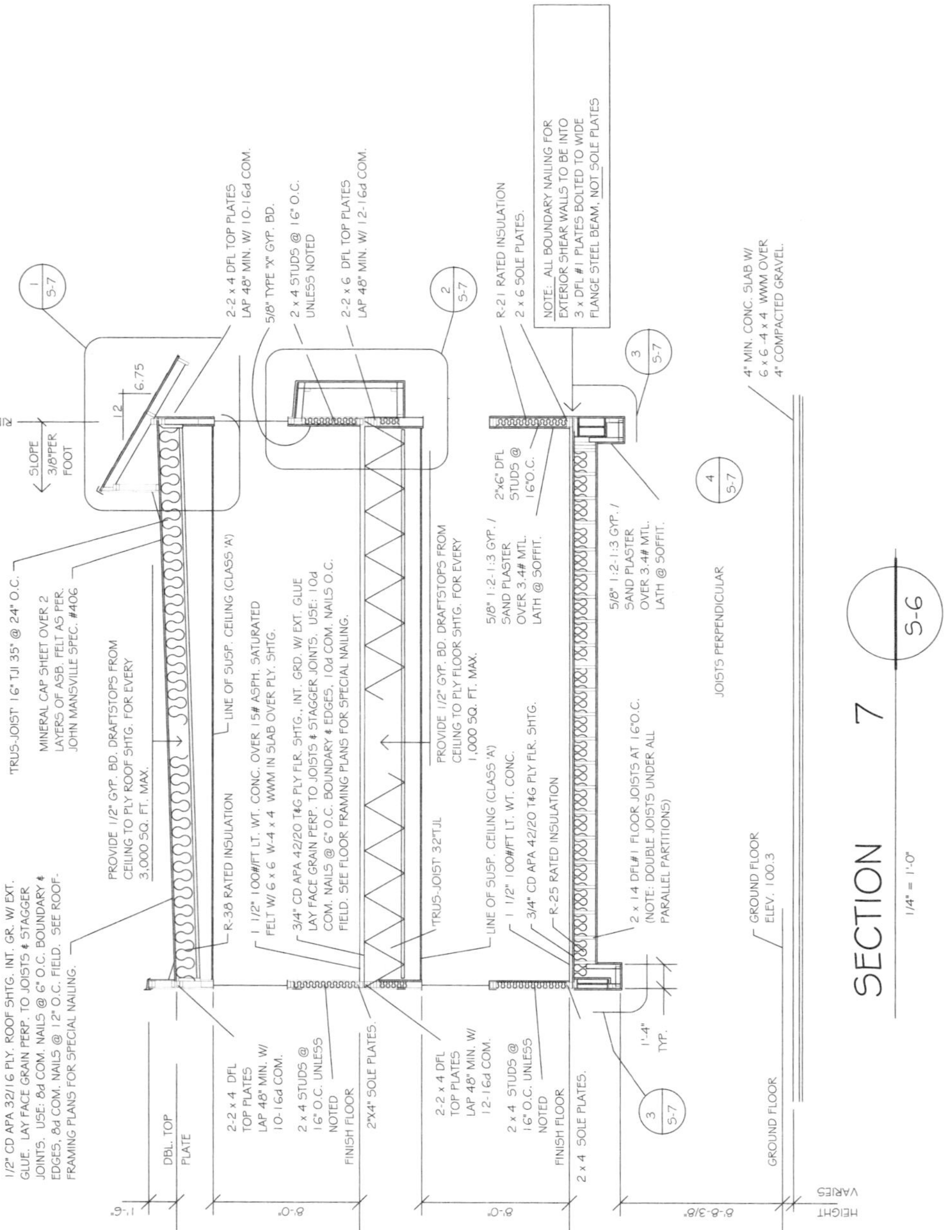

Figure 16.57 *A completed section using stock notes that were inserted in the drawing and moved to the desired location.*

CHAPTER 16 EXERCISES

1. Start a new drawing and type the following text using left justified text with a height of .125" and a TrueType font.

   All sheathing should be 1/2" standard grade 32/16-ply interior type with exterior glue. Lay perpendicular to rafters and stagger all joints. Nail with 8d's @ 6" o.c. at edge, blocking and beams, and 8d's @ 12" o.c. @ field. Use common wire nails.

   Save the drawing as E-16-1.

2. Open a new drawing and make ROMANS the active font, with an obliquing angle of 10° and a height of 0.125".

   Use 8x8x16 grade "A" concrete block units with a triple score. Use # 5 diameter rebar @ 24" o.c. each way solid-grout all steel cells.

   Save the drawing as E-16-2.

3. Open a new drawing and make CIBT the active font and a height of 0.125". Use a rotation angle of 35°. Write the following note:

   5 1/8" x 13 1/2" exposed glu-lam ridge beam f:2200

   Save the drawing as E-16-3.

4. Start a new drawing and make ROMANT the active font with a height of .125". Use a width factor of 1.25. Use the center option to type the following text:

   ATTENTION ALL CADD

   OPERATORS.

   DUE TO

   BUDGETARY PROBLEMS

   YOUR MOM DOES NOT WORK HERE.

   PLEASE CLEAN UP

   YOUR OWN MESS.

   Save the drawing as E-16-4.

5. Start a new drawing and create a style for text using ROMANS as the font with a height of 0.125". Use a width factor of 1.00. Create a second style named TITLES using ROMANT as the font, with a height of 0.25 and a width factor of 1.5. Assign each style a layer named for the style and provide a separate color for each. Use **Mtext** to create the following text:

   INSULATION NOTES

   Insulate all exterior walls with 5 1/2" high-density fiberglass batt insulation, R-21 min.

Insulate all flat ceiling joist with 12" R-38 batts (no paper facing required).

Insulate all vaulted ceiling with 10 1/4" high-density, paper-faced fiberglass R-38 min. with 2" air space above.

Insulate all wood floors with 8" fiberglass batts, R-25 with paper face. Install plumbing on heated side of insulation.

CAULKING NOTES

Caulk the following openings with expanded foam or backer rods. Elastomatic, copolymer, siliconized acrylic latex caulks may also be used where appropriate.

Any space between window and door frames.

Between all exterior wall sole plates and plywood sheathing.

On top of rim joist prior to plywood floor application.

Wall sheathing to top plate.

Joints between wall and foundation.

Joints between wall and roof.

Joints between wall panels.

Save the drawing as E-16-5.

6. Open drawing E-16-5 and make the following changes to the insulation notes.

   Change wall insulation to reflect 6" fiberglass batt insulation, R-19 min.

   Change all flat ceiling to 12" R-38 batts with paper facing.

   Change all vaulted ceiling to 10" paper-faced fiberglass R-30 min. with 2" air space above.

   Change wood floors to 6" fiberglass batts, R-19 with paper face.

   Save the drawing as E-16-6.

7. Open drawing E-11-9. Use a text font other than TXT and label the drawing using drawing E-12-8 as a guide. Complete the drawing and add all hatch patterns. Place the required text box on a layer separate from other text. Use the **Line** command to add all leader lines. Save the drawing as E-16-7.

8. Open drawing E-14-10 and label the materials that have been drawn using a suitable font. Use **Qtext** to disable the text. Create a separate layer for all text and save the drawing as E-16-8.

   E-16-9. Open drawing E-14-9. Create separate layers, styles, and fonts for titles and text and label the drawing. Use straight lines to connect text to the object being described. Save the drawing as E-16-9.

10. Open drawing FLOOR14 and design a 2" wide title block along the right side of the page. Design space for page number, date, revision date, client information, and designer information. Since you are the designer, provide a company name and logo and your name, address, and phone number. Create new layers for the title block lines and text that will remain constant, and a separate layer for text that will vary for each job. Use a minimum of three different text fonts. Save the drawing as FLOOR14.
11. Start a new drawing and type the following note: PROVIDE A 5° +/–. 5° BEND AT ALL #6 DIA. STEEL REINFORCING. STEEL TO BE WITHIN 1% OF SPECIFIED LENGTH. Save it as E-16-11.
12. Start a new drawing and create a line of text and make a copy of the text. Mirror one of the lines using a variable of 1 and one of the lines with a variable of 0. Save the drawing as E-16-12.
13. Start a new drawing and use the **Mtext** command to make a list of a minimum of three guidelines that describe the creation of multiline text. Use a second text style and justification point and type a second list of guidelines for describing text created with the **Text** command. Save the drawing as E-16-13.
14. Open drawing E-16-13 and make a copy of each set of text. Edit the new text to add additional words in the existing sentences and change the color, style and width of text. Save the drawing as E-16-14.

CHAPTER 16 QUIZ

1. What text option will allow text to be placed between two selected points, at any angle?
2. What text option can be used to place multiple lines of text in a column?
3. If text is to be 0.125 high on a drawing that is to be plotted at a scale of 1"=20'–0", what should the text height be?
4. Explain the difference between rotation angle and oblique angle.
5. What is the difference between a style and a font?
6. What is the process for producing a degree symbol?
7. What commands are used to correct errors in text?
8. What effect will setting the style height option have on future text?
9. Explain how to correct the following problem: VERIFY EXATC hEIGHT AT JOBSITE.
10. When are lowercase letters appropriate for text on construction drawings?
11. Write the required entry to correctly specify three bolts that are three quarters of an inch in diameter at six inches on center. Include all special characters required if the style is set to be Romans.
12. A beam is located in an opening in a wall that is at 45° to horizontal. How should the text be oriented to describe the beam?
13. List and describe the two types of notes generally found on construction drawings.
14. What should be considered when choosing a font?
15. What are the effects of setting the text height in the **Style** command as opposed to setting the height in the **Text** command?
16. When would **Qtext** be used?
17. What is **Mtext** best suited for?
18. What should be the value for **Mirrtext** if you want to mirror the text with the object?
19. Your instructor wants you to create a body of text that starts in the upper right corner of a template drawing. The notes will be 4" wide and be 1/2" down and 4 1/2" from the border. What are the commands and options, and what is the proper sequence for placing these notes?
20. What is the process required to save a paragraph created using **Mtext**?
21. A paragraph of multiline text has been created using ROMANS but it should be in Italics. How can this be fixed?
22. Explain the major differences between **Ddedit** and **Properties** on multiline text.
23. What does the .CUS extension designate?

24. What is the advantage in placing stock notes on a template drawing?
25. A client has decided to change from floor joist to open-web trusses after all of the written specifications have been added to the drawing. How can this change be easily included?

CHAPTER 17

Introduction to Dimensions

In addition to the visual representation and the text used to describe a feature, dimensions are needed to describe the size and location of each member of a structure. Figure 17.1 shows a floor plan for a dormitory and the dimensions used to describe the location of structural members. In this chapter you will be introduced to

- Basic principles of dimensioning
- Guidelines for placing dimensions on plan views and on drawings showing vertical relationships

Commands to be introduced in this chapter include

- **Dimlinear**
- **Dimaligned**
- **Dimordinate**
- **Dimradius**
- **Dimdiameter**
- **Dimcenter (Center Mark)**
- **Dimangular**
- **Dimbaseline**
- **Dimcontinue**
- **Quick Dimension**
- **Qleader**

Several methods will be used to enter the dimension commands. Because so many options are available to place dimensions, this chapter will examine only the default values for dimensioning. Chapter 18 will introduce methods for creating different styles of dimensions and describe how to adjust the variables that control the size and

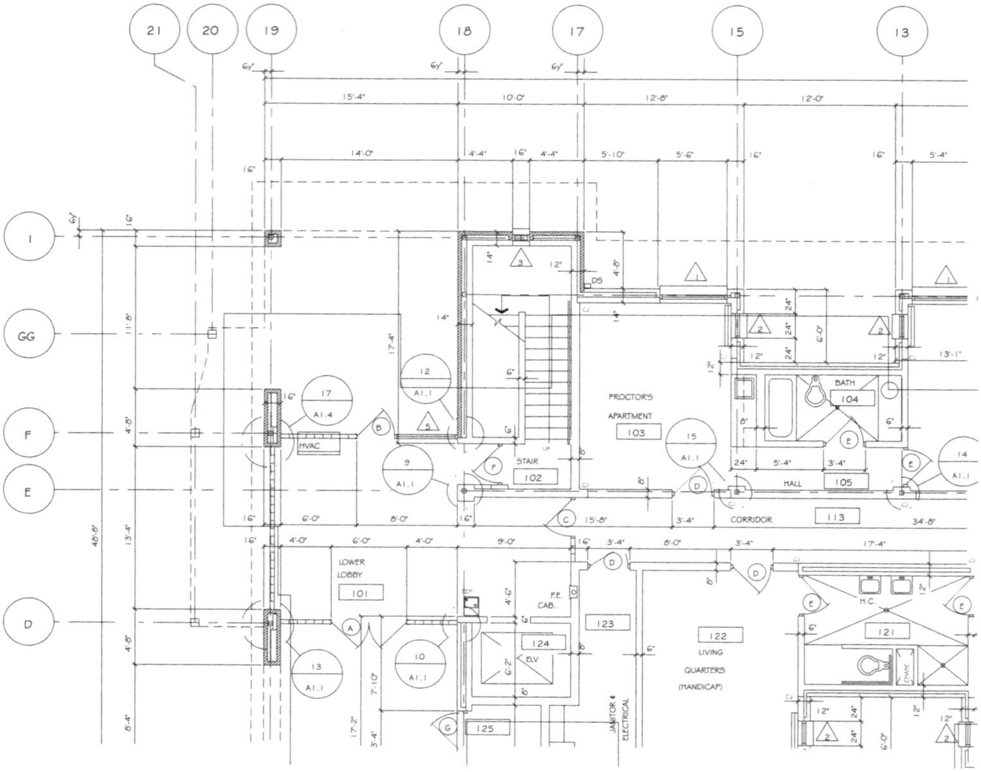

Figure 17.1 *Dimensions are used throughout construction drawings to show the location of each system to be installed. (Courtesy G. William Archer, AIA, Archer and Archer, P.A.)*

placement of the dimensions. Because so many options are included in the software, some options that do not apply to the construction industry will be skipped entirely. Feel free to explore these options on your own, or consult the Help menu.

BASIC PRINCIPLES OF DIMENSIONING

To make better use of AutoCAD's potential, you need to understand several basic dimensioning concepts and terms before you explore the dimension commands. These principles include linetypes, placement of dimension features, and locating exterior and interior drawing features.

DIMENSIONING COMPONENTS

Dimensioning features include extension and dimension lines, text, and line terminators. Each component can be seen in Figure 17.2.

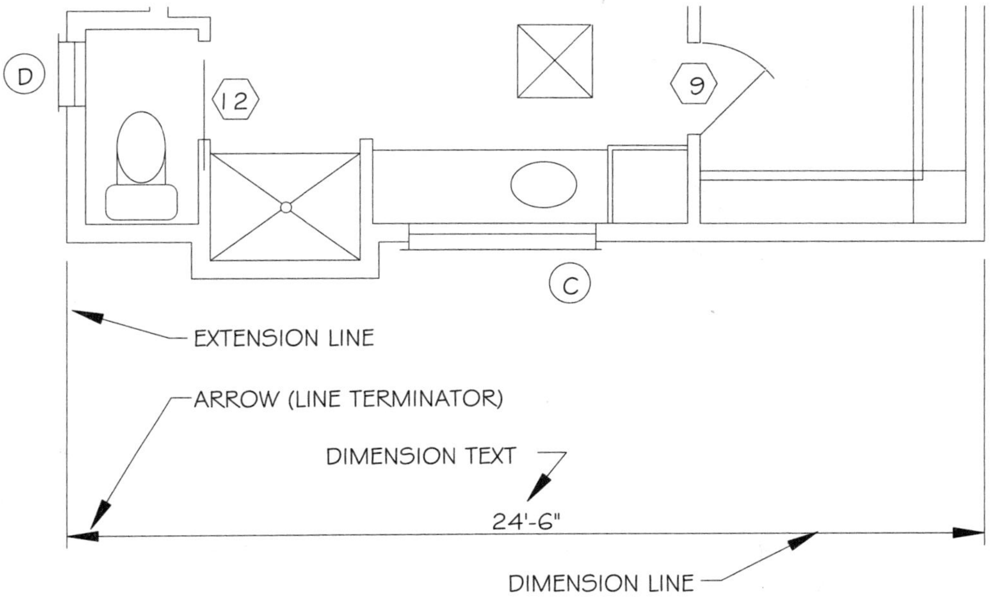

Figure 17.2 *Dimensions are composed of the dimension text, a dimension line, extension lines, and arrows (line terminators).*

Extension Lines

Extension lines are thin lines that extend out from the object being described and set the limits for the dimensions. Extension lines are usually offset about 1/8" from the object, and they extend past the last dimension line between 1/16" and 1/8". The offsets and extensions can be seen in Figure 17.3.

Two different types of linetypes are typically used for extension lines. Solid lines are used to dimension to the exterior face of an object, such as a wall or footing. A centerline is used to dimension to the center of a wood wall or the center of other objects. Figure 17.4 shows examples of each.

Dimension Lines

Dimension lines are thin lines used to show the extent of the object being described. Exact location will vary with each office, but dimension lines should be placed in such a way as to leave room for notes, but still close enough to the features being described so that clarity will not be hindered. Guidelines for placement will be discussed later in this chapter.

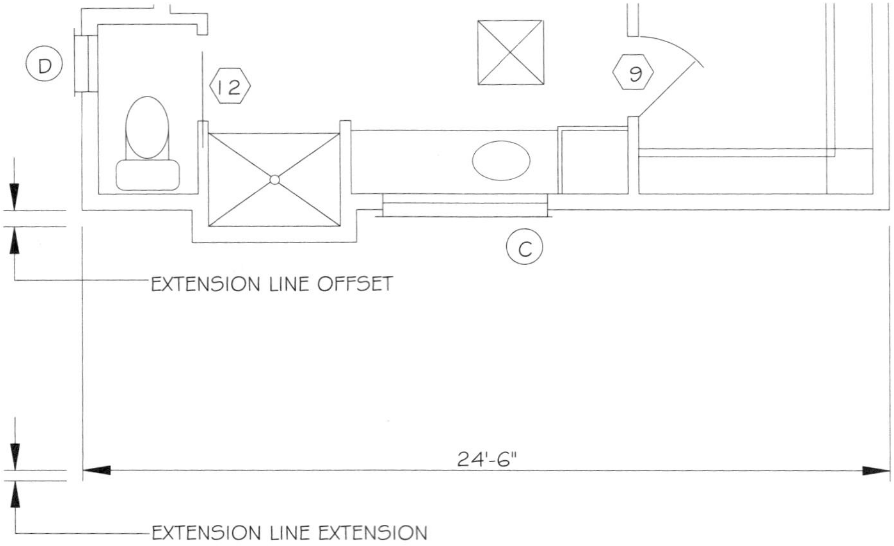

Figure 17.3 *Extension lines are typically placed about 1/8" from the object, and extend about 1/8" past the dimension line.*

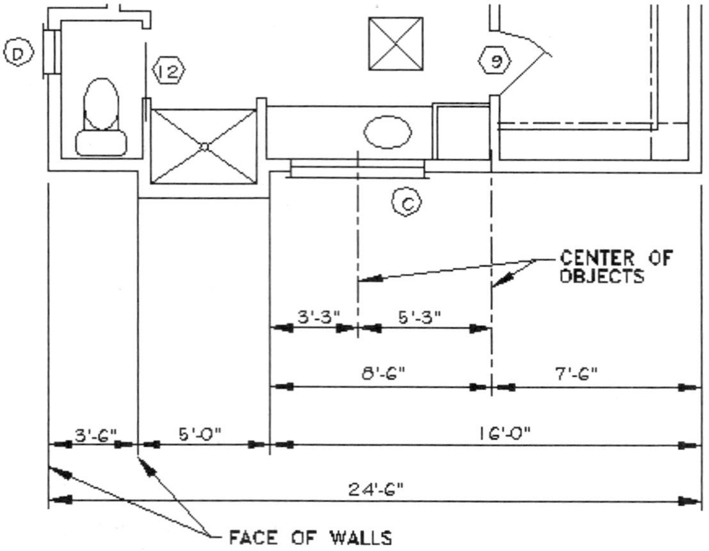

Figure 17.4 *Continuous extension lines are used to dimension to the edge of a surface; centerlines are used to dimension to the center of objects. Professionals often use a continuous line in place of a centerline to save time.*

Text

The text for dimensioning is placed above the dimension line, and centered between the two extension lines. On the left and right sides of the structure, text is placed above the dimension line and rotated so that the text can be read from the right side of the drawing page. Examples of each placement can be seen in Figure 17.5. On objects placed at an angle other than horizontal or vertical, dimension lines and text are placed parallel to the oblique object, as seen in Figure 17.6. Often not enough space is available for the text to be placed between the extension lines when small areas are dimensioned. Although options vary with each office, several alternatives for placing dimensions in small spaces can be seen in Figure 17.7.

Terminators

The default method for terminating dimension lines at an extension line is with an arrowhead. Many offices use a tick mark or dot in an effort to duplicate their manual drawing styles. A third option is a thickened tick mark, which offers good con-

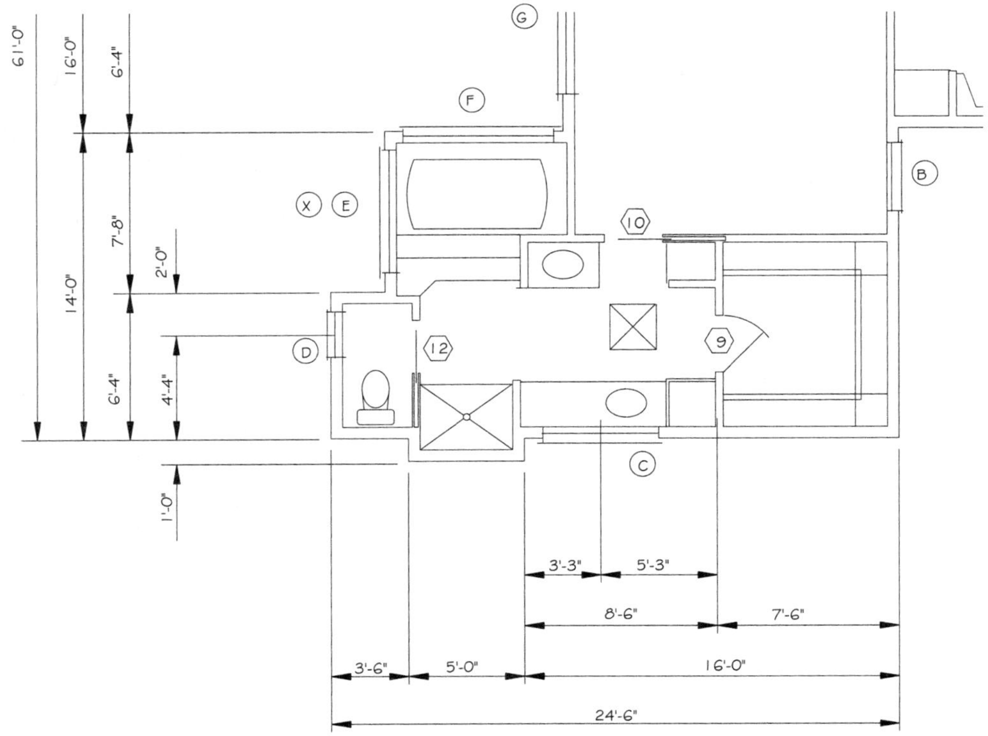

Figure 17.5 *Text should be placed above the dimension line and read from the bottom or right side of the drawing.*

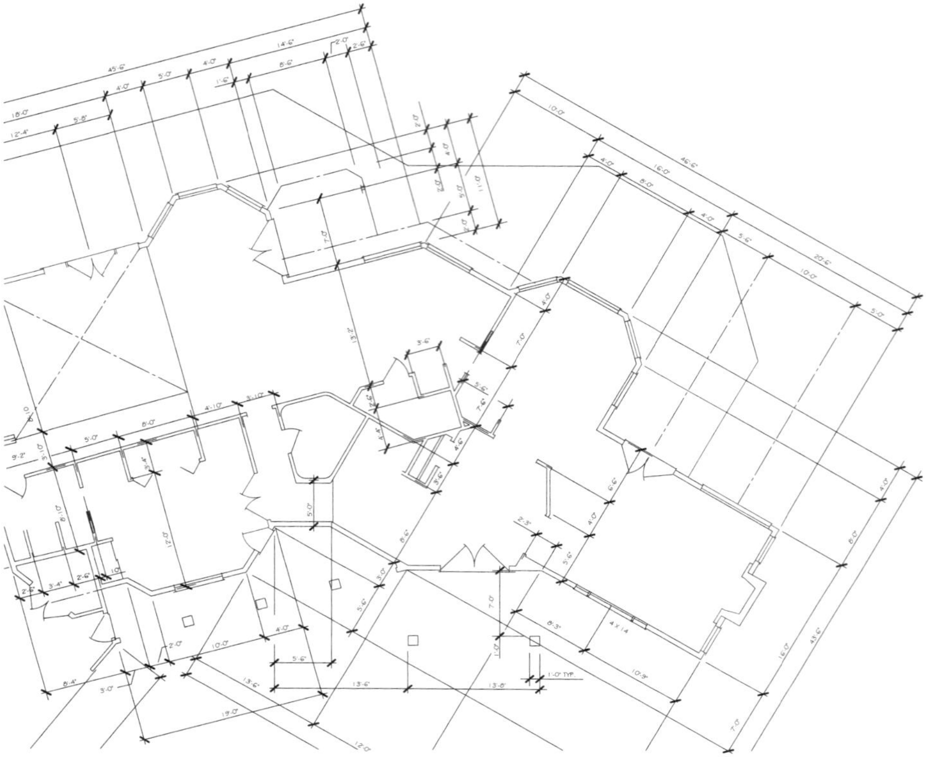

Figure 17.6 *When objects on an angle are dimensioned, dimension lines and text are placed parallel to the inclined surface. (Courtesy Residential Designs.)*

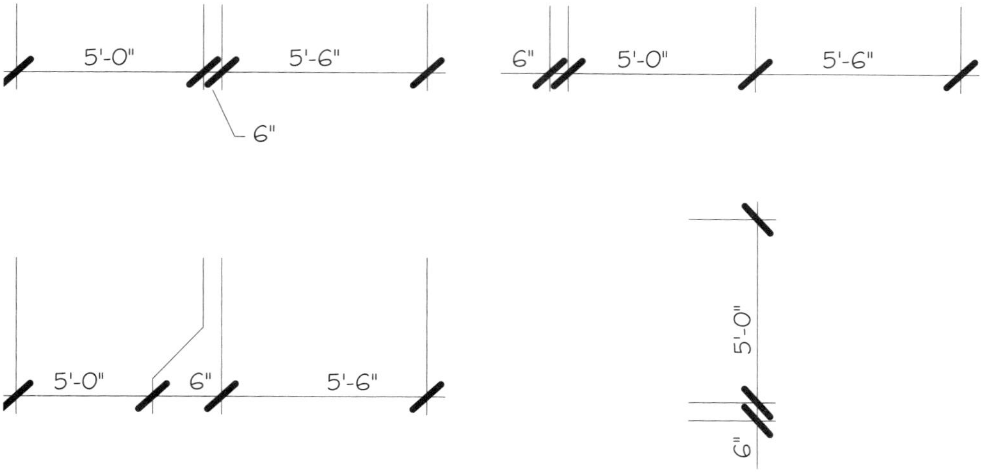

Figure 17.7 *Common professional alternatives for placing dimensions in small spaces.*

trast between lines, closely resembles its manual counterpart, and can be plotted quickly. All four options can be seen in Figure 17.8. AutoCAD also offers several other options that will be introduced in the next chapter.

DIMENSION PLACEMENT

Construction drawings requiring dimensions typically consist of plan views, such as the floor, foundation, and framing plans, and drawings showing vertical relationships, such as exterior and interior elevations, sections, details, and cabinet drawings. Each type of drawing has its own set of challenges to be overcome in placing drawings. No matter the drawing type, dimensions should be placed on a layer titled *-ANNO-DIMS. The * represents the letter of the proper originator such as A or S, ANNO represents annotation, and DIMS represents dimensions.

Note: Each of the examples that follow in this chapter has been created by professional designers and architects by altering the dimension style. This chapter will introduce each of the options for placing dimension in their default settings. Chapter 18 will introduce methods for altering the dimension qualities.

PLAN VIEWS

A discussion about placing dimensions in plan view can be divided into the areas of interior and exterior dimensions. Whenever possible, dimensions should be placed outside the drawing area.

Exterior Dimensions

Most offices start by placing an overall dimension on each side of the structure that is approximately 2" from the exterior wall. Moving inward, with approximately 1/2" between lines, are dimension lines used to describe major jogs in exterior walls, the distance from wall to wall, and the distance from wall to window or door to wall. Examples of placing these four dimension lines can be seen in Figure 17.9.

Two different systems are used to represent the dimensions between exterior and interior walls. Architects tend to represent the distance from edge to edge of walls as seen in Figure 17.10. Engineering firms tend to represent the distance from exterior edge to center of interior wood walls using methods shown in Figure 17.11. Concrete walls are dimensioned to the edge by both disciplines, as seen in Figure 17.12. Concrete foot-

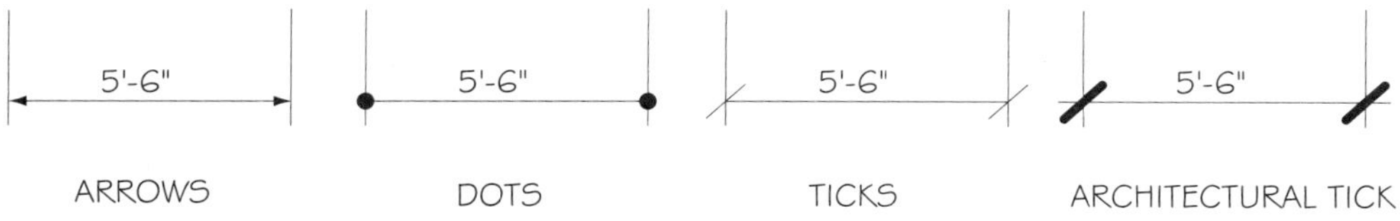

Figure 17.8 *Alternatives for terminating dimension lines.*

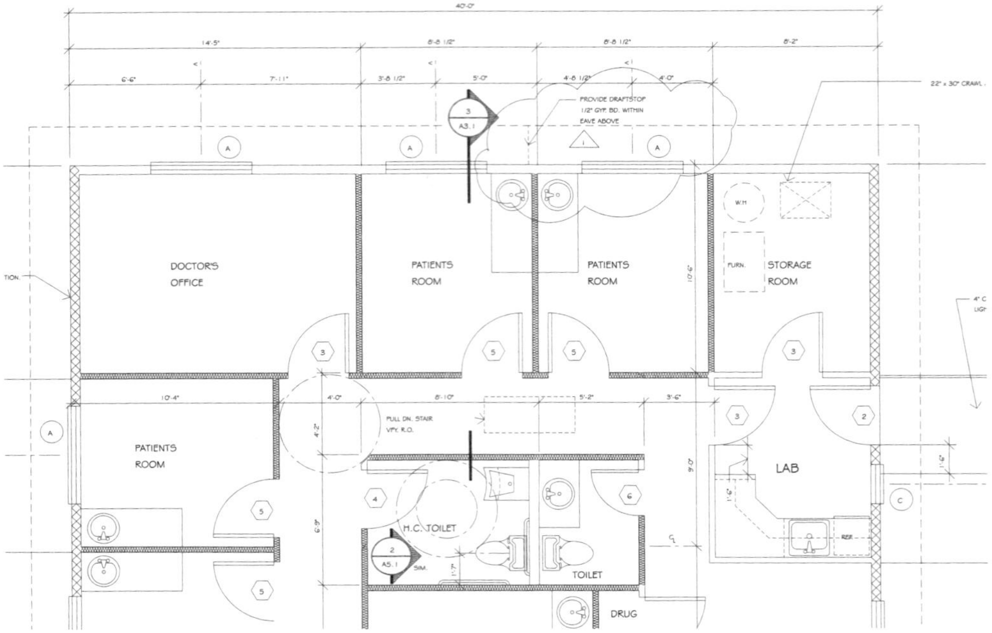

Figure 17.9 *Floor plans are typically dimensioned by placing an overall dimension, dimensions to locate major jogs, wall-to-wall dimensions, and wall-to-opening dimensions. (Courtesy Scott R. Beck, Architect.)*

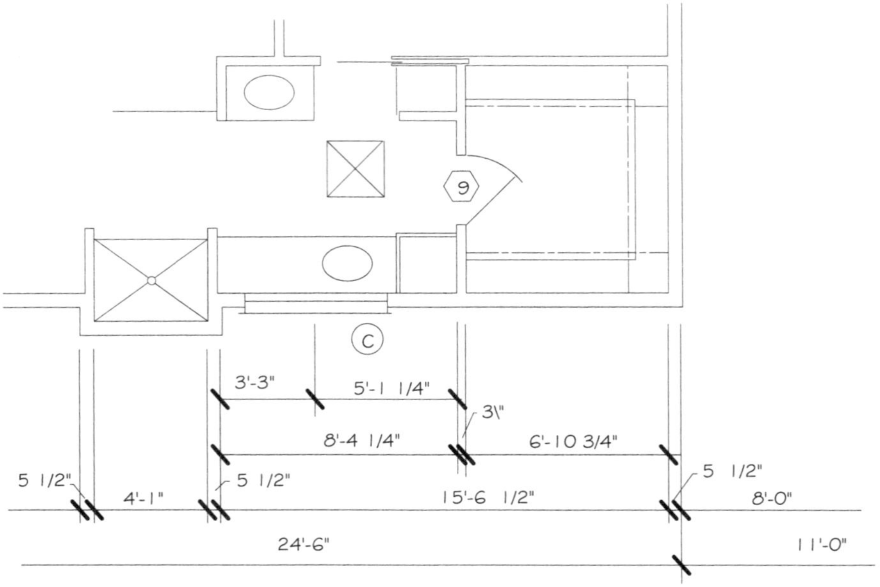

Figure 17.10 *Some architectural firms dimension from edge to edge of interior walls.*

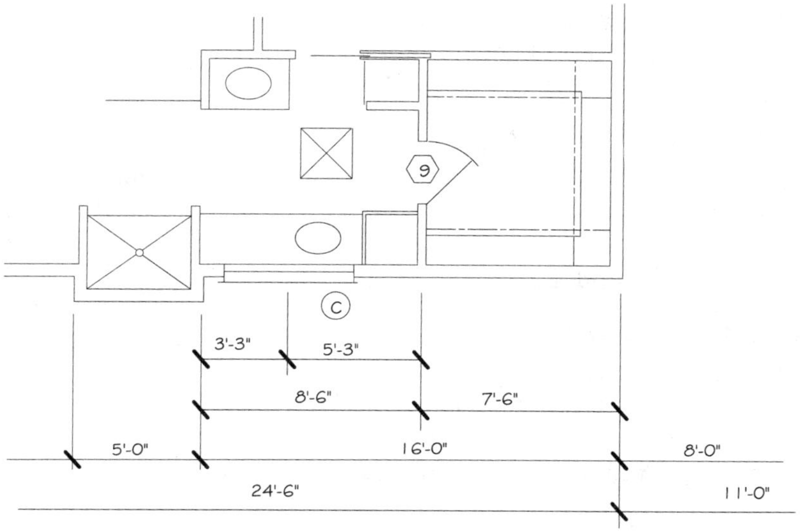

Figure 17.11 *Engineering firms often dimension to the center of interior walls.*

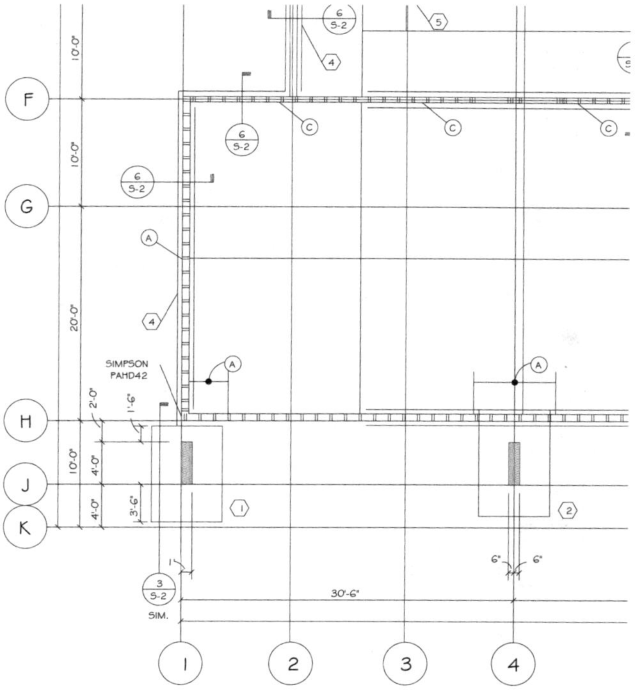

Figure 17.12 *Concrete walls are dimensioned from edge to edge. (Courtesy Van Domelen/Looijenga/McGarrigle/Knauf Consulting Engineers.)*

ings normally are dimensioned to the center. Examples of dimensioning concrete footings and walls can be seen in Figure 17.13.

Interior Dimensions

The two main considerations in placing interior dimensions are clarity and grouping. Dimension lines and text must be placed so that they can be read easily and so that neither interferes with other information that must be placed on a drawing. Information should also be grouped together as seen in Figure 17.14, so that construction workers can find dimensions easily.

VERTICAL DIMENSIONS

Unlike the plan views, which show horizontal relationships, the elevations, sections, and details require dimensions that show vertical relationships, as seen in Figure 17.15. Typically, these dimensions originate at a line that represents a specific point, such as the finish grade, a finish floor elevation, or a plate height.

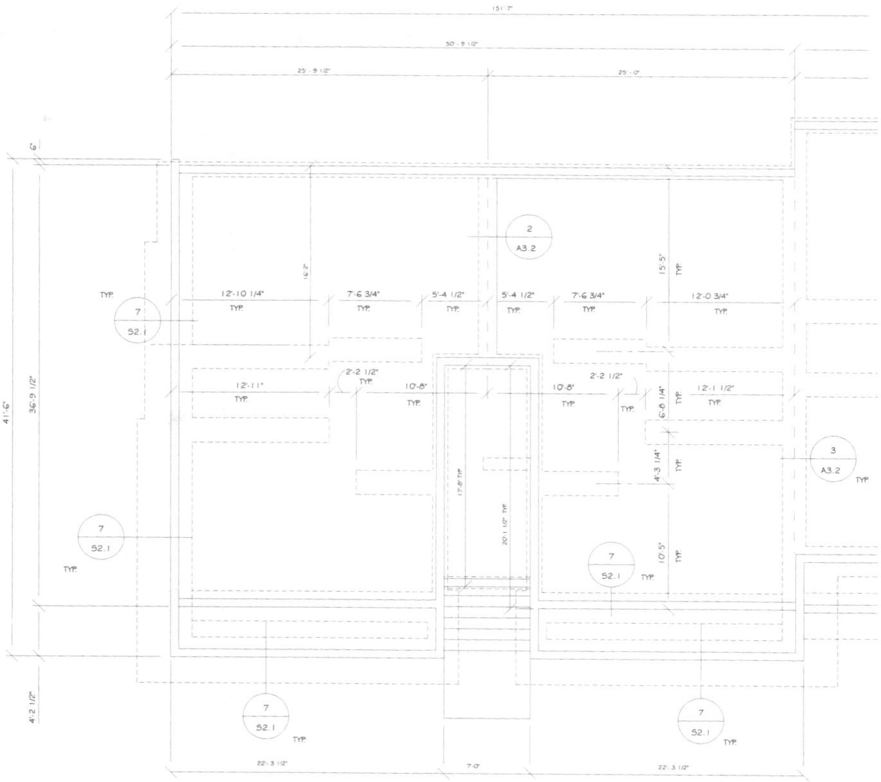

Figure 17.13 *Interior concrete piers and footings are dimensioned to their centers. (Courtesy Scott R. Beck, Architect.)*

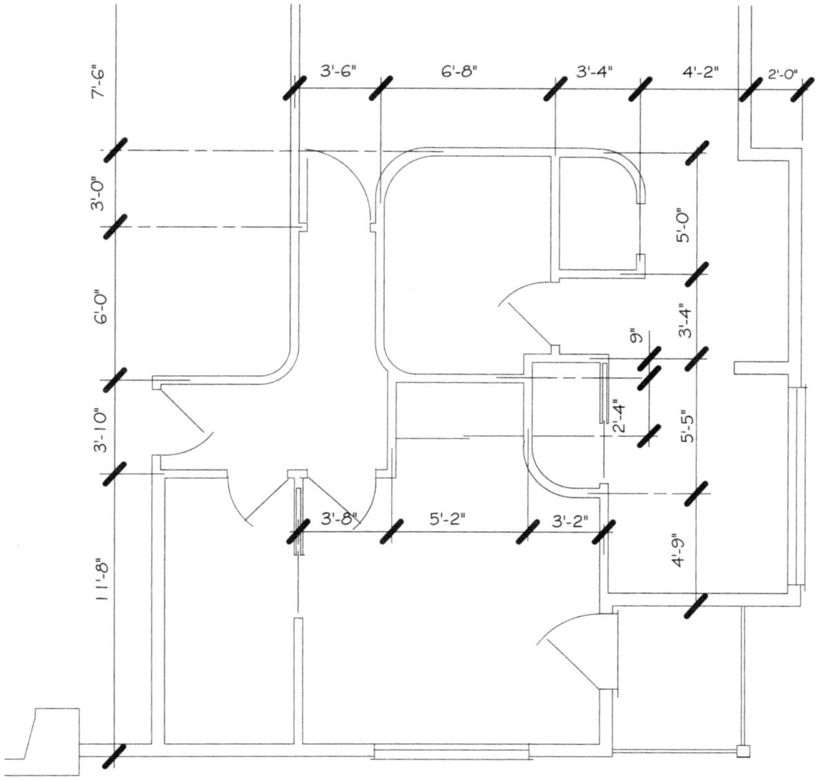

Figure 17.14 *Dimensions placed on the inside of a drawing should be grouped together and be clearly visible. (Courtesy Residential Designs.)*

ACCESSING DIMENSIONING COMMANDS

Several methods are available for accessing the commands that control dimensions. Commands can be accessed by selecting buttons on the **Dimension** toolbar or by selecting an option from the **Dimension** menu. The menu and toolbar can be seen in Figure 17.16. Each of the options displayed on the menu and toolbar will be presented in this chapter. Dimensions can also be entered at the Command prompt. Options are also available in dialog boxes, which will be presented in Chapter 18.

DIMENSIONING OPTIONS

Eleven methods are available for placing dimensions on a drawing. The **Dimlinear** and **Dimaligned** commands are used for providing linear dimensioning. **Dimdiameter**, Dimradius, Dimangular, and **Dimcenter** are used to describe circular features. Three other commands, **Dimbaseline**, **Dimcontinue**, and **Dimordinate**, are used to place dimensions and can be combined with other commands. The **Qleader** command can

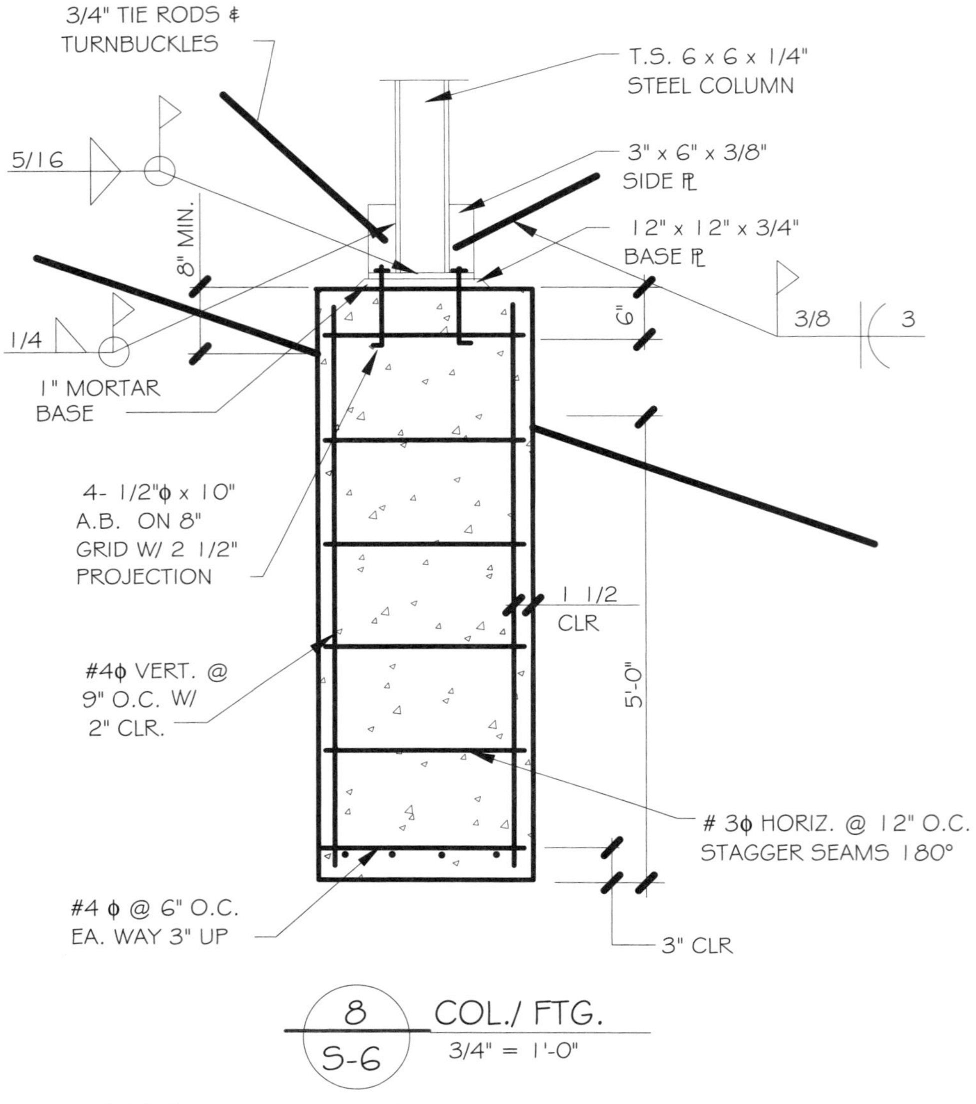

Figure 17.15 *Elevations, sections, and details each require dimensions to show vertical relationships. (Courtesy Residential Designs.)*

be used for referencing text and dimensions to a drawing. The **Tolerance** command can be used to access geometric dimensioning and tolerancing symbols. These symbols are rarely used and will not be covered in detail. **Style** can be used to access the **Dimension Style Manager** dialog boxes that are used to save dimensioning styles. Styles will be discussed in Chapter 18.

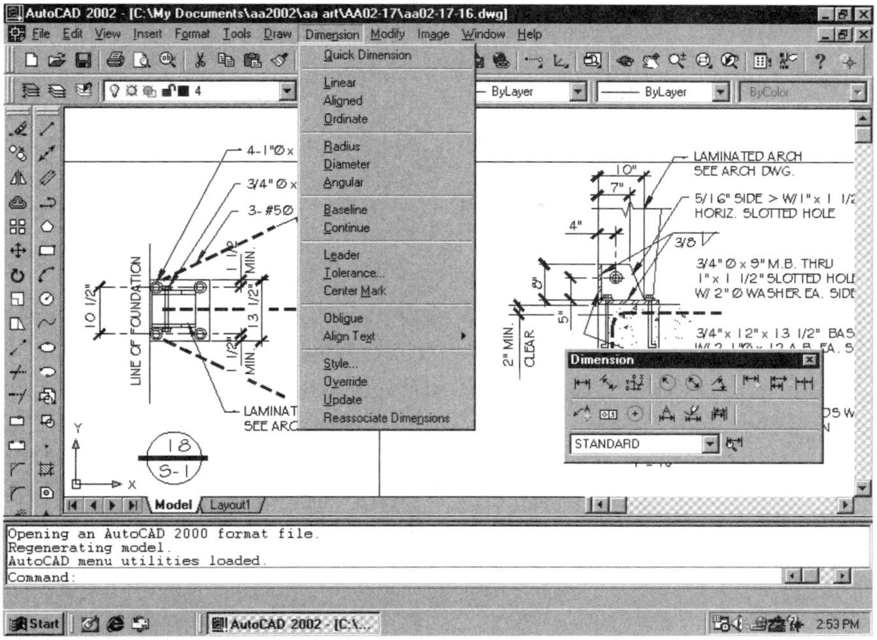

Figure 17.16 *The commands for placing dimensions can be selected on the* **Dimension** *toolbar or from the* **Dimension** *menu, or entered by keyboard.*

Note: In addition to the dimensioning commands listed above, AutoCAD allows dimensions to be placed using the **Quick Dimension** command. It can be used to create a series of dimensions using baseline or continued dimensions. The following portion of this chapter will introduce each dimensioning method. The final portion of this chapter will help you combine options using **Quick Dimension**. Chapter 18 will introduce methods of altering the features of each dimensioning method.

LINEAR DIMENSIONS

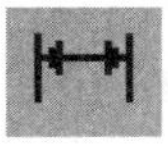

Linear dimensions are used to dimension straight surfaces common to construction features. Selecting the **Linear Dimension** button on the **Dimension** toolbar or **Linear** from the **Dimension** menu will produce the Command prompt, which will be discussed shortly. You can also start the command by typing **DLI** (ENTER) at the Command prompt. This will produce the prompt:

Command: **DLI** (ENTER) *(Or click the Linear Dimension button.)*

DIMLINEAR

Specify first extension line origin or <select object>:

AutoCAD now offers two options to place a dimension. The default method is to specify two points to define the dimension. A second method allows selection of the object to be dimensioned. Once the object is selected, the end points are automatically defined.

Select Extension Line Origin

The start and end of a line or distance to be dimensioned can be specified. The command sequence for specifying each endpoint is as follows:

> Specify first extension line origin or <select object>: *(Select point.)*
>
> Specify second extension line origin: *(Select point.)*
>
> Specify dimension line location or [Mtext/Text/Angle/Horizontal/ Vertical/Rotated]: *(Select dimension line location.)*

The default requires you to move the cursor and select the desired location to place the dimension line. Be sure to activate OSNAP settings before selecting dimension locations. As the second extension line location is selected, the dimension and extension lines are dragged into position to allow for visual inspection, but the coordinate display can be useful in spacing dimension lines. The dimension line location should be selected to allow room for all future notes. Place the dimension line in the desired location and click the select button. The process can be seen in Figure 17.17.

Mtext

Selecting the Mtext option allows the dimension value to be replaced by a body of text in a similar manner as the Text option. Typing **M** (ENTER) at the dimension prompt will produce a display similar to Figure 17.18 with the cursor in front of the brackets. The brackets represent the default dimension value. With the cursor preceding the brackets, text will be written followed by the dimension. An entry of **F.O.S.** <> would be displayed as F.O.S. 40'–0". An entry of <>**F.O.S.** will be displayed as 40'–0" F.O.S. Using the cursor to remove the brackets will remove the default dimension value and allow only the text to be placed.

Text

One of the strengths of AutoCAD is also one of its weaknesses. AutoCAD is extremely accurate. Sometimes the dimension text that will be assigned is too accurate for the drawing purpose. Chapter 18 will introduce methods of controlling dimension accurately. If you haven't been accurate as you created the drawing, the text assigned to the dimension will not produce the desired results. Choosing the Text option before the dimension line location is selected allows you to override the dimension text value by entering the value by keyboard. Use the following sequence to change the value from 3 1/2" to 4":

> Specify dimension line location or [Mtext/Text/Angle/Horizontal/ Vertical/Rotated]: **T** (ENTER)

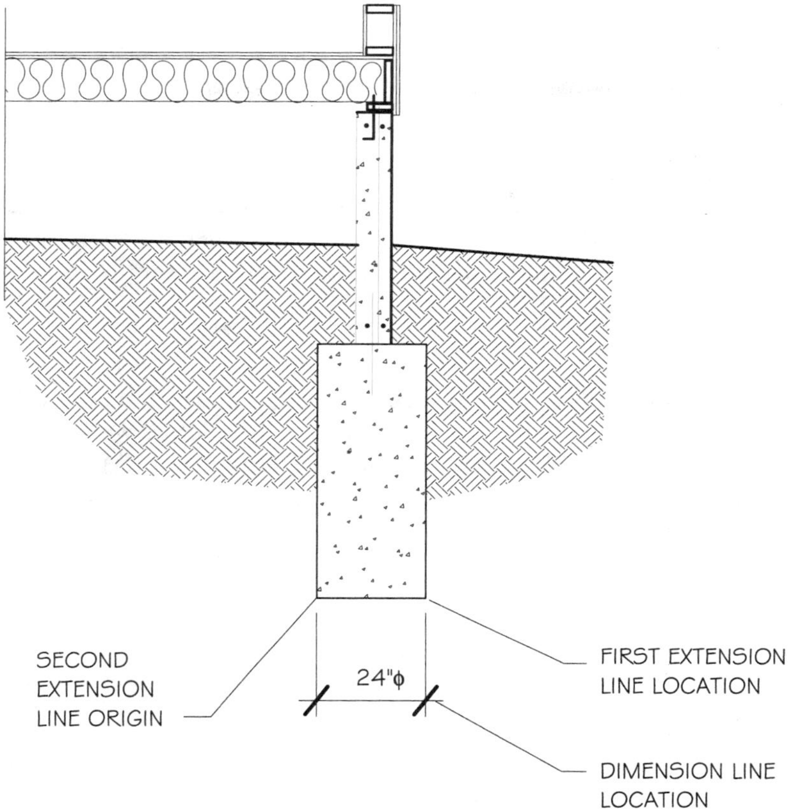

Figure 17.17 *Horizontal and vertical dimensions can be placed using the Dimlinear command. The dimension is placed by selecting the starting location for each extension line and then selecting the location for the dimension line. As the location for the second extension line is selected, the dimension is displayed and can be dragged into position.*

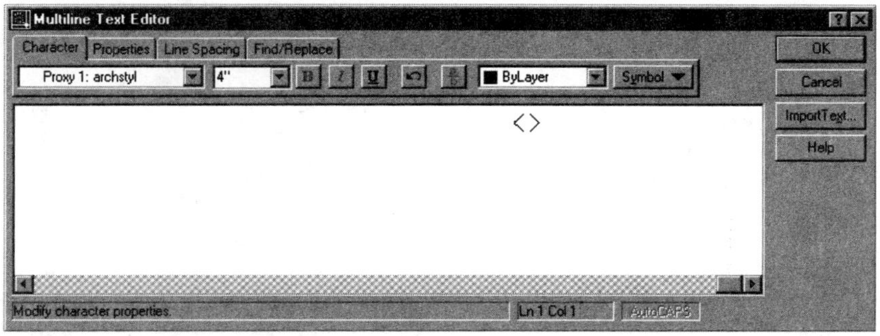

Figure 17.18 *Typing* **M** *(ENTER) at the Command prompt will produce the* **Multiline Text Editor**.

Enter dimension text <3.5>: **4** (ENTER)

Specify dimension line location or [Mtext/Text/Angle/Horizontal/Vertical/Rotated]: *(Select the dimension location.)*

Command:

Pressing (ENTER) returns the dimension and allows it to be dragged into position. Pressing SPACEBAR and then (ENTER) rather than assigning a value at the prompt for dimension text displays the extension lines, dimension lines, and terminators but no text is assigned. This option can be useful for specifying the limits of a specific product, as seen in Figure 17.19.

Angle

Type **A** (ENTER) at the line location prompt to alter the angle used to display the text. Although this is not often done, it could be especially helpful to make a dimension for a small space stand out. Figure 17.20 shows an example of text rotated at 45° to the dimension line. The command sequence, once the two endpoints are selected, is as follows:

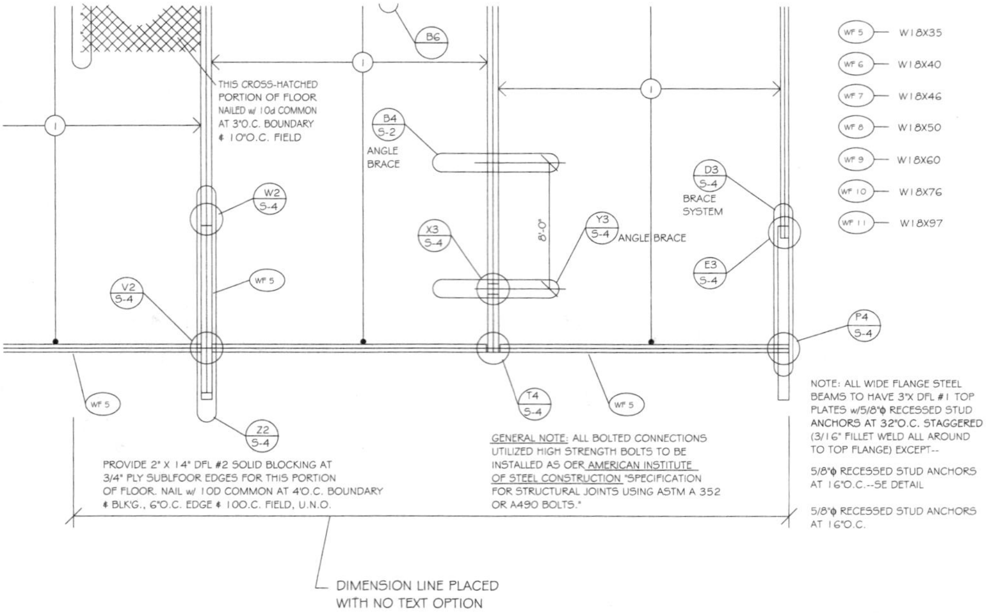

Figure 17.19 *Rather than entering a text value, pressing SPACE followed by ENTER allows the dimension line to be placed with no dimension.*

Specify dimension line location or [Mtext/Text/Angle/Horizontal/Vertical/Rotated]: **A** (ENTER)

Specify angle of dimension text: **45** (ENTER)

Specify dimension line location or [Mtext/Text/Angle/Horizontal/Vertical/Rotated]: *(Select the dimension location.)*

Dimension text = 20'-6"

Command:

Horizontal and Vertical

Typing **H** (ENTER) or **V** (ENTER) option at the prompt allows the specified dimension direction to be entered. AutoCAD automatically places a vertical dimension if two vertical endpoints are selected or a horizontal dimension line if two horizontal endpoints are selected.

Rotated

Typing R (ENTER) at the prompt rotates the extension and dimension lines to a specified angle from the desired surface. This option places the dimension in a rotated position, and can be used to draw attention to a dimension describing a small space. Use this option to place the extension and dimension lines for referencing a note to a specific area. The sequence to place the lines at 20° to a vertical surface is as follows:

Specify dimension line location or [Mtext/Text/Angle/Horizontal/Vertical/Rotated]: **R** (ENTER)

Specify angle of dimension line <0>: **20** (ENTER)

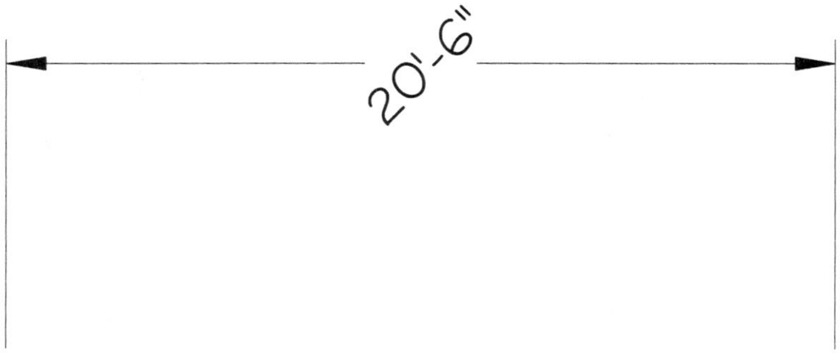

Figure 17-20 *The text angle can be rotated from the dimension line using the A option. Although not a common method of dimensioning, rotated text can be used to help draw attention to the dimension.*

Specify dimension line location or [Mtext/Text/Angle/Horizontal/
Vertical/Rotated]: *(Select the dimension location.)*

Dimension text = 0'-6"

Command:

Figure 17.21 shows an example of the rotated option.

Select Object Option

Up to this point, each of the options discussed was made available once the two origins for the extension line endpoints were selected. By choosing the Select Object option, a specific line can be selected, and each endpoint will be determined automatically. The command sequence is as follows:

Command: **DLI** (ENTER) *(Or click the Linear Dimension button.)*

DIMLINEAR

Specify first extension line origin or <select object>: (ENTER)

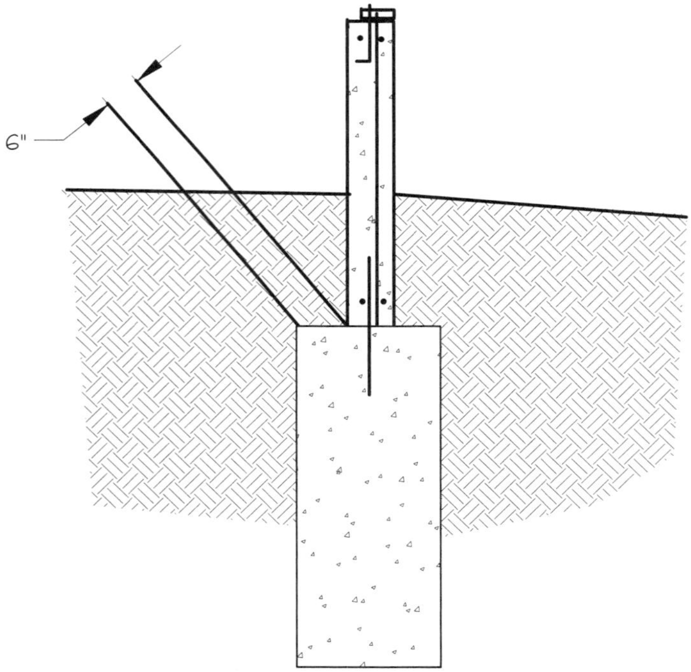

Figure 17.21 *The extension lines, dimension lines, and text can be rotated from the surface to be dimensioned using the* **R** *option.*

Once (ENTER) is pressed, the crosshairs are changed to a pick box and a specific object can be selected:

Select object to dimension: *(Specify object to be dimensioned)*

Specify dimension line location or [Mtext/Text/Angle/Horizontal/ Vertical/ Rotated]: *(Select the desired location.)*

Dimension text = 24

Command:

This process can be seen in Figure 17.22. The effect of the **Dimlinear** command on vertical dimensions is shown in Figure 17.23. If the area to be dimensioned is smaller than the length required to place the arrows and text, each will be placed outside the extension lines. Methods of altering placement will be discussed in Chapter 18.

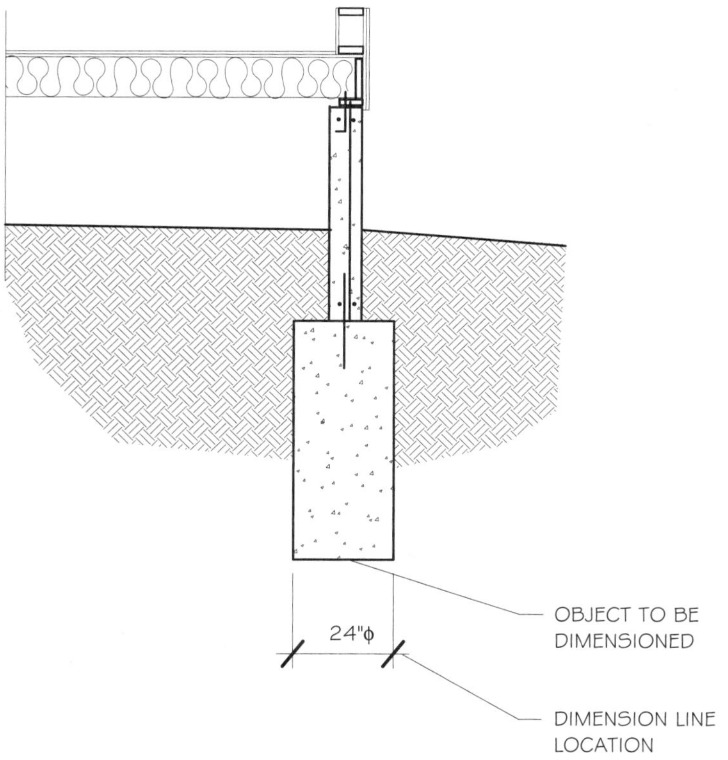

Figure 17.22 *Choosing the Select Object option rather than selecting an extension line location allows an object to be selected. Extension lines will automatically be placed at each end of the selected object.*

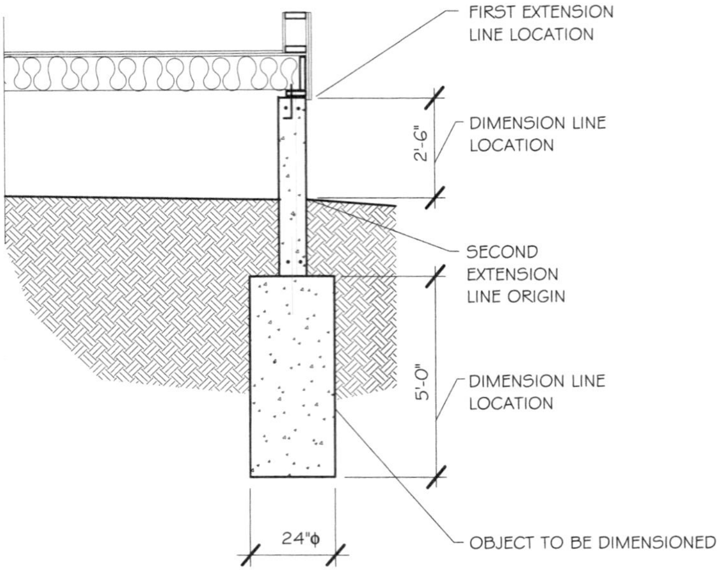

Figure 17.23 *Placement of vertical dimension lines using* **Dimlinear** *with the Select Object option to place the lower dimension and the two-point selection method on the upper dimension.*

ALIGNED DIMENSIONS

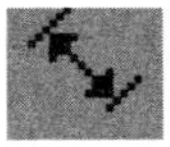

This method of placing dimensions is useful when the object being described is not parallel to the drawing borders. An example of aligned dimensions can be seen on the foundation plan shown in Figure 17.24. The command sequence is similar to the **Dimlinear** command. Start the **Dimaligned** command by selecting the **Aligned Dimension** button on the Dimension toolbar, selecting **Aligned** from the **Dimension** menu, or by typing **DAL** (ENTER) at the Command prompt:

Command: **DAL** (ENTER) *(Or click the Aligned Dimension button.)*

DIMALIGNED

Specify first extension line origin or <Select object>: *(Select first point.)*

Specify second extension line origin: *(Select second point.)*

Specify dimension line location or [Mtext/Text/Angle]:

Dimension text <45'-0">:

Command:

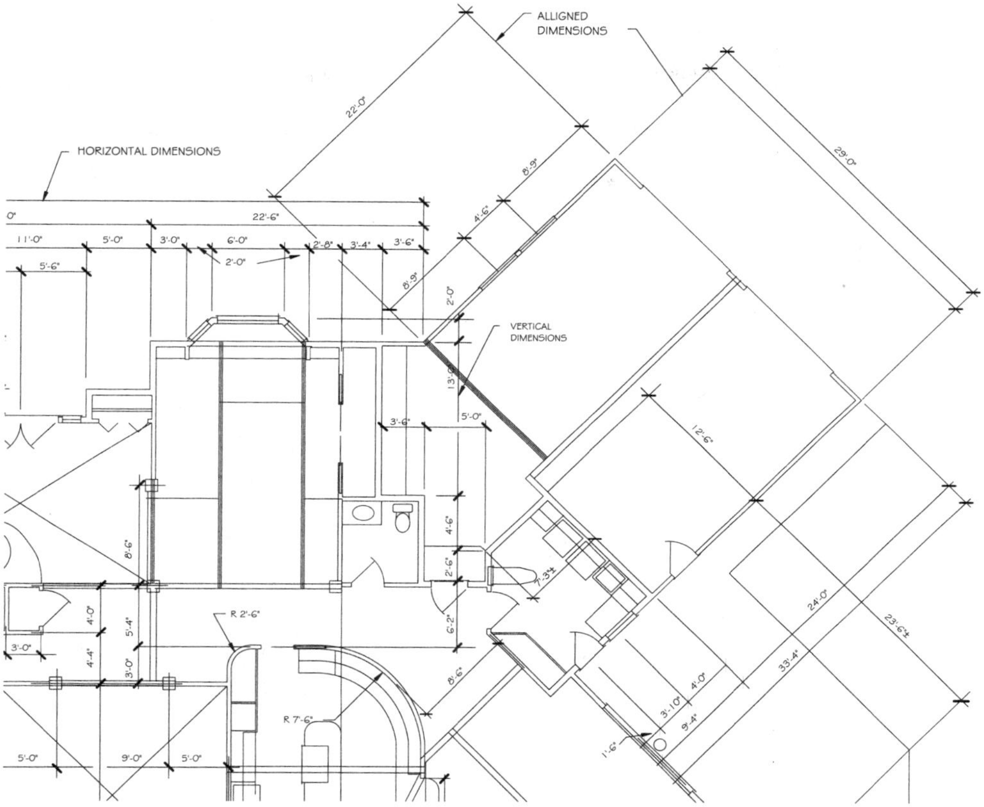

Figure 17.24 *When a surface is not horizontal or vertical, the* **Dimaligned** *command can be used to place dimensions. (Courtesy Residential Designs.)*

If a location is selected for the dimension line, the line will automatically be placed parallel to the surface defined by the endpoints. Selecting the Mtext or Text option allows the default text value to be altered. Selecting the Angle option allows the angle of the text to be altered. Each of the options is the same as the corresponding options for the **Dimlinear** command. Selecting the Select Object option automatically places the extension lines at the ends of the selected line and allows the dimension to be dragged into position.

The effects of the **Dimaligned** command can be seen in Figure 17.25.

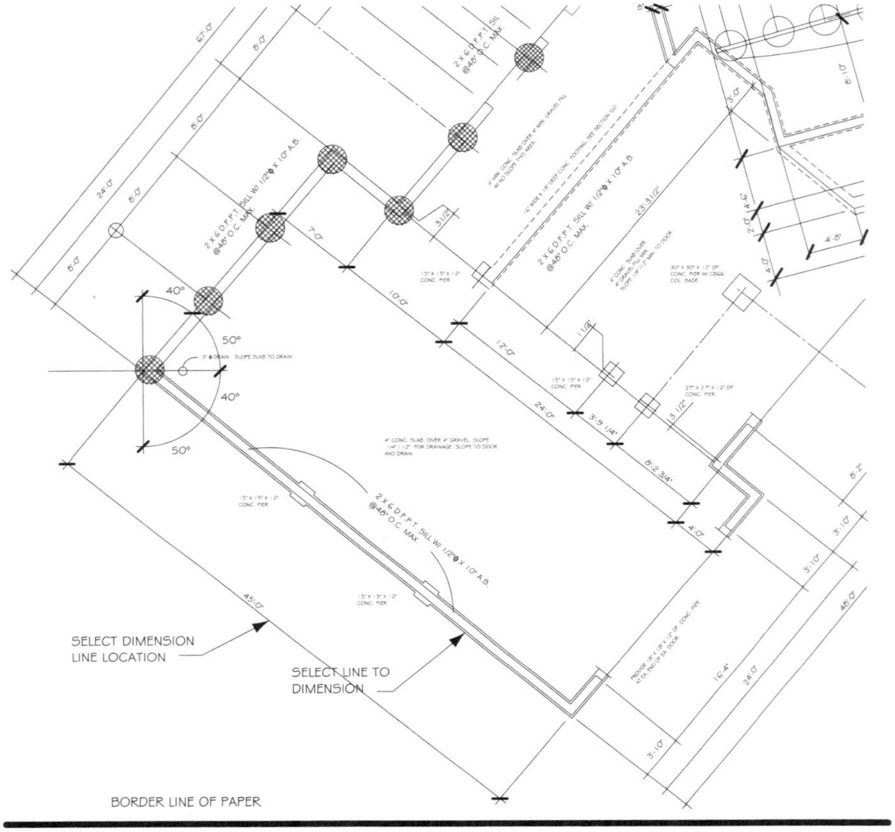

Figure 17.25 *Placing dimensions using the* **Dimaligned** *command is similar to using the* **Dimlinear** *command.*

ORDINATE DIMENSIONS

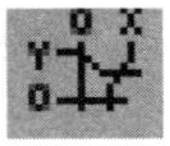

Ordinate dimensioning consists of placing dimensions without the use of dimension lines and arrows. An example of ordinate dimensions can be seen in Figure 17.26. They are not typically used in construction, but are occasionally used in placing vertical dimensions on details and sections. Start the **Dimordinate** command by selecting the **Ordinate Dimension** button on the **Dimension** toolbar or **Ordinate** from the **Dimension** menu or by typing **DOR** (ENTER) at the Command prompt:

Command: **DOR** (ENTER) *(Or click the Ordinate Dimension button.)*

DIMORDINATE

Select feature location: *(Select the surface to be dimensioned.)*

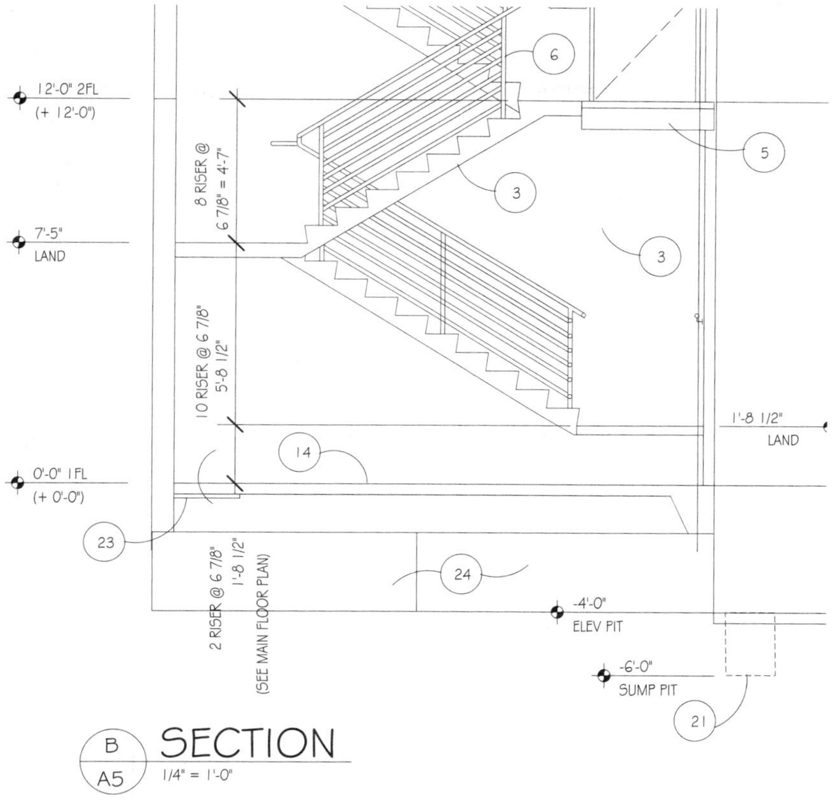

Figure 17.26 *Placement of dimensions using the* **Dimordinate** *command. All dimensions are placed using only a single extension line to represent each finish floor. All dimensions are referenced from the main floor level. (Courtesy Russ Hanson, HDN Architects, PC.)*

AutoCAD is prompting for the location that will serve as a base. On a section, this is often the finished floor elevation. It will be helpful to move the 0,0 location of the world coordinate system from the lower corner of the drawing to a corner of the object being dimensioned. The ORTHO mode should also be set to ON. Once the initial edge has been selected, the prompt will be altered to read

> Specify leader endpoint or [Xdatum/Ydatum/Mtext/Text/Angle]:
> *(Select the endpoint of the extension line.)*
>
> Dimension text: <10'-0"> (ENTER)

The leader is a single horizontal line when the Ydatum ordinate is used and a single vertical line for the Xdatum.

RADIUS DIMENSIONS

The **Dimradius** command is useful for many engineering and architectural applications. The way the information is placed will vary depending on the size of the circle or arc to be dimensioned. Figure 17.27 shows each option. The type of dimension also depends on the **Dimvar** settings. Start the command by selecting the **Radius Dimension** button on the **Dimension** toolbar, selecting **Radius** from the **Dimension** menu, or by typing **DRA** (ENTER) at the Command prompt:

Command: **DRA** (ENTER) *(Or click the Radius Dimension button.)*

DIMRADIUS

Select arc or circle: *(Select desired arc or circle.)*

Dimension text <10'-6">:

Specify dimension line location or [Mtext/Text/Angle]: *(Select location for dimension.)*

Command:

When you select the circle or arc to be dimensioned, the default dimension is dragged as you move the cursor. Entering **M**, **T** or **A** allows the text or angle value to be altered. The Mtext, Text, and Angle options function exactly as discussed with other

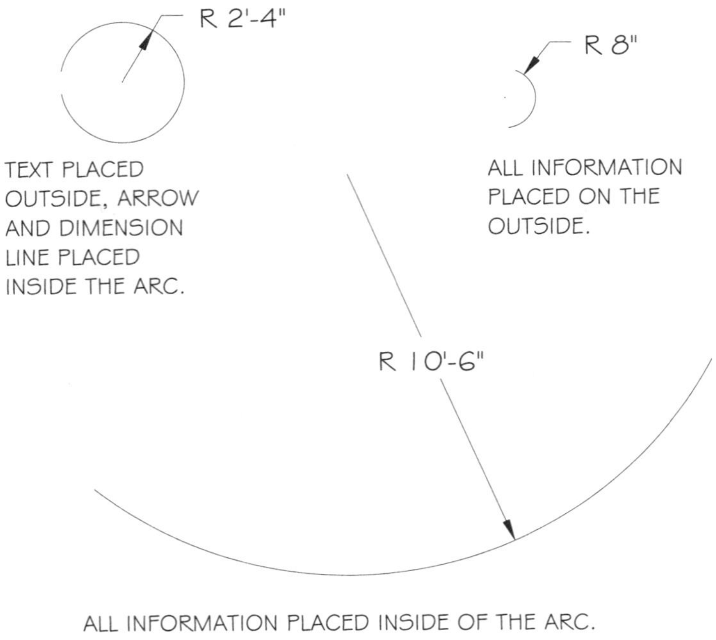

Figure 17.27 *The* **Dimradius** *command allows a radius to be placed using three different techniques.*

commands. Once the desired Text and Angle prompts have been responded to, the prompt will be returned to locate the dimension line. Figure 17.28 shows an example of the use of the **Dimradius** command.

DIAMETER DIMENSIONS

The **Dimdiameter** command will allow you to dimension the diameter of circular objects. Start the command by selecting the **Diameter Dimension** button on the **Dimension** toolbar or **Diameter** from the **Dimension** menu or by typing **DDI** (ENTER) at the Command prompt. Three methods of placing the diameter specifications on a drawing can be seen in Figure 17.29. The default method is to use a leader line to display the dimension. The command sequence is as follows:

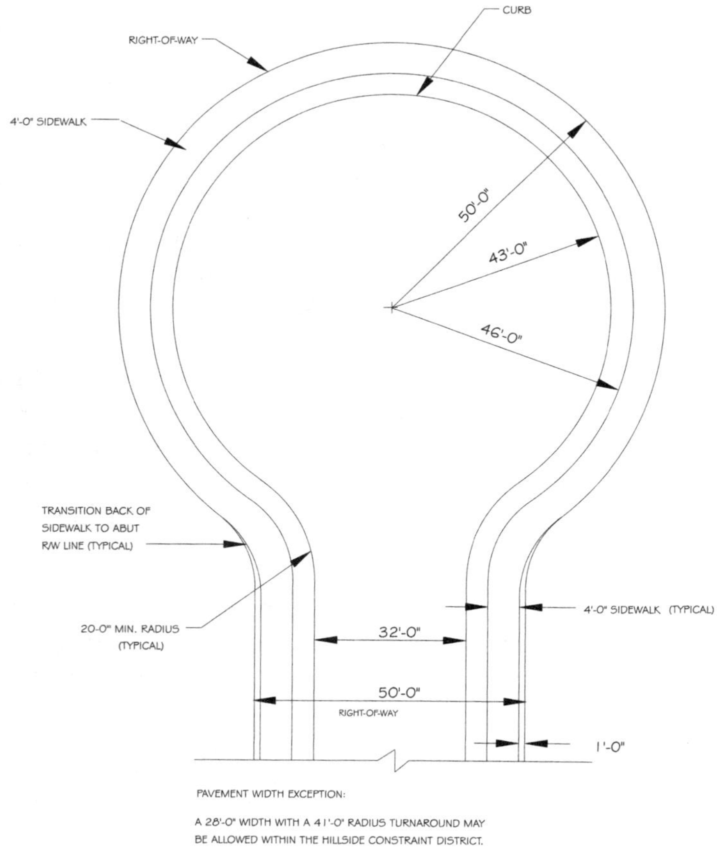

Figure 17.28 *Placing of dimensions using the* **Dimradius** *command. (Courtesy Department of Environmental Services, City of Gresham, Oregon.)*

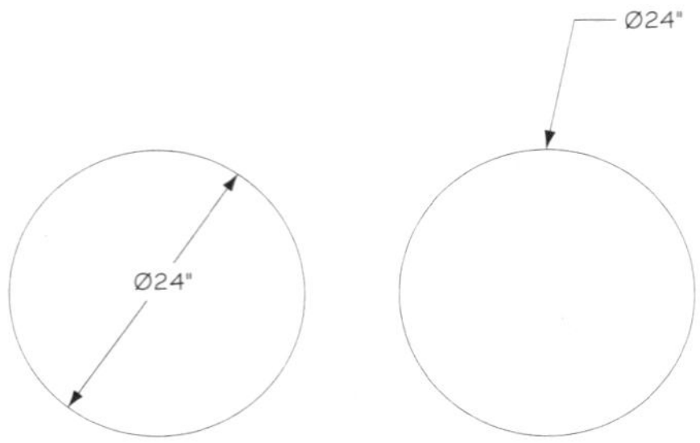

Figure 17.29 *The* **Dimdiameter** *command allows the diameter to be placed using two different methods.*

Command: **DDI** (ENTER) *(Or click the Diameter Dimension button.)*

DIMDIAMETER

Select arc or circle:

Dimension text <24.00>:

Specify dimension line location or [Mtext/Text/Angle]: *(Specify a point or enter an option.)*

Command:

When you select the circle to be dimensioned, the default dimension is dragged as you move the cursor. Depending on the size of the circle, the dimension toggles between being placed inside or outside the circle. Placing the dimension inside or outside the circle depends entirely on the size of the circle and the room available. Chapter 19 will discuss methods for controlling where the text is placed. Typing **M** (ENTER), **T** (ENTER), or **A** (ENTER) allows the text or angle value to be altered. Mtext, Text and Angle function exactly as discussed earlier for the **Dimlinear** command. Once the desired prompts have been responded to, the prompt will be returned for locating the dimension line.

CREATING A CENTER MARK

The **Center Mark** command will provide a center mark for a circle or an arc. This command, together with the **Dimcen** command, can be used to control the center mark. The **Dimcen** command allows the size of the mark to be altered. The command sequence is as follows:

Command: **DIMCEN** (ENTER)

Enter new value for DIMCEN <0'-0 1/16">: *(Provide desired value)* **6** (ENTER)

Command:

As the Command prompt is returned, nothing is changed. Selecting **Center Mark** on the **Dimension** toolbar or from the **Dimension** menu will now allow a circle or arc to be selected. The command sequence is as follows:

> Command: **DCE** (ENTER) *(Or click the Center Mark button.)*
>
> DIMCENTER
>
> Select arc or circle: *(Select desired circle to receive center mark.)*
>
> Command:

ANGULAR DIMENSIONS

The **Dimangular** command can be used to dimension angles formed by selecting two straight nonparallel lines, an arc, a circle, and another point or three points. You can start the command by selecting the **Angular Dimension** button on the **Dimension** toolbar, or by selecting **Angular** from **Dimension** menu. To enter the command at the keyboard, type **DAN** (ENTER) at the Command prompt. This will produce the following prompt:

> Command: **DAN** (ENTER) *(Or click the Angular Dimension button.)*
>
> DIMANGULAR
>
> Select arc, circle, line, or <specify vertex>:

Choosing Two Lines

One of the most common situations for dimensioning angles is describing the angle formed between two intersecting lines. Respond to the prompt by selecting a line. The command sequence continues with

> Select arc, circle, line, or <specify vertex>: *(Select first line.)*
>
> Select second line: *(Select second point.)*
>
> Specify dimension arc line location or [Mtext/Text/Angle]: *(Select location.)*
>
> Dimension text <44.00>:
>
> Command:

As the dimension arc line location (Mtext/Text/Angle) prompt is displayed, you will be able to choose the location of the dimension placement relative to the angle. Figure 17.30 shows alternatives for placing the angle dimensions.

Angles Based on an Arc

When you respond to the initial angle prompt by choosing an arc, dimensions can be placed to describe the tangent points of an arc. This could be especially useful when labeling subdivision maps or other large parcels of land. The command sequence is as follows:

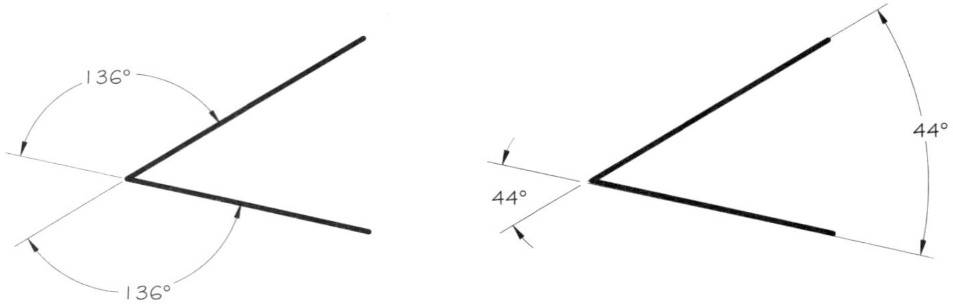

Figure 17.30 *Alternatives for placement of dimensions using the* **Dimangular** *command.*

Command: **DAN** (ENTER) *(Or click the Angular Dimension button.)*

DIMANGULAR

Select arc, circle, line, or <specify vertex>: *(Select first line.)*

Select second line: *(Select second point.)*

Specify dimension arc line location or [Mtext/Text/Angle]: *(Select location.)*

Dimension text <156.00>:

Command:

As you select the second line, the angle is displayed. Depending on the location of the cursor, the display can be located in one of four positions. The Text and Angle options are the same as with other dimensioning commands. Figure 17.31 shows an example of how text will be placed.

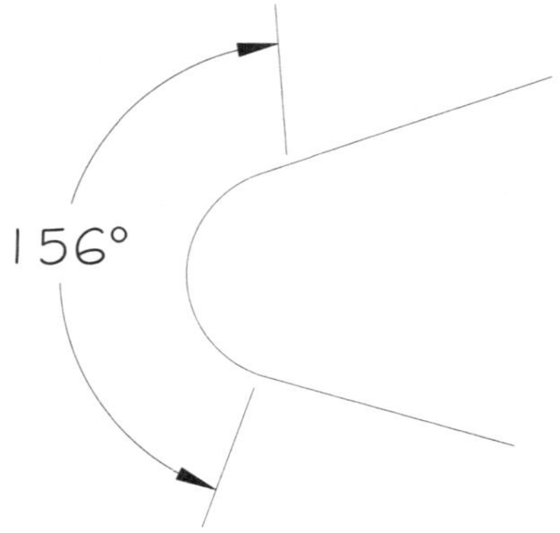

Figure 17.31 *Dimensioning of an arc using the* **Dimangular** *command.*

Angles Based on a Circle

When you respond to the initial angle prompt by choosing a circle, dimensions can be placed to describe angular patterns formed within a circle. This type of dimensioning is often used in specifying steel placement in circular columns and footings. The command sequence is as follows:

> Command: **DAN** (ENTER) *(Or click the Angular Dimension button.)*
>
> DIMANGULAR
>
> Select arc, circle, line, or <specify vertex>: *(Select desired object.)*
>
> Specify dimension arc line location or [Mtext/Text/Angle]: *(Select location.)*
>
> Dimension text <67.00>:
>
> Command:

Options for placing the dimension line can be seen in Figure 17.32.

Describing Three Points

When you respond to the original angular prompt with (ENTER), you will be allowed to select three different points that can be used to describe an angle. This option can be used to describe the centerpoint of bolts around a centerpoint in a steel connector strap. The two options for placing the angle can be seen in Figure 17.33. The command sequence is as follows:

> Command: **DAN** (ENTER) *(Or click the Angular Dimension button.)*
>
> DIMANGULAR
>
> Select arc, circle, line, or <specify vertex>: (ENTER)

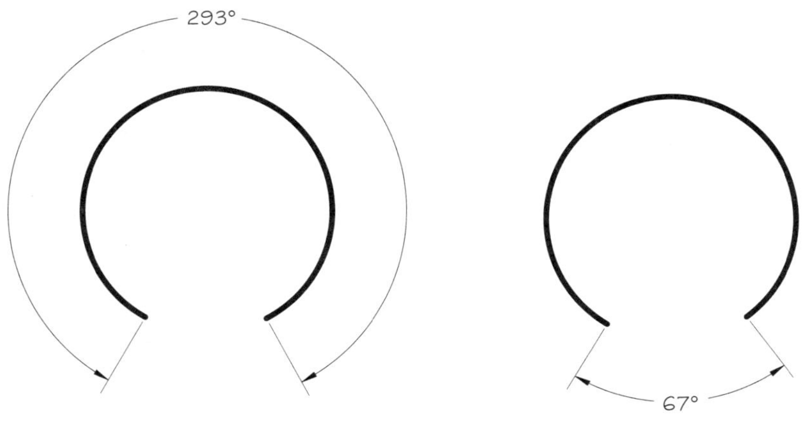

Figure 17.32 *Placement of angular dimensions using the circle option.*

Specify angle vertex: *(Select a circle center point.)*

Specify first angle endpoint: *(Select a point.)*

Specify second angle endpoint: *(Select a point.)*

Specify dimension arc line location or [Mtext/Text/Angle]: *(Select desired text location.)*

Dimension text <69>:

Command:

BASELINE DIMENSIONS

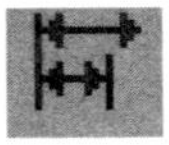

Occasionally, when the spacing of objects is extremely critical, the baseline system of placing dimensions can be used. Baseline dimensions (also called datum dimensions) assume one edge to be perfect and reference all dimensions back to that surface. Figure 17.34 shows an example of baseline dimensions. Although this type of dimensioning is not used in residential design, it is common in some areas of commercial construction where a high degree of accuracy must be achieved. The **Dimbaseline** command sequence must be started by using a linear, angular or ordinate existing dimension. Start the command by selecting the **Baseline Dimension** button on the **Dimension** toolbar or **Baseline** from the **Dimension** menu. The command can also be started by typing **DBA** (ENTER) at the Command prompt. AutoCAD will use the first extension line from the existing dimension as the base point for other dimensions. The command sequence is as follows:

Command: **DLI** (ENTER) *(Or click the Linear Dimension button.)*

DIMLINEAR

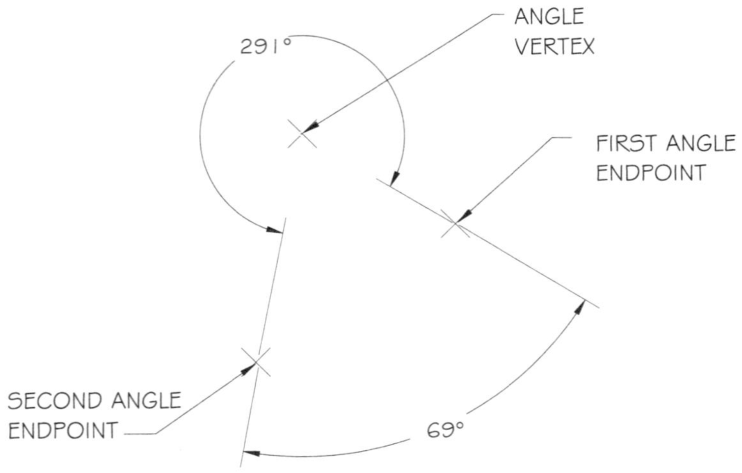

Figure 17.33 *Placement of dimensions using the three point option.*

Specify first extension line origin or <select object>: (ENTER)

Select object to dimension:

Specify dimension line location or [Mtext/Text/Angle/Horizontal/Vertical/Rotated]: *(Select the desired dimension location.)*

Dimension text <30'-0">:

Command:

Command: **DBA** (ENTER) *(Or click the Baseline Dimension button.)*

DIMBASELINE

Specify a second extension line origin or [Undo/Select] <Select>: *(Select base for second point.)*

Dimension text <50'-0">:

Specify a second extension line origin or [Undo/Select] <Select>: *(Select base point for the next dimension.)*

Dimension text <60'-0">:

Specify a second extension line origin or [Undo/Select] <Select>: *(Select base point for the next dimension.)*

Dimension text <70'-0">:

Specify a second extension line origin or [Undo/Select] <Select>: (ENTER)

Select base dimension: (ENTER)

Command:

CONTINUOUS DIMENSIONS

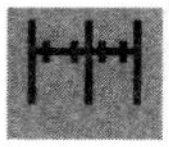

A series of dimensions is usually needed to dimension a structure. In Figure 17.1, dimension lines extend from the exterior walls, to interior walls, and finally to the exterior wall on the opposite side. These dimensions can be placed with the **Dimcontinue** command. The command sequence starts by using the **Dimlinear** command to place the first dimension. The **Dimcontinue** command must be based on a linear, angular or ordinate dimension. Start the **Dimcontinue** command by selecting the **Continue Dimension** button on the **Dimension** toolbar or by selecting **Continue** from the **Dimension** menu. You can also start the command by typing **DCO** (ENTER) at the Command prompt:

Command: **DCO** (ENTER) *(Or click the Continue Dimension button.)*

DIMCONTINUE

Select continued dimension: *(Select a dimension to continue from.)*

Specify a second extension line origin or [Undo/Select] <Select>:

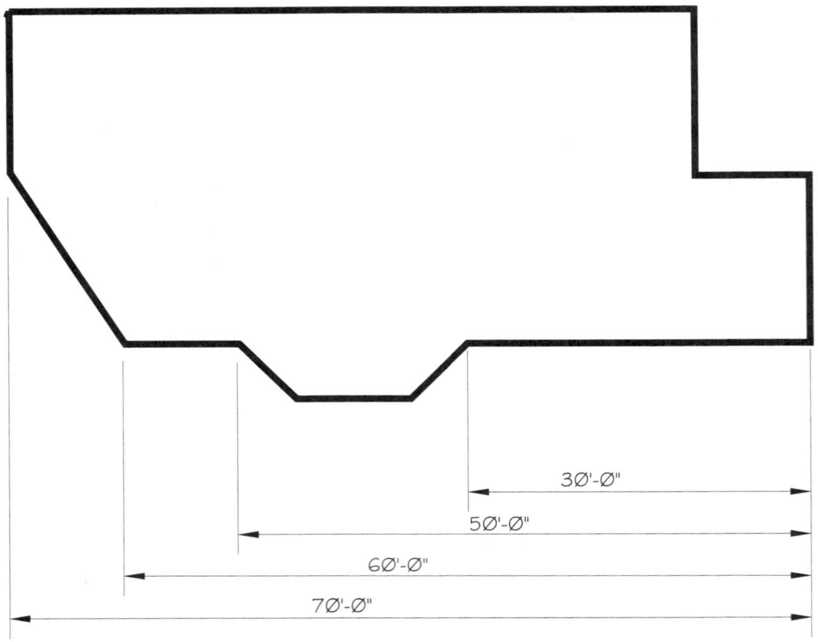

Figure 17.34 *Placement of dimensions using the* **Dimbaseline** *command.*

After you select the starting point for the second extension line, AutoCAD uses the second extension line of the **Dimlinear** command sequence as the first extension line for the **Dimcontinue** sequence. Press ESC to end the **Dimcontinue** command sequence. Figure 17.35 shows an example of placing dimensions using the **Dimcontinue** command with the default settings. The command sequence to place the dimensions across the bottom is as follows:

Command: **DLI** (ENTER) *(Or click the Linear Dimension button.)*

DIMLINEAR

Specify first extension line origin or <select object>: (ENTER)

Select object to dimension:

Specify dimension line location or [Mtext/Text/Angle/Horizontal/
Vertical/Rotated]: *(Select the desired dimension location.)*

Dimension text <10'-0">:

Command:

Command: **DCO** (ENTER) *(Or click the Continue Dimension button.)*

DIMCONTINUE

Specify a second extension line origin or [Undo/Select]: <Select>
(Select second point.)

Dimension text = 10'-0"

Specify a second extension line origin or [Undo/Select]: <Select>
(Select next point.)

Dimension text = 20'-0"

Specify a second extension line origin or [Undo/Select]: <Select>
(Select next point.)

Dimension text = 30'-0"

Specify a second extension line origin or [Undo/Select]: <Select>
(ENTER)

Select continued dimension: (ENTER)

Command:

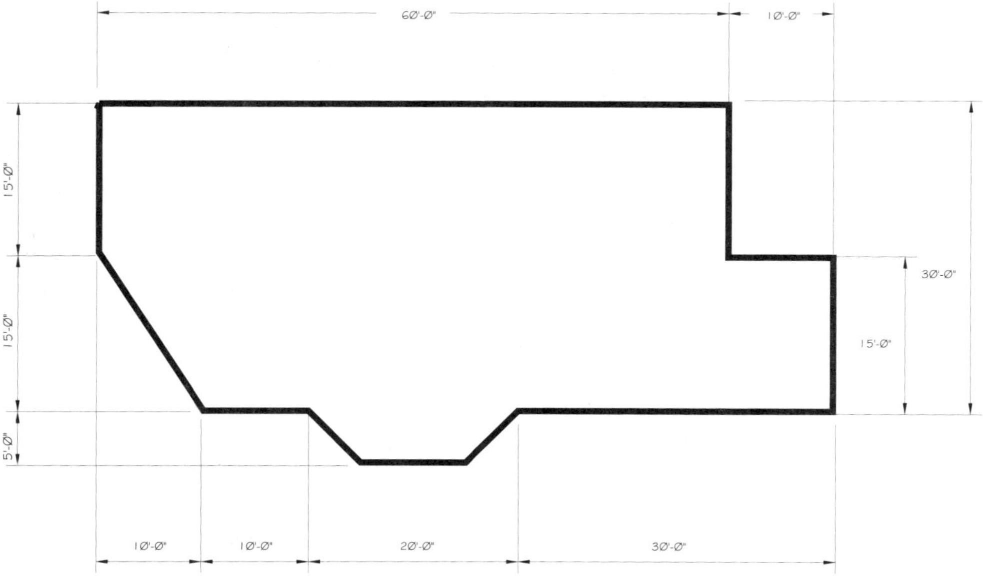

Figure 17.35 *Placement of dimensions using the* **Dimlinear** *and* **Dimcontinue** *commands. The 10'–0" dimension on the bottom left was placed using the* **Dimlinear** *command. The next 10'–0" dimension was placed using the* **Dimcontinue** *command. The* **Dimcontinue** *command uses the second extension line of the preceeding sequence as the first extension line of the* **Dimcontinue** *sequence.*

QUICK DIMENSIONS

You've been exposed to several different methods for placing dimensions throughout this chapter. Each works well for a specific need. AutoCAD provides an additional method for placing dimensions that combines features of many of the previous options. As the name implies, the **Quick Dimension** command allows a single dimension or a series of dimensions to be placed quickly. **Quick Dimension** excels in placing strings of continuous or baseline dimensions with very few prompts required. Start the command by selecting the **Quick Dimension** button on the **Dimension** toolbar, by typing **QDIM** (ENTER), or by selecting **Quick Dimension** from the **Dimension** menu. Each will produce the following prompt at the command line:

> Command: **QDIM** (ENTER) *(Or click the Quick Dimension button.)*
>
> Select geometry to dimension: *(Select object or objects to be dimensioned.)*

AutoCAD is waiting for you to select an object to be dimensioned. Press (ENTER) to end the selection process. The Command prompt will now change to display the following:

> Select Geometry to dimension: (ENTER)
>
> Specify dimension line position, or [Continuous/Staggered/Baseline/Ordinate/Radius/Diameter/datumPoint/Edit]
>
> <Continuous>: *(Select the desired dimension location.)*

If only one object is selected, the command will function similarly to the Select Objects option of the **Dimlinear** command. An object to be dimensioned is selected, followed by a location for the dimension to be placed. As (ENTER) is pressed, the dimension will be displayed, and can then be dragged into position as the cursor is moved. When the dimension is in the desired position, left-click to add the dimension to the drawing. The process can be seen in Figures 17.36a and 17.36b.

The best features of **Quick Dimension** can be seen if multiple objects are selected to be dimensioned. In Figure 17.37a, four objects have been selected to be dimensioned. Once the selection set is completed and (ENTER) is pressed, dimensions to describe each feature will be displayed. A dimension to describe each feature can be placed by identifying one location point. The process can be seen in Figures 17.37a and 17.37b. The command process to place the four dimensions is as follows:

> Command: **QDIM** (ENTER) *(Or click the Quick Dimension button.)*
>
> Select Geometry to dimension: *(Select object or objects to be dimensioned.)*

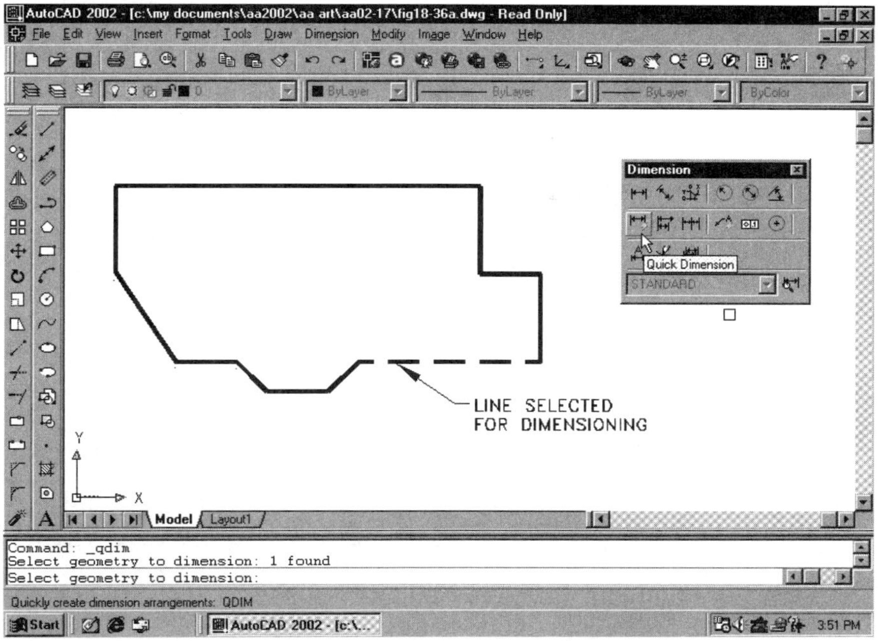

Figure 17.36a *The* **Quick Dimension** *command functions similarly to the Select Objects option of the* **Dimlinear** *command if only one object is selected to be dimensioned.*

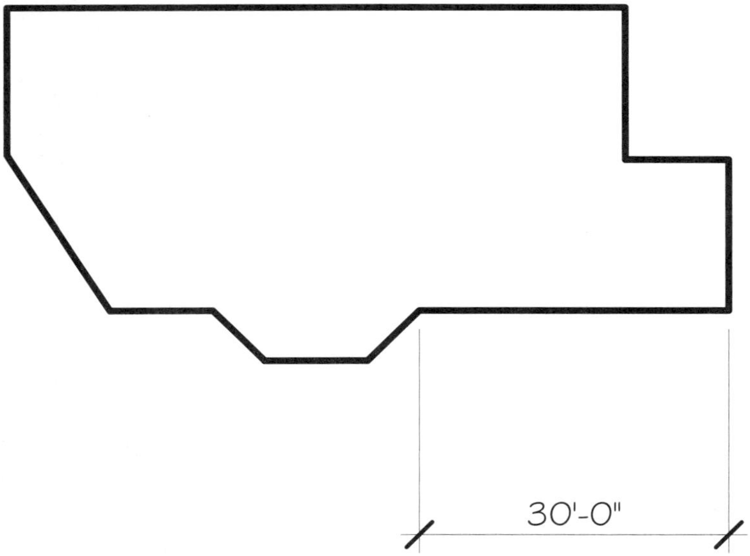

Figure 17.36b *With the* **Quick Dimension** *command, only an object and location are required to place a dimension.*

Select Geometry to dimension: 1 found,

Select Geometry to dimension: 1 found, 2 total

Select Geometry to dimension: 1 found, 3 total

Select Geometry to dimension: 1 found, 4 total

Select Geometry to dimension: (ENTER)

Specify dimension line position, or [Continuous/Staggered/Baseline/
Ordinate/Radius/Diameter/datumPoint/Edit]

<Continuous>: *(Select location to place the dimension string.)*

Command:

Note: Keep in mind that when defining a selection set with multiple objects, you need to hold down SHIFT to add selections. If you forget to hold SHIFT down, the objects to be dimensioned will be selected, but the previous selection will be deleted from the selection set.

QUICK DIMENSION PLACEMENT OPTIONS

The Baseline option of **Quick Dimension** combines the features of the **Dimbaseline** dimension option and the Continuous process to place multiple dimensions. The command functions in a method similar to the Continuous option in that each object must

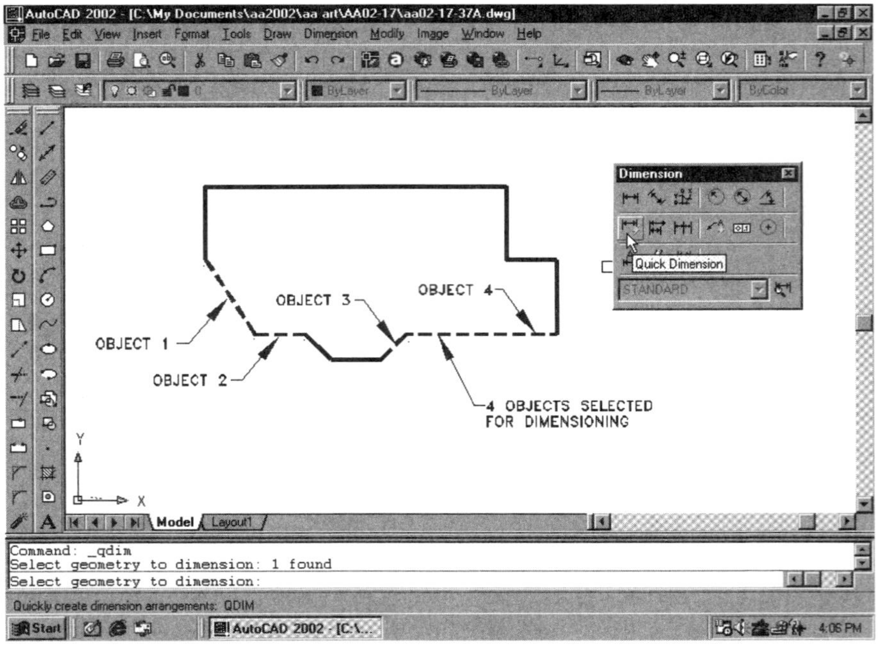

Figure 17.37a *The best features of* **Quick Dimension** *can be seen if multiple objects are selected to be dimensioned. In this example, four objects have been selected to be dimensioned.*

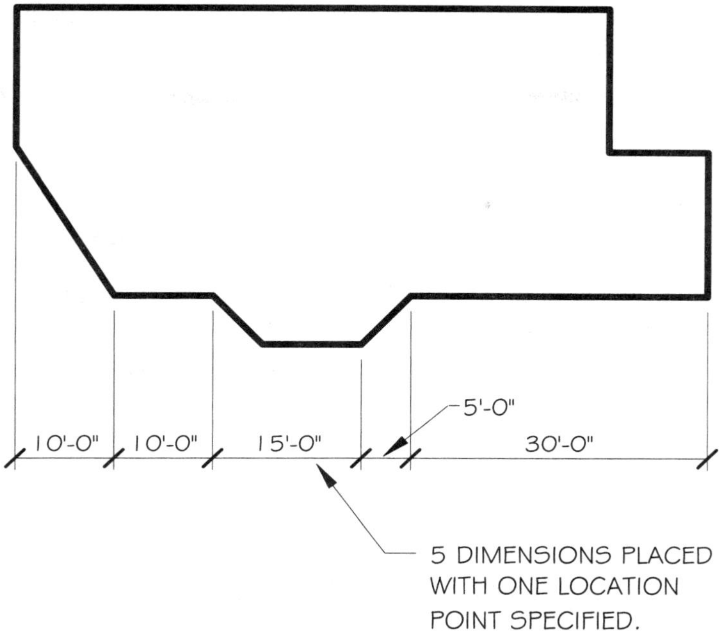

Figure 17.37b *A dimension to describe each feature is placed by identifying one location point.*

be selected, followed by the location. Figure 17.38 shows an example of placing dimensions using the Baseline option of **Quick Dimension**. The command sequence is as follows:

Command: **QDIM** (ENTER) *(Or click the Quick Dimension button.)*

Select Geometry to dimension: *(Select object or objects to be dimensioned.)*

Select Geometry to dimension: 1 found,

Select Geometry to dimension: 1 found, 2 total

Select Geometry to dimension: 1 found, 3 total

Select Geometry to dimension: 1 found, 4 total

Select Geometry to dimension: (ENTER)

Specify dimension line position, or [Continuous/Staggered/Baseline/Ordinate/Radius/Diameter/datumPoint/Edit]

<Continuous>: **B** (ENTER)

Specify dimension line position, or [Continuous/Staggered/Baseline/
Ordinate/Radius/Diameter/datumPoint/Edit]

<Baseline>: *(Select location to place the dimension string.)*

Command:

The Staggered option of **Quick Dimension** can be used to place staggered dimensions. Once objects to be dimensioned are selected, type **S** (ENTER) when prompted for an option. Figure 17.39 shows an example of staggered dimensions.

Typing **O** (ENTER) at the prompt allows ordinate dimensions to be placed using methods for selection similar to other **Quick Dimension** options.

The Radius and Diameter options of **Quick Dimension** allow you to select and dimension multiple circles or arcs of various sizes in one command sequence rather that having to repeat a dimension sequence for each object.

The datumPoint option will place dimensions in a pattern similar to the Ordinate option. The major difference between the two options is that datumPoint allows any point to be set as the base point. This option could be used to establish a finish floor level as 00, and allow dimensions to be placed above and below the base point (the finish floor). Figure 17.40 shows an example of dimensions placed using the datumPoint

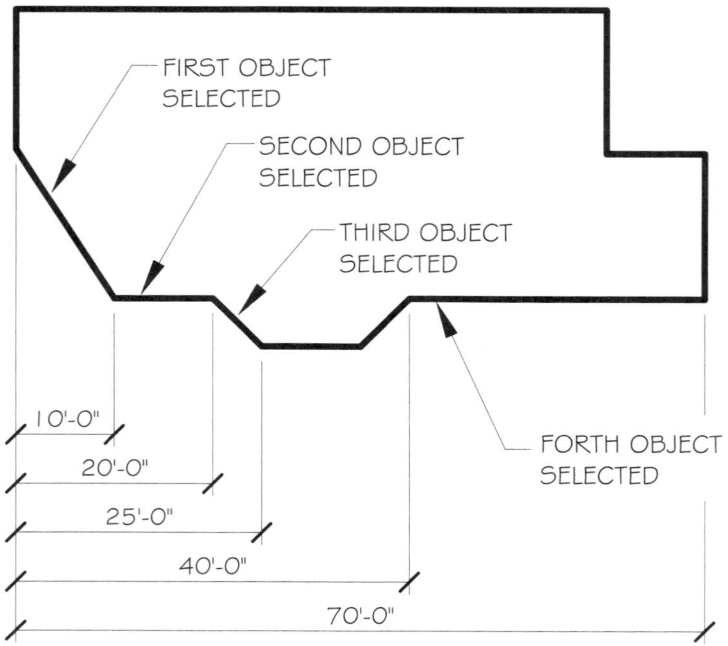

Figure 17.38 *The Baseline option of the* **Quick Dimension** *command functions similarly to the Continuous option in that each object must be selected, followed by the location.*

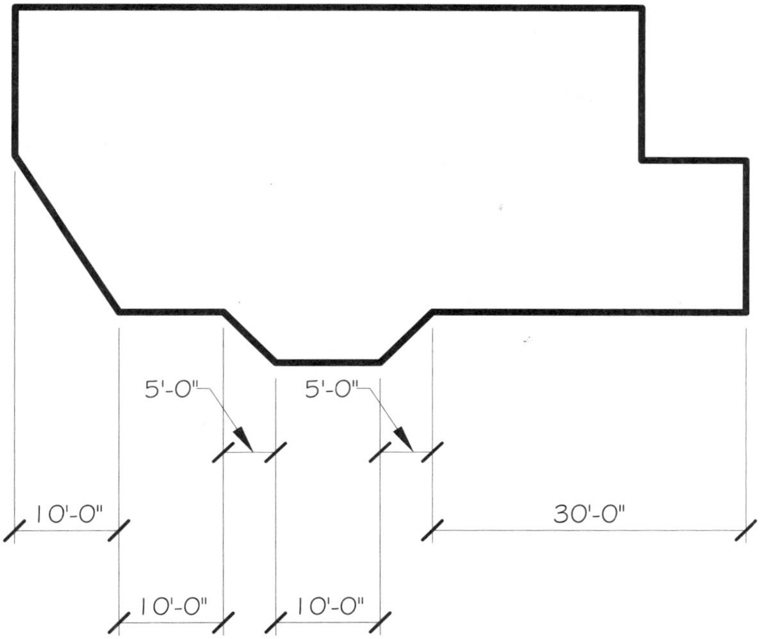

Figure 17.39 *Typing* **S** *(ENTER) when prompted for an option allows staggered dimensions to be placed.*

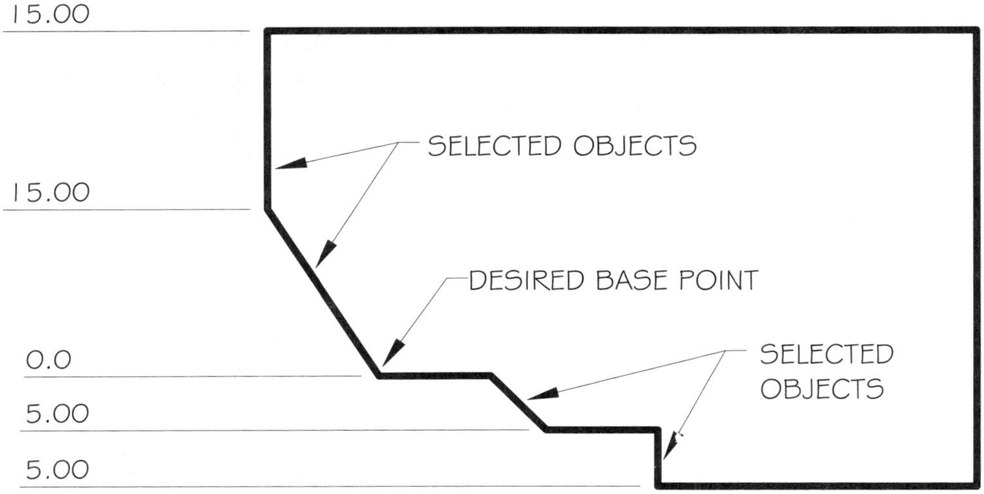

Figure 17.40 *The datumPoint option allows ordinate dimensions to be placed using any point to be set as the base point.*

option. The option is started by typing **P** (ENTER) at the option prompt. The command sequence to place datumPoint dimensions is as follows:

Command: **QDIM** (ENTER) *(Or click the Quick Dimension button.)*

Select Geometry to dimension: *(Select object or objects to be dimensioned)*

Select Geometry to dimension: 1 found,

Select Geometry to dimension: 1 found, 2 total

Select Geometry to dimension: 1 found, 3 total

Select Geometry to dimension: 1 found, 4 total

Select Geometry to dimension: (ENTER)

Specify dimension line position, or [Continuous/Staggered/Baseline/Ordinate/Radius/Diameter/datumPoint/Edit]

<Continuous>: **P** (ENTER)

Select new datum point: *(Select the point to be the base of the dimensions.)*

Specify dimension line position, or [Continuous/Staggered/Baseline/Ordinate/Radius/Diameter/datumPoint/Edit]

<Ordinate>: *(Select location to place the dimension string.)*

Command:

LEADER LINES

The **QLeader** command can be used to place a leader line to connect notes or dimensions to a specific portion of the drawing. Figure 17.41 shows examples of the use of leader lines. Leaders are generally placed at any angle except horizontal or vertical. (Some architectural offices use horizontal and vertical leaders with a custom arrow to duplicate manual practice.) A horizontal leg is placed at the end of the leader line to tie the text to the leader. The horizontal leg should extend from the left side of the first line of text or from the right side of the last line of text. Figure 17.42 shows common methods of placing leader lines.

USING QUICK LEADERS

The **Qleader** command should be used when several local notes need to be referenced to the drawing, or when you want to control the width of the space where the notes will be placed. Start the command by selecting the **Quick Leader** button on the **Dimension** toolbar, by typing **LE** (ENTER) at the Command prompt, or by selecting **Leader** from the **Dimension** menu. Each will produce the prompt:

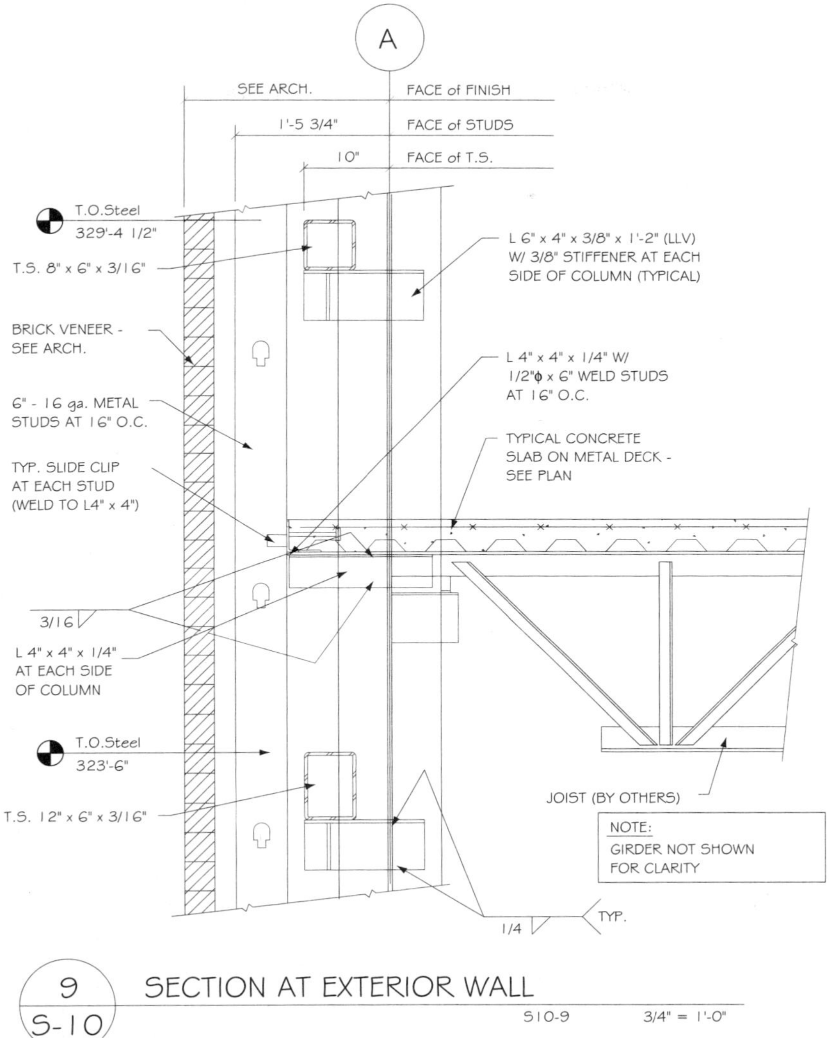

Figure 17.41 *Leader lines can be used to attach text, dimensions and welding symbols to a specific portion of the drawing. (Courtesy Van Domelen/Looijenga/McGarrigle/Knauf Consulting Engineers.)*

Command: **LE** (ENTER) *(Or click the Quick Leader button.)*

QLEADER

Specify first leader point, or [Settings] <Settings>: *(Select desired start point for leader.)*

Specify next point: *(Select desired end point of leader.)*

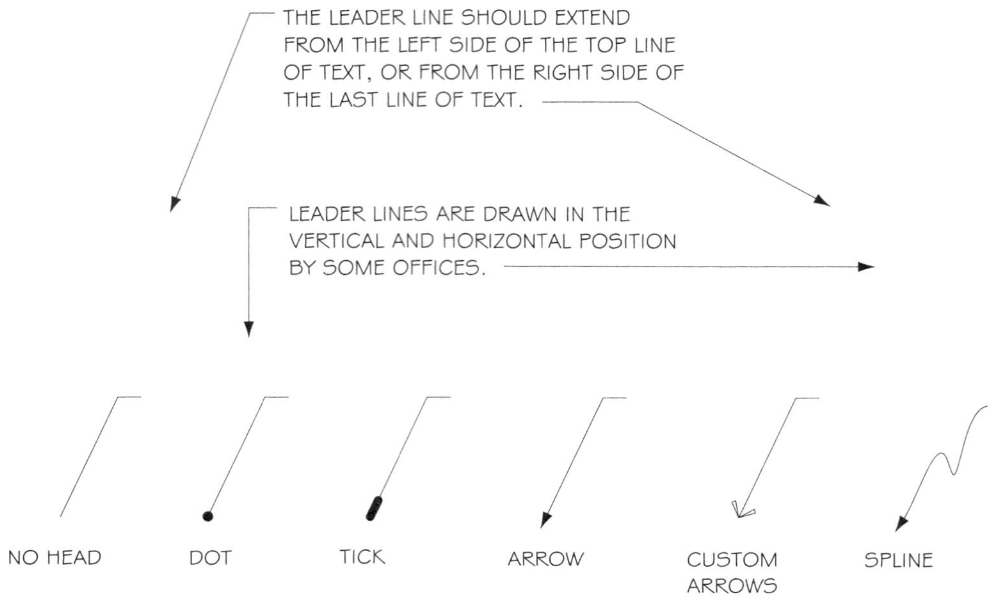

Figure 17.42 *Common methods of placing and terminating leader lines.*

Specify next point: *(Select desired end point for horizontal leader tail.)*

Specify text width <0">: (ENTER)

Enter first line of annotation text <Mtext>:

The first option to be considered is Settings; it allows several features of the leader to be altered. Settings will be explored after the default option of **Qleader** has been explored. For now, select a starting point for the leader. The "first" and "next" point prompts are like those of the Line command. AutoCAD will place a line with the tip of an arrow extending from the "first" point to the "next" point. The Settings option allows alternatives to the arrow to be used. A prompt for a second "next" point is also provided. As shown in Figure 17.43, this second "next" point produces a horizontal line.

Text Width

Once the location of the leader is specified, you will be prompted to provide the text width. The Text Width option allows the space to be specified where the annotation will be placed. The default setting for width is 0". A line of text will be of unspecified length with this setting. Text will be placed in one line until (ENTER) is pressed to start a new line of text. If a value is provided for the text width, that value will specify the width of each line of text. Typing **4** (ENTER) will allow the text line to be 4" long. Once the specified width has been achieved, a new line of text will be started automatically.

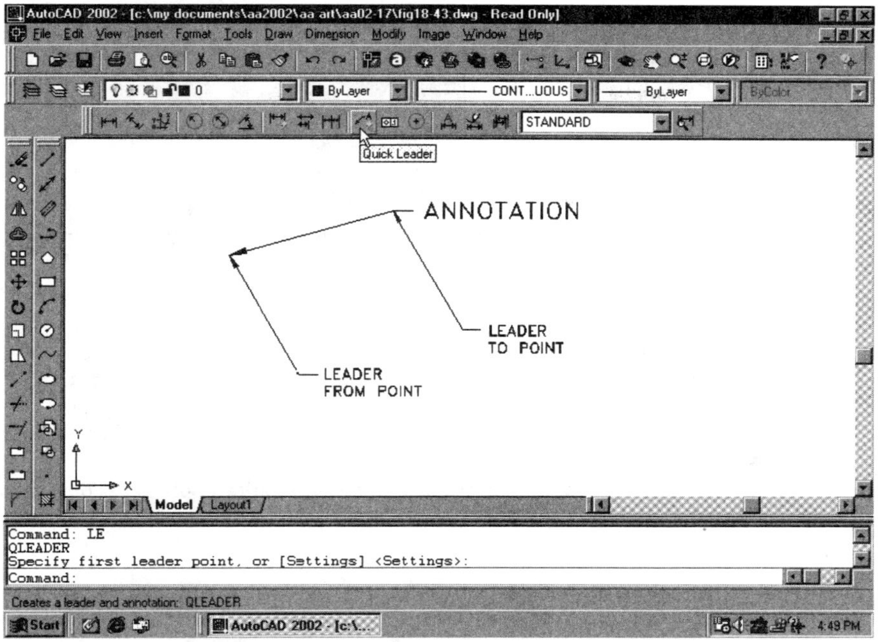

Figure 17.43 *When* **Qleader** *is used, the "first" and "next" point prompts are like those of the* **Line** *command.*

Entering Annotation

The final prompt of the **Qleader** command is

Enter first line of annotation text <Mtext>:

Three options are available for entering text beside the leader. Pressing ESC will place the leader line and close the command without placing any text beside the leader. This option works well if stock notes have been stored in a drawing template, and they just need to be moved into the proper position. The notes can be placed and then leader lines can be placed to reference the note to the drawing.

A second option as the text prompt is displayed is to begin entering the text at the Command prompt. As text is entered at the keyboard, it will not be displayed in the drawing area. Figure 17.44 shows how text will be displayed on the Command prompt. Pressing (ENTER) will display a prompt for additional lines of text. When the desired text has been entered, press (ENTER) to end the command and display the text. Figure 17.45 shows the leader and the placement of the text.

The third option of placing text at the leader line is to press (ENTER) at the text prompt. This option allows multiple lines of text, numbers or symbols to be placed at the end

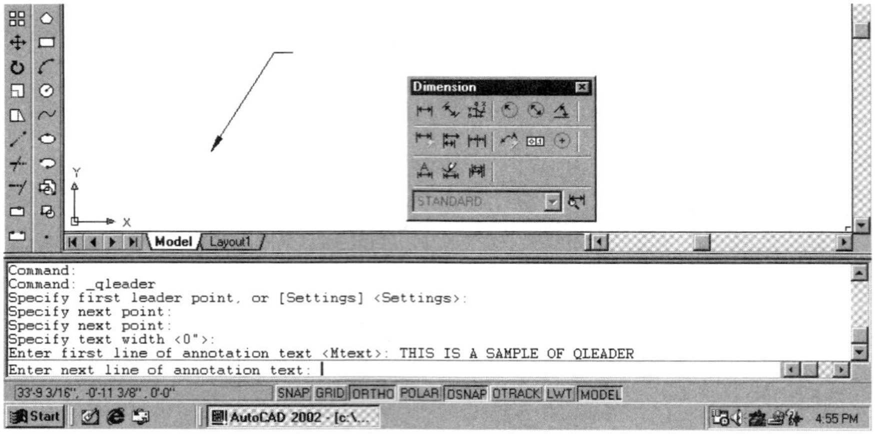

Figure 17.44 Qleader *allows the text to be entered at the Command prompt. As text is entered at the keyboard, it will be displayed at the Command prompt and not beside the leader line.*

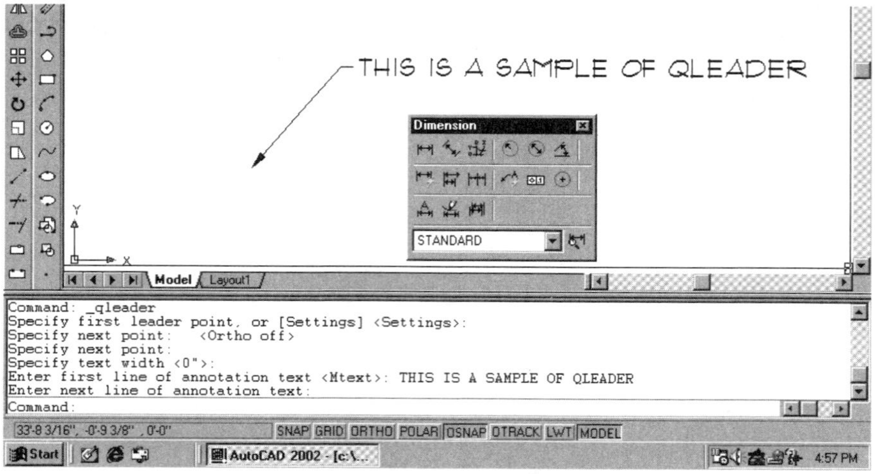

Figure 17.45 *When the text has been entered, pressing* ENTER *will close the command and display the text beside the leader line.*

of the leader line using the **Multiline Text Editor**. Text entered by keyboard will be displayed in the edit box and will not be placed by the leader line until the **OK** button is clicked. Figure 17.46 shows an example of the edit box. The **Multiline Text Editor** was explored as multiline text was examined. See Chapter 16 for a review of the **Multiline Text Editor**.

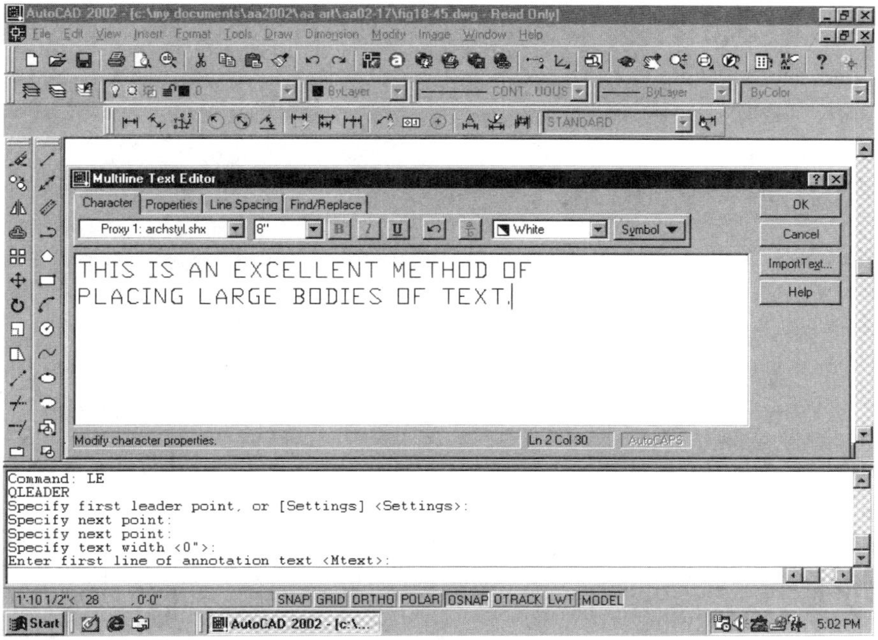

Figure 17.46 *Pressing* ENTER *at the text prompt allows text, numbers, or symbols to be placed at the end of the leader line using the* **Multiline Text Editor** *dialog box. Text entered by keyboard will be displayed in the edit box and not placed beside the leader line until the* **OK** *button is clicked.*

ADJUSTING QLEADER SETTINGS

Earlier you were introduced to the Settings option of the **Qleader** command. Rather than adjust the settings, methods of placing text were explored. The Settings option will present controls for annotation, leader line and arrow types, and methods of attaching the leader to the leader lines. **Qleader** settings can be altered by pressing (ENTER) at the first prompt.

> Command: **LE** (ENTER) *(Or click the Quick Leader button.)*
>
> QLEADER
>
> Specify first leader point, or [Settings] <Settings>: (ENTER)

Pressing (ENTER) will produce the **Leader Settings** dialog box shown in Figure 17.47.

Annotation Tab

The **Annotation** tab allows controls for **Annotation Type**, **MText options**, and **Annotation Reuse**.

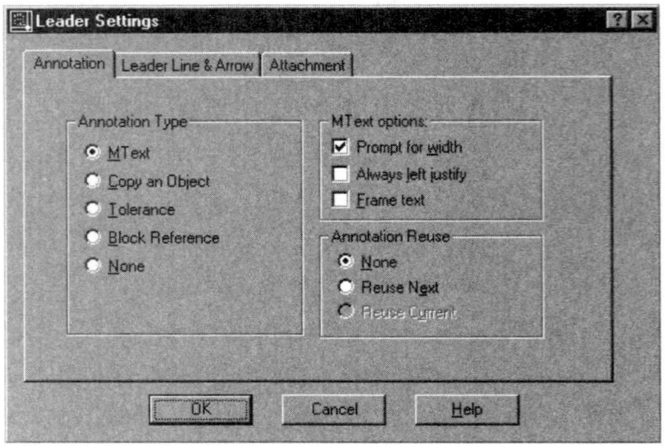

Figure 17.47 *Pressing* ENTER *at the first* **Qleader** *prompt will produce the* **Leader Settings** *dialog box.*

Annotation Type

The default **Annotation Type** setting of this tab will place text using **Mtext**. Remember, it's the default setting because it is meets a large variety of needs. Other options include the following:

The **Copy an Object** option allows existing text to be copied to the new leader line. With this option active, once the leader line is placed, a prompt to select text to be copied is displayed. As the text body is selected, it will automatically be copied to the new location. This option could be used on a drawing such as the exterior elevations, where the same material is often referenced in each elevation. The **Copy** command will do the same thing.

The **Tolerance** option will display the **Tolerance** dialog box and allow the creation of a feature control frame. These are tolerance symbols that are used to control the accuracy of a dimension in relation to a specific surface or condition. Tolerances are rarely added to architectural dimensions.

The **Block Reference** option allows predefined groups of text to be inserted by the leader line. Blocks will be introduced in later chapters.

The **None** option will place a leader with no text. Uses for this option have been introduced earlier in this chapter. The benefit of using this option is that it quickly produces the leader line with no prompt for text and no need to remember if you press ESC or (ENTER).

MText Options

The **MText** options area is activated only if **MText** is the selected annotation type. The **Prompt for width** check box, if deactivated, can be used to suppress the width prompt. If you're placing text in a drawing and don't care how long the lines of text are, suppress the prompt. The **Always left justify** check box allows you to specify aligned text. Most professionals pride themselves on their neatness, and aligned text will add to a professional appearance. You've explored justification as multiline text was introduced. Further options will explore methods of altering text to leader line relationship. An option that you should explore is **Frame text**. With this option active, text will be displayed as seen in Figure 17.48.

Annotation Reuse

This portion of the **Annotation** tab sets options for reusing text. The default option is not to reuse text at a leader line. The **Reuse Next** option will reuse the text that is created with the current command sequence for all future leaders. The current text will be placed by each leader until the setting is altered. The **Reuse Current** option will reuse current annotation. With this option active, AutoCAD will automatically select this option when you reuse text after selecting the **Reuse Next** option.

Leader Line & Arrow Tab

The **Leader Line & Arrow** tab contains options that can be used to control the appearance of the leader line and arrow. The tab can be seen in Figure 17.49.

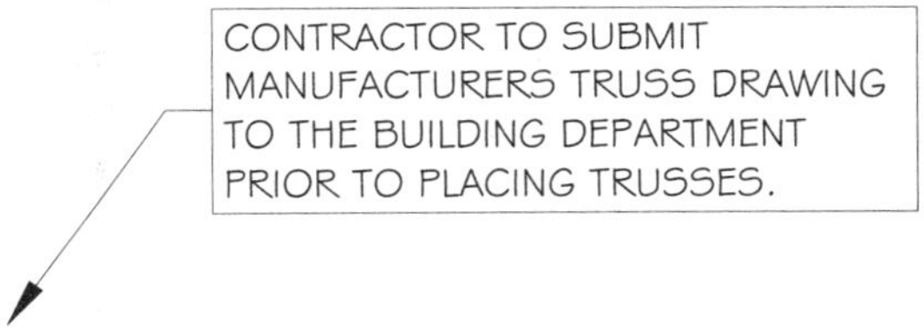

Figure 17.48 *The* **Frame text** *option will enclose annotation text in a box. The option provides an excellent method of drawing attention to important notes.*

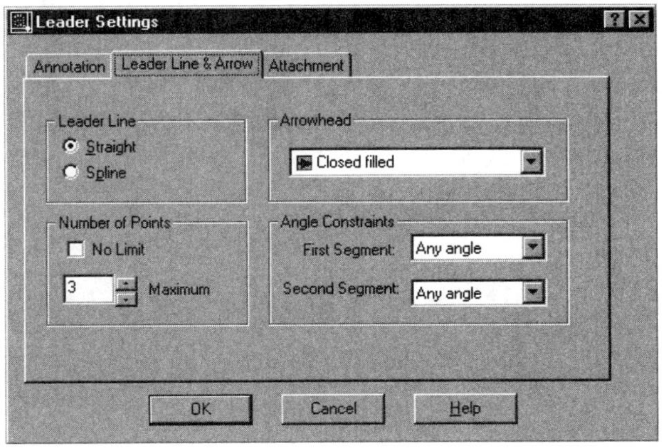

Figure 17.49 *The* **Leader Line & Arrow** *tab of the* **Leader Settings** *dialog box.*

Leader Line

Two options are available for placing the leader line. Throughout this chapter, each example has been created using the **Straight** option. Figure 17.50 shows an example of a **Spline** leader line.

Number of Points

This option sets the number leader points that you will be prompted to provide. With the **No Limit** option selected, prompts for points will be provided until (ENTER) is pressed twice. Setting a number for the maximum number of prompts will limit the number of prompts displayed as you proceed through the command. Set the value for one more than the number of leader lines you desire. In the default setting of 3, you will be prompted to place text after specifying the two lines formed by the three points.

Note: Adjusting the number of points will be a powerful tool when you create **Spline** leader lines.

Arrowhead

This portion of the tab allows the leader line terminator to be set. Figure 17.51 shows the options that are available. These options can be used to replace the filled arrowhead. To replace the filled arrow, select the desired terminator. As the selection is made, the list will be closed and the **Leader Line & Arrow** tab will be displayed. Figure 17.52 shows a detail with Right Angle arrows. Notice that the last option in the list is User Arrow. This option will display the **Select Custom Arrow Block** dialog box that can be used to select predefined arrows that have been saved as a block. Chapter 19 will introduce the **Block** command.

Angle Constraints

These options will allow the angle of each segment of the leader line to be controlled. Values for each option include Any angle, Horizontal, 90, 45, 30 and 15 degrees. Options are provided for the first and second lines of a leader line. Using the Any angle option for the **First Segment** provides freedom in placing text around a drawing. Selecting an angle provides uniformity in appearance, but it might make placing text more challenging. Specifying the Horizontal option for the **Second Segment** will provide a horizontal tail without requiring ORTHO to be active.

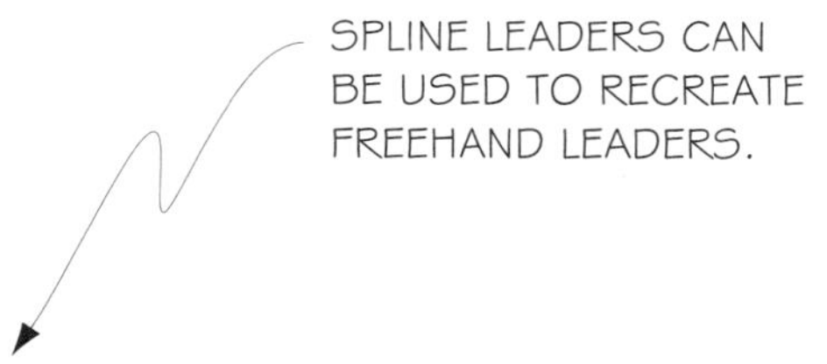

Figure 17.50 *A spline leader line.*

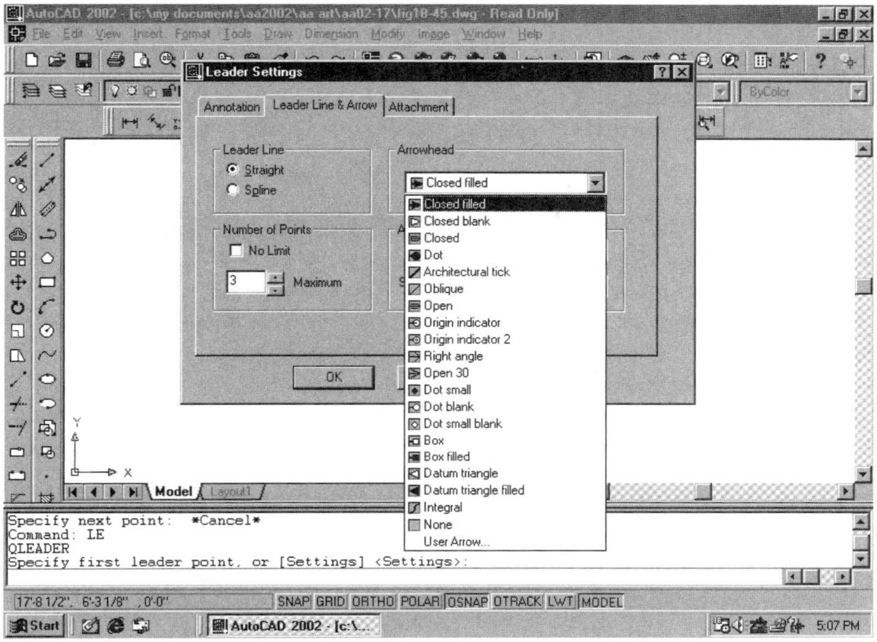

Figure 17.51 *Options that are available for leader line terminators.*

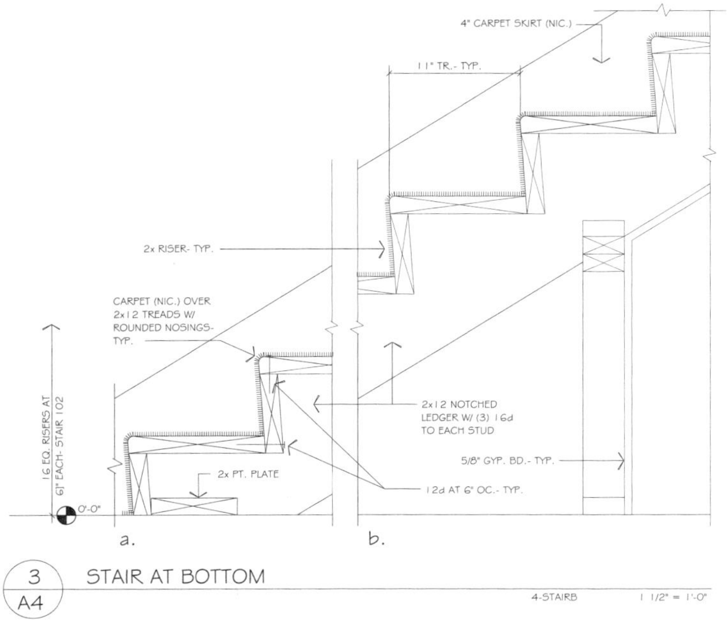

Figure 17.52 *Many architectural firms use right angle arrows. (Courtesy Barrentine, Bates and Lee, AIA.)*

Attachment Tab

The **Attachment** tab shown in Figure 17.53 can be used to set the attachment of multiline text and the leader line. Settings are available to control the leader to text relationship depending on which side of the text the leader is placed. Options can be seen in Figure 17.54.

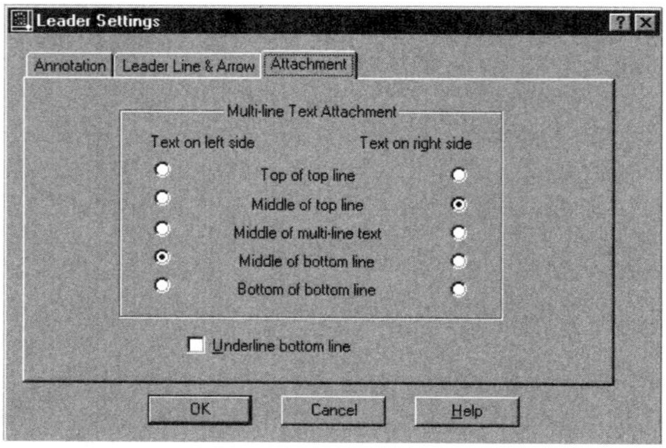

Figure 17.53 *The* **Attachment** *tab can be used to set the attachment of multiline text and the leader line.*

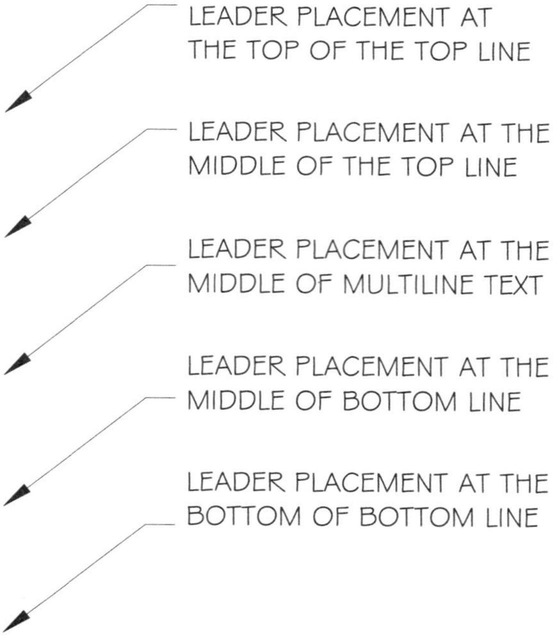

Figure 17.54 *Options for placing the leader line in relationship to the text.*

CHAPTER 17 EXERCISES

1. Start a new drawing and set the units to architectural with 1/16" accuracy. Draw a hexagon with a horizontal top and base inscribed in a 4" diameter circle. Use the **Dimlinear** command to provide overall dimensions. Save the drawing as E-17-1.
2. Open drawing E-17-1. Change the hexagon to a layer titled BASE. Change the dimensions to a layer titled DIMEN and turn the layer off. Create a layer titled ANGLE and make it current. Provide dimensions to indicate the angle of the sides. Save the drawing as E-17-2.
3. Open drawing E-6-9 and dimension objects a, c, e, and g using two-place decimal dimensions. Place the remaining objects on a layer titled EXTRA and freeze the layer. Once the dimensions are complete, freeze the existing dimensions and change the units to architectural with 1/16" accuracy. Dimension the objects again using alternatives to the methods already used. Save the drawing as E-17-3.
4. Open drawing E-9-3 and scale the drawing using a scale factor of .5. Create a layer for the dimensions. Completely dimension the drawing so the steel beam can be fabricated. Save the drawing as E-17-4.
5. Open drawing E-9-5 and create a layer titled DIM1. Rescale the drawing by a factor of .25 and completely dimension the drawing so that the metal hanger can be fabricated. Save the drawing as E-17-5.
6. Open drawing E-15-4 and create a layer titled DIM1. Change the lines representing the one-story footing to .5 wide polylines. Place the two-story footing on a new layer and set that layer to OFF. Rescale the drawing using a scale factor of .25. Completely dimension the one-story footing and save the drawing as E-17-6.
7. Start a new drawing and draw the following object. Starting at @ 2,2:

   B. @ 1.5,0
   C. @ 0,1
   D. @ 1.25,0
   E. @ .75,.75
   F. @ 1.5,0
   G. @ 0,.75
   H. @ 1.5,0
   J. @ 0,1.75
   K. @ –2,0
   L. @ 0,–.75
   M. @ –2,0
   N. @ 0,–1.25
   P. @ –2.5,0
   Q. @ .5,–1.5
   R. AND THEN CLOSE.

   Change the line segments to polylines. Completely dimension the drawing, using the **Dimcontinue** command where possible. Save the drawing as E-17-7.

8. Open drawing E-17-7 and create a new layer for the existing dimensions. Change the dimensions to the new layer that was just created, and set it to OFF. Create another layer for dimensions and completely dimension the object using baseline dimensioning. Assume the lower left corner as the starting point for horizontal and vertical dimensions.

9. Open the architectural template, and draw an object similar to Figure 17.35. Create a layer for placing dimensions. Dimension every surface using the Continuous option of **Quick Dimension**. Save the drawing as E-17-9.

10. Open drawing E-17-9 and create a new layer for dimensions. Dimension each surface using the Baseline option of **Quick Dimension**. Save the drawing as E-17-10.

11. Open drawing E-17-10 and create another layer for dimensions. Dimension each surface using the Staggered option of **Quick Dimension**. Save the drawing as E-17-11.

12. Use Figure 14.15 as a guide and draw a detail for a concrete block wall. Use varied lineweight and colors to represent the CMUs, steel and mortar. Create separate layers for CMU, steel, mortar, dimensions, text, and detail reference bubble. Do not place any dimension on the drawing. Place leaders so that text will be placed using an architectural font. Assume that the drawing will be plotted at a scale of 3/4"=1'–0". Save the drawing as E-17-12.

13. Use Figure 16.5 as a guide and draw a detail for a roof to wall connection. Use varied lineweight and colors to represent materials that have been cut by the cutting plane. Create separate layers for drawing, dimensions, and text. Do not place any dimensions on the drawing. Place leaders so that text will be placed using an architectural font in 4.5" wide columns. Use right angle arrows. Assume that the drawing will be plotted at a scale of 1/2"=1'–0". Save the drawing as E-17-13.

14. Use Figure 11.1 as a guide and draw detail 5/S-7. Use varied lineweight, colors, and layers to represent the concrete, steel, dimensions, text, and detail reference bubble. Do not place any dimension on the drawing. Place leaders so that text will be placed using an architectural font. Use right angle arrows. Assume that the drawing will be plotted at a scale of 1"=1'–0". Save the drawing as E-17-14.

15. Use Figure 11.2 as a guide and draw detail of a truss to wall connection. Use varied lineweight, colors, and layers to represent the concrete, steel, dimensions, text, and detail reference bubble. Do not place any dimension on the drawing. Place leaders so that text will be placed using an architectural font. Set the angle control for the first leader line segment to 90° and use a right angle arrow. Assume that the drawing will be plotted at a scale of 1"=1'–0". Save the drawing as E-17-15.

CHAPTER 17 QUIZ

1. What four components comprise dimensioning?
2. How should text be placed on architectural drawings to ease the job of the print reader?
3. List four options typically used for the intersections of dimension and extension lines.
4. Describe how to place dimensions relative to a concrete wall.
5. List four groups of commands that relate to placing dimensions on a drawing.
6. Describe two methods of locating extension lines for **Dimlinear** dimensions.
7. List four methods to describe angular dimensions.
8. Open one of the drawing exercises that were completed for this chapter, and provide the name of the layer that was added to the drawing base by AutoCAD.
9. List three methods of placing linear dimensions.
10. Explain the difference between aligned and rotated dimensions.
11. What is the first prompt for the **Dimlinear** command asking for?
12. Should the default text value always be accepted? Explain your answer.
13. What does the Angle option alter in the **Dimlinear** command?
14. What is the effect of pressing (ENTER) at the first **Dimlinear** command prompt rather than selecting a point?
15. Should a dimension be placed inside or outside a circular object?
16. What is the process for placing a 6" center mark in the center of a circular steel tube?
17. List and explain the options for leader annotation.
18. Four lines of text need to be placed by a leader line. What option should be used?
19. Explain the options for **Dimordinate** dimensioning.
20. List five options of leader lines, besides the text, that can be altered.

CHAPTER 18

Placing Dimensions on Drawings

In Chapter 17 you were introduced to dimension requirements of different drawings and the methods of describing linear, angular, circular, continuous, and baseline dimensions. Each system was described using the default values. If all of your drawings could fit on a computer screen, everything would be fine. To allow for drawings of various sizes, the spacing of extension and dimension lines and the size of text and arrows must be altered. In this chapter you'll be introduced to

- Altering dimension variables
- Establishing dimension styles
- Editing existing dimensions

Commands to be explored include

- **Dimstyle**
- **Dimoverride**
- **Dimedit**
- **Dimtedit**
- **Dimension Update**

CONTROLLING DIMENSION VARIABLES WITH STYLES

As text was explained, you were introduced to the **Style** command. **Style** allows you to group variables such as the font, text height, width, oblique angle, and orientation so that several different styles of text can easily be created within one drawing. AutoCAD also allows dimension styles to be created within a drawing to meet the needs of various situations. To save time, individual styles as well as each dimensioning variable should be set on template drawings on the ANNO-DIMS layer.

INTRODUCTION TO DIMENSION STYLES

A dimension style is a set of dimension variables that control various aspects of placing the extension lines, dimension lines, and annotation in relation to these lines and line terminators. The type of drawing you are working on will affect how the variables are to be set.

- Dimensions that are placed on a site plan are often written in engineering units.
- Dimensions that locate the property lines are placed with no extension or dimension lines.
- Dimensions placed to locate utilities in an easement are placed using baseline dimensions expressed in feet and inches.
- Dimensions written on architectural drawings are written in feet and inches using continuous dimensions. Dimensions on elevations and sections can be placed using ordinate dimensioning.

You can create styles that incorporate the requirements for each of these dimensions and save them in a template drawing. You will be introduced to creating various styles later in this chapter.

Before a style can be created, you must have a thorough understanding of the dimension variables. Dimension variables are the qualities that control how dimensions will appear. You will be introduced to dimensioning variables through the default style of Standard. Later you'll be introduced to methods of creating specific styles to meet the needs of each drawing type.

CONTROLLING DIMENSION VARIABLES

Dimension variables are altered using the **Dimension Style Manager** dialog box, displayed in Figure 18.1. Access the dialog box by selecting **Dimension Style** on the Dimension toolbar, by selecting **Dimension Style** from the **Format** menu, by selecting **Style** from the **Dimension** menu or by typing **D** (ENTER) at the Command prompt.

DIMENSION STYLE MANAGER

The main areas of the **Dimension Style Manager** dialog box include the **Styles** box, a **Preview** window, and the buttons **Set Current**, **New**, **Modify**, **Override**, and **Compare**.

Current Dimension Style

Notice that in Figure 18.1 the current style is named Standard, the name of the default settings for AutoCAD. As styles are created, the name of each style will be listed in the **Styles** box. The current style will be highlighted. Once multiple styles have

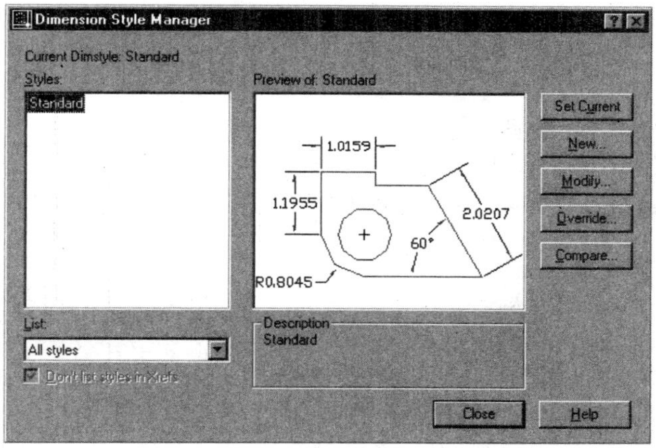

Figure 18.1 *Dimension variables are created and altered using the* **Dimension Style Manager** *dialog box. Access the dialog box by selecting* **Dimension Style** *on the* **Dimension** *toolbar, by selecting* **Dimension Style** *from the* **Format** *menu, by selecting* **Style** *from the* **Dimension** *menu, or by typing* **D** *(ENTER) at the Command prompt.*

been created, a different style can be made current by highlighting the name of the style and then selecting the **Set Current** button.

List

The **List** box displays the options that control the display of dimension styles. Options include All Styles and Styles in Use. With All Styles selected, all of the dimension styles contained in the drawing will be displayed. With Styles in Use selected, the display will display only dimension styles that are referenced by dimensions in the drawing.

Just below the **List** box is the check box for displaying dimension styles in externally referenced drawings. XREF drawings will be discussed in future chapters.

Control Buttons

Five buttons are available for creating and altering the dimension style. Each will be introduced now and explored in detail throughout the chapter.

Set Current

This option has already been briefly introduced. Once a style is created, you can set it as the current style by highlighting the style and selecting the **Set Current** button. Each of the settings of the drawing style will be applied to the new dimensions.

New

Selecting the **New** button will display the **Create New Dimension Style** dialog box shown in Figure 18.2. The box is used to name new dimension styles, to select the style that will be used as a base to build the new style, and indicate the dimension types that the new style will be applied to. Components include the following:

New Style Name—By default, AutoCAD will assign Copy of Standard as the new style name. With the name highlighted, change the style name to provide a useful description.

Start With—This list box determines what will be used as the base to form the new style. Rather than starting from scratch, AutoCAD will use the Standard style as a base and allow these options to be adjusted to meet the drawing needs. As new styles are created, each style name will be listed, allowing it to be selected to be the base for a new style.

Use for—Selecting the **Use for** edit arrow will display the list shown in Figure 18.3, which allows a style to be created that will be applied only to certain types of dimensions. This option could be used if you would like to use the Standard style with black text as a base but allow all radius dimensions to be red. Selecting Standard as the **Start With** option, and Radius Dimensions as the **Use** for selection will begin the process of creating a sub-style of the Standard style. With these options active, each time the Standard style is used radius dimensions will be red, and all other dimensions will be black.

Continue—Selecting the **Continue** button will display the **New Dimension Style** dialog box. The display will be explored in the next section of this chapter.

Modify

Selecting the **Modify** button in the **Dimension Style Manager** dialog box allows an existing style to be modified. As the **Modify** button is selected, the **Dimension Style Manager** dialog box is removed and the **Modify Dimension Style** dialog box is displayed. The dialog box is identical to the **New Dimension Style** dialog box. Methods of altering an existing style will be explored later in this chapter.

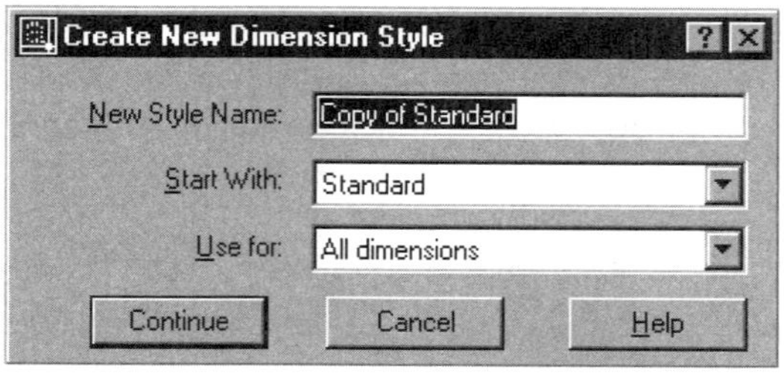

Figure 18.2 *Selecting the* **New** *button will display the* **Create New Dimension Style** *dialog box.*

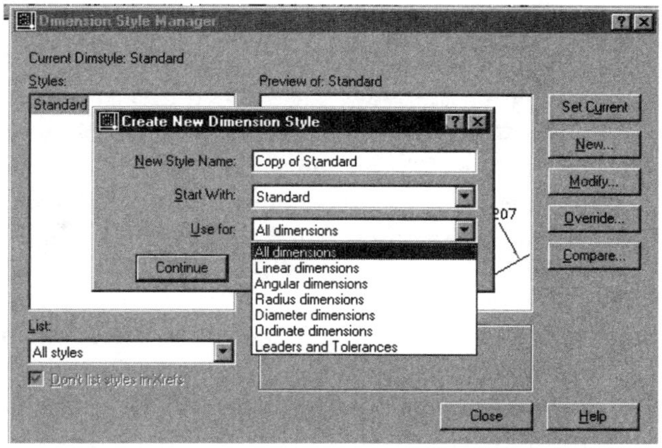

Figure 18.3 *Selecting the* **Use for** *edit arrow will display a list of dimension types that the style will be applied to.*

Override

Selecting the **Override** button will produce the **Override Dimension Style** dialog box. The box is identical to the **New Dimension Style** dialog box. This option allows a temporary override to a dimension style. Methods of overriding an existing style will be explored later in this chapter.

Compare

Selecting this option displays the **Compare Dimension Styles** dialog box shown in Figure 18.4. (The contents of the display will depend on the styles being compared.) This dialog box can be very helpful in comparing the differences between two styles or to display all of the properties of one style.

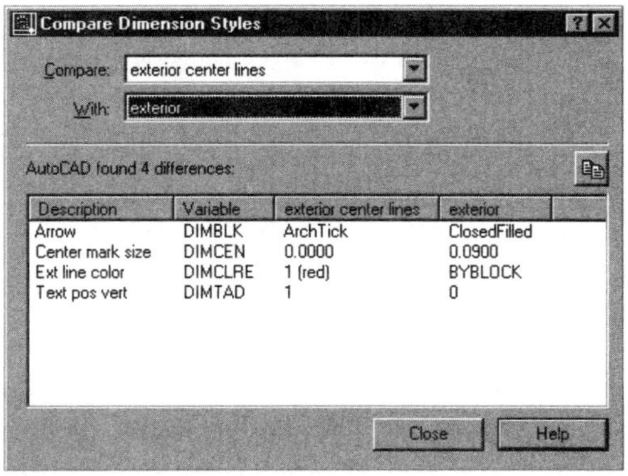

Figure 18.4 *The* **Compare Dimension Styles** *dialog box can be used to compare the differences between two styles or to display all of the properties of one style.*

CREATING A NEW STYLE

You've just explored the **Dimension Style Manager** and the **Create New Dimension Style** dialog boxes. Selecting the **New** button in the **Dimension Style Manager** dialog box and then the **Continue** button in the **Create New Dimension Style** dialog box will display the **New Dimension Style** dialog box shown in Figure 18.5. (The exact title will vary depending on the name of the style being created.) The dialog box contains six tabs for controlling the dimension style properties.

LINES AND ARROWS TAB

The **Lines and Arrows** tab contains five major areas for adjusting the lines and arrowheads to be used with the new style. These areas are the display window, **Dimension Lines**, **Extension Lines**, **Arrowheads**, and **Center Marks**. A display tab is in each tab and will demonstrate the effect of the current tab settings.

Display Window

The display window presents a visual representation of how the dimension properties will be displayed in the drawing area using the current settings. Each time a property is modified in the **Line and Arrows** tab, the display will be updated to reflect the new property setting.

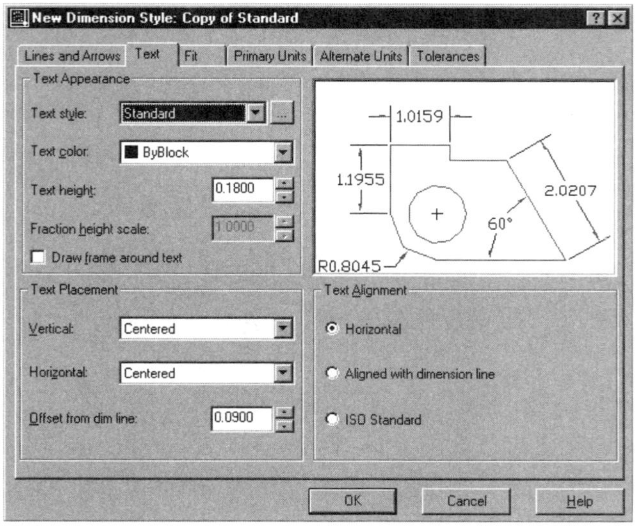

Figure 18.5 *Selecting the* **New** *button in the* **Dimension Style Manager** *dialog box and then the* **Continue** *button in the* **Create New Dimension Style** *dialog box will display a* **New Dimension Style** *dialog box and allow the properties of a dimension style to be set.*

Dimension Lines

The **Dimension Lines** area contains settings to control **Color**, **Lineweight**, **Extend beyond ticks**, **Baseline spacing**, and **Suppress**.

Color

The **Color** list displays and sets the color for dimension leader lines. Selecting the **Color** edit arrow displays the seven basic colors. Selecting the Other option displays the **Select Color** dialog box. Although a different color can be set for each aspect of a dimension, colors are best controlled by the layer that they are to be placed on.

Lineweight

Selecting the **Lineweight** edit arrow displays the lineweight list, allowing the lineweight of dimension lines to be set.

Extend Beyond Ticks

By default, this option is inactive. It will become active once the arrowhead is altered to be Oblique, Architectural Tick, Integral or None. This option controls how far the dimension line will extend past the extension line. Figure 18.6 shows examples of adjusting the extension. Most architectural offices set the extension value to equal the height of the text to be used.

Baseline Spacing

Selecting the **Baseline spacing** edit box allows the distance between lines of dimensions to be set. AutoCAD refers to this increment as the baseline setting. The default setting for architectural units is 0'–0 3/8". Most professionals use a distance of 1/2". Figure 18.7 shows an example of the baseline spacing.

Figure 18.6 *The default placement of dimension line extension on the left is often altered so that the dimension line extends past the extension line to simulate manual drafting methods.*

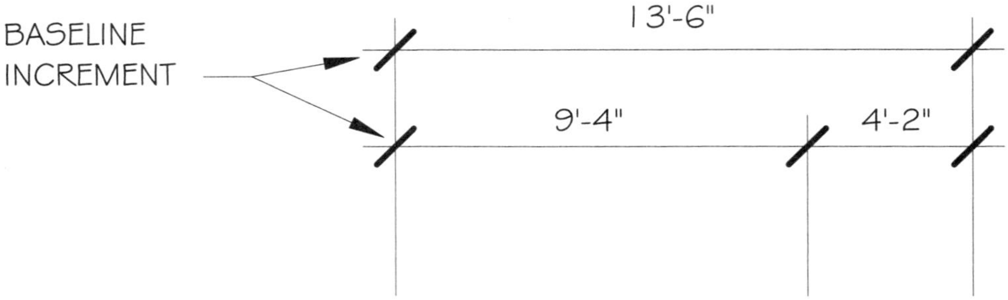

Figure 18.7 *The baseline increment controls the spacing between stacked lines of dimensions.*

Suppress

These options work as toggles allowing parts of the dimension line to be omitted. The **Dim Line1** and **Dim Line2** check boxes allow the first or second portion of the dimension line to be omitted from the drawing. Figure 18.8 shows an example of the first or second dimension line suppressed. The first or second dimension line is relative to the selection of the first and second extension line. These options are generally not set active for construction drawings.

Extension Lines

This area has features similar to those in the **Dimension Lines** area, except these settings control each aspect of the extension line in relation to the object being dimensioned.

Color

This option can be used to set the color of the extension lines. Selecting the **Color** edit arrow displays the color list. Although a different color can be set for extension lines, colors are best controlled by the layer that they are to be placed on.

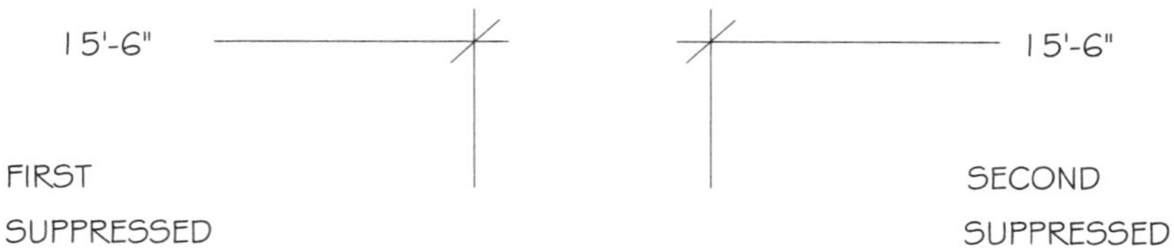

Figure 18.8 *Suppression of dimension lines is generally not required in architectural drafting.*

Lineweight
Selecting the **Lineweight** edit arrow displays the lineweight list allowing the lineweight of extension lines to be set.

Extend Beyond Dim Lines
The **Extend beyond dim lines** option allows the distance that the extension line extends past the dimension line to be controlled. The default value is 3/16". An extension offset of 1/8" is typical on most professional drawings. The effect of setting **Extend beyond dim lines** can be seen in Figure 18.9.

Offset from Origin
The **Offset from origin** option is used to control the gap between the extension lines and the object being dimensioned. The default value is 1/16" but a space of 1/8" is common in most offices. The effect of setting the **Offset from origin** is shown in Figure 18.10.

Suppressing Extension Lines
This option allows for the suppression of the first and second extension lines, as shown in Figure 18.11. By default, extension lines will be displayed at each end of the dimension line. This option works well for establishing overall sizes. The first extension line can be omitted by selecting the **Ext Line1** box. With this option active, as the location for extension lines is selected, the first location will not receive an extension line. Selecting both boxes suppresses the extension lines at each end of the dimension line. This is useful if a dimension line is to be placed between two exist-

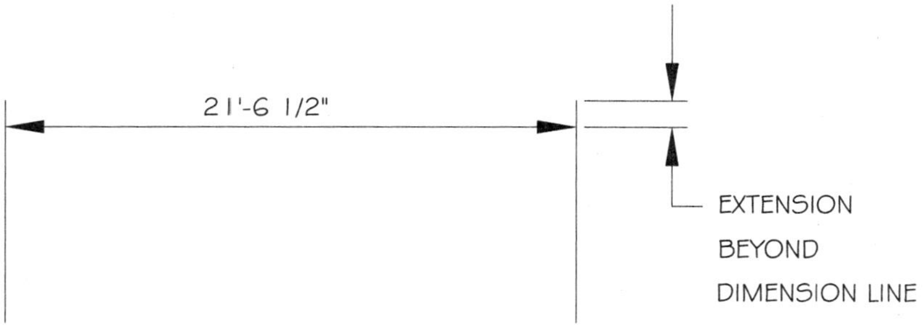

Figure 18.9 *The extension line is usually set to extend above the dimension line by a distance equal to the text height.*

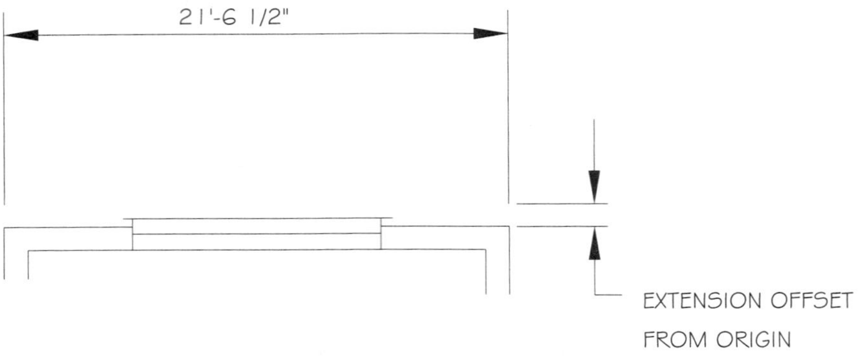

Figure 18.10 *The extension line should be offset from the object being described. A space of 1/8" is typical.*

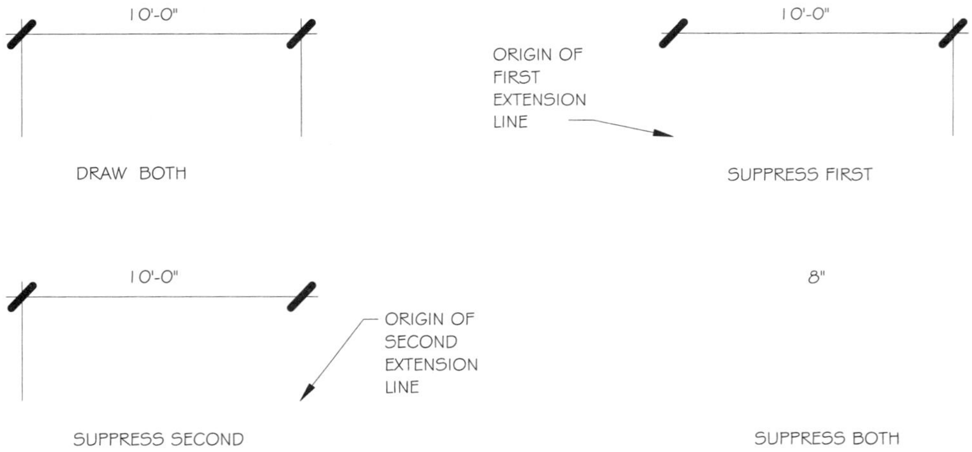

Figure 18.11 *Suppression of extension lines.*

ing extension lines. This option also works well when the object is being used as the extension lines. This is typically done when dimensioning between floor and ceiling levels on sections or elevations. Figure 18.12 shows examples of suppressing extension lines.

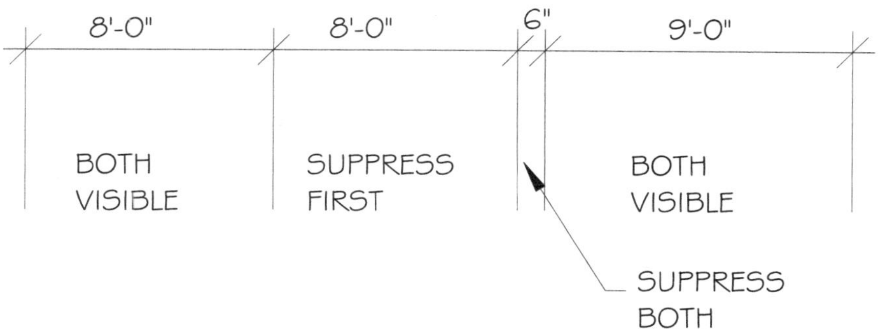

Figure 18.12 *Practical applications for suppressing extension lines.*

Arrowheads

This area controls the type of dimension line terminator to be used. The current selection for the first and second terminator, the closed filled arrow, is displayed in the edit box. By default, AutoCAD uses the terminator selected for the first dimension line used for the second line. The type of terminator can be selected by selecting the button or the arrow on the right side of the first list box. Selecting the button or the down arrow displays the list (Figure 18.13 shows a sample). Making the selection of

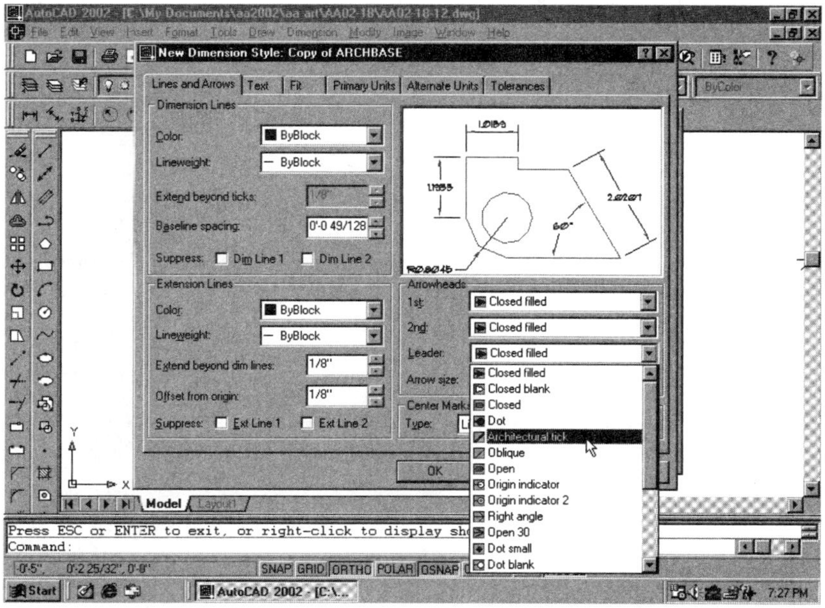

Figure 18.13 *Selecting the button or the down arrow displays the terminator list. The balance of the list can be seen by using the down arrow.*

Architectural Tick automatically alters the display of each arrow to tick marks. You can alter the second terminator to be different from the first by using the list for second option. Rarely is this done on professional drawings, and you should not decide to start a new trend! Examples of the variations of dimension line terminators can be seen in Figure 18.14.

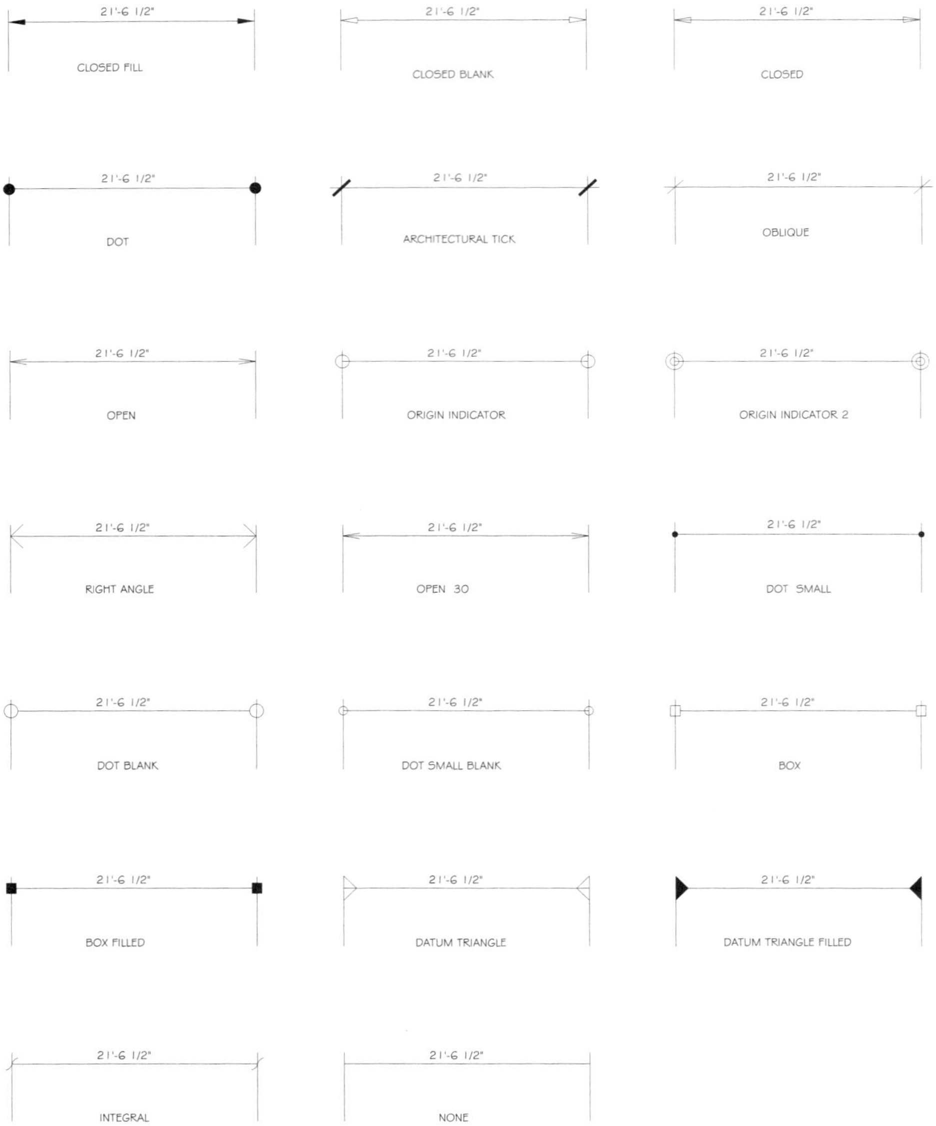

Figure 18.14 *Dimension line terminators available in AutoCAD.*

User Arrow

Most architectural and engineering offices use the Oblique, Architectural Tick, or User Arrow for a terminator. The user-defined tick (User Arrow) can be created using the **Block** command (discussed in Chapter 19). Once the desired block is drawn, it can be named and saved to the drawing base. Selecting User Arrow displays the dialog box shown in Figure 18.15. The name that was assigned to the block can be assigned in the **Select Custom Arrow Block** dialog box. The display window will now display the name of the block rather than showing an icon.

Arrow Size

The **Arrow size** edit box can be used to determine the size of the terminator relative to other dimensioning features. Selecting a tick value that is the same height as the text is a safe guideline to follow. If 3/8" high text is to be used, change the value to .375.

Center Marks

The **Center Marks** box is used to control how and when the center mark for radial dimensions will be displayed. The center mark is drawn only if the dimensions for **Dimcenter**, **Dimradius** or **Dimdiameter** are placed outside the circle or arc being described. Figure 18.16 shows the options for placement.

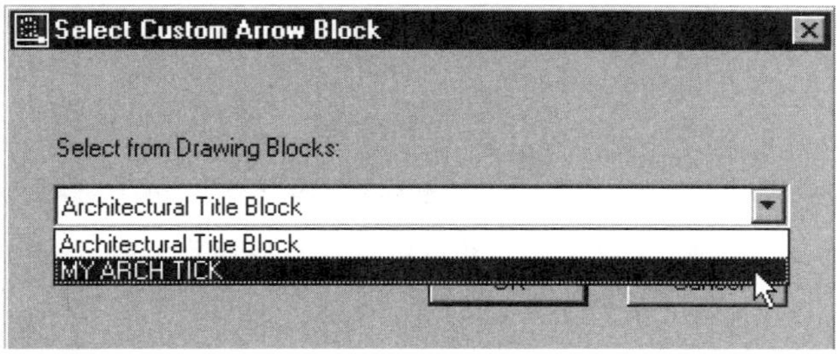

Figure 18.15 *A user-defined arrow can be created as a block and added into the drawing. Chapter 19 will introduce the* **Block** *command.*

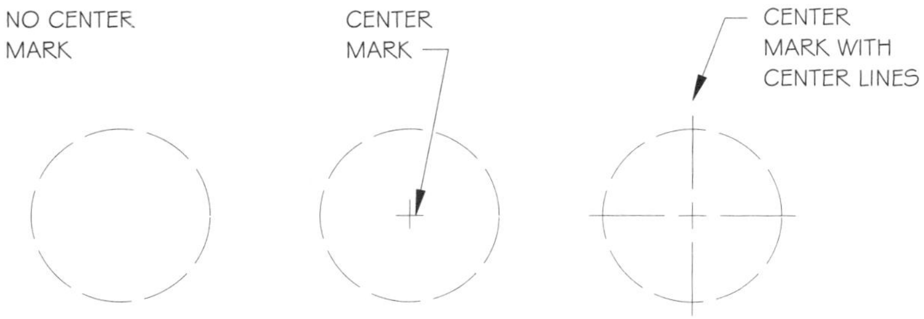

Figure 18.16 *Options for displaying center marks.*

Displaying the Center Mark

The default setting of Mark displays the center mark using the center Size value. Selecting the Line option displays centerlines that project beyond the limits of the circle. Selecting the **None** radio button will display no markings for the centerpoint of a circle or arc. The **Size** edit box can be used to control the size of the center mark. Many professionals adjust the default setting of 1/16" to 1/8".

TEXT TAB

The **Text** tab is used to set the appearance, placement, and alignment of the text that is used with dimensions. Each of the four major areas of the tab can be seen in Figure 18.17.

Text Appearance

The **Text Appearance** portion of the **Text** tab contains controls for the dimension text style, color, height, size, and frame.

Text Style

The **Text style** edit box displays the name of the current text style. It also allows one of the defined text styles to be used for dimension text. Selecting the [...] ellipsis button beside the edit box displays the **Text Style** dialog box, which allows a new text style to be created.

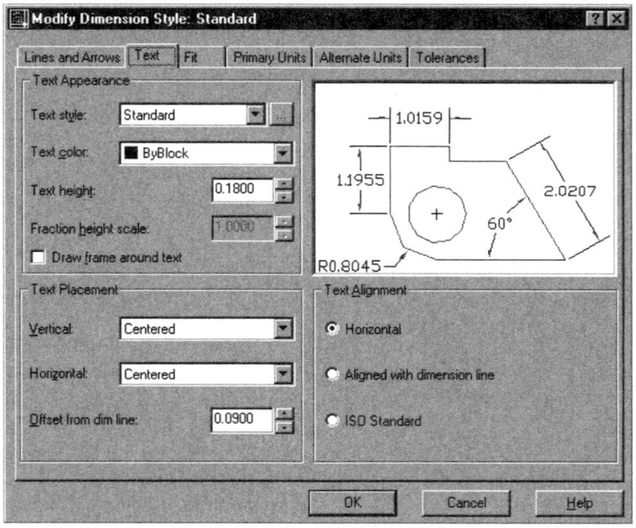

Figure 18.17 *The* **Text** *tab is used to set the appearance, placement, and alignment of the text that is used with dimensions.*

Text Color

The **Text color** box displays and sets the color for dimension text. Selecting the **Color** edit arrow displays the seven basic colors. Selecting the Other option displays the **Select Color** dialog box.

Text Height

The **Text height** box displays the current text height and allows the height of the dimension text to be altered. The height entered here must be coordinated with the height that is entered in the **Text Style** dialog box. If a text height is entered in the **Text Style** dialog box, that height will override the text height entered here in the **Text** tab. For the value of the **Text** tab to be active, a text value of 0 must be entered in the **Text Style** dialog box.

Fraction Height Scale

The **Fraction height scale** option sets the scale of fractions relative to the dimension text. In its current state, the option is inactive. For this option to be active, the **Precision** must be set on the **Primary Units** tab. Once a precision has been set, the scale factor can be altered. With the default value of 1.0000, the height of fractions will be the same as the dimension text height.

Framing Text

The **Draw frame around text** check box is currently inactive. Selecting this option will place a box around the text. With the current settings, selecting the box will do nothing. Once values are altered in the **Text Placement** area, selecting this option box will display a box around a dimension similar to Figure 18.18.

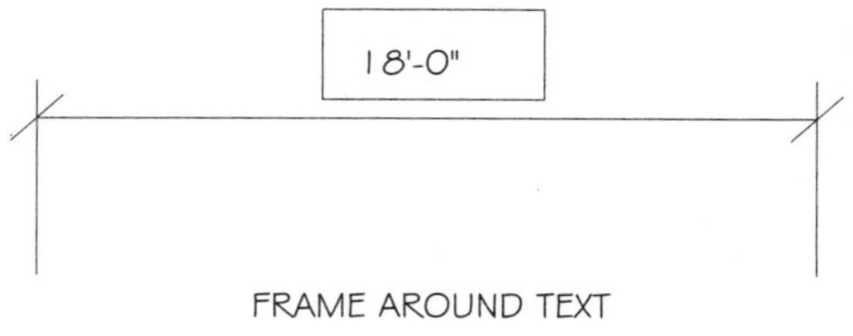

Figure 18.18 *Once values are altered in the* **Text Placement** *area, selecting the* **Draw frame around text** *check box will display a box around a dimensions. This option can be used to draw attention to a dimension or to distinguish between new and existing construction.*

Figure 18.19 *Two common settings of the* **Vertical Text Placement** *list.*

Text Placement

The **Text Placement** area of the **Text** tab controls the vertical and horizontal placement of the dimension text, as well as the offset distance.

Vertical

This option controls the vertical justification of dimension text relative to the dimension line. Options include Centered, Above Outside, and JIS. Figure 18.19 shows the two options that relate to construction drawing. Neither the Outside or the JIS (Japanese Industrial Standards) are used on construction drawings.

Centered—This option breaks the dimension lines so that text is centered between the dimension lines.

Above—This option is often used for architectural and engineering drawings. The option places the dimension text above the dimension line. The distance from the dimension line to the bottom of text is controlled by the Offset from dim line option.

Outside—The Outside option places the dimension text on the side of the dimension farthest away from the defining point.

JIS—This option places dimension text to conform to Japanese Industrial Standards (JIS) representation.

Horizontal

The **Horizontal** box allows the location of text along the dimension line to be set. The box contains a button and an edit box that can be used to control where the text is placed. Selecting the down arrow allows the options to be altered. As the arrow is selected, the **Horizontal** list is displayed. Selecting an option makes it current and removes the list. Options include the following:

Centered—The default value places text centered along the dimension line.

At Ext Line 1—This option places the text close to the first extension line.

At Ext Line 2—This option places the text close to the second extension line.

Over Ext Line 1—This option places the text parallel and above the first extension line.

Over Ext Line 2—This option places the text parallel and above the second extension line.

Examples of each option can be seen in Figure 18.20.

Offset from the Dimension Line

The **Offset from dim line** edit box displays the current gap between the text and the dimension line. The setting also allows the gap between the dimension line and the text to be altered. In construction drawings, the text is placed above the dimension line. The gap value should be set to equal half of the text height value.

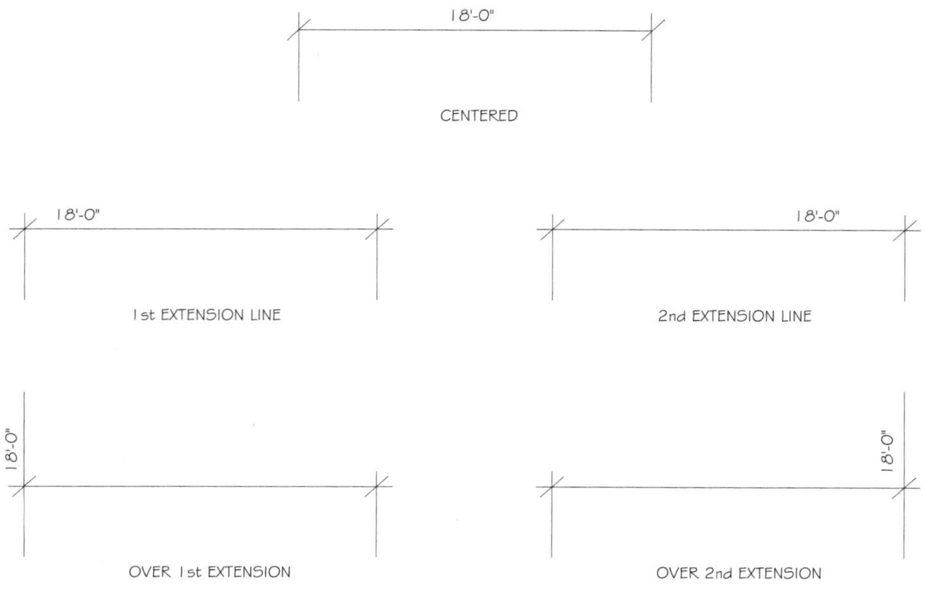

Figure 18.20 *Placement of the dimension text using the* **Horizontal Text Placement** *list.*

Text Alignment

The **Text Alignment** area of the **Text** tab can be used to control the orientation of dimension text to the dimension line. Choosing one of the inactive options alters the selection. Options include the default option of **Horizontal**, **Aligned with dimension line**, and **ISO Standard**. Figure 18.21 shows an example of each type of text placement.

Horizontal

With this box active, dimension text is placed horizontal, regardless of the angle of the dimension line. The option should be set to OFF for most construction drawings so that the text remains parallel to the dimension line.

Aligned with Dimension Line

This option is most typically associated with construction drawings. Selecting this option will align dimension text to be parallel with the dimension line.

ISO Standard

This option aligns text with the dimension line if the space between the extension lines is wide enough to display the text. Text will be placed in a horizontal position outside the extension lines if the text will not fit between the extension lines.

FIT TAB

The **Fit** tab provides controls for the placement of text in relationship to the dimension line. The **Fit** tab can be used to set the values that control the placement of text, arrows, and leaders for dimension placement. Figure 18.22 shows the **Fit** tab of the **Modify Dimension Style** dialog box. Figure 18.23 shows common text placements that might be seen on a floor plan.

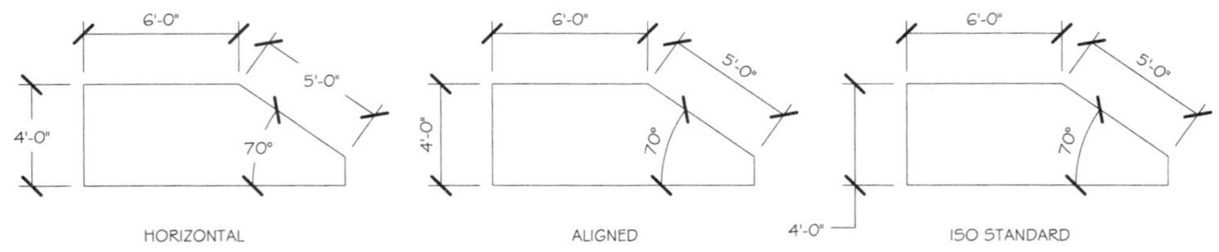

Figure 18.21 *The* **Text Alignment** *area of the* **Text** *tab can be used to control the orientation of dimension text to the dimension line. Most construction drawings use aligned dimensions.*

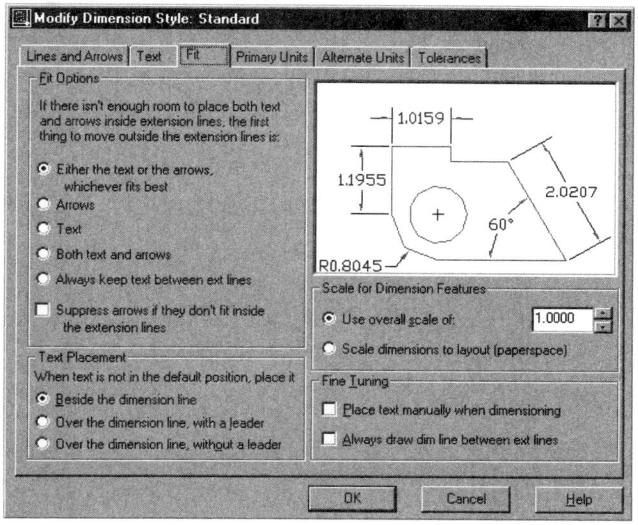

Figure 18.22 *The* **Fit** *tab of the* **Dimension Style** *dialog box.*

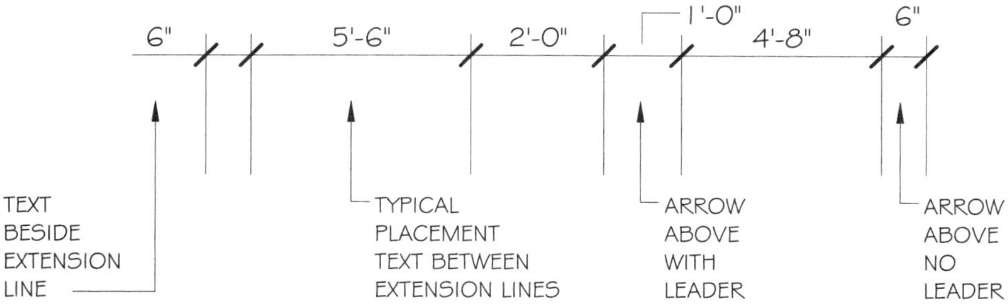

Figure 18.23 *Common dimension text placements used on floor plans.*

Fit Options

This portion of the **Fit** tab controls the placement of text and arrowheads if there is not enough space to place both inside the extension lines. If you are using Architectural Ticks, you'll have more space between the extension lines than if you use arrows. Figure 18.24 shows examples of placement options.

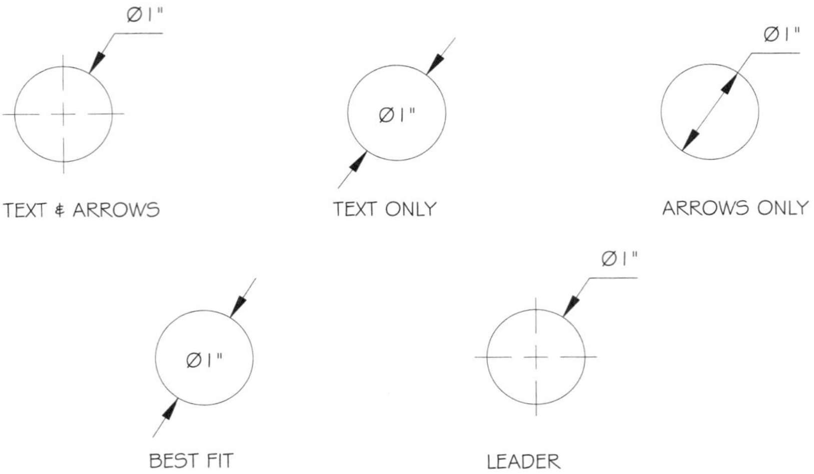

Figure 18.24 *Examples of* **Fit** *options. Notice that depending on the size of the space to be dimensioned, some options produce the same results.*

Either the Text or the Arrows, Whichever Fits Best

Selecting the default option will display the arrows and the text inside the extension lines when space is available. If sufficient space is not available between the extension lines, the text is placed between the extension lines and the arrows are placed outside the extension lines. If no space is available for text, only the arrows are placed between the extension lines. If space is not available for either text or arrows, both will be placed outside the extension lines.

Arrows

This option displays the arrows and the text inside the extension lines when space is available. If both will not fit between the extension lines, the arrows will be placed inside and the text will be placed outside the extension lines. If sufficient space is not available between the extension lines, the arrows will be placed outside the extension lines.

Text

This option also displays the arrows and the text inside the extension lines when space is available. If sufficient space is not available between the extension lines, the text is placed between the extension lines and the arrows are placed outside the extension lines. If no space is available for text, both will be placed outside the extension lines.

Both Text and Arrows

This option always displays the arrows and the text outside the extension lines. This option works well with small dimensions placed at the end of a dimension string.

Always Keep Text Between Ext Lines
This option always displays the text between the extension lines. This option works well with small dimensions placed in the middle of a dimension string.

Suppressing Arrows
The **Suppress arrows if they don't fit inside the extension line** option keeps the text between the extension lines and removes arrowheads when space is not available.

Text Placement

The **Text Placement** portion of the **Fit** tab allows for the control of the placement of text when it is not in the default position. Options include **Beside the dimension line**, **Over the dimension line, with a leader**, and **Over the dimension line, without a leader**.

Beside the Dimension Line
This option displays the text and the arrows outside the extension lines. This option works well in combination with using the **Either the text or the arrows, whichever fits best** option of the **Fit Options** area with small dimensions placed at the end of a dimension string.

Over the Dimension Line, with a Leader
This option displays the text above the normal position of text and to the side of the dimension line and uses an arrow to reference the text to the dimension line. This option works well in combination with using the **Either the text or the arrows, whichever fits best** option of the **Fit Options** area with small dimensions placed in the middle of a dimension string, when several small dimensions will be placed near each other.

Over the Dimension Line, without a Leader
This option displays the text above the normal position of text above the dimension line but does not use an arrow to reference the text to the dimension line. This option works well in combination with using the **Either the text or the arrows, whichever fits best** option of the **Fit Options** area with small dimensions placed in the middle of a dimension string.

Scale for Dimension Features

Up to this point of planning dimensions, each dimension will only be legible on a 12" x 9" screen. This area of the **Fit** tab sets the overall scale value or the paper space scaling for all dimensions. Options include **Use overall scale of** and **Scale dimensions to layout (paperspace)**.

Use Overall Scale of

The **Use overall scale of** edit box allows the scale of all dimensioning features to be increased or decreased based on the scale a drawing is to be plotted at. Typically, you will be drawing in model space with actual sizes. Text that is to be 1/8" high when plotted at a scale of 1/4"=1'–0" will need to be set to 6" in height. To have legible arrows, offsets, and text you'll have to convert every variable. If you actually have a life, you might just want to set the scale factor to 48. This will control all dimension variables and leave you a little free time. To find the desired scale factor, take the reciprocal of the drawing scale. A scale of 1/4"=1'–0" is 12/.25, a scale factor of 48. Other common scale factors include the following:

ARCHITECTURAL VALUES		ENGINEERING VALUES	
Plotting Scale	**Scale Factor**	**Plotting Scale**	**Scale Factor**
3/4"=1'–0"	16	1"=1'–0"	12
1/2"=1'–0"	24	1"=10'–0"	120
3/8"=1'–0"	32	1"=100'–0"	1200
1/4"=1'–0"	48	1"=20'–0"	240
3/16"=1'–0"	64	1"=200'–0"	2400
1/8"=1'–0"	96	1"=30'–0"	360
3/32"=1'–0"	128	1"=40'–0"	480
1/16"=1'–0"	182	1"=50'–0"	600
		1"=60'–0"	720

Scale Dimensions to Layout

Selecting the **Scale dimensions to layout (paperspace)** box will apply a scaling factor to dimensions plotted in paper space. The option sets a scale factor based on the scaling between the current model space viewport and paper space.

Fine Tuning

The final portion of the **Fit** tab allows minor text adjustments to be made: Options include **Place text manually when dimensioning** and **Always draw dim lines between ext lines**.

Place Text Manually when Dimensioning

Activating this check box allows the location for dimension text to be specified as you place each dimension regardless of the justification. With this box active, AutoCAD ignores other justification settings and allows you to specify where the text will be placed. Adjusting the variable can be useful on radial dimensions, as shown in Figure 18.25.

Always Draw Dimension Line

When a small space is dimensioned the text will be placed outside the extension lines. Selecting the **Always draw dim line between ext lines** option will place a dimension line between the extension lines even when the text and arrows are outside the extension lines. Figure 18.26 shows the effect of using this option on a radial dimension.

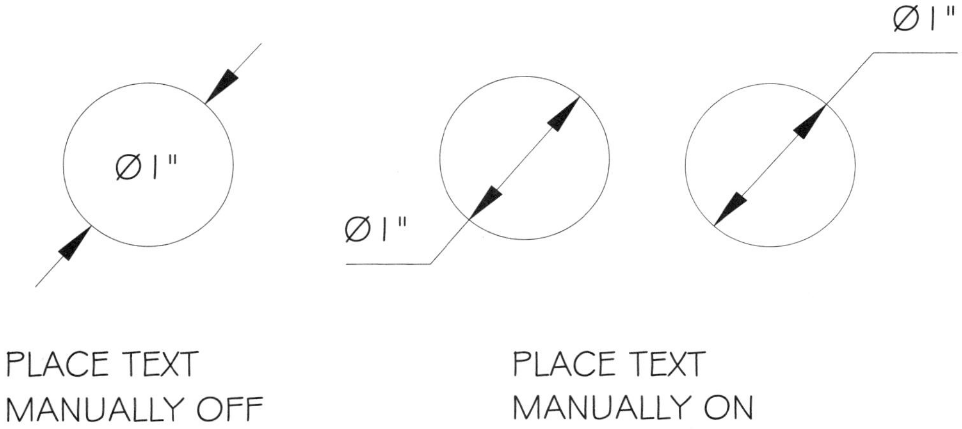

Figure 18.25 *With the* **Place text manually when dimensioning** *check box active AutoCAD ignores other justification settings and allows you to specify where the text will be placed.*

Figure 18.26 *Selecting the* **Always draw dim line between ext lines** *option will place a dimension line between the extension lines even when the text and arrows are outside the extension lines.*

PRIMARY UNITS TAB

A dimension of 14'–0" is considered a primary dimension unit. Most construction-related dimensions consist of primary units, with no other information required. Occasionally a tolerance might be added to the primary units. This portion of the **Modify Dimension Style** dialog box controls the display of the primary measurement units and any prefixes or suffixes for the dimension text. Selecting the **Primary Units** tab displays a tab similar to Figure 18.27. This tab is used to set the variables that control the appearance of text for dimension and leader lines. The tab includes the areas of **Linear Dimensions**, **Measurement Scale**, **Angular Dimensions**, and **Zero Suppression**.

Linear Dimensions

This portion of the **Primary Units** tab displays and sets the format and precision for linear dimensions. Key elements include the **Unit format**, **Precision**, **Fraction format**, **Decimal separator**, **Round off**, **Prefix**, and **Suffix**.

Unit Format

Even though the units of the drawing are set to Architectural, the default **Unit format** setting is decimal. Select the down arrow to display the list, which includes Scientific units, Decimal units, Engineering units, Architectural units, Fractional units, and Windows Desktop. The Architectural units will meet the needs for most drawings, although Engineering units are used on site-related drawings.

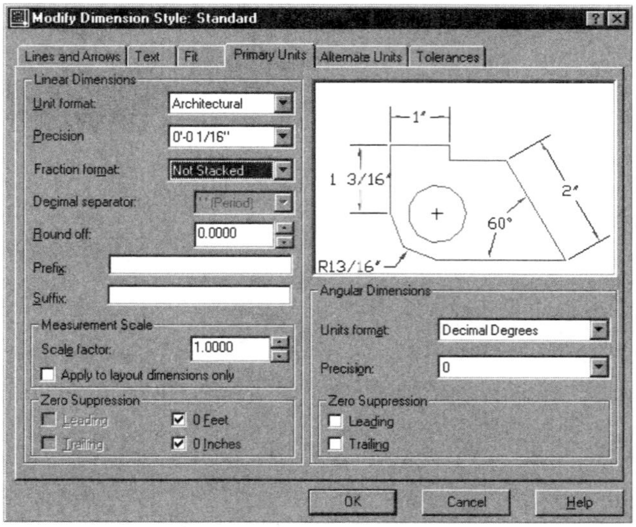

Figure 18.27 *The* **Primary Units** *tab.*

Precision

Selecting the down arrow of the **Precision** edit box displays the precision of the dimension. Options for the Architectural format are shown in Figure 18.28. For most dimensions, a setting of 0'–0" will provide needed accuracy. Most engineers and architects determine the fractions based on calculations rather than relying on drawing accuracy. If a fraction needs to be added to the dimension, it can be added by editing the text once the dimension is displayed.

Fraction Format

The **Fraction format** edit box sets the format for the display of fractions. Options include Diagonal, Horizontal, and Not Stacked. Each option can be seen in Figure 18.29.

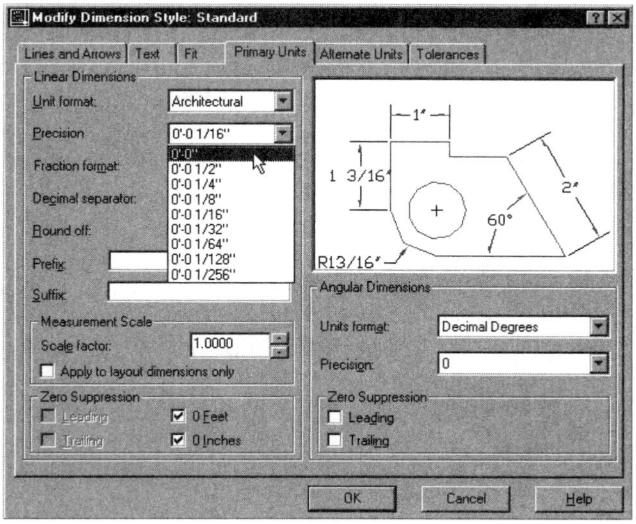

Figure 18.28 *The* **Precision** *list controls the accuracy assigned to a dimension.*

Figure 18.29 *The* **Fraction format** *options.*

Decimal Separator

This option is inactive when the **Unit format** is set to Architectural units. If the **Unit format** is set to Decimal, the option will be activated. Selecting the edit arrow displays the period, comma and space options.

Round Off

This option determines the value for rounding measurements for all dimensions types except for Angular. Entering a value of 1/8 rounds all measurement values to the nearest eighth inch.

Prefix and Suffix

A prefix or suffix can be added by typing in the appropriate box. Common suffixes might include TYP (typical), MM (millimeter), or MAX (maximum). This is one area that the control codes that were introduced in Chapter 17 might come in useful. The %%C code can be used to enter the diameter symbol as a prefix or suffix. Other control codes include %%D (degree symbol) and %%P (plus/minus symbol).

Measurement Scale

The **Measurement Scale** area of the **Primary Units** tab provides options for setting the scale factor and a method for limiting the scale values.

Scale Factor

The **Scale** box specifies the scale factor of a dimension without affecting the component, angles, or tolerance values. The option affects all linear distances except Angular. Alter the scale factor by using the up/down arrows or by highlighting the existing value and then typing a new value. A scale factor of 2 doubles the value of the dimension that is entered on the drawing. With a scale factor of 2, a dimension of 8" will be written when a 4" long object is dimensioned. For most architectural drawings the default value of 1 does not need to be altered.

Apply to Layout Dimensions Only

The **Apply to layout dimensions only** check box will apply a scaling factor to dimensions plotted in paper space. The layout scale factor will be adjusted to reflect the zoom scale factor for objects created in model space viewports.

Zero Suppression

This area of the **Primary Units** tab controls the display of zeros for leading, trailing, feet and inches. When a check mark is in the box, it is activated.

Leading

With this box active, the zero is suppressed preceding a decimal point in units less than one. Seven tenths will be written .7 rather than 0.7. This option will not be available when Architectural units are current.

Trailing
This box controls the display of zeros after a decimal point. With the box active, no zeros will be displayed after the decimal. Seven tenths will be written .7 rather than .70. This option will not be available when Architectural units are current.

0 Feet
This option will be available only when Architectural units are current. A dimension less than one foot can be displayed as 0'–8" or 8". Professional practice is mixed, and the setting for the **0 Feet** box will vary from office to office. Selecting the **0 Feet** box causes the zero representing feet to be suppressed when the dimension is less than twelve inches.

0 Inches
This option will be available only when Architectural units are current. Most professionals display the zero after the foot symbol. Writing 5'–0" is preferred to writing 5'. With this box active, zeros will be suppressed. This box should be inactive.

Controlling the Display of Angles

The **Angular Dimensions** area controls the format of angular dimensions. Options include **Units format**, **Precision**, and **Zero Suppression**. Options are similar to their linear counterparts.

Units Format
The **Units format** for angular dimensions includes the options of Decimal Degrees, Degrees/Minutes/Seconds, Gradians, and Radians. The Decimal Degrees option will meet the needs for most architectural drawings. The Degrees/Minutes/Seconds option will work best for most site-related drawings.

Precision
The Precision list will vary depending on the **Units format** setting. When units are measured in decimals, the precision will be measured in decimal places from zero to eight places.

Zero Suppression
This area suppresses the leading or trailing zero of angular dimensions. With the **Leading** option activated, 0.75 will be displayed as .75. The **Trailing** option, when activated, suppresses zeros that follow the decimal. .7500 will be displayed as .75.

ALTERNATE UNITS TAB

This portion of the **Modify Dimension Style** dialog box is not used to create a style for most construction drawings. As you display the tab, each option is inactive. Selecting the **Display alternate units** check box activates the options. The **Alternate**

Units tab is shown in Figure 18.30. Major components include the **Alternate Units**, **Zero Suppression**, and the **Placement** areas. Features of this tab function in a manner similar to their counterparts on the **Primary Units** tab. One option that should be considered on this tab is the ability to create and display metric units behind the architectural primary units. Selecting Decimal as the **Unit format**, a multiplier for all units of 25.4, and a suffix of mm will display dimensions similar to Figure 18.31.

TOLERANCES TAB

A tolerance is a range that is allowed to compensate for human error. If a note specifies that bolts in a steel plate are to be 3" o.c., and the holes are placed at 3 1/16" o.c., should the plate be scrapped? A tolerance specification will define what is an acceptable range of variation. Most areas of construction do not use a tolerance zone, although some steel and concrete detailing might require tolerances. The **Tolerances** tab of the **Modify Dimension Style** dialog box, shown in Figure 18.32, controls the display and precision of the tolerance. Major components include the **Tolerance Format**, **Zero Suppression**, and the **Alternate Unit Tolerance** areas. Features of this tab function in a manner similar to their counterparts on the **Primary Units** tab. With the default **Method** value of None, most of the box is inactive. Once a method of tolerance specification is selected, the balance of options on the tab will be activated. Figure 18.33 shows examples of the available tolerance methods.

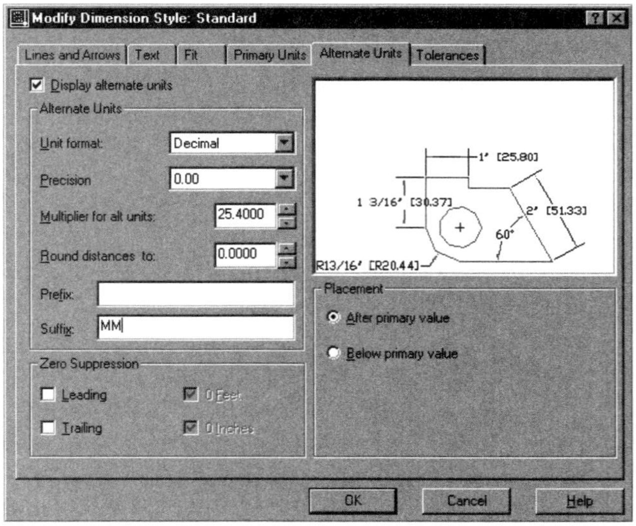

Figure 18.30 *The* **Alternate Units** *tab options are inactive as it it first displayed. Selecting the* **Display alternate units** *check box activates this tab, allowing alternate units to be displayed behind the primary units.*

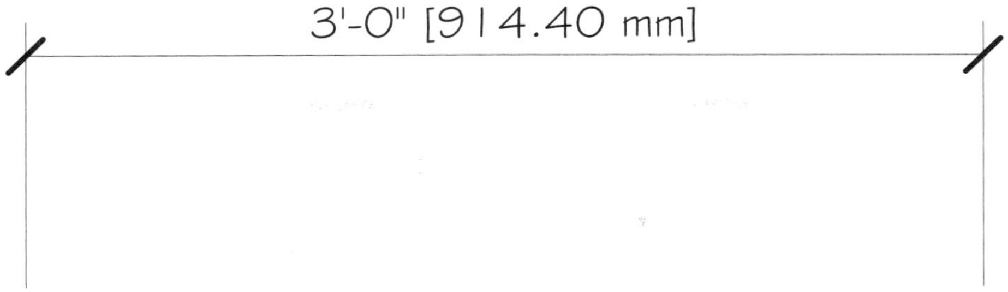

Figure 18.31 *Selecting Decimal as the alternate unit format, a multiplier for all units of 25.4, and a suffix of mm will display architectural units followed by metric dimensions.*

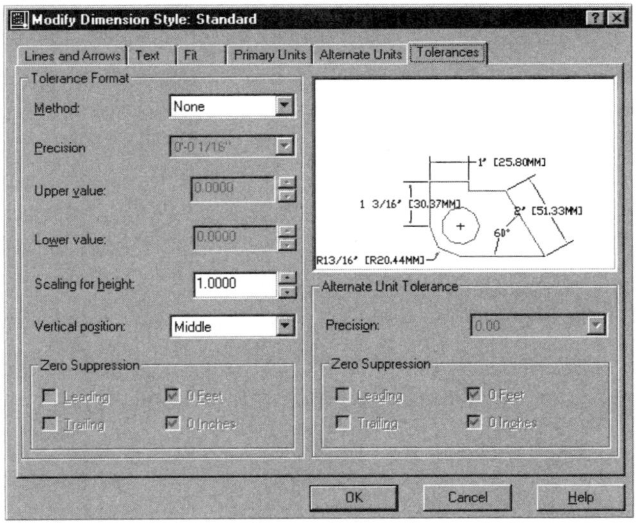

Figure 18.32 *The Tolerances tab of the* **Modify Dimension Style** *dialog box controls the display and precision of the tolerances*

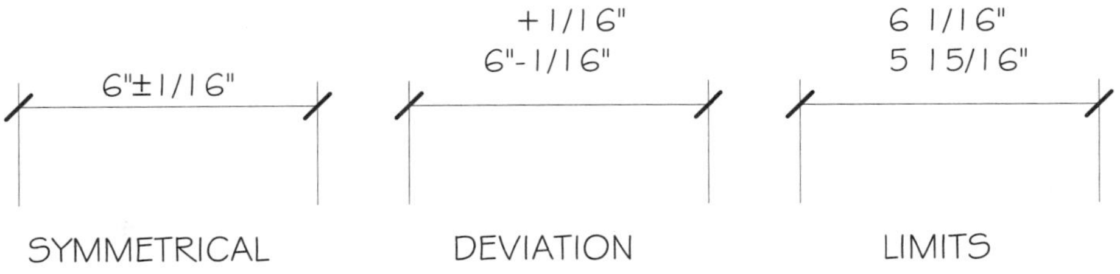

Figure 18.33 *Tolerance methods available in AutoCAD include symmetrical, deviation, and limits.*

COMBINING VARIABLES TO CREATE A STYLE

If you're feeling overwhelmed by styles, variables, and values, don't be discouraged. Remember, most of these values will be set on a template drawing and left alone. Throughout this chapter comparisons have been made using the **Modify Dimension Style** dialog box and its tabs. This section will help you combine these values to create a style. Dimensioning variables can be combined to create a particular style in much the same way that options are combined with **Text** to create different text styles. Styles can be created, saved, modified and quickly loaded by using the **Dimension Style Manager** dialog box shown in Figure 18.1.

CREATING DIMENSIONING STYLES

Dimensioning styles allow you to save groups of settings for dimension variables. If an office does both architectural and civil engineering projects, two distinct dimensioning styles could be created and saved in template drawings to speed the dimensioning process. Even in a firm that does only architectural drawings, varied styles can be useful to group option settings. For instance, on a multilevel structure, some dimensions will be shown on each floor plan and the framing and foundation plans. These might be placed in a style titled ARCHBASE that uses a certain size, color and style of text with two extension lines. Styles might be created for dimensions that will be used to dimension to the center of interior walls and have the left or right extension line suppressed, with another style created for a style that has both extension lines suppressed. Once created, these styles can be quickly loaded and interchanged.

Use the following steps to create a dimension style suitable for a plan view to be plotted at a scale of 1/4"=1'–0":

1. Select the **Dimension Style** button on the **Dimension** toolbar. This will display a **Dimension Style Manager** dialog box similar to Figure 18.1.
2. Select the **New** button. This will display a **Create New Dimension Style** dialog box similar to Figure 18.2.
3. Enter the desired style name. In this case PLAN VIEW BASE will be used.
4. Select an existing style to use as a base. In this case the Standard style will be used.
5. Select what dimensions that this style will be applied to in the **Use for** edit box. In this case, All Dimensions will be used.
6. Select the **Continue** button. This will display a **New Dimension Style: PLAN VIEW BASE** dialog box. Proceed through each tab and set the options to meet your specific situation. The next section will recommend values that you might want to consider using.

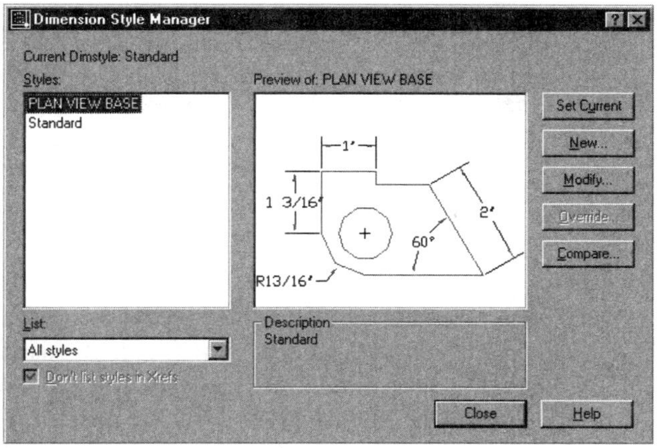

Figure 18.34 *In the* **Dimension Style Manager** *dialog box, select PLAN VIEW BASE from the* **Styles** *box, select the* **Set Current** *button, and then select the* **Close** *button to return to the drawing area.*

7. Once each option has been selected, click the **OK** button. This will return the **Dimension Style Manager** dialog box, similar to Figure 18.34.
8. Select PLAN VIEW BASE from the **Styles** box.
9. Select the **Set Current** button.
10. Select the **Close** button.

This process will create the PLAN VIEW BASE style and set it as the current dimension style.

Common Style Settings

Although offices often have individual dimensioning standards, several common options can be set to meet most needs. Use the Standard style as a base to create the desired settings to update your template:

Lines and Arrows Tab (Figure 18.13)

Color and Lineweight: By Block	Arrowheads: Architectural Tick
Extend beyond ticks: 1/8"	Arrow size: 1/8"
Baseline spacing: 1/2"	Center Mark: Line
Extend beyond dim lines: 1/8"	Center Mark Size: 1/8"

Text Tab (Figure 18.17)

Text style: Stylus BT	Offset from dim line: 1/8"
Text color: By BLock	Vertical: Above
Text height: 1/8"	Aligned with dimension line: ON

Fit Tab (Figure 18.22)

Fit Options: Text | Use overall scale of: 48

Always draw dim line between ext lines: ON

Primary Units Tab (Figure 18.27)

Unit format: Architectural | Precision: 0'–0 1/2"

Fraction format: Not Stacked

Following the guidelines to create PLAN VIEW BASE will produce dimensions similar to Figure 18.35a. The dimension style works well for the overall dimensions, but it is somewhat limited when an object has a jog in it. If the style is used to place dimension A, the desired results will be obtained, but two extension lines will be placed on the left side of space A. The style will reproduce with the desired results if you're careful to always use OSNAP when you select where the extension lines will be placed. Failure to use OSNAP might produce results similar to Figure 18.35b. Even if OSNAP or **Quick Dimension** is used to place the dimensions for space B and C, the desired results will not be achieved. The extension line for B will extend to be within 1/8" of surface B, leaving no offset for surface A. The same problem will occur at B as surface C is dimensioned.

CREATING A NEW STYLE BASED ON AN EXISTING STYLE

A better method of placing the interior dimensions is to create a style that will suppress one or both of the extension lines. This can easily be done using the following procedure:

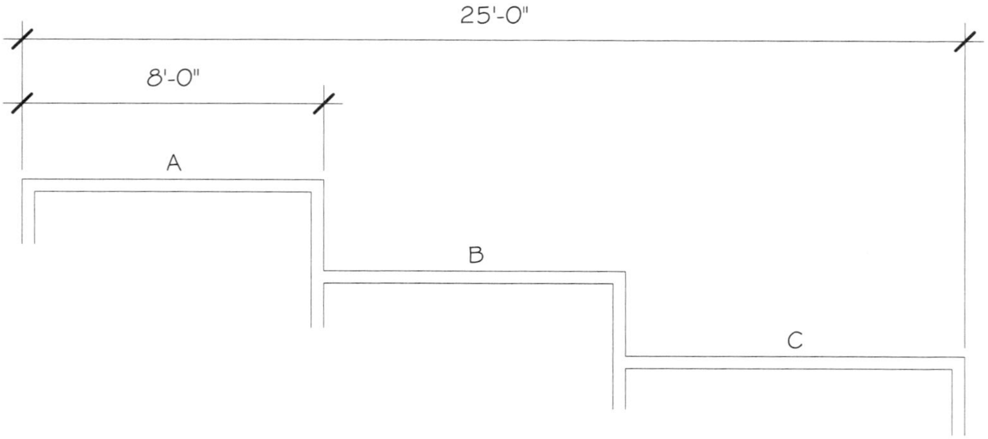

Figure 18.35a *Using the PLAN VIEW BASE style will produce dimensions suitable for many architectural applications.*

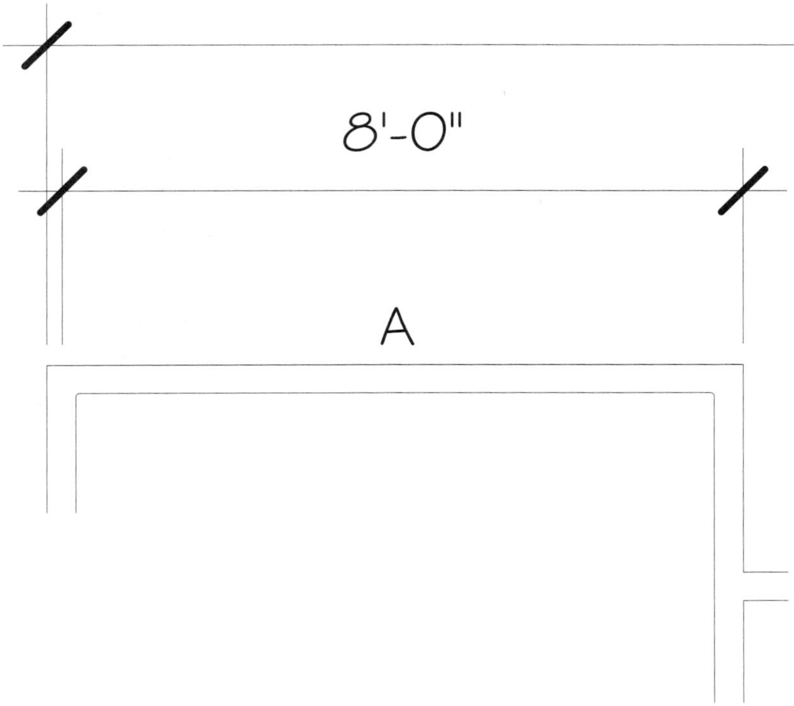

Figure 18.35b *Failure to use OSNAP might produce extension lines that do not align.*

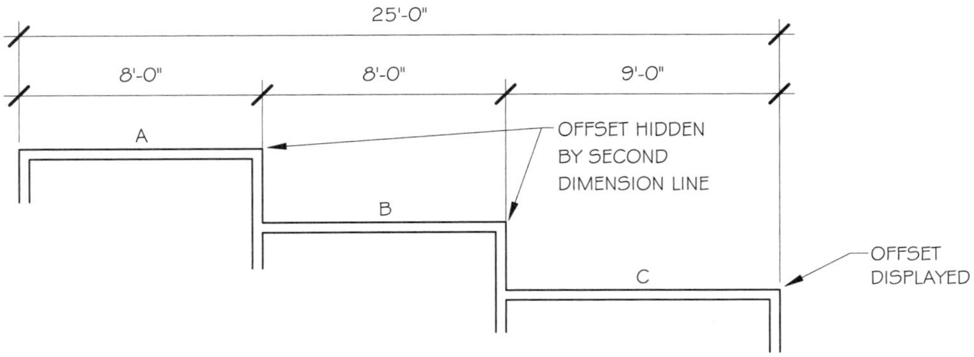

Figure 18.35c *Even if OSNAP or* **Quick Dimension** *is used to place the dimensions for space B and C, the desired results will not be achieved. The extension line for B will extend to be within 1/8" of surface B, leaving no offset for surface A. The same problem will occur at B as surface C is dimensioned.*

1. Display the **Dimension Style Manager** dialog box and select the **New** button.
2. In the **Create New Dimension Style** dialog box, enter a style name of INTERIOR.
3. Use PLAN VIEW BASE as the **Start With** style and use it with All Dimensions.
4. Click **Continue** to advance to the **New Dimension Style** dialog box.
5. Use the **Lines and Arrows** tab to suppress Line 1.
6. Click the **OK**, **Set Current**, and **Close** buttons to set the new style as the current style and return to the drawing area.

Using the PLAN VIEW BASE and the INTERIOR styles will allow dimensions to be placed accurately for the overall length and for spaces A, B, and C. Figure 18.35d shows the result of using the PLAN VIEW BASE and the INTERIOR styles.

MODIFYING A STYLE

Once a style has been created, you might find that it is not all you hoped it would be. The **Modify** option in the **Dimension Style Manager** dialog box allows an existing dimension style to be altered. The process for altering an existing style is similar to the process used to create a new style using an existing style as a base. An existing style can easily be modified using the following procedure:

1. Display the **Dimension Style Manager** dialog box and select the style to be modified from the **Styles** display box.
2. With the desired style highlighted, select the **Modify** button. This will display the **Modify Dimension Style** dialog box. The box is exactly the same as the

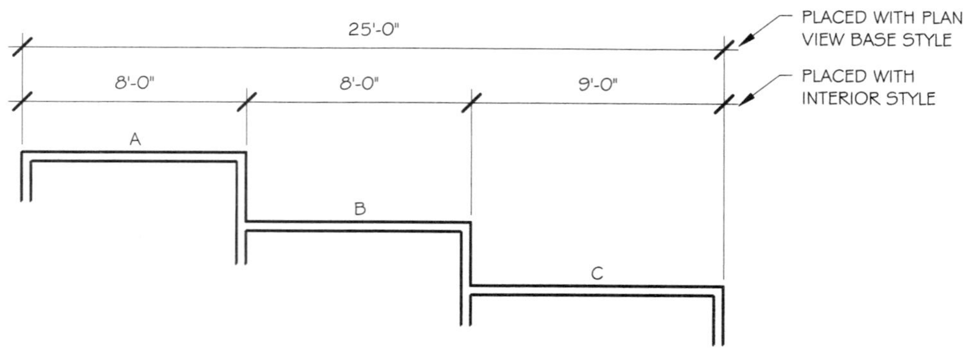

Figure 18.35d *Using the PLAN VIEW BASE and the INTERIOR styles will allow dimensions to be placed accurately for the overall length and for spaces A, B, and C.*

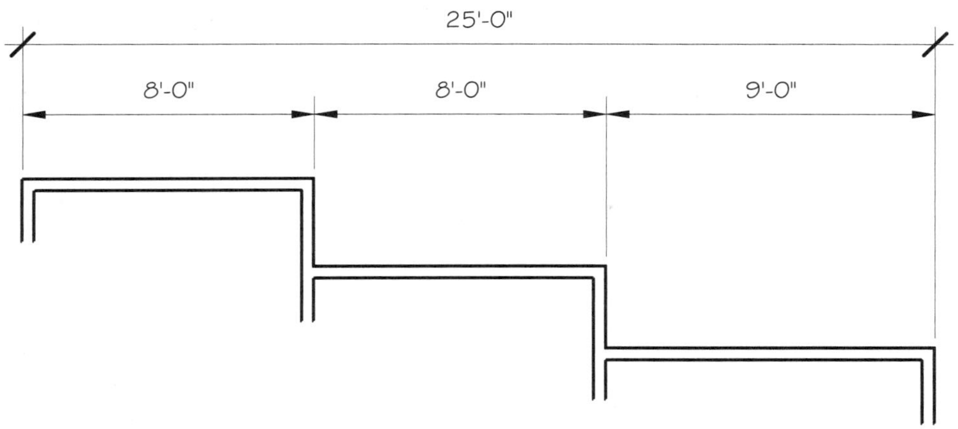

Figure 18.36 *Using the* **Modify** *option on the INTERIOR style will alter the display shown in Figure 18.35d to resemble Figure 18.36.*

New Dimension Style dialog box.

3. Use the necessary tabs to alter the existing values. For the example, the **Lines and Arrows** tab was used to alter the arrowheads.
4. Click the **OK** and the **Close** buttons to modify the existing style and return to the drawing area.

Using the **Modify** option on the INTERIOR style will alter the display to resemble Figure 18.36.

OVERRIDING EXISTING STYLES

Using the **Modify** option of the **Dimension Style Manager** dialog box changes an entire style to the specified parameters. Using the **Override** option allows an existing style to stay intact while altering new features that are added to the style. Use the following steps to override an existing style:

1. Display the **Dimension Style Manager** dialog box and select the style to override from the **Styles** display box. In this example, the INTERIOR style will be used as the style to be altered.
2. With the desired style highlighted, select the **Override** button. This will display the **Override Current Style** dialog box. The box is exactly the same as the **New Dimension Style** dialog box.
3. Use the necessary tabs to alter the existing values. For the example, the **Lines and Arrows** tab was used to alter the arrowheads, and the **Text** tab was used to center the text between the dimension line.

4. Click the **OK** and the **Close** buttons to override the existing style and return to the drawing area.

Using the **Override** option on the INTERIOR style will alter the display to resemble Figure 18.37. Styles can also be overridden at the Command prompt by typing **DOV** (ENTER). This will produce the prompt:

Enter dimension variable name to override [or Clear overrides]:

Enter the name of the desired variable to be altered followed by the name of the new style.

COMPARING STYLES

The final option of the **Dimension Style** dialog box is to compare styles. This option can be used to list the properties of a style or to compare the properties of two

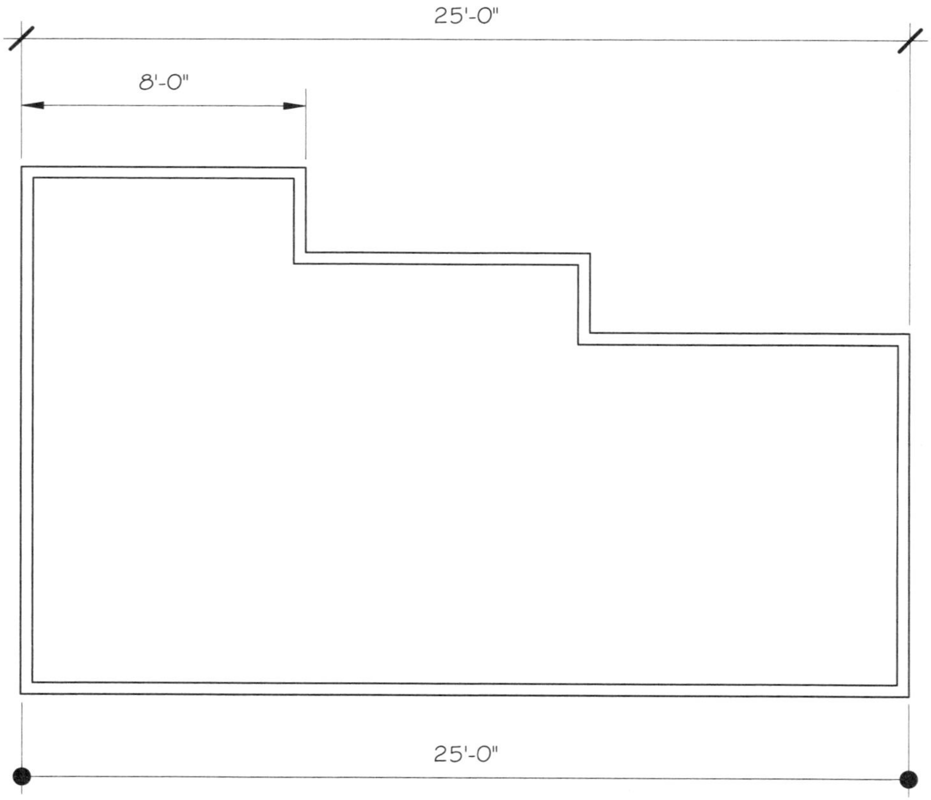

Figure 18.37 *Using the* **Override** *option on the INTERIOR style to alter the display to show dot arrowheads and centered text.*

dimension styles. The function performed depends on the settings of the **Compare** and **With** edit boxes. To list the properties of a style, select the name of the desired style in the **Compare** edit box, and select <None> in the **With** list. Figure 18.38 displays a partial listing of the properties for the PLAN VIEW BASE style. Entering a style name in the **Compare** edit box and a different style name in the **With** edit box will display a listing of the property differences similar to Figure 18.39. The list of properties and the list of differences can be copied to the Windows Clipboard. Once copied to the Clipboard, the listings can be pasted into other Windows applications.

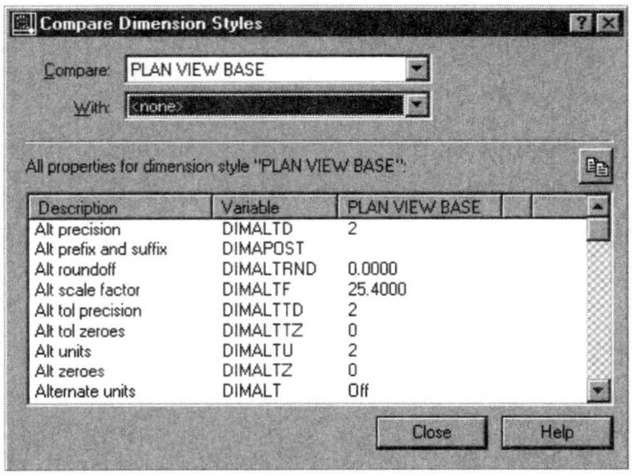

Figure 18.38 *A partial listing of the properties for the PLAN VIEW BASE style.*

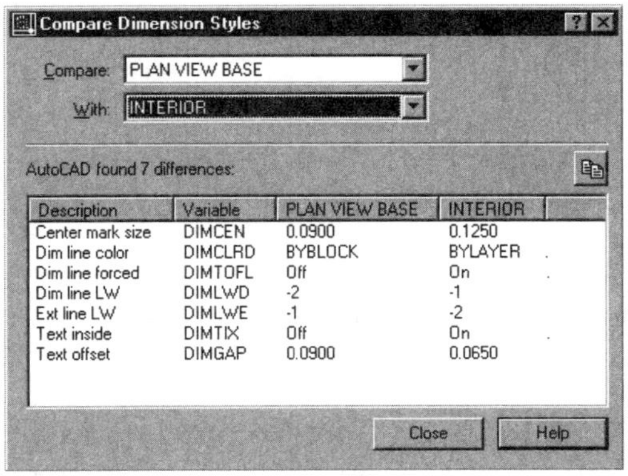

Figure 18.39 *Entering a style name in the* **Compare** *edit box and a different style name in the* **With** *edit box will display a listing of the property differences.*

RENAMING OR DELETING STYLES

A style can be renamed or deleted using the **Style** portion of the **Dimension Style Manager** dialog box. Use the following procedure to rename a style:

1. Display the **Dimension Style Manager** dialog box and move the cursor to the **Styles** portion.
2. Highlight a style in the **Styles** portion of the dialog box.
3. With the style highlighted, right-click to produce a shortcut menu.
4. Select the **Rename** option and enter the desired new name.
5. Continue renaming other menu listings or select **Close** to remove the dialog box and return to the drawing area.

Similarly, deleting a dimension style is done in the same dialog box, using the following procedure:

1. Once in the style to be deleted is highlighted, right-click and select the **Delete** option from the shortcut menu.
2. Continue deleting other styles or select **Close** to remove the dialog box and return to the drawing area.

APPLYING STYLES

Each dimension you create has a style. If you started to place dimensions prior to reading this chapter, the dimensions were created using the Standard style. Throughout this chapter, you've learned to create, modify, and override dimension styles. In addition to selecting the **Set Current** button in the **Dimension Style Manager** dialog box, two other methods are available to alter the dimension style. A new style can be applied by using the shortcut menu or by using the **Dim Style Control** box on the **Object Properties** toolbar.

Using the Shortcut Menu

Use the following steps to assign a new dimension style to an existing dimension with the shortcut menu:

1. Select the desired dimension and right-click.
2. Choose **Dim Style** from the shortcut menu.
3. Choose the name of the desired style to be applied to the selected dimension and press (ENTER).

The process can be seen in Figure 18.40.

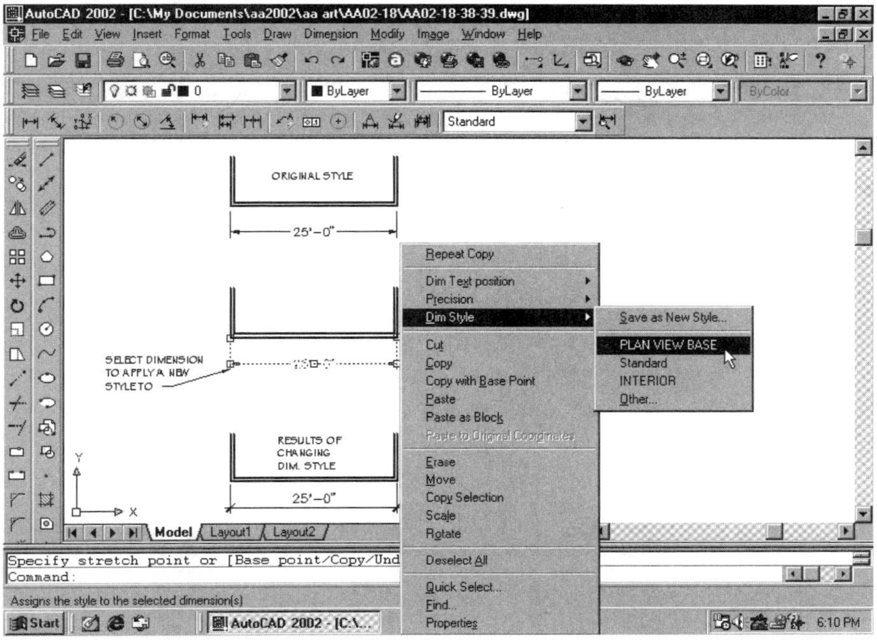

Figure 18.40 *A dimension style can be altered by activating the grips and right-clicking to produce the shortcut menu, and then selecting the name of the desired style.*

Using the Dim Style Control Box

Use the following steps to assign a new dimension style to an existing dimension with the **Dim Style** edit box:

1. Select the desired dimension.
2. Display the **Dim Style Control** menu on the **Object Properties** toolbar.
3. Choose the name of the desired style to be applied to the selected dimension and press (ENTER).

EDITING DIMENSION PLACEMENT

The placement of dimensions can be edited using grips or the editing commands. The editing commands can be selected from the shortcut menu or by the methods presented in earlier chapters. To use the shortcut menu, select a dimension with the cursor. Each of the grips will be activated. Right-click to obtain the editing menu. Notice in Figure 18.41 that a grip is displayed for the dimension text, at each intersection of the dimension and extension lines, and at what appears to be the end of the extension line. Closer examination will show the grips that appear to be at the end of the extension lines actually lie beyond the lines. These grips mark the definition points or Defpoints

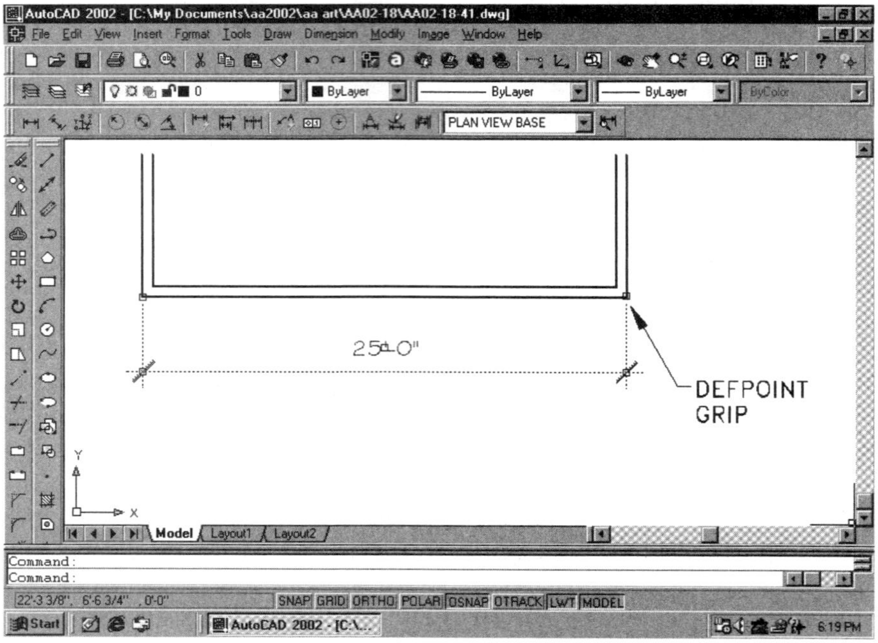

Figure 18.41 *A grip is displayed for the dimension text, at each intersection of the dimension and extension lines, and at the definition points (DefPoints) for the dimensions. AutoCAD automatically assigns definition points for every dimension to define the dimension location.*

for the dimensions. AutoCAD automatically assigns definition points for every dimension to define the dimension location. Use the **Layers** button to examine the layers for the current drawing and you'll find that the program created a layer titled Defpoints to display the definition points. The layer will not be plotted. Dimensions can be edited using grips without knowing the location of the definition points. The Defpoints will automatically be displayed. If you alter dimensions using the edit commands, the Defpoints must be included in the selection set. Using the Node mode of Object Snap will aid in Defpoint selection.

STRETCHING DIMENSIONS

Dimensions can be stretched to alter their position using either grips or the **Stretch** command. The process of stretching a dimension can be seen in Figure 18.42. Use the following steps to alter the dimension position using grips:

1. Select the dimension to be stretched.
2. Make one of the grips on the dimension line the hot grip.
3. Drag the grip to the desired location. OSNAP will aid in accurate placement.

This process will stretch all features of the dimension. To move just the text, select the text grip. Make it the hot grip, and move the text to the desired location.

The **Stretch** command can be used to stretch dimensions using the same procedure that is used to move drawing objects. Use the Crossing option to select the dimension to be relocated. **Stretch** can't be used to relocate the text.

EDITING OBJECTS AND DIMENSIONS WITH THE MODIFY MENU

Objects and dimensions can be edited using the editing commands contained on the **Modify** menu as well as the grip editing functions. As an object is edited, the dimensions are also edited.

Stretch

Dimensions can be stretched in the same way that drawing objects are stretched. If the dimension is selected with the object, as seen in Figure 18.43, both will be stretched, and the dimension text will be revised automatically.

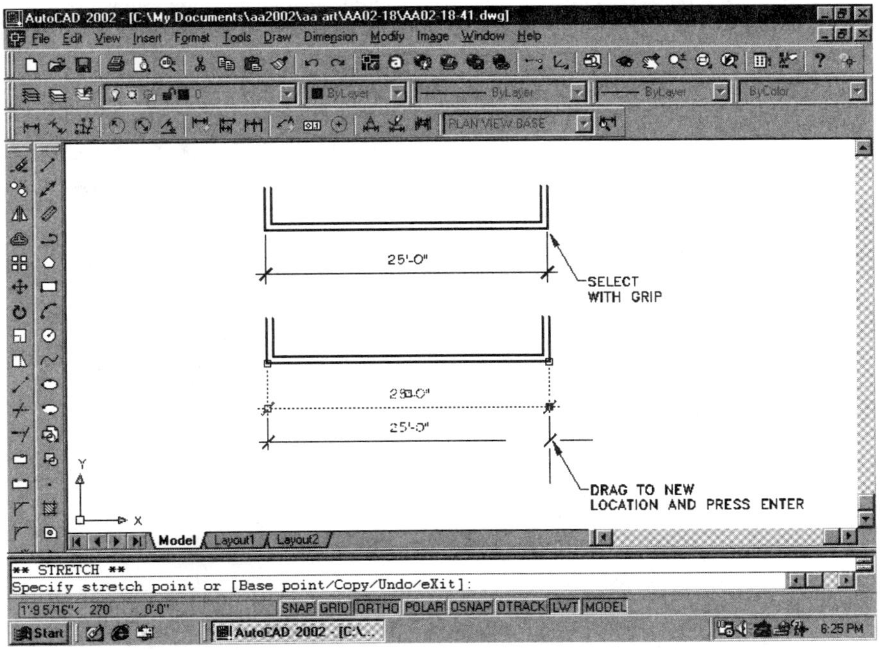

Figure 18.42 *Select the dimension to be stretched and then make one of the grips on the dimension line the hot grip. Drag the hot grip to the desired location. OSNAP will aid in accurate placement when you are aligning with other dimension lines.*

Extend

The **Extend** command sequence also can be used to extend a dimension line and adjust the text accordingly. Figure 18.44 shows the effects of the **Extend** command on dimensions.

Trim

The **Trim** command sequence can be used to shorten a dimension line and adjust the text accordingly. Figure 18.45 shows the effects of **Trim** on dimensions.

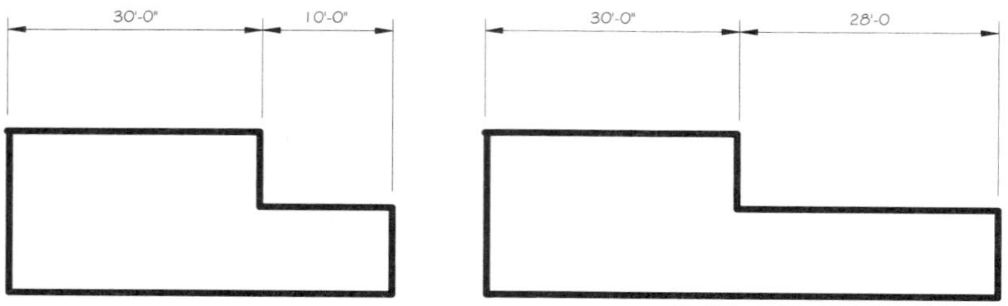

Figure 18.43 *Using the* **Stretch** *command to alter an object with dimensions.*

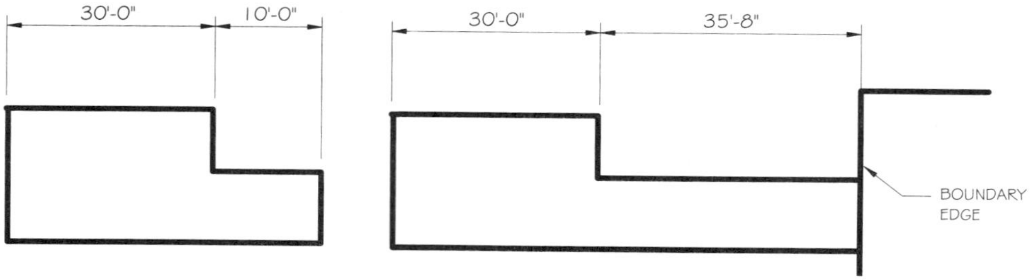

Figure 18.44 *The effect of the* **Extend** *command on an object with dimensions.*

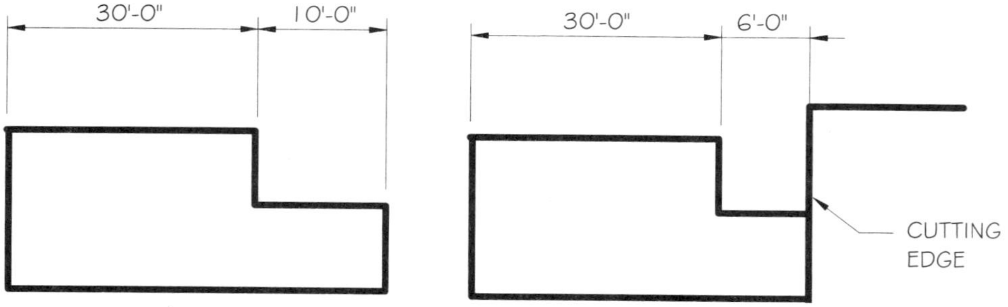

Figure 18.45 *The* **Trim** *command also affects an object, the dimension lines, and text.*

EDITING DIMENSION TEXT

Dimensions that have been placed on a drawing can be edited using the **Properties**, **Dimedit** and **Dimtedit** commands.

EDITING DIMENSION TEXT USING PROPERTIES

The dimension text can easily be edited using the **Properties** command. Use the following steps to alter dimension text using the **Properties** command:

1. Select the dimension to be edited.
2. Select the **Properties** button on the **Standard** toolbar or by one of the other access methods.
3. From the **Alphabetic** tab, choose **Text Override**.
4. Enter the desired dimension in the **Text Override** edit box.
5. Continue to alter properties or close the **Properties** dialog box.

Text entered in the **Text Override** edit box will override the measured dimension of the feature. The true measurement will continue to be displayed in the **Measurement** box of the **Properties** dialog box. Dimension text properties can also be altered by selecting the text, right-clicking and then selecting **Properties** from the shortcut menu. Other options for altering the location and text will be introduced later in this section.

Note: This feature can be especially helpful to dimension an object that must be a certain size, but you were unsure of the exact size when the part was drawn and dimensioned. If you know bolts are typically placed in steel plates at a 3" spacing, you can draw and dimension a detail prior to receiving the final calculation from the engineer. If the engineer determines that this application needs the bolts at 3 1/2" spacing, the dimension can be altered without editing the bolt locations on the drawing. The bolt locations will not be drawn to scale, but if the math works, so does the detail, and you've saved time.

EDITING DIMENSION TEXT USING DIMEDIT

The **Dimedit** command allows dimension text to be edited. The existing text can be edited, rotated, moved, or restored to its original position. The angle of the extension lines can also be altered from perpendicular to oblique. Start the command by selecting the **Dimension Edit** button on the **Dimension** toolbar, or by typing **DED** (ENTER) at the Command prompt:

Command: **DED** (ENTER) *(Or click the Dimension Edit button.)*

Enter type of dimension editing [Home/New/Rotate/Oblique]
<Home>:

Changing the Dimension Text

To alter existing text, select the New option:

> Enter type of dimension editing [Home/New/Rotate/Oblique]
> <Home>: **N** (ENTER)

This will produce the **Multiline Text Editor** and allow the desired text to be entered. Remove the brackets to replace the existing text. If the new text is placed in front of the brackets, the new text will be placed in front of the existing dimension text. Enter the new text and select the **OK** button to remove the editor. The prompt will continue:

> Select objects: *(select dimension text to be altered)* 1 found
>
> Select objects: (ENTER)
>
> Command:

The result of the command can be seen in Figure 18.46.

Rotating Existing Text

Choosing the Rotate option allows existing text to be rotated. Selecting the text in Figure 18.29 to be rotated to a 45° angle would require the following command sequence:

> Command: **DED** (ENTER) *(Or click the Dimension Edit button.)*
>
> Enter type of dimension editing [Home/New/Rotate/Oblique]
> <Home>: **R** (ENTER)
>
> Specify angle for dimension text: **45** (ENTER)
>
> Select object: *(Select text to be altered.)*
>
> Select object: (ENTER)
>
> Command:

This would produce text as shown in Figure 18.47.

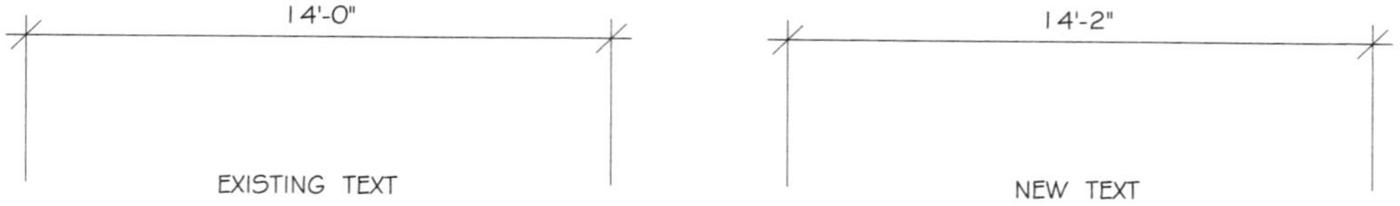

Figure 18.46 *The New option of the* **Dimedit** *command can be used to edit existing dimension text.*

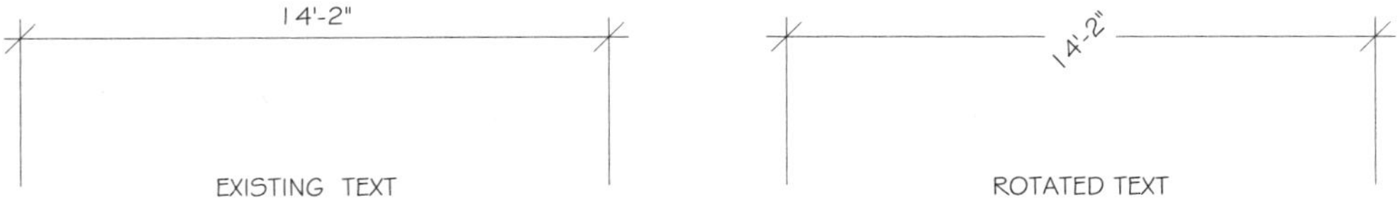

Figure 18.47 *The Rotate option of the* **Dimedit** *command can be used to rotate existing dimension text.*

Altering the Extension Line Angle

The angle of existing extension lines can be altered using the Oblique option of **Dimedit**. The command sequence to alter the angle of the extension lines in Figure 18.48 is as follows:

> Command: **DED** (ENTER) *(Or click the Dimension Edit button.)*
>
> Enter type of dimension editing [Home/New/Rotate/Oblique] <Home>: **O** (ENTER)
>
> Select object: *(Select dimension to be altered.)* 1 found
>
> Enter obliquing angle *(press (ENTER) for none)*: **-75** (ENTER)
>
> Command:

The results of the sequence can be seen in Figure 18.48.

Returning Home

Using the default option of Home will return the text to its default position.

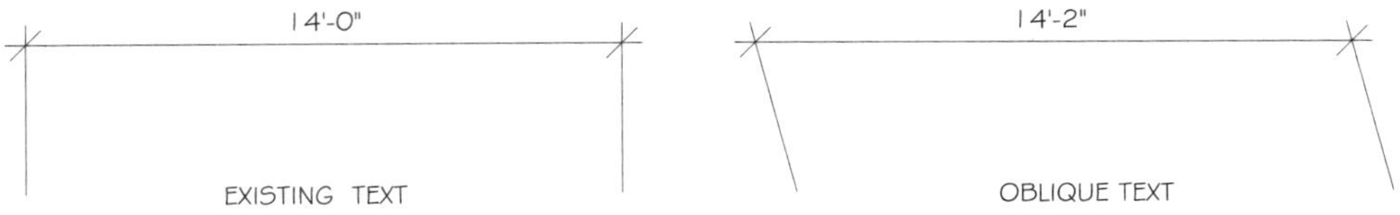

Figure 18.48 *The Oblique option of the* **Dimedit** *command can be used to alter the oblique angle of existing extension lines.*

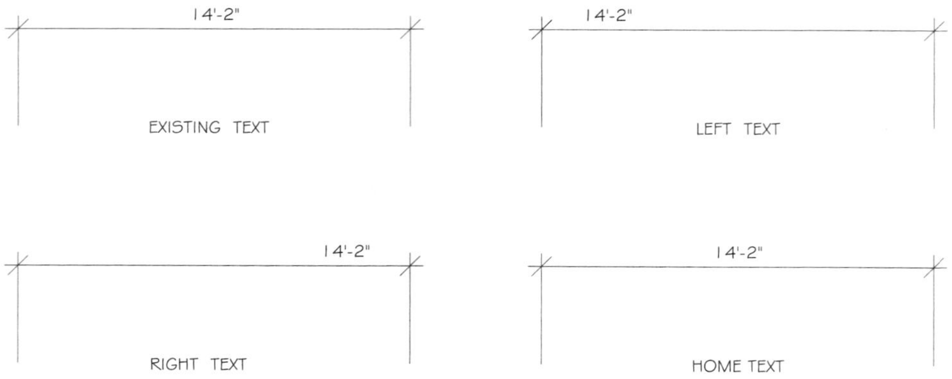

Figure 18.49 *The options of* **Dimtedit** *will alter the location of existing dimension text.*

EDITING DIMENSION TEXT USING DIMTEDIT

The **Dimtedit** command is used to position the text relative to the dimension line. Start the command by selecting the **Dimension Text Edit** button on the **Dimension** toolbar, by selecting **Align Text** from the **Dimension** menu, or by typing **DIMTED** (ENTER) at the Command prompt. Any access method will prompt you to select a dimension to be edited. Once a dimension is selected, the following prompt will be displayed.

> Specify new location for dimension text or [Left/Right/Center/
> Home/Angle):

Choosing the Left option will move the text toward the left end of the dimension line. Choosing Right will move the text toward the right end of the dimension line. Choosing Angle will allow the text to be rotated, choosing Center will center the text on the dimension line, and choosing Home will return the text to its default position. Each option can be seen in Figure 18.49.

UPDATING DIMENSIONS

The **Dimension Update** command allows existing dimensions to be selected and altered to a different style. The command is started by selecting the **Dimension Update** button on the Dimension toolbar or by selecting **Update** from the **Dimension** menu. The command duplicates the action of the **Dimension Text Edit** command but automatically applies the Apply option. The command sequence is as follows:

> Command: (Select the Dimension Update button.)
>
> DIMSTYLE

Enter a dimension style option

Current dimension style: Standard

Enter dimension style option (Enter the name of the desired style or press (ENTER) to update the dimension to the current style)

[Save/Restore/Status/Variable/Apply/?] <Restore>: apply

Select objects: *(Select any dimension(s) whose settings are to be updated to conform to the current dimension style.)* 1 found

Select objects: (ENTER)

Command:

WHAT IT ALL MEANS

More so than any area of AutoCAD, dimensioning seems to cause new users the most anxiety. You've read through what must seem like an unlimited number of commands and options, and you might be having a hard time getting all of this information organized. The good news is that once you set the variables using the dialog boxes and store them in template drawings, most of your work is done.

ALTERNATIVE METHODS

Now that you've heard the good news, you might have guessed that there is some bad news. As good as the software is, it is really intended for mechanical drafters. They're nice people, but they do things so differently. You've probably noticed that all of the extension lines throughout the last two chapters have been created with continuous lines. This is totally unacceptable for many architectural offices. Continuous extension lines have traditionally been used to represent dimensions that extend to the face of a surface such as a stud wall. Center extension lines are traditionally used to represent the center of objects. This shortcoming can be overcome, but at a cost to other dimensioning values.

To take best advantage of the editing qualities, dimensions typically are placed as associative dimensions. Dimensions could be set as normal dimensions, which would allow the extension lines to be edited using the Change command. Another option is to suppress the extension lines for interior objects, and place them in the drawing using the **Line** command using a centerline type.

An alternative to giving up the qualities of associative dimensions is to use the **Explode** command; this command will be covered in detail in Chapter 20. Basically, the command will allow you to select a group of objects that function as one object, and then allow you to separate that one group. This will allow the majority of dimensions to retain their editing qualities, while allowing the selected dimensions to be edited as individual lines, arrows, and text.

CHAPTER 18 EXERCISES

1. Open drawing Floor14. Use the appropriate dialog box and a style suitable for dimensions with both extension lines, a style with one dimension line, and a style with no extension lines. Set the text offset from the dimension line for all styles to be 1/16", with a baseline increment of 1/2". Assign each style a different color. Mark circles with a 1/8" center mark and provide for centerlines. Use 1/8" for the feature offset and extension above line settings. Use 1/8" long tick marks and a 1/8" long tick extension. Set the text height as 1/8" with the text forced inside, and above the extension line. Set the zero suppression controls so that a zero will be shown following feet, but not preceding inch sizes. Save these styles and then save the drawing as FLOOR14.
2. Use appropriate dimension techniques to label the floor to plate height of drawing E-15-8. Save the drawing as ELEV.
3. Use appropriate dimension techniques to label the base footings created for drawing E-13-4. Save the drawing as FOOTINGS.
4. Use appropriate dimension techniques to label the two-story footing created for drawing E-13-4. Show the concrete as 8" above grade, with an 8" stem wall, a 7"x15" footing that extends 18" below the finish grade. Save the drawing as FOOTINGS.
5. Open drawing 14-6. Use baseline dimension methods with dot line terminators to locate each tree center. Place all dimensions separate from all other information.
6. Open drawing E-11-7 and provide the needed dimensions to describe the structure. Use the required commands to create the linetypes and text of the original problem. Hatch the 8" wide wall with ANSI31. Save the drawing as E-18-7.
7. Open drawing E-9-7 and provide the needed dimensions and notes to describe the foundation plan created for E-9-7. Change the lines representing the beams to 3" wide lines. Label the main structural members to represent post and beam methods of your region. Save the drawing as E-18-7.

8. Start a new drawing and use the drawing below as a guide to show the support for two 6 3/4" x 37 1/2" glu-lam beams. Establish separate layers for the beam, steel, text, and dimensions. Completely label and dimension the drawing. Set the required scale factors for plotting at a scale at 1"=1'–0".

   The beams will be supported on a 5/16" x 6 7/8" x 24" base plate with two 5/16" x 10" x 24" side plates welded to the base plate with a 1/4" fillet weld. The base plate will be supported by a 6"x6"x1/4" steel column and welded with a 5/16" fillet weld all around. Each beam will be bolted to the side plate with four 5/8" diameter machine bolts. Bolts will be 1 1/2" in from the top and sides and be 3" on center.

   A second connector plate will be located with the center of the 3" x 22" x 5/16" steel plate 8" down from the top of the beam. Two bolts will be used to join the plate to the beam. Bolts will be 2" from the plate end, and 3" on center. Save the drawing as E-18-8.

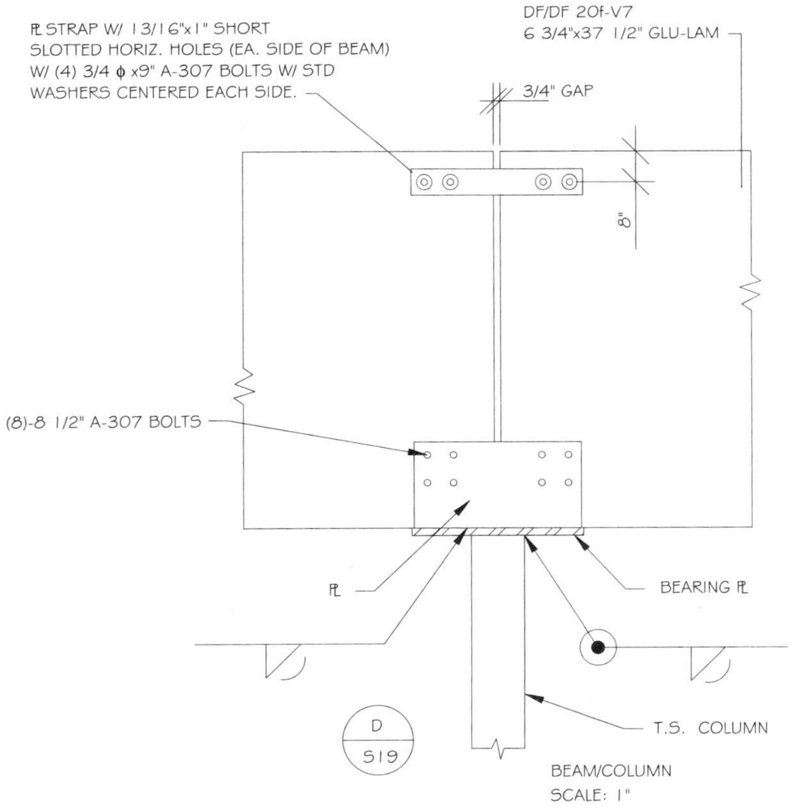

9. Start a new drawing and use the drawing below as a guide to show the column-to-base-plate connection. Establish separate layers for the steel, concrete, text, hatch, and dimensions. Completely label and dimension the drawing. Set the required scale factors for plotting at a scale of 1 1/2"=1'–0".

   The column will be a 6"x6"x3/8" t.s. welded to a 10"x10"x7/8" base plate with a 3/16" fillet weld all around. Use four 18" long anchor bolts set in an 8 1/2" square grid around the column. Support the base plate on 1" mortar. Save the drawing as E-18- 9.

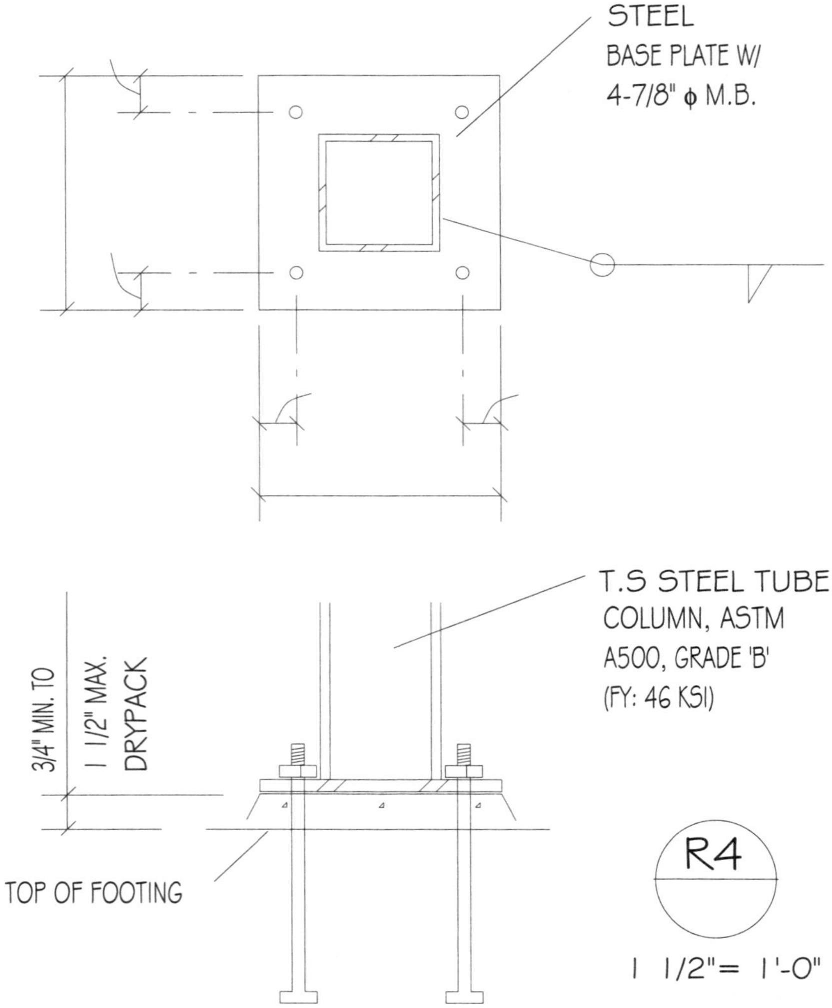

10. Start a new drawing and use the drawing below as a guide to show the beam-to-beam connection. Establish separate layers for the steel, text, hatch, and dimensions. Completely label and dimension the drawing. Set the required scale factors for plotting at a scale of 1 1/2"=1'–0".

    The support beam is a W18x76 (18 1/4" dp.) steel wide flange with a W18x46 (18" dp.) on the left and a W18x40 (17 7/8" dp.) on the right. Use two 4"x12"x5/16" steel connection plates welded to the W18x76 with 5/16"x12" fillet each side. Use four machine bolts through the plate and beam @ 3" o.c., 1 1/2" from the edge. Save the drawing as E-18-10.

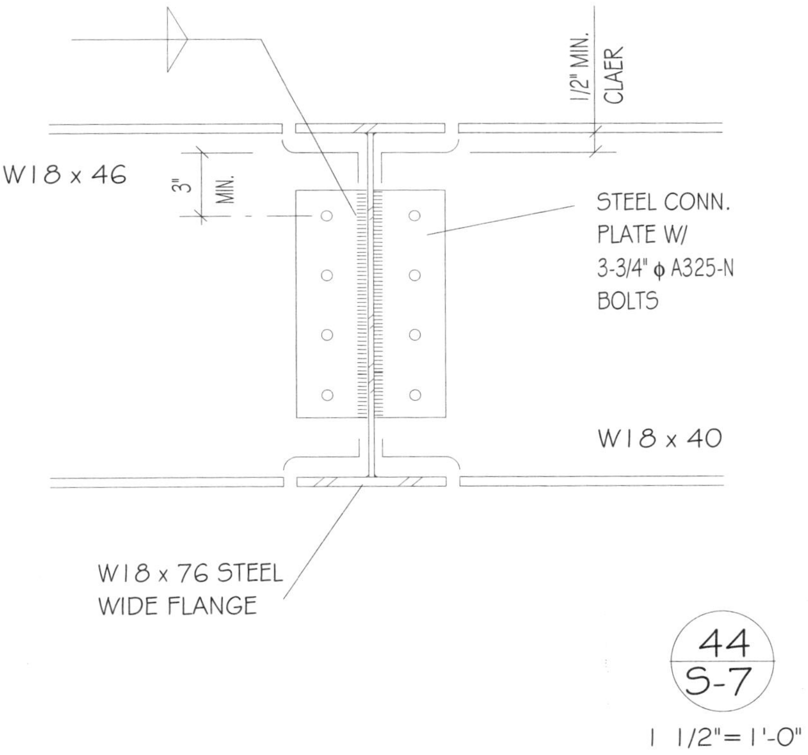

11. Open drawing FLOOR14 and use the attached drawing as a guide to draw the slab-on-grade plan. Establish separate layers for the concrete, text, and dimensions. Completely label and dimension the drawing. Save the drawing as E-18-11.

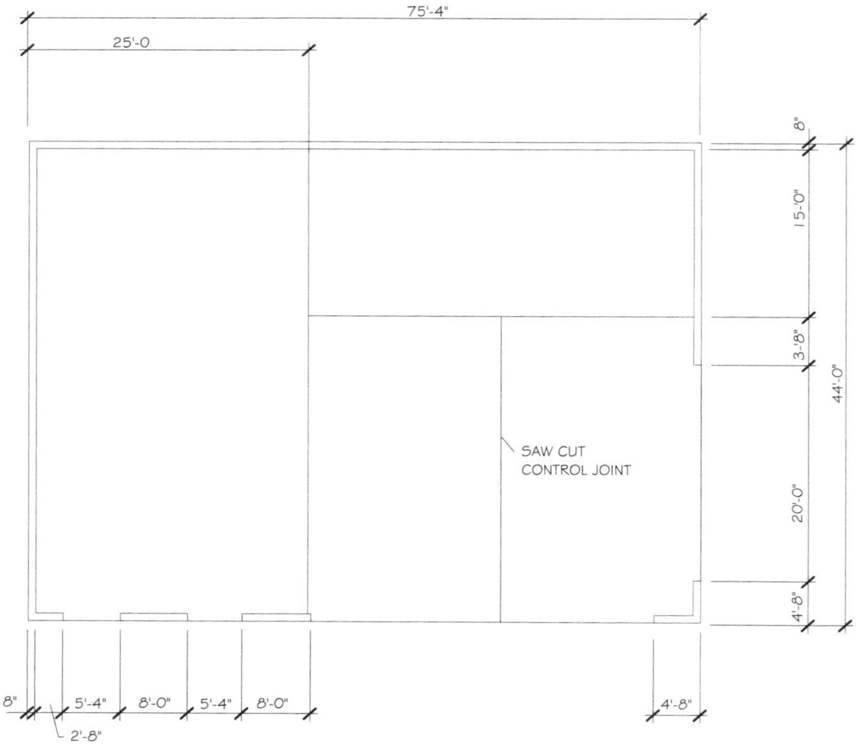

SLAB PLAN

1/8" = 1'-0"

12. Start a new drawing and use the drawing below as a guide to steel framing elevation. Establish separate layers for the steel, text, and dimensions. Completely label and dimension the drawing. Set the required scale factors for plotting at a scale of 3/8"=1'–0". Establish the eave height as 24 feet. Locate the first Z brace at 7 feet above the floor, with the balance at 72" o.c. Vertical supports will be set 12" in from each end of the 100'-0" structure with interior supports at 25'–0" o.c. Save the drawing as E-18-12.

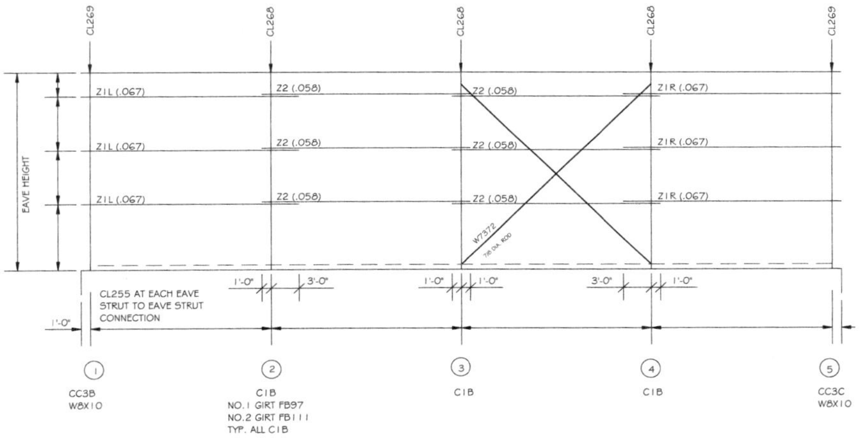

13. Start a new drawing and use the drawing below as a guide to draw the wall and floor detail. Establish separate layers for the wood, concrete, steel, text, and dimensions. Completely label and dimension the drawing. Set the required scale factors for plotting at a scale of 1/2"=1'–0". Establish the upper floor 30" above the lower floor, and the bottom of the 4x10 that supports the lower floor is to be 18" clear to finish grade. Dimension the height of the rail as 42". Save the drawing as E-18-13.

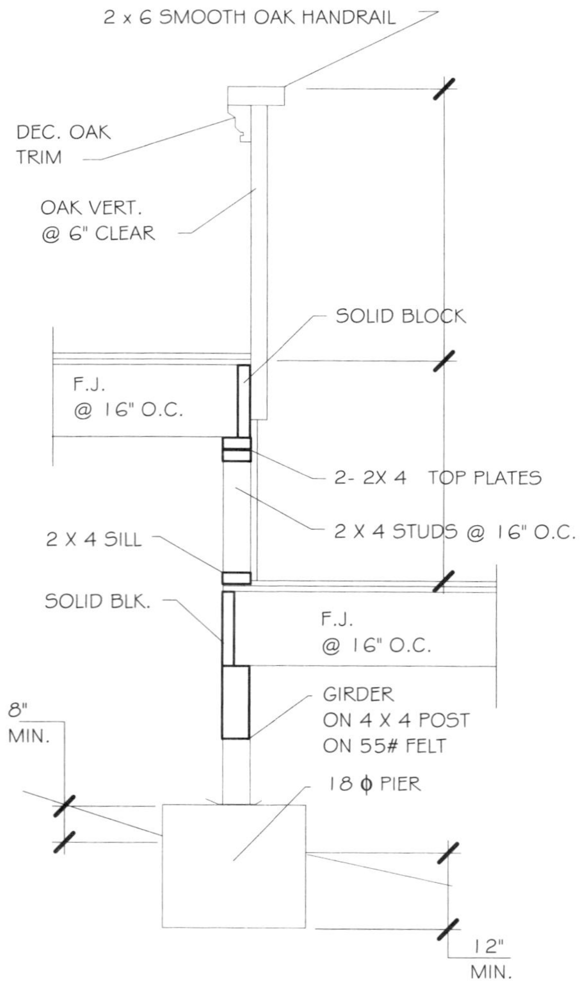

CHAPTER 18 QUIZ

1. A group of dimensioning controls that can be saved for future use is called a ______.
2. A ________________________________ is used to control individual aspects of dimensioning.
3. What command can be typed to access a listing of dimensioning features?
4. How can styles be used in an engineering office that does both civil and structural projects?
5. What value controls the size of each dimension feature and how should this value be set for a drawing that will be reproduced at a scale of 1/4"=1'–0"?
6. What option will remove both extension lines from a dimension?
7. The spacing between two dimension lines is controlled by what option?
8. Use the **Help** menu and determine what the **Dimgap** controls.
9. List two features that can be displayed to describe round objects.
10. List five options for terminating a dimension line that are suitable for an architectural office.
11. List and describe two features of the extension line as it relates to the dimension line and the object being described, and give the value that controls each.
12. What are the options for controlling the height of text?
13. Name the dialog box that can be used to make changes to dimensioning text.
14. What style will be used as a base when dimensioning styles are created?
15. What option will allow for changing an attribute without changing the entire style?
16. What does the **Dimtedit** command affect?
17. Describe how associative dimensions will affect editing.
18. How can the Leader command be used to provide two arrows pointing to the same text?
19. What dimensioning command is used if the text is to be placed so that it is not parallel, perpendicular, or aligned with a surface?
20. What command can be used to change existing dimensions to the current style?
21. Sketch examples of right angle and closed terminators.
22. Explain the process for creating a user-defined terminator.
23. Explain the options to display a centerline in each circle to be dimensioned.
24. Give the proper scale factor for drawing a site plan to be plotted at a scale of 1/8"=1'–0" and explain how the value is determined.

25. Explain the process to place text above the dimension line.
26. List and explain the options for controlling zeros in dimensions.
27. Explain the difference between associative and normal dimensions.
28. Explain the options of the **Dimedit** command.
29. Explain the effect of selecting an extension line to be erased.
30. Explain the effect of stretching a window that is dimensioned 3'–0" long to a length of 6'–0".

SECTION 4

GROUPING AND PRESENTING INFORMATION

- CREATING BLOCKS AND WBLOCKS
- ADDING ATTRIBUTES TO ENHANCE BLOCKS
- OBLIQUE AND ISOMETRIC DRAWINGS
- WORKING WITH MULTIPLE DRAWINGS
- COMBINING DRAWINGS USING XREF
- WORKING WITH LAYOUTS AND VIEWPORTS
- CONTROLLING OUTPUT
- AUTOCAD AND THE INTERNET

CHAPTER 19

Creating Blocks and Wblocks

You might have been wondering, while working through the first 18 chapters, why you haven't started drawing some really neat projects. Since Chapter 2 you've had the basic drawing skills to draw most projects you could imagine. Subsequent chapters have helped you to develop skills needed to ease the drawing and design process but still have not equipped you to master a complicated project easily. This chapter will

- Introduce you to creating blocks for repetitive use
- Explore methods of inserting blocks into a drawing
- Examine methods for editing blocks that have been inserted in a drawing
- Explore methods to redefine blocks that have been inserted in a drawing
- Examine the benefits of using wblocks in multiple windows

Commands to be introduced in this chapter include

- **Block**
- **Insert**
- **Minsert**
- **Base**
- **DesignCenter**
- **Explode**
- **Refedit**
- **Wblock**

BLOCKS

Drawings are composed of symbols. To show a door on a floor plan, you don't draw a door; instead you use a line and an arc to represent a door. A block is a group of objects that are treated as one object. The objects used to represent a door symbol can be saved as a block and inserted throughout the drawing as one object. The **Block** command allows symbols that are used repeatedly to be created, saved, and inserted easily in the drawing in which they were created, which provides many benefits for the user.

BENEFITS OF BLOCKS

Symbols that are used frequently can be saved as a block, which will greatly increase drawing speed, uniformity, and efficiency.

Speed

If you're drawing a typical residential floor plan, doors will need to be represented 19 or 20 times. Rather than draw each door, you can draw one door and save it as a block. As a block, a door can be inserted quickly and the size altered or the position rotated, as seen in Figure 19.1. With a little more effort, these same operations can also be accomplished with the **Copy**, **Rotate**, **Mirror**, **Array**, and **Scale** commands, but why use several commands when you could use one?

Editing

Throughout the evolution of construction drawings, many changes and revisions will take place. Blocks can greatly aid in this revision process. If a client decides that all of the doors in a building are to be 6 inches wider, several hours might be required to change all of the doors on the various floor plans. If the doors are drawn as a block, the block can be redefined and all references to that block will be updated automatically.

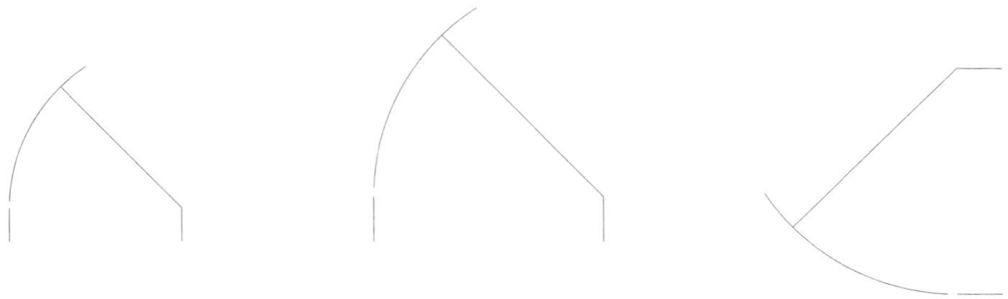

Figure 19.1 *Blocks can be inserted in a drawing, altered in size, and rotated to meet specific needs.*

Attributes

An attribute is text that is associated with a block. Many blocks, such as a door, require text to explain variations. Size, material, quantity, and supplier are examples of information that can be saved as an attribute with a door block. This text is called an attribute and can vary with each application of the block. Attributes can be displayed on the screen as text or they can be stored with the drawing but made invisible. Attributes associated with blocks can also be extracted from a drawing and transferred to a database form, such as a door schedule. Chapter 20 will explore adding attributes to blocks.

Efficiency

In addition to saving time and providing a quick method for revisions, the **Block** command combines the storage of groups of objects into one unit. In drawing a door, information must be stored about the start and end points of the original line and arc, the radius of the arc, the scale factor as the symbol is enlarged or reduced for various uses, and the location of each unit as it is displayed throughout the drawing. When the symbol is installed as a block, only a block reference is stored, saving valuable storage space.

USES FOR BLOCKS

Before a possible listing of blocks is considered, it should be remembered that the list is endless. Any symbol that is used repeatedly should be stored in a template drawing as a block. A sample listing for architectural blocks would include items for the site, floor, and foundation plans, elevations, and sections as listed below.

Site plans—North arrow, trees, shrubs, spas, pools, fountains, curbcuts and driveways, utility symbols, and general notes. Plots for subdivisions might include structure shapes, sidewalks, and decks or patios.

Floor plans—Appliances, door and window symbols, heating equipment, heating and cooling symbols, electrical symbols, and plumbing symbols.

Framing plans—Joist and rafter direction indicators, section tags, and general notes.

Foundation plans—Piers, post and piers, beam and column symbols, general notes, and foundation details.

Elevations—Door styles, window styles, shutters, trim, eave molding, decorative posts, garage doors, gable wall vents, trees and shrubs, lighting fixtures, and people.

Sections—Structural shapes such as plates, sills, double sills, beams, columns and piers, foundation components, eave components, insulation, standard dimensions, and common notes.

In addition to drawing simple symbols as a block, you can combine several individual blocks to form a larger block. Figure 19.2 shows an example of a typical truss-to-plate intersection. The block could be inserted in a section drawing, mirrored, and then the truss shape completed by the **Fillet** and **Extend** commands, as seen in Figure 19.3.

DRAWING OBJECTS FOR BLOCKS

Because a block is a group of objects that is saved as one object, there is no special method for drawing the block. The components that will comprise the block can be created using any of the drawing or editing commands. Some consideration should be given to the layer the block is created on and the size of the block while it is being constructed, to avoid some problems when the block is inserted in the drawing.

The Effects of Layer on Blocks

A block will take on the characteristics of the layer on which it is created. A block can also consist of objects that have been drawn on several layers with different linetypes and colors. When the block is inserted in a drawing, the block will retain its original characteristics. If a lavatory is created on a layer titled MPLUMB and inserted in a drawing with the UPPLUMB layer current, the lavatory will be inserted in the drawing on the MPLUMB layer. If the block was created on a layer that does not exist in the current drawing, the layer containing the block will be created. There are, however, three exceptions to this guideline.

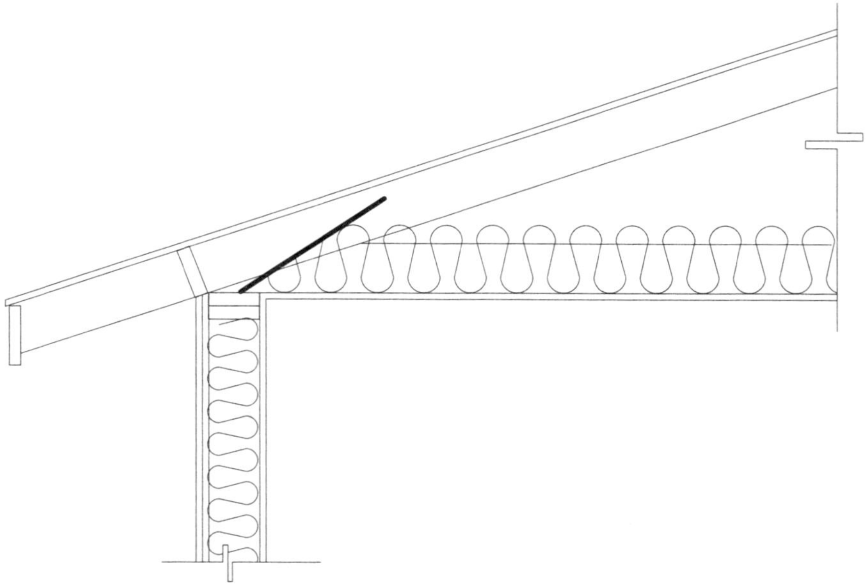

Figure 19.2 *Several objects can be combined to form a block.*

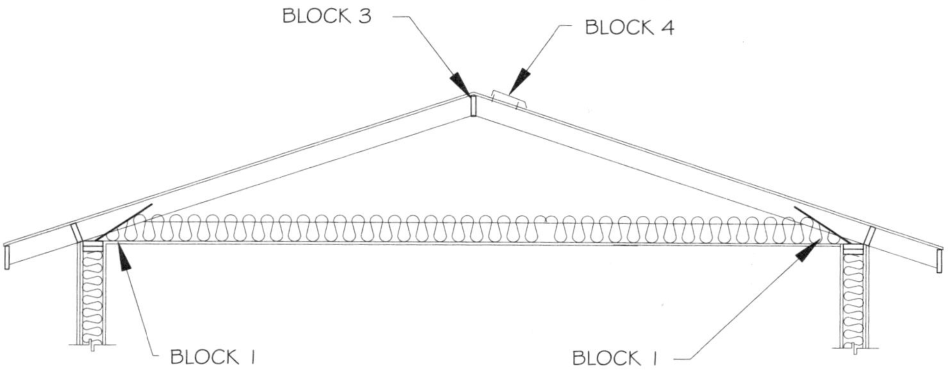

Figure 19.3 *Using the* **Mirror, Fillet,** *and* **Extend** *commands, this section was created by inserting and connecting blocks.*

- Blocks that are created on the 0 layer will be placed on the current layer when placed in a drawing.
- Blocks drawn with the color created BYLAYER are drawn with the color of the current layer, but will reflect the color of the layer into which the block is inserted.
- Blocks drawn with the linetype created BYBLOCK are drawn with the linetype of the current layer, but will reflect the linetype of the layer in which the block is inserted.

Block Size

Once the block is saved, it can be inserted throughout a drawing. When the **Insert** command is introduced later in this chapter, you will have an opportunity to alter the size of the block in the X and Y directions. Altering the size of the block during the insertion process can be easier if the size of the block is considered during creation.

Some blocks, such as toilets, electrical symbols, and appliances, are a standard size. These blocks can be drawn at their actual size and inserted. Other blocks, such as doors, windows, tubs, and showers, vary in size with each application. Drawing these blocks in one-unit squares will allow the size to be adjusted each time it is inserted. Rather than having separate blocks for 4', 5', and 6' windows, one window can be created, and the X value controlling its length can be altered.

The size of the unit can vary depending on the drawing the block will be inserted in. On floor plans, you might use 1" as the unit and scale up, or use 1' as the unit and scale down. Figure 19.4 shows a window created using a 12" unit, and the effects that can be achieved with one block.

Figure 19.4 *Blocks that vary in size should be created using one unit, such as a foot, so the size of the block can be adjusted easily each time the block is inserted.*

CREATING A BLOCK

Figure 19.5 shows the drawing process for creating a toilet, which can then be saved as a block. Keep in mind that although the toilet has been drawn, it is not a block yet. Once the drawing is ready to be reproduced, start the **Block** command by selecting the **Make Block** button on the **Draw** toolbar, by typing **B** (ENTER) at the Command prompt or by selecting **Make** from the **Block** cascading menu on the **Draw** menu. Each method will produce the **Block Definition** dialog box shown in Figure 19.6.

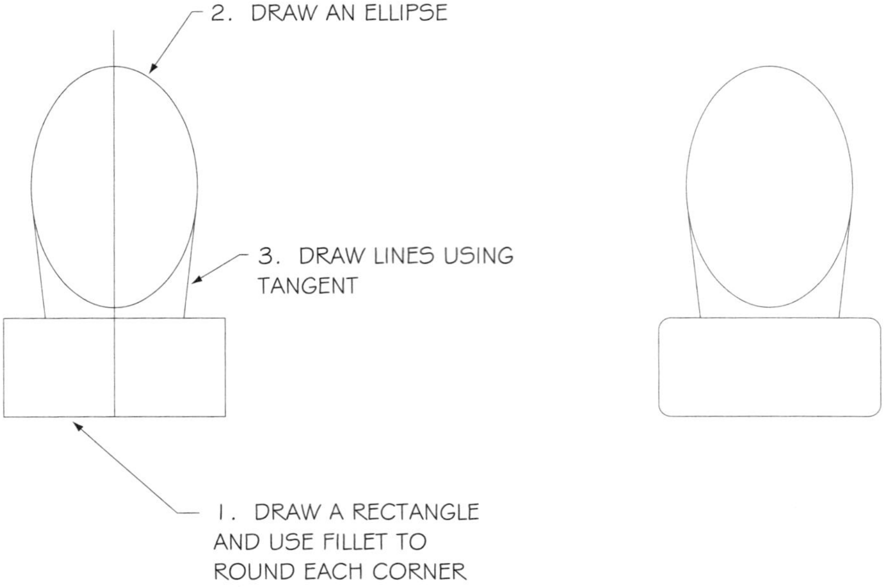

Figure 19.5 *A toilet can be created by drawing an ellipse and a rectangle and connecting the two with straight lines. The* **Fillet** *command was used to round each corner.*

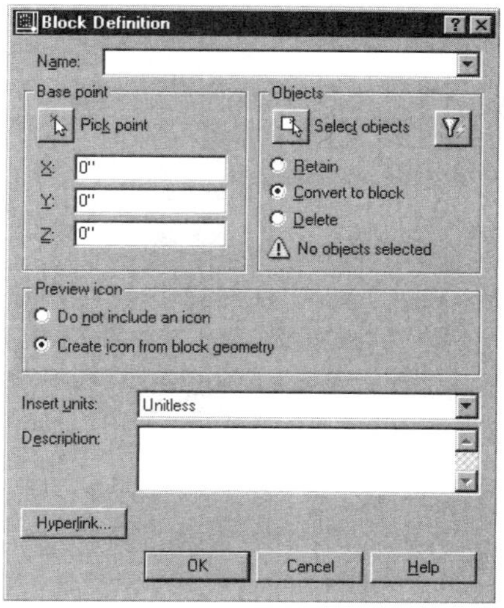

Figure 19.6 *Once the drawing is ready to be reproduced, start the* **Block** *command by selecting the* **Make Block** *button on the* **Draw** *toolbar, by typing* **B** *(ENTER) at the Command prompt, or by selecting* **Make** *from* **Block** *on the* **Draw** *menu.*

Choosing a Name

Start the process by entering the name of the block to be created in the **Name** edit box. A name of up to 255 characters can be used to describe the block. The block name can contain letters, digits, and special characters such as $, -, and _. Any character not used by Microsoft Windows or AutoCAD, including blank spaces can be used. Use a name that will describe the contents. For instance, DOOR-SWG-32 would be an appropriate name to describe a 2'–8" swinging door. If a name for an existing block is provided for the new block, the old block will be destroyed and the new block will take its place. Before it is vaporized, you will be given an alert message by AutoCAD warning that the block is about to be redefined.

Selecting the Objects

Once the name has been provided, click either the **Pick point** or **Select objects** button. Each option must be selected, but the order will not affect the outcome of the command. For this discussion, the object will be selected first. Select the **Select objects** or **Quick Select** button to define the objects that will be in the block. (The use of the **Quick Select** dialog box will be explained shortly.) Selecting the **Select objects** button temporarily removes the dialog box from the screen and allows the objects that will make up the block to be selected. As the dialog box is removed, the command line will prompt to select objects. Objects to comprise the block can be selected using any object selection method. Once the desired objects have been

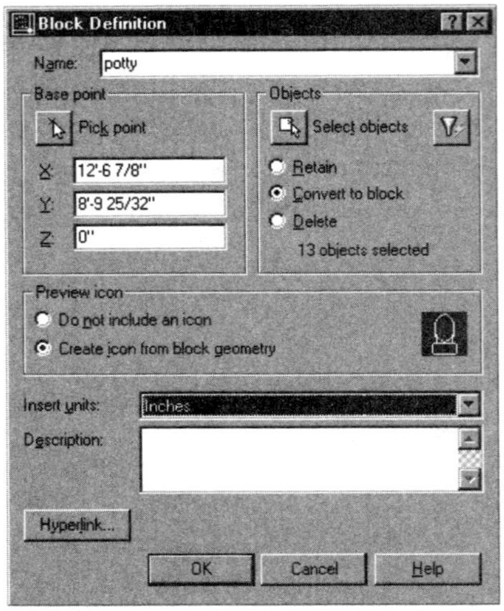

Figure 19.7 *Once the objects to comprise the block have been selected, pressing (ENTER) will return the* **Block Definition** *dialog box with an image of the selected objects displayed in the Preview icon display box.*

selected, pressing (ENTER) will return the **Block Definition** dialog box. An image of the selected objects will be displayed in the **Preview icon** display similar to Figure 19.7. The option for no icon can be activated if you do not want to see the image you'll be using as a block.

Note: The **Blockicon** command can be used to display images of blocks stored in drawings created using AutoCAD 2000 or previous releases. The command can be entered at the Command prompt and will provide prompts for what blocks you would like to update. Pressing (ENTER) will update all blocks.

Using Quick Select

If the **Quick Select** button is selected, a **Quick Select** dialog box similar to Figure 19.8 will be displayed. The **Quick Select** dialog box can be used to define the filtering criteria that will be used to select the objects for the block. Although simple blocks such as the toilet can easily be selected using a window, **Quick Select** can be used to select complex shapes or multiple objects. Notice that the current filter is defined as the entire drawing in the **Apply to** edit box. If there is no current selection set, the entire drawing will be used as a block; this option will be explored later. To select an object for the block, click the **Select Objects** button to the right of the **Apply to** edit box. This will temporarily remove the dialog box from the screen and allow the objects to be selected. The process is the same as used with the **Select objects** button in the **Block Definition** dialog box. Once the desired objects

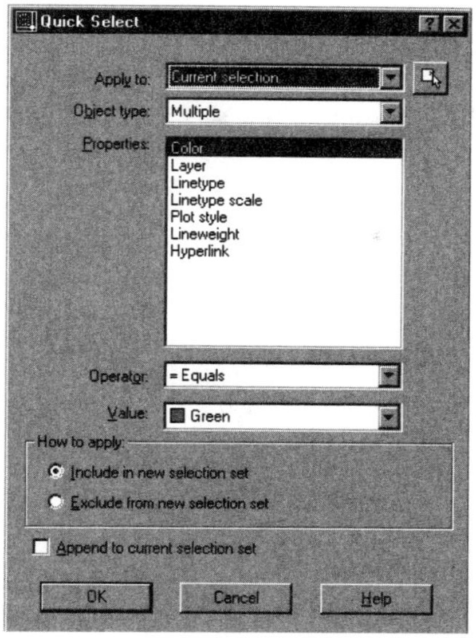

Figure 19.8 *The* **Quick Select** *dialog box can be used to define the filtering criteria that will be used to select the objects for the block.*

have been selected, pressing (ENTER) will return the **Quick Select** dialog box. Other key elements of the **Quick Select** dialog box include the following:

- **Object type** specifies the type of object to use to define the filter. The current choices are Multiple, and Block Reference. Since you're defining a block, choose Block Reference.
- The **Properties** box can be used to specify properties to be used as a filter. Objects to form the block can be defined by the color, lineweight, linetype, or any of the other listed properties.
- **Operator** controls the range of the selected property filter. Depending on the selected property, options can include Equals, Not Equal, Greater Than, Less Than, and Select All. Selecting the edit arrow displays a list to allow the operator value to be altered.
- **Value** defines the filter property value. If the selected property is a color, the value will specify a color. If the property is a layer, the 0 layer is the default value. Selecting the edit arrow displays a list to allow the value to be altered.
- The **How to apply** area of the dialog box specifies if the selected objects will form or be excluded from the selected set to form the block. Choosing **Exclude from new selection set** will create a block of all objects that do not match the selection filter.

Note: To create a block such as a toilet, use **Select objects**. **Quick Select** can be used to select all of the walls of a floor plan to form a block without having to select each one of them. If a filter set is defined that will include all of the walls, or exclude everything except the walls, the block can be defined much faster than if each wall has to be selected.

Object Selection Options

In addition to methods for selecting objects to define the block, the **Objects** area in the **Block Definition** dialog box contains options to retain, convert, and delete the selected objects.

Retain—Selecting **Retain** will keep the selected objects in their original state in the selection set. Once the object is selected to be a block, the original objects will remain in their current location.

Convert to block—Selecting this option will remove the selected objects from the selection set and replace them with a previously defined block.

Delete—Selecting this option will remove the objects that define the block from the drawing. This option works well if you create the toilet (or any other object) that is to become the block in the middle of the living room. With **Delete** active, once the object is selected to be a block, the original object will be removed from the drawing. It will still exist as a block in the drawing base, but it can't be seen until the block is inserted.

SPECIFYING THE BASE POINT

The insertion point of the block will align with a point that will be designated as the block is inserted into other parts of a drawing. The insertion point can be selected by entering X and Y coordinates or by selecting any point of the object. The **Pick point** button temporarily removes the dialog box from the screen and allows the insertion point to be selected. Figure 19.9 shows the effect of the insertion point of a block with the insertion point at the destination. The insertion point should be chosen using OSNAP based on the block's use. Blocks such as a toilet, a lavatory, or a sink will be inserted most easily using a centerpoint. Other objects such as a sill or beam might best be located from an edge or intersection. Figure 19.10 shows some typical insertion points for common architectural features. The insertion point also can be used as a rotation point for the block as the block is inserted in its new location. Figure 19.11 shows how a block can be rotated about its insertion point. Other options will be discussed as the Insert command is explored. Using a centerpoint or the lower left corner of a block will help to keep track of where to insert the block.

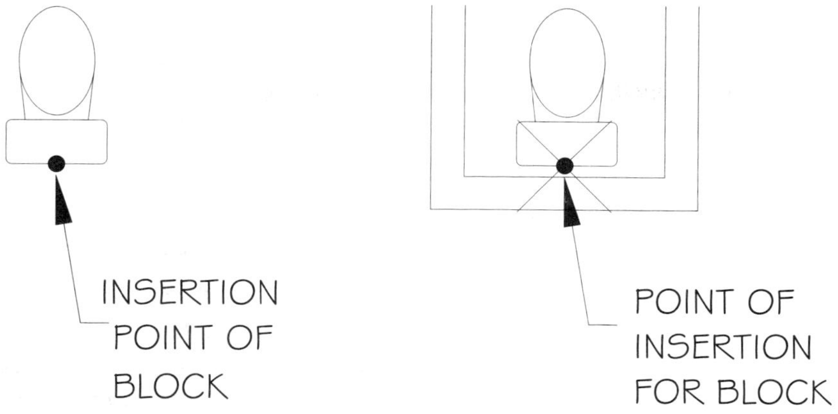

Figure 19.9 *An insertion point defines the location of the block to be used as a handle.*

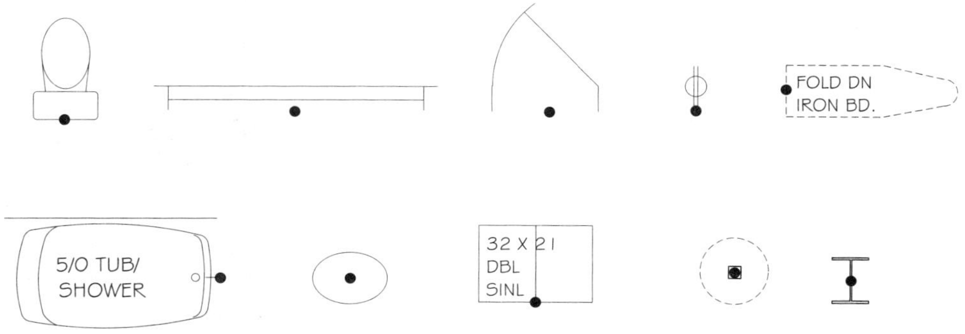

Figure 19.10 *Typical insertion points for common architectural features.*

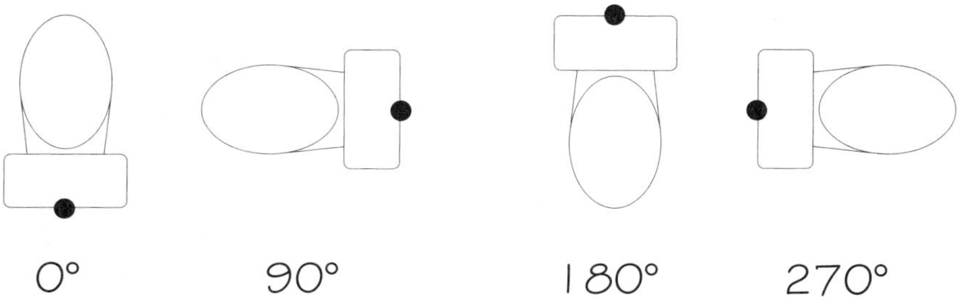

Figure 19.11 *Rotating a block around its insertion point.*

Description

The **Description** edit box allows a description to be assigned to the block. A block can be created without providing a description, but on a complex block, the description might help you distinguish it from a similar block.

Summary of Steps in Creating a Block

It's taken several pages to explain what a block is. Remember that a block is just a collection of objects that can be used over and over. Because they are so important to your drawing future, use the summary as you create blocks. Use the following steps to create blocks:

1. Display the **Block Definition** dialog box by selecting the **Make Block** button on the **Draw** toolbar, by typing **B** (ENTER) at the Command prompt, or by selecting **Make** from **Block** on the **Draw** menu.
2. Provide a name that will define the block.
3. Select the drawing objects to be reproduced as blocks.
4. Select a base point to be used the block is inserted in the drawing.
5. Select modifiers in the **Objects** and **Preview** portions of the dialog box.
6. Click **OK** to save the block to the drawing base.

The object has now been saved as part of the drawing base and is ready to be reused throughout the drawing. If another toilet, or what ever object is used as a block, needs to be added to the drawing, you can add it to the drawing by inserting the block rather than redrawing the objects.

Note: You've just read through the process to create a block. As you create a block, it might leave you waiting for something exciting to happen. The creation of a block is nothing special. It's what you can do with it that will revolutionize your drawing process.

PREPARING BLOCKS FOR INSERTION

Once a block has been created and saved, it can be placed in a drawing using the **Insert** command. The command can be accessed by selecting the **Insert Block** button on the **Draw** toolbar, by typing **I** (ENTER) at the Command prompt, or by selecting **Block** from the **Insert** menu. Each method will produce an **Insert** dialog box similar to Figure 19.12.

Choosing a Block

The **Name** edit box is used to select the block to be inserted. The name of the block to be inserted can be typed in the edit box or selected from a list of blocks contained in the drawing. Selecting the edit arrow beside the box will display a listing of all blocks contained in the drawing. Selecting the **Browse** button will open the

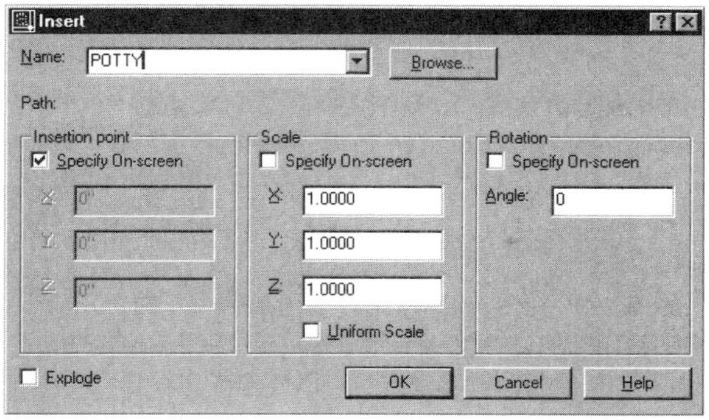

Figure 19.12 *Selecting the* **Insert Block** *button on the* **Draw** *toolbar, typing* **I** *(ENTER) at the Command prompt, or selecting* **Block** *on the* **Insert** *menu will display the* **Insert** *dialog box.*

Select Drawing File dialog box. The dialog box is the same as the file selection dialog boxes that you've used as drawing files are opened. The dialog allows a block or file to be selected for insertion. Inserting files will be introduced later in this chapter.

Specifying an Insertion Point

Activating the **Insertion Point Specify On-screen** check box allows an insertion point to be used to specify where the block will be located. With the **Specify On-screen** box inactive, the X and Y coordinate locations (Z is the vertical reference for 3D) can be used to locate the insertion point. As the block was created, an insertion point was determined. Once the desired insertion parameters are adjusted, a prompt will be displayed to provide a location to place the original insertion point. Figure 19.13 shows the relationship of the block insertion point to the destination insertion point. Once the insertion point is selected, the block will be dragged into the indicated position, and the scale can be altered. The parameters to control the insertion can each be adjusted before removing the dialog box to insert the block, or at the Command prompt before the block is inserted.

Scale

This area allows you to define scale factors for the block to be inserted. With the **Specify On-screen** check box active, a prompt will be displayed at the Command prompt before the block is inserted. With the box inactive, the X, Y, and Z edit boxes are activated. Unless a different value is entered, the value entered for X will become the default value for Y. Any dimensions contained in the block will be multiplied by the scale factor as the block is inserted. If the toilet in Figure 19.13 was drawn as full size, the scale for inserting the toilet would be 1. Inserting a window created as a 1' block to be a 6' long window in a 6" exterior wall would require an X scale factor of

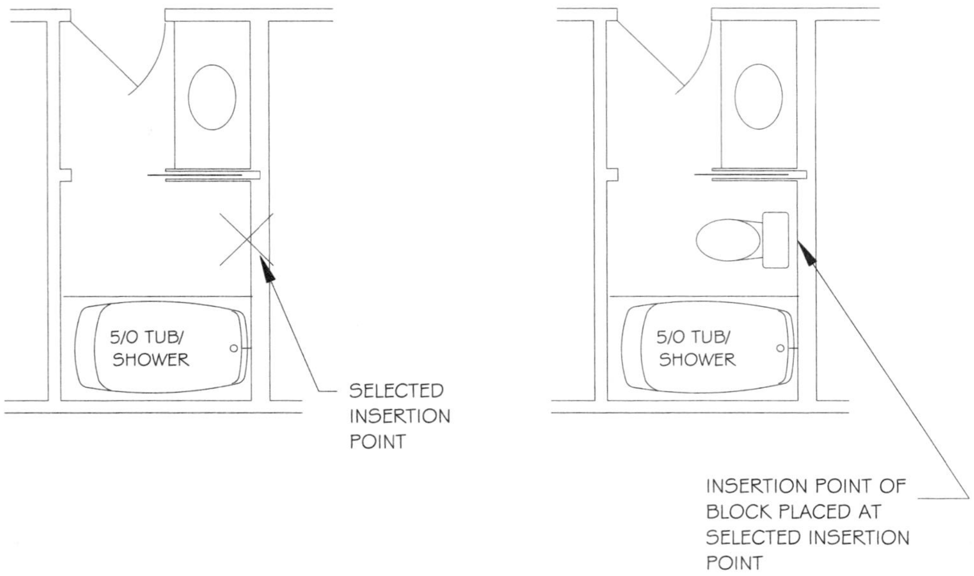

Figure 19.13 *The effect of the selected insertion point of a drawing and the insertion point of a block.*

6, and a Y scale factor of .5. With the **Uniform Scale** check box active, the Y value will equal the X value. The effects of the scale factor can be seen in Figure 19.14.

Negative Scale Factors

The location of a block can be altered by using negative X and Y values. Using negative values combines the Insert and Mirror commands. Figure 19.15 shows an example of negative value alternatives.

Rotation

The **Rotation** area allows you to supply a rotation angle. With the **Specify On-screen** box active, a prompt will be displayed at the Command prompt before the block is inserted. With the box inactive, the **Angle** edit box is activated and an angle can be entered to control the rotation of the block at insertion. Figure 19.16 shows an object rotated around its insertion point. Any angle can be provided for the rotation angle.

Explode

Blocks function as one object when they are inserted in a drawing. If the **Explode** check box is activated, as the block is inserted in the drawing, each object of the block will function as a separate object.

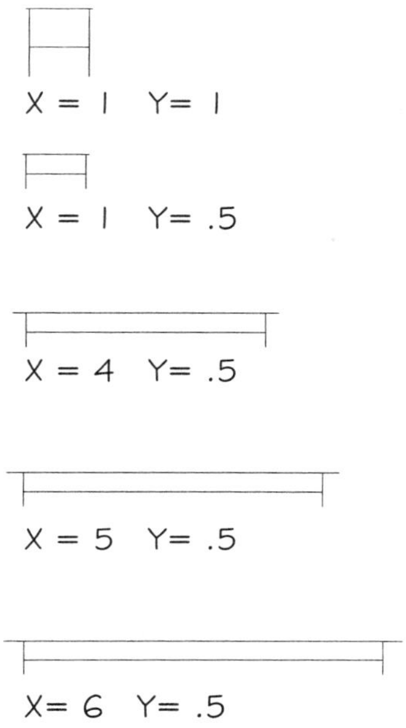

Figure 19.14 *The effect of X/Y scale factors on a block.*

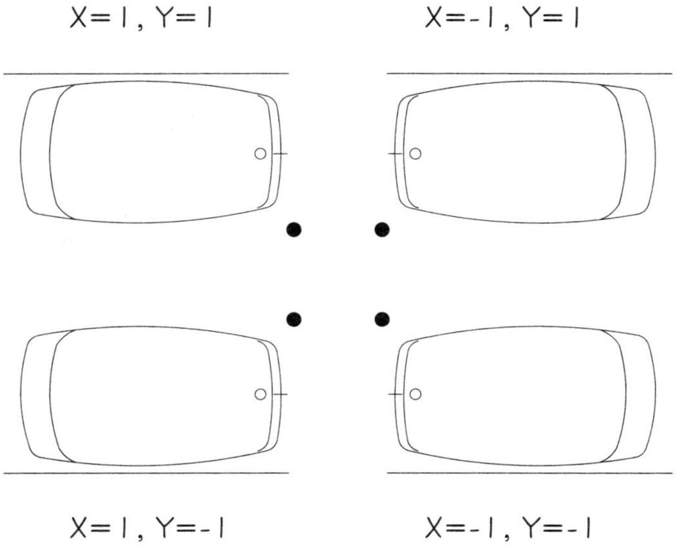

Figure 19.15 *Altering a block using negative scale factors.*

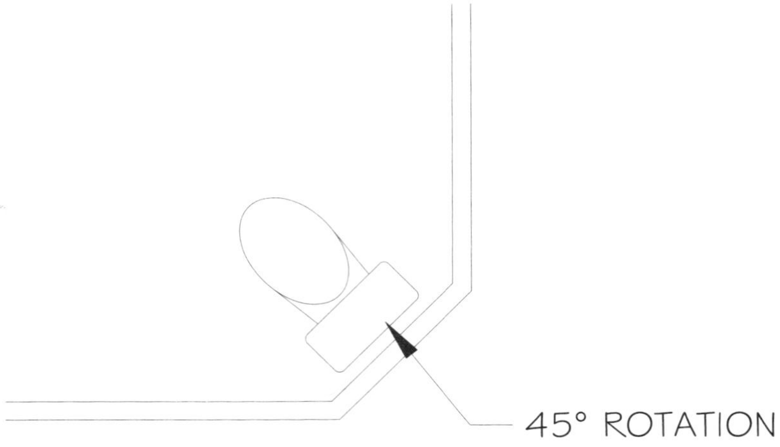

Figure 19.16 *Specifying an angle of rotation allows a block to be rotated relative to its insertion point.*

INSERTING BLOCKS

Once the parameters of the **Insert** dialog box have been set, selecting the **OK** button will close the dialog box and allow a selection point to be specified. With the **Insert** dialog box in the default settings, the following prompt will be displayed at the Command prompt:

> Command: **I** (ENTER) *(Or click the Insert Block button. Clicking the OK button will close the Insert dialog box and continue the Command prompt.)*
>
> Specify insertion point or [Scale/X/Y/Z/Rotate/PScale/PX/PY/PZ/PRotate]:

Selecting an insertion point will insert the block at the specified parameters. The scale, insertion coordinates, and rotation values that were set in the **Insert** dialog box can be altered at the Command prompt. Additional prompts are also shown at the Command prompt.

WORKING WITH BLOCKS

In addition to inserting blocks, several other options can be used to control the use of blocks. Useful tools include the **Minsert** command for multiple inserts, using an entire drawing as a block, building blocks by using other blocks in a process known as nesting, exploding blocks, and controlling blocks with the **DesignCenter**.

MINSERT—MAKING MULTIPLE COPIES

The **Minsert** command combines the features and the command sequences of **Insert** and **Array**. **Minsert** will allow blocks to be arrayed in a rectangular pattern typical of what might be found in the interior column supports for multilevel structures, or the supports of a post-and-beam foundation. Start the **Minsert** command by typing **MINSERT** (ENTER) at the Command prompt. Figure 19.17 shows an example of a multiple insert. The drawing was created using the following command sequence:

Command: **MINSERT** (ENTER)

MINSERT

Enter block name or [?]: <toilet> **PIER** (ENTER)

Specify insertion point or [Scale/X/Y/Z/Rotate/PScale/PX/PY/PZ/PRotate]: (Select the insertion point.)

Enter X scale factor, specify opposite corner, or [Corner/XYZ] <1>: (ENTER)

Enter Y scale factor, < use X scale factor>: (ENTER)

Specify rotation angle: <0>: (ENTER)

Enter number of rows (—-) <1>: **3** (ENTER)

Enter number of columns (|||) <1>: **6** (ENTER)

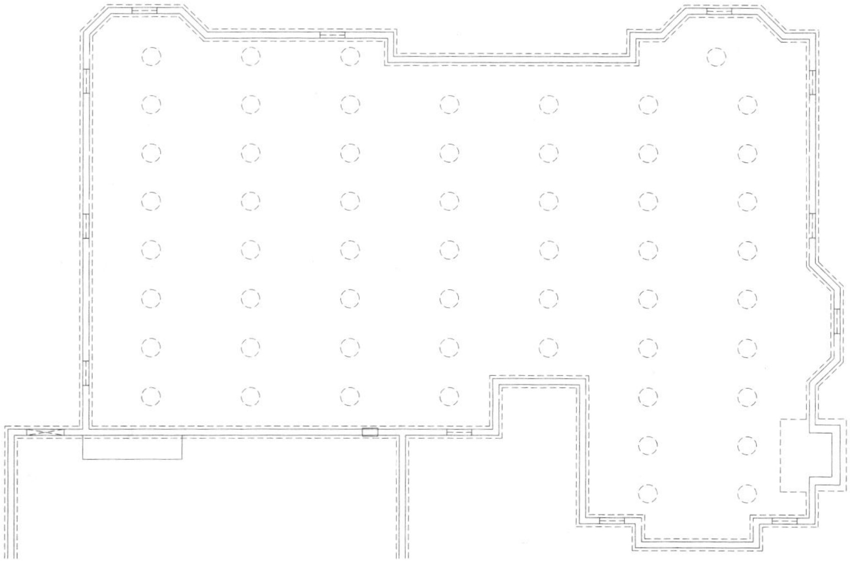

Figure 19.17 *The* **Minsert** *command allows blocks to be arrayed in a rectangular pattern. (Courtesy Piery & Barclay Designers, Inc., A.I.B.D.)*

Enter distance between rows or specify unit cell (---): **96** (ENTER)

Specify distance between columns (|||): **48** (ENTER)

Command:

The array pattern created with **Minsert** has the qualities of a block, except that it cannot be exploded.

MAKING AN ENTIRE DRAWING A BLOCK

An entire drawing can be used to form a block. When the entire drawing is to be a block, an insertion point other than 0,0 might be desired. The **Base** command can be used to define a new base point. Access the command by typing **BASE** (ENTER) at the Command prompt or by selecting **Base** from **Block** on the **Draw** menu. Each will produce the following prompt:

Command: **BASE** (ENTER)

BASE

Enter base point: *(Select a point.)*

The base point will now become the insertion point.

A better alternative to this method of using an entire drawing as a block will be introduced later, as wblocks are introduced, and in Chapter 23, as externally referenced drawings (Xrefs) are explored.

NESTING BLOCKS

So far you've been introduced to creating blocks to represent common symbols such as a toilet, sink, door, or any number of repetitive drawing elements. Blocks can also be combined with other objects or blocks to form a nested block. A kitchen for a large apartment complex could be created as a block and inserted in each unit. The kitchen block will contain other blocks such as the dishwasher, sink, stove, refrigerator, and pantry. Joining blocks of individual units can create entire buildings of the complex. Uses for nested blocks are almost endless for construction drawings. Figure 19.18 shows an example of nested blocks used to create a foundation plan.

Care must be taken in the use of nested blocks to avoid confusion in naming and layering. Use the following guidelines to avoid confusion:

- Assign properties individually rather than ByBlock if all instances of a block require the same color, layer, linetype and lineweight properties.
- To control the properties of each block using the properties of the layer the block is inserted on, draw the objects that form each block, including nested blocks, on the 0 layer. Set properties such as color, layer, linetype and lineweight properties ByLayer.

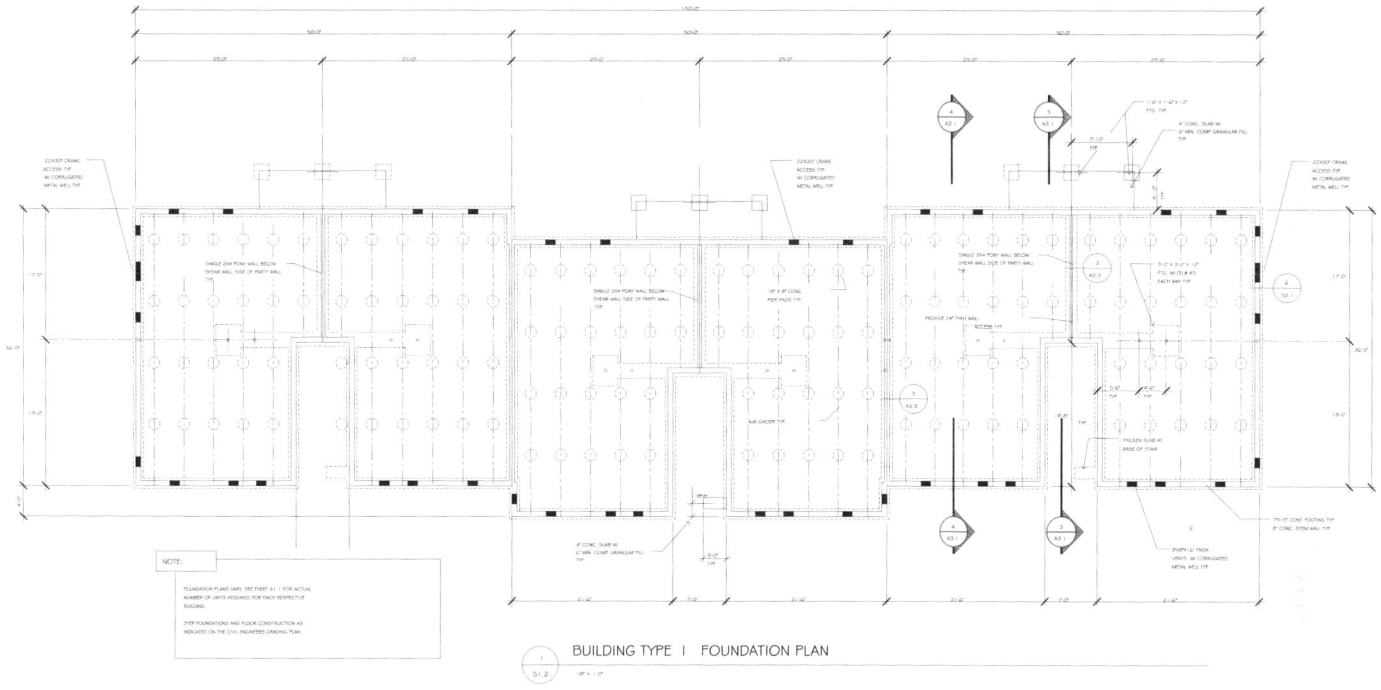

Figure 19.18 *Blocks for the vents, piers, girders are combined to create a foundation for each unit. A block was then created for each unit and mirrored to form the adjoining unit. A block was then made of each pair of units, and then blocks of each pair were inserted to form the building foundation.*

USING BLOCKS IN MULTIPLE DRAWINGS

One of the benefits of working with multiple drawings is that you can move or copy blocks in one drawing to a different drawing. This can be done with opened or closed drawings. Use the following steps to copy blocks between two open drawings.

1. Select the block to be copied.
2. Right-click to display the shortcut menu.
3. Select **Copy**.
4. Move to the other drawing and make it the current drawing by selecting the title bar.
5. Select **Paste** from the shortcut menu or the **Edit** menu.
6. Insert the block in the desired location.

CONTROLLING BLOCKS WITH DESIGNCENTER

The DesignCenter allows blocks to be inserted from drawings that are not open. If you work with consulting firms, you can copy a block from a drawing in their office, if you have the proper access to their network. For now, we'll work on copying a block from one drawing to another. Future chapters will provide information on working with networked machines and consulting firms using the Internet. To copy a block from a drawing file on your computer, display the DesignCenter by selecting the **DesignCenter** button on the **Standard** toolbar. Chapter 22 will introduce additional methods for opening the DesignCenter. The DesignCenter will resemble Figure 19.19. (The exact appearance will depend on what drawings are opened.) Click the **Desktop** button in the upper left corner. This will produce a display exactly like

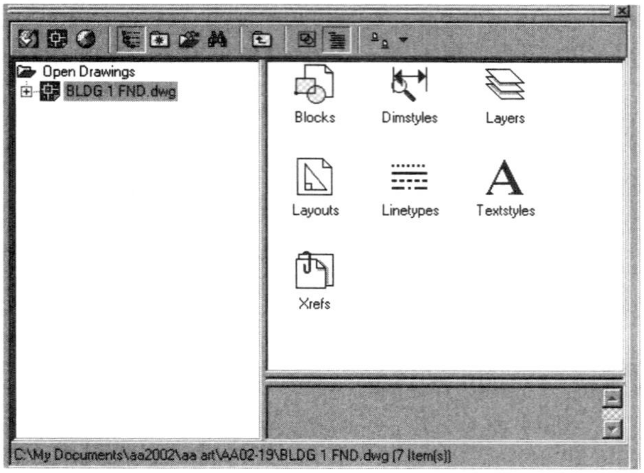

Figure 19.19 *The DesignCenter.*

that produced by Windows Explorer. Figure 19.20 shows a portion of this display. Double-clicking the listing of FLOOR PLAN will produce a display of the contents of the drawing similar to Figure 19.21. Double-clicking the Blocks listing will provide a listing of the blocks contained in the drawing similar to Figure 19.22. Highlighting the name of the desired block displays the block and the display window and allows that block to be dragged into the current drawing.

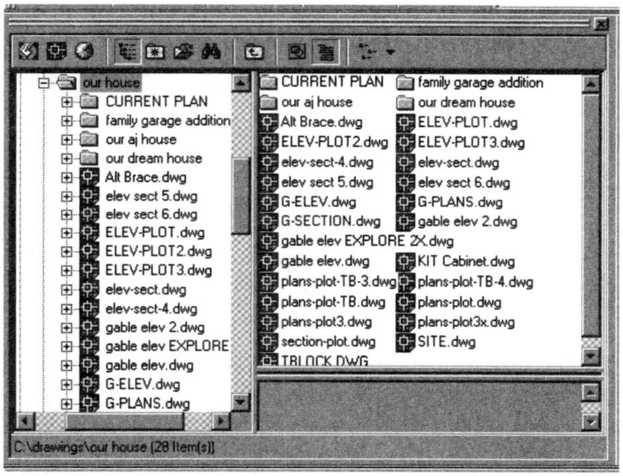

Figure 19.20 *Selecting the* **Desktop** *button in the DesignCenter will produce a display similar to Windows Explorer.*

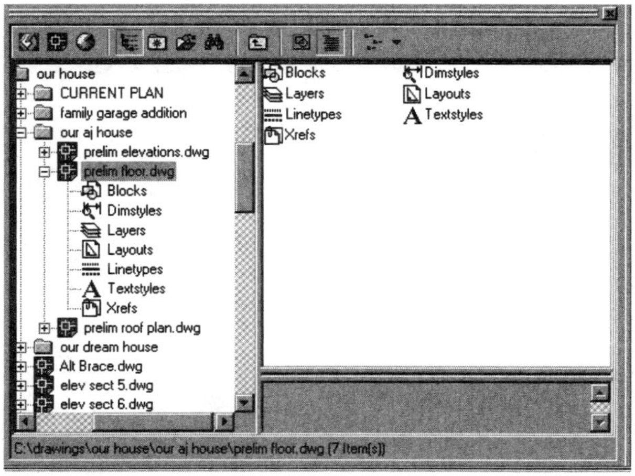

Figure 19.21 *Selecting a drawing from the list will display a list of the contents.*

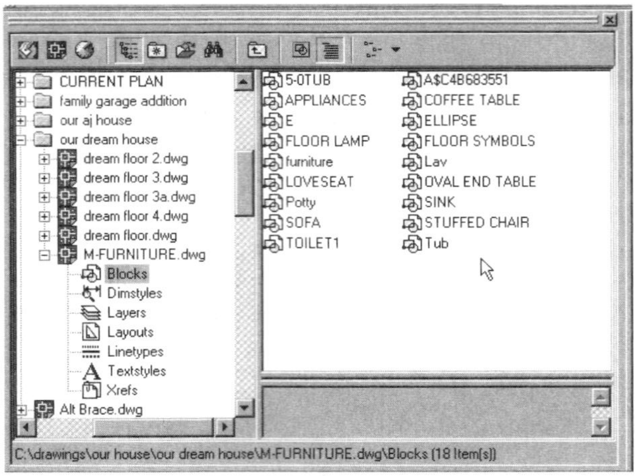

Figure 19.22 *Selecting blocks from the list in Figure 19.21 will display a list of blocks in the selected drawing. One or all of the listed blocks can be copied to a new drawing.*

USING HYPERLINKS

Selecting the **Insert Hyperlink** button on the **Standard** toolbar and selecting an object opens the **Insert Hyperlink** dialog box, allowing a hyperlink to be associated with the block definition. Figure 19.23 shows the **Insert Hyperlink** dialog box. You can also display the dialog box by selecting **Hyperlink** from the **Insert** menu or by typing **HYPERLINK** (ENTER) at the Command prompt. Hyperlinks are

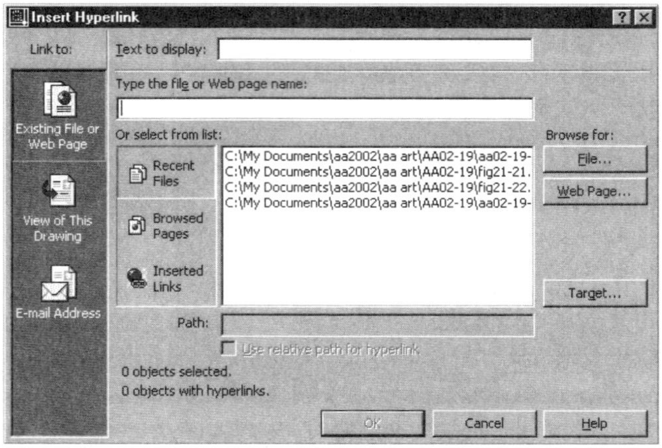

Figure 19.23 *Selecting the* **Insert Hyperlink** *button will display the* **Insert Hyperlink** *dialog box, allowing a hyperlink to be associated with a block definition.*

created in AutoCAD drawings as pointers to associated files. Hyperlinks can be used for varied tasks such as launching a word processing program, opening a specific file, pointing to a named location in a file, or activating your Web browser and loading a specified HTML page. Hyperlinks can be used to point to locally stored files, files on a network drive, or files on the Internet. Cursor feedback is automatically provided to indicate when the crosshairs are over a graphical object that has an attached hyperlink. You can then select the object and use the **Hyperlink** shortcut menu to open the file associated with the hyperlink. Chapter 26 will also explore the **Hyperlink** command.

EDITING BLOCKS

Blocks can be edited by changing the description, by using the **Explode** command, or by using the Redefine option of the **Block** command.

EDITING BLOCK DESCRIPTIONS

A block description can be altered in the **Block Definition** dialog box. Display the dialog box by selecting the **Make Block** button on the **Draw** toolbar or using one of the methods mentioned earlier in this chapter. Select the edit arrow beside the **Name** box. This will display a listing of the current blocks. Select the name of the block that will be altered. Enter the changes that will be made. Typically, this might include adding objects to the block. Click **OK** when you're satisfied with the description.

REDEFINING A BLOCK

Even if a block has been inserted in a drawing in several locations, it is possible to update all of the blocks without editing them individually. Blocks are edited by redefining a copy of one of the blocks, then saving the edited copy of the block using the existing block name. It's also important to remember that this process will update ALL blocks of the specified name. There is no "I only wanted a few of them" option. Use the following sequence to redefine a block. For the following example, the TOILET block will be altered.

1. Select the **Edit Block or Xref** button on the **Refedit** toolbar, select **In-place Xref and Block Edit Reference** from the **Modify** menu or type **REFEDIT** (ENTER) at the command line.
2. Select the block to be edited. Selecting a block to be altered will produce a **Reference Edit** dialog box similar to Figure 19.24.
3. Select the name of the block from the **Reference name** edit box.
4. Select the **Enable unique layer and symbol names** check box.
5. Select the **Display attribute definitions for editing** check box.
6. Click the **OK** button. The command line will prompt to select nested objects.

7. Select the block to be edited. Other uses of the selected block might appear to be altered. Don't panic! The blocks will be restored shortly. As a block is selected, the **Refedit** toolbar will be displayed, if has not been previously opened.
8. Edit the block as desired.
9. Select the **Save Back Changes to Reference** button on the **Refedit** toolbar. This will display a warning box indicating that all of the selected blocks are about to be updated.
10. Click the **OK** button in the warning box. The box will be closed and all of the selected blocks will be redisplayed to reflect the indicated changes.

Each toilet used throughout the drawing will now be updated to reflect the current changes. Figure 19.25 shows an example of a redefined block.

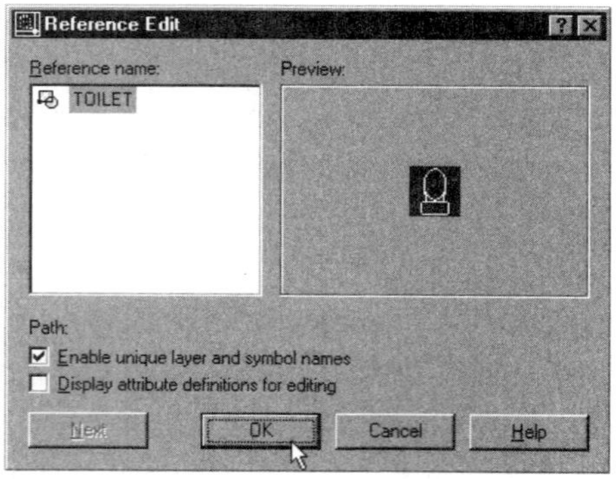

Figure 19.24 *The* **Reference Edit** *dialog box can be used to select a block to be edited.*

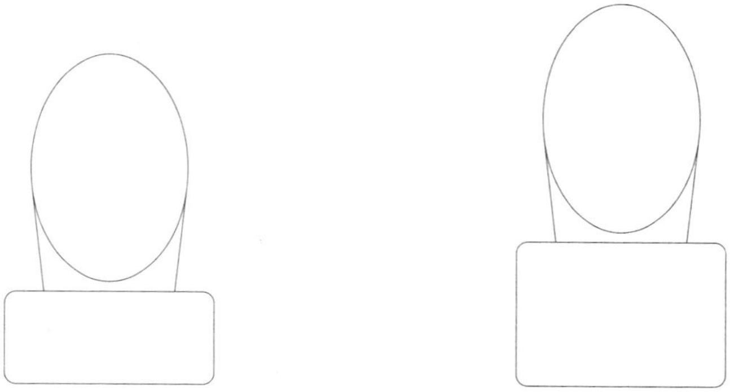

Figure 19.25 *When a block is redefined, all occurrences of the block in the drawing will be updated.*

EXPLODING BLOCKS

If the **Explode** option is activated in the **Insert** dialog box, each object in the block will function independently of other objects in the block. By default, as blocks are inserted in a drawing, all objects in the block will function as one object for editing purposes. Combining all of the objects of a block as one object is great if you want to **Move**, **Rotate**, or **Copy** a block. Only one object has to be selected and the entire block will respond to the command. Using other editing commands such as **Erase** to remove one object also will cause the entire block to be erased. To erase just one object in the block, you must use the **Explode** command (first introduced in Chapter 12). Access the command by selecting the **Explode** button on the **Modify** toolbar, by typing **X** (ENTER) at the Command prompt, or by selecting **Explode** from the **Modify** menu. The command sequence is as follows:

Command: **X** (ENTER) (Or click the Explode button.)

Select objects: *(Select the desired blocks.)*

Select object: *(This prompt will continue until all of the desired blocks are selected.)* (ENTER)

Command:

The block will be displayed with no apparent change. Individual components of the block can now be edited. Blocks with equal X, Y, and Z values will keep their shape when exploded. Blocks with unequal X, Y, and Z values may change shapes when exploded. Blocks that are inserted with **Minsert** can't be exploded.

WBLOCKS

Blocks are a great way to make repeated copies of a symbol throughout a drawing. A symbol that is used on a foundation for one structure can be used on the foundation of several other structures. Blocks are saved in the drawing file they are created in, but with the aid of the AutoCAD DesignCenter, a block can be used on other drawings. If a block is saved on a template drawing, the block can be used each time the template drawing is used. The **Wblock** command provides another option for blocks to be used in multiple drawings no matter where the drawing was created, by saving the block as a .DWG file. A wblock will be displayed as a .DWG file each time a directory of drawings is made, but it will act as a block when inserted in a drawing. Figures 19–26a and 19-26b show examples of details and sections created using blocks and combined using the **Wblock** command.

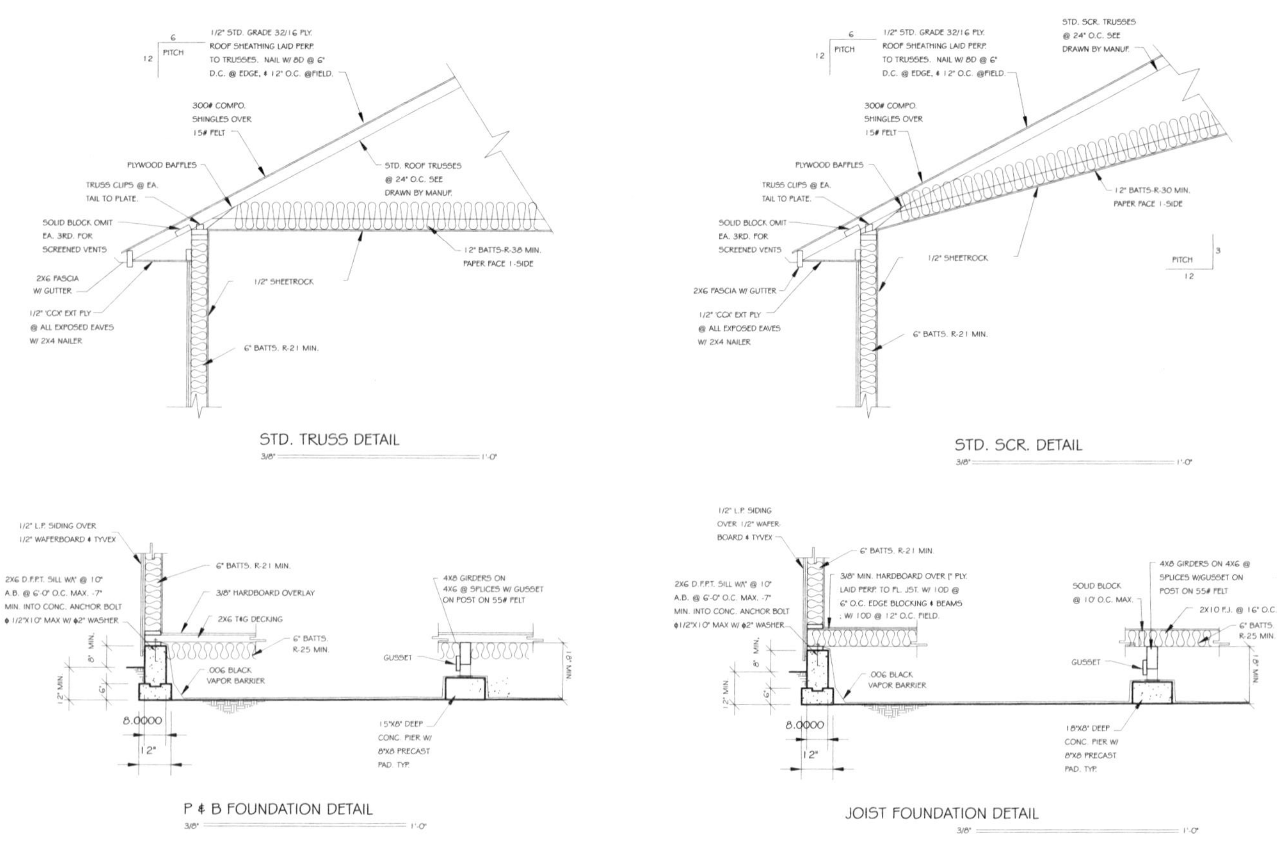

Figure 19.26a *Common details can be stored using the* **Wblock** *command.*

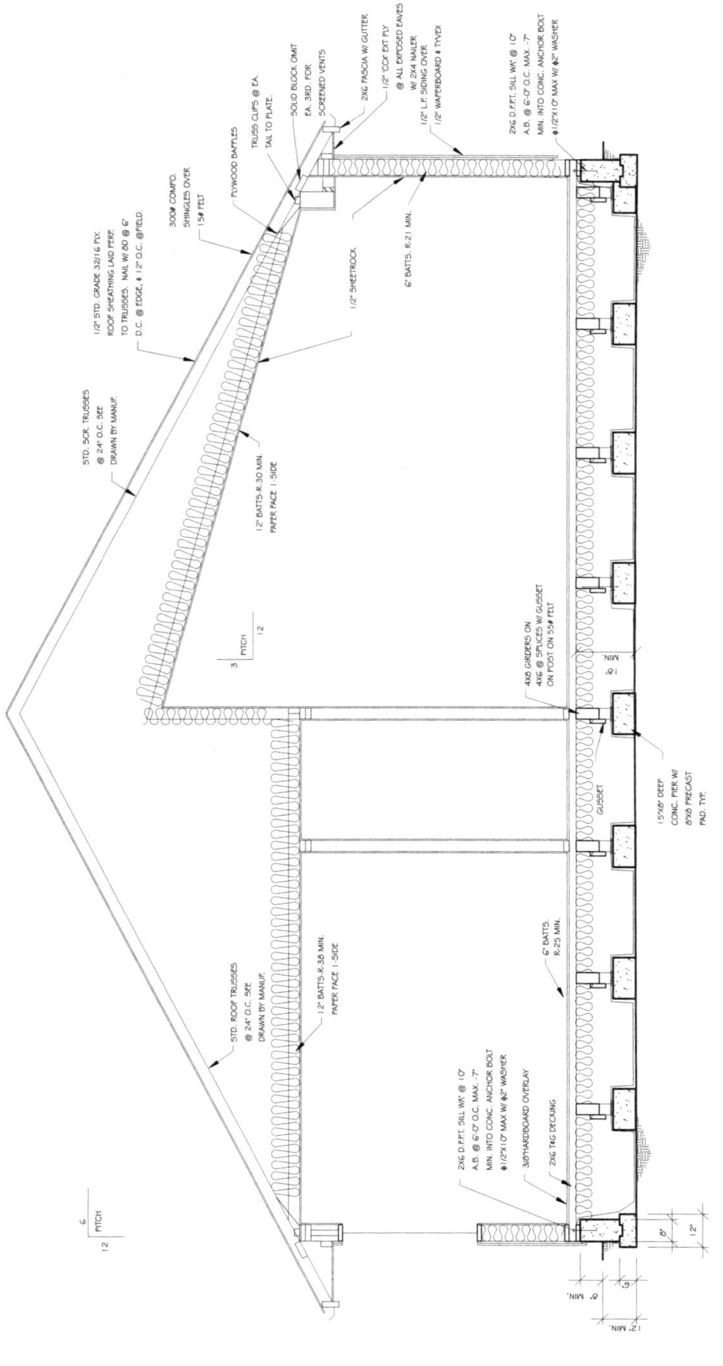

Figure 19.26b *Blocks saved using the* **Wblock** *command can be combined to make other drawings.*

USING WBLOCKS

Symbols and construction components that are saved as a wblock can be used to save hours of drawing time on future drawings. Using an existing drawing as a wblock can also save hours on drawings such as elevations or sections that have features that must be located based on their position on the floor plan. By freezing all unnecessary materials and thawing only materials such as the roof outline, walls, windows and exterior doors, a wblock of the floor plan can be created and inserted in the ELEVATION.DWG file. The wblock can then be copied and rotated for use in projecting each elevation, as shown in Figure 19.27. Projecting irregular shaped structures from a wblock of the floor plan allows each surface to be accurately projected. Sizes can be projected to determine both vertical and horizontal relationships. The roof and wall locations are projected vertically from the plan to the elevation, and heights are projected from one elevation to the next.

CREATING WBLOCKS

Typing **W** (ENTER) at the Command prompt will start the command and display a **Write Block** dialog box similar to Figure 19.28. The dialog box is used to create and save blocks to a file for future use. The two major areas of the **Write Block** dialog box include the **Source** and **Destination** areas. The areas that are active will depend on the selections made in the **Source** portion of the dialog box.

Source

The **Source** area of the **Write Block** dialog box selects existing blocks or objects, writes them as a file, and specifies an insertion point for when they are inserted. Options are provided for selecting an existing block or an entire drawing, or specifying objects to form the wblock.

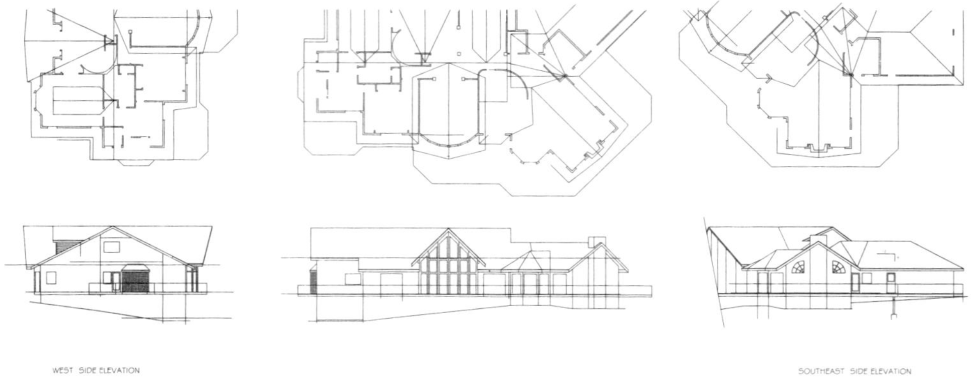

Figure 19.27 *A block was made of the floor plan; then it was reinserted and rotated to project each elevation. (Courtesy Residential Designs.)*

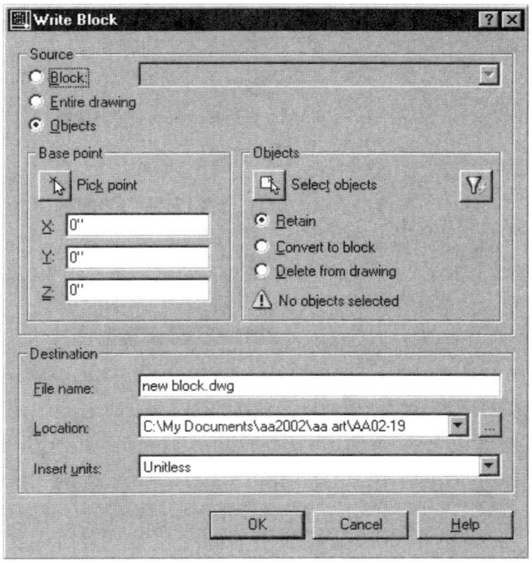

Figure 19.28 *Typing* **W** *(ENTER) at the Command prompt will display the* **Write Block** *dialog box, allowing blocks, drawings, or objects to be saved for future use.*

Block

Selecting the **Block** option allows an existing block to be selected to be stored as wblock file. With the **Block** setting active, the **Block** edit box is activated and the **Base point** and **Objects** portions of the dialog box are inactive. Selecting the edit arrow will display a listing of blocks contained in the current drawing. Selecting a name from the list will allow that block to be inserted. The **Destination** portion of the block displays where the wblock file will be stored. If the **OK** button is clicked, the selected block will be saved as a drawing file in the indicated destination. Methods of altering the destination will be examined in the next section of this chapter.

Entire Drawing

Selecting the **Entire drawing** option allows an entire drawing to be saved as a wblock. This option will have the same results as when **Saveas** is used. The benefit of using the option is that unreferenced objects such as blocks, layers, and linetypes that were not used will be removed from the drawing file. This is an excellent method for reducing the size of a drawing and for saving a drawing at the end of a work session.

Objects

Using the default option of **Objects** allows drawing objects to be selected to form the wblock. The process is very similar to the procedure used to create a block. With the **Objects** button active, the **Base point** and **Objects** portions of the dialog box are activated so that objects and a base point can be selected. The order in which the base point or the objects are selected can be altered. Each of these functions is exactly the same as their counterparts in the **Block** command. As each button is selected, the dialog box is removed, allowing for a selection to be made. Once the selection is made, the dialog box is restored, allowing for other options to be altered.

Destination

The box allows the name, location, and units of the drawing file to be specified.

File Name

The **File name** display will vary depending on which of the three **Source** options is active. If the **Block** option is active, the name of the block entered in the **Block** edit box will be displayed in the **File name** edit box. If the **Entire drawing** box is active, the current name of the drawing will be displayed in the **File name** edit box. With the **Objects** option active, the default name of New Block.dwg will be displayed. Best results will be achieved by providing a more descriptive name.

Location

Because blocks stored using the **Wblock** command are treated as a drawing file, a storage location must be given. The **Location** edit box displays the current drive and path that will be used to store the drawing. Selecting the **Browse for Folder** button allows the drive and folder to be altered. Selecting the button produces a display similar to Windows Explorer. Once a folder name has been selected, the **Create Drawing** dialog box will allow for selection of a file name and a location for the file.

Note: Great care should be given in the naming and storing of wblocks. The name should clearly describe the contents of the drawing. Wblocks should be stored in folders and subfolders that will define the use. For instance, a name of PROTOS could be used for the folder. Subfolder names would include names such as SITE, FLOOR, ELEV, FND, or other common drawings to be used. Each subfolder would contain only wblocks for that specific type of drawing. Figure 19.29 shows a sample listing of wblocks for an architectural firm.

Summary of Steps in Creating a Wblock

Remember that a wblock is just a collection of objects that can be used over and over in a multitude of drawings. Use the following steps to create wblocks:

1. Display the **Write Block** dialog box by typing **W** (ENTER) at the Command prompt
2. Choose the source of the block to be created.
3. If the wblock will be comprised of a pre-existing block, provide a name for the block.

SITE PLAN
- NORTH ARROW
- GENERAL NOTES
- PARKING NOTES
- PAVING NOTES
- HANDIC PARKING
- PERP. SPACE
- 30^ PARKING
- FIRE HYDRANT
- CURB CUT
- LIGHT STANDARD

FLOOR PLAN
- PLUMB
  - DBL. SINK
  - TRIPLE SINK
  - VEGGIE SINK
  - BATH SINK
  - LAUNDRY SINK
  - TOILET
  - BIDET
  - 60" TUB
  - 72" TUB
  - SPA
  - 36" SHOWER
  - 48" SHOWER
  - 60" SHOWER
  - WASHER
  - WATER HTR.
- ELECTRICAL
  - DISHWASHER
  - WASHER
  - WATER HTR.
  - F.A.U.
  - IRON BR.
  - VACUUM
  - RANGE
  - COOK TOP
  - DBL. OVEN
  - REFR.
  - TRASH COMP.
- DOOR
  - SINGL. SW.
  - DBL. SW.
  - SLIDING
  - FRENCH
  - BI-PASS
  - BI-FOLD
  - OVERHEAD
- WINDOW
  - STANDARD
  - BAY
  - BOW
  - GARDEN
- SKYLIGHT

FOUNDATION
- GENERAL NOTES
- VENTS
- PIER
- GIRDER
- FIREPLACE
- BAY

ELEVATIONS
- ROOF
  - VENT
  - SKYLIGHT
  - METAL CAP
  - MASON CAP
  - CUPOLA
- WALLS
  - DOOR-FR
  - DOOR-SLDG
  - DR-6 PANEL
  - DR- 10 LITE
  - DR-FRONT-1 LT
  - DR-FRT-SING LT.
  - DR-FRT-DBL. LT.
  - WIN-SLDG
  - WIN-FIXED
  - WIN-FIX/AWN
  - WIN-CASE
  - WIN-DBL. HNG
  - WALL LITE
  - SHUTTER
  - WALL VENT
  - RIDGE
  - VENT
- DECKS
  - FENCE
  - WOOD VERT
  - WOOD HORIZ
  - IRON RAIL
  - STAIR
- LANDSCAPING
  - LOW BUSH
  - BAMBOO
  - BERRY
  - VINE
  - TREE-FIR
  - TREE-OAK
  - TREE-PALM
  - TREE-ALDER
  - TREE-MAPLE

SECTIONS
- ROOF
  - TRUSS
    - STAND
    - SCISSOR
  - STICK
    - STD RAFT.
    - VAULTED
    - OPEN BM.
- UP FLOOR
  - TJI's
  - JOIST
  - TRUSS
- GND. FLOOR
  - TJI's
  - FLOOR JST.
  - POST / BM.
  - CONC. SLAB

ELECTRICAL
- PLUGS
  - 110 C.O.
  - 110 GFCI
  - 110 WP
  - 110 - 4 PLEX
  - 110- FLOOR
  - 220 C. O.
  - J-BOX
- LIGHTS
  - CEILING
  - WALL
  - CAN
  - FLOOD
  - FLUOR-SHOP
  - FLUOR-SURFACE
  - FLUOR-RECESSED
- LOW VOLT
  - YARD
  - FLOOR
  - STAIR
- SWITCHES
  - SING POLE
  - DIMMER
  - TIMER
  - THREE WAY
- SPECIALITIES
  - SMOKE DET.
  - DOOR BELL
  - CHIME
  - CIRCUIT BOX
  - CEIL. FAN
  - BATH FAN
  - PHONE
  - STEREO
  - TV
  - VACUUM
  - INTERCOM SP
- YARD

Figure 19.29 *Wblocks should be stored in folders and subfolders so that they can be easily accessed. Most offices name folders for the drawings that will use the wblock.*

4. Select the drawing objects to be reproduced as blocks.
5. Select a base point to be used at the block is inserted in the drawing.
6. Select a name to designate the file.
7. Specify the drive, folder, and subfolder to be used as the destination.
8. Click the **OK** button to save the block as a drawing file.

The object has now been saved as a drawing file and is ready to be reused throughout the existing drawing or in a new drawing. If the toilet from Figure 19.9 needs to be added to the existing drawing or to a totally new drawing file, it can be added to the drawing using the **Insert** command.

INSERTING WBLOCKS

Once a symbol or drawing has been saved as a wblock, it can be used just as any other drawing. It can be retrieved using the drawing **Startup** dialog box as a new drawing session is started, or it can be inserted in an existing drawing using the **Insert** command. The **Insert** command will bring an existing drawing into the current drawing from the Command prompt. Each aspect of inserting a wblock is the same as for inserting a block.

EDITING WBLOCKS

One of the strengths of using an entire drawing as a wblock is that when it is inserted, it will move as one object. Insert a wblock of a floor/roof plan, and all of the components will move as one object. A wblock can be altered using the same commands used to edit a block.

CHAPTER 19 EXERCISES

1. Open the FOOTINGS drawing that was updated in Chapter 18. Save the drawing as a block titled FND1STORY.
2. Open drawing FLOOR14 and use the drawing to create a wblock titled BASE14.
3. Open drawing E-18-7. Save the drawing as a wblock titled FNDPLN. Insert FNDPLN in drawing FND1STORY, rescale the footing and save the entire drawing as FNDDLT.
4. Open drawing FNDDLT and use it to create a wblock titled FNDPLN1. Insert FNDPLN1 in BASE14.
5. Use the attached drawing as a guide to draw the following symbols and save them as a wblock in a folder titled \PROTO\FLOOR using the designated block name. Create all blocks on the 0 layer and use 4" high text.

DESCRIPTION	SYMBOL	BLOCK NAME
BATH		
60" x 32" tub	A	TUB60
72" x 32" tub	B	TUB72
19" x 16" lavatory	C	LAV
Toilet	D	POTTY
36" shower	E	SHOWER36
42" shower	F	SHOWER42
60" shower	G	SHOWER60
32" x 21" double sink	H	DBLSINK with a garbage disposal on either side
36" x 21" double sink	J	XDBLSINK with a garbage disposal on small side
43" x 21" triple sink	K	XXSINK with a garbage disposal in the middle
16" x 16" veggie sink	L	VEGSINK
Dishwasher	M	DW
Trash compactor	N	TC
Lazy susan	P	LS
Range	Q	RANGE

Double oven	R	DBLOVEN
18" pantry	S	PAN

UTILITY

Washing machine	T	WASH
Dryer	U	DRYER
Built-in ironing board	V	IRON
21"x21" laundry tray	W	LT
Water heater	X	WH
Hose bibb	Y	HB

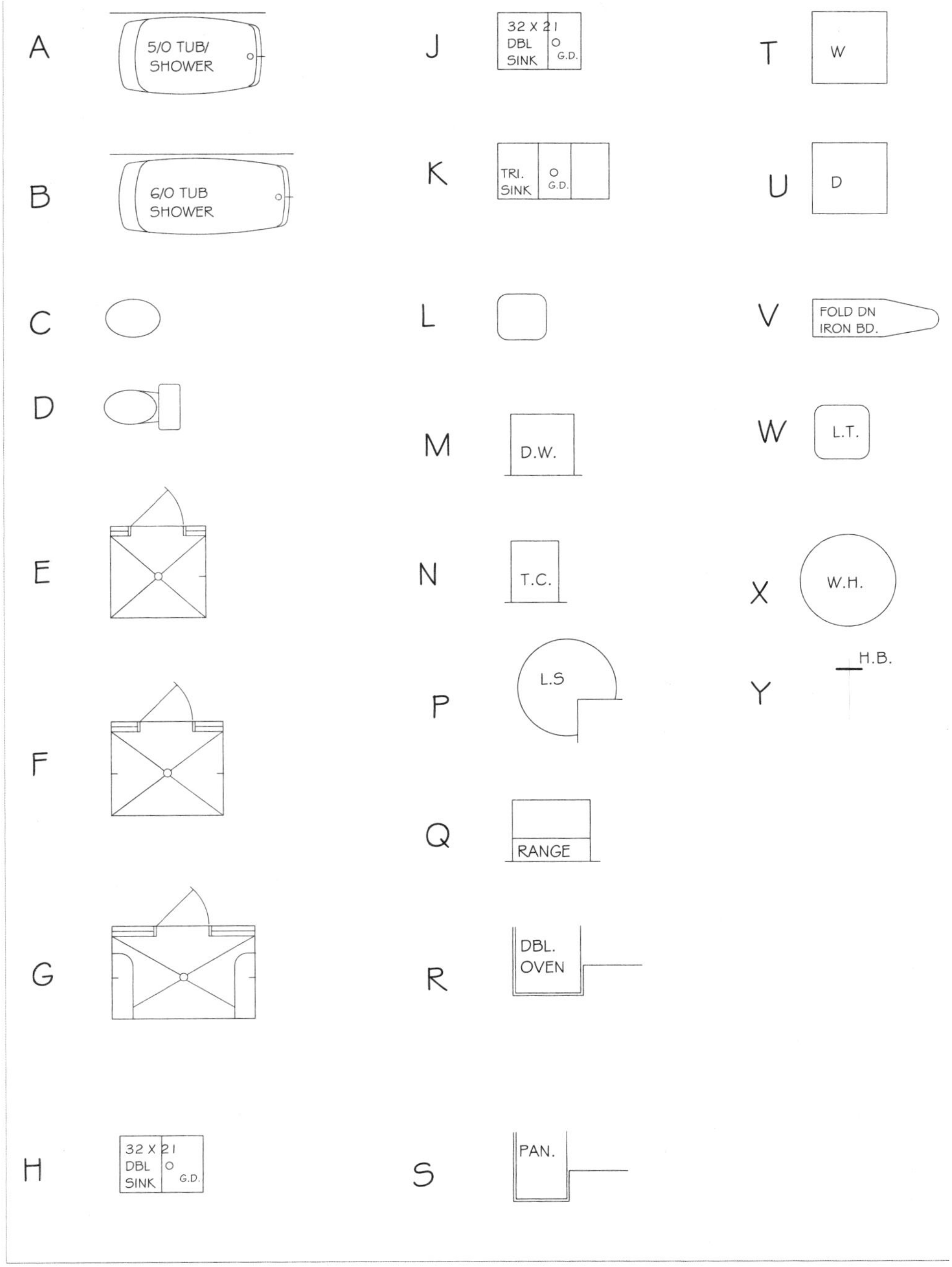
A
5/0 TUB/
SHOWER
B
6/0 TUB
SHOWER
C
D
E
F
G
H
32 X 21
DBL
SINK
G.D.
J
32 X 21
DBL
SINK
G.D.
K
TRI.
SINK
G.D.
L
M
D.W.
N
T.C.
P
L.S
Q
RANGE
R
DBL.
OVEN
S
PAN.
T
W
U
D
V
FOLD DN
IRON BD.
W
L.T.
X
W.H.
Y
H.B.

6. Use the drawing below as a guide to draw the following symbols, and save them as a wblock in a folder titled \PROTO\ELECT using the designated block name. Use a 6" diameter circle for all electrical symbols with no text. Use an 8" diameter symbol with 4" high text for symbols that require text to be placed inside the symbol. Create all blocks on the 0 layer.

DESCRIPTION	SYMBOL	BLOCK NAME
PLUGS		
110 c.o.	A	110CO
Multi 110 c.o.	B	110CO2
219 c.o.	C	219CO
110 half-hot c.o.	D	110HH
110 waterproof	E	110WP
110-ground fault	F	110GFI interrupter
Floor-mounted 110 c.o.	G	110F
Junction box	H	JBOX
LIGHT FIXTURES		
Ceiling-mounted light	J	CEILLITE fixture
Wall-mounted light	K	WALLLITE fixture
Can light fixture	L	CANLITE
Recessed ceiling light	M	RECESSLITE fixture
Wall-mounted spotlights	N	WALLSPOT
24" x 48" surface-	P	24FLUOR mounted fluorescent fixture
48"x48" surface-	Q	48FLUOR mounted fluorescent fixture
48" shop	R	48SHOP fluorescent fixture
48" track light	S	TRACK
SWITCHES		
Single-pole switch	T	SWITCH
Three-way switch	U	SWITCH3
Switch with dimmer	V	DIMSWITCH
SPECIAL SYMBOLS		
Vacuum	W	VACUUM

Intercom	X	
INTERCOM		
Phone	Y	PHONE
Stereo speakers	Z	SPEAKERS
Smoke detector	AA	SMOKE
Light, heat, fan	BB	LHF

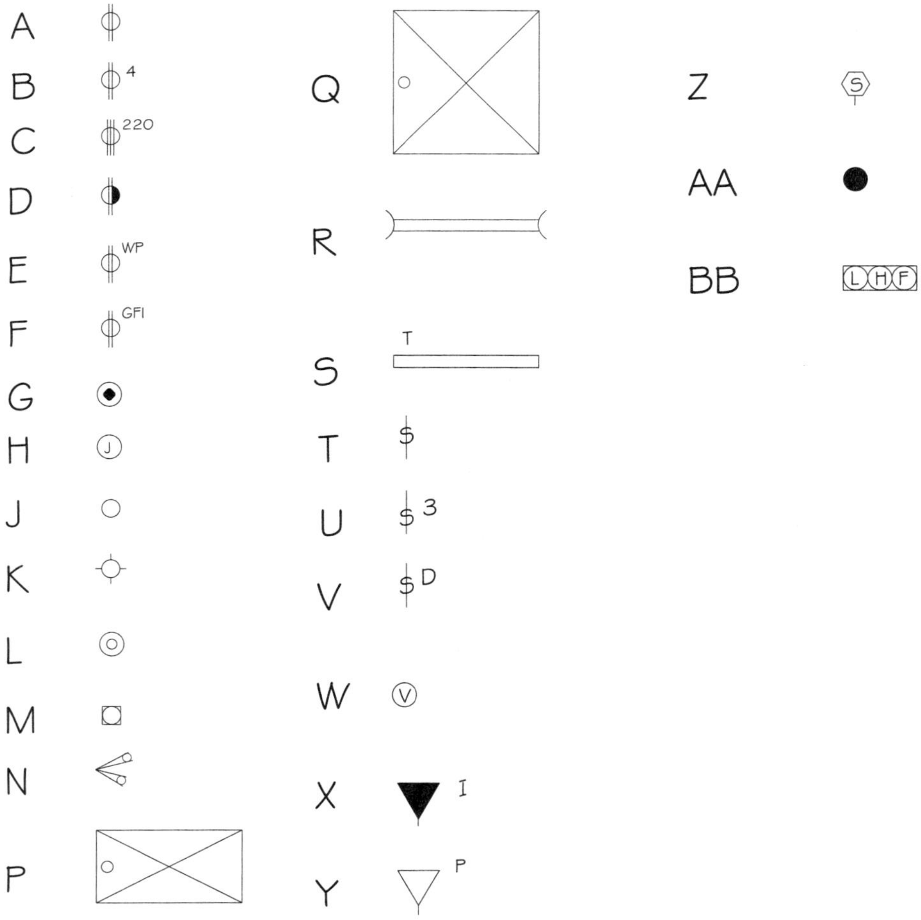

7. Use the drawing below as a guide to draw the following symbols, and save them as a wblock in a folder titled \PROTO\FLOOR\DOORS using the designated block name. Create all blocks on the 0 layer using a 12" block unit.

DESCRIPTION	SYMBOL	BLOCK NAME
Exterior door	A	XDOOR
Exterior door with 1 sidelight	B	XDOOR1LITE
Exterior door with two sidelights	C	XDOOR2LITE
Pair of exterior doors with 2 sidelights	D	PR2DOOR2LITE
Interior door	E	INDOOR
Exterior sliding glass door	F	XSLDOOR
Interior by-pass door	G	BYPASS
Interior bi-fold door	H	BYFOLD
Interior folding door	J	FOLD
Window	K	WINDOW

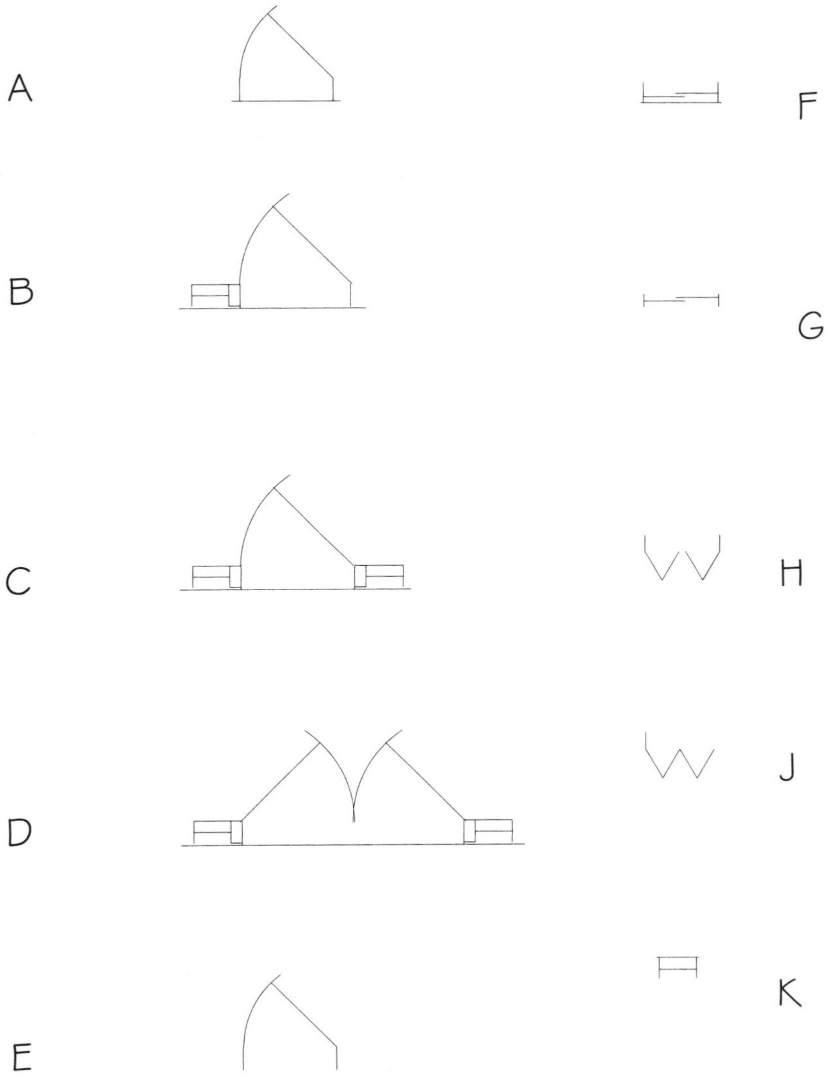
A
B
C
D
E
F
G
H
J
K

8. Open the appropriate template and use the drawing below as a guide to drawing the bathroom. Draw the exterior windows 6 feet wide, except for a 2-foot wide window at the toilet room, a 4-foot window over the counter, and 3-foot wide windows in the bedroom. Use a pair of 2-foot wide doors in the bedroom, a pair of 3-foot doors from the bedroom to the exterior, and a 28" pocket door into the toilet room. All other doors to be 30" wide. Create the following layers: WALLS, PLUMB, APPLIANCES, DIMEN, NOTES, DOOR-WIN and CABS. Create and insert the required blocks to complete the drawing. Provide dimensions and label all fixtures. Save the drawing as a wblock titled BATH.

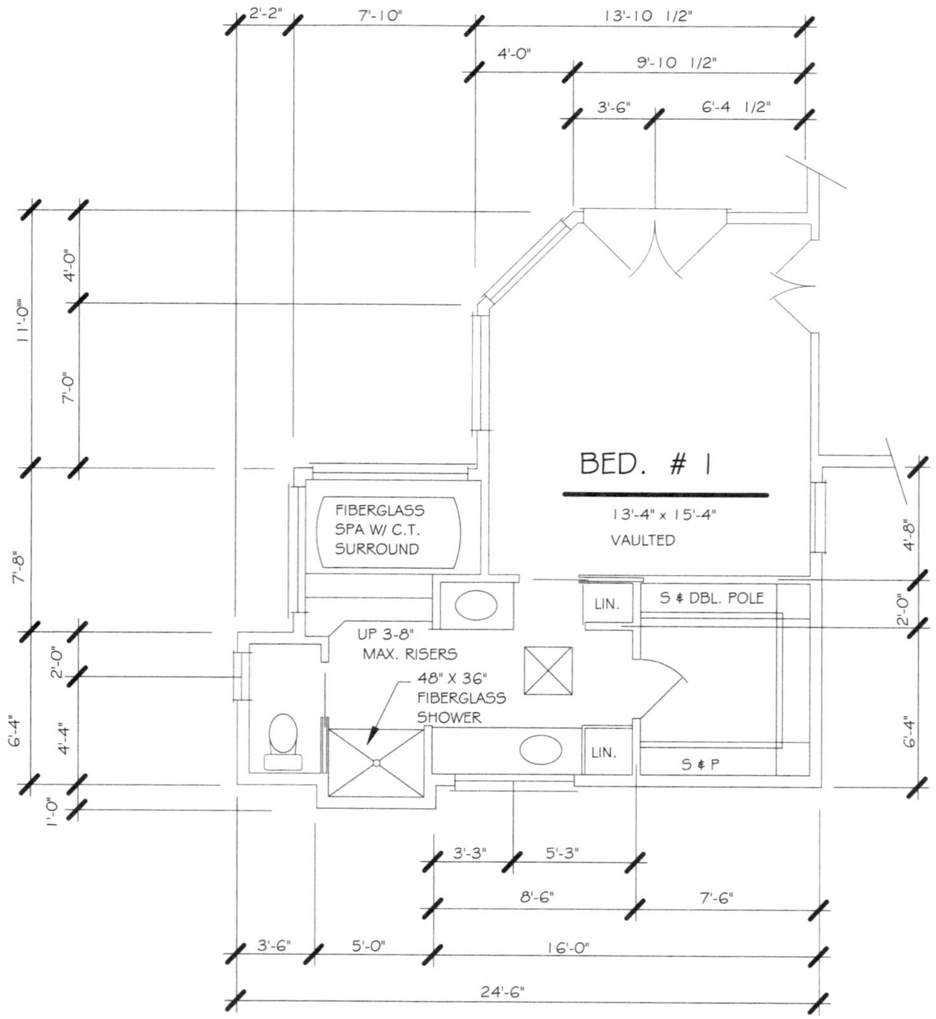

9. Open the appropriate template and use the attached drawing to draw the kitchen. Design door and window sizes to fit the space provided. Create layers that will logically separate information for future printing on floor, framing, and electrical plans. Insert the required blocks to complete the drawing. Align the bathroom with the utility room and provide a 42" fiberglass shower unit. Provide dimensions, determine the square footage of each room, and label the drawing. Turn off all layers containing notes and dimensions and design an electrical plan that assumes all appliances are electrical. Use 2 ceiling-mounted fixtures in the nook, a surface-mounted fixture in the utility room, 8 can lights in the kitchen and 2 in the hallway, a wall-mounted fixture over the bathroom sink, and a ceiling-mounted fixture in the center of the bath. Save the drawing as a wblock titled KITCHEN.

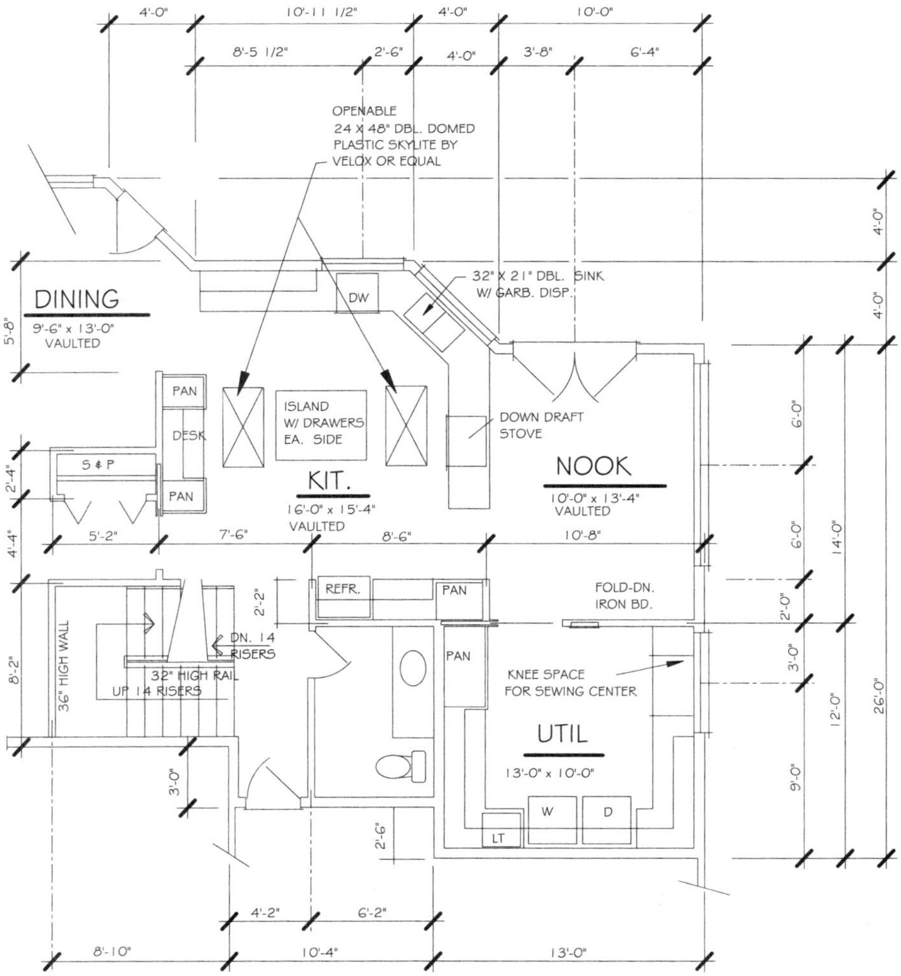

10. Open the FOOTING drawing and use it to create a wblock drawing titled BASEFTG. Save the file with an appropriate title.

CHAPTER 19 QUIZ

1. How can a wblock increase drawing speed?
2. Why is using blocks more efficient than copying an object?
3. If a block is created on the 0 layer, where will the block be located when it is inserted?
4. When the color of a block is set "BY BLOCK," how will the block be affected?
5. What is the advantage of drawing blocks such as a window in a 12" box rather than full size?
6. List the process to create a block.
7. How does the Insertion Point specified when the block is defined affect the insertion point requested when the block is inserted?
8. What command will allow for multiple blocks to be inserted?
9. What command will allow for editing blocks?
10. What is the quickest method for editing the same mistake in several blocks?
11. Explain the difference between a block and a wblock.
12. List the process for inserting a wblock in a drawing.
13. How can a block be mirrored before it is inserted?
14. What method can be used to insert a block by selecting two points?
15. What is the effect of activating the **Explode** check box in the **Insert** dialog box?

CHAPTER 20

Adding Attributes to Enhance Blocks

Even with all the qualities that blocks and wblocks add to a drawing, they can still be enhanced. Many blocks need text to explain or provide a specification. Text can be added to a block using three methods. In the last chapter, text was added to the drawings using Text, and saved as part of the block. This works well if the text remains constant, such as the specification for a double kitchen sink. If text is added to the block, the block must be exploded if the text is to be edited. The Explode command allows the text to be edited, but characteristics of the block might be lost. Once a block has been inserted, text can be added to the drawing base to supplement the block by using the **Text** or **Mtext** command. This option works well only if a few blocks need written specifications.

- This chapter will introduce a new method of adding text to a block by using attributes.

Commands to be introduced in this chapter include

- **Attdef**
- **Attdisp**
- **Attedit**
- **Attreq**
- **Battman**
- **Eattedit**
- **Global**
- **Eattext**

THE BENEFITS OF ATTRIBUTES

An attribute is a grouping of objects that contains text. An example of a block with attributes can be seen in Figure 20.1. These text groupings are similar to a block, which is then attached to a drawing block. Attributes have several features that make them superior to text created with **Text** or **Mtext** and are well worth the effort to master. Once the benefits have been explored, thought must be given to defining the attribute text, controlling how the attributes will be displayed, changing the values, and extracting attributes from a drawing to compile a written listing of attributes.

The two major benefits of adding text through the use of attributes are control and retrieval of information.

CONTROLLING TEXT

Using attributes will allow for assigning prompts to a block to request information each time that the block is displayed. For instance, each time the door block is inserted in a drawing, these prompts could also be provided: size, thickness, type, manufacturer, finish, rough opening and hardware requirements.

Default values can be provided, but altered with each insertion. Using attributes also allows for correcting the text that is displayed without altering the block characteristics. These prompts might be hidden during the display or plotting process, even though the block they are attached to is displayed.

EXTRACTING ATTRIBUTE INFORMATION

In the example above, much of the information that will comprise the attribute is displayed in a plant schedule. Attributes can be extracted from the drawing, stored in a separate file, and printed using a third party database text editor, such as the Microsoft Notepad. A list of attributes also can be compiled and printed within the drawing.

PLANNING THE DEFINITION

It is easiest to add attributes to a drawing before the drawing has been saved as a block. Once the symbols have been drawn, a decision must be made about what specifica-

Figure 20.1 *Attributes can be added to a block to explain qualities that could vary with each insertion.*

tions will be associated with the block. For best efficiency this should be done in a planning session prior to entering the **Attdef** (ATTribute DEFinition) command. For a block of a window, this might include the size, type, frame material, manufacturer, glazing, and rough opening. Other than the actual space limitations within the drawing, there is no limit to the amount of information that can be associated with a block. As the attributes are being defined, an option is even available to keep the attributes invisible.

After the material to be listed with the block has been determined, decide how the prompts for each attribute will be displayed. Attribute prompts are usually in the form of a question, but statements can also be used. For instance, the prompt might read "What is the window size?" or "Provide the window size." Determine the prompts for each attribute that you will be listing.

ENTERING THE ATTRIBUTE COMMANDS

Several commands will be used as you work with attributes. Once the block has been drawn and you've decided what attributes are to be assigned, the process of attaching text to the block is started by using the **Attdef** command. Commands for controlling attributes include the **Attdef** (ATTribute DEFinition), **Attdisp** (ATTribute DISPlay), **Attedit** (ATTribute Edit), and **Attreq** (ATTribute REQuest) commands.

CREATING ATTRIBUTES

Each of the qualities of an attribute can be controlled using the **Attribute Definition** dialog box or at the command line. Consult the **Help** menu (or professional help), if you would rather use the slower, harder Command prompt method. Access the dialog box by typing **ATTDEF** (ENTER) at the Command prompt or by selecting **Define Attributes** from **Block** on the **Draw** menu. Each method will display the dialog box shown in Figure 20.2. Major elements of the dialog box include the **Mode**, **Attribute**, **Insertion Point**, and **Text Options** areas.

ATTRIBUTE DISPLAY

The **Mode** area allows the display options for the attribute to be set. These include **Invisible**, **Constant**, **Verify**, and **Preset**, with a default setting of Normal. If none of the options is selected, the Normal mode will be kept. The Normal mode will prompt you for each attribute and the response will be displayed.

Invisible

Selecting the **Invisible** check box will make the attributes invisible. Attributes will still be attached to the block when it is inserted, but they will not be displayed. If a separate window schedule is to be created, values can be invisible so they are not displayed twice. The **Attdisp** command, which will be introduced shortly, can be used to override the **Invisible** mode setting.

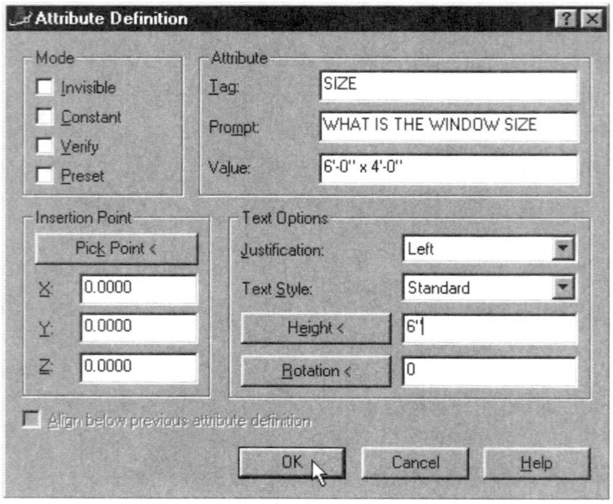

Figure 20.2 *Typing* **ATTDEF** *(ENTER) at the Command prompt or selecting* **Define Attributes** *from* **Block** *on the* **Draw** *menu will produce the* **Attribute Definition** *dialog box. The box can be used to design the tag, prompt, value, and insertion point of each attribute, as well as controlling how each will be displayed.*

Constant

Selecting the **Constant** check box indicates that all uses of the block will have the same attribute value. The value is entered as the attribute is being defined and set throughout the use of the block. No prompt for new values will be given when the block is inserted with this box active. If this is not selected, the default **Normal** mode will allow variables to be changed at a later time.

Verify

By defining an attribute with this mode active, you will be given an opportunity to verify that the attribute value is correct during the insertion process.

Preset

This mode will allow you to create attributes that are variable but not requested each time the block is inserted. When a block containing a preset attribute is inserted, the attribute value is not requested and is automatically set to its default value.

DEFINING ATTRIBUTES

Once the mode has been set, attributes can be defined. The three **Attribute** area edit boxes are **Tag**, **Prompt**, and **Value**, and they can be seen in Figure 20.3.

Attribute Tag

The tag is used to identify an attribute definition in a similar way that a specific name is used to identify a block. The tag will identify each occurrence of the attribute

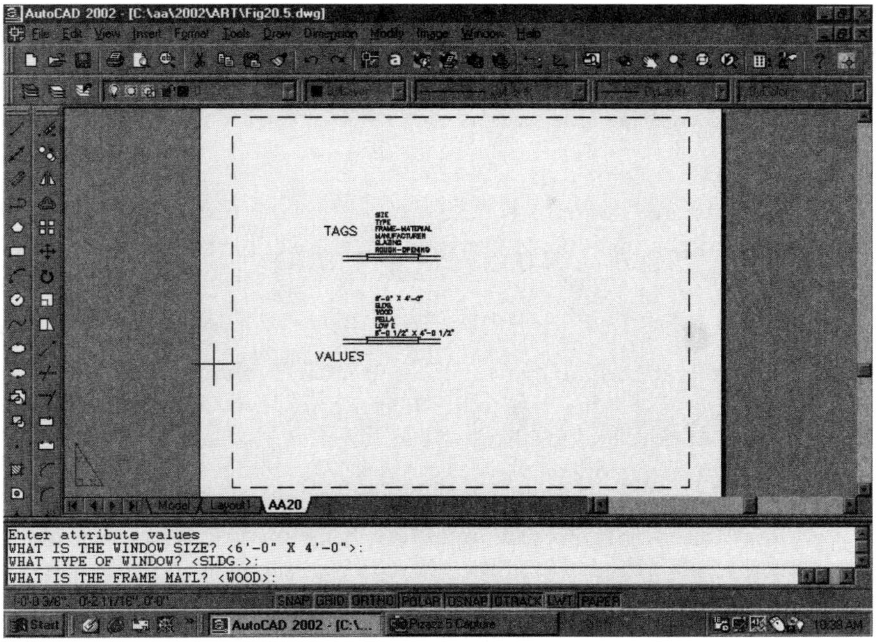

Figure 20.3 *The attribute tags are displayed as the block is being inserted. The values are displayed once the block has been inserted.*

throughout the drawing. During the initial planning stage for the windows, you might decide that information regarding the size, type, frame material, manufacturer, glazing, and rough opening should be associated with each window block. Each of these listings can be used as an attribute tag, but each tag must be entered individually with the prompt that follows used to control only this tag. As the tag is entered, it can be composed of any letters or characters except a blank space. A hyphen can be used between words to create a space. To name a tag, type a tag such as **SIZE** in the **Tag** edit box.

Attribute Prompt

The attribute prompt is the prompt you will see when the block is inserted. As you plan the prompts, questions or statements should be used. To add the prompt, type a prompt such as **WHAT IS THE WINDOW SIZE?** in the **Prompt** edit box. If the **Prompt** edit box is left blank, the tag (SIZE) will be used as the attribute prompt. If the **Constant** mode is active, the **Prompt** option will be inactive.

Attribute Value

Once the tag and the prompt have been specified, provide a value in the **Value** edit box. The value that is entered at this prompt will be displayed as the default value as the block is inserted. A specific value such as 6'-0"x4'-0" can be entered, or the null response can be used. If the **Constant** mode for the attribute has been selected, a request for the prompt will not be made, and the prompt "Attribute Value:" will be displayed instead. To enter a value such as 6'-0"x4'-0", type it in the **Value** edit box.

TEXT OPTIONS

The **Text Options** area contains settings to control the text appearance. Options include edit lists for justification and style and edit boxes for height and rotation values.

Justification

Selecting the **Justification** edit arrow will display the list shown in Figure 20.4. The options are the same as those used with the **Text** command. To replace the Left option, select the desired option, and the list will be closed.

Style

Selecting the **Text** style edit arrow will produce a list of current drawing text styles. To alter the style, select the name of the desired style. The selected style name will be placed in the text box, and the new style will become current.

Height

The **Height** edit box allows the text height to be set. You can edit the existing value or select the **Height<** button. If the **Height<** button is selected, the **Attribute Definition** dialog box is removed and a prompt is displayed to allow points to be selected to indicate the text height. You can enter a value at the Command prompt instead

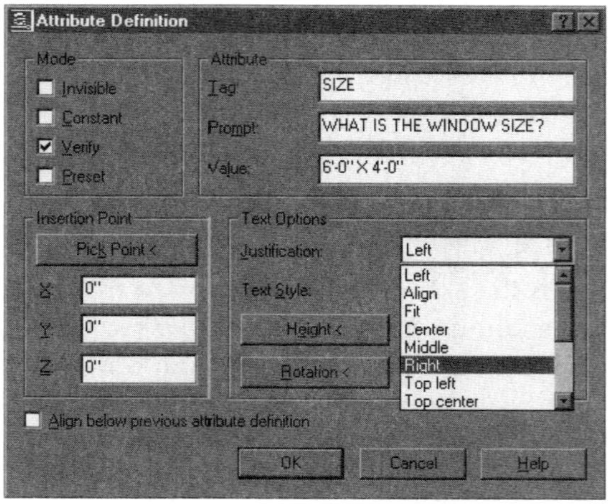

Figure 20.4 *Options for justification can be displayed by selecting the* **Justification** *edit arrow. The options are the same as those of the* **Text** *command.*

of selecting two points. When the points or value have been provided, the dialog box will be restored and the indicated height will be displayed in the **Height** edit box.

Rotation

The rotation angle for attributes is set in the same manner as the height. Enter the desired angle in the text box or select the **Rotation<** button to select the angle with the cursor.

SELECTING THE INSERTION POINT

The **Insertion Point** area allows the location of the attribute to be selected. Selecting the **Pick Point<** button removes the dialog box and allows the insertion point to be selected using the cursor. If the exact location is known, coordinates can be entered using the **X**, **Y**, and **Z** text boxes.

ALIGNING ADDITIONAL ATTRIBUTES

The **Align below previous attribute definition** check box will allow additional attributes to be placed below previous attributes, using the same justification. The option is not active when the first attribute is written. With the option active, the **Text Options** and **Insertion Point** areas will be inactive.

Once all of the options have been selected, clicking the **OK** button will close the dialog box and display the tag at the selected insertion point, similar to the display shown in Figure 20.5. Don't be discouraged at seeing the tags displayed as the command is completed. The tags will not be displayed as the block is inserted. If you are not satisfied with the tags, they can be altered using the **Properties** or **Ddedit** command.

REPEATING THE PROCESS

To add additional attributes to the block, press (ENTER) to repeat the previous command, and the entire process can be repeated. By repeating the process five more times, the tags for type, frame material, manufacturer, glazing, and rough opening can be added to the block. Activating the **Align below previous attribute definition** check box will automatically place additional tags directly below existing tags with the proper spacing. Repeating the entire sequence will produce a display similar to the one in Figure 20.6.

Figure 20.5 *The window block with the SIZE attribute tag displayed.*

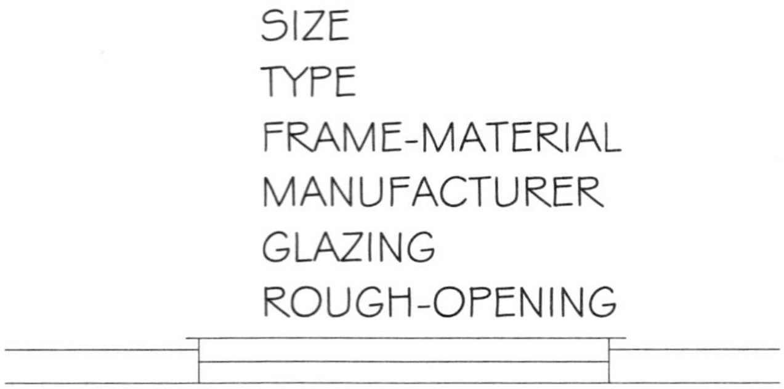

Figure 20.6 *Repeating the procedure used to create the SIZE tag allows an unlimited number of tags to be attached to a block.*

ADDING ATTRIBUTES TO EXISTING BLOCKS

If a block has already been created, attributes can still be added to it. The original block name must be changed, however, to avoid an error message. If the existing POTTY block is inserted in a drawing and attributes are added to it, it cannot be saved using the original name of POTTY. Enter a new name to use the existing block with the desired attributes. Many companies will use a name such as POTTY-A, using the -A to indicate that the block contains attributes. It's much easier to add the attribute prior to saving the block.

CONTROLLING ATTRIBUTE DISPLAY

Rather than wasting disk space by storing POTTY and POTTY-A to distinguish between blocks with and without attributes, the **Attdisp** (ATTribute DISPlay) command can be used to hide attributes if you decide that they should not be displayed. This command sequence is as follows:

Command: **ATTDISP** (ENTER)

Enter attribute visibility setting [Normal/ON/OFF] <Normal>:

Normal will display all attributes, ON will display the variables, and OFF will make all attributes invisible.

The **Attreq** (ATTribute REQuest) command will allow attributes to be displayed but will suppress the attribute prompts. Some blocks, such as kitchen sink, typically retain their default values with very little variation. When the prompt is not required for attributes, the **Attreq** value can be toggled from 1 to 0. The 0 setting will display all attributes in their default setting.

ALTERING ATTRIBUTE VALUES PRIOR TO INSERTION

An **Edit Attribute Definition** dialog box similar to Figure 20.7 can be displayed during the **Insert** process by selecting the **Edit Text** button on the **Modify II** toolbar, by typing **ED** (ENTER) at the Command prompt, or by selecting **Text** from the **Modify** menu. This dialog box can be used to display and alter attribute values. The default values are listed to the right side of the current prompts. Alter values using methods similar to the methods used for other dialog boxes.

The **Properties** dialog box can also be used to edit attributes. Display the dialog box selecting the **Properties** button on the **Object Properties** toolbar, by entering **MO** at the Command prompt, or by selecting **Properties** from the **Modify** menu. The **Properties** dialog box can be used to edit the color, layer, linetype, and each of the options defined as the attribute was created.

ATTACHING ATTRIBUTES TO A BLOCK

Attributes can be attached to a block when a block is defined or redefined (review Chapter 19). Select the attributes and the object when prompted to select the objects that will comprise the block. The order that attributes are selected when the block is created will determine the order of display when the block is inserted.

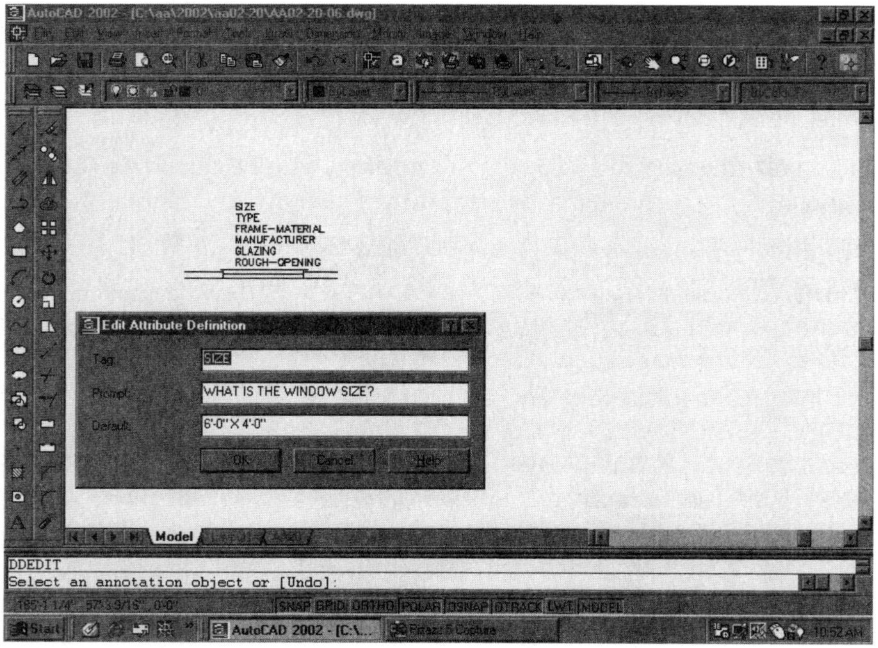

Figure 20.7 *The* **Edit Attribute Definition** *dialog box can be used to alter the tag, prompt, and default value for an attribute before it is saved with a block. Access the dialog box by typing* **ED** *(ENTER) at the Command prompt.*

Inserting a Block with Attributes

The **Insert** command introduced in Chapter 19 can be used to insert blocks with attributes in a drawing. Blocks with attributes are inserted in a method similar to inserting blocks with no attributes. Once the block to be inserted is selected, the prompt will be given for the insertion point. Each of the options for a block can now be adjusted from the command line. The final aspect of inserting the block will be the display of the prompts defined as the attributes were assigned. Figure 20.8 shows the tags and the values for a window block.

EDITING ATTRIBUTES

Once the values for an attribute have been assigned to a block, they can be altered using the **Edit Attributes** dialog box. Display the dialog box by selecting the **Edit Attribute** button on the **Modify II** toolbar, by selecting **Single** from **Attribute** on the **Modify** menu, or by typing **ATTEDIT** (ENTER) at the Command prompt. As the command is entered, you'll be prompted to first select a block reference to be edited. Once the block is selected, a dialog box similar to Figure 20.9a is displayed to allow each attribute value to be altered. Alter the text by typing in the desired edit box. Clicking the **OK** button will alter the values displayed on the screen similar to Figure 20.9b. Clicking the **Cancel** button will remove the dialog box, with no changes made to the drawing base.

There are two tools available that will allow modification of the attribute once it has been placed in a block. The **Block Attribute Manager** (**Battman**), and the **Enhanced Attribute Manager** (**Eattedit**). The **Block Attribute Manager** should be used if you want to modify the block definition of the attribute. The **Enhanced Attribute Editor** is the choice if you wish to edit the attributes in an individual block.

THE BLOCK ATTRIBUTE MANAGER

The **Block Attribute Manager** will allow you to easily modify the attribute definition within a block without having to explode or redefine the block. All changes are

Figure 20.8 *The window on the left shows how the tags will be displayed as the block is created. The window on the right shows how the values will be displayed once the block is inserted in a drawing.*

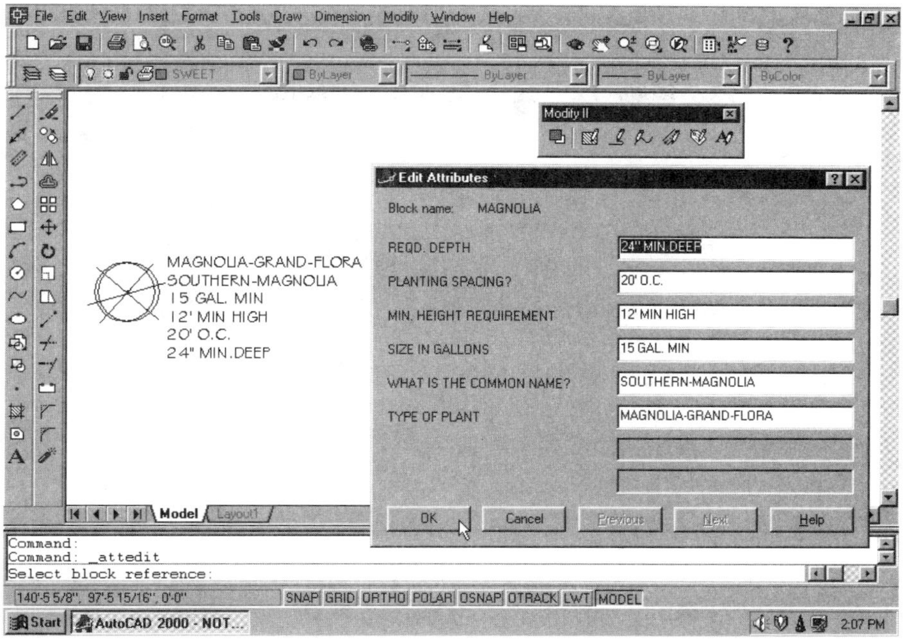

Figure 20.9a *Attributes can be edited using the* **Edit Attributes** *dialog box. Access the dialog box by selecting the* **Edit Attribute** *button on the* **Modify II** *toolbar, by selecting* **Single** *from* **Attribute** *on the* **Modify** *menu, or by typing* **ATTEDIT** *(ENTER) at the Command prompt.*

immediately reflected in the existing block insertions. This is useful when a minor change is needed and you do not wish to completely redefine the block containing the attribute. Access the **Block Attribute Manager** by first selecting **Attribute** from **Object** on the **Modify** menu, and then selecting **Block Attribute Manager**. You can also type **BATTMAN** (ENTER) at the Command prompt. Figure 20.10 shows an example of the **Block Attribute Manager** dialog box.

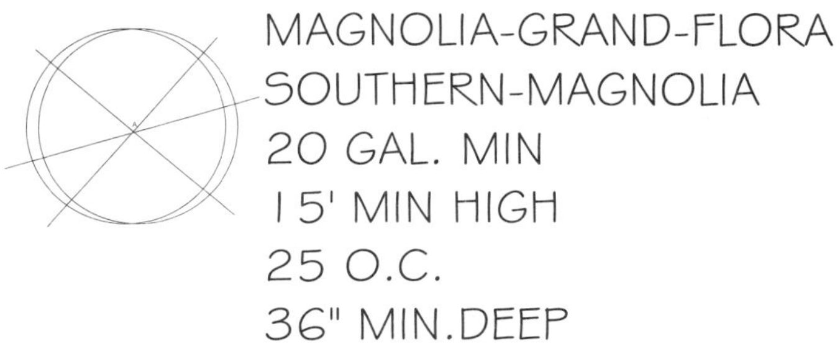

Figure 20.9b *The effects of altering the values in the* **Edit Attributes** *dialog box in Figure 20.9a.*

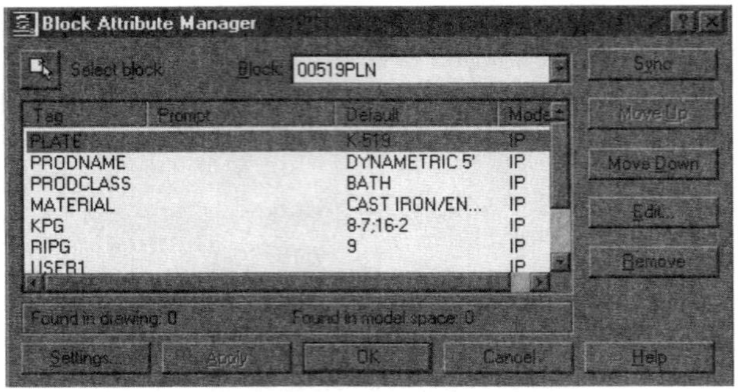

Figure 20.10 *The* **Block Attribute Manager** *dialog box allows you to easily modify the attribute definition within a block without having to explode or redefine the block.*

The **Block Attribute Manager** dialog box allows you to easily manipulate block attribute properties without redefining the block in the following ways:

- Remove an attribute completely from a block definition.
- Change the order of attributes. This affects the order in which the attributes appear when using the **Edit Attributes** dialog box (**DDATTE** or **ATTEDIT**) and the order in which the attribute prompts appear for input during the block insertion.
- Force the attributes in existing block references to be updated to match their block table definition. This updating process does not affect the attribute value.
- Change an attribute's tag name, prompt string, default value, or mode.
- Change an attribute's text settings, such as the size or font.
- Change an attribute's properties, such as color or layer.
- Customize the main attribute list to display selected columns of information and highlight any redundant tag names.

By default, all changes made to a block with attributes using the **Block Attribute Manager** dialog box are then assigned to all the existing references of that block in the drawing file.

Select Block

The **Select block** button enables you to select blocks with attributes in the current drawing for editing. This name is then highlighted in the **Block** list with the attribute information from the block. The information will then be displayed in the attribute list. If you selected an attribute in the block object, that attribute will be the one that is highlighted in the attribute list.

Block

The **Block** list will display the names of all the blocks with attributes in the block table. (This list will not include any blocks that do not contain attributes.) When a block is selected, the attribute list is updated to display the attributes that are contained in the block that is selected. If the **Block Attribute Manager** dialog box has previously been accessed in the current drawing, the last block and attribute edited will be displayed by default. If the last block edited no longer exists, the selection will default to the first block in the list.

Attribute List

The attribute list displays information about all the attributes defined in the block selected in the **Block** list, including the tag, prompt, default value, and the modes. This list can be customized to display specific columns of information and to highlight redundant tag names by using the **Settings** button.

Sync

The **Sync** button forces all of the existing references to the block displayed in the **Block** list to be updated to the new settings to match the block table definition. This update does not affect attribute values. This operation will automatically occur when the block is edited.

Edit

The **Edit** button will display the **Edit Attribute** dialog box for the highlighted attribute and allow the information to be modified.

THE EDIT ATTRIBUTE DIALOG BOX

The **Edit Attribute** dialog box consists of three tabs for editing different aspects of the highlighted attributed (see Figure 20.11). When **Auto preview changes** is checked, all edits that affect the appearance of the attribute will automatically be displayed in the editor. This permits you to see how edits being made will actually affect the attributed block. Tabs include the following:

Attribute—On the Attribute tab, you can edit the attributes tag, prompt, default value, and mode.

Text Options—Allows changes to values that affect the text-based properties of the attribute, as in figure 20.12.

Properties—Allows editing of values that affect the general appearance of the attribute, either in the editor or during plotting. See figure 20.13.

THE ENHANCED ATTRIBUTE EDITOR

The **Enhanced Attribute Editor** dialog box will allow editing of the attributes in the block that is selected. You can change the attribute value, any text-related settings like

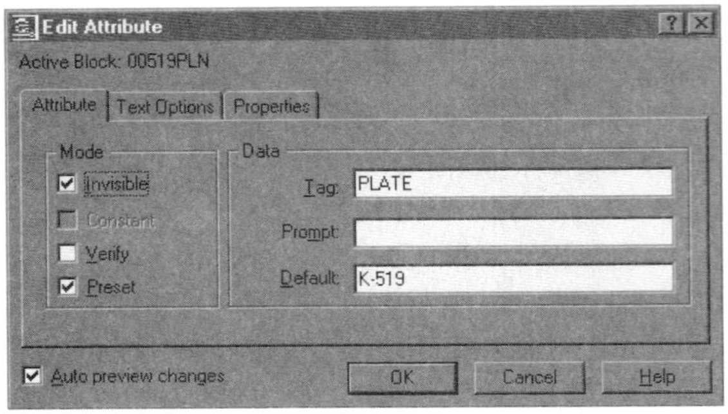

Figure 20.11 *The* **Edit Attribute** *dialog box shows the three tabs available for editing the highlighted attribute.*

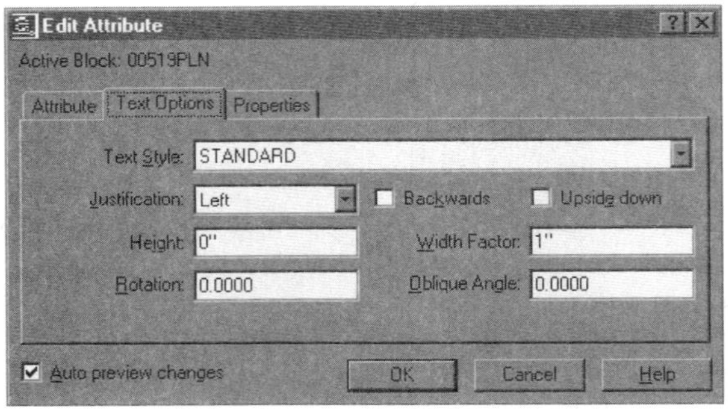

Figure 20.12 *The* **Text Options** *tab allows editing of the text properties of the attribute.*

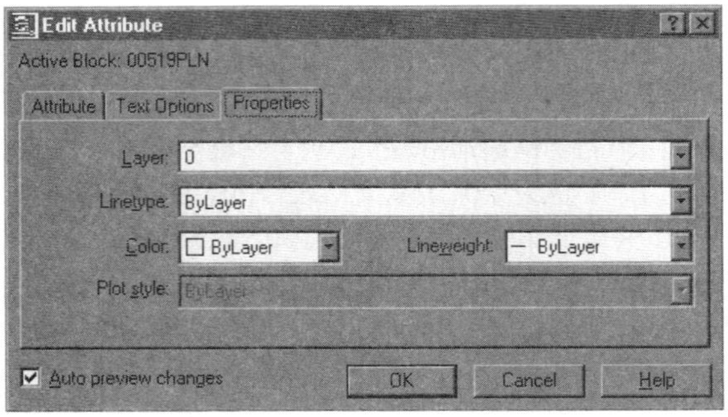

Figure 20.13 *The Properties tab will change the appearance of the attribute.*

size and orientation, and the properties of the attribute. All of the changes will be displayed automatically in the editor as they are made. To access the **Enhanced Attribute Editor**, select **Attribute** from **Object** on the **Modify** menu and then **Single**. You can also type **EATTEDIT** (ENTER) at the Command prompt. The **Enhanced Attribute**

Editor dialog box has three tabs available to allow editing the different features of the selected attribute in a block, as well as a button for selecting the block, as shown in Figure 20.14.

Select block—The **Select block** button allows you to select a block with attributes in the current drawing for editing. The name of the block appears at the top of the dialog box, and the attribute information assigned to that block will be added to the attribute list.

Attribute—This tab shows the information about all of the attributes that are a part of the selected block. When an attribute is selected from the list, the value can be modified.

Text Options—This tab allows editing of those values that will apply to the text properties of the attribute (see Figure 20.15).

Properties—This tab allows you to modify the appearance of the attribute either in the editor itself or at the time of plotting (see Figure 20.16).

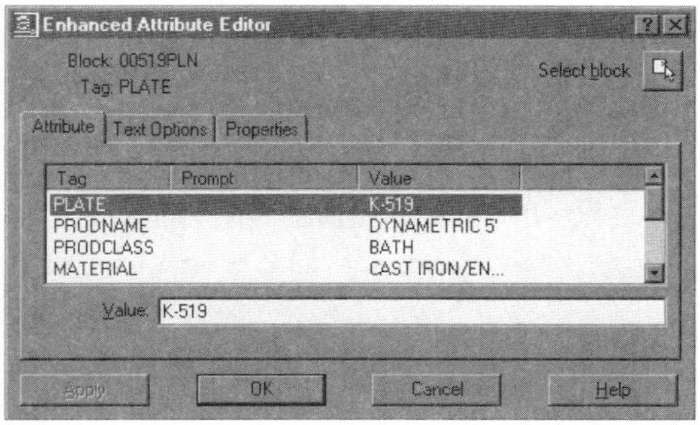

Figure 20.14 *The* **Enhanced Attribute Editor** *dialog box contains three tabs for changing features of the attribute.*

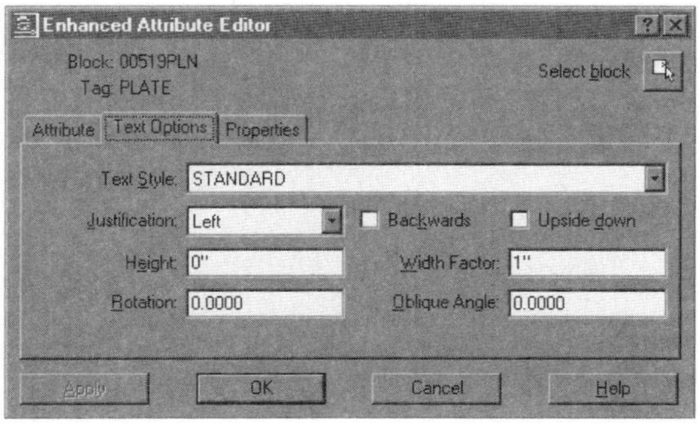

Figure 20.15 *The* **Text Options** *tab will modify the attribute text properties.*

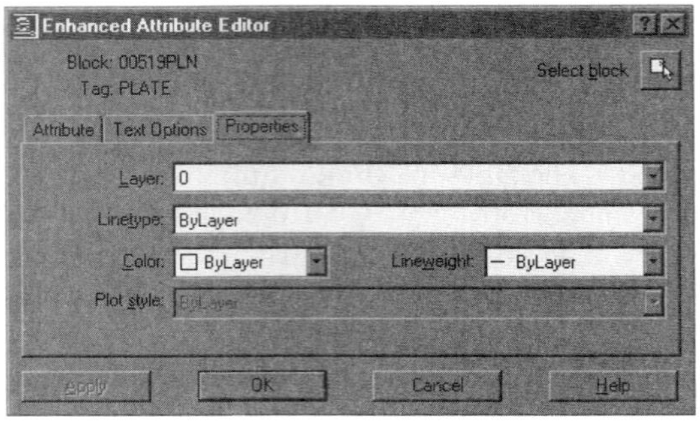

Figure 20.16 *The* **Properties** *tab will change the appearance of the attribute.*

GLOBAL EDITING

Selecting **Attribute** from **Object** on the **Modify** menu and then selecting **Global** allows the same attribute to be edited in all of its applications throughout a drawing. This can be useful if you find an error in a block value that has been inserted several times. Rather than changing each attribute individually, you can change all of them using the global option. The command sequence is as follows:

Command: *(Select Global from Attribute, Object on the Modify menu.)*

Edit attributes one at a time? [Yes/No] <Y>:

Accepting the default value will allow you one last chance to verify each attribute before it is altered. Typing **N** (ENTER) at the prompt will change all of the specified attributes, once other parameters have been set. The prompt will continue:

Performing global editing of attribute values

Edit only attributes visible on screen? [Yes/No] <Y>:

Typing **N** (ENTER) will display the AutoCAD Text window with the following message:

Drawing must be regenerated afterward.

Any changes that are made to attributes will not be reflected until the drawing is regenerated. If **Autoregen** is set to ON, a drawing regeneration will be performed when this command is finished.

Editing Visible Attributes

If a Y is accepted at the prompt and only the attributes shown on the screen are to be edited, four prompts will be displayed. These include the following:

Enter block name specification <*>:

Enter attribute tag specification <*>:

Enter attribute value specification <*>:

Select Attributes:

Specific sets of block names, attribute tags, and attribute values, including wild card characters, can be entered to define the selection set for editing. For an attribute to be selected for editing, the block name, attribute tag, and attribute value must all match the name, tag, and value entered for the selection set. Be sure to match the case, uppercase or lowercase, used for the selection sets to the case used when the attributes were entered originally. Matching uppercase titles with lowercase titles will not be accepted as matching sets. Attributes can be selected using Box, Window, Crossing, WPolygon, or CPolygon. Once selected, attributes will appear highlighted. When the selection process is complete, the prompt will change to

Select attribute:

Enter string to change:

The "string to change" prompt initiates a search of all eligible attribute strings for matching strings. Each matching string will be changed to reflect the entry for the "new string" prompt. A block titled FOOTING with a tag of DIA. and a value of 17" diameter x 8" could be changed to 18" by entering the following values:

Enter string to change: **17** (ENTER)

Enter new string: **18** (ENTER)

Entering * at the "string to change" prompt will cause the entry at the "new string" prompt to be placed ahead of all eligible attribute value strings. If a block is specified as SINK with an attribute value of DBL. SINK, it can be changed to 32x21 DBL SINK by entering the following:

Enter string to change: *

Enter new string: **32x21** (SPACE) (ENTER)

Be sure to use (SPACE) to provide a space after the 21, or the new display will read 32x21DBL SINK.

EXTRACTING ATTRIBUTES

One of the most useful benefits of using attributes is the ability to extract information stored as an attribute and assemble it in a written list. Window, door, and finish schedules are typically associated with architectural drawings. Pier, steel, and beam schedules are but a few of the types of lists in engineering drawings. Figure 20.17 shows portions of a floor plan and the doors that will be extracted. The doors shown are a fraction of the total number of doors that would be represented on a complete multilevel structure. Each door was inserted in the floor plan as a block, with one visible and four invisible attributes. Attributes have been assigned to represent symbol, size, style, manufacturer, price, and quantity. Because these values are assigned as attributes, an accurate schedule can be maintained and quickly updated. With the Eattext command, a complete listing of all of the doors used throughout the project can be written to a file.

USING A DATABASE

A database program can be used to manipulate the data stored in block attributes. A listing of doors or windows for a multilevel structure is a database that has essential data associated with it—such as size, type, rough opening, etc. Two terms generally associated with each database are *record* and *field*. A list of each of the construction components related to doors would be a record. The door, frame material, hinge type, and finishing method are each an example of a record. A door might be swinging, sliding, folding, or overhead. Each listing of a door is a record. The swing information about a door is a field. Swinging doors can be divided by interior, exterior, raised panel, slab, 1-lite, or 10-lite. It is possible to take a listing of doors that are listed by size, and by using a database, reorder the listing by type. AutoCAD allows a list to be generated that contains all of the objects to be manipulated by the database.

EXTRACTING THE ATTRIBUTES OF A BLOCK

The **Attribute Extraction Wizard** allows the easy extraction of block attribute data to comma-separated text (.csv), Excel (.xls), and Access (.mdb) formats. You can attach alias names to blocks and attributes, publish data from

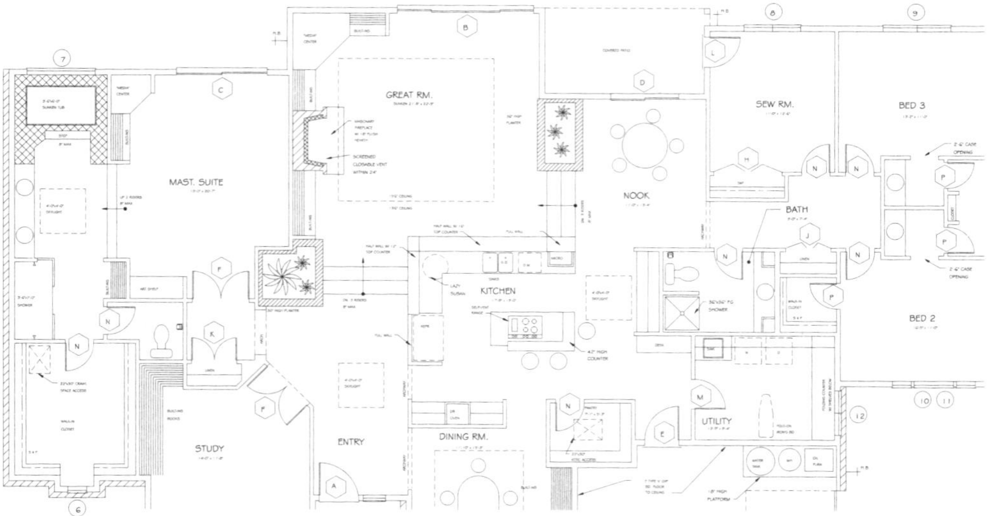

Figure 20.17 *Extracting attributes can be very useful on drawings containing multiple blocks with attributes such as this floor plan.*

multiple drawings and xref attachments, and save templates of selected blocks, attributes, and alias information for reuse with any drawing or set of drawings. The **Attribute Extraction Wizard** will guide you through the process of attribute extraction. Select the **Attribute Extract** button on the **Modify II** toolbar or type **EATTEXT** (ENTER) at the command line to begin the wizard.

The **Attribute Extraction - Select Drawing** dialog box determines the source of the attribute data. It will allow you to create a selection set of objects from the current drawing, select all of the blocks in the current drawing, or select multiple drawings to extract the information from. Figure 20.18 shows a sample of how the source data is to be selected.

The **Attribute Extraction - Settings** dialog box allows you to include external reference files (xrefs) and nested blocks in your search. (Nested blocks were introduced in Chapter 19.) Figure 20.19 shows how the **Attribute Extraction - Settings** dialog box appears.

The **Attribute Extraction - Use Template** dialog box allows the selection of a template file. Template files store information related to which attributes and blocks are to be included in the output and the view from which data is extracted. The template will be created as the criteria are assigned with the wizard. You can then save the template for use in future attribute extractions. Figure 20.20 shows how to choose an existing template file.

The **Attribute Extraction - Select Attributes** dialog box consists of two sections:

- The **Blocks** list will show the list of blocks from the selected drawing or drawings. The toggle boxes will allow selection of a block to be included in the

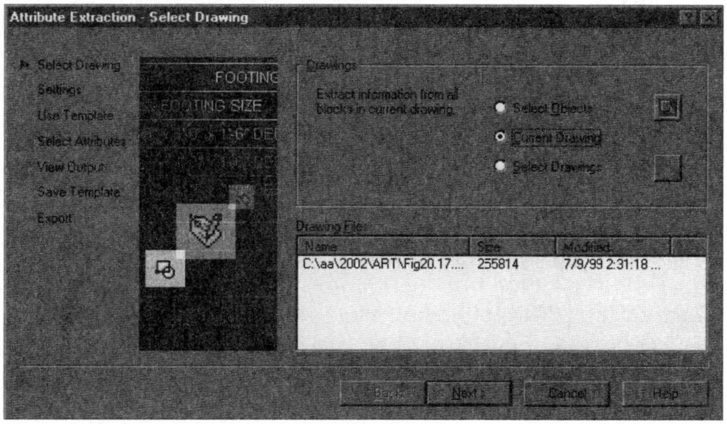

Figure 20.18 *The* **Attribute Extraction - Select Drawings** *dialog box determines the source of the attribute data.*

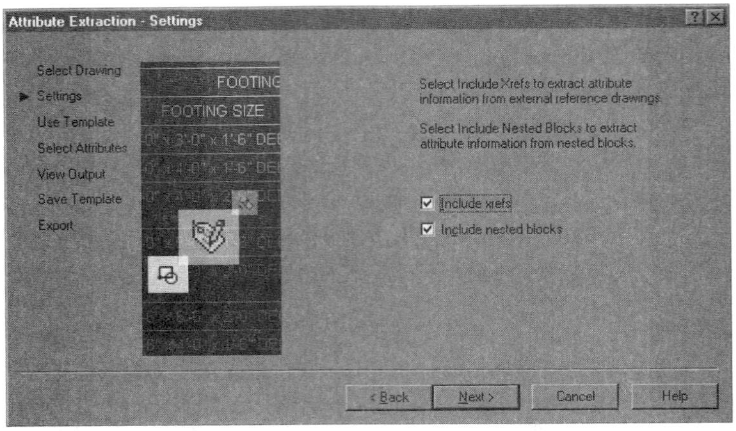

Figure 20.19 *External references and nested blocks can be included in the search.*

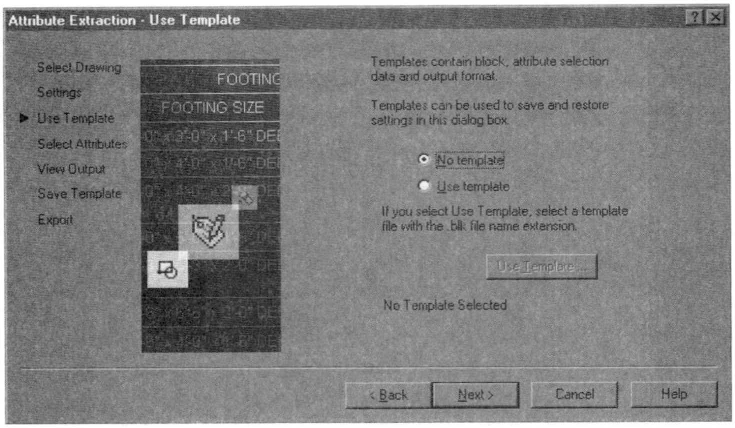

Figure 20.20 *A template can be selected from a previous extraction.*

extraction. The **Alias** column allows assignment of a name to each block name. The **Number** column displays the number of occurrences of each block.

- The **Attributes for block** list shows the list of attributes that relate to the block selected. The toggle box allows selection of attributes that are to be included in the output. The Alias column allows using a different name for each attribute. Then the value of each attribute will also be displayed.

Figure 20.21 shows an example of the **Attribute Extraction - Select Attributes** dialog box.

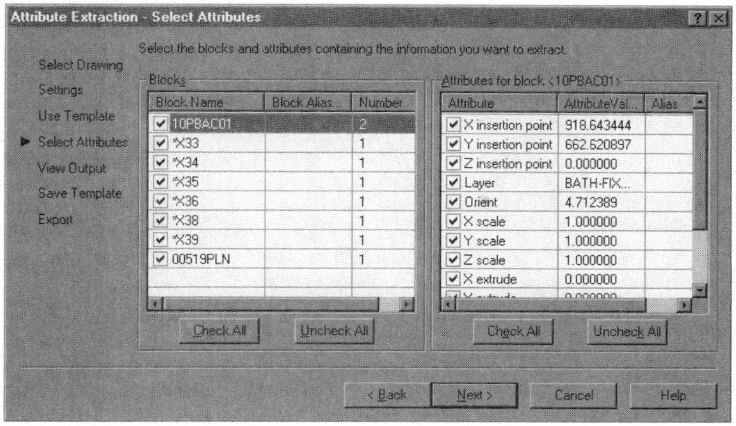

Figure 20.21 *The* **Attribute Extraction - Select Attributes** *dialog box contains a block list and an attribute list.*

The **Attribute Extraction - View Output** dialog box shown in Figure 20.22 will now show a list of the information that has been extracted. A list of the extracted information can be copied to the clipboard. This allows the information to be viewed and printed with a variety of applications.

The **Attribute Extraction - Save Template** dialog box allows a copy of the template to be stored in a file that can be used again in an attribute extraction (see Figure 20.23). The file will have a file extension of .blk. Saving the template is a good idea if the attribute extraction needs to be repeated, with the same information included.

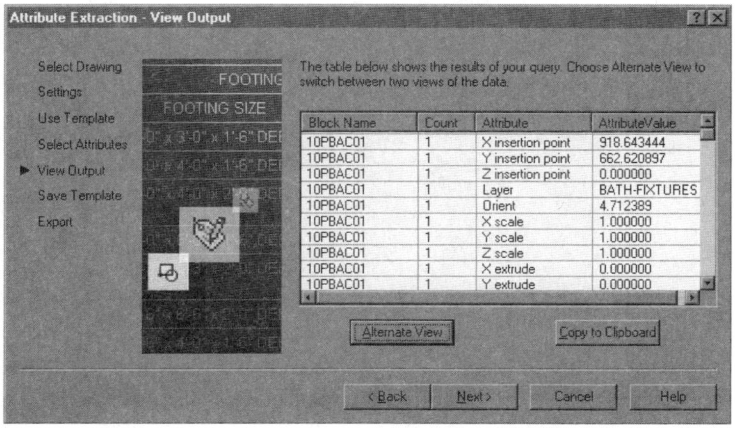

Figure 20.22 *The* **Attribute Extraction - View Output** *dialog box displays a sample of the information that will be extracted.*

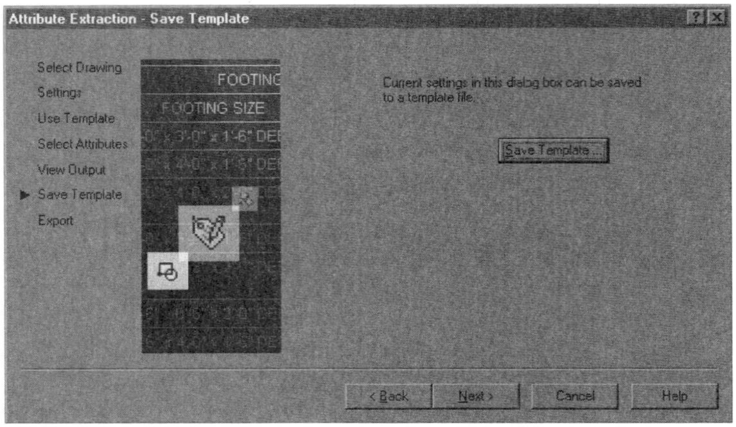

Figure 20.23 *The* **Attribute Extraction - Save Template** *dialog box allows the information from the template to be saved and given a name for future use.*

The **Attribute Extraction - Export** dialog box allows you to export the extracted data to a comma-separated text file (.csv) that can be used by some database programs. In addition, if you have Microsoft Excel or Microsoft Access, you can also export the extracted data set to .xls (Excel) or .mdb (Access) format. Figure 20.24 is an example of the **Attribute Extraction - Export** dialog box. Simply choose the type of data to match the software you will be using and name the file with the proper file extension.

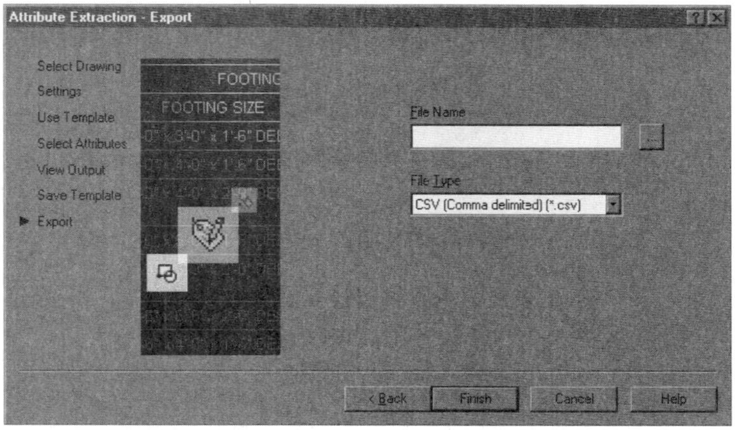

Figure 20.24 *The* **Attribute Extraction – Export** *dialog box allows the extracted data to be shared with other software programs.*

CHAPTER 20 EXERCISES

1. Draw an 18" diameter footing using dashed lines and show a 4x4 post in the center of the pier. Use 5" high text to install the following attributes.

TAG	PROMPT	VALUE
Dia.	What is the diameter?	18" dia.
Depth	What is the pier depth?	12" min. into grade
Strength	What is the conc. strength?	2500 p.s.i.
Rebar	Will the pier be reinforced?	10x10-4x4 wwm 3" up
Post	Size of post to be supported?	4x4 D.F.L. #1
Anchor	Will an anchor be used?	Simpson CB44 base

Save the drawing as a wblock on a diskette as FNDPIER.

2. Draw an elevation of a 3'–0" x 4'–6" double-hung window and use 5" high text to install the following attributes. Set the mode to invisible display.

TAG	PROMPT	VALUE
Manuf.	List the manuf.	POZZI
Frame	List the frame size	3'–0" x 4'–6"
Unit	List the unit size	UN=3'–2 5/8" x 4'–8 7/8"
R.O.	List the rough opening	RO=3'–0 1/2" x 4'–6 1/2"
Frame	Frame material	Kiln-dried fir
Exterior	Exterior finish	Prime ext. face
Sash	Sash material	1 11/16" thick sash
Glaze	Glazing material	9/16" air space
Weather	Weather stripping	foam-backed
Hardware	Hardware color	Bronze
Screen	Screen material	Alum.-bronze tint
Assembly	Where assembled	Site assembled
Shipping	Destination	Seattle, WA.

Prior to creating a wblock of this material, change the Frame material to Kiln-dried Western pine. Change the hardware to WHITE. Ship from Bend, OR. Change the Assembly point to Factory. Save the drawing as a wblock with a name of EL3-46DH.

3. Open a new drawing and draw a 24" x 24" skylight in plan view. Add the following attribute tags, prompts, and values.

TAG	PROMPT	VALUE
Size	Unit size	24"x24"
Manuf.	Who is the manuf?	Velux
Model	Model number	SF
Frame	Frame material	Alum.-clad frame
Finish	Finish material	Lacquered finish
Glass	Glazing	Double-tempered glazing
Flashing	Flash material	32ga. lacquered alum.

Save the drawing on a diskette as a wblock titled SKYLIGHT. Insert the skylight in a drawing three different times. Change the size so that the skylight will be 24x24, 24x48, and 48x48. Change the size value of each skylight to reflect the proper size. Use global editing to change the model value to FS, change the frame to Alum-clad Frame, and change the finish to 22ga. Save the drawing as E-20-3.

4. Open drawing E-19-5 and insert the POTTY block. EXPLODE the block and add the following attribute tags, prompts, and values.

TAG	PROMPT	VALUE
Manuf.	Who is the manuf?	Am. Std.
Cat-no	What is the Catalog number?	2006.014
Descrip.	What is the description?	Elon. Lexington
Color	What color will be used?	White
Supplier	Who will provide fixture?	General cont.
Installer	Who will install fixture?	Plumber
Hardware	Who will supply hardware?	Owner

Save the drawing as a wblock titled POTTY.

CHAPTER 20 QUIZ

1. How are attributes saved?
2. What commands can be used to add a block or wblock to a drawing?
3. What must be done to add attributes to an existing block?
4. Describe what an attribute prompt is.
5. What is an attribute tag?
6. List the four modes for attribute display and give a brief explanation of each.
7. What is the initial step to assign attributes to blocks?
8. List the commands that produce dialog boxes for use with attributes.
9. List two methods to edit an attribute.
10. What **Attdia** setting is required to produce dialog box displays?
11. What two commands allow for attribute editing?
12. How can global editing of attributes be limited?
13. How do the prompts "Enter string to change" and "Enter new string" affect each other?
14. What options are provided when attributes are edited individually?

CHAPTER 21

Oblique and Isometric Drawings

The drawings that have been examined throughout this text have been orthographic drawings. These are drawings that show an object as you would look directly at it. Objects are assumed to be located in a glass box, with the sides of the object parallel to the sides of the glass box. Orthographic drawings comprise the majority of construction drawings. Pictorial drawings, including oblique and isometric drawings, are used in some disciplines of engineering and architectural drawings to show specific details. This chapter will introduce

- The theory of oblique and isometric drawings
- The creation of oblique and isometric drawings

Commands to be explored in this chapter include

- **Snap**
- **Isoplane**
- **Ellipse**
- **Dimali**
- **Dimedit**

ABOUT OBLIQUE AND ISOMETRIC DRAWINGS

Before examining what oblique and isometric drawings are, it is important to understand what they are not. These drawings appear to be three-dimensional, but are really two-dimensional drawings using lines that are created using X and Y coordinates.

3D DRAWINGS

One of the major features of AutoCAD is its ability to create true 3D drawings. Three-dimensional drawings require creation of a 3D model. For instance, in drawing a floor plan, heights are provided in addition to the normal information that would be drawn. Once the model is created, AutoCAD can rotate the model in space, allowing the viewing point to be altered. You've most likely seen commercials on television with structures growing out of a floor plan and progressing through the entire construction process. These drawings could be created starting with 3D models. Individual points, lines, and planes comprising the drawings are stored and become part of the drawing base. Figure 21.1 shows an example of a 3D drawing created using AutoCAD and a third party software program. Although these drawings are both beautiful and very useful, how they are created will not be discussed in this text.

OBLIQUE DRAWINGS

Oblique drawings show height, width, and depth of an object. Three common methods of oblique drawings include cavalier, cabinet, and general drawings. Examples of each can be seen in Figure 21.2. Each system places one or more surfaces parallel to the viewing plane and shows the receding surface at an angle. Each drawing method can be used at any scale.

Cavalier Drawings

Cavalier drawings use the same scale for the front and receding surfaces. The front side is drawn parallel to the viewing plane. The receding surface is drawn at a 45° angle to show depth. See Figure 21.3.

Figure 21.1 *3D drawings rendered in AutoCAD often resemble a photograph.*

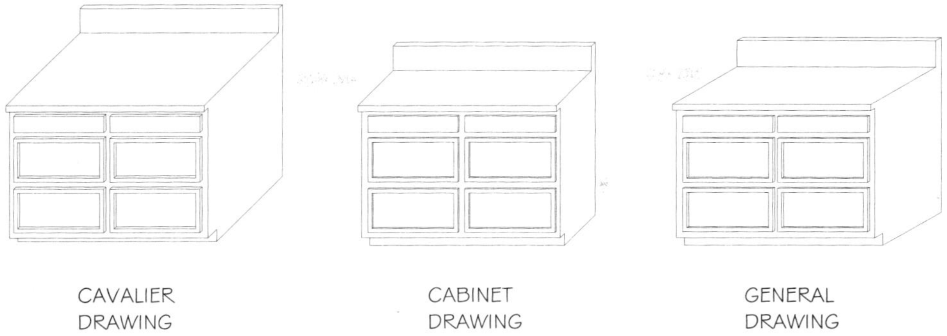

Figure 21.2 *Although lacking the details of 3D drawings, three types of oblique drawings are often used to show height, width, and depth of an object.*

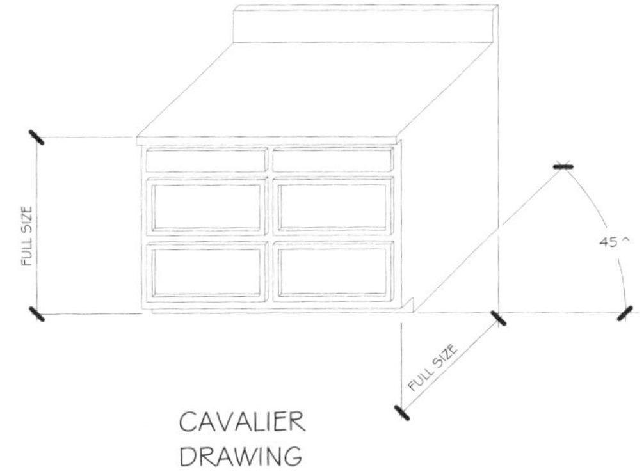

Figure 21.3 *Cavalier drawings show all surfaces in full scale, with the receding side drawn at a 45° angle.*

Cabinet Drawings

Cabinet drawings are similar to cavalier drawings but show the receding surfaces using a scale that is half of the scale used on the front surfaces. The receding surfaces are also shown at a 45° angle. See Figure 21.4.

General Drawings

General drawings represent the receding surfaces using a scale that is three-quarters of the scale used to draw the front surfaces. The angle used to create the receding edge can be any angle except 45°. Figure 21.5 shows an example of a general oblique drawing.

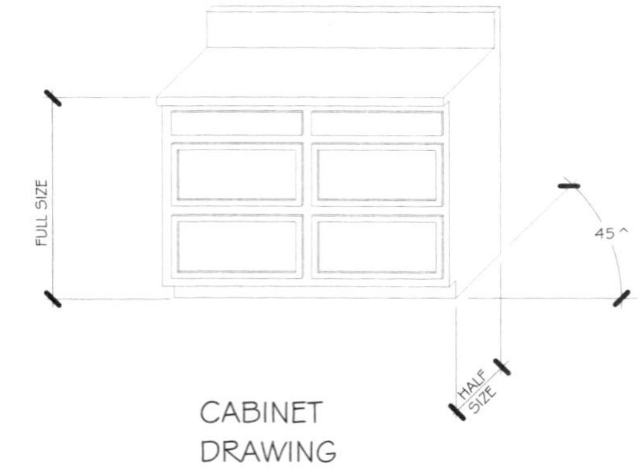

Figure 21.4 *Cabinet drawings show the front view at full scale, with the receding side shown at half scale and a 45° angle.*

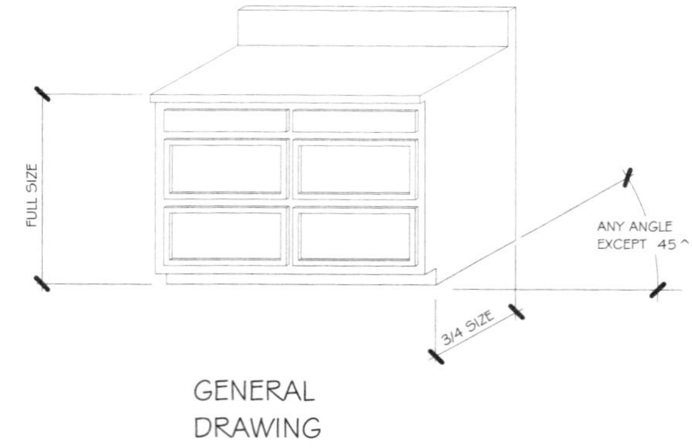

Figure 21.5 *General drawings show the front view at full scale, with the receding side shown at three-quarter scale and any angle except 45°.*

ISOMETRIC DRAWINGS

Isometric drawings are the most common form for showing three surfaces of an object. Rather than having one surface resting squarely on a base, the object is rotated at 35° 16' from a horizontal plane. This results in a view of the object from the front where each edge is drawn at 30° to a horizontal base. All vertical lines remain vertical, and an equal scale is used to represent all surfaces. Figure 21.6 shows the basic construction methods for an isometric drawing. Lines that are parallel to one of the three axes will remain parallel. Inclined surfaces that are not parallel to an axis will need to be projected from a point that lies on a line that is parallel to an axis. Figure 21.7 shows the steps for blocking out an irregularly shaped detail.

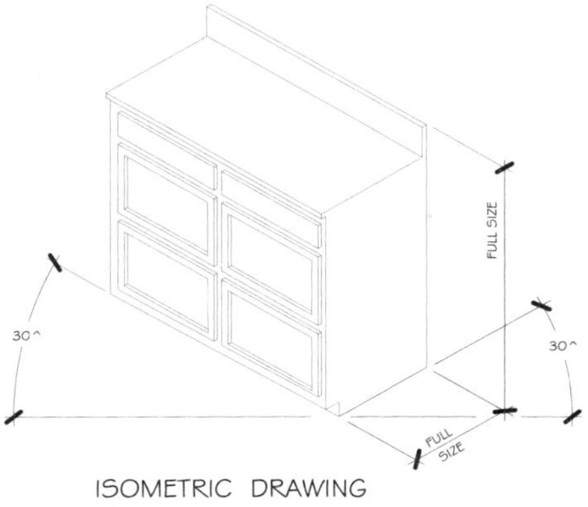

Figure 21.6 *Isometric drawings show all sides at full scale, and each side at a 30° angle.*

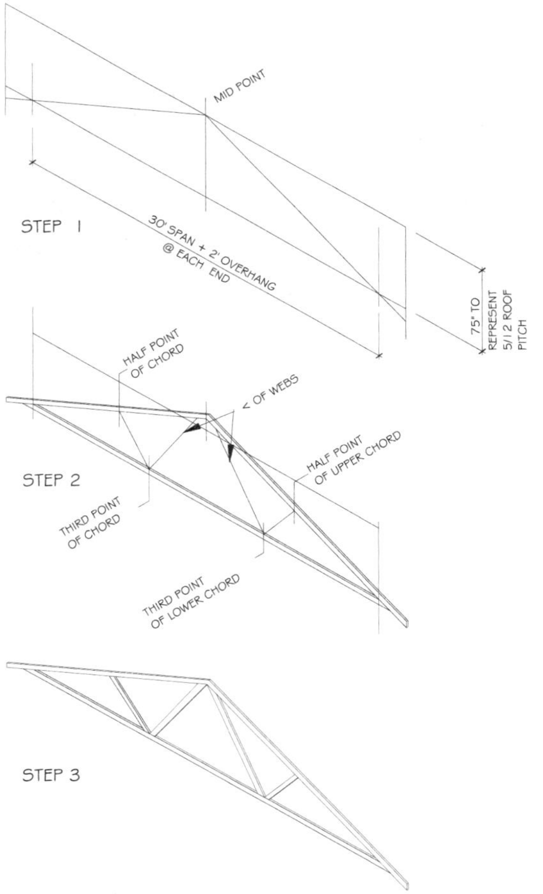

Figure 21.7 *Blocking irregularly shaped objects in rectangles will aid in drawing layout.*

DRAWING TOOLS FOR PICTORIAL DRAWINGS

The drawing tools used to create two-dimensional drawings also can be used to create pictorial drawings. Commands such as **Snap**, **Grid**, **Isoplane**, and **Ellipse** can be used to aid in drawing setup.

SNAP

Pictorial drawings can be created easily by altering the snap setting in the **Drafting Settings** dialog box (choose **Drafting Settings** from the **Tools** menu). Select **Isometric Snap** from the **Snap Type & Style** area of the **Snap and Grid** tab. Clicking the **OK** button will close the dialog box, alter the snap setting to isometric, and restore the drawing area. Returning to the drawing display will display an isometric grid and cursor, as seen in Figure 21.8. Activating SNAP and GRID will greatly aid in creating isometric drawings.

CROSSHAIRS ORIENTATION

Three surfaces are seen in pictorial drawings and are referred to by AutoCAD as Isoplane Left, Isoplane Top, and Isoplane Right. These three orientations are also available for the crosshairs to ease drawing construction. Figure 21.9 shows the three iso-

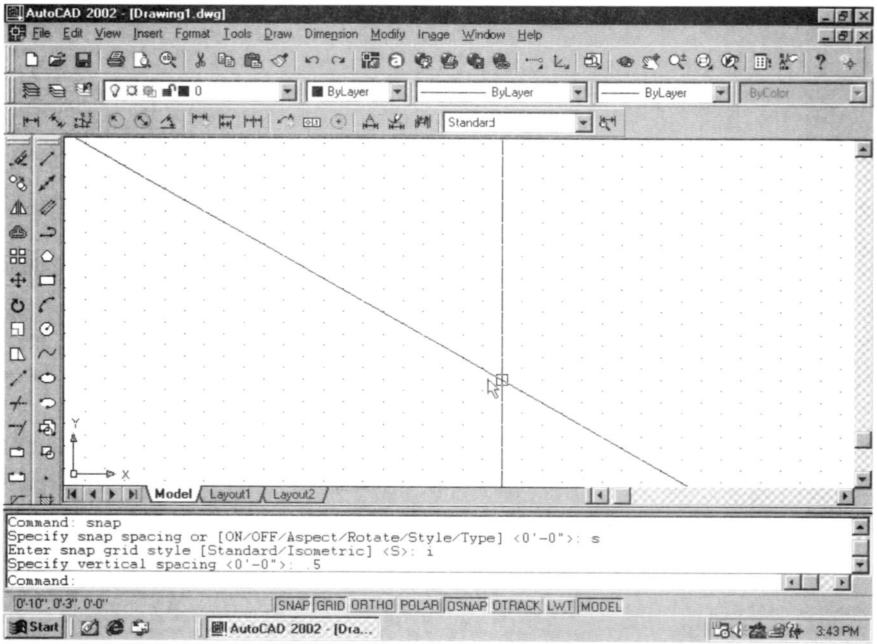

Figure 21.8 *An isometric grid can be established in the* **Drafting Settings** *dialog box (select* **Drafting Settings** *from the* **Tools** *menu). Select* **Isometric Snap** *from the* **Snap Type & Style** *area of the* **Snap and Grid** *tab.*

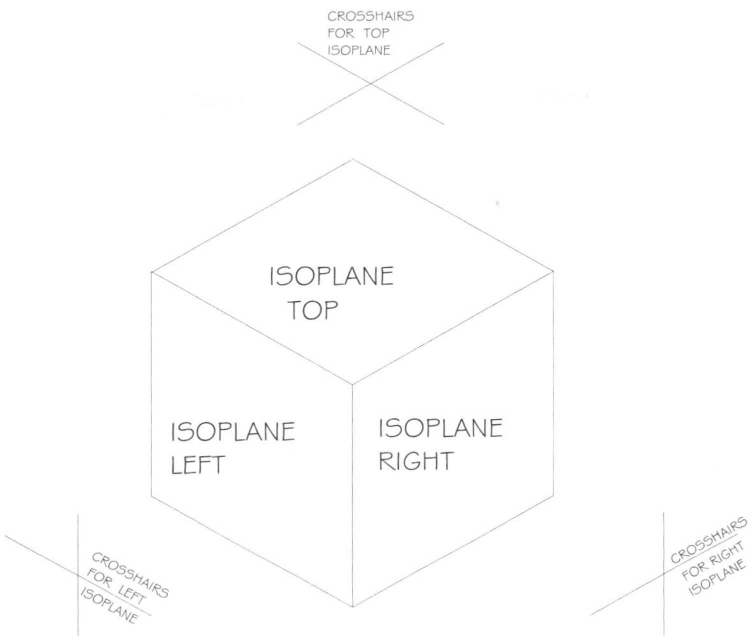

Figure 21.9 *AutoCAD refers to the three isometric surfaces as Isoplane Left, Isoplane Top, and Isoplane Right.*

metric planes and the options for crosshairs positioning. The crosshairs orientation can be changed using one of several methods, including a function key, the **Isoplane** command, and a CTRL combination.

Function Key

The easiest toggle between the three isoplane options is to use the function key F5. Each time F5 is pressed, the crosshairs will toggle to the next isoplane and the current setting will be displayed at the Command prompt.

The Isoplane Command

The crosshairs position can also be controlled using the **Isoplane** command. The command sequence is as follows:

Command: **ISOPLANE** (ENTER)

Enter isometric plane setting [Left/Top/Right] <top>: **L** (ENTER)
(Enter either option.)

Current Isometric plane is: LEFT

Command:

The letter representing an option can be typed, followed by (ENTER). Pressing (ENTER) at the toggle prompt will automatically change the crosshairs to the next option and restore the Command prompt.

Control Key Control

The crosshairs can also be controlled by pressing the CTRL+E. Pressing CTRL+E with the crosshairs in the current setting will display a prompt of

Command: <Isoplane Top>:

Repeated pressing of CTRL+E will transparently toggle the crosshairs between the Left, Top, and Right settings.

DRAWING ISOMETRIC CIRCLES

Circles in pictorial drawings are represented by ellipses. With the **Isometric Snap** set, start the **Ellipse** command and select the Isocircle option; an ellipse will be drawn automatically in the current Isoplane setting, as seen in Figure 21.10. The command sequence is as follows:

Command: **EL** (ENTER) *(Or click the Ellipse button.)*

Specify axis endpoint of ellipse or [Arc/Center/Isocircle]: **I** (ENTER)

Specify center of isocircle: *(Specify desired point.)*

Specify radius of isocircle or [Diameter]: *(Specify desired diameter.)*

Command:

A centerpoint can be selected or a Radius or Diameter specified. One of the drawbacks to using pictorial drawings to show construction details is that circular objects often appear distorted. Round objects will appear as an ellipse in an isometric drawing. Figure 21.11 shows the proper orientation of circles for each plane of an isometric drawing. Notice that the centerline of the circle is parallel to the edge of one of the three surfaces.

ISOMETRIC TEXT

No special commands are required to place text on pictorial drawings. On small details, text is best placed in its normal position and connected to the detail with leader lines. On larger drawings, text can be placed parallel to the isoplane on which it will appear by adjusting the Obliquing Angle in the **Text** command. Figure 21.12 shows an example of text placed at 30°, –30°, 90°, and –90°. This can best be done by setting up text styles for each angle.

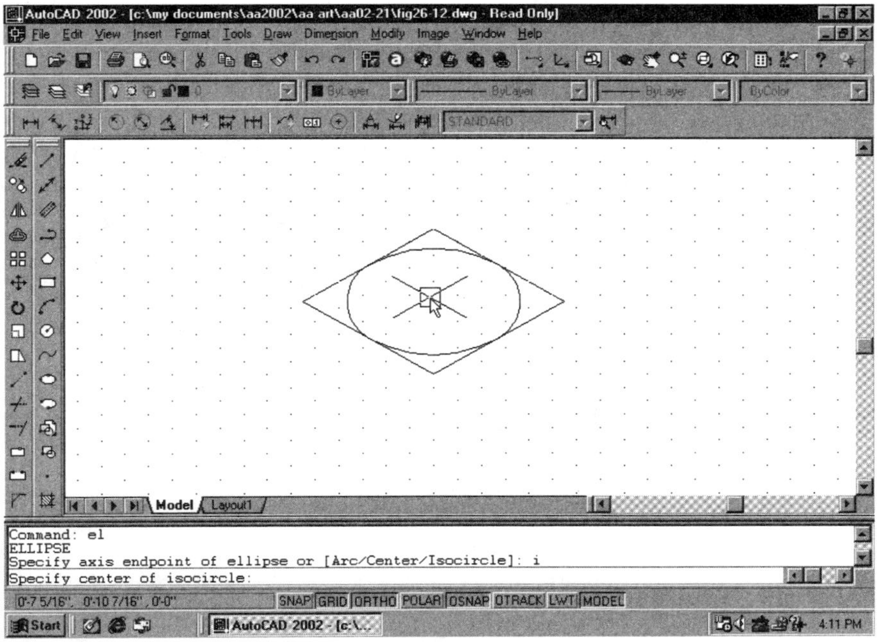

Figure 21.10 *An isometric ellipse will be drawn automatically in the current isoplane setting using the Isocircle option of the* **Ellipse** *command.*

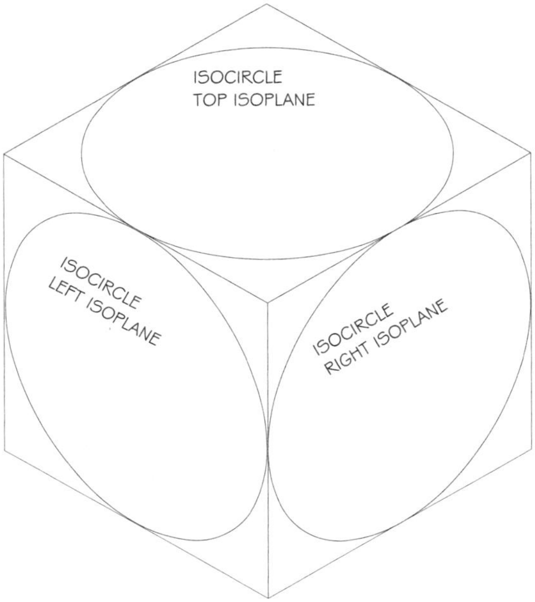

Figure 21.11 *Proper orientation of isometric circles.*

ISOMETRIC DIMENSIONS

Although no specific command exists to create isometric dimensions, dimensions created using the **Dimali** (**Aligned Dimension**) command can be adjusted quickly to conform to an isometric drawing. Aligned dimensions can be placed by selecting the **Aligned Dimension** button on the **Dimension** toolbar or by typing **DIMALI** (ENTER) at the Command prompt. Figure 21.13 shows an example of dimensions added to an isometric drawing using the **Dimali** command. Once dimension text is created, start the **Dimedit** command by typing **DIMEDIT** (ENTER) or selecting the **Dimension Edit** button on the **Dimension** toolbar. Select Oblique at the **Dimedit** prompt. You will be asked to select a dimension to be edited, and then prompted to specify an obliquing angle. Use either 30° or –30° for the angle. The command sequence is as follows:

Command: **DIMALI** (ENTER) *(Or click the Aligned Dimension button.)*

Specify first extension line origin or <select object>: *(Select first point.)*

Specify second extension line origin: *(Select second point.)*

Specify dimension line location or [Mtext/Text/Angle]: *(Select point.)*

Dimension text <2 3/4">: (ENTER)

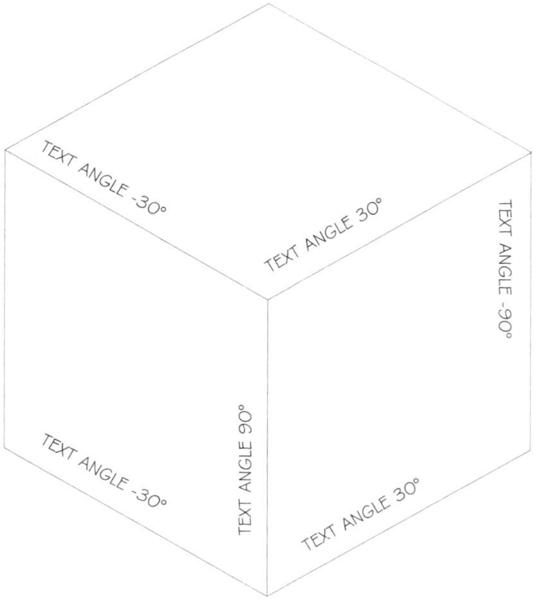

Figure 21.12 *Text can be placed in each plane of an isometric drawing using the Text command and rotating the angle.*

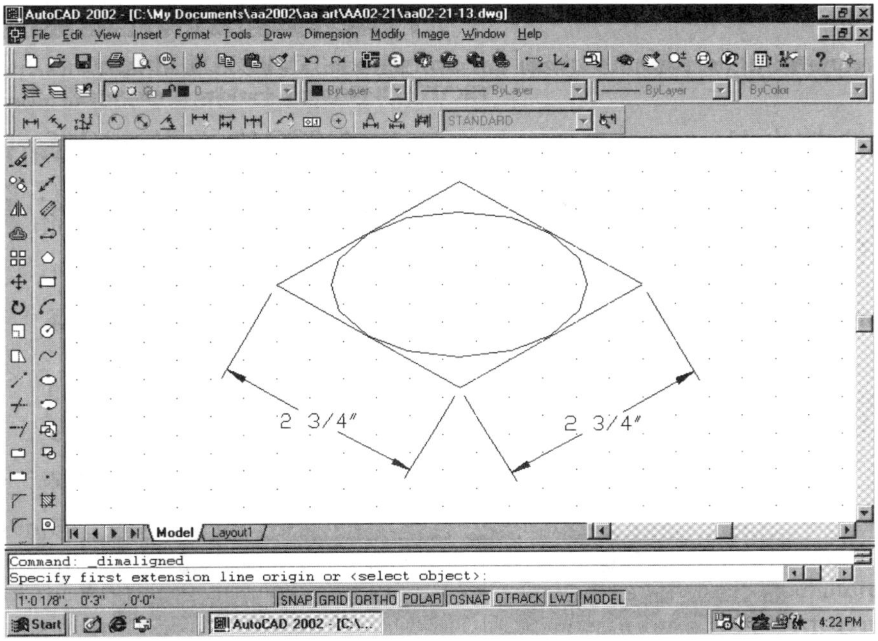

Figure 21.13 *Dimensions can be placed on an isometric drawing using the Dimali command.*

Command: **DIMEDIT** (ENTER) *(Or click the Dimension Edit button.)*

Enter type of dimension editing [Home/New/Rotate/Oblique]
<Home>: **O** (ENTER)

Select objects: *(Select dimension to edit.)*

1 found

Select object: (ENTER)

Enter obliquing angle (RETURN for none): **30** (ENTER)

Command:

The result of this sequence can be seen in Figure 21.14.

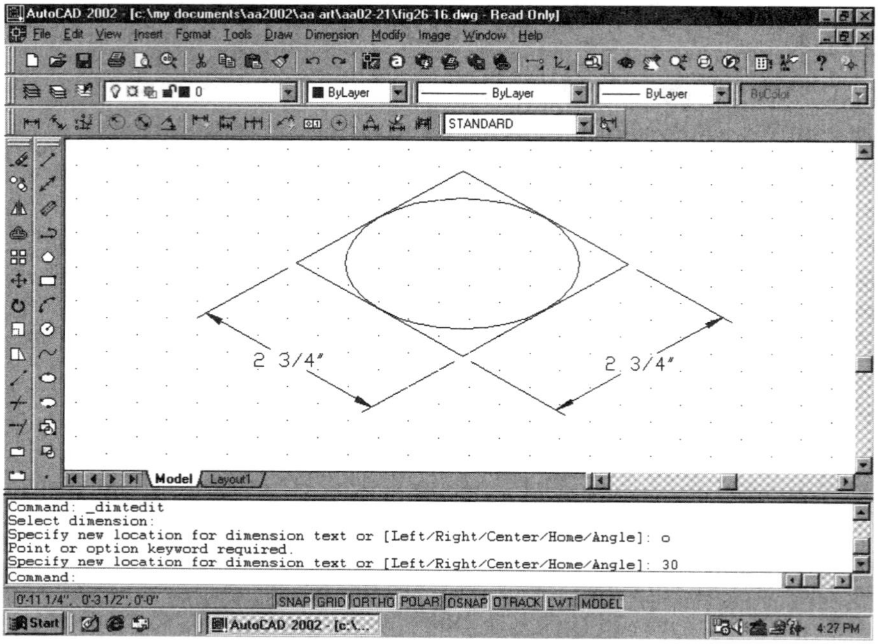

Figure 21.14 *Entering* **O** *(ENTER) for Oblique at the* **Dimedit** *prompt will alter aligned dimensions to isometric dimensions.*

CHAPTER 21 EXERCISES

1. Start a new drawing and draw a cube that is 8' wide, 6' high, and 4'–6" deep using each of the oblique drawing methods. Place each drawing on a separate layer. Save the drawing as E-21-1.
2. Start a new drawing and draw a cube that is 8' in each direction using isometric techniques. Place an isometric circle in each plane so that the circle is tangent to the edges of each plane. Save the drawing as E-21-2.
3. Start a new drawing and draw an isometric cube that is 6' in each direction. Show the bottom, left, and right planes. Place an isometric circle in each plane so that the circle is tangent to the edges of each plane. Save the drawing as E-21-3.
4. Use the drawing below and draw an isometric drawing showing the framing of a 5'–0" x 3'–6" window. Use 2x6 studs. Label the top plates, a 4x8 header, 2x6 sill, and 2x6 studs, king studs, trimmers, and cripples. Save the drawing as E-21-4.

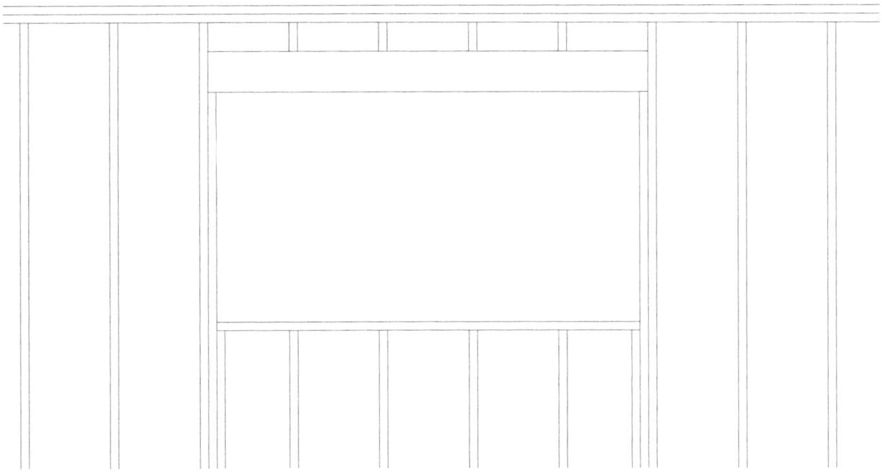

5. A window supplier would like to update its catalogue. Provide an isometric drawing of a 4'–0" wide x 3'–6" high vinyl window with a half-round window above. Show the window with a grid in each slider, and a radial pattern in the upper window. Draw the frame as 1" aluminum, and the grid as 1/2" wide material. Place the grid on a layer separate from the window layer. Save the drawing as E-21-5.

6. Use the floor plan below and draw an isometric drawing showing the left wall with the refrigerator and double oven. Design and show drawers and cabinet drawers. Use glass in some of the doors and show the shelving. Consult vendor catalogues for specific sizes of appliances and cabinets. Omit the refrigerator. Show 8'–0" high ceilings. Save the drawing as E-21-6.

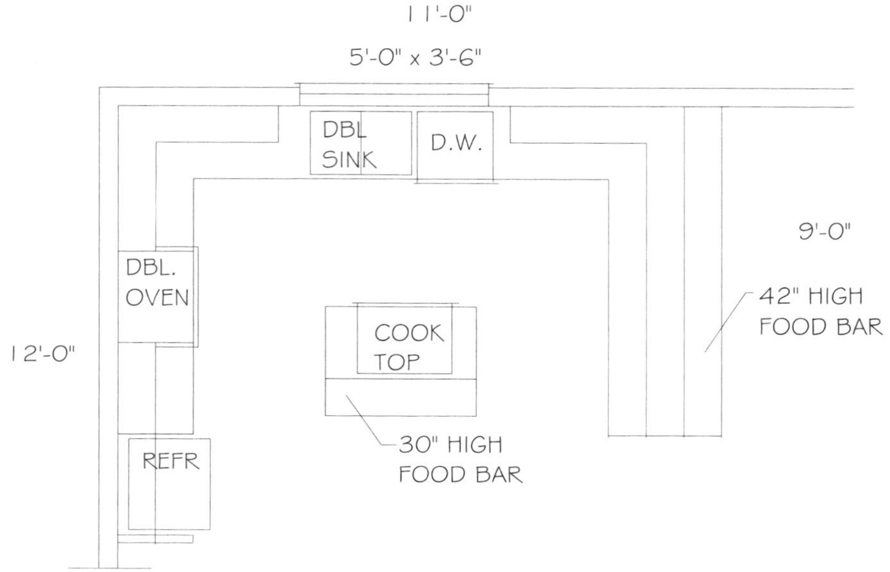

7. Use the floor plan for drawing E-21-6 and the drawing started in Exercise 6 to draw an isometric of the complete kitchen. Draw the cabinets over the food bar as 30" high. Save the drawing as E-21-7.

8. Use the drawing below to draw an isometric drawing showing the metal beam seat. Show the bracket, but omit the beam. Make the side plates from 1/2" steel and the base plate from 7/8" x 7" wide steel plates. Place bolts at 2" in from each edge and at 3" o.c. Dimension the size of the steel plates. Save the drawing as E-21-8.

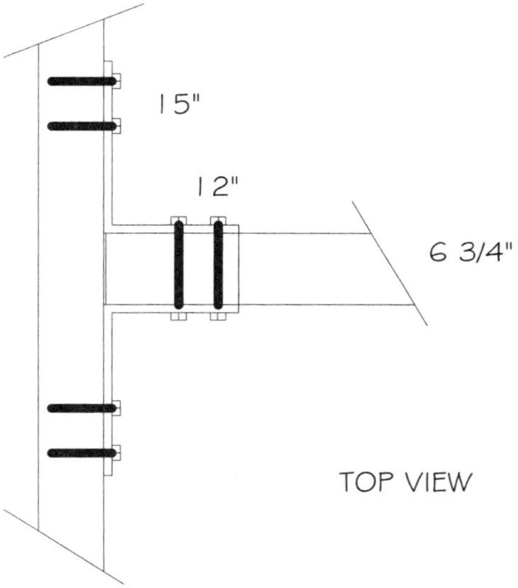

TOP VIEW

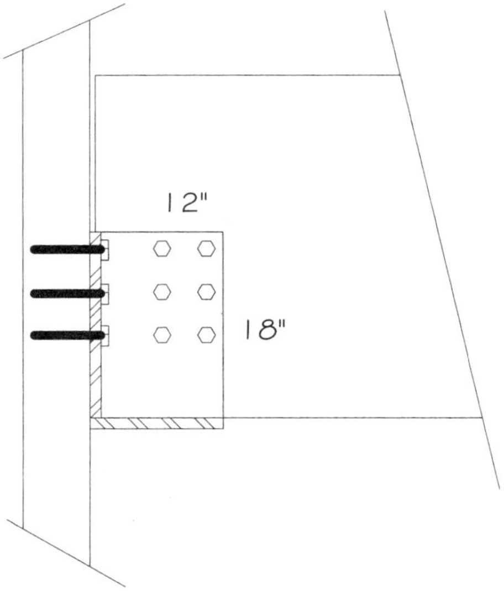

SIDE VIEW

9. Use the drawing below as a guide to draw an isometric view of the truss. Assume that the top and bottom chords are made from two 2x3s, and the webs are 1" in diameter. Show the total distance from the top of the top chord to the bottom of the metal bracket as 3 1/2". Show the total depth as 36 inches, and lay out the webs at 45° from the top chord. Save the drawing as E-21-9.

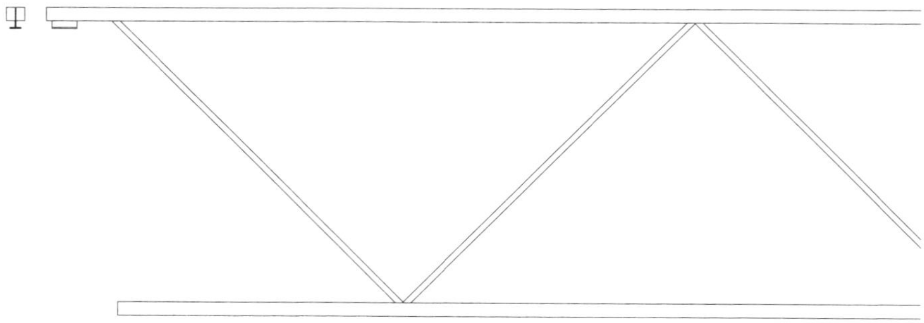

10. Use the drawing below as a guide to draw an isometric view of the 6 3/4" x 13 1/2" glu-lam beam supported on two 6x6 posts at 10 feet o.c. on 18" diameter concrete piers. Show piers extending 8" minimum above the finish grade. Show the wall framed with 2x6 studs at 16" o.c. Frame the floor with 2x10 floor joists at 12" o.c. 12' above the grade covered with 3/4" plywood floor sheathing and an 18" cantilever. Use a 13" x 8" metal hanger with 2" x 10" legs to post. Show two bolts through the hanger into each beam, and two bolts through the hanger legs and post. Place bolts 1 1/2" from steel edges and 3" o.c. Use 1/2" steel cables with turnbuckles forming an "X" between each post. Attach the cable to each post with eye bolts 6" up and from the bottom. Place the top eyebolt 3" below the bottom hanger's leg bolt. Label major materials and dimension the cantilever and post spacing. Because your drawing will be used to assemble this project and not to manufacture each item, great accuracy is not required on items such as the turnbuckle and eyebolts. Save the drawing as E-21-10.

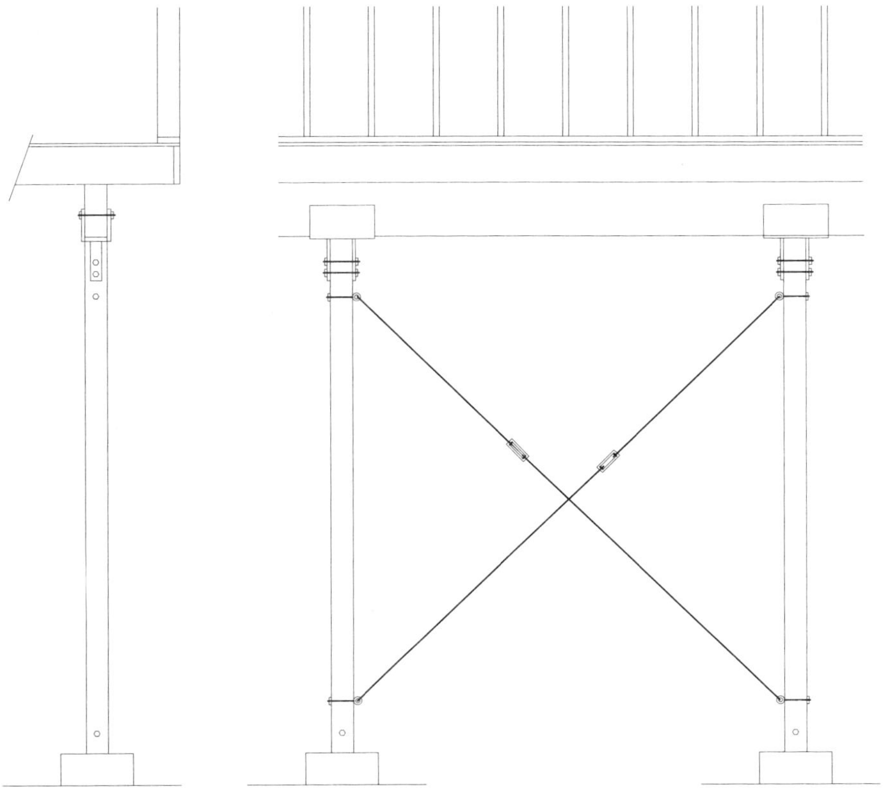

CHAPTER 21 QUIZ

1. What are the differences between cabinet and general oblique drawings?
2. Compare the differences between cavalier and cabinet drawings.
3. Describe the relationship of horizontal and vertical lines in an isometric drawing.
4. What command and option adjust the grid for isometric drawings?
5. List the three options for selecting the isometric snap options.
6. Explain the easiest way to construct isometric circles.
7. List common angles for placing text on isometric drawings.
8. Sketch and label the three positions of the crosshairs in isometric drawings.
9. What is the major difference between oblique and 3D drawings?
10. Explain the process for placing dimensions on an isometric drawing.

CHAPTER 22

Working with Multiple Drawings

This chapter will introduce

- General considerations for multiple document environments
- Methods of moving information between two or more drawing files
- Features of DesignCenter

Commands to be introduced include

- **Cut**
- **Copy**
- **Paste**
- **Copybase**
- **Pasteblock**
- **Pasteorig**
- **Adcenter**

THE MULTIPLE DOCUMENT ENVIRONMENT

Template drawings have been used throughout this text to store frequently used styles, settings, and objects for use in similar drawings. This chapter will introduce methods for moving objects and information between two or more drawing files. Objects or drawing properties can be moved between drawings using cut/copy and paste, object dragging, the **Match Properties** command, and concurrent command execution.

GENERAL CONSIDERATIONS FOR MULTIPLE DOCUMENT ENVIRONMENTS

AutoCAD is by default set to work with multiple documents. The toggle for switching from multiple to single document interface is the **Single Drawing Compatibility**

Mode button in the **General options** area of the **System** tab in the **Options** dialog box. (Access the dialog box by selecting **Options** from the **Tools** menu.) With the box inactive, multiple drawings can be opened in one AutoCAD session. Don't gloss over the concept of multiple drawings and one drawing session. Just as you can have several programs open on your desktop at once and rapidly switch from one program to another, AutoCAD allows you to have several drawings open at the same time. Chapter 1 introduced the techniques for displaying multiple drawings. Start AutoCAD and select **Open a Drawing**. Select an existing drawing to be opened. While holding down CTRL, select three other drawings to be opened, and then click the **Open** button. You have now used one of the most powerful tools of AutoCAD and have four drawings opened in one drawing session.

Now select **Cascade** from the **Window** menu. Your drawing display will resemble Figure 22.1. The title bar of each drawing will be displayed. The drawing with the blue title bar on top of the stack is the active drawing. Apart from being on top of the pile, it is the only drawing that can currently be used. Return to the **Window** menu and select **Tile Vertical**. Now select **Tile Horizontal** instead from the **Window** menu. The drawing display will now resemble Figure 22.2. With four drawings open, the horizontal and vertical displays will appear alike. With an odd number of drawings open, they will be displayed in vertical rows with the **Tile Vertical** option. Important considerations to keep in mind when multiple drawings are open are as follows:

- In any **Window** display selection, any one of the four drawings can be made active. Notice that each small drawing area has all of the components of a normal AutoCAD drawing.
- Any one of the four drawings can be maximized or minimized to ease viewing, or all can be left open and you can switch from drawing to drawing as needed.
- You can start a command in one drawing, switch to another drawing and perform a different command, and then return to the original drawing and complete the command in progress. As you return to the original drawing, click anywhere in the drawing area, and the original command will be continued.
- Each drawing is saved as a separate drawing file, independent of the other open files. You can close a file without affecting the contents of the remaining open files.

Opening Multiple Drawings from Windows Explorer

In addition to being opened from within AutoCAD, multiple drawings can also be opened from Windows Explorer. To start the process, open Windows Explorer and double-click the drawing to be opened. Once the drawing is opened, minimize the drawing and reduce the size of the AutoCAD drawing session by clicking the **Restore** button at the top right side of the toolbar so that the drawings in Windows Explorer can still be seen. With a drawing open, additional drawings can be dragged

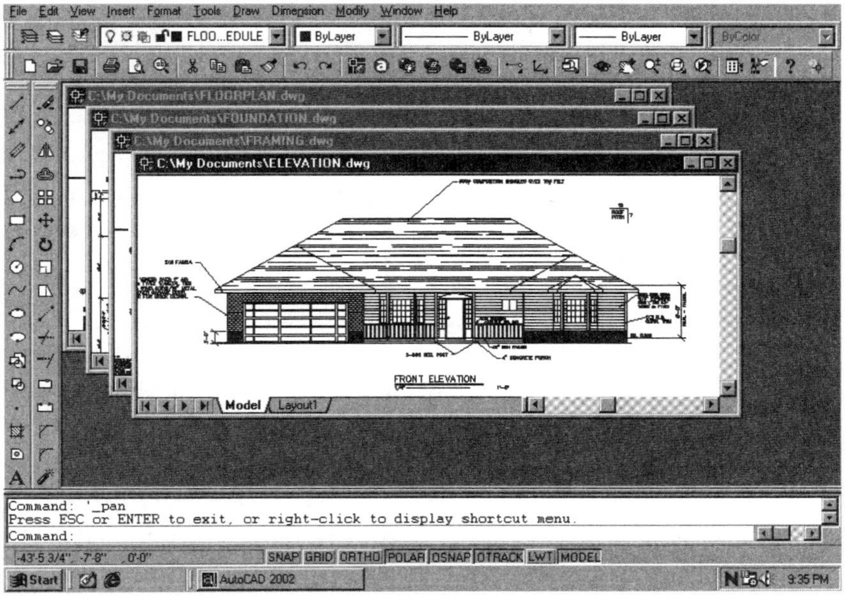

Figure 22.1 *Selecting* **Cascade** *from the* **Window** *menu will display the title bar of each drawing. The drawing with the blue title bar on top of the stack is the active drawing.*

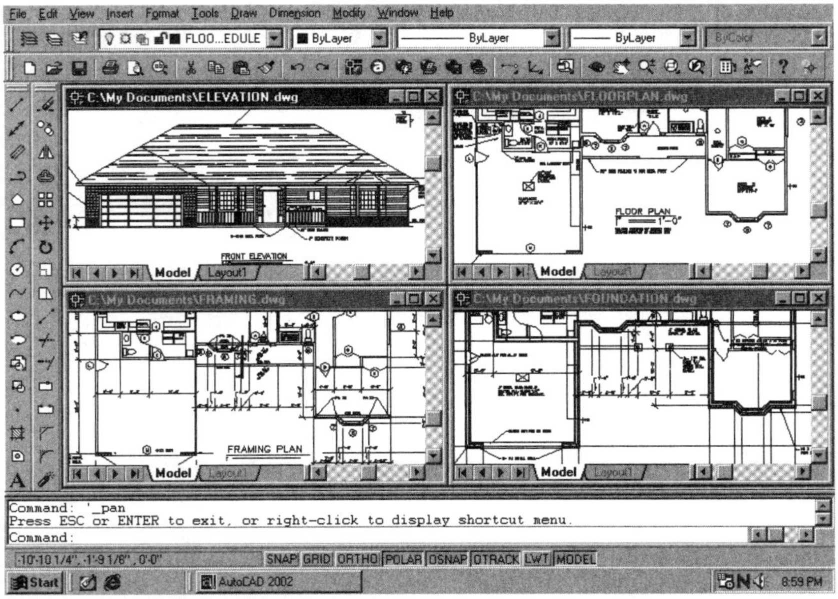

Figure 22.2 *Selecting* **Tile Horizontal** *from the* **Window** *menu will display each open drawing. In this orientation, any one of the four drawings can be made active. Notice that each small drawing area has all of the components of a normal AutoCAD drawing.*

from Windows Explorer and dropped into the current drawing session. Drag Files by pressing and holding CTRL and the left mouse button while the desired drawing is selected. Once the Drawing icon is in the drawing area, release the CTRL and select button to open the drawing. Drawings will be placed in a cascading display similar to Figure 22.1.

Note: If a drawing is dragged from Windows Explorer into a current drawing, the **Insert** command will be activated. Instead of opening multiple drawings, you'll be inserting a drawing in the current drawing.

USING THE CUT, COPY, AND PASTE COMMANDS

In addition to standard procedures that are allowed using the **Cut**, **Copy**, and **Paste** commands of any Windows software, AutoCAD allows drawing objects and properties to be transferred directly from one drawing to another using these same commands. Each command can be performed using the **Edit** menu, by keyboard, or by shortcut menu. **Cut** will remove objects from a drawing and move them to the Clipboard, from where they can be placed in a new drawing using **Paste**. **Copy** allows an object to be reproduced in another location. The object to be copied is reproduced in its new location using the **Paste** command. As the object is pasted to the new drawing, so are any new layers, linetypes, or other properties associated with the object. Figures 22.3a and 22.3b show examples of using **Copy** to place the footing detail into the second drawing. The footing was copied by using the following sequence:

From the Source Drawing:

Command: *(Select Copy from the shortcut menu.)*

_copyclip

Select objects: *(Select object to be copied using any selection method.)*

Select objects: (ENTER)

Command:

In the Second Drawing:

Command: *(Select Paste from the shortcut menu.)*

_pasteclip

Select insertion point: *(Select the location for the new object.)*

Command:

Copy With Base Point

The **Copy with Base Point** command on the shortcut menu is similar to the **Copy** command. You can also start the command by typing **COPYBASE** (ENTER) at the Command prompt. With this option, before you select objects to copy, you'll be

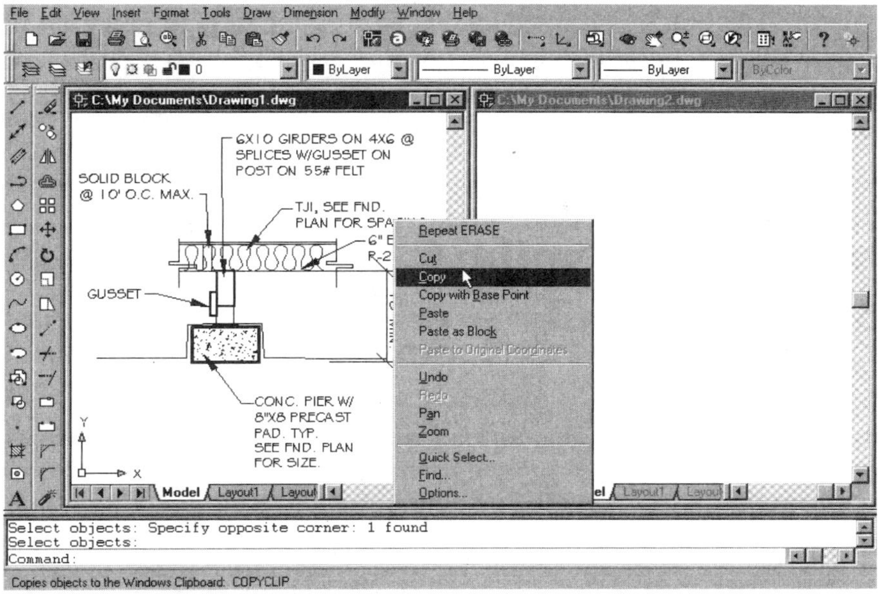

Figure 22.3a *The footing in the left drawing can be copied into the right drawing by selecting* **Copy** *from the shortcut menu.*

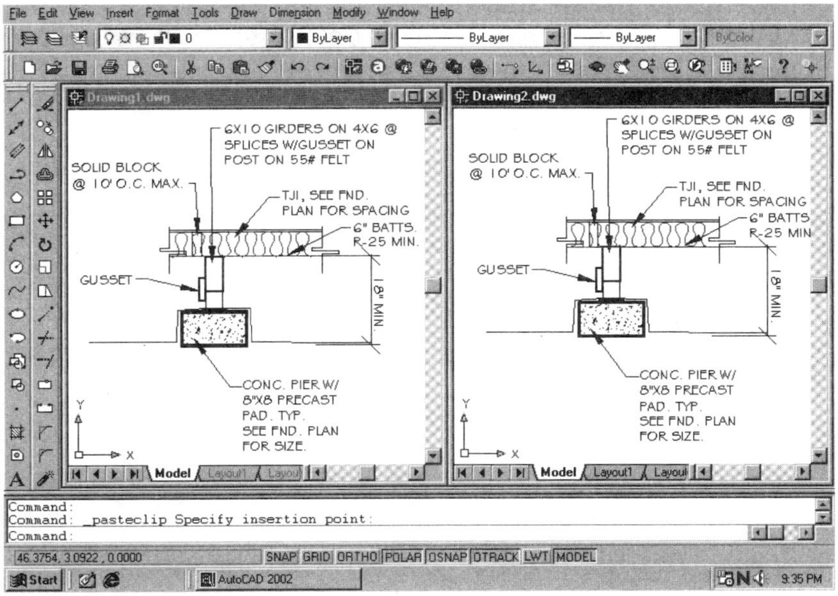

Figure 22.3b *Once the object selected to be copied, move the cursor to the desired drawing, and single-click to make that drawing active. Right-click and choose* **Paste** *from the shortcut menu, and then select an insertion point.*

prompted to provide a base point. This option works well when objects need to be inserted accurately. The roof detail in Figure 22.4 was copied into the new drawing using the **Copy with Base Point** command. The roof detail was copied using the following sequence:

From the Source Drawing:

Command: *(Select Copy with Base Point from the shortcut menu.)*

_copybase

Specify base point: *(Select the base point for the objects to be copied.)*

Select objects: *(Select object to be copied using any selection method.)*

Select objects: (ENTER)

Command:

In the Second Drawing:

Command: *(Select Paste from the shortcut menu.)*

_pasteclip

Select insertion point: *(Select the location for the new object.)*

Command:

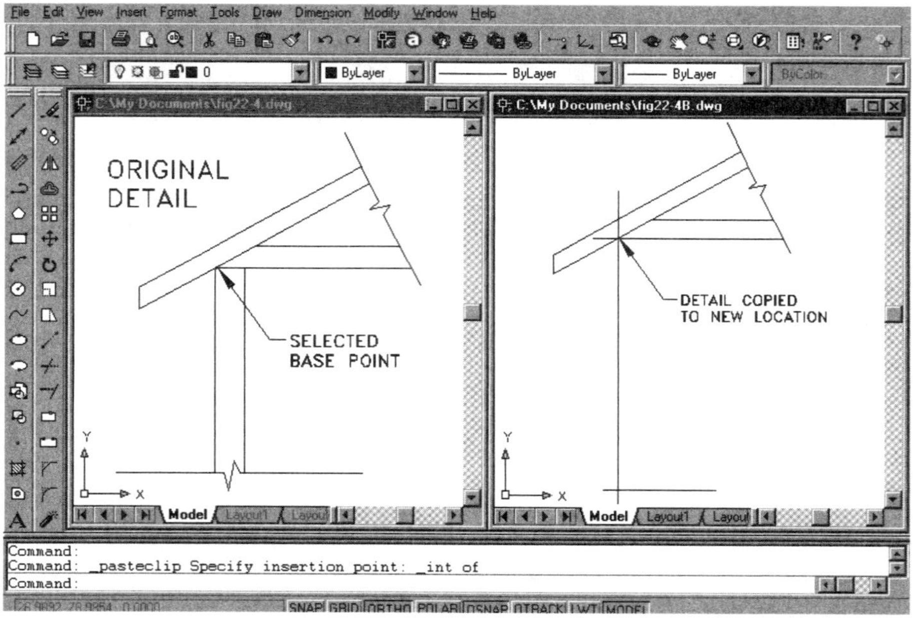

Figure 22.4 *The* **Copy with Base Point** *command, similar to the* **Copy** *command, works well when objects need to be inserted accurately. With this option, before you select objects to copy, you'll be prompted to provide a base point.*

Paste as Block

The **Paste as Block** command can be used to paste a block into a different drawing. It functions similarly to the **Paste** command. Select the objects to be copied in the original drawing and select the **Copy** command. Selected objects do not need to be a block, because the **Paste as Block** command will turn them into a block. Next activate the new drawing and right-click to select the **Paste as Block** command. The copied objects are now a block. To be edited, they must be exploded. You can also start the command by typing **PASTEBLOCK** (ENTER) at the Command prompt. The command sequence is the same as for the **Paste** command. You will notice that blocks that are copied will be renamed. A name such as ROOFWALL will be assigned a name such as A$C2AA9F67C4 by AutoCAD. The new block can be renamed using the **Rename** command.

Paste to Original Coordinates

This command can be used to copy an object in one drawing to the exact same location in another drawing. The command could be useful in copying an object from one apartment unit to the same location in another unit. The command can be started by using the shortcut menu or by typing **PASTEORIG** (ENTER) at the Command prompt. The command is active only when the Clipboard contains AutoCAD data from a drawing other than the current drawing. Selecting an object to be copied starts the command sequence, and it is completed by making another drawing active and then selecting the **Paste to Original Coordinates** command from the shortcut menu.

DRAGGING OBJECTS

AutoCAD allows objects to be moved between drawings using a drag and drop sequence. Dragging allows the following operations:

- Move or copy objects to a new location in the same drawing.
- Move or copy objects to another drawing in the same drawing session.
- Copy objects to a drawing in another session.
- Copy objects to another application.

Dragging can be carried out using either the left or right mouse buttons. Slightly different results and procedures will be required using each method.

Selecting Objects to Drag with Left-Click

The procedure to drag using the left mouse button begins by selecting the objects to be moved or copied. Any object selection method can be used to select objects. Once the set is selected, move the cursor over one of the selected objects and press and hold down the left mouse button. Moving the cursor while the left button is held down will show the effect of moving the selected objects. Release the mouse button when the objects are in the desired location. If you're working with a single drawing, the selected objects will be moved. Objects can be copied in a single drawing by pressing CTRL prior to pressing the left mouse button to drag the object. (Objects will always be copied when dragged into a different drawing.) You are not allowed to drag objects into a drawing that has an active command. The unavailable cursor—a circle with a diagonal line through it—is displayed if you attempt to drag objects into a drawing with an active command. Figure 22.5 shows the steps to copy a detail into another drawing using the left mouse button.

Selecting Objects to Drag with Right-Click

The procedure to drag using the right mouse button begins by selecting the objects to be moved or copied. Any object selection method can be used to select objects. Once the set is selected, move the cursor over one of the selected objects and press the right mouse button. Moving the cursor while the button is held down will show the effect of moving the selected objects. Release the mouse button when the objects are in the desired location. As the right button is released, a menu will be displayed. The menu will depend on what type of drawing you're working in. The menu options are as follows:

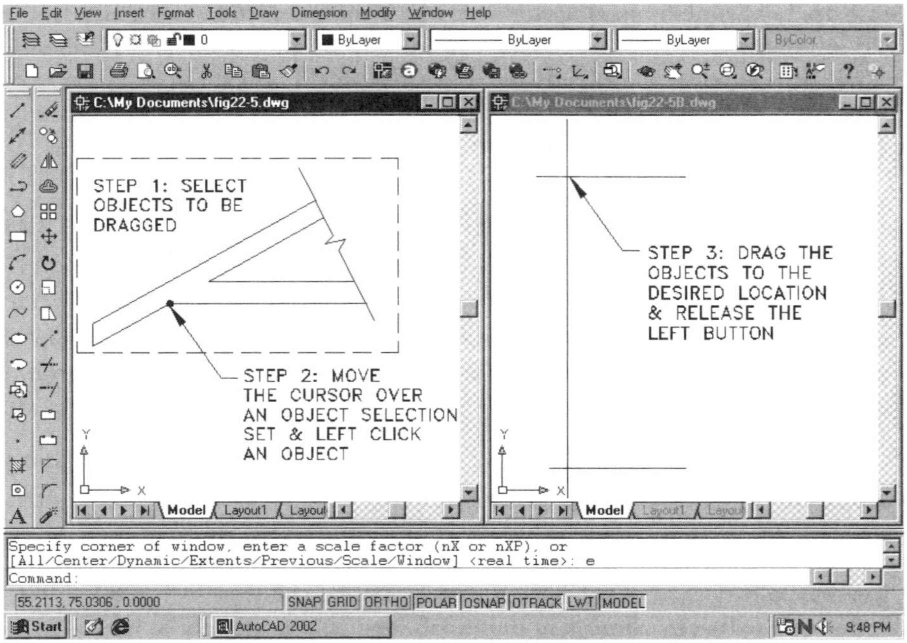

Figure 22.5 *Objects can be moved or copied easily using the drag and drop technique.*

Drag in the Same Drawing	**Drag in a Different Drawing**
•Move Here	•Copy Here
•Copy Here	•Paste as Block
•Paste as Block	•Paste to Orig Coords
•Cancel	•Cancel

Select a menu option to end the drag and drop sequence.

MATCHING PROPERTIES

The **Match Properties** command cannot be used to move or copy objects, but it can be used to copy properties from one object to another. The **Match Properties** command was introduced in Chapter 13 for transferring properties within a drawing. The command can also be used to transfer properties from an active drawing to other drawings. The process to change the layer and linetype of objects in a drawing to match another drawing can be completed using the following process:

1. Select the source object using any selection method.
2. Select **Match Properties** on the Standard toolbar.
3. Move to the target drawing and make it the current drawing. The **Match Properties** cursor will be displayed by the selection box.
4. Select the target objects. The command can be continued within the current drawing, or another drawing can be made current, and the properties of the original object can be transferred to additional objects.

EXPLORING THE AUTOCAD DESIGNCENTER

The DesignCenter can be used to locate, organize, and customize drawing information. It can be used to directly transfer data such as blocks, layers, and external references from one drawing to another without using the clipboard. Drawing content can be moved from other open files in AutoCAD or from files stored on your machine, on a network drive, or on an Internet site. Access the DesignCenter by selecting the **AutoCAD DesignCenter** button on the **Standard** toolbar, or by typing **ADCENTER** (ENTER) at the Command prompt. Either method will display a DesignCenter similar to Figure 22.6. The DesignCenter can also be displayed or closed using CTRL+2.

The default position of DesignCenter is docked on the left side of the drawing screen. You can resize the DesignCenter by clicking a border, the palette sidebar or by grabbing a corner and moving the cursor in a diagonal direction. If you are using the DesignCenter, the **Properties** dialog box, and a drawing, you might be tempted to shrink the size of the DesignCenter window. Resist shrinking the DesignCenter past the point where two rows of icons can't be seen in the palette display. The display can be reduced so no palette is displayed, but then you've limited the usefulness of the display. Clicking and dragging the title bar will float the DesignCenter, allowing it to be moved to any location in the drawing area. Figure 22.7 shows an example of a floating DesignCenter. You can dock the DesignCenter by double-clicking the title bar of the DesignCenter display or by dragging it to the left or right side of the drawing area.

FEATURES OF DESIGNCENTER

The DesignCenter provides direct access to any drawing file that you have on your machine, network, or the Internet. Files stored on diskette, the hard drive, network or Internet can be accessed and reused as a library for retrieving or coping blocks, referenced drawings, layouts, layers, linetypes, text styles or dimension styles. Other operations of the DesignCenter include the following:

- Create a Favorites folder to create a shortcut to frequently accessed drawings, folders, or Internet locations.
- View, attach, insert, or copy and paste layer and block definitions of any file you have access to into the current drawing. This feature could be used to update a drawing supplied by a subcontractor to match current standards.

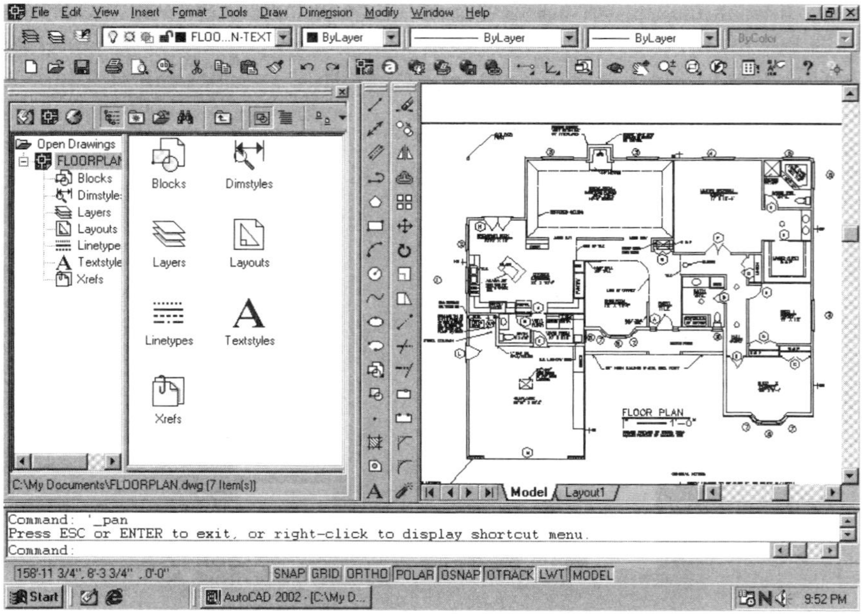

Figure 22.6 *Access the DesignCenter by selecting the* **AutoCAD DesignCenter** *button on the* **Standard** *toolbar or by typing* **ADCENTER** *(ENTER) at the Command prompt. The DesignCenter can be used to directly transfer data such as blocks, layers, and external references from one drawing to another.*

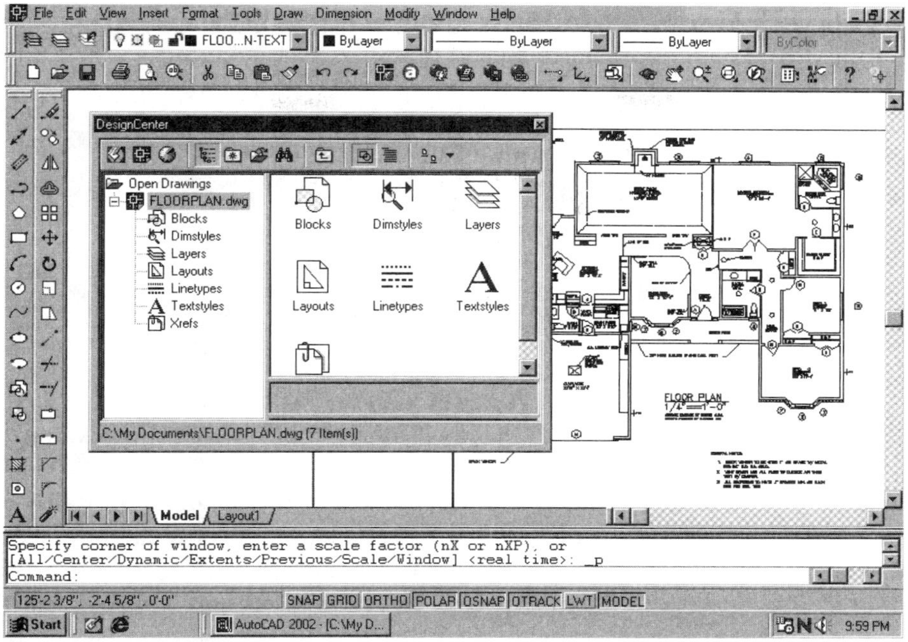

Figure 22.7 *Clicking and dragging the title bar will float the DesignCenter, allowing it to be moved to any location in the drawing area. You can dock the DesignCenter by double-clicking the title bar of the DesignCenter or by dragging it to the left or right side of the drawing area.*

- Open drawing files by dragging the file from the tree view into the current drawing area.
- Find drawing contents stored in any folder, drive, network, or from any Internet site that you have access to, based on search criteria for the source. Search tools such as a name or date can be used to locate materials to be dragged into the current drawing file.
- Use the DesignCenter palette to view aspects of the object, block, or drawing content.

EXAMINING THE DESIGNCENTER TOOLBAR

The **DesignCenter** toolbar is located across the top of the DesignCenter. Before examining the contents, you might need to adjust the size of the DesignCenter window to be sure that all of the contents are displayed. The toolbar consists of eleven buttons for controlling the contents of the DesignCenter. These buttons, from left to right, are **Desktop**, **Open Drawings**, **History**, **Tree View Toggle**, **Favorites**, **Load**, **Find**, **Up**, **Preview**, **Description**, and **Views**. Each can be seen in Figure 22.8.

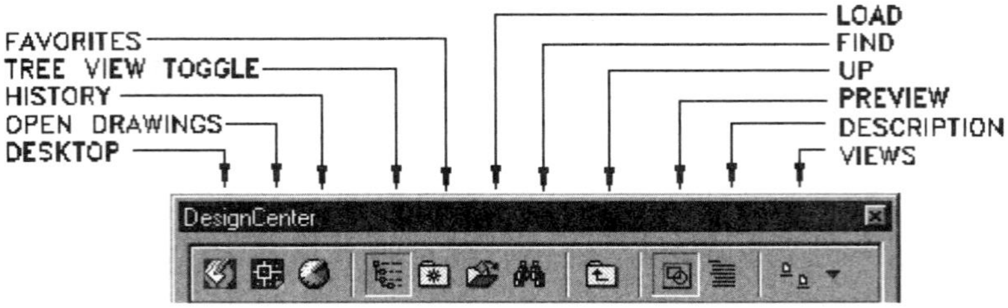

Figure 22.8 *The buttons on the DesignCenter toolbar.*

Desktop and Open Drawings

These buttons serve as toggles to control what is displayed in the DesignCenter. When **Desktop** is selected, the display will resemble Windows Explorer. All resources available to your workstation will be displayed, including internal drives, local network, and Internet files. The exact display will depend on the contents of the selected drive and the setting for the **Tree View Toggle** button. Selecting the **Open Drawings** button will display a list of drawings that are currently opened, similar to Figure 22.9. Notice that the drawings have either a + or – symbol preceding the title. The + symbol indicates that additional information can be displayed about the drawing file or folder. Double-clicking the drawing name will display the drawing contents and double-clicking a drawing with a full display (–) will close the contents of that file.

History

Selecting this button will display a listing of the last twenty locations accessed through the DesignCenter. The history can also be displayed by right-clicking in the palette and then selecting **History** from the shortcut menu.

Tree View Toggle

This button displays or hides the tree view in much the same way that **Details** and **Large Icons** of the **View** menu of Windows Explorer alter that display. Tree view displays open drawings, history, and files or folders for all of the drives available to your computer. Figure 22.9 shows a display using the tree view. Figure 22.10 shows a display with the tree view hidden. You can hide the tree view by clicking the **Tree View Toggle** button or by right-clicking in the palette and selecting **Tree** from the shortcut menu. With the tree view hidden, you can then restore the tree by clicking the button or by right-clicking in the palette and then clicking Tree from the shortcut menu.

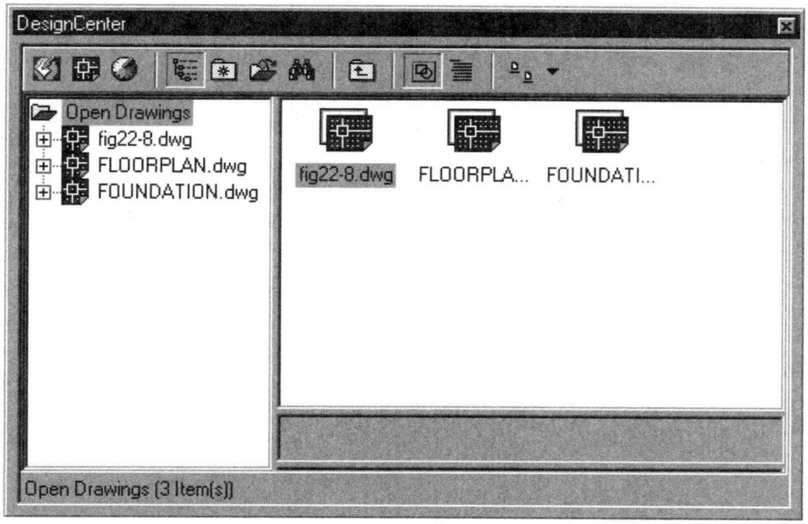

Figure 22.9 *Selecting the* **Open Drawings** *button will display a list of drawings that are currently opened. Notice that the drawings have either a + or – symbol preceding the title. The + symbol indicates that additional information can be displayed about the drawing file or folder. Double-clicking the drawing name will display the drawing contents and double-clicking a drawing with a full display (–) will close the contents of that file.*

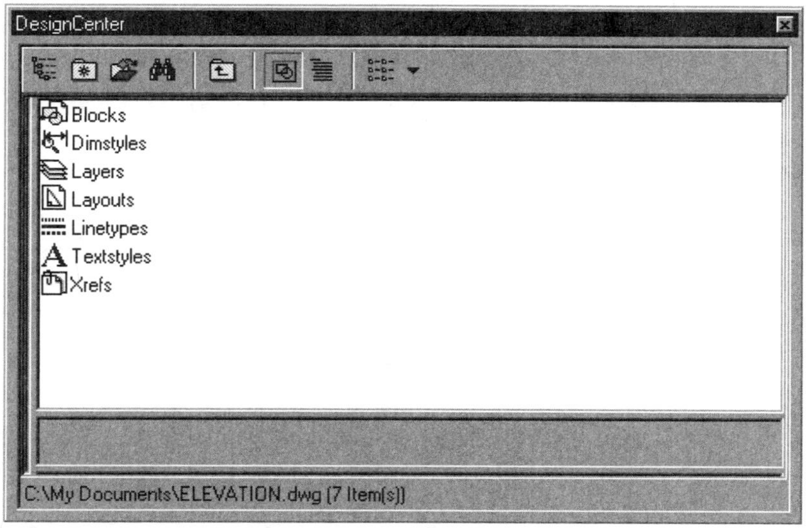

Figure 22.10 *The DesignCenter display with the tree view hidden.*

Favorites

This option of the DesignCenter displays shortcuts to files or folders that you need to access quickly on a regular basis. The tree, palette, and shortcut menus each include menus for accessing the Favorites folder.

Load

Selecting the **Load** button displays the **Load DesignCenter Palette** dialog box. The dialog box is similar to the **Select File** dialog box used to open drawings. You can also display it by right-clicking in the palette and then selecting **Load** from the shortcut menu. You can choose any available drive from the **Look in** edit box. The initial step in using the DesignCenter is to load the palette. Use the following steps to load the palette:

1. Select the **Load** button.
2. In the **Load DesignCenter Palette** dialog box, choose one of the following options:

    Look in—This source displays the files and folders on local and network drives.

    Favorites—The **Favorites** button displays the contents of the Favorites folder.

3. Select the listing that is to be loaded.

Find

In addition to selecting information from the desktop, you can search for drawings or parts of drawings using **Find**. Selecting the **Find** button displays a **Find** dialog box similar to Figure 22.11. The dialog box can be used to narrow the search to find drawings, blocks, layers, layouts, text styles, dim styles, and referenced drawings. Currently the search will be conducted in all drawings. Selecting the **Look for** edit box allows the search to be limited to Blocks, Dimstyles, Drawings, Drawing and Blocks, Hatch pattern files, Hatch patterns, Layers, Layouts, Linetypes, Text Styles, or Xrefs. Depending on the field, the tabs in the dialog box will be altered. The tabs for finding drawings include **Drawings**, **Date Modified**, and **Advanced**.

Drawings

This tab can be used to specify the name or text that will be the object of the search. In Figure 22.12, the name ELECT is entered in the **Search for the word(s)** edit box. The **In the field(s)** edit box can be used to limit the search criteria. In the current setting, the search will be conducted in file names. Alternatives include looking in titles, subjects, author, and keywords.

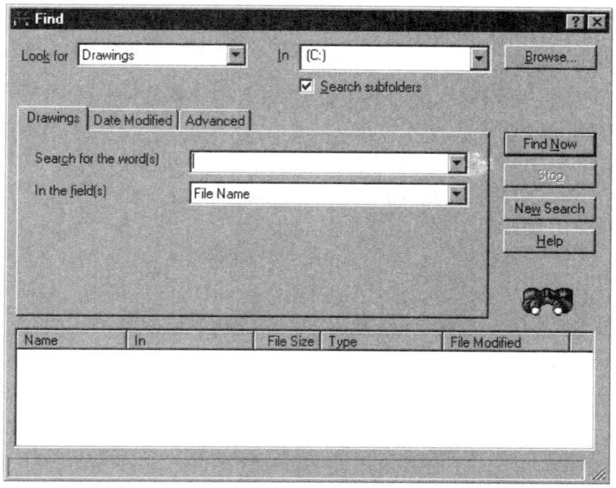

Figure 22.11 *Selecting the* **Find** *button displays a* **Find** *dialog box. The dialog box can be used to find drawings, blocks, dimstyles, layers, layouts, linetypes, text styles, and referenced drawings.*

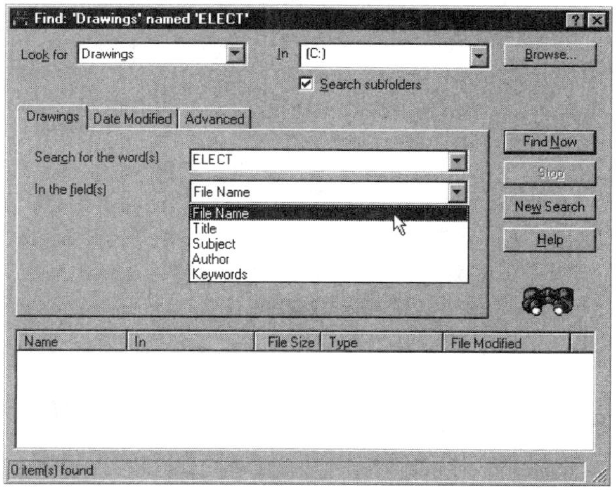

Figure 22.12 *Entering ELECT in the* **Search for** *edit box and clicking the* **Find Now** *button will conduct a search for all file names containing the word ELECT.*

Date Modified

The **Date Modified** tab allows a search to be conducted based on the date that the file was created or last modified. A date range specified in days or months can be used if you're not sure of the exact date the drawing was created.

Advanced

The **Advanced** tab allows additional search parameters to be set based on text contained in a portion of a file or by the file size. A search can be defined using the Containing edit box to search block name, block and drawing descriptions, attribute tag, or attribute value for text containing, but not limited to, a specified text. The search parameters can also be limited to looking in files of a minimum or maximum size.

Up

The **Up** button is used to move through the contents of the DesignCenter. The **Up** button performs the same function here as it does in the **Select File** dialog box. Each time the button is selected, the display is moved up one level from the current location.

Preview

The **Preview** window displays an image of a selected file or block.

Description

If a description was created for objects such as a block, this portion of the dialog box will display the description. The description will be given for the file or object that is displayed in the **Preview** window. A description can also be displayed by right-clicking in the palette, and then selecting **Description** from the shortcut menu.

Views

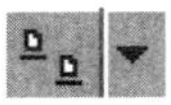

Selecting **Views** presents one of four options for displaying the format of the contents of the DesignCenter. Each time the button is selected, a different method of presentation will be displayed. Selecting the edit arrow beside the button will present a menu of the four options, so that the desired display method can be selected without having to scroll through each option. Display options include **Large icons**, **Small icons**, **List**, and **Details**.

Shortcut Menus

Right-clicking in the palette produces a shortcut menu similar to Figure 22.13. The shortcut menu contains the same options as found on the **DesignCenter** toolbar.

OPENING DRAWINGS USING THE DESIGNCENTER

The DesignCenter can be used to open drawings. Start AutoCAD using any of the options presented throughout this text. The DesignCenter can now be used to drag a drawing into an empty drawing area with the following steps:

1. Display the DesignCenter.
2. Select the drawing source from the Desktop.

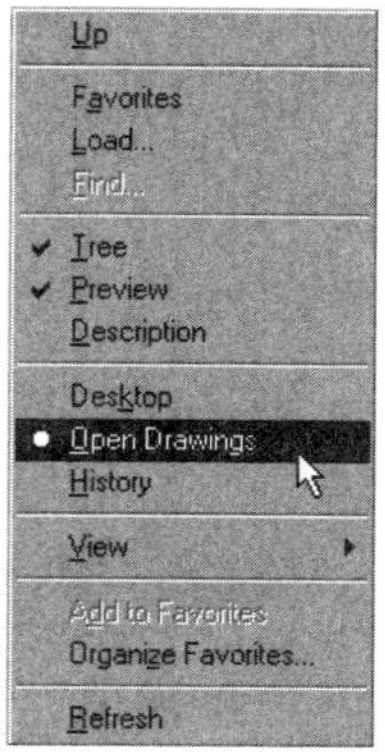

Figure 22.13 *Right-clicking in the palette produces a shortcut menu that contains the same options as found on the* **DesignCenter** *toolbar.*

3. Drag the icon of the drawing to be opened and drop it in the drawing area.
4. Provide the insertion point and responses to the scaling prompts.
5. Resize the drawing window as needed to display the drawing.

A drawing can also be placed in another drawing by using the shortcut menu. From the DesignCenter, select the file to be opened, right-click, and select the **Open to Drawing** option from the shortcut menu.

ADDING CONTENT TO DRAWINGS USING THE DESIGNCENTER

One of the best uses for the DesignCenter is adding content from the palette to a new or existing drawing. Information can be selected from the palette or from **Find** and dragged directly into a drawing without opening the drawing containing the original. Chapter 19 introduced methods of moving blocks using DesignCenter. DesignCenter can also be used to attach external referenced drawings and copy layers between drawings.

Inserting Blocks

The DesignCenter provides two methods for inserting blocks in a drawing: using the default scale and rotation of the block, or altering the block parameters as the block is dragged into the new drawing.

Providing Parameters for Block Insertion

For most purposes this will be the preferred method of inserting a block using the DesignCenter. Use the following sequence to specify the insertion point, scale, and rotation angle as a block is inserted:

1. Display the DesignCenter.
2. Use the palette to select the block to be copied.

3. Use the right mouse button to drag the desired block to the new drawing.
4. Release the right button to drop the block into the drawing area. This will display a shortcut menu.
5. Choose **Insert Block** from the shortcut menu. This will display the **Insert** dialog box.
6. Enter the values for each parameter or choose the **Specify On-screen** option for each parameter.
7. Click the **OK** button to insert the block at the specified location.

Using the Default Scale

When blocks are inserted using the default scaling and rotation option, the block is inserted using Autoscaling. Autoscaling compares the units of the drawing with those of the block. As the block is inserted, AutoCAD scales the block as needed, based on the ratio of the block to the new drawing. Blocks inserted using this method might not retain their true size. Use the following sequence to insert a block using the default parameters:

1. Display the DesignCenter.
2. Use the palette to select the block to be copied.
3. Drag the desired block to the new drawing.
4. Drop the block at the desired location.

Attaching Referenced Drawings

The DesignCenter can be used to attach a referenced drawing using steps similar to those used to attach a block and provide the parameters. Use the following procedure to bring an xref into a new drawing.

1. Display the DesignCenter.
2. Use the palette to select the xref to be attached.
3. Use the right mouse button to drag the desired xref to the new drawing.
4. Release the right button to drop the xref into the drawing area. This will display a shortcut menu.
5. Choose **Attach Xref** from the shortcut menu. This will display the **External Reference** dialog box.
6. Select **Attach** or **Overlay** from the **Reference Type** box.
7. Enter the values for each parameter or choose the **Specify On-screen** option for each parameter.
8. Click the **OK** button to complete the command.

Working with Layers

The DesignCenter can be used to copy layers from one drawing to another. Typically, a template drawing containing stock layers can be used when you create new drawing files. This option of the DesignCenter is useful when you're working with drawings created by a consulting firm and the drawing needs to conform to office standards. The DesignCenter can be used to drag layers from a template, or any other drawing, into the new drawing and ensure drawing consistency. Drag layers into a new drawing using the following command sequence:

1. Display the DesignCenter.
2. Use the palette or **Find** to select the drawing that contains the layers to be copied.
3. Double-click the name of the drawing to display the options of DesignCenter.
4. Double-click Layers.
5. Highlight the names of the layers to be added to the new drawing (holding down the CTRL allows you to select several layers at once).
6. Press the right mouse button and drag the layer names to the new drawing.
7. When the cursor is in the new drawing area, release the right button to display the shortcut menu.
8. Select **Add Layers(s)** from the shortcut menu.

ACCESSING FAVORITE CONTENTS

As with most programs, the AutoCAD DesignCenter allows frequently used sources to be stored for easy retrieval. The final section of this chapter will explore methods of adding shortcuts and displaying and organizing your favorite drawing tools.

Adding Shortcuts to Favorites

Folders, files, drawings, or drawing content can be added to the Favorites list. Selecting a drawing or other option to be placed in the Favorites list adds a shortcut to the AutoCAD Favorites folder. The selected item is not moved to the folder, only the directions for quickly fetching it. Use the following procedure for adding items to the AutoCAD favorites:

1. Right-click the item to be listed as a favorite.
2. Select **Add to Favorites.** The selected item will be listed the next time the Favorites listing is displayed.

Display the contents of the Favorites folder by selecting the **Favorites** button on the DesignCenter toolbar or by right-clicking in the palette background and selecting **Favorites**. The shortcut saved in the Favorites folder can be copied, moved, or deleted using the Autodesk Explorer. The explorer is accessed by right-clicking in the palette and choosing **Organize Favorites** from the shortcut menu. Figure 22.14 shows an example of the Autodesk Explorer.

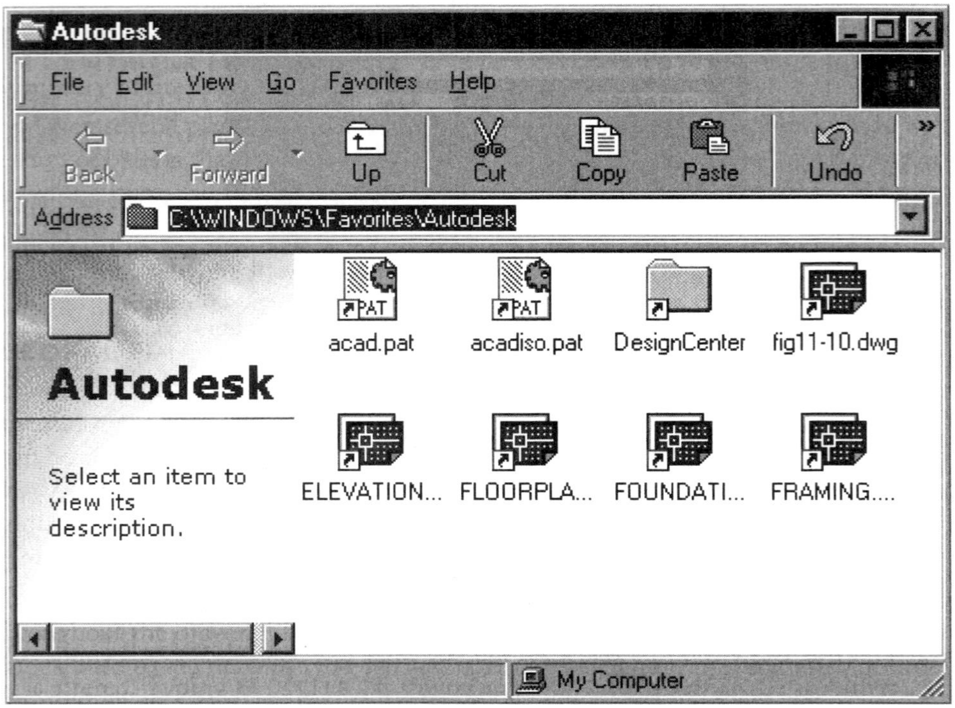

Figure 22.14 *Right-clicking in the palette and choosing* **Organize Favorites** *from the shortcut menu will display the Autodesk Explorer.*

CHAPTER 22 EXERCISES

1. Open any three drawings and copy features from one drawing to the other two. Save each drawing as E22-1a, 1b, and 1c.
2. Open the drawings created in Exercise 1. Copy one object from one of the drawings using the base point option so that the new object aligns with an existing intersection. Save the drawing as E-22-2.
3. Open one of the drawings from Exercise 2. Use the **Paste as Block** option to place two of the blocks created in Chapter 19 in the drawing.
4. Open a template drawing and draw a 10' long x 5' wide bathroom. Show all walls as 4" wide. Use the DesignCenter and move blocks for a toilet, sink, and tub into the bathroom.
5. Open the floor plan that you've created and move five electrical symbols into the drawing. Save the drawing as ELECT.

CHAPTER 22 QUIZ

1. Describe the options for inserting blocks using the DesignCenter.
2. Having multiple drawings opened in one drawing session is a powerful tool. Describe the different options used to display your drawings, and explain how to toggle from one drawing to another.
3. Explain the process to copy an object from one drawing to another in a multiple drawing session.
4. Explain the difference between **Paste with Original Coordinates** and **Paste as Block**.
5. How can the **Match Properties** command be used in a multiple drawing environment?
6. List two methods for opening the DesignCenter.
7. What determines whether an object will be moved or copied using drag and drop methods with the left mouse button?
8. Explain the effects of right-click on drag and drop methods.
9. How can the **History** button of the DesignCenter be used?
10. Describe the four displays that can be used in the DesignCenter.
11. You need to find a site drawing created in early September. Explain the process.
12. What do the + and – symbols mean when they are displayed by a folder or file name.
13. Explain the process to place a drawing in the Favorites folder.
14. A drawing template contains five of the best layer names ever created. How can these layers be transferred into a new drawing?
15. You've come to your senses and want to remove three of the layers that were just added to the drawing. How can this be done?

CHAPTER 23

Combining Drawings Using Xref

This chapter will introduce

- Merging a drawing into another drawing
- Externally referenced drawing options
- Methods for editing externally referenced drawings

Commands to be introduced include

- **Xref**
- **Xbind**
- **Xclip**
- **Refedit**
- **Refset**
- **Refclose**

Most of the drawings for a structure are worked on by teams of CAD drafters representing several firms. In addition to the architectural and engineering firms responsible for the design of a structure, CAD drafters often work for landscape and interior architects, as well as plumbing, mechanical, electrical, and civil firms. Each of these firms must have up-to-date drawings to ensure an error-free construction process. One way to provide these drawings to subcontractors is by the use of externally referenced drawings, or xrefs. The use of the **Xref** command is instrumental in assuring quality drawings. The **Xref** command can also be used to compile a drawing sheet containing details or drawings completed at varied scales.

EXTERNAL REFERENCED DRAWINGS

In Chapter 19 it was suggested that a wblock of a floor plan could be inserted in a drawing and used to form the base of that drawing. For example, the floor plan can form the base drawing for the electrical, mechanical, plumbing, and framing plans. This works well, but it does take up lots of disk space. A more efficient method than inserting a wblock is to use an externally referenced drawing or xref. An externally referenced drawing is similar to a wblock, in that it is displayed each time the master drawing is accessed. These drawings, however, are not stored as part of the master drawing file. Each time the base drawing is updated, all drawings that it is attached to will also be updated. Externally referenced drawings are created through the **Xref** command. Figure 23.1 shows a drawing that serves as the base drawing. Figures 23.2a and 23.2b show drawings that make use of the base drawing.

SAVING SPACE

An externally-referenced drawing can be brought into the current drawing for viewing, but it does not become part of the current drawing base. Depending on controls set by the originator, the xref source file may or may not be able to be altered by subcontractors working in the drawing. An xref is similar to a block, except that no drawing objects are recorded in the drawing base. Only the drawing name and a small amount of information needed to access the drawing are stored in the new drawing file.

AUTOMATIC UPDATING

Externally referenced drawings provide excellent benefits for drawing projects that are being developed by a team of engineers, architects, contractors, and drafters. Because of time constraints, subcontractors often need part of the drawings before they are complete. When xref drawings are used, every time the host drawing is accessed, the most recently saved version of the external drawing is loaded. Changes made to the xref source file will be updated automatically in every drawing where it is referenced. For example, if an engineering firm is working off a network and if one team member is working on the plan views and another is working on exterior elevations, every time the xref floor plans are loaded for reference for the elevations, they will be updated.

ASSEMBLING MULTI-SCALED DETAILS

A common use for xref drawings is to assemble details for a stock detail sheet. In most professional offices, stock libraries of details have been created for every possible construction alternative that is likely to be encountered. You can assemble a sheet of the appropriate details using externally referenced drawings with the knowledge that the most up-to-date version of the details is being provided. The next chapter will explore the process for assembling drawings to be plotted at different scales within the same drawing.

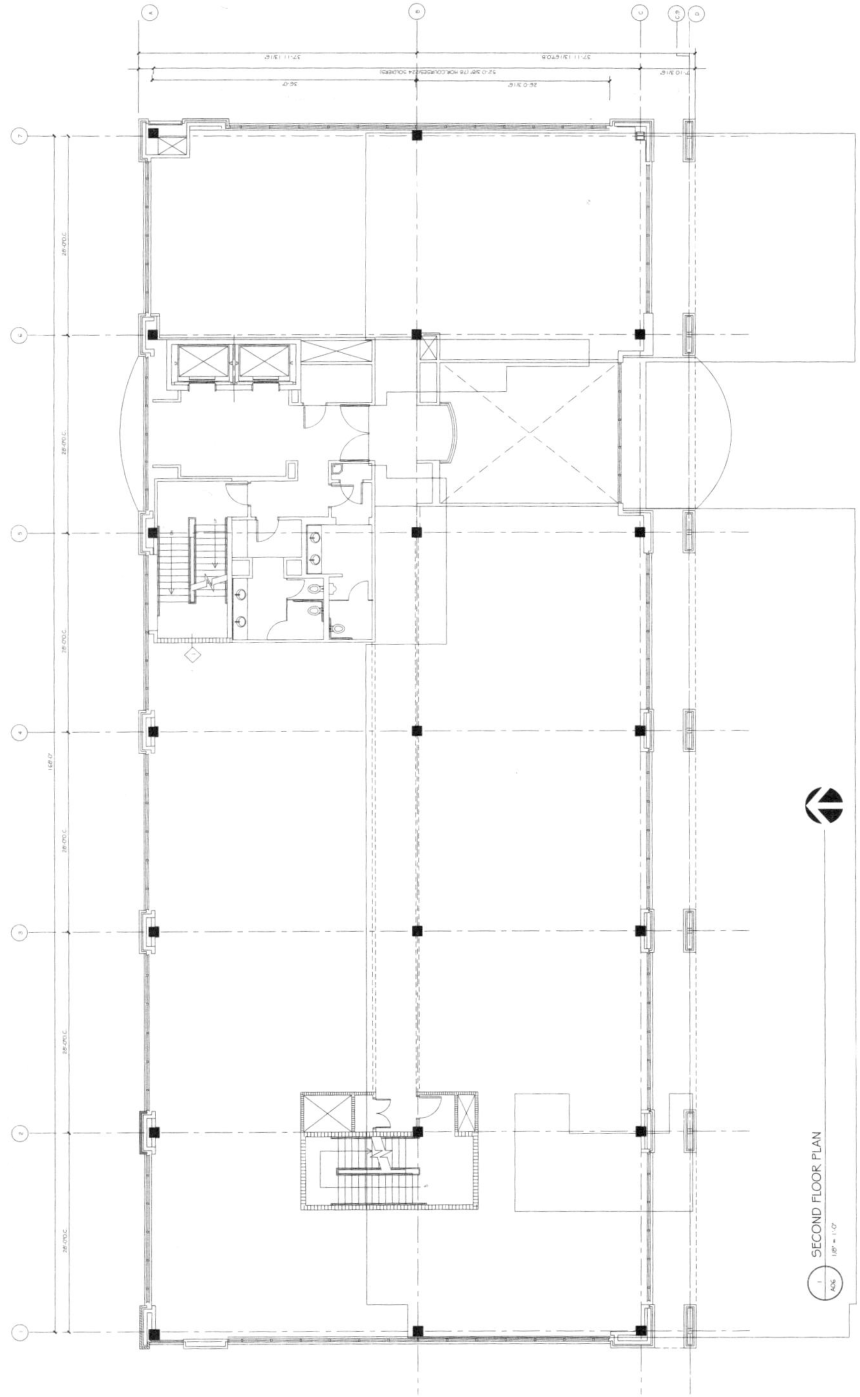

Figure 23.1 *The* **Xref** *command can be used to attach a drawing to another drawing. The information contained in the attached drawing is not stored in the current drawing. Each time the base drawing is updated, all drawings that it is attached to will also be updated. (Courtesy Peck, Smiley, Ettlin Architects.)*

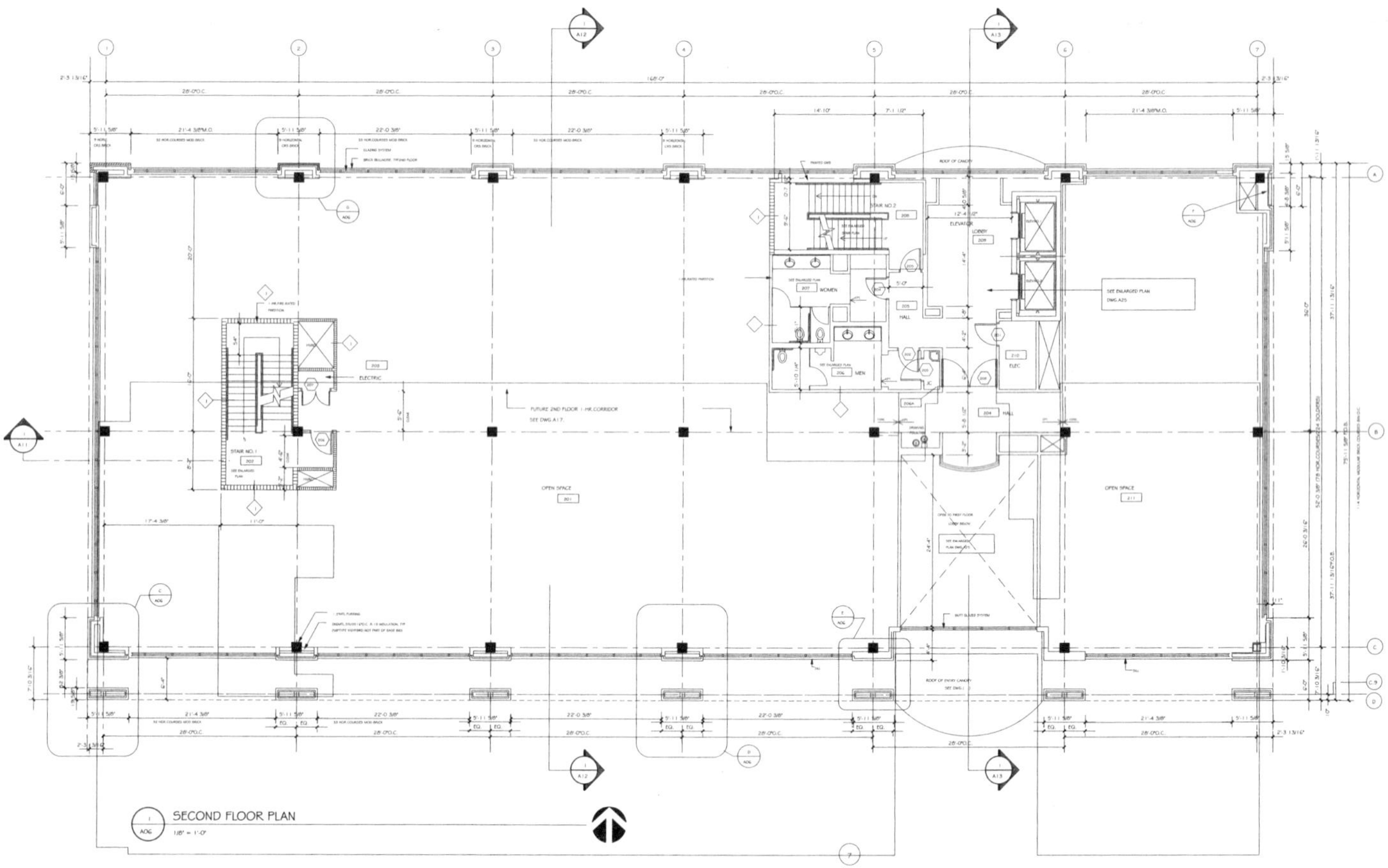

Figure 23.2a *The drawing shown in Figure 23.1 is used as the base for the floor plan shown here and the drawings prepared by the mechanical, electrical, and plumbing subcontractors. (Courtesy Peck, Smiley, Ettlin Architects.)*

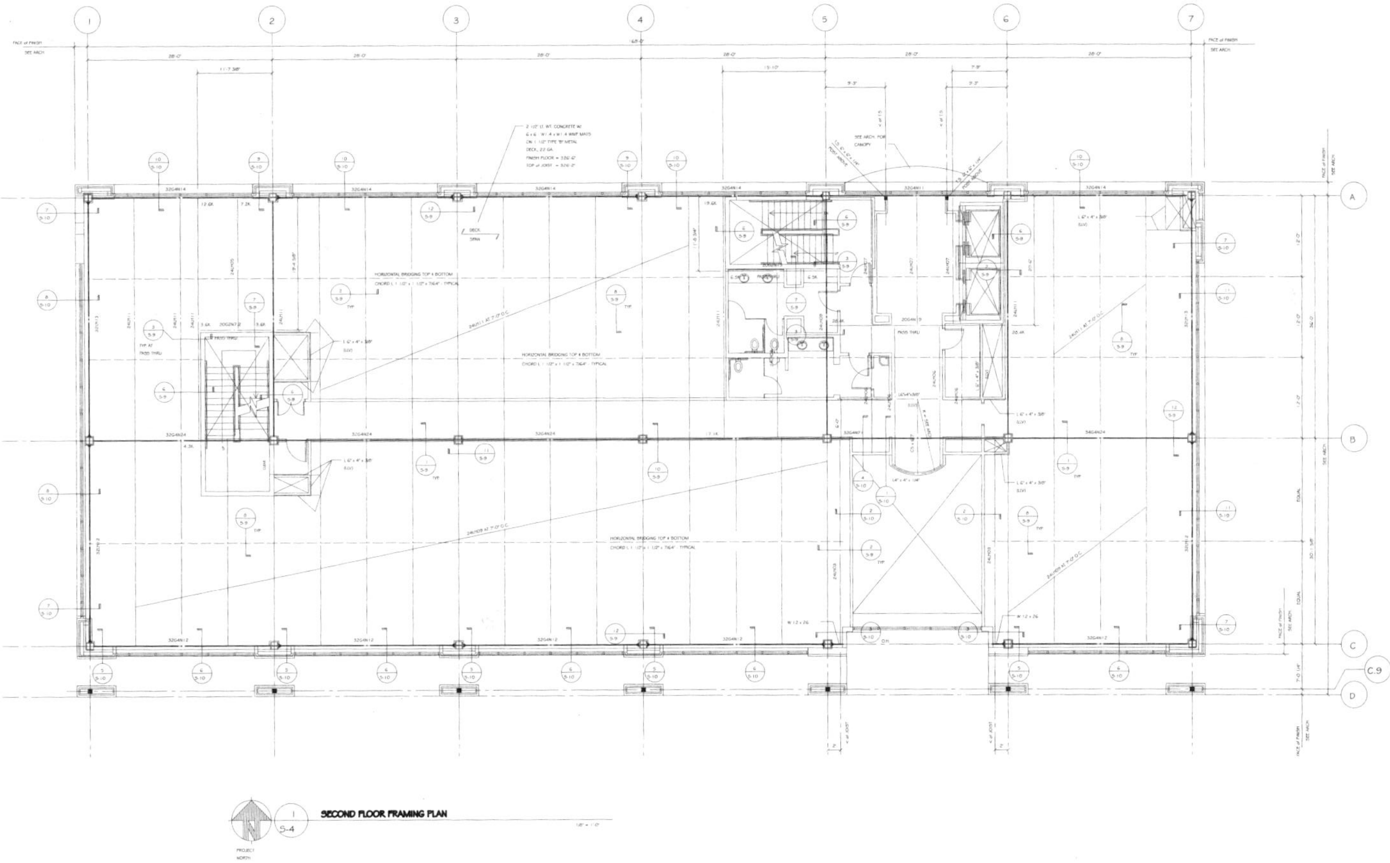

Figure 23.2b *The framing plan is completed by an engineering firm that uses the base drawing created by the architectural team. The drawing is attached using* **Xref**, *and then the needed material is added to the drawing. (Courtesy Van Domelen/Looijeng/McGarrigle/Knauf Consulting Engineers.)*

ATTACHING AN XREF TO A DRAWING

Although the **Xref** command can be completed from the command line, using the **Xref Manager** dialog box will greatly aid the use of the command. Access the dialog box by selecting the **External Reference** button on either the **Insert** or the **Reference** toolbar, or by typing **XR** (ENTER) at the Command prompt. Each method will produce the **Xref Manager** dialog box shown in Figure 23.3. The dialog box allows a drawing to be referenced to the current drawing. It also allows the status of each xref and the relationship of xrefs to one another to be managed. The dialog box is used to do the following:

- Attach an xref
- Change the xref path
- Detach an existing xref
- Reload or unload an existing xref
- Bind an entire xref definition to the current drawing

Once a drawing has been referenced to the drawing, display the **Xref Manager** dialog box using the shortcut menu: select an xref, right-click in the drawing area, and then select **Xref Manager.**

Selecting **External Reference** from the **Insert** menu or selecting **Attach** in the **Xref Manager** dialog box will display the **Select Reference File** dialog box. Once a file to be referenced is selected, an **External Reference** dialog box similar to Figure 23.4 is displayed. The dialog box allows a drawing to be selected for external reference to the current file, much the way the Insert command works.

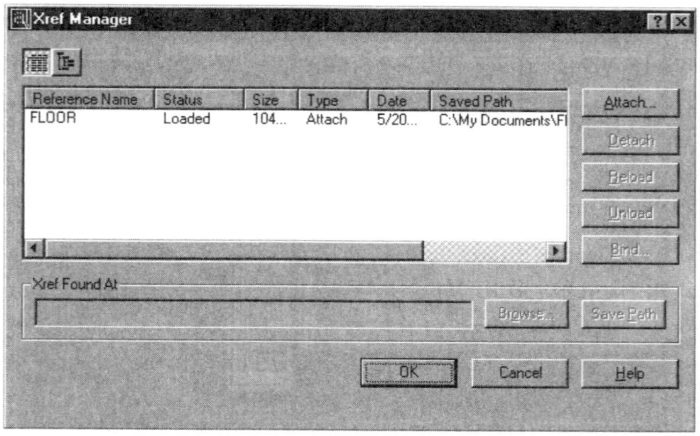

Figure 23.3 *The* **Xref Manager** *dialog box can be used to reference a drawing into a base drawing. Access the dialog box by selecting the* **External Reference** *button from either the* **Insert** *or* **Reference** *toolbar, or by typing* **XR** *(ENTER) at the Command prompt.*

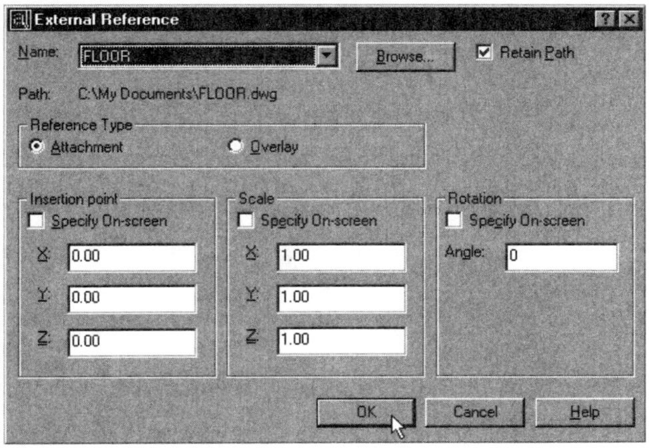

Figure 23.4 *The* **External** *Reference dialog box can be used to reference a drawing to another drawing using a procedure similar to the* **Insert** *command. Access the dialog box by selecting* **External Reference** *from the* **Insert** *menu.*

Figure 23.5a shows a floor plan that will be used as the base drawing in the following example. The base drawing is titled C:\DRAWINGS\SCHOOL\ARCH\FLOOR.DWG. Figure 23.5b shows the electrical information that will be attached

Figure 23.5a *This portion of a floor plan will serve as a base for the electrical plan shown in Figure 23.5b. The information on the referenced drawing will be displayed as the floor plan is accessed, but it will not be stored in the FLOOR.DWG drawing. (Courtesy Michael Jefferis.)*

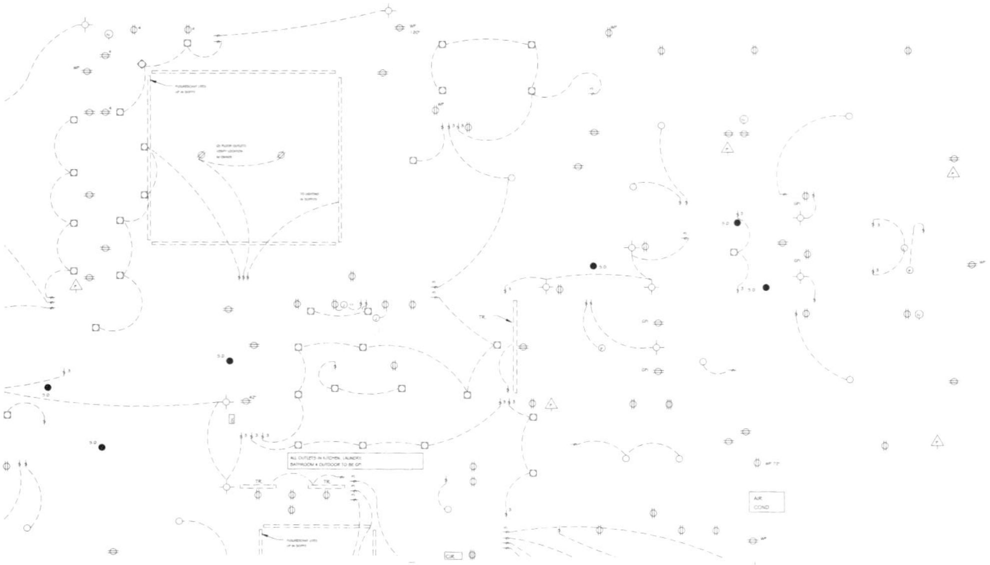

Figure 23.5b *By using referenced drawings, subcontractors from various firms can each complete their work without having to wait for others to finish. This electrical information was added to the base drawing shown in Figure 23.5a.*

to the base drawing to make the electrical plan. This drawing is stored as C:\DRAWINGS\SCHOOL\ARCH\ELECT.DWG.

To attach the electrical information to the floor drawing, display the **Xref Manager** dialog box and select the **Attach** button. This will display the **Select Reference File** dialog box. The box is similar to each dialog box that is used to open a new drawing. Select the desired folder and file of the drawing to be attached. With the desired file highlighted, click the **Open** button to display the **External Reference** dialog box shown in Figure 23.4. This dialog box allows you to control how and where the attached drawing will be located in the base drawing by using the **Reference Type** and parameters areas.

EXTERNAL REFERENCE DIALOG BOX OPTIONS

So far, the **External Reference** dialog box shown in Figure 23.4 has been used only to attach one reference drawing to a base drawing. Several parameters can be adjusted to alter its use.

NAME AND PATH OPTIONS

The **Name** edit box was empty as the dialog box was first displayed. Once a drawing has been attached, the name of the drawing will be displayed. When multiple drawings are attached, selecting the edit arrow will display a list of the referenced drawings.

Selecting an attached drawing from the list will display the drawing name in the **Name** edit box, and its location will be displayed in the **Path** display. Selecting **Browse** will display the **Select Reference File** dialog box, allowing new xrefs to be selected for the current drawing.

Retain Path

The **Retain Path** check box dictates if the full path of the referenced drawing will be saved. By default, this option is active so that the path to the referenced drawing will be saved in the database of the host drawing. If the option is deactivated, the referenced drawing is saved in the database without a path. However, AutoCAD will search paths associated with the current drawing name. The inactive option is often used when referenced drawings are shared with subcontractors who will work with the drawings. With this check box inactive, as AutoCAD loads the host drawing, it will only search for the referenced drawing in the path specified in the **Files** tab of the **Options** dialog box. If the **Retain Path** check box is active, it will search through only the file folders specified in the path. This will usually be the desired method. If you are working on several jobs at the same time, your drawing will not be overwritten by the wrong job. For example, if you are working on several jobs requiring an ELECT.dwg file to be attached, if you have the **Retain Path** box inactive, it will only search C:\PROGRAM FILES\PARKER\ELECT, rather than C:\PROGRAM FILES\ACAD2002\ELECT.dwg. If you complete this search, AutoCAD may find several different ELECT plans, but it would grab the first file found. It is best to be as specific on the path as possible. Make sure, as you create new drawings, that they are placed in the correct location.

REFERENCE TYPE

This portion of the **External Reference** dialog box provides the options of **Attachment** and **Overlay**, for deciding how an external drawing will be attached to the base drawing. Attached drawings allow you to build a drawing using other drawings.

Attaching a Drawing

With the default setting of **Attachment**, clicking the **OK** button will remove the dialog box, return you to the drawing area, and provide a prompt for an insertion point. As the point is selected with the mouse, the drawing will be attached and the command will be closed. Figure 23.6 shows the merged drawings from Figure 23.5a and 23.5b. The insertion point of the lower right corner was used with an Object Snap of Intersection to assure proper alignment of the two drawings.

If a second drawing is to be attached to the host drawing, selecting the **Attach** button will produce the **Select Reference File** dialog box. Once one drawing has been selected, when the **Open** button is selected, the **External Reference** dialog box is redis-

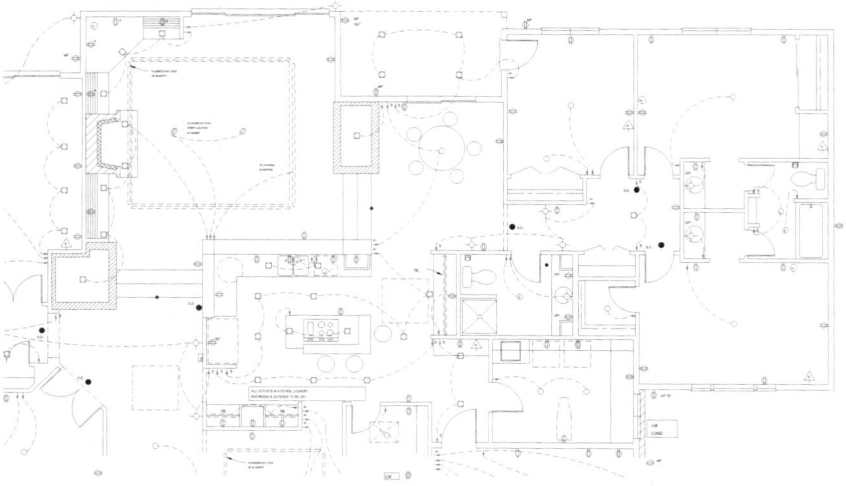

Figure 23.6 *The result of attaching the electrical drawing to the base drawing. (Courtesy Michael Jefferis.)*

played. Once a drawing is attached, AutoCAD adds dependent symbols in the xref to the base drawing. Dependent symbols are named items such as blocks, dimension styles, layers, linetypes, and text styles. Although these dependent symbols can be viewed, they can't be altered unless you preset values to allow the user to alter the drawing. This is one of the benefits of an xref—you can give a referenced drawing to a subcontractor and not have to worry about the drawing being altered. When a drawing is referenced to a base drawing, you'll notice that the layer names of the xref will be altered. Layer names for attached drawings will be displayed showing the name of the drawing, followed by a vertical bar and the layer name.

A name such as 72PLUMB|UPPLUMB would indicate a referenced drawing of 72PLUMB and a layer title of UPPLUMB.

Overlaying an External Drawing

Choosing the **Overlay** option will produce a drawing that looks like an attached drawing. The major difference between an attached and an overlaid drawing occurs when the referenced drawing contains nested reference drawings. A nested drawing is a drawing within another referenced drawing. When a referenced drawing containing nested drawings is overlaid, the nested drawings within the drawing will not be displayed. When a nested drawing is attached, all objects will be displayed. The **Overlay** option is typically used when you need to share information. An overlaid drawing will allow others to view how their portion of the project relates to the drawing. Rather than showing all of the drawings like the Mechanical and Electrical Xrefs, with the **Overlay** option, you can still benefit from the automatically updated files.

SETTING PARAMETERS

As the drawing in Figure 23.5b was attached to the drawing in Figure 23.5a, the insertion point was selected prior to attachment. Using the parameters portions of the **External Reference** dialog box allows the insertion point, scale factors, and rotation angle to be controlled for a referenced drawing.

Insertion Point

In the default setting, the **Specify On-screen** check box is active, allowing the insertion point to be selected with the cursor or by entering coordinates at the Command prompt. With the **Specify On-screen** check box deactivated, the text boxes for each option are activated. These coordinates allow specific X and Y coordinates to be specified for the insertion point of the referenced drawing. These options correspond to their counterparts in the Insertion Points for the **Insert** command.

Note: For most firms, it is common to use an insertion point of 0,0. With an insertion point of 0,0, all drawings created by subcontractors will align perfectly. This is a much easier method than worrying about what corner of a structure might have been used as a base point.

Scale

This option specifies the scale factors for the selected referenced drawing as it is inserted in the host drawing. The default values for the X, Y, and Z scale factors are set as 1. Selecting the **Specify On-screen** check box for scale factors will deactivate the scale factor values and allow the values to be adjusted from the Command prompt as the drawing is attached.

Rotation

The **Rotation** portion of the dialog box is similar to that of the **Insert** dialog box. Angle allows the rotation angle of the referenced drawing to be controlled as it is inserted in the base drawing. In the default setting, a rotation angle can be entered in the edit box. With the **Specify On-screen** check box active, the rotation angle can be set at the Command prompt as the drawing is attached.

XREF MANAGER DIALOG BOX OPTIONS

So far, the **Xref Manager** dialog box shown in Figure 23.3 has been used only to attach one reference drawing to a host drawing. This dialog box can be used to list, attach, load, and modify referenced drawings contained in the current drawing. The display provided in the box will depend on the active button in the upper left corner of the dialog box. The buttons toggle between **List View** and **Tree View**. Change the display by selecting the inactive option or by pressing F3 or F4.

LIST VIEW OPTION

In the default setting, the dialog box is displayed in **List View**, with information related to the reference name, status, size, type, date, and save path. Information can be provided in an alphabetical list of the referenced drawings contained in the current drawing. Notice in Figure 23.3, that not all of the information is visible. The path scrolls off the edge of the display field. Move the cursor to the line that divides the **Reference Name** area and the **Status** area. As the line is touched, the cursor turns to a double arrow, allowing the size of the box to be altered. This can be done to each box to allow a complete display. (See Figure 23.7a.) Click a list column heading to sort the referenced drawings by the column name. To sort referenced files by date, click the reference **Date** column.

Reference Name

This column lists the names of the referenced drawings stored in the definition table of the drawing.

Status

This column shows the current status of referenced drawings. Options include the following:

Loaded—The attached drawing is displayed.

Not Found—The referenced drawing cannot be found.

Orphaned—Attached to another referenced drawing that is unreferenced, unresolved, or not found.

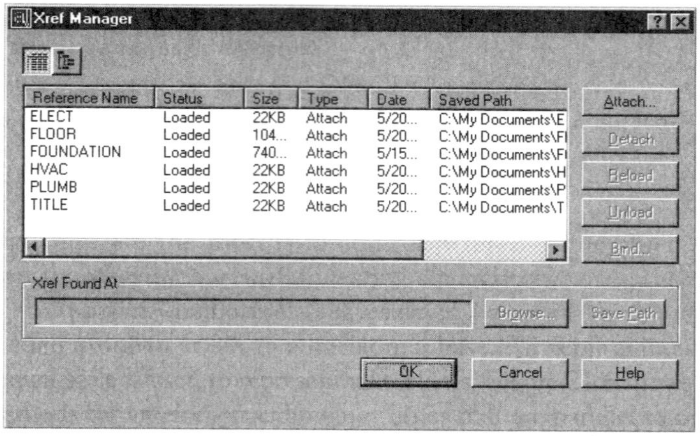

Figure 23.7a *The space between listings can be reduced to allow longer listings to be completely viewed. Place the cursor on the lines that divide titles, to reduce or enlarge each display space.*

Reload—The referenced drawing has been marked to be reloaded from the drawing once the **Xref Manager** dialog box is closed.

Unload—The referenced drawing has been marked to be unloaded from the drawing once the **Xref Manager** dialog box is closed.

Unloaded—The referenced drawing is not displayed and will not be displayed again until the reload option is selected.

Unreferenced—An unreferenced drawing is attached to the drawing but erased.

Unresolved—The referenced drawing cannot be read by AutoCAD.

Size

The **Size** column displays the size of the referenced drawing.

Type

The **Type** column indicates if the referenced drawing was placed using the **Attachment** or the **Overlay** option.

Date

The **Date** column lists the last date that the referenced drawing was modified.

Saved Path

The **Saved Path** column shows the path of the referenced drawing.

TREE VIEW OPTION

With the **Tree View** active, a hierarchical representation of referenced drawings is shown similar to the manner that listings in Windows Explorer are displayed. The **Tree View** method displays nested drawings in their relationship to the attached drawing, as well as the attachment method and the status. Figure 23.7b shows a listing using the **Tree View** option.

EXTERNALLY REFERENCED DRAWING OPTIONS

Once the drawing is attached, each time the base drawing is opened, the attached drawing will be loaded and displayed as long as the path to the attached drawing is accessible. In addition to loading the referenced drawing, the other options of the **External Reference** dialog box will become active. Options include **Attach**, **Detach**, **Reload**, **Unload**, and **Bind**.

ATTACH

You've already explored the **Attach** option. Use the following summary of the steps to attach a drawing to your current drawing:

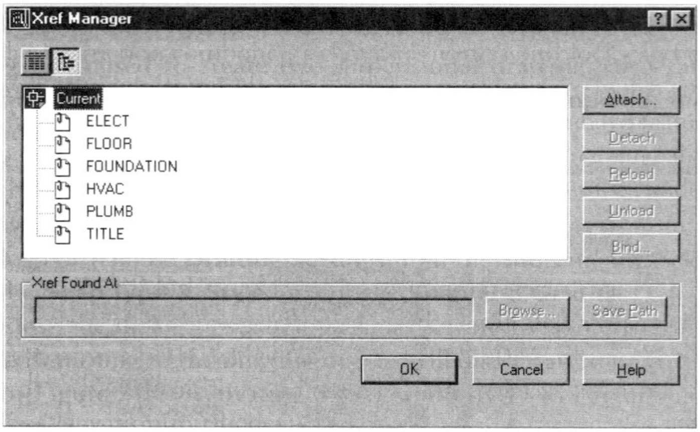

Figure 23.7b *Listings in the* **Xref Manager** *dialog box displayed using the Tree View option.*

1. Choose the **External Reference** button on the **Insert** or **Reference** toolbar, or type **XR** (ENTER) at the Command prompt.
2. Select **Attach** and choose a drawing to be attached to the current drawing from the **Select Reference File** dialog box. Select the **Open** button to continue. Selecting **Open** will display the **External Reference** dialog box.
3. Select the reference type to be used. Typically, this will be **Attachment**.
4. Specify the insertion point, scale, and rotation angle, or select the **Specify On-screen** option for each parameter.
5. Click the **OK** button to attach the referenced drawing to the current drawing.

DETACH

The **Detach** option will remove unneeded referenced drawings from the host drawing. Similar to the way **Erase** removes a block, **Detach** removes copies of the referenced drawing and its definition. Detach a drawing by first selecting the name of the unneeded xref to highlight it and then selecting the **Detach** button, which will remove the xref listing from the dialog box. If the **OK** button is clicked, the selected xref will be removed from the screen and all references to the drawing will be removed from the base drawing. Detach an attached drawing file using the following procedure:

1. Choose the **External Reference** button on the **Insert** or **Reference** toolbar. To use the shortcut menu, select the referenced drawing to detach, right-click in the drawing area, and choose **Xref Manager** from the menu.

2. Select the drawing to be detached from the host drawing.
3. Select the **Detach** button. This will detach the drawing file from the current file.
4. Click the **OK** button to return to the drawing area.

RELOAD

This option allows an attached referenced drawing to be reloaded and updated in the middle of a drawing session. If a coworker revises the referenced drawing while you're working on a drawing that contains the xref, your xref is out of date. The **Reload** option allows the most current referenced drawing to be used without your having to exit the drawing session and reopening the master drawing. The process to reload a drawing is similar to the **Detach** option.

UNLOAD

The **Unload** option allows a referenced drawing to be temporarily removed from the screen display without removing the xref from the drawing base. The **Unload** option is similar to using the Freeze option of the **Layer** command. The referenced drawing is not displayed or considered in regeneration, increasing the drawing speed. The drawing can be reactivated by using the **Reload** option. The process to reload a drawing is similar to the Detach option. The **Unload** option is very handy if you are working on a very large drawing and you don't need certain information at the current time but will need the information at a later time.

BIND

The **Bind** option allows a referenced drawing to become a permanent part of the base drawing. When this option is used, referenced drawings function as a wblock rather than an xref. When a drawing is referenced to a host drawing, the drawing is displayed, but you can't modify the referenced drawing. Using the **Bind** option allows an external drawing to be edited, but the xref will no longer be updated as the original drawing is edited. The **Bind** option adds dependent symbols to the drawing base so that they can be used just like any other drawing object. Layer names such as 72PLUMB|UPPLUMB in an attached drawing will be altered to read 72PLUMB0UPPLUMB. This naming system allows you to quickly identify attached and bound layers. The **Bind** option should be used only when you know for sure that the review process for the referenced drawing is complete and the drawing is no longer subject to change. Because construction drawings are subject to change, **Bind** might not be an acceptable option. The **Xbind** command, introduced later in this chapter, allows specific symbols of a drawing to be permanently attached.

Binding a Drawing

Selecting the **Bind** button will produce the **Bind Xrefs** dialog box shown in Figure 23.8, providing the options of **Bind** and **Insert**.

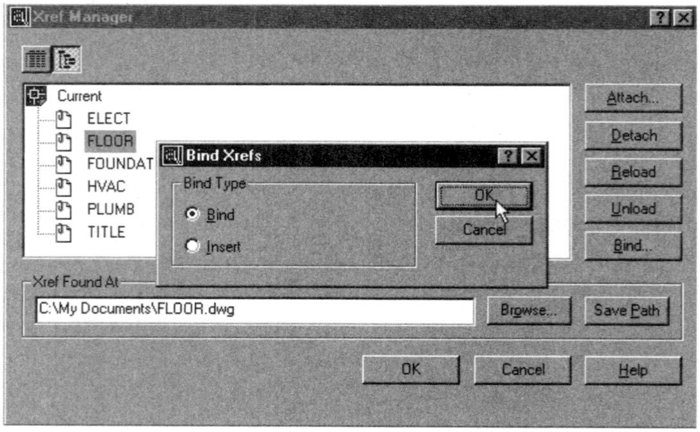

Figure 23.8 *The* **Bind** *option allows a referenced drawing to be permanently attached to a base drawing. Selecting the name of an attached drawing from the* **Xref Manager** *dialog box activates each of the* **Xref** *options. Selecting the* **Bind** *option displays the* **Bind Xrefs** *dialog box.*

Bind

The default option of **Bind** will permanently attach a drawing to the base. Clicking the **OK** button removes the dialog box, binds the attached drawing to the host drawing, and updates the names of dependent symbols. Symbols in the bound drawing can now be edited just as any object created in the host drawing.

Insert

If the **Insert** option of **Bind** is used, an attached drawing is bound to the base drawing as if a wblock had been inserted in a drawing. **Insert** will bind the referenced drawing to the current drawing in a method similar to detaching and inserting the reference drawing. Clicking the **OK** button removes the dialog box, inserts the attached drawing in the host drawing, and updates the names of dependent symbols. A layer of 72PLUMB|UPPLUMB would be renamed to UPPLUMB.

THE XBIND COMMAND

The **Bind** option adds dependent symbols to the drawing base, so that they can be used. Automatic updating and other qualities of xref drawings are lost. This can be overcome with the **Xbind** command, which allows a portion of a referenced drawing—such as a block, layer, or linetype—to be permanently added to a current drawing, with the balance of the external drawing retaining its qualities. Access the **Xbind** command by selecting **External Reference Bind** on the Reference toolbar, by selecting **Object**, **External Reference**, **Bind** from the **Modify** menu, or by typing **XB** (ENTER) at the Command prompt. Each option will produce the **Xbind** dialog box, shown in Figure 23.9, which lists referenced drawings that are attached to the

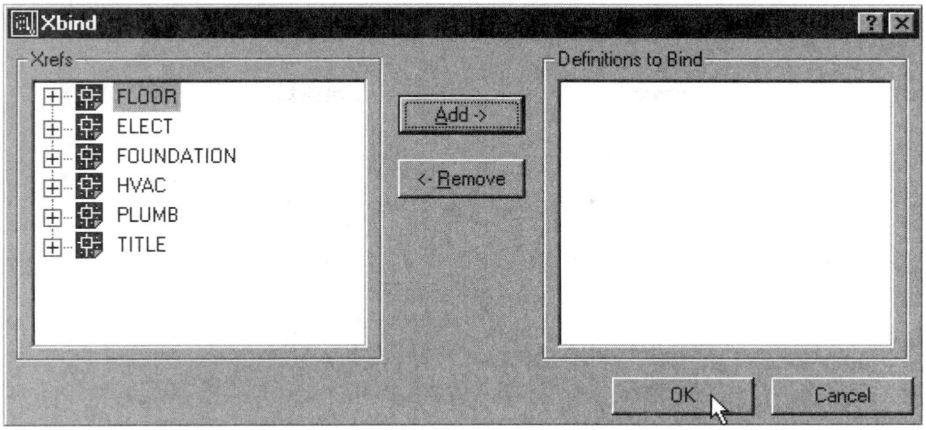

Figure 23.9 *The* **Xbind** *dialog box displays a listing of drawings that are attached to the current drawing. The + symbol preceding AutoCAD 2002 icon indicates that the listing can be expanded.*

base drawing. In Figure 23.9, six drawings are attached to the base drawing. Each drawing is preceded by the + symbol and the AutoCAD 2002 icon. The + indicates that the listing can be expanded. Selecting one of these options and double-clicking will display the contents of the drawing. In Figure 23.10a, the FLOOR listing was expanded, revealing that the drawing contains nested Blocks, Dimstyles, Layers, Linetypes, and text styles. Notice in Figure 23.10b that the block FLOOR|10PBAC01 has been selected to be added, using the **Xbind** command. If the desired dependent symbol to bind, FLOOR|10PBAC01, is selected and the **Add>** button is clicked, the name of the block will be moved from the **Xref** listing to the **Definitions to Bind** display. (See Figure 23.10c.) Clicking the **OK** button will bind the block to the base drawing and close the **Xbind** dialog box. Selecting a listing from the **Definitions to Bind** display and selecting the **<Remove** button will unbind the definition and restore it to the **Xrefs** listing.

WORKING WITH XCLIP

The **Xclip** command allows portions of an attached xref drawing to be hidden. The command allows a boundary to be created in a referenced drawing so that all material that is outside the boundary is invisible. Access the command by selecting **External Reference Clip** on the **Reference** toolbar or by typing **XC** (ENTER) at the Command prompt. Each access method will produce a Select object prompt. Any object selection method can be used to select xrefs to be clipped. The command sequence is as follows:

Command: **XC** (ENTER) *(Or click the Xclip button.)*

Select object: *(Select xrefs to be clipped.)*

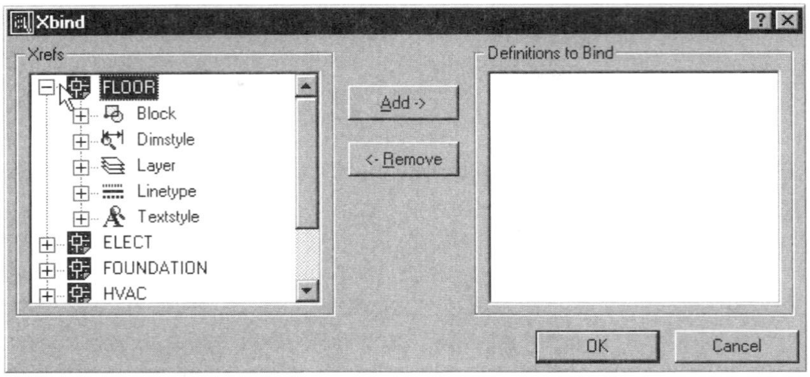

Figure 23.10a *Selecting the FLOOR listing reveals nested items in the Block, Dimstyle, and line Textstyle.*

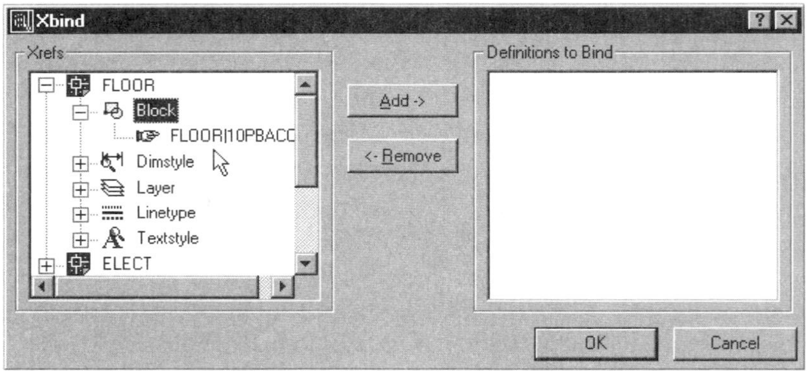

Figure 23.10b *Selecting the FLOOR|10PBAC01 block for binding to the current drawing.*

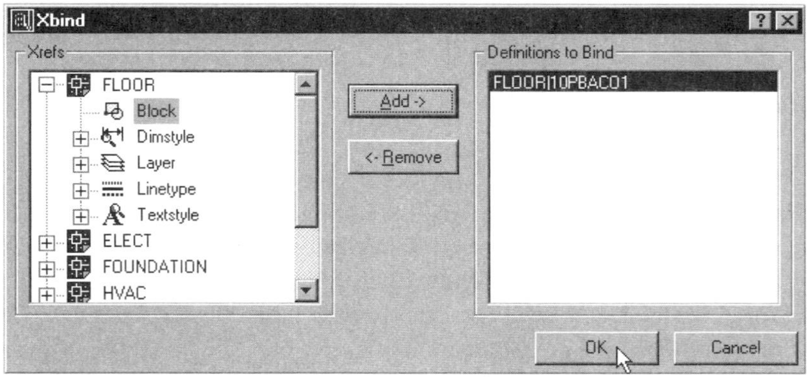

Figure 23.10c *Selecting the name and selecting the* **Add>** *button will bind the selected definition to the base drawing.*

I found

Select objects: (ENTER)

[ON/OFF/Clipdepth/Delete/generate Polyline/New boundary]
<New>: (ENTER)

Accepting the default value allows the clipping boundary to be selected and produces the following prompts:

Specify Clipping boundary:

[Select polyline/Polygonal/Rectangular] <Rectangular>: (ENTER)

Specify first corner: *(Select first window corner.)*

Specify opposite corner: *(Select opposite window corner.)*

Command:

Options for setting the boundary include the default method of forming a rectangular boundary with a selection window, selecting a polyline, or by selecting Polygonal. Selecting Polygonal allows a boundary to be set by specifying points for the vertices of a polygon. Accepting the default by pressing (ENTER) produces prompts to select the First and Other corners. As the corners are selected, objects in the xref that are outside the boundary will become invisible. This will be helpful if there are items on certain layers that you need visible and others that need to be invisible. Rather than altering the layers, you can exclude some objects with **Xclip**.

Figure 23.11a is an example of a footing detail that has been referenced to a host drawing. Figure 23.11b shows the detail on the right side of Figure 23.11a, hidden using the Xclip command.

EDITING REFERENCED DRAWINGS

AutoCAD 2002 allows externally referenced drawings that you've created to be edited, using in-place editing. In-place editing allows the referenced drawing to be revised. If the referenced drawing selected for editing contains attached xrefs or block definitions, the reference and nested references are displayed in the **Reference Edit** dialog box. Display this dialog box by selecting the **Edit Block or Xref** button on the **Refedit** toolbar, or by selecting **In-place Xref and Block Edit** from the **Modify** menu. The toolbar and the dialog box can be seen in Figure 23.12. The toolbar presents five options for editing referenced drawings.

EDIT BLOCK OR XREF

A referenced drawing will respond like a block if you attempt to edit it. Using the **Refedit** command allows individual objects in the referenced drawing to be altered. You can edit objects by selecting the **Edit Block or Xref** button on the **Refedit** toolbar and then choosing the referenced drawing to be edited.

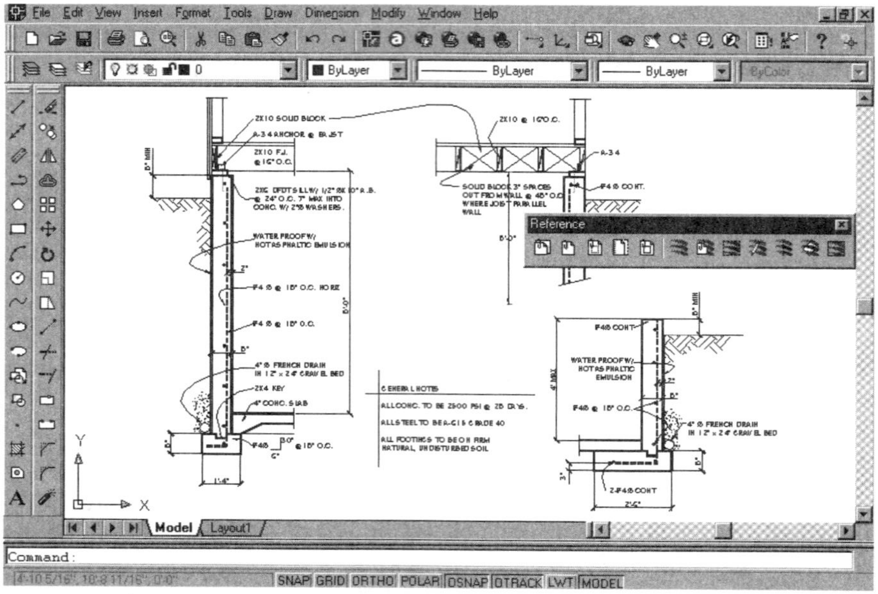

Figure 23.11a *The* **Xclip** *command allows information that has been attached to a base drawing to be hidden. The command requires that a boundary be selected. Anything outside the boundary will be made invisible.*

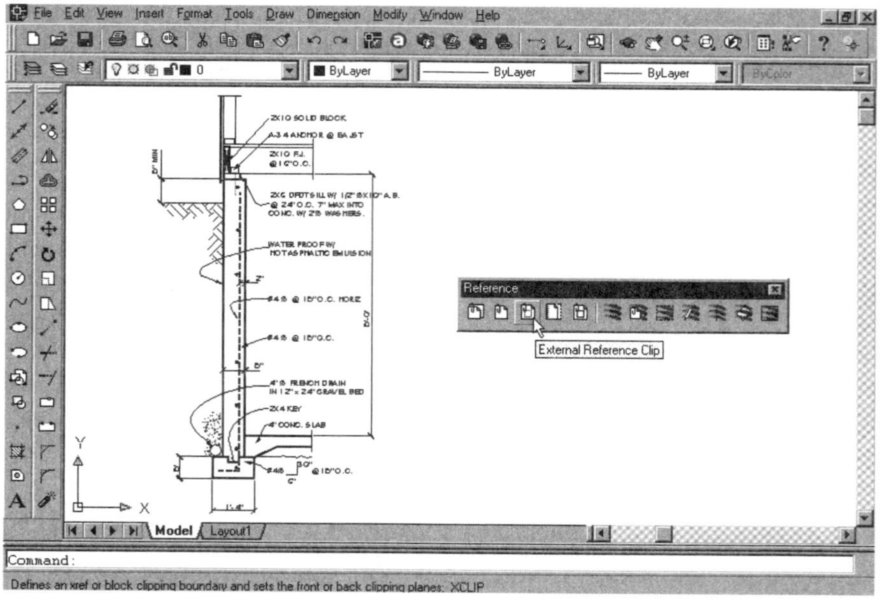

Figure 23.11b *The detail on the left side of Figure 23.11a was placed inside the boundary. With the* **Xclip** *command, the detail on the right side is still attached to the drawing base but is removed from the display set.*

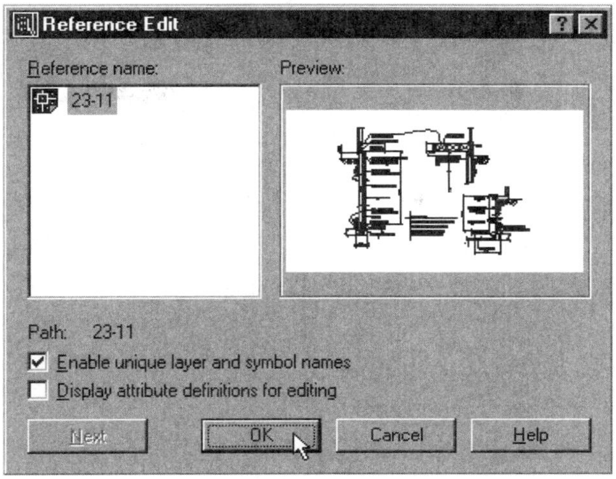

Figure 23.12 *AutoCAD allows referenced drawings to be edited using in-place editing with the* **Reference Edit** *dialog box. Display the box by selecting the* **Edit Block or Xref** *button on the* **Refedit** *toolbar, or by selecting* **In-place Xref and Block Edit** *from the* **Modify** *menu.*

Using the Reference Edit Dialog Box

The **Reference Edit** dialog box allows a referenced drawing to be edited. Key elements of the dialog box include the **Reference Name** box, **Preview** window, **Next**, and editing options.

Reference Name—This portion of the dialog box displays the referenced drawing that has been selected for in-place editing and any nested drawings or blocks within the selected drawing.

Preview—The **Preview** window displays an image of the drawing selected for editing.

Next—This option can be used to cycle through selectable references.

Path—The file location of the selected referenced drawing.

Enable unique layer and symbol names—This option controls whether layer and symbol names of objects extracted from the referenced drawing will be unique or altered. In the active mode, symbol and layer names will be displayed with a prefix of $#$. The altered prefix allows for easy identification. With the option inactive, layer and symbols will appear as they do in the original drawing.

Display attribute definitions for editing—This option controls whether variable attribute definitions in a block reference will be extracted and displayed during editing. Referenced drawings and blocks without definitions are not affected by this option. When active, attributes will be invisible but the object and the attribute definitions are available for editing. If changes are saved back to the block reference, the original reference remains unchanged. The new attribute definition will affect only future insertions of the block.

Once each option is set and you've selected a reference to edit, you'll be prompted to select specific objects in the referenced drawing to edit.

Editing Referenced Drawings Using Refedit

The **Refedit** command is used to select and edit a referenced drawing. Use the following procedure to edit individual objects of a referenced drawing:

1. Select the **Edit Block or Xref** button on the **Refedit** toolbar, or type **REFEDIT** (ENTER) at the Command prompt.
2. Select the reference to be edited. This will highlight the referenced drawing and display the **Reference Edit** dialog box. If nested blocks or drawings are contained in the selected xref, all references available for selection are displayed in the dialog box.
3. If multiple drawings are listed in the dialog box, highlight the drawing to be edited.
4. Set the options of **Enable unique layer and symbol names** and **Display attribute definitions for editing to meet drawing needs**.
5. Click the **OK** button.
6. Select objects to be edited.

This is your opportunity to select objects in the referenced drawing that will be edited. Once selected, objects in the referenced drawing can be edited just like any other drawing.

Adding or Removing Objects from the Working Set

To add objects to the working set or remove them from it as the referenced drawing is being edited, you can use the **Refset** command. The working set consists of what will be extracted, edited, and saved back to the drawing. You'll be editing and saving changes to the original drawing—even though you're working on a referenced drawing through a host drawing. You can supply a referenced floor plan to the HVAC subcontractor, and they can alter the plan as needed to allow for ducts and chases to be installed. As they save changes to the plan, you can now supply an up-to-date plan to other consultants. Start the command by selecting the appropriate button on the **Refedit** toolbar, by selecting **Add to Workset** or **Remove from Workset** from **In-place**

Xref and Block Edit on the **Modify** menu, or by typing **REFSET** (ENTER) at the Command prompt.

Adding Objects to the Working Set

To add objects to the working set, start the **Refset** command by selecting the **Add** button on the **Refedit** toolbar.

This will display the prompt:

> Transfer objects between the Refedit working set and host drawing...
>
> Enter an option [Add/Remove] <Add>: _Add
>
> Select objects: *(Select objects to be added to the working set.)*

The command will vary slightly when entered by keyboard. You'll have to select Add or Remove at the prompt. (When using the toolbar, you've already made the decision by selecting the button.) Once objects are selected, a prompt will be displayed giving the number of objects added to the working set. The prompt will resemble the following:

> 3 Added to working set.

Removing Objects from the Working Set

To remove objects from the working set, start the **Refset** command by selecting the **Remove** button on the **Refedit** toolbar.

This will display the prompt:

> Transfer objects between the Refedit working set and host drawing...
>
> Enter an option [Add/Remove] <Add>: _Rem
>
> Select objects: *(Select objects to be removed to the working set.)*

The command will vary slightly when entered by keyboard. Once objects are selected, a prompt will be displayed giving the number of objects removed from the working set. The prompt will resemble the following:

> 3 Removed from working set.

Objects removed from the working set will appear in a gray tone referred to as *dithered.* The intensity of the line can be controlled using the **Display** tab of the **Options** dialog box. (Select **Options** from the **Tools** menu.) Use the **Reference Edit Fading Intensity** setting to control the display of removed objects. A higher value provides greater contrast.

Saving and Discarding Changes with Refclose

Up to this point, you've referenced an external drawing to a drawing and edited a portion of the referenced drawing. If you try to edit other portions of the referenced drawing, AutoCAD will warn you:

> **Command not allowed, (drawing name) already checked out for editing**

This is AutoCAD's way of not allowing two subcontractors to attempt to use and update a drawing at the same time. If you save the drawing, the new drawing will still contain your revisions, but the original drawing will remain unaltered. The **Refclose** command is used to either save back or discard changes made during the editing process of a referenced drawing. Once you've used **Refclose**, you can edit other referenced drawings or blocks. The command can be selected on the **Refedit** toolbar, shortcut menu, menu, or by keyboard. As the command is entered by keyboard, you'll be prompted to enter a letter for the Discard or Save option.

Discard

Selecting Discard will vaporize the working set so that the source drawing remains in its original state. Changes that you've made will remain in the current drawing, but the original drawing remains unchanged. You'll be given a warning, but proceeding will remove changes to the working set. **Undo** will restore the reference editing session if you change your mind after using Discard.

Save

Selecting the Save option will save the changes made to the working set back to the original drawing.

DENYING ACCESS TO REFERENCED DRAWINGS

Throughout the discussion of in-place editing, it has been assumed that you don't mind if someone alters your drawing. The idea of someone working for the HVAC contractor making changes to your floor plan might sound pretty good. The CAD operator for the HVAC is saving you the work of incorporating their required changes into the base drawing. If for some reason you need to provide drawings to another firm, but you want to ensure accuracy, you can restrict the ability of others to edit the referenced drawing. Security can be maintained on the **Open/Save** tab of the **Options** dialog box (select **Options** from the **Tools** menu). By default, the **Allow other users to Refedit current drawing** check box in the **External References (Xrefs)** area is active. Making this option inactive will remove other users' ability to edit your source drawing.

CHAPTER 23 EXERCISES

1. Open the FOOTING drawing and use it to create an xref drawing titled BASEFTG. Open drawing FLOOR14 and attach BASEFTG to the drawing. Save the drawings as XREFTEST.
2. Open a drawing template containing the border and title block. Attach any four drawings in your library. Provide a title and the scale used by the original drawing.
3. Open any three drawings and attach them to a drawing base. Save the drawing as XREF.

CHAPTER 23 QUIZ

1. Describe how a referenced drawing can be used in an office to complete a project.
2. List and briefly define the six **Xref** options.
3. List the command, the options required, and the prompts to attach a drawing to another drawing.
4. What is a dependent symbol and how does it affect a drawing?
5. Explain the effects of overlaying a drawing.
6. The Bind option of **Xref** has been used. How will the option affect the external drawing?
7. Explain the effects of **Xbind** on a referenced drawing.
8. How does the **Xclip** command affect an attached drawing?
9. How does the **Refedit** command affect a referenced drawing?
10. What is the difference between **Insert** and **Xref**?

CHAPTER 24

Working with Layouts and Viewports

This chapter will introduce

- Comparisons between model space and paper space
- Creating and using layouts
- Working with tiled viewports
- Working with floating viewports

Commands to be introduced include

- **Layout**
- **Vports**
- **Mspace**
- **Pspace**
- **Vpclip**

Chapter 2 introduced the layout tabs of the AutoCAD template drawings. A layout is a paper space drawing environment that represents how your drawing created in model space will appear on a sheet of paper when plotted. Throughout this text you've used the model space tab of the Architectural template. The template has been edited to include your favorite linetypes, lineweights, text fonts, and dimensioning styles. This chapter will introduce you to using the layout tabs to provide a predictable plotting setup. Methods of using existing layouts and creating new layouts will be explored. By completing the steps in this chapter, you'll be able to take drawings that have been created in previous chapters, place them in a viewport within the layout, and prepare them for plotting. Methods for creating multiple viewports within a drawing will also be explored. Multiple viewports allow objects that must be displayed at different scales within the same drawing to be plotted easily.

COMPARING MODEL AND PAPER SPACE

As you begin to prepare a drawing for plotting it's important that you understand the use of the terms model space, paper space, and layout. In Chapter 2 you were introduced to drawing templates that were created in paper space. Throughout the text you've been drawing in model space. Model space is where you create the drawing; paper space is where you plot. A layout is a paper space tool that contains one or more viewports to aid in plotting. If you open the Architectural template and select the Architectural Title Block tab, your drawing is displayed in paper space. The paper space UCS icon is displayed in the lower left corner, and the architectural title block surrounds your drawing. If you click the PAPER button on the status bar below the Command prompt, the UCS icon is changed to reflect model space and is placed inside the drawing template. Your drawing is now in a viewport or window that is in model space within a paper space border. Open a template or refer to Chapter 2, and you'll notice that the architectural template contains a black line that overlays the border. This box is the viewport (see Figure 24.1). You can edit the drawing in the model space viewport without having to return to the model space tab. Later in this chapter you'll be introduced to methods for dividing model space into tiled viewports to represent different views of the drawing. Switch back to paper space by clicking the MODEL button, and the drawing is displayed in the paper space layout with viewports, ready for plotting.

Note: As a general rule, NEVER alter the location of the drawing in paper space. When the drawing location is altered in paper space, it can affect layouts by other subcontractors who might be working on the project. Alter the drawing in model space and then return to paper space to display the drawing with the title block and borders.

AN OVERVIEW OF PAPER SPACE

To understand the process of viewing the floor plan drawing in Figure 24.2, visualize a sheet of 24" x 36" vellum with a title block and border printed on the sheet. Imagine a hole, the viewport, cut in the vellum that allows you to look through the paper and see the floor plan. This is a simplified version of what is required to display a drawing for plotting. Figure 24.3 shows the theory of displaying a drawing in model space to a layout in a template drawing created in paper space. The floor plan shown in Figure 24.1 is 90'–0" wide in model space. If you were to hold a sheet of D size vellum in front of the plan, the paper would seem minute. To make the floor plan fit inside the viewport on the paper, you're going to have to hold the paper a great distance away from the floor plan until the drawing is small enough to be seen through the hole. AutoCAD will automatically figure the distance. By entering the desired scale factor for plotting, you can reduce the floor plan to fit inside the viewport and maintain a scale typically used in the construction trade. Multiple viewports can be placed in a single layout, and multiple layouts can be created within a single drawing file.

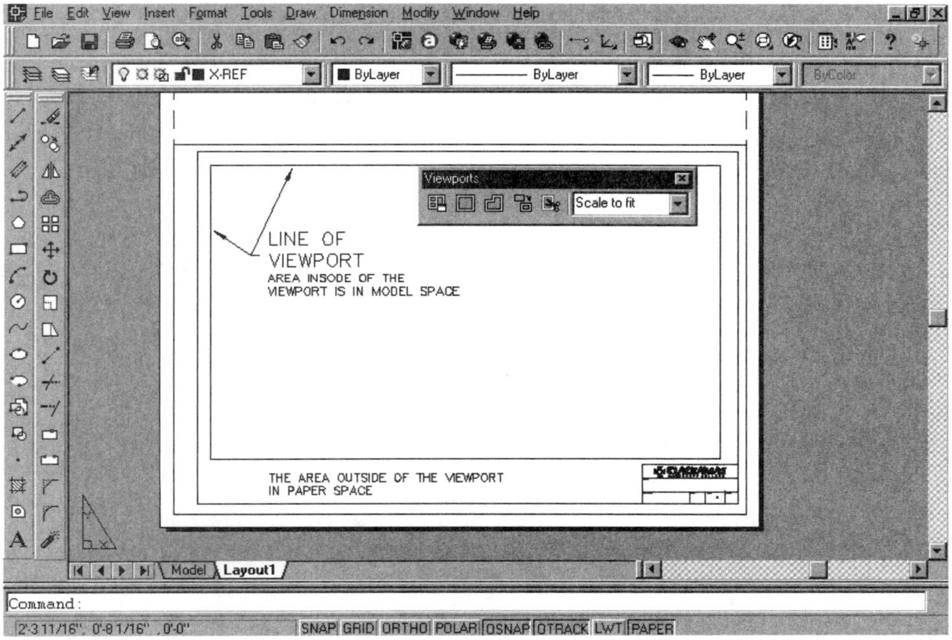

Figure 24.1 *The architectural template contains a black line that overlays the border. This box is the viewport.*

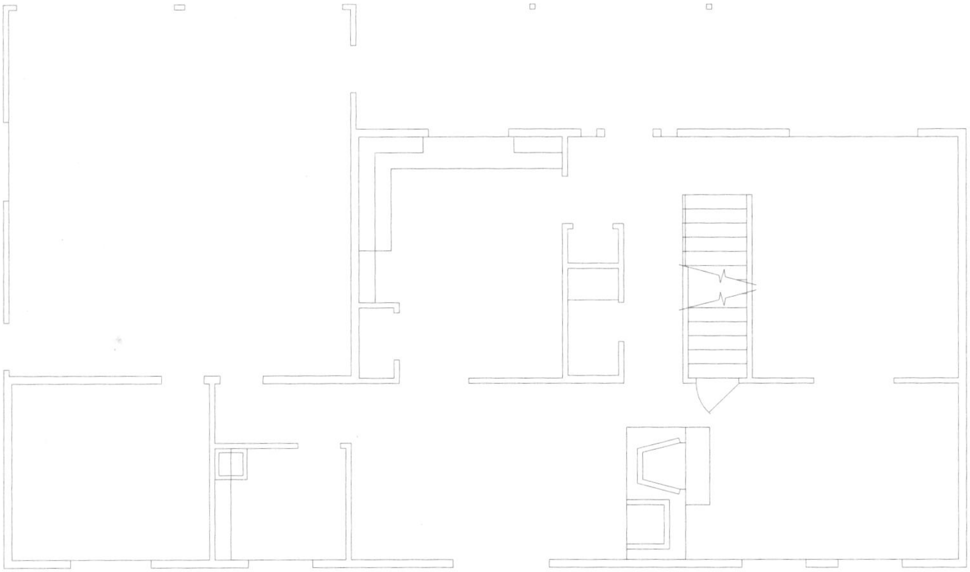

Figure 24.2 *A floor plan that is 90' long can be displayed in paper space using a paper space layout. (Courtesy Kyle Jones.)*

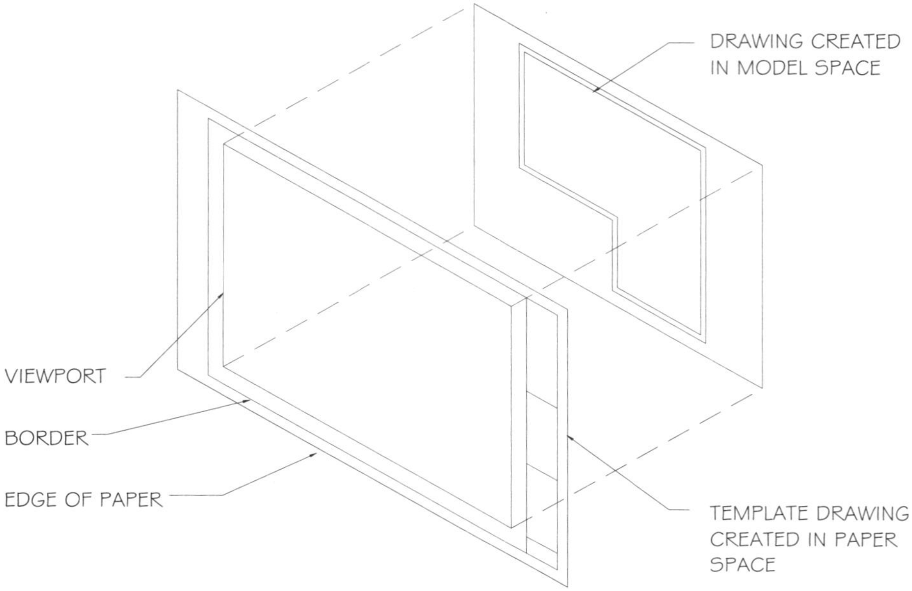

Figure 24.3 *A viewport is provided in each drawing layout created in paper space. Additional layouts can be created to allow for different plotting displays.*

WORKING WITH LAYOUTS

Throughout this text you've been using the Architectural template or a template that you've created to meet your needs. When your drawing is complete, you can select the layout tab to begin preparing the drawing for reproduction. If you're using the Architectural template, the layout tab is titled Architectural Title Block. As the tab is displayed for the first time, the **Page Setup** dialog box, introduced in an earlier chapter, is displayed. (Chapter 25 will deal with page setup in depth.) Layout settings for plotting can be set once and stored with the drawing. For now, the goal is to help you create drawing displays that are ready for plotting. When your drawing is complete, use the following steps to adjust a drawing in the viewport in paper space:

1. Select the Architectural tab.
2. Click the PAPER button on the bottom of the screen to enter model space in the viewport.
3. Use the All or Extents option of **Zoom** to view the drawing.
4. Use **Pan** to center the drawing in the layout.
5. Click the MODEL button on the bottom of the screen to return to paper space.

The procedure to view the drawing in paper space is simple and should be familiar by now. In addition to using layouts for plotting, you can create several layouts for the same drawing so that different parameters can be emphasized in the plotting process. Different scales and paper sizes can be assigned to different layouts, or different plotting devices can be assigned to the layouts.

CREATING NEW LAYOUTS

The existing template and layout work well for displaying a single drawing for plotting. Additional paper space layouts can be created to plot various layers, or in some other way customize a drawing for a specific plotting need. Layouts can be assigned a unique name, and multiple paper space viewports can be assigned to one or more of the layouts. Independent settings can be assigned and stored in each layout for plotter and page setup. Once created, layouts can be copied, deleted, moved, or renamed. Selecting **Create Layout from Wizards** on the **Tools** menu allows a new layout to be created. The **Create Layout Wizard** contains a series of pages that walk you through the process of creating a new layout. Options are similar to the plotting options introduced in Chapter 5. Each setting will be further discussed in the next chapter. Figure 24.4 shows an example of the beginning page of the **Create Layout Wizard**. This page is used to assign a name to the layout being created. Once the desired name is entered, selecting the **Next** button will produce a display for choosing a printer from a list of configured printing devices. Using multiple layouts with multiple plotter configurations will allow you to create check prints with a laser printer, or full size plots with a plotter without having to adjust plotter settings each time you make a plot. Once the plotter to be assigned to the layout is selected,

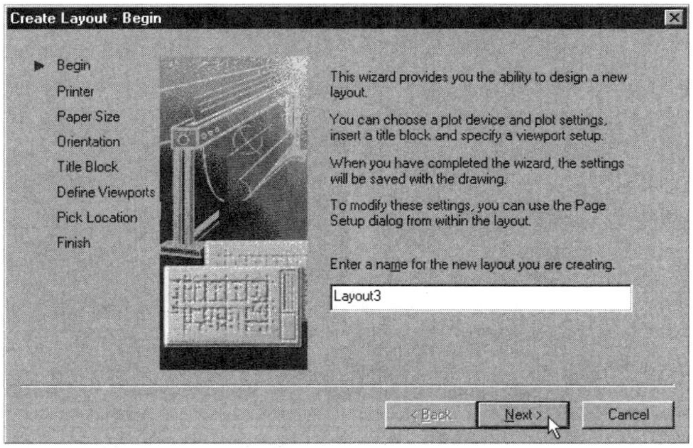

Figure 24.4 *The* **Create Layout Wizard** *contains a series of pages that walk you through the process of creating a new layout. Options are similar to the plotting options introduced in Chapter 5. The beginning page of the* **Create Layout Wizard** *is used to assign a name to the layout being created.*

choosing **Next** will display a **Create Layout - Paper Size** dialog box similar to Figure 24.5. Depending on the configured plotter, this dialog box allows the paper size and drawing units to be selected.

Selecting the **Next** button allows the orientation of the plot to the paper to be configured. Options include **Portrait** or **Landscape**. When finished, selecting the **Next** button allows a predefined template to be assigned to the layout. By default, the name of the template assigned to the previous layout will be highlighted and displayed in the **Preview** area. By using multiple layouts, you can display various title blocks with the drawing. This could be helpful for displaying a floor plan drawn by the architectural team and the framing plan for the same structure drawn by the structural team.

Once a title block is selected, clicking the **Next** button displays a **Create Layout - Define Viewports** dialog box similar to Figure 24.6. This dialog box can be used to define a viewport and select the scale for presenting the information in the viewport. By default, a single viewport is displayed. Creating multiple viewports will be discussed in the next section of this chapter. The default setting for the viewport scale factor is **Scaled to Fit**. In this setting, a 90' structure will be crammed into the viewport. Most construction drawings need to be plotted at a measurable scale. Selecting the **Next** button after selecting the desired scale will display the **Create Layout - Pick Location** dialog box, which prompts you to select the corner locations of the viewport. In the Architectural layout, the viewport edges align with the top, right, and bottom borders. This option allows a different viewport size to be selected for the new layout. The size of the viewport will depend on the size of the drawing to be displayed, the desired scale, and the number of viewports to be created in the layout. As the size is selected, the final page of the wizard will be displayed. This page will display the name of the lay-

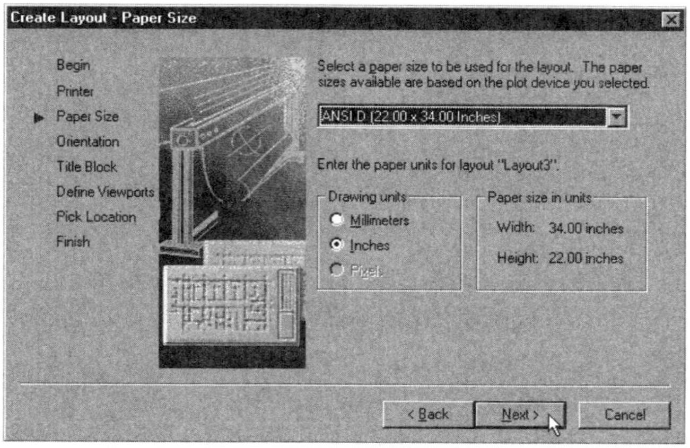

Figure 24.5 *Once the plotter to be assigned to the layout is selected, choosing* **Next** *will display a* **Create Layout - Paper Size** *dialog box that allows the paper size and drawing units to be selected.*

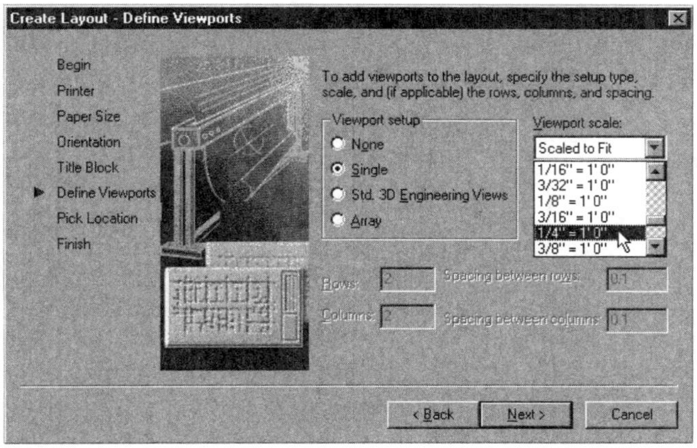

Figure 24.6 *The* **Define Viewports** *page allows the scale of the drawing in the viewport to be set and maintained in paper space.*

out you've created and methods for altering the layout. Selecting the **Finish** button will place the layout in the drawing base as the current layout.

Summary of Steps in Creating a New Layout

In summary, the steps to creating a new layout with a wizard are as follows:

1. Select **Create Layout from Wizard** on the **Tools** menu.
2. Provide a name for the layout being created.
3. Select the plotting device.
4. Select the paper size and the drawing units.
5. Select the orientation.
6. Select the title block to be used.
7. Select the number of viewports and the scale to be used in the viewport.
8. Select the corners of the viewport.
9. Click the **Finish** Button to return to the drawing area.

The new layout will now be displayed as the current layout, similar to Figure 24.7a. Notice that the viewport that was created is not large enough to display the full floor plan. Selecting the viewport border will display the grips for the viewport. Make one of the grips hot, and then stretch the viewport to the desired size, as shown in Figure 24.7b. Pressing ESC will end the **Stretch** command and restore the viewport. Figure 24.7c shows the results of stretching the viewport. Other modifications can be made to the layout by using the **Page Setup** dialog box. Access this dialog box by

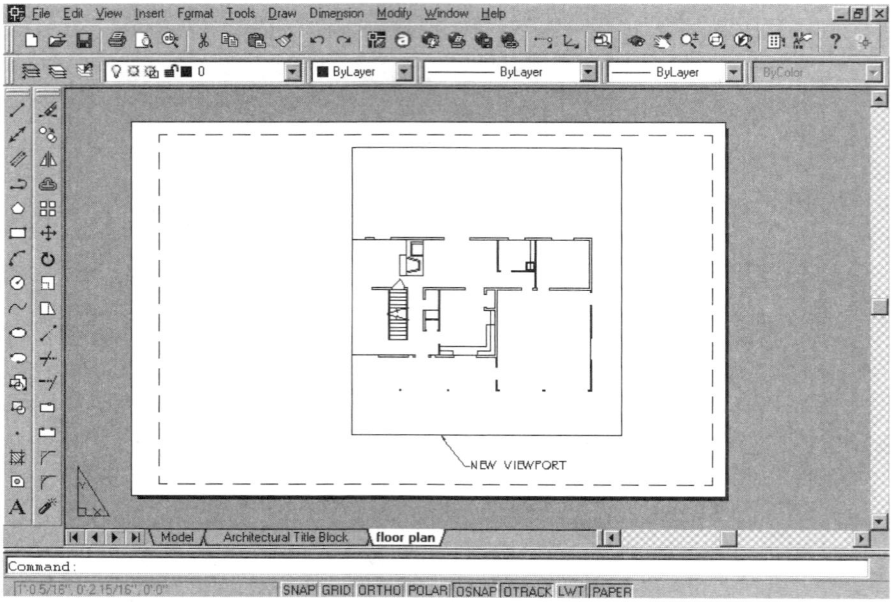

Figure 24.7a *The viewport that was created might not be large enough to display the entire drawing.*

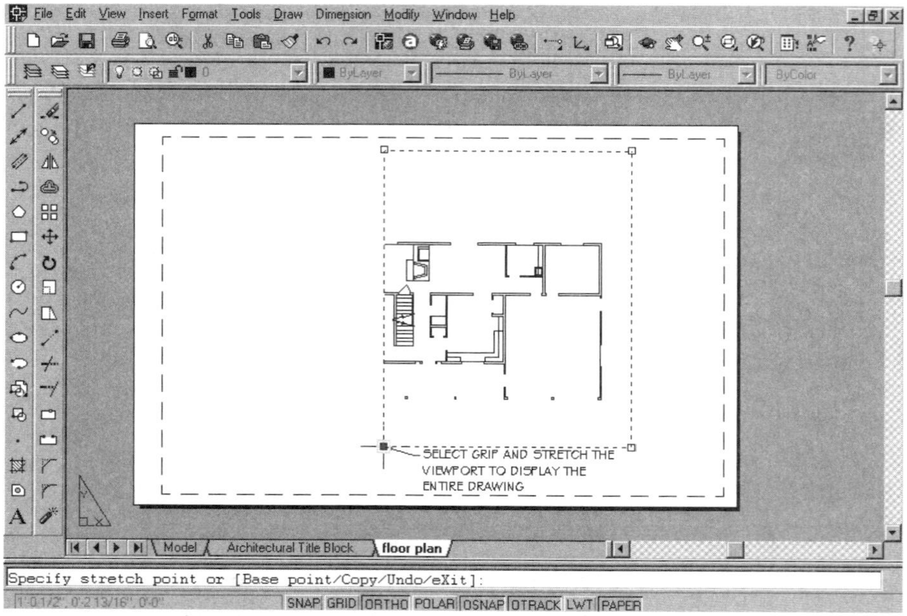

Figure 24.7b *Selecting the viewport border will display the grips for the viewport. Make one of the grips hot, and then stretch the viewport to the desired size.*

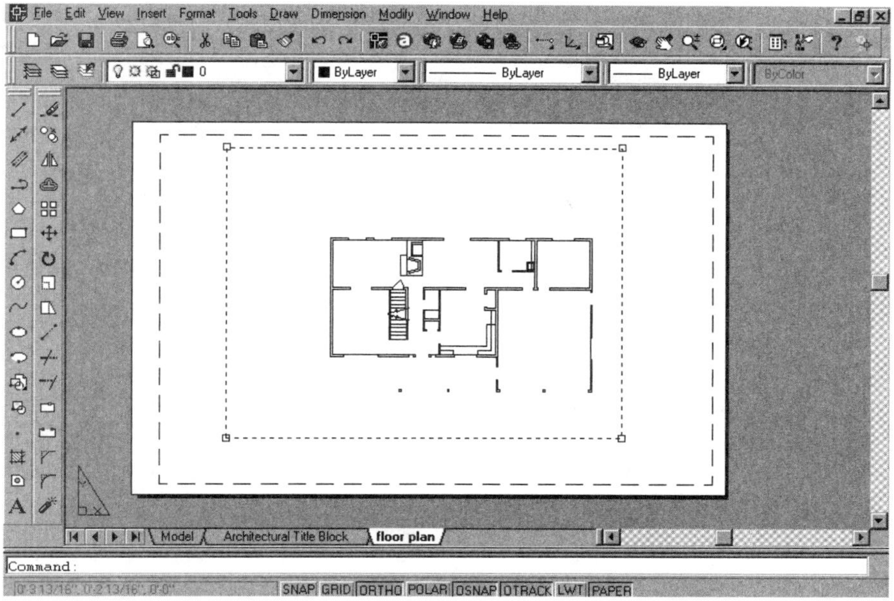

Figure 24.7c *A viewport can be stretched to display the full drawing. Pressing* ESC *will end the* **Stretch** *command and restore the viewport.*

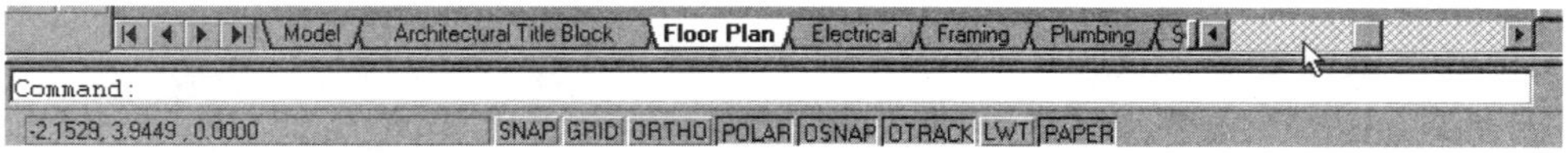

Figure 24.8a *Depending on the size of the slide bar, not all the layout tabs might be displayed.*

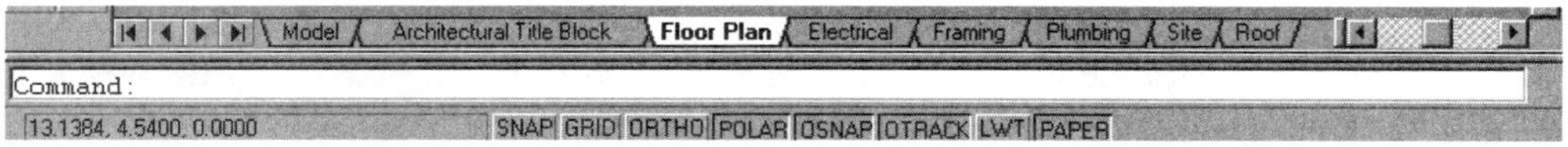

Figure 24.8b *Decrease the size of the slide bar to display more layout titles.*

selecting the **Page Setup** button on the **Layouts** toolbar or by right-clicking the Model tab or a layout tab and choosing **Page Setup**.

Note: As several layouts are created, all of the layout names might not appear. To display more of the layout tabs, select and hold the edge of the slide bar, and then move the cursor to the right. Figures 24.8a and 24.8b show the process. You can also display the other layouts by using the arrow keys directly to the left of the Model tab. This allows you to scroll through the layout tabs and select the layout you need.

ALTERING LAYOUTS

You can create a new layout or alter an existing one by typing **LAYOUT** (ENTER) at the Command prompt. This method will present the following prompt at the command line:

Enter layout option [Copy/Delete/New/Template/
Rename/SAveas/Set/?] <Set>:

These options can be used to create, define, and save a new drawing layout.

Copy

Selecting Copy will create a new layout by making a copy of an existing layout. A prompt for the name of the layout to copy will be displayed at the Command prompt. The default name will be the name of the current layout—in this case, the Architectural Title Block is the default name. If a new layout name is not provided, the new layout assumes the name of the copied layout, with an incremental number added to the name. The command sequence is as follows:

Enter layout option [Copy/Delete/New/Template/
Rename/SAveas/Set/?] <Set>: C (ENTER)

Enter layout to copy <Architectural Title Block>: (ENTER)

Enter layout name for copy <Architectural Title Block (2)>: (ENTER)

Layout "Architectural Title Block" copied to "Architectural Title
Block (2)".

Delete

This option will delete a layout. By default, the current layout will be deleted. (The Model tab cannot be deleted.) One or more layouts can be deleted. If all layouts are removed, an empty layout titled Layout1 will be created and displayed.

New

This option creates a new layout tab without the use of the **Create Layout Wizard**. Selecting the New option will display the following prompts:

Enter layout option [Copy/Delete/New/Template/
Rename/SAveas/Set/?] <Set>: N (ENTER)

Enter new Layout name <Architectural Title Block>: COOL STUFF
(ENTER)

Command:

Template

Selecting the Template option creates a new layout based on an existing template (.DWT) or drawing file (.DWG). The layouts from an existing template or drawing file can be inserted in a new drawing. In addition to the layout, objects contained in the layout can also be made part of the layout being created. The command sequence is as follows:

Enter layout option [Copy/Delete/New/Template/
  Rename/SAveas/Set/?] <Set>: T (ENTER)

Pressing (ENTER) will display the **Select Template from File** dialog box, allowing a template to be selected. Selecting the **File type** edit box allows a drawing file to be selected. Once the template or file is selected, it will be displayed in the preview area. Selecting the **Open** button will close the **Select Template from File** dialog box and provide the **Insert Layout** dialog box that verifies the template to be added to the layout. Selecting **OK** will return the layout display with the template. You can also select a template by choosing the **Layout From Template** button on the **Layouts** toolbar.

Rename

This option can be used to rename a layout. The layout that was last used is shown as the default. Prompts will be displayed to select the layout to be renamed, and to select the new name. Prompts to rename the JUNK layout as COOL STUFF2 are as follows:

Enter layout option [Copy/Delete/New/Template/
  Rename/SAveas/Set/?] <Set>: R (ENTER)

Enter Layout to rename <Architectural Title Block>: JUNK (ENTER)

Enter new layout name: COOL STUFF2 (ENTER)

Layout "JUNK" renamed to "COOL STUFF2".

Command:

Layout names are not case sensitive and can be up to 255 characters long. Only 32 characters are displayed on the layout tab.

Saveas

As you complete the **Create Layout Wizard**, the new layout is stored as part of the current drawing. Any existing drawing can be saved and used as a template drawing. You can save all of the objects, properties and settings as a .DWT file (template) by selecting the Saveas option from the **Layout** command. New drawings can be created from drawing templates or new layouts can be created in the current drawing based on a layout in the template file.

Set

The Set option makes an existing layout the current layout. Clicking the tab of a layout does the same thing.

The ? Option

Selecting this option will list all of the layouts defined in a drawing. This can be helpful when all layout names cannot be displayed below the drawing. Pressing F2 will also display a list of the current layouts.

WORKING WITH VIEWPORTS

AutoCAD allows two types of viewports to be created. Floating viewports are used to create plotting layouts in paper space. Tiled viewports can be created in model space to aid in the display of the drawing. Up to four tiled viewports can be created, but only one viewport can be active at a time. Tiled viewports allow you to divide the screen into multiple areas so that you have multiple views of different parts of the same object. Figure 24.9 shows an example of a sheet of details. Creating multiple viewports would allow you to zoom in on a beam-to-column connection in one viewport, a column-to-footing detail in another viewport, and a display of the whole sheet of details in a third viewport, as shown in Figure 24.10. This multiple use of viewports allows for drawing and editing within the enlarged portions of the drawing without having to zoom in and out to move from one detail to another. It is important to remember that even though you are seeing multiple images on the screen, you're working with only one database. An object that is drawn or edited in one viewport will also affect the other viewports.

DISPLAYING MULTIPLE TILED VIEWPORTS

Multiple tiled viewports can be displayed in several configurations that are controlled by the **Viewports** dialog box. Display the **Viewports** dialog box by selecting the **Display Viewports Dialog** button on the **Viewports** toolbar, by selecting **New Viewports** from **Viewports** on the **View** menu, or by typing **VPORTS** (ENTER) at the Command prompt. Each method will produce the display shown in Figure 24.11. In the current setting, one viewport is displayed. Selecting a listing from the viewport listing will alter the display in the preview to reflect the change. Figure 24.12 shows the display for each option. Use the following steps to display multiple tiled viewports in a new configuration:

1. Display the **Viewports** dialog box.
2. Select the name of the desired viewport configuration from the **Standard viewports** list.
3. Select Display as the current setting for the **Apply to** list.
4. Select 2D as the current setting for the **Setup** list.

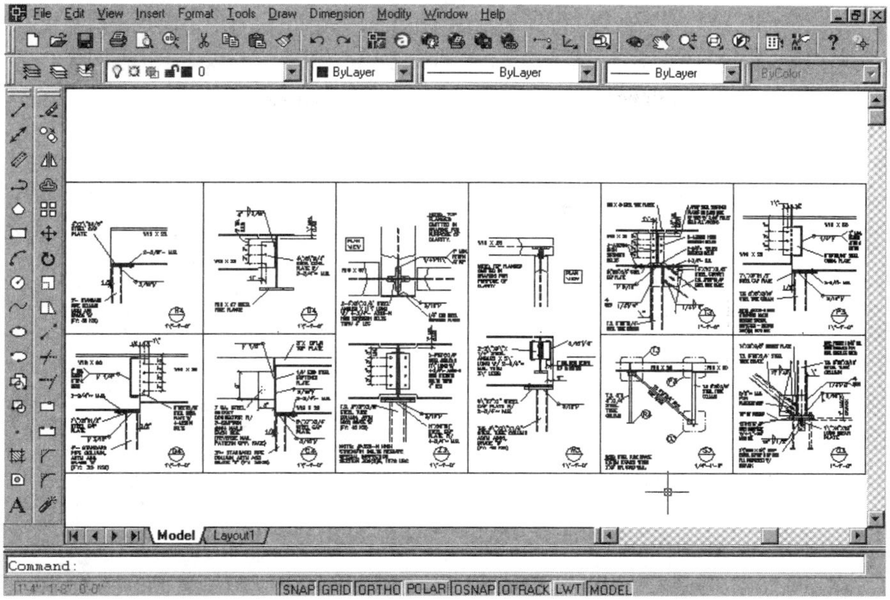

Figure 24.9 *When you work on a sheet of details, it is often hard to coordinate information from one detail to another.*

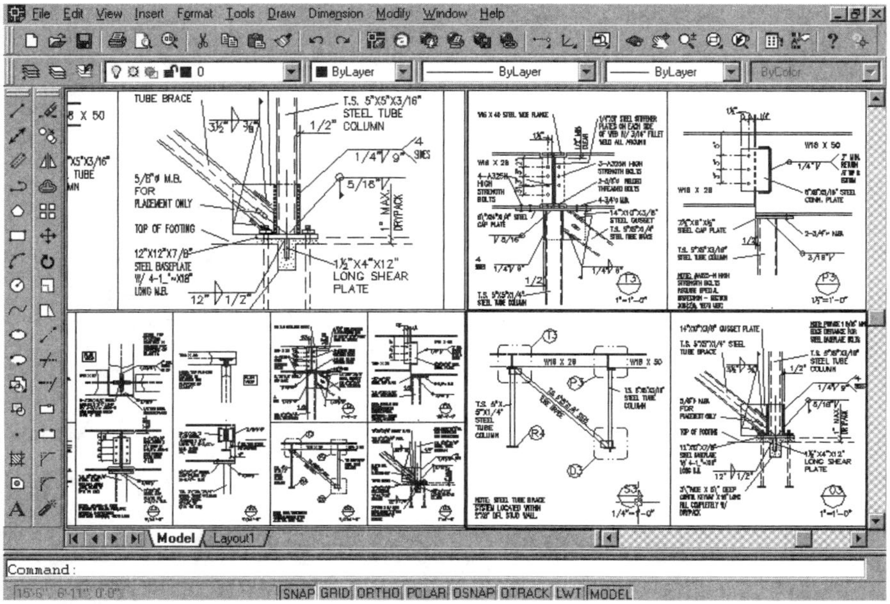

Figure 24.10 *Tiled viewports allow you to divide the screen into multiple images so that you have multiple zoom images of different parts of one drawing file. The details shown in Figure 24.9 can be easily viewed using multiple viewports.*

5. Click the **OK** button to apply the configuration and close the dialog box.

Notice, in Figure 24.10, that a bold black line surrounds the viewport on the right side. This viewport also contains the crosshairs. As you move the cursor around the screen, you'll notice that the crosshairs only work in the drawing on the right side. This view is called the active viewport. You can draw or edit in this viewport just as in any other drawing. As the cursor crosses the center of the drawing into the viewport on the left, the crosshairs change to an arrow. Notice, in comparing Figure 24.13 with Figure 24.14, that information related to the steel column was removed from the lower left corner of the upper left viewport, which also removed the information from the database and the total view. Although the total detail is not legible, you can see that information is missing. The fact that the information is not legible is the point. Viewports make multiple enlarged views of one drawing that are legible for easy drawing or editing, while leaving a view of the overall drawing for perspective.

To make a different viewport active, move the cursor to the desired viewport and click. To convert a drawing like Figure 24.13 to one similar to Figure 24.14, change between active viewports and use the **Zoom** command. Remember, once active, all drawing and editing commands function normally. First make the upper left viewport active. Select the detail in the upper right quadrant for a **Zoom** window and enlarge to fill the viewport. Finally, make the viewport on the lower left active and select the detail in the lower right quadrant for a **Zoom**, using the Window option. With these details enlarged, you can make changes to each without having to zoom from one detail to another.

Working in Multiple Viewports

In addition to easing viewing, viewports can ease working. A command such as **Line** can be started in one viewport and completed in another viewport. This could be especially helpful on large drawings, such as exterior elevations. Exterior elevations are typically created side by side so that heights can be projected from one view to another. If all views are to be displayed, each view will be tiny. Creating a viewport for each elevation allows for easy viewing. Being able to draw lines from one viewport to another will speed production. Use the following steps to draw a line between viewports:

1. Select the viewport that contains the start point, making it the active viewport.
2. Start the **Line** command sequence.
3. Select the viewport that will contain the ending point, making it the active viewport.
4. Select the ending point for the line.

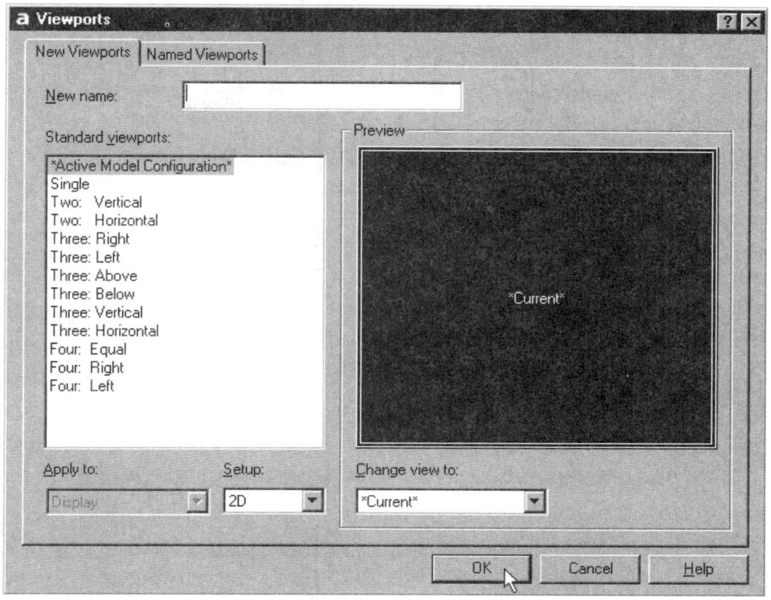

Figure 24.11 *Display the* **Viewports** *dialog box by selecting the* **Display Viewports Dialog** *button on the* **Viewports** *toolbar, by selecting* **New Viewports** *from* **Viewports** *on the* **View** *menu, or by typing* **VPORTS** *(ENTER) at the Command prompt.*

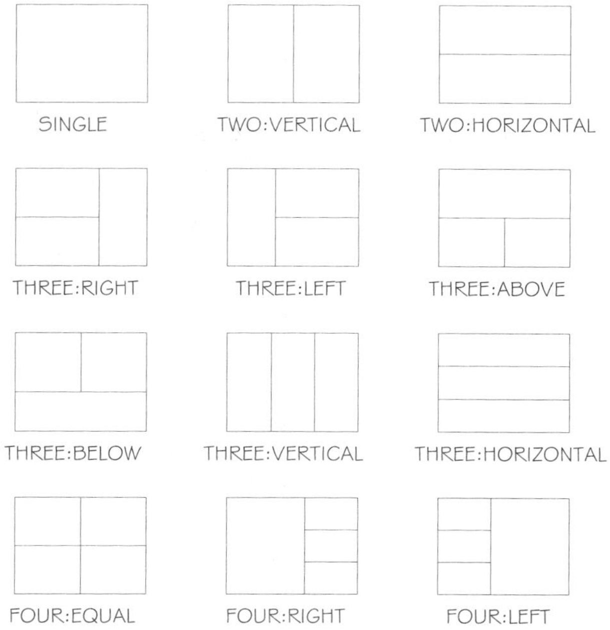

Figure 24.12 *Possible display options for tiled viewports.*

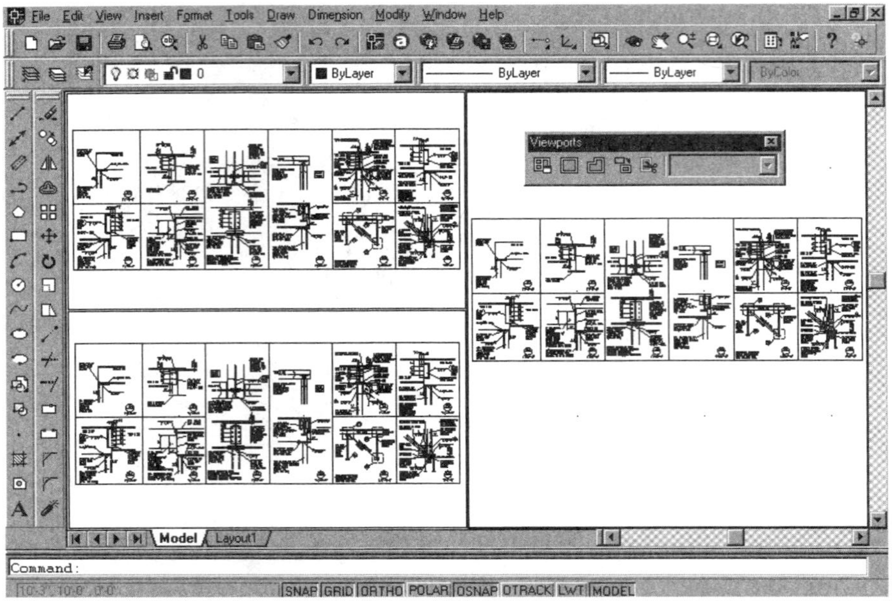

Figure 24.13 *Selecting Three: Right followed by the* **OK** *button will display the drawing shown in Figure 24.10 in three viewports. (Courtesy StructureForm Masters.)*

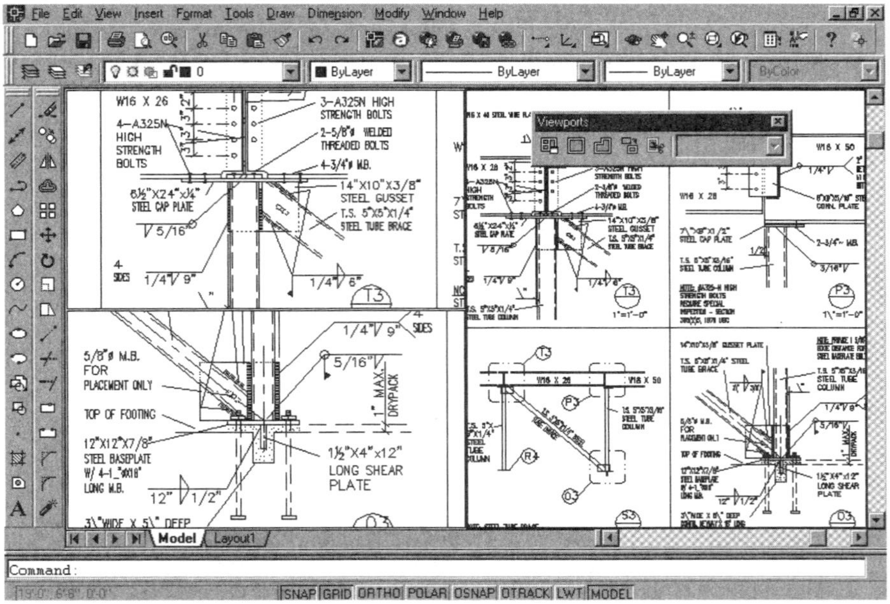

Figure 24.14 *The viewports in Figure 24.13 have each been edited to show this display. With the upper left viewport active, the detail was enlarged using the* **Zoom** *command to display the beam-to-beam connection. The same process was used in the lower left viewport to enlarge the column detail. (Courtesy StructureForm Masters.)*

A line will now be displayed between the two selected points, similar to Figure 24.15.

WORKING WITH FLOATING VIEWPORTS

Each time you display a layout, you're working with a floating viewport. The viewport is referred to as floating because it can be moved and the size can be altered, just like a floating dialog box. The viewport allows you to look through the paper and see the drawing created in model space. The floating viewport is used to display the drawing in the layout for plotting. An unlimited amount of viewports can be created in paper space, but only 64 viewports can be visible at one time. The balance of this chapter will explore floating viewports and how they can be used to prepare a drawing layout for plotting. A floating viewport can be considered as a drawing object that provides a view into model space. Several floating viewports can be created within a layout, and the viewports can be overlapped or separated from each other. A floating viewport can also be an irregular shape. Because floating viewports are considered objects, drawing objects displayed in the viewport cannot be edited. To edit objects within the viewport,

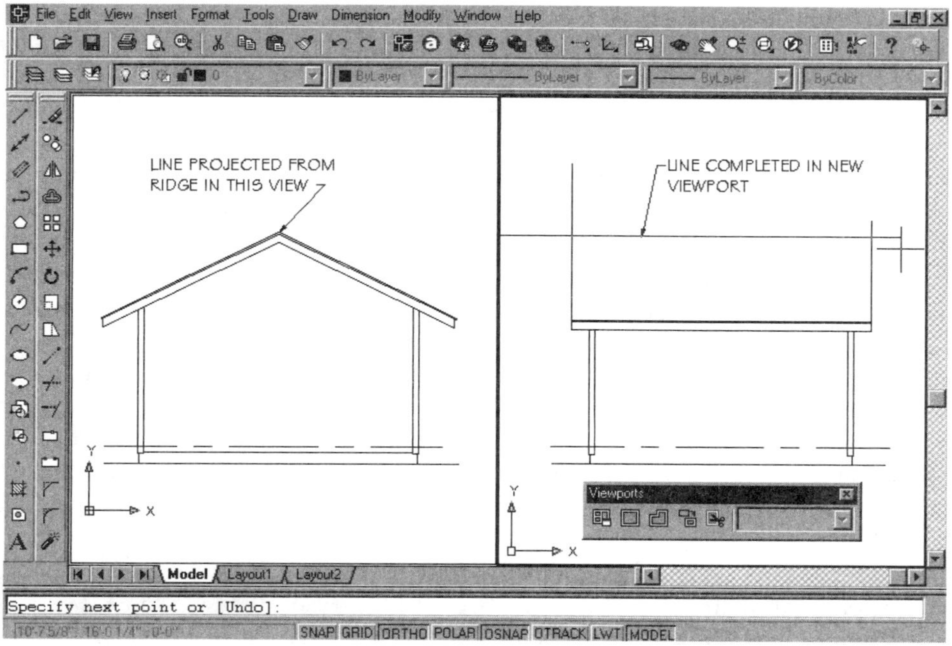

Figure 24.15 *A command can be started in one viewport and completed in another viewport. Select the viewport that contains the start point, making it the active viewport, and start the command sequence. Next select the viewport that will contain the endpoint, making it the active viewport, and select the endpoint for the line.*

model space must be restored. You can toggle between model space and paper space by choosing either the Model or layout tab, depending on which one is inactive. Other methods for toggling between model and paper space include the following:

- Double-clicking in a floating viewport will switch the display to model space.

Note: Remember, while editing in model space in a layout, you risk changing the drawing setup. This should only be used by experienced AutoCAD users or when in the initial setup of the view inside the viewport.

- Select the PAPER or MODEL button on the status bar. Remember, when you're in a layout, the page is in paper space, but the viewport can be toggled to model space or paper space. When PAPER button is displayed, the viewport is a paper space display of the drawing in a paper layout. Selecting PAPER will change the button to MODEL and display the drawing in a model space viewport within a paper space layout.
- Type **MS** (ENTER) for **Mspace** or **PS** (ENTER) for **Pspace** at the Command prompt.

Most professionals work in model space and arrange the final drawing for output in paper space. While you work in a layout, with the viewport in paper space, any material added to the drawing will be added to the layout, but not shown in the model display. This will prove useful as the final plotting layout is constructed.

ALTERING VIEWPORT PROPERTIES

Because viewports are objects created by AutoCAD, they have properties such as color, layer, linetype, lineweight, and plot style. You can alter each of these properties using the **Properties** dialog box. As you create a viewport, a new layer can be created and displayed for the viewport. This layer should be frozen before the drawing is saved so that it is not plotted. Linetype and lineweight are ignored in rectangular viewports but are displayed in nonrectangular viewports.

VIEWPORT SCALING

Most users of AutoCAD plot their drawings displayed in a layout at a scale of 1:1. To do this, the drawing must be inserted or attached to the viewport at a predetermined scale factor. AutoCAD has created an aid to do this. Earlier it was pointed out that the scale can be set as a drawing is placed in a viewport. The scale can be locked using the **Properties** dialog box. Locking the viewport scaling will prevent the scale from being altered if the drawing is zoomed once it is displayed in the viewport. The scale can be locked by the following procedure:

1. Display the layout containing the viewport to be locked.
2. Select the edge of the viewport so that it is highlighted.

3. Display the **Properties** dialog box by selecting the **Properties** button on the **Standard** toolbar.
4. Select **Display locked** from the **Properties** dialog box.
5. Select Yes from the **Display locked** edit list.
6. Close the **Properties** dialog box and return to the drawing area.

The scale of the selected viewport is now locked. If you attempt to alter the scale while in model space, AutoCAD displays "warning: Viewport is view-locked." If you change the zoom factor in the viewport, only paper space objects are affected.

Note: It's been stressed not to edit objects in viewports while in paper space. Once you've placed a lock on the viewport, it is safe to edit.

CREATING MULTIPLE FLOATING VIEWPORTS

The **Viewports** toolbar shown in Figure 24.16 can be used to create floating viewports. Buttons include **Display Viewports Dialog**, **Single Viewport**, **Polygonal Viewport**, **Convert Objects to a Viewport**, **Clip Existing Viewport**, and **Viewport Scale**. Up to this point, you've used layouts with viewports that take up the entire drawing area. Floating viewports can be created in a layout that fills the entire drawing area or only takes up a small portion of the drawing area. Floating viewports can be created using a process that is similar to the process for creating tiled viewports. Floating viewports are created using the **Viewports** dialog box shown in Figure 24.17. Compare the **Viewports** dialog box in Figure 24.17 with the **Viewports** dialog box in Figure 24.11. Although each dialog box is very similar, they create different types of viewports, depending on how they are accessed. The dialog box shown in Figure 24.11 was accessed while in model space. Use the following steps to create a floating viewport.

1. Display a layout by selecting an existing layout tab or by creating a new layout.
2. Display the **Viewports** dialog box by selecting the **Display Viewports Dialog** button on the **Viewports** toolbar, by selecting **New Viewports** from **Viewports** on the **View** menu, or by typing **VPORTS** (ENTER) at the Command prompt.

Figure 24.16 *The* **Viewports** *toolbar can be used to control floating viewports. Buttons include* **Display Viewports Dialog**, **Single Viewport**, **Polygonal Viewport**, **Convert Objects to a Viewport**, **Clip Existing Viewport**, *and* **Viewport Scale**.

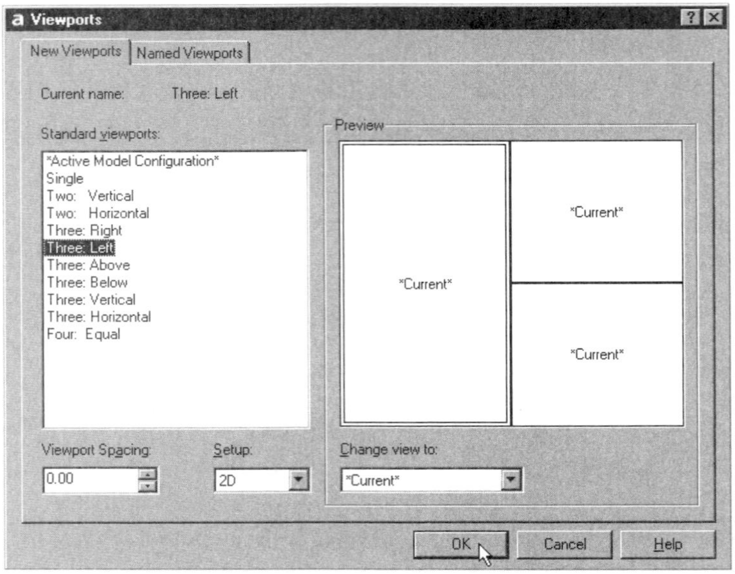

Figure 24.17 *The* **Viewports** *dialog box. Although this dialog box is very similar to the dialog box shown in Figure 24.11, they create different types of viewports, depending on how they are accessed.*

3. Select the desired viewport configuration from the **Standard viewports** list.
4. Enter the desired spacing to be maintained between viewports in the **Viewports Spacing** edit box.
5. Select 2D in the **Setup** edit box.
6. Click the **OK** button to apply the configuration and close the dialog box.

A process similar to placing tiled viewports in a layout can be used to place multiple floating viewports in an existing layout.

CREATING A SINGLE FLOATING VIEWPORT

Floating viewports provide an excellent means of assembling multi-scaled drawings for plotting. New floating viewports can be created without using the standard viewport configurations. Figure 24.18 shows a sheet of details assembled for plotting using floating viewports. Construction projects typically comprise drawings that are drawn at a variety of scales. These details range in scale from 3/8"=1'–0" to 1 1/2"=1'–0". This sheet of details cannot be easily drawn or plotted without the use of multiple viewports and inserting or referencing each detail to a base drawing. As with drawings where a single object is attached, the use of scale factors to control text and dimension variables will be critical to the success of the final outcome of the drawing. The process for creating multiple viewports will be similar to the creation of one viewport.

The following example will demonstrate the steps used to prepare the section and three details shown in Figure 24.19 for plotting. The section will be displayed at a scale of 1/4"=1'–0", two of the details will be displayed at a scale of 3/4"=1'–0", and the stair section will be displayed at a scale of 1/2"=1'–0". For most users, drawings such as a floor plan or section will be created in a drawing template using model space. The following guidelines will walk you through the process of creating multiple viewports in a template and inserting or attaching multiple drawings. For this discussion, the **Insert** command will be used. The drawings could also be attached using **Xref** command, depending on the drawings needs.

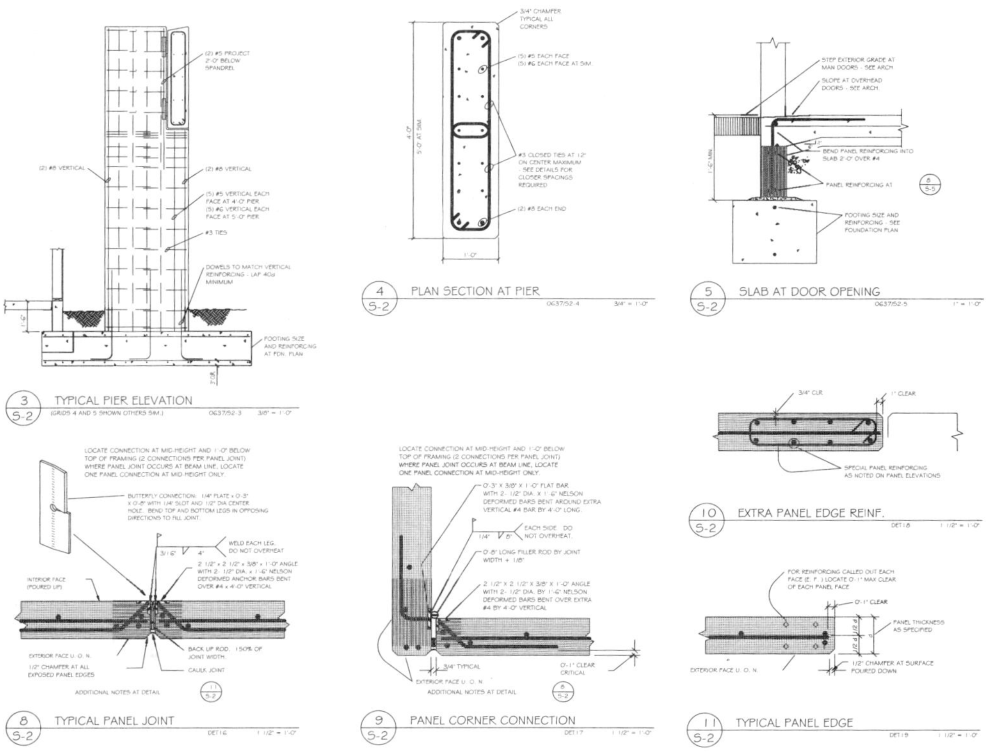

Figure 24.18 *Details drawn at varied scales can be assembled on one sheet for plotting using the* **Vports** *command. (Courtesy Van Domelen/Looijenga/McGarrigle/Knauf Consulting Engineers.)*

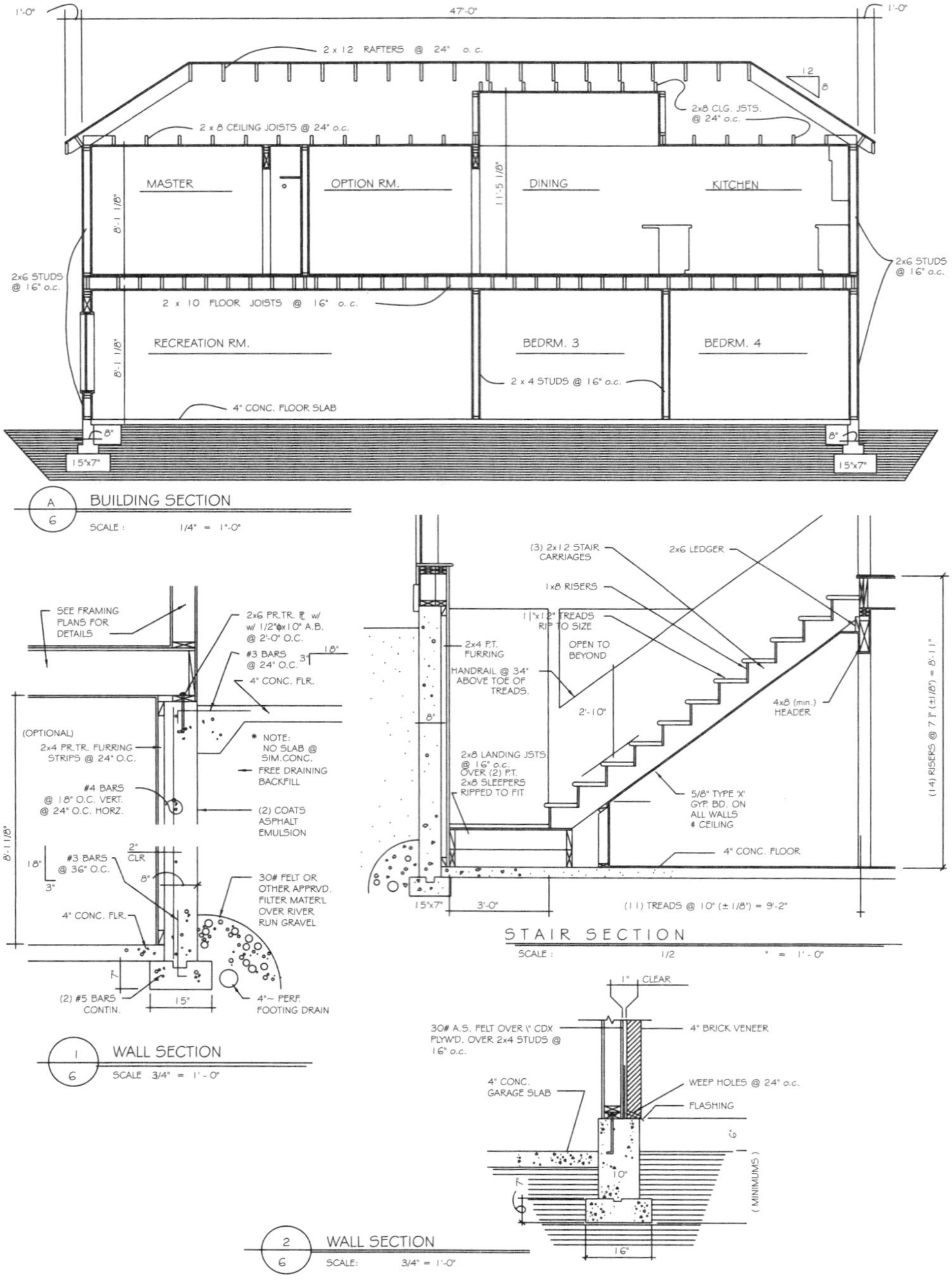

Figure 24.19 *Drawings to be attached to the base sheet. (Courtesy Piercy & Barclay Designers, Inc., A.I.B.D.)*

Displaying Model Space Objects in the Viewport

The easiest method for displaying multiple drawings at multiple scales is by inserting each of the drawings to be plotted in the template in model space. With the details arranged in one drawing similar to Figure 24.19, switch to the Architectural Title Block layout. Use the following steps to display the drawings in multiple viewports:

1. In the Architectural Title Block layout, click the PAPER button in the status bar to toggle the viewport to model space. Use the **Zoom All** option to enlarge the area of model space to be displayed in the existing viewport. Each of the four drawings will be displayed, but each will be at an unknown scale. Your drawing will resemble Figure 24.20.
2. Select what will be the largest drawing to work with first. In our example, the viewport for displaying the section will be adjusted first.
3. Activate the **Viewports** toolbar.
4. Set the scale of the viewport to the appropriate value for displaying the section using the **Viewport Scale** menu on the **Viewports** toolbar. For this example a scale of 1/4"=1'–0" was used.
5. Click the MODEL button on the status bar to return to paper space in the Architectural Title Block layout.

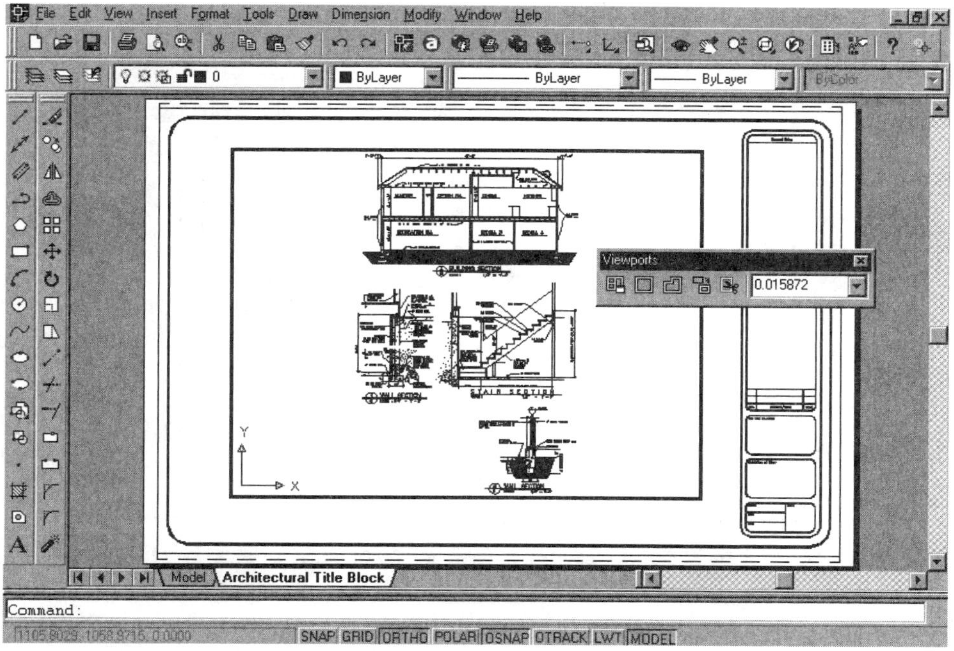

Figure 24.20 *Toggle to model space and use the* **Zoom All** *option to view all four drawings in your viewport, but each will be at an unknown scale.*

Adjusting the Existing Viewport

Once the desired drawing is inserted or attached to the viewport, the size of the viewport can be altered if the entire drawing can't be seen. To alter the viewport size, move the cursor to touch the viewport and activate the viewport grips. Select one of the grips to make it hot and then use the hot grip to drag the window to the desired size. The section has now been inserted in the template and is ready to be plotted at a scale of 1/4"=1'–0". The display would resemble Figure 24.21.

Note: All of the drawing commands can be used inside the viewport, but try to limit editing to changes that will affect only this plot. Changes to the drawing that will be needed on future drawings should be made to the original drawing. Remember, if the original is attached to this drawing by **Xref**, the changes made to the original drawing will be reflected in all uses of the referenced drawing. Enter model space prior to making changes to the drawing in the current viewport. This will allow you to add detail numbers and drawing titles that change with each use of the drawing.

Creating Additional Viewports

Use the following steps to prepare for additional viewports:

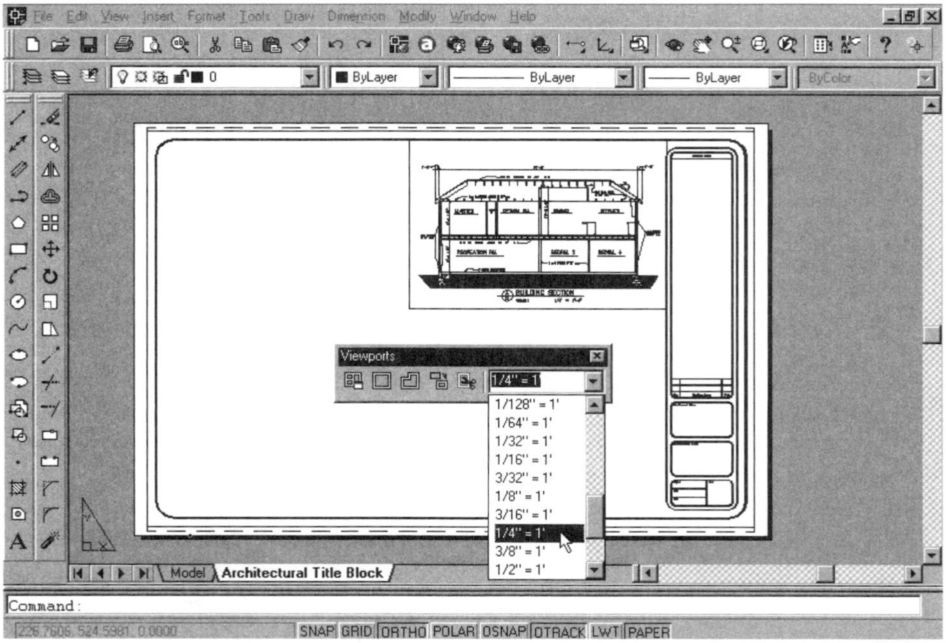

Figure 24.21 *Once the desired drawing is inserted or attached to the viewport, set the scale of the viewport to the appropriate value for displaying the drawing using the* **Viewport Scale** *menu on the* **Viewports** *toolbar. Next, alter the size of the viewport if the entire drawing can't be seen. The section is now ready to be plotted at a scale of 1/4"=1'–0".*

1. Set the current layer to Viewport. This layer is part of each template.
2. Select **Single Viewport** on the **Viewports** toolbar.
3. Select the corners for the new viewport. The coordinates for the viewport can be entered or specified by indicating the corners of the viewport—just as if it were a window used for determining a selection set.

The results of this command can be seen in Figure 24.22. Although you should try to size the viewport accurately, you can stretch it to enlarge or reduce its size once the scale has been set.

Altering the Second Viewport

In its current state, if a new viewport is created, the display from the existing viewport will be shown in the new viewport as well. Use the following steps to alter the display.

1. Click the PAPER button on the status bar to toggle the viewport to model space.
2. Make the new viewport the current viewport by placing the cursor in the viewport and clicking.

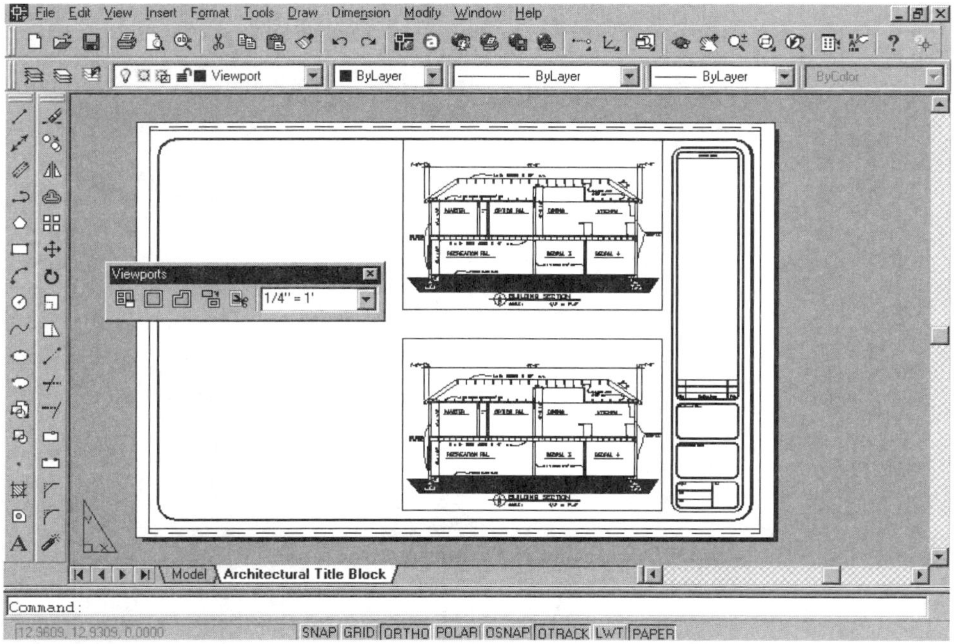

Figure 24.22 *If a new viewport is created, the display from the existing viewport will be shown in the new viewport as well.*

3. Use the **Zoom All** option to display all of the model space contents in the second viewport.
4. Set the viewport scale to the appropriate value for displaying the section. For this example, a scale of 3/4"=1'–0" will be used to display the wall section.
5. Use the **Pan** command to center the wall section in the viewport.
6. Click the MODEL button in the status bar to toggle the drawings to paper space.
7. Select the edge of the viewport to display the viewport grips.
8. Make a grip hot and shrink the viewport so that only the wall section is shown.
9. Use the **Move** command to move the viewport to the desired position in the template.
10. Save the drawing for plotting.

The drawing should now resemble Figure 24.23

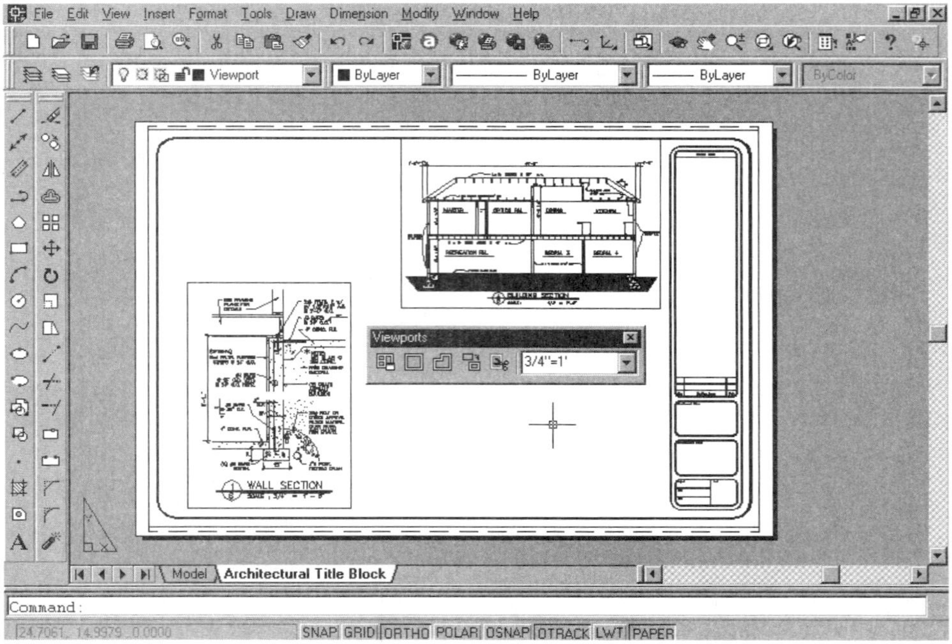

Figure 24.23 *Adjust the viewport scale to the desired value, and use the* **Pan** *command to move the drawings so that the desired information is centered in the viewport. Once it is centered, adjust the size of the viewport so only the desired material is displayed.*

Repeating the Process for Additional Viewports

Once the viewport for the wall section been adjusted, additional viewports can be created. The process to create the viewports to display the BRICK1ST and STAIR 8FTWALL drawings is the same process as used to create the viewport for the WALL SECTION drawing. If you have planned the sheet contents well, the viewports for BRICK1ST and STAIR can also be created. Placing all of the viewports at once will eliminate having to keep switching from paper space to model space. Once all of the desired details have been added to the layout, the viewport outlines can be frozen. Start the **Layer** command, select VIEWPORT as the layer to be edited, and then select the Freeze option. Freezing the viewports will eliminate the line of the viewport being produced as the drawing is reproduced. Any viewports that are still visible were created on a layer other than VIEWPORT. Change any existing viewports to the correct layer. The finished drawing, with the viewports frozen, will resemble Figure 24.24

CONTROLLING THE DISPLAY IN FLOATING VIEWPORTS

As you placed drawings in viewports, the **Pan** and **Zoom** commands were used to alter what was displayed in each viewport. AutoCAD allows displayed objects to be altered by screening, turning viewports ON or OFF, and hiding lines. Hiding lines is typically associated with 3D drawings and will not be covered in this chapter. Consult the Help menu for additional information.

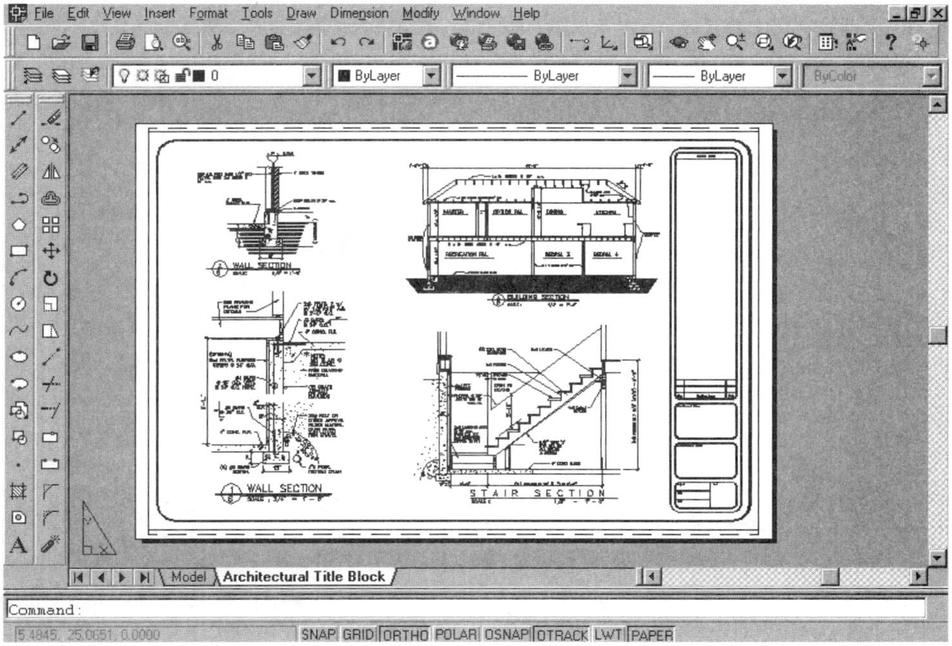

Figure 24.24 *The finished drawing, with the Viewports layer frozen, is ready for plotting.*

Screening Objects in Viewports

Screened objects use less ink when plotted, so they appear dimmer on screen and when plotted. Screening could be used to distinguish between objects, without plotting in color. On remodeling projects, existing objects can be screened to easily distinguish between new and existing work. To screen an object, you need to assign it a plot style and then assign the screening value to that plot style. (Plot styles will be discussed in the next chapter.) Values ranging from 0 to 100 can be assigned, with 100 being the default value. A value of 100 assigns no screening and will provide normal plotting values. A value of 0 would assign no ink to the object and have the same effect as freezing the viewport contents. Figure 24.25 shows an existing footing and new wood floor framing.

Displaying a Floating Viewport

By default, a viewport is ON as it is created. The performance of your machine can be affected if a large number of complex viewports are created. Turning viewports OFF can save regeneration time for large drawings. Toggle viewports ON or OFF using the **Properties** dialog box and the following steps:

1. Select the viewport to alter.

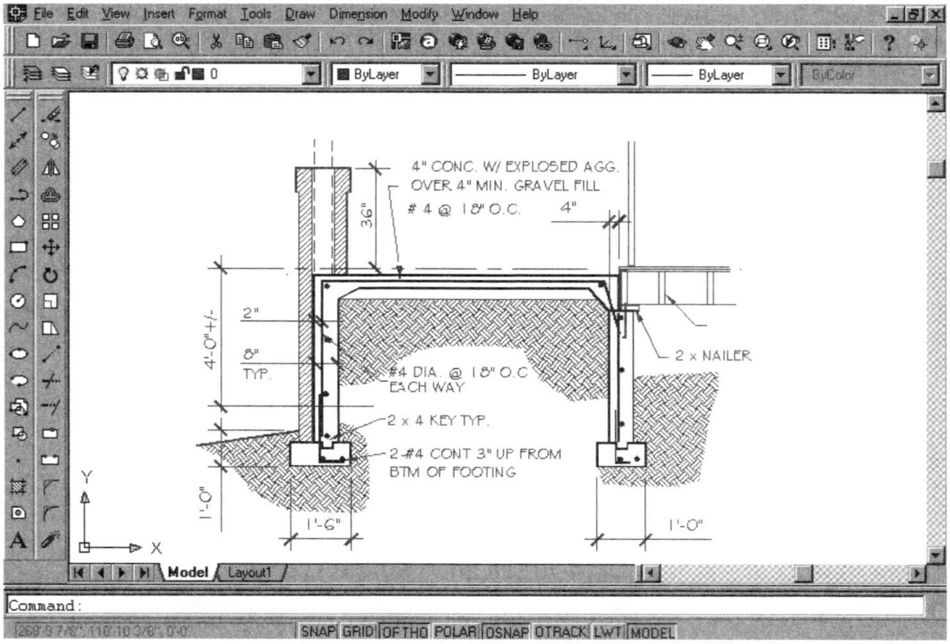

Figure 24.25 *Screened objects use less ink when plotted so they appear dimmer when plotted. To screen an object, assign it a plot style, and then assign the screening value to that plot style. The concrete in this detail was assigned a value of 50.*

2. Open the **Properties** dialog box.
3. Select **On** from the **Misc** listing and then select No from the edit list.
4. Close the **Properties** dialog box.

As the viewport is toggled to No, the contents of the viewport will be removed from the drawing display. The line representing the viewport can be removed from the display by using the **Layer Properties Manager** dialog box to freeze the layer. Figure 24.26 shows the drawing from Figure 24.24 with two of the viewports in the No setting.

Controlling Layer Display with the Layer Properties Manager

The contents of specific layers in each viewport can be controlled using the **Layer Properties Manager**. Using the layer options works well when an object will be displayed in two or more views. Specific aspects of the drawings, such as layers containing text, can be frozen in one viewport and displayed in another. Options for freezing layers in floating viewports include **Freeze in All Viewports**, **Freeze in Active Viewport**, and **Freeze in New Viewport**. These options are inactive as the **Layer Properties Manager** dialog box is displayed, but they are activated as a layer is highlighted. The **Layer Properties Manager** can also be used to turn plotting ON or OFF for visible

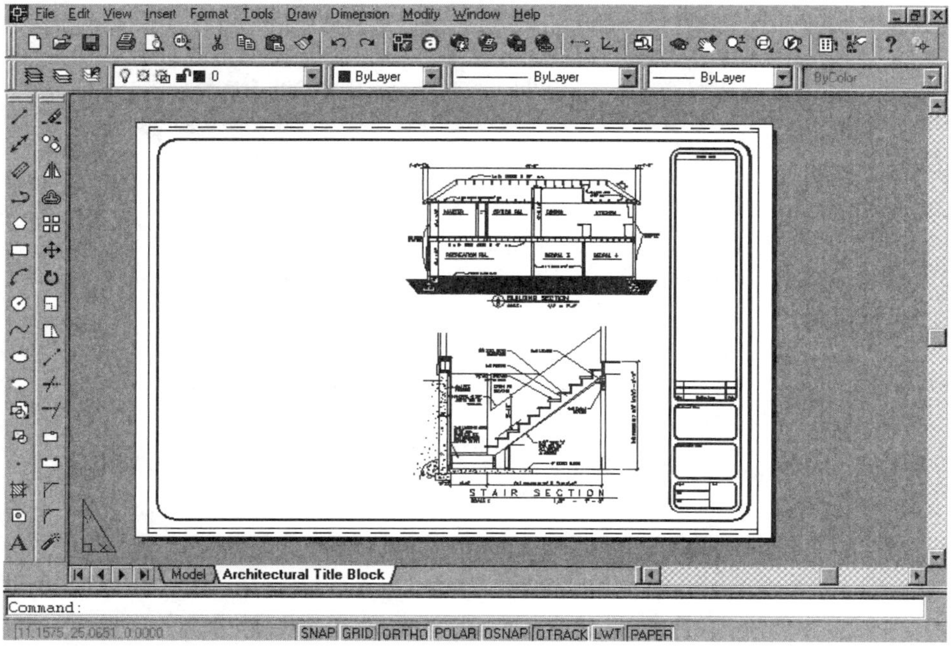

Figure 24.26 *Two of the viewports from Figure 24.24 have been toggled to the NO setting. As the viewport is toggled to NO, the contents of the viewport will be removed from the drawing display.*

layers. A layer created for interoffice notes can be displayed on the screen, but frozen for plotting. Layers are frozen for plotting by selecting the **Plot** icon of the desired layer.

Controlling Scaling in Viewports

Altering the size of a floating viewport does not change the scale of the drawing within the viewport. When the scale was adjusted for the viewport, you set a consistent scale for accurate display and plotting. Earlier in the text, selecting a scale for linetypes was discussed. If a hidden line is to be 1/8" long when plotted at 1/4"=1'–0", a scale factor must be applied to the line to create the desired result. The scale for linetypes can be based on the drawing units of the space where the object was created, or based on paper space units. The **Psltscale** system variable can be used to maintain a uniform linetype scaling for objects displayed at different zoom factors. The **Psltscale** variable can be set by the following steps:

1. Select **Linetype** from the **Format** menu.
2. In the **Linetype Manager** dialog box, select **Show Details**.
3. Enter the desired global scale factor to apply to the linetypes in the **Global Scale Factor** edit box.
4. Click **OK** to return to the drawing area.

CREATING VIEWPORTS WITH IRREGULAR SHAPES

AutoCAD allows viewports to be created in an irregular shape that best suits the object to be displayed. Create a nonrectangular viewport by selecting the **Polygonal Viewport** button, by clipping an existing viewport, or by converting an existing object to a viewport.

CREATING POLYGONAL VIEWPORTS

A polygonal viewport is created by selecting the **Polygonal Viewport** button on the **Viewports** toolbar, by selecting **Polygonal Viewport** from **Viewports** on the **View** menu, or by typing **-VPORTS** (ENTER) and then **P** (ENTER) at the Command prompt. The command is completed by selecting a series of points using a command sequence similar to the sequence for creating a polyline. The command sequence is as follows:

Command: **-VPORTS** (ENTER) *(Click the Polygonal Viewport button.)*

_-vports

Specify corner of viewport or

[ON/OFF/Fit/Hideplot/Lock/Object/Polygonal/Restore/2/3/4/ <Fit>:
_P

Specify start point: *(Specify viewport starting point.)*

Specify next point or [Arc/Length/Undo]: *(Select points to define polygon.)*

Specify next point or [Arc/Close/Length/Undo]: *(Select points until the polygon is defined. As the polygon is closed, the command will end with the following prompt.)* Regenerating model

Command:

Figure 24.27 shows an example of an irregularly shaped viewport.

USING VPCLIP TO ADJUST THE VIEWPORT

The **Vpclip** command can be used to clip or redefine the shape of an existing viewport. Access the command by selecting the **Clip Existing Viewport** button on the **Viewports** toolbar or by typing **VPCLIP** (ENTER) at the Command prompt. Selecting the viewport to be accessed, right-clicking to display the shortcut menu, and then selecting **Viewport Clip** will also access the command. The command sequence is as follows:

Command: **–VPCLIP** (ENTER) *(Or click the Clip Existing Viewport button.)*

Select viewport to clip: *(Select desired viewport.)*

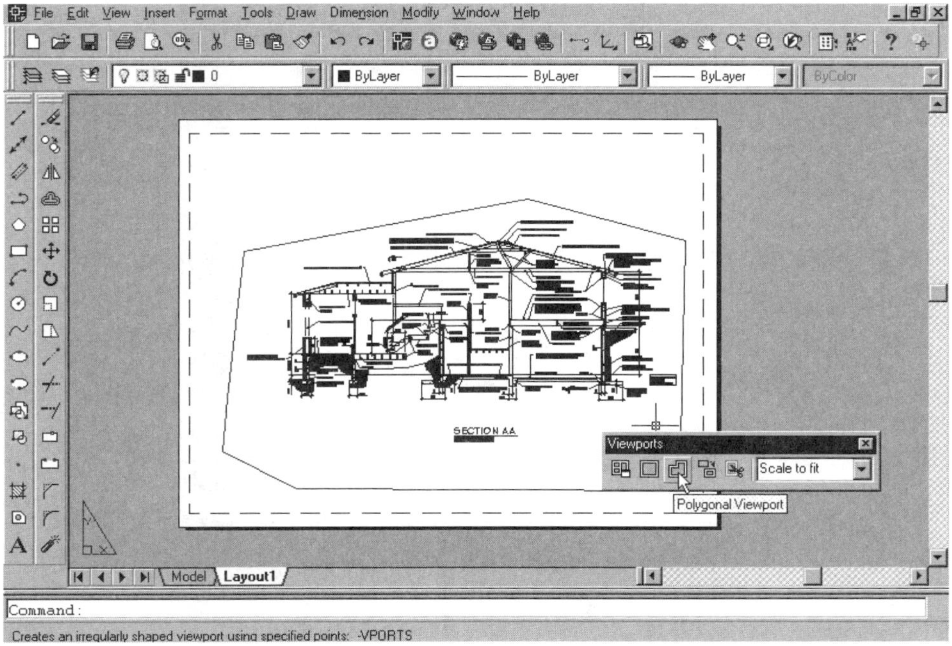

Figure 24.27 *A polygonal viewport can be created by selecting the* **Polygonal Viewport** *button on the* **Viewports** *toolbar.*

Select Clipping object or [Polygonal/Delete] <Polygonal>: (ENTER)

Specify start point: *(Specify viewport starting point.)*

Specify next point or [Arc/Length/Undo]: *(Select a minimum of three points to define polygon.)*

Specify next point or [Arc/Close/Length/Undo]: *(Select points until the polygon is defined.)*

Command:

Objects that lie within the polygon will be displayed. Objects outside the polygon will be removed from the viewport and will not be displayed. The viewport in Figure 24.28 has been edited using **Vpclip**. Irregular viewports can be edited by selecting the edge of the viewport. With the viewport grips displayed, the viewport can be edited in the same way as any other object is altered with grips. (See Figure 24.29.)

CONVERTING EXISTING OBJECTS TO A VIEWPORT

AutoCAD allows existing shapes to be converted to viewports. The object can be a circle, ellipse, region, spline, or closed polyline comprised of lines and arc segments. Access the command by selecting the **Convert Object to a**

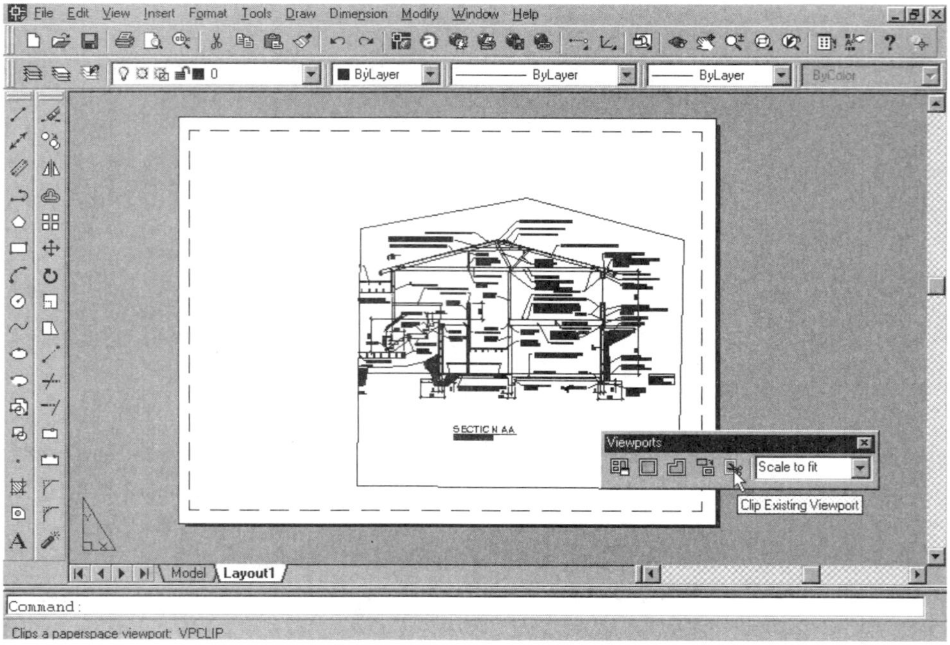

Figure 24.28 *The* **Vpclip** *command can be used to clip or redefine the shape of an existing viewport. You can also access the command by selecting the* **Clip Existing Viewport** *button on the* **Viewports** *toolbar.*

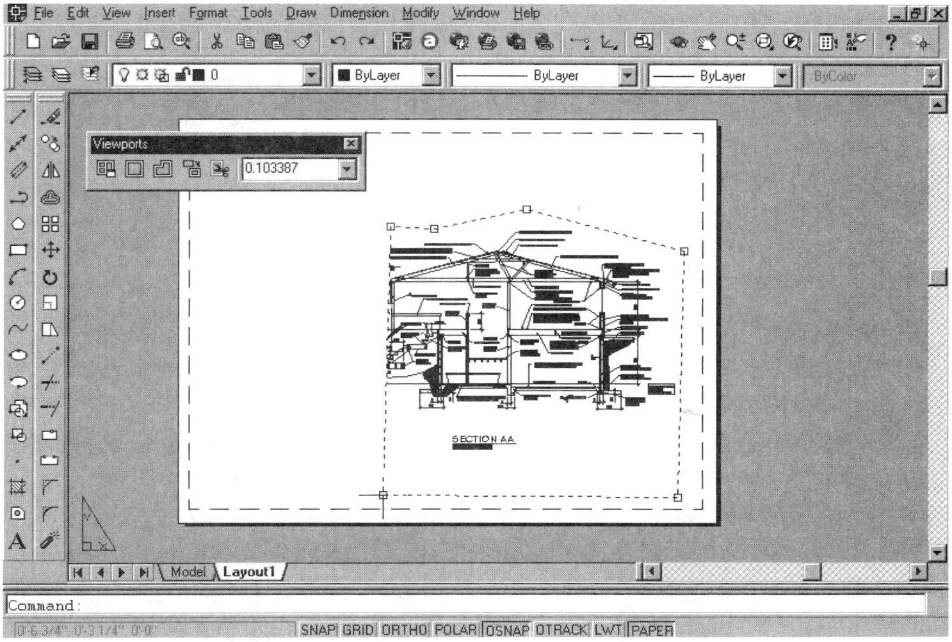

Figure 24.29 *A viewport can be edited by selecting a viewport grip.*

Viewport button on the **Viewports** toolbar. The command is completed using the following command sequence:

> Command: *(Click the Convert Objects to a Viewport button on the Viewports toolbar.)*
>
> Specify corner of viewport or [ON/OFF/ Fit/Hideplot/Lock/ Object/Polygonal/Restore/2/3/4] <Fit>:
>
> Select object to clip viewport: *(Select desired object)* Regenerating model

Options include the following:

> **ON and OFF**—These serve as toggle switches to display or hide the contents of the individual viewports. Selecting either option will provide a prompt to select objects so the desired viewport can be controlled. This option can be very useful when you plot stock sheets of details, if one or two attached details do not apply to the current use. Setting these viewports to OFF will eliminate the viewport from the plot. This option is also very useful for controlling **Regen** time. Inactive viewports can be set to OFF while the current viewport is edited. If all viewports are set to OFF, you will not be allowed to work in model space of a layout until a new viewport is created.

Fit—This option will create a viewport to fit the current screen display. The size of viewport will be determined by the dimensions of the paper space view. In the example used throughout this chapter, the objects were created first in base drawings and then drawings were inserted in a template. The Fit option will allow a viewport to be created around an existing drawing or a portion of a drawing.

Hideplot—This option will allow hidden lines to be removed from selected viewports when plotting in paper space. This option generally is not required in architectural and engineering drawings.

Lock—Locks the current viewport in a similar manner to locking a layer.

Object—Selects a circle, closed polyline, ellipse, region, or spline to be used as a viewport.

Polygonal—Creates an irregularly shaped viewport using the same process that is used when the **Polygonal Viewport** button is selected.

Restore—This option can be used to restore saved viewport layouts. The prompt is

?/Name of window configuration to insert <default name>:

The name of a saved view can be entered. Typing the question symbol (?) will provide a list of view names.

2, 3, 4—Selecting one of these options will create two, three, or four viewports without your having to reissue the command. The required number of viewports is entered, followed by specifications for how to arrange the viewport. Each option is similar to the arrangements used for tiled viewports.

CHAPTER 24 EXERCISES

1. Open the Architectural template and create a new layout titled COOL STUFF. Use the **Viewports** dialog box and create viewports using the Three: Above layout. Insert a drawing in model space. Adjust each viewport so that a different aspect of the drawing is centered in each viewport. Save the drawing as E-24-1.
2. Open the Architectural template and use the **Create Layout Wizard** to create a new layout titled VIEWPORT TEST. Set the printer as the default system printer, and if necessary select an appropriate paper size. Set the orientation to portrait. Assign a single viewport suitable for displaying a floor plan at 1/4"=1'–0". Select the corners of the viewport so that it will only fill half of the page. Save the drawing as E-24-2.
3. Open any three drawings and insert them in a template drawing. Create a floating viewport to display the drawings at a suitable scale. Save the drawing as E-24-3
4. Open the Architectural drawing template. Attach any four drawings from your library. Each drawing is to be shown at a different scale when plotted. Provide a title and the scale for each detail. Save the drawing as E-24-4.
5. Open drawing E-24-4 and clip two of the drawings so that they are not displayed. Save the drawing as E-24-5.

CHAPTER 24 QUIZ

1. Describe how a referenced drawing can be used in an office to complete a project.
2. Use the **Help** menu and determine how and what the **Tilemode** setting affects.
3. What layer is added to the drawing base when you work with viewports? How will it affect the display?
4. What command or process is used to create a viewport?
5. What are the benefits of using layouts?
6. How can the display in the second viewport be altered so that it does not display the same drawing as another viewport?
7. How does paper space differ from model space?
8. A drawing has been inserted in a template in model space but it can't be seen. Describe how to show it.
9. List the steps to create a layout using the **Create Layout Wizard**.
10. Explain the difference between floating and tiled viewports.
11. How is a scale applied to the display in a viewport?
12. List three options for controlling the display of the viewport.
13. A drawing is displayed at the proper scale in a viewport, but the entire drawing can't be seen. Describe how the problem can be fixed.
14. Three drawings are displayed in a viewport, but you would like to hide one from a meeting with a client. Describe how to hide the drawing.
15. Describe the process to create an irregularly shaped viewport.

CHAPTER 25

Controlling Output

Chapter 24 walked you through the process of taking a drawing created in model space and placing it in a layout for plotting. As computers become more common in contractor and subcontractor offices, more construction projects will be transmitted between offices electronically. AutoCAD allows an electronic plot to be created that is accessible over the Internet. This allows the designers for the subcontractors to have access to areas of the project that require contributions from them. In addition to providing access to the drawings, electronic communication can save many hours by eliminating the need to produce paper copies. Unfortunately, some offices have not made the transition to CAD and still require paper copies. This chapter will introduce methods for

- Controlling plotting
- Controlling the plotting device
- Assigning plot styles
- Configuring the drawing to be plotted

The commands to be introduced in this chapter include

- **Plot**
- **Stylemanager**

AN OVERVIEW OF PLOTTING

The commands that control plot devices and settings can be accessed from the **Plot** dialog box, which consists of the **Plot Device** and **Plot Settings** tabs. The **Plot Device** tab is used to assign plotters and plot style tables. The **Plot Settings** tab is used to set plotting parameters. Access the **Plot** dialog box by selecting the **Print** button on the **Standard** toolbar, by typing **PLOT** (ENTER) at the Command prompt, or by selecting **Print** from the **File** menu. Each method will produce a **Plot** dialog box similar to Figure 25.1. The **Plot** dialog box can be used to control either a printer or a plotter. (Printing was introduced in Chapter 5.)

Each time a new layout was entered, the **Page Setup** dialog box was displayed. The **Page Setup** dialog box was first introduced when layout tabs were first examined in Chapter 4. Although it was dealt with quickly when the initial drawing process was started, as you gain experience, it should be used to define plotting parameters. When the drawing is complete, it will be plotted based on the specified parameters during the page setup. As you gain experience, plotting parameters can be specified and saved within a layout. The **Plot** dialog box is similar to the **Page Setup** dialog box. Plotting parameters assigned in the **Plot** dialog box will override those set in the **Page Setup** dialog box.

Note: It's important to keep sight of the goal as you begin this chapter. Your goal is to produce a plot at a useable scale. Since Chapter 5, you've been making scaled plots using the **Scaled to Fit** option. This produced scaled drawings, but the scale was most likely a scale known only to AutoCAD and not the rest of the known world. In the industry, most drawings are plotted at a scale such as 1/4"=1'–0". AutoCAD produces a warning as the **Plot**

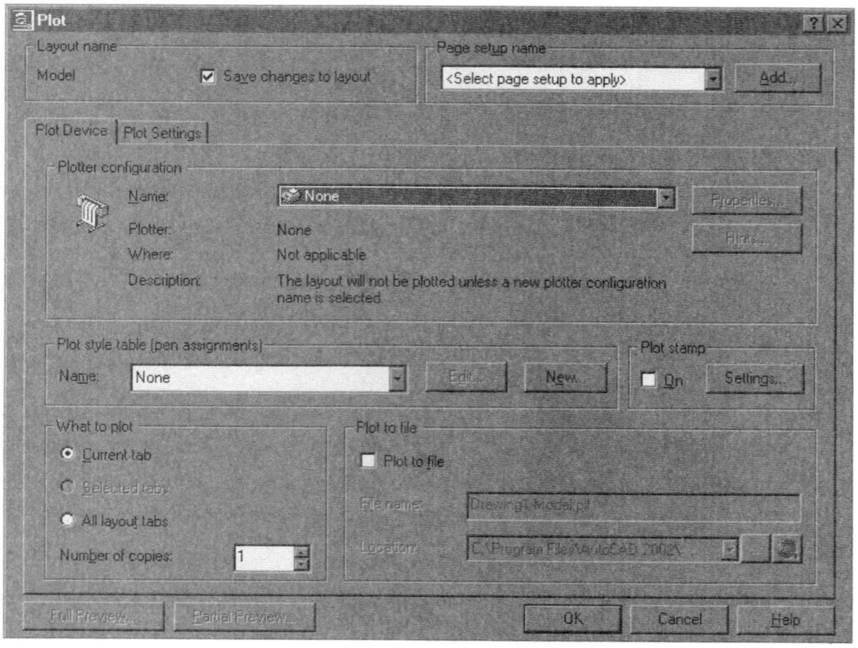

Figure 25.1 *The commands to control plotting can be accessed from the Plot dialog box. The dialog box consists of the* **Plot Device** *and* **Plot Settings** *tabs. The* **Plot Device** *tab is used to select plotters and plot style tables. The* **Plot Settings** *tab is used to set plotting parameters. Access the* **Plot** *dialog box by selecting the* **Print** *button on the* **Standard** *toolbar, by typing* **PLOT** *(ENTER) at the Command prompt, or by selecting* **Print** *from the* **File** *menu.*

dialog box is displayed that might seem a bit intimidating for new users. AutoCAD has added many aids and wizards to help make plotting easier, but at the same time has made the initial process seem more complex. Think of yourself in a new, top-of-the-line, bright red, convertible sports car. All you want to do is drive—you'll read the manual when the thrill wears off. The purpose of Chapter 5 was to get you driving. You've taken AutoCAD plotting on a test spin. This chapter will introduce the concepts needed to produce a drawing at a standard scale. Once the process to produce a plot has been explored, you'll be introduced to the many bells and whistles that have been provided by AutoCAD to enhance plotting.

REVIEWING THE PLOTTING PROCESS

A scaled plot can be created easily using the **Plot** dialog box with the same procedure that was introduced in Chapter 5. As the dialog box is displayed, you'll be prompted to specify the parameters that will define the plot. Once the **Plot** dialog box is displayed, the plotting process will include the following steps:

1. If a named page setup has been previously defined and saved, it can be selected from the Page Setup Name list.
2. Using the **Plot Device** tab, verify that the desired plotter is current.
3. Verify that the correct plot style table (pen assignments) is attached to the drawing.
4. Use the **Plot Settings** tab to define the plotting parameters. Settings include **Paper size**, **Plot area**, **Drawing orientation**, **Plot scale**, **Plot offset**, and **Plot options**.
5. Select a preview option.
6. Press (ENTER). If the preview does not provide the expected results, redefine the plotting parameters on each tab as needed and repeat the preview process.
7. When the plot preview meets your expectation, select the **OK** button to produce the plot.

EXPLORING THE PLOT DIALOG BOX

The **Plot** dialog box consists of two tabs and three areas that are common to each tab. At the top of the dialog box are areas that describe the **Layout name** and **Page setup name**. At the bottom of the display are buttons to control preview options and buttons to continue or cancel the plotting process. The tabs are used to control the plotting devices and the plot settings.

Layout Name Display

The **Layout name** area of the **Plot** dialog box displays the name of the layout to be plotted. In Figure 25.1 the layout displayed is the Model. The selected layout names will be displayed in this area if multiple layouts are selected to be plotted. Model will be displayed if the Model layout is the current viewport. Model space will produce a plot of the drawing, without displaying the border and title block. For best results, select a layout that will display the drawing with a title block.

Saving Plotting Specifications
AutoCAD allows the plotting parameters that are defined in the **Plot** dialog box to be attached to the layout. The next time you need a plot of the selected drawing, instead of setting the parameters of the two tabs, you only need to click the **OK** button of the **Plot** dialog box. With the **Save changes to layout** check box active, changes made to each tab will be saved in the current layout. This option is not available if multiple layouts are selected for plotting in the **What to plot** area.

Page Setup Name
This portion of the **Plot** dialog box displays a listing of named and saved page setups. Attempting to plot a drawing displayed in the Architectural title block would include the following options: Previous Plot, Architectural Title Block, and Architectural English-no output device.

Add
Selecting the **Add** button will display the **User Defined Page Setups** dialog box, which is used to name the plot setup that is to be saved to the layout. In addition to naming the current setup, this option can be used to rename, delete, or import user-defined setups and apply them to the current drawing.

Preview Options
Once all of the plotting parameters have been set, the preview options can be used to view the outcome of the plot before the plot is produced. Options include the full and partial previews. Each will be explored in detail after all of the settings have been explored. A full preview will display the drawing just as it will appear when plotted. A partial preview will show a box that defines the limits of the objects to be plotted inside a box that defines the plotting area. The **Partial Plot Preview** dialog box displays the **Paper size**, **Printable area**, **Effective area**, and **Warnings**. The **Warnings** area displays a list of warnings that will affect the effective plotting area. The other three categories can each be set on the **Plot Settings** tab.

EXPLORING THE PLOT DEVICE TAB

As the **Plot** dialog box is displayed, the **Plot Settings** tab is displayed by default. Once a plotter is configured to your workstation, the plot device will not need to be accessed for each plot. The key elements of the **Plot Device** tab include settings for the current plotter and settings for using a plot style table. The plotting devices that are configured to your workstation or network can be used by AutoCAD to plot drawings.

As you've made check prints from model space, AutoCAD has used the plotter specified in the **Options** dialog box to determine the default plotter. As you plot using a layout, the plotting device specified on the **Layout** tab is used. The layout stores settings for plot devices, the plot style table, print area, rotation, paper size, and scale. These settings can be provided as the drawing is started, and stored as a named page setup. When the drawing is ready to be plotted, these values can used to control the plot production. The **Plotter Manager** controls existing plotters, modifies existing configurations, and adds new plotter configurations. To access this feature choose the **Plotter Manager** from the **File** menu. Users of previous releases of AutoCAD will appreciate the elimination of command line prompts and the addition of the **Add-A-Plotter Wizard** to configure plotters.

Defining Plotting Parameters

The **Plot Device** tab displays the settings and controls for the plotter to use, a plot style table, the layout to be plotted, and information for saving the drawing to a file. The major areas of the **Plot Device** tab include **Plotter configuration**, **Plot styles table (pen assignment)**, **What to plot**, **Plot to file**, and five buttons common to both tabs.

Plot Configuration

The **Plotter configuration** area of the tab displays the current plotting device and its operating parameters. It also displays the port it is configured to. Selecting the **Name** edit box will display a list similar to Figure 25.2. The list shows each available plotter that is connected to either the workstation or the network. Notice that an icon is displayed by each listing to distinguish between system devices and those stored with a PC3 file name. A system printer or plotter is a device that is configured in the Printer folder of My Computer, or added by selecting **Printers** from the **Settings** menu. Selecting the **Add Printer** option from the Windows **Printers** Menu will produce a prompt for the **Add Printer Wizard** and allow a printer to be added and accessed by other programs. You can select previously configured printers using the **Device and Default Information** box of the **Print/Plot Configuration** dialog box. Methods of configuring a printer or plotter and adding it to this list will be introduced later in this chapter. Selecting the **Properties** button will display a **Plotter Configuration (PC3)** editor. The editor is used to modify or view the current plotter parameters. Selecting the **Hints** button will display information on improving the plot quality specific to the configured plotting device.

A PC3 file is a file created by AutoCAD that is drawing independent. PC3 files are just like any other file in that they are portable and can be shared between different workstations. A PC3 file contains all of the information on what to plot and how to plot, including the specified plotting device, paper size, and pen information. For users of previous releases of AutoCAD, PCP and PC2 files can be imported into AutoCAD 2002. These older formats will automatically be converted to PC3 format.

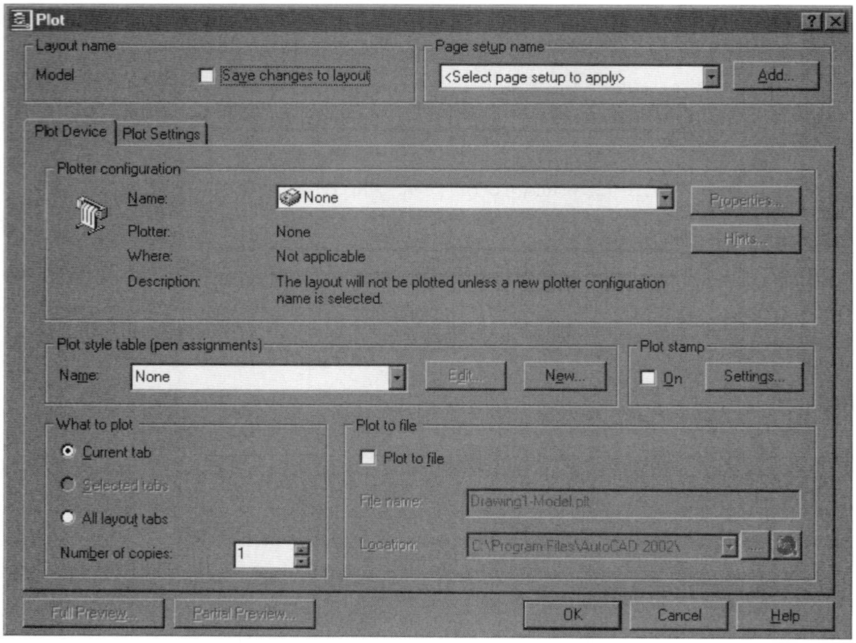

Figure 25.2 *Selecting the* **Name** *edit box will display a list of available plotting devices that are connected to either the workstation or the network.*

Plot Style Table (Pen Assignments)

If you've used an older release of AutoCAD, part of configuring the plotter was to configure pen assignments. AutoCAD 2002 replaces pen assignments with plot styles. Just as a style can be assigned to text or dimension, AutoCAD allows a style to be assigned to each layout and model space view. This portion of the **Plot Device** tab is used to create, edit, or set the plot style table assigned to the current model and layout tabs. Selecting the **Name** edit box displays a list of available styles. Later in this chapter you'll be introduced to the procedure to create and plot with plot styles. Selecting the **New** button will start the **Add Color-Dependent Plot Style Table Wizard**. The wizard can be used to create a new plot style table. Once a plot style table has been created, it will be listed in the **Name** box. Notice that in Figure 25.3, the listing shows several plot styles, even though none have been defined. These are plot styles defined by AutoCAD. Most architectural drawings can be plotted by selecting either the monochrome.ctb, Default R14 pen assignment.ctb, or the R14 cfg-1-Default System Printer.ctb.

What to Plot

This portion of the **Plot Device** tab allows you decide if you will plot using model space or one of the layout tabs of the drawing. The default setting is the **Current tab**.

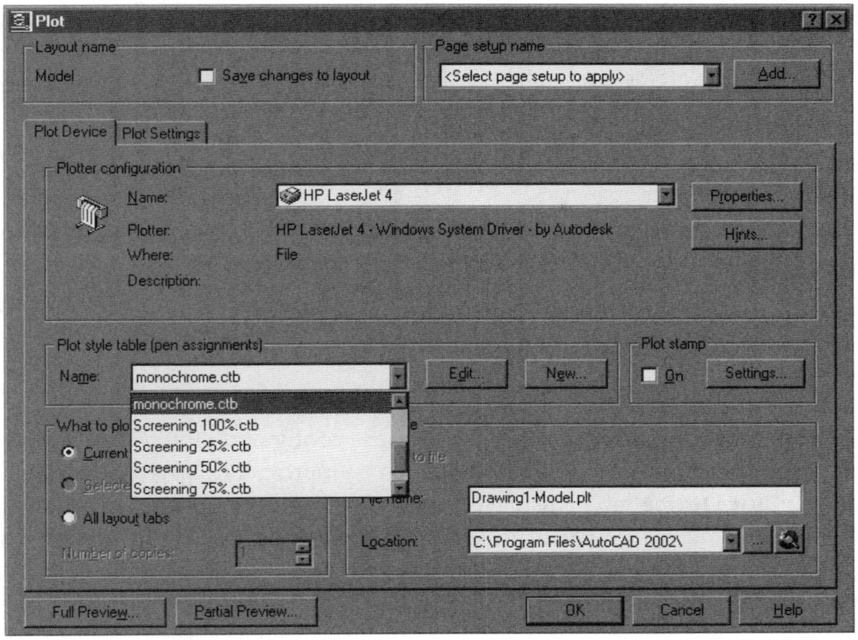

Figure 25.3 *The plot style* **Name** *listing shows several plot style tables defined by AutoCAD. As you create new styles, they will be added to the list.*

If the model space view of the drawing is displayed on the screen, this is what will be plotted. Setting a layout as the current display will make it the current tab, selected as the drawing to be plotted. If **All Layout Tabs** is selected, all of the layouts contained in the current drawing will be plotted. The **Selected Tabs** option is currently inactive when only one tab is selected for plotting. To make this option active, hold down (CTRL) and select the tabs to be plotted prior to displaying the **Plot** dialog box. The **Number of copies** setting determines the number of plots to be made of each layout.

Plot to File

When you complete all of the options on both tabs of the **Plot** dialog box, the information will be sent to the plotting device and the information will be processed. The time required to produce the plot will depend on the size of the drawing file and the type of plotter. If the **Plot to file** check box is selected, the plot will be stored as a .PLT file instead of being sent to the plotting device. Saving a plot file allows the drawing to be plotted at a more convenient time. As the **Plot to file** check box is selected, the other options in this area will become active. A name such as Floor-Layout1 will be displayed, indicating the drawing and layout names. The **Location** edit box shows the location of the current drawing. The PLT file will be stored in the same location as the drawing to be plotted. Selecting the [...] button will display the **Browse**

for Folder dialog box, with a display similar to the Explorer display. The dialog box allows the storage location of the plot file to be specified using methods similar to that used by other dialog boxes for saving files. The file will be stored to disk with the indicated name followed by the .PLT extension. Any storage device in your machine or accessible to the network can be used to store the plot file. Selecting the **Browse the Web** button allows the drawing to be stored at a specified Web site.

EXPLORING THE PLOT SETTINGS TAB

The **Plot Settings** tab is used to control the paper size and paper units, plot area, drawing orientation, plot scale, plot offset, and plot options. Each area can be seen in Figure 25.4.

Paper Size and Paper Units

This portion of the **Plot Settings** tab displays the configured plotter, the current paper size, the printable area, and toggles from inch to metric. The plotter that was specified on the **Plot Device** tab is displayed as the **Plot device**. This display indicates the standard plotting sizes that can be accommodated by the current plotting device. Selecting the edit arrow will display a list similar to Figure 25.5. (The list will vary, based on the configured plotter.) The paper size to be used for the plot can be select-

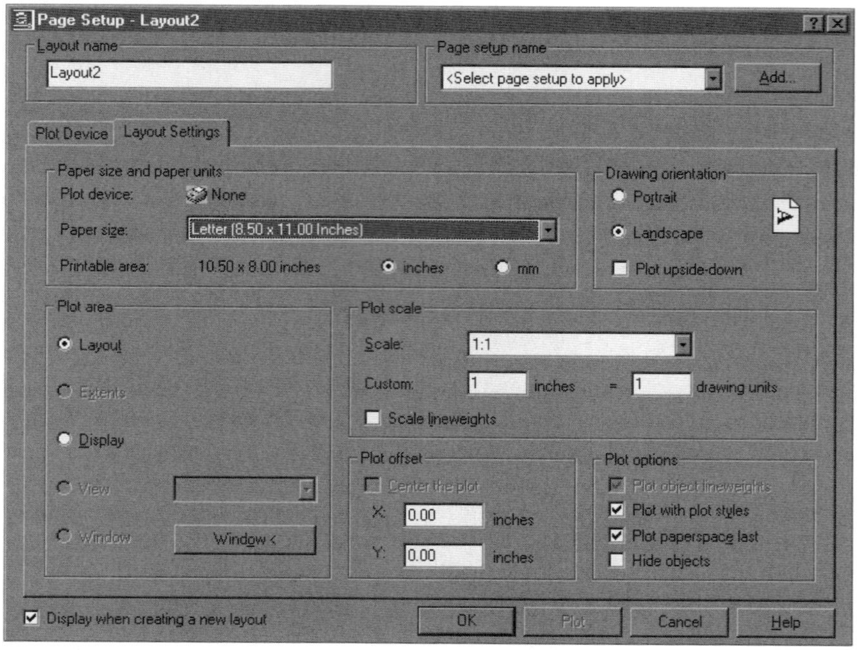

Figure 25.4 *The* **Plot Settings** *tab is used to control the paper size and paper units, plot area, drawing orientation, plot scale, plot offset, and plot options.*

ed from the list. The sizes displayed are the actual paper sizes. Sizes are listed as width (X) x height (Y). Once the desired value is selected, the **Printable area** will be updated. This is the maximum area available based on the current paper size. The final option of this area of the tab is to select the plotting units. Options include inches and mm.

The Plot Area

This area of the **Plot Settings** tab provides options for describing what portion of the current drawing will be plotted. Because the entire drawing often does not need to be plotted, five options are given to decide what portion of the drawing will be plotted.

Layout

This option will plot everything within the margins of the selected paper size. This will be the option of choice when the entire contents of the layout need to be plotted.

Extents

Choosing this option will plot the current drawing with the extent of the drawing objects as the maximum limits that will be displayed in the plot. The option is sim-

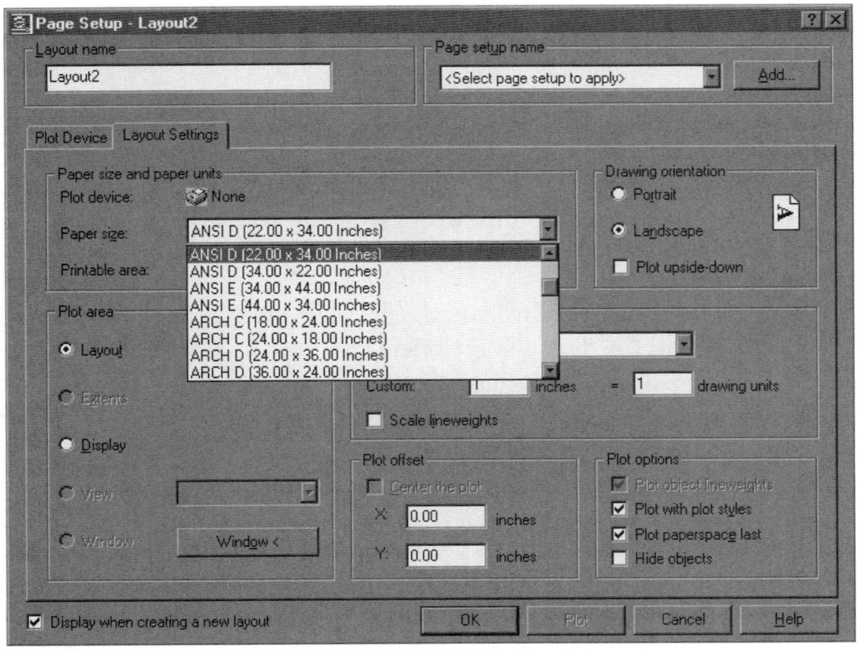

Figure 25.5 *The plotter that was specified in the* **Plot Device** *tab is displayed as the* **Plot device**. *The* **Paper size** *list indicates the standard plotting sizes that can be accommodated by the current plotting device. The list will vary, based on the configured plotter.*

ilar to the Extents option of the **Zoom** command. Using this option might eliminate objects at the perimeter of the drawing because the plotter cannot reach the edge of the plotting surface.

Display
This setting plots only the material that is currently displayed on the screen. If you have zoomed into a drawing, only that portion of the drawing will be plotted.

View
If specific views have been saved using the **View** command, the **View** option for plotting will be activated. This option will allow specific views to be plotted as they were saved. In the initial setting, the **View** radio button is inactive. When a drawing contains named views, the **View** button will be active and a listing showing each saved view name can be displayed. Selecting the desired view to be plotted will highlight that view, place the name in the edit box, close the list, and allow the selections to continue.

Window
This option will allow a specific area of a drawing to be defined for plotting. In the same way that a window is used to zoom or select objects for editing, a window can be specified here to define the selection plot. Initially the **Window** radio box is inactive. If the **Window<** button is selected, the **Plot** dialog box will be removed from the display area, and prompts for defining the window will be displayed. The prompts are as follows:

> Specify first corner: *(Select first window corner.)*
>
> Specify other corner: *(Select opposite window corner.)*

Once the window is defined, the **Plot** dialog box will be redisplayed.

Drawing Orientation

This area of the **Plot Settings** tab allows the drawing orientation to be specified. The box on the right side of the display is the orientation icon and not a selection box. The current orientation of the box is based on the configured plotter and indicates the drawing will be plotted in the normal orientation for construction drawing. This setting is referred to as *landscape*. When **Landscape** is used, the A will be placed on its side, indicating the print will be read from the long edge. When the plot is rotated so that it is read from the title block end, the plot is referred to as a portrait plot. The A will be in the vertical position indicating the print will be read looking from the title block end. A portrait plot is more likely to be used with laser printers. The orientation of the drawing to the paper can be adjusted by selecting the appropriate radio button. A check box is also available for plotting the drawing upside-down. Selecting this option will rotate the portrait or landscape layout 180°.

Plot Scale

The **Plot scale** area of the **Plot Settings** tab controls settings for the scale and lineweight scale. As drawings have been created throughout this text, they've been drawn at real size rather than at a reduced scale, as would be done in manual drawing. In Chapter 24 you were introduced to methods of creating layouts for plotting. If you're plotting a layout, the scale was set as the layout was created. This will allow the scale factor to be at 1:1 here. If you want to make a plot of a drawing created in model space, you'll need to provide a scale factor for plotting. Select the **Scale** edit arrow and the desired scale from the list to provide a plotting scale. Figure 25.6 shows a portion of the **Scale** list showing common scale factors that can be used to produce a plot at a given size. Dimensions and text were both multiplied by a scale factor in anticipation of the scale at which the drawing would be plotted.

An alternative to selecting a precise scale is to activate the **Scaled to Fit** box on the **Plot Settings** tab of the **Options** dialog box. This option will allow the desired plot to be sized to fit within the parameters of the paper. The resulting scale will be displayed in the **Custom** display box. Drawings that are scaled to fit are useful as check prints to determine completeness, but they are not suitable for most projects within the construction industry. The **Custom** box is better suited for entering a scale that

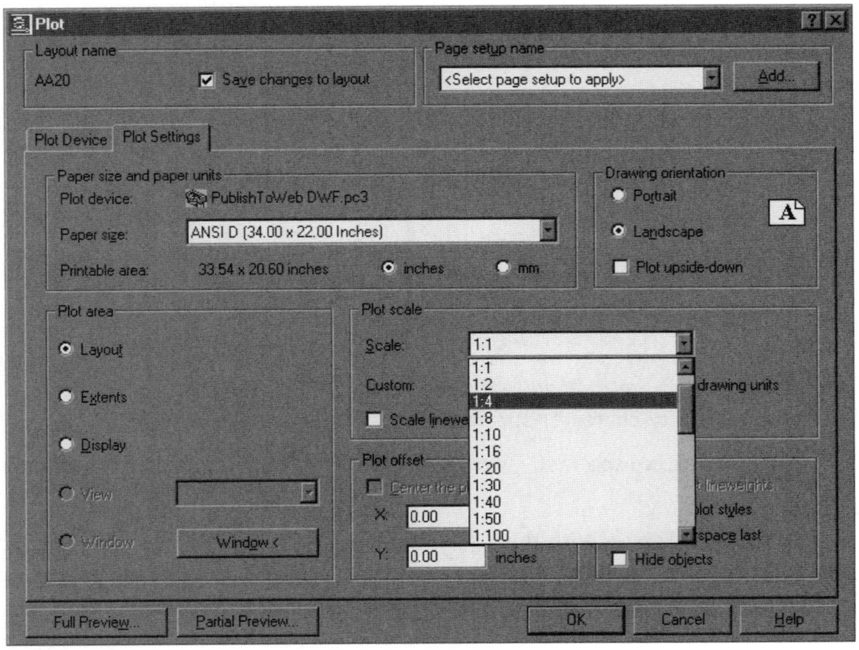

Figure 25.6 *Select the* **Scale** *edit arrow to display a list of common scale factors that can be used to produce a plot at a given size.*

is not listed. The scale is created by entering the number of inches equal to the number of drawing units.

Use the **Scale lineweights** check box to scale lineweights in proportion to the plotted scale. Lineweights selected during drawing construction normally specify the line width of printed objects and are plotted with the specified width regardless of the plot scale.

Plot Offset

The **Plot** offset portion of the **Plot Settings** tab contains the **Center the plot** check box and a display for X and Y coordinates. Activating the **Center the plot** option will automatically center the plot in the center of the plotting paper. The plot offset values specify an offset distance of the drawing to be plotted from the lower left corner of the paper. If you're plotting a layout, the lower left corner of the plotting area is positioned at the lower left edge of the plotting material. You can offset the origin of the plot from the paper edge by entering a positive or negative value. The plotter unit values are in inches or millimeters on the paper. Entering a value in the **X** edit box specifies the plot origin in the X direction. Entering a value in the **Y** edit box specifies the plot origin in the Y direction. The default setting for plotting origin is X=0–Y=0. For most plotters, this will be the lower left corner of the paper. By selecting the X or Y origin box, new values can be entered. Entering an X value of 12 and a Y value of 10 will place the new origin 12 units to the right and 10 units above the original plotting origin.

Plot Options

This portion of the **Plot Settings** tab contains four check boxes for controlling the outcome of the plot. The options control lineweight, plot styles, and the current plot style table. Plot styles will be introduced in the next portion of this chapter.

Plot object lineweights—With this box active, lineweights specified in the drawing will be plotted.

Plot with plot styles—With this option active, plot styles will be applied to objects that are defined in the plot style table. All style definitions with different property characteristics are stored in the plot style tables.

Plot paperspace last—With this option active, model space geometry will be plotted first.

Hide objects—With this option active, hidden lines will be removed from layouts. Viewports can be used to remove hidden lines for model space drawings.

Previewing the Plot

You've now explored each tab of the **Plot** dialog box. Once values have been entered for each option, the drawing can be plotted. Selecting either preview button allows the outcome of the current plot settings to be viewed prior to the actual plot. Two options are available for previewing the plot: **Partial Preview** and **Full Preview**.

Partial Preview
Selecting the **Partial Preview** button will produce a display similar to the one seen in Figure 25.7. A white box represents the paper size and the limits of the effective plotting area are specified on the screen by a dashed line. A blue box represents the area that the drawing will fill. A warning also will be given of any problems that may be encountered during the plot. Common warnings displayed in the warning box include the following:

Effective area too small to display

Origin forced effective area off display

Plotting area exceeds plotter maximum

A small red triangular Rotation Icon is displayed in the corner to represent the current rotation. The triangle will be moved, depending on the **Landscape**, **Portrait**, or **Upside-down** selection. Clicking **OK** will restore the **Plot** dialog box.

Full Preview
Choosing this option will produce a graphic display of the drawing as it will appear when plotted. Figure 25.8 shows an example of a full preview. A few seconds will elapse between the selection of the **Full Preview** button and the display of the proposed plot. An outline of the paper size will be drawn on the screen, which displays the portion of the drawing to be plotted. Once the drawing is displayed, the **Realtime Zoom** but-

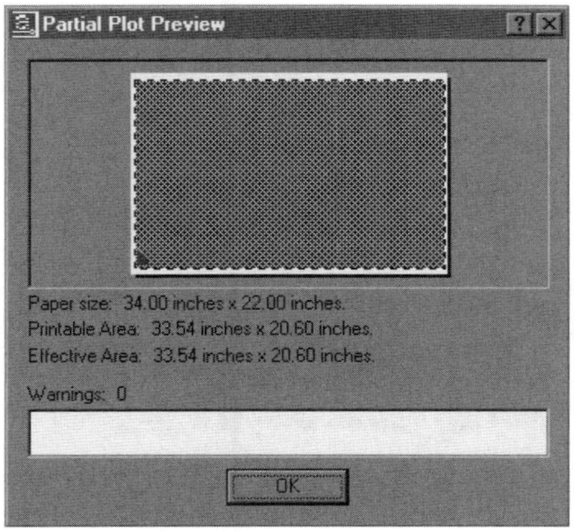

Figure 25.7 *Selecting the* **Partial Preview** *button will produce a display of the area to be plotted. The white box represents the paper size, and the limits of the effective plotting area are specified by a dashed line. A blue box represents the area that the drawing will fill.*

ton is displayed. The **Pan** and **Zoom** options can be quite useful in viewing detailed areas of the drawing a final time prior to plotting. Once the indicated **Pan** or **Zoom** is performed, press (ENTER) to restore the **Plot** dialog box.

Producing the Plot

The two tabs of the **Plot** dialog box have been examined. If you change your mind on the need for a plot, click the **Cancel** button. The dialog box will be terminated, and the drawing will be redisplayed. If you are satisfied with the plotting parameters that have been established, click the **OK** button. The dialog box will be removed from the screen and the Command prompt will display the effective plotting area. With older pen plotters, you might be given a prompt to remind you to position paper in the plotter. Newer plotters use rolls of paper, but they also need to be reloaded occasionally. Don't laugh! After you labor through the **Plot** dialog box the first few times, simple things like turning the plotter on and installing the paper might tend to be forgotten. If you need to set up the plotter, you can do it before you click **OK**. Your plotter might have additional features, such as pen pressure and speed, to be adjusted. The completed plot can be seen in Figure 25.9.

WORKING WITH PLOT STYLES

Rather than setting pen assignments like past releases, AutoCAD 2002 allows a plot style table to be created and saved as a file. Plot styles are a new object property assigned to all drawing objects to control the appearance of the finished plot. Plot styles can be used as the drawing is reproduced to override the object properties

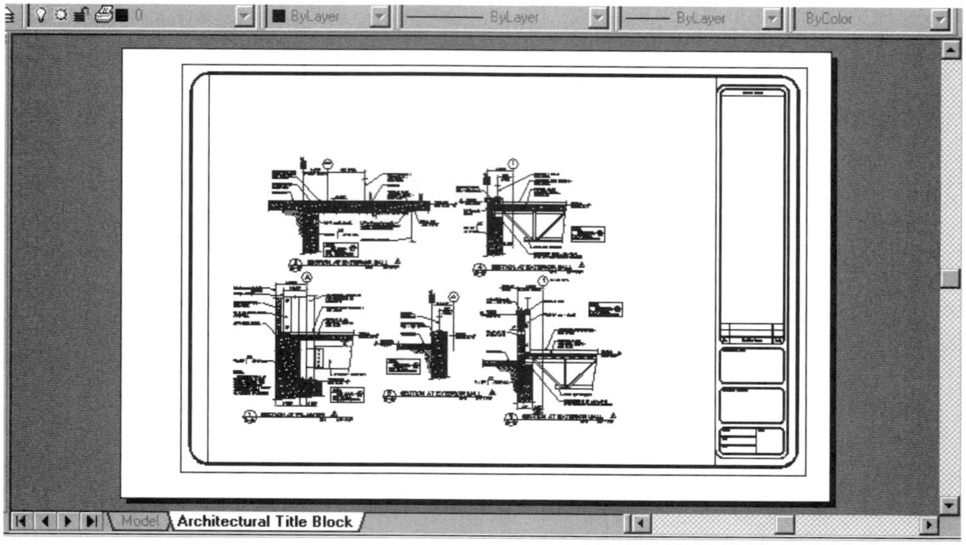

Figure 25.8 *A full preview shows each drawing object as it will appear in the finished plot.*

assigned as the drawing was created. They can be used as overrides for color, dithering, gray scale, linetype, lineweight, pen assignments, screening, end styles, join styles, and fill styles. They can be assigned layer by layer or object by object. Each aspect of the style is defined in a plot style table that can be attached to the model or layout tab. By assigning different plot styles to the same layout, you can reproduce the drawing in different ways to meet the needs of different clients or subcontractors. The three steps required to use a plot style are as follows:

1. Define the plot style using a plot style table.
2. Attach the plot style table to the layout.
3. Assign the plot style to an object or layer in the drawing.

The effects of the attached plot style can be viewed on the **Layout Settings** tab of the **Page Setup** dialog box.

EXPLORING PLOT STYLE MODES

The two types of modes used for all drawings in AutoCAD 2002 are **Color Dependent** and **Named plot styles**. The mode is controlled using the **Plotting** tab

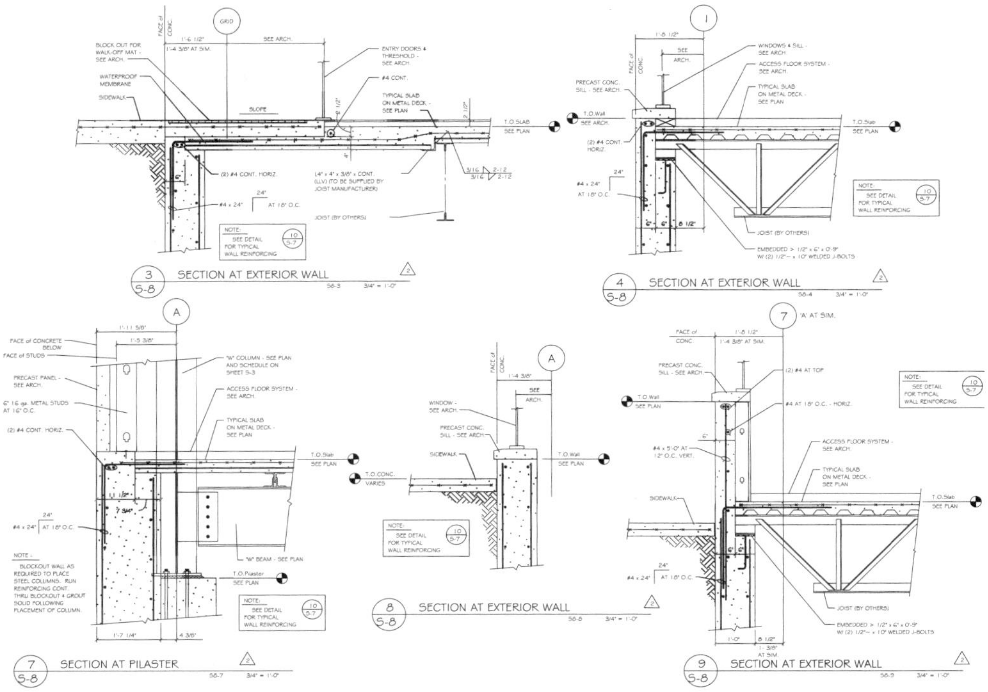

Figure 25.9 *The plot of the drawing previewed in Figure 25.8. (Courtesy Van Domelen/Loojenga/McGarrigle/Knauf Consulting Engineers.)*

of the **Options** dialog box. (Select **Options** from the **Tools** menu.) The mode is controlled in the **Default plot style behavior for new drawings** area of the **Plotting** tab. Users of previous releases of AutoCAD will be most familiar with the **Color Dependent** mode.

Color-Dependent Plot Styles

A color-dependent plot style controls the plotting based on the color assigned to objects in the drawing. There are 255 color-dependent plot styles based on the 255 colors in the AutoCAD color index. Color-dependent plot styles are very similar to the pen assignments introduced with AutoCAD Release 14. Key characteristics of a color dependent plot style include the following:

- Plot characteristics are assigned to each color to determine how the drawing will be plotted.
- Color-dependent plot styles cannot be added, deleted or renamed in a style table.

Using a color-dependent plot style, you can alter the appearance of all red drawing objects by changing the plot style that corresponds to RED. For most of the drawings that you've done in this text, a color-dependent plot style will work well. All red lines could be assigned to a specific pen, all blue lines could be assigned to a different pen width. However, as your drawings become more complex, using the object color could limit the use of color in a drawing. By associating color with a specific pen, you lose the ability to work with a color independent of lineweight or linetype. Creating color-dependent plot styles will be explored as the use of the **Add Plot Style Table Wizard** is explored.

Named Plot Styles

A named plot style depends on an assigned name rather than the color of an object. Named styles are independent of object colors in the style. Key characteristics of a named plot style include the following:

- The Normal style assigned by AutoCAD cannot be modified.
- Each defined style in the plot style table has a name and specific plotting characteristics.
- A named plot style can be assigned to layers or objects.

CREATING A PLOT STYLE TABLE

One of the most common uses of a plot style table is to plot a layout containing a wide variety of colors on a plotter using black lines. The drawing can be plotted without using a plot style table, but each line might be displayed in a gray tone rather than black. By creating a plot style table, you can assign each color the color black for plotting, with the same color intensity. A plot style table can be created using the **Stylemanager** command. Access the command by selecting **Plot Style Manager** from **File** on the **Standard** toolbar or by typing **STYLEMANAGER** (ENTER) at the Command prompt. Each method will display the **Plot Styles** dialog box shown in Figure 25.10. You can also access the **Plot Styles** dialog box by selecting the **Add or Edit Plot Styles Tables** button on the **Plotting** tab of the **Options** dialog box. Each of the available plot style tables is displayed in the **Plot Styles** dialog box by a name and icon. Selecting the **Add A Plot Style Table** option will start the wizard shown in Figure 25.11. After reading the introductory text, click the **Next** button.

Figure 25.12 shows the second dialog box prompt of the **Add Plot Style Table Wizard**. By default, you'll be starting the plot table from scratch. The dialog box also allows options for the following:

> **Use My R14 Plotter Configuration**—this option allows a new plot style table to be created using pen assignments stored in an R14 drawing, but you don't have access to the PCP or PCP2 files associated to the drawing.

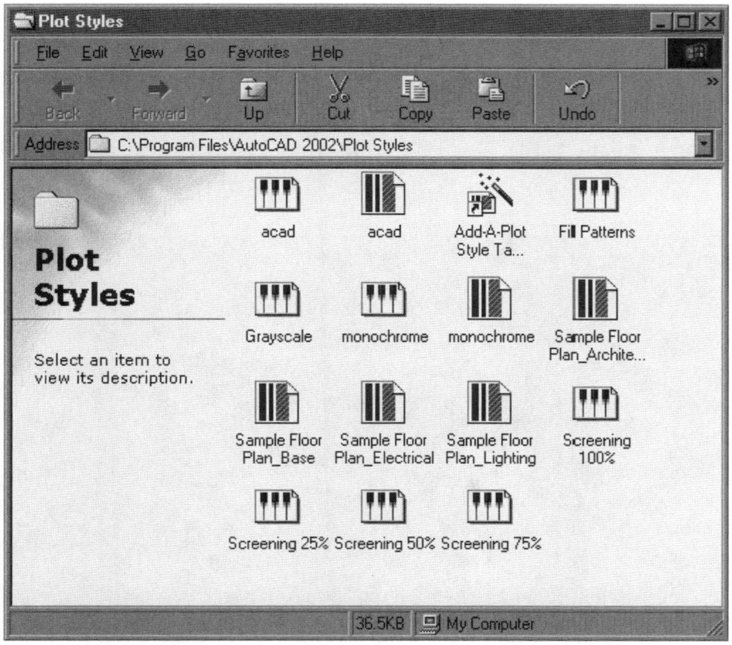

Figure 25.10 *A plot style table can be created using the* **Stylesmanager** *command. Access the command by selecting* **Plot Style Manager** *from* **File** *on the* **Standard** *toolbar or by typing* **STYLESMANAGER** *(ENTER) at the Command prompt. Each method will display the* **Plot Styles** *dialog box.*

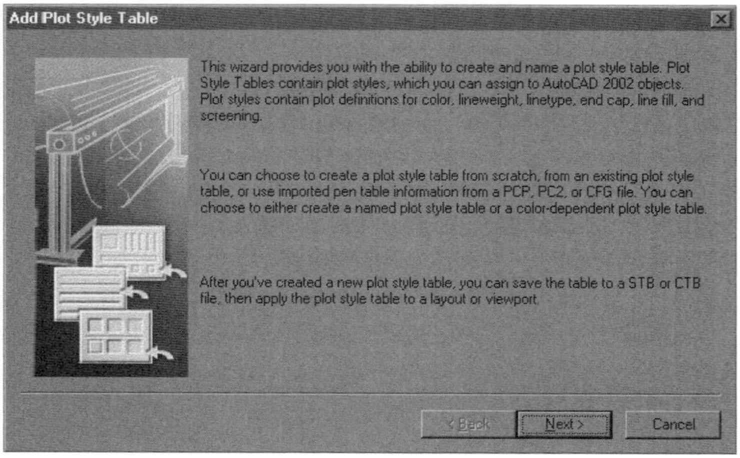

Figure 25.11 *Selecting the* **Add A Plot Style Table** *option will start the wizard. After reading the introductory text, click the* **Next** *button to proceed through the setup.*

> **Use a PCP or PCP2 file**—This option allows a new plot style table to be created using pen assignments stored in acadr14.cfg files.

As you proceed through the initial viewing of the wizard, accept the default option and select the **Next** button. The next dialog box of the **Add Plot Style Table Wizard** will display a prompt for the type of table to be used. Options include **Color-Dependent Plot Style Table** and **Named Plot Style Table**. Choose the **Named**

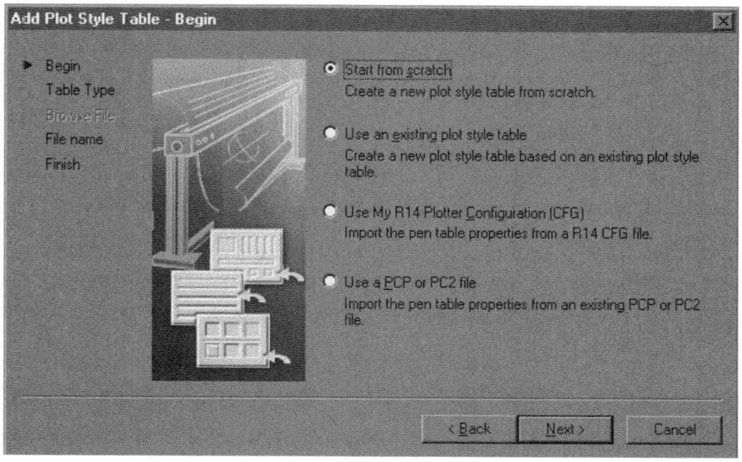

Figure 25.12 The **Add Plot Style Table - Begin** *dialog box shows the options for selecting a starting point for creating a plot style table. By default, you'll be starting the plot table from scratch.*

Plot Style Table button. Selecting the **Next** button will display the **File Name** dialog box. Enter a name to describe the plot style in the **Name** edit box. For this example a name of MONO will be used. Once a name has been entered in the edit box, select the **Next** button to display the final page of the wizard.

Up to this point, you've started a plot style table named MONO. The final page of the setup wizard allows the naming process to be completed or the plot style to be edited. Selecting the **Plot Style Editor** button will display a **Plot Style Table Editor** dialog box similar to Figure 25.13. This table can be used to alter the properties of the drawing to be plotted. Later sections of this chapter will deal with altering drawing properties. Select the **Finish** button to create a named plot style table and exit the wizard. Selecting **Finish** will create a .STB file that is listed in the **Plot Styles** dialog box. Selecting the **Use this plot style table for layouts in new drawings and pre-AutoCAD 2002 drawings** option will make the newly created plot style the default plot style for future drawings.

Creating a Color-Dependent Plot Style Table

The process to create a color-dependent plot style is similar to the process of creating a named style. The steps to create a color-dependent include the following:

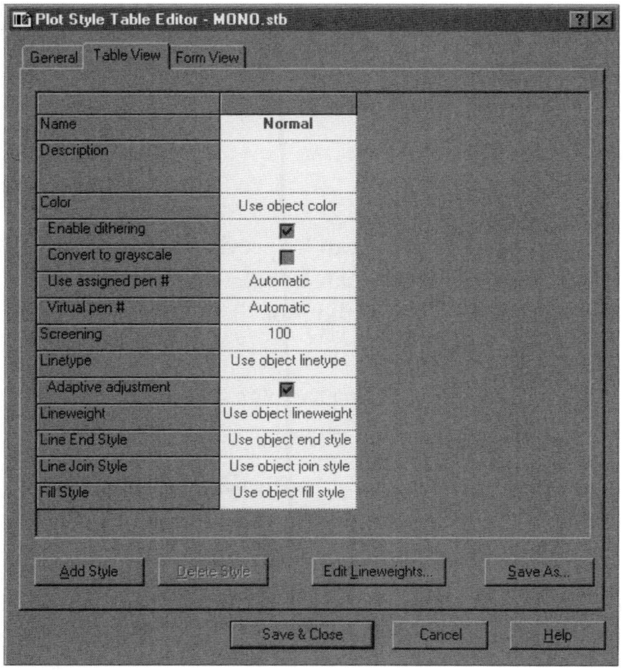

Figure 25.13 *Selecting the* **Plot Style Editor** *button will display a Plot Style Table Editor dialog box. This dialog box can be used to alter the properties of the drawing to be plotted.*

1. Select the **Add A Plot Style Table** button in the **Plot Styles** dialog box.
2. Select the **Next** button after reading the introductory material.
3. Select the option to be used to start the plotting style table and select the **Next** button.
4. Choose the type of plot style table to be created and select the **Next** button.
5. Enter the name of the plot style table and choose the **Next** button.
6. Set the options on the **Finish** page and click the **Finish** button.

The color-dependent plot style will be stored as a .CTB file and listed in the **Plot Styles** dialog box.

ATTACHING A PLOT STYLE TABLE TO A LAYOUT

A plot style created with the **Add Plot Style Table Wizard** can be attached to the model or layout tab using the **Page Setup** dialog box. Use the following steps to attach a plot style to a layout tab:

1. Select the desired layout tab to apply the plot style.
2. Select **Page Setup** from the **File** menu.
3. Using the **Page Setup** dialog box, select the **Plot Device** tab.
4. Select the desired plot style to be applied in the **Plot Styles** portion of the **Plot Device** tab.
5. Click the **OK** button.

The qualities assigned to the plot style can now be applied to the selected layout.

Viewing a Plot Style

Once a plot style has been created, it can be previewed before it is plotted. Changes made to object properties in a layout can be seen when a drawing is regenerated, or you can use the **Full Preview** option in the **Plot** dialog box. Plot styles can be displayed in a drawing using the following steps:

1. Select the layout tab containing the plot style to be used.
2. Select **Page Setup** from the **File** menu.
3. Select Display Plot Styles from the **Plot Style Table (Pen Assignments) in a Layout** on the **Plot Device** tab.
4. Click the **OK** button.

The properties of the plot style will now be displayed. If the changes are not displayed, type **REGENALL** (ENTER) at the Command prompt to display plot styles in a layout.

Adding Named Plot Styles

Once you've created your own plot style, it must be added to a plot style table before it can be applied to the layout to be plotted. Named plot styles are added to the plot style table using the **Plot Style Table Editor**. Plot styles are added using the following sequence:

1. Select **Plot Style Manager** from **File** on the Standard toolbar.
2. Use one of the following methods to display the **Plot Style Table Editor**:
   - Double-click a .STB file.
   - Right-click a .STB file and select **Open** from the shortcut menu.
3. In the **Active Plot Style Table** of the **Current Plot Style Table** dialog box or the **Select Plot Style Table** dialog box, select the desired plot style and choose **Editor**.
4. Select the **Table View** tab or **Form View** tab in the **Plot Style Table Editor** dialog box to view the style settings.
5. Select the **Add Style** button.

A style titled Style1 will be added to the table. While the new style is still highlighted, change the name to better describe the style contents. Figure 25.14 shows an example of a style that was added. Adjust the options as needed, and select the **Save & Close** button to return to the **Plot Styles** dialog box.

Deleting a Named Plot Style

As plotting requirements change throughout the life of a project, a plot style in a named plot style table can be deleted. The procedure to delete a style is similar to the procedure used to add the style. Use the following steps to delete a file:

1. Select **Plot Style Manager** from **File** on the Standard toolbar.
2. Double-click the .STB file to be edited.
3. Use the **Form View** tab of the **Plot Style Table Editor** dialog box to select the plot style to be deleted.
4. With the style highlighted, click the **Delete Style** button.
5. Select the **Save & Close** button or the **Cancel** button to return to the **Plot Styles** dialog box.

DEFINING A PLOT STYLE

Figure 25.14 shows an example of the MY LINES plot style. Properties of a style can be defined or edited, using similar methods used to alter a layer color in the **Layer Properties Manager**. These settings will provide overrides to be used while plotting for values defined in the drawing. Settings can be defined using the **Table View** or

Form View tab of the **Plot Style Table Editor** dialog box. The **Table View** tab shown in Figure 25.14 works well if only a few styles are assigned to the table. Figure 25.15 shows an example of the **Form View** tab. The **Form View** tab works well if several styles have been assigned to the table. Notice that the style names are listed on the left side, and the values for the current style are listed on the right side.

To define the color for the plot style requires the following steps:

1. Using the **Table View** tab of the **Plot Style Table Editor** dialog box, select the current color setting for the plot style to be altered. As Use object color is selected, the standard color list will be displayed.
2. Select the desired color. As the color is selected, the list will be closed, allowing other options to be defined.

Other options can be defined using similar methods.

Enable Dithering

Dithering is used to approximate colors using a dot pattern. If the plotter does not support dithering, this option will be inactive. If the plotter supports dithering, dither-

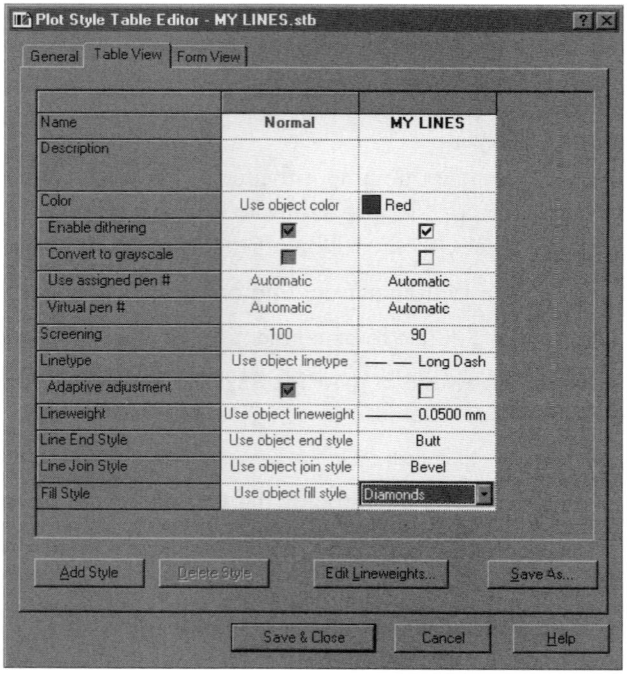

Figure 25.14 *The* **Table View** *tab of the* **Plot Style Table Editor** *dialog box can be used to add a plot style. Notice that the color, screening, linetype, lineweight, and line styles have been altered.*

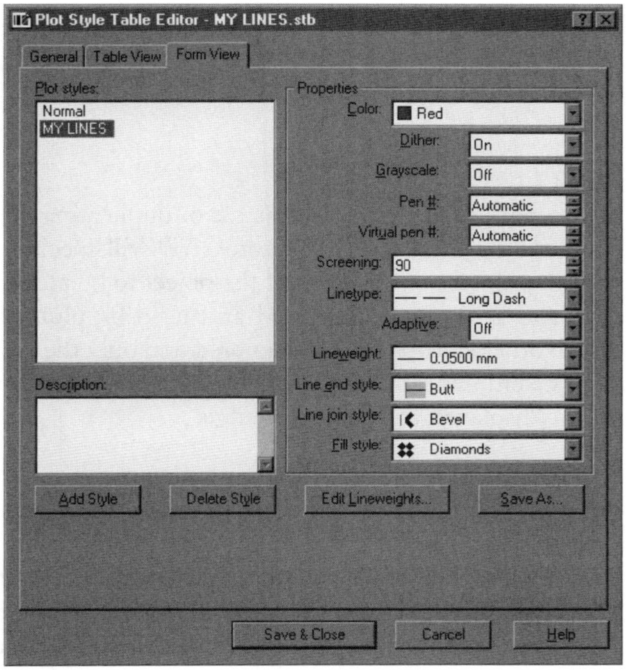

Figure 25.15 *The* **Form View** *tab works well if several styles have been assigned to the table. Notice that the style names are listed on the left side, and the values for the current style are listed on the right side.*

ing can be turned on to provide dim colors more visibility. With dithering OFF, AutoCAD maps colors to the nearest color.

Convert to Grayscale

With this option active and a plotter that supports grayscale, AutoCAD will convert colors to grayscale.

Use Assigned Pen Numbers

This option specifies which pen will be used with each plot style. The default setting is automatic. In the **Plotter Configuration Editor**, you can specify that a certain object will be plotted using pen #1 as black and .02" , and pen #2 as blue and .031". With this option, pen #1 can be assigned to one plot style and pen #2 can be assigned a different plot style. This can be a very useful tool if a subcontractor has used pen assignments different from the desired standard. Using the Automatic or 0 setting, AutoCAD will select the pen closest in color to the color of the object you are plotting. The value can't be altered if a color-dependent plot table is being used or if the plot style color setting of Use object color is active. Values can be entered by keyboard or by using the range arrow. As the box is selected, arrows are displayed to adjust the range. With a setting of 0, only the up arrow is active. A pen value from 1 to 32 can be specified.

Virtual Pen Number

Most new plotters do not use pens but are capable of duplicating pen plotter output by using virtual pens. In its current setting, AutoCAD will specify a color from the AutoCAD Color Index closest to the color of the object to be plotted. A virtual pen number between 1 and 255 can be provided and used by plotters without pens. Using virtual pens, all other style settings are ignored and only the virtual pen settings are used. Values can be entered by keyboard or by using the range arrow. As the box is selected, arrows are displayed to adjust the range.

Screening

This setting allows a color intensity to be specified by controlling the amount of ink to be placed during the plot. The default setting of 100 plots at full intensity. A value between 0 and 100 can be entered. Entering 0 will have the same effect as freezing the assigned objects. Values can be entered by keyboard or by using the range arrow. As the box is selected, arrows are displayed to adjust the range. With a setting of 100, only the down arrow is active.

Linetype

Selecting the **Linetype** option will display a linetype list similar to Figure 25.16. The default setting for plot style linetype is Use object linetype. Selecting a linetype from the list will override the linetypes specified in the drawing.

Adaptive Adjustment

This setting adjusts the linetype scale to be used to display the linetype pattern. This setting should be OFF when the display of the linetype is important.

Lineweight

The **Lineweight** box displays a sample of the selected lineweight and the numeric value of the lines to be used with the style. An override value can be selected from the drop-down list, displayed as the box is selected.

Line End Style

In addition to using the line end style specified for drawing objects, AutoCAD allows four options to be assigned to override the line end style of drawing objects. Selecting this field will produce a list with four options in addition to the default Use object end style: Butt, Diamond, Round, and Square.

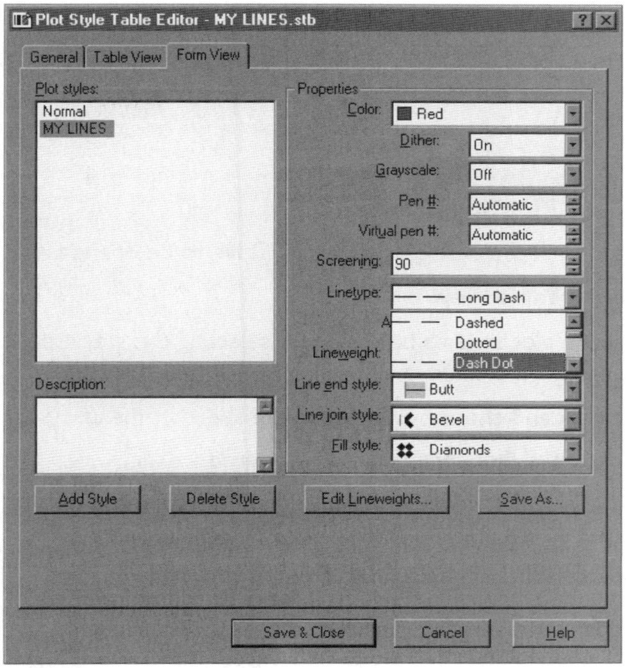

Figure 25.16 *Selecting the* **Linetype** *option will display a linetype list. The default setting for plot style linetype is Use object linetype.*

Line Join Style

AutoCAD allows four options to be assigned to override the line join style specified for drawing objects. Selecting this field will produce a menu with the four options in addition to the default Use object join style: Bevel, Diamond, Miter, and Round.

Assigning Fill Style

AutoCAD allows fill patterns to be assigned to override the fill style specified for drawing objects. The patterns will be applied to donuts, plines, and solids. Selecting this field will produce a list with the options. In addition to the default of Use object fill style, options include Solid, Checkerboard, Crosshatch, Diamond, Horizontal Bars, Slant Left, Slant Right, Square Dots, and Vertical Bars.

EDITING PLOT STYLES

A plot style can be edited using the same methods that were used to define it. Either the **Table View** or the **Form View** tab can be used to edit the plot style. Qualities in a plot style can be edited using the same methods that were used to define the style. In addition to altering the qualities, the **Plot Style Table Editor** dialog box can be used to add, delete, copy, paste, and modify plot styles.

Copying a Plot Style

A plot style can be copied using the **Table View** tab of the **Plot Style Table Editor** dialog box. A plot style can be moved from one table to another table by using the drag and drop technique. In addition, one or more settings from a plot style can be copied to another style using the following steps:

1. Open the **Plot Style Manager**.
2. Double-click the desired .CTB or .STB file to be opened.
3. Make the **Table View** tab current.
4. Select the plot style to be copied by selecting the gray area above the plot style name. (You can copy individual settings by selecting individual fields.)
5. Right-click and choose **Copy** from the shortcut menu.
6. Switch to the plot style table to be altered.
7. Select the plot style to be replaced by selecting the gray area above the plot style name.
8. Right-click and choose **Paste** from the shortcut menu.

Renaming Plot Styles

A named plot style can be renamed using the **Rename** command. Before a style can be renamed, it must exist in the plot style table associated with the layout. The new plot style name must be a name that is not currently used by any object or layer. Use the following steps to rename a named plot style:

1. Select **Rename** from the **Format** menu.
2. Select **Plot Style** from the **Named Objects** list.
3. Select the name of the plot style to be renamed from the **Items** list. The name will now be displayed in the **Old Name** box.
4. Provide a new name in the **Rename To** box.
5. Select the **Rename To** button. The new name will now be displayed in the list box.
6. Click the **OK** button to return to the drawing area.

CHAPTER 25 EXERCISES

1. Use a laser printer or plotter and set the required parameters to plot E-19-8 at a scale of 1/4"=1'–0".
2. Open drawing E-14-4. Plot the drawing at a scale of 3/8"=1'–0".
3. Open drawing E-15-10. Use the **Copy** command and create a second copy of the elevation. Design a second elevation using a different architectural style and materials. Save the drawing file as E-25-3. Plot the drawing containing your border and title block to a C size sheet of vellum.
4. Plot your drawing template containing the border and title block on a D size sheet of paper. Create a new base drawing that will fit within the limits of the borders. Attach any four drawings in your library. Each drawing is to be shown at a different scale when plotted. Provide a title and the scale used by the original drawing prior to plotting.

CHAPTER 25 QUIZ

1. How many colors can be configured for plotting?
2. Is it possible to have more than one plotting device connected to a single computer? Explain your answer.
3. What options are available to preview the information to be plotted prior to plotting, and how are they accessed?
4. Describe the meaning of the display used in a partial plot preview.
5. What tab contains a display of the current plotting scale?
6. What can be used to control how much of the current drawing will be displayed?
7. List five options for controlling what portion of the drawing will be plotted.
8. Describe the typical origin location for a plotter and a printer.
9. What is the advantage of saving the plot information with **Plot to file**?
10. Which takes precedence, the page setup or the plot style?
11. How does AutoCAD deal with pen assignments?
12. Explain the difference between portrait and landscape.
13. What plot style should be used to plot a drawing with 46 different colors using only black?
14. What command is used to create a plot style?
15. List two types of plot style tables and explain which will most typically meet your needs.
16. Explain the steps to attach a plot style to a layout.
17. What is the purpose of a plot style?
18. You've inserted five details in a drawing template. Each of the details was created for plotting at a different scale. Each was inserted in the template using the proper methods. What scale should be used to plot this drawing?
19. How are plotters configured using AutoCAD 2002?
20. What does the **What to plot** area control?

CHAPTER 26

AutoCAD and the Internet

The Internet is an important way to exchange files and information. AutoCAD allows you to access and store drawings and related files and to interact through the Internet. The AutoCAD Internet features allow drawings to be opened and saved to an Internet location. The features in AutoCAD allow hyperlinks to be added to drawings, so that a variety of related documents can easily be accessed. AutoCAD allows drawings to be displayed in Web format (DWF) files that can be viewed with an Internet browser. DWF files can be viewed by anyone with an Internet browser using a free WHIP!® plug-in or Volo View Express provided by Autodesk. This chapter will introduce methods for

- Accessing and working with drawings on the Internet
- Launching a Web browser
- Using a hyperlink, or Uniform Resource Locator (URL)
- Using the Drawing Web Format (DWF) to display drawings on the Internet
- Publish drawings to the Web

Commands to be introduced include

- **Browser**
- **Inetlocation**
- **Hyperlink**
- **Pasteashyperlink**
- **Publishtoweb**
- **Etransmit**

INTERNET ACCESS

You are probably familiar with the best-known uses for the Internet: email (electronic mail) and the Web (short for the World Wide Web). Email lets users exchange messages and data at very low cost. The Web brings together text, graphics, audio, and movies in an easy-to-use format. To use the features of AutoCAD requires Internet access and an Internet browser to be installed on your workstation. To view and plot AutoCAD DWF files from an Internet browser requires the WHIP! 4.0 plug-in or the Volo Viewer Express from autodesk.com to be downloaded. An Internet browser must also be installed on your computer. With the required software loaded, AutoCAD can open, insert, and save drawings to and from the Internet. To save files to an Internet location requires access rights to the directory where the files are stored.

AutoCAD is able to launch a Web browser from within AutoCAD using the Browser command. The browser is used for viewing drawings in 2D format on Web pages. The **Hyperlink** command can be used to link a drawing with other documents on your computer or the Internet. For example, a window or door schedule created on a spreadsheet can be linked to a floor plan created in AutoCAD. The electronic plot option of the **Plot** command can create DWF files for viewing 2D drawing on Web pages. With the **Open**, **Insert** and **Saveas** commands, AutoCAD can import or export drawings to and from the Internet.

WORKING WITH UNIFORM RESOURCE LOCATORS

A uniform resource locator (URL) is the file naming system of the Internet. This system allows files such as a text file, a Web page, a program file, an audio file, or movie clip to be located on the Internet. A typical URL would resemble http://www.autodesk.com. The prefix http:// is not typically required to be entered. Most Web browsers automatically add in the routing prefix, which saves you a few keystrokes. Common URLs you should be familiar with include the following:

URL	Meaning
http://www.autodesk.com	Autodesk Primary Web site
http://www.autodeskpress.com	Autodesk Press Web site

URLs can access several different kinds of resources such as Web sites, email, or news groups. No matter the resource, the URL always take on the same general format: "scheme://networkaddress."

The scheme accesses the specific resource on the Internet such as

file://	File located on your computer's hard drive or local network
ftp://	File Transfer Protocol (used for downloading files)
http://	HyperText Transfer Protocol (the basis of Web sites)

mailto://	Electronic mail (email)
news://	Usenet news (news groups)
telnet://	Telnet protocol
gopher://	Gopher protocol

The characters :// indicate a network address. Autodesk recommends these formats for specifying URL-style file names:

Web Site	http://servername/pathname/filename
FTP Site	ftp://servername/pathname/filename
Local File	file:///drive:/pathname/filename
or	file:///drive/pathname\filename
or	file://\\\\localPC\pathname\filename
or	file://\\\\localPC\pathname\filename
or	file:////localPC/pathname/filename
Network File	file://localhost/drive:/pathname/filename
or	\\\\localhost\drive:\pathname\filename
or	file://localhost/drive|/pathname/filename

The following definitions will help to clarify this confusing terminology.

Term	**Meaning**
Servername	A name or location of a computer on the Internet such as www.autodesk.com
Pathname	The same as a subdirectory or folder name
Drive	Your local drive
localPC	A file located on your computer
localhost	The name of the network host computer

If you are not sure of the name of the network host computer, use Windows Explorer to check the Network Neighborhood for the network names of computers.

OPENING AND SAVING INTERNET FILES

The **Browser** command lets you start a Web browser while inside AutoCAD. The two most common browsers are Microsoft Internet Explorer and Netscape Navigator. By default, the **Browser** command uses the Web browser program that is registered in your computer's Windows operating system. AutoCAD lists the name of the browser before prompting you for the uniform resource locator (URL). The URL is the Web site address, such as http://www.autodesk.com. The **Browser** command is used to launch a Web browser from within AutoCAD and requires that Netscape Navigator or Microsoft Internet Explorer be installed on your computer. The command can also be used in scripts, toolbar or menu macros, and AutoLISP routines to automatically access the Internet. The command can be accessed using the following command sequence:

Command: **BROWSER** (ENTER)

Enter Web location (URL) < http://www.autodesk.com> :*(Enter URL.)*

Pressing (ENTER) will accept the default URL, an HTML file added to your computer during the installation of AutoCAD. This URL is Autodesk's own Web site. After you type the URL and press (ENTER), AutoCAD launches the Web browser and contacts the Web site, such as the Autodesk Web site shown in Figure 26.1.

CHANGING THE DEFAULT WEB SITE

Altering the **Inetlocation** system variable setting can change the default Web page that your browser starts with from within AutoCAD. The variable stores the URL used by the **Browser** command and the **Browse the Web** dialog box. Make the change as follows:

Command: **INETLOCATION** (ENTER)

Figure 26.1 *Display the Autodesk Web site by typing* **BROWSER** (ENTER) *at the command prompt and then entering the name of the URL.*

Enter new value for INETLOCATION <"http://www.autodesk.
com">:*(Type URL.)*

DRAWINGS ON THE INTERNET

When a drawing is stored on the Internet, you access it from within AutoCAD using the standard **Open**, **Insert**, and **Save** commands. Instead of specifying the file's location with the usual drive-subdirectory-file name format, such as c:\acad14\file-name.dwg, use the URL format. The URL is the universal file naming system used by the Internet to access any file located on any computer hooked up to the Internet.

OPENING DRAWINGS FROM THE INTERNET

Use the **Open** command from the **File** menu to open a drawing from the Internet. The options available are to search the Web, open from the saved favorites, browse my files at Point A, or browse the Buzzsaw or Redspark sites shown in figure 26.2.

Search the Web—Display the **Browse the Web** dialog box, a simplified Web browser. This is the button next to the look in box that is the image of the world.

Look in Favorites—Open the Favorites folder, which is the equivalent to bookmarks that store Web addresses.

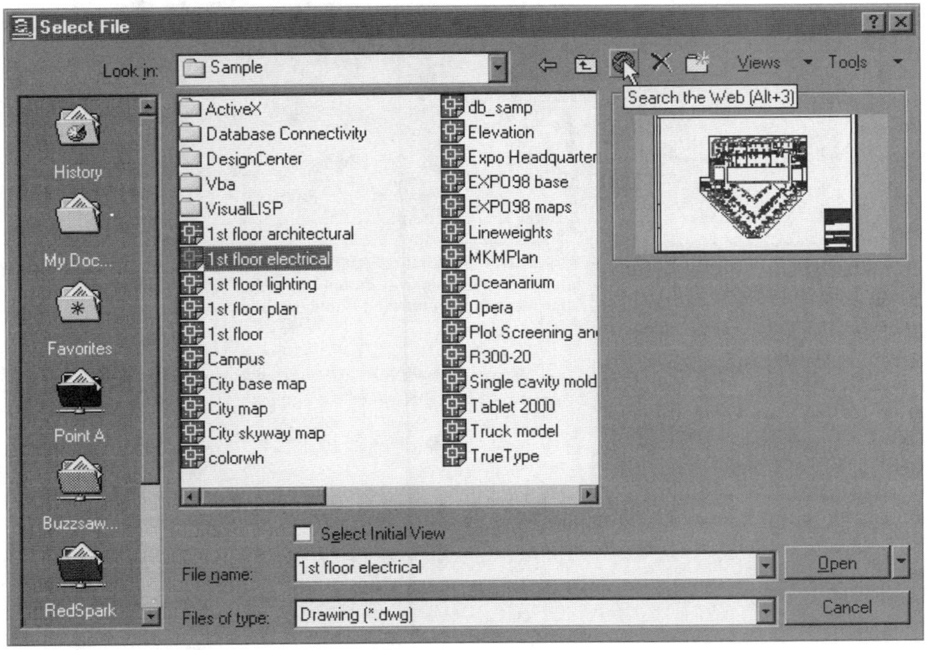

Figure 26.2 *The* **Select File** *dialog box contains the options to aid in working with the Internet.*

Point A—Browse for files at Point A Autodesk Web site

Buzzsaw—Browse for files at the Buzzsaw Web site

Redspark—Browse for files at the Redspark Web site

Add to Favorites—Save the current URL (hyperlink) to the Favorites folder. Access this option from the **Tools** menu in the upper right corner of the dialog box.

When you click the **Search the Web** button, AutoCAD opens the **Browse the Web** dialog box. This dialog box is a simplified version of a Web browser. The purpose of this dialog box is to allow you to browse files at a Web site. By default, the **Browse the Web** dialog box displays the contents of the URL stored in the **Inetlocation** system variable (See Figure 26.3a). You can easily change this to another folder or Web site, as noted earlier. (See Figure 26.3b).

The **Browse the Web** dialog box has six buttons across the top to aid in working with files. These include the following:

Back—Go back to the previous URL.

Forward—Go forward to the next URL.

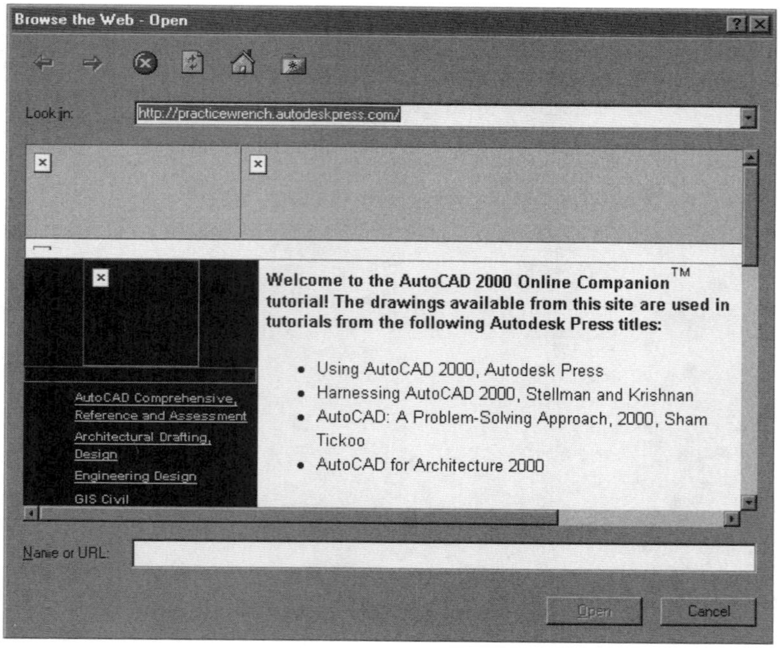

Figure 26.3a *By default, the* **Browse the Web - Open** *dialog box displays the contents of the URL stored in the* **Inetlocation** *system variable.*

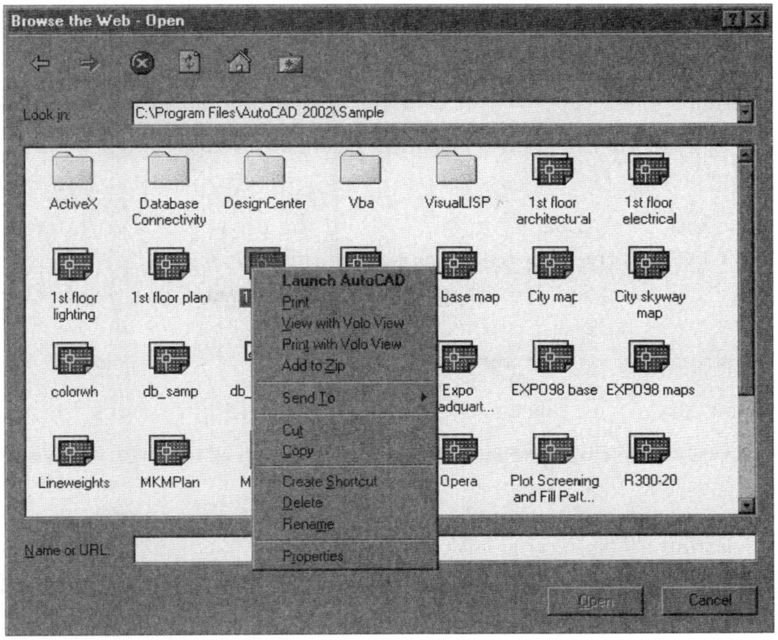

Figure 26.3b *The* **Browse the Web - Open** *dialog box and shortcut menu.*

Stop—Halt displaying the Web page. This option is useful if the connection is slow or the page is very large.

Refresh—Redisplay the current Web page.

Home—Return to the location specified by the Inetlocation system variable.

Favorites—List stored URLs (hyperlinks) or bookmarks. If you have previously used Internet Explorer, you will find all your favorites listed here. Favorites are stored in the \Windows\Favorites folder on your computer.

The **Look in** edit box allows you to type the URL. Alternatively, click the down arrow to select a previous destination. If you have stored Web site addresses in the Favorites folder, then select a URL from that list. You can either double-click a file name in the window or type a URL in the edit box. The following table gives the templates and examples for typing the URL to open a drawing file:

Drawing Location	Template URL
Web or HTTP Site	http://servername/pathname/filename.dwg
	http://practicewrench.autodeskpress.com/wrench.dwg
FTP Site	ftp://servername/pathname/filename.dwg
	ftp://ftp.autodesk.com

Local File	drive:\pathname\filename.dwg
	c:\acad 2000\sample\tablet2000.dwg
Network File	\\localhost\drive:\pathname\filename.dwg
	\\upstairs\e:\install\sample.dwg

When you open a drawing over the Internet, it will probably take much longer than opening a file found on your computer. During the file transfer, AutoCAD displays a dialog box to report the progress. If your computer uses a 28.8 Kbps modem, you should allow about five to ten minutes per megabyte of drawing file size. If your computer has access to a faster T1 connection to the Internet, you should expect a transfer speed of about one minute per megabyte.

The **Open** command will place the drawing in the current drawing file.

Note: The **Locate** and **Find File** buttons in the **Select File** dialog box do not work for locating files on the Internet.

Tutorial: Opening a Drawing at a Web Site

The Autodesk Press Web site has an area that allows you to practice using the Internet with AutoCAD. In this tutorial, you open a drawing file located at that Web site.

1. Start AutoCAD.
2. Ensure that you have a live connection to the Internet. If you normally access the Internet via a telephone (modem) connection, dial your Internet service provider now.
3. Select **Open** from the **File** menu or click the **Open** button on the **Standard** toolbar. The **Select File** dialog box is displayed.
4. Click the **Search the Web** button. The **Browse the Web** dialog box is displayed, as shown in Figure 26.4.
5. In the **Look in** edit box, type **PRACTICEWRENCH.AUTODESK-PRESS.COM** (ENTER). After a few seconds, the Browse the Web dialog box displays the Web site.
6. In the **Name or URL** edit box, type **WRENCH.DWG** (ENTER).

AutoCAD begins transferring the file. Depending on the speed of your Internet connection, this will take between a couple of seconds and half a minute. Notice the drawing of the wrench in AutoCAD that is shown in Figure 26.5.

INSERTING A BLOCK FROM THE INTERNET

When a block (symbol) is stored on the Internet, you can access it from within AutoCAD using the **Insert** command. When the Insert dialog box appears, click the

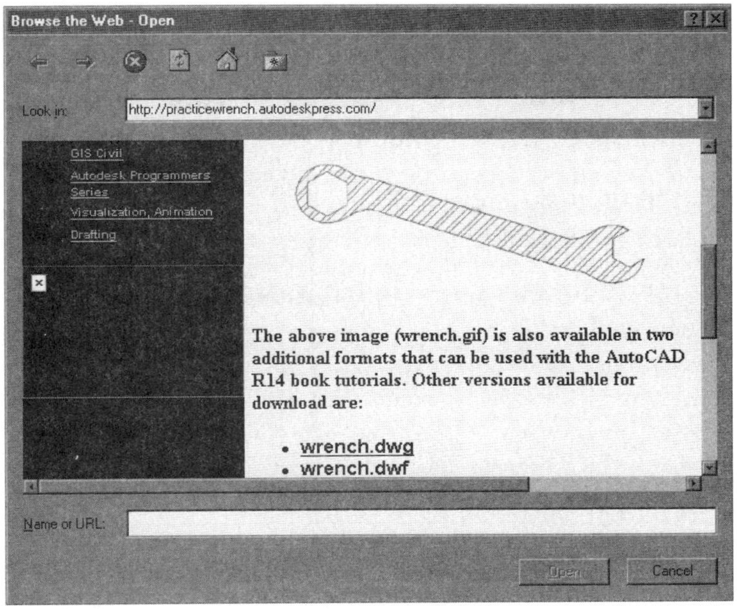

Figure 26.4 *The* **File Download** *display box.*

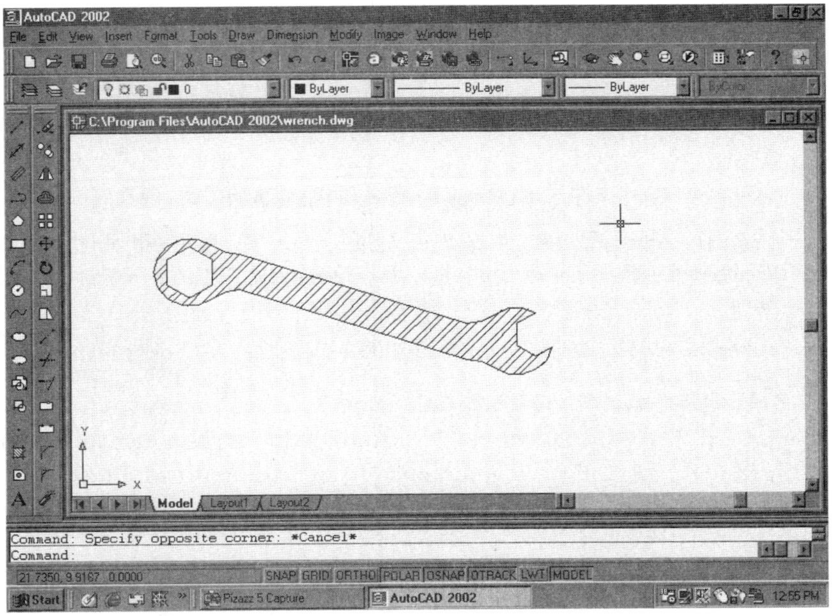

Figure 26.5 *The Practice Wrench is placed in a drawing.*

Browse button to display the **Select Drawing File** dialog box. This is identical to the dialog box discussed above. After you select the file, AutoCAD downloads the file and continues with the **Insert** command's familiar prompts.

The process is identical for accessing external reference (xref) and raster image files. Other files that AutoCAD can access over the Internet include 3D Studio, SAT (ACIS solid modeling), DXB (drawing exchange binary), WMF (Windows metafile), and EPS (encapsulated PostScript). All of these options are found on the **Insert** menu.

ACCESSING OTHER FILES ON THE INTERNET

Most other file-related dialog boxes allow you to access files from the Internet. This allows your firm or agency to have a central location that stores drawing standards. When you need to use a linetype or hatch pattern, for example, you can access the LIN or PAT file over the Internet. More than likely, you would have the location of those files stored in the Favorites list.

Some examples include the following:

Linetypes—Select **Linetype** from the **Format** menu. In the **Linetype Manager** dialog box, click the **Load**, **File**, and **Look in Favorites** buttons.

Hatch Patterns—Use the Web browser to copy .PAT files from a remote location to your computer.

Layer Name—Select **Layer Properties Manager** from **Layers**. In the **Layer Manager** dialog box, click the **Import** and **Look in Favorites** buttons.

LISP and ARX Applications—Select **Load Applications** from the **Tools** menu.

Scripts—Select **Run Scripts** from the **Tools** menu.

Menus—Select **Customize Menus** from the **Tools** menu. In the **Menu Customization** dialog box, click the **Browse** and **Look in Favorites** buttons.

Images—Select **View** from **Image Displays** on the **Tools** menu.

SAVING THE DRAWING TO THE INTERNET

When you are finished editing a drawing in AutoCAD, you can save it to a file server on the Internet with the **Save** command. If you inserted the drawing from the Internet in the default drawing using **Insert**, AutoCAD insists you first save the drawing to your computer's hard drive.

When a drawing of the same name already exists at that location, AutoCAD warns you just as it does when you use the **Saveas** command. Recall from the **Open** command that AutoCAD uses your computer system's temporary subdirectory, hence the reference to it in the dialog box.

USING HYPERLINKS WITH AUTOCAD

AutoCAD allows you to employ URLs in two ways: directly within an AutoCAD drawing, and indirectly in DWF files displayed by a Web browser. (URLs are also known as hyperlinks, the term we will use from now on.) Hyperlinks are created, edited, and removed with the **Hyperlink** command, using a dialog box, and the **-Hyperlink** command, using prompts at the command line. The **Hyperlink** command (select **Hyperlink** from the **Insert** menu) prompts you to "Select objects" and then displays the **Insert Hyperlink** dialog box. (See Figure 26.6.) As a shortcut, you can press CTRL+K or click the **Insert Hyperlink** button on the **Standard** toolbar.

As an alternative, you can use the **-Hyperlink** command. This command displays its prompts at the command line, and is useful for scripts and AutoLISP routines. The **-Hyperlink** command has the following syntax:

Command: **-HYPERLINK** (ENTER)

Enter an option [Remove/Insert] <Insert>: (ENTER)

Enter hyperlink insert option [Area/Object] <Object>: (ENTER)

Select objects: *(Pick an object.)*

1 found Select objects: (ENTER)

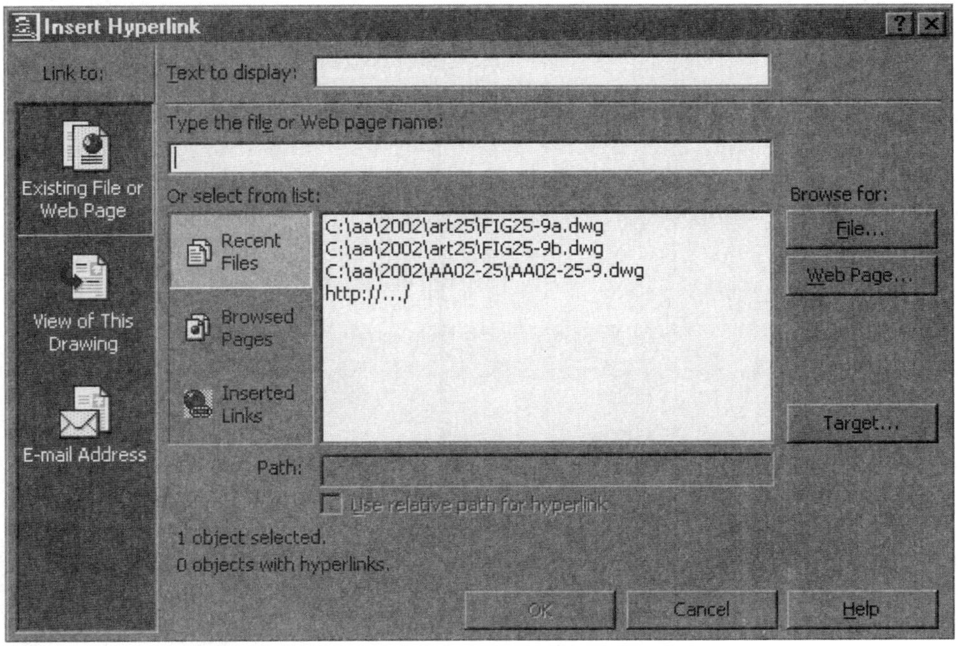

Figure 26.6 *The* **Insert Hyperlink** *dialog box.*

Enter hyperlink <current drawing>: *(Enter the name of the document or Web site.)*

Enter named location <none>: *(Enter the name of a bookmark or AutoCAD view.)*

Enter description <none>: *(Enter a description of the hyperlink.)*

Notice that the command also allows you to remove a hyperlink. It does not, however, allow you to edit a hyperlink. To do this, use the Insert option and re-specify the hyperlink data.

In addition, this command allows you to create hyperlink areas, which are rectangular areas that can be thought of as a 2D hyperlink. The dialog box-based **Hyperlink** command does not create hyperlink areas. When you select the Area option, the rectangle is placed automatically on layer URLLAYER and colored red.

In the following sections, you'll learn how to apply and use hyperlinks in an AutoCAD drawing and in a Web browser through the dialog box **Hyperlink** command.

HYPERLINKS INSIDE AUTOCAD

AutoCAD allows you to add a hyperlink to any object in the drawing. An object is permitted just a single hyperlink. On the other hand, a single hyperlink can be applied to a selection set of objects.

You can tell an object has a hyperlink by passing the cursor over it. The cursor displays the "linked Earth" icon, as well as a tooltip describing the link, similar to Figure 26.7.

If, for some reason, you do not want to see the hyperlink cursor, you can turn it off. Select **Options** from the **Tools** menu and then click the **User Preferences** tab shown in Figure 26.8. The **Display hyperlink cursor and shortcut menu** check box toggles the display of the hyperlink cursor, and you can toggle the **Display hyperlink tooltip** option.

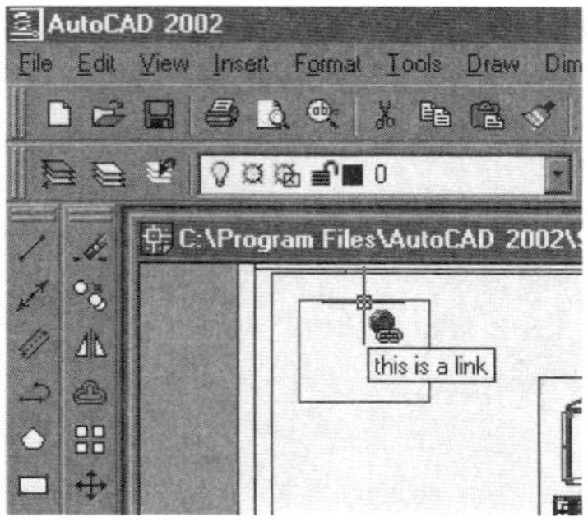

Figure 26.7 *An example of a hyperlink.*

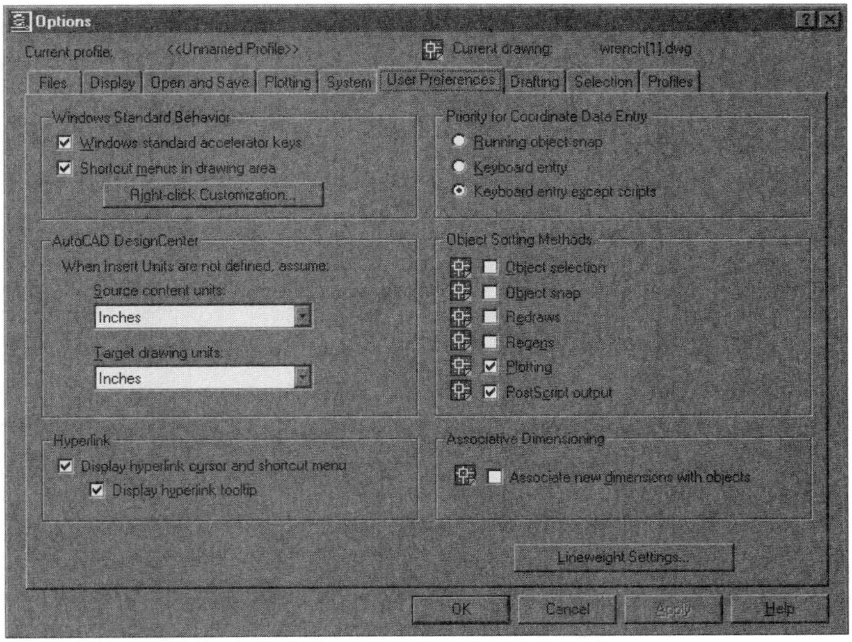

Figure 26.8 *The* **Options** *dialog box controls the display of hyperlinks.*

Tutorial: Attaching Hyperlinks

Let's see how that works by trying an example. In this example, we have the drawing of a floor plan. To this drawing, we will add hyperlinks to another AutoCAD drawing, a Word document, and a Web site. Hyperlinks must be attached to objects. For this reason, we will place some text in the drawing and then attach the hyperlinks to the text.

1. Start AutoCAD
2. Open the 1st floor plan.dwg file found in the AutoCAD Sample folder. (If necessary, click the Model tab to display the drawing in model space.)
3. Start the **Text** command and place the text in the drawing as shown in the following command sequence:

   Command: **TEXT** (ENTER)

   Current text style: "Standard" Text height: 0.20000

   Specify start point of text or [Justify/Style]: *(Pick a point in the drawing.)*

   Specify height <0.2000>: 2 (ENTER)

Specify rotation angle of text <0>: (ENTER)

Enter text: Site Plan (ENTER)

Enter text: Lighting Specs (ENTER)

Enter text: Electrical (ENTER)

Enter text: (ENTER)

The result up to this point can be seen in Figure 26.9

4. Click the Site Plan text to select it.
5. Click the **Insert Hyperlink** button on the **Standard** toolbar. Notice the **Insert Hyperlink** dialog box.
6. Select the **File** button in the **Browse for** area.
7. Go to AutoCAD's Sample folder and select the City base map.dwg file. Click **Open**. AutoCAD does not open the drawing; rather, it copies the file's name to the **Insert Hyperlink** dialog box. Notice in Figure 26.10 the name of the drawing you selected in the **Type the File or Web page name** edit box.
8. Click **OK** to close the dialog box. Move the cursor over the Site Plan text. Notice in Figure 26.11 that the display of the "linked Earth" icon appears; a moment later, the tooltip displays the URL.

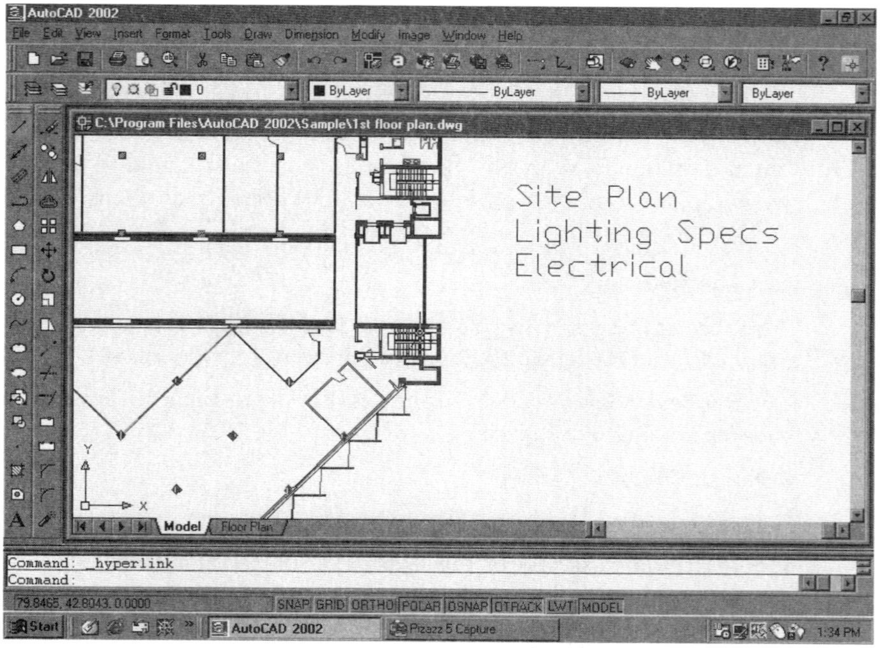

Figure 26.9 *Text is added to the drawing to be used for hyperlinks.*

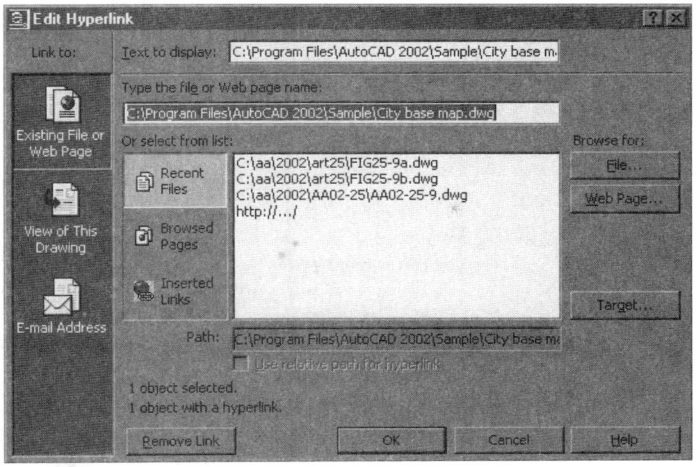

Figure 26.10 *The* **Edit Hyperlink** *dialog box.*

Figure 26.11 *The hyperlink cursor and tooltip.*

9. To try out the link click "Site Plan" to select it. Right-click and select **Open "C:Program Files\AutoCAD 2002\Sample\City base map.dwg"** from **Hyperlink** on the shortcut menu, as shown in Figure 26.12. Notice that AutoCAD opens "City base map.dwg". To see both drawings, select **Tile Vertically** from the **Window** menu. The screen will now appear as shown in Figure 26.13.

10. Hyperlinks can also be applied to a text document from a word processor. The process to create the hyperlink is the same process you have just seen. Select **Open from the hyperlink** in the shortcut menu (to access this menu right-click the hyperlink) and select the object. Now right click and the menu will appear. A word processor will start and open the document. Figure 26.14 is an example of a document that is opened by a hyperlink. If the Word processor has a Web toolbar, open it and click the **Back** button. This will send the computer back to AutoCAD.

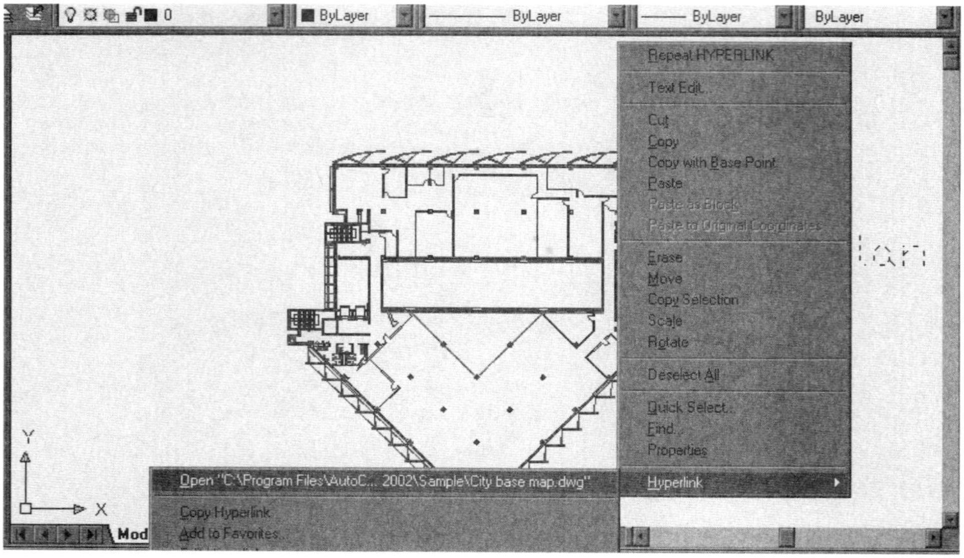

Figure 26.12 *The hyperlink shortcut menu.*

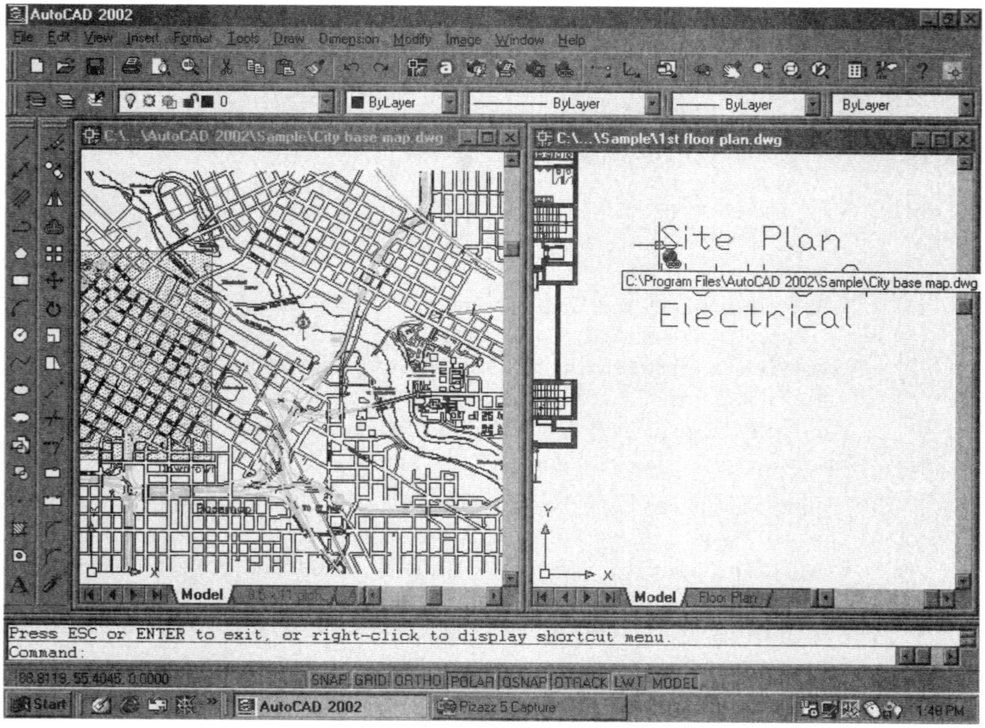

Figure 26.13 *Viewing two drawings.*

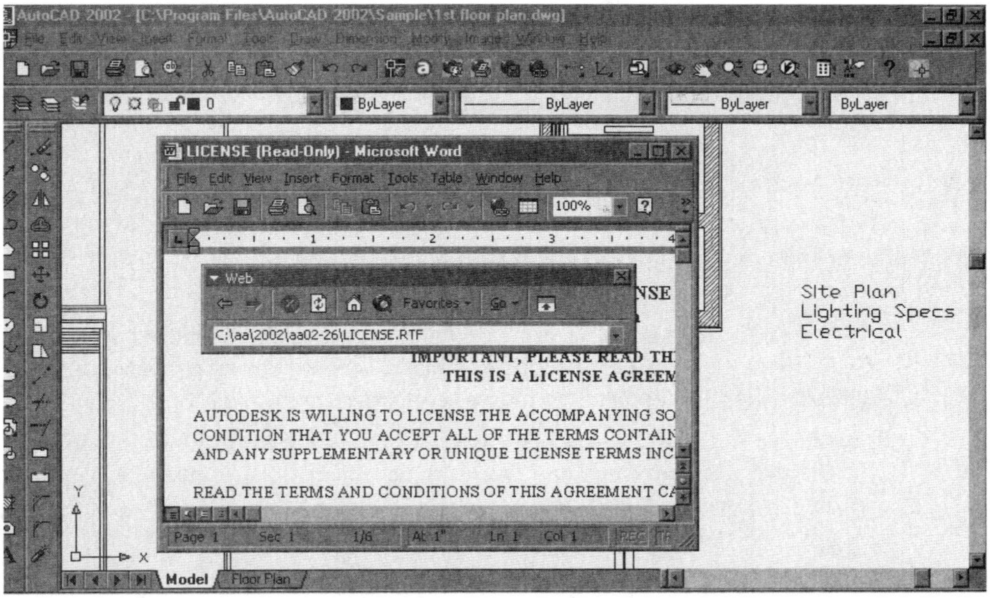

Figure 26.14 *Viewing a Word document.*

11. Back in AutoCAD, right-click any toolbar and select the **Web** toolbar. Notice in Figure 26.15 that the **Web** toolbar has four buttons. From left to right, these are as follows:

Button	**Command**	**Meaning**
Go Back	**Hyperlinkback**	Move back to the previous hyperlink.
Go Forward	**Hyperlinkfwd**	Move to the next hyperlink.
Stop Navigation	**Hyperlinkstop**	Stops the hyperlink access action.
Browse the Web	**Browser**	Launches the Web browser.

When you work with hyperlinks in AutoCAD, you might come across these three limitations:

- AutoCAD does not check that the URL you type is valid.

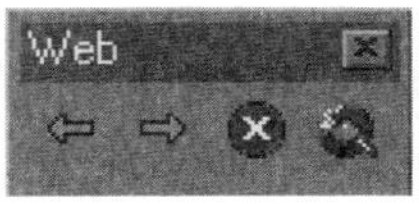

Figure 26.15 *Display AutoCAD's* **Web** *toolbar by moving the cursor to any open toolbar, right-clicking, and selecting* **Web***.*

- If you attach a hyperlink to a block, be aware that the hyperlink data is lost when you scale the block unevenly, stretch the block, or explode it.
- Wide polylines and rectangular hyperlink areas are only "sensitive" on their outline.

Pasting as Hyperlink

AutoCAD has a shortcut method for creating hyperlinks in the drawing. The **Pasteashyperlink** command pastes any text in the Windows Clipboard as a hyperlink to any object in the drawing. Here is how it works:

1. In a word processor, select some text and copy it to the Clipboard using either CTRL+C or the **Edit Copy** command. The text can be a URL such as http://www.autodeskpress.com or any other text.
2. Switch to AutoCAD and select **Paste as Hyperlink** from the **Edit** menu. Note that this command does not work (grayed out) if anything else is in the Clipboard, such as a picture.
3. Select one or more objects, as prompted:

   Command: _**PASTEASHYPERLINK** (ENTER)

   Select objects: *(Select an object.)*

   1 found Select objects: (ENTER)
4. Pass the cursor over the object and note the hyperlink cursor and tooltip. The tooltip displays the same text that you copied from the document.

If the text you copy to the Clipboard is very long, AutoCAD displays only portions of it in the tooltip, using ellipses (...) to shorten the text. You cannot select text in the AutoCAD drawing to paste as a hyperlink. You can, however, copy the hyperlink from one object to another. Select the object, right-click, and select **Copy Hyperlink** from the **Hyperlink** shortcut menu. The hyperlink is copied to the Clipboard. You can now paste the hyperlink into another document, or use AutoCAD's **Paste as Hyperlink** command to attach the hyperlink to another object in the drawing.

Note: The **Match Properties** command does not work with hyperlinks.

Highlighting Objects with URLs

Although you can see the rectangle of area URLs, the hyperlinks themselves are invisible. To see them, you can use the **Qselect** command, which highlights all objects that match specifications. Select **Quick Select** from the **Tools** menu. AutoCAD displays a **Quick Select** dialog box similar to Figure 26.16

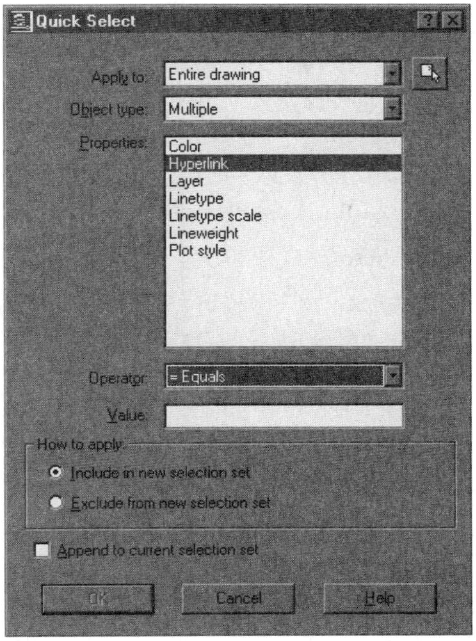

Figure 26.16 *Selecting* **Quick Select** *from the* **Tools** *menu displays the* **Quick Select** *dialog box.*

Enter these specifications: Select Entire Drawing in the **Apply to** edit box. Select Multiple for **Object type** and Hyperlink for **Properties**. Select * Wildcard as the **Operator** and enter * in the **Value** edit box.

Click **OK**, and AutoCAD highlights all objects that have a hyperlink. Depending on your computer's display system, the highlighting shows up as dashed lines or as another color.

Editing Hyperlinks

Now that you know where the objects with hyperlinks are located, you can use the **Hyperlink** command to edit the hyperlinks and related data. Select the hyperlinked object and start the **Hyperlink** command. When the **Edit Hyperlink** dialog box appears (it looks identical to the **Insert Hyperlink** dialog box), make the changes and click **OK**.

Removing Hyperlinks from Objects

To remove a hyperlink from an object, use the **Hyperlink** command with the object selected. When the **Edit Hyperlink** dialog box appears, click the **Remove Hyperlink** button.

To remove a rectangular area hyperlink, you can simply use the **Erase** command. Start the command and select the rectangle, and AutoCAD erases the rectangle.

HYPERLINKS OUTSIDE AUTOCAD

The hyperlinks you place in the drawing are also available for use outside AutoCAD. The Web browser makes use of the hyperlinks when the drawing is exported in DWF format. To help make the process clearer, here are the steps that you need:

1. Open a drawing in AutoCAD.
2. Attach hyperlinks to objects in the drawing with the **Hyperlink** command. To attach hyperlinks to areas, use the **-Hyperlink** command's Area option.
3. Export the drawing in DWF format using the **Plot** command's DWF ePlot PC2 plotter configuration.
4. Copy the DWF file to your Web site.
5. Start your Web browser with the **Browser** command.
6. View the DWF file and click on a hyperlink spot.

THE DRAWING WEB FORMAT (DWF)

To display AutoCAD drawings on the Internet, Autodesk developed a new file format called drawing Web format (DWF). The DWF file has several benefits over DWG files and some drawbacks. The DWF file is compressed to make it as much as eight times smaller than the original DWG drawing file, so that it takes less time to transmit over the Internet, particularly with the relatively slow telephone modem connections. The DWF format is more secure, because the original drawing is not being displayed—another user cannot tamper with the original DWG file.

However, the DWF format has some drawbacks:

- You must go through the extra step of translating from DWG to DWF.
- DWF files cannot display rendered or shaded drawings.
- DWF is a flat 2D-file format; therefore, it does not preserve 3D data, although you can export a 3D view.
- AutoCAD itself cannot display DWF files.
- DWF files cannot be converted back to DWG format without using file translation software from a third-party vendor.
- Earlier versions of DWF did not handle paper space objects (version 2.x and earlier) or linewidths and non-rectangular viewports (version 3.x and earlier).

To view a DWF file on the Internet, your Web browser needs a *plug-in*, which is a software extension that lets a Web browser handle a variety of file formats. Autodesk makes the DWF plug-in freely available from its Web site at http://www.autodesk.com/whip. It's a good idea to regularly check for updates to the DWF plug-in, which is updated about twice a year.

OTHER DWF VIEWING OPTIONS

In addition to WHIP! with a Web browser, Autodesk provides two other options for viewing DWF files.

- CADViewer Light is designed to be a DWF viewer that works on all operating systems and computer hardware because it is written in Java. As long as your Windows, Mac, or Unix computer has access to Java (included with most Web browsers), you can view DWF files.
- Volo View Express is a stand-alone viewer that views and prints DWG, DWF, and DXF files.

Both products can be downloaded free from the Autodesk Web site.

VIEWING DWG FILES

To view AutoCAD DWG and DXF files on the Internet, your Web browser needs a DWG-DXF plug-in. At the time of publication, these plug-ins were available from the following vendors:

Autodesk: http://www.autodesk.com/whip

SoftSource: http://www.softsource.com/

California Software Labs: http://www.cswl.com

Cimmetry systems: http://www.cimmetry.com

CREATING A DWF FILE

Use the following steps to create a DWF file from AutoCAD.

1. Start the **Plot** command by typing **PLOT** (ENTER) at the Command prompt or selecting **Plot** from the **File** menu. The **Plot** dialog box is displayed.
2. In the **Name** list box of the **Plotter Configuration** area, select DWF ePlot pc3.
3. Click the **Properties** button. This will display the **Plotter Configuration Editor** dialog box similar to Figure 26.17.
4. Click Custom Properties in the tree view. Notice the **Custom Properties** button that appears in the lower half of the dialog box.
5. Click the **Custom Properties** button. The **DWF Properties** dialog box similar to Figure 26.18 is displayed.

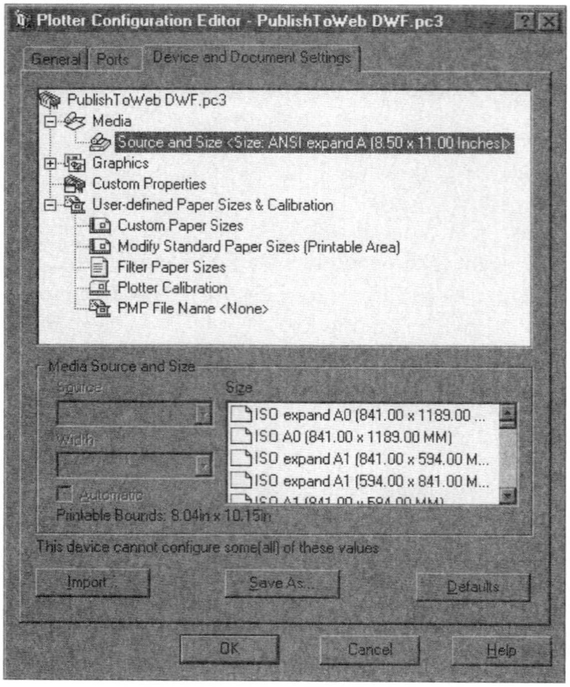

Figure 26.17 *Selecting the* **Properties** *button in the* **Plot** *dialog box displays the* **Plotter Configuration Editor** *dialog box.*

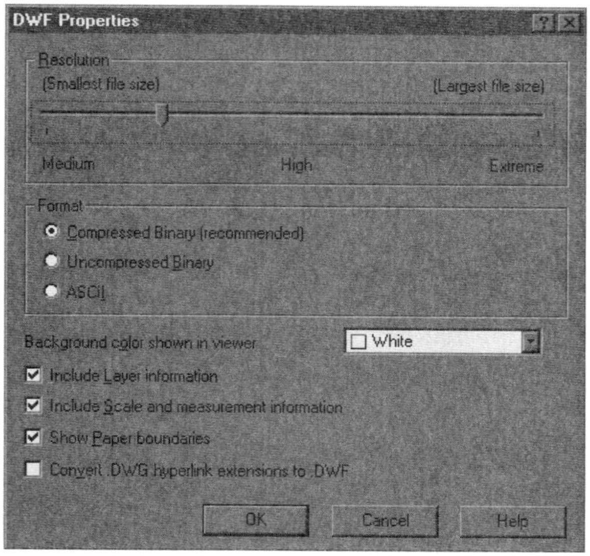

Figure 26.18 *Selecting the* **Custom Properties** *button displays the* **DWF Properties** *dialog box.*

6. Select the desired options, as described below, and click **OK** to exit the dialog boxes back to the **Plot** dialog box.
7. Accept the DWF file name listed in the **File name** text box, or type a new name. If necessary, change the location that the file will be stored in. Note the two buttons: one has an ellipsis [...] that displays the **Browse for Folder** dialog box. The second button brings up AutoCAD's internal Web browser, a simplified version of the Microsoft browser.
8. Click **OK** to save the drawing in DWF format.

DWF Properties

You can use the **DWF Properties** dialog box to set **Resolution** and **Format**, and you can determine other display options as well.

Resolution

Unlike AutoCAD DWG files that are based on real numbers, DWF files are saved using integer numbers. The **Medium** precision setting saves the drawing using 16-bit integers, which is adequate for all but the most complex drawings. **High** resolution saves the DWF file using 20-bit integers, while **Extreme** resolution saves using 32-bit integers. More significant is the difference in file size. When a DWF file from the "1st floor plan.dwg" file was created (342 KB), the medium resolution DWF file was 27 KB, while the extreme resolution file was 1,100 KB—some 40 times larger. The medium resolution DWF file would be transmitted over the Internet 40 times faster, a significant saving in time.

Format

Compression further reduces the size of the DWF file. You should always use compression, unless you know that another application cannot decompress the DWF file. Compressed binary format is seven times smaller than ASCII format (7,700 KB). Therefore, a compressed binary DWF file transmits over the Internet seven times faster than an ASCII DWF file.

Other Options

In most cases, you would use all these options:

Background color shown in viewer—While white is probably the best background color, you can choose any of AutoCAD's 255 colors.

Include Layer information—This option includes layers, which allows you to toggle layers off and on when the drawing is viewed by the Web browser.

Include Scale and measurement information—This allows you to use the Location option in the Web browser's WHIP! plug-in to show scaled coordinate data.

Show Paper boundaries—Includes a rectangular boundary at the drawing's extents.

Convert .DWG hyperlink extensions to .DWF—Includes hyperlinks in the DWF file.

VIEWING DWF FILES

To view a DWF file, you need to use a Web browser with a special plug-in that allows the browser to correctly interpret the file. (Remember: you cannot view a DWF file with AutoCAD.) Autodesk named their DWF plug-in WHIP!, short for Windows HIgh Performance.

Autodesk updates the DWF plug-in approximately twice a year. Each update includes some new features. In summary, all versions of the DWF plug-in perform the following functions:

- DWF files created by AutoCAD are displayed within a browser.
- Right-clicking the DWF image displays a cursor menu with commands.
- Realtime pan and zoom let you change the view of the DWF file as quickly as a drawing file in AutoCAD.
- Embedded hyperlinks let you display other documents and files.
- File compression means that a DWF file appears in your Web browser much faster than the equivalent DWG drawing file would.
- Print the DWF file alone or along with the entire Web page.
- Works with Netscape Navigator or Microsoft Internet Explorer. A separate plug-in is required, depending on which of the two browsers you use.
- Allows you to drag a DWG file from a Web site into AutoCAD as a new drawing or as a block.
- Displays a named view stored in the DWF file.
- Allows you to specify a view using X,Y coordinates.
- Toggles layers off and on.

If you don't know whether the DWF plug-in is installed in your Web browser, select **About Plug-ins** from **Help** in the browser's menu bar. You might need to scroll through the list of plug-ins to find a display similar to Figure 26.19. If the plug-in is not installed or is an older version, then you need to download it from Autodesk's Web site at http://www.autodesk.com

The WHIP! Viewer is in the product list of the Product area on this site. A better alternative is to use the Volo View Express from the same location.

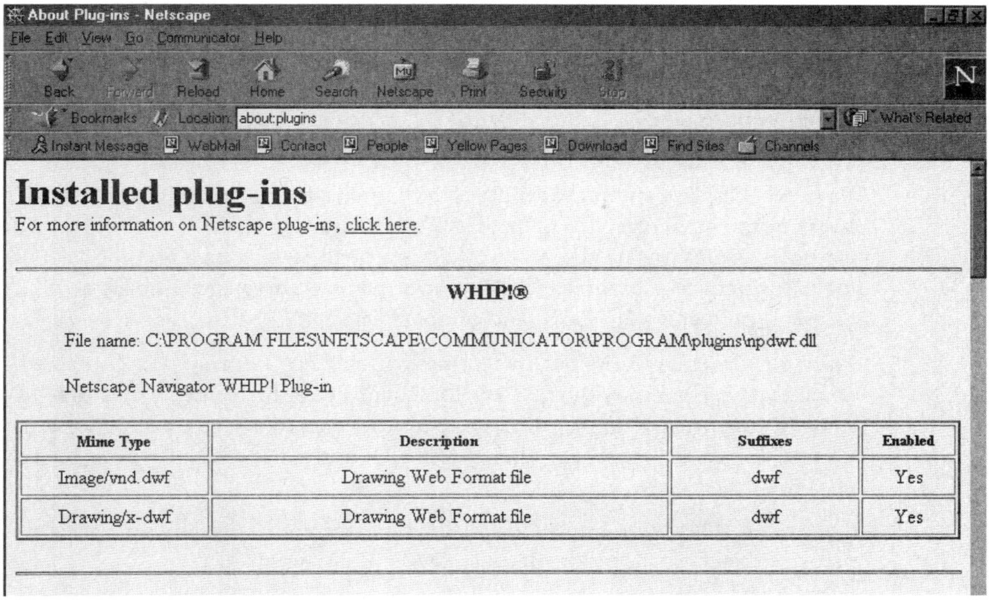

Figure 26.19 *Checking the status of browser plug-ins.*

For Netscape users, the file is quite large at 3.5 MB and it takes about half an hour to download it, using a typical 28.8 Kbaud modem. The file you download from the Autodesk Web site is either a self-installing file (with a .JAR extension) or a self-extracting installation file with a name such as Whip4.Exe. After the download is complete, follow the instructions on the screen. For Microsoft Internet Explorer users, the DWF plug-in is an ActiveX control.

DWF PLUG-IN COMMANDS

To display the DWF plug-in commands, move the cursor over the DWF image and right-click. This displays a shortcut menu with commands, such as **Pan**, **Zoom**, and **Named Views** similar to Figure 26.20. Options include the following:

Pan—The default command. Click the select button and move the mouse. The cursor changes to an open hand, signaling that you can pan the view around the drawing. This is exactly the same as realtime panning in AutoCAD. Panning works only when you are zoomed in; it does not work in **Full View** mode.

Zoom—Like the **Zoom** command in AutoCAD. The cursor changes to a magnifying glass. Hold down the select button and move the cursor up to zoom in and down to zoom out.

Zoom Rectangle—The same as **Zoom** Window in AutoCAD. The cursor changes to a plus sign. Click the select button at one corner, then drag the rectangle to specify the size of the new view.

Fit to Window—The same as AutoCAD's **Zoom Extents** command. You see the entire drawing.

Layers—Displays a non-modal dialog box similar to Figure 26.23 that lists all layers in the drawing. (Unlike AutoCAD's modal dialog boxes, a non-modal dialog box remains on the screen—you do not need to dismiss a non-modal dialog box to continue working.) Click a layer name to toggle its visibility between on (yellow lightbulb) and off (blue lightbulb).

Named Views—Works only when the original DWG drawing file contained named views created with the View command. Selecting Named Views **displays a non-modal dialog box that allows you** to select a named view. Click a named view to see it; click the small x in the upper-right corner to dismiss the dialog box.

Location—Displays a non-modal dialog box that shows the x, y, z coordinates of the cursor location. Because DWF files are only 2D, the Z coordinate is always 0.

Full View—Causes the Web browser to display the DWF image as large as possible. This is useful for increasing the physical size of a small DWF image. When you have finished viewing the large image, right-click and select the **Back** command or click the browser's **Back** button to return to the previous screen.

Highlight URLs—Displays a flashing gray effect on all objects and areas with a hyperlink. Figure 26.24 shows the text highlighted. This helps you see where the hyperlinks are. To read the associated URL, pass the cursor over a highlighted area and look at the URL on the browser's status line. A shortcut is to hold down SHIFT, which causes the URLs to highlight until you release the key. Clicking the URL opens the document with its associated application.

Copy—Copies the DWF image to the Windows Clipboard in EMF (Enhanced Metafile or "Picture") format.

Print—Prints only the DWF image. To print the entire Web page, including the DWF image, use the browser's **Print** button.

Save As—saves the DWF file in three formats to your computer's hard drive:

- DWF
- BMP (Windows bitmap)
- DWG (AutoCAD drawing file)—works only when a copy of the DWG file is available at the same subdirectory as the DWF file.

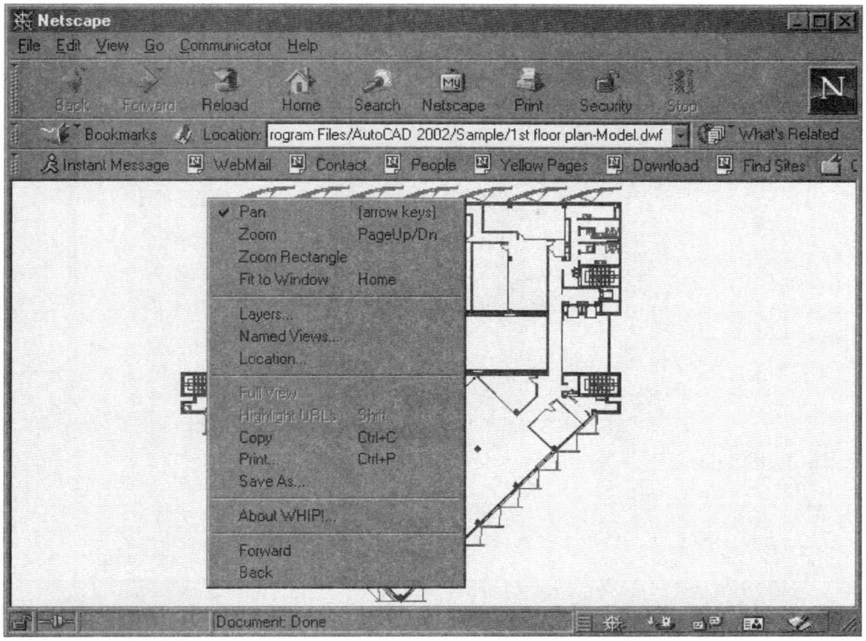

Figure 26.20 *Moving the cursor over the DWF image and right-clicking displays the WHIP!'s shortcut menu.*

You cannot use the **Save As** command until the entire DWF file has been transmitted to your computer.

About WHIP—Displays information about the DWF file similar to Figure 26.21, including DWF file revision number, description, author, creator, source file name, creation time, modification time, source creation time, source modification time, current view left, current view right, current view bottom, and current view top.

Forward—Almost the same as clicking the browser's **Forward** button. When the DWF image is in a frame, only that frame goes forward.

Back—Almost the same as clicking the browser's **Back** button. It works differently when the DWF image is displayed in a frame.

Dragging from Browser to Drawing

The DWF plug-in allows you to perform several drag and drop functions. Hold down the key to drag a DWF file from the browser into AutoCAD. Remember that AutoCAD cannot translate a DWF file to DWG drawing format. For this reason, this form of drag and drop works only when the originating DWG file exists in the same subdirectory as the DWF file.

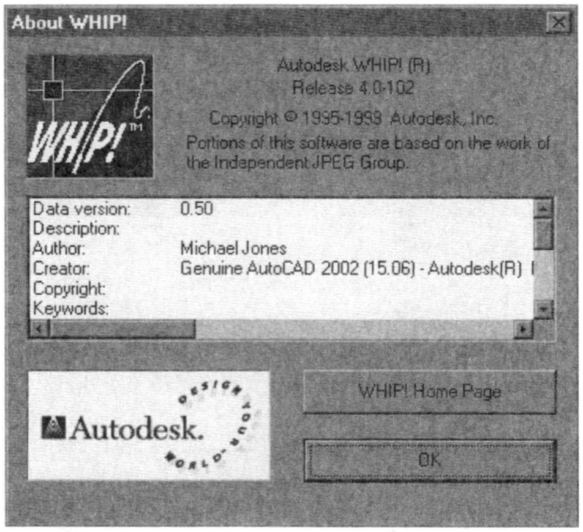

Figure 26.21 *Display of the WHIP! version number.*

AutoCAD executes the **Insert** command using the following sequence:

Command: _.-insert Enter block name or [?]: C:\CAD\ACAD\SAMPLE\1st floor plan.dwg

Specify insertion point or [Scale/X/Y/Z/Rotate/PScale/PX/PY/PZ/PRotate]: 0,0 (ENTER)

Enter X scale factor, specify opposite corner, or [Corner/XYZ] <1>: (ENTER)

Enter Y scale factor <use X scale factor>: (ENTER)

Specify rotation angle <0>: (ENTER)

To see the inserted drawing, you might need to use the Extents option of the **Zoom** command.

Another drag and drop function is to drag a DWF file from the Windows Explorer into the Web browser. This causes the Web browser to load the DWF plug-in, and then display the DWF file. Once it is displayed, you can execute all of the commands listed in the previous section.

Finally, you can also drag a DWF file from Windows Explorer (Windows 95/98) or File Manager (Windows NT) into AutoCAD. This causes AutoCAD to launch another program that is able to view the DWF file. This does not work if you do not have other software on your computer system capable of viewing DWF files.

PUBLISHING DRAWINGS TO THE WEB

Publishing drawings to the Web allows the files to be accessed across the Internet. The drawings can be posted on a Web site in a number of different file formats. A wizard is available in AutoCAD to make this a simple process for any user. To begin to use the wizard, choose **Publish to Web** from the **File** menu. Other AutoCAD features that will allow files to be shared across the Web include the I-drop technology and **eTransmit**.

PUBLISH TO THE WEB

The **Publish to Web** command provides a method for creating a Web page that contains images of AutoCAD drawings. This tool eliminates the need to write programming code, which means that no previous Web development knowledge will be necessary with this method. To begin publishing your drawings to the Web, choose **Publish to Web** from the **File** menu. The following figures and instructions will walk you through the process.

The **Publish to Web - Begin** dialog box as seen in Figure 26.22 has two options available. You can create a new Web page to contain the drawings or edit an existing Web page. You should be familiar with using wizards from previous examples, so simply choose the option needed and select the Next button.

In the **Publish to Web - Create Web Page** dialog box, there are three areas to complete. Specify the name of your Web page, choose the storage location, and type in a description if needed.

The next step is the option to choose the type of image to use. This is done using the **Publish to Web - Select Image Type** dialog box. There are three image types available in the list. A file that is stored as a .DWF, one that is JPEG, or a file format of PNG. The dialog box in Figure 26.24 gives a description of each file type. If the file is an image file type, the image size can be adjusted.

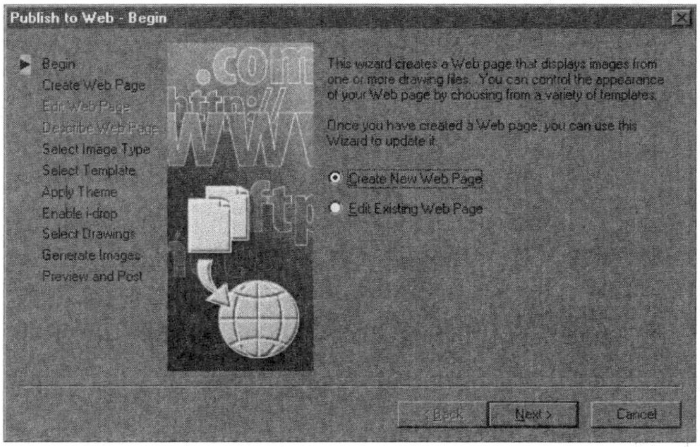

Figure 26.22 *The* **Publish to Web Wizard**.

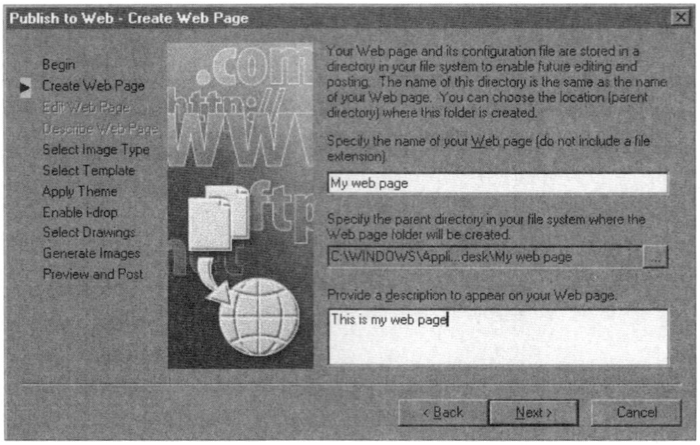

Figure 26.23 *Name and provide a description for your Web page.*

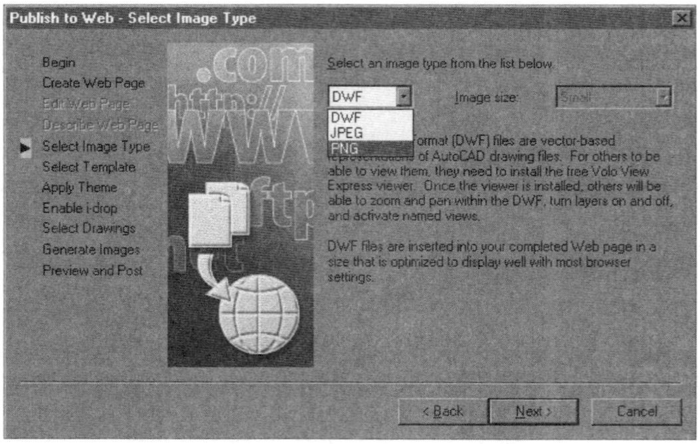

Figure 26.24 *Select the image type for the Web page.*

The **Publish to Web - Select Template** dialog box controls the display of the images in the Web page. A preview window shows how the images will appear in your Web page with the title on the top. See Figure 26.25.

The **Publish to Web - Apply a Theme** dialog box will control the color scheme to be used and improve the appearance of the Web page. The list shown in Figure 26.26 has several options, and the preview window shows how your Web page will look on the Web.

With the **Publish to Web - Enable i-drop** dialog box in Figure 26.27 you can enable the feature that will allow the files in your Web page to be dragged from the Web into

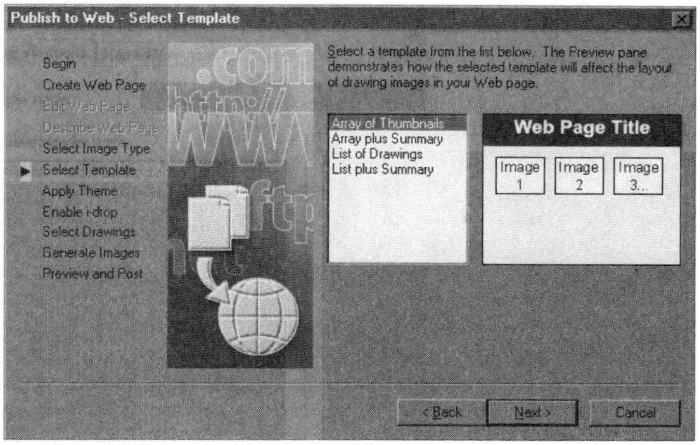

Figure 26.25 *Select a template for the Web page layout.*

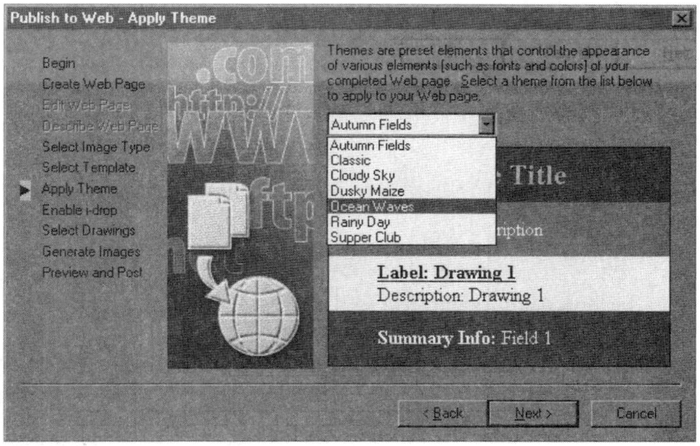

Figure 26.26 *Apply a theme for the color scheme of the Web page.*

an AutoCAD drawing file. This will post .DWG files along with the images, to allow easy access to the drawings.

Now that the setup of the Web page is complete, it is time to select the drawings to be posted on the Web page. The **Publish to Web - Select Drawings** dialog box in Figure 26.28 is the way to make your drawing selections. The current drawing is displayed in the **Drawing** box and a browse [...] button is available to open a file dialog box to add additional files. When the file is selected, the layout list contains the name of the layout to be included from the drawing. To add additional layouts from the same file, you will need to repeat the process and use the **Add** button to include each one.

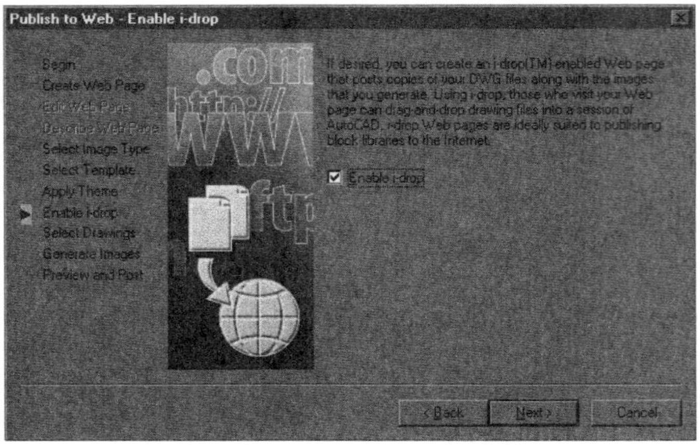

Figure 26.27 *Enable I-drop to allow dragging from the Web page.*

A description can be included for each drawing to make a comment about the drawing on the Web page. Figure 26.29 shows three layouts selected. Notice you can add drawings to the Web page, update the drawings if there have been modifications since the page was made, remove drawings from the Web page, or control the location of the drawing with the **Move Up** and **Move Down** buttons.

Now AutoCAD will finish the Web page drawings by creating all the images necessary for the Web page. The next step is to decide if all of the images need to be regenerated or only those that have changed since the page was created. Refer to Figure

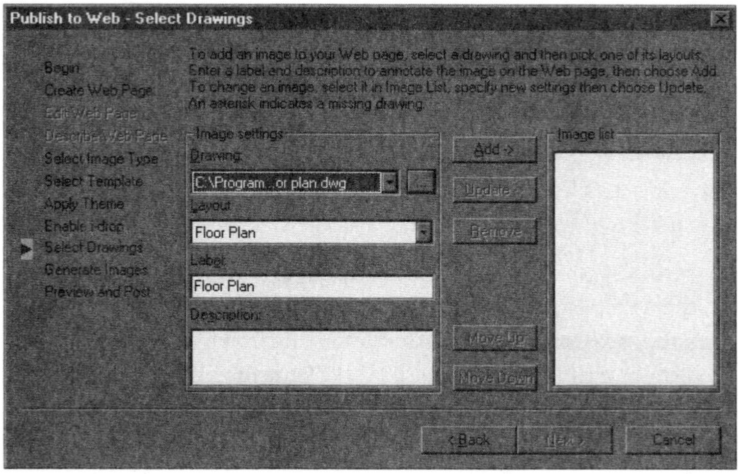

Figure 26.28 *Select the drawings to publish in the Web page.*

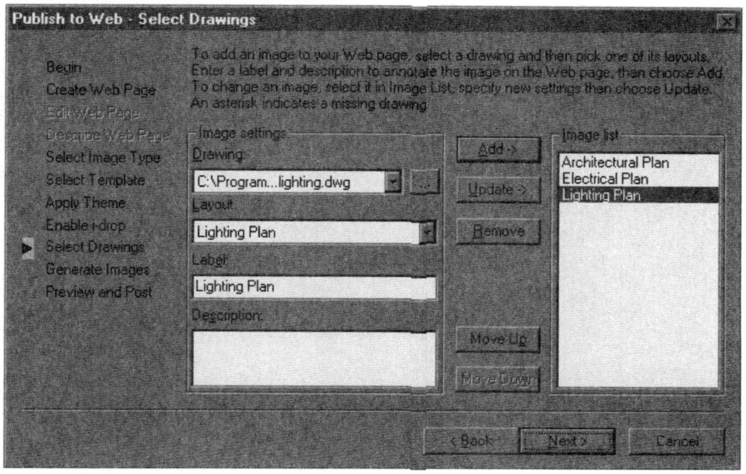

Figure 26.29 *The image list that is added to the Web page.*

26.30. The reason for the two settings is that this wizard is the tool used to edit existing Web pages as well. When you select the **Next** button, AutoCAD will generate all the images requested. This might take a few minutes, depending on how much information is included.

The final display of the wizard is **Publish to Web - Preview and Post** dialog box, shown in Figure 26.31. It is used to notify you that your Web page is now available in the listed directory. You can either examine the new page using the **Preview** button or post the page by selecting the **Post Now** button. Once the page is posted, you can generate an email message that will include a hyperlinked URL to the page.

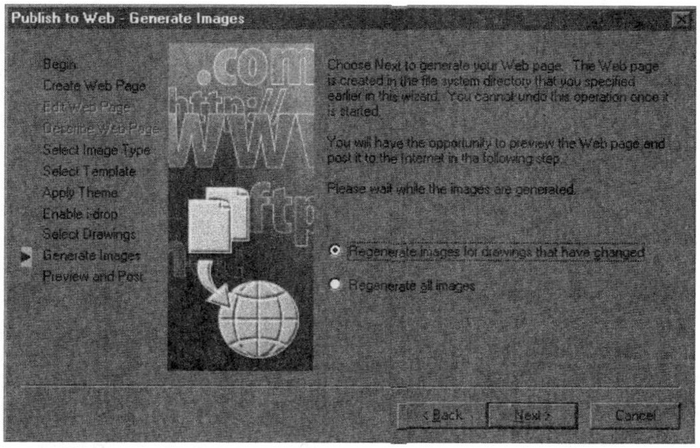

Figure 26.30 *Clicking* **Next** *in the* **Publish to Web - Generate Images** *dialog box will convert the files to Web file format.*

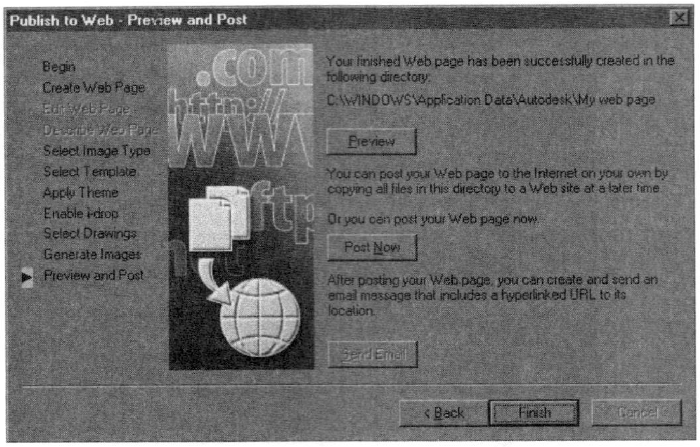

Figure 26.31 *Preview and post the images to the Web.*

ETRANSMIT

The **eTransmit** feature in AutoCAD enables you to pack currently open files with all of the associated files and x-refs into a transmittal set that can be sent to the desired location. This is very similar to sending email messages across the Web. This will allow files to be shared both internally in a company and externally to any location on the Web. To begin a transmittal, select **eTransmit** from the **File** menu. Figure 26.32 is an example of the **eTransmit** dialog box.

The following is a list of the features available:

Notes

If an explanation of the enclosed file is needed, type it in this box.

Type

The file can be compressed to speed up the transfer rate in a self-extracting executable file (.exe) or a .zip file.

Security

You can protect the transmittal by applying a password. This will keep intruders on the Web from viewing the files. Select the **Password** button to protect the transmittal.

Email Notification

An automatic email message can be sent to let the recipient know that the transmittal is available for use. This is a good way to inform the receiving party about the drawings, the location of the information, or password needed to view the designs.

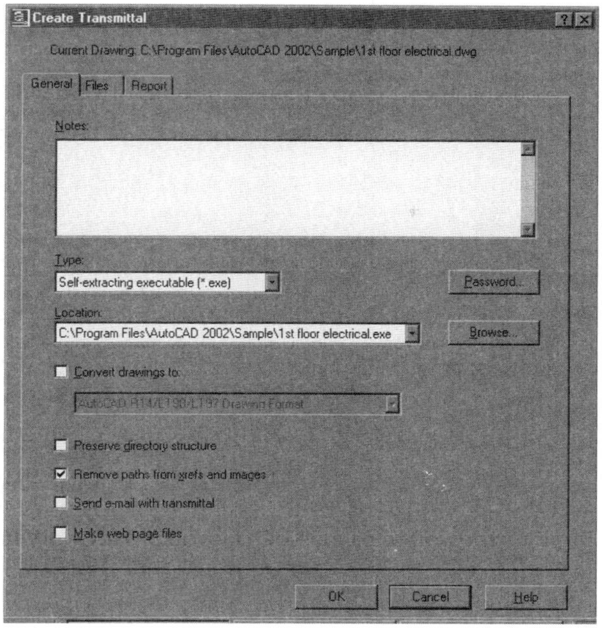

Figure 26.32 *Create a Transmittal to send a file to a client through the Web.*

Web Posting

This allows the posting of the information to be handled across the Internet without tying up the email system. This would be useful if there area lot of people involved with the project.

Drawing Conversion

Both AutoCAD 2000 and Release 14 drawing file formats are supported. Select the **Convert drawings to** check box and choose a format from the list.

Report Generation

This feature on the **Report** tabwill generate a complete log about the transmittal set. This is a tracking system that is available to all the recipients to track transmittal activities, manage any problems, and determine which files have been included.

You should also activate the **Remove paths from xrefs and images** check box so there are no problems with sending the files that contain these objects.

MEET NOW

Meet now is a feature that allows hosting interactive meetings over the Internet using Microsoft's Net Meeting Technology. If you are familiar with chat sessions on the Internet, this feature is very similar. It allows voice, data, and video to be utilized in design team collaboration. Because it lies beyond the scope of what a new user would be expected to do in an office, it will not be further explored in this text. Refer to Microsoft and Autodesk documentation to learn more about this feature.

CHAPTER 26 QUIZ

1. Can you launch a Web browser from within AutoCAD?
2. What does DWF mean?
3. What is the purpose of DWF files?
4. What is URL short for?
5. Which of the following URLs are valid?

   www.autodesk.com

   http://www.autodesk.com

   Both of the above

   Neither of the above
6. What is the purpose of a URL?
7. What is FTP short for?
8. What is a "local host"?
9. Can URLs be used in an AutoCAD drawing?
10. The purpose of URLs is to let you create ______________________ between files.
11. When you attach a URL to a block, the URL data is ____________ when you scale the block unevenly, stretch the block, or explode it.
12. Can you attach a URL to rays and xlines?
13. The **Attachurl** command allows you to attach a URL to _________________ and _________________.
14. To see the location of URLs in a drawing, use the ___________________ command.
15. Rectangular URLs are stored on layer______________________________.
16. The Listurl command tells you ____________________________________.
17. The **Detachurl** command removes a _________________ from an object.
18. Compression in the DWF file causes it to take (less, more, the same) time to transmit over the Internet.
19. A DWF is created from a _______________ file using the ________________ or _______________ command.
20. What does a plug-in allow a Web browser do?
21. Can a Web browser view DWG drawing files over the Internet?
22. _______ is an HTML tag for embedding graphics in a Web page.

23. Are the WHIP! plug-in and the WHIP! display driver the same?
24. A file being transmitted over the Internet via a 28.8 Kbps modem takes about _____________ minutes per megabyte.
25. To open a drawing located on the Internet, use the ___________________ command.

INDEX

LICENSE AGREEMENT FOR AUTODESK PRESS

A Thomson Learning Company

Educational Software/Data

You the customer, and Autodesk Press incur certain benefits, rights, and obligations to each other when you open this package and use the software/data it contains. BE SURE YOU READ THE LICENSE AGREEMENT CAREFULLY, SINCE BY USING THE SOFTWARE/DATA YOU INDICATE YOU HAVE READ, UNDERSTOOD, AND ACCEPTED THE TERMS OF THIS AGREEMENT.

Your rights:

1. You enjoy a non-exclusive license to use the enclosed software/data on a single microcomputer that is not part of a network or multi-machine system in consideration for payment of the required license fee, (which may be included in the purchase price of an accompanying print component), or receipt of this software/data, and your acceptance of the terms and conditions of this agreement.
2. You own the media on which the software/data is recorded, but you acknowledge that you do not own the software/data recorded on them. You also acknowledge that the software/data is furnished "as is," and contains copyrighted and/or proprietary and confidential information of Autodesk Press or its licensors.
3. If you do not accept the terms of this license agreement you may return the media within 30 days. However, you may not use the software during this period.

There are limitations on your rights:

1. You may not copy or print the software/data for any reason whatsoever, except to install it on a hard drive on a single microcomputer and to make one archival copy, unless copying or printing is expressly permitted in writing or statements recorded on the diskette(s).
2. You may not revise, translate, convert, disassemble or otherwise reverse engineer the software/data except that you may add to or rearrange any data recorded on the media as part of the normal use of the software/data.
3. You may not sell, license, lease, rent, loan, or otherwise distribute or network the software/data except that you may give the software/data to a student or and instructor for use at school or, temporarily at home.

Should you fail to abide by the Copyright Law of the United States as it applies to this software/data your license to use it will become invalid. You agree to erase or otherwise destroy the software/data immediately after receiving note of Autodesk Press' termination of this agreement for violation of its provisions.

Autodesk Press gives you a LIMITED WARRANTY covering the enclosed software/data. The LIMITED WARRANTY can be found in this product and/or the instructor's manual that accompanies it.

This license is the entire agreement between you and Autodesk Press interpreted and enforced under New York law.

Limited Warranty

Autodesk Press warrants to the original licensee/ purchaser of this copy of microcomputer software/ data and the media on which it is recorded that the media will be free from defects in material and workmanship for ninety (90) days from the date of original purchase. All implied warranties are limited in duration to this ninety (90) day period. THEREAFTER, ANY IMPLIED WARRANTIES, INCLUDING IMPLIED WARRANTIES OF MERCHANTABILITY AND FITNESS FOR A PARTICULAR PURPOSE ARE EXCLUDED. THIS WARRANTY IS IN LIEU OF ALL OTHER WARRANTIES, WHETHER ORAL OR WRITTEN, EXPRESSED OR IMPLIED.

If you believe the media is defective, please return it during the ninety day period to the address shown below. A defective diskette will be replaced without charge provided that it has not been subjected to misuse or damage.

This warranty does not extend to the software or information recorded on the media. The software and information are provided "AS IS." Any statements made about the utility of the software or information are not to be considered as express or implied warranties. Delmar will not be liable for incidental or consequential damages of any kind incurred by you, the consumer, or any other user.

Some states do not allow the exclusion or limitation of incidental or consequential damages, or limitations on the duration of implied warranties, so the above limitation or exclusion may not apply to you. This warranty gives you specific legal rights, and you may also have other rights which vary from state to state. Address all correspondence to:

AutodeskPress
3 Columbia Circle
P. O. Box 15015
Albany, NY 12212-5015